OTHER FUNCTIONS

Absolute Value Function
$$f(x) = |x|$$

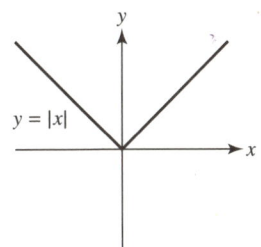

Greatest Integer Function
$$f(x) = [\![x]\!]$$

D0138708

Rational Function
$$f(x) = \frac{1}{x} \ (x \neq 0)$$

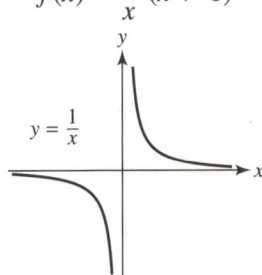

Square Root Function
$$f(x) = \sqrt{x}$$

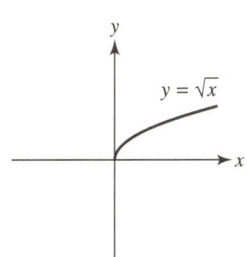

Cube Root Function
$$f(x) = \sqrt[3]{x}$$

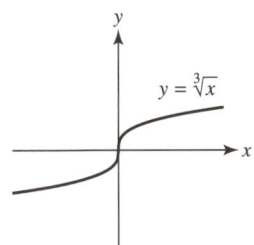

Exponential Function, $a > 1$
$$f(x) = a^x, a > 1$$

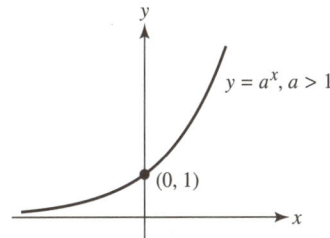

Exponential Function, $0 < a < 1$
$$f(x) = a^x, 0 < a < 1$$

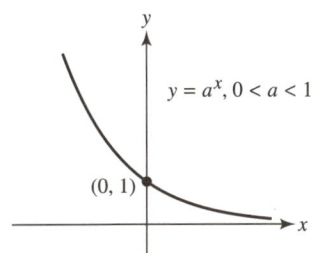

Logarithmic Function, $a > 1$
$$f(x) = \log_a x, a > 1$$

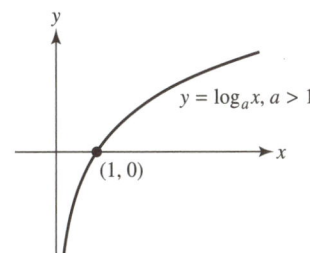

Precalculus
through Modeling and Visualization

Gary K. Rockswold
Minnesota State University, Mankato

John Hornsby
University of New Orleans

Margaret L. Lial
American River College

 **ADDISON-WESLEY**

An imprint of Addison Wesley Longman, Inc.

Reading, Massachusetts • Menlo Park, California • New York • Harlow, England
Don Mills, Ontario • Sydney • Mexico City • Madrid • Amsterdam

Sponsoring Editor: Bill Poole
Senior Editorial Project Manager: Christine O'Brien
Project Manager: Elka Block
Editorial Assistant: Jennifer Kerber
Managing Editor: Karen Guardino
Production Supervisor: Rebecca Malone
Text Design and Project Management: Elm Street Publishing Services, Inc.
Executive Marketing Manager: Brenda L. Bravener
Manufacturing Buyer: Evelyn Beaton
Cover Design: Barbara T. Atkinson
Cover Photograph: VCG/FPG International LLC.
Art Editing Supervisor: Meredith Nightingale
Compositor: Typo-Graphics, Inc.
Technical Art Illustration: Techsetters, Inc.
Reflective Illustration: Jim Bryant, Todd Bonita

Photo Credits: pp. 1, 101, 182, 296, 386, 491, 562, 634, 728 © 1998 PhotoDisc, Inc.

Library of Congress Cataloging-in-Publication Data
Rockswold, Gary.
    Precalculus through modeling and visualization / Gary
Rockswold, John Hornsby, Margaret L. Lial.—1st ed.
    p.   cm.
Includes bibliographical references and index.
ISBN 0-321-05777-5 (alk. paper)
   1. Functions.   I. Hornsby, John.   II. Lial, Margaret L.
III. Title.   IV. Title: Precalculus
QA331.3.R63   2000   99-35868
512'.13—dc21   CIP

Reprinted with corrections, May 2000

2 3 4 5 6 7 8 9 10—QWT—020100

This book is dedicated to the students.

CONTENTS

CHAPTER 9
FURTHER TOPICS IN ALGEBRA   728

PREFACE

This text offers an innovative approach to the material in this course. It concentrates on broad mathematical themes and places emphasis on the unifying concepts and skills that are used repeatedly throughout mathematics. It also consistently links mathematical concepts to real applications in order to promote student intuition and understanding. By demonstrating the importance of mathematics in an accessible manner, this text helps breathe meaning and life into the course.

The concept of function is the unifying theme of the text. Models, applications, visualization, and technology are used consistently to introduce and investigate families of functions. Real data, graphs, and tables also play an important role in the course, giving meaning to the numbers and concepts that students encounter. In exploring functions, symbolic, graphical, verbal, and numerical approaches are all heavily emphasized, and students are frequently asked to discuss and interpret their results to demonstrate that they have mastered the concepts.

Approach

When introducing mathematical ideas, this text moves from the concrete to the abstract, rather than the reverse. It is our philosophy that learning is increased when students can relate a concept to something in their lives. Hence, we often introduce and motivate mathematical concepts using applications to help make the ideas accessible to students. For example, graphing parabolas is introduced by means of a model of AIDS-related deaths; robotics are used to motivate trigonometric functions of angles; the concept of translations of graphs is related to a discussion of computer graphics; the concept of direct variation is applied to space requirements for storing music on compact discs; and a model based on waiting in line at a parking garage motivates the concept of vertical asymptotes. In this way, students see the importance of a topic from a practical and intuitive point of view, with models and applications playing a central part in the learning process. This approach increases both student interest and motivation. Once students are interested and motivated, symbolic concepts and abstractions become more meaningful. Students will see this approach as offering an opportunity to succeed in mathematics.

Features

- **Applications**

  It is our belief that students become more effective problem-solvers by being exposed to applications throughout the course. We have included a wide variety of unique, up-to-date applications in nearly every section. They are designed so that they can be introduced with a minimum expenditure of class time. The following are some of the types of applications included in the text: (See Index of Applications, p. I-1.)

Indoor air pollution	Weather	Global Positioning System
Tuition and fees	Smoking and stress	Tides
AIDS	Internet	Daylight hours
Skin cancer	Medical transplants	Music
Ozone layer	Pollution	Electronics
Global warming	CDs and music	Kinesiology
Urban heat island	Gender equity	Touch-tone phones
Computer graphics	Crime and prisons	Surveying
Digital photography	Credit cards and debt	Solar power
Highway design	Cellular technology	Aerial photography
Acid rain	Robotics	Fractals

- **Section Exercises**

  The exercise sets are the heart of any mathematics text, and we have included a wide variety of exercises within each exercise set.

- **Skill-Building Exercises**

  Skill-building exercises are placed at the beginning of each exercise set. The questions are grouped under headings that identify the skill or concept being tested.

- **Applied Exercises**

  Applied exercises appear in every section and are labeled according to subject matter to make the assignment easier. Many of the applied exercises emphasize modeling and interpreting graphs, tables, and data, and nearly every section contains numerous application exercises that are unique to this text.

- **"Writing about Mathematics"**

  Each exercise set contains exercises that require students to write about and interpret mathematics. These writing exercises often ask the students to synthesize or generalize mathematical concepts, demonstrating their mastery of the concepts.

- **"Checking Basic Concepts"**

  This feature includes a small set of exercises provided after every two sections. They can be used by students for review purposes or as group activities. They require 10–15 minutes to complete and can be used during class if time is available.

- **Chapter Review Exercises**

  Chapter review exercises contain both skill-building and applied exercises, stress graphical, numerical, and algebraic techniques, and provide students with the review necessary to successfully pass a chapter test.

- **Extended and Discovery Exercises**

  Extended and discovery exercises occur at the end of each chapter. These exercises are more involved, often require discovery, and can be used for either collaborative learning or homework assignments.

- **Chapter Introductions and Summaries**

  Many college algebra students have little or no understanding of the goals and motivation behind mathematics. Chapter introductions provide insight into mathematics, while the chapter summaries tie together fundamental concepts and themes.

- **"Putting It All Together"**

  This helpful feature occurs at the end of each section and summarizes important techniques and reinforces the mathematical concepts presented in that section. It is presented in an easy-to-follow grid format.

- **"Making Connections"**

  This feature occurs throughout the text and helps students see how concepts presented in the course are interrelated.

- **Critical Thinking**

  This feature is included in each section and poses questions that can be used for either classroom discussion or homework. The questions ask students to apply mathematical concepts beyond what has already been discussed.

- **Review Notes**

  Rather than present much of the material from intermediate algebra in a review chapter, review notes occur as they are needed. They briefly review a specific mathematical concept.

- **Technology Notes**

  This feature occurs throughout the text and offers students guidance, suggestions, and cautions on the use of technology. It is identified within the text by the icon .

- **Sources**

  Since hundreds of application examples and exercises are presented, hundreds of sources are included. These sources are shown throughout the text and demonstrate the importance of mathematics in a wide variety of areas and disciplines.

- **Web Site**

  A new Web site, awlonline.com/Precalculus, has been created to increase student success in the course by offering section-by-section tutorial help, downloadable programs for TI graphing calculators, author tips, and more. The icon in the text alerts students to instances when this site would be helpful. The site will also be useful to instructors by providing dynamic resources for use in their courses.

Supplements for the Student

Printed Supplements

Student's Solutions Manual

ISBN 0-321-06666-9

Gary Rockswold, Minnesota State University, Mankato, and Terry A. Krieger, South Dakota State University

- Complete solutions to all odd-numbered exercises.
- Ask your bookstore about ordering.

Graphing Calculator Manual

ISBN 0-321-06667-7

Stuart Moskowitz, Humboldt State University

- Instruction on graphing calculator usage.
- Keystroke operations for the following calculator models: TI-82 ®, TI-83 ®, TI-85 ®, TI-86 ®, Casio Color Graphics 9850 Plus ® , HP38G ®.
- Worked-out examples taken directly from the text.

Media Supplements
InterAct Math Tutorial Software

Throughout the text, the icon indicates when this software would be helpful to students.

Windows/Macintosh (Dual-Platform CD-ROM) ISBN 0-321-06829-7
InterAct Math Tutorial Software has been developed and designed by professional software engineers working closely with a team of experienced math educators. InterAct Math Tutorial Software includes exercises that are linked with every objective in the textbook and require the same computational and problem-solving skills as their companion exercises in the text. Each exercise has an example and an interactive guided solution that are designed to involve students in the solution process and to help them identify precisely where they are having trouble. The software recognizes common student errors and provides students with appropriate customized feedback. With its sophisticated answer recognition capabilities, InterAct Math Tutorial Software recognizes appropriate forms of the same answer for any kind of input. It also tracks student activity and the scores for each section, which can then be printed out.

The software is free to qualifying adopters or can be bundled with books for sale to students.

Videotape Series

Throughout the text, the icon [icon] indicates when this software would be helpful to students.

ISBN 0-321-06833-5
- Keyed specifically to the text.
- An engaging team of lecturers provide comprehensive coverage of each section and every topic.
- Utilizes worked-out examples, visual aids, and manipulatives to reinforce concepts.
- Emphasizes the relevance of material to the real world and relates mathematics to students' everyday lives.
- Can be ordered by mathematics instructors or departments.

Supplements for the Instructor
Printed Supplements
Instructor's Solutions Manual
ISBN 0-321-06665-0
Gary Rockswold, Minnesota State University, Mankato, and Terry A. Krieger, South Dakota State University
- Complete solutions to all exercises.
- Consult your Addison-Wesley sales representative for details.

Instructor's Testing Manual
ISBN 0-321-06668-5
- Provides prepared tests for each chapter.
- Consult your Addison-Wesley sales representative for details.

Media Supplements
TestGen-EQ with QuizMaster-EQ
Windows/Macintosh (Dual-Platform CD-ROM) ISBN 0-321-06831-9

TestGen-EQ's friendly graphical interface enables instructors to easily view, edit, and add questions, transfer questions to tests, and print tests in a variety of fonts and forms. Search and sort features let the instructor quickly locate questions and arrange them in a preferred order. Six question formats are available, including short-answer, true-false, multiple-choice, essay, matching, and bi-modal formats. A built-in question editor gives the user power to create graphs, import graphics, insert mathematical symbols and templates, and insert variable numbers or text. Computerized testbanks include algorithmically defined problems organized according to each textbook. An "Export to HTML" feature lets instructors create practice tests for the Web.

QuizMaster-EQ enables instructors to create and save tests using TestGen-EQ so students can take them for practice or a grade on a computer network. Instructors can set preferences for how and when tests are administered. QuizMaster-EQ automatically grades the exams, stores results on disk, and allows the instructor to view or print a variety of reports for individual students, classes, or courses.

Consult your Addison-Wesley sales representative for details.

InterAct Math Plus Software

InterAct Math Plus combines course management and on-line testing with the features of the basic InterAct Math tutorial software to create an invaluable teaching resource. Consult your Addison-Wesley representative for details.

Acknowledgments

Many people contributed to the development of this textbook. We would like to express our appreciation to the following reviewers whose comments and suggestions were invaluable in preparing this text:

Ignacio Alarcón	*Santa Barbara City College*
Fred Alderman	*Valencia Community College*
Chris Burditt	*Napa Valley College*
Tim Carroll	*Eastern Michigan University*
Charles Curtis	*Missouri Southern State College*
Arunas Dagys	*St. Xavier University*
Madelyn Gould	*DeKalb College*
Linda Green	*Santa Fe Community College*
William Grimes	*Central Missouri State University*
Margaret D. Hovde	*Grossment College*
Heidi Howard	*Florida Community College—Jacksonville*
Annita W. Hunt	*Clayton College and State University*
Terry A. Krieger	*South Dakota State University*
Adam Lewenberg	*University of Akron*
JoAnn Lewin	*Edison Community College*
Joanne Manville	*Bunker Hill Community College*
Joe May	*North Hennepin Community College*
Johnny Moore	*Indian River Community College*
Kimberly Myers	*University of Cincinnati*
Nancy H. Olson	*Johnson County Community College*
Wing Park	*College of Lake County*
Sallie Paschal	*DeKalb College*
Doug Proffer	*Collin County Community College*
Joseph W. Rody	*Arizona State University*
Richard Semmler	*Northern Virginia Community College*

Sandy Spears	*Jefferson Community College*
Joe A. Stickles, Jr.	*Marshall University*
Eddie Warren	*University of Texas—Arlington*

We would like to recognize all the work and effort of Terry Krieger, Georgia Mederer, and Deana Richmond in their accuracy checking and proofreading of the text and its ancillaries. We are thankful for the excellent cooperation and support received from everyone at Addison Wesley Longman. Thanks go to Greg Tobin for giving his support to the project. Particular recognition is due Bill Poole, Christine O'Brien, and Donna Bagdasarian, who gave invaluable advice, encouragement, and assistance. We express our special gratitude to Elka Block, who provided guidance, insight, and many helpful suggestions. We also appreciate the outstanding contributions of Becky Malone, Karen Guardino, Brenda Bravener, Barbara Atkinson, Tricia Mescall, Joe Vetere, Caroline Fell, Jennifer Kerber, Meredith Nightingale, and Alex Levering. Thanks go to Wendy Rockswold, who proofread the manuscript and was instrumental in its development, and to Michele Heinz of Elm Street Publishing Services for her valuable help. We would also like to thank the many students who participated in classroom testing of this edition. Their suggestions were insightful. Please feel free to write to us with your comments. Your opinion is important to us. You can write to the attention of Gary Rockswold, Department of Mathematics, Minnesota State University, Mankato, MN 56002.

Gary K. Rockswold
John Hornsby
Margaret L. Lial

CHAPTER 1 Introduction to Functions and Graphs

"Why do I need to learn math?" This is a question commonly asked by students across the country. A historical perspective provides some reasons for studying mathematics.

All of us have ancestors who were not only illiterate but also unable to perform even basic arithmetic. This was not due to a lack of intelligence, but a lack of education. Around 400 A.D., Saint Augustine declared that people who could add and subtract were in conspiracy with the devil. One hundred years ago many Americans were educated in rural schools. It was not uncommon for school boards to have difficulty finding teachers who could multiply and divide. In the 1940s and 1950s, many mathematics teachers had no training beyond algebra. Today, many high schools offer calculus.

Knowledge of mathematics provides access into today's world. Without mathematical skills and understanding, doors are closed and opportunities are lost. Most vocations and professions require a higher level of mathematical understanding than in the past. Seldom are the mathematical expectations of employees lowered, as the work place becomes more technical—not less. The Department of Education cites

It is not enough to have a good mind; the main thing is to use it well.

— **René Descartes**

mathematics as a key to achievement in our society. Students with a solid mathematics background earned, on average, 38 percent more per hour than their peers without such a background.

Mathematics is vital to understanding and interpreting the data that technology creates. Our society is in transition from an industrial to an informational society. An informational society's primary social and economic activity is the production, storage, and communication of information. Communication skills are essential, and mathematics is the language of technology. Mathematics allows society to quantify its experiences.

Decisions in high school or college to avoid mathematics classes can affect life-long vocations, incomes, and lifestyles. Switching majors simply to elude a mathematics prerequisite may prevent a person from pursuing his or her real vocational dreams. Neither illiteracy nor innumeracy are desirable attributes for attaining one's full potential. Students today are preparing for the twenty-first century. What skills will be needed in society during the next 50 years? Although no one can answer this question with certainty, history tells us that the importance of mathematics will not diminish but likely increase.

Sources: A. Toffler and H. Toffler, *Creating a New Civilization; USA Today.*

1.1 Data, Numbers, and Algorithms

Natural Numbers • Integers and Rational Numbers • Real Numbers • Scientific Notation • Algorithms

Introduction

Throughout history, humans have had a desire to communicate information and knowledge. Languages evolved to satisfy this desire. The need for numbers existed in nearly every society. Numbers first occurred in the measurement of time, currency, goods, and land. As the complexity of a society increased, so did the data and information that was communicated. One tribe from the Torres Strait, located between Australia and New Guinea, counted to only six. Any number higher than six was referred to as "many." This number system met the needs of their society. On the other hand, a highly technical society could not function with so few numbers. As a result, number systems have evolved over millennia.

This section discusses sets of numbers that are vital to our technological society. It also shows how numbers can be used to represent different types of data. With the aid of algorithms, we can draw conclusions about important information contained in data. (Reference: *Historical Topics for the Mathematics Classroom, Thirty-first Yearbook,* National Council of Teachers of Mathematics (NCTM).)

Natural Numbers

One important set of numbers found in most societies is the set of **natural numbers.** These numbers are comprised of the *counting numbers* 1, 2, 3, 4, 5, ... Natural numbers can be used when data is not broken down into fractional parts.

EXAMPLE 1   *Describing image resolution*

The screens for computer terminals or graphing calculators are made up of tiny units called pixels. (The word "pixel" is a contraction of "picture element.") A rectangular screen on a graphing calculator might be 95 pixels across and 63 pixels high, whereas a high-resolution computer terminal could be 2048 by 2048 pixels. Figure 1.1 shows a graphing calculator screen. If we look closely, each number is comprised of several rectangular pixels. Fractional parts of a pixel do not occur. As a result, natural numbers are appropriate to describe them. As the number of pixels increases, so do the screen resolution and clarity.

(a) Find the number of pixels in a graphing calculator screen that is 95 by 63 pixels.

(b) The photograph in Figure 1.2 shows an image of Jupiter and two of its moons, Io and Europa, taken by *Voyager 1*. This photograph is 800 by 800 pixels. How many pixels are there?

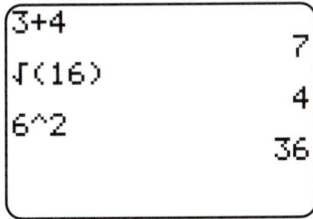

FIGURE 1.1

FIGURE 1.2 Source: NASA.

Solution

(a) There are $95 \times 63 = 5985$ pixels.

(b) There are $800 \times 800 = 640{,}000$ pixels.   ∎

Integers and Rational Numbers

The **integers**

$$\{\ldots, -3, -2, -1, 0, 1, 2, 3, \ldots\}$$

are a set of numbers that contains the natural numbers. Historically, negative numbers were not readily accepted. They did not appear to have real meaning. However, today when a person opens a personal checking account for the first time, negative numbers quickly take on meaning. There is a significant difference between a positive and a negative balance.

A **rational number** is any number that can be expressed as the ratio of two integers $\frac{p}{q}$, where $a \neq 0$. Rational numbers include the integers. Some examples of rational numbers are

$$\frac{2}{1}, \frac{1}{3}, -\frac{1}{4}, \frac{-50}{2}, \frac{22}{7}, 0, \text{ and } 1.2.$$

Note that 0 and 1.2 are both rational numbers. They can be represented by the fractions $\frac{0}{1}$ and $\frac{12}{10}$. Because two fractions that look different can be equivalent, rational numbers have more than one form. Rational numbers also can be expressed in a decimal form that either *repeats* or *terminates*. For example, $\frac{2}{3} = 0.\overline{6}$, a repeating decimal, and $\frac{1}{4} = 0.25$, a terminating decimal. The overbar indicates that $0.\overline{6} = 0.6666666....$

Critical Thinking

The number 0 was invented well after the natural numbers. Many societies did not have a zero—for example, there is no Roman numeral for 0. Discuss some possible reasons for this.

The next example shows how integers and rational numbers are used in measuring temperature.

EXAMPLE 2   *Comparing temperature scales*

The Fahrenheit temperature scale is commonly used in the United States. A temperature of 0°F is approximately the coldest temperature possible for a mixture of salt and water to remain liquid. On the Celsius temperature scale, pure water freezes at 0°C and boils at 100°C. Table 1.1 shows the relationship between the two temperature scales.

TABLE 1.1

°F	°C	Observation
−89	$-67\frac{2}{9}$	Alcohol freezes.
−40	−40	Mercury freezes.
0	$-17\frac{7}{9}$	Snow and salt mixture freezes.
32	0	Water freezes.
100	$37\frac{7}{9}$	People are feverish.
212	100	Water boils.

(a) If the outside temperature decreases by 40°C, find the corresponding decrease in degrees Fahrenheit.

(b) One degree Fahrenheit is equivalent to how many degrees Celsius?

Solution

(a) As seen in Table 1.1, there is a 40°C decrease from water freezing to mercury freezing. The corresponding change on the Fahrenheit scale is $32 - (-40) = 72$°F.

(b) Since a 40° change on the Celsius scale is equivalent to a 72° change on the Fahrenheit scale, 1°F is equivalent to $\frac{40}{72} = \frac{5}{9}$°C. ∎

Real Numbers

Real numbers can be represented by decimal numbers. Since every fraction has a decimal form, real numbers include rational numbers. However, some real numbers cannot be expressed by fractions. These numbers are called **irrational numbers.** The numbers $\sqrt{2}$, $\sqrt{15}$, and π are examples of irrational numbers.

They can be represented by decimals, but not by decimals that either repeat or terminate. However, any real number can be *approximated* by a terminating decimal. Examples of real numbers include

$$2, -10, 151.6, -131.3337, \frac{1}{3} = 0.\overline{3}, -\sqrt{5} \approx -2.2361, \text{ and } \sqrt{11} \approx 3.3166.$$

Review Note   *Approximately Equal*

The symbol "≈" means **approximately equal.** This symbol will be used in place of an equals sign whenever two unequal quantities are close in value. For example, $\frac{1}{4} = 0.25$, whereas $\frac{1}{3} \approx 0.3333$. It is often wise to use approximately equal when the exact value is not clearly evident, such as when we estimate a value from a graph.

EXAMPLE 3   *Classifying numbers*

Classify each real number as a natural number, integer, rational number, or irrational number:

$$5, -1.2, \frac{13}{7}, -\sqrt{7}, -12, \sqrt{16}.$$

Solution

The numbers 5 and $\sqrt{16} = 4$ are natural numbers. We classify 5, $\sqrt{16}$, and -12 as integers. The rational numbers include 5, $\sqrt{16}$, -12, -1.2, and $\frac{13}{7}$. The only irrational number is $-\sqrt{7}$. ∎

Even though a data set may be comprised of only integers, decimals often are used to analyze it. One common way to analyze data is to find the *arithmetic average,* or more simply, the *average.* The average of a set of integers need not be an integer. Sometimes it will be a fraction or decimal.

EXAMPLE 4   *Analyzing deaths from AIDS*

Table 1.2 lists the annual deaths caused by AIDS in the United States from 1990 to 1993. Calculate the average number of deaths over this four-year period.

TABLE 1.2				
Year	1990	1991	1992	1993
Deaths	31,269	36,165	39,307	41,920

Source: Department of Health and Human Services.

Solution

The average number of deaths from AIDS between 1990 and 1993 was equal to

$$\frac{31,269 + 36,165 + 39,307 + 41,920}{4} = \frac{148,661}{4} = 37,165.25$$

This average of four integers is a rational number *and* a real number, but not an integer. ∎

Except for imaginary and complex numbers, which you may have studied in a previous course, all of the numbers used in this course are real numbers. Real numbers have applications in many areas such as interest rates, grade point averages, and the Consumer Price Index (CPI).

The CPI is often referred to as the "cost of living index" and is the numerical scale most commonly used to measure inflation. It tracks the prices of basic consumer goods. If the CPI changes from c_1 to c_2, then the **percent change** is given by $\frac{c_2 - c_1}{c_1} \times 100$. Over 40 economists calculate the CPI annually, at a cost of $26 million. The time from 1982 to 1984 is called the *reference period,* where the CPI is defined to be 100.0.

EXAMPLE 5   *Interpreting the Consumer Price Index*

Table 1.3 lists the CPI for selected years from 1950 to 1995. A CPI of 24.1 indicates that a typical sample of consumer goods costing $100 in 1982–1984 would have cost $24.10 in 1950.

TABLE 1.3

Year	1950	1955	1960	1965	1970	1975	1980	1985	1990	1995
CPI	24.1	26.8	29.6	31.5	38.8	53.8	82.4	107.6	130.7	153.5

Source: Bureau of the Census.

(a) Discuss the change in the CPI over this 45-year period.
(b) Approximate the percent change in prices from 1970 to 1980.
(c) From 1980 to 1990 the average tuition and fees at public colleges and universities rose from $804 to $1908. Determine if the percent change in tuition and fees was more or less than the percent change in the CPI for the same time period. (Source: The College Board.)

Solution

(a) The CPI increased during every five-year period between 1950 and 1995. This reflects the fact that prices have consistently risen.
(b) In 1970 the CPI was 38.8 and in 1980 it was 82.4. The percent change was approximately $\frac{82.4 - 38.8}{38.8} \times 100 \approx 112\%$.
(c) In 1980 the CPI was 82.4 and in 1990 it was 130.7. The CPI rose by $\frac{130.7 - 82.4}{82.4} \times 100 \approx 59\%$, while tuition and fees rose by $\frac{1908 - 804}{804} \times 100 \approx 137\%$. The rise in tuition and fees outpaced the CPI. ∎

Scientific Notation

Numbers that are large or small in absolute value occur frequently in applications. For simplicity these numbers are often expressed in scientific notation. Table 1.4 lists examples of numbers in *standard form* and in *scientific notation.*

TABLE 1.4

Standard Form	Scientific Notation	Application
93,000,000 mi	9.3×10^7 mi	Distance to Sun
360,000	3.6×10^5	Dots in 1 sq. in. of a laser print
10,000,000,000	1×10^{10}	Estimated world population in 2050
0.00000538 sec	5.38×10^{-6} sec	Time for light to travel 1 mile
0.000005 cm	5×10^{-6} cm	Size of a typical virus

To write 0.00000538 in scientific notation, start by moving the decimal point to the right of the first nonzero digit 5 to obtain 5.38. Since the decimal point was moved six places to the *right,* the exponent of 10 is −6. Thus, $0.00000538 = 5.38 \times 10^{-6}$. When the decimal point is moved to the *left,* the exponent of 10 is positive, rather than negative. Here is a formal definition of scientific notation.

Scientific notation

A real number r is in **scientific notation** when r is written as $c \times 10^n$, where $1 \leq |c| < 10$ and n is an integer.

The accuracy of a measurement may be shown by the use of **significant digits.** For example, scientists might measure the average distance from Earth to the sun to be $93,000,000 = 9.3 \times 10^7$ miles. A more accurate measurement would be $92,960,000 = 9.296 \times 10^7$ miles. The first measurement has two significant digits, while the second has four significant digits. In Table 1.4 the estimated world population in the year 2050 is given to one significant digit, while the time for light to travel one mile is given to three significant digits.

EXAMPLE 6 *Applying scientific notation*

A compact disc (CD) is an optically readable device that is capable of storing large amounts of data. Some CDs can store 600 million bytes of information. (A **byte** is the amount of memory necessary to store a single character, such as one letter or a blank space. The phrase "math test" requires nine bytes of storage.) (Source: Grolier Electronic Publishing, Inc.)
(a) Express in scientific notation the number of bytes available on some CDs.
(b) Approximate how many sentences like this one could be stored on a CD. Express your answer in scientific notation using two significant digits.
(c) When playing a CD, sound channels are sampled (or read) at a rate of 44,100 times per second per channel. If two channels are played for one minute of stereo music, find the total number of times the sound channels are sampled. Express your answer in scientific notation using three significant digits.

Solution

(a) Move the decimal point in 600,000,000 to the left 8 places to obtain 6. Then 600,000,000 can be written as 6×10^8.

(b) The sentence

"Approximate how many sentences like this one could be stored on a CD."

is composed of 69 characters including the blank spaces and the period.

$$\frac{600{,}000{,}000}{69} \approx 8{,}700{,}000 = 8.7 \times 10^6$$

The CD could store the sentence approximately 8.7 million times.

(c) Two channels are being sampled $2 \times 44{,}100 = 88{,}200$ times per second. In 60 seconds, they would be sampled $60 \times 88{,}200 = 5{,}292{,}000 = 5.292 \times 10^6$ times, or 5.29×10^6 times rounded to three significant digits. ∎

EXAMPLE 7   *Computing in scientific notation*

Use a calculator to compute the expressions in scientific notation. Round each result to two significant digits.

(a) $(3.1 \times 10^{11})(5.6 \times 10^2)$

(b) $\left(\dfrac{4.55 \times 10^{-3}}{5.991 \times 10^8}\right)(1.25 \times 10^6)$

Solution

(a) Calculators often express powers of ten using "E." For example, 3.1×10^{11} is displayed as "3.1E11" on a calculator screen. A calculator may display the result of $(3.1 \times 10^{11})(5.6 \times 10^2)$ as 1.736E14. See Figure 1.3. Rounding to two significant digits, the answer is 1.7×10^{14}.

(b) A calculator shows that $\dfrac{4.55 \times 10^{-3}}{5.991 \times 10^8}(1.25 \times 10^6) \approx 9.493406777 \times 10^{-6}$.

See Figure 1.4. The result, rounded to two significant digits, is 9.5×10^{-6}. ∎

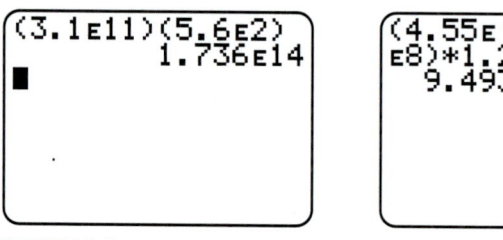

FIGURE 1.3   FIGURE 1.4

Algorithms

The concept of an algorithm is fundamental to human thought. In order to complete a task or calculate a result, a process or algorithm must be followed. When a consumer needs to assemble an item, there are typically directions included in the package. Directions provide a step-by-step procedure. Frequently, the order of the steps is essential. Algorithms also occur in mathematics. An **algorithm** is a finite sequence of well-defined instructions or steps, receiving input and producing output. For example, the division algorithm receives the divisor and dividend, and outputs the quotient and remainder. If an algorithm receives the same input on two different occasions, it must produce identical outputs. In an algorithm, instructions involving chance, such as the rolling of dice, are not allowed.

Temperatures are often displayed in both degrees Fahrenheit and degrees Celsius. Algorithm 1.1 can be used to convert a Celsius temperature to a Fahrenheit temperature.

ALGORITHM 1.1 Converting degrees Celsius to degrees Fahrenheit

Step 1:   Input the Celsius temperature x.

Step 2:   Multiply x by $\frac{9}{5}$.

Step 3:   Add 32 to the answer in STEP 2. Let this result be y.

Step 4:   Output y, the equivalent Fahrenheit temperature.

These steps or instructions satisfy the requirements of an algorithm. There are four well-defined steps that do not involve chance. This algorithm accepts one input x and produces one output y. If the algorithm receives the same Celsius temperature twice, it outputs the same Fahrenheit temperature both times.

EXAMPLE 8   *Converting degrees Celsius to degrees Fahrenheit*

Use Algorithm 1.1 to compute the equivalent Fahrenheit temperature for 25°C.

Solution

Step 1:   Let $x = 25$.

Step 2:   Multiply $\frac{9}{5} \times 25 = 45$.

Step 3:   Add $45 + 32 = 77$. Let $y = 77$.

Step 4:   Output the Fahrenheit temperature of 77°F.

Thus, 25°C is equivalent to 77°F.   ■

Algorithms occur in a wide variety of applications. Some algorithms require more than one input. Finding the average of two numbers is an example. Algorithm 1.2 can be used to calculate a student's grade point average, where a 4.00 represents a perfect A average. It requires five inputs, namely the number of credit hours earned for each of the five possible grades.

ALGORITHM 1.2 Computing grade point average (GPA)

Step 1:   Input the number of credit hours earned of grades A, B, C, D, and F, as a, b, c, d, and f, respectively.

Step 2:   Evaluate the expression $4a + 3b + 2c + 1d$. Call this result s.

Step 3:   Add $a + b + c + d + f$. Call the sum n.

Step 4:   Calculate the quotient $\frac{s}{n}$. Let the answer be y.

Step 5:   Output y, the grade point average.

EXAMPLE 9   *Using an algorithm to calculate grade point average*

Use Algorithm 1.2 to compute the grade point average for a student who has earned 20 credits of A, 28 credits of B, 20 credits of C, 8 credits of D, and 4 credits of F grades.

Solution

Step 1:   Let $a = 20$, $b = 28$, $c = 20$, $d = 8$, and $f = 4$.

Step 2:   Evaluate $4a + 3b + 2c + 1d = 4(20) + 3(28) + 2(20) + 1(8) = 212$. Let $s = 212$.

Step 3:   Add $a + b + c + d + f = 20 + 28 + 20 + 8 + 4 = 80$.
Let $n = 80$.

Step 4:   Calculate $\dfrac{s}{n} = \dfrac{212}{80} = 2.65$. Let $y = 2.65$.

Step 5:   Output the student's GPA of 2.65.   ■

Critical Thinking

Give an example of something that cannot be computed by an algorithm.

1.1 PUTTING IT ALL TOGETHER

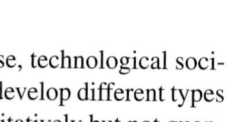

Data, numbers, and algorithms play central roles in a diverse, technological society. Because of the variety of data, it has been necessary to develop different types of numbers. Without numbers, data could be described qualitatively but not quantitatively. For example, we might say that the day seems hot, but would not be able to give an actual number for the temperature. Accurate comparisons with temperatures on other days would be difficult, if not impossible.

Algorithms are used in almost every facet of our lives, providing the procedures used to systematically complete tasks and perform computations. They determine how to construct buildings and cars, evaluate our transcripts to see if we have met graduation requirements, and predict tomorrow's weather.

The following table summarizes some of the concepts found in this section.

Concept	Examples	Comments
Natural numbers	1, 2, 3, 4, 5, . . .	Sometimes referred to as the *counting numbers*
Integers	. . . , −2, −1, 0, 1, 2, . . .	Includes the natural numbers
Rational numbers	$\dfrac{1}{2}, -3, \dfrac{128}{6}, -0.335, 0$ $0.25 = \dfrac{1}{4}, 0.\overline{33} = \dfrac{1}{3},$ and all fractions	Includes integers; all fractions $\dfrac{p}{q}$, where p and q are integers with $q \ne 0$; all repeating and all terminating decimals
Real numbers	$\pi, \sqrt{7}, -\dfrac{4}{7}, 0, -10$ $0.\overline{3} = \dfrac{1}{3}, 1000, \sqrt{15}$	Any number that can be expressed in decimal form Includes the rational numbers
Scientific notation	3.12×10^4 -1.4521×10^{-2} 5×10^9 -3.98715×10^9 1.5987×10^{-6}	Writes a number in the form $c \times 10^n$, where $1 \le |c| < 10$ and n is an integer Used to represent numbers that are large or small in absolute value
Algorithm	Installing software, multiplication, or assembling a bicycle	A finite sequence of well-defined instructions—cannot involve chance or random events

1.1 EXERCISES

   Tape 1

Classifying Numbers

Exercises 1–8: Classify each number as one or more of the following: natural number, integer, rational number, or real number.

1. 400,000 (Minimum cost in dollars to start a McDonald's franchise)

2. $\frac{23}{50}$ (Fraction of 1994 civilian labor force who are female)

3. −13 (Percent change in the release of toxic substances into the water between 1988 and 1993)

4. 1,800,000,000 (Sales in dollars for TLC Beatrice International Holding, Inc.)

5. 7.5 (Average number of gallons of water used each minute while taking a shower)

6. 21.9 (Nielsen rating of the TV show *Home Improvement* during 1993–1994)

7. $90\sqrt{2}$ (Distance in feet from home plate to second base in baseball)

8. −100 (Wind chill when the temperature is −30°F and the wind speed is 40 mph)

Exercises 9 and 10: Classify each number as one or more of the following: natural number, integer, rational number or irrational number.

9. $\pi, -3, \frac{2}{9}, \sqrt{9}, 1.\overline{3}, -\sqrt{2}$

10. $\frac{3}{1}, -\frac{5}{8}, \sqrt{7}, 0.\overline{45}, 0, 5.6 \times 10^3$

Exercises 11–16: For each measured quantity, state the set of numbers that is most appropriate to describe it. Choose from the natural numbers, integers, or rational numbers. Explain your answer.

11. Shoe sizes

12. Populations of states

13. Gallons of gasoline

14. Speed limits

15. Temperatures given in a winter weather forecast in Montana

16. Numbers of compact disc sales

Scientific Notation

Exercises 17–20: Write the number in scientific notation.

17. 186,300 (New lung cancer cases reported in 1995)

18. 71,934 (New AIDS cases reported in New York City during 1994)

19. 0.05333 (Fraction of accidental U.S. deaths attributed to drowning in 1993)

20. 0.001471 (Fraction of U.S. farmland found in New Hampshire in 1994)

Exercises 21–24: Write the number in standard form.

21. 1×10^{-6} (Approximate wavelength in meters of visible light)

22. 9.11×10^{-31} (Weight in kilograms of an electron)

23. 2×10^8 (Number of years for the sun to complete an orbit in our galaxy)

24. 4.86×10^{12} (Federal debt in dollars in March 1995)

Exercises 25–28: Use a calculator to evaluate the expression. Write your result in scientific notation with two significant digits.

25. $(9.87 \times 10^6)(34 \times 10^{11})$

26. $\dfrac{8.947 \times 10^7}{0.00095}$

27. $(56 \times 10^{-7}) + \sqrt{3.4 \times 10^{-2}}$

28. $\sqrt[3]{(2.5 \times 10^{-8})} + 10^{-7}$

Exercises 29–34: Use a calculator to evaluate the expression. Round your result to the nearest thousandth.

29. $\sqrt[3]{192}$

30. $\sqrt{(32 + \pi^3)}$

31. $|\pi - 3.2|$

32. $\dfrac{1.72 - 5.98}{35.6 + 1.02}$

33. $\dfrac{0.3 + 1.5}{5.5 - 1.2}$

34. $3.2(1.1)^2 - 4(1.1) + 2$

Real Data Computation

35. *Consumer Price Index*   (Refer to Example 5.) From 1985 to 1994 the average tuition and fees at private colleges and universities rose from $6,121 to $11,709, while the CPI changed from 107.6 to 148.2. Compare the percent change in tuition and fees to the percent change in the CPI. (Sources: Bureau of the Census, The College Board.)

36. *Size of the Milky Way*   The speed of light is about 186,000 miles per second. The Milky Way galaxy has an approximate diameter of 6×10^{17} miles. Estimate the number of years it takes for light to travel across the Milky Way. (Source: C. Ronan, *The Natural History of the Universe.*)

37. *Federal Debt*   The amount of federal debt has changed dramatically during the past 25 years. (Sources: Department of the Treasury, Bureau of the Census.)
    (a) In 1970 the population of the United States was 203,000,000 and the federal debt was $370 billion. Find the debt per person.
    (b) In 1995 the population of the United States was approximately 260,000,000 and the federal debt was $4.86 trillion. Find the debt per person.
    (c) If this trend were to continue, estimate the debt per person in the year 2020. (*Hint:* There may be more than one way to do this estimate.)

38. *Discharge of Water*   The Amazon River discharges water into the Atlantic Ocean at an average rate of 4,200,000 cubic feet per second, which is the greatest of any river in the world. Is this more or less than 1 cubic mile of water per day? Explain your calculations. (*Hint:* 1 cubic mile = 5280^3 cubic feet.) (Source: *The Guinness Book of Records 1993.*)

39. *Thickness of an Oil Film*   A drop of oil measuring 0.12 cubic centimeter is spilled onto a lake. The oil spreads out in a circular shape having a diameter of 23 centimeters. Approximate the thickness of the oil film. (*Hint:* The thickness of the oil film is equal to its volume divided by its area.)

40. *Thickness of Gold Foil*   A flat, rectangular sheet of gold foil measures 20 centimeters by 30 centimeters and has a mass of 23.16 grams. If one cubic centimeter of gold has a mass of 19.3 grams, find the thickness of the gold foil. (*Hint:* The thickness of the foil is equal to its volume divided by its area.) (Source: U. Haber-Schaim, *Introductory Physical Science.*)

41. *Analyzing Debt*   A one-inch high stack of $100 bills contains about 250 bills. In 1995 the federal debt was approximately 5 trillion dollars.
    (a) If the entire federal debt were converted into a stack of $100 bills, how many feet high would it be?
    (b) The distance between Los Angeles and New York is approximately 2500 miles. Could this stack of $100 bills reach between these two cities?

42. *Toll-Free Numbers*   In 1995 the toll-free 800 telephone numbers that were available for businesses and individuals were nearly depleted. In November 1993, 3.5 million numbers had been assigned and by November 1994, this number reached 5.5 million. The total number could not exceed 7.8 million without making changes. (Source: Database Services Management.)
    (a) Determine how many numbers were still available in November 1994.
    (b) If no changes had been made, estimate the number of months after November 1994, in which toll-free 800 numbers would have been available.

43. *Credit Card Debt*   The table lists total consumer installment debt for selected years. This type of debt includes outstanding credit card balances but not home mortgages. (Source: Federal Reserve Board.)

Year	1975	1980	1985	1990	1994
Debt ($ billions)	$168.7	$302.1	$526.3	$751.9	$925.0

    (a) Discuss the trend in the amount of consumer debt over this time period.
    (b) Find the arithmetic average of the consumer debt in 1980 and 1990. How does this average compare to the actual debt in 1985?

44. *Military Personnel*   The table lists the number of female personnel on active duty in the military for selected years. (Source: Department of Defense.)

Year	Female Personnel
1970	41,479
1980	171,418
1985	211,606
1990	227,018
1994	199,688

    (a) Discuss the change in the number of active-duty female military personnel between 1970 and 1994.
    (b) Use the concept of an arithmetic average to estimate the number of female personnel in 1992.

45. *Alcohol Consumption*   The table lists the average annual U.S. consumption in gallons of alcohol per person (14 years old or over) from 1940 to 1992. (Sources: Department of Health and Human Services, Bureau of the Census.)

Year	1940	1950	1960	1970	1980	1990	1992
Alcohol (gal)	1.56	2.04	2.07	2.52	2.76	2.46	2.31

(a) Discuss the trend in the per capita consumption of alcohol.

(b) In 1990 there were 199,609,000 people 14 years old or over. Estimate the total gallons of alcohol consumed in the United States in 1990. Express this value in scientific notation with two significant digits.

46. *Social Security*   The table lists the predicted percent differences between how much is collected by the Social Security Administration in payroll taxes and how much is paid out to its recipients. A positive percentage indicates that more money is being collected than is being paid out, whereas a negative percentage indicates the opposite. (Source: Social Security Administration.)

Year	1995	2005	2015	2025	2035	2045
Percentage	+1.1	+0.8	−0.2	−3.5	−4.4	−4.2

(a) Describe the trend in this percentage. Is there any cause for concern?

(b) The baby boomers begin to retire in 2013. Interpret how their retirements influence the data in this table.

Executing Algorithms

Exercises 47–50: *Temperature Conversion*   Use Algorithm 1.1 to convert the Celsius temperature to an equivalent Fahrenheit temperature.

47. 20°C

48. 100°C

49. −40°C; What is interesting about your answer?

50. −273°C; This is the coldest temperature possible.

Exercises 51 and 52: *GPA Calculation*   Use Algorithm 1.2 to compute the GPA for the following credits and grades. Round your answer to the nearest hundredth.

51. 15 credits of A, 32 of B, 43 of C, 8 of D, and 2 of F

52. 85 credits of A, 23 of B, 17 of C, 4 of D, and 1 of F

Exercises 53–56: *Body Mass Index*   Many studies have tried to find a recommended relationship between a person's height and weight. The given algorithm computes the body mass index (BMI). This algorithm requires two inputs, height in inches and weight in pounds. It outputs a single real number y. Federal guidelines suggest that $19 \leq y \leq 25$ is desirable. (Source: Associated Press.)

ALGORITHM Finding body mass index

Step 1: Input a person's weight W in pounds and height H in inches.

Step 2: Multiply W by 0.455.

Step 3: Multiply H by 0.0254.

Step 4: Square the result in STEP 3.

Step 5: Divide the answer in STEP 2 by the answer in STEP 4.
Let the result be y.

Step 6: Output y, the BMI.

Compute the BMI for each individual.

53. 300 pounds, 7 feet 1 inch (Shaquille O'Neal, professional basketball player) (Source: The Topps Company, Inc.)

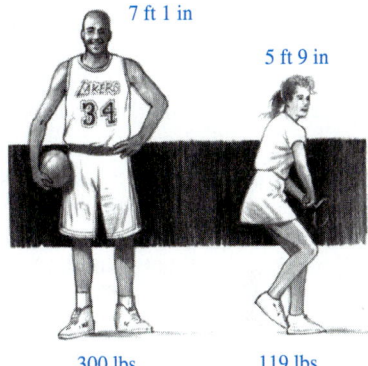

7 ft 1 in

5 ft 9 in

300 lbs

119 lbs

54. 119 pounds, 5 feet 9 inches (Steffi Graf, professional tennis player) (Source: J. Monroe, *Steffi Graf.*)

55. 153 pounds, 5 feet 10 inches (Jackie Joyner-Kersee, track and field athlete) (Source: M. Goldstein and J. Larson, *Jackie Joyner-Kersee Superwoman.*)

56. 195 pounds, 6 feet 1 inch (Deion Sanders, professional athlete) (Source: The Topps Company, Inc.)

Exercises 57–60: *Indoor Air Pollution Risk*   Indoor air pollution, such as tobacco smoke, formaldehyde in building materials, radon, and asbestos, presents a serious health risk for many Americans. Researchers often estimate the lifetime individual risk R for a particular pollutant. If R is known, then the given algorithm can be used to estimate the number of deaths caused by a pollutant each year in the United States. (Source: A. Hines, *Indoor Air Quality & Control.*)

ALGORITHM Estimating risk from indoor air pollution

Step 1: Input R.

Step 2: Divide R by 72, the average life expectancy in the United States.
Let the result be A.

Step 3: Multiply A times 260 million, the approximate population of the United States. Let the result be y.

Step 4: Output y, the estimated number of annual deaths due to an indoor air pollutant with a lifetime individual risk R.

57. *Radon Gas Risk*   Radon gas occurs naturally in the soil when uranium disintegrates into lead. It enters homes through cracks in basement walls and floors. It can cause lung cancer. The Environmental Protection Agency estimates that $R = 3.9 \times 10^{-3}$ for radon gas. Estimate the number of deaths caused by radon gas annually in the United States.

58. *Passive Smoking Risk*   Environmental tobacco smoke has recently become a concern for non-smokers. Some studies estimate that $R = 2 \times 10^{-3}$ for environmental tobacco smoke in the United States. Estimate the number of deaths caused by environmental tobacco smoke annually.

59. *Smoking Risk*   Smoking increases a person's risk of lung cancer dramatically. For smokers it is estimated that $R = 0.59$. Currently, there are approximately 50 million people who smoke. Modify the above algorithm to estimate the number of annual deaths caused by smoking.

60. *Formaldehyde Risk*   Exposure to formaldehyde is blamed for approximately 7000 deaths in the United States annually. Estimate the lifetime individual risk R for formaldehyde.

Writing Algorithms

61. An algorithm is written to calculate the area of a square. State what this algorithm will need for input.

62. An algorithm is written to calculate the gas mileage for an automobile. State what this algorithm will need for input.

63. Suppose that tuition is $95.25 per semester credit and student fees are fixed at $31.50 per semester for each student. Write an algorithm to calculate the tuition and fees for taking x credits.

64. Write an algorithm to calculate the area of a circle in square feet given the radius x in inches. Your algorithm should receive x as input and output the area y.

65. Write an algorithm that will calculate the sale price of an item whose regular price is reduced by $x\%$. Your algorithm should receive the regular price P and the percentage discount x as input. Output the sale price y.

66. Write an algorithm that will compute the number of fluid ounces held in a cylindrical can having a radius r and height h measured in inches. Your algorithm should receive r and h as input, and output the number of fluid ounces y. (*Hint:* $V = \pi r^2 h$; 1 cubic inch $= \dfrac{128}{231}$ fluid ounce.)

Writing about Mathematics

1. Describe some basic sets of numbers that are used in mathematics.

2. Explain what an algorithm is. Give one example of a step-by-step procedure that is an algorithm and one that is not an algorithm.

1.2 Visualization of Data

One-Variable Data • Two-Variable Data • Visualizing Data with a Graphing Calculator

Introduction

Computer technology, the Internet, and electronic communication are creating large amounts of data. The challenge is to convert this data into meaningful information that can be used to solve important problems and create new knowledge. Before conclusions can be drawn about data, it must be analyzed. A powerful tool in this step is visualization. Visual presentations have the capability of communicating vast quantities of information in short periods of time. A typical page of graphics contains one hundred times more information than a page of text. The map in Figure 1.5 shows the annual average precipitation in the United States.

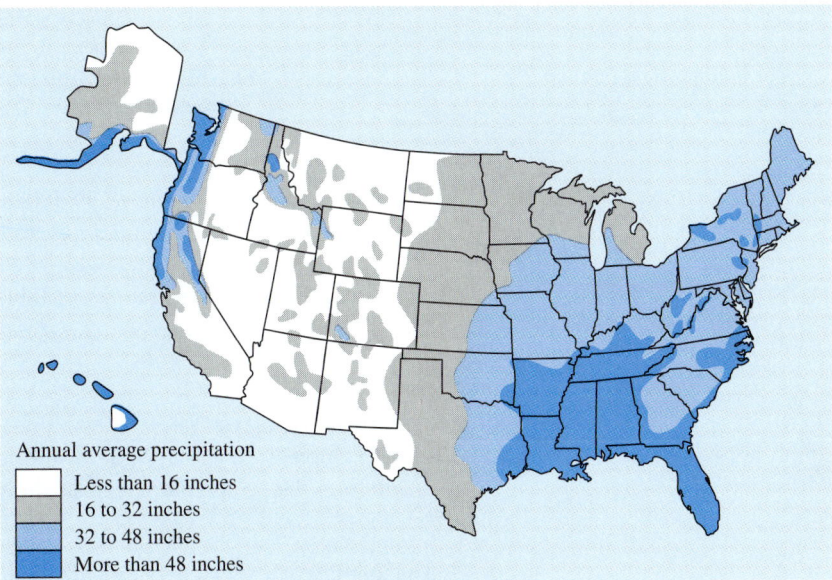

FIGURE 1.5   Source: J. Williams, *The Weather Almanac 1995.*

Imagine trying to describe this map using only words. It would be an enormous task, and the final result would not be nearly as clear as the visual map. Through visualization, it is possible to quickly recognize important trends and draw conclusions. This section discusses how different types of data can be visualized using various mathematical techniques.

One-Variable Data

Data often occurs in the form of a list. A list of test scores without names is an example. The only variable involved is the score. Data of this type, involving only one variable, is referred to as **one-variable data.** If the values in a list are unique, they can be represented visually on a number line.

EXAMPLE 1   *Determining the maximum, minimum, and average of a list of data*

Table 1.5 lists the monthly average temperatures in degrees Fahrenheit at Mould Bay, Canada. (Source: A. Miller and J. Thompson, *Elements of Meteorology.*)

TABLE 1.5												
Temperature (°F)	−27	−31	−26	−9	12	32	39	36	21	1	−17	−24

(a) Plot these temperatures on a number line.
(b) Find the maximum and minimum temperatures.
(c) Determine the average of these twelve temperatures.

Solution

(a) In Figure 1.6 on the next page, the numbers −27, −31, −26, −9, 12, 32, 39, 36, 21, 1, −17, and −24 are plotted on a number line.

FIGURE 1.6 Monthly Average Temperatures

(b) The maximum temperature of 39°F is plotted farthest to the right in Figure 1.6. Similarly, the minimum temperature of −31°F is plotted farthest to the left.

(c) The sum of the 12 temperatures in Table 1.5 equals 7. The average of these temperatures is $\frac{7}{12} \approx 0.6°F$. ■

Critical Thinking

In Example 1(c), the average of the temperatures is approximately 0.6°F. Interpret this temperature. Explain your reasoning.

The **median** of a sorted list is equal to the value that is located in the middle of the list. If there is an odd number of data items, the median is the middle data item. If there is an even number of data items, the median is the average of the two middle items. The **range** of a list of data is the difference between the maximum and minimum values. The range is a measure of the *spread* or *dispersion* in the data. Averages, medians, and ranges can be found for one-variable data sets.

EXAMPLE 2   *Determining the median and range for a list of data*

The monthly average precipitations in inches for Seattle are 5.7, 4.2, 3.7, 2.4, 1.7, 1.4, 0.8, 1.1, 1.9, 3.5, 5.9, and 5.9. (Source: J. Williams.)
(a) Sort this data in increasing order and display it using a table.
(b) Find the median and the range of this data set.
(c) Interpret the median and the range.

Solution

(a) The sorted data is displayed in Table 1.6.

TABLE 1.6

Precipitation (in.)	0.8	1.1	1.4	1.7	1.9	2.4	3.5	3.7	4.2	5.7	5.9	5.9

(b) Since there is an even number of data items, the median is the average of the two middle values, 2.4 and 3.5. This is $\frac{2.4 + 3.5}{2} = 2.95$. The largest value is 5.9 and the smallest value is 0.8. The range is equal to their difference, $5.9 − 0.8 = 5.1$.

(c) In Seattle, half the months have an average precipitation of less than 2.95 inches (the median) and half the months have an average precipitation of greater than 2.95 inches. For any two months, the average precipitations vary by at most 5.1 inches (the range). ◼

Technology Note

Some graphing calculators can analyze one-variable data using a *box plot* or *box-and-whisker plot.* Figure 1.7 shows a box plot for the data in Example 2. The median is given as 2.95. Other information can be obtained from this type of plot. Consult your manual if you are interested.

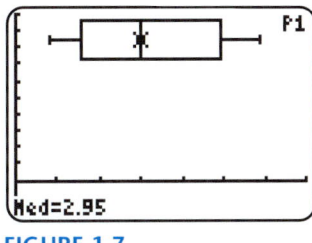

FIGURE 1.7

Two-Variable Data

It may be possible for a relationship to exist between two lists of data. In Example 2, each monthly average precipitation could be associated with a month. By forming a relationship between the month and the precipitation, more information is communicated. Table 1.7 lists the monthly average precipitation in inches for Portland, Oregon. In this table, 1 corresponds to January, 2 to February, and so on, until 12 represents December. Showing the relationship between a month and its average precipitation is accomplished by joining the two lists together, so that corresponding months and precipitations are visually paired.

TABLE 1.7												
Month	1	2	3	4	5	6	7	8	9	10	11	12
Precipitation (in.)	6.2	3.9	3.6	2.3	2.0	1.5	0.5	1.1	1.6	3.1	5.2	6.4

If x is the month and y is the precipitation, then the ordered pair (x, y) represents the average amount of precipitation y during the month x. For example, the ordered pair $(5, 2.0)$ indicates that in May the average precipitation is 2.0 inches, whereas the ordered pair $(2, 3.9)$ indicates that the average precipitation in February is 3.9 inches. *Order is important* in an ordered pair.

Since the data Table 1.7 involves two variables, the month and precipitation, we refer to it as **two-variable data.** It is important to realize that a relation established by two-variable data is between two lists rather than within a single list. January is not related to August and a precipitation of 6.2 inches is not associated with 1.1 inches. Instead January is paired with 6.2 inches, while August is paired with 1.1 inches. We now define the mathematical concept of a relation.

> ### Relation
>
> A **relation** is a set of ordered pairs.

If we denote the ordered pairs in a relation by (x, y), then the set of all x-values is called the **domain** of the relation and the set of all y-values is called the **range.** In Table 1.7 the domain is

$$D = \{1, 2, 3, 4, 5, 6, 7, 8, 9, 10, 11, 12\},$$

and the range is

$$R = \{6.2, 3.9, 3.6, 2.3, 2.0, 1.5, 0.5, 1.1, 1.6, 3.1, 5.2, 6.4\}.$$

Notice that the word "range" has different meanings for one-variable and two-variable data. For one-variable data, the range is a *real number* that describes the spread of the data in a single list. For two-variable data, the range is a *set of numbers* associated with a set of ordered pairs. As is often the case in English, we can determine the correct meaning of "range" from the context in which it is used.

To visualize two-variable data, a number line for each list of data is needed. We often use the **Cartesian coordinate plane** or *xy***-plane.** The horizontal axis is the *x***-axis** and the vertical axis is the *y***-axis.** The axes intersect at the **origin** and determine four regions called **quadrants,** numbered I, II, III, and IV, counterclockwise as shown in Figure 1.8. We can plot the ordered pair (x, y) using the x- and y-axes. The point $(1, 2)$ is located in quadrant I, $(-2, 3)$ in quadrant II, $(-4, -4)$ in quadrant III, and $(1, -2)$ in quadrant IV. A point lying on a coordinate axis does not belong to any quadrant. The point $(-3, 0)$ is located on the x-axis, whereas the point $(0, -3)$ lies on the y-axis.

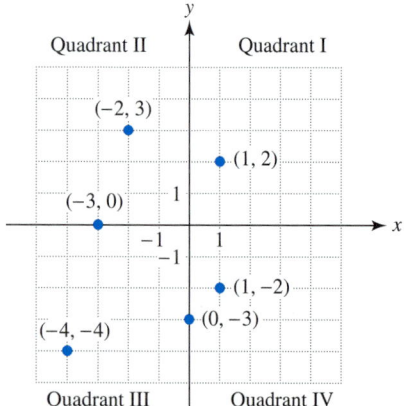

FIGURE 1.8 The *xy*-plane

The term **scatterplot** is given to a graph in the *xy*-plane, where distinct data points are plotted. A scatterplot is shown in Figure 1.9.

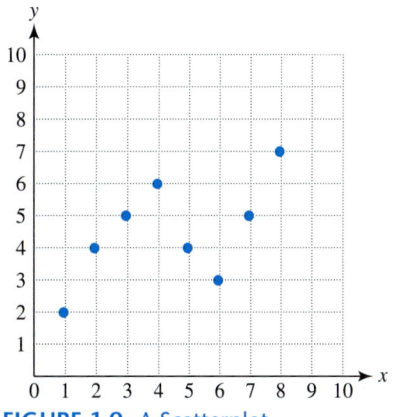

FIGURE 1.9 A Scatterplot

EXAMPLE 3   *Making scatterplots*

Use Table 1.7 to make a scatterplot of the monthly average precipitations in Portland, Oregon.

Solution

Use the *x*-axis for the months and the *y*-axis for the precipitation amounts. To make a scatterplot, simply graph the ordered pairs $(1, 6.2)$, $(2, 3.9)$, $(3, 3.6)$, $(4, 2.3)$, $(5, 2.0)$, $(6, 1.5)$, $(7, 0.5)$, $(8, 1.1)$, $(9, 1.6)$, $(10, 3.1)$, $(11, 5.2)$ and $(12, 6.4)$ in the *xy*-plane as shown in Figure 1.10.

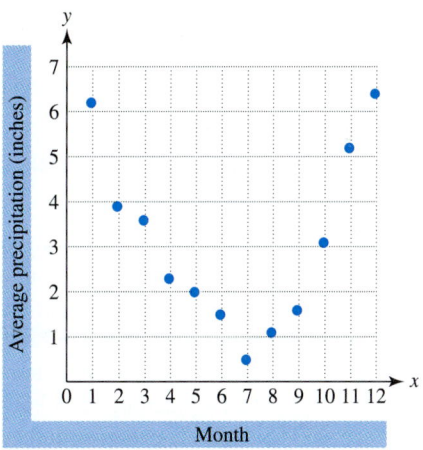

FIGURE 1.10 Monthly Average Precipitation

Sometimes it is helpful to connect the data points in a scatterplot with straight-line segments. This type of graph visually emphasizes changes in the data. It is called a **line graph.**

EXAMPLE 4   *Interpreting a line graph*

The line graph in Figure 1.11 on the following page shows the number of college graduates in thousands at ten-year intervals between 1900 and 1990. (Sources: Department of Education, Center for Educational Statistics.)

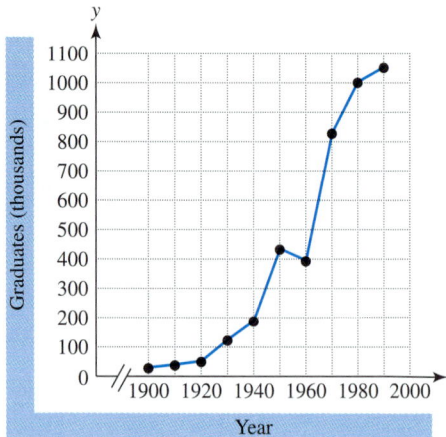

FIGURE 1.11 College Graduates

(a) Did the number of graduates ever decrease during this time period? Explain.
(b) Approximate the number of college graduates in the year 1970.
(c) Determine the ten-year period when there was the greatest increase in the number of college graduates.

Solution

(a) Yes, the number decreased slightly between 1950 and 1960. For this time period, the line segment slopes downward from left to right.
(b) According to the graph, the number of college graduates in 1970 was approximately 830,000.
(c) The greatest increase corresponds to the steepest line segment. This occurs between 1960 and 1970. ∎

Visualizing Data with a Graphing Calculator

Graphing calculators provide several features beyond those found on scientific calculators. The bottom rows of keys on graphing calculators are often similar to those found on scientific calculators. However, graphing calculators have additional keys that can be used to create tables, scatterplots, line graphs, and other types of graphs. Another difference between graphing and scientific calculators is that a graphing calculator has a larger screen along with graphing capabilities.

The *viewing rectangle* or *window* on a graphing calculator is similar to the view finder in a camera. A camera cannot take a picture of an entire scene. It must be centered on some object, photographing only a portion of the available scenery. A camera can capture different views of the same scene by zooming in and out. Graphing calculators have similar capabilities. The *xy*-plane is infinite. The calculator screen can show only a finite, rectangular region of the *xy*-plane. The viewing rectangle must be specified by setting minimum and maximum values for both the *x*- and *y*-axes before a graph can be drawn. Graphing calculators also can zoom in and out. Zooming in shows more detail in a smaller region of a graph, whereas zooming out gives a better overall picture of the graph.

We will use the following terminology regarding the size of a viewing rectangle. **Xmin** is the minimum *x*-value and **Xmax** is the maximum *x*-value along the *x*-axis. Similarly, **Ymin** is the minimum *y*-value and **Ymax** is the maximum

y-value along the y-axis. Most graphs show an x-scale and y-scale using tick marks on the respective axes. Sometimes the distance between consecutive tick marks is one unit, while other times it might be five units. The distance represented by consecutive tick marks on the x-axis is called **Xscl,** while the distance represented by consecutive tick marks on the y-axis is called **Yscl.** See Figure 1.12. This information about the viewing rectangle can be written concisely as [Xmin, Xmax, Xscl] by [Ymin, Ymax, Yscl]. For example, $[-10, 10, 1]$ by $[-10, 10, 1]$ means that Xmin $= -10$, Xmax $= 10$, Xscl $= 1$, Ymin $= -10$, Ymax $= 10$, and Yscl $= 1$. This setting is commonly referred to as the *standard viewing rectangle.*

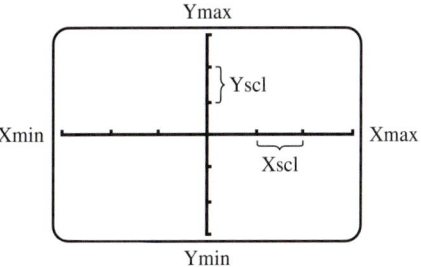

FIGURE 1.12

EXAMPLE 5 *Setting the viewing rectangle*

Show the standard viewing rectangle and the viewing rectangle $[-2, 3, 0.5]$ by $[-100, 200, 50]$ on your calculator.

Solution

The window settings and viewing rectangles are displayed in Figures 1.13–1.16. Notice that in Figure 1.14, there are 10 tick marks on the positive x-axis, since its length is 10 and the distance between consecutive tick marks is 1.

FIGURE 1.13

$[-10, 10, 1]$ by $[-10, 10, 1]$

FIGURE 1.14

FIGURE 1.15

$[-2, 3, 0.5]$ by $[-100, 200, 50]$

FIGURE 1.16 ■

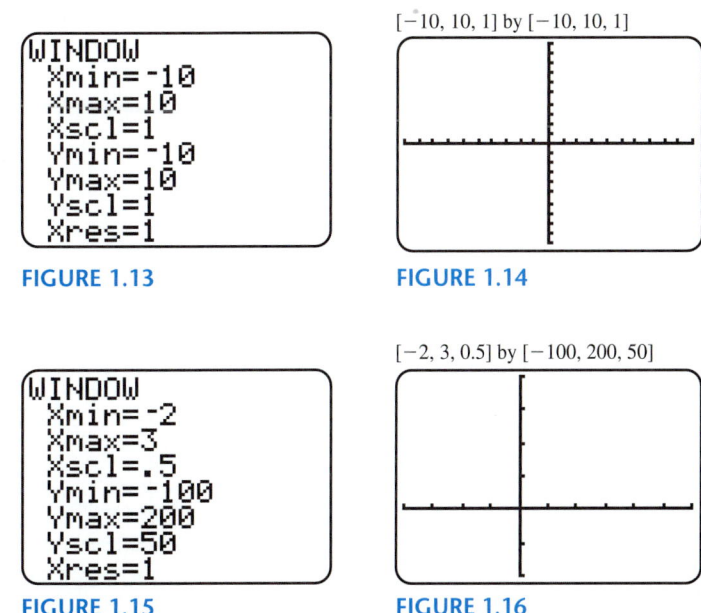

EXAMPLE 6   *Making a scatterplot with a graphing calculator*

Plot the points $(-5, -5)$, $(-2, 3)$, $(1, -7)$, and $(4, 8)$ in the standard viewing rectangle.

Solution

The standard viewing rectangle is $[-10, 10, 1]$ by $[-10, 10, 1]$. The points $(-5, -5)$, $(-2, 3)$, $(1, -7)$, and $(4, 8)$ are plotted in Figure 1.17.

$[-10, 10, 1]$ by $[-10, 10, 1]$

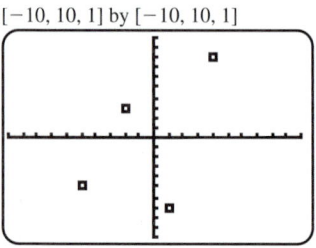

FIGURE 1.17

EXAMPLE 7   *Creating a line graph with a graphing calculator*

Table 1.8 lists the number of compact discs (CDs) in millions sold between 1987 and 1994. Make a line graph of these sales in the viewing rectangle $[1986, 1995, 1]$ by $[0, 700, 100]$.

TABLE 1.8

Year	1987	1988	1989	1990	1991	1992	1993	1994
CDs (millions)	102.1	149.7	207.2	286.5	333.3	407.5	495.4	662.1

Source: Recording Industry Association of America.

Solution

Enter the points $(1987, 102.1)$, $(1988, 149.7)$, $(1989, 207.2)$, $(1990, 286.5)$, $(1991, 333.3)$, $(1992, 407.5)$, $(1993, 495.4)$, and $(1994, 662.1)$. A line graph can be created by selecting this option on your graphing calculator. The graph is shown in Figure 1.18.

$[1986, 1995, 1]$ by $[0, 700, 100]$

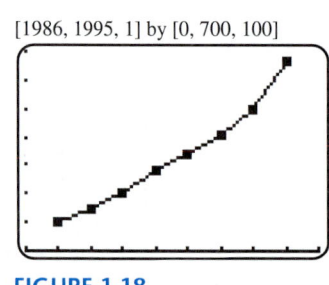

FIGURE 1.18

1.2 PUTTING IT ALL TOGETHER

The following table summarizes some concepts found in this section.

Type of Data	Methods of Visualization	Comments
One-variable data	Number line, list, one-column or one-row table	The data items are the same type, and can be described using x-values Computations of averages, medians, and ranges are performed on one-variable data
Two-variable data	Two-column or two-row table, scatterplot, line graph, or other types of graphs in the xy-plane	Two types of data are related, and can be described using ordered pairs (x, y) A relation is a set of ordered pairs (x, y), where the set of x-values is the domain and the set of y-values is the range

1.2 EXERCISES   Tape 1

Data Involving One Variable

Exercises 1–4: Sort each list of numbers from smallest to largest and display them in a table.
  (a) *Determine the maximum and minimum values.*
  (b) *Calculate the average, median, and range. Round each result to the nearest hundredth when appropriate.*

1. $-10, 25, 15, -30, 55, 61, -30, 45, 5$
2. $-1.25, 4.75, -3.5, 1.5, 2.5, 4.75, 1.5$
3. $\sqrt{15}, 2^{2.3}, \sqrt[3]{69}, \pi^2, 2^{\pi}, 4.1$
4. $\frac{22}{7}, 3.14, \sqrt[3]{28}, \sqrt{9.4}, 4^{0.9}, 3^{1.2}$

Exercises 5–8: Geography Each set of numbers contains data about geographic features of the world.
  (a) *Plot the numbers on a number line.*
  (b) *Calculate the average, median, and range for the set of numbers. Express each answer using three significant digits. Interpret your results.*
  (c) *Try to identify the geographic feature associated with the largest number in each set.*

5. {840, 87.8, 227, 280, 84.2, 196, 306, 165} (Areas of largest islands in thousands of square miles) (Source: National Geographic Society.)

6. {11.2, 9.91, 31.7, 22.3, 12.3, 26.8, 12.2, 12.0, 11.0, 24.9, 23.0} (Areas of largest freshwater lakes in thousands of square miles) (Source: U.S. National Oceanic and Atmospheric Administration.)

7. {19.3, 18.5, 29.0, 7.31, 16.1, 22.8, 20.3} (Highest elevations of the continents in thousands of feet) (Source: *National Geographic Atlas of the World.*)

8. {4145, 3720, 3360, 3740, 3990, 3650, 3590, 3030} (Lengths of longest rivers in miles) (Source: U.S. National Oceanic and Atmospheric Administration.)

9. *Baseball Salaries* In 1995 the average major league baseball salary was $1,168,263, while the minimum salary was $109,000. Suppose that one player earned $6,465,000, while five other players earned the minimum salary. (Source: Associated Press.)
  (a) Determine the average salary for the six players.
  (b) Find their median salary.
  (c) Discuss why "average salary" can be misleading.

10. *Tennis Earnings* Martina Navratilova had the following earnings for the first six years of her professional tennis career: $173,668, $128,535,

$300,317, $450,757, $747,548, and $749,250. (Source: J. Leder, *Martina Navratilova.*)

(a) Determine her average and median winnings during this time period.
(b) Find the range of these earnings.

11. *Airline Complaints*   The table lists the number of baggage complaints per 1000 passengers for ten popular airlines in November 1995. (Source: Department of Transportation.)

Complaints	5.78	5.55	4.60	4.95	5.12
	4.14	5.63	3.78	5.91	5.27

(a) Find the range of these values.
(b) If each airline surveyed 1,000,000 passengers, approximate the maximum and minimum number of complaints.

12. *Traffic Deaths and Alcohol*   The numbers of alcohol-related traffic deaths in the United States from 1990 to 1995 are listed in the table. Compare the average and median numbers of deaths. (Source: Mothers Against Drunk Driving.)

Deaths	22,084	19,887	17,859	17,473	16,589	17,274

Data Involving Two Variables

Exercises 13–18: Complete the following.
(a) *Find the domain and range of the relation.*
(b) *Determine the maximum and minimum of the x-values; of the y-values.*
(c) *Label appropriate scales on the xy-axes.*
(d) *Plot the relation by hand in the xy-plane.*

13. $\{(0, 5), (-3, 4), (-2, -5), (7, -3), (0, 0)\}$
14. $\{(1, 1), (3, 0), (-5, -5), (8, -2), (0, 3)\}$
15. $\{(10, 50), (-35, 45), (20, -55), (75, 25), (-40, 60), (-25, -25)\}$
16. $\{(11, 15), (2, -13), (-5, -14), (-7, 19), (-17, -4), (-1, 1), (5, -11)\}$
17. $\{(0.1, -0.3), (0.5, 0.4), (-0.7, 0.5), (0.8, -0.1), (0.9, 0.9)\}$
18. $\{(-1.2, 1.5), (1.0, 0.5), (-0.3, 1.1), (-0.8, -1.3), (1.4, -1.6)\}$

Exercises 19–24: Show the viewing rectangle on your graphing calculator. Predict the number of tick marks on the positive x-axis and the positive y-axis.

19. Standard viewing rectangle
20. $[-12, 12, 2]$ by $[-8, 8, 2]$
21. $[0, 100, 10]$ by $[-50, 50, 10]$
22. $[-30, 30, 5]$ by $[-20, 20, 5]$
23. $[1980, 1995, 1]$ by $[12,000, 16,000, 1000]$
24. $[1900, 1990, 10]$ by $[1650, 28,650, 1000]$

Exercises 25–28: Match the settings for a viewing rectangle with the correct figure.

25. $[-5, 5, 1]$ by $[-5, 5, 1]$
26. $[-5, 5, 1]$ by $[-2, 2, 1]$
27 $[-100, 100, 50]$ by $[-100, 100, 10]$
28. $[-2, 8, 1]$ by $[-3, 7, 1]$

(a)

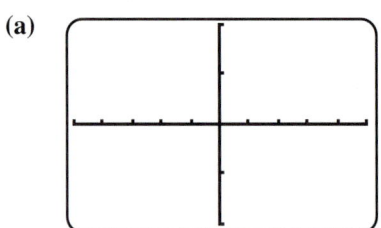

(b)

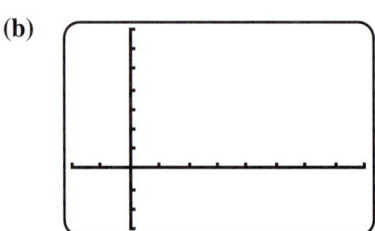

(c)

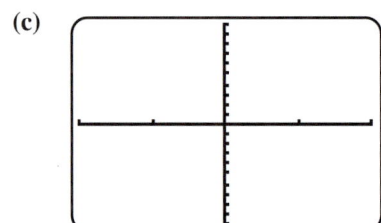

(d)

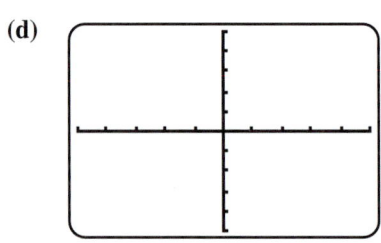

Exercises 29–34: Make a scatterplot of the relation.

29. $\{(4, 3), (-2, 1), (-3, -3), (5, -2)\}$

30. $\{(5, 5), (2, 0), (-2, 7), (2, -8), (-1, -5)\}$

31. $\{(20, 40), (-25, -15), (-20, 25), (15, -25)\}$

32. $\{(-13, 12), (3, 10), (-15, -4), (12, -9)\}$

33. $\{(3.1, 6.2), (-5.1, 10.1), (-0.7, -1.4), (1.8, 3.6), (-4.9, -9.8)\}$

34. $\{(-1.2, 0.6), (1.0, -0.5), (-0.4, 0.2), (-2.8, 1.4), (2.8, -1.4)\}$

Exercises 35–38: Each table contains real data involving two variables.

   (a) *Determine the maximum and minimum values for each variable in the table.*

   (b) *Use your results from part (a) to determine an appropriate viewing rectangle.*

   (c) *Make a scatterplot of the data.*

   (d) *Make a line graph of the data.*

35. Sales of cassette tapes in millions of dollars

x (year)	1988	1989	1990	1991	1992	1993	1994
y (sales)	3385	3345	3472	3020	3116	2916	2976

Source: Recording Industry Association of America.

36. Sales of compact discs in millions of dollars

x (year)	1988	1989	1990	1991	1992	1993	1994
y (sales)	2090	2588	3452	4338	5327	6511	8465

Source: Recording Industry Association of America.

37. Projected population for Asian-Americans in millions

x (year)	1996	1998	2000	2002	2004
y (population)	9.7	10.5	11.2	12.0	12.8

Source: U.S. Census projections.

38. Estimated usage of the Internet in millions of users

x (year)	1989	1991	1993	1995
y (users)	1.6	7.5	20.1	49.6

Source: The Internet Society.

39. Refer to Exercises 35 and 36. Discuss the trends in cassette and CD sales from 1988 to 1994. What would be your prediction for the future sales of cassette tapes and CDs?

40. Refer to Exercise 37. Make a prediction of the Asian-American population in the year 2006.

Exercises 41 and 42: Graduate Degrees *Create a table that contains the same data as the line graph.*

41. Numbers of doctorate degrees in thousands conferred in the United States from 1950 to 1995

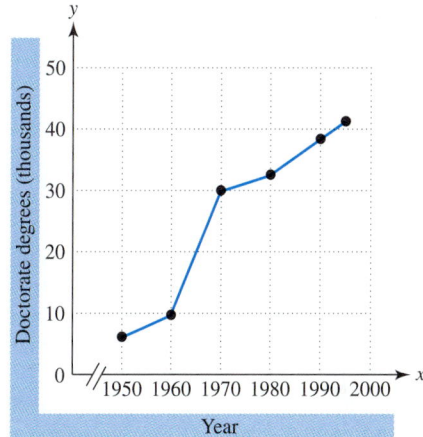

Source: Department of Education.

42. Numbers of master's degrees in thousands conferred in the United States from 1950 to 1995

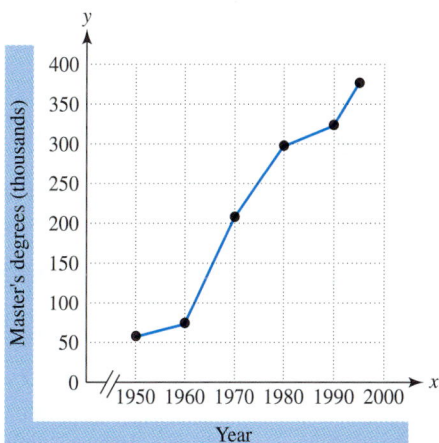

Source: Department of Education.

43. *Handguns* The line graph on the next page shows the number of handguns in millions manufactured in the United States annually from 1981 to 1991.

   (a) Has the number of handguns manufactured increased each year? Explain.

   (b) Determine the years in which the largest three-year increase in handgun production took place.

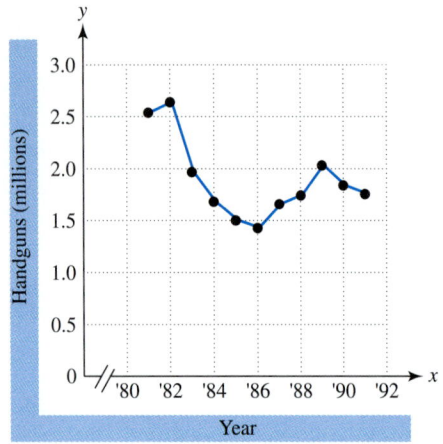

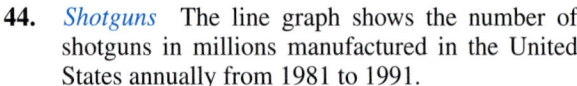

Source: Bureau of Alcohol, Tobacco, and Firearms.

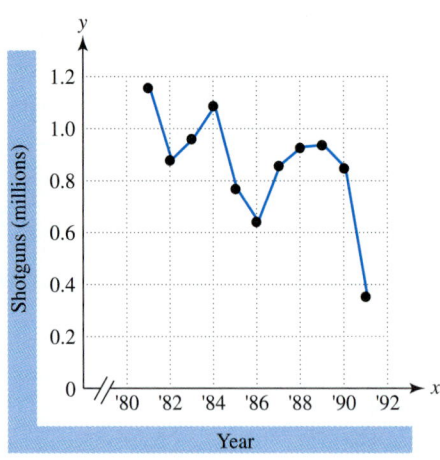

Source: Bureau of Alcohol, Tobacco, and Firearms.

44. *Shotguns* The line graph shows the number of shotguns in millions manufactured in the United States annually from 1981 to 1991.
  (a) When did the largest yearly decrease in the manufacture of shotguns take place?
  (b) When did the largest three-year decrease in shotguns occur?

Writing about Mathematics

1. Describe how one- and two-variable data can be represented. Give one example of each type of data.

2. If a person works at an hourly rate of R dollars for H hours, then the pay P is computed by the equation $P = R \cdot H$. Discuss ways to represent the pay P for various values of R and H using a table. Give examples.

CHECKING BASIC CONCEPTS FOR SECTIONS 1.1 AND 1.2

1. Approximate each expression to three significant digits.
  (a) $\sqrt{4.2(23.1 + 0.5^3)}$
  (b) $\dfrac{23 + 44}{85.1 - 32.9}$

2. Write each number using scientific notation.
  (a) 348,500,000
  (b) -1237.4
  (c) 0.00198

3. Write an algorithm that outputs the number of seconds y in x years.

4. The average depths in feet of the four oceans are 13,215, 12,881, 13,002, and 3953. Calculate the average, median, and range of these depths. Interpret your results.

5. Make a scatterplot and a line graph with the points $(-5, -4)$, $(-1, 2)$, $(2, -2)$, and $(3, 6)$. State the quadrant in which each point lies.

1.3 Functions and Their Representations

Algorithms and Functions • Representations of Functions • Formal Definition of a Function • Graphing Calculators and Functions • Identifying Functions

Introduction

Any concept can be made more or less complicated by how it is represented. Many advances throughout history were the result of inventing new representations. For example, the theory of germs provided a new representation of disease that greatly improved medical science.

Similar events have occurred in mathematics. Historically, fractions were developed much earlier than decimals. The introduction of decimals greatly simplified arithmetic involving fractions. For example, sums like $\frac{1}{2} + \frac{1}{4}$ caused difficulty for people because of unlike denominators. However, their decimal equivalents could be easily combined as $0.50 + 0.25 = 0.75$. By developing a new representation, an old problem was made easier.

In this section the concept of a function is introduced. A function has different representations that are closely related to both algorithms and relations. We begin by discussing a special type of algorithm that produces exactly one output for each valid input. (Reference: L. Motz and J. Weaver, *The Story of Mathematics*.)

Algorithms and Functions

An algorithm receives data as input and then performs a finite sequence of instructions. The result is called the output. This can be represented visually by the following diagram.

$$\text{Input Data} \quad \rightarrow \quad \boxed{\text{Algorithm}} \quad \rightarrow \quad \text{Output Data}$$

As an example, consider Algorithm 1.3 that converts x yards to y feet.

ALGORITHM 1.3 Converting yards to feet
Step 1: Input x, the number of yards.
Step 2: Multiply x by 3. Let the result be y.
Step 3: Output y, the equivalent number of feet.

This algorithm establishes a relation between two sets of numbers. Each valid input x in yards determines *exactly one* output y in feet, which can be represented by the ordered pairs (x, y). If an algorithm produces *exactly one output for each valid input,* then the algorithm *computes a function.* Algorithm 1.3 computes a function that converts yards to feet. We say *y is a function of x* because the output y is determined by and depends on the input x. As a result, y is called the **dependent variable,** and x is the **independent variable.** Since Algorithm 1.3 requires only one input, it computes a **function of one input** or a **function of one variable.** To emphasize that y is a function of x, the notation $y = f(x)$ is used. The symbol $f(x)$ does not represent multiplication of a variable f and a variable x. The notation $y = f(x)$ is called **function notation,** is read "y equals f of x," and denotes that function f with input x produces output y. The function computed by Algorithm 1.3 can be summarized by the following diagram.

$$\text{Input } x \text{ yards} \quad \rightarrow \quad \boxed{\text{Compute } y = f(x) \text{ using function } f \text{ with input } x} \quad \rightarrow \quad \text{Output } y \text{ feet}$$

For example, if $x = 4$ yards, then $y = f(4) = 12$ feet. The input is 4 and the output is 12. Similarly, if $x = 5$, then $y = f(5) = 15$.

The set of valid or meaningful inputs x is called the **domain** of the function and the set of corresponding outputs y is the **range.** For example, suppose a function f computes the height of a ball thrown into the air after x seconds. Then the domain of f might consist of all times while the ball was in flight, and the range would include all heights attained by the ball.

The following can be computed by functions because they result in one output for each valid input.

- Calculating the square of a number x
- Finding the sale price of an item discounted 25% with regular price x
- Naming the biological mother of person x

At first, one might think that every algorithm computes a function, but the next example exhibits an algorithm that does not compute a function.

EXAMPLE 1   *Determining whether an algorithm is a function*

Algorithm 1.4 outputs the students in a class having a particular birthday. Determine if it computes a function.

ALGORITHM 1.4 Listing students in a class with a particular birthday
Step 1:   Input x, a day of the year.
Step 2:   Output the name of each student in the class with birthday x.

Solution

For an algorithm to compute a function, each meaningful input must produce *exactly one* output. Algorithm 1.4 does not compute a function, because it is possible to have a day (input) that is the birthday for more than one student.   ■

Other examples of algorithms that do not represent functions are listed below. In each case, there are inputs that may produce more than one output.

- Naming the children of parent x
- Listing the biological grandparents of person x
- Finding all prime numbers between 1 and x

Representations of Functions

A function f forms a relation between inputs x and outputs y that can be expressed as a set of ordered pairs (x, y). As is the case with any relation, the set of x-values is called the domain of f and the set of y-values is called the range of f. A function involves two-variable data and can be represented using verbal descriptions, tables, diagrams, symbols, and graphs.

Verbal Representation.   Algorithm 1.3 can be thought of as a *verbal representation* of a function. Another verbal representation for this function would be "Multiply x yards by 3 to obtain y feet."

Numerical Representation.   A function that converts yards to feet is shown in Table 1.9. A table might be appropriate for someone who has not learned how to multiply.

TABLE 1.9							
Yards	1	2	3	4	5	6	7
Feet	3	6	9	12	15	18	21

A table is an example of a *numerical representation* of a function because it lists numeric values associated with the function. One difficulty with a numerical representation is that it is often either inconvenient or impossible to list all possible inputs x. For this reason we sometimes refer to a table of this type as a *partial numerical representation* as opposed to a *complete numerical representation,* which would include all elements from the domain of a function. For example, many valid inputs do not appear in Table 1.9, such as $x = 10$ or $x = 0.75$.

Diagrammatic Representation.   Functions are sometimes represented using *diagrammatic representations* or *diagrams.* Figure 1.19 is a diagram of a function with domain $D = \{1, 2, 3, 4\}$ and range $R = \{3, 6, 9, 12\}$. An arrow is used to show that input x produces output y. For example, input 2 results in output 6 or $f(2) = 6$. On the other hand, Figure 1.20 shows a relation, but not a function, because input 2 results in two different outputs, 5 and 6. Other types of diagrams also are possible to illustrate functions.

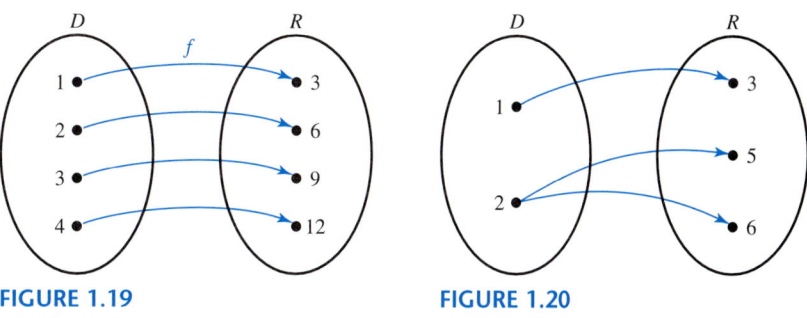

FIGURE 1.19   **FIGURE 1.20**

Symbolic Representation.   A formula provides a *symbolic representation* of a function. The computation performed by Algorithm 1.3 is expressed by the equation $y = 3x$. If f is this function, then the formula becomes $f(x) = 3x$, where $f(x)$ equals y. We say that function f is *represented by* or *given by* $f(x) = 3x$. It follows that $f(6) = 3 \cdot 6 = 18$.

Similarly, if a function g computes the square of a number x, then g can be represented by $g(x) = x^2$. A formula is an efficient but less visual way to define a function.

Graphical Representation.   Leonhard Euler (1707–1783) invented the function notation $f(x)$. He also was the first to allow a function to be represented by a graph, rather than only by a symbolic expression.

A *graphical representation* or *graph* visually associates an x-input with a y-output. In a graph of a function, the ordered pairs (x, y) are plotted in the xy-plane. The ordered pairs

$$(1, 3), (2, 6), (3, 9), (4, 12), (5, 15), (6, 18), \text{ and } (7, 21)$$

from Table 1.9 are plotted in Figure 1.21 on the next page. This scatterplot suggests a line for the graph of the function that converts yards to feet.

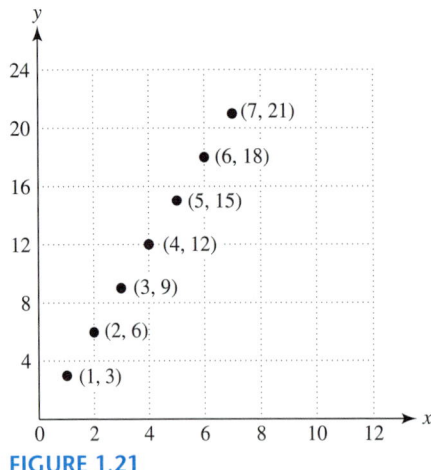

FIGURE 1.21

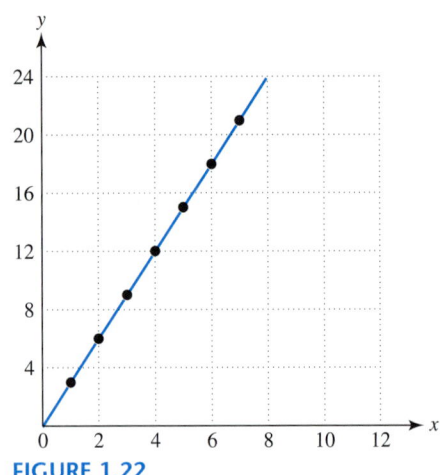

FIGURE 1.22

We can use the formula $y = 3x$ to find ordered pairs that include noninteger values of x and y, such as (0.5, 1.5). Plotting all such points results in a line with no breaks as shown in Figure 1.22.

Although there are many ways to represent a function, there is no *best* representation. It depends on both the individual and the application. Sometimes a formula works best—other times a graph or a table is preferable.

Formal Definition of a Function

Because the idea of a function is a fundamental concept in mathematics, it is important that we define a function precisely. This definition should allow for all representations of a function. The commonality with every representation is the concept of an ordered pair.

A relation is a set of ordered pairs. Would it be sufficient to say that a function is a relation? No, because a function must produce only one output for each valid input.

Function

A **function** is a relation where each element in the domain corresponds to exactly one element in the range.

The set of ordered pairs for a function can be either finite or infinite. The function $f = \{(1, 2), (3, 4), (5, 6)\}$ is a finite set of ordered pairs. On the other hand, the function represented by $g(x) = x^2$ with all real numbers as its domain generates an infinite set of ordered pairs.

EXAMPLE 2   *Computing average income as a function of educational attainment*

The function f computes the average 1992 annual earnings in dollars by educational attainment. This function is defined by $f(N) = 12{,}809$; $f(H) = 18{,}737$; $f(B) = 32{,}629$; $f(G) = 48{,}653$, where N denotes no diploma, H a high school diploma, B a bachelor's degree, and G a graduate degree.

(a) Write f as a set of ordered pairs.
(b) Give the domain and range of f.
(c) Discuss the relationship between education and income. (Sources: Bureau of the Census, Department of Commerce.)

Solution

(a) $f = \{(N, 12{,}809), (H, 18{,}737), (B, 32{,}629), (G, 48{,}653)\}$.
(b) The domain of f is $D = \{N, H, B, G\}$ and the range is $R = \{12{,}809, 18{,}737, 32{,}629, 48{,}653\}$.
(c) The greater the educational attainment the greater the annual earnings. On average, a person with a bachelor's degree earns approximately 2.5 times as much as someone without a high school diploma. ∎

Unless stated otherwise, the domain of a function f is the set of all real numbers for which its symbolic representation is defined. This is called the **implied domain** of a function. The implied domain can be thought of as the set of all valid inputs that make sense in the expression for $f(x)$.

EXAMPLE 3   *Evaluating a function and determining its domain*

Let a function f be represented symbolically by $f(x) = \dfrac{x}{x^2 - 1}$.

(a) Evaluate $f(3)$ and $f(a)$.
(b) Find the implied domain of f.

Solution

(a) To evaluate $f(3)$ substitute 3 for x in the formula.

$$f(3) = \frac{3}{3^2 - 1} = \frac{3}{8}$$

Evaluate $f(a)$ in a similar manner.

$$f(a) = \frac{a}{a^2 - 1}$$

(b) The expression for $f(x)$ is not defined when the denominator $x^2 - 1 = 0$. Therefore, the implied domain of f is all real numbers such that $x \neq \pm 1$. ∎

EXAMPLE 4   *Finding the domain and range of a function*

A graph of $f(x) = \sqrt{x - 2}$ is shown in Figure 1.23 on the next page. Use both the symbolic and graphical representations to find the domain and range of f.

Solution

The expression $\sqrt{x - 2}$ is defined and equal to a real number whenever $x - 2 \geq 0$. Therefore, the domain of f is all real numbers x such that $x \geq 2$. This conclusion is supported by the graph of f. Notice that points occur on the graph only where $x \geq 2$.

The range of f is the set of y-values satisfying $y = \sqrt{x - 2}$ when $x \geq 2$. The graph of f never lies below the x-axis. The arrow on this graph indicates that the y-values continue to increase without a maximum. Thus, the range consists of all nonnegative numbers. ∎

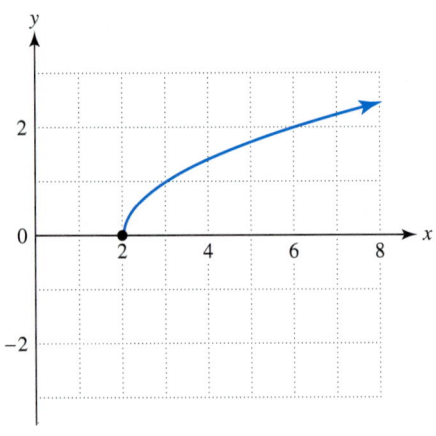

FIGURE 1.23

Critical Thinking

Suppose a ball is thrown into the air. Sketch an approximate graph of a function f that outputs the height y (in feet) after x seconds. Use your graph to describe the domain and range of f.

Graphing Calculators and Functions

Graphing calculators can create graphical and numerical representations of a function—usually more efficiently and reliably than pencil-and-paper techniques. However, a graphing calculator uses the same basic method that we might use to draw a graph. For example, one way to sketch a graph of $y = x^2$ is to first make a table of values. See Table 1.10.

TABLE 1.10

x	-3	-2	-1	0	1	2	3
y	9	4	1	0	1	4	9

We can plot these points in the xy-plane as shown in Figure 1.24. Next we might connect the points with a smooth curve, as shown in Figure 1.25.

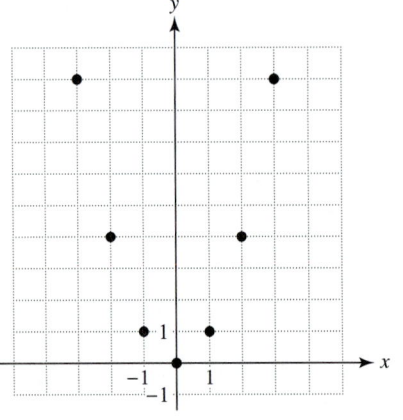

FIGURE 1.24

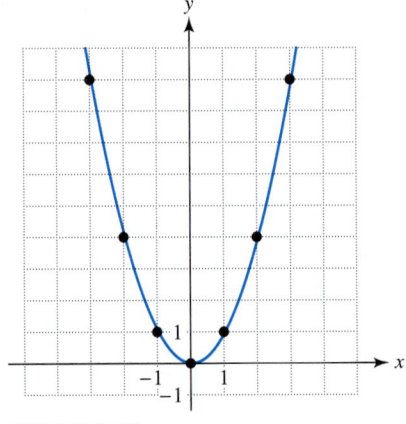

FIGURE 1.25

A graphing calculator typically follows the same process to draw a graph. It plots numerous points and connects them to make a graph.

EXAMPLE 5   *Representing and evaluating a function*

When the relative humidity is less than 100%, air cools at a rate of 5.4°F for every 1000-foot increase in altitude. (Source: L. Battan, *Weather in Your Life.*)

(a) Give verbal, symbolic, graphical, and numerical representations of a function f that computes this change in temperature for an increase in altitude of x-thousand feet. Let the domain of f be $0 \le x \le 6$.

(b) Evaluate $f(3)$ using each representation.

Solution

(a) *Verbal*   Multiply the input x by -5.4 to obtain the change in temperature.

Symbolic   Let $f(x) = -5.4x$.

Graphical   Since $f(x) = -5.4x$, enter $Y_1 = -5.4X$ as shown in Figure 1.26. Graph Y_1 in a viewing rectangle such as $[0, 6, 1]$ by $[-35, 10, 5]$. See Figure 1.27.

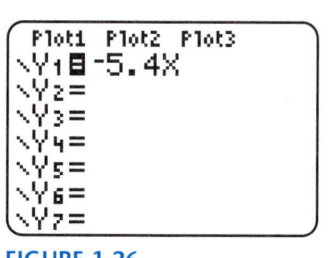

FIGURE 1.26

$[0, 6, 1]$ by $[-35, 10, 5]$

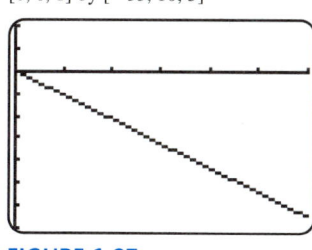

FIGURE 1.27

Numerical   It is impossible to list all inputs x, since $0 \le x \le 6$. However, Table 1.11 is a partial numerical representation of f at $x = 0, 1, 2, \ldots, 6$.

TABLE 1.11

x	0	1	2	3	4	5	6
$f(x)$	0	-5.4	-10.8	-16.2	-21.6	-27	-32.4

$[0, 6, 1]$ by $[-35, 10, 5]$

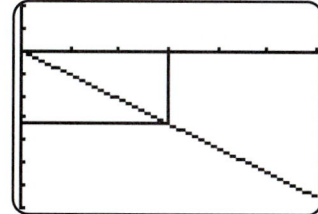

FIGURE 1.28

(b) *Verbal*   Multiply 3 by -5.4 to obtain -16.2. This means that there is a 16.2°F *decrease* in temperature associated with a 3000-foot increase in altitude.

Symbolic   Since $f(x) = -5.4x$, $f(3) = -5.4(3) = -16.2$.

Graphical   To evaluate $f(3)$ graphically, start by visualizing the graph in Figure 1.28. Estimate the y-value on the graph of f when $x = 3$. Since each tick mark on the y-axis represents 5, this value is approximately -16. (Note that answers may vary slightly when performing estimations graphically.)

Numerical   Using Table 1.11, $f(3)$ evaluates to -16.2. ∎

EXAMPLE 6 *Evaluating representations of a function*

People who sustain leg injuries often require crutches. It is possible to estimate the proper crutch length without using trial and error. The function given by $f(x) = 0.72x + 2$ can compute the appropriate crutch length in inches for a person with a height of x inches. (Source: *Journal of the American Physical Therapy Association.*)

(a) Graph f in the viewing rectangle [60, 90, 10] by [40, 80, 10]. Estimate $f(60)$.
(b) Evaluate $f(60)$. Interpret your result.
(c) Represent f numerically using a table for the values $x = 60, 61, 62, \ldots, 75$. Determine the proper crutch length for a person six feet tall.

Solution

(a) To graph f let $Y_1 = .72X + 2$ as shown in Figure 1.29. Then set the viewing rectangle to [60, 90, 10] by [40, 80, 10]. The graph of f is shown in Figure 1.30.

 The horizontal tick marks on the x-axis represent $x = 60, 70, 80, 90$. The vertical tick marks represent $y = 40, 50, 60, 70, 80$. From the graph, $f(60) \approx 45$.

[60, 90, 10] by [40, 80, 10]

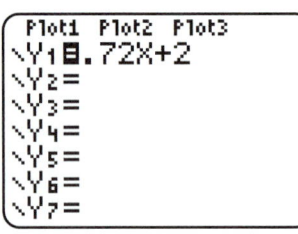

FIGURE 1.29

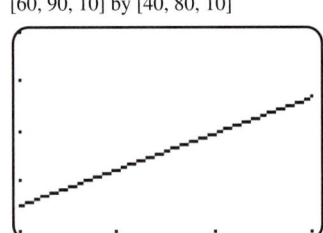

FIGURE 1.30

(b) Since $f(x) = 0.72x + 2$, $f(60) = 0.72(60) + 2 = 45.2$. This means that the appropriate crutch height for a person 60 inches tall is about 45 inches.

(c) To create a table on a graphing calculator, we retain the formula for Y_1 and specify both the *starting value* for x (TblStart) and the *increment* between successive x-values (ΔTbl). Here we start the table at $x = 60$ and increment by 1. (There is no viewing rectangle to set.) Figures 1.31 and 1.32 illustrate these steps.

FIGURE 1.31

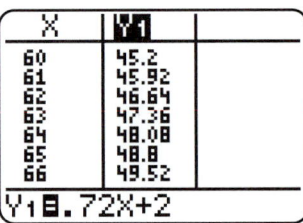

FIGURE 1.32

By scrolling down, the lower portion of the table may be accessed. To find the crutch length for a person six feet tall, evaluate f at $x = 72$ inches. Figure 1.33 shows that $f(72) = 53.84$. Thus, a person 72 inches tall requires a crutch that is about 54 inches.

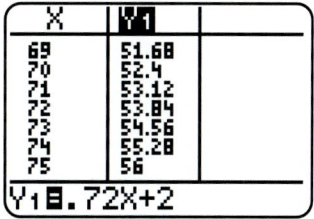

FIGURE 1.33   ∎

In the previous example a table of f was made for $x = 60, 61, 62, \ldots, 75$. We will refer to these steps using the sentence "Table $f(x) = 0.72x + 2$ starting at $x = 60$, incrementing by 1." Sometimes the phrase "until $x = 75$" will be included to specify the last x-value in the table.

Identifying Functions

By applying the definition of a function, we can determine if a relation represents a function.

EXAMPLE 7   *Determining if a set of ordered pairs is a function*

The set S of ordered pairs (x, y) represents the monthly average temperature y in degrees Fahrenheit for the month x in Washington, D. C. Determine if S is a function. (Source: A. Miller and J. Thompson, *Elements of Meteorology.*)

$S = \{$(January, 33), (February, 37), (March, 45), (April, 53), (May, 66), (June, 73), (July, 77), (August, 77), (September, 70), (October, 51), (November, 48), (December, 37)$\}$

Solution

The input x is the month and the output y is the monthly average temperature. The set S is a function, since each month is paired with one and only one monthly average temperature. Each x-value is paired with exactly one y-value.   ∎

Vertical Line Test. To conclude that a graph represents a function, we must be convinced that it is impossible for two distinct points with the same x-coordinate to lie on the graph. For example, the ordered pairs $(4, 2)$ and $(4, -2)$ are distinct points with the same x-coordinate. These two points could not lie on the graph of the same function because input 4 would result in two outputs, ± 2. A function has exactly one output for each valid input. When the points (4, 2) and (4, −2) are plotted, they lie on the same vertical line as shown in Figure 1.34. A graph passing through these points intersects the line twice as illustrated in Figure 1.35.

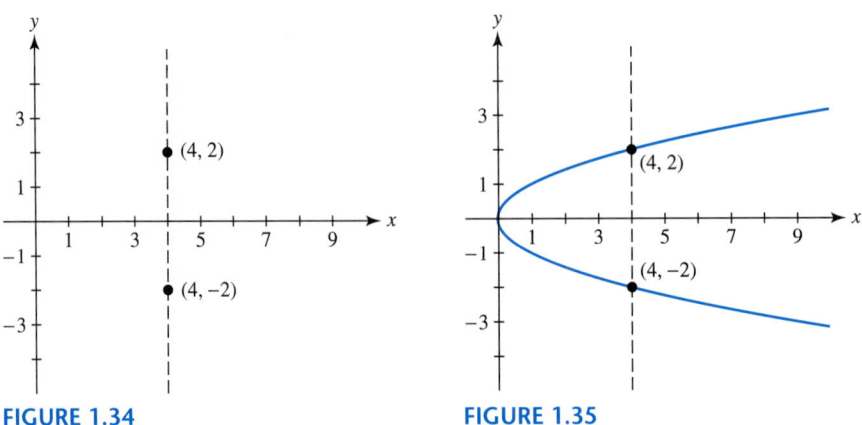

FIGURE 1.34   **FIGURE 1.35**

To determine if a graph represents a function, simply visualize vertical lines in the *xy*-plane. If each vertical line intersects a graph at no more than one point, then it is a graph of a function. This is called the **vertical line test** for a function.

EXAMPLE 8   *Determining if a graph represents a function*

Figure 1.36 shows a graph of the world population in billions from 1950 to 1995. Does this graph represent a function?

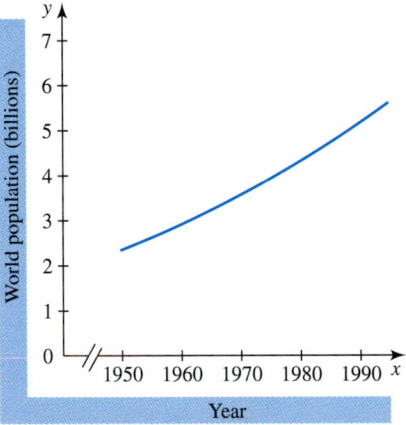

FIGURE 1.36   Sources: United Nations, Bureau of the Census.

Solution

Any vertical line will cross the graph at most once. See Figure 1.37. Therefore, the graph does represent a function.

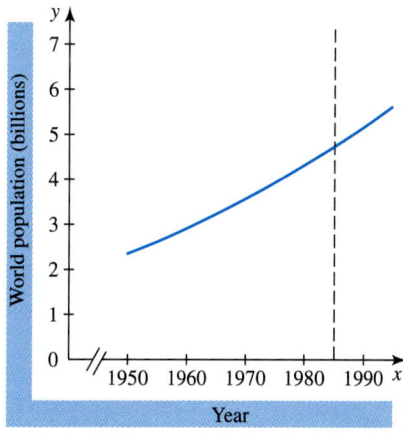

FIGURE 1.37

■

EXAMPLE 9   *Representing the orbit of Mercury*

A graph of the path of the planet Mercury around the sun is shown in Figure 1.38. The sun is located at the origin.

(a) Could the graph of a function model this path?

(b) Determine the minimum number of functions that would be necessary to create the graph of Mercury's orbit.

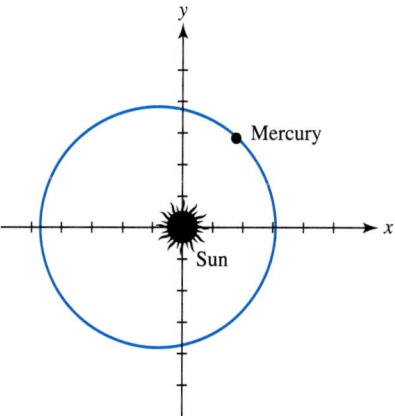

FIGURE 1.38 Mercury's Orbit

Solution

(a) The path of Mercury cannot be modeled by the graph of a function because vertical lines can intersect the path twice. See Figure 1.39 on the following page.

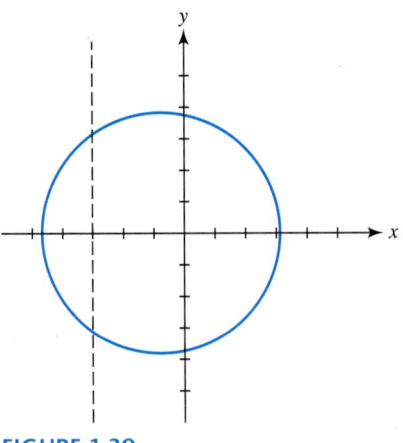

FIGURE 1.39

(b) Since a vertical line cannot intersect the graph more than twice, two functions could produce the graph. The graph of one function could represent the portion of the orbit above the x-axis, while the graph of a second function could represent the portion of the orbit below the x-axis. For these functions, a vertical line would intersect either graph at most once. ∎

1.3 PUTTING IT ALL TOGETHER

One of the most important concepts in mathematics is that of a function. A function always produces exactly one output for each element in its domain. The domain of a function f consists of the set of all valid x-inputs, while the range is the set of corresponding y-outputs. The following table summarizes different representations of a function.

Type of Representation	Explanation	Comments
Verbal, algorithmic	Words describe precisely what is computed	May be oral or written May be in algorithmic form where each step is precisely stated
Symbolic	Mathematical formula	Is an efficient and concise way of representing a function, e.g., $f(x) = 2x^2 - 3$
Numerical, tabular	Lists of individual inputs and their outputs	May be in the form of a table or an explicit set of ordered pairs
Graphical, diagrammatic	Graphs and diagrams; shows inputs and outputs visually	Uses no words, formulas, or tables Many types of graphs and diagrams are possible

1.3 EXERCISES   Tape 1

Representation and Evaluation of Functions

Exercises 1–6: Graph f by hand. Support your result using a graphing calculator.

1. $f(x) = 2x$
2. $f(x) = \sqrt{x}$
3. $f(x) = \frac{1}{2}x^2$
4. $f(x) = -2$
5. $f(x) = |x - 1|$
6. $f(x) = 4 - x$

Exercises 7–10: Express the verbal representation for the function f using symbolic and graphical representations. Let $y = f(x)$ and $0 \le x \le 100$ when graphing.

7. To convert x gallons to y liters, multiply x by 3.785.
8. To convert x pounds to y kilograms, divide x by 2.205.
9. To convert x kilometers to y miles, divide x by 1.609.
10. To convert x acres to y square feet, multiply x by 43,560.

Exercises 11–16: Use f(x) to determine verbal, graphical, and numerical representations. For the numerical representation use a table with $x = -2, -1, 0, 1, 2$.

11. $f(x) = x^2$
12. $f(x) = 2x - 5$
13. $f(x) = \frac{1}{x}$
14. $f(x) = 8$
15. $f(x) = |x|$
16. $f(x) = 5 - x$

Exercises 17 and 18: Express a function f with the specified representation.

17. *Counterfeit Money* It is estimated that nine out of every one million bills are counterfeit. Give a numerical representation that computes the predicted number of counterfeit bills in a sample of x million bills for $x = 0, 1, 2, \ldots, 6$. (Source: Department of the Treasury.)

18. *Cost of Driving* In 1997 the average cost of driving a new car was 41 cents per mile. Give symbolic, graphical, and numerical representations that compute the cost in dollars of driving x miles. For the numerical representation, let $x = 1, 2, 3, 4, 5, 6$. (Source: Associated Press.)

19. *Classified Ad Costs* The cost of running a classified advertisement for two days in a newspaper is given by the following algorithm.

ALGORITHM Computing the cost of a classified advertisement

Step 1: Input x, the number of lines.
Step 2: For the first three lines or less, charge $9.60.
Step 3: For each additional line over three, add $3.20 to the result in STEP 2.
Step 4: Output the answer in STEP 3 as y, the total cost in dollars.

(a) Describe the function computed by this algorithm using numerical and graphical representations. Let $x = 1, 2, 3, 4, 5, 6$.
(b) Find the cost of a five-line advertisement.

20. *Global Warming* A rough approximation to the average global surface temperature in degrees Fahrenheit from 1856 to 1995 is given by $f(x) = 0.01(x - 1856) + 57.3$, where x is the year. (Source: *The New York Times.*)
(a) Determine the average temperature in 1995.
(b) Table f starting at $x = 1990$, incrementing by 1.
(c) Use the table to find the yearly change in the average global temperature given by f.

21. *Music Videos* The function f computes the number y in millions of music videos sold during year x. (Source: Recording Industry Association of America.)

$$f = \{(1990, 9.2), (1991, 6.1), (1992, 7.6), (1993, 11.0)\}$$

(a) Evaluate $f(1993)$.
(b) Identify the domain and range of f.
(c) Suppose this relation is updated to include the ordered pair (1994, 11.2), while the ordered pair (1990, 9.2) is removed. Let this new function be g. Find the domain and range of g.

22. *Motorcycle Registrations* The table lists millions of motorcycle registrations in the United States by year. Let this table represent a function f, where $y = f(x)$ computes millions of registrations y in year x. (Source: Federal Highway Administration.)

x (year)	y (millions)
1980	5.7
1985	5.4
1990	4.3
1994	3.9

(a) Represent f with a diagram.
(b) Identify the domain and range of f.
(c) Use the diagram to evaluate $f(1990)$.

23. *Average Temperatures* The monthly average high temperatures in degrees Fahrenheit at Daytona Beach can be approximated by $f(x) = 0.0151x^4 - 0.438x^3 + 3.60x^2 - 6.49x + 72.5$, where $x = 1$ corresponds to January and $x = 12$ to December.
    (a) Graph f in $[0.5, 12.5, 1]$ by $[65, 95, 5]$. Interpret the graph.
    (b) Table f for $x = 1, 2, 3, \ldots, 12$. Round values to the nearest degree.
    (c) Use the table to evaluate f for November.

24. *Unhealthy Air Quality* The Environmental Protection Agency (EPA) monitors air quality in U.S. cities. The function f, represented by the table, gives the annual number of days with unhealthy air quality in Los Angeles.

x	1988	1989	1990	1991	1992	1993
$f(x)$	226	212	163	157	169	131

   (a) Find $f(1992)$.
   (b) Determine when the greatest decrease in the number of unhealthy days occurred.
   (c) Discuss the trend of air pollution in Los Angeles.

Exercises 25 and 26: Refer to the graph of f in the figure.
   (a) *Evaluate $f(0)$ and $f(2)$.*
   (b) *Find all x such that $f(x) = 0$.*

25.

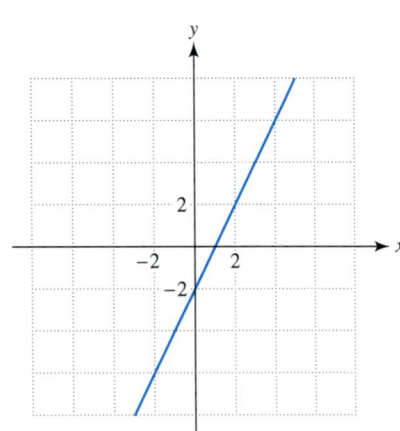

26.

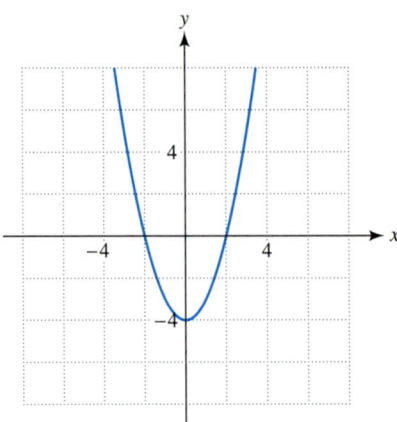

Exercises 27–30: Graph f in $[-4.7, 4.7, 1]$ by $[-3.1, 3.1, 1]$.
   (a) *Use the graph to estimate $f(2)$.*
   (b) *Evaluate $f(2)$ symbolically.*

27. $f(x) = 0.25x^2$

28. $f(x) = 3 - 1.5x^2$

29. $f(x) = \sqrt{x + 2}$

30. $f(x) = |1.6x - 2|$

Exercises 31–40: Complete the following.
   (a) *Evaluate $f(x)$ at the indicated values of x.*
   (b) *Find the implied domain of f.*

31. $f(x) = x^3$ at $x = -2, 5$

32. $f(x) = 2x - 1$ at $x = 8, -1$

33. $f(x) = \sqrt{x}$ at $x = 4, a$

34. $f(x) = \sqrt{1 - x}$ at $x = -2, a$

35. $f(x) = \dfrac{1}{x - 1}$ at $x = -1, a + 1$

36. $f(x) = \dfrac{3x - 5}{x + 5}$ at $x = -1, a$

37. $f(x) = -7$ at $x = 6, a$

38. $f(x) = x^2 - x + 1$ at $x = 1, -2$

39. $f(x) = \dfrac{1}{x^2}$ at $x = 4, -7$

40. $f(x) = \sqrt{x - 3}$ at $x = 4, a + 4$

Exercises 41–44: Use the graph of the function f to esti-mate its domain and range.

41.

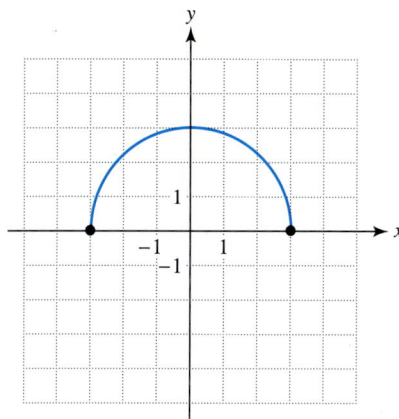

42.

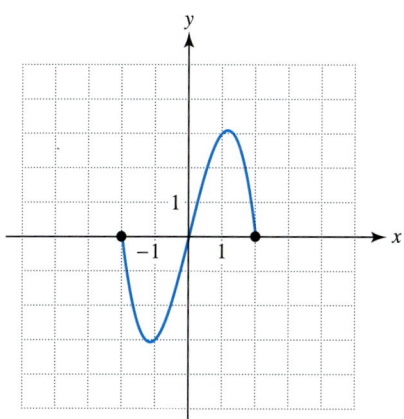

43.

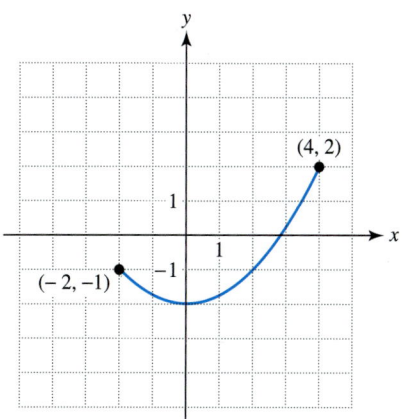

44.

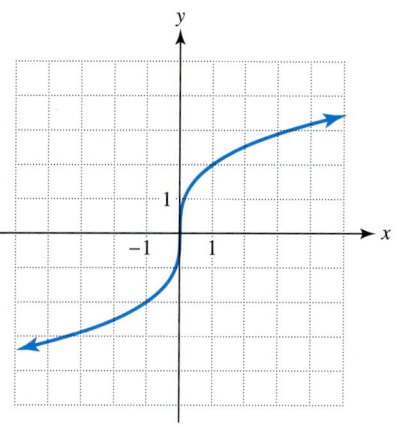

Identifying Functions

45. *Average Precipitation*   The table lists the monthly average precipitation P in Las Vegas, Nevada, where $x = 1$ corresponds to January and $x = 12$ to December. (Source: J. Williams, *The Weather Almanac 1995.*)

x (month)	1	2	3	4	5	6
P (in.)	0.5	0.4	0.4	0.2	0.2	0.1

x (month)	7	8	9	10	11	12
P (in.)	0.4	0.5	0.3	0.2	0.4	0.3

  (a) Determine the value of P during May.
  (b) Is P a function of x? Explain.
  (c) If $P = 0.4$, determine x.

46. *Wind Speeds*   The table lists the monthly average wind speed W in miles per hour in Louisville, Kentucky, where $x = 1$ corresponds to January and $x = 12$ to December. (Source: J. Williams.)

x	1	2	3	4	5	6
W	10.4	12.7	10.4	10.4	8.1	8.1

x	7	8	9	10	11	12
W	6.9	6.9	6.9	8.1	9.2	9.2

  (a) Determine the month with the greatest average wind speed.
  (b) Is W a function of x? Explain.
  (c) If $W = 6.9$, determine x.

Exercises 47–52: Determine if the graph represents a function. If it does not represent a function, determine the minimum number of functions necessary to create the graph. If it represents a function determine its domain and range.

47.

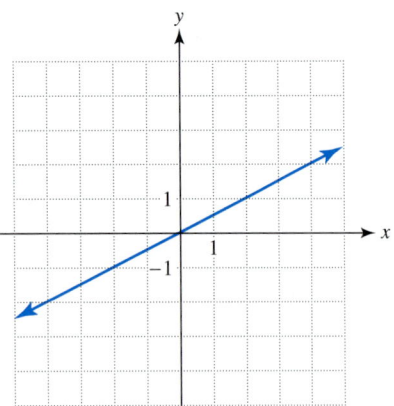

48.

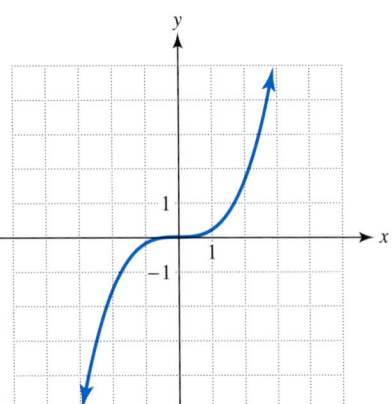

49.

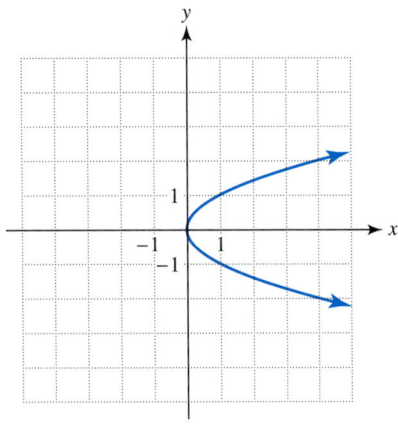

50.

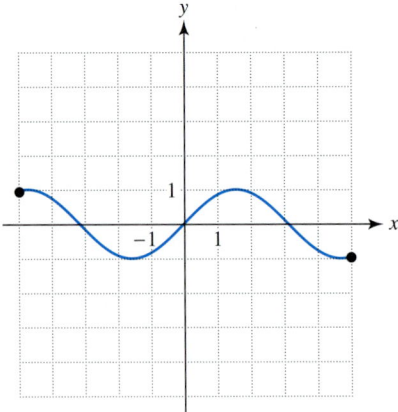

51.

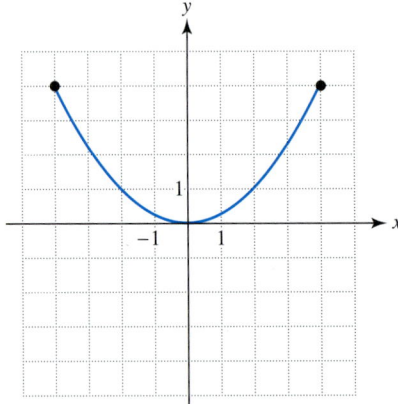

52.

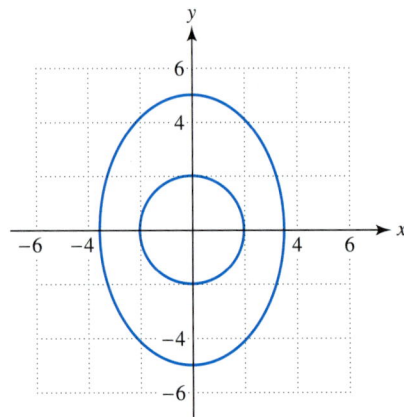

Exercises 53–56: Determine if the following describes a function. Explain your answer.

53. Calculating the cube root of a number

54. Calculating your age (to the nearest whole number) on a given day

55. Listing the students who passed a given English exam

56. Finding the *x*-values in the domain of a relation

Exercises 57–62: Determine if S defines a function.

57. $S = \{(1, 2), (2, 3), (4, 5), (1, 3)\}$

58. $S = \{(-3, 7), (-1, 7), (3, 9), (6, 7), (10, 0)\}$

59. $S = \{(a, 2), (b, 3), (c, 3), (d, 3), (e, 2)\}$

60. $S = \{(a, 2), (a, 3), (b, 5), (-b, 7)\}$

61. *S* is given by the table.

x	1	3	1
y	10.5	2	−0.5

62. *S* is given by the table.

x	1	2	3
y	1	1	1

Writing about Mathematics

1. Explain how you could use a graph or a complete numerical representation of a function to determine its domain and range.

2. Explain in your own words what a function is. How is a function different from a relation?

1.4 Types of Functions and Their Rates of Change

Constant Functions • Linear Functions • Slope • Nonlinear Functions • Average Rate of Change • Increasing and Decreasing Functions

Introduction

A central theme of applied mathematics is modeling real-world phenomena with functions. Because applications involving real data are diverse, mathematicians have created a wide assortment of functions. In fact, professional mathematicians design new functions every day for use in business, education, and government. Mathematics is not static—it is dynamic. It requires both ingenuity and creativity to analyze data and make predictions about the future. This section demonstrates how different types of data require a variety of functions to describe them. These functions can be categorized as constant, linear, or nonlinear.

Constant Functions

The monthly average wind speeds in miles per hour at Hilo, Hawaii, from May through December are listed in Table 1.12.

TABLE 1.12

Month	May	June	July	Aug	Sept	Oct	Nov	Dec
Wind Speed (mph)	7	7	7	7	7	7	7	7

Source: J. Williams, *The Weather Almanac 1995.*

It is apparent that the monthly average wind speed is constant between May and December. This data can be described by a set *f* of ordered pairs (x, y), where *x* is the month and *y* is the wind speed. The months have been assigned the standard numbers.

$$f = \{(5, 7), (6, 7), (7, 7), (8, 7), (9, 7), (10, 7), (11, 7), (12, 7)\}$$

A scatterplot of *f* is shown in Figure 1.40 on the next page. By the vertical line test *f* is a function.

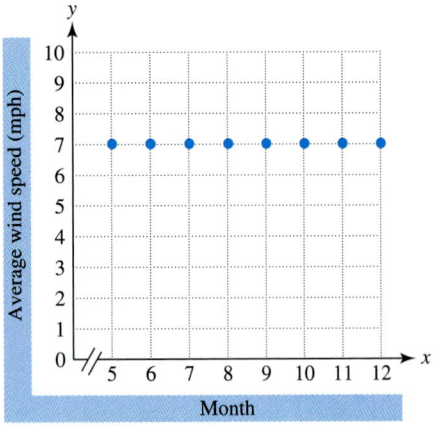

FIGURE 1.40 A Discrete Constant Model

The function f is given by $f(x) = 7$, where $x = 5, 6, 7, \ldots, 12$. The output of f never changes. We say f is a *constant function*.

Constant function

A function f represented by $f(x) = b$, where b is a fixed number, is a **constant function.**

In the previous example $f(x) = 7$, so $b = 7$. The range of f is $R = \{7\}$. The range of a constant function contains a single element. The domain of f is $D = \{5, 6, 7, 8, 9, 10, 11, 12\}$. Since f is defined only at individual or discrete values of x, f is called a **discrete function.** The graph of a discrete function suggests a scatterplot.

Sometimes it is more convenient to describe discrete data with a continuous graph. If the domain of $f(x) = 7$ is expanded to $D = \{x \mid 5 \le x \le 12\}$, its graph

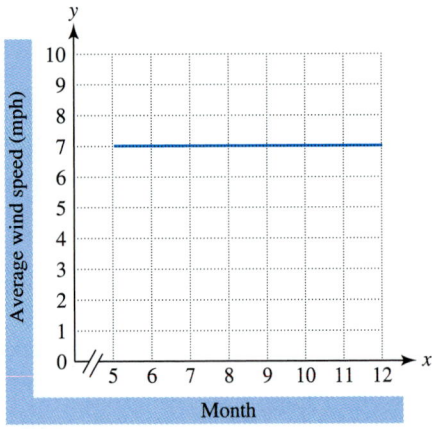

FIGURE 1.41 A Continuous Constant Model

becomes a continuous horizontal line without any breaks. See Figure 1.41. The graph of a **continuous function** can be sketched without picking up the pencil.

If the domain includes $x = 6.5$, what does $f(6.5) = 7$ represent? The expression $f(6)$ computes the average wind speed in June, while $f(7)$ gives the average wind speed in July. We might interpret $f(6.5)$ to represent the average wind speed from June 15 to July 15. Other interpretations are possible.

Review Note   *Set Builder Notation* ─────────────

The expression $\{x \mid 5 \le x \le 12\}$ is written in *set builder notation*. It represents the set of all real numbers x, such that x is greater than or equal to 5 and less than or equal to 12.

───

As simple as constant functions might appear, they occur frequently in applications. The following are examples that can be modeled by constant functions. In each case the independent variable is time.

- A thermostat computes a constant function regardless of the weather outside by maintaining a set temperature.
- A cruise control in a car computes a constant function by maintaining a fixed speed, regardless of the type of road or terrain.
- A compact disc player spins a CD at a constant number of revolutions each minute in order to play properly.

Linear Functions

A car is initially located 30 miles north of the Texas border, traveling north on Interstate 35 at 60 miles per hour. The distances between the automobile and the border are listed in Table 1.13 for various times.

TABLE 1.13

Elapsed Time (h)	0	1	2	3	4	5
Distance (mi)	30	90	150	210	270	330

It can be seen that the distance increases by 60 miles every hour. This data is modeled by a set f of ordered pairs (x, y), where x is the elapsed time and y is the distance from the border.

$$f = \{(0, 30), (1, 90), (2, 150), (3, 210), (4, 270), (5, 330)\}$$

The set f is a function, but not a constant function. A scatterplot of f is shown in Figure 1.42 on the following page. The scatterplot suggests a line that rises from left to right.

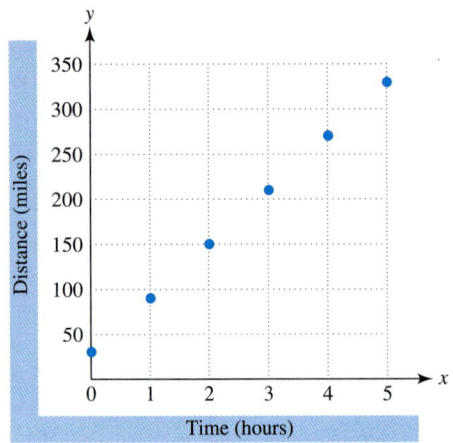

FIGURE 1.42 A Discrete Linear Model

If the car travels for x hours, the distance traveled can be computed by multiplying 60 times x and adding the initial distance of 30 miles. This computation can be expressed as $f(x) = 60x + 30$. The formula is valid for nonnegative values of x. For example, $f(1.5) = 60(1.5) + 30 = 120$ means that the car is 120 miles from the border after 1.5 hours. The graph of $f(x) = 60x + 30$ is shown in Figure 1.43. We call f a *linear function*.

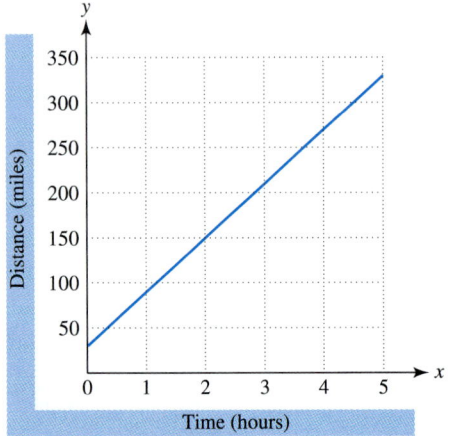

FIGURE 1.43 A Continuous Linear Model

Linear function

A function f represented by $f(x) = ax + b$, where a and b are constants, is a **linear function.**

If $a = 0$ then $f(x) = b$, which defines a constant function. Thus, any constant function is also a linear function. In the example of the moving car, $f(x) = 60x + 30$, so $a = 60$ and $b = 30$. The value of a represents the speed of the car, while b corresponds to the initial distance of the car from the border. Other examples of linear functions include the following. The values of the constants a and b are also given.

i. $f(x) = 1.5x - 6$ $a = 1.5,\quad b = -6$

ii. $f(x) = 8x$ $a = 8,\quad\ \ b = 0$

iii. $f(x) = 72$ $a = 0,\quad\ \ b = 72$

iv. $f(x) = 1.9 - 3x$ $a = -3,\quad b = 1.9$

A distinguishing feature of a linear function is that each time x increases by one unit, the value of $f(x)$ always changes by an amount equal to a. This type of behavior occurs in the following applications that are modeled by linear functions. Try to determine the value of the constant a.

• The wages earned by an individual working x hours at \$6.25 per hour
• The amount of tuition and fees when registering for x credits, if each credit costs \$75 and the fees are fixed at \$56
• The distance traveled by light in x seconds, if the speed of light is 186,000 miles per second

Slope

The graph of a linear function is a line. Slope is a real number that measures the "tilt" of a line in the xy-plane. If the input x to a linear function increases by one unit, then the output y changes by a constant amount that is equal to the slope of its graph. In Figure 1.44 a line passes through the points (x_1, y_1) and (x_2, y_2). The *change in y* is $y_2 - y_1$, while the *change in x* is $x_2 - x_1$. The ratio of the change in y to the change in x is called the *slope*.

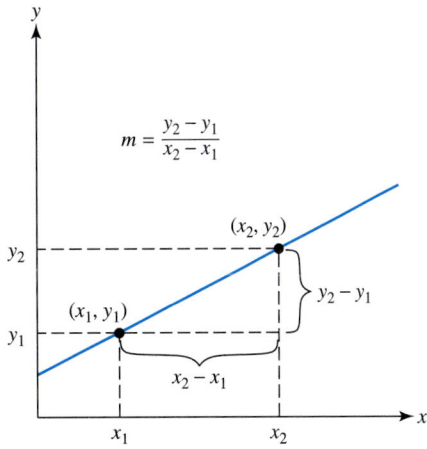

FIGURE 1.44

Slope

The **slope** m of the line passing through the points (x_1, y_1) and (x_2, y_2) is

$$m = \frac{y_2 - y_1}{x_2 - x_1}$$

where $x_1 \neq x_2$.

If the slope of a line is positive, the line *rises* from left to right. If the slope is negative, the line *falls* from left to right. Slope 0 indicates that the line is horizontal. When $x_1 = x_2$, the line is vertical and the slope is undefined. Figures 1.45–1.48 illustrate these situations. Slope 2 indicates that a line rises 2 units for every unit increase in x.

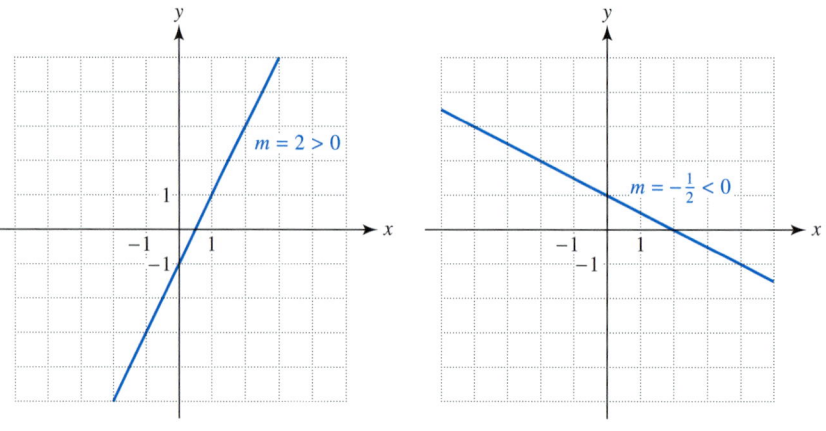

FIGURE 1.45   **FIGURE 1.46**

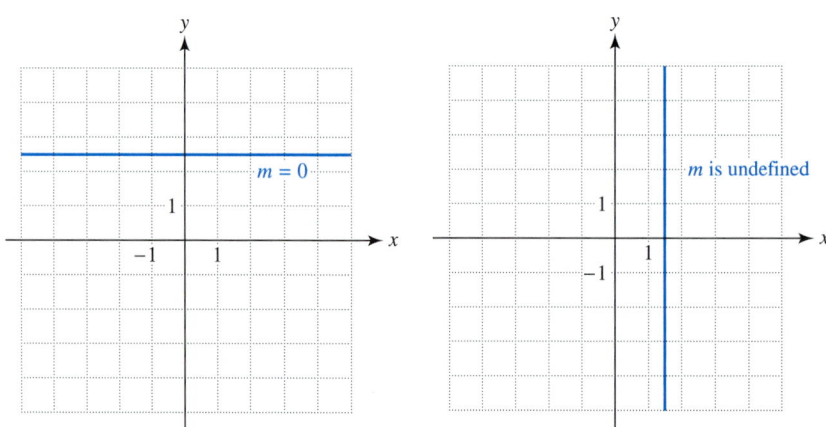

FIGURE 1.47   **FIGURE 1.48**

EXAMPLE 1   *Calculating the slope of a line*

Find the slope of the line passing through the points $(-4, 1)$ and $(2, 4)$. Plot these points together with the line. Interpret the slope.

Solution

The slope is

$$m = \frac{y_2 - y_1}{x_2 - x_1} = \frac{4 - 1}{2 - (-4)} = \frac{1}{2}.$$

A graph of the line passing through these two points is shown in Figure 1.49. A slope of $\frac{1}{2}$ means that for every unit increase in x, there is half a unit increase in y.   ■

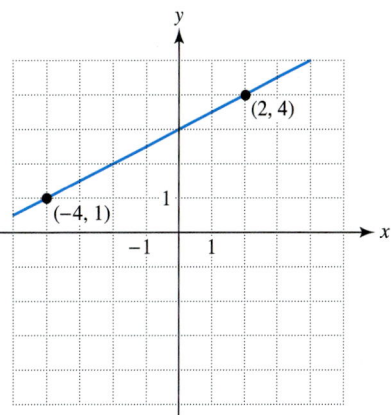

FIGURE 1.49

The graph of $f(x) = ax + b$ is a line. Since $f(0) = b$ and $f(1) = a + b$, the graph of f passes through the points $(0, b)$ and $(1, a + b)$. The slope of this line is

$$\frac{y_2 - y_1}{x_2 - x_1} = \frac{a + b - b}{1 - 0} = \frac{a}{1} = a.$$

In applications involving linear functions, slope sometimes is interpreted as a *rate of change*. In Example 6 of Section 1.3, the recommended crutch length for a person x inches tall was given by $f(x) = 0.72x + 2$. The slope of its graph is 0.72. One interpretation of this slope is that for each one-inch increase in the height of a person, the crutch length should be increased by 0.72 inch.

Squaring a Viewing Rectangle.   The graph of the line $y = x$ in the standard viewing rectangle is shown in Figure 1.50. The visual distance between consecutive tick marks on the axes are not equal. As a result, the line $y = x$ does not appear to have slope 1. Notice that although the box has five units on each side, it appears as a rectangle and not as a square. The visual change in y is less than the visual change in x. The reason for this is that the graphing calculator screen is wider than it is high. If we graph $y = x$ in $[-9, 9, 1]$ by $[-6, 6, 1]$, the length of the y-axis in the viewing rectangle is $\frac{2}{3}$ of the length of the x-axis. See Figure 1.51. There is a uniform spacing between consecutive tick marks on both axes. In this case, the line $y = x$ makes a $45°$ angle with the x-axis. The rectangle in Figure 1.50 now appears as a *square* in Figure 1.51. This process of setting the viewing rectangle is called *squaring a viewing rectangle*. The ratio of $\frac{2}{3}$ may be different on other calculators. Consult your manual.

$[-10, 10, 1]$ by $[-10, 10, 1]$

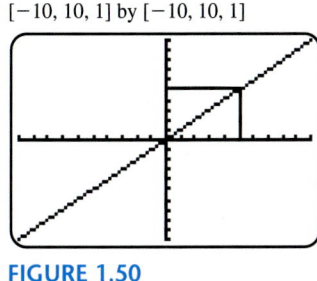

FIGURE 1.50

$[-9, 9, 1]$ by $[-6, 6, 1]$

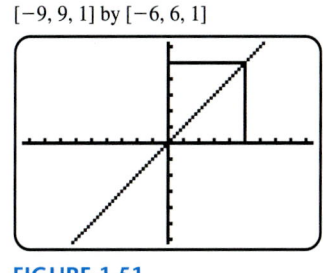

FIGURE 1.51

Nonlinear Functions

Table 1.14 shows the population of the United States at twenty-year intervals from 1800 to 1980. Between 1800 and 1820 the population increased by 5 million, between 1820 and 1840 it increased by 7 million, and between 1840 and 1860 the increase was 14 million. If these increases in population had been equal, this data could be modeled by a linear function.

TABLE 1.14

Year	Population (millions)	Year	Population (millions)
1800	5	1900	76
1820	10	1920	106
1840	17	1940	132
1860	31	1960	179
1880	50	1980	226

The actual population data is nonlinear. In Figure 1.52 the data together with a nonlinear function f have been plotted. The graph of f is curved rather than straight.

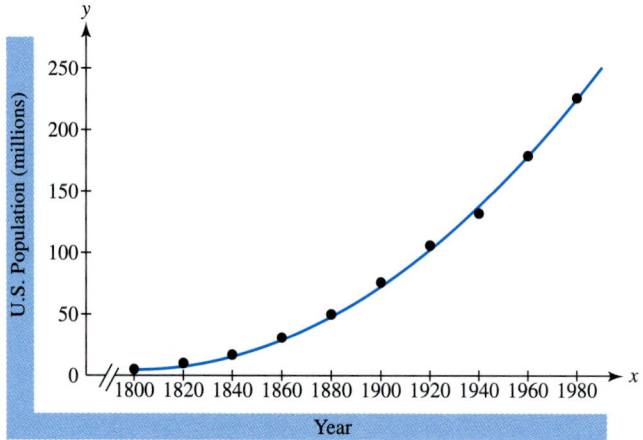

FIGURE 1.52 A Nonlinear Model

If a function is not linear, then it is called a **nonlinear function.** The graph of a nonlinear function is not a line. With a nonlinear function, it is possible for the input x to increase by one unit and the output y to change by different amounts. Nonlinear functions *cannot* be written in the form $f(x) = ax + b$. Examples of nonlinear functions include the following.

   i. $f(x) = x^2 - 3x + 2$

  ii. $f(x) = \sqrt{x}$

 iii. $f(x) = x^3$

Real-world phenomena often are modeled using nonlinear functions. The following are some examples of quantities that can be described by nonlinear functions.

- The total number of people who have contracted AIDS beginning in 1980
   (The increase in the number of AIDS cases is not the same each year.)
- The monthly average temperature in Chicago
   (Monthly average temperatures increase and decrease throughout the year.)
- The height of a child between the ages of 2 and 18
   (A child grows faster at certain ages.)

Critical Thinking

The time required to drive a distance of 100 miles depends on the average speed. Let the function f compute this time, given the average speed x as input. For example, $f(50) = 2$, since it would take two hours to travel 100 miles at an average speed of 50 miles per hour. Make a numerical representation of f. Is f linear or nonlinear?

Polynomial functions are frequently used to approximate data. They can be either linear or nonlinear. The implied domain of a polynomial function is all real numbers, and its graph is continuous.

Polynomial function

A **polynomial function** f **of degree** n **in the variable** x can be represented by

$$f(x) = a_n x^n + \cdots + a_2 x^2 + a_1 x + a_0$$

where each coefficient a_k is a real number, $a_n \neq 0$, and n is a nonnegative integer. The **leading coefficient** is a_n and the **degree** is n.

Symbolic representations, degrees, and leading coefficients of some polynomial functions are given.

i. $f(x) = 10$ Degree 0, $a_0 = 10$

ii. $f(x) = 2x - 3.7$ Degree 1, $a_1 = 2$

iii. $f(x) = 3x^2 - 1.4x + 1$ Degree 2, $a_2 = 3$

iv. $f(x) = -\frac{1}{2}x^6 + 4x^4 + x$ Degree 6, $a_6 = -\frac{1}{2}$

A polynomial function of degree 2 or higher is a nonlinear function. Functions (i) and (ii) are linear, whereas functions (iii) and (iv) are nonlinear. As a result, polynomial functions are used to model both linear and nonlinear data. For example, in Figure 1.52 the degree 2 polynomial function given by $f(x) = \frac{221}{32,400}x^2 - \frac{221}{9}x + 22,105$ was used to model the population of the United States.

Average Rate of Change

The graphs of nonlinear functions are not straight, so there is no notion of a single slope. The slope of the graph of a linear function gives its rate of change. With a nonlinear function we speak of an average rate of change. Suppose the points

(x_1, y_1) and (x_2, y_2) lie on the graph of a nonlinear function f. See Figure 1.53. The slope of the line L passing through these two points represents the *average rate of change of f from x_1 to x_2*. The line L is referred to as a *secant line*. If different values for x_1 and x_2 are selected, then a different secant line and a different average rate of change usually result.

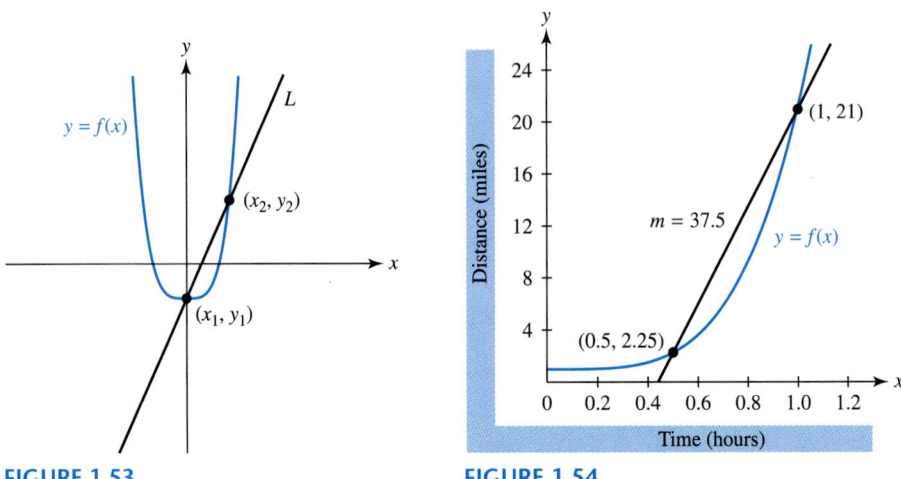

FIGURE 1.53   **FIGURE 1.54**

In applications the average rate of change measures how fast a quantity is changing over an interval of its domain, on average. For example, suppose the graph of the function f in Figure 1.54 represents the distance y in miles that a car has traveled on a straight highway after x hours. The points $(0.5, 2.25)$ and $(1, 21)$ lie on this graph. Thus, after 0.5 hour the car has traveled 2.25 miles and after 1 hour the car has traveled 21 miles. The slope of the line passing through these two points is

$$m = \frac{y_2 - y_1}{x_2 - x_1} = \frac{21 - 2.25}{1 - 0.5} = 37.5.$$

This means that during the half-hour period from 0.5 to 1 hour the average rate of change, or average velocity, was 37.5 miles per hour. These ideas lead to the following definition.

Average rate of change

Let (x_1, y_1) and (x_2, y_2) be distinct points on the graph of a function f. The **average rate of change of f from x_1 to x_2** is

$$\frac{y_2 - y_1}{x_2 - x_1}.$$

If f is a constant function, its average rate of change is zero. For a linear function defined by $f(x) = ax + b$, its average rate of change is equal to a, the slope of its graph. The average rate of change for a nonlinear function varies.

In Figure 1.55, the average rates of change of the nonlinear function f are not constant, but depend on the values of x_1 and x_2. The average rate of change from 1 to 5 is equal to the slope of the line segment connecting the points $(1, 1)$ and $(5, 4)$. This is given by $m_1 = \dfrac{4 - 1}{5 - 1} = \dfrac{3}{4}$. From 5 to 9, the average rate of change is equal to the slope of the line segment connecting the points $(5, 4)$ and $(9, 9)$, or $m_2 = \dfrac{9 - 4}{9 - 5} = \dfrac{5}{4}$.

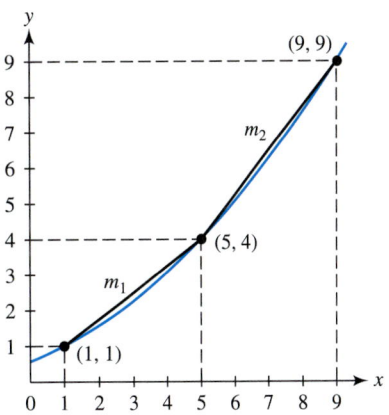

FIGURE 1.55 A Nonlinear Function

EXAMPLE 2   *Reporting cases of tetanus*

Table 1.15 lists the number of reported cases of tetanus for selected years.

TABLE 1.15

Year	1950	1960	1970	1980	1990
Cases of Tetanus	486	368	148	95	64

Source: Department of Health and Human Services.

(a) Make a line graph of the data. Let the line graph represent a function f.
(b) Find the average rate of change of f for each ten-year period.
(c) Interpret these average rates in terms of tetanus cases.

Solution

(a) A line graph using the points $(1950, 486)$, $(1960, 368)$, $(1970, 148)$, $(1980, 95)$, and $(1990, 64)$ is shown in Figure 1.56 on the following page.

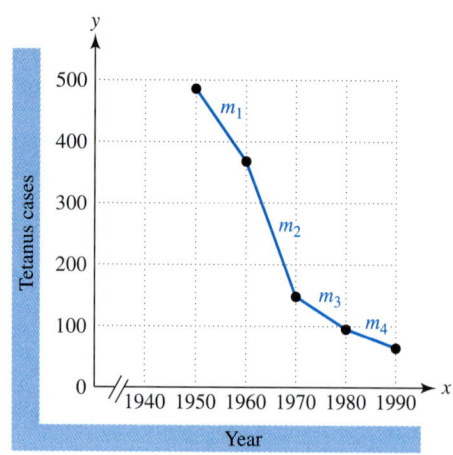

FIGURE 1.56 Tetanus Cases

(b) The average rates of change are equal to the slopes of the four line segments in the line graph.

$$m_1 = \frac{368 - 486}{1960 - 1950} = -11.8$$

$$m_2 = \frac{148 - 368}{1970 - 1960} = -22.0$$

$$m_3 = \frac{95 - 148}{1980 - 1970} = -5.3$$

$$m_4 = \frac{64 - 95}{1990 - 1980} = -3.1$$

(c) The value $m_1 = -11.8$ means that, on average, the number of tetanus cases between 1950 and 1960 *decreased* by 11.8 cases per year. The other three slopes can be interpreted in a similar manner. ∎

EXAMPLE 3   *Modeling solid waste with a polynomial function*

The gross weight in millions of tons of aluminum waste generated each year in the United States can be modeled by $f(x) = -\frac{7}{60,000}x^3 + \frac{13}{2000}x^2 - \frac{1}{75}x + \frac{2}{5}$. The domain of f is $0 \le x \le 30$, where $x = 0$ corresponds to 1960, $x = 1$ to 1961, and so on, until $x = 30$ represents 1990. (Source: Environmental Protection Agency.)
(a) Evaluate $f(20)$ and $f(30)$. Interpret the results.
(b) Compute the average rate of change of f from 20 to 30. Interpret the result.

Solution

(a) $f(20) = -\frac{7}{60,000}(20)^3 + \frac{13}{2000}(20)^2 - \frac{1}{75}(20) + \frac{2}{5} = 1.8$. In 1980 about 1.8 million tons of aluminum waste was generated. Similarly, $f(30) = 2.7$ indicates that in 1990 the aluminum waste was 2.7 million tons.

(b) The points (20, 1.8) and (30, 2.7) lie on the graph of f. The average rate of change of f from 20 to 30 is equal to the slope of the line passing through these points.

$$\frac{y_2 - y_1}{x_2 - x_1} = \frac{2.7 - 1.8}{30 - 20} = \frac{0.9}{10} = 0.09.$$

During the ten-year period from 1980 to 1990, there was an average increase in aluminum waste of 0.09 million tons per year.   ■

Increasing and Decreasing Functions

The concepts of increasing and decreasing relate to whether the graph of a function rises or falls. Intuitively, if we could walk along the graph of an increasing function *from left to right,* it would be uphill. For a decreasing function, we would walk downhill. We sometimes speak of a function f increasing or decreasing over an interval of its domain. For example, in Figure 1.57 the function is decreasing (the graph falls) when $-2 \le x \le 0$ and increasing (the graph rises) when $0 \le x \le 2$.

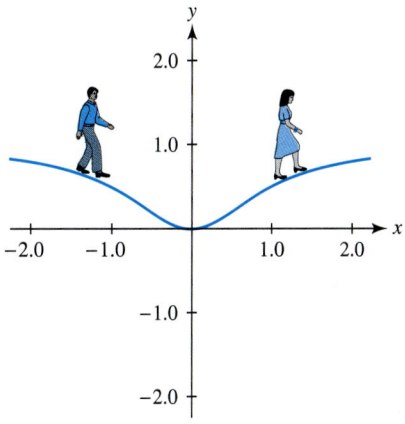

FIGURE 1.57

Increasing and decreasing functions are defined as follows.

Increasing and decreasing functions

Suppose that a function f is defined over an interval I on the number line. If x_1 and x_2 are in I,

(a) f **increases** on I if, whenever $x_1 < x_2, f(x_1) < f(x_2)$;

(b) f **decreases** on I if, whenever $x_1 < x_2, f(x_1) > f(x_2)$.

Figures 1.58 and 1.59 on the following page illustrate these concepts.

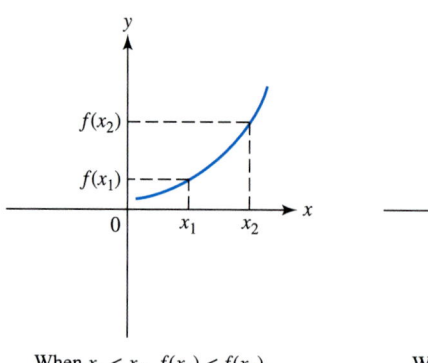

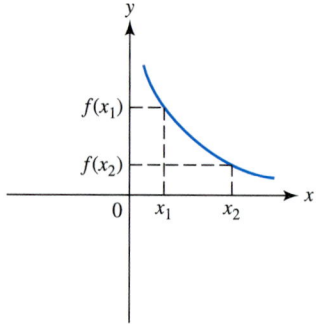

When $x_1 < x_2$, $f(x_1) < f(x_2)$.
f is increasing.

FIGURE 1.58

When $x_1 < x_2$, $f(x_1) > f(x_2)$.
f is decreasing.

FIGURE 1.59

EXAMPLE 4   *Determining where a function is increasing or decreasing*

Graph $f(x) = 4x - \dfrac{1}{3}x^3$ in $[-9.4, 9.4, 1]$ by $[-6.2, 6.2, 1]$. Use the trace feature to approximate the x-intervals where f is increasing or decreasing.

Solution

The graph of f is shown in Figure 1.60. Since x-intervals where f is increasing or decreasing are determined by moving from left to right, move the cursor to the left side of the graph.

$[-9.4, 9.4, 1]$ by $[-6.2, 6.2, 1]$

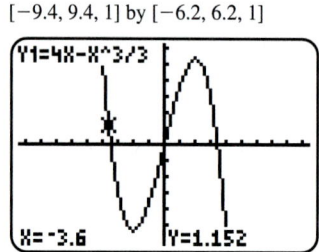

FIGURE 1.60

$[-9.4, 9.4, 1]$ by $[-6.2, 6.2, 1]$

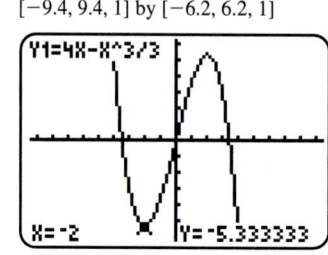

FIGURE 1.61

$[-9.4, 9.4, 1]$ by $[-6.2, 6.2, 1]$

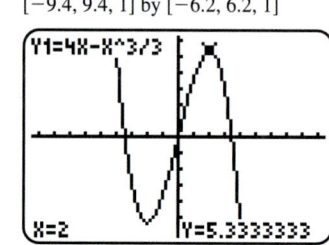

FIGURE 1.62

Notice that, as the cursor moves to the right, the x-values increase, while the y-values decrease, until $x = -2$. Thus, f is decreasing on the interval $x \le -2$. See Figure 1.61. Then the y-values increase from $x = -2$ to $x = 2$, so f is increasing on the interval $-2 \le x \le 2$. To the right of $x = 2$ the y-values decrease. See Figure 1.62. It follows that f is decreasing when either $x \le -2$ or $x \ge 2$, and f is increasing when $-2 \le x \le 2$.   ■

Critical Thinking

How is the slope of the graph of a linear function related to whether the function is increasing, decreasing, or neither? Explain.

Technology Note ──────────────────────────────

The viewing rectangle $[-9.4, 9.4, 1]$ by $[-6.2, 6.2, 1]$ is an example of a *decimal* viewing rectangle. When tracing, the cursor stops on points with convenient x-values. Consult your manual.

──

The next example illustrates that the concepts of increasing and decreasing can be extended to include discrete functions.

EXAMPLE 5   *Determining where a discrete function is increasing or decreasing*

A complete numerical representation of a function f is given in Table 1.16. This function computes the annual budgets for national parks from 1994 to 1998. The input x is the year and the output $f(x)$ is the amount of the budget in billions of dollars. Determine where f is increasing or decreasing.

TABLE 1.16

x (year)	1994	1995	1996	1997	1998
$f(x)$ ($ billions)	1.45	1.37	1.36	1.42	1.60

Source: National Park Service.

Solution

From Table 1.16 we can see that the amount spent on national parks decreased from 1994 to 1996 and then increased from 1996 to 1998. The scatterplot in Figure 1.63 supports this conclusion.

[1993, 1999, 1] by [1.3, 1.7, 0.1]

FIGURE 1.63

1.4 PUTTING IT ALL TOGETHER

The following table summarizes important concepts related to constant, linear, and nonlinear functions.

Concept	Constant Function	Linear Function	Nonlinear Function
Slope of graph	Always zero	Always constant	No notion of one slope
Average rate of change	Always zero	Always constant	Can vary
Increasing or decreasing	Neither increasing nor decreasing	May be increasing or decreasing but not both	May be both increasing and decreasing
Graph	Horizontal line	Nonvertical line	Not a line
Examples	$f(x) = 2$	$f(x) = 2x - 2$	$f(x) = x^2 - 3$
Examples	$f(x) = -3$	$f(x) = -x + 3$	$f(x) = x^3 - 4x$

1.4 EXERCISES

   Tapes 1, 3

Constant and Linear Functions

1. *Thermostat*   Let $y = f(x)$ describe the temperature y of a room that is kept at 68°F after x hours.
   (a) Represent f symbolically and graphically over a 24-hour period for $0 \le x \le 24$.
   (b) Table f starting at $x = 0$, incrementing by 4, until $x = 24$.
   (c) What type of function is f?

2. *Cruise Control*   Let $y = f(x)$ describe the speed y of an automobile after x minutes if the cruise control is set at 55 miles per hour.

   (a) Represent f symbolically and graphically over a 15-minute period for $0 \le x \le 15$.
   (b) Table f starting at $x = 0$, incrementing by 3, until $x = 15$.
   (c) What type of function is f?

3. *Average Wind Speed*   The table lists the average wind speed in miles per hour at Myrtle Beach, South Carolina. The months are assigned the standard numbers. (Source: J. Williams.)

Month	1	2	3	4	5	6	7	8	9	10	11	12
Wind (mph)	7	8	8	8	7	7	7	7	7	6	6	6

(a) Could this data be represented exactly by a constant function?

(b) Determine a continuous, constant function f that models this data.

(c) Graph f and the data.

4. *Income Tax Rates* The table lists the lowest personal income tax rate in percent between 1991 and 1996. (Source: Internal Revenue Service.)

Year	1991	1992	1993	1994	1995	1996
Rate (%)	15	15	15	15	15	15

(a) Determine a discrete, constant function f that describes this data. Represent f symbolically.

(b) What is the domain of f?

(c) Graph f.

5. *Circles* In the table the functions f, g, and h all compute quantities related to circles, where x is the radius.

(a) Identify each function as constant, linear, or nonlinear.

(b) Determine what quantity each function computes.

x	$f(x)$	$g(x)$	$h(x)$
1	2π	π	2
2	4π	4π	4
3	6π	9π	6
4	8π	16π	8
5	10π	25π	10

6. *Income Tax* The federal income tax rate for single individuals in 1995 was 15% of taxable income between $0 and $23,500, and 28% of taxable income between $23,500 and $56,550. (Source: Internal Revenue Service.)

(a) The linear function given by $f(x) = 0.15x$ computes the tax on taxable incomes x between $0 and $23,500. Determine the income tax on a taxable income of $9,600.

(b) The linear function given by $f(x) = 0.28x - 3055$ computes the tax on taxable incomes x between $23,500 and $56,550.

Determine the income tax on a taxable income of $25,600.

7. Suppose that a car's distance from a service center along a straight highway can be described by a constant function of time. Discuss what can be said about the car's velocity.

8. Suppose that a car's distance in miles from a rest stop along a straight stretch of an interstate highway after x hours can be modeled by $f(x) = ax$. Discuss what can be said about the car's velocity.

Exercises 9–14: Write a symbolic representation for a function f that computes the following.

9. The number of pounds in x ounces

10. The number of dimes in x dollars

11. The distance traveled by a car moving at 50 miles per hour for x hours

12. The monthly electric bill in dollars for using x kilowatt-hours at 6 cents per kilowatt-hour plus a fixed fee of $6.50

13. The cost of downhill skiing x times with a $500 season pass

14. The total number of hours in day x

Slope

Exercises 15–22: If possible, find the slope of the line passing through each pair of points.

15. $(4, 6), (2, 5)$

16. $(-8, 5), (-3, -7)$

17. $(-0.5, 9.2), (-0.3, 7.6)$

18. $(1.6, 12), (1.6, 5)$

19. $(1997, 5.6), (1994, 7.9)$

20. $(1824, 108), (1900, 380)$

21. $(-5, 6), (-5, 8)$

22. $(17, 7), (19, 7)$

23. *Age in the U.S.* The median age of the U.S. population for each year x between 1820 and 1995 can be approximated by $f(x) = 0.09x - 147.1$. (Source: Bureau of the Census.)

(a) Graph f in $[1820, 1995, 20]$ by $[0, 40, 10]$.

(b) Compute the median ages of the U. S. population in 1820 and 1995.

(c) What is the slope of the graph of f? Interpret the slope.

24. *Commercial Banks* From 1987 to 1997 the numbers of federally insured commercial banks could be modeled by $f(x) = -458x + 923,769$, where x is the year. (Source: Federal Deposit Insurance Corporation.)

(a) Graph f in [1987, 1997, 1] by [8000, 15,000, 1000].

(b) Estimate the number of commercial banks in 1987 and in 1997.

(c) What is the slope of the graph of f? Interpret the slope.

25. *Women's Olympic Times*   The winning times in seconds for the women's 200-meter dash at the Olympic Games can be approximated by $f(x) = -0.0635x + 147.9$, where x is the year with $1948 \le x \le 1996$. (Source: United States Olympic Committee.)

(a) Graph f in [1948, 1996, 4] by [21, 25, 1].

(b) What is the slope of the graph of f? Interpret the slope.

(c) Approximate the average change in the winning times over a four-year period.

26. *Men's Olympic Times*   The winning times in seconds for the men's 400-meter hurdles at the Olympic Games can be approximated by $f(x) = -0.0975x + 240.8$, where x is the year with $1900 \le x \le 1996$. (Source: United States Olympic Committee.)

(a) Graph $f(x)$ in [1900, 1996, 4] by [45, 60, 1].

(b) What is the slope of the graph of f? Interpret the slope.

(c) Approximate the average change in the winning times over a four-year period.

Increasing and Decreasing Functions

Exercises 27–38: Graph f in an appropriate viewing rectangle. Use both the trace feature and the formula for $f(x)$ to complete the following.

(a) *Decide whether f is constant, linear, or nonlinear.*

(b) *Estimate the x-intervals where f is increasing or decreasing.*

27. $f(x) = -2x + 5$

28. $f(x) = -0.01x - 2$

29. $f(x) = 2x^{1.01} - 1$

30. $f(x) = 5.123$

31. $f(x) = 1$

32. $f(x) = \sqrt{(-x)}$

33. $f(x) = 0.2x^3 - 2x + 1$

34. $f(x) = 0.5x^2 - 2x - 3$

35. $f(x) = 5x - x^2$

36. $f(x) = -\dfrac{1}{2}x^3 + 4x - 3$

37. $f(x) = x^4 - 5x^2 + 4$

38. $f(x) = x^4 - 2x^3 - 5x^2 + 6x$

Exercises 39–42: Analyzing Real Data   *For each data set complete the following.*

(a) *Make a line graph of the data in an appropriate viewing rectangle. Let this graph represent a function f.*

(b) *If possible, determine where f is increasing or decreasing.*

(c) *Decide whether f is linear or nonlinear.*

39. Interest income earned in one year on x dollars invested at 7% per year

x (dollars)	500	1000	2000	3500
Interest	$35	$70	$140	$245

40. World energy consumption in quadrillion Btu (*Note*: 1 quadrillion $= 1 \times 10^{15}$)

Year	1990	1991	1992	1993
Consumption	345	345	345	345

Source: Energy Information Administration, *International Energy Annual.*

41. Median incomes of full-time female workers

Year	1970	1980	1990	1993
Income	$5,440	$11,591	$20,591	$22,469

Source: Bureau of the Census, *Current Population Reports.*

42. Median incomes of full-time male workers

Year	1970	1980	1990	1993
Income	$9,184	$19,173	$28,979	$31,077

Source: Bureau of the Census, *Current Population Reports.*

43. Use the table to classify each discrete function as increasing, decreasing, or neither.

x	$f_1(x)$	$f_2(x)$	$f_3(x)$	$f_4(x)$
8	-2	3	4	-1
1	20	1	-8	4
3	15	-3	-2	5
11	-7	-10	10	8
5	4	-11	0	9

44. *Interest on Car Loans* The table gives monthly payments on four-year car loans for various dollar amounts x and interest rates. For example, a $10,000 loan at 10% would have monthly payments of $254 over four years.

x	5%	10%	13%
$5000	$115	$127	$134
$10,000	$230	$254	$268
$15,000	$345	$381	$402
$20,000	$460	$508	$536

(a) Determine if the payment is a decreasing or increasing function of x at a fixed-interest rate.
(b) Determine if the payment is a linear function of x at a fixed-interest rate. Interpret your answer.

Curve Sketching

Exercises 45–50: Critical Thinking   Assume that each function is continuous. Do not use a graphing calculator.

45. Sketch a graph of an increasing linear function f with $f(3) = -1$.

46. On the same coordinate axes, sketch the graphs of two linear functions f and g that never intersect and have f decreasing. What must be true about g?

47. On the same coordinate axes, sketch the graphs of a constant function f and a nonlinear function g that intersect exactly twice. What can be said about whether g is increasing or decreasing?

48. Sketch a graph of a linear function f that intersects a constant function g exactly once.

49. Sketch a graph of an increasing nonlinear function f. What can be said about its average rates of change?

50. Sketch a graph of a decreasing nonlinear function f. What can be said about its average rates of change?

Average Rates of Change

Exercises 51–54: Use the graph of $f(x)$ to discuss the average rates of change of f.

51. $f(x) = 4$

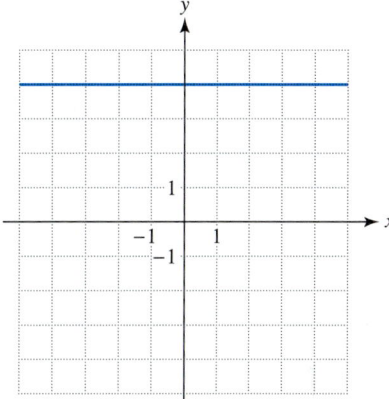

52. $f(x) = 2x - 1$

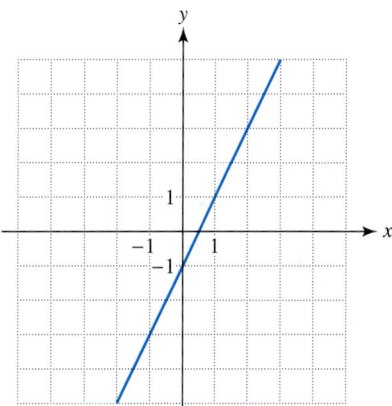

53. $f(x) = -0.3x^2 + 4$

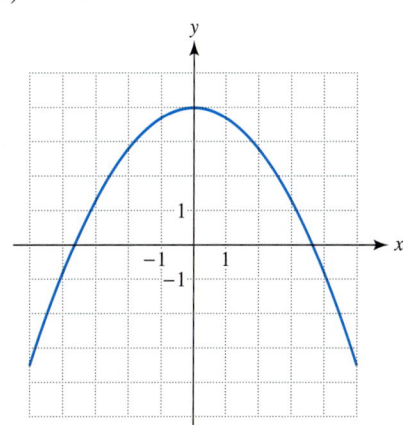

54. $f(x) = 0.3x^2 - 4$

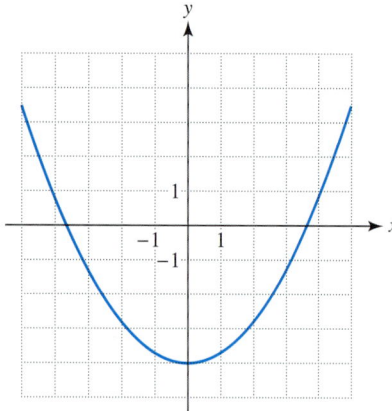

Exercises 55–60: Compute the average rate of change of f from x_1 to x_2. Round your answer to two decimal places when appropriate. Interpret your result graphically.

55. $f(x) = 7x - 2$, $x_1 = 1$, and $x_2 = 4$

56. $f(x) = -8x + 5$, $x_1 = -2$, and $x_2 = 0$

57. $f(x) = x^3 - 2x$, $x_1 = 0.5$, and $x_2 = 4.5$

58. $f(x) = 0.5x^2 - 5$, $x_1 = -1.3$, and $x_2 = 3.7$

59. $f(x) = \sqrt{2x - 1}$, $x_1 = 1$, and $x_2 = 3$

60. $f(x) = \sqrt[3]{x + 1}$, $x_1 = 0.3$, and $x_2 = 1.2$

61. *Remaining Life Expectancy*   The accompanying table lists the average *remaining* life expectancy E in years for females at age x. (Source: Department of Health and Human Services.)

x (yr)	0	10	20	30	40	50	60	70	80
E (yr)	72.3	69.9	60.1	50.4	40.9	31.6	23.1	15.5	9.2

(a)   Make a line graph of the data.
(b)   Assume that the graph represents a function f. Compute the average rates of change of f during each ten-year period. Interpret the results.
(c)   Determine the life expectancy (not the remaining life expectancy) of a female whose age is 20. What is the life expectancy of a female who is 70 years old? Discuss one reason why these two expectancies are not equal.

62. *Average Life Expectancy*   The accompanying table lists the average life expectancy E in years for males born in year x. (Source: Department of Health and Human Services.)

x (yr)	1920	1930	1940	1950	1960	1970	1980	1990
E (yr)	53.6	58.1	60.8	65.6	66.6	67.1	70.0	71.8

(a)   Make a line graph of the data.
(b)   Assume that the graph represents a function f. Compute the average rates of change of f during each ten-year period. Interpret the results.

63. *AIDS Deaths*   The table lists reported numbers of AIDS deaths in the United States. (Source: Department of Health and Human Services.)

Year	1981	1985	1990	1995
AIDS Deaths	128	6948	31,269	45,765

(a)   Calculate the average rates of change between consecutive data points in the table.
(b)   Interpret these average rates of change.
(c)   What does your answer from (b) indicate about the spread of the disease?

64. *Temperature Change*   The table gives the outside temperature in degrees Fahrenheit on a summer afternoon.

Time (P.M.)	12:00	1:00	3:15	4:30	6:00
Temperature (°F)	78	82	88	93	91

(a)   Calculate the average rate of change between consecutive data points in the table.
(b)   Determine the time interval(s) when the thermometer was rising the fastest, on average.

Writing about Mathematics

1.   Describe two methods to determine if a data set can be modeled by a linear function. Create or find a data set that consists of four or more ordered pairs. Determine if your data set can be modeled by a linear function.

2.   Suppose you are given a graphical representation of a function f. Explain how you would determine whether f was constant, linear, or nonlinear. How would you determine the type if you were given a numerical or symbolic representation? Give examples.

CHECKING BASIC CONCEPTS FOR SECTIONS 1.3 AND 1.4

1. Create symbolic, numerical, and graphical representations of a function f that computes the number of feet in x miles. For the numerical representation let $x = 1, 2, 3, 4, 5$.

2. Sketch the graphs of two relations—one that represents a function and one that does not represent a function. Explain your answer.

3. Find the slope of the line passing through the points $(-2, 4)$ and $(4, -5)$. If the graph of $f(x) = ax + b$ passes through these two points, what is the value of a?

4. Identify each representation of a function f as constant, linear, or nonlinear. Support your answer graphically. If possible, determine the x-intervals where f is increasing or decreasing.

(a) $f(x) = -1.4x + 5.1$
(b) $f(x) = 2x^2 - 5$
(c) $f(x) = 25$

5. The table shows corporate profits y in billions of dollars for three years x.
(a) Make a line graph of the data. Let this graph represent a function f.
(b) Compute the average rate of change of f from 1990 to 1993 and from 1993 to 1995. Interpret your results.

x (yr)	1990	1993	1995
y ($ billions)	236	384	463

1.5 Modeling Data with Functions

Exact and Approximate Models • Quadratic Models • Interpreting Mathematical Models

Introduction

Throughout history, people have attempted to explain natural phenomena by creating models. A model is based on observed data. It can be a diagram, an equation, a verbal expression, or some other form of communication. Models are used in diverse areas such as economics, physics, chemistry, astronomy, psychology, religion, and mathematics. Regardless of where it is used, a model is an *abstraction* that has the following two characteristics.

(1) A model is able to explain present phenomena. It should not contradict data and information that is already known to be correct.
(2) A model is able to make predictions about future data or results. It should be able to use current information to forecast future phenomena or create new information.

Mathematical models are used to forecast business trends, design the shapes of airplanes, estimate ecological trends, control highway traffic, describe epidemics, predict weather, and discover new information when human knowledge is inadequate. In this section mathematical models are used to describe several different sets of data.

Exact and Approximate Models

Not all mathematical models are exact representations of data. Data might appear to be nearly linear, but not exactly linear. In this case a linear function can be used to provide an *approximate model* of the data. In Figure 1.64 on the following page the data is modeled exactly by a linear function, whereas in

Figure 1.65 the data is modeled approximately. In real applications, an approximate model is much more likely to occur than an exact model.

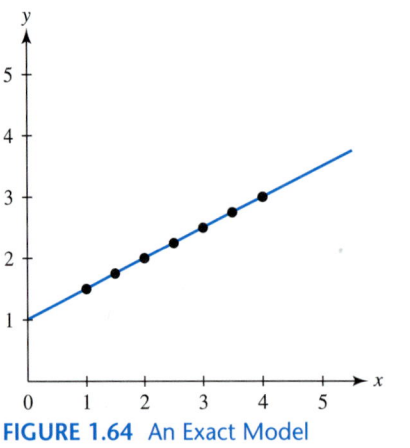

FIGURE 1.64 An Exact Model

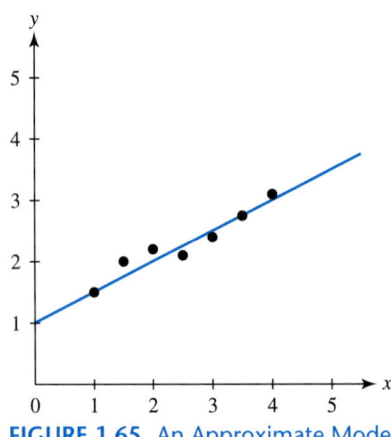

FIGURE 1.65 An Approximate Model

EXAMPLE 1   *Modeling data with a linear function*

Per capita personal incomes in the United States from 1990 to 1994 are listed in Table 1.17.

TABLE 1.17

Year	1990	1991	1992	1993	1994
Income	$18,635	$19,091	$20,105	$20,800	$21,809

Source: Department of Commerce.

This data can be modeled by $f(x) = 805.7x - 1,584,866$, where x is the year.
(a) Graph f and the data in [1989, 1995, 1] by [17,000, 23,000, 1000]. Use the graph to decide whether this model is exact or approximate.
(b) Find the slope of the graph of f. Interpret the slope.
(c) Use f to estimate per capita personal income in 1996.

Solution
(a) Graph $Y_1 = 805.7X - 1584866$ along with the data points

$$(1990, 18,635), (1991, 19,091), (1992, 20,105), (1993, 20,800),$$
$$\text{and } (1994, 21,809)$$

as shown in Figure 1.66. The model is approximate, since the line does not pass through each point.

[1989, 1995, 1] by [17,000, 23,000, 1000]

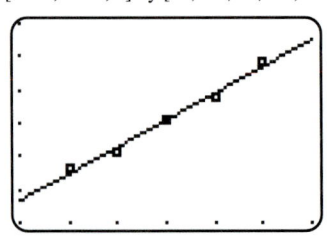

FIGURE 1.66

(b) Since f is expressed in the form $f(x) = ax + b$, where $a = 805.7$ and $b = -1,584,866$, the slope of the graph of f is 805.7. This slope suggests that per capita personal income has risen by approximately \$806 per year.

(c) To estimate per capita personal income in 1996, evaluate $f(1996)$.

$$f(1996) = 805.7(1996) - 1,584,866 = \$23,311.20 \qquad ▪$$

Quadratic Models

Many mathematical models use nonlinear functions to describe data. One important nonlinear function is a *quadratic function.* A quadratic function is a polynomial function with degree 2. Its implied domain includes all real numbers.

Quadratic function

Let a, b, and c be real numbers with $a \neq 0$. A function f represented by $f(x) = ax^2 + bx + c$ is a **quadratic function.**

The following are symbolic representations of quadratic functions.

$$f(x) = 2x^2 - 4x - 1$$
$$g(x) = 4 - x^2$$
$$h(x) = \frac{1}{3}x^2 + \frac{2}{3}x + 1$$

Graphs of these three functions are shown in Figures 1.67–1.69, respectively, below and on the following page. A graph of a quadratic function is a **parabola.** A parabola opens upward if a is positive and opens downward if a is negative. For example, since $a = -1 < 0$ for $g(x) = 4 - x^2$, the graph of g opens downward. The highest point on a parabola that opens downward or the lowest point on a parabola that opens upward is called the **vertex.** The vertical line passing through the vertex is called the **axis of symmetry.** The vertex and axis of symmetry are shown in each figure, except for Figure 1.68 where the axis of symmetry corresponds to the y-axis.

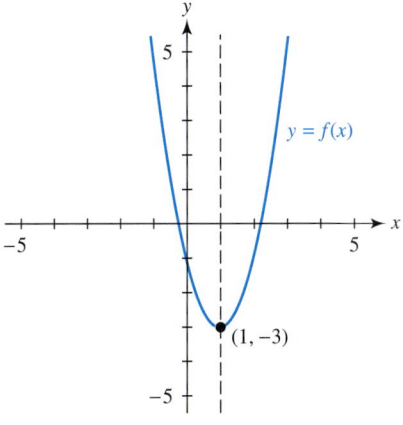

FIGURE 1.67

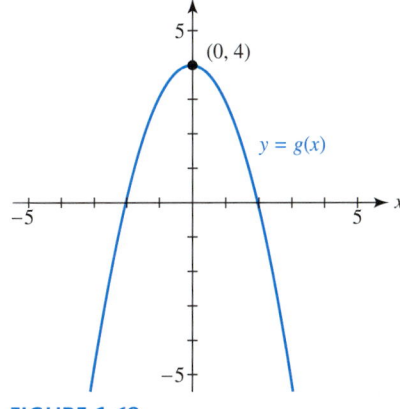

FIGURE 1.68

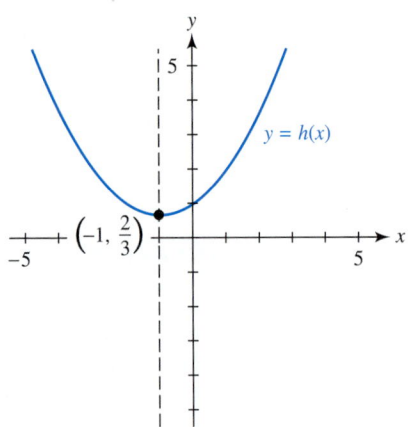

FIGURE 1.69

The leading coefficient a of a quadratic function not only determines whether its graph opens upward or downward, but it also controls the width of the parabola. The leading coefficients of $f(x)$, $g(x)$, and $h(x)$ are 2, -1, and $\frac{1}{3}$, respectively. Larger values for $|a|$ result in a parabola becoming narrower, while smaller values for $|a|$ result in a parabola becoming wider.

Given a symbolic representation for a quadratic function f, the following formula can be used to find the vertex symbolically.

Vertex formula

The *vertex* of the graph of $f(x) = ax^2 + bx + c$ with $a \neq 0$ is the point
$$\left(-\frac{b}{2a}, f\left(-\frac{b}{2a}\right)\right).$$

EXAMPLE 2   *Using the vertex formula*

Find the vertex of the graph of $f(x) = 1.5x^2 - 6x + 4$ symbolically. Support your answer graphically and numerically.

Solution

Symbolic   If $f(x) = 1.5x^2 - 6x + 4$, then $a = 1.5$, $b = -6$, and $c = 4$. The x-coordinate of the vertex is

$$x = -\frac{b}{2a} = -\frac{(-6)}{2(1.5)} = 2.$$

The y-coordinate of the vertex can be found by evaluating $f(2)$.

$$y = f(2) = 1.5(2)^2 - 6(2) + 4 = -2$$

Thus, the vertex is $(2, -2)$.

Graphical   This result can be supported by graphing $Y_1 = 1.5X^{\wedge}2 - 6X + 4$ as shown in Figure 1.70. The vertex is the lowest point on the graph.

Numerical   Numerical support is shown in Figure 1.71. The minimum y-value of -2 occurs when $x = 2$.

$[-4.7, 4.7, 1]$ by $[-3.1, 3.1, 1]$

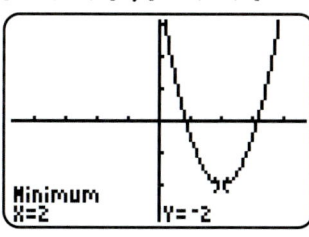

FIGURE 1.70

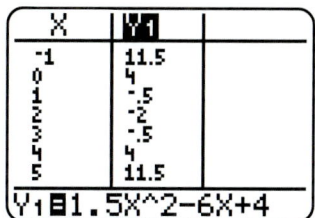

FIGURE 1.71   ∎

In the next example a quadratic function models research data involving photosynthesis.

EXAMPLE 3   *Maximizing a quadratic function graphically and symbolically*

In one study the efficiency of photosynthesis in an Antarctic species of grass was investigated. Table 1.18 lists results for various temperatures. The temperature x is in degrees Celsius and the efficiency y is given as a percent. The purpose of the research was to determine the temperature at which photosynthesis is most efficient. (Source: D. Brown and P. Rothery, *Models in Biology: Mathematics, Statistics and Computing.*)

TABLE 1.18

x (°C)	−1.5	0	2.5	5	7	10	12	15	17	20	22	25	27	30
y (%)	33	46	55	80	87	93	95	91	89	77	72	54	46	34

(a) Plot the data. Discuss reasons why a quadratic function might model this data.
(b) The quadratic function given by $f(x) = -0.249x^2 + 6.77x + 46.37$ was used to model the data. Plot f and the data in the same viewing rectangle.
(c) Determine graphically the temperature at which f predicts photosynthesis is most efficient.
(d) Solve part (c) symbolically.

Solution

(a) A plot of the data is shown in Figure 1.72. The y-values first increase and then decrease as the temperature x increases. The data suggests a parabolic shape opening downward.

$[-5, 35, 5]$ by $[20, 110, 10]$

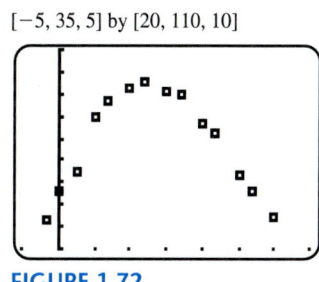

FIGURE 1.72

(b) The graph of $Y_1 = -.249X^2 + 6.77X + 46.37$ with the data is shown in Figure 1.73. Although the model is not exact, the parabola describes the general trend in the data. Notice that the parabola is opening downward since $a = -0.249 < 0$.

(c) The vertex is the highest point on the graph. Its coordinates are approximately (13.6, 92.4) as shown in Figure 1.74. This quadratic model predicts that the highest efficiency is about 92.4% and it occurs near 13.6°C. Although there are percentages in the table higher than 92.4%, 13.6°C is a reasonable estimate for the optimum temperature. The function f is attempting to model the general trend in the data and predict future results.

$[-5, 35, 5]$ by $[20, 110, 10]$ $[-5, 35, 5]$ by $[20, 110, 10]$

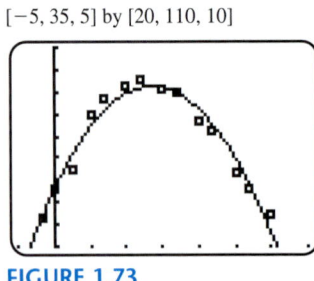   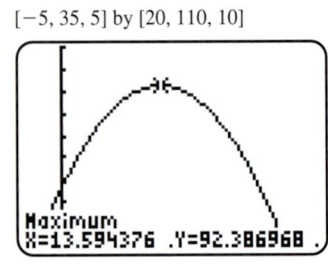

FIGURE 1.73   **FIGURE 1.74**

(d) The coordinates of the vertex of the graph of $f(x) = -0.249x^2 + 6.77x + 46.37$ can be found using the vertex formula with $a = -0.249$ and $b = 6.77$.

$$x = -\frac{b}{2a} = -\frac{6.77}{2(-0.249)} \approx 13.6°C$$

$$y = f\left(-\frac{b}{2a}\right) \approx -0.249(13.6)^2 + 6.77(13.6) + 46.37 \approx 92.4\%$$

The graphical and symbolic solutions provide similar results.   ■

 Technology Note *Regression*

In Example 3, the researchers used a technique called *quadratic regression* to determine $f(x) = -0.249x^2 + 6.77x + 46.37$. *Linear regression* was used in Example 1 to find $f(x) = 805.7x - 1,584,866$. Regression is discussed later in the text. However, you may wish to consult the manual for your graphing calculator to determine these formulas.

EXAMPLE 4   *Selecting a function to model data*

The number of women gainfully employed in the work force has changed significantly since 1900. Table 1.19 lists these numbers in millions for selected years.

TABLE 1.19

Year	1900	1910	1920	1930	1940	1950	1960	1970	1980	1990
Work Force (millions)	5.3	7.4	8.6	10.8	12.8	18.4	23.2	31.5	45.5	56.6

Source: Department of Labor.

[−10, 100, 10] by [0, 60, 10]

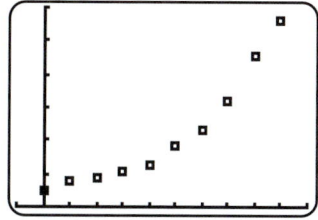

FIGURE 1.75

[−10, 100, 10] by [0, 60, 10]

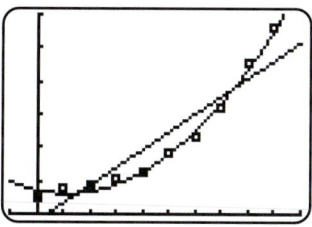

FIGURE 1.76

(a) Make a scatterplot of the data in $[-10, 100, 10]$ by $[0, 60, 10]$. Let $x = 0$ correspond to 1900, $x = 10$ to 1910, and so on.

(b) Conjecture whether the function f_1 or f_2 models the data better.

$$f_1(x) = 0.00797x^2 - 0.180x + 7.41$$
$$f_2(x) = 0.537x - 2.15$$

Graph the two polynomial functions together with the data to test your conjecture.

(c) Estimate the number of women in the work force in 1989.

Solution

(a) A scatterplot of the data is shown in Figure 1.75.

(b) The data does not lie on a line—it is *nonlinear.* Thus, we suspect that the quadratic function f_1, whose graph opens upward, may model the data. Graph $Y_1 = .00797X^2 - .18X + 7.41$ and $Y_2 = .537X - 2.15$, together with the data as in Figure 1.76. The quadratic function f_1 models the data better than the linear function f_2.

(c) To estimate the work force in 1989, evaluate $f_1(89)$.

$$f_1(89) = 0.00797(89)^2 - 0.180(89) + 7.41 \approx 54.5 \text{ million}$$ ■

Interpreting Mathematical Models

EXAMPLE 5 *Interpreting a graphical model*

Figure 1.77 shows a graph of a function f that computes the approximate length in millimeters of a small fish, called *Lebistes reticulatus,* during the first 14 weeks of its life. (Source: D. Brown.)

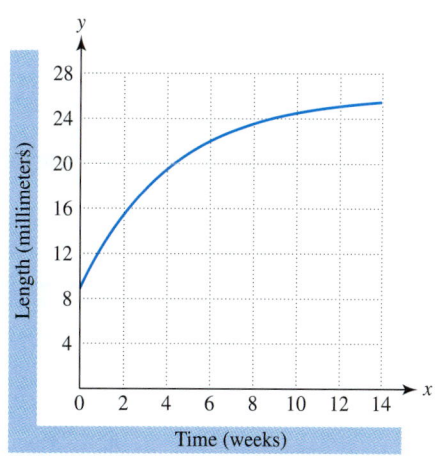

FIGURE 1.77 Growth of a Small Fish

(a) Estimate the length of the fish when it hatches, at six weeks, and at twelve weeks.

(b) Approximate the average rates of change of f from hatching to six weeks and from six weeks to twelve weeks. During which time period is the fish growing faster?

Solution

(a) At $x = 0$ the fish hatches and the graph intersects the y-axis at approximately 9 millimeters. At six weeks its length is about 22 millimeters, and at twelve weeks it is about 25 millimeters.

(b) From part (a), the points $(0, 9)$, $(6, 22)$, and $(12, 25)$ lie on the graph of f. See Figure 1.78. To estimate the average rate of change of f from hatching to six weeks, find the slope m_1 of the line passing through the points $(0, 9)$ and $(6, 22)$.

$$m_1 = \frac{y_2 - y_1}{x_2 - x_1} = \frac{22 - 9}{6 - 0} = \frac{13}{6} \approx 2.2$$

This means that the fish is growing at an average rate of 2.2 millimeters per week between hatching and six weeks. The average rate of change of f from six weeks to twelve weeks is equal to the slope m_2 of the line passing through the points $(6, 22)$ and $(12, 25)$.

$$m_2 = \frac{y_2 - y_1}{x_2 - x_1} = \frac{25 - 22}{12 - 6} = \frac{1}{2}$$

From six weeks to twelve weeks the fish grows at an average rate of 0.5 millimeter per week. Therefore, the fish grows faster from hatching to six weeks.

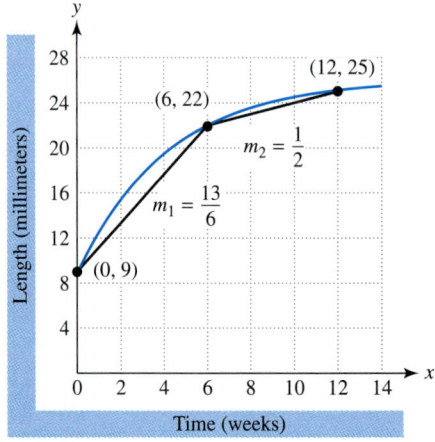

FIGURE 1.78 Average Rates of Change

Critical Thinking

During the years from 1875 to 1905 the Hudson Bay Company in Canada logged trap records of both lynx and snowshoe hares. Line graphs representing these records are shown.

(a) Describe in general terms how the numbers of lynx and snowshoe hares changed during this time period.

(b) The lynx is a predator of the snowshoe hare. Use this fact to help explain why the numbers of lynx and hares varied as they did.

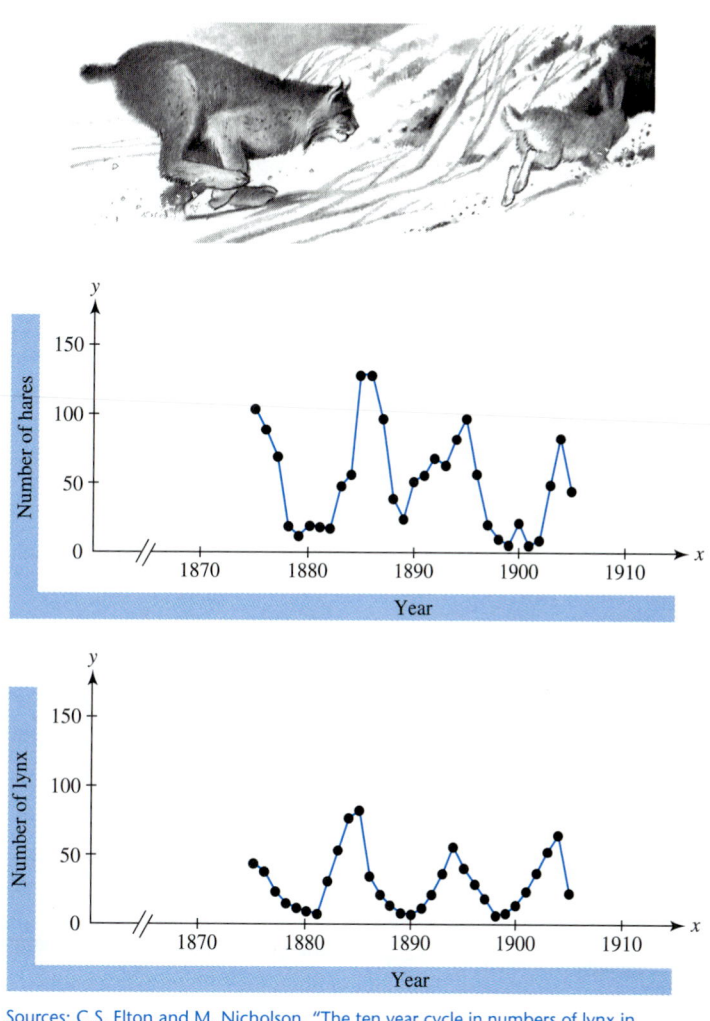

Sources: C.S. Elton and M. Nicholson, "The ten year cycle in numbers of lynx in Canada"; D. Brown.

1.5 PUTTING IT ALL TOGETHER

Mathematics plays a central role in our attempt to describe the world around us. Mathematical models are created every day to develop new products, explain current trends, and make predictions about the future. Linear and quadratic functions are used to model data. The following table summarizes important concepts related to linear and quadratic models.

Concept	Symbolic Representation	Explanation
Linear model	$f(x) = ax + b$, where a and b are constants	Models data that changes at a constant rate Data may be increasing or decreasing, but not both
Quadratic model	$f(x) = ax^2 + bx + c$, where a, b, and c are real numbers with $a \neq 0$	Models nonlinear data Data may be increasing, decreasing, or both
Vertex formula	The vertex for $f(x) = ax^2 + bx + c$ is the point $\left(-\dfrac{b}{2a}, f\left(-\dfrac{b}{2a}\right)\right)$	

1.5 EXERCISES   Tapes 1, 3

Basics of Quadratic Functions

Exercises 1–6: Sketch a graph of f by hand. Use a graphing calculator to support your result.

1. $f(x) = x^2$

2. $f(x) = -2x^2$

3. $f(x) = -\dfrac{1}{2}x^2$

4. $f(x) = 4 - x^2$

5. $f(x) = x^2 - 3$

6. $f(x) = x^2 - 2x + 1$

Exercises 7–12: Complete the following.

(a) *Determine if f is a linear or quadratic function. Support your answer by graphing f in an appropriate viewing rectangle.*

(b) *Evaluate f(a) for the given value of a without a calculator.*

7. $f(x) = 2x^2 - 5$, $a = 3$

8. $f(x) = 3x + 1$, $a = -8$

9. $f(x) = -x + 5$, $a = -\dfrac{3}{2}$

10. $f(x) = 3 - x - \dfrac{1}{2}x^2$, $a = 2$

11. $f(x) = 0.5x^2 - 4x + 5$, $a = 2$

12. $f(x) = 3^2 x - \dfrac{1}{3}$, $a = \dfrac{1}{3}$

Exercises 13–16: Approximate f(a) for the given value of a to two significant digits.

13. $f(x) = 1.3x^2 - 5.2x + 1.8$, $a = 1.7$

14. $f(x) = 1.2x^2 - x$, $a = 0.35$

15. $f(x) = 3.45x^2 + 9.6$, $a = 3.2$

16. $f(x) = 1 - \dfrac{1}{3}x + \dfrac{1}{6}x^2$, $a = 1.9$

Exercises 17–24: Graph the quadratic function f in an appropriate viewing rectangle.

(a) *Find the vertex symbolically. Support your answer graphically and numerically.*

(b) *Identify the x-intervals where f is increasing or decreasing.*

17. $f(x) = 6 - x^2$

18. $f(x) = 2x^2 - 2x + 1$

19. $f(x) = x^2 - 6x$

20. $f(x) = -2x^2 + 4x + 5$

21. $f(x) = x^2 - 3.8x - 2$

22. $f(x) = -4x^2 + 16x$

23. $f(x) = 1.5 - 3x - 6x^2$

24. $f(x) = 0.25x^2 - 1.5x + 1$

Mathematical Models

25. *Modeling Growth* One of the earliest studies (1913) about population growth was done using

yeast plants. A small amount of yeast was placed in a container with a fixed amount of nourishment. The units of yeast were recorded every two hours. The data is listed in the table. (Sources: T. Carlson, *Biochem*; D. Brown.)

Time	0	2	4	6	8
Yeast	9.6	29.0	71.1	174.6	350.7

Time	10	12	14	16	18
Yeast	513.3	594.8	640.8	655.9	661.8

(a) Make a scatterplot of the data.
(b) Determine the two-hour time period when the yeast population increased the most.
(c) Use the graph to describe the growth of the yeast plants.

26. *Stopping Distance*   Faster driving speeds require greater stopping distances. If a car doubles its speed, the stopping distance more than doubles. Graph each of the three functions in [20, 70, 10] by [0, 500, 100]. Determine which function best models the stopping distance in feet for a car traveling x miles per hour.

$f_1(x) = 450$ (constant model)

$f_2(x) = 6x$ (linear model)

$f_3(x) = 0.07x^2 + 1.2x - 22$ (quadratic model)

27. *Seedling Growth*   The line graph shows the heights y in centimeters of melon seedlings growing at different temperatures x in degrees Celsius. (Source: R. Pearl, "The growth of *Cucumis melo* seedlings at different temperatures.")

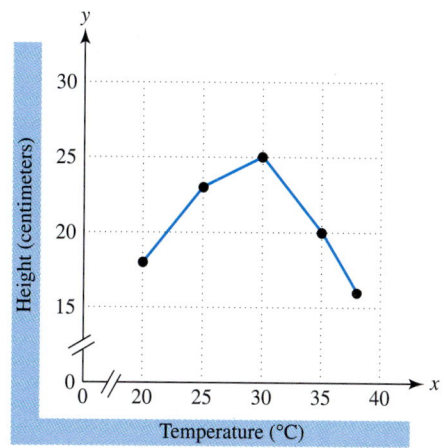

(a) Describe the effect that temperature appears to have on the growth of the seedlings.
(b) Use graphing to determine which function models the data better,

$$f_1(x) = -0.095x^2 + 5.4x - 52.2 \quad \text{or}$$
$$f_2(x) = 0.002x^2 - 0.5x + 34.$$

(c) Use a function to estimate the optimum temperature for the growth of melon seedlings.

28. *Survival Rates of Birds*   The survival rates of sparrowhawks vary according to their age. The table summarizes the results of one study by listing the age in years and the percent in decimal form of birds that survived the previous year. For example, 52% of sparrowhawks that reached age six lived to be seven years old. (Source: D. Brown.)

Age	1	2	3	4	5
Percent	0.45	0.60	0.71	0.67	0.67

Age	6	7	8	9
Percent	0.61	0.52	0.30	0.25

(a) Describe the relationship between age and the likelihood of survival for one year.
(b) Make a scatterplot of the data together with the functions

$$f_1(x) = -0.0357x + 0.711 \quad \text{and}$$
$$f_2(x) = -0.0207x^2 + 0.171x + 0.330$$

in [0, 10, 1] by [0, 0.8, 0.1]. Decide which function models the data better.
(c) Use a function to estimate the age when a sparrowhawk has the best chance for surviving the next year.

29. *Blood Pressure*   In one study, the relationship between age and average systolic blood pressure was analyzed. The female participants were grouped in ten-year intervals of 30 to 39, 40 to 49, and so on. The age listed in the table is the midpoint of the interval. (Source: D. Brown.)

Age	35	45	55	65	75
Systolic Pressure	114	124	143	158	166

(a) Does there appear to be a relationship between age and blood pressure? Explain.

(b) Make a scatterplot of the data together with the functions

$$f_1(x) = 1.38x + 65.1 \text{ and}$$
$$f_2(x) = 43 + 12\sqrt{x}$$

in [30, 80, 5] by [110, 170, 5]. Decide which function models the data better.

30. *Distance from Home*   At noon two friends start driving on a straight road from home to a movie theater 25 miles away. Traffic is slow and they travel at a constant speed of 25 miles per hour until they reach the theater. The movie lasts two hours, after which they drive home at a constant speed of 50 miles per hour. Let f be a function that represents the distance between the car and home at time x in hours. Assume that $x = 0$ corresponds to noon.

(a) Sketch a graph of f for $0 \leq x \leq 3.5$.

(b) Compute the average rates of change of f from noon to 1:00 P.M., from 1:00 P.M. to 3:00 P.M., and from 3:00 P.M. to 3:30 P.M. Interpret these average rates of change.

Exercises 31–34: Quadratic Models   Match the physical situation with the graph of the quadratic function that models it best.

31. The height y of a stone thrown from ground level after x seconds

32. The number of people attending a popular movie x weeks after its opening

33. The temperature after x hours in a house where the furnace quits and a repairperson fixes it

34. The number of reported AIDS cases from 1980 to year x

a.

b.

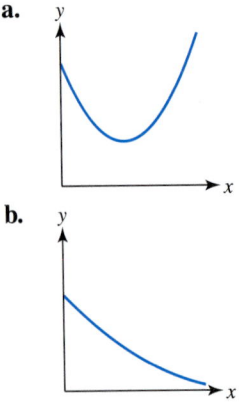

c.

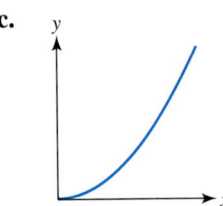

d.

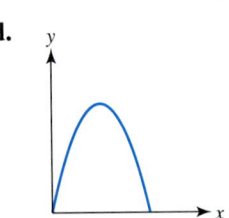

Exercises 35–38: Maximizing Altitude   If air resistance is ignored, the height h of a projectile above the ground after x seconds is given by $h(x) = -\frac{1}{2}gx^2 + v_0x + h_0$. In this formula g is the acceleration due to gravity, v_0 is the projectile's initial velocity, and h_0 is its initial height. The leading coefficient is negative since gravity pulls downward. If v_0 is positive the projectile is propelled upward, and if v_0 is negative it is propelled downward. This formula is valid not only for Earth, but also for other celestial bodies. A ball is thrown straight up at 88 feet per second from a height of 25 feet.

(a) For each value of g, graphically estimate both the maximum height and the time when this occurs.

(b) Solve part (a) symbolically.

35. $g = 32$ (Earth)

36. $g = 5.1$ (Moon)

37. $g = 13$ (Mars)

38. $g = 88$ (Jupiter)

Interpretation of Graphical Models

39. *Smoking and Stress*   Smoking coupled with significant stress is an additional risk factor associated with lung cancer. The accompanying figure models the results of a study that shows the health effect that smoking and stress have on individuals. The graph computes the percentage y of people who died from lung cancer both with stress (solid curve) and without stress (dashed curve), while smoking x cigarettes per day. (Adapted from: H. Eysenck, *Smoking, Personality, and Stress.*)

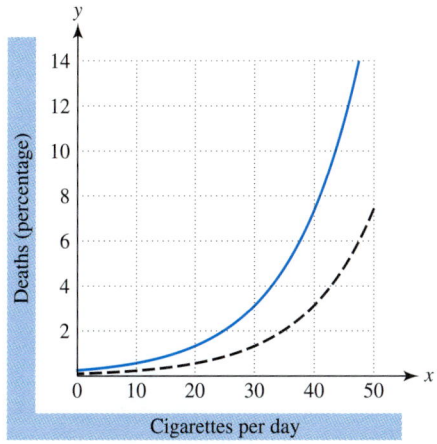

(a) Estimate the percentages of individuals who died from lung cancer and smoked 40 cigarettes per day with and without stress.

(b) What effect does increasing the number of cigarettes smoked per day have on the likelihood of dying from cancer?

(c) Discuss the combined effect of smoking and stress.

40. *Identifying a Model*   Two identical cylindrical tanks A and B each contain 100 gallons of water. Tank A has a pump removing water from it at a constant rate of eight gallons per minute. Tank B has a plug removed from its bottom and water begins to flow out. The accompanying figure shows a graph of the amount of water in each tank after x minutes.

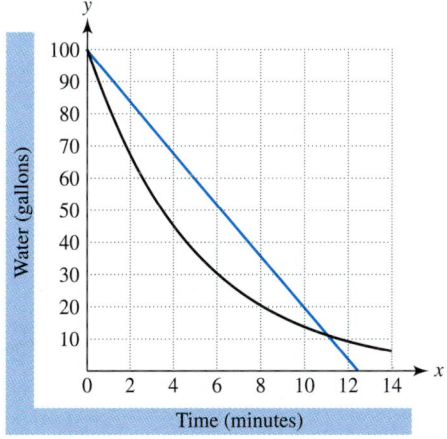

(a) Identify which graph is associated with each tank. Explain your answer.

(b) Use the graphs to determine which tank is half-empty first and which tank is empty first.

(c) Estimate how many minutes are required for each tank to become half-empty.

41. (Refer to Example 5.) The figure shows a graph of a function f that models the weight in milligrams of

a small fish, called *Lebistes reticulatus,* during the first 14 weeks of its life. (Source: D. Brown.)

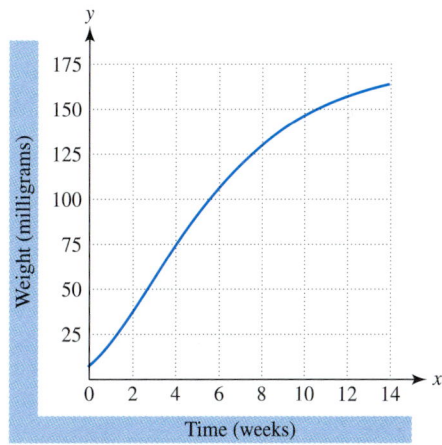

(a) Estimate the weight of the fish when it hatches, at six weeks, and at twelve weeks.

(b) Approximate the average rates of change of f from hatching to six weeks and from six weeks to twelve weeks. Interpret these rates of change. During which time period is the fish gaining weight the fastest?

42. *Average Temperature*   The graph shows monthly average temperature in degrees Fahrenheit at a location in Canada. The variable x is in months, with $x = 1$ corresponding to January. (Source: A. Miller and J. Thomson, *Elements of Meteorology.*)

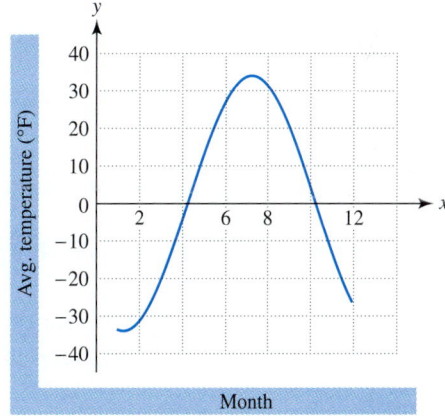

(a) Determine the month with the highest average temperature.

(b) Estimate the lowest monthly average temperature.

(c) Conjecture what the *yearly* average temperature might be. Explain your reasoning.

43. *Carbon Dioxide Levels*   At Mauna Loa, Hawaii atmospheric carbon dioxide levels in parts per million (ppm) have been measured regularly since 1958. The accompanying figure shows a graph of the carbon dioxide levels between 1960 and 1995. (Source: A. Nilsson, *Greenhouse Earth.*)

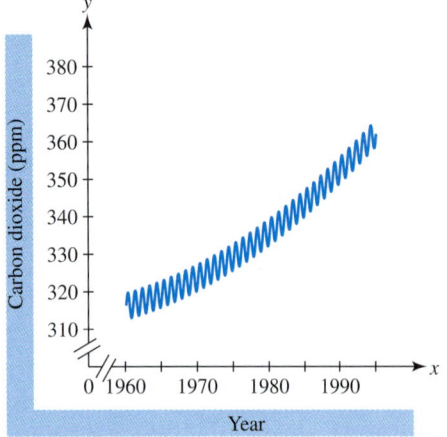

(a) What is the overall trend in the carbon dioxide levels at Mauna Loa?

(b) Discuss what happens to the carbon dioxide levels during each year.

(c) Give a possible explanation for the shape of this graph.

44. (Refer to the previous exercise.) The atmospheric carbon dioxide levels at Barrow, Alaska, in parts per million (ppm) from 1970 to 1995 are shown in the figure. (Source: M. Zeilik, *Introductory Astronomy and Astrophysics.*)

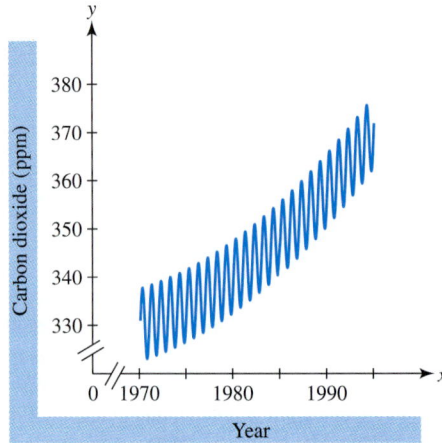

(a) Compare this graph with the graph from the previous exercise.

(b) Discuss possible reasons for their similarities and differences.

Writing about Mathematics

1. Create an example of a scatterplot that could be modeled by a linear function f. Explain what the slope of the graph of f represents.

2. Create an example of a scatterplot that could be modeled by a quadratic function f. Explain why the leading coefficient and vertex are important concepts when graphing a parabola.

1.6 Transformations of Graphs

Translating Graphs of Functions • Fitting Data Visually • Dynamically Displayed Data

Introduction

In the previous section the functions used to model data were given. In real applications functions are not provided—they must be created. This section presents techniques based on visualization, to determine functions that describe data.

Data can be presented either *statically* or *dynamically.* Static displays of data do not demonstrate movement or change in the data. Data printed on a page is typically static. Dynamic displays of data can show movement of data and how it changes over time. A typical weather forecast on television exhibits data both ways. The listing of the high and low temperatures is a static representation of data. The movement of a storm or changes in cloud patterns are dynamic displays of data. Movies and video games consist of a sequence of individual pictures, displayed at a rapid speed that the eye cannot detect. Each individual picture can be thought of as static data, but as a sequence, they become dynamic.

Before we can create functions to model data, we need to develop the mathematical techniques necessary for translating graphs of functions. (References: S. Hoggar, *Mathematics for Computer Graphics;* A. Watt, *3D Computer Graphics.*)

Translating Graphs of Functions

Graphs of three functions defined by $f(x) = x$, $g(x) = x^2$, and $h(x) = \sqrt{x}$ will be used to demonstrate translations in the xy-plane. Symbolic, numerical, and graphical representations of these functions are shown in Figures 1.79–1.81, respectively, below and on the following page. Points listed in the table are plotted on the graph.

x	-2	-1	0	1	2
y	-2	-1	0	1	2

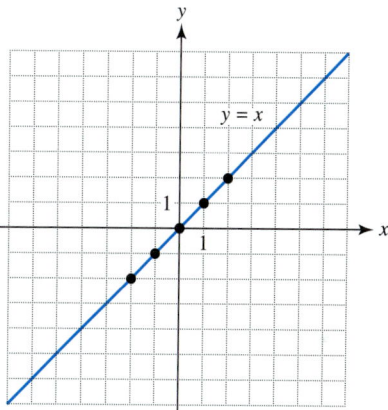

FIGURE 1.79

x	-2	-1	0	1	2
y	4	1	0	1	4

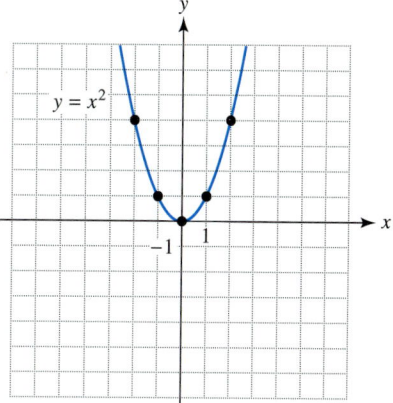

FIGURE 1.80

x	0	1	4
y	0	1	2

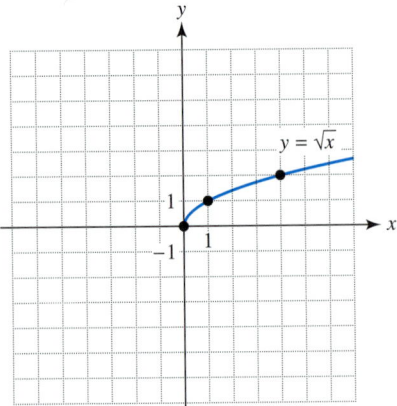

FIGURE 1.81

Vertical Shifts.   If 2 is added to the formula of each function, the original graphs are shifted upward 2 units. The graphs of $y = x + 2$, $y = x^2 + 2$, and $y = \sqrt{x} + 2$ are shown in Figures 1.82–1.84. A table of points is included, together with the graph of the original function. Notice how the y-values in both the graphical and numerical representations increase by 2 units.

x	−2	−1	0	1	2
y	0	1	2	3	4

x	−2	−1	0	1	2
y	6	3	2	3	6

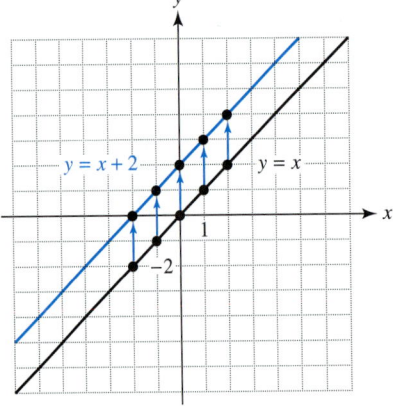

FIGURE 1.82

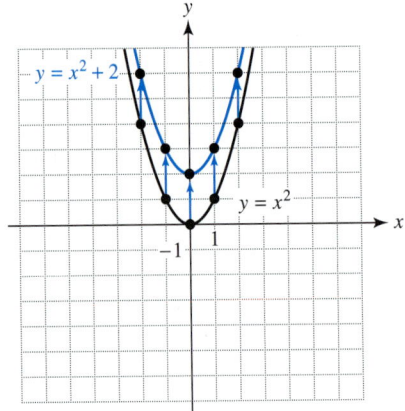

FIGURE 1.83

x	0	1	4
y	2	3	4

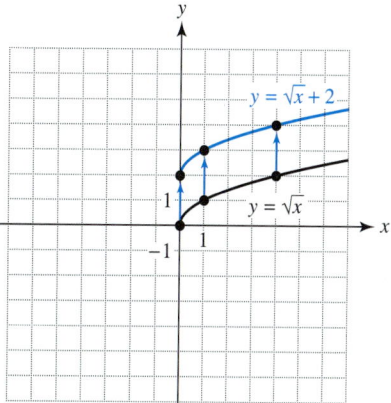

FIGURE 1.84

If 2 is subtracted from each of the original formulas, the graphs would translate down 2 units. Verify this by graphing $y = x - 2$, $y = x^2 - 2$, and $y = \sqrt{x} - 2$. Translations of this type are called *vertical shifts*. They do not alter the shape of the graph—only its position. A vertical shift is an example of an *isometry*, because distance between any two points on the graph does not change. The original and shifted graphs are congruent.

Horizontal Shifts. If the variable x is replaced by $(x - 2)$ in each of the formulas for f, g, and h, a different type of shift results. In Figures 1.85–1.87 below and on the following page, the graphs and tables of $y = x - 2$, $y = (x - 2)^2$, and $y = \sqrt{(x - 2)}$ are shown, together with the graphs of the original functions.

x	0	1	2	3	4
y	−2	−1	0	1	2

x	0	1	2	3	4
y	4	1	0	1	4

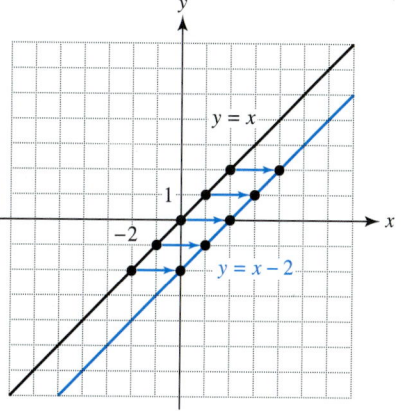

FIGURE 1.85

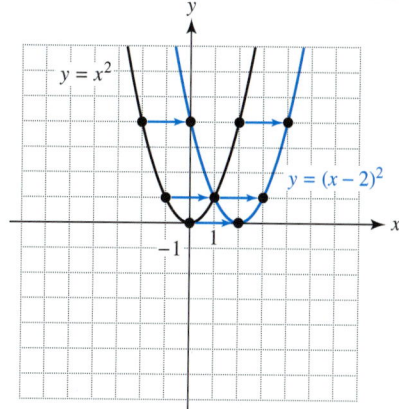

FIGURE 1.86

x	2	3	6
y	0	1	2

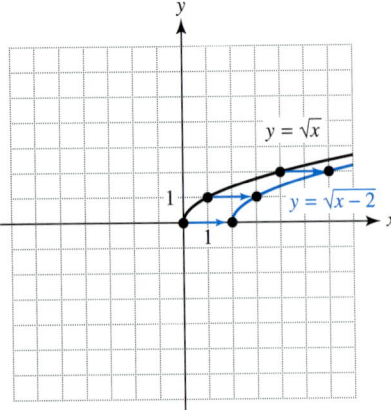

FIGURE 1.87

Each new graph suggests a shift of the original graph to the *right* by 2 units. Notice that a table for a shifted graph can be obtained from the original table by *adding* 2 to each *x*-value.

If the variable x is replaced by $(x + 3)$ in each equation, the original graphs are translated to the *left* 3 units. The graphs of $y = x + 3$, $y = (x + 3)^2$, and $y = \sqrt{(x + 3)}$ and their tables are shown in Figures 1.88–1.90. This type of translation is a *horizontal shift*. Horizontal shifts are also examples of isometries. The table for the shifted graph is obtained from the original table by *subtracting* 3 from each *x*-value.

x	-5	-4	-3	-2	-1
y	-2	-1	0	1	2

x	-5	-4	-3	-2	-1
y	4	1	0	1	4

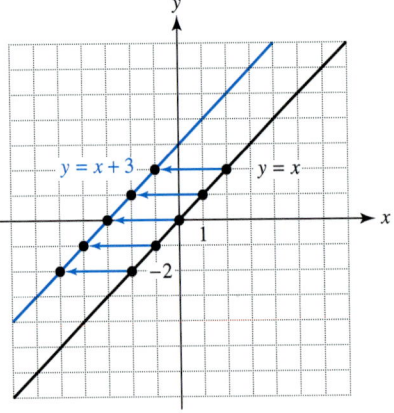

FIGURE 1.88

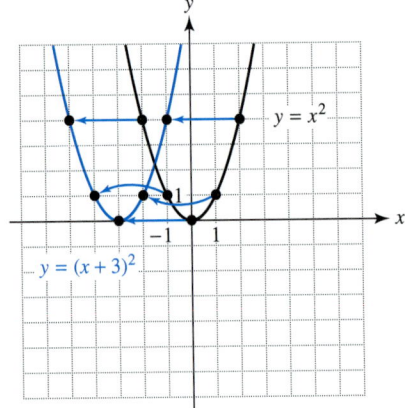

FIGURE 1.89

x	-3	-2	1
y	0	1	2

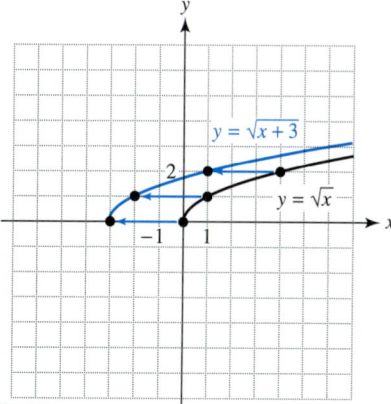

FIGURE 1.90

These ideas are summarized in the following box.

Vertical and horizontal shifts

Let f be a function, and let k be a positive number.

To graph	Shift the graph of $y = f(x)$ by k units
$y = f(x) + k$	upward
$y = f(x) - k$	downward
$y = f(x - k)$	right
$y = f(x + k)$	left

Shifts can be combined to translate a graph of $y = f(x)$ both vertically and horizontally. For example, to shift the graph of $y = f(x)$ to the right 2 units and downward 4 units, we graph $y = f(x - 2) - 4$. If $y = |x|$, then $y = f(x - 2) - 4$ $= |x - 2| - 4$ as shown in Figure 1.91 on the following page. This technique is illustrated in the next example.

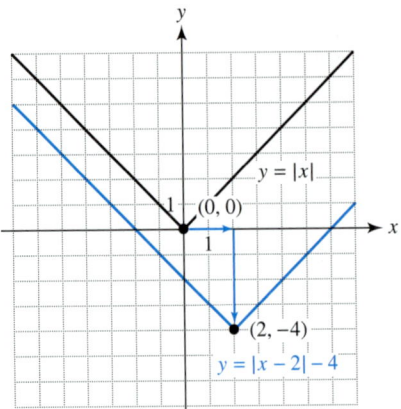

FIGURE 1.91

EXAMPLE 1   *Combining vertical and horizontal shifts*

Find an equation that shifts the graph of $f(x) = \frac{1}{2}x^2 - 4x + 6$ left 8 units and upward 4 units. Support this result by graphing.

Solution

To shift the graph of f left 8 units, replace x with $(x + 8)$ in the formula for $f(x)$. This results in

$$y = f(x + 8) = \frac{1}{2}(x + 8)^2 - 4(x + 8) + 6.$$

To shift the graph of this new equation upward 4 units, add 4 to the formula to obtain

$$y = f(x + 8) + 4 = \frac{1}{2}(x + 8)^2 - 4(x + 8) + 10.$$

The graphs of $Y_1 = .5X^2 - 4X + 6$ and $Y_2 = .5(X + 8)^2 - 4(X + 8) + 10$ are shown in Figure 1.92. The vertex of Y_2 is 8 units left and 4 units upward compared to the vertex of Y_1.

$[-9, 9, 2]$ by $[-6, 6, 2]$

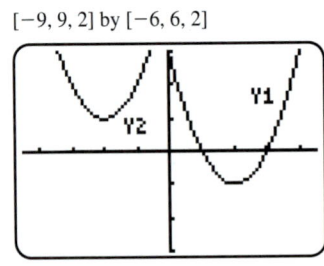

FIGURE 1.92   ■

Fitting Data Visually

Our knowledge about transformations of graphs can be used to create modeling functions. We can transform the line $y = x$ so that it passes through a point (h, k) with a slope m by adjusting the slope and then using horizontal and vertical shifts. This results in the equation $y = m(x - h) + k$. For example, to obtain the line passing through the point (2, 3) with slope $\frac{1}{2}$, perform the following transformations. Figure 1.93 shows these transformations.

$$y = x \qquad y = \frac{1}{2}x \qquad y = \frac{1}{2}(x - 2) \qquad y = \frac{1}{2}(x - 2) + 3$$

Original equation   Adjust slope   Horizontal shift   Vertical shift
to $\frac{1}{2}$ right 2 units   upward 3 units

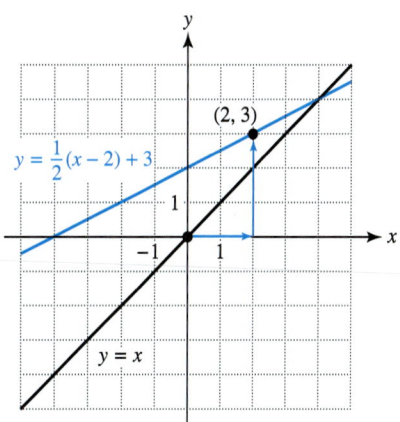

FIGURE 1.93

These concepts are now summarized.

Equation of a line

An equation of the line passing through the point (h, k) with slope m is

$$y = m(x - h) + k.$$

EXAMPLE 2   *Using translations to model linear data*

Tuition has risen at private colleges. Table 1.20 lists the average tuition for selected years.

TABLE 1.20

Year	1980	1985	1990	1995
Tuition	$3,617	$6,121	$9,340	$12,432

Source: The College Board.

(a) Make a scatterplot of this data in [1978, 1997, 1] by [0, 13,000, 1000]. What type of model does the scatterplot suggest?
(b) Use transformations of the line $y = x$ to find a linear function given by $f(x) = m(x - h) + k$ that models this data. Interpret the slope m.
(c) Use f to estimate tuition in 1987 and in 2000.

Solution

(a) The scatterplot in Figure 1.94 suggests a linear model.

[1978, 1997, 1] by [0, 13,000, 1000]

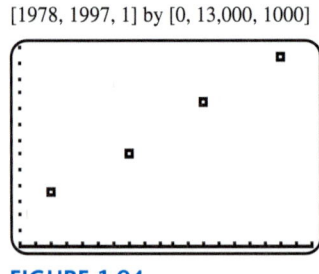

FIGURE 1.94

(b) Let the slope be m and select a data point in Figure 1.94 that appears to lie on a line that models the data. For example, suppose we choose the point (1980, 3617). Translate the line $y = x$ right 1980 units and up 3617 units by replacing x by $(x - 1980)$ and adding 3617. This new line with slope m has the equation $y = m(x - 1980) + 3617$.

To find an appropriate value for the slope m, graph the equations $y = m(x - 1980) + 3617$ with $m = 10$, 100, and 1000. This is shown in Figure 1.95. Notice how a larger slope m creates a line with a greater tilt.

[1978, 1997, 1] by [0, 13,000, 1000]

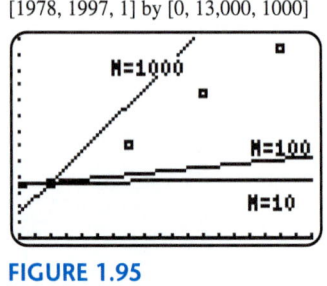

FIGURE 1.95

From Figure 1.95, the slope m that we desire is between 100 and 1000. Using trial and error, $m \approx 575$. A graph of $y = 575(x - 1980) + 3617$ and the data are shown in Figure 1.96. There is no "correct" answer for m. However, most values would fall between 500 and 700. A value of $m = 575$ indicates that between 1980 and 1995 tuition at private colleges rose approximately $575 per year.

[1978, 1997, 1] by [0, 13,000, 1000]

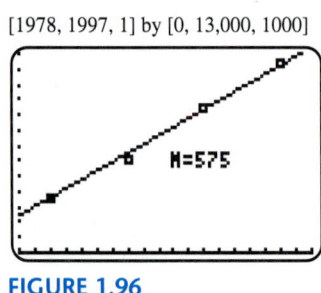

FIGURE 1.96

(c) To estimate average tuition in 1987 and 2000, evaluate $f(1987)$ and $f(2000)$, respectively.

$$f(1987) = 575(1987 - 1980) + 3617 = \$7{,}642$$

$$f(2000) = 575(2000 - 1980) + 3617 = \$15{,}117 \qquad ■$$

In 1981 the first case of AIDS was reported in the United States. Since then, AIDS has become one of the most devastating diseases of our time. Table 1.21 lists the total *cumulative* deaths from AIDS in the United States.

TABLE 1.21

Year	1982	1984	1986	1988	1990	1992	1994
AIDS Deaths	619	5605	24,593	61,911	120,811	196,283	270,533

Source: Department of Health and Human Services.

By examining Table 1.21, it can be seen that the data is nonlinear. The numbers of AIDS deaths have not increased by a constant amount during each two-year period. Therefore, modeling this data requires a nonlinear function, such as a quadratic function.

We can use transformations of the parabola $y = x^2$ to model nonlinear data. The vertex of the graph of $y = x^2$ is (0, 0). By adjusting the leading coefficient and applying horizontal and vertical shifts, we can transform the graph of $y = x^2$ into a parabola that has a desired shape with vertex (h, k). The resulting equation is $y = a(x - h)^2 + k$. For example, to obtain a parabola with a vertex of $(-2, -1)$ and a leading coefficient of -3 perform the following steps. Figure 1.97 illustrates this process.

$y = x^2$	$y = -3x^2$	$y = -3(x + 2)^2$	$y = -3(x + 2)^2 - 1$
Original equation	Adjust leading coefficient	Shift vertex left 2 units	Shift vertex downward 1 unit

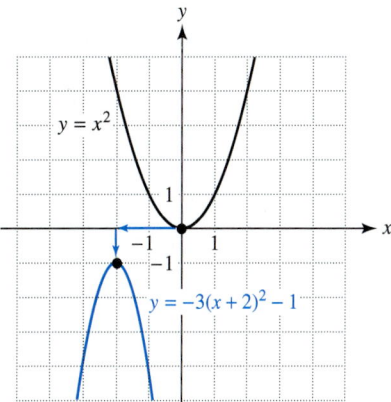

FIGURE 1.97

The equation $y = a(x - h)^2 + k$ is referred to as the *standard equation of a parabola.* The axis of symmetry is vertical.

> ### Standard equation of a parabola
>
> The **standard equation of a parabola** with vertex (h, k) is
>
> $$y = a(x - h)^2 + k$$
>
> where $a \neq 0$ is a constant. If $a > 0$ the parabola opens upward, and if $a < 0$ the parabola opens downward.

In the next example a function is created that models the numbers of AIDS deaths.

EXAMPLE 3   *Modeling AIDS deaths*

Use the data in Table 1.21 to complete the following.
(a) Make a scatterplot of the data. What type of model does the scatterplot suggest?
(b) Determine a quadratic function defined by $f(x) = a(x - h)^2 + k$ that models this data. Graph f together with the data points.
(c) Use f to predict the number of AIDS deaths by 2000.

Solution

(a) A scatterplot of the data in Table 1.21 is shown in Figure 1.98. The data is nonlinear and increasing. The scatterplot suggests the right half of a parabola opening upward.

[1980, 1996, 2] by [−50,000, 300,000, 50,000]

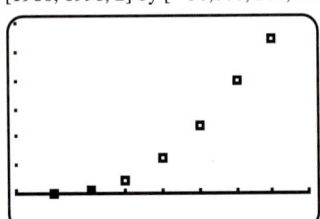

FIGURE 1.98

(b) Since the parabola is opening up, it follows that $a > 0$ and the vertex is the lowest point on the parabola. The known minimum number of deaths is 619 in 1982. One possibility for the vertex (h, k) is $(1982, 619)$. Translate the graph of $y = x^2$ right 1982 units and up 619 units. If the leading coefficient is a, this results in $f(x) = a(x - 1982)^2 + 619$. To estimate a value for a, graph $y = a(x - 1982)^2 + 619$ with $a = 500, 1000,$ and 5000 in the same viewing rectangle with the data points. The resulting graphs are shown in Figure 1.99. The larger the value of a, the faster the graph of f increases.

[1980, 1996, 2] by [−50,000, 300,000, 50,000]

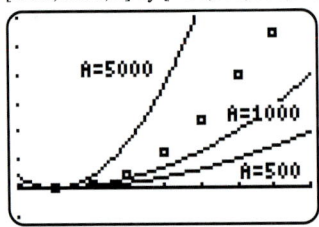

FIGURE 1.99

From the graph it can be seen that the desired value of a is between 1000 and 5000. The fact that a is quite large indicates that the number of deaths is increasing rapidly. With a little experimentation, a reasonable value for a near 1900 is found. A scatterplot of the data and the graph of $f(x) = 1900(x - 1982)^2 + 619$ are shown in Figure 1.100.

[1980, 1996, 2] by [−50,000, 300,000, 50,000]

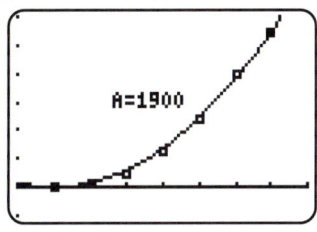

A=1900

FIGURE 1.100

(c) To predict the number of deaths from AIDS by 2000, evaluate $f(2000)$.

$$f(2000) = 1900(2000 - 1982)^2 + 619 = 616{,}219$$ ■

Critical Thinking

Find current data on AIDS deaths. Test the model presented in Example 3 to see if it is accurate. If it is not, try to improve the model.

Dynamically Displayed Data

With the introduction of computer graphics, dynamic displays of data are routine. Translations of graphs and figures play an important role in this new technology. In early motion pictures, it was common to have the background move to create the appearance that the actors and actresses were moving. This same technique is used today in two-dimensional graphics.

In video games, the background often is translated to give the illusion that the player in the game is moving. A simple scene of a mountain and an airplane is shown in Figure 1.101. In order to make it appear as though the airplane is flying to the right, the image of the mountain could be translated horizontally to the left as shown in Figure 1.102. (Reference: C. Pokorny and C. Gerald, *Computer Graphics.*)

FIGURE 1.101

FIGURE 1.102

EXAMPLE 4   *Using translations to model movement*

Suppose the mountain in Figure 1.101 can be described by the quadratic function represented by $f(x) = -0.4x^2 + 4$ and that the airplane is located at the point $(1, 5)$.

(a) Graph f in $[-4, 4, 1]$ by $[0, 6, 1]$, where the units are in kilometers. Plot a point to mark the position of the airplane.

(b) Assume that the airplane is moving horizontally to the right at 0.4 kilometer per second. To give a video player the illusion that the airplane is moving, graph the image of the mountain and the position of the airplane after five seconds and after ten seconds.

Solution

(a) The graph of $y = f(x) = -0.4x^2 + 4$ and the position of the airplane at $(1, 5)$ are shown in Figure 1.103. The "mountain" has been shaded to emphasize its position.

(b) Five seconds later, the airplane has moved $5(0.4) = 2$ kilometers right. In ten seconds it has moved $10(0.4) = 4$ kilometers right. To graph these new positions, translate the graph of the mountain 2 and 4 kilometers to the left. Replace x with $(x + 2)$ and graph

$$y = f(x + 2) = -0.4(x + 2)^2 + 4$$

together with the point $(1, 5)$. In a similar manner graph

$$y = f(x + 4) = -0.4(x + 4)^2 + 4.$$

The results are shown in Figures 1.104 and 1.105. The position of the airplane at $(1, 5)$ has not changed. However, it has the appearance that it has moved to the right.

$[-4, 4, 1]$ by $[0, 6, 1]$

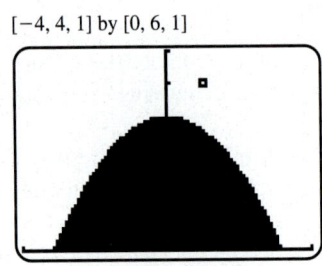

FIGURE 1.103

$[-4, 4, 1]$ by $[0, 6, 1]$

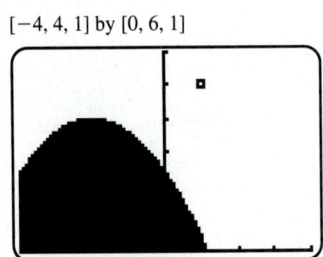

FIGURE 1.104

$[-4, 4, 1]$ by $[0, 6, 1]$

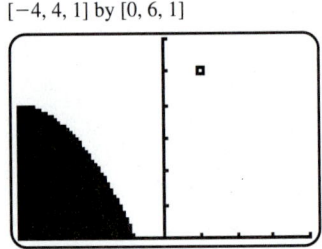

FIGURE 1.105   ■

Critical Thinking

Discuss how one might create the illusion of the airplane moving to the left and *gaining altitude* as it passes over the mountain.

1.6 PUTTING IT ALL TOGETHER

Visualization is an important technique that can be used to find a function that models a data set. The linear function given by $f(x) = x$ can be transformed to model any linear data set. Similarly, the quadratic function represented by $f(x) = x^2$ can be transformed to describe any data set whose scatterplot resembles a portion of a parabola. By understanding the form of a few basic graphs and implementing simple transformations, we can model a wide variety of data. The following table summarizes several results.

Type of Equation	$y = m(x - h) + k$	$y = a(x - h)^2 + k$
Characteristics of its graph	Line passing through point (h, k) with slope m	Parabola with vertex (h, k); $a > 0$ opens upward; $a < 0$ opens downward
Examples		
Equation (basic):	$y = x$	$y = x^2$
Examples	$m > 0$	$a > 0$
Equation:	$y = m(x - h) + k$	$y = a(x - h)^2 + k$
Examples	$m < 0$	$a < 0$
Equation:	$y = m(x - h) + k$	$y = a(x - h)^2 + k$

1.6 EXERCISES   Tape 1

Transforming Graphical Representations

Exercises 1–8: Use transformations of the graphs of $y = x$ and $y = x^2$ to sketch a graph of f by hand. Support your answer using a graphing calculator.

1. $f(x) = x - 5$

2. $f(x) = 2x + 1$

3. $f(x) = (x + 4)^2$

4. $f(x) = (x - 5)^2 + 3$

5. $f(x) = 2(x - 1) + 1$

6. $f(x) = 2(x - 1)^2 + 1$

7. $f(x) = -x^2 + 4$

8. $f(x) = -3(x + 4) - 3$

Exercises 9–14: Use transformations of the graphs of
$y = \sqrt{x}$ *and* $y = |x|$ *to sketch a graph of f by hand.*
Support your answer using a graphing calculator.

9. $f(x) = \sqrt{x + 1}$
10. $f(x) = \sqrt{x} + 1$
11. $f(x) = |x| - 4$
12. $f(x) = |x - 4|$
13. $f(x) = \sqrt{x - 3} + 2$
14. $f(x) = |x + 2| - 3$

Exercises 15–20: Use the accompanying graph of
$y = f(x)$ *to sketch the graphs of each equation.*

15. **(a)** $y = f(x) + 2$
    (b) $y = f(x + 3)$

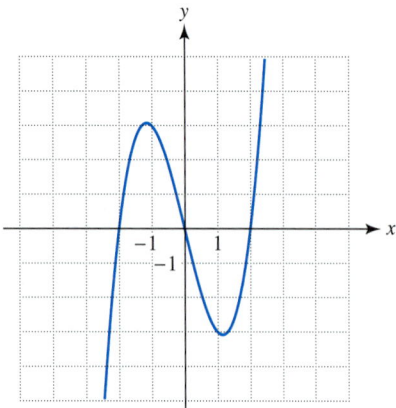

16. **(a)** $y = f(x - 1)$
    (b) $y = f(x) - 4$

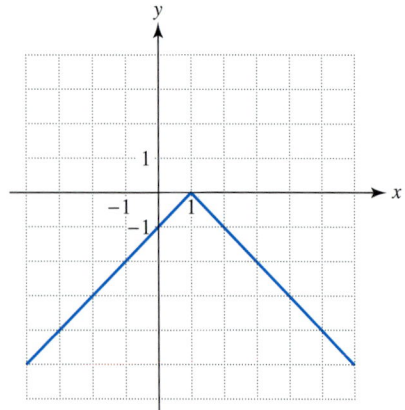

17. **(a)** $y = f(x + 3) - 2$
    (b) $y = f(x - 2) + 1$

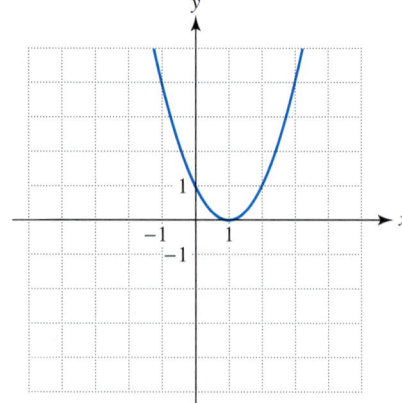

18. **(a)** $y = f(x - 1) - 2$
    (b) $y = f(x + 2) + 1$

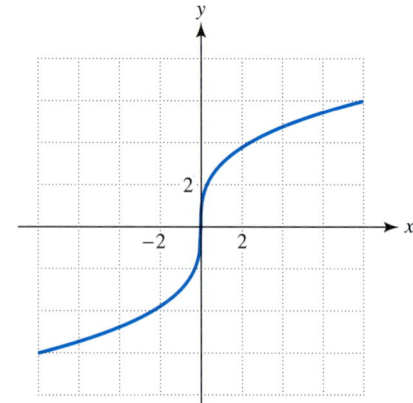

19. **(a)** $y = f(x) - 2$
    (b) $y = f(x - 1) + 2$

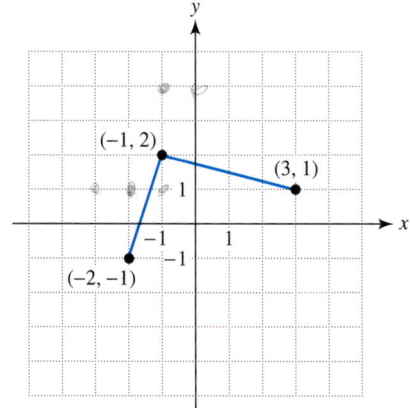

20. **(a)** $y = f(x - 2)$
    (b) $y = f(x + 1) + 1$

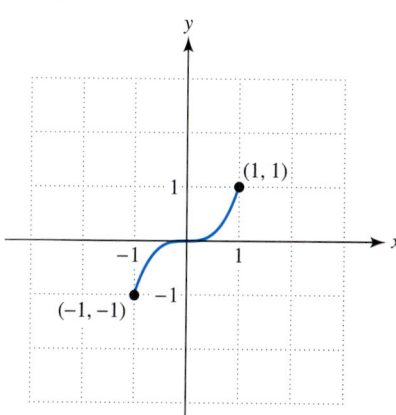

21. *National Debt*   The line graph shows annual federal deficits in billions of dollars from 1990 to 1994. In 1995 the federal deficit was $192 billion. Sketch a line graph that shows the deficits from 1991 to 1995. (Source: Office of Management and Budget.)

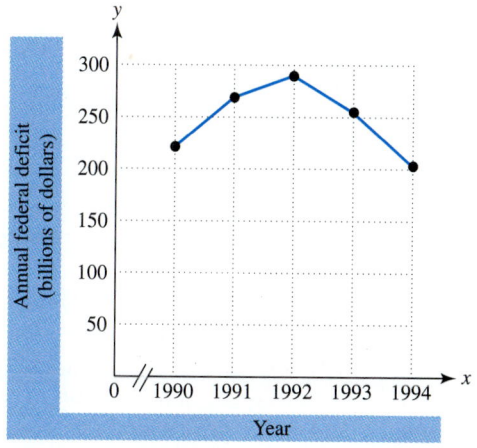

22. *Fuel Efficiency*   The line graph shows the fuel efficiency in miles per gallon of domestic cars at five-year intervals from 1955 to 1985. In 1990 and 1995 the fuel efficiencies were 26.9 and 27.3, respectively. Sketch a line graph that shows fuel efficiency from 1965 to 1995. (Source: Department of Transportation.)

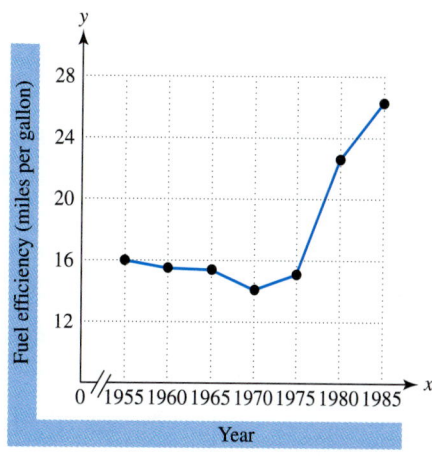

Transforming Symbolic Representations

Exercises 23–28: Find a formula for the expression.

23. $f(x + 3)$ if $f(x) = 2x$
24. $f(x) + 3$ if $f(x) = \sqrt{x}$
25. $f(x) - 7$ if $f(x) = |x| + 1$
26. $f(x - 5)$ if $f(x) = x^2 - x$
27. $f(x + 1) - 3$ if $f(x) = x^3 - 2$
28. $f(x - 2) + 1$ if $f(x) = x^2 - 2x + 2$

Exercises 29–32: (Refer to Example 1.) Find an equation that shifts the graph of f by the desired amounts. Graph f and the shifted graph in the same viewing rectangle.

29. $f(x) = x^3$;   right 2 units, downward 3 units
30. $f(x) = 3x - 4$;   left 3 units, upward 1 unit
31. $f(x) = x^2 - 4x + 1$;   left 6 units, upward 4 units
32. $f(x) = 5 - 3x - \frac{1}{2}x^2$;   right 5 units, downward 8 units

33. *Consumer Spending*   The numerical representation for f computes the average consumer spending in dollars on media and entertainment for the year x.

x	1990	1991	1992	1993	1994	1995	1996
$f(x)$	$360	$380	$410	$430	$460	$490	$520

Source: Veronis, Suhler & Associates.

(a) Give a numerical representation for a function g that computes the average consumer spending in dollars on media and entertainment during the year x, where $x = 0$ corresponds to 1990 and $x = 6$ to 1996.
(b) What are the domains of f and g?
(c) Write an equation that shows the relationship between $f(x)$ and $g(x)$.

34. *Public College Tuition* The function f, represented by $f(x) = 122.8x + 786.2$, models the average tuition at public four-year colleges from 1981 to 1994. (Source: The College Board.) The value $x = 1$ corresponds to 1981, $x = 2$ to 1982, and so on. Use transformations to determine a function g that computes the average tuition at four-year colleges during the year x, where $x = 1981$, $1982, \ldots ,1994$.

Transforming Numerical Representations

Exercises 35–40: Two functions f and g are related by the given equation. Use the numerical representation of f to make a numerical representation of g.

35. $g(x) = f(x) + 7$

x	1	2	3	4	5
$f(x)$	5	1	6	2	7

36. $g(x) = f(x) - 10$

x	0	5	10	15	20
$f(x)$	-5	11	21	32	47

37. $g(x) = f(x - 2)$

x	-4	-2	0	2	4
$f(x)$	5	2	-3	-5	-9

38. $g(x) = f(x + 50)$

x	-100	-50	0	50	100
$f(x)$	25	80	120	150	100

39. $g(x) = f(x + 1) - 2$

x	1	2	3	4	5
$f(x)$	2	4	3	7	8

40. $g(x) = f(x - 3) + 5$

x	-3	0	3	6	9
$f(x)$	3	8	15	27	31

Linear Models

Exercises 41–44: Find the equation of the line satisfying the given conditions. Graph the line.

41. Slope $\frac{1}{2}$ passing through $(1, -3)$

42. Slope -2 passing through $(0, 3)$

43. Slope -165.1 passing through $(1995, 4582)$

44. Slope 65 passing through $(1985, 112.5)$

Exercises 45–48: The data set can be modeled exactly by $f(x) = m(x - h) + k$.
  (a) *Make a scatterplot of the data.*
  (b) *Visually fit f to the data by determining values for the constants m, h, and k.*
  (c) *Check your answer by making a table of f.*

45.

x	1	2	3	4
y	3	5	7	9

46.

x	3	4	5	6
y	4	2	0	-2

47.

x	-6	-3	0	3
y	1	-8	-17	-26

48.

x	-5	5	15	·25
y	-15	25	65	105

Exercises 49–52: Modeling Real Data Each table contains real data that can be modeled approximately by $f(x) = m(x - h) + k$.
  (a) *Make a scatterplot of the data. (Do not try to plot the undetermined point in the table.)*
  (b) *Visually fit f to the data by approximating values for the constants m, h, and k. Graph f together with the data on the same coordinate axes.*
  (c) *Interpret the slope m.*
  (d) *Use f to approximate the undetermined value in the table.*

49. Municipal solid waste in millions of tons

Year	1960	1965	1970	1975
Waste	87.8	103.4	121.9	128.1

Year	1980	1985	1990	1995
Waste	151.5	164.4	195.7	?

Source: Environmental Protection Agency.

50. Population in millions of the western region of the United States

Year	1950	1960	1970	1980	1990	2000
Population	20.2	28.1	34.8	43.2	52.8	?

Source: Bureau of the Census.

51. Projected Asian-American population in millions

Year	1996	1997	1998	1999
Population	9.7	10.1	10.5	10.9

Year	2000	2001	2002	2005
Population	11.2	11.6	12.0	?

Source: Bureau of the Census.

52. State and federal prison inmates in thousands

Year	1988	1989	1990	1991
Inmates	628	712	774	826

Year	1992	1993	1994	1995
Inmates	884	970	1054	?

Source: Bureau of Justice *Statistics Bulletin*.

Quadratic Models

Exercises 53–56: Find the standard equation of a parabola satisfying the given conditions. Graph the parabola.

53. Leading coefficient 2, vertex $(2, -3)$

54. Leading coefficient -0.4, vertex $(-1, 4)$

55. Leading coefficient -2.59, vertex $(1990, 35)$

56. Leading coefficient 1850, vertex $(1980, 125)$

Exercises 57–60: Each data set can be modeled exactly by $f(x) = a(x - h)^2 + k$, where a is an integer.
  (a) *Make a scatterplot of the data.*
  (b) *Visually fit f to the data by determining values for the constants a, h, and k. (Hint: Use the first data point as the vertex.)*
  (c) *Check your answer by tabling f.*

57.

x	2	3	4	5
y	3	5	11	21

58.

x	-2	-1	0	1
y	-5	-7	-13	-23

59.

x	-5	-4	-3	-2
y	3	0	-9	-24

60.

x	5	6	8	11
y	-10	-9	-1	26

Exercises 61–64: Modeling Real Data Each table contains real data that can be modeled approximately by $f(x) = a(x - h)^2 + k$.
  (a) *Make a scatterplot of the data. (Do not try to plot the undetermined point(s) listed in the table.)*
  (b) *Visually fit the quadratic function f to the data by approximating values for the constants a, h, and k. Graph f together with the data in the same viewing rectangle.*
  (c) *Use f to approximate the undetermined value(s) in the table.*

61. Poverty income thresholds in dollars for one person

Year	1960	1965	1970	1975	1980	1985	1990	1995
Income	1490	1582	1954	2724	4190	5469	6652	?

Source: Bureau of the Census.

62. Poverty income thresholds in dollars for two people

Year	1960	1965	1970	1975	1980	1985	1990	1995
Income	1924	2048	2525	3506	5363	6998	8509	?

Source: Bureau of the Census.

63. Total cumulative AIDS cases in the United States

Year	1982	1984	1986	1988
Cases	1586	10,927	41,910	106,304

Year	1990	1992	1994	1998
Cases	196,576	329,205	441,528	?

Source: Department of Health and Human Services.

64. U.S. population in millions

Year	1800	1820	1840	1860	1870	1880
Population	5	10	17	31	?	50

Year	1900	1920	1940	1960	1980	2000
Population	76	106	132	179	226	?

Source: Bureau of the Census.

65. Find a quadratic function given by $f(x) = a(x - h)^2 + k$ that models the data in Table 1.18 from Section 1.5.

66. Find a quadratic function given by $f(x) = a(x - h)^2 + k$ that models the data in Table 1.19 from Section 1.5.

Using Transformations to Represent Data Dynamically

67. (Refer to Example 4.) Suppose that the airplane in Figure 1.101 is flying at 0.2 kilometer per second to the left, rather than to the right. If the position of the airplane is fixed at $(-1, 5)$, graph the image of the mountain and the position of the airplane after 15 seconds.

68. (Refer to Example 4.) Suppose that the airplane in Figure 1.101 is traveling to the right at 0.1 kilometer per second and gaining altitude at 0.05 kilometer per second. If the airplane's position is fixed at $(-1, 5)$, graph the image of the mountain and the position of the airplane after 20 seconds.

69. *Modeling a Weather Front* Suppose a cold front is passing through the United States at noon with a shape described by the function $y = \frac{1}{20}x^2$. Each unit represents 100 miles. Des Moines, Iowa, is located at $(0, 0)$, and the positive y-axis points north.

  (a) If the cold front is moving south at 40 miles per hour and retains its present shape, graph its location at 4 P.M.

  (b) Suppose that by midnight the vertex of the front has moved 250 miles south and 210 miles east of Des Moines, maintaining the same shape. Columbus, Ohio, is located approximately 550 miles east and 80 miles south of Des Moines. Plot the locations of Des Moines and Columbus together with the new position of the cold front.

Determine whether the cold front has reached Columbus.

70. *Modeling Motion in Computer Graphics* The first figure contains a picture that is composed of lines and curves. In this exercise we will model only the semicircle that outlines the top of the silo. In order to make it appear that the person is walking to the right, the background must be translated horizontally to the left as shown in the second figure.

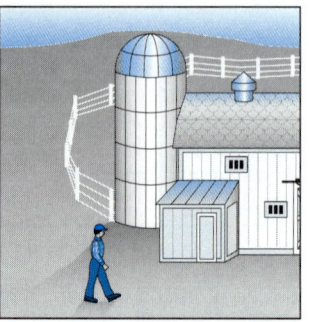

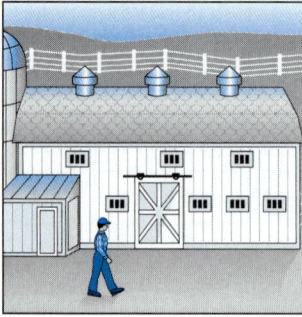

The semicircle at the top of the silo in the first figure is described by $f(x) = \sqrt{9 - x^2} + 12$.

  (a) Graph f in the (square) viewing rectangle $[-12, 12, 1]$ by $[0, 16, 1]$.

  (b) To give the illusion that the walker is moving to the right at 2 units per second, graph the top of the silo after 1 second, and after 4 seconds.

Writing about Mathematics

1. Compare the graphs of $y_1 = a(x - h) + k$ and $y_2 = a(x - h)^2 + k$. Explain what the constants a, h, and k represent in each equation.

2. Let k be a positive number. Explain how to shift the graph of $y = f(x)$ upward, downward, left, or right k units. Give examples.

CHECKING BASIC CONCEPTS FOR SECTIONS 1.5 AND 1.6

1. Graph $f(x) = 2x^2 - 6x + 2$.
   (a) Evaluate $f(3)$.
   (b) Find the vertex symbolically and graphically.

2. Graph $f(x) = a(x - h)^2 + k$ for the given values of $a, h,$ and k. Explain how each constant influences the graph of f.

   (a) $a = -\dfrac{3}{2}, h = -2,$ and $k = 4$

   (b) $a = \dfrac{1}{2}, h = 1,$ and $k = -3$

3. Graph $f(x) = x^3$. Then, conjecture how the graph of each equation will appear compared to the graph of f. Test your conjecture by graphing f and the equation in the same viewing rectangle.

   (a) $y = (x + 4)^3$
   (b) $y = x^3 - 3$
   (c) $y = (x - 5)^3 + 3$

4. The table lists the amount of memory x in megabytes (MB) needed to record y seconds of music. Find a linear function given by $f(x) = m(x - h) + k$ that models this data. Graph f and the data together. Interpret the slope m.

x (MB)	0.5	0.9	1.5	1.8
y (sec)	23.5	42.3	70.5	84.6

CHAPTER 1 Summary

Numbers are essential to mathematics. Without them it would be impossible to compare and quantify physical objects and events. Today, vast amounts of data are being created. Chapter 1 demonstrates how mathematics uses functions and their representations to model and summarize real data. In order to obtain information from data, algorithms, graphs, tables, and formulas are used. Four common representations of a function are verbal, graphical, numerical, and symbolic. There is no *best* representation of a function, but rather each representation gives unique insight into what a function computes. Any representation of a function provides exactly one output y for each valid input x. The set of all valid inputs is the *domain* and the set of all possible outputs is the *range*.

Three basic types of functions are constant, linear, and nonlinear. A *constant function* outputs only one y-value regardless of the input x and can be represented by $f(x) = b$. Its graph is a horizontal line. Constant functions are simple examples of linear functions. A *linear function* has a nonvertical line as its graph and is represented by $f(x) = ax + b$. The output from a linear function is predictable. If the input x increases by one unit, then the output y always changes by the same amount. This amount is equal to the slope of its graph, which is given by the constant a. For example, if $a = -3$, then the output y decreases by 3 units for every unit increase in x. The third type of function is a *nonlinear function*. Its graph is not a line. Therefore, its output y does not necessarily change by the same amount for each unit increase in x. Instead of slope, average rate of change may be used to help describe how outputs change for nonlinear functions.

By applying transformations of graphs, linear functions defined by $f(x) = m(x - h) + k$ can model linear data, while quadratic functions given by $f(x) = a(x - h)^2 + k$ can model some types of nonlinear data.

Review Exercises

Exercises 1 and 2: Classify each number listed as one or more of the following: natural number, integer, rational number, or real number.

1. $-2, \frac{1}{2}, 0, 1.23, \sqrt{7}, \sqrt{16}$

2. $55, 1.5, \frac{104}{17}, 2^3, \sqrt{3}, -1000$

Exercises 3–6: Write each number in scientific notation.

3. 1,891,000

4. $-13,850$

5. 0.0000439

6. 0.0001001

Exercises 7 and 8: Write each number in standard form.

7. 1.52×10^4

8. -7.2×10^{-3}

9. *Speed of Light*   The average distance between the planet Mars and the sun is approximately 228 million kilometers. Estimate the time required for sunlight, traveling at 300,000 kilometers per second, to reach Mars. (Source: C. Ronan, *The Natural History of the Universe.*)

10. *Flow Rate*   Niagara Falls has an average of 212,000 cubic feet of water passing over it each second. Is this rate more or less than 10 cubic miles per year? Explain your calculations. (*Hint:* 1 cubic mile $= 5280^3$ cubic feet.) (Source: National Geographic Society.)

11. *Body Mass Index*   (Refer to Exercises 53–56, Section 1.1.) Compute the body mass index for a person who is 5 feet tall and weighs 120 pounds.

12. *Volume of a Sphere*   Write an algorithm to calculate the volume of a sphere in cubic feet given the radius r in inches. Your algorithm should receive r as input and output the volume V. (*Hint:* The formula for the volume of a sphere is $V = \frac{4}{3}\pi r^3$.)

Exercises 13 and 14: Sort each list of numbers from smallest to largest and display them in a table.
   (a) Determine the maximum and minimum values in the list.
   (b) Calculate the average, median, and range.

13. $-5, 8, 19, 24, -23$

14. $8.9, -1.2, -3.8, 0.8, 1.7, 1.7$

Exercises 15 and 16: Complete the following.
   (a) Find the domain and range of the relation.
   (b) Determine the maximum and minimum of the x-values; of the y-values.
   (c) Label appropriate scales on the xy-axes.
   (d) Plot the relation by hand in the xy-plane.

15. $\{(-1, -2), (4, 6), (0, -5), (-5, 3), (1, 0)\}$

16. $\{(10, 20), (-35, -25), (60, 60), (-70, 35), (0, -55)\}$

Exercises 17 and 18: Make a scatterplot of the relation in an appropriate viewing rectangle. Determine if the relation is a function.

17. $\{(10, 13), (-12, 40), (-30, -23), (25, -22), (10, 20)\}$

18. $\{(1.5, 2.5), (0, 2.1), (-2.3, 3.1), (0.5, -0.8), (-1.1, 0)\}$

19. *CD Usage*   CD-ROM has become a widely used data storage technology for computers. The line graph describes the numbers in millions of CD-ROM sales in 1991, 1992, 1993, and 1994. Assume that this graph represents a function f. (Source: Dataquest, *Multimedia Market Trends.*)

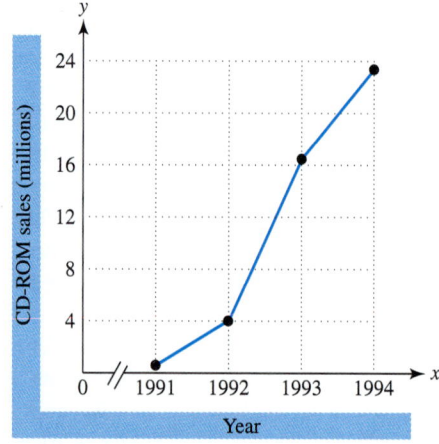

   (a) Evaluate $f(1993)$.

(b) Estimate the value of x when $f(x) = 4$. Interpret this answer.

(c) When did the greatest increase in sales occur? Approximate this increase.

(d) Represent f in numerical form.

20. *Stock Exchange* The table shows the highest New York Stock Exchange membership price for various years. (Source: New York Stock Exchange, Fact Book.)

Year	1955	1970	1980	1990	1994
Price (dollars)	90,000	320,000	275,000	430,000	830,000

(a) Make a scatterplot of the data. Let x-values represent years and y-values represent membership prices.

(b) Make a line graph of the data. Discuss how prices changed.

Exercises 21 and 22: Use the verbal representation to express the function f symbolically, graphically, and numerically. Let $y = f(x)$ with $0 \le x \le 100$. For the numerical representation, use $x = 0, 25, 50, 75, 100$.

21. To convert x pounds to y ounces, multiply x by 16.

22. To find the area y of a square, multiply the length x of a side by itself.

23. *ACT Scores* The function f defined by

$$f(1991) = 20.6, f(1992) = 20.6, f(1993) = 20.7,$$
$$\text{and } f(1994) = 20.8$$

computes the national average ACT composite score in the year x. (Source: The American College Testing Program.)

(a) Use ordered pairs to describe f.

(b) Identify the domain and range of f.

(c) Find all x such that $f(x) = 20.6$.

24. (a) Use the graph of f to evaluate $f(-3)$ and $f(1)$.

(b) Find all x such that $f(x) = 0$.

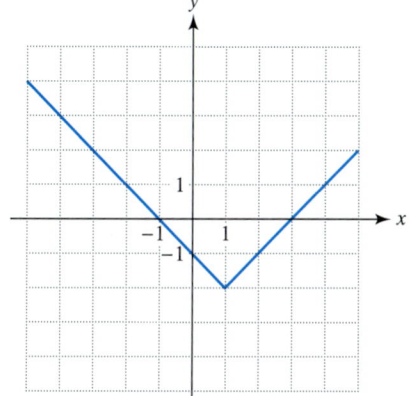

Exercises 25–28: Complete the following for the function f.

(a) *Evaluate $f(a)$ at the indicated values of a.*

(b) *Find the implied domain of f.*

25. $f(x) = \sqrt[3]{x}$ at $a = -8, 1$

26. $f(x) = 3x + 2$ at $a = -2, 5$

27. $f(x) = \dfrac{1}{x^2 - 4}$ at $a = -3, 2$

28. $f(x) = \sqrt{x + 3}$ at $a = -3, 1$

29. Suppose that for each input x an algorithm outputs all y satisfying the equation $x = y^2$. Does this algorithm compute a function? Explain.

30. Give an example of a relation that involves real data. List some or all of the ordered pairs in the relation.

Exercises 31 and 32: Determine if the graph represents a function. If it does not represent a function, find the minimum number of functions necessary to create the graph.

31.

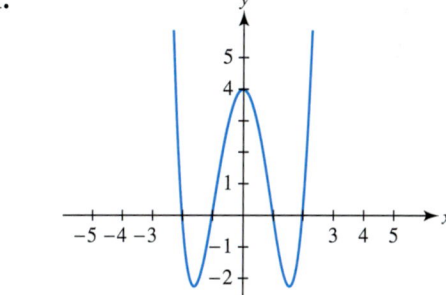

32.

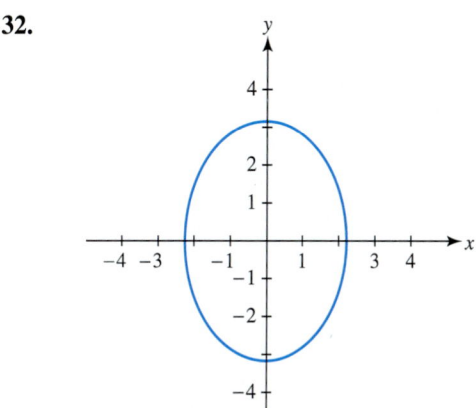

Exercises 33–36: If possible, find the slope of the line passing through each pair of points.

33. $(-1, 7), (3, 4)$ **34.** $(1, -4), (2, 10)$

35. $(8, 4), (-2, 4)$ **36.** $(-2, -6), (-2, 7)$

Exercises 37–42: Decide whether the function f is constant, linear, or nonlinear. Support your answer graphically.

37. $f(x) = 8 - 3x$ **38.** $f(x) = 2x^2 - 3x - 8$

39. $f(x) = |x + 2|$ **40.** $f(x) = 2x$

41. $f(x) = 6$ **42.** $f(x) = x^3 - x + 1$

Exercises 43 and 44: Complete the following.
   (a) *Graph f in the standard viewing rectangle.*
   (b) *Estimate the x-intervals on which f is increasing or decreasing.*

43. $f(x) = x^2 - x$ **44.** $f(x) = x^3 - 4x$

45. Use the complete numerical representations for the functions *f*, *g*, and *h* to classify each as increasing, decreasing, or neither.

x	f(x)	g(x)	h(x)
1	−1	−3	−8
7	10	−7	−2
4	3	−5	4
−2	−6	−1	5

46. Sketch a graph of a function *f* with $f(-2) = 3$, $f(2) = -3$, and $f(4) = 4$.

47. *Survival Rates*   The survival rates for song sparrows from 100 eggs are shown in the table. The values listed are the numbers of song sparrows that attain a given age. For example, six song sparrows reach an age of two years from 100 eggs laid in the wild. (Source: S. Kress, *Bird Life*.)

Age	0	1	2	3	4
Number	100	10	6	3	2

   (a) Make a line graph of the data. Interpret the data.
   (b) Does this line graph represent a function?

48. Compute the average rate of change of $f(x) = x^2 - x + 1$ from 1 to 3.

Exercises 49 and 50: Complete the following for the quadratic function f.
   (a) *Evaluate f(a) at the specified value of a without a calculator.*
   (b) *Find the vertex symbolically. Support your result graphically and numerically.*

49. $f(x) = -x^2 + 4x + 1, a = 4$

50. $f(x) = 0.2x^2 - 5x + 2, a = -2$

Exercises 51–54: Use transformations of basic graphs to sketch a graph of f by hand. Support your result using a graphing calculator.

51. $f(x) = x^2 - 4$ **52.** $f(x) = 3(x - 2) + 1$

53. $f(x) = (x + 4)^2 - 2$ **54.** $f(x) = \sqrt{x + 4}$

Exercises 55–58: Find the formula for the expression.

55. $f(x - 2)$ if $f(x) = 2x$

56. $f(x) - 3$ if $f(x) = \sqrt{x} + 1$

57. $f(x + 1) - 2$ if $f(x) = x^2 - 3$

58. $f(x - 4) + 1$ if $f(x) = -3x^3$

Exercises 59 and 60: Two functions f and g are related by the given equation. Use the numerical representation of f to construct a numerical representation of g.

59. $g(x) = f(x) + 2$

x	1	2	3	4
f(x)	−3	3	4	7

60. $g(x) = f(x - 5)$

x	0	5	10	15
f(x)	−5	11	21	32

61. Find a function given by $f(x) = m(x - h) + k$ that models the data exactly.

x	1	2	3	4
y	7	4	1	−2

62. Find a function given by $f(x) = a(x - h)^2 + k$ that models the data exactly.

x	2	3	4	5
y	4	7	16	31

63. *Children on the Internet*   The table shows the estimated number of children who have access to the Internet at home. (Source: Jupiter Communications.)

Year	1995	1996	1997	1998	1999	2000
Children (millions)	9	12	15	19	23	27

(a) Make a scatterplot of the data. Discuss any trends.
(b) Find a linear function represented by $f(x) = m(x - h) + k$ that models this data.
(c) Interpret the slope m.

64. *Average Rate of Change* Let $f(x) = 0.5x^2 + 50$ compute the outside temperature in degrees Fahrenheit at x P.M. where $1 \le x \le 5$.
(a) Graph f in $[1, 5, 1]$ by $[40, 70, 5]$. Is f linear or nonlinear?
(b) Calculate the average rate of change of f from 1 P.M. to 4 P.M.
(c) Interpret this average rate of change verbally and graphically.

65. *World Population* The function given by $f(x) = 0.000478x^2 - 1.813x + 1720.1$ models the world population in billions between 1950 and 1995, where x is the year.
(a) Evaluate $f(1985)$ and interpret the result.
(b) Use f to estimate world population in the year 2000.

66. *Electrical Circuits* The figure shows the voltage y in a household electrical circuit over a time interval of $\frac{1}{60}$ second.

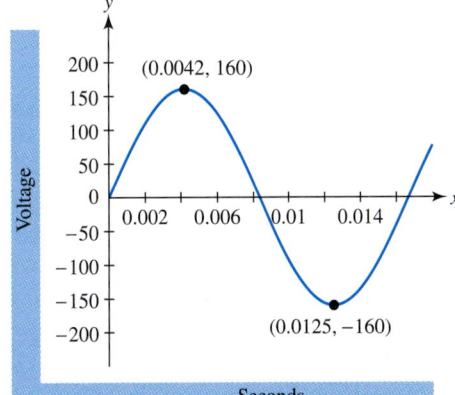

(a) What is the maximum voltage?
(b) Determine the times when the voltage is decreasing.
(c) Use your results from part (b) to determine when the voltage is increasing. Let $0 \le x \le \frac{1}{60}$.

Extended and Discovery Exercises

1. *Global Warming* If the global climate were to warm significantly, as a result of the greenhouse effect or other climatic change, the Arctic ice cap would start to melt. It is estimated that this ice cap contains the equivalent of 680,000 cubic miles of water. Over 200 million people currently live on soil that is less than 3 feet above sea level. In the United States, several large cities have low average elevations, such as Boston (14 feet), New Orleans (4 feet), and San Diego (13 feet). In this exercise you will estimate the rise in sea level if this cap were to melt and determine whether this event would have a significant impact on people. (Sources: Department of the Interior, Geological Survey.)
(a) The surface area of a sphere is given by the formula $4\pi r^2$ where r is its radius. Although the shape of the earth is not exactly spherical, it has an average radius of 3960 miles. Estimate the surface area of the earth.

(b) Oceans cover approximately 71% of the total surface area of the earth. How many square miles of the earth's surface are covered by oceans?
(c) Approximate the potential rise in sea level by dividing the total volume of the water from the ice cap by the surface area of the oceans.
(d) Discuss the implications of your calculation. How would cities like Boston, New Orleans, and San Diego be affected?
(e) The Antarctic ice cap contains 6,300,000 cubic miles of water. Estimate how much sea level would rise if this ice cap melted.

2. *Daylight Hours* The graph of the function f in the accompanying figure approximates the number of daylight hours at Columbus, Ohio, where x is measured in days. Assume that $x = 1$ corresponds to January 1, $x = 2$ to January 2, and so on, until $x = 365$ represents December 31.

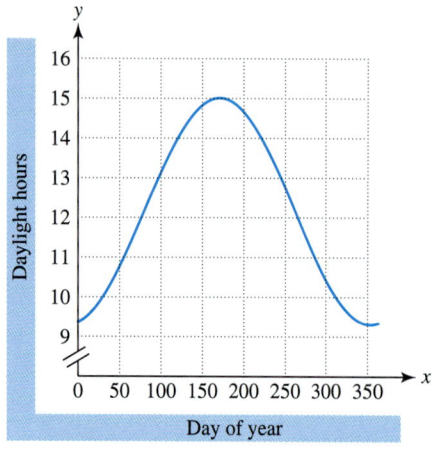

Day of year

(a) Use the graph and your knowledge about the seasons to determine the days when daylight hours are increasing at Columbus.
(b) Use your answer from part (a) to determine the days when daylight hours are decreasing.
(c) Discuss how the graph would change for a city that is located either north or south of Columbus. How would the graph appear for a city located at the equator?

3. *Modeling Water Flow*   A cylindrical container measuring 16 centimeters high and 12.7 centimeters in diameter with a 0.5 centimeter hole at the bottom was completely filled with water. As water leaked out, the height of the water level inside the container was measured every 15 seconds. The results of the experiment appear in the table.

Time (sec)	0	15	30	45	60	75
Height (cm)	16	13.8	11.6	9.8	8.1	6.6

Time (sec)	90	105	120	135	150	165	180
Height (cm)	5.3	4.1	3.1	2.3	1.4	0.8	0.5

(a) Make a scatterplot of the data. Is the data linear or nonlinear?
(b) Find a function f that models this data.
(c) Graph f together with the data.
(d) At first, water rushed out the hole in the container. Gradually the water began to flow more slowly. Discuss how this event affects whether the data is linear or nonlinear. Describe how water would flow if it could be modeled accurately by a linear function.

4. *Modeling a Cold Front*   A weather map of the United States on April 22, 1996, is shown in the figure. There was a cold front roughly in the shape of a circular arc passing north of Dallas and west of Detroit. The center of the arc was located near Pierre, South Dakota, with a radius of about 750 miles. If Pierre has the coordinates $(0, 0)$ and the positive y-axis points north, then the equation of the front can be modeled by $f(x) = -\sqrt{750^2 - x^2}$, where $0 \le x \le 750$. (Source: AccuWeather, Inc.)

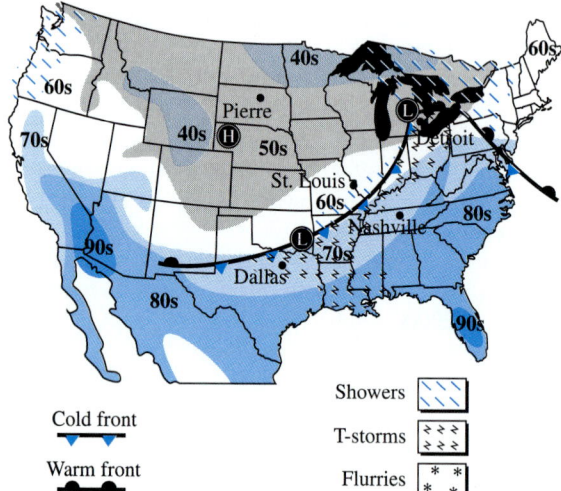

(a) St. Louis is located at $(535, -400)$ and Nashville is at $(730, -570)$. Plot these points and graph f in $[0, 1200, 100]$ by $[-800, 0, 100]$. Did the cold front reach these cities?
(b) During the next 12 hours, the center of the front moved approximately 110 miles south and 160 miles east. Assuming the cold front did not change shape, use transformations of graphs to determine an equation that models its new location.
(c) Use graphing to determine visually if the cold front reached both cities.

CHAPTER 2

Linear Functions and Equations

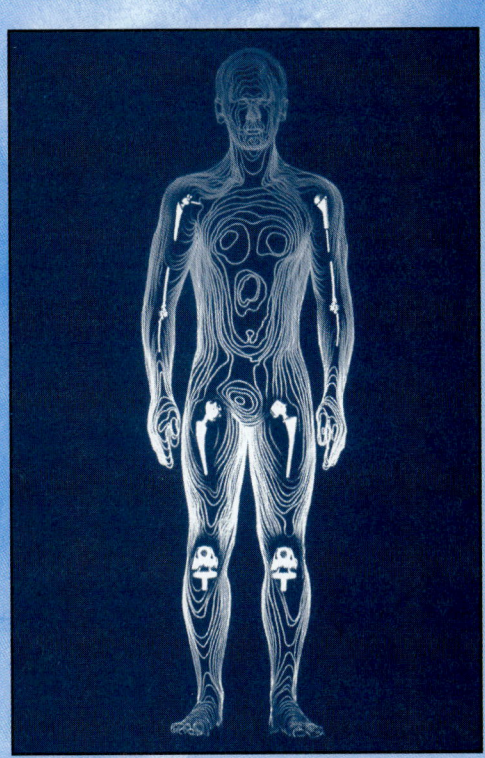

*T*o learn how to pound a nail, we might start by watching a skilled carpenter. Next, we would pick up a hammer and try it for ourselves. If we had never done it before, it could take practice before we became proficient. Acquiring skills in mathematics is similar. The best way to learn mathematics is to *do* mathematics.

In this chapter we discuss tools and techniques used in mathematics for solving linear equations and inequalities. In addition to symbolic and numerical methods, visual approaches are also introduced. Mathematical concepts can be communicated visually. A graph on a calculator screen contains approximately the same amount of information as a page of text. However, many people find the information in a graph easier to absorb. Much of the brain is devoted to processing visual information. Sophisticated scientific visualization, which you may have seen in movies or on television, is still in its infancy. The following table lists the approximate age in years of various forms of communication. Notice that eyesight has had a considerably longer time to develop than any other form.

*S*ay it, I'll forget.
Demonstrate it, I may recall.
But if I'm involved, I'll understand.
— **Old Chinese Proverb**

Form of Communication	Age (years)
Eyesight	500,000,000
Oral communication	50,000
Written words	5000
Printed books	500
Television	60
Scientific visualization	20

Mathematics makes technology possible. Technology makes mathematics easier to learn. But regardless of how advanced technology becomes, learning mathematics still requires the involvement of both the student and the instructor.

Sources: M. Friedhoff and W. Benzon, *The SECOND Computer Revolution VISUALIZATION.* G. Nielson and B. Shriver, Editors, *Visualization in Scientific Computing.*

2.1 Linear Equations

Equations • The *x*-Intercept Method • The Intersection-of-Graphs Method • Solving Linear Equations Symbolically

Introduction

A primary objective of both theoretical and applied mathematics is solving equations. Billions of dollars are spent each year on computers and personnel to solve equations that hold the answers for creating better products. Equations occur in a wide variety of forms. Some equations can be solved easily, whereas others require enormous amounts of resources. Without graphical, numerical, and symbolic techniques for solving equations, our society would not have televisions, CD players, satellites, fiber optics, CAT scans, computers, or accurate weather forecasts.

Chapter 1 introduced the concept of a function and its representations. In this chapter we see that applications involving functions lead to equations. For example, the function represented by $f(x) = 575(x - 1980) + 3617$ models the average tuition at private colleges from 1980 to 1994. One way to predict the year when the average tuition is $14,000 would be to solve the equation

$$575(x - 1980) + 3617 = 14,000.$$

Equations

An **equation** is a statement that two mathematical expressions are equal. Equations always contain an equals sign. Some examples of equations include

$$x + 15 = 9x - 1, \qquad x^2 - 2x + 1 = 2x, \qquad z + 5 = 0,$$
$$xy + x^2 = y^3 + x, \qquad \text{and} \qquad 1 + 2 = 3.$$

The first three equations involve one variable, the fourth equation has two variables, and the fifth equation contains only constants. For now, our discussion will concentrate on equations involving one variable.

To **solve** an equation means to find all values for the variable that make the equation a true statement. Such values are called **solutions.** The set of all solutions is the **solution set.** The solutions to the equation $x^2 - 1 = 0$ are -1 or 1, written as $x = \pm 1$. Either value for x **satisfies** the equation. The solution set is $\{-1, 1\}$. Two equations are **equivalent** if they have the same solution set. For example, the equations $x + 2 = 5$ and $x = 3$ are equivalent.

If an equation has no solutions, then its solution set is empty and the equation is called a **contradiction.** The equation $x + 2 = x$ has no solutions and is a contradiction. On the other hand, if every (meaningful) value for the variable is a solution, then the equation is an **identity.** The equation $x + x = 2x$ is an identity since every value for x makes the equation true. Any equation that is satisfied by some, but not all, values of the variable is a **conditional equation.** The equation $x^2 - 1 = 0$ is a conditional equation. Only the values -1 and 1 for x make this equation a true statement.

As with functions, equations can be either *linear* or *nonlinear.* A linear equation is one of the simplest types of equations.

Linear equation in one variable

A **linear equation** in one variable is an equation that can be written in the form

$$ax + b = 0,$$

where a and b are real numbers with $a \neq 0$.

There is a close relationship between functions and equations. If h is a linear function, then it can be represented symbolically by $h(x) = ax + b$. Therefore, every linear equation in one variable can be written in the form $h(x) = 0$, where h is a linear function. If an equation is not linear, then it is **nonlinear,** and can also be expressed as $h(x) = 0$, where h is a nonlinear function. This is illustrated in the next example.

EXAMPLE 1   *Identifying equations as linear or nonlinear*

Write each equation in the form $h(x) = 0$ and give a symbolic representation for the function h. Identify h and the equation as either linear or nonlinear. Graph h to support your answer.
(a) $3x - 17 = 16 + x$
(b) $x^3 - 5x^2 + 6 = 1 - x^2$

Solution
(a) We begin by transposing terms on the right side of the equation to the left side.

$$3x - 17 = 16 + x \qquad \text{Original equation}$$
$$3x - 17 - 16 - x = 0 \qquad \text{Subtract 16 and } x \text{ from both sides.}$$
$$2x - 33 = 0 \qquad \text{Simplify.}$$

The original equation is equivalent to $2x - 33 = 0$. A symbolic representation for h is $h(x) = 2x - 33$. This is in the form $h(x) = ax + b$, where $a = 2$ and $b = -33$. Therefore, both the original equation and h are linear. The linear graph of h is shown in Figure 2.1.

$[-60, 60, 5]$ by $[-40, 40, 5]$

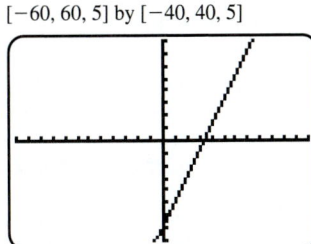

FIGURE 2.1

$[-10, 10, 1]$ by $[-10, 10, 1]$

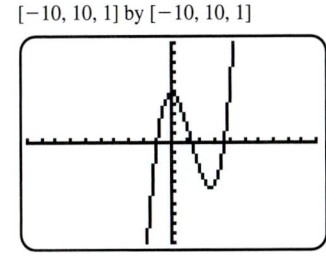

FIGURE 2.2

(b) Transpose the terms to obtain a zero on the right side.

$$x^3 - 5x^2 + 6 = 1 - x^2 \qquad \text{Original equation}$$
$$x^3 - 5x^2 + 6 - 1 + x^2 = 0 \qquad \text{Subtract 1 and add } x^2.$$
$$x^3 - 4x^2 + 5 = 0 \qquad \text{Simplify.}$$

Let $h(x) = x^3 - 4x^2 + 5$. Since $h(x)$ cannot be written in the form $ax + b$, both h and the original equation are nonlinear. The nonlinear graph of h is shown in Figure 2.2. ∎

The *x*-Intercept Method

Every equation in one variable—linear or nonlinear—can be written in the form $h(x) = 0$ for some function h. A solution to the equation $h(x) = 0$ corresponds to a point where the graph of $y = h(x)$ intersects the *x*-axis. An ***x*-intercept** is the *x*-coordinate of a point where a graph intersects the *x*-axis. As a result, solving the equation $h(x) = 0$ graphically reduces to the task of locating all *x*-intercepts. This method of solving an equation is called the ***x*-intercept method.**

The graph of $y = h(x)$, where $h(x) = -2x + 4$, is shown in Figure 2.3. This graph has one *x*-intercept, 2. If 2 is substituted for x in the formula for h, then $h(2) = -2(2) + 4 = 0$. Therefore, 2 is a **zero** of the function h and is a solution to the equation $-2x + 4 = 0$. The graph of another function f, having two *x*-intercepts, -3 and 2, is shown in Figure 2.4. The solutions to the equation $f(x) = 0$ are -3 and 2. Moreover, the zeros of f are -3 and 2. In general, the *x*-intercepts on the graph of a function are zeros of the function.

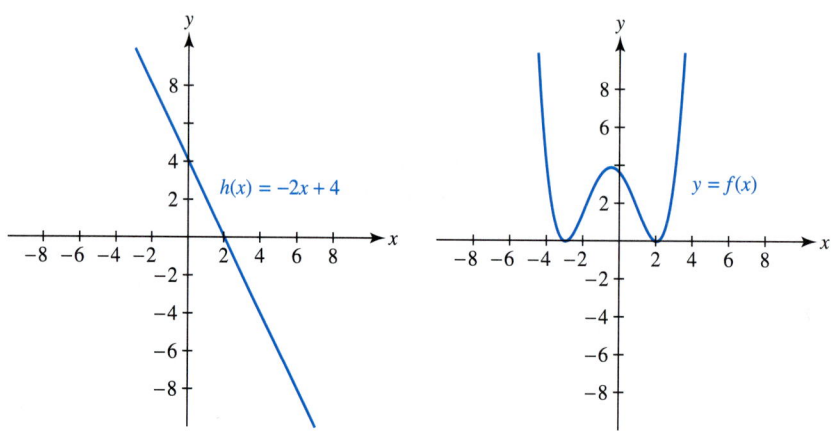

FIGURE 2.3   FIGURE 2.4

Critical Thinking

Any linear equation can be expressed as $h(x) = 0$, where h is given by $h(x) = ax + b$ with $a \neq 0$. Sketch possible graphs of h. How many solutions to the linear equation $ax + b = 0$ are there? Explain.

EXAMPLE 2   *Solving a linear equation by the x-intercept method*

The linear function defined by $f(x) = 575(x - 1980) + 3617$ models the average tuition at private colleges in the United States between 1980 and 1994. Use the *x*-intercept method to estimate when average tuition is $14,000.

Solution

The solution of the equation $575(x - 1980) + 3617 = 14{,}000$ corresponds to when tuition is \$14,000. Start by writing this equation in the form $h(x) = 0$.

$$575(x - 1980) + 3617 = 14{,}000 \qquad \text{Original equation}$$

$$575(x - 1980) - 10{,}383 = 0 \qquad \text{Subtract 14,000 from both sides.}$$

We now have $h(x) = 575(x - 1980) - 10{,}383$. Graph h as shown in Figure 2.5 and locate the x-intercept. This intercept can be approximated by either using zoom and trace features or built-in utilities for finding zeros. This x-intercept is located near 1998.1 as shown in Figure 2.6. Rounded to the nearest year, this linear model predicts the average tuition at private colleges to be \$14,000 in 1998.

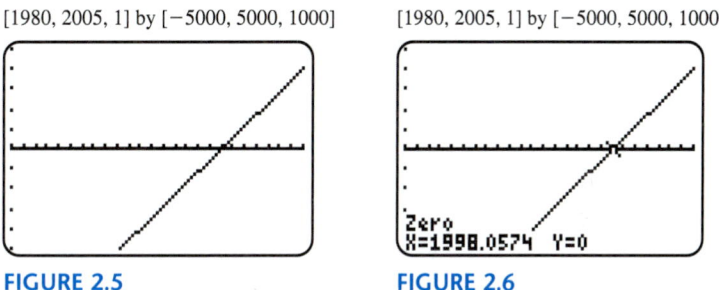

[1980, 2005, 1] by [−5000, 5000, 1000]   [1980, 2005, 1] by [−5000, 5000, 1000]

FIGURE 2.5   **FIGURE 2.6**   ■

The Intersection-of-Graphs Method

The equation $f(x) = g(x)$ results whenever the symbolic representations of two functions f and g are set equal to each other. A solution to this equation corresponds to the x-coordinate of a point where the graphs of f and g intersect. This technique is called the **intersection-of-graphs method.** If the graphs of f and g are lines with different slopes, then their graphs intersect once. For example, if $f(x) = 2x + 1$ and $g(x) = -x + 4$, then the equation $f(x) = g(x)$ becomes $2x + 1 = -x + 4$. To apply the intersection-of-graphs method, we graph $y_1 = 2x + 1$ and $y_2 = -x + 4$ as shown in Figure 2.7.

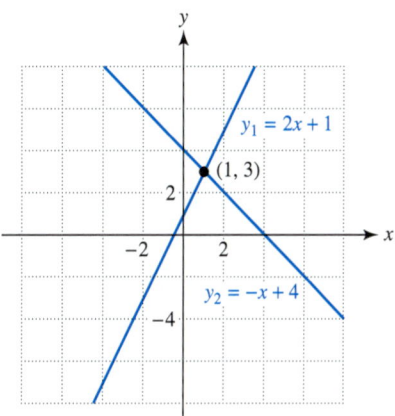

FIGURE 2.7

Their graphs intersect at the point (1, 3). Since the variable in the equation $2x + 1 = -x + 4$ is x, the solution is the x-coordinate of the point of intersection. When $x = 1$, the functions f and g both assume the value 3, that is, $f(1) = 2(1) + 1 = 3$ and $g(1) = -1 + 4 = 3$. The value 3 is the y-coordinate of the point of intersection (1, 3).

EXAMPLE 3   *Solving a linear equation by the intersection-of-graphs method*

From 1985 to 1990, sales of compact discs in millions can be modeled by $f(x) = 51.6(x - 1985) + 9.1$, whereas sales of vinyl LP records in millions can be modeled by $g(x) = -31.9(x - 1985) + 167.7$. Use the intersection-of-graphs method to approximate the year x when sales of LP records and compact discs were equal. What were their sales at that time? (Source: Recording Industry Association of America.)

Solution

Sales of CDs and LP records were equal when the equation $f(x) = g(x)$ is satisfied. To solve this equation for x using the intersection-of-graphs method, graph $Y_1 = 51.6(X - 1985) + 9.1$ and $Y_2 = -31.9(X - 1985) + 167.7$ as shown in Figure 2.8.

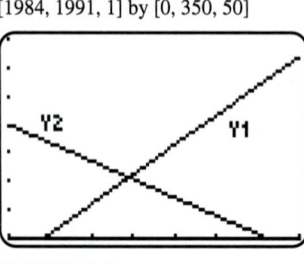

[1984, 1991, 1] by [0, 350, 50]   [1984, 1991, 1] by [0, 350, 50]

FIGURE 2.8   **FIGURE 2.9**

The point of intersection can be estimated using either zoom and trace features or other built-in utilities. From Figure 2.9, the graphs of f and g intersect near the point (1986.9, 107.1). The x-value corresponds to the year that CD and LP record sales were equal. The y-value represents the number (in millions) of each that were sold. Rounded to the nearest year, in 1987 both CD and LP record sales were about 107 million.   ■

EXAMPLE 4   *Solving a linear equation by the intersection-of-graphs method*

Formaldehyde is an indoor air pollutant found in plywood, foam insulation, and carpeting. When concentrations in the air reach 33 micrograms per cubic foot $(\mu g/ft^3)$ eye irritation can occur. One square foot of new plywood can emit 140 μg per hour. (Source: A. Hines, *Indoor Air Quality & Control.*)

(a) A room has 100 square feet of new plywood flooring. Find a linear function f that computes the amount of formaldehyde in micrograms emitted in x hours.

(b) The room contains 800 cubic feet of air and has no ventilation. Use the inter-section-of-graphs method to determine how long it takes for concentrations to reach 33 μg/ft^3.

Solution

(a) Each square foot of plywood emits 140 μg/hr. One hundred square feet would emit $140 \cdot 100 = 14{,}000$ micrograms each hour. Thus, $f(x) = 14{,}000x$.

(b) The formaldehyde concentration reaches 33 μg/ft^3 when the plywood has emitted a total of $33 \cdot 800 = 26{,}400$ μg. To find this time, solve the equation $f(x) = 26{,}400$ by graphing $Y_1 = 14000X$ and $Y_2 = 26400$. Their graphs intersect near $(1.89, 26{,}400)$ as shown in Figure 2.10. Thus, it will take about 1.9 hours for formaldehyde concentrations to reach 33 μg/ft^3.

[0, 3, 0.5] by [0, 40,000, 10,000]

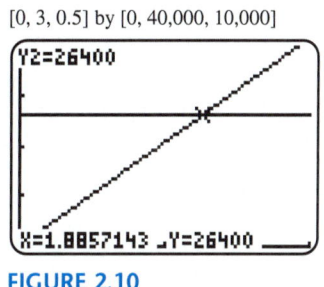

FIGURE 2.10

Solving Linear Equations Symbolically

In the previous examples, linear equations were solved graphically. Linear equations can also be solved symbolically. A symbolic approach provides an exact solution.

EXAMPLE 5 *Solving a linear equation symbolically*

In Example 4, the equation $14{,}000x = 26{,}400$ is solved graphically. Solve this equation symbolically.

Solution

To solve an equation symbolically, we write a sequence of equivalent equations.

$$14{,}000x = 26{,}400 \qquad \text{Original equation}$$

$$x = \frac{26{,}400}{14{,}000} \qquad \text{Divide both sides by 14,000.}$$

$$x = \frac{66}{35} \qquad \text{Simplify.}$$

The symbolic solution for the linear equation is exact, while the graphical solution found in Example 4 is approximate.

Linear equations can be solved symbolically, graphically, and numerically as illustrated in the next example.

EXAMPLE 6   *Solving a linear equation symbolically, graphically, and numerically*

The Celsius and Fahrenheit temperature scales have the same temperature reading at a unique value x that satisfies the linear equation $x = \frac{5}{9}(x - 32)$. Solve this equation symbolically. Support the result graphically and numerically.

Solution

Symbolic Solution   Start by multiplying both sides of the equation by 9.

$x = \dfrac{5}{9}(x - 32)$	Original equation
$9x = 5(x - 32)$	Multiply both sides by 9.
$9x = 5x - 160$	Distributive property
$4x = -160$	Subtract 5x from both sides.
$x = \dfrac{-160}{4}$	Divide both sides by 4.
$x = -40$	Simplify.

Thus, $-40°$F is equivalent to $-40°$C.

Graphical and Numerical Solutions   This result can be supported graphically and numerically by letting $Y_1 = X$ and $Y_2 = (5/9)(X - 32)$. See Figures 2.11 and 2.12.

$[-60, -10, 10]$ by $[-60, 10, 10]$

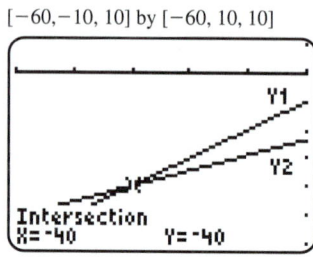

FIGURE 2.11   FIGURE 2.12   ■

Review Note   ***The Distributive Property*** ─────────────

In Example 6 the distributive property is used to help simplify an equation. The distributive property states that $a(b + c) = ab + ac$ for all real numbers a, b, and c. When applying this property, it is important to remember that a is multiplied with *both* b and c. Some examples are given.

$$5(x + 6) = 5x + 5(6) = 5x + 30$$
$$-(2x - 3) = -1(2x - 3) = -2x + 3$$
$$2x(7 - 4x) = 14x - 8x^2$$

EXAMPLE 7   *Solving a linear equation symbolically*

Solve $3(2x - 5) = 10 - (x + 5)$. Check your answer.

Solution

$$3(2x - 5) = 10 - (x + 5) \qquad \text{Original equation}$$

$$6x - 15 = 10 - x - 5 \qquad \text{Distributive property}$$

$$6x - 15 = 5 - x \qquad \text{Simplify.}$$

$$7x - 15 = 5 \qquad \text{Add } x \text{ to both sides.}$$

$$7x = 20 \qquad \text{Add 15 to both sides.}$$

$$x = \frac{20}{7} \qquad \text{Divide both sides by 7.}$$

To check this answer let $x = \dfrac{20}{7}$ in the original equation and simplify.

$$3(2x - 5) = 10 - (x + 5) \qquad \text{Original equation}$$

$$3\left(2 \cdot \frac{20}{7} - 5\right) \stackrel{?}{=} 10 - \left(\frac{20}{7} + 5\right) \qquad \text{Let } x = \frac{20}{7}.$$

$$\frac{15}{7} = \frac{15}{7}\surd \qquad \text{The answer checks.} \qquad \blacksquare$$

EXAMPLE 8   *Solving a linear equation graphically, symbolically, and numerically*

Figure 2.13 shows the graph of a linear function f that models the number of radio stations on the air from 1950 to 1995. (Source: National Association of Broadcasters.)

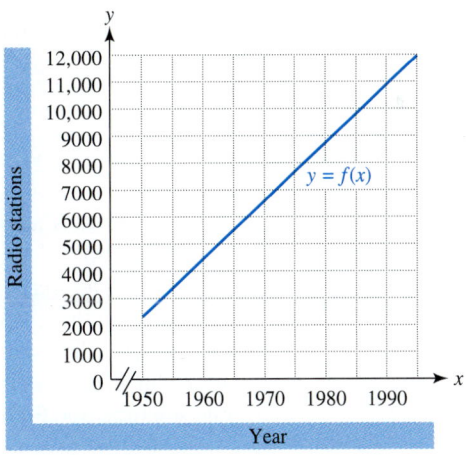

FIGURE 2.13 Radio Stations on the Air

(a) Visually apply the intersection-of-graphs method to estimate when 10,000 stations were on the air.

(b) A symbolic representation of f is $f(x) = 214.2(x - 1950) + 2322$. Determine an equation whose solution is the year when there were 10,000 radio stations on the air. Solve this equation symbolically.

(c) Solve the equation from part (b) numerically.

(d) Compare the graphical, symbolic, and numerical solutions.

Solution

(a) *Graphical Solution* Visually estimate the *x*-value where the graphs of $y = 10,000$ and $y = f(x)$ intersect as shown in Figure 2.14. This *x*-value is approximately 1986.

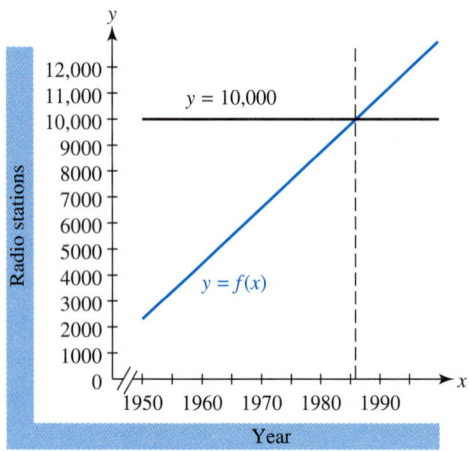

FIGURE 2.14

(b) *Symbolic Solution*   Solve the linear equation $f(x) = 10,000$.

$$214.2(x - 1950) + 2322 = 10,000 \qquad \text{Original equation}$$

$$214.2(x - 1950) = 7678 \qquad \text{Subtract 2322 from both sides.}$$

$$x - 1950 = \frac{7678}{214.2} \qquad \text{Divide both sides by 214.2.}$$

$$x = \frac{7678}{214.2} + 1950 \qquad \text{Add 1950 to both sides.}$$

$$x \approx 1985.845 \qquad \text{Approximate.}$$

According to this model, there were approximately 10,000 radio stations on the air in 1986.

(c) *Numerical Solution*   Table $Y_1 = 214.2(X - 1950) + 2322$, starting at 1950, incrementing by 1. Approximate the value of *x* where Y_1 is equal to 10,000. Figure 2.15 shows that the number of radio stations was approximately 10,000 in 1986. If desired, we could increment by 0.1 to obtain a more accurate solution. See Figure 2.16.

X	Y₁
1983	9390.6
1984	9604.8
1985	9819
1986	10033
1987	10247
1988	10462
1989	10676

X=1986

FIGURE 2.15

X	Y₁
1985.5	9926.1
1985.6	9947.5
1985.7	9968.9
1985.8	9990.4
1985.9	10012
1986	10033
1986.1	10055

X=1985.8

FIGURE 2.16

(d) The graphical, symbolic, and numerical solutions all give similar results. ∎

In Example 8, the number of radio stations on the air is modeled by a continuous linear function f. In part (c), it is assumed that since $Y_1 < 10{,}000$ when $x = 1985$ and $Y_1 > 10{,}000$ when $x = 1986$, there is an x-value between 1985 and 1986 where $Y_1 = 10{,}000$. This is true because the graph of Y_1 is *continuous* with no breaks. This basic concept is referred to as the *intermediate value property*.

Intermediate value property

Let (x_1, y_1) and (x_2, y_2) with $y_1 \neq y_2$ be two points on the graph of a continuous function f. Then, on the interval $x_1 \leq x \leq x_2$, f takes on every value between y_1 and y_2 at least once.

The intermediate value property is illustrated in Figure 2.17. The points $(2, -2)$ and $(7, 6)$ lie on the graph of a function f. The value 3 lies between the y-values of -2 and 6. Because f is continuous, $f(x)$ must equal 3 for some x on the interval $2 \leq x \leq 7$. This x-value is 4, since the point $(4, 3)$ lies on the graph of f. Loosely speaking, the intermediate value property is saying that we cannot draw a continuous curve that connects the points $(2, -2)$ and $(7, 6)$ without crossing the line $y = 3$. The only way not to cross this line would be to pick up the pencil. However, this creates a discontinuous graph, rather than a continuous one.

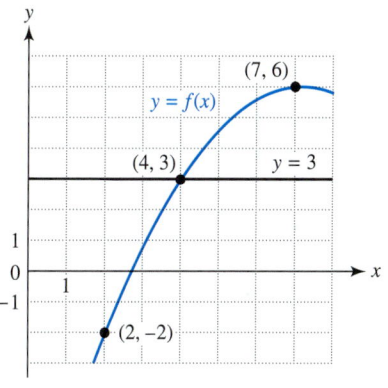

FIGURE 2.17

There are many examples of the intermediate value property. Physical motion is usually considered to be continuous. Suppose at one time a car is traveling at 20 miles per hour and at another time it is traveling at 40 miles per hour. It is logical to assume that the car traveled 30 miles per hour at least once between these times. In fact, by the intermediate value property, the car must have assumed all speeds between 20 and 40 miles per hour at least once. Similarly, if a jet airliner takes off and flies at an altitude of 30,000 feet, then by the intermediate value property, we may conclude that the airliner assumed all altitudes between ground level and 30,000 feet at least once.

Percentages. Applications involving percentages often result in linear equations, because percentages can be computed by linear functions. Taking P percent of x is performed by $f(x) = Px$, where P is in decimal form. For example, to calculate 35% of x, let $f(x) = 0.35x$. Then 35% of $150 can be computed by $f(150) = 0.35(150) = 52.5$ or $52.50.

EXAMPLE 9   *Solving an application involving percentages*

A survey found that 76% of bicycle riders do not wear helmets. (Source: Opinion Research Corporation for Glaxo Wellcome, Inc.)

(a) Find a symbolic representation for a function that computes the number of people who do not wear helmets among x bicycle riders.

(b) There are approximately 38.7 million riders of all ages who do not wear helmets. Write a linear equation whose solution gives the total number of bicycle riders. Solve this equation.

Solution

(a) A linear function f that computes 76% of x is given by $f(x) = 0.76x$.

(b) We must find the x-value for which $f(x) = 38.7$ million. The linear equation is $0.76x = 38.7$. Solving gives $x = \dfrac{38.7}{0.76} \approx 50.9$ million bike riders. ∎

2.1   PUTTING IT ALL TOGETHER

Two basic types of equations are linear and nonlinear. Any linear equation in one variable can be written in the form $ax + b = 0$ where $a \neq 0$. It has exactly one solution. Linear equations can be solved symbolically, graphically, and numerically.

Figure 2.18 visually summarizes some of the steps and decisions involved with solving a linear equation. Although there are exceptions, this diagram provides a general overview of the process.

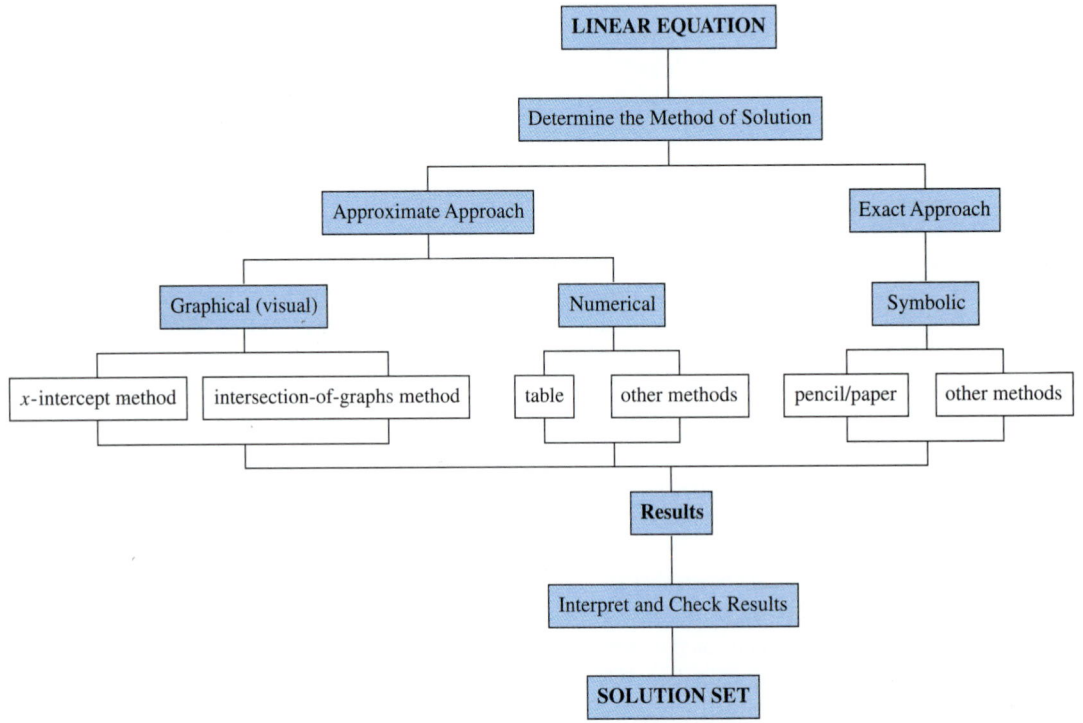

FIGURE 2.18

2.1 EXERCISES   Tape 2

Identifying Linear and Nonlinear Equations

Exercises 1–6: (Refer to Example 1.) Complete the following.
  (a) *Write the equation in the form $h(x) = 0$, and give a symbolic representation for the function h.*
  (b) *Identify both h and the equation as either linear or nonlinear.*
  (c) *Graph h to support your response to part (b).*

1. $3x - 1.5 = 7$

2. $100 - 23x = 20x$

3. $1.2x^2 - 3x = x^2 + 1$

4. $x^2 - 3x = x^3$

5. $7x - 55 = 3(x - 8) + x$

6. $2x^3 - 8x = 2(x^3 - 8) + 2x$

Solving Linear Equations Graphically

Exercises 7 and 8: A linear equation is solved using both the x-intercept and intersection-of-graphs methods. Find the solution by interpreting each graph.

7. **a.**

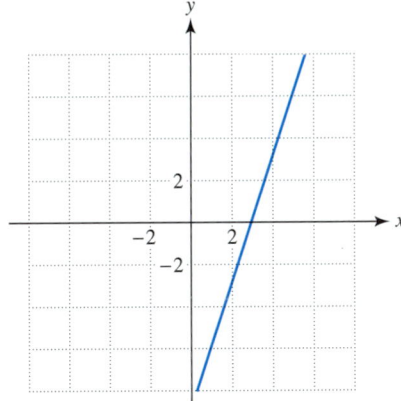

  b.

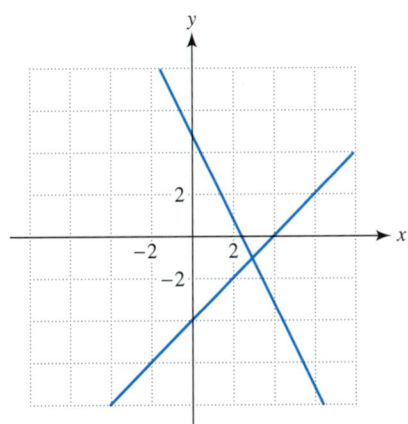

8. **a.**

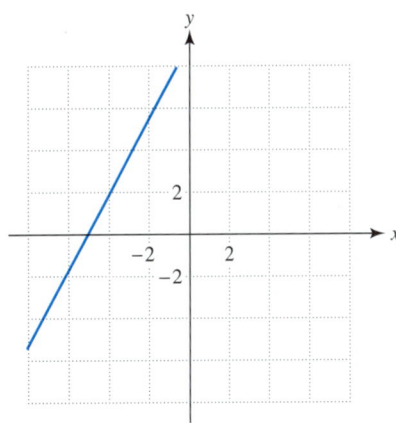

  b.

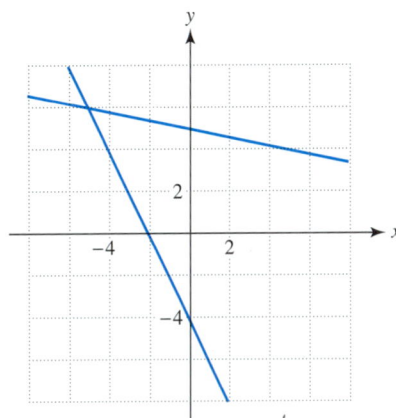

Exercises 9–18: Solve the linear equation using
  (a) *the x-intercept method, and*
  (b) *the intersection-of-graphs method.*
Approximate the solution to four significant digits whenever appropriate.

9. $5x - 1.5 = 5$

10. $8 - 2x = 1.6$

11. $3x - 1.7 = 1 - x$

12. $\sqrt{2}x = 4x - 6$

13. $3.1(x - 5) = \dfrac{1}{5}x - 5$

14. $65 = 8(x - 6) - 5.5$

15. $\dfrac{6 - x}{7} = \dfrac{2x - 3}{3}$

16. $\pi(x - \sqrt{2}) = 1.07x - 6.1$

17. $\sqrt{5}(3 - 2.1x) = 18$

18. $3(x - 4.7) + 5(x + 1.1^2) = 15$

Solving Linear Equations Numerically

Exercises 19 and 20: Let $Y_1 = f(x)$ *and* $Y_2 = g(x)$, *where* f *and* g *are linear functions. Use the table to approximate the solution to the given linear equation. Round answers to the nearest integer.*

19. $f(x) = 0$

X	Y₁	
-5	-3.95	
-4.5	-2.95	
-4	-1.95	
-3.5	-.95	
-3	.05	
-2.5	1.05	
-2	2.05	

X=-5

20. $f(x) = g(x)$

X	Y₁	Y₂
0	5.5	6.01
.5	5.25	5.51
1	5	5.01
1.5	4.75	4.51
2	4.5	4.01
2.5	4.25	3.51
3	4	3.01

X=0

Exercises 21–26: Use tables to solve the equation numerically to the nearest tenth.

21. $2x - 7.2 = 10$

22. $5.8x - 8.7 = 0$

23. $5x - 16 = 5 - x$

24. $\dfrac{3 - x}{2} + 2x = 96$

25. $6x + \dfrac{3}{4} = \dfrac{x - 5}{9}$

26. $8 - 2.1x = 3(x - \sqrt{2}) + \pi$

Solving Linear Equations Symbolically

Exercises 27–40: Complete the following.
  (a) *Solve the equation symbolically.*
  (b) *Check your answer whenever possible.*
  (c) *Classify each equation as being a contradiction, an identity, or a conditional equation.*

27. $3x - 7 = 8$

28. $12 = 13 - 17x$

29. $5x - 1 = 5x + 4$

30. $7 - 9x = 1.5x$

31. $3(x - 1) = 5$

32. $22 = -2(2x + 1.4)$

33. $0.5(x - 2) + 5 = 0.5x + 4$

34. $\dfrac{1}{2}x - 2(x - 1) = 0$

35. $\dfrac{x + 1}{2} = x - 1$

36. $\dfrac{2x + 1}{3} = \dfrac{2x - 1}{3}$

37. $\dfrac{1 - 2x}{4} = \dfrac{3x - 1.5}{-6}$

38. $3(x - 2) + x = 2(x + 3)$

39. $0.5(x - 1980) + 5 = 10$

40. $5(x - 1995) - 15 = 65$

Solving Linear Equations by More than One Method

Exercises 41–44: Solve the equation
  (a) *symbolically,*
  (b) *graphically, and*
  (c) *numerically.*

41. $6x - 8 = -7x + 18$

42. $5 - 8x = 3(x - 7) + 37$

43. $5.5x - 16 = 2.3x + 8$

44. $\dfrac{x + 1}{3} = 3x$

Applications

45. *U.S. Median Income* Median family income between 1980 and 1993 can be modeled by $f(x) = 1321.7(x - 1980) + 21,153$, where x is the year. (Source: Department of Commerce.)
  (a) Determine graphically when the median income was $27,735.
  (b) Solve part (a) numerically.
  (c) Solve part (a) symbolically.

46. *Worker Wages* The average weekly earnings of full-time workers in the lower two-thirds of the pay scale in the year x can be modeled by $f(x) = -1.3256x + 3012.27$. The function f computes wages in 1994 dollars and its domain is $D = \{x \mid 1979 \leq x \leq 1995\}$. (Source: Department of Labor.)
  (a) Determine graphically when the average weekly earnings were $367.70.
  (b) Solve part (a) symbolically.
  (c) What happened to the real earning power of American workers in the lower two-thirds of the pay scale from 1979 to 1995?

47. *Classroom Ventilation* Ventilation is an effective method for removing indoor air pollutants. According to the American Society of Heating, Refrigerating and Air-Conditioning Engineers (ASHRAE), a classroom should have a ventilation rate of 900 cubic feet per hour for each person in the classroom. (Source: ASHRAE.)

(a) Find a function V that computes the ventilation rate in cubic feet per hour that is necessary for a classroom containing x people.

(b) Determine the hourly ventilation rate necessary for a classroom containing 50 people.

(c) If a classroom contains 10,000 cubic feet with 50 people, how many times in one hour should all of the air in the classroom be replaced?

(d) In areas like bars and lounges that allow smoking, the ventilation rate should be increased to 3000 cubic feet per hour for each person. Compared to classrooms, by what factor should the ventilation rate be increased in smoking areas?

48. *Celestial Orbits* To escape the gravity of any celestial body, such as Earth or the sun, a spacecraft must attain a certain velocity, called the *escape velocity*, denoted by v_e. In order to go into a circular orbit, a slower velocity v_c is necessary. The relation between v_e and v_c is described by $v_e = \sqrt{2}v_c$. (Source: H. Karttunen, *Fundamental Astronomy.*)

(a) The velocity for a circular orbit around Earth is 17,700 miles per hour. Approximate the escape velocity for Earth.

(b) A spacecraft from Earth requires a velocity of 94,000 miles per hour in order to escape the solar system. What velocity is needed to travel in a circular orbit around the sun?

49. *Lead Poisoning* Lead is a neurotoxin found in drinking water, old paint, and polluted air. It is particularly hazardous to people because it is not easily eliminated from the body. As directed by the "Safe Drinking Water Act" of 1974, the Environmental Protection Agency (EPA) proposed a maximum lead level in public drinking water of 0.05 milligram per liter. The maximum amount of lead in milligrams that can be ingested in x years under these guidelines is given by $f(x) = 36.5x$. (Source: N. Nemerow and A. Dasgupta, *Industrial and Hazardous Waste Treatment.*)

(a) Determine graphically the number of years before a person could ingest one gram of lead under these guidelines.

(b) Solve part (a) symbolically.

50. *Fat Grams* Some slices of pizza contain 17 grams of fat.

(a) Find a linear function f that computes the number of fat grams in x slices of this type of pizza.

(b) It is recommended that a person requiring 2000 calories daily consume 65 grams of fat or less per day. Graph f together with $y_1 = 65$, $y_2 = 130$, and $y_3 = 195$ in [0, 15, 1] by [0, 210, 10]. Use the intersection-of-graphs method to find how many pizza slices contain 1, 2, and 3 daily allowances of fat.

(c) Solve part (b) symbolically and compare answers.

51. *Height of a Tree* In the accompanying figure, a person five feet tall casts a shadow four feet long. A nearby tree casts a shadow that is 33 feet long. Find the height of the tree by solving a linear equation.

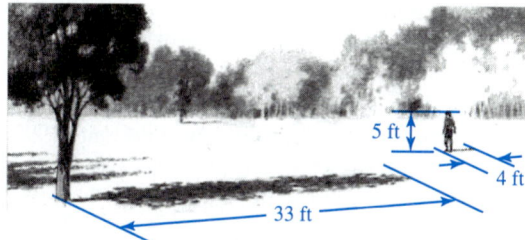

52. *Two-Cycle Engines* Two-cycle engines, used in snowmobiles, chain saws, and outboard motors, require a mixture of gasoline and oil. For certain engines the amount of oil in pints that should be added to x gallons of gasoline is computed by $f(x) = 0.16x$. (Source: Johnson Outboard Motor Company.)

(a) Explain why it is reasonable to expect that f is linear.

(b) How much gasoline should be mixed with one quart of oil?

53. *Grades* In order to receive an A in a college course it is necessary to obtain an average of 90% correct on three one-hour exams of 100 points each and on one final exam of 200 points. If a student scores 82, 88, and 91 on the one-hour exams, what is the minimum score on the final exam that the person can receive and still earn an A?

54. *Auto Maintenance* The cost of auto maintenance as a percentage of total operating expenses has risen since 1984. The function given by $f(x) = 0.2025x - 396.5$ models the percentage of the total operating expenses that were spent on maintenance

in the year x, where $1984 \leq x \leq 1996$. (Source: Runzheimer International.)
(a) Determine graphically the year when this percentage first reached 6%.
(b) Solve part (a) numerically.
(c) Solve part (a) symbolically.

55. *Deaths in the U.S.* There were 2.28 million deaths in the United States during the year 1994. This number represented 8.8 deaths per 1000 people. Approximate the population of the United States in 1994. (Source: Bureau of the Census.)

56. *Skin Cancer* In 1995 the population of the United States was approximately 261 million and there were 34,100 cases of skin cancer reported. (Sources: American Cancer Society, Bureau of the Census.)
(a) Determine the rate of skin cancer per 1000 people.
(b) Find a linear function f that outputs the approximate number of cases of skin cancer in a population of x thousand.

57. *Cellular Phone Survey* In a survey of 100 motorists who had been in accidents, a newspaper reported that 13.7% owned cellular phones. Do you believe that the mathematics in this survey was presented correctly? Explain. (Source: Associated Press.)

58. *Corporate Downsizing* In 1991 and 1992, millions of workers lost their jobs due to corporate downsizing. In a 1994 survey of these workers, 32%, or 1.76 million workers, found full-time work for equal or higher pay, 24% were jobless, while 28% were working full time for less pay. The remaining were either self-employed or working part time. How many of these workers were jobless? (Source: Department of Labor.)

59. *French Fry Production* The United States, the top producer of frozen french fries, increased its exports in 1995 by 36% over the 1994 level. In 1995, 48.5% of the 360,184 tons of french fries exported were sold to Japan. (Source: Department of Agriculture.)

(a) How many tons were exported to all countries during 1994?
(b) Determine the total amount of french fries sold to Japan in 1995.
(c) Suppose that an order of french fries contains four ounces. Use scientific notation to express the total number of orders of fries exported to Japan in 1995.
(d) In 1995 the population of Japan was approximately 125 million. How many orders of fries is this for each person?

60. *Salty Snacks* The average American ate 22 pounds of salty snacks in 1994, up 25.7% from 1988. What was the average consumption in 1988? (Source: Snack Foods Association.)

61. *Sale Price* A store is discounting all regularly priced merchandise by 25%. Find a function f that computes the sale price of an item having a regular price of x. If an item normally costs $56.24, what is its sale price?

62. To continue Exercise 61, use f to find the regular price of an item that costs $19.62 on sale.

Intermediate Value Property

63. The points $(1, -4)$ and $(5, 6)$ lie on the graph of $f(x) = 2.5x - 6.5$. Find a point $(x, 3.5)$ that lies on the graph of f. Sketch a graph to illustrate how the intermediate value property applies in this situation.

64. Sketch a graph of a function f that passes through the points $(-2, 3)$ and $(1, -2)$ but never takes on a value of 0. What must be true about the graph of f?

Writing about Mathematics

1. Describe two basic graphical methods used to solve a linear equation. Give examples.

2. Describe verbally how to solve the linear equation $ax + b = 0$. What assumptions have you made about the value of a?

2.2 Linear Inequalities

Inequalities • Techniques for Solving Inequalities • Interval Notation

Introduction

If a person weighing 143 pounds needs to purchase a life preserver for whitewater rafting, it is doubtful that there is one designed exactly for this weight. Life preservers are manufactured to support a range of body weights. A vest that is approved for weights between 120 and 160 pounds would be appropriate. Every airplane has a maximum weight allowance. It is important that this weight limit is accurately determined. However, most people feel more comfortable at takeoff if that maximum has not been reached. This is because any weight that is less than the maximum is also safe and allows for a greater margin of error. Both of these situations involve the concept of inequality.

In mathematics much effort is expended toward solving equations and determining equality. One reason for this is that equality is frequently a boundary between *greater than* and *less than*. The solution to an inequality often can be found by first locating where two expressions are equal. Since equality and inequality are closely related, many of the techniques used to solve equations also can be applied to inequalities.

Inequalities

Inequalities result whenever the equals sign in an equation is replaced with any one of the symbols $<$, $\leq$, $>$, or $\geq$. Some examples of inequalities include

$$x + 15 < 9x - 1, \qquad x^2 - 2x + 1 \geq 2x, \qquad z + 5 > 0,$$
$$xy + x^2 \leq y^3 + x, \qquad \text{and} \qquad 2 + 3 > 1.$$

The first three inequalities involve one variable, the fourth inequality contains two variables, and the fifth inequality has only constants. As with linear equations, our discussion focuses on inequalities in one variable.

To **solve** an inequality means to find all values for the variable that make the inequality a true statement. Such values are **solutions** and the set of all solutions is the **solution set** of the inequality. Two inequalities are **equivalent** if they have the same solution set. It is common for an inequality to have an infinite number of solutions. For instance, the inequality $x - 1 > 0$ has an infinite number of solutions because any real number x satisfying $x > 1$ is a solution. The solution set is $\{x \mid x > 1\}$.

Review Note   *Properties of Inequalities* ——————————————

Let a, b, and c be real numbers.

1. $a < b$ and $a + c < b + c$ are equivalent.
(The same number may be added to both sides of an inequality.)
2. If $c > 0$, then $a < b$ and $ac < bc$ are equivalent.
(Both sides of an inequality may be multiplied by the same positive number.)
3. If $c < 0$, then $a < b$ and $ac > bc$ are equivalent.
(Both sides of an inequality may be multiplied by the same negative number provided the inequality symbol is reversed.)

Note: Replacing $<$ with $\leq$ and $>$ with $\geq$ results in similar properties. The word "multiplied" may be replaced by "divided" in Properties 2 and 3.

Like functions and equations, inequalities in one variable can be classified as linear or nonlinear.

Linear inequality in one variable

A **linear inequality** in one variable is an inequality that can be written in the form

$$ax + b > 0,$$

where $a \neq 0$. (The symbol $>$ may be replaced by $\geq$, $<$, or $\leq$.)

Examples of linear inequalities include

$$3x - 4 < 0, \qquad 7x + 5 \geq x, \qquad x + 6 > 23, \qquad \text{and} \qquad 7x + 2 \leq -3x + 6.$$

Using techniques from algebra, each of these inequalities can be transformed into one of the forms $ax + b > 0$, $ax + b \geq 0$, $ax + b < 0$, or $ax + b \leq 0$. For example, by subtracting x from both sides of $7x + 5 \geq x$, we obtain the equivalent inequality $6x + 5 \geq 0$.

Techniques for Solving Inequalities

In residential areas, a typical speed limit is 30 miles per hour. A driver traveling x miles per hour is obeying the speed limit whenever $x \leq 30$, and breaking the speed limit if $x > 30$. The equation $x = 30$ represents the boundary between obeying the limit and breaking it. It is unnecessary for highway workers to list every legal speed. By posting the speed limit or boundary, people can easily determine whether or not they are speeding.

A similar situation occurs in the solution of mathematical inequalities. One technique to solve an inequality is first to determine where equality occurs. When solving linear inequalities, equality is always the boundary between *greater than* and *less than*. For example, to solve the inequality $4 - 2x > 0$, we can replace inequality with equality and solve the equation $4 - 2x = 0$ to obtain $x = 2$. **Test values** can be used to find the solution set for the related inequality. If we substitute a test value that is less than 2 into the inequality, such as $x = 0$, then the inequality is true. Any test value for x that is greater than 2 makes the inequality false. Therefore, the solution set for $4 - 2x > 0$ is $\{x \mid x < 2\}$. The table in the margin contains several test values, showing that the expression $4 - 2x$ is positive when $x < 2$ and negative when $x > 2$. In this example, $x = 2$ is the *boundary number*. Creating a table is an efficient numerical method to evaluate test values in an expression.

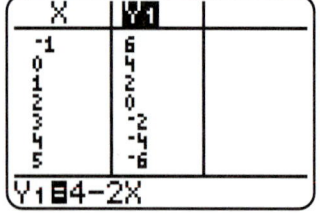

The three basic methods used to solve a linear equation were graphical, symbolic, and numerical. They can also be applied to linear inequalities. For example, the intersection-of-graphs method can be extended to solve inequalities. Figure

2.19 shows the velocity of two cars in miles per hour after x minutes. V_A denotes the velocity of Car A, while V_B denotes the velocity of Car B.

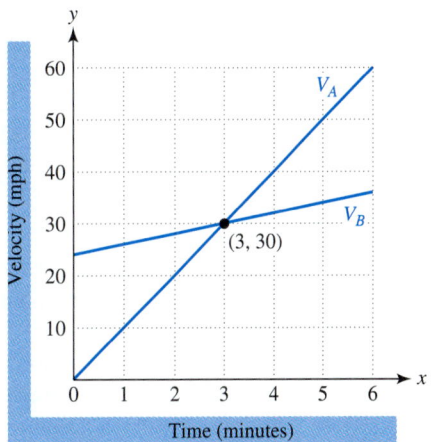

FIGURE 2.19 Velocities of Two Cars

At three minutes $V_A = V_B$, and both cars are traveling at 30 miles per hour. To the left of $x = 3$, the graph of V_A is below the graph of V_B so Car A is traveling slower than Car B. Thus,

$$V_A < V_B \text{ when } 0 \le x < 3.$$

To the right of $x = 3$ the graph of V_A is above the graph of V_B, so Car A is traveling faster than Car B. Thus,

$$V_A > V_B \text{ when } 3 < x \le 6.$$

This technique is used in the next example.

EXAMPLE 1   *Using the intersection-of-graphs method*

When the air temperature reaches the dew point, fog may form. This phenomenon also causes clouds to form at higher altitudes. Both the air temperature and the dew point decrease at a constant rate as the altitude above ground level increases. If the ground-level temperature and dew point are T_0 and D_0, the air temperature can be approximated by $T(x) = T_0 - 29x$ and the dew point by $D(x) = D_0 - 5.8x$ at an altitude of x miles.

(a) If $T_0 = 75°F$ and $D_0 = 55°F$, determine the altitudes where clouds will not form. See Figure 2.20.

(b) The slopes of the graphs for the functions T and D are called *lapse rates*. Interpret their meanings. Explain how these two slopes ensure a strong likelihood of clouds forming. (Source: A. Miller and R. Anthes, *Meteorology*.)

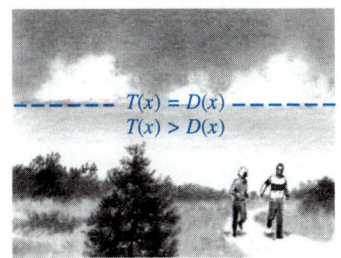

FIGURE 2.20

Solution

(a) Since $T_0 = 75$ and $D_0 = 55$, let $T(x) = 75 - 29x$ and $D(x) = 55 - 5.8x$. Graph $Y_1 = 75 - 29X$ and $Y_2 = 55 - 5.8X$ as in Figure 2.21 on the following page.

[0, 2, 1] by [0, 80, 10]

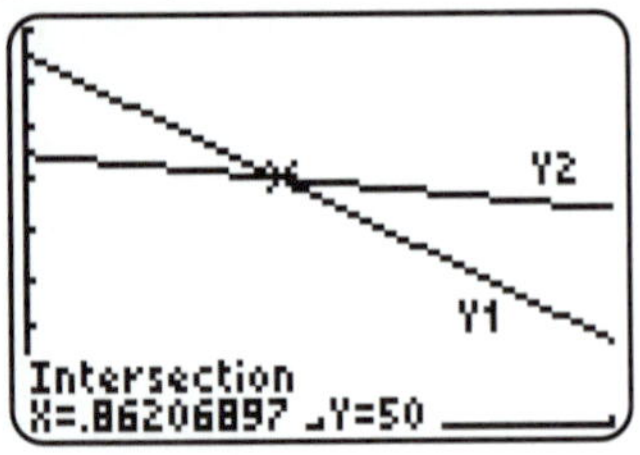

FIGURE 2.21

The graphs intersect at $(k, 50)$ where $k \approx 0.86$. This means that the air temperature and dew point are both 50°F at about 0.86 mile above ground level. Clouds will not form below this altitude or when the graph of Y_1 is above the graph of Y_2. The solution set is $\{x \mid 0 \le x < k\}$, where $k \approx 0.86$.

(b) The slope of the graph of T is -29. This means that for each one-mile increase in altitude, the air temperature *decreases* by 29°F. Similarly, the slope of the graph of D is -5.8. The dew point *decreases* by 5.8°F for every one-mile increase in altitude. As the altitude increases, the air temperature decreases at a faster rate than the dew point. As a result, the air temperature typically cools to the dew point at higher altitudes. Above this altitude clouds may form. ∎

Critical Thinking

How does the difference between the air temperature and the dew point at ground level affect the altitudes at which clouds may form? Explain.

The x-intercept method can be applied to a linear inequality in the form $ax + b > 0$ by graphing $h(x) = ax + b$. The linear graph of h has one x-intercept k. In this case, k is the boundary number. The solution set includes x-values where $h(x) > 0$. This corresponds to where the graph of h is above the x-axis. The solution set to $ax + b > 0$ in Figure 2.22 is $\{x \mid x > k\}$, and in Figure 2.23 it is $\{x \mid x < k\}$. This technique also may be applied to linear inequalities in the forms $ax + b < 0$, $ax + b \ge 0$, and $ax + b \le 0$.

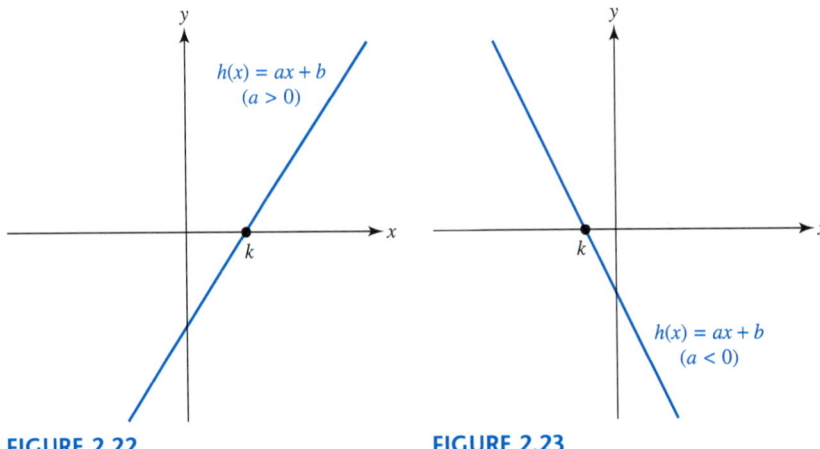

$h(x) = ax + b$
$(a > 0)$

$h(x) = ax + b$
$(a < 0)$

FIGURE 2.22   **FIGURE 2.23**

EXAMPLE 2 *Solving an inequality by the x-intercept method*

Solve the inequality $\sqrt{2}x + 1.7 > -3.3x + 5.1$ using the *x*-intercept method.

Solution

To apply the *x*-intercept method, write the inequality as $h(x) > 0$ for some linear function h.

$$\sqrt{2}x + 1.7 > -3.3x + 5.1 \qquad \text{Original inequality}$$
$$\sqrt{2}x + 1.7 + 3.3x - 5.1 > 0 \qquad \text{Subtract 5.1 and add 3.3}x.$$
$$(\sqrt{2} + 3.3)x - 3.4 > 0 \qquad \text{Simplify.}$$

Graph $h(x) = (\sqrt{2} + 3.3)x - 3.4$ as shown in Figure 2.24.

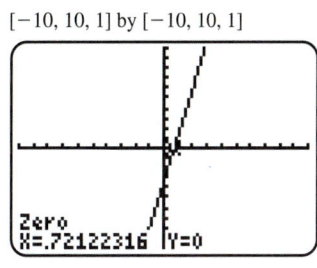

$[-10, 10, 1]$ by $[-10, 10, 1]$

Zero
X=.72122316 Y=0

FIGURE 2.24

Let the *x*-intercept be *k*, where $k \approx 0.72$ as shown in Figure 2.24. The *x*-intercept represents the boundary between *greater than zero* and *less than zero*. The graph of *h* is above the *x*-axis, and hence $h(x) > 0$, for *x*-values to the right of the *x*-intercept. Therefore, the solution set is $\{x \mid x > k\}$, where $k \approx 0.72$. ∎

EXAMPLE 3 *Solving an inequality symbolically, graphically, and numerically*

Solve the inequality $-2x + 5 < 12$ symbolically. Give graphical and numerical support to your answer.

Solution

Symbolic Solution To solve an inequality symbolically we apply the properties of inequalities.

$$-2x + 5 < 12 \qquad \text{Original inequality}$$
$$-2x < 7 \qquad \text{Subtract 5.}$$
$$x > -\frac{7}{2} \qquad \text{Divide by } -2. \text{ Reverse inequality.}$$

The solution set is $\left\{ x \mid x > -\frac{7}{2} \right\}$.

Graphical and Numerical Solution Since $-2x + 5 < 12$ is equivalent to $-2x - 7 < 0$, graph $Y_1 = -2x - 7$. In Figure 2.25 on the following page, $Y_1 < 0$ to the right of the *x*-intercept of -3.5. The solution set is $\{x \mid x > -3.5\}$. Numerical support is shown in Figure 2.26 on the following page, where several test values are evaluated. $Y_1 = 0$ when $x = -3.5$, and $Y_1 < 0$ for *x*-values greater than -3.5.

[−10, 10, 1] by [−10, 10, 1]

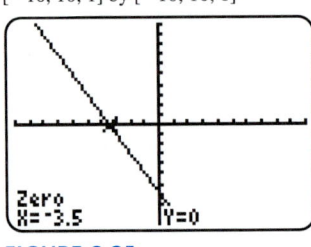

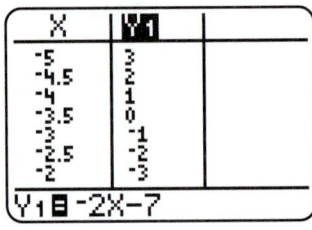

FIGURE 2.25

FIGURE 2.26 ■

EXAMPLE 4 *Solving a compound inequality graphically and symbolically*

As the altitude above ground level increases, the air temperature becomes cooler. The function given by $T(x) = T_0 - 29x$ models the Fahrenheit temperature x miles high, where T_0 is the ground temperature. (Source: A. Miller.)

(a) Suppose $T_0 = 70°F$. Use the intersection-of-graphs method to determine the altitudes where the air temperature is from 32°F to 50°F. See Figure 2.27.

(b) Solve part (a) symbolically.

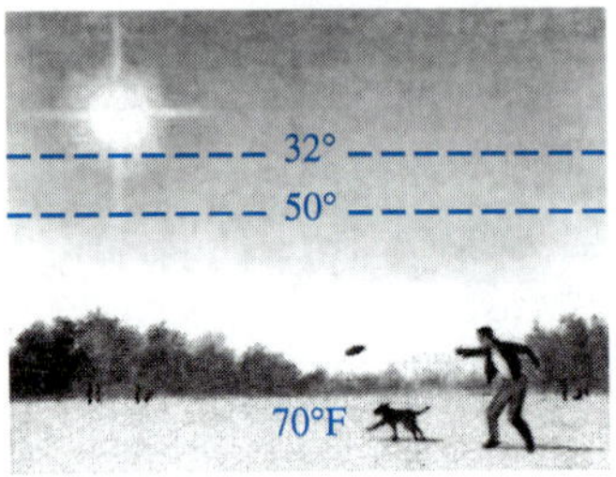

FIGURE 2.27

Solution

(a) Let $T(x) = 70 - 29x$. The solution set consists of x-values where the compound inequality $32 \leq T(x) \leq 50$ is true. Graph $Y_1 = 32$, $Y_2 = 70 - 29X$, and $Y_3 = 50$. Their graphs intersect near the points (0.69, 50) and (1.31, 32) as shown in Figures 2.28 and 2.29. The air temperature is from 32°F to 50°F whenever the graph of $Y_2 = 70 - 29X$ is between the graphs of $Y_1 = 32$ and $Y_3 = 50$. Thus, the solution set is $\{x \mid k_1 \leq x \leq k_2\}$, where $k_1 \approx 0.69$ and $k_2 \approx 1.31$. This means that the air temperature is between 32°F and 50°F from about 0.69 mile to 1.31 miles above ground level.

[0, 3, 1] by [0, 80, 10]

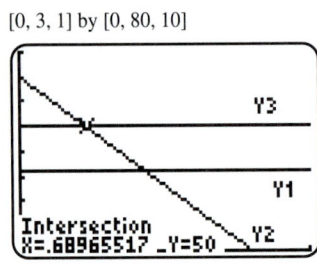

FIGURE 2.28

[0, 3, 1] by [0, 80, 10]

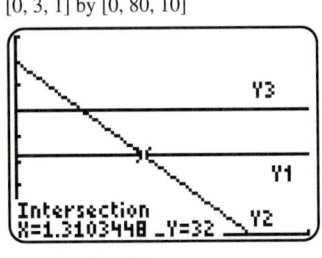

FIGURE 2.29

(b) The air temperature is from 32°F to 50°F for altitudes x satisfying the following.

$$32 \leq 70 - 29x \leq 50 \qquad \text{Compound inequality}$$

$$-38 \leq -29x \leq -20 \qquad \text{Subtract 70 from each expression.}$$

$$\frac{-38}{-29} \geq x \geq \frac{-20}{-29} \qquad \text{Divide by } -29. \text{ Reverse each inequality.}$$

$$\frac{38}{29} \geq x \geq \frac{20}{29} \qquad \text{Simplify.}$$

$$\frac{20}{29} \leq x \leq \frac{38}{29} \qquad \text{Rewrite the inequality.}$$

The solution set is $\left\{ x \mid \dfrac{20}{29} \le x \le \dfrac{38}{29} \right\}$. Both the graphical and symbolic methods produce similar results since $\dfrac{20}{29} \approx 0.69$ and $\dfrac{38}{29} \approx 1.31$. Notice that the exact values of $k_1 = \dfrac{20}{29}$ and $k_2 = \dfrac{38}{29}$ were found using a symbolic approach rather than a graphical approach. ∎

EXAMPLE 5   *Solving an inequality numerically*

For the 55- to 64-year age group, the percentage of women in the work force is given by $f(x) = \dfrac{13}{15}(x - 1950) + 28$, where x is between the years 1950 and 1965.

For men over 65 years, the percentage who were in the work force is computed by $g(x) = -\dfrac{19}{15}(x - 1950) + 47$. The domain for both f and g is $D = \{1950, 1951, 1952, \ldots, 1965\}$. Solve the inequality $f(x) > g(x)$ numerically. (Source: J. Schulz, *The Economics of Aging.*)

Solution

Let $Y_1 = (13/15)(X - 1950) + 28$ and $Y_2 = (-19/15)(X - 1950) + 47$. We must find x-values where $Y_1 > Y_2$. Table Y_1 and Y_2, starting at 1950, incrementing by 1 as shown in Figure 2.30.

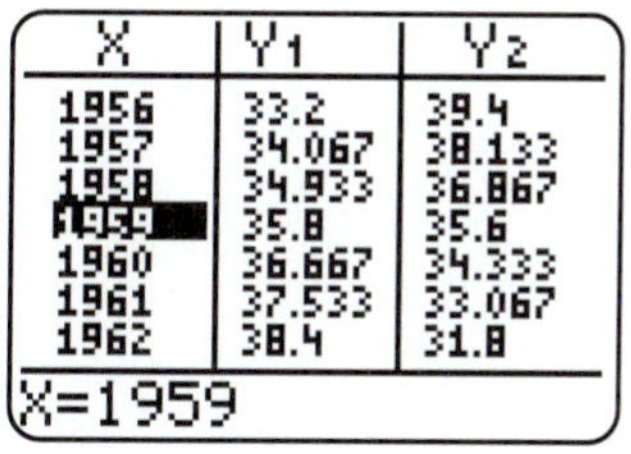

X	Y₁	Y₂
1956	33.2	39.4
1957	34.067	38.133
1958	34.933	36.867
1959	35.8	35.6
1960	36.667	34.333
1961	37.533	33.067
1962	38.4	31.8

X=1959

FIGURE 2.30

There is no year in D where $Y_1 = Y_2$. In 1958, $Y_1 < Y_2$ and in 1959, $Y_1 > Y_2$. For the given age categories, the percentage of women in the work force first exceeded the percentage of men in 1959. Since Y_1 represents an increasing linear function, while Y_2 represents a decreasing linear function, this trend continues each year after 1959. The solution set is $\{1959, 1960, 1961, \ldots, 1965\}$. ∎

Interval Notation

The solution set in Example 4 consists of all real numbers x satisfying $\dfrac{20}{29} \le x \le \dfrac{38}{29}$. This solution set can be graphed using a number line as shown in Figure 2.31.

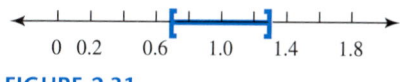

0   0.2   0.6   1.0   1.4   1.8

FIGURE 2.31

A convenient notation for number line graphs is called **interval notation.** Instead of drawing the entire number line as in Figure 2.31, the solution set can be expressed as $\left[\dfrac{20}{29}, \dfrac{38}{29}\right]$. Because the solution set includes the endpoints $\dfrac{20}{29}$ and $\dfrac{38}{29}$, the interval is a **closed interval** and brackets are used. A solution set that includes all real numbers satisfying $-2 < x < 3$ would be expressed as the **open interval** $(-2, 3)$. Parentheses indicate that the endpoints are not included in the solution set. An example of a **half-open interval** is $[0, 4)$, which represents the inequality $0 \leq x < 4$.

Table 2.1 provides some examples of interval notation. The symbol ∞ refers to infinity. It does not represent a real number. The notation $(1, \infty)$ means $\{x \mid x > 1\}$ or simply $x > 1$. Since this interval has no maximum x-value, ∞ is used in the position of the right endpoint. A similar interpretation holds for the symbol $-\infty$, which represents negative infinity.

TABLE 2.1   Interval Notation

Inequality	Interval Notation	Graph
$-2 < x < 2$	$(-2, 2)$ open interval	
$-1 < x \leq 3$	$(-1, 3]$ half-open interval	
$-3 \leq x \leq 2$	$[-3, 2]$ closed interval	
$x > -3$	$(-3, \infty)$ infinite interval	
$x \leq 1$	$(-\infty, 1]$ infinite interval	
$-\infty < x < \infty$ (entire number line)	$(-\infty, \infty)$ infinite interval	

MAKING CONNECTIONS
Points and Intervals

The expression $(2, 5)$ has two possible meanings. The first is that $(2, 5)$ is an ordered pair, which can be plotted as a point on the xy-plane. The second is that $(2, 5)$ represents an open interval. To alleviate confusion, phrases like "the point $(2, 5)$" or "the interval $(2, 5)$" may be used.

EXAMPLE 6   *Solving inequalities symbolically*

Solve the linear inequalities symbolically. Express the solution set using interval notation.

(a) $-\dfrac{x}{2} + 1 \leq 3$

(b) $-8 < \dfrac{3x - 1}{2} \leq 5$

(c) $5(x - 6) < 2x - 2(1 - x)$

Solution

(a) Simplify the inequality as follows.

$$-\frac{x}{2} + 1 \le 3 \qquad \text{Original inequality}$$

$$-\frac{x}{2} \le 2 \qquad \text{Subtract 1.}$$

$$x \ge -4 \qquad \text{Multiply by } -2. \text{ Reverse the inequality.}$$

In interval notation the solution set is $[-4, \infty)$.

(b) The two parts of this compound inequality can be solved simultaneously.

$$-8 < \frac{3x - 1}{2} \le 5 \qquad \text{Original inequality}$$

$$-16 < 3x - 1 \le 10 \qquad \text{Multiply by 2.}$$

$$-15 < 3x \le 11 \qquad \text{Add 1.}$$

$$-5 < x \le \frac{11}{3} \qquad \text{Divide by 3.}$$

The solution set is $\left(-5, \dfrac{11}{3} \right]$.

(c) Start by applying the distributive property on both sides of the inequality.

$$5(x - 6) < 2x - 2(1 - x) \qquad \text{Original inequality}$$

$$5x - 30 < 2x - 2 + 2x \qquad \text{Distributive property}$$

$$5x - 30 < 4x - 2 \qquad \text{Simplify.}$$

$$x - 30 < -2 \qquad \text{Subtract } 4x.$$

$$x < 28 \qquad \text{Add 30.}$$

The solution set is $(-\infty, 28)$. ∎

Critical Thinking

When humans breathe, carbon dioxide is emitted. In one study of college students, the emission rates of carbon dioxide in grams per hour were measured during both lectures and exams. The average emission rate L during lectures satisfied $25.33 \le L \le 28.17$, whereas the average emission rate E during exams was in the range of $36.58 \le E \le 40.92$. Interpret these results and discuss some reasons that might account for them. (Source: T. Wang, *ASHRAE Trans.*)

2.2 PUTTING IT ALL TOGETHER

Applications involving linear functions result in both linear equations and inequalities. Like a linear equation, the solution set for a linear inequality can be found graphically, numerically, and symbolically. One common strategy for solving a linear inequality is first to locate the x-value that results in equality. This boundary number represents the boundary between *greater than* and *less than*. Using this value, the solution set for the linear inequality can be found.

The following table lists methods to solve a linear inequality in the form $h(x) > 0$ or $f(x) > g(x)$. Inequalities involving $<$, $\leq$, and $\geq$ are solved in a similar manner.

Method	Description
x-Intercept Method	To solve the inequality $h(x) = ax + b > 0$, graph h. The graph has one x-intercept. *Greater than* occurs where the graph is above the x-axis. This is either to the right or to the left of the x-intercept.
Intersection-of-Graphs Method	For an inequality in the form $f(x) > g(x)$, graph $y_1 = f(x)$ and $y_2 = g(x)$. Find the point of intersection. The solution set includes x-values where the graph of f is above the graph of g. This is either to the right or to the left of the point of intersection.
Numerical Method I	To solve the inequality $h(x) = ax + b > 0$, table h. The solution set includes x-values where $h(x) > 0$. This is either to the right or to the left of the x-value that satisfies the equation $h(x) = 0$.
Numerical Method II	Solve $f(x) > g(x)$ by tabling f and g. Approximate the x-value where $f(x) = g(x)$. The solution set includes all x-values where the table values of $f(x)$ are greater than the table values of $g(x)$.
Symbolic Method I	Use the properties of inequalities to simplify $ax + b > 0$ to either $x > k$ or $x < k$ for some real number k.
Symbolic Method II	Solve the equation $ax + b = 0$ to obtain $x = k$. Use test values to determine if the solution set for $ax + b > 0$ includes $x > k$ or $x < k$.

2.2 EXERCISES   Tape 2

Solving Linear Inequalities Graphically

1. *Interest* The function f computes the annual interest y on a loan of x dollars with an interest rate of 10%. The graphs of f and the horizontal line $y = 100$ are shown in the figure. Determine the loan amounts that result in the following. Express your answers verbally.
   (a) An annual interest equal to $100
   (b) An annual interest of more than $100
   (c) An annual interest of less than $100
   (d) An annual interest of $100 or more

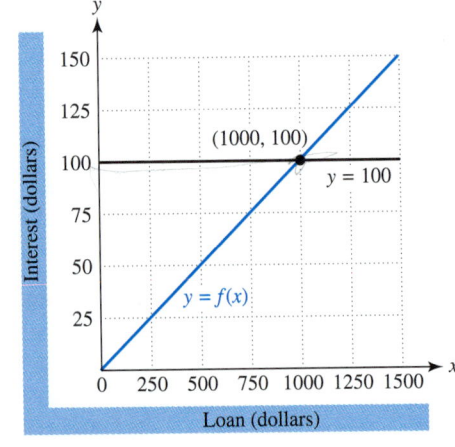

2. *U.S. Population*   The function f models the population of the United States from 1970 to 1995. The graphs of f and the horizontal line $y = 226$ are shown in the figure. Use the graphs to determine when each of the following were satisfied. Express your answers verbally.

(a)   A population equal to 226 million

(b)   A population of 226 million or less

(c)   A population of 226 million or more

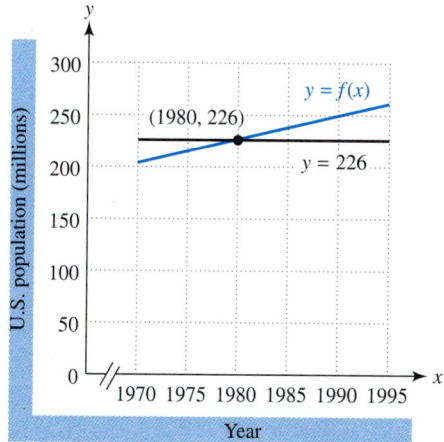

3.   Use the figure for Exercise 1 to determine the x-values that satisfy the equation or inequality. Express your answers symbolically.

(a) $f(x) = 100$

(b) $f(x) > 100$

(c) $f(x) < 100$

(d) $f(x) \geq 100$

4.   Use the figure for Exercise 2 to determine the x-values that satisfy the equation or inequality. Express your answers symbolically.

(a) $f(x) = 226$

(b) $f(x) \leq 226$

(c) $f(x) \geq 226$

5. *Distance Between Cars*   Cars A and B are both traveling in the same direction. Their distances in miles north of St. Louis after x hours are computed by the functions f_A and f_B, respectively. The graphs of f_A and f_B are shown in the figure for $0 \leq x \leq 10$.

(a)   Which car is traveling faster? Explain.

(b)   How many hours elapse before the two cars are the same distance from St. Louis? How far are they from St. Louis when this occurs?

(c)   During what time interval is Car B farther from St. Louis than Car A?

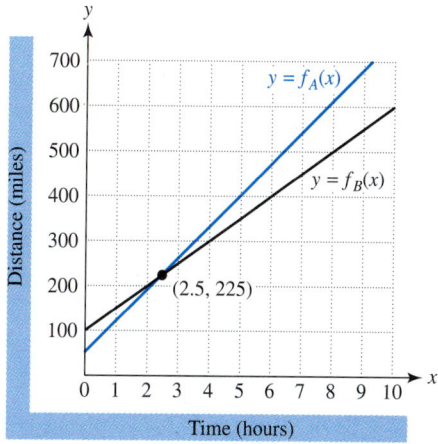

6. *Distance*   The linear function f computes the distance y in miles between a car and the city of Omaha after x hours, where $0 \leq x \leq 6$. The graphs of f and the horizontal lines $y = 100$ and $y = 200$ are shown in the figure. Use the graphs to answer the following.

(a)   Is the car moving toward or away from Omaha? Explain.

(b)   Determine the times when the car is 100 miles or 200 miles from Omaha.

(c)   When is the car from 100 to 200 miles from Omaha?

(d)   When is the car's distance from Omaha greater than 100 miles?

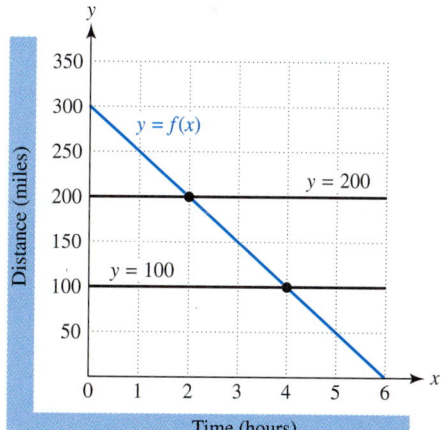

7. The graphs of two linear functions f and g are shown in the figure.
   (a) Solve the equation $g(x) = f(x)$.
   (b) Solve the inequality $g(x) > f(x)$.

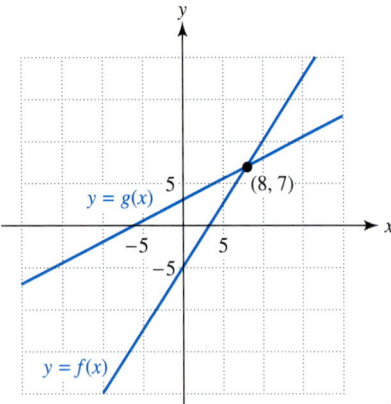

8. Use the figure to solve each equation or inequality.
   (a) $f(x) = g(x)$
   (b) $g(x) = h(x)$
   (c) $f(x) < g(x) < h(x)$
   (d) $g(x) > h(x)$

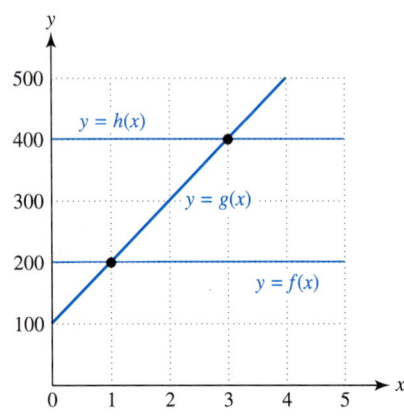

Exercises 9–14: Solve the linear inequality using
   (a) *the x-intercept method, and*
   (b) *the intersection-of-graphs method.*
9. $5x - 4 > 10$
10. $-3x + 6 \le 9$
11. $-2(x - 1990) + 55 \ge 60$
12. $\sqrt{2}x > 10.5 - 13.7x$
13. $\sqrt{5}(x - 1.2) - \sqrt{3}x < 5(x + 1.1)$
14. $1.238x + 0.998 \le 1.23(3.987 - 2.1x)$

Exercises 15–18: Solve the compound linear inequality using the intersection-of-graphs method.
15. $3 \le 5x - 17 < 15$

16. $-4 < \dfrac{55 - 3.1x}{4} < 17$

17. $1.5 \le 9.1 - 0.5x \le 6.8$

18. $0.2x < \dfrac{2x - 5}{3} < 8$

Solving Linear Inequalities Numerically

Exercises 19–22: Assume Y_1 and Y_2 represent linear functions with the set of real numbers for their domains. Use the table to solve the inequalities.

19. $Y_1 > 0$, $Y_1 \le 0$

X	Y1	
0	220	
1	165	
2	110	
3	55	
4	0	
5	-55	
6	-110	
X=0		

20. $Y_1 < 0$, $Y_1 \ge 0$

X	Y1	
-5	-32	
-4	-16	
-3	0	
-2	16	
-1	32	
0	48	
1	64	
X= -5		

21. $Y_1 \ge Y_2$, $Y_1 < Y_2$

X	Y1	Y2
0	200	-300
10	150	-250
20	100	-200
30	50	-150
40	0	-100
50	-50	-50
60	-100	0
X=0		

22. $Y_1 > Y_2$, $Y_1 \le Y_2$

X	Y1	Y2
8	140	110
9	145	125
10	150	140
11	155	155
12	160	170
13	165	185
14	170	200
X=8		

Exercises 23–26: Complete the following.
   (a) *Write each linear inequality using one of the forms: $h(x) > 0$, $h(x) \ge 0$, $h(x) < 0$, or $h(x) \le 0$, where h is a linear function.*

(b) *Use a table to approximate to the nearest tenth the x-value where h(x) = 0. Solve the inequality.*

23. $-2x - 4 > 2x + 2$

24. $\sqrt{11}x \leq 5.5 + \pi x$

25. $\dfrac{3x - 1}{5} < 15$

26. $1.5(x - 0.7) + 1.5x < 1$

Interval Notation

Exercises 27–38: Express each of the following in interval notation.

27. $x \geq 5$ **28.** $x < 100$

29. $4 \leq x < 19$ **30.** $-4 < x < -1$

31. $x \leq -37$ **32.** $\{x \mid x \leq -3\}$

33. $\{x \mid -1 \leq x\}$ **34.** $17 > x \geq -3$

35.

36.

37.

38.

Solving Linear Inequalities Symbolically

Exercises 39–50: Solve each inequality symbolically. Express the solution set in interval notation.

39. $2x + 6 \geq 10$

40. $-4x - 3 < 5$

41. $-2(x - 10) + 1 > 0$

42. $3(x + 5) \leq 0$

43. $\dfrac{x + 2}{3} \geq 5$

44. $\dfrac{2 - x}{6} < 0$

45. $-1 \leq 2x \leq 4$

46. $5 < 4x - 1 \leq 11$

47. $3 \leq 4 - x \leq 20$

48. $-5 < 1 - 2x < 40$

49. $3x - 1 < 2(x - 3) + 1$

50. $5x - 2(x + 3) \geq 4 - 3x$

Solving Linear Inequalities by More than One Method

Exercises 51–54: Solve the inequality
(a) *symbolically,*
(b) *graphically, and*

(c) *numerically.*

51. $6x - 8 > 16$

52. $2.5x \leq 4.5x - 3$

53. $3(x - 2) < -2(x - 1) + 7$

54. $6.5x - (2x - 1) \geq -8$

55. Solve Example 1(a) numerically and symbolically.

56. Solve Example 5 graphically and symbolically.

You Decide the Method

Exercises 57–60: Solve the inequality. Write the solution set in interval notation. Approximate the endpoints of intervals to three significant digits when appropriate.

57. $2x - 8 > 5$

58. $\pi x - 5.12 \leq \sqrt{2}x - 5.7(x - 1.1)$

59. $5.1x - \pi \geq \sqrt{3} - 1.7x$

60. $5 < 4x - 2.5$

Applications

61. *Temperature and Altitude*   (Refer to Example 4.) Suppose the temperature x miles above ground level is given by $T(x) = 85 - 29x$.
(a) Use the intersection-of-graphs method to estimate the altitudes where the temperature is below freezing. Assume that the domain of T is $0 \leq x \leq 6$.
(b) What does the x-intercept on the graph of T represent?
(c) Solve part (a) symbolically.

62. *Clouds and Temperature*   (Refer to Example 1.) Suppose the ground-level temperature is 65°F and the dew point is 50°F.
(a) Use the intersection-of-graphs method to estimate the altitudes where clouds will not form.
(b) Solve part (a) symbolically.

63. *Prices of Homes*   The median prices of a single-family home from 1980 to 1990 can be approximated by $P(x) = 3421x + 61,000$, where $x = 0$ corresponds to 1980 and $x = 10$ to 1990. (Source: Department of Commerce.)
(a) Interpret the slope of the graph of P.
(b) Estimate graphically the years when the median price range was from $71,000 to $88,000.
(c) Solve part (b) symbolically.

64. *Motorcycles*   The number of Harley-Davidson motorcycles manufactured between 1985 and 1995 can be approximated by $N(x) = 6409(x - 1985) + 30,300$, where x is the year. (Source: Harley-Davidson.)
(a) Has the demand for Harley-Davidson motorcycles increased or decreased over this time period? Explain your reasoning.

(b) Estimate graphically the years when production was between 56,000 and 75,000.

(c) Solve part (b) symbolically.

65. *Paved Roads* The total length in thousands of miles of surfaced roads owned by county and local governments can be modeled by $M(x) = 36.1(x - 1900) - 331$, where x is the year and $x \leq 1994$. (Source: American Automobile Manufacturers Association (AAMA).)

(a) Find the x-intercept of the graph of M symbolically. Round this value to the nearest year.

(b) Explain the significance of the x-intercept in this problem.

66. *Ozone and Pollution* At ground level, ozone is a toxic gas to both plants and animals, and can cause respiratory problems and eye irritation in humans. Automobiles are a major source of this type of harmful ozone. Ozone can enter buildings through ventilation systems and should not exceed 50 parts per billion (ppb) indoors. It can be removed from the air using filters. In one scientific study, an air filter was able to remove 43% of the ozone entering a building. Approximate the range of ozone concentrations in outside air that this filter will reduce to an acceptable level. (Source: Parmar and Grosjean, "Removal of Air Pollutants from Museum Display Cases.")

67. *Indoor Air Pollution* Kitchen gas ranges are a source of indoor pollutants such as carbon monoxide and nitrogen dioxide. One of the most effective ways to remove contaminants from the air while cooking is to operate a range hood that is vented. If a range hood is capable of removing x liters of air per second from a kitchen, then the percentage of the contaminants that are also removed may be calculated by $f(x) = 1.06x + 7.18$ where $10 \leq x \leq 75$. (Source: R. L. Rezvan, "Effectiveness of Local Ventilation in Removing Simulated Pollutants from Point Sources.")

(a) Graph f in [10, 75, 5] by [0, 100, 5].

(b) How does increasing the amount of ventilation x affect the percentage of pollutants removed from the kitchen? Interpret the slope of the graph of f.

(c) Use the intersection-of-graphs method to estimate the x-values where 50% to 70% of the pollutants are removed.

68. *Medicare Costs* Based on current trends, estimates of future Medicare costs in billions of dollars can be modeled by $f(x) = 18x - 35,750$, where $1995 \leq x \leq 2007$. (Source: Office of Management and Budget.)

(a) Estimate graphically and numerically the years when Medicare costs will range from 250 to 340 billion dollars.

(b) Interpret the slope of the graph of f.

69. *Oil Consumption* From 1973 to 2005, oil consumption in million metric tons of oil equivalent (Mtoe) by the Middle East can be modeled by $f(x) = 10.1x - 19,904$, where x is the year. Similarly, oil consumption by Eastern Europe can be modeled by $g(x) = 2.89x - 5622$. (Source: International Energy Agency, *Global Energy: The Changing Outlook.*)

(a) Numerically determine the year when the oil consumption by the Middle East and Eastern Europe were approximately equal.

(b) Determine the years when oil consumption by the Middle East exceeds that of Eastern Europe.

70. *Long-Distance Calls* The function represented by $f(x) = 0.4(x - 1989) + 4.6$ models the total time in billions of hours of long-distance telephone calls made in the United States during the year x. The domain of f is $D = \{1989, 1990, \ldots, 1995\}$. (Source: USA Today.)

(a) Write an inequality whose solution is the years when the total long-distance calling time exceeded 6 billion hours.

(b) Solve the inequality numerically.

Properties of Inequalities

Exercises 71–78: Assuming that a, b, and c are real numbers, answer the question. Then, illustrate each situation by choosing actual values for a, b, and c.

71. What is the solution set for $ax + b > 0$ if $a < 0$?

72. If $a \geq b$ and $c < 0$, how do ac and bc compare?

73. If $a \geq b$ and $c > 0$, how do ac and bc compare?

74. If $a \leq b$, how do $a - c$ and $b - c$ compare?

75. If $0 < a < b$, how do a^2 and b^2 compare?

76. If $a < b < 0$, how do a^2 and b^2 compare?

77. If $a < b < 0$, how do $\frac{1}{a}$ and $\frac{1}{b}$ compare?

78. If $0 < a < b$, how do $\frac{1}{a}$ and $\frac{1}{b}$ compare?

Writing about Mathematics

1. Suppose the solution to the equation $ax + b = 0$ with $a > 0$ is $x = k$. Discuss how the value of k may be used to help solve the inequalities $ax + b > 0$ and $ax + b < 0$. Illustrate this graphically. How would the solution sets change if $a < 0$?

2. Describe how to numerically solve the linear inequality $ax + b \leq 0$. Give an example.

CHECKING BASIC CONCEPTS FOR SECTIONS 2.1 AND 2.2

1. Solve the linear equation $4(x - 2) = 2(5 - x) - 3$ using each technique. Compare your results.
   (a) Graphically, using both the x-intercept and intersection-of-graphs methods
   (b) Numerically
   (c) Symbolically

2. Solve the inequality $2(x - 4) > 1 - x$ symbolically, graphically, and numerically.

3. Solve the compound inequality $-2 \leq 1 - 2x \leq 3$ symbolically, graphically, and numerically. Express the solution set using interval notation.

2.3 Equations of Lines

Forms for Equations of Lines • Horizontal, Vertical, Parallel, and Perpendicular Lines • Direct Variation (Optional)

Introduction

Lines are a fundamental geometric concept that have applications in a variety of areas such as computer graphics, business, and science. Any quantity that experiences growth at a constant rate can be modeled by the graph of a linear function, which is a line. In this section we discuss how equations of lines can be determined, and some of their applications.

Forms for Equations of Lines

In Section 1.6 it was shown that an equation of the line passing through the point (h, k) with slope m is $y = m(x - h) + k$. This was accomplished using translations of graphs. This equation can be found using a different approach.

Suppose that a nonvertical line passes through the point (h, k) with slope m. If (x, y) is any point on this line with $x \neq h$, then $m = \dfrac{y - k}{x - h}$. See Figure 2.32.

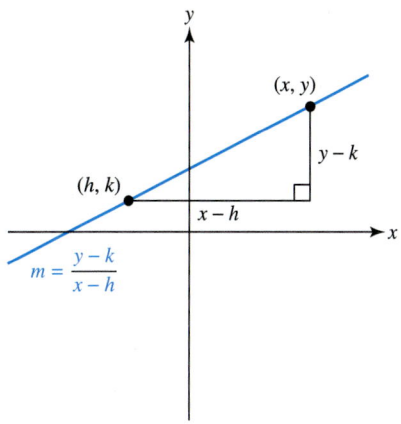

FIGURE 2.32

Using this slope formula, the equation of the line can be found.

$$m = \frac{y - k}{x - h} \qquad \text{Slope formula}$$

$$y - k = m(x - h) \qquad \text{Cross multiply.}$$

$$y = m(x - h) + k \qquad \text{Add } k \text{ to both sides.}$$

The equation $y - k = m(x - h)$ is traditionally called the **point-slope form** of the equation of a line. Since we think of y as being a function of x, written $y = f(x)$, the equivalent form $y = m(x - h) + k$ will also be referred to as the point-slope form. The point-slope form is not unique, since any point on the line can be used for (h, k). However, these point-slope forms are *equivalent*—their graphs are identical.

EXAMPLE 1   *Estimating growth and investment in cellular communication*

Cellular phone use has grown dramatically in the United States. In New York City when there were 25,000 customers, the investment cost per cellular site was $12 million. (A cellular site would include such things as a relay tower to transmit signals between cellular phones.) When the number of customers rose to 100,000, the investment cost rose to $96 million. Although cost usually decreases with additional customers, this was not the case for early cellular technology. Instead, cost increased due to the need to purchase expensive real estate and to establish communication between a large number of cellular sites. The relationship between customers and investment costs per site was approximately linear as shown in Figure 2.33. (Source: M. Paetsch, *Mobile Communications in the U.S. and Europe.*)

(a) Find a point-slope form of the line passing through the points (25,000, 12) and (100,000, 96).

(b) Use this equation to estimate the investment cost per cellular site when there were 70,000 customers in New York City.

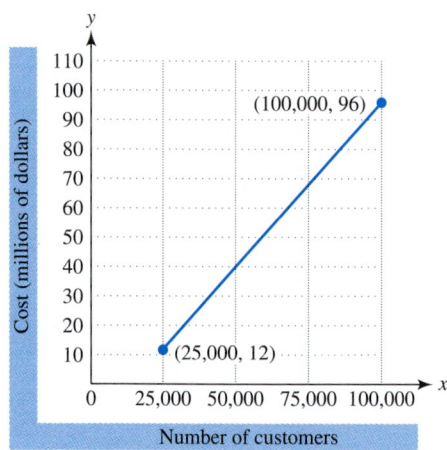

FIGURE 2.33   Cost of a Cellular Site

Solution

(a) The slope m of the line segment passing through the points (25,000, 12) and (100,000, 96) is

$$m = \frac{96 - 12}{100,000 - 25,000} = 0.00112.$$

A point-slope form of the line passing through (25,000, 12) with slope 0.00112 is found as follows.

$$y = m(x - h) + k \qquad \text{Point-slope form}$$

$$y = 0.00112(x - 25{,}000) + 12 \qquad \text{Substitute for } (h, k) \text{ and } m.$$

In this equation y represents the investment cost for each cellular site in millions of dollars, when there were x customers.

(b) For $x = 70{,}000$ customers, the investment cost was

$$y = 0.00112(70{,}000 - 25{,}000) + 12 = 62.4$$

or $62.4 million per cellular site.   ■

In Example 1 the data point (100,000, 96) could be used for (h, k) instead of (25,000, 12). The resulting equation would be $y = 0.00112(x - 100{,}000) + 96$. This form can be simplified.

$$y = 0.00112(x - 100{,}000) + 96$$

$$y = 0.00112x - 112 + 96 \qquad \text{Distributive property}$$

$$y = 0.00112x - 16 \qquad \text{Simplify.}$$

The first equation can be simplified similarly.

$$y = 0.00112(x - 25{,}000) + 12$$

$$y = 0.00112x - 28 + 12 \qquad \text{Distributive property}$$

$$y = 0.00112x - 16 \qquad \text{Simplify.}$$

Both point-slope forms simplify to the same equation.

The form $y = mx + b$ is a special form for the equation of a line called the *slope-intercept form.* It is a convenient form because it is unique. When $x = 0$, $y = m(0) + b = b$. Thus, the point $(0, b)$ lies on the graph of $y = mx + b$. The real number b represents the y-intercept. A **y-intercept** is the y-coordinate of a point where a graph intersects the y-axis. See Figure 2.34. The slope m and y-intercept b determine a line. In Example 1, the slope-intercept form of the line is $y = 0.00112x - 16$, where $m = 0.00112$ and $b = -16$.

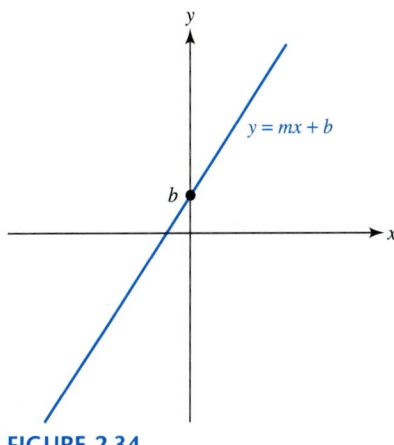

FIGURE 2.34

> ### Slope-intercept form
>
> The line with slope m and y-intercept b has an equation
>
> $$y = mx + b,$$
>
> the **slope-intercept form** of the equation of a line.

EXAMPLE 2   *Finding equations of lines*

Find a point-slope form for the line that satisfies the conditions. Then convert this equation into slope-intercept form.
(a) Slope 4 passing through the point $(-3, -7)$
(b) x-intercept -4, y-intercept 2

Solution

(a) Let $m = 4$ and $(h, k) = (-3, -7)$ in the point-slope form.

$$y = m(x - h) + k \qquad \text{Point-slope form}$$
$$y = 4(x + 3) - 7 \qquad \text{Substitute.}$$

The slope-intercept form can be found by simplifying.

$$y = 4(x + 3) - 7 \qquad \text{Point-slope form}$$
$$y = 4x + 12 - 7 \qquad \text{Distributive property}$$
$$y = 4x + 5 \qquad \text{Slope-intercept form}$$

(b) The line passes through the points $(-4, 0)$ and $(0, 2)$. Its slope is

$$m = \frac{2 - 0}{0 - (-4)} = \frac{1}{2}.$$

Thus, a point-slope form is $y = \frac{1}{2}(x + 4) + 0$, where the point $(-4, 0)$ is used for (h, k). The slope-intercept form is $y = \frac{1}{2}x + 2$. ∎

EXAMPLE 3   *Interpreting slope-intercept form*

The distance y in miles that a bicycle rider is from home after x hours is shown in Figure 2.35.
(a) Estimate the y-intercept. What does the y-intercept represent?
(b) The graph passes through the point $(2, 6)$. Discuss the meaning of this point.
(c) Find the slope-intercept form of this line. Interpret the slope.
(d) Determine the x-intercept symbolically. What does the x-intercept represent?

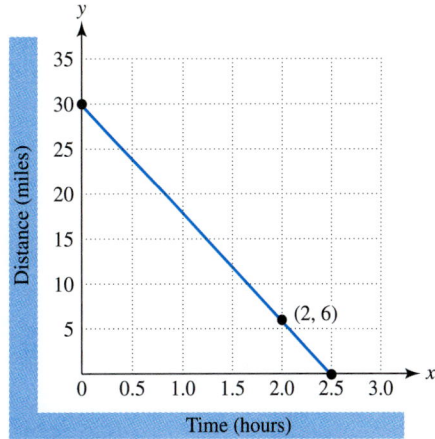

FIGURE 2.35 Distance from Home

Solution

(a) From the graph, the y-intercept is 30. This indicates that the bicyclist is initially 30 miles from home.

(b) The point $(2, 6)$ means that after two hours the bicyclist is six miles from home.

(c) The line passes through the points $(0, 30)$ and $(2, 6)$. Its slope is

$$m = \frac{6 - 30}{2 - 0} = -12.$$

The slope-intercept form is $y = -12x + 30$. A slope of -12 means that the bicyclist is traveling at 12 miles per hour toward home. The negative sign indicates that the distance between the rider and home is decreasing.

(d) The x-intercept is found by letting $y = 0$ in the slope-intercept equation.

$$0 = -12x + 30 \qquad \text{Substitute } y = 0.$$
$$12x = 30 \qquad \text{Add } 12x \text{ to both sides.}$$
$$x = 2.5 \qquad \text{Divide by 12 and simplify.}$$

(The graph supports this result). An x-intercept of 2.5 indicates that the bicyclist arrives home after 2.5 hours. ∎

Horizontal, Vertical, Parallel, and Perpendicular Lines

The graph of a constant function f, represented by $f(x) = b$, is the horizontal line $y = b$. This line has slope 0 and y-intercept b.

A vertical line cannot be represented by a function, since distinct points on a vertical line have the same x-coordinate. In fact, this is the distinguishing feature about points on a vertical line—they all have the same x-coordinate. The vertical line in Figure 2.36 is $x = 3$. The equation of a vertical line with x-intercept k is expressed symbolically as $x = k$ as shown in Figure 2.37 on the following page. Horizontal lines have a slope equal to zero, while vertical lines have an undefined slope.

Critical Thinking

Why do you think that a vertical line sometimes is said to have "infinite slope"? What are some problems with taking this phrase too literally?

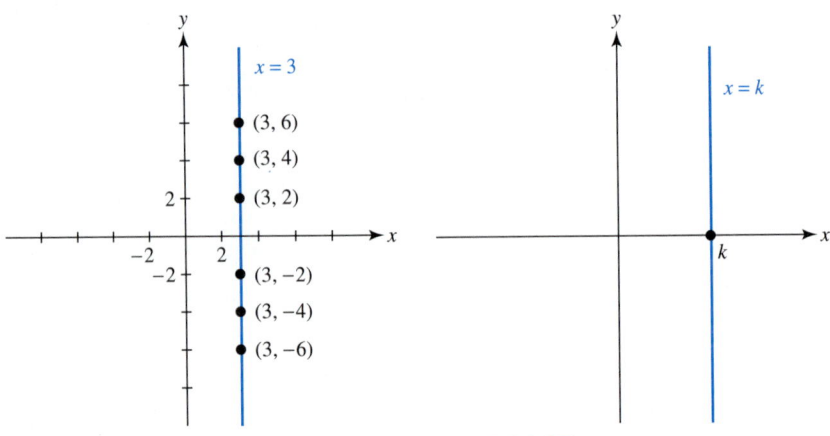

FIGURE 2.36   **FIGURE 2.37**

Equations of horizontal and vertical lines

An equation of the horizontal line with y-intercept b is $y = b$. An equation of the vertical line with x-intercept k is $x = k$.

EXAMPLE 4   *Finding equations of horizontal and vertical lines*

Find equations of vertical and horizontal lines passing through the point $(8, 5)$. Graph the lines in $[0, 15, 1]$ by $[0, 10, 1]$.

Solution

The x-coordinate of the point $(8, 5)$ is 8. The vertical line $x = 8$ passes through every point in the xy-plane with an x-coordinate of 8, including the point $(8, 5)$. Similarly, the horizontal line $y = 5$ passes through every point with a y-coordinate of 5, including $(8, 5)$. The lines $x = 8$ and $y = 5$ are graphed in Figure 2.38.

$[0, 15, 1]$ by $[0, 10, 1]$

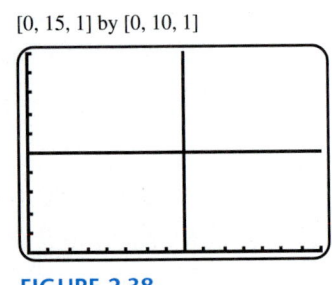

FIGURE 2.38   ■

Slope is an important concept when determining if two lines are either parallel or perpendicular. Two nonvertical parallel lines have equal slopes.

> ### Parallel lines
>
> Two lines with slopes m_1 and m_2, neither of which is vertical, are parallel if and only if $m_1 = m_2$.

Review Note   *If And Only If*

The phrase "if and only if" is used when two statements are mathematically equivalent. If two nonvertical lines are parallel, then it is true that $m_1 = m_2$. Conversely, if two nonvertical lines have equal slopes, then they are parallel. Either condition implies the other.

EXAMPLE 5   *Finding parallel lines*

Find the slope-intercept form of a line parallel to $y = -2x + 5$, passing through the point $(-2, 3)$. Support your result graphically.

Solution

Since the line $y = -2x + 5$ has slope -2, any parallel line also has slope -2. The line passing through $(-2, 3)$ with slope -2 is determined as follows.

$$y = -2(x + 2) + 3 \qquad \text{Point-slope form}$$
$$y = -2x - 4 + 3 \qquad \text{Distributive property}$$
$$y = -2x - 1 \qquad \text{Slope-intercept form}$$

Graphical support is shown in Figure 2.39 where the lines $Y_1 = -2X + 5$ and $Y_2 = -2X - 1$ are parallel.

$[-9, 9, 1]$ by $[-6, 6, 1]$

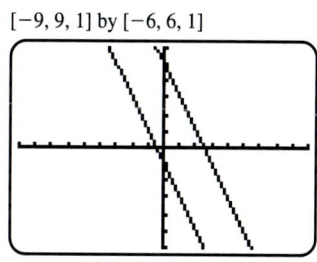

FIGURE 2.39   ■

Cross hairs, found in telescopes, view finders, and video games, are examples of perpendicular lines. Graphing calculators often use perpendicular line segments for a cursor as shown in Figure 2.40 on the following page. In video games provisions sometimes are made for cross hairs to tilt in order to simulate things such as a craft rolling from side to side. In this situation, the cross hairs remain perpendicular, but they are not always in a horizontal-vertical position. Two lines with nonzero slopes are perpendicular if the product of their slopes is equal to -1. The following result is given without proof.

[−10, 10, 1] by [−10, 10, 1]

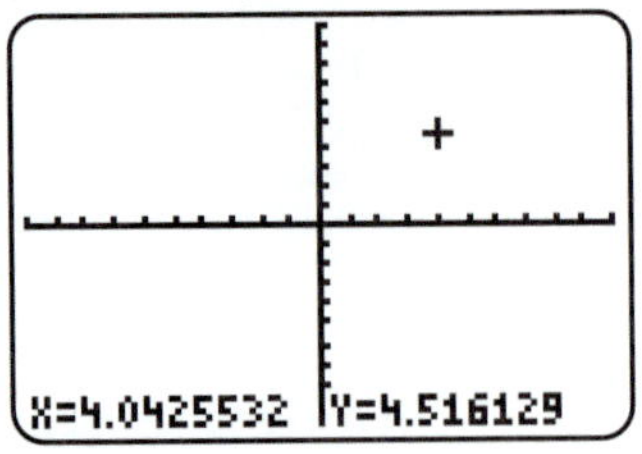

X=4.0425532   Y=4.516129

FIGURE 2.40

Perpendicular lines

Two lines with nonzero slopes m_1 and m_2 are perpendicular if and only if $m_1 m_2 = -1.$

EXAMPLE 6   *Finding perpendicular lines*

Suppose one cross hair is given by the equation $y = -\dfrac{1}{2}x + 9.$ If the cross hairs are centered on the point (8, 5), find the equation of the other cross hair. Graph the cross hairs in the viewing rectangle [0, 15, 1] by [0, 10, 1].

Solution

The line $y = -\dfrac{1}{2}x + 9$ has slope $-\dfrac{1}{2}.$ The slope of the perpendicular line is 2, since

$$m_1 m_2 = -\frac{1}{2} \cdot 2 = -1.$$

We find the equation of the line passing through the point (8, 5) with slope 2. A point-slope form is $y = 2(x - 8) + 5,$ which simplifies to $y = 2x - 11.$

Graph $Y_1 = -.5X + 9$ and $Y_2 = 2X - 11$ as shown in Figure 2.41. The perpendicular lines intersect at (8, 5) as expected. Why is it important to use a square viewing rectangle?

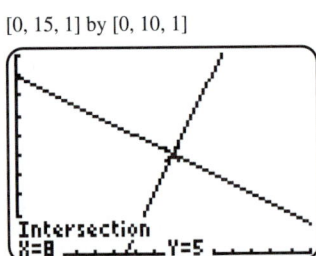

[0, 15, 1] by [0, 10, 1]

Intersection
X=8   Y=5

FIGURE 2.41

EXAMPLE 7   *Graphing a rectangle*

One side of a rectangle has vertices (0, 0) and (5, 3). Graph four lines that outline the boundary of the rectangle, if the point (0, 5) lies on one side of this rectangle.

Solution

We begin by sketching the rectangle as shown in Figure 2.42. One side of the rectangle is determined by the points $(0, 0)$ and $(5, 3)$. The slope of this line is $m = \dfrac{3 - 0}{5 - 0} = \dfrac{3}{5}$ and its y-intercept is 0. The equation of this line is $y = \dfrac{3}{5}x$. A second line passes through $(0, 0)$ and is perpendicular to the first line. The slope of the second line is $-\dfrac{5}{3}$, since $\dfrac{3}{5}\left(-\dfrac{5}{3}\right) = -1$. The equation of this line is $y = -\dfrac{5}{3}x$.

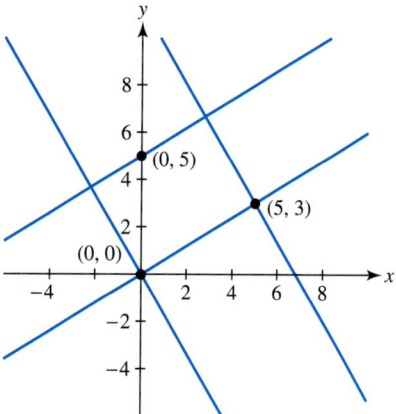

FIGURE 2.42

A third line passes through the point $(5, 3)$ and is parallel to the line $y = -\dfrac{5}{3}x$. Its equation is $y = -\dfrac{5}{3}(x - 5) + 3$. These three lines are graphed in Figure 2.43 as $Y_1 = (3/5)X$, $Y_2 = -(5/3)X$, and $Y_3 = -(5/3)(X - 5) + 3$.

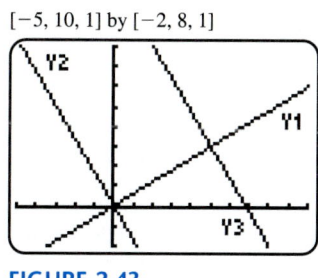

$[-5, 10, 1]$ by $[-2, 8, 1]$

FIGURE 2.43

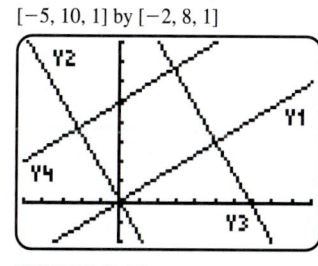

$[-5, 10, 1]$ by $[-2, 8, 1]$

FIGURE 2.44

The fourth line of the rectangle must pass through the point $(0, 5)$ and be parallel to Y_1. This line has slope $\dfrac{3}{5}$, y-intercept 5, and equation $y = \dfrac{3}{5}x + 5$. The four lines are graphed in Figure 2.44. ■

Critical Thinking

Find four equations whose graphs would result in the rectangle in Figure 2.44 being translated two units to the right and one unit down. Graph your equations to test them.

Direct Variation (Optional)

When a change in one quantity causes a proportional change in another quantity, the two quantities are said to *vary directly* or to *be directly proportional*. For example, if we work for $8 per hour, our pay is proportional to the number of hours that we work. Doubling the hours doubles the pay, tripling the hours triples the pay, and so on. This is stated more precisely as follows.

> ### Direct variation
>
> Let x and y denote two quantities. Then y is **directly proportional** to x, or y **varies directly** with x, if there exists a nonzero number k such that
>
> $$y = kx.$$
>
> The number k is called the **constant of proportionality** or the **constant of variation.**

If a person earns $57.75 working for seven hours, the constant of proportionality k is the hourly pay rate. If y represents the pay in dollars and x the hours worked, then k is found by substituting values for x and y into the equation $y = kx$ and solving for k. That is,

$$57.75 = k(7) \qquad \text{or} \qquad k = \frac{57.75}{7} = 8.25,$$

so the hourly pay rate is $8.25.

Given a set of data points, one method to determine if y is directly proportional to x is to graph the ordered pairs (x, y). If the points lie on a line that passes through the origin, then y varies directly with x and the constant of proportionality k is equal to the slope of the line. A second method is to compute the ratio $\frac{y}{x}$ for each ordered pair. The equation $y = kx$ implies that $k = \frac{y}{x}$, so each ratio $\frac{y}{x}$ will equal k.

Both techniques are demonstrated in the next example.

EXAMPLE 8   *Modeling storage requirements for recording music*

Recording music requires an enormous amount of storage. A compact disc (CD) can hold approximately 600 million bytes. One million bytes is commonly referred to as a **megabyte** (MB). (See Example 6, Section 1.1 for an explanation of a byte.) Table 2.2 lists the megabytes x needed to record y seconds of music.

TABLE 2.2

x (MB)	0.129	0.231	0.415	0.491	0.667	1.030	1.160	1.260
y (sec)	6.010	10.74	19.27	22.83	31.00	49.00	55.25	60.18

Source: Gateway 2000 System CD.

(a) Make a scatterplot of the data.
(b) How does y vary with x? Explain why this relationship seems reasonable.

(c) Compute the ratio $\frac{y}{x}$ for each musical segment in the table. Interpret these ratios.

(d) Approximate a constant of proportionality k satisfying the equation $y = kx$. Graph the data and the equation together.

(e) Estimate the maximum number of seconds of music that can be placed on a 1.44-megabyte floppy disc and on a 600-megabyte CD.

Solution

(a) A scatterplot of the data is shown in Figure 2.45.

$[-0.1, 1.5, 0.25]$ by $[0, 70, 10]$

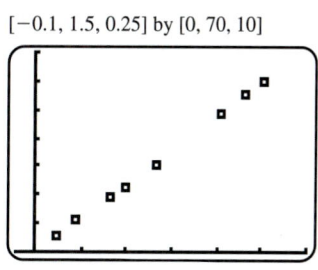

FIGURE 2.45

(b) The points in the scatterplot suggest a line that passes through the origin. Time y appears to vary directly with storage x. This seems reasonable. If the length of a song doubles, the number of megabytes needed to record it should also double.

(c) The ratios are shown in Table 2.3. For example, $\frac{6.010}{0.129} \approx 46.6$. These ratios represent the number of seconds that can be recorded on one megabyte. They are approximately equal indicating direct variation.

TABLE 2.3

x (MB)	0.129	0.231	0.415	0.491	0.667	1.030	1.160	1.260
y (sec)	6.010	10.74	19.27	22.83	31.00	49.00	55.25	60.18
y/x	46.6	46.5	46.4	46.5	46.5	47.6	47.6	47.8

(d) From Table 2.3 it appears that approximately 47 seconds of music can be recorded on one megabyte. Therefore, let the constant of proportionality be $k = 47$. The data points and the equation $y = 47x$ are graphed in Figure 2.46.

$[-0.1, 1.5, 0.25]$ by $[0, 70, 10]$

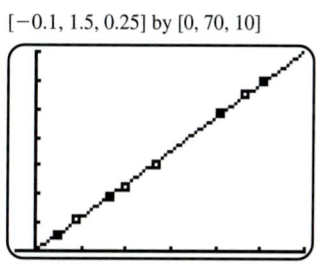

FIGURE 2.46

(e) Since $y = 47x$, a 1.44-megabyte floppy disc could store at most $47 \cdot 1.44 = 67.68 \approx 68$ seconds of music. On a 600-megabyte CD, at most $47 \cdot 600 = 28{,}200$ seconds, or 470 minutes, of music can be recorded. ■

2.3 PUTTING IT ALL TOGETHER

The following table summarizes two important forms for the equation of a line.

Concept	Comments	Example
Point-Slope Form $y = m(x - h) + k$ or $y - k = m(x - h)$	Is used to find the equation of a line, given two points or one point and the slope	Given two points $(1, 2)$ and $(3, 5)$, first compute $m = \frac{5 - 2}{3 - 1} = \frac{3}{2}$. An equation of this line is $y = \frac{3}{2}(x - 1) + 2$.
Slope-Intercept Form $y = mx + b$	Is a unique equation for a line, determined by the slope m and the y-intercept b	An equation of the line with slope $m = 3$ and y-intercept $b = -5$ is $y = 3x - 5$.

The following table summarizes the important concepts involved with special types of lines.

Concept	Equation(s)	Example
Horizontal line	$y = b$, where b is a constant	A horizontal line with y-intercept 5 has the equation $y = 5$.
Vertical line	$x = k$, where k is a constant	A vertical line with x-intercept -3 has the equation $x = -3$.
Parallel lines	$y = m_1 x + b_1$ and $y = m_2 x + b_2$, where $m_1 = m_2$	The lines $y = 2x - 1$ and $y = 2x + 5$ are parallel because they both have slope 2.
Perpendicular lines	$y = m_1 x + b_1$ and $y = m_2 x + b_2$, where $m_1 m_2 = -1$	The lines $y = 3x - 5$ and $y = -\frac{1}{3}x + 2$ are perpendicular because $m_1 m_2 = 3\left(-\frac{1}{3}\right) = -1$.

2.3 EXERCISES   Tape 2

Graphical Interpretation

1. A person is driving a car on a straight road. The graph shows the distance y in miles that this individual is from home after x hours.
   (a) Is the person traveling toward or away from home?

   (b) The graph passes through $(1, 35)$ and $(3, 95)$. Discuss the meaning of these points.
   (c) Find the slope-intercept form of the equation of the line. Interpret the slope.
   (d) Use the graph to estimate the y-coordinate of the point $(4, y)$ lying on the graph. Then, find this coordinate symbolically.

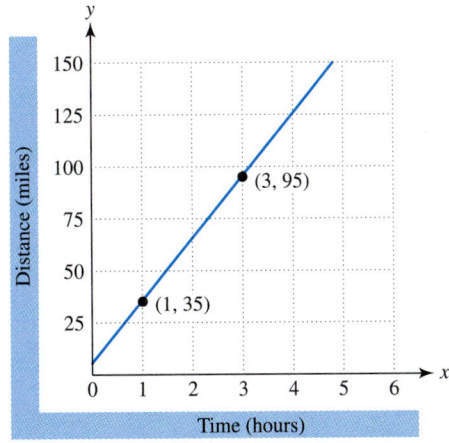

2. The graph shows the amount of water y in a 100-gallon tank after x minutes have elapsed.
   (a) Is water entering or leaving the tank? How much water is in the tank after three minutes?
   (b) Find both the x- and y-intercepts. Interpret their meanings.
   (c) Find the slope-intercept form of the equation of the line. Interpret the slope.
   (d) Use the graph to estimate the x-coordinate of the point $(x, 50)$ that lies on the line. Then, find this coordinate symbolically.

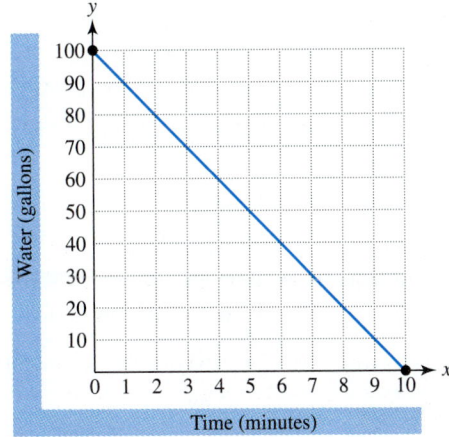

Exercises 3–8: Match the situation to the graph (a–f) that models it best.

3. The approximate distance y that Earth is from the sun at time x

4. The approximate federal debt y from 1985 to 1990

5. The remaining hours y that a light bulb will work after it has been turned on for x hours

6. The amount of money y earned working for x hours at a fixed hourly rate

7. The annual average temperature y in degrees Celsius at Nome, Alaska, in the year x

8. The profit (revenue minus cost) from selling x items

a.

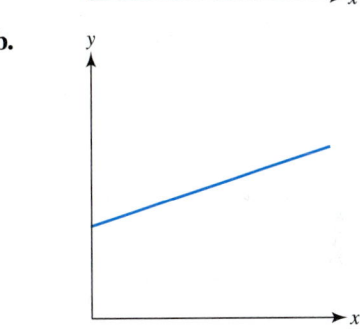

b.

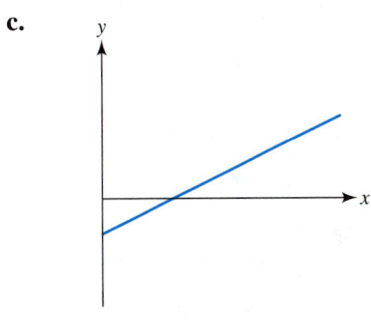

c.

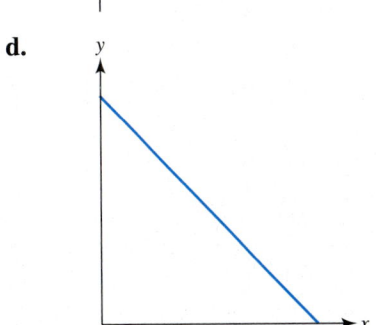

d.

e.

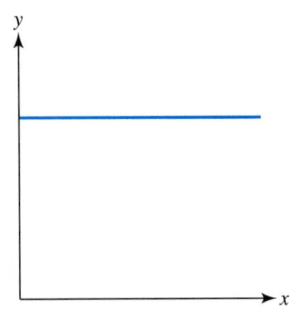

c.

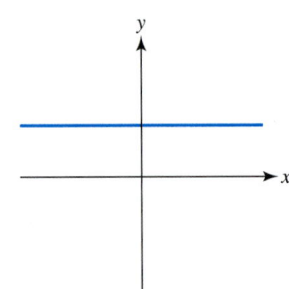

f.

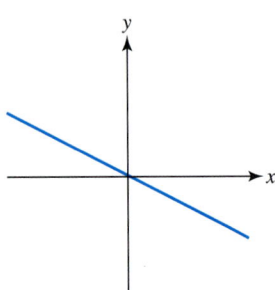

d.

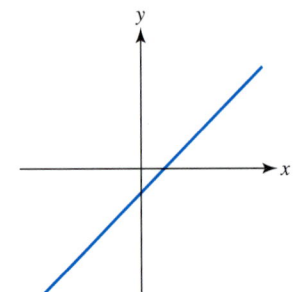

Exercises 9–12: Match the equation with its graph, where a and b are constants.

9. $y = ax + b, a > 0$

10. $y = ax + b, a < 0$ and $b \neq 0$

11. $y = ax, a < 0$

12. $y = b$

a.

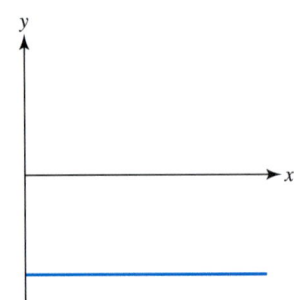

b.

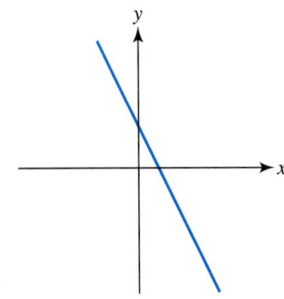

Equations of Lines

Exercises 13–26: Find a point-slope form for the equation of the line satisfying the conditions.

13. Slope -2.4, passing through $(4, 5)$

14. Slope 1.7, passing through $(-8, 10)$

15. Passing through $(1, -2)$ and $(-9, 3)$

16. Passing through $(-6, 10)$ and $(5, -12)$

17. Passing through $(1980, 5)$ and $(1990, 25)$

18. Passing through $(1990, -3)$ and $(1996, 19)$

19. y-intercept 5, slope -7.8

20. y-intercept -155, slope 5.6

21. y-intercept 45, x-intercept 90

22. x-intercept -6, y-intercept -8

23. Parallel to $y = 4x + 16$, passing through $(-4, -7)$

24. Parallel to $y = -\dfrac{3}{4}(x - 100) - 99$, passing through $(1.5, \sqrt{3})$

25. Perpendicular to $y = -\dfrac{2}{3}(x - 1980) + 5$, passing through $(1980, 10)$

26. Perpendicular to $y = 6x - 1000$, passing through $(15, -7)$

Exercises 27–34: Find an equation of the line satisfying the following conditions.

27. Vertical, passing through $(-5, 6)$

28. Vertical, passing through $(1.95, 10.7)$

29. Horizontal, passing through $(-5, 6)$

30. Horizontal, passing through $(1.95, 10.7)$

31. Perpendicular to $y = 15$, passing through $(4, -9)$

32. Perpendicular to $x = 15$, passing through $(1.6, -9.5)$

33. Parallel to $x = 4.5$, passing through $(19, 5.5)$

34. Parallel to $y = -2.5$, passing through $(1985, 67)$

Determining Intercepts

Exercises 35–38: Determine symbolically the x- and y-intercepts on the graph of the equation. Support your results graphically.

35. $y = 8x - 5$

36. $y = -1.5x + 15$

37. $y = 3(x - 2) - 5$

38. $y = -2(x + 1) + 7$

Applications

39. *Tuition and Fees*   The graph models average tuition and fees in dollars at public four-year colleges from 1981 to 1995. (Source: The College Board.)

   (a) The graph passes through $(1984, 1225)$ and $(1987, 1621)$. Interpret these two points.

   (b) Find a point-slope form for this line. Interpret the slope.

   (c) Use the graph to estimate the year when tuition and fees were \$2,000. Then, determine this year symbolically.

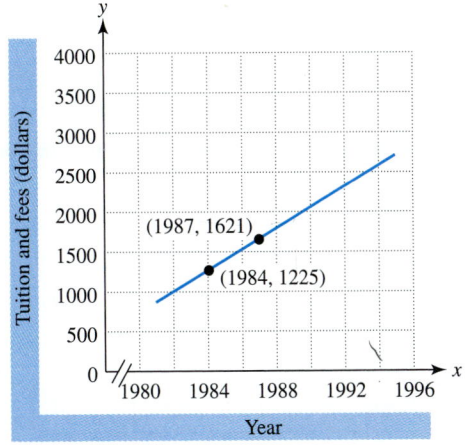

40. *Antibiotic Resistance*   The graph approximates the percentage of gonorrhea cases with antibiotic resistance diagnosed in the United States from 1985 to 1990. (Source: S. Teutsch and R. Churchill, *Principles and Practice of Public Health Surveillance.*)

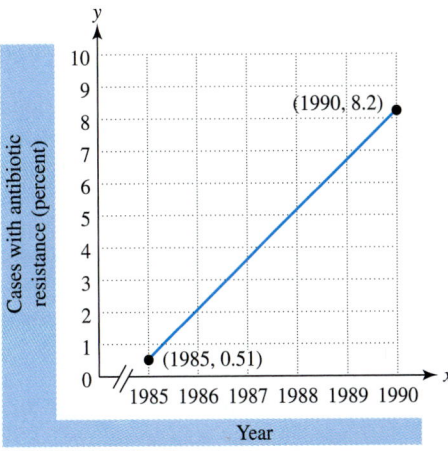

   (a) The points $(1985, 0.51)$ and $(1990, 8.2)$ lie on the graph. Interpret each point.

   (b) Determine a point-slope form for the equation of this line.

   (c) Estimate the percentage of gonorrhea cases with antibiotic resistance diagnosed in 1987 graphically and symbolically.

41. *Antarctic Ozone Layer*   Stratospheric ozone occurs in the atmosphere between altitudes of 12 and 18 miles and is an important filter of ultraviolet light from the sun. Ozone in the stratosphere is frequently measured in Dobson units. The Dobson scale is linear where 300 Dobson units is a midrange value that corresponds to an ozone layer three millimeters thick. In 1991 the reported minimum in the Antarctic *ozone hole* was about 110 Dobson units. (Source: R. Huffman, *Atmospheric Ultraviolet Remote Sensing.*)

   (a) The equation $y = kx$ describes the thickness y in millimeters of the ozone layer that is x Dobson units. Find k.

   (b) How thick was the ozone layer over the Antarctic in 1991?

   (c) Was the ozone hole actually a *hole* in the ozone layer?

42. *Farm Pollution*   In 1988 the number of farm pollution incidents reported in England and Wales was 4000. This number had increased roughly at a rate of 280 per year since 1979. (Source: C. Mason, *Biology of Freshwater Pollution.*)

   (a) Find an equation of a line $y = m(x - h) + k$ that describes this data, where y represents the number of pollution incidents during the year x.

   (b) Estimate graphically and symbolically the year when there were approximately 2300 incidents.

43. *Cost of Driving*   The cost of driving a car includes both fixed costs and mileage costs. Assume that it

costs $350 per month for insurance and car payments and it costs $0.29 per mile for gasoline, oil, and routine maintenance.
  (a) Determine a linear function that computes the annual cost of driving this car x miles.
  (b) What does the y-intercept on the graph of f represent?

44. *Average Wages* The average hourly wage (adjusted to 1982 dollars) was $7.81 in 1986 and $7.41 in 1992. (Source: Department of Commerce.)
  (a) Find an equation of a line that passes through (1986, 7.81) and (1992, 7.41).
  (b) Interpret the slope.
  (c) Approximate the hourly wage in 1990. Compare it to the actual value of $7.52.

45. *Safer in Jail?* The murder rate in the United States remained roughly constant from 1984 to 1990 at 10 murders per 100,000 people each year. During the same time period, the murder rate among state and federal inmates decreased linearly from 30 to 8 per 100,000 inmates. (Source: Department of Justice.)
  (a) Find the equation of a line that passes through (1984, 30) and (1990, 8).
  (b) Estimate when the murder rate in state and federal prisons was first less than the rate for all Americans.

46. *HIV Infection Rates* It was estimated in 1996 that there were 22 million HIV infections worldwide with an annual infection rate of 3.1 million. (Source: Centers for Disease Control and Prevention.)
  (a) Find a symbolic representation of a linear function that models the number of worldwide HIV infections during year x.
  (b) Estimate graphically, numerically, and symbolically the year when there might be 30 million HIV infections.

Perspectives and Viewing Rectangles

47. Graph $y = \frac{1}{1024}x + 1$ in [0, 3, 1] by [−2, 2, 1].
  (a) Is the graph a horizontal line?
  (b) Conjecture why the calculator screen appears as it does.

48. Graph $y = 1000x + 1000$ in the standard viewing rectangle.
  (a) Is the graph a vertical line?
  (b) Explain why the calculator screen appears as it does.

49. *Square Viewing Rectangle* Graph the lines $y = 2x$ and $y = -\frac{1}{2}x$ in the standard viewing rectangle.
  (a) Do the lines appear to be perpendicular?
  (b) Graph the lines in the following viewing rectangles.

  i.   [−15, 15, 1] by [−10, 10, 1]
  ii. [−10, 10, 1] by [−3, 3, 1]
  iii. [−3, 3, 1] by [−2, 2, 1]

  Do the lines appear to be perpendicular in any of these viewing rectangles?
  (c) Determine the viewing rectangles where perpendicular lines will appear perpendicular. (Answers may vary depending on the model of graphing calculator used.)

50. Continuing with Exercise 49, conjecture which viewing rectangles result in the graph of a circle with radius 5 and center at the origin appearing circular.

  i.   [−9, 9, 1] by [−6, 6, 1]
  ii. [−5, 5, 1] by [−10, 10, 1]
  iii. [−5, 5, 1] by [−5, 5, 1]
  iv. [−18, 18, 1] by [−12, 12, 1]

  Test your conjecture by graphing this circle in each viewing rectangle. (*Hint:* Graph the equations
  $$y_1 = \sqrt{25 - x^2} \text{ and } y_2 = -\sqrt{25 - x^2}$$
  to create the circle.)

Graphing a Rectangle

Exercises 51–54: (Refer to Example 7.) Graph the rectangle that satisfies the stated conditions.

51. Vertices (0, 0), (2, 2), and (1, 3)
52. Vertices (1, 1), (5, 1), and (5, 5)
53. Vertices (4, 0), (0, 4), (0, −4), and (−4, 0)
54. Vertices (1, 1), (2, 3), and the point (3.5, 1) lies on a side of the rectangle

Direct Variation

Exercises 55–58: Find the constant of proportionality k and the undetermined value in the table if y is directly proportional to x. Support your answer by graphing the equation y = kx and the data points.

55.
x	3	5	6	8
y	7.5	12.5	15	?

56.
x	1.2	4.3	5.7	?
y	3.96	14.19	18.81	23.43

57. Sales tax y on a purchase of x dollars

x	$25	$55	?
y	$1.50	$3.30	$5.10

58. Cost y of buying x compact discs having the same price

x	3	4	5
y	$41.97	$55.96	?

59. *Cost of Tuition*   The cost of tuition is directly proportional to the number of credits taken. If 11 credits cost $720.50, find the cost of taking 16 credits. What is the constant of proportionality?

60. *Strength of a Beam*   The maximum load that a horizontal beam can carry is directly proportional to its width. If a beam 1.5 inches wide can support a load of 250 pounds, find the load that a beam of the same type can support if its width is 3.5 inches.

61. *Hooke's Law*   The distance that a spring stretches is directly proportional to the size of the weight hung on the spring. Suppose a 15-pound weight stretches a spring eight inches as shown in the figure.
(a)   The equation $y = kx$ models this situation. Find k.
(b)   How far will a 25-pound weight stretch this spring?

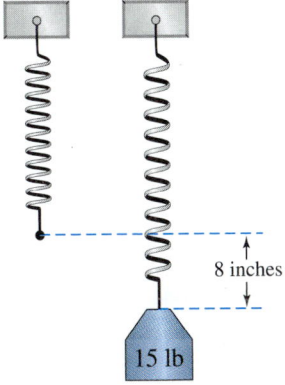

8 inches

15 lb

62. *Distance Traveled*   Suppose a car travels at 60 miles per hour for x hours. Then the distance y that the car travels is directly proportional to x. Find the constant of proportionality. How far does the car travel in 3 hours?

63. *Stopping Distance*   The approximate stopping distances y in feet for a car traveling at x miles per hour are listed in the table. Determine if stopping distance is directly proportional to speed. Interpret your results. (Source: L. Haefner, *Introduction to Transportation Systems.*)

x (mph)	0	20	40	60	80
y (ft)	0	118	324	620	1004

64. *Electrical Resistance*   The electrical resistance of a wire varies directly with its length. If a 255-foot wire has a resistance of 1.2 ohms, find the resistance of 135 feet of the same type of wire. Interpret the constant of proportionality in this situation.

Writing about Mathematics

1.   Compare and contrast the slope-intercept form and the point-slope form. Give examples of each.

2.   Give an example of two quantities in real life that vary directly. Explain your answer. Use an equation to describe the relationship between the two quantities.

2.4 Piecewise-Defined Functions

Evaluating and Graphing Piecewise-Defined Functions • The Greatest Integer Function • The Absolute Value Function • Equations and Inequalities Involving Absolute Values

Introduction

When a function f models real data, it is possible that no single formula can conveniently represent f. An example is first-class postage. In 1997 a first-class letter weighing up to one ounce cost $0.32 to mail, whereas a letter weighing 1.1 ounces cost $0.55 to mail. It is common to define a postage function in pieces that depend on the weight of the letter. Functions defined in pieces are called **piecewise-defined functions.** They typically use different formulas on various intervals of their domains. Our discussion in this section centers on piecewise-defined functions where each piece is linear.

Evaluating and Graphing Piecewise-Defined Functions

In 1997 postage rates in dollars for first-class letters weighing up to five ounces could be computed by the piecewise-defined function f.

$$f(x) = \begin{cases} 0.32 & \text{if } 0 < x \le 1 \\ 0.55 & \text{if } 1 < x \le 2 \\ 0.78 & \text{if } 2 < x \le 3 \\ 1.01 & \text{if } 3 < x \le 4 \\ 1.24 & \text{if } 4 < x \le 5 \end{cases}$$

The function f is not a constant function because the cost varies according to the weight x of a letter. However, each piece of f is a constant. Therefore, f is a **piecewise-constant function.** Since each constant piece is linear, f also can be called a **piecewise-linear function.**

The graph of f is composed of horizontal line segments. When $0 < x \le 1$, $f(x) = 0.32$ as shown in Figure 2.47. A small circle occurs at the point (0, 0.32) because $x = 0$ is not in the domain of f. The other pieces of f are graphed similarly. Since there are breaks in the graph, f is not a continuous function. It is *discontinuous* at $x = 1, 2, 3,$ and 4. This type of function is called a **step function.**

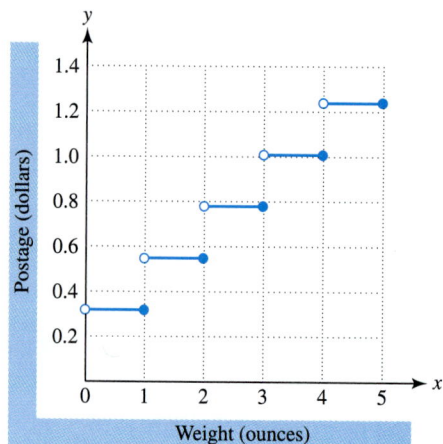

FIGURE 2.47 First-Class Postage in 1997

EXAMPLE 1   *Evaluating a graphical representation*

Figure 2.48 depicts a graph of a piecewise-linear function f. It models the amount of water in thousands of gallons in a swimming pool after x hours have elapsed.
(a) Use the graph to evaluate $f(0), f(25),$ and $f(40)$. Interpret the results.

(b) Discuss how the amount of water in the pool changed. Is f a continuous function on the interval [0, 50]?

(c) Interpret the slope of each line segment in the graph.

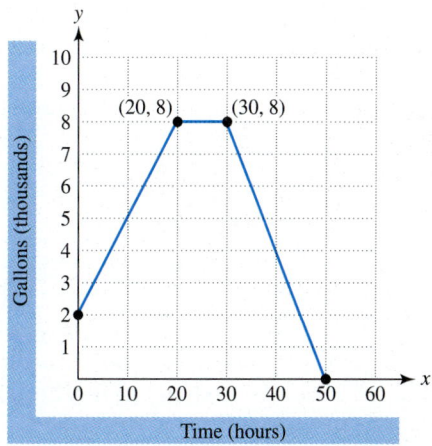

FIGURE 2.48 Water in a Swimming Pool

Solution

(a) From Figure 2.48, when $x = 0$, $y = 2$ so $f(0) = 2$. Initially, there are two thousand gallons of water in the pool. The points (25, 8) and (40, 4) lie on the graph of f. Therefore, $f(25) = 8$ and $f(40) = 4$. After 25 hours there were eight thousand gallons in the pool, and after 40 hours there were four thousand gallons.

(b) During the first 20 hours, water in the pool increases from two thousand to eight thousand gallons. For the next 10 hours, the water level is constant. Between 30 and 50 hours, the pool is drained. Since there are no breaks in the graph, f is a continuous function on the interval [0, 50].

(c) Slope indicates the rate at which water is entering or leaving the pool. The first line segment connects the points (0, 2) and (20, 8). Its slope is $m = \dfrac{8 - 2}{20 - 0} = 0.3$. Since the units are thousands of gallons and hours, water is entering the pool during this time at a rate of 300 gallons per hour. Between 20 and 30 hours the slope of the line segment is 0. Water is neither entering nor leaving the pool. The slope of the line segment connecting the points (30, 8) and (50, 0) is $m = \dfrac{0 - 8}{50 - 30} = -0.4$. On this interval, the amount of water is *decreasing* at a rate of 400 gallons per hour. ∎

Critical Thinking

Suppose that a function g models the amount of water in a swimming pool. Discuss whether g must be continuous. If g were discontinuous at some point in time, what must happen to the amount of water in the pool?

EXAMPLE 2   *Analyzing pollution and brown trout*

Due to acid rain, the percentage of lakes in Scandinavia that lost their population of brown trout increased dramatically between 1940 and 1975. Based on a sample of 2850 lakes, this percentage can be approximated by the piecewise-linear function f. (Source: C. Mason, *Biology of Freshwater Pollution.*)

$$f(x) = \begin{cases} \dfrac{11}{20}(x - 1940) + 7 & \text{if } 1940 \le x < 1960 \\[2ex] \dfrac{32}{15}(x - 1960) + 18 & \text{if } 1960 \le x \le 1975 \end{cases}$$

(a) Determine the percentage of lakes that lost brown trout by 1947 and by 1972.
(b) Sketch a graph of f.

Solution

(a) Since $1940 \le 1947 < 1960$, $f(1947)$ is calculated using the first formula.

$$f(1947) = \frac{11}{20}(1947 - 1940) + 7 = 10.85 \text{ (percent)}.$$

Similarly, $f(1972)$ is found using the second formula.

$$f(1972) = \frac{32}{15}(1972 - 1960) + 18 = 43.6 \text{ (percent)}.$$

Therefore, by 1947 approximately 11% of the lakes had lost their population of brown trout. By 1972 this percentage had increased to about 44%.

(b) Because f is a piecewise-linear function, the graph of f consists of two line segments. The first segment is determined by $y = \dfrac{11}{20}(x - 1940) + 7$. If $x = 1940$ then $y = 7$, and if $x = 1960$ then $y = 18$. Place a dot at $(1940, 7)$ and a small circle at $(1960, 18)$, since the year 1960 is not included in the interval $1940 \le x < 1960$. These points are endpoints of the line segment.

The second line segment is determined by $y = \dfrac{32}{15}(x - 1960) + 18$. If $x = 1960$ then $y = 18$, and if $x = 1975$ then $y = 50$. This segment includes $(1960, 18)$ and $(1975, 50)$, so place a dot at both endpoints. Notice that this fills in the small circle at $(1960, 18)$ from the first line segment. The graph of f is shown in Figure 2.49. ∎

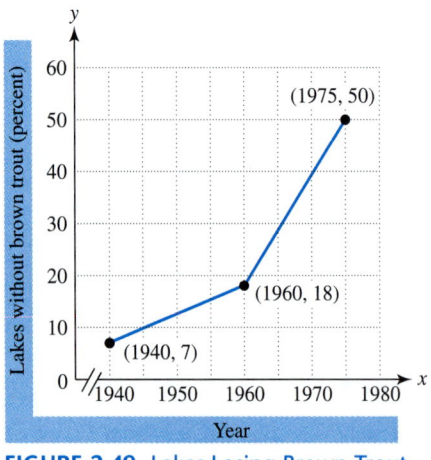

FIGURE 2.49 Lakes Losing Brown Trout

The Greatest Integer Function

A common piecewise-defined function is the greatest integer function, denoted by $f(x) = [x]$. The **greatest integer function** is defined as follows.

$[x]$ is the greatest integer less than or equal to x.

Some examples that evaluate $[x]$ include

$$[6.7] = 6, \quad [3] = 3, \quad [-2.3] = -3, \quad [-10] = -10, \quad \text{and} \quad [-\pi] = -4.$$

The graph of $y = [x]$ is shown in Figure 2.50. A symbolic representation is also given. The greatest integer function is both a piecewise-constant function and a step function.

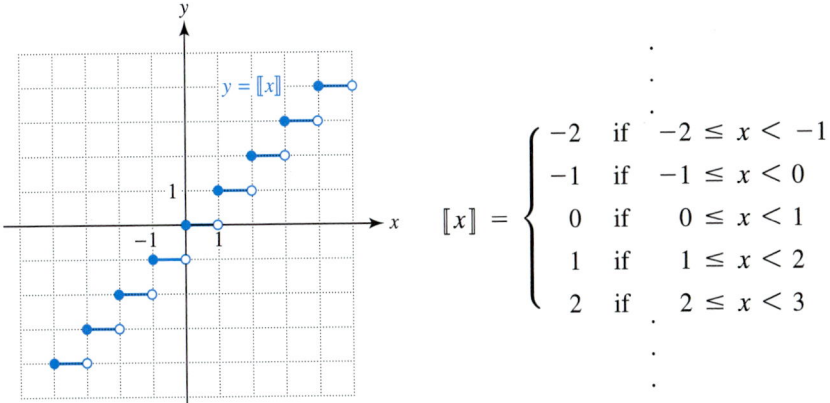

$$[x] = \begin{cases} \vdots \\ -2 & \text{if} \quad -2 \le x < -1 \\ -1 & \text{if} \quad -1 \le x < 0 \\ 0 & \text{if} \quad 0 \le x < 1 \\ 1 & \text{if} \quad 1 \le x < 2 \\ 2 & \text{if} \quad 2 \le x < 3 \\ \vdots \end{cases}$$

FIGURE 2.50 The Greatest Integer Function

In many applications, fractional parts are either not allowed or ignored. Long-distance telephone calls often are timed to the minute, framing lumber for houses is measured in two-foot multiples, and mileage charges for rental cars may be calculated to the mile.

Suppose a car rental company charges $31.50 per day plus $0.25 for each mile driven, where fractions of a mile are ignored. The function given by $f(x) = 0.25[x] + 31.50$ calculates the cost of driving x miles in one day. For example, the cost of driving 100.4 miles is

$$f(100.4) = 0.25[100.4] + 31.50 = 0.25(100) + 31.50 = \$56.50.$$

On some calculators and computers, the greatest integer function is denoted by int(X). (Check your manual.) A graph $Y_1 = 0.25*\text{int}(X) + 31.5$ is shown in Figure 2.51.

[0, 10, 1] by [31, 35, 1]

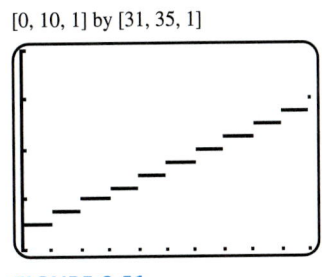

FIGURE 2.51

Technology Note *Connected and Dot Modes*

It is common for graphing calculators to connect points in order to make a graph look continuous. However, if a graph has breaks in it, a graphing calculator may connect points where there should be breaks. In dot mode, points are plotted but not connected. Figure 2.52 is the same graph shown in Figure 2.51, except that it is plotted in connected mode. Notice that connected mode generates an inaccurate graph.

[0, 10, 1] by [31, 35, 1]

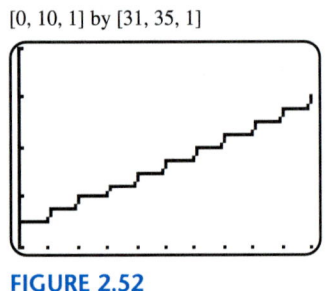

FIGURE 2.52

The Absolute Value Function

The graph of $y = |x|$ is shown in Figure 2.53. It is V-shaped and cannot be represented by a single linear function. However, it can be represented by the lines $y = x$ (when $x \geq 0$) and $y = -x$ (when $x < 0$.) This suggests that the absolute value function can be defined symbolically using a piecewise-linear function.

$$|x| = \begin{cases} -x & \text{if } x < 0 \\ x & \text{if } x \geq 0 \end{cases}$$

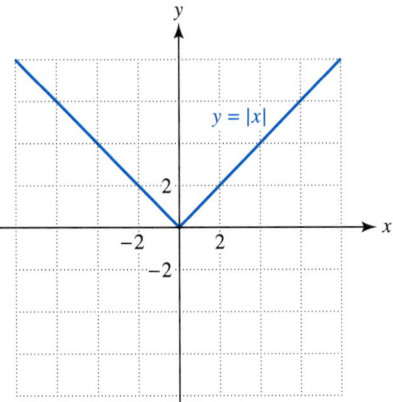

FIGURE 2.53 The Absolute Value Function

EXAMPLE 3 *Analyzing the graph of $y = |ax + b|$*

For each linear function f, graph $y = f(x)$ and $y = |f(x)|$ separately. Discuss how the absolute value affects the graph of f.

(a) $f(x) = x + 2$

(b) $f(x) = -2x + 4$

Solution

(a) On some graphing calculators, the absolute value function is denoted by abs(X). (Check your manual.) The graphs of $Y_1 = X + 2$ and $Y_2 =$ abs(X + 2) are shown in Figures 2.54 and 2.55, respectively. The graph of Y_1 is a line with x-intercept -2. The graph of Y_2 is V-shaped. The graphs are identical to the right of the x-intercept. To the left of the x-intercept, the graph of $Y_1 = f(x)$ passes below the x-axis, while the graph of $Y_2 = |f(x)|$ is the reflection of $Y_1 = f(x)$ across the x-axis. The graph of $Y_2 = |f(x)|$ does not dip below the x-axis, because absolute value is never negative.

$[-9, 9, 1]$ by $[-6, 6, 1]$ $[-9, 9, 1]$ by $[-6, 6, 1]$

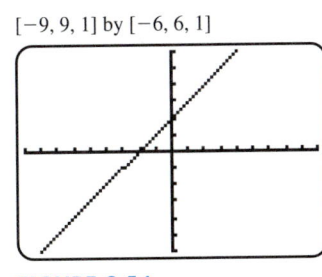

FIGURE 2.54   **FIGURE 2.55**

(b) The graphs of $Y_1 = -2X + 4$ and $Y_2 =$ abs(-2X + 4) are shown in Figures 2.56 and 2.57, respectively. Again, the graph of Y_2 is V-shaped. The graph of $Y_2 = |f(x)|$ is the reflection of f across the x-axis whenever the graph of $Y_1 = f(x)$ is below the x-axis.

$[-9, 9, 1]$ by $[-6, 6, 1]$ $[-9, 9, 1]$ by $[-6, 6, 1]$

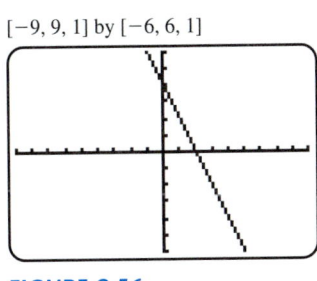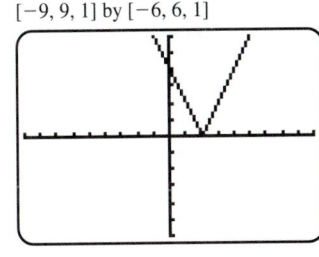

FIGURE 2.56   **FIGURE 2.57**   ■

Critical Thinking

Graph $f(x) = \sqrt{x^2}$ in $[-15, 15, 1]$ by $[-10, 10, 1]$. Find a second symbolic representation for f. Explain your reasoning.

Equations and Inequalities Involving Absolute Values

The equation $|x| = 5$ has two solutions, $x = \pm 5$. This is shown visually in Figure 2.58 on the next page, where the graph of $y = |x|$ intersects the horizontal line $y = 5$ at the points $(\pm 5, 5)$.

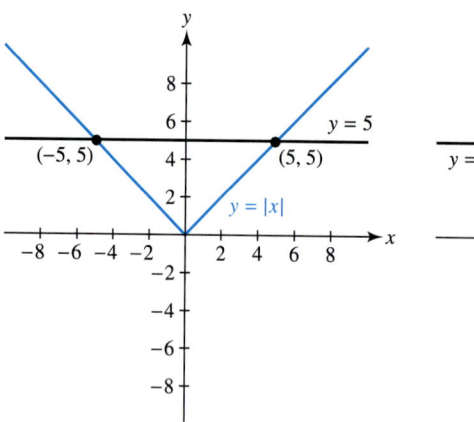

FIGURE 2.58

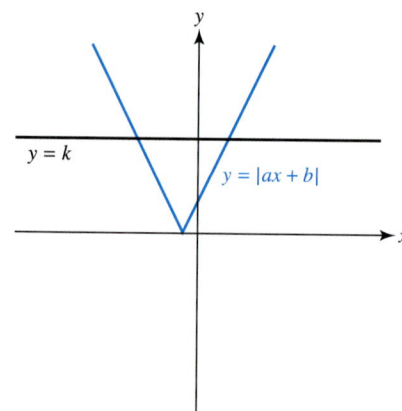

FIGURE 2.59

In general, the graph of $y = |ax + b|$ with $a \neq 0$ is V-shaped. It intersects the horizontal line $y = k$ twice whenever $k > 0$ as illustrated in Figure 2.59. Thus, there are two solutions to the equation $|ax + b| = k$.

Absolute value equations

Let k be a positive number.

$|ax + b| = k$ is equivalent to $ax + b = \pm k$.

EXAMPLE 4 *Solving an equation involving absolute value*

Solve the equation $|2x + 5| = 2$ graphically, numerically, and symbolically.

Solution

Graphical Solution Graph $Y_1 = \text{abs}(2X + 5)$ and $Y_2 = 2$. The V-shaped graph of Y_1 intersects the horizontal line at the points $(-3.5, 2)$ and $(-1.5, 2)$ as shown in Figures 2.60 and 2.61. The solutions are -3.5 and -1.5.

Numerical Solution Table $Y_1 = \text{abs}(2X + 5)$ and $Y_2 = 2$ as in Figure 2.62. The solutions to $Y_1 = Y_2$ are -3.5 and -1.5.

[−9, 9, 1] by [−6, 6, 1]

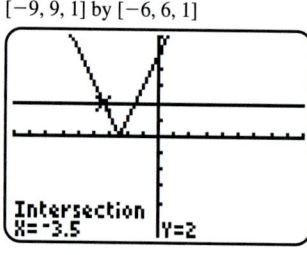

FIGURE 2.60

[−9, 9, 1] by [−6, 6, 1]

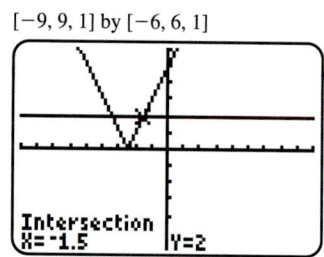

FIGURE 2.61

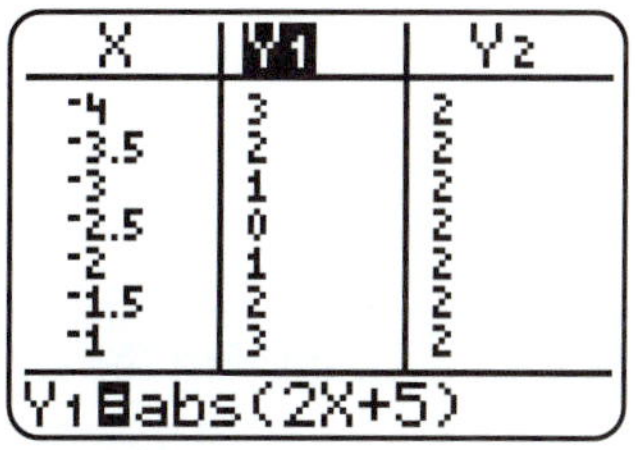

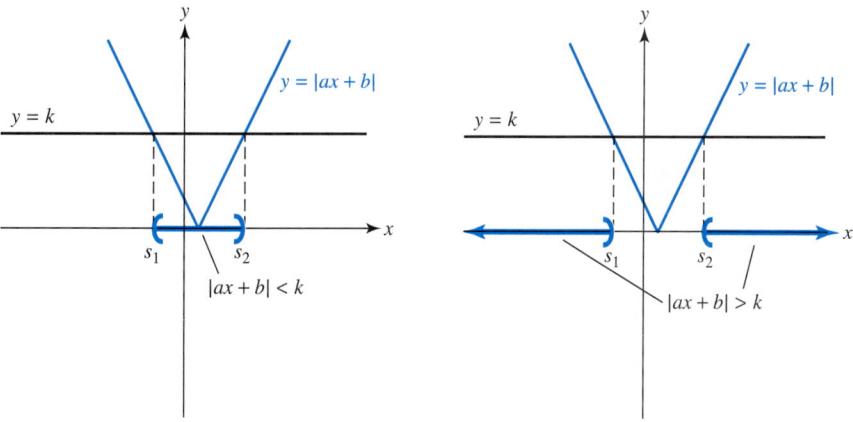

FIGURE 2.62

Symbolic Solution   The equation $|2x + 5| = 2$ is satisfied when $2x + 5 = \pm 2$.

$$2x + 5 = 2 \qquad \text{or} \qquad 2x + 5 = -2$$

$$2x = -3 \qquad \text{or} \qquad 2x = -7$$

$$x = -\frac{3}{2} \qquad \text{or} \qquad x = -\frac{7}{2} \qquad \blacksquare$$

In Figure 2.63 the solutions to $|ax + b| = k$ are labeled s_1 and s_2. The V-shaped graph of $y = |ax + b|$ is below the horizontal line $y = k$ between s_1 and s_2 or when $s_1 < x < s_2$. The solution set for the inequality $|ax + b| < k$ is shaded on the x-axis. In Figure 2.64 the V-shaped graph is above the horizontal line $y = k$ left of s_1 or right of s_2. That is, when $x < s_1$ or $x > s_2$. The solution set for the inequality $|ax + b| > k$ is shaded on the x-axis. Notice that in both figures, equality (determined by s_1 and s_2) is the boundary between *greater than* and *less than*. For this reason, s_1 and s_2 are called *boundary numbers*.

FIGURE 2.63   **FIGURE 2.64**

These results are generalized.

Absolute value inequalities

Let the solutions to $|ax + b| = k$ be s_1 and s_2, where $s_1 < s_2$ and $k > 0$.

1. $|ax + b| < k$ is equivalent to $s_1 < x < s_2$.

2. $|ax + b| > k$ is equivalent to $x < s_1$ or $x > s_2$.

Similar statements can be made for inequalities involving $\le$ or $\ge$.

Note: The *union symbol* ∪ may be used to write $x < s_1$ or $x > s_2$ in interval notation. For example, $x < -2$ or $x > 5$ is written as $(-\infty, -2) \cup (5, \infty)$ in interval notation. This indicates that the solution set includes all real numbers in
either $(-\infty, -2)$ or $(5, \infty)$.

EXAMPLE 5   ***Analyzing the temperature range in Santa Fe***

The inequality $|T - 49| \leq 20$ describes the range of monthly average tempera-
tures T in degrees Fahrenheit for Santa Fe, New Mexico. (Source: A. Miller and J.
Thompson, *Elements of Meteorology.*)
(a) Solve this inequality graphically and symbolically.
(b) The high and low monthly average temperatures satisfy the absolute value
equation $|T - 49| = 20$. Use this fact to interpret the results from part (a).

Solution

(a) *Graphical Solution*   Graph $Y_1 = \text{abs}(X - 49)$ and $Y_2 = 20$ as in Figure 2.65.
The V-shaped graph of Y_1 intersects the horizontal line at the points (29, 20)
and (69, 20). See Figures 2.66 and 2.67. The graph of Y_1 is below the graph
of Y_2 between these two points. Thus, the solution set consists of all temperatures T satisfying $29 \leq T \leq 69$.

[20, 80, 5] by [0, 35, 5]

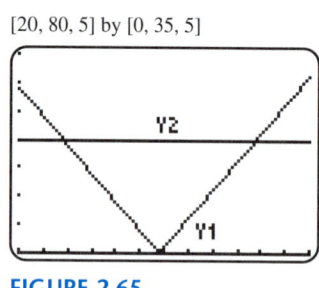

FIGURE 2.65

[20, 80, 5] by [0, 35, 5]

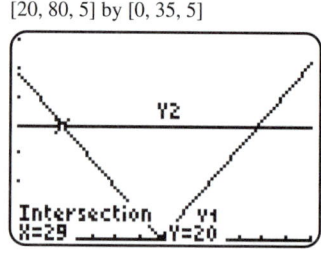

FIGURE 2.66

[20, 80, 5] by [0, 35, 5]

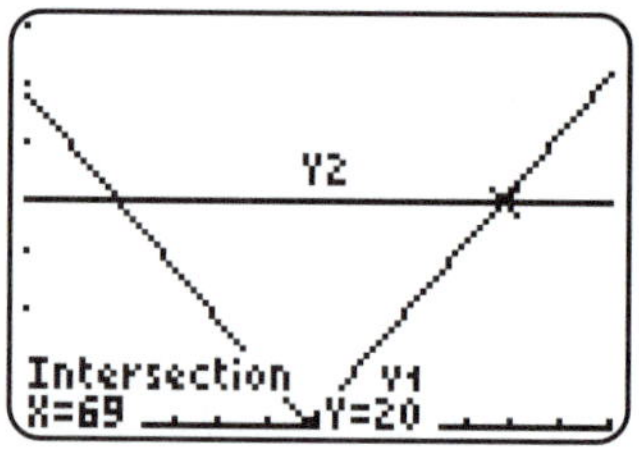

FIGURE 2.67

Symbolic Solution   First solve the related equation $|T - 49| = 20$.

$$T - 49 = -20 \qquad \text{or} \qquad T - 49 = 20$$
$$T = 29 \qquad \text{or} \qquad T = 69$$

Thus, by our previous discussion, the inequality $|T - 49| \leq 20$ is equivalent
to $29 \leq T \leq 69$.
(b) The solutions to $|T - 49| = 20$ are 29 and 69. Therefore, the monthly average
temperatures in Sante Fe vary between a low of 29°F (January) and a high of
69°F (July). The monthly averages are always within 20 degrees of 49°F. ■

EXAMPLE 6 *Solving equations and inequalities involving absolute values*

Solve the equation symbolically. Then, solve the related inequality. Support your results graphically.

(a) $|2x - 5| = 6$, $|2x - 5| \le 6$

(b) $|5 - x| = 3$, $|5 - x| > 3$

Solution

(a) Begin by solving $2x - 5 = \pm 6$.

$$2x - 5 = 6 \qquad \text{or} \qquad 2x - 5 = -6$$
$$2x = 11 \qquad \text{or} \qquad 2x = -1$$
$$x = \frac{11}{2} \qquad \text{or} \qquad x = -\frac{1}{2}$$

The solutions to $|2x - 5| = 6$ are $-\frac{1}{2}$ and $\frac{11}{2}$. The solution set for the inequality $|2x - 5| \le 6$ includes all real numbers x satisfying $-\frac{1}{2} \le x \le \frac{11}{2}$. Graphical support is shown in Figures 2.68 and 2.69.

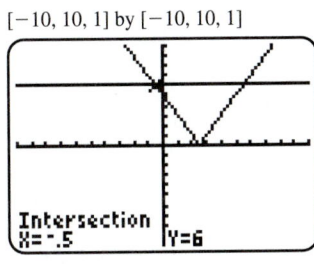

$[-10, 10, 1]$ by $[-10, 10, 1]$

FIGURE 2.68

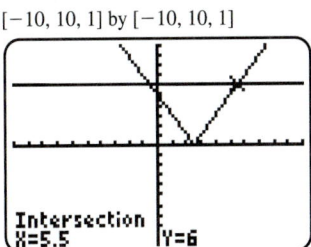

$[-10, 10, 1]$ by $[-10, 10, 1]$

FIGURE 2.69

(b) Begin by solving $5 - x = \pm 3$.

$$5 - x = 3 \qquad \text{or} \qquad 5 - x = -3$$
$$-x = -2 \qquad \text{or} \qquad -x = -8$$
$$x = 2 \qquad \text{or} \qquad x = 8$$

The solutions to $|5 - x| = 3$ are 2 and 8. By our previous discussion, the solution set for $|5 - x| > 3$ includes all real numbers x left of 2 and right of 8. Thus, $|5 - x| > 3$ is equivalent to $x < 2$ or $x > 8$. Graphical support is shown in Figures 2.70 and 2.71.

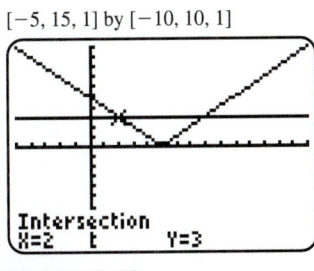

$[-5, 15, 1]$ by $[-10, 10, 1]$

FIGURE 2.70

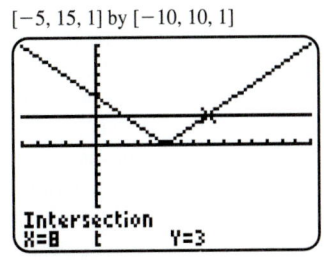

$[-5, 15, 1]$ by $[-10, 10, 1]$

FIGURE 2.71

2.4 PUTTING IT ALL TOGETHER

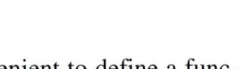

Piecewise-defined functions are used when it is not convenient to define a function using one formula. The greatest integer function and the absolute value function are two examples of piecewise-defined functions.

The following table summarizes methods for solving equations and inequalities involving absolute value. (Inequalities containing the symbols $\leq$ and $\geq$ are solved in a similar manner.)

Problem	Symbolic Solution	Graphical Solution
Solve $\lvert ax + b \rvert = k$ with $k > 0$.	Solve the two equations $ax + b = k$ and $ax + b = -k$.	Graph $Y_1 = \lvert ax + b \rvert$ and $Y_2 = k$. Find the x-values of the two points of intersection.
Solve $\lvert ax + b \rvert < k$ with $k > 0$.	Solve $\lvert ax + b \rvert = k$. Let these values be s_1 and s_2, where $s_1 < s_2$. The solutions to the inequality satisfy $s_1 < x < s_2$.	Graph $Y_1 = \lvert ax + b \rvert$ and $Y_2 = k$. Find the x-values of the two points of intersection. The solutions are between these values on the number line, where the graph of Y_1 lies below the graph of Y_2.
Solve $\lvert ax + b \rvert > k$ with $k > 0$.	Solve $\lvert ax + b \rvert = k$. Let these values be s_1 and s_2, where $s_1 < s_2$. The solutions to the inequality satisfy $x < s_1$ or $x > s_2$.	Graph $Y_1 = \lvert ax + b \rvert$ and $Y_2 = k$. Find the x-values of the two points of intersection. The solutions lie "outside" these values on the number line, where the graph of Y_1 lies above the graph of Y_2.

2.4 EXERCISES   Tape 2

Evaluating and Graphing Piecewise-Defined Functions

1. *Speed Limits*   The graph of $y = f(x)$ gives the speed limit y along a rural highway after traveling x miles.
   (a) What are the maximum and minimum speed limits along this stretch of highway?
   (b) Estimate the miles of highway with a speed limit of 55 miles per hour.
   (c) Evaluate $f(4), f(12),$ and $f(18)$.
   (d) At what x-values is the graph discontinuous? Interpret each discontinuity.

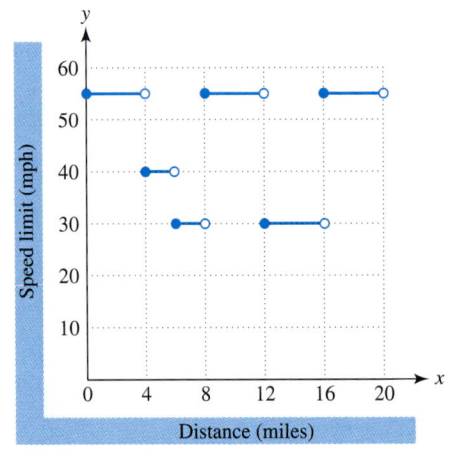

2. *ATM*   The graph of $y = f(x)$ depicts the amount of money y in dollars in an automatic teller machine (ATM) after x minutes.
   (a) Determine the initial and final amounts of money in the ATM.
   (b) Evaluate $f(10)$ and $f(50)$. Is f continuous?
   (c) How many withdrawals occurred during this time period?
   (d) When did the largest withdrawal occur? How much was it?
   (e) How much was deposited into the machine?

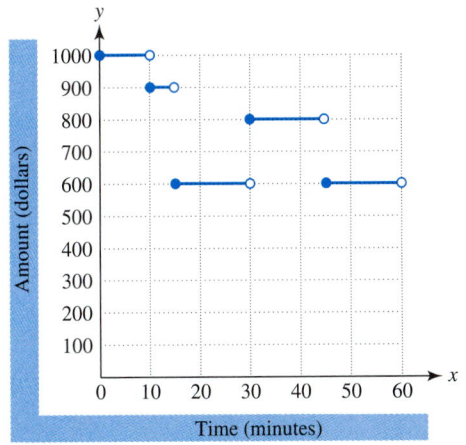

3. *Swimming Pool Levels*   The graph of $y = f(x)$ represents the amount of water in thousands of gallons remaining in a swimming pool after x days.
   (a) Estimate the initial and final amounts of water contained in the pool.
   (b) When did the amount of water in the pool remain constant?
   (c) Approximate $f(2)$ and $f(4)$.
   (d) At what rate was water being drained from the pool when $1 \le x \le 3$?

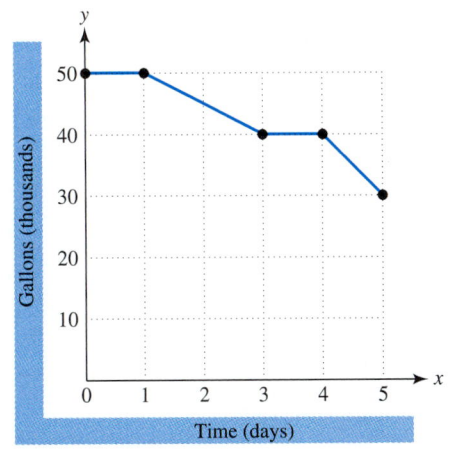

4. *Gasoline Usage*   The graph shows the gallons of gasoline y in the gas tank of a car after x hours.
   (a) Estimate how much gasoline was in the gas tank when $x = 3$.
   (b) Interpret the graph.
   (c) During what time interval did the car burn gasoline at the fastest rate?

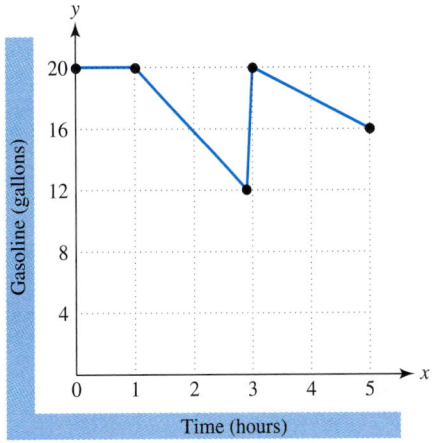

Exercises 5–8: Use $f(x)$ to complete the following.

$$f(x) = \begin{cases} 3x - 1 & \text{if } -5 \le x < 1 \\ 4 & \text{if } 1 \le x \le 3 \\ 6 - x & \text{if } 3 < x \le 5 \end{cases}$$

5. Evaluate f at $x = -3, 1, 2,$ and 5.
6. On what interval is f constant?
7. Sketch a graph of f. Is f continuous?
8. Find the x-value(s) where $f(x) = 2$.

Exercises 9–12: Use $g(x)$ to complete the following.

$$g(x) = \begin{cases} -2x - 6 & \text{if } -8 \le x \le -2 \\ x & \text{if } -2 < x < 2 \\ 0.5x + 1 & \text{if } 2 \le x \le 8 \end{cases}$$

9. Evaluate g at $x = -8, -2, 2,$ and 8.
10. For what x-values is g decreasing?
11. Sketch a graph of g. Is g continuous?
12. Find the x-value(s) where $g(x) = 0$.

Greatest Integer Function

Exercises 13–16: Complete the following.
   (a) *Use dot mode to graph the function f in the standard viewing rectangle.*
   (b) *Evaluate $f(-3.1)$ and $f(1.7)$.*

13. $f(x) = [\![2x - 1]\!]$
14. $f(x) = [\![x + 1]\!]$
15. $f(x) = 2[\![x]\!] + 1$

16. $f(x) = [-x]$

17. *Lumber Costs* Lumber that is used to frame up walls of houses is frequently sold in multiples of two feet. If the length of a board is not exactly a multiple of two feet, there is often no charge for the additional length. For example, if a board measures at least eight feet but less than ten feet, then the consumer is charged for only eight feet.

(a) Suppose that the cost of lumber is $0.80 every two feet. Find a symbolic representation of a function f that computes the cost of a board x feet long for $6 \le x \le 18$.

(b) Graph f.

(c) Determine the costs of boards with lengths of 8.5 feet and 15.2 feet.

18. *Long-Distance Phone Bills* Suppose that the charges for a long-distance call are $0.50 for the first minute and $0.25 for each additional minute. Assume that a fraction of a minute is rounded up.

(a) Determine the cost of a phone call lasting 3.5 minutes.

(b) Find a symbolic representation for a function f that computes the cost of a telephone call x minutes long, where $0 < x \le 5$. (*Hint:* Express f as a piecewise-constant function.)

(c) The greatest integer function is not convenient to use in part (b). Explain why.

Exercises 19–24: (Refer to Example 3.)

(a) *Graph* $y = f(x)$ *in the standard viewing rectangle.*

(b) *Use the graph of* $y = f(x)$ *to sketch a graph of the equation* $y = |f(x)|$. *Verify your graph with a calculator.*

(c) *Determine the x-intercept for the graph of* $y = |f(x)|$.

19. $y = 2x$ **20.** $y = 0.5x$

21. $y = 3x - 3$ **22.** $y = 2x - 4$

23. $y = 6 - 2x$ **24.** $y = 2 - 4x$

Exercises 25 and 26: Let $f(x) = |ax + b|$ *where* $a \ne 0$ *and* $g(x) = k$, *where* $k > 0$ *is a constant. The points where the graphs of* f *and* g *intersect are shown in the viewing rectangles. Solve each equation and inequality.*

25. (a) $f(x) = g(x)$

(b) $f(x) < g(x)$

(c) $f(x) > g(x)$

[−10, 10, 1] by [−10, 10, 1]

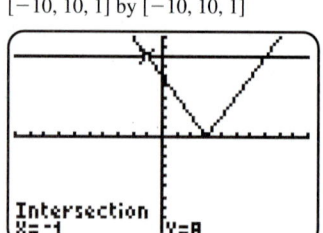

[−10, 10, 1] by [−10, 10, 1]

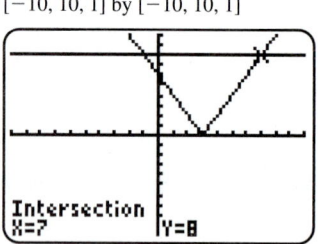

26. (a) $f(x) = g(x)$

(b) $f(x) \le g(x)$

(c) $f(x) \ge g(x)$

[−6, 12, 1] by [−6, 6, 1]

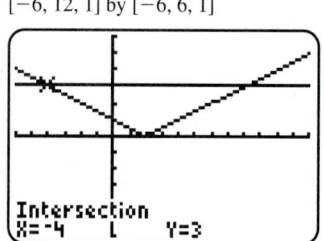

[−6, 12, 1] by [−6, 6, 1]

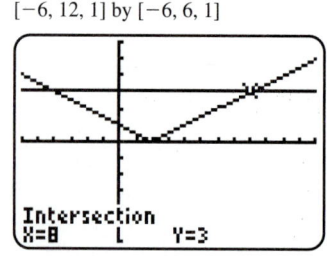

Exercises 27–32: Solve the equation
  (a) *graphically,*
  (b) *numerically, and*
  (c) *symbolically.*
Then solve the related inequality.

27. $|x| = 5$, $|x| \leq 5$
28. $|x| = 15$, $|x| > 15$
29. $|2x - 5| = 10$, $|2x - 5| < 10$
30. $|3x - 4| = 8$, $|3x - 4| \leq 8$
31. $|5 - 3x| = 2$, $|5 - 3x| > 2$
32. $|4x - 7| = 5$, $|4x - 7| \geq 5$

Exercises 33–38: Solve the equation graphically and symbolically. Then solve the related inequality.

33. $|2.1x - 0.7| = 2.4$, $|2.1x - 0.7| \geq 2.4$
34. $\left|\frac{1}{2}x - \frac{3}{4}\right| = \frac{7}{4}$, $\left|\frac{1}{2}x - \frac{3}{4}\right| \leq \frac{7}{4}$
35. $|3x| + 5 = 6$, $|3x| + 5 > 6$
36. $|x| - 10 = 25$, $|x| - 10 < 25$
37. $\left|\frac{2}{3}x - \frac{1}{2}\right| = -\frac{1}{4}$, $\left|\frac{2}{3}x - \frac{1}{2}\right| \leq -\frac{1}{4}$
38. $|5x - 0.3| = -4$, $|5x - 0.3| > -4$

You Decide the Method

Exercises 39–50: Solve the inequality. Write the solution in interval notation.

39. $|3x - 1| < 8$
40. $|15 - x| < 7$
41. $|7 - 4x| \leq 11$
42. $|-3x + 1| \leq 5$
43. $|0.5x - 0.75| < 2$
44. $|2.1x - 5| \leq 8$
45. $|2x - 3| > 1$
46. $|5x - 7| > 2$
47. $|-3x + 8| \geq 3$
48. $|-7x - 3| \geq 5$
49. $|0.25x - 1| > 3$
50. $|-0.5x + 5| \geq 4$

Sketching Graphs

51. Sketch a graph that depicts the amount of water in a 100-gallon tank. The tank is initially empty and then filled at a rate of five gallons per minute. Immediately after it is full, a pump is used to empty the tank at two gallons per minute.

52. Sketch a graph showing the mileage that a person is from home after x hours if that individual drives at 40 mph to a park 20 miles away, remains at the park

for two hours, and then returns home at a speed of 20 mph.

53. Sketch a graph showing the amount of money y that a person makes after working x hours in one day, where $0 \leq x \leq 12$. Assume that the worker earns \$8 per hour and receives time-and-half pay for all hours over eight.

54. Suppose that two cars, both traveling at a constant speed of 60 miles per hour, approach each other on a straight highway. If they are initially four miles apart, sketch a graph of the distance between the cars after x minutes, where $0 \leq x \leq 4$.

Applications

Exercises 55–60: Average Temperatures (Refer to Example 5.) The inequality describes the range of monthly average temperatures T in degrees Fahrenheit at a certain location.
  (a) *Solve the inequality.*
  (b) *If the high and low monthly average temperatures satisfy equality, interpret the inequality.*

55 $|T - 43| \leq 24$, Marquette, Michigan
56. $|T - 62| \leq 19$, Memphis, Tennessee
57. $|T - 50| \leq 22$, Boston, Massachusetts
58. $|T - 10| \leq 36$, Chesterfield, Canada
59. $|T - 61.5| \leq 12.5$, Buenos Aires, Argentina
60. $|T - 43.5| \leq 8.5$, Punta Arenas, Chile

61. *Shoe Size* Professional basketball player Shaquille O'Neal is 7 feet 1 inch tall and weighs 300 pounds. The table lists his shoe sizes at certain ages. (Source: *USA Today.*)

Age	20	21	22	23
Shoe Size	19	20	21	22

  (a) Write an expression that gives his shoe size at age $x = 20, 21, 22$, and 23.
  (b) Suppose that after age 23 his shoe size did not change. Sketch a graph of a continuous, piecewise-linear function f that models his shoe size from ages 20 to 26.

62. *ACT Scores* The table lists the average composite scores on the American College Test (ACT) for selected years. (Source: The American College Testing Program.)

Year	1970	1975	1980	1985	1990	1995
Score	19.9	18.6	18.5	18.6	20.6	20.9

(a) Make a line graph of the data in [1968, 1997, 5] by [15, 36, 1].
(b) If this graph represents a piecewise-linear function f, find a symbolic representation for the piece of f located on the interval [1990, 1995].
(c) Evaluate $f(1993)$. (The actual average composite score in 1993 was 20.7.)

63. *Violent Crimes* The table lists victims of violent crime per 1000 people in 1993 by age group using interval notation. (Source: Department of Justice.)

Age	[12, 16)	[16, 19)	[19, 24)	[24, 35)
Crime Rate	125	121	98	61

Age	[35, 50)	[50, 65)	[65, 90)
Crime Rate	45	18	8

(a) Sketch the graph of a piecewise-constant function that models the data, where x represents age.
(b) Discuss the impact that age has on the likelihood of being a victim of a violent crime.

64. *New Homes* The table lists the numbers in millions of houses built for various time intervals from 1940 to 1990 using interval notation. (Source: Bureau of the Census, *American Housing Survey*.)

Year	[1940, 1950)	[1950, 1960)	[1960, 1970)
Houses	8.5	13.6	16.1

Year	[1970, 1980)	[1980, 1990)
Houses	11.6	8.1

(a) Sketch the graph of a piecewise-constant function that models this data, where x represents the year.
(b) Discuss the trends in housing starts between 1940 and 1990.

65. *Perspectives and Viewing Rectangles* The table lists the average math scores on the Scholastic Aptitude Test (SAT) for selected years. (Source: The College Board.)

Year	1970	1975	1980	1985	1990	1995
Score	488	472	466	475	476	482

(a) Make a line graph of the data in [1968, 1997, 5] by [460, 490, 5].
(b) If this graphical representation of a piece-wise-linear function appeared in a newspaper, how would many people interpret the college entrance math scores?
(c) The math scores may vary between 200 and 800. Remake the line graph in [1968, 1997, 5] by [200, 800, 100]. Discuss how interpretation of data can be influenced by changing the scale on the y-axis.

66. *Hourly Wages* The average hourly wage in dollars from 1938 to 1996 can be modeled by $f(x)$. (Source: Department of Labor.)

$$f(x) = \begin{cases} 0.084(x - 1938) + 0.25 & \text{if } 1938 \le x < 1967 \\ 0.312(x - 1967) + 2.69 & \text{if } 1967 \le x \le 1996 \end{cases}$$

(a) Use f to estimate the average hourly wages in 1956, 1967, and 1995.
(b) Interpret the slope of each line segment in the graph of f.

67. *Cesarean Births* The percentages of babies delivered by Cesarean birth are listed in the table. (Source: S. Teutsch.)

Year	1970	1975	1980	1985	1990
Cesarean Births	5%	11%	17%	23%	23%

(a) Make a line graph of the data. Interpret the graph.
(b) Find values for the constants a, b, and c so that $f(x)$ models the data.

$$f(x) = \begin{cases} a(x - 1970) + b & \text{if } 1970 \le x < 1985 \\ c & \text{if } 1985 \le x \le 1990 \end{cases}$$

(c) Evaluate $f(1978)$ and $f(1988)$. Interpret the results.

68. *Student Loans* The table lists the amounts in millions of dollars of government-guaranteed student loans from 1990 to 1994. (Source: *USA Today*.)

Year	1990	1991	1992	1993	1994
Loans ($ millions)	12.3	13.5	14.7	16.5	18.3

(a) Make a line graph of the data. Can a linear function model the data exactly?
(b) Find values for the constants a, b, c, and d so that $f(x)$ models the data.

$$f(x) = \begin{cases} a(x - 1990) + b & \text{if } 1990 \le x < 1992 \\ c(x - 1992) + d & \text{if } 1992 \le x \le 1994 \end{cases}$$

(c) Evaluate $f(1991)$ and $f(1993)$. Do the results agree with the actual values listed in the table?

69. *Internet Accounts*   In 1994 there were 6 million online, dial-up accounts on the Internet. This amount increased at a rate of approximately 4 million per year until 1997. Because of new technologies such as direct broadcast satellite, the number of accounts may remain constant from 1997 to 2000. (Sources: Forrester Research, Globe Research, *New York Times* Service.)
(a) Sketch a graph of a piecewise-linear function f that models this situation in the year x, where $1994 \le x \le 2000$.
(b) Find a symbolic representation of f.

70. *Flow Rates*   A water tank has an inlet pipe with a flow rate of five gallons per minute and an outlet pipe with a flow rate of three gallons per minute. A pipe can be either closed or completely open. The graph shows the number of gallons of water in the tank after x minutes have elapsed. Use the concept of slope to interpret the graph.

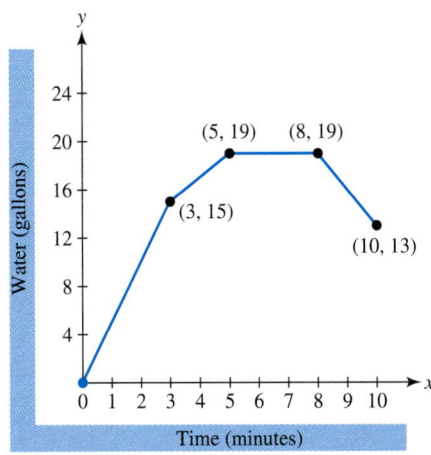

Writing about Mathematics

1. Find a piecewise-constant function that describes movie prices for children, adults, and senior citizens. Represent your function verbally, graphically, and symbolically. Explain your results.

2. Explain why some functions are defined piecewise. Give examples.

CHECKING BASIC CONCEPTS FOR SECTIONS 2.3 AND 2.4

1. Find an equation of the line passing through the points $(-3, 4)$ and $(5, -2)$. Give equations of lines that are parallel and perpendicular to this line.

2. Find equations of horizontal and vertical lines that pass through the point $(-4, 7)$.

3. Sketch a graph of

$$f(x) = \begin{cases} 2 - x & \text{if } -5 \le x < 0 \\ 2x - 1 & \text{if } 0 \le x \le 5 \end{cases}$$

(a) Evaluate $f(-2)$, $f(0)$, and $f(3)$.
(b) Is f continuous?

4. (a) Solve the equation $|2x - 1| = 5$ graphically, numerically, and symbolically.
(b) Use part (a) to solve the linear inequalities $|2x - 1| \le 5$ and $|2x - 1| > 5$.

2.5 Linear Approximation

Midpoint Formula • Distance Formula • Interpolation and Extrapolation • Linear Regression (Optional)

Introduction

In previous sections, several instances were presented where real data was approximated using linear functions and graphs. The approximation of data with lines and linear functions is called **linear approximation.** It is an essential technique used in both elementary and advanced mathematics. In this section we discuss some of the many types of linear approximation.

Midpoint Formula

A common way to make estimations is to average data values. For example, in 1980 the average cost of tuition and fees at public colleges and universities was $800, whereas in 1982 it was $1000. One might estimate the cost of tuition and fees in 1981 to be $900. This averaging is referred to as finding the midpoint. If a line segment is drawn between two data points, then its *midpoint* is the unique point on the line segment that is equidistant from the endpoints.

On a real number line, the midpoint M of two data points x_1 and x_2 is calculated by averaging their coordinates as shown in Figure 2.72. For example, the midpoint of -3 and 5 is $M = \dfrac{-3 + 5}{2} = 1.$

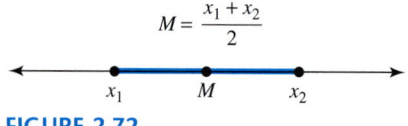

$$M = \frac{x_1 + x_2}{2}$$

$x_1 \qquad M \qquad x_2$

FIGURE 2.72

The midpoint formula in the xy-plane is similar to the formula for the real number line, except that both coordinates are averaged. Figure 2.73 shows midpoint M located on the line segment connecting the two data points (x_1, y_1) and (x_2, y_2). The x-coordinate of the midpoint M is located halfway between x_1 and x_2 and is $\dfrac{x_1 + x_2}{2}$. Similarly, the y-coordinate of M is the average of y_1 and y_2.

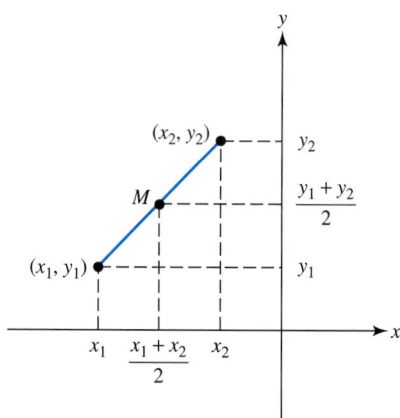

FIGURE 2.73

> ### Midpoint formula in the *xy*-plane
>
> The **midpoint** of the line segment with endpoints (x_1, y_1) and (x_2, y_2) in the *xy*-plane is $\left(\dfrac{x_1 + x_2}{2}, \dfrac{y_1 + y_2}{2} \right)$.

The population of the United States from 1800 to 1990 is modeled by the nonlinear function f in Figure 2.74. The line segment connecting the points (1800, 5) and (1990, 249) does not accurately describe the population over the time period from 1800 to 1990. However, over a shorter time interval a linear approximation may be more accurate.

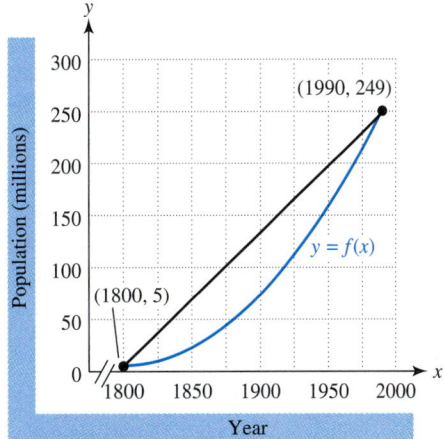

FIGURE 2.74 U.S. Population

EXAMPLE 1 *Using the midpoint formula*

In 1970 the population of the United States was 203 million and in 1990 it was 249 million. (Source: Bureau of the Census.)
(a) Use the midpoint formula to estimate the population in 1980.
(b) Describe this approximation graphically.

Solution

(a) The U.S. populations in 1970 and 1990 can be represented by the data points (1970, 203) and (1990, 249). The midpoint M of the line segment connecting these points is

$$M = \left(\frac{1970 + 1990}{2}, \frac{203 + 249}{2} \right) = (1980, 226).$$

The midpoint formula estimates a population of 226 million in 1980. (The actual population was 226.5 million.)

(b) Figure 2.75 on the following page shows the U.S. population and a line segment with midpoint M connecting the data points (1970, 203) and (1990, 249). Notice that the line segment appears to be an accurate linear approximation over a relatively short period of 20 years.

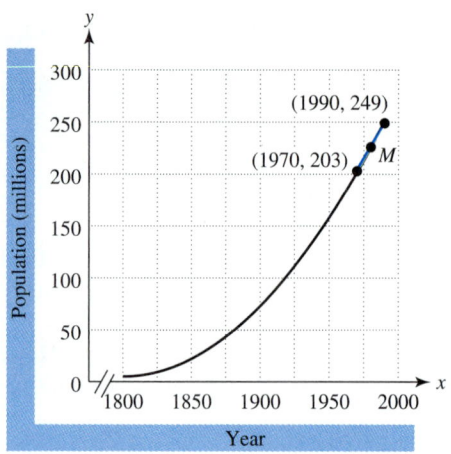

FIGURE 2.75 A Linear Approximation

Distance Formula

In the xy-plane, the length of a line segment with endpoints (x_1, y_1) and (x_2, y_2) can be calculated using the Pythagorean Theorem. See Figure 2.76. The lengths of the legs of the right triangle are $|x_2 - x_1|$ and $|y_2 - y_1|$. The distance d is the hypotenuse of the right triangle. Applying the Pythagorean Theorem, $d^2 = (x_2 - x_1)^2 + (y_2 - y_1)^2$. Since distance is nonnegative, it follows that $d = \sqrt{(x_2 - x_1)^2 + (y_2 - y_1)^2}$.

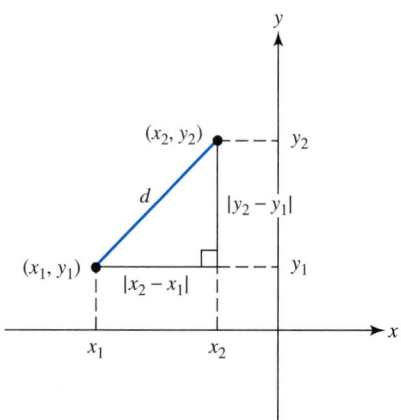

FIGURE 2.76

Distance formula

The **distance** d between the points (x_1, y_1) and (x_2, y_2) in the xy-plane is

$$d = \sqrt{(x_2 - x_1)^2 + (y_2 - y_1)^2}.$$

Figure 2.77 shows a line segment connecting the points $(-1, 3)$ and $(4, -3)$. Its length is

$$d = \sqrt{(4 - (-1))^2 + (-3 - 3)^2} = \sqrt{61} \approx 7.81.$$

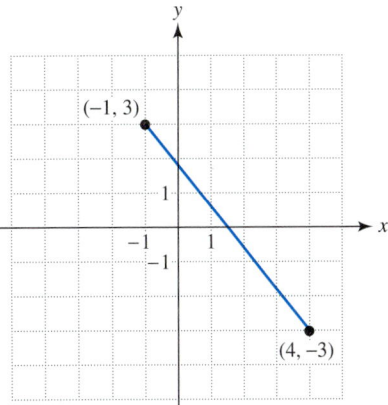

FIGURE 2.77

Curves designed by engineers for highways and railroads frequently have parabolic, rather than circular, shapes. The reason for this is that a parabolic curve becomes sharper gradually. See Figure 2.78. If railroad tracks changed abruptly from straight to circular, the momentum of the locomotive could cause a derailment. Figure 2.79 illustrates straight track connecting to a circular curve. (Source: F. Mannering and W. Kilareski, *Principles of Highway Engineering and Traffic Analysis.*)

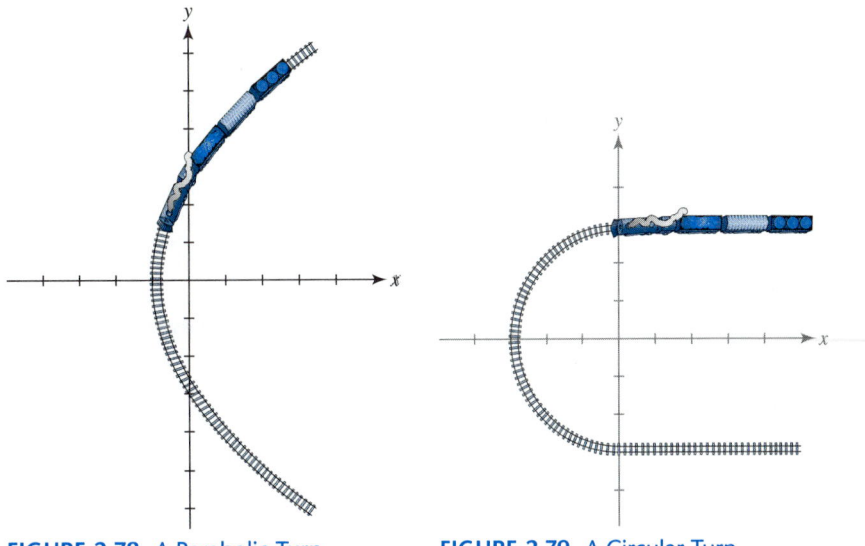

FIGURE 2.78 A Parabolic Turn   **FIGURE 2.79** A Circular Turn

In order to design a curve and estimate its cost, engineers determine the distance around a curve before it is built. In Figure 2.80 on the following page the distance along a parabolic curve from *A* to *C* is approximated by two line segments *AB* and *BC*. The distance formula can be used to calculate the length of each segment. The sum of these two lengths gives a crude estimate for the length of the curve.

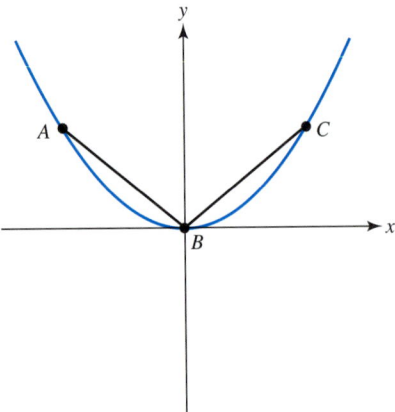

FIGURE 2.80 An Estimate of Curve Length

A better estimate would result from four line segments as shown in Figure 2.81. As the number of segments increases so does the accuracy of the linear approximation.

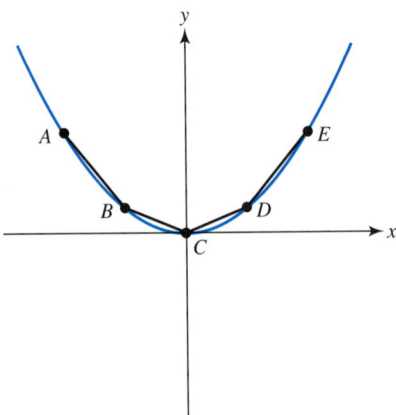

FIGURE 2.81 A Better Estimate

EXAMPLE 2   *Approximating the length of a curve*

Suppose that a curve designed for railroad tracks is represented by the equation $y = 0.2x^2$, where the units are in kilometers.

(a) The points $(-3, 1.8)$, $(-1.5, 0.45)$, $(0, 0)$, $(1.5, 0.45)$, and $(3, 1.8)$ lie on the graph of $y = 0.2x^2$. Approximate the length of the curve from $x = 3$ to $x = -3$ using line segments connecting these points.

(b) Show this approximation graphically.

Solution

(a) Let the ordered pairs $(-3, 1.8)$, $(-1.5, 0.45)$, $(0, 0)$, $(1.5, 0.45)$, and $(3, 1.8)$ correspond to the points *A, B, C, D,* and *E,* respectively. As illustrated in Figure 2.81, the length of the curve can be approximated by calculating the sum of the distances between these points. Using symmetry we can double the lengths of AB and BC to obtain a linear approximation for the length of the curve.

$$AB: \quad \sqrt{(-1.5 - (-3))^2 + (0.45 - 1.8)^2} \approx 2.018$$
$$BC: \quad \sqrt{(0 - (-1.5))^2 + (0 - 0.45)^2} \approx 1.566$$

Thus, the distance along the curve is approximately $2(2.018 + 1.566) \approx 7.17$ kilometers. (The actual distance is about 7.23 kilometers.)

(b) Figure 2.82 shows a line graph connecting the five data points. The sum of the lengths of these four line segments is about 7.17 kilometers. Figure 2.83 shows the line graph and the curve $y = 0.2x^2$.

$[-3, 3, 1]$ by $[-1, 3, 1]$ $[-3, 3, 1]$ by $[-1, 3, 1]$

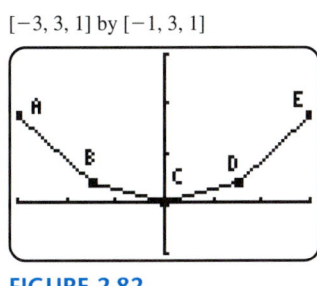 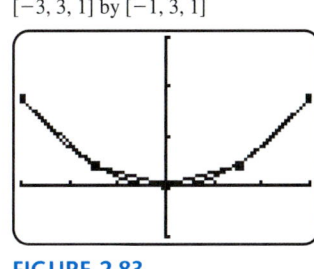

FIGURE 2.82   **FIGURE 2.83** ∎

Critical Thinking
If the distance along any curve is estimated using line segments, how will this estimation compare to the actual distance along the curve?

Interpolation and Extrapolation

Total energy production in the Asia-Pacific region of the world is predicted to increase substantially between 1990 and 2005. Table 2.4 lists this energy production in million metric tons of oil equivalent (Mtoe) for selected years. (Source: R. Andre-Pascal, *Global Energy: The Changing Outlook.*)

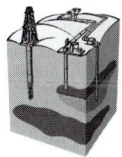

TABLE 2.4				
Year	1990	1995	2000	2005
Energy (Mtoe)	415	530	645	760

If the data in Table 2.4 is used to estimate energy production during 1998, this would be an example of **interpolation** since 1998 lies between 1990 and 2005. If this data is used to estimate energy production in either 1980 or 2010, this would be an example of **extrapolation** because these years do not lie between 1990 and 2005. Interpolation usually provides more reliable results than extrapolation. Extrapolation should be used cautiously.

EXAMPLE 3   *Estimating past and future energy production*

(a) Use Table 2.4 to determine an equation of a line that models the data. Graph both the data and this line. Interpret the slope.

(b) Use this line to estimate energy production in 1998 and 1970. Discuss the results.

Solution

(a) The data increases by exactly 115 Mtoe every five years, or by $\frac{115}{5} = 23$ Mtoe per year. Thus, the line with slope 23 passing through $(1990, 415)$ models this data exactly. Its equation is $y = 23(x - 1990) + 415$. The graph of the data and the line are shown in Figure 2.84. The slope indicates that energy production in the Asia-Pacific region is expected to increase at a rate of 23 Mtoe per year.

[1988, 2007, 1] by [350, 820, 100]

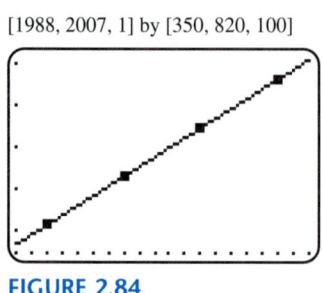

FIGURE 2.84

(b) To estimate energy production in 1998 and 1970, substitute 1998 and 1970 for x.

$$y = 23(1998 - 1990) + 415 = 599 \text{ Mtoe}$$
$$y = 23(1970 - 1990) + 415 = -45 \text{ Mtoe}$$

The interpolated value for 1998 is reasonable, since it lies between the 1995 and 2000 values of 530 and 645 Mtoe, respectively. The extrapolated value for 1970 is not reasonable because it is negative. It might be acceptable to use extrapolation for estimating energy production in 1988 or 2007, since these years lie relatively close to the domain of the data. However, extrapolating 20 years back provides a meaningless result. ∎

Linear Regression (Optional)

Throughout Chapters 1 and 2, linear functions and lines have been used to model data involving the variables x and y. Unknown values for y were predicted at given values of x. Problems where one variable is used to predict the behavior of a second variable are called **regression** problems. If a line is used to approximate the data, then this is referred to as **linear regression.**

We have already solved numerous problems using linear regression by selecting a line that *visually* fits the data in a scatterplot. However, this technique has some disadvantages. First, it does not produce a unique line. Different people may arrive at different lines to fit the same data. Secondly, the line is not determined automatically by a calculator or computer. A person must view the data and adjust the line until it "fits." By contrast, a statistical method used to determine a unique regression line is based on *least squares.*

Most graphing calculators have the capability to calculate the least-squares regression line automatically after the data points have been entered as seen in the following example.

EXAMPLE 4   *Predicting airline passenger growth*

Table 2.5 lists the estimated numbers in millions of airline passengers at some of the fastest growing airports in 1992 and 2005. (Source: Federal Aviation Administration.)

TABLE 2.5	Airline Passengers (millions)		
Airport	1992	2005	
Harrisburg International	0.7	1.4	
Dayton International	1.1	2.4	
Austin Robert Mueller	2.2	4.7	
Milwaukee General Mitchell	2.2	4.4	
Sacramento Metropolitan	2.6	5.0	
Fort Lauderdale–Hollywood	4.1	8.1	
Washington Dulles	5.3	10.9	
Greater Cincinnati	5.8	12.3	

(a) Make a scatterplot of the data using the 1992 data for x-values and the corresponding 2005 data for y-values.

(b) Use a calculator to find the least-squares regression line.

(c) Raleigh-Durham International had 4.9 million passengers in 1992. Assuming Raleigh-Durham International follows the same trend in growth as the other airports, estimate the number of passengers using this airport in 2005. Compare this result with the Federal Aviation Administration (FAA) estimation of 10.3 million passengers.

Solution

(a) Plot the data as shown in Figure 2.85.

[0, 6, 1] by [0, 14, 1]

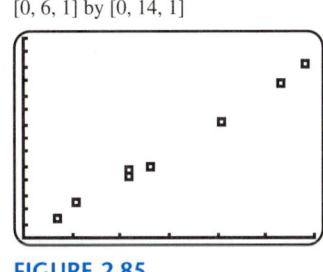

FIGURE 2.85

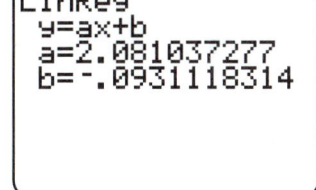

FIGURE 2.86

[0, 6, 1] by [0, 14, 1]

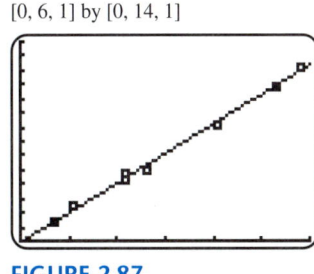

FIGURE 2.87

(b) The equation of the regression line is shown in Figure 2.86. The line $y = 2.081x - 0.0931$ and the data are graphed together in Figure 2.87.

(c) When $x = 4.9$, $y = 2.081(4.9) - 0.0931 \approx 10.1$ million passengers. This is quite close to the FAA prediction of 10.3 million.

2.5 PUTTING IT ALL TOGETHER

Linear approximation, a fundamental tool used frequently in mathematics, usually becomes more accurate as the interval of approximation becomes smaller. One important type of linear approximation is a regression line. A regression line can be found by at least three different techniques. One is using trial and error to visually fit a line to a set of data. A second is to choose two data points and find the equation of the line passing through these points. The third is to find the least-squares regression line using technology.

The following table outlines some important ideas presented in this section.

Concept	Comments	Example
Midpoint formula $\left(\dfrac{x_1 + x_2}{2}, \dfrac{y_1 + y_2}{2} \right)$	Computes a linear approximation	Given $(2, -3)$ and $(4, 5)$, then $M = \left(\dfrac{2 + 4}{2}, \dfrac{-3 + 5}{2} \right) = (3, 1).$
Distance formula $\sqrt{(x_2 - x_1)^2 + (y_2 - y_1)^2}$	Finds the distance between the points (x_1, y_1) and (x_2, y_2) in the xy-plane	Given $(2, -1)$ and $(-1, 3)$, then $d = \sqrt{(-1 - 2)^2 + (3 - (-1))^2}$ $= \sqrt{9 + 16} = \sqrt{25} = 5.$
Interpolation	Estimates values that are between two or more known data values	If the data points $(0, 1)$ and $(2, 3)$ are used to estimate the value of y when $x = 1$, then this would involve interpolation.
Extrapolation	Estimates values that are not between two known data values	If the data points $(0, 1)$ and $(2, 3)$ are used to estimate the value of y when $x = 4$, then this would involve extrapolation.

2.5 EXERCISES

Midpoint Formula

Exercises 1–6: Use the midpoint formula to complete the following.

1. *Charge Card Balances* The outstanding balances on VISA and MasterCard totaled $254 billion in 1992 and $444 billion in 1996. Estimate this debt in 1994. (Actual amount was $335 billion.) (Sources: BankCard Holders of America, Federal Reserve Board.)

2. *State and Federal Inmates* In 1992 there were 883,656 inmates in state and federal prisons and in 1994 there were 1,053,738. Estimate the number of inmates in 1993. (Actual number was 970,444.) (Source: Department of Justice.)

3. *State and Federal Inmates* In 1980 there were 329,821 inmates in state and federal prisons and in 1983 there were 436,885. Estimate the number of inmates in 1986. (Actual number was 544,972.) (Source: Department of Justice.)

4. *Population Estimates* The population of the United States was 151 million in 1950 and 179 million in 1960. Estimate the population in 1970. (Actual population was 203 million.) (Source: Bureau of the Census.)

5. *Olympic Times* In the Olympic Games, the 200-meter dash is run in approximately 20 seconds. Estimate the time to run the 100-meter dash.

6. Between any two real numbers a and b there is always another real number. How could such a number be found?

Exercises 7–14: Find the midpoint of the line segment connecting the points.

7. $(1, 2), (5, -3)$

8. $(-6, 7), (9, -4)$

9. $(1.5, 2.9), (-5.7, -3.6)$

10. $(9.4, -4.5), (-7.7, 9.5)$

11. $(\sqrt{2}, \sqrt{5}), (\sqrt{2}, -\sqrt{5})$

12. $(\sqrt{7}, 3\sqrt{3}), (-\sqrt{7}, -\sqrt{3})$

13. $(a, b), (-a, 3b)$

14. $(-a, b), (3a, b)$

Distance Formula

Exercises 15–20: Find the distance in the xy-plane between the two points. Round approximate results to the nearest hundredth.

15. $(2, -2), (5, 2)$

16. $(0, -3), (12, -8)$

17. $(7, -4), (9, 1)$

18. $(-1, -6), (-8, -5)$

19. $(-6.5, 2.7), (3.6, -2.9)$

20. $(3.6, 5.7), (-2.1, 8.7)$

Exercises 21–24: Curve Length (Refer to Example 2.) Use three line segments connecting the four points to estimate the length of the curve on the graph of f from $x = -1$ to $x = 2$. Graph f and a line graph of the four points in the indicated viewing rectangle.

21. $f(x) = x^2$; $(-1, 1), (0, 0), (1, 1), (2, 4)$; $[-4.5, 4.5, 1]$ by $[-1, 5, 1]$

22. $f(x) = \sqrt[3]{x}$; $(-1, -1), (0, 0), (1, 1), (2, \sqrt[3]{2})$; $[-3, 3, 1]$ by $[-2, 2, 1]$

23. $f(x) = 0.5x^3 + 2$; $(-1, 1.5), (0, 2), (1, 2.5), (2, 6)$; $[-4.5, 4.5, 1]$ by $[0, 6, 1]$

24. $f(x) = 2 - 0.5x^2$; $(-1, 1.5), (0, 2), (1, 1.5), (2, 0)$; $[-3, 3, 1]$ by $[-1, 3, 1]$

25. The distance along the curve of $y = x^2$ from $(0, 0)$ to $(3, 9)$ is approximately 9.747. Use this fact to estimate the distance along the curve of $y = 9 - x^2$ from $(0, 9)$ to $(3, 0)$.

26. Estimate the distance along the curve of $y = \sqrt{x}$, from $(1, 1)$ to $(4, 2)$. (The actual value is approximately 3.168.)

Interpolation and Extrapolation

Exercises 27–30: The table lists data that is exactly linear.

   (a) Find the slope-intercept form of the line that passes through these data points.

   (b) Predict y when $x = -2.7$ and 6.3. Decide whether these calculations involve interpolation or extrapolation.

27.

x	-3	-2	-1	0	1
y	-7.7	-6.2	-4.7	-3.2	-1.7

28.

x	-2	-1	0	1	2
y	10.2	8.5	6.8	5.1	3.4

29.

x	5	23	32	55	61
y	94.7	56.9	38	-10.3	-22.9

30.

x	-11	-8	-7	-3	2
y	-16.1	-10.4	-8.5	-0.9	8.6

Linear Regression

Exercises 31–34: Complete the following.

   (a) Use a calculator to find the least-squares regression line in the form $y = ax + b$. Give approximations for a and b to five significant digits.

   (b) Use the regression line to estimate y when $x = 2.4$.

31.

x	-3	-2	-1	0	1	2	3
y	-11.1	-8.4	-5.7	-2.6	1.1	3.9	7.3

32.

x	-4	-2	0	2	4
y	1.2	2.8	5.3	6.7	9.1

33.

x	1	3	5	7	10
y	5.8	-2.4	-10.7	-17.8	-29.3

34.

x	-4	-3	-1	3	5
y	37.2	33.7	27.5	16.4	9.8

Linear Approximation Applications

35. *Jail Population* The table lists the average daily jail population in thousands for three selected years. (Source: Department of Justice.)

Year	1983	1988	1993
Population (thousands)	228	336	466

(a) Make a line graph of this data in [1982, 1994, 1] by [200, 500, 100].
(b) Conjecture whether the midpoint formula gives an estimate for 1988 that is high or low.
(c) Compute the midpoint estimate. Was your conjecture correct?

36. *Juvenile Crime*   The table lists the numbers of juveniles charged in court from 1990 to 1995 in Blue Earth County, Minnesota. (Source: Blue Earth County Attorney's Office.)

Year	1990	1991	1992	1993	1994	1995
Juveniles	346	379	453	566	681	713

(a) Use the midpoint formula to conjecture whether this data is exactly linear.
(b) Make a line graph of this data in [1989, 1996, 1] by [300, 800, 100]. Was your conjecture correct? Describe the trend in juvenile crime in Blue Earth County.

37. *Never Married*   The number of females in their early thirties who have never married rose from 6 to 20% between 1970 and 1994. Use the midpoint method to estimate this percentage in 1982.

38. *Never Married*   The number of males in their early thirties who have never married rose from 9 to 30% between 1970 and 1994. Use the midpoint method to estimate this percentage in 1982.

39. *Federal Debt*   The table lists the total federal debt in billions of dollars for three different years.

Year	1985	1987	1989
Debt ($ billions)	1828	2354	2881

Source: Office of Management and Budget.

(a) Use the midpoint method to predict whether the data is exactly linear.
(b) Find a linear function f whose graph passes through the points (1985, 1828) and (1989, 2881).
(c) Evaluate $f(1987)$. How does this value compare to the midpoint estimate? Explain.

40. Continuing with Exercise 39,
(a) Use f to estimate the federal debt in 1984, 1988, and 1996.
(b) The actual debts in 1984, 1988, and 1996 were 1572, 2602, and 5020 billion dollars,

respectively. Compare the accuracies of extrapolation and interpolation.

41. *Long-Distance Revenue*   The table lists total long-distance telephone revenues in billions of dollars for three different years. (Source: *USA Today.*)

Year	1991	1992	1993
Revenue ($ billions)	60.9	63.9	66.9

(a) Is the data exactly linear?
(b) Conjecture what revenues were in 1990.

42. *Corporate Downsizing*   Due to corporate downsizing, companies hire temporary workers rather than full-time employees because of the cost savings. As a result, employment through temporary work agencies increased significantly during the 1990s. In 1992 there were 1.35 million workers placed by temporary agencies and in 1994 there were 1.97 million. (Source: National Association of Temporary Staffing Services.)
(a) Find a linear function f whose graph passes through the data points (1992, 1.35) and (1994, 1.97).
(b) Use f to estimate the number of workers placed in 1990 and 1993. (The actual numbers were 1.17 and 1.64 million, respectively.)
(c) Compare the accuracies of interpolation and extrapolation.

43. *Water Pollution*   At one time, the Thames River in England supported an abundant community of fish. By 1915, however, pollution destroyed all the fish in a 40-mile stretch near its mouth for a 45-year period. Since then, improved sewage treatment facilities and other ecological steps have resulted in a dramatic increase in the number of different fish present. The number of species present from 1967 to 1978 can be modeled by $f(x) = 6.15x - 12,059$, where x is the year.
(a) Using a table, find the year when the number of species first exceeded 70.
(b) The domain of f is $1967 \leq x \leq 1978$. Discuss whether f would continue to give good approximations beyond 1978. What would likely happen to the number of species of fish present after time passes and pollution disappears? (Source: C. Mason, *Biology of Freshwater Pollution.*)

44. *Ozone Layer*   Ozone is the only gas capable of screening out lethal ultraviolet light. Without ozone, ultraviolet light would kill terrestrial life. Although the ozone layer can vary in thickness, it is typically only three millimeters thick. Worldwide, it has been estimated that for each 0.03 millimeter decrease in the thickness of the ozone layer, there would be an additional 100,000 cases of blindness annually due to cataracts. (Source: R. K. Turner, *Environmental Economics.*)

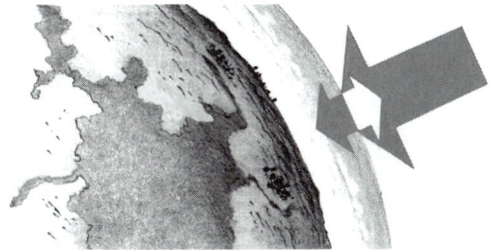

(a) Find a linear function that computes the estimated number of additional cases of blindness each year from cataracts for a decrease of x millimeters in the ozone layer.

(b) Approximate the additional annual cases of blindness that may occur due to a decrease in the ozone layer of 0.05 millimeter.

45. *Distant Galaxies*   In the late 1920s the famous observational astronomer Edwin P. Hubble (1889–1953) determined both the distance to several galaxies and the velocity at which they were receding from Earth. Four galaxies with their distances in light-years and velocities in miles per second are listed in the table. (Sources: A. Acker and C. Jaschek, *Astronomical Methods and Calculations; A. Sharov and I. Novikov, Edwin Hubble, The Discoverer of the Big Bang Universe.*)

Galaxy	Distance	Velocity
Virgo	50	990
Ursa Minor	650	9300
Corona Borealis	950	15,000
Bootes	1700	25,000

(a) Let x represent distance and y velocity. Plot the four data points in $[-100, 1800, 100]$ by $[-1000, 28{,}000, 1000]$.

(b) Find the least-squares regression line that models this data.

(c) If the galaxy Hydra is receding at a speed of 37,000 miles per second, estimate its distance.

46. *Cellular Phones*   One technological problem with cellular phones is the delay time involved with placing a call when the system is busy. One study analyzed this delay. The table shows that as the number of calls increased by P percent, the average delay time D to put through a call also increased.

(Source: A. Mehrotra, *Cellular Radio: Analog and Digital Systems.*)

P (%)	0	20	40	60	80	100
D (minutes)	1	1.6	2.4	3.2	3.8	4.4

(a) Let P correspond to x-values and D to y-values. Find the least-squares regression line that models this data. Plot the data and the regression line.

(b) Estimate the delay for a 50% increase in the number of calls.

47. *Women in Politics*   The table lists percentages of women in state legislatures for past years. (Source: National Women's Political Caucus.)

Year	1975	1977	1979	1981	1983	1985
Percent	8.0	9.1	10.3	12.1	13.3	14.8

Year	1987	1989	1991	1993	1995
Percent	15.7	17.0	18.3	20.5	20.7

(a) Make a scatterplot of the data.

(b) Find the least-squares regression line that models this data.

(c) Interpret the slope.

(d) Assuming that trends continue, use this line to estimate the percentage of women in state legislatures in 1998.

48. *Juvenile Crime*   Find a least-squares regression line that models the data in Exercise 36. Estimate the number of juveniles charged in court during 1996.

Calculator Experiment

Exercises 49 and 50: Graph the function f in the standard viewing rectangle.

(a) *Choose any curved portion of the graph of f and repeatedly zoom in. Describe how the graph appears. Repeat this on different portions of the graph.*

(b) *Under what circumstances could a linear approximation be used to accurately model a nonlinear graph?*

49. $f(x) = 4x - x^3$

50. $f(x) = x^4 - 5x^2$

Writing about Mathematics

1. Explain the difference between interpolation and extrapolation. Sketch a graph of $y = x^2$ and the line passing through the points $(0, 0)$ and $(1, 1)$. Use the graph to explain why interpolation tends to be more accurate than extrapolation.

2. What type of data can the midpoint formula be used to approximate? Explain your answer with the aid of a graph.

CHECKING BASIC CONCEPTS FOR SECTION 2.5

1. Find the length of the line segment connecting the points $(-2, 3)$ and $(4, 2)$. Calculate the midpoint of this line segment. Illustrate your calculations graphically.

2. According to the 1996 federal *Dietary Guidelines for Americans*, the recommended minimum weight for a person 58 inches tall is 91 pounds, while for a person 64 inches tall it is 111 pounds.
   (a) Find an equation of the line that passes through the points $(58, 91)$ and $(64, 111)$.

   (b) Use this line to estimate the minimum weight for someone 61 inches tall. Then use the midpoint formula to find this minimum weight. Are the results the same? Why?

3. Find the least-squares regression line for the data from Example 1 in Section 1.5. Estimate per capita personal income in 1995. Did this calculation involve interpolation or extrapolation?

CHAPTER 2 Summary

A linear function f can be expressed in the form $f(x) = ax + b$, where a and b are constants. When a linear function is used in an application, a linear equation frequently results. A linear equation, written as $ax + b = 0$ where $a \neq 0$, can be solved graphically, numerically, and symbolically. Two important graphical techniques for solving linear equations are the x-intercept method and the intersection-of-graphs method.

Any linear equation can be converted into a linear inequality by replacing the equals sign with one of the following: $<, \leq, >,$ or $\geq$. Many of the techniques used to solve linear equations can be used to solve linear inequalities because equality represents the boundary between *greater than* and *less than*.

The graph of a linear function is a line. A line is completely determined by either a point and a slope, or by two points. Two important forms of an equation of a line are the point-slope form and the slope-intercept form. The point-slope form is $y = m(x - h) + k$. In this equation m is the slope and (h, k) is a point that lies on the line. The slope-intercept form is $y = mx + b$, where m is the slope and b is the y-intercept. The slope-intercept form is unique.

Piecewise-defined functions occur when it is difficult to express a function symbolically with one formula. This type of function uses different formulas on different intervals of its domain. Piecewise-defined functions are not always continuous. Two examples of piecewise-linear functions are the greatest integer function and the absolute value function.

The midpoint formula can be used for linear approximations. It finds the midpoint of a line segment connecting two points. The distance formula calculates the distance between two points in the xy-plane. It can be used to estimate the length of a curve. Interpolation and extrapolation are two methods of approximation. Interpolation is usually more accurate than extrapolation. Data can be modeled using least-squares regression lines. These lines are usually computed by a calculator or computer.

Review Exercises

Exercises 1–4: Solve the linear equation graphically and symbolically.

1. $5x - 22 = 10$

2. $5(4 - 2x) = 16$

3. $-2(3x - 7) + x = 2x - 1$

4. $\pi x + 1 = 6$

Exercises 5 and 6: Use a table to solve each linear equation numerically to the nearest tenth.

5. $3.1x - 0.2 = 2(x - 1.7)$

6. $\sqrt{7} - 3x = 2.1(1 + x)$

7. *Median Income* The median U.S. family income between 1980 and 1993 can be modeled by $f(x) = 1321.7(x - 1980) + 21{,}153$, where x is the year. (Source: Department of Commerce.)
  (a) Solve the equation $f(x) = 25{,}000$ graphically. Interpret the solution.
  (b) Solve part (a) symbolically.

8. *Course Grades* In order to receive a B grade in a college course, it is necessary to have an overall average of 80% correct on two one-hour exams of 75 points each and one final exam of 150 points. If a person scores 55 and 72 on the one-hour exams, what is the minimum score on the final exam that the person can receive and still earn a B?

9. The graphs of two linear functions f and g are shown in the figure. Solve each equation or inequality.
  (a) $f(x) = g(x)$
  (b) $f(x) < g(x)$
  (c) $f(x) > g(x)$

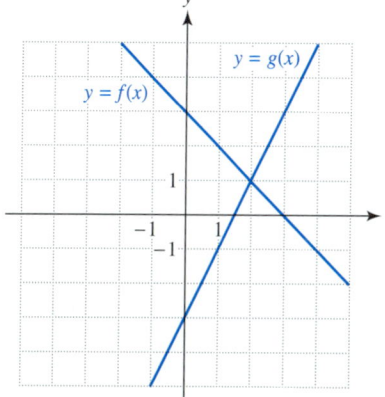

10. The graphs of three linear functions f, g, and h with domains $D = \{x \mid 0 \le x \le 7\}$ are shown in the figure. Solve each equation or inequality.
  (a) $f(x) = g(x)$
  (b) $g(x) = h(x)$
  (c) $f(x) < g(x) < h(x)$
  (d) $g(x) > h(x)$

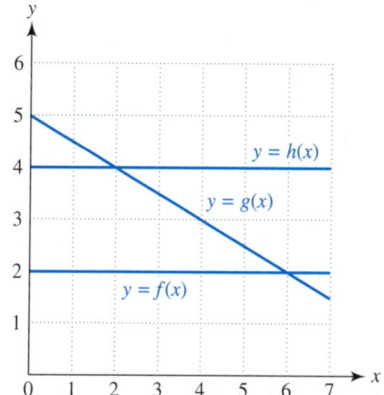

*Exercises 11–14: Solve the linear inequality graphi-
cally and symbolically. Write the solution set in interval
notation.*

11. $3x - 4 \leq 2 + x$

12. $-2x + 6 \leq -3x$

13. $-2 \leq 5 - 2x < 7$

14. $-1 < \dfrac{3x - 5}{-3} < 3$

15. *Medicare Costs* Future estimates for Medicare
costs in billions of dollars can be modeled by
$f(x) = 18x - 35{,}750$, where $1995 \leq x \leq 2007$.
(Source: Office of Management and Budget.)
   (a) Determine symbolically when Medicare costs
   are expected to be from 268 to 358 billion
   dollars.
   (b) Support your answer graphically and numerically.

16. *Temperature Scales* The table shows equivalent
temperatures in degrees Celsius and degrees
Fahrenheit.

°F	−40	32	59	95	212
°C	−40	0	15	35	100

   (a) Plot the data by having the x-axis correspond
   to Fahrenheit temperature and the y-axis to
   Celsius temperature. What type of relation
   exists between the data?
   (b) Find a function C that receives the Fahrenheit
   temperature x as input and outputs the corresponding Celsius temperature. Interpret the
   slope.
   (c) If the temperature is 83°F, what is this temperature in degrees Celsius?

*Exercises 17–20: Find the slope-intercept form of the
equation of a line satisfying the following conditions.*

17. Slope 7, passing through $(-3, 9)$

18. Passing through $(2, -4)$ and $(7, -3)$

19. Passing through $(1, -1)$, parallel to $y = -3x + 1$

20. Passing through $(-2, 1)$, perpendicular to $y = 2(x + 5) - 22$

*Exercises 21–24: Find an equation of the line satisfying
the following conditions.*

21. Parallel to the y-axis, passing through $(6, -7)$

22. Parallel to the x-axis, passing through $(-3, 4)$

23. Horizontal, passing through $(1, 3)$

24. Vertical, passing through $(1.5, 1.9)$

25. *Distance from Home* The graph depicts the
mileage y that a person driving a car on a straight
road is from home after x hours. Interpret the graph.
What speeds did the car travel?

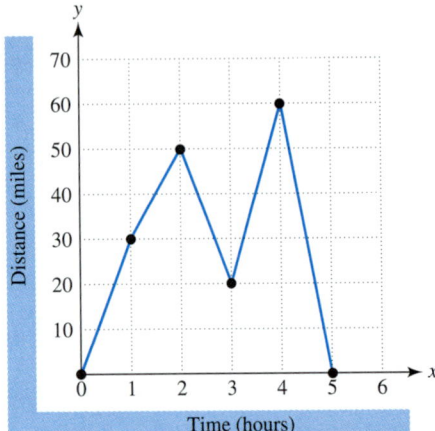

26. *Distance between Bicyclists* The graph shows the
mileage after x hours between two bicyclists traveling in opposite directions along a straight road.
   (a) At what time did the bicycle riders meet?
   (b) When were they 20 miles apart?
   (c) Find the times when they were less than 20
   miles apart.
   (d) Estimate the sum of the speeds of the two
   bicyclists.

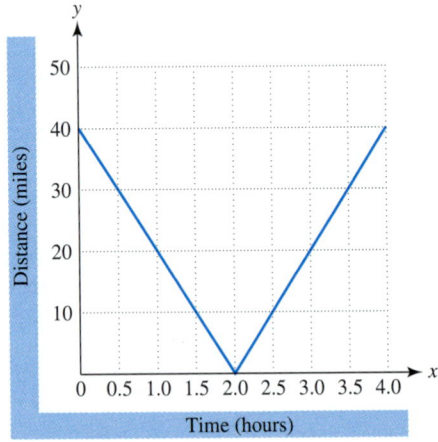

27. Use $f(x)$ to complete the following.

$$f(x) = \begin{cases} 8 + 2x & \text{if } -3 \leq x \leq -1 \\ 5 - x & \text{if } -1 < x \leq 2 \\ x + 1 & \text{if } 2 < x \leq 5 \end{cases}$$

   (a) Evaluate f at $x = -2, -1, 2,$ and 3.
   (b) Sketch a graph of f. Is f continuous?
   (c) Determine the x-value(s) where $f(x) = 3$.

28. *Graphical Model*   A 500-gallon water tank is initially full, then emptied at a constant rate of 50 gallons per minute. Then the tank is filled by a pump that outputs 25 gallons of water per minute. Sketch a graph that depicts the amount of water in the tank after x minutes.

29. Graph $f(x) = [\![x]\!]$ without a calculator on the interval $-4 \le x \le 4$.

30. If $f(x) = [\![2x - 1]\!]$, evaluate $f(-3.1)$ and $f(2.5)$.

Exercises 31–34: Solve the equation graphically, numerically, and symbolically. Use the solutions to help solve the related inequality.

31. $|x| = 3,\quad |x| > 3$

32. $|-3x + 1| = 2,\quad |-3x + 1| < 2$

33. $|3x - 7| = 10,\quad |3x - 7| > 10$

34. $|4 - x| = 6,\quad |4 - x| \le 6$

35. *ACT Scores*   The table lists the average composite ACT scores for selected years. (Source: The American College Testing Program.)

Year	1989	1990	1991	1992	1993	1994
Score	20.6	20.6	20.6	20.6	20.7	20.8

(a) Make a line graph of the data.
(b) Let this graph represent a piecewise-linear function f. Find a symbolic representation for f.

36. Find the midpoint of the line segment connecting the points $(-1, 3)$ and $(13, 23)$.

37. *Population Estimates*   In 1992 the population of a city was 143,247 and in 1996 it was 167,933. Estimate the population in 1994 using the midpoint formula.

38. Find the distance between the points $(-3, 0)$ and $(5, -6)$.

39. *HIV Infections*   In 1996 there were between 650 and 900 thousand HIV infections in the United States with an annual infection rate of 40,000. Assume that this trend continues. (Source: Centers for Disease Control and Prevention.)

(a) Find linear functions f and g that model the maximum and minimum number of HIV infections in the United States in future years.
(b) Graph f and g for $1996 \le x \le 2000$. Explain why their graphs are parallel.
(c) Does the vertical distance between their graphs change for different values of x? What does this distance represent?

40. The table lists data that is exactly linear.
(a) Determine the slope-intercept form of the line that passes through these data points.

(b) Predict y when $x = -1.5$ and 3.5. State whether these calculations involve interpolation or extrapolation.

x	-3	-2	-1	0	1	2	3
y	6.6	5.4	4.2	3	1.8	0.6	-0.6

41. *SAT Scores*   The table lists the average verbal SAT scores for selected years. (Source: The College Board.)

Year	1991	1993	1995
Score	422	424	428

(a) Find a point-slope form of the equation of the line passing through the points (1991, 422) and (1995, 428). Interpret the slope of this line.
(b) Use the line found in part (a) to approximate the average SAT verbal score in 1996. Did this estimate involve interpolation or extrapolation?

42. *Curve Length*   Estimate the length of the curve between the points $(-1, -1)$ and $(1, 1)$ on the graph of $y = x^3$.

43. *Flow Rates*   (Refer to Exercise 70 in Section 2.4.) Suppose the tank is modified so that it has a second inlet pipe, which flows at a rate of two gallons per minute. Interpret the graph by determining when each inlet and outlet pipe is open or closed.

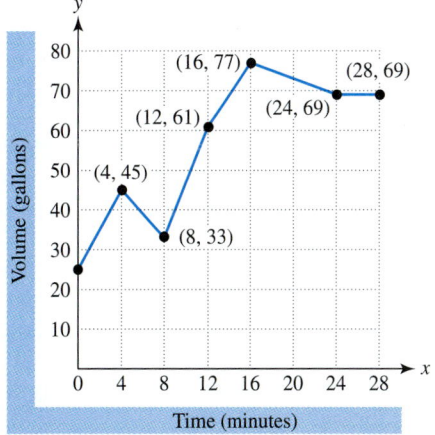

44. *Air Temperature*   For altitudes up to 4 kilometers, moist air will cool at a rate of about 6°C per kilometer. If the ground temperature is 25°C, at what altitudes would the air temperature be from 5°C to 15°C? Solve this graphically and symbolically. (Source: A. Miller and R. Anthes, *Meteorology.*)

Extended and Discovery Exercises

1. *Indoor Air Pollution*   Today, indoor air pollution has become more hazardous as people spend 80 to 90% of their time in tightly sealed, energy-efficient buildings that often lack proper ventilation. Many contaminants such as tobacco smoke, formaldehyde, radon, lead, and carbon monoxide often are allowed to increase to unsafe levels. Mathematics plays a central role in risk assessment of pollutants. If x people are exposed to a contaminant, then the resulting number of cancer cases can sometimes be estimated by $f(x) = Cx$. In this formula, C is a constant that represents the excess cancer risk per year. For example, if $C = 0.01$ then an exposed individual has a one percent greater chance of developing cancer during a period of one year. (Source: A. Hines, *Indoor Air Quality & Control.*)

(a) Environmental tobacco smoke is a health hazard. The EPA has estimated that the annual excess cancer risk is $C = 0.00002$. The population of the United States was approximately 265 million in 1996. Assuming that everyone is exposed to tobacco smoke to some degree, estimate the annual number of cancer cases caused by environmental tobacco smoke.

(b) Formaldehyde is a volatile organic compound that has come to be recognized as a highly toxic indoor air pollutant. Its source is found in building materials such as fiberboard, plywood, foam insulation, and carpeting. Some studies have estimated that $C = 0.0007$ for formaldehyde. In a population of 100,000 people exposed to formaldehyde, how many are likely to develop cancer during a year?

(c) Uranium occurs in small traces in most soils throughout the world. Radon gas occurs naturally in homes and is produced when uranium decays radioactively into lead. Radon enters buildings through basements, water supplies, and natural gas lines. Exposure to radon gas is a known lung cancer risk. According to the EPA, radon causes between 5000 and 20,000 lung cancer cases each year in the United States. Approximate a range for the constant C, if the population of the United States was 250 million when the EPA study was performed.

(d) Discuss why it is reasonable that f is a linear function.

2. *Archeology*   It is possible for archeologists to estimate the height of an adult based only on the length of the humerus, a bone located between the elbow and the shoulder. The approximate relationship between the height y of an individual and the length x of the humerus is shown in the table for both males and females. All measurements are recorded in inches. Tables like this are the result of measuring bones from many skeletons. Individuals may vary from these values.

x	8	9	10	11	12	13	14
y (females)	50.4	53.5	56.6	59.7	62.8	65.9	69.0
y (males)	53.0	56.0	59.0	62.0	65.0	68.0	71.0

(a) Find the estimated height of a female with a 12-inch humerus.

(b) Plot the ordered pairs (x, y) for both sexes. What type of relation exists between the data?

(c) For each one-inch increase in the length of the humerus, what are the corresponding increases in the heights of females and of males?

(d) Determine linear functions f and g that model this data for females and males, respectively.

(e) Suppose part of a broken humerus is estimated to be between 9.7 and 10.1 inches in length. If the sex is not known, use f and g to approximate the range for the height of an individual of each sex.

3. Continuing with Exercise 2, have members of the class measure their heights and the lengths of their humeri (plural of humerus) in inches.

(a) Make a table of the results.

(b) Find regression lines that fit the data points for males and females.

(c) Compare your results with the table in the previous exercise.

4. *Comparing Ages*   The age of Earth is approximately 4.45 billion years old. The earliest evidence of dinosaurs dates back 200 million years, whereas the earliest evidence for *Homo sapiens* dates back 300,000 years. If the age of Earth was condensed into one year, determine the approximate times when dinosaurs and *Homo sapiens* first appeared.

200 million years ago

300,000 years ago

Exercises 5 and 6: *Error Tolerances* Applications involving absolute values occur frequently in error analysis. When a product is manufactured to meet a certain requirement, there is typically an error tolerance specified. For example, if a wheel is required to have a circumference C of 200 inches, then any value for C from 199 to 201 inches might be acceptable. In this case, the circumference C must not differ from 200 by more than ±1 inch. (The ±1 is used to denote that C could be either high or low by 1 inch.) This can be written concisely using the inequality $|C - 200| \leq 1$.

5. The perimeter P of a square having a side with a length of x feet is given by $P = 4x$. Determine the x-values that satisfy the inequality $|P - 1000| < 4$ graphically, numerically, and symbolically. What values of P satisfy the inequality? Interpret your results.

6. The circumference of a circular garden with a radius r is to have a 150-foot fence around it. The length of the fence can vary from 150 feet by at most ±2 feet. What range of values are possible for r?

CHAPTER 3 Nonlinear Functions and Equations

*T*he location of a parked car can be described by a constant function, while the position of a car traveling at a constant velocity can be modeled by a linear function. However, the location of a vehicle moving with a changing velocity must be modeled by a nonlinear function. Neither a linear nor a constant function accurately characterizes this situation. In real-world applications, quantities frequently do not change at a constant rate. As a result, nonlinear functions are necessary to describe and predict events.

Both this chapter and the next present different types of phenomena that are modeled by nonlinear functions. Some examples are the amount of air in a person's lungs, weather, concentrations of a drug in the bloodstream, the spread of AIDS, a person's heart rate, growth of *E. coli* bacteria, use of marijuana among high school seniors, numbers of endangered species, and growth in computer technology. Applications involving nonlinear functions and equations exist in almost every subject area.

Until the development of calculators and computers, many nonlinear problems were either too difficult or impossible to solve by hand. Only recently has technology been able to provide approximate solutions to

> *T*he purpose of computing is insight, not numbers.
>
> — **Richard Hamming**

many complicated and important questions. While technology and visualization provide a new way for us to investigate mathematical problems and the world around us, they are not replacements for mathematical understanding. Fundamental concepts change little over time. The human mind is capable of mathematical insight and decision making, but does not usually perform long arithmetic calculations proficiently. On the other hand, computers and calculators are incapable of possessing mathematical insight, but are excellent at performing arithmetic and other routine computation. In this way, technology complements the human mind.

Reference: R. Wolff and L. Yaeger, *Visualization of Natural Phenomena.*

3.1 Quadratic Equations and Inequalities

Solving Quadratic Equations • Solving Quadratic Inequalities

Introduction

In Chapter 2 applications involving linear functions led to linear equations and inequalities. In a similar manner, applications involving nonlinear functions often result in nonlinear equations and inequalities. Linear equations can always be solved symbolically. This is not the case with some types of nonlinear equations. Their solutions must be approximated using numerical and graphical techniques.

Quadratic equations are one of the simplest types of nonlinear equations. They can be solved symbolically, as well as numerically and graphically.

Solving Quadratic Equations

Any quadratic function f can be represented by $f(x) = ax^2 + bx + c$, where a, b, and c are constants with $a \neq 0$. For example, the quadratic function given by $D(x) = 1986x^2 + 6751x + 5271$ approximates the cumulative number of deaths from AIDS in the United States from 1984 to 1994. In this formula, $a = 1986$, $b = 6751$, and $c = 5271$. The variable x represents the year, where $x = 0$ corresponds to 1984, $x = 1$ to 1985, and so on, until $x = 10$ corresponds to 1994. Once the function D has been found, it is natural to use D to obtain information about AIDS. For example, to estimate the year when the total number of AIDS deaths might reach 500,000, we could solve the following equation.

$$D(x) = 500,000$$

$$1986x^2 + 6751x + 5271 = 500,000 \quad \text{Substitute for } D(x).$$

$$1986x^2 + 6751x - 494,729 = 0 \quad \text{Subtract 500,000 from both sides.}$$

The equation $1986x^2 + 6751x - 494,729 = 0$ is a *quadratic equation.* We solve this equation in Example 6.

Quadratic equation

A **quadratic equation** in one variable is an equation that can be written in the form

$$ax^2 + bx + c = 0,$$

where a, b, and c are real numbers with $a \neq 0$.

MAKING CONNECTIONS
Functions and Equations

Any linear equation can be written as $g(x) = 0$, where g is a linear function given by $g(x) = ax + b$ with $a \neq 0$. Any quadratic equation can be written as $f(x) = 0$, where f is a quadratic function given by $f(x) = ax^2 + bx + c$ with $a \neq 0$.

Quadratic equations can be solved graphically, numerically, and symbolically. Four basic symbolic strategies include factoring, the square root property, completing the square, and the quadratic formula. We begin by discussing factoring.

Factoring.   Factoring is a common technique used to solve equations. It is based on the **zero-product property,** which states that if $ab = 0$, then either $a = 0$ or $b = 0$. It is important to remember that this property only works for 0. For example, if $ab = 1$, then this does *not* imply that either $a = 1$ or $b = 1$, since $a = \dfrac{1}{2}$ and $b = 2$ satisfy $ab = 1$.

EXAMPLE 1   *Solving a quadratic equation with factoring*

Solve the quadratic equation $2x^2 + 2x - 11 = 1$. Check your results.

Solution
Start by writing the equation in the form $ax^2 + bx + c = 0$.

$2x^2 + 2x - 11 = 1$	Given equation
$2x^2 + 2x - 12 = 0$	Subtract 1 from both sides.
$x^2 + x - 6 = 0$	Divide both sides by 2. (Optional step)
$(x + 3)(x - 2) = 0$	Factor.
$x + 3 = 0$ $\quad$ or $\quad$ $x - 2 = 0$	Zero-product property
$x = -3$ $\quad$ or $\quad$ $x = 2$	Solve.

These solutions can be checked by substituting them into the original equation.

$$2(-3)^2 + 2(-3) - 11 \stackrel{?}{=} 1 \qquad 2(2)^2 + 2(2) - 11 \stackrel{?}{=} 1$$
$$1 = 1\sqrt{} \qquad\qquad\qquad 1 = 1\sqrt{}$$

The solutions are -3 and 2. ∎

EXAMPLE 2   *Solving a quadratic equation symbolically, graphically, and numerically*

On the shoreline surrounding certain types of recreational lakes, the Department of Natural Resources (DNR) restricts the sizes of buildings. Storage buildings built within 100 feet of the shoreline are sometimes limited to a maximum area of 120 square feet. Suppose a person is building a rectangular shed that has an area of 120 square feet and is seven feet longer than it is wide. See Figure 3.1 on the next page. Determine its dimensions symbolically, graphically, and numerically.

FIGURE 3.1

Solution

Symbolic Solution Let x be the width of the shed. Then $x + 7$ represents the length. Since area is equal to width times length, we solve the following equation.

$$x(x + 7) = 120 \qquad \text{Area is equal to 120 square feet.}$$
$$x^2 + 7x = 120 \qquad \text{Distributive property}$$
$$x^2 + 7x - 120 = 0 \qquad \text{Subtract 120 from both sides.}$$
$$(x + 15)(x - 8) = 0 \qquad \text{Factor.}$$
$$x + 15 = 0 \quad \text{or} \quad x - 8 = 0 \qquad \text{Zero-product property}$$
$$x = -15 \quad \text{or} \quad x = 8 \qquad \text{Solve.}$$

Since dimensions cannot be negative, the solution that has meaning is $x = 8$. The length is 7 feet longer than the width, so the dimensions of the shed are 8 feet by 15 feet.

Graphical Solution The graphs of $Y_1 = X(X + 7)$ and $Y_2 = 120$ intersect at the points $(-15, 120)$ and $(8, 120)$ as shown in Figures 3.2 and 3.3. The solutions to the equation are -15 and 8.

Numerical Solution Table $Y_1 = X(X + 7)$ and $Y_2 = 120$. The solution of 8 is shown in Figure 3.4, while the solution of -15 may be found by scrolling through the table. The equation $Y_1 = Y_2$ is satisfied when $x = -15$ or 8. The symbolic, graphical, and numerical solutions agree.

$[-20, 20, 5]$ by $[-20, 150, 10]$ $[-20, 20, 5]$ by $[-20, 150, 10]$

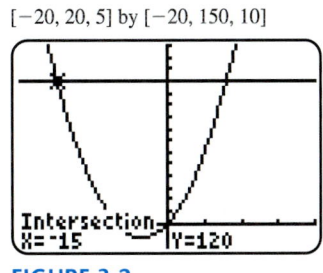 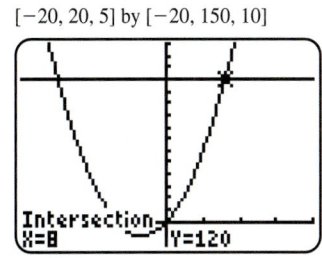

FIGURE 3.2   **FIGURE 3.3**

X	Y₁	Y₂
5	60	120
6	78	120
7	98	120
8	120	120
9	144	120
10	170	120
11	198	120

X=8

FIGURE 3.4

The Square Root Property. Some types of quadratic equations can be written as $x^2 = k$, where k is a positive number. The solutions to this equation are $\pm\sqrt{k}$. For example, $x^2 = 16$ has two solutions, ± 4. We refer to this as the *square root property.*

EXAMPLE 3 *Using the square root property*

If a metal ball is dropped 100 feet from a water tower, its distance above the ground after x seconds is given by $h(x) = 100 - 16x^2$. See Figure 3.5. Determine how long it takes the ball to hit the ground. Support your result graphically and numerically.

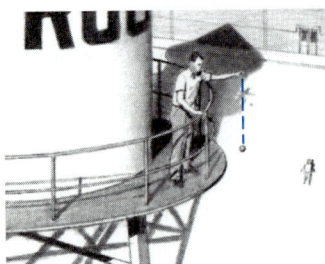

FIGURE 3.5

Solution

The ball strikes the ground when the equation $100 - 16x^2 = 0$ is satisfied.

$$100 - 16x^2 = 0$$

$$100 = 16x^2 \qquad \text{Add } 16x^2 \text{ to both sides.}$$

$$\frac{100}{16} = x^2 \qquad \text{Divide both sides by 16.}$$

$$\pm\frac{10}{4} = x \qquad \text{Square root property}$$

$$x = \pm\frac{5}{2} \qquad \text{Simplify.}$$

In this example only positive values for time are valid. Thus, after 2.5 seconds the ball strikes the ground.

Graphical support is shown in Figures 3.6 and 3.7. The x-intercepts for $Y_1 = 100 - 16X\text{^}2$ are ± 2.5. For numerical support, table $Y_1 = 100 - 16X\text{^}2$ as shown in Figure 3.8. The equation $Y_1 = 0$ is satisfied when $x = \pm 2.5$.

[−5, 5, 1] by [−30, 150, 10]

FIGURE 3.6

[−5, 5, 1] by [−30, 150, 10]

FIGURE 3.7

X	Y1
-3.5	-96
-2.5	0
-1.5	64
-.5	96
.5	96
1.5	64
2.5	0

Y1 = 100−16X²

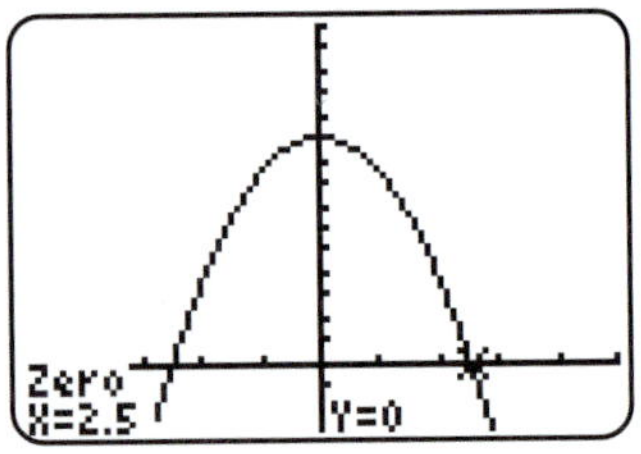

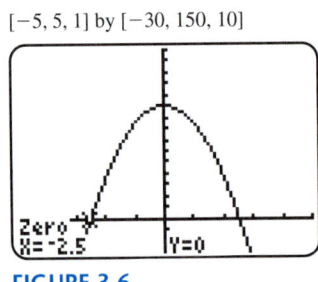

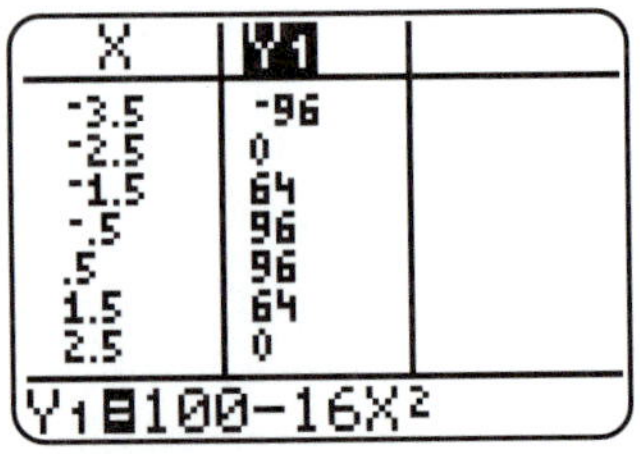

FIGURE 3.8 ■

Completing the Square. Another technique that may be used to solve a quadratic equation is *completing the square*. If a quadratic equation can be written in the form $x^2 + kx = d$, where k and d are constants, then the equation can be solved using the fact that

$$x^2 + kx + \left(\frac{k}{2}\right)^2 = \left(x + \frac{k}{2}\right)^2.$$

For example, $k = 6$ in the equation $x^2 + 6x = 7$, so add $\left(\frac{k}{2}\right)^2 = \left(\frac{6}{2}\right)^2 = 9$ to both sides.

$x^2 + 6x = 7$	Original equation
$x^2 + 6x + 9 = 7 + 9$	Add 9 to both sides.
$(x + 3)^2 = 16$	Factor.
$x + 3 = \pm 4$	Square root property
$x = -3 \pm 4$	Subtract 3 from both sides.
$x = -7 \quad$ or $\quad x = 1$	Simplify.

Completing the square can be applied to quadratic equations that do not factor easily.

EXAMPLE 4   *Completing the square*

Solve the equation $x^2 - 8x + 9 = 0$.

Solution
Start by writing the equation in the form $x^2 + kx = d$.

$x^2 - 8x + 9 = 0$	Given equation
$x^2 - 8x = -9$	Subtract 9 from both sides.
$x^2 - 8x + 16 = -9 + 16$	Add $\left(\frac{k}{2}\right)^2 = \left(\frac{-8}{2}\right)^2 = 16.$
$(x - 4)^2 = 7$	Factor.
$x - 4 = \pm\sqrt{7}$	Square root property
$x = 4 \pm \sqrt{7}$	Solve. ∎

The Quadratic Formula. The quadratic formula can be used to find the solutions to any quadratic equation. Its derivation is based on completing the square and left as an exercise.

Quadratic formula

The solutions of the quadratic equation $ax^2 + bx + c = 0$, where $a \neq 0$, are

$$x = \frac{-b \pm \sqrt{b^2 - 4ac}}{2a}.$$

EXAMPLE 5 *Using the quadratic formula*

Solve the equation $3x^2 - 6x + 2 = 0$.

Solution

Let $a = 3$, $b = -6$, and $c = 2$.

$$x = \frac{-b \pm \sqrt{b^2 - 4ac}}{2a}$$ Quadratic formula

$$x = \frac{6 \pm \sqrt{(-6)^2 - 4(3)(2)}}{2(3)}$$ Substitute for a, b, and c.

$$x = \frac{6 \pm \sqrt{12}}{6}$$ Simplify.

$$x = 1 \pm \frac{1}{6}\sqrt{12}$$ Divide.

$$x = 1 \pm \frac{1}{3}\sqrt{3}$$ $\sqrt{12} = \sqrt{4 \cdot 3} = 2\sqrt{3}$ ■

Critical Thinking

Use the results of Example 5 to evaluate each expression mentally.

$$3\left(1 + \frac{1}{3}\sqrt{3}\right)^2 - 6\left(1 + \frac{1}{3}\sqrt{3}\right) + 2$$

$$3\left(1 - \frac{1}{3}\sqrt{3}\right)^2 - 6\left(1 - \frac{1}{3}\sqrt{3}\right) + 2$$

EXAMPLE 6 *Estimating AIDS deaths*

Solve the quadratic equation $1986x^2 + 6751x - 494{,}729 = 0$ (as discussed earlier) to estimate when the total number of AIDS deaths may reach 500,000. Use graphical, numerical, and symbolic methods.

Solution

Graphical Solution Graph $Y_1 = 1986X^2 + 6751X - 494729$. Two x-intercepts, located at $x \approx -17.6$ and $x \approx 14.2$, are shown in Figures 3.9 and 3.10. Since $x = 0$ corresponds to 1984, $x = -17.6$ represents $1984 - 17.6 = 1966.4$ and $x = 14.2$ represents $1984 + 14.2 = 1998.2$. AIDS was unknown in 1966, so this model estimates that about 500,000 AIDS deaths may be reported by 1998.

$[-30, 30, 10]$ by $[-6 \times 10^5, 6 \times 10^5, 10^5]$ $[-30, 30, 10]$ by $[-6 \times 10^5, 6 \times 10^5, 10^5]$

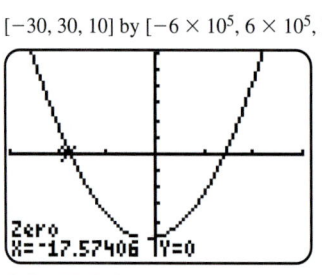 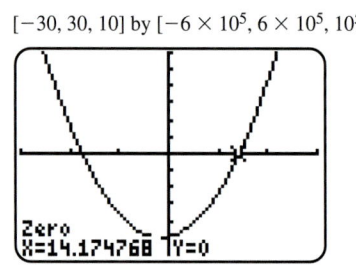

FIGURE 3.9   **FIGURE 3.10**

Numerical Solution Table Y_1 incrementing by 0.5 as shown in Figure 3.11. The values for Y_1 change sign from positive to negative between $x = -18$ and -17.5. By the intermediate value property, there is a solution between these values. Similarly, in Figure 3.12 the values of Y_1 change from negative to positive between $x = 14$ and $x = 14.5$.

X	Y1
-19.5	128803
-19	93948
-18.5	60086
-18	27217
-17.5	-4659
-17	-35542
-16.5	-65432

X=-17.5

FIGURE 3.11

X	Y1
12.5	-1E5
13	-71332
13.5	-41642
14	-10959
14.5	20717
15	53386
15.5	87048

X=14

FIGURE 3.12

Symbolic Solution Apply the quadratic formula with $a = 1986$, $b = 6751$, and $c = -494{,}729$.

$$x = \frac{-b \pm \sqrt{b^2 - 4ac}}{2a}$$

$$x = \frac{-6751 \pm \sqrt{6751^2 - 4(1986)(-494{,}729)}}{2(1986)}$$

$$x = \frac{-6751 \pm \sqrt{3{,}975{,}703{,}177}}{3972}$$

$$x \approx 14.2 \ \text{ or } \ -17.6$$

Again we obtain two solutions. The solution in the future is $1984 + 14.2 \approx 1998$. ■

The Discriminant. One important difference between linear and nonlinear equations is that nonlinear equations may have more than one solution. If the quadratic equation $ax^2 + bx + c = 0$ is solved graphically, the parabola $y = ax^2 + bx + c$ can intersect the x-axis 0, 1, or 2 times as illustrated in Figures 3.13–3.15, respectively. Each x-intercept is a real solution to the quadratic equation $ax^2 + bx + c = 0$.

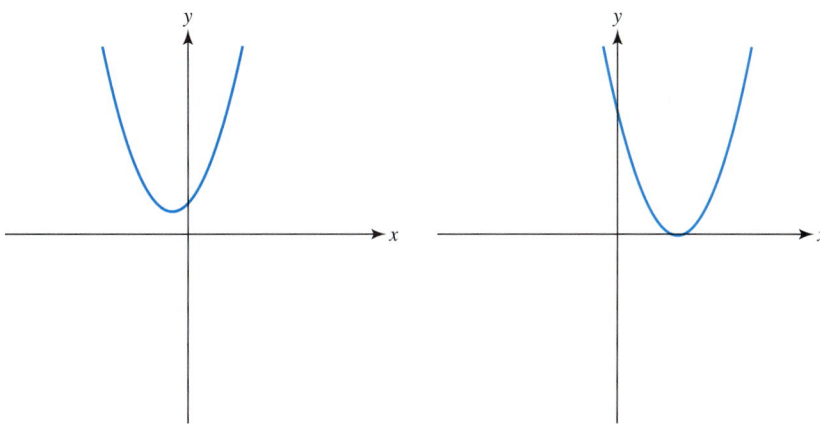

FIGURE 3.13 No Real Solutions

FIGURE 3.14 One Real Solution

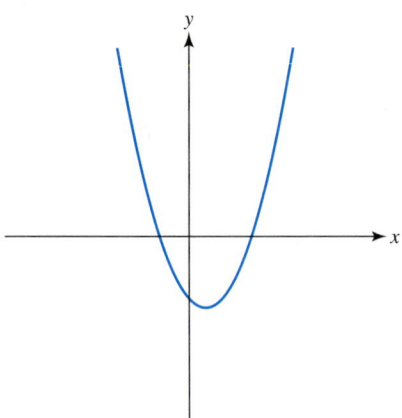

FIGURE 3.15 Two Real Solutions

The quantity $b^2 - 4ac$ in the quadratic formula is called the **discriminant.** It provides information about the number of solutions to a quadratic equation.

> ## Quadratic equations and the discriminant
>
> To determine the number of real solutions to $ax^2 + bx + c = 0$, with $a \neq 0$, evaluate the discriminant $b^2 - 4ac$.
>
> **1.** If $b^2 - 4ac > 0$, there are two real solutions.
> **2.** If $b^2 - 4ac = 0$, there is one real solution.
> **3.** If $b^2 - 4ac < 0$, there are no real solutions.
>
> *Note:* When $b^2 - 4ac < 0$, the solutions to a quadratic equation may be expressed as complex numbers. This is discussed in Section 3.5.

In Example 6 the discriminant is

$$b^2 - 4ac = 6751^2 - 4(1986)(-494,729) = 3,975,703,177 > 0.$$

Since the discriminant is positive, there are two real solutions.

EXAMPLE 7   *Using the discriminant*

Use the discriminant to determine the number of solutions to the quadratic equation $9x^2 - 12.6x + 4.41 = 0$. Then solve the equation using the quadratic formula. Support your result graphically.

Solution
Let $a = 9$, $b = -12.6$, and $c = 4.41$. The discriminant is given by

$$b^2 - 4ac = (-12.6)^2 - 4(9)(4.41) = 0.$$

Since the discriminant is 0, there is one solution.

$$x = \frac{-b \pm \sqrt{b^2 - 4ac}}{2a} \qquad \text{Quadratic formula}$$

$$x = \frac{-(-12.6) \pm \sqrt{0}}{18} \qquad \text{Substitute.}$$

$$x = 0.7 \qquad \text{Simplify.}$$

The only solution is 0.7. A graph of $y = 9x^2 - 12.6x + 4.41$ is shown in Figure 3.16. Notice that the graph suggests that there is one x-intercept corresponding to 0.7.

[0, 1.5, 0.1] by [−0.5, 1, 0.1]

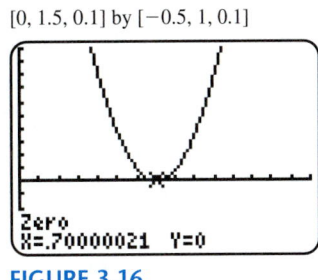

FIGURE 3.16   ■

Solving Quadratic Inequalities

A quadratic equation can be written as $ax^2 + bx + c = 0$. If the equals sign is replaced by $>$, $\geq$, $<$, or $\leq$ a quadratic inequality results. Since equality often is the boundary between *greater than* and *less than*, a first step in solving a quadratic inequality is to determine the x-values where equality occurs. These x-values are the *boundary numbers* (see Section 2.2). We begin by discussing numerical and graphical solutions of quadratic inequalities.

A well-conditioned athlete's heart rate can reach 200 beats per minute (bpm) during strenuous physical activity. Upon quitting, a typical heart rate decreases rapidly at first and then gradually levels off as shown in Table 3.1.

TABLE 3.1

Time (min)	0	2	4	6	8
Heart Rate (bpm)	200	150	115	90	80

Adapted from: V. Thomas, *Science and Sport.*

EXAMPLE 8   *Solving quadratic inequalities numerically and graphically*

The heart rate shown in Table 3.1 can be modeled by $R(x) = \frac{15}{8}(x - 8)^2 + 80$, where x represents time and $0 \leq x \leq 8$.

(a) Numerically estimate when the heart rate was from 125 to 160.

(b) Solve part (a) graphically.

Solution

(a) Table $Y_1 = (15/8)(X - 8)^2 + 80$ as shown in Figure 3.17. Use the test values in the table to estimate when $125 \leq Y_1 \leq 160$. The heart rate is 160 bpm after about 1.5 minutes and 125 bpm after about 3 minutes. Thus, the heart rate varies from 125 to 160 bpm when $k_1 \leq x \leq k_2$, where $k_1 \approx 1.5$ minutes and $k_2 \approx 3$ minutes.

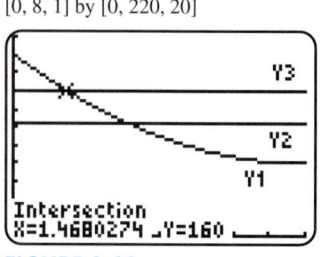

[0, 8, 1] by [0, 220, 20]

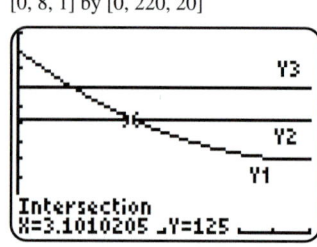

[0, 8, 1] by [0, 220, 20]

FIGURE 3.17   **FIGURE 3.18**   **FIGURE 3.19**

(b) Graph $Y_1 = (15/8)(X - 8)^2 + 80$, $Y_2 = 125$, and $Y_3 = 160$. See Figures 3.18 and 3.19. Their graphs intersect near the points (1.47, 160) and (3.10, 125). The graph of Y_1 lies between the graphs of Y_2 and Y_3 when $k_1 \leq x \leq k_2$, where $k_1 \approx 1.47$ and $k_2 \approx 3.10$. ∎

In Section 2.4 we solved inequalities involving $|ax + b|$. Quadratic inequalities can be solved similarly. Any quadratic inequality can be written as $ax^2 + bx + c > 0$, $ax^2 + bx + c \geq 0$, $ax^2 + bx + c < 0$, or $ax^2 + bx + c \leq 0$, where $a > 0$. In each case the graph of $y = ax^2 + bx + c$ is a parabola opening upward. For example, Figure 3.20 shows the parabola $y = x^2 - x - 6$.

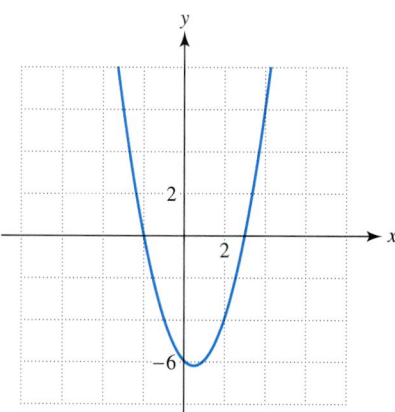

FIGURE 3.20

Since $a = 1 > 0$, this parabola opens upward. It has x-intercepts -2 and 3, which satisfy the equation $x^2 - x - 6 = 0$. The parabola lies below the x-axis between the intercepts (or boundary numbers), so the solution set to the inequality $x^2 - x - 6 < 0$ is $\{x \mid -2 < x < 3\}$. Similarly, the solutions to

$x^2 - x - 6 > 0$ include x-values either left of $x = -2$ or right of $x = 3$, where the parabola is above the x-axis. This solution set is $\{x \mid x < -2 \text{ or } x > 3\}$.

This is supported by the accompanying figure. When $x = -2$ or 3, the expression $x^2 - x - 6$ equals 0. It is negative for test values between these zeros and positive for test values less than $x = -2$ or greater than $x = 3$.

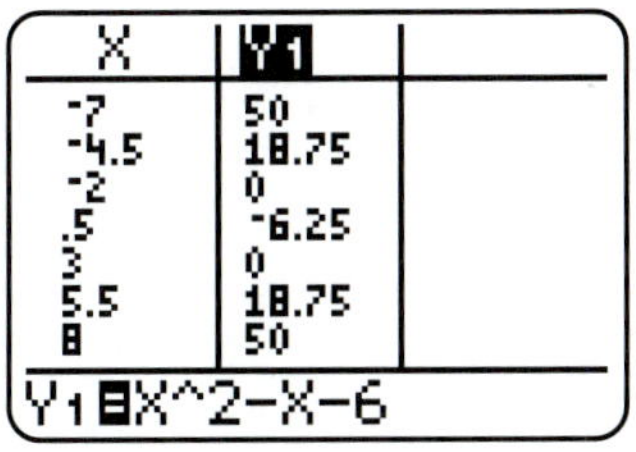

X	Y1
-7	50
-4.5	18.75
-2	0
.5	-6.25
3	0
5.5	18.75
8	50

Y1⬛X^2-X-6

Interval	Test Value	$f(x) = x^2 - x - 6$
$x < -2$	$x = -7$	$f(-7) = 50 > 0$
$-2 < x < 3$	$x = 0.5$	$f(0.5) = -6.25 < 0$
$x > 3$	$x = 8$	$f(8) = 50 > 0$

This discussion is summarized.

Solving quadratic inequalities

Let $ax^2 + bx + c = 0$ with $a > 0$ have distinct solutions s_1 and s_2, where $s_1 < s_2$.

$$ax^2 + bx + c < 0 \text{ is equivalent to } s_1 < x < s_2.$$
$$ax^2 + bx + c > 0 \text{ is equivalent to } x < s_1 \text{ or } x > s_2.$$

Quadratic inequalities involving $\le$ or $\ge$ can be solved similarly. The next example illustrates this technique.

EXAMPLE 9 *Solving quadratic inequalities symbolically*

Solve each inequality. Support your results graphically.
(a) $3x^2 + 8x - 3 < 0$
(b) $x(4 - x) \le 3$
(c) $2 + 3x^2 > x$

Solution
(a) Begin by solving the equation $3x^2 + 8x - 3 = 0$.

$$3x^2 + 8x - 3 = 0$$
$$(x + 3)(3x - 1) = 0$$
$$x = -3 \qquad \text{or} \qquad x = \frac{1}{3}$$

The x-values satisfying $3x^2 + 8x - 3 < 0$ lie between $x = -3$ and $x = \frac{1}{3}$. The solution set is $\left\{ x \mid -3 < x < \frac{1}{3} \right\}$. To support this result, graph $Y_1 = 3X^2 + 8X - 3$ as in Figure 3.21. Notice that the parabola lies below the x-axis between the x-intercepts.

(b) Write the inequality so that $a > 0$.

$$x(4 - x) \leq 3 \qquad \text{Given inequality}$$
$$4x - x^2 \leq 3 \qquad \text{Distributive property}$$
$$-x^2 + 4x - 3 \leq 0 \qquad \text{Subtract 3 from both sides.}$$
$$x^2 - 4x + 3 \geq 0 \qquad \text{Multiply by } -1. \text{ Reverse the inequality.}$$

Next, solve the equation $x^2 - 4x + 3 = 0$.

$$x^2 - 4x + 3 = 0$$
$$(x - 1)(x - 3) = 0$$
$$x = 1 \qquad \text{or} \qquad x = 3$$

The solution set for $x^2 - 4x + 3 \geq 0$ is $\{x \mid x \leq 1 \text{ or } x \geq 3\}$. The test values shown in the accompanying figure support this result. The expression $Y_1 = X(4 - X)$ is less than or equal to $Y_2 = 3$ when $x \leq 1$ or $x \geq 3$.

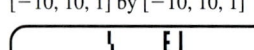

Graphical support is shown in Figure 3.22. Notice that the parabola $y = x^2 - 4x + 3$ lies above the x-axis left of $x = 1$ or right of $x = 3$.

(c) Write the inequality $2 + 3x^2 > x$ as $3x^2 - x + 2 > 0$. If we attempt to solve the equation $3x^2 - x + 2 = 0$ by factoring we are unsuccessful. The discriminant is

$$b^2 - 4ac = (-1)^2 - 4(3)(2) = -23 < 0.$$

Since the discriminant is negative, there are no real solutions. The graph of $y = 3x^2 - x + 2$ is a parabola opening upward, with no x-intercepts. See Figure 3.23. It always lies above the x-axis. Therefore, the solution set for $3x^2 - x + 2 > 0$ includes all real numbers.

$[-10, 10, 1]$ by $[-10, 10, 1]$

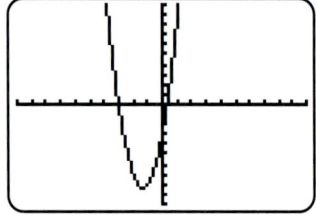

FIGURE 3.21

$[-10, 10, 1]$ by $[-10, 10, 1]$

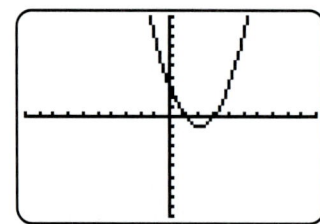

FIGURE 3.22

$[-10, 10, 1]$ by $[-10, 10, 1]$

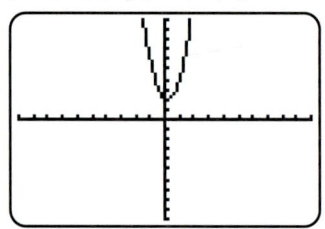

FIGURE 3.23   ∎

Critical Thinking —————————————————————————————

The graph of $y = ax^2 + bx + c$ with $a < 0$ is a parabola opening downward. Suppose this parabola has x-intercepts of -4 and 3. Visualize this graph and solve the following quadratic inequalities.

    i. $ax^2 + bx + c > 0$ **ii.** $ax^2 + bx + c < 0$

3.1 PUTTING IT ALL TOGETHER

Quadratic equations are nonlinear equations. Four basic symbolic methods for solving quadratic equations are factoring, the square root property, completing the square, and the quadratic formula. Graphical and numerical methods also can be applied to quadratic equations.

Quadratic inequalities can be solved by first determining the boundary numbers (the x-values where equality occurs). The graph of $y = ax^2 + bx + c$ is a parabola opening upward if $a > 0$ and opening downward if $a < 0$. Visualizing the general shape of a parabola together with its x-intercepts can be a valuable aid when solving quadratic inequalities.

The following table summarizes some important concepts regarding quadratic equations and inequalities.

Concept	Example	Comments
Quadratic equation $ax^2 + bx + c = 0$	$2x^2 + 3x - 2 = 0$ $(x + 2)(2x - 1) = 0$ $x = -2$ or $x = \frac{1}{2}$	May have 0, 1, or 2 real solutions. Can be solved graphically, numerically, and symbolically.
Square root property $x^2 = k$ is equivalent to $x = \pm\sqrt{k}$.	$x^2 = 9$ is equivalent to $x = \pm 3$.	Can be used to solve a quadratic equation that does not have an x-term.
Quadratic formula $x = \dfrac{-b \pm \sqrt{b^2 - 4ac}}{2a}$	To solve the equation $-x^2 + 4x + 5 = 0$ let $a = -1$, $b = 4$, $c = 5$, and simplify.	Although factoring can sometimes be faster, the quadratic formula always provides the exact solution set.
Discriminant $b^2 - 4ac$	$2x^2 + 5x - 6 = 0$ $5^2 - 4(2)(-6) = 73$	$b^2 - 4ac > 0$ implies 2 real solutions. $b^2 - 4ac = 0$ implies 1 real solution. $b^2 - 4ac < 0$ implies 0 real solutions.
Quadratic inequality $ax^2 + bx + c > 0$ (or $\geq$, $<$, $\leq$)	$x^2 - 5x + 4 \leq 0$	Quadratic inequalities can be solved symbolically, graphically, and numerically. Begin by solving the equation $ax^2 + bx + c = 0$.

3.1 EXERCISES

   Tape 3

Quadratic Equations

Exercises 1–4: The graph of $f(x) = ax^2 + bx + c$ is shown in the figure.
- **(a)** *State whether $a > 0$ or $a < 0$.*
- **(b)** *Solve the equation $ax^2 + bx + c = 0$.*
- **(c)** *Determine if the discriminant is positive, negative, or zero.*

1.

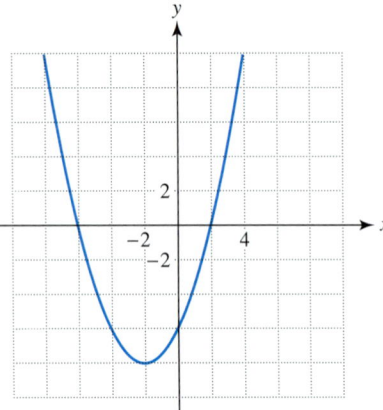

2.

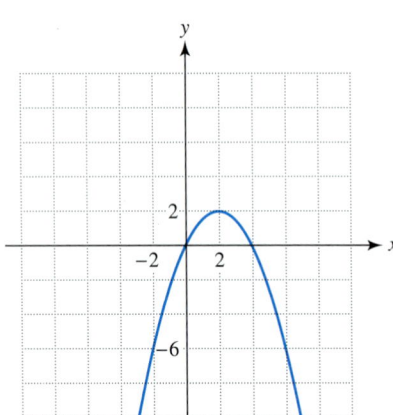

3.

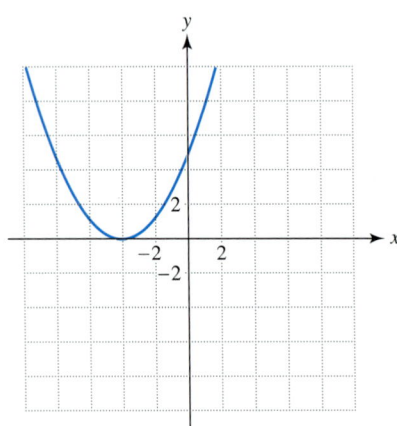

4.

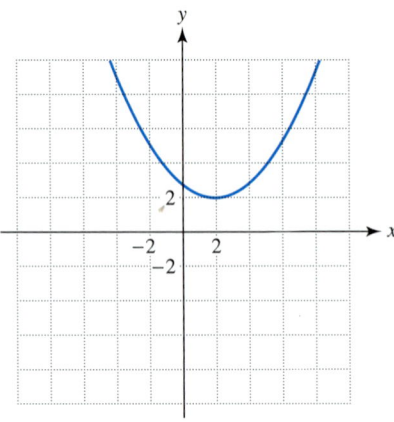

Exercises 5–14: Solve each quadratic equation
- **(a)** *graphically,*
- **(b)** *numerically, and*
- **(c)** *symbolically.*

Express graphical and numerical solutions to the nearest tenth when appropriate.

5. $x^2 - x - 6 = 0$ **6.** $2x^2 + 5x - 3 = 0$

7. $2x^2 = 6$ **8.** $x^2 - 225 = 0$

9. $4x^2 - 12x = -9$ **10.** $-4x(x - 1) = 1$

11. $4x - 2 = x^2$ **12.** $2x^2 - 3x - 3 = 0$

13. $x^2 + 1 = x$ **14.** $2x^2 + x = 2$

Exercises 15–20: Solve the quadratic equation graphically.

15. $20x^2 + 11x = 3$

16. $x^2 - 3.1x = 0.32$

17. $2.5x^2 = 4.75x - 2.1$

18. $-2x^2 + 4x = 1.595$

19. $x(x + 24) = 6912$

20. $x(31.4 - 2x) = -348$

Exercises 21–32: Complete the following.
- **(a)** *Write each equation as $ax^2 + bx + c = 0$.*
- **(b)** *Calculate the discriminant $b^2 - 4ac$ and determine the number of real solutions.*
- **(c)** *Solve each equation symbolically. Support your answer graphically.*

21. $3x^2 = 12$ **22.** $8x^2 - 2 = 14$

23. $x^2 - 2x = -1$ **24.** $6x^2 = 4x + 2$

25. $2x^2 + 3x = 12 - 2x$ **26.** $3x^2 - 2 = 5x$

27. $\frac{1}{4}x^2 + 3x = x - 4$ **28.** $9x(x - 4) = -36$

29. $x\left(\dfrac{1}{2}x + 1\right) = -\dfrac{13}{2}$

30. $4x = 6 + x^2$

31. $3x^2 = 1 - x$

32. $x(5x - 3) = 1$

Exercises 33–36: Solve the equation by completing the square.

33. $x^2 + 4x = 6$

34. $x^2 - 10x = 1$

35. $x^2 + 5x - 4 = 0$

36. $2x^2 + 4x - 5 = 0$

Applications Involving Quadratic Equations

37. *Safe Runway Speed* The road (or taxiway) used by aircraft to exit from the runway should not contain sharp curves. The safe radius for any curve depends on the taxiing speed of an airplane. The table lists the recommended minimum radius R of these exit curves, where the taxiing speed of the airplane is x miles per hour. (Source: Federal Aviation Administration.)

x (mph)	10	20	30	40	50	60
R (ft)	50	200	450	800	1250	1800

  (a) If the taxiing speed x of the plane doubles, what happens to the minimum radius R of the curve?

  (b) The Federal Aviation Administration (FAA) used $R(x) = ax^2$ to compute the values in the table. Determine the constant a.

  (c) If $R = 500$, find x. Interpret your results.

38. *Falling Object* The table lists the velocity and distance traveled by a falling object.

Elapsed Time (sec)	0	1	2	3	4	5
Velocity (ft/sec)	0	32	64	96	128	160
Distance (ft)	0	16	64	144	256	400

  (a) Make a scatterplot of the ordered pairs (time, velocity) and (time, distance) in the same viewing rectangle $[-1, 6, 1]$ by $[-10, 450, 50]$.

  (b) Find a function v that models the velocity.

  (c) The distance is modeled by $d(x) = ax^2$. Find the constant a.

  (d) Use the table to mentally estimate the elapsed time when the distance traveled by the falling object is 200 feet. Determine this time symbolically using d. Find the velocity at this time.

39. *Airline Passengers* The number of worldwide airline passengers y between 1950 and 1990 in millions is approximated in the table. The variable x represents the year, where $x = 0$ corresponds to 1950 and $x = 40$ to 1990. (Source: International Civil Aviation Organization.)

x	0	10	20	30	40
y	31	106	386	734	1273

This data may be modeled by $f(x) = 0.68x^2 + 3.8x + 24$.

  (a) Graph f and the data in $[-5, 45, 5]$ by $[0, 1500, 100]$.

  (b) Determine graphically the year when there were 1 billion passengers.

  (c) Solve part (b) symbolically.

40. *Pollution* Air pollution has not only occurred in the United States, but also in European countries. The table lists the emissions of sulfur S and nitrogen N in millions of tons for various years in the United Kingdom. (Source: C. Mason, *Biology of Freshwater Pollution.*)

Year	1950	1960	1970	1980
S	2.3	2.8	3.0	2.4
N	0.3	0.4	0.5	0.6

  (a) Two functions are given, where x represents the year.

$$f_1(x) = -0.00275x^2 + 10.8125x - 10,625.2$$
$$f_2(x) = 0.01x - 19.2$$

One function models sulfur emissions, while the other models nitrogen emissions. Mentally match the functions with the correct data.

  (b) Support your answers by graphing each function with the appropriate data.

  (c) Estimate the year(s) when sulfur emissions totaled 2.5 million tons.

41. *Biology* Some types of worms have a remarkable capacity to live without moisture. The table shows the number of worms y surviving after x days in one study. (Source: D. Brown and P. Rothery, *Models in Biology: Mathematics, Statistics and Computing.*)

x (days)	0	20	40	80	120	160
y (worms)	50	48	45	36	20	3

(a) Find a quadratic function in the form $f(x) = a(x - h)^2 + k$ that models this data.

(b) Solve the quadratic equation $f(x) = 0$. Do both solutions have real meaning? Explain.

42. *Stopping Distance* Braking distance for cars may be approximated by $D(x) = \dfrac{x^2}{30k}$. The input x is the car's velocity in miles per hour and the output $D(x)$ is the braking distance in feet. The positive constant k is a measure of the traction of the tires. Small values of k indicate a slippery road. (Source: L. Haefner, *Introduction to Transportation Systems*.)

(a) Let $k = 0.3$. Table D starting at $x = 10$, incrementing by 10, until $x = 70$.

(b) Use the table to evaluate $D(60)$. Interpret the result.

(c) If the speed of a car doubles, does its braking distance double? Explain.

43. *Screen Dimensions* The width of a rectangular computer screen is 2.5 inches more than its height. If the area of the screen is 93.5 square inches, determine its dimensions

(a) graphically,

(b) numerically, and

(c) symbolically.

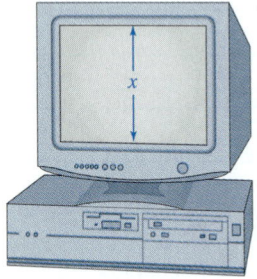

44. *Maximizing Area* A rectangular pen for a pet is under construction using 100 feet of fence.

(a) Determine graphically, numerically, and symbolically the dimensions that result in an area of 576 square feet.

(b) Find the dimensions that give maximum area.

Quadratic Inequalities

Exercises 45–48: The graph of f is shown in the figure. Solve each inequality.

45. **(a)** $f(x) < 0$
   (b) $f(x) \geq 0$

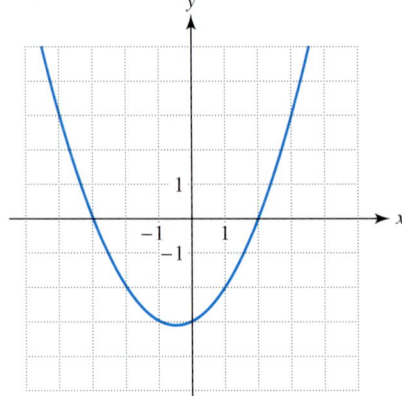

46. **(a)** $f(x) > 0$
   (b) $f(x) < 0$

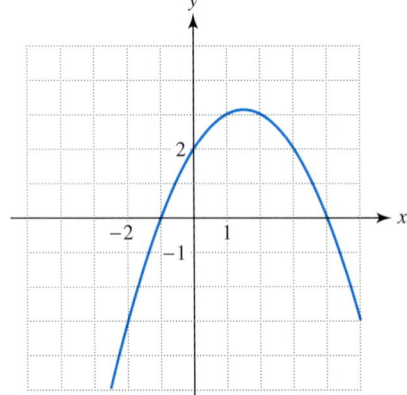

47. **(a)** $f(x) \leq 0$
   (b $f(x) > 0$

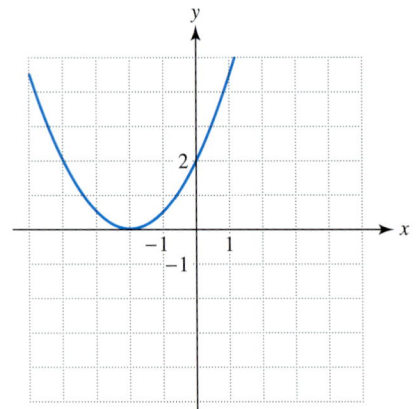

48. (a) $f(x) \geq 0$
(b) $f(x) \leq 0$

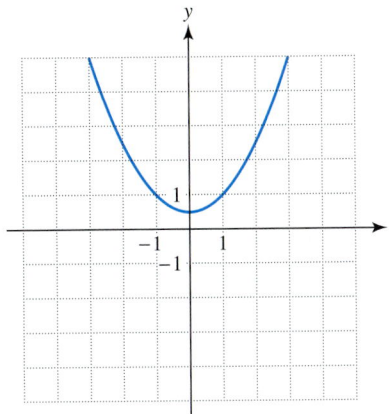

Exercises 49–52: Solve the quadratic inequality
(a) *graphically,*
(b) *numerically, and*
(c) *symbolically.*

49. $x^2 - 3x - 4 < 0$ **50.** $2x^2 + 5x + 2 \leq 0$

51. $x^2 + x > 6$ **52.** $-3x \geq 9 - 12x^2$

Exercises 53–60: Solve each quadratic inequality symbolically. Support your results graphically.

53. $x^2 \leq 4$ **54.** $2x^2 > 16$

55. $x(x - 4) \geq -4$ **56.** $x^2 - 3x - 10 < 0$

57. $-x^2 + x + 6 \leq 0$ **58.** $-x^2 - 2x + 8 > 0$

59. $6x^2 - x < 1$ **60.** $5x^2 \leq 10 - 5x$

You Decide the Method

Exercises 61–64: Solve the quadratic inequality.

61. $x^2 - 5 \leq 0$

62. $0.5x^2 - 3.2x > -0.9$

63. $7x^2 + 515.2 \geq 179.8x$

64. $-10 < 3x - x^2$

Applications Involving Quadratic Inequalities

65. *AIDS*   Use the quadratic function D presented in this section to estimate graphically the years when the total number of AIDS deaths was from 150,000 to 225,000.

66. *Photosynthesis*   (Refer to Example 3 in Section 1.5.) Use the function f to estimate graphically the temperatures where the rate of photosynthesis was 80% or more.

67. *Cellular Phones*   Our society is in transition from an industrial to an informational society. Cellular communication has played an increasingly large role in this transition. The number of cellular sub-

scribers in thousands from 1985 to 1991 can be modeled by $f(x) = 163x^2 - 146x + 205$, where x is the year and $x = 0$ corresponds to 1985. (*Source:* M. Paetsch, *Mobile Communication in the U.S. and Europe.*)
(a) Write a quadratic inequality whose solution set represents the years when there were over 2 million subscribers.
(b) Solve this inequality graphically.

68. *Air Density*   As the altitude increases, air becomes thinner or less dense. An approximation to the density of air at an altitude of x meters above sea level is given by

$$d(x) = (3.32 \times 10^{-9})x^2 - (1.14 \times 10^{-4})x + 1.22.$$

The output is the density of air in kilograms per cubic meter. The domain of d is $0 \leq x \leq 10{,}000$. (*Source:* A. Miller and J. Thompson, *Elements of Meteorology.*)
(a) Denver is sometimes referred to as the mile-high city. Compare the density of air at sea level and in Denver. (*Hint:* 1 ft ≈ 0.305 m)
(b) Determine graphically the altitudes where the density ranges from 0.5 to 1 kilogram per cubic meter.

69. *Modeling Water Flow*   A cylindrical container measuring 16 centimeters high and 12.7 centimeters in diameter with a 0.5-centimeter hole at the bottom was completely filled with water. As water leaked out, the height of the water level inside the container was measured every 15 seconds. The results of the experiment appear in the accompanying table.

Time (sec)	0	15	30	45	60	75	90
Height (cm)	16	13.8	11.6	9.8	8.1	6.6	5.3

Time (sec)	105	120	135	150	165	180
Height (cm)	4.1	3.1	2.3	1.4	0.8	0.5

(a) Explain why this data cannot be modeled by a linear function.
(b) Use the table to estimate when the height of the water was from 5 to 10 centimeters.
(c) The data can be modeled by $f(x) = 0.0003636x^2 - 0.1511x + 15.92$, where x represents the time. Graph f and the data in the same viewing rectangle.
(d) Solve part (b) using $f(x)$.

70. *Quadratic Formula*   Prove the quadratic formula by completing the following.

(a) Write $ax^2 + bx + c = 0$ as $x^2 + \dfrac{b}{a}x = -\dfrac{c}{a}$.

(b) Complete the square to obtain

$$\left(x + \frac{b}{2a}\right)^2 = \frac{b^2 - 4ac}{4a^2}.$$

(c) Use the square root property and solve for x.

Writing about Mathematics

1. Discuss three different symbolic methods for solving a quadratic equation. Make up a quadratic equation and use each method to find the solution set.

2. Explain how to determine the solution set for the inequality $ax^2 + bx + c < 0$, where $a > 0$. How would the solution set change if $a < 0$?

3.2 Nonlinear Functions and Their Graphs

Extrema of Nonlinear Functions • Symmetry • Power Functions

Introduction

In Chapter 1 two basic types of data, linear and nonlinear, were introduced. When linear data is plotted, it lies on a straight line. Unlike linear data, nonlinear data can increase over one interval and decrease on another. It has been necessary for mathematicians to invent a wide variety of nonlinear functions to model the different types of nonlinear data. In fact, new functions are being created by the hundreds every year.

In this section we discuss nonlinear functions. We begin by describing the graphs of nonlinear functions.

Extrema of Nonlinear Functions

In Section 1.4, polynomial functions of degree 2 or higher were introduced as examples of nonlinear functions. Symbolic representations of two polynomial functions are

$$f(x) = \frac{1}{2}x^2 - 2x - 4 \qquad \text{and} \qquad g(x) = -\frac{1}{4}x^4 + \frac{2}{3}x^3 + \frac{5}{2}x^2 - 6x.$$

Their graphs are shown in Figures 3.24 and 3.25. Some values have been rounded.

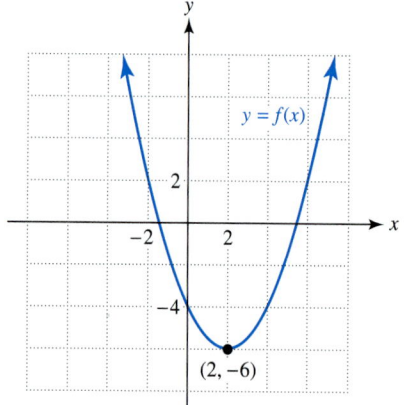

FIGURE 3.24

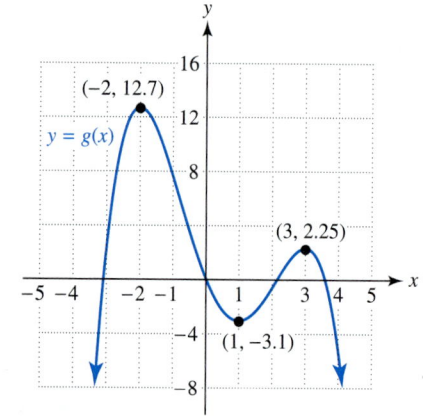

FIGURE 3.25

Figure 3.24 shows the graph of a parabola. If one traces along this graph from left to right, the y-values decrease until the vertex $(2, -6)$ is reached. To the right of the vertex, the y-values increase. The minimum y-value on the graph of f is -6. It is called the *absolute minimum* of f. The function f has no *absolute maximum* because there is no largest y-value on a parabola opening upward.

In Figure 3.25 the highest "peak" on the graph of g is located at $(-2, 12.7)$. Therefore, the absolute maximum of g is 12.7. There is a smaller peak located at the point $(3, 2.25)$. In a small interval near $x = 3$, the y-value of 2.25 is locally the largest. We say that g has a *local maximum* of 2.25. Similarly, a "valley" occurs on the graph of g. The lowest point is $(1, -3.1)$. The value -3.1 is not the smallest y-value on the entire graph of g. Therefore, it is not an absolute minimum. Rather, -3.1 is a *local minimum*.

Maximum and minimum values that are either absolute or local are called **extrema** (plural of extremum). A function may have several local extrema, but at most one absolute maximum and one absolute minimum. However, it is possible for a function to assume an absolute extremum at two different values of x. In Figure 3.26 the absolute maximum is 11. It occurs at $x = \pm 2$.

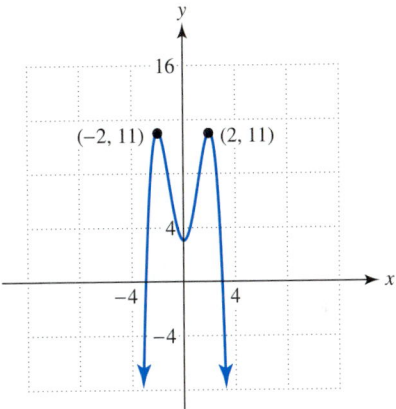

FIGURE 3.26

This discussion is summarized in the following.

Absolute and local extrema

Let c be in the domain of f.

    $f(c)$ is an **absolute maximum** if $f(c) \geq f(x)$ *for all x in the domain of f.*
    $f(c)$ is an **absolute minimum** if $f(c) \leq f(x)$ *for all x in the domain of f.*
    $f(c)$ is a **local maximum** if $f(c) \geq f(x)$ when x is *near c.*
    $f(c)$ is a **local minimum** if $f(c) \leq f(x)$ when x is *near c.*

Note: The expression "near c" means that there is an open interval in the domain of f containing c, where $f(c)$ satisfies the inequality.

EXAMPLE 1 *Identifying and interpreting extrema*

Figure 3.27 shows the graph of a function *f* that models the volume of air in a person's lungs measured in liters after *x* seconds. (Adapted from: V. Thomas, *Science and Sport.*)

(a) Determine the absolute maximum and the absolute minimum of *f*. Interpret the results.

(b) Identify two local maxima (plural of maximum) and two local minima (plural of minimum) of *f*. Interpret the results.

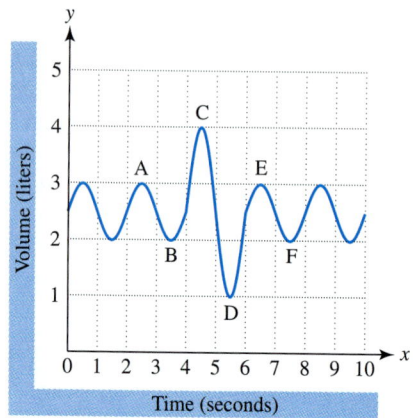

FIGURE 3.27 Volume of Air in a Person's Lungs

Solution

(a) The absolute maximum is four liters and occurs at C. The absolute minimum is one liter and occurs at D. At C a deep breath has been taken and the lungs are more inflated. After C, the person exhales beyond normal breathing until the lungs contain only one liter of air at D.

(b) One local maximum is three liters. It occurs at A and E and represents the amount of air in a person's lungs after inhaling normally. One local minimum is two liters. It occurs at B and F and represents the amount of air after exhaling normally. Another local maximum is four liters, which was also an absolute maximum. Similarly, one liter is a local minimum. ∎

EXAMPLE 2 *Modeling ocean temperatures*

The monthly average ocean temperature in degrees Fahrenheit at Bermuda can be modeled by $f(x) = 0.0215x^4 - 0.648x^3 + 6.03x^2 - 17.1x + 76.4$, where $x = 1$ corresponds to January and $x = 12$ to December. The domain of *f* is $D = \{x \mid 1 \leq x \leq 12\}$. (Source: J. Williams, *The Weather Almanac 1995.*)

(a) Graph *f* in [1, 12, 1] by [50, 90, 10].

(b) Estimate the absolute extrema. Interpret the results.

Solution

(a) The graph of $Y_1 = .0215X^4 - .648X^3 + 6.03X^2 - 17.1X + 76.4$ is shown in Figure 3.28.

(b) Many graphing calculators have the capability to find maximum and minimum y-values. The points associated with absolute extrema are shown in Figures 3.28 and 3.29. An absolute minimum of about 61.5 corresponds to the point $(2.01, 61.5)$. This means that the average ocean temperature is coldest during the month of February $(x = 2)$ when it reaches a minimum of about $61.5°F$.

An absolute maximum of approximately 82 corresponds to the point $(7.61, 82.0)$. The warmest average ocean temperature occurs during August $(x \approx 8)$ when it reaches a maximum of $82°F$. We might also say that this maximum occurs in late July, since $x \approx 7.61$.

[1, 12, 1] by [50, 90, 10] [1, 12, 1] by [50, 90, 10]

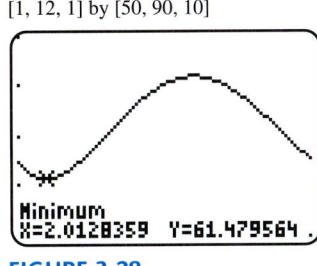 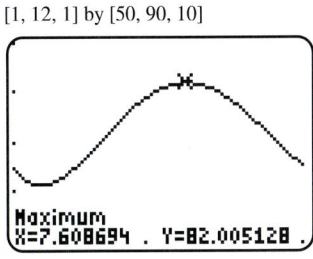

FIGURE 3.28   **FIGURE 3.29**   ■

MAKING CONNECTIONS
Local and Absolute Extrema

Extrema can either be local or absolute. When using a graph to identify either local or absolute extrema, it is important to remember that an extremum is a *y-value*—not a point (x, y). It is possible for an extremum to be both local and absolute.

Symmetry

Symmetry is used frequently in art, mathematics, and computer graphics. Many objects are symmetric along a vertical line so that the left and right sides are mirror images of each other. If an automobile is viewed from the front, the left side is typically a mirror image of the right side. Similarly, animals and people usually have an approximate left-right symmetry. Graphs of functions may also exhibit this type of symmetry as shown in Figures 3.30–3.32.

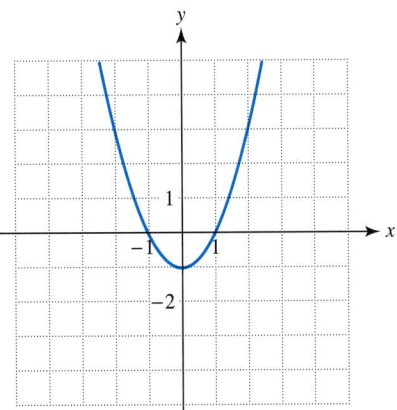 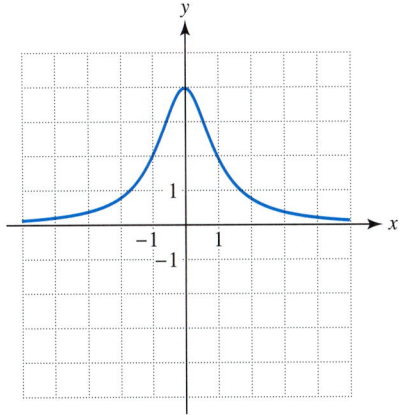

FIGURE 3.30   **FIGURE 3.31**

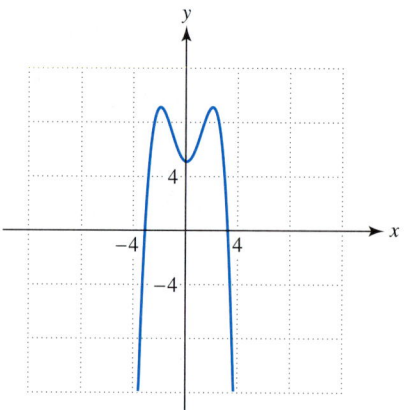

FIGURE 3.32

If each graph were folded along the *y*-axis, the left and right halves would match. These graphs are **symmetric with respect to the *y*-axis.** A function whose graph satisfies this characteristic is called an *even function*.

Figure 3.33 shows a graph of an even function *f*. Since the graph is symmetric with respect to the *y*-axis, the points (x, y) and $(-x, y)$ both lie on the graph of *f*. Thus, $f(x) = y$ and $f(-x) = y$, and so $f(x) = f(-x)$ for an even function. This means that if we change the sign of the input, the output does not change. For example, if $g(x) = x^2$ then $g(2) = g(-2) = 4$. Since this is true for every input, *g* is an even function.

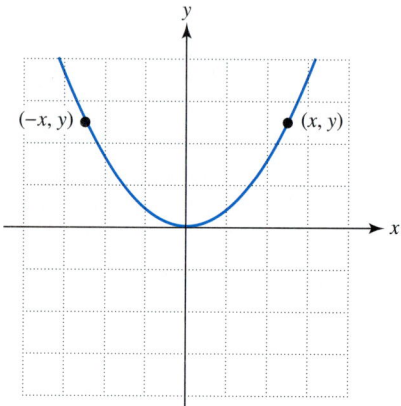

FIGURE 3.33 An Even Function

Even function

A function *f* is an **even function** if $f(-x) = f(x)$ for every *x* in its domain. The graph of an even function is symmetric with respect to the *y*-axis.

A second type of symmetry is shown in Figures 3.34–3.36. If we could spin the graph about the origin, the original graph would reappear after half a turn. These graphs are **symmetric with respect to the origin** and represent *odd functions*.

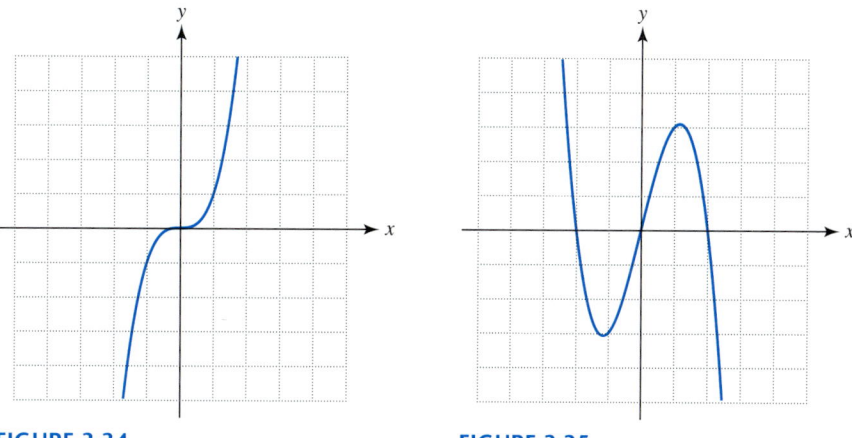

FIGURE 3.34   **FIGURE 3.35**

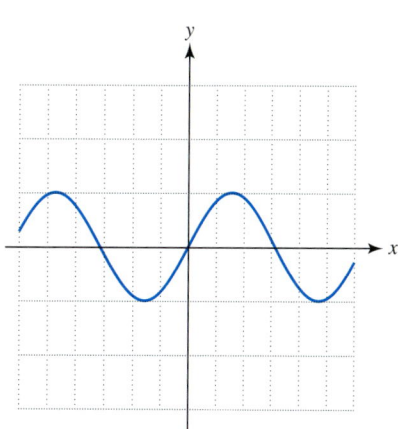

FIGURE 3.36

In Figure 3.37 the point (x, y) lies on the graph of an odd function f. If this point spins half a turn or $180°$ around the origin, its new location is $(-x, -y)$. Thus, $f(x) = y$ and $f(-x) = -y$. It follows that $f(-x) = -y = -f(x)$ for any odd function f. This means that changing the sign of the input only changes the sign of the output. For example, if $g(x) = x^3$ then $g(3) = 27$ and $g(-3) = -27$. Since this is true for every input, g is an odd function.

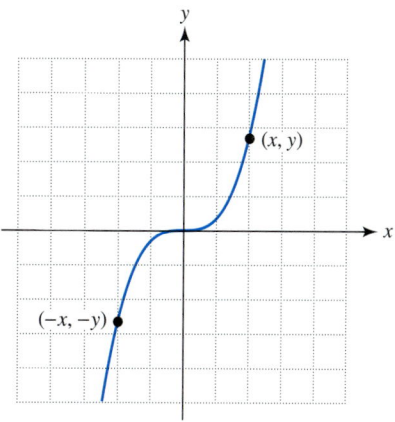

FIGURE 3.37 An Odd Function

> ### Odd function
>
> A function f is an **odd function** if $f(-x) = -f(x)$ for every x in its domain. The graph of an odd function is symmetric with respect to the origin.

The terms *odd* and *even* have special meaning when they are applied to a polynomial function f. If $f(x)$ contains terms that only have odd powers of x, then f is an odd function. Similarly, if $f(x)$ contains terms that only have even powers of x (and possibly a constant term), then f is an even function. For example, $f(x) = x^6 - 4x^4 - 2x^2 + 5$ is an even function, whereas $g(x) = x^5 + 4x^3$ is an odd function. This can be shown symbolically as follows.

$$
\begin{aligned}
f(-x) &= (-x)^6 - 4(-x)^4 - 2(-x)^2 + 5 && \text{Substitute } -x \text{ for } x. \\
&= x^6 - 4x^4 - 2x^2 + 5 && \text{Simplify.} \\
&= f(x) && f \text{ is an even function.} \\
g(-x) &= (-x)^5 + 4(-x)^3 && \text{Substitute } -x \text{ for } x. \\
&= -x^5 - 4x^3 && \text{Simplify.} \\
&= -g(x) && g \text{ is an odd function.}
\end{aligned}
$$

Critical Thinking

If 0 is in the domain of an odd function f, what point must lie on its graph? Explain your reasoning.

It is important to remember that the graphs of many functions exhibit no symmetry with respect to either the y-axis or the origin. As a result, these functions are neither odd nor even.

EXAMPLE 3 *Identifying odd and even functions*

For each representation of a function f, identify whether f is odd, even, or neither.

(a)

TABLE 3.2

x	-3	-2	-1	0	1	2	3
$f(x)$	10.5	2	-0.5	-2	-0.5	2	10.5

(b)

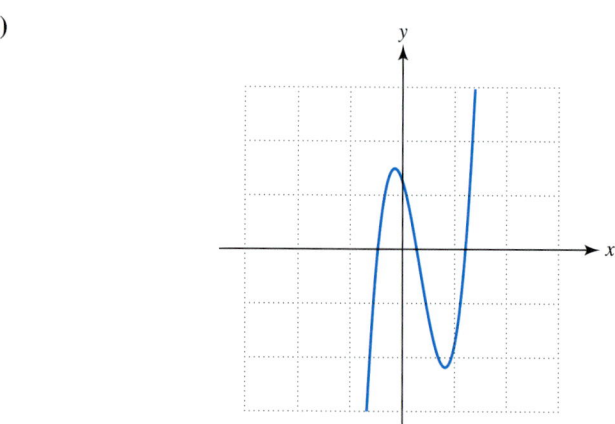

FIGURE 3.38

(c) $f(x) = x^3 - 5x$

(d) f is the cube root function.

Solution

(a) The function defined by Table 3.2 has domain $D = \{-3, -2, -1, 0, 1, 2, 3\}$. Notice that $f(-3) = 10.5 = f(3)$ and $f(-2) = 2 = f(2)$. The function f satisfies the statement $f(-x) = f(x)$ for every x in D. Thus, f is an even function.

(b) If we fold the graph in Figure 3.38 on the y-axis, the two halves do not match, so f is not an even function. Similarly, f is not odd since spinning its graph half a turn about the origin does not result in the same graph. The function f is neither odd nor even.

(c) Since f is a polynomial containing only odd powers of x, it is odd. This also can be shown symbolically as follows.

$$f(-x) = (-x)^3 - 5(-x) \qquad \text{Substitute } -x \text{ for } x.$$
$$= -x^3 + 5x \qquad \text{Simplify.}$$
$$= -(x^3 - 5x) \qquad \text{Distributive property}$$
$$= -f(x) \qquad f \text{ is an odd function.}$$

This result is supported graphically in Figure 3.39.

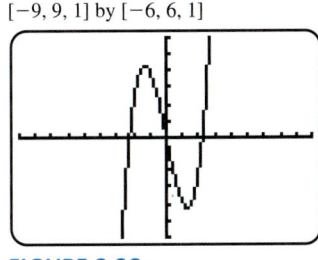

$[-9, 9, 1]$ by $[-6, 6, 1]$ $[-6, 6, 1]$ by $[-4, 4, 1]$

FIGURE 3.39   **FIGURE 3.40**

(d) Since $f(x) = \sqrt[3]{x}$, graph $Y_1 = \sqrt[3]{(X)}$ or equivalently $Y_1 = X^{\wedge}(1/3)$ as shown in Figure 3.40. Spinning the graph half a turn about the origin results in the same graph. The cube root function is odd. If desired, this can be shown symbolically.

$$f(-x) = \sqrt[3]{-x} \qquad \text{Substitute } -x \text{ for } x.$$
$$= \sqrt[3]{(-1)x} \qquad \text{Let } -x = (-1)x.$$
$$= -1\sqrt[3]{x} \qquad \sqrt[3]{-1} = -1$$
$$= -f(x) \qquad f(x) = \sqrt[3]{x} \qquad \blacksquare$$

Critical Thinking

Discuss the possibility of the graph of a function being symmetric with respect to the x-axis.

Power Functions

Power functions are examples of nonlinear functions.

> ### Power function
>
> A function f given by $f(x) = x^a$, where a is a constant, is a **power function.**
> If $a = \dfrac{1}{n}$ for some integer $n \geq 2$, then f is a **root function** given by $f(x) = x^{1/n}$, or equivalently, $f(x) = \sqrt[n]{x}$.

Symbolic representations of power functions include

$$f_1(x) = x^2, \qquad f_2(x) = x^{\pi}, \qquad f_3(x) = x^{0.4}, \qquad \text{and} \qquad f_4(x) = \sqrt[3]{x^2}.$$

Frequently, the domain of a power function f is restricted to nonnegative numbers. Suppose the rational number $\dfrac{p}{q}$ is written in lowest terms. Then the domain of $f(x) = x^{p/q}$ is all real numbers whenever q is odd and all nonnegative real numbers whenever q is even. If a is an irrational number, the domain of $f(x) = x^a$ is all nonnegative real numbers.

For example, the domain of $f(x) = x^{1/3}$ $\left(f(x) = \sqrt[3]{x}\right)$ is all real numbers, whereas the domain of $g(x) = x^{1/2}$ $\left(g(x) = \sqrt{x}\right)$ is all nonnegative numbers.

EXAMPLE 4   *Graphing power functions*

Graph $f(x) = x^a$ for $a = 0.3$, 1, and 1.7 for $x \geq 0$. Discuss the effect that a has on the graph of f.

Solution
The graphs of $y = x^{0.3}$, $y = x^1$, and $y = x^{1.7}$ are shown in Figure 3.41. Larger values of a cause the graph of f to increase faster.

[0, 9, 1] by [0, 6, 1]

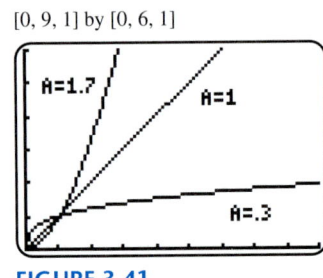

FIGURE 3.41   ■

Review Note   *Properties of Exponents*
Let m and n be positive integers. Assume that b is a real number and $\sqrt[n]{b}$ is defined.

Property	Example
1. $b^{m/n} = (b^m)^{1/n} = (b^{1/n})^m$	$4^{3/2} = (4^3)^{1/2} = (4^{1/2})^3 = 2^3 = 8$
2. $b^{m/n} = \sqrt[n]{b^m} = (\sqrt[n]{b})^m$	$8^{2/3} = \sqrt[3]{8^2} = (\sqrt[3]{8})^2 = 2^2 = 4$
3. $(b^m)^n = b^{mn}$	$(2^3)^2 = 2^6 = 64$
4. $b^{-m} = \dfrac{1}{b^m}$	$4^{-2} = \dfrac{1}{4^2} = \dfrac{1}{16}$

EXAMPLE 5 *Solving equations involving rational exponents*

Write $f(x)$ using radical notation and determine the domain of f. Then solve the equation $f(x) = b$ for the given value of b. Support your results graphically.

(a) $f(x) = x^{1/4}$, $b = 3$

(b) $f(x) = x^{2/3}$, $b = 4$

Solution

(a) Since $x^{1/4} = \sqrt[4]{x}$, let $f(x) = \sqrt[4]{x}$. This computation involves an even root, so the domain of f is all nonnegative real numbers. Next we solve the equation $f(x) = 3$.

$$x^{1/4} = 3 \qquad \text{Substitute for } f(x).$$
$$(x^{1/4})^4 = 3^4 \qquad \text{Raise to the fourth power.}$$
$$x = 81 \qquad \text{Properties of exponents}$$

Graphical support is shown in Figure 3.42, where $Y_1 = X^{\wedge}(1/4)$ and $Y_2 = 3$.

[0, 120, 20] by [0, 4, 1]

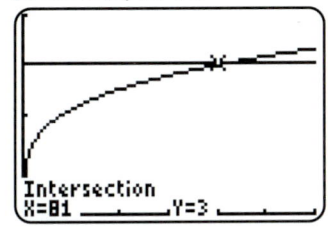

FIGURE 3.42

[−12, 12, 1] by [−8, 8, 1]

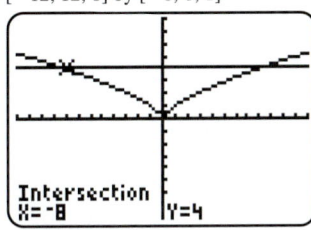

FIGURE 3.43

[−12, 12, 1] by [−8, 8, 1]

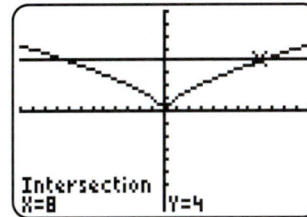

FIGURE 3.44

(b) Since $x^{2/3} = (x^2)^{1/3} = \sqrt[3]{x^2}$, write $f(x) = \sqrt[3]{x^2}$. An odd root can be taken of any real number. Therefore, the domain of f is all real numbers.

$$x^{2/3} = 4 \qquad \text{Original equation}$$
$$(x^{2/3})^3 = 4^3 \qquad \text{Cube both sides.}$$
$$x^2 = 64 \qquad \text{Properties of exponents}$$
$$x = \pm 8 \qquad \text{Square root property}$$

Graphical support is shown in Figures 3.43 and 3.44, where $Y_1 = X^{\wedge}(2/3)$ and $Y_2 = 4$. ∎

EXAMPLE 6 *Solving an equation involving a radical*

Solve $x = \sqrt{15 - 2x}$. Check your solutions.

Solution

Begin by squaring both sides of the equation.

$$x = \sqrt{15 - 2x} \qquad \text{Given equation}$$
$$x^2 = (\sqrt{15 - 2x})^2 \qquad \text{Square both sides.}$$
$$x^2 = 15 - 2x \qquad \text{Simplify.}$$
$$x^2 + 2x - 15 = 0 \qquad \text{Add } 2x \text{ and subtract 15.}$$
$$(x + 5)(x - 3) = 0 \qquad \text{Factor.}$$
$$x = -5 \quad \text{or} \quad x = 3 \qquad \text{Solve.}$$

Now substitute these values in the original equation $x = \sqrt{15 - 2x}$.

$$-5 \neq \sqrt{15 - 2(-5)} = 5, \qquad 3 = \sqrt{15 - 2(3)} \checkmark$$

Thus, 3 is the only solution. This result is supported in Figure 3.45, where $Y_1 = X$ and $Y_2 = \sqrt{(15 - 2X)}$. Notice that no point of intersection occurs when $x = -5$.

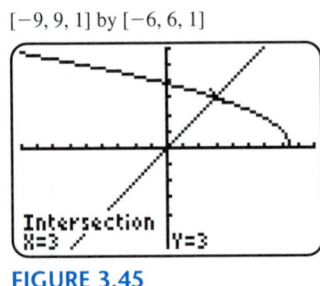

$[-9, 9, 1]$ by $[-6, 6, 1]$

FIGURE 3.45

The value -5 in Example 6 is called an *extraneous solution* because it does not satisfy the original equation. It is important to check results whenever *squaring* has been used to solve an equation.

EXAMPLE 7   *Determining the period of planetary orbits*

Johannes Kepler (1571–1630) was the first to recognize that the orbits of planets are elliptical, rather than circular. He also found that a power function models the relationship between a planet's distance from the sun and its period of revolution. Table 3.3 lists the average distance x from the sun and the time y in years for several planets to orbit the sun. The distance x has been normalized so that Earth is one unit away from the sun. For example, Jupiter is 5.2 times farther from the sun than Earth and requires 11.9 years to orbit the sun. (Source: C. Ronan, *The Natural History of the Universe.*)

TABLE 3.3

Planet	x (distance)	y (period)
Mercury	0.387	0.241
Venus	0.723	0.615
Earth	1.00	1.00
Mars	1.52	1.88
Jupiter	5.20	11.9
Saturn	9.54	29.5

(a) Make a scatterplot of the data. Estimate graphically a value for a so that $f(x) = x^a$ models the data.

(b) Numerically check the accuracy of f.

(c) The average distances of Uranus, Neptune, and Pluto from the sun are 19.2, 30.1, and 39.5, respectively. Use f to estimate the periods of revolution for these planets. Compare these answers to the actual values of 84.0, 164.8, and 248.5 years.

Solution

(a) Make a scatterplot of the data and then graph $y = x^a$ for different values of a. By viewing the graphs of $y = x^{1.4}$, $y = x^{1.5}$, and $y = x^{1.6}$ in Figures 3.46–3.48, it can be seen that $a \approx 1.5$.

[0, 10, 1] by [0, 30, 10]
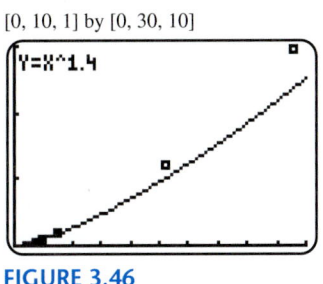
FIGURE 3.46

[0, 10, 1] by [0, 30, 10]
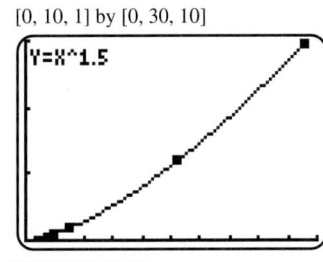
FIGURE 3.47

[0, 10, 1] by [0, 30, 10]
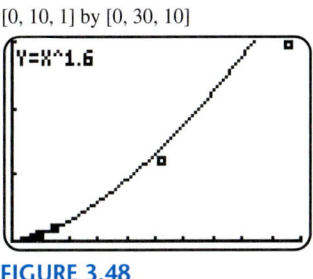
FIGURE 3.48

(b) Let $f(x) = x^{1.5}$ and table $Y_1 = X^{\wedge}1.5$. The values shown in Figure 3.49 model the data remarkably well.

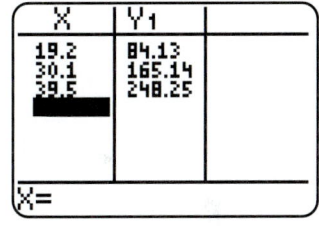

FIGURE 3.49   FIGURE 3.50

(c) To approximate the number of years for Uranus, Neptune, and Pluto to orbit the sun, evaluate f at $x = 19.2$, 30.1, and 39.5 as shown in Figure 3.50. These values are close to the actual values.   ■

3.2 PUTTING IT ALL TOGETHER

Nonlinear functions have graphs that can increase or decrease over their domains. Absolute and local extrema and symmetry are used to describe nonlinear graphs. Polynomial, power, and root functions are examples of nonlinear functions. Power functions, which include root functions, can be represented by $f(x) = x^a$, where a is a constant. Root functions can be represented by $g(x) = x^{1/n}$, where $n \geq 2$ is an integer. The square root and cube root functions are examples. The following table summarizes some important concepts related to the graphs of nonlinear functions.

Concept	Comments	Graphical Example
Even function	$f(-x) = f(x)$ Graph is symmetric with respect to the y-axis.	
Odd function	$f(-x) = -f(x)$ Graph is symmetric with respect to the origin.	
Absolute maximum (minimum)	*The* maximum (minimum) y-value on the graph $y = f(x)$	
Local maximum (minimum)	A maximum (minimum) y-value on the graph $y = f(x)$ in an open interval of the domain of f	

3.2 EXERCISES   Tape 3

Nonlinear Functions

Exercises 1–6: Determine if f is a polynomial function, power function, or both. If f is a polynomial function, state its degree. If f is a power function, decide if it is a root function.

1. $f(x) = \sqrt[5]{x}$
2. $f(x) = x^{1.58}$
3. $f(x) = 5x^2 - 4x + 1$
4. $f(x) = 7 - 6x + 3x^4$
5. $f(x) = x^{5/3}$
6. $f(x) = x^5$

Exercises 7 and 8: Complete the following.

(a) *Plot the data listed in the table. Without evaluating the functions f or g, conjecture which models the data better. Explain your reasoning.*

(b) *Verify your conjecture by graphing f and g together with the data.*

7. $f(x) = 1.5x + 2, \quad g(x) = x^{2.3}$

x	2	4	6	8
y	5	8	11	14

8. $f(x) = 0.03(x - 1850) + 1,$
   $g(x) = 0.00021(x - 1850)^2 + 1$

Estimated world population

x (year)	1850	1930	1975	1995
y (billions)	1	2	4	5.7

Absolute and Local Extrema

9. *Wiretaps* The line graph shows the approximate number of state and federal wiretaps approved from 1968 to 1992. If the line graph represents a function *f*, find the local extrema. (*Hint:* Local extrema do not occur at endpoints.) (Source: Administrative Office of the U.S. Courts.)

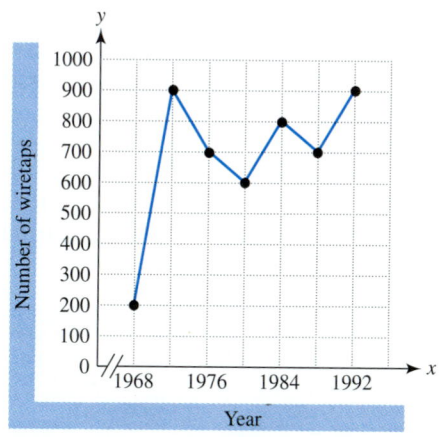

10. *Sunspots* The table shows the number of sunspots during the month of January from 1950 to 1969. Let the line graph of this data represent a function *f*, where the input is the year and the output is the number of sunspots. Determine the local and absolute extrema of *f*. (Source: R. Wolff and L. Yaeger, *Visualization of Natural Phenomena.*)

Year	1950	1951	1952	1953	1954
Sunspots	102	60	41	27	0

Year	1955	1956	1957	1958	1959
Sunspots	23	74	165	203	217

Year	1960	1961	1962	1963	1964
Sunspots	146	58	39	20	15

Year	1965	1966	1967	1968	1969
Sunspots	18	28	111	122	104

Exercises 11–14: Use the graph of f to estimate the
(a) *local extrema, and*
(b) *absolute extrema.*

11.

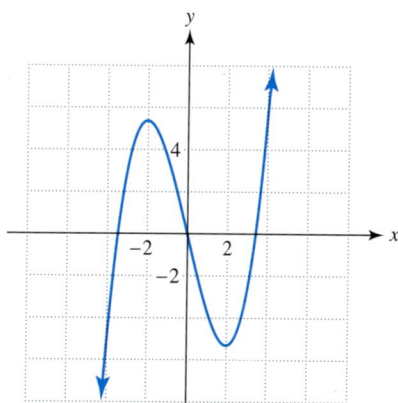

12.

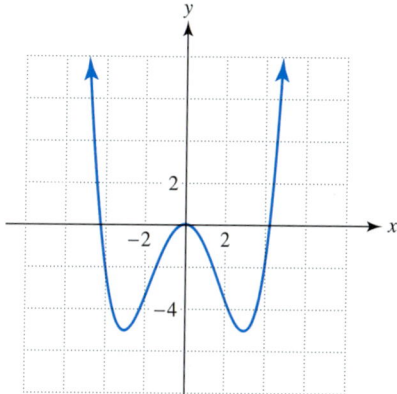

13.

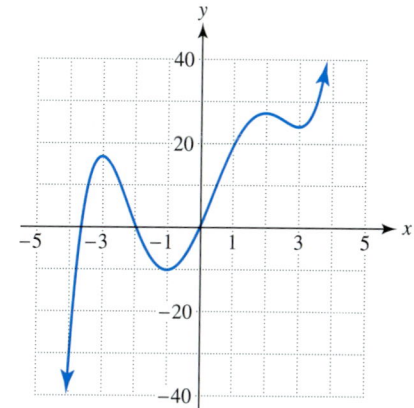

14.

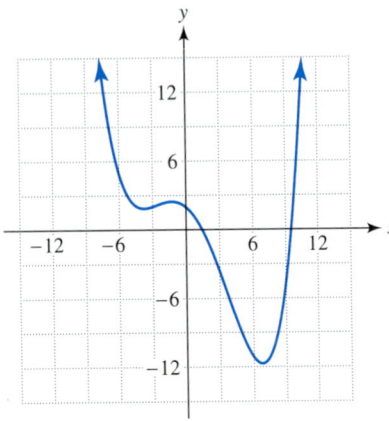

Exercises 15–20: Determine graphically any
(a) *local extrema, and*
(b) *absolute extrema.*

15. $f(x) = -x^2 + 3x + 3$

16. $f(x) = x^{2/3}$

17. $f(x) = 0.5x^4 - 5x^2 + 4.5$

18. $f(x) = 0.01x^5 + 0.02x^4 - 0.35x^3 - 0.36x^2 + 1.8x$

19. $f(x) = \dfrac{8}{1 + x^2}$

20. $f(x) = x^{1/3}$

21. *Energy* The U.S. consumption of energy by residential and commercial users from 1950 to 1980 can be modeled by the polynomial function represented by

$$f(x) = -0.00113x^3 + 0.0408x^2 - 0.0432x + 7.66,$$

where $x = 0$ corresponds to 1950 and $x = 30$ to 1980. Consumption is measured in quadrillion Btu. (Source: Department of Energy.)
(a) Graph f in [0, 30, 5] by [6, 16, 1]. Describe the energy usage during this time period.
(b) Approximate the local maximum and interpret the result.

22. *Natural Gas* The U. S. consumption of natural gas from 1965 to 1980 can be modeled by the quartic function f, represented by

$$f(x) = 0.0001234x^4 - 0.005689x^3 + 0.08792x^2 - 0.5145x + 1.514,$$

where $x = 5$ corresponds to 1965 and $x = 20$ to 1980. Consumption is measured in trillion cubic feet. (Source: Department of Energy.)
(a) Graph f in [5, 20, 5] by [0.4, 0.8, 0.1]. Describe the energy usage during this time period.
(b) Determine the local extrema and interpret the results.

23. *Heating Costs* In colder climates of the United States the cost for natural gas to heat homes can vary dramatically from one month to the next. The polynomial function given by

$$f(x) = -0.1213x^4 + 3.462x^3 - 29.22x^2 + 64.68x + 97.69$$

models the monthly cost in dollars of heating a typical home. The input x represents the month, where $x = 1$ corresponds to January and $x = 12$ to December. (Source: Minnegasco, A NORAM Energy Company.)
(a) Conjecture where absolute extrema might occur for $1 \le x \le 12$.
(b) Graph f in [1, 12, 1] by [0, 150, 10]. Identify the absolute extrema in this graph and interpret the results.

24. *Blow Fly Experiment* In one experiment, blow flies (*Lucilia cuprina*) were kept in a laboratory for roughly 20 generations. Adequate food and water were provided. However, there was a limited supply of ground liver. Adult flies need a nutrient found in liver in order to lay eggs. Both the number of adult flies and the number of eggs laid per day are shown in the figures at the top of the next column. Interpret how these two graphs and their local extrema relate to each other. (Sources: A. J. Nicholson, "An Outline of the Dynamics of Animal Populations"; E. Pielou, *Population and Community Ecology, Principles and Methods.*)

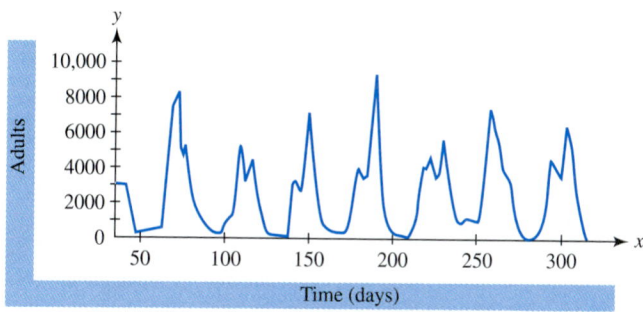

Time (days)

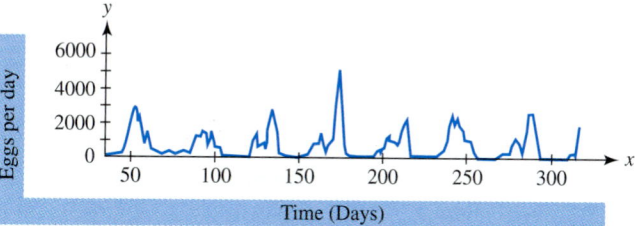

Time (Days)

Symmetry

Exercises 25 and 26: The accompanying figure shows a graph of either an odd or even function f together with a point on its graph. Use this point to determine a second point on the graph of f.

25.

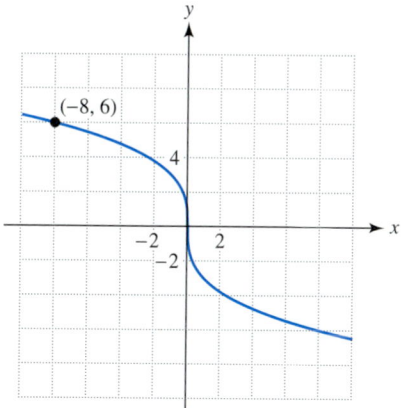

26.

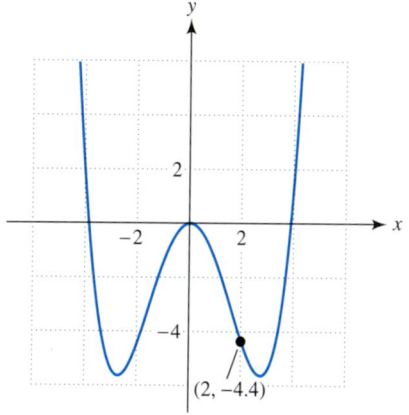

27. The table is a complete representation of f. Decide if f is even, odd, or neither.

x	-100	-10	-1	0	1	10	100
$f(x)$	56	-23	5	0	-5	23	-56

28. The table is a complete representation of f. Decide if f is even, odd, or neither.

x	-5	-3	-1	1	2	3
$f(x)$	-4	-2	1	1	-2	-4

29. Complete the table if f is an even function.

x	-3	-2	-1	0	1	2	3
$f(x)$	21	?	-25	?	?	-12	?

30. Complete the table if f is an odd function.

x	-5	-3	-2	0	2	3	5
$f(x)$	13	?	-5	?	?	-1	?

31. Complete the table so the function f is neither odd nor even.

x	-2	-1	0	1	2
$f(x)$	5	?	?	3	?

32. Complete the table if the function f is odd.

x	-2	0	2
$f(x)$	5	?	?

33. A partial graph of an odd function with domain $D = \{x \mid -2 \le x \le 2\}$ is shown in the figure. Make a sketch of the complete graph.

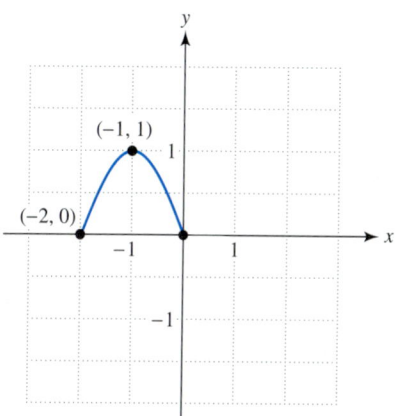

34. A partial graph of an even function f with domain $D = \{x \mid -10 \le x \le 10\}$ is shown in the figure. Make a sketch of the complete graph.

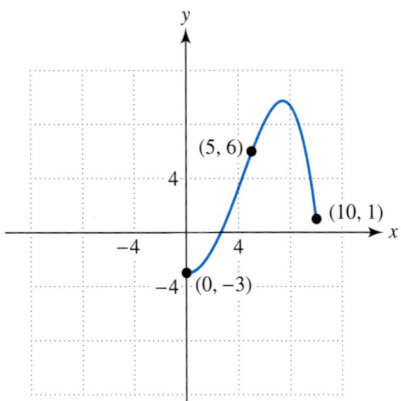

35. *Average Temperature* The graph approximates the monthly average temperatures in degrees Fahrenheit in Austin, Texas. In this graph x represents the month, where $x = 0$ corresponds to July.
- **(a)** Is this a graph of an odd or even function?
- **(b)** June corresponds to $x = -1$ and August to $x = 1$. The average temperature in June is $83°F$. What is the average temperature in August?
- **(c)** March corresponds to $x = -4$ and November to $x = 4$. According to the graph, how do their average temperatures compare?
- **(d)** Interpret what this type of symmetry implies about the average temperatures in Austin.

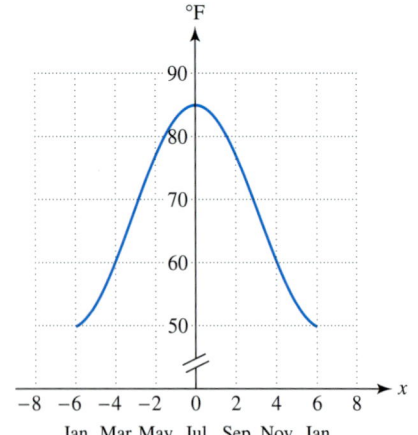

36. *Average Temperature* The accompanying graph shows the deviation of the monthly average temperatures from the yearly average temperature in Anchorage, Alaska. In this graph x represents the month, where $x = 0$ corresponds to October. For example, the average temperature in October is equal to the average yearly temperature, so its deviation is $0°F$. The deviation in July ($x = -3$) is $22°F$ above the yearly average.
- **(a)** October and what other month have a temperature deviation of $0°F$ from the yearly average?
- **(b)** Is this a graph of an odd or even function?
- **(c)** September corresponds to $x = -1$ and November to $x = 1$. The average temperature in November is $11°F$ below the average yearly temperature. What does this indicate about the average temperature in September?
- **(d)** August corresponds to $x = -2$ and December to $x = 2$. How do their temperatures compare to the average yearly temperature?
- **(e)** Interpret what this type of symmetry implies about the average monthly temperatures in Anchorage.

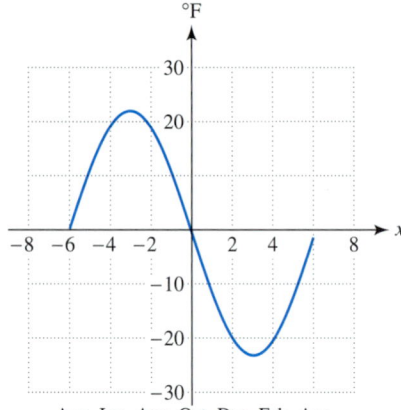

37. *Height of a Projectile* If a projectile is shot into the air, it attains a maximum height and then falls back to the ground. Suppose that $x = 0$ corresponds to the time when the projectile's height is maximum. If air resistance is ignored, its height h above the ground at any time x may be modeled by $h(x) = -16x^2 + h_{max}$, where h_{max} is the projectile's maximum height above the ground. Height is measured in feet and time in seconds. Let $h_{max} = 400$ feet.
- **(a)** Evaluate $h(-2)$ and $h(2)$. Interpret these results.
- **(b)** Evaluate $h(-5)$ and $h(5)$. Interpret these results.
- **(c)** Graph h in $[-5, 5, 1]$ by $[0, 500, 100]$. Is h an even or odd function?
- **(d)** How do $h(x)$ and $h(-x)$ compare when $-5 \le x \le 5$? What does this indicate about the flight of a projectile?

38. (Refer to the previous exercise.) The velocity in feet per second of a projectile shot into the air is given by $v(t) = -32t$, where $-5 \leq t \leq 5$ and $t = 0$ corresponds to the time when the maximum height is attained. A positive velocity indicates that the projectile is traveling up, whereas a negative velocity indicates that it is falling.

    (a) Evaluate $v(-2)$ and $v(2)$. Interpret each result.

    (b) Graph v in $[-5, 5, 1]$ by $[-200, 200, 100]$. Is v an even or odd function?

    (c) How do $v(t)$ and $v(-t)$ compare when $-5 \leq t \leq 5$? What does this mean about the velocity of a projectile?

Exercises 39–48: Determine if f is even, odd, or neither. Support your answer graphically.

39. $f(x) = 3x^4 - 1.5x^2 - 9$

40. $f(x) = 4x^3 - 18x$

41. $f(x) = 3x^5 + 6x$

42. $f(x) = 3x^6 + x^4 - 4$

43. $f(x) = 7x^2 - 4x + 1$

44. $f(x) = 3x^3 - 2x + 1$

45. $f(x) = \sqrt{1 - x^2}$

46. $f(x) = \dfrac{1}{x^3}$

47. $f(x) = x^{2/3}$

48. $f(x) = -7x$

Concepts

49. Sketch a graph of a linear function that is an odd function.

50. Sketch a graph of a linear function that is an even function.

51. Does there exist a continuous odd function that is always increasing and whose graph passes through the points $(-3, -4)$ and $(2, 5)$? Explain.

52. Is there an even function whose domain is all real numbers and is always decreasing? Explain.

Power Functions

Exercises 53–60: Evaluate each expression by hand. Then check your result using a calculator.

53. $8^{2/3}$

54. $16^{3/2}$

55. $16^{-3/4}$

56. $25^{-3/2}$

57. $81^{0.5}$

58. $32^{1/5}$

59. $64^{1/6}$

60. $16^{-0.25}$

Exercises 61–64: (Refer to Example 5.)

    (a) Write $f(x)$ using radicals.

    (b) State the domain of f.

    (c) Solve the equation $f(x) = k$ for the given value of k. Support your answer graphically.

61. $f(x) = x^{3/2}$, $k = 27$

62. $f(x) = x^{3/4}$, $k = 8$

63. $f(x) = x^{4/3}$, $k = 10$

64. $f(x) = x^{7/5}$, $k = 7$

Exercises 65–72: Solve the equation. Check each solution.

65. $x^3 = 8$

66. $x^4 = \dfrac{1}{81}$

67. $x^{1/4} = 3$

68. $x^{1/3} = \dfrac{1}{5}$

69. $\sqrt{x + 2} = x - 4$

70. $\sqrt{2x + 1} = 13$

71. $\sqrt{3x + 7} = 3x + 5$

72. $\sqrt{x} = \sqrt{x - 5} + 1$

Exercises 73–76: Evaluate f(b) at the given b. Approximate each result to five significant digits.

73. $f(x) = x^{1.62}$, $b = 1.2$

74. $f(x) = x^{-0.71}$, $b = 3.8$

75. $f(x) = x^{3/2} - x^{1/2}$, $b = 50$

76. $f(x) = x^{5/4} - x^{-3/4}$, $b = 7$

Exercises 77 and 78: Match f(x) with its graph. Assume that a and b are constants with $0 < a < 1 < b$.

77. $f(x) = x^a$

78. $f(x) = x^b$

    a.

b.

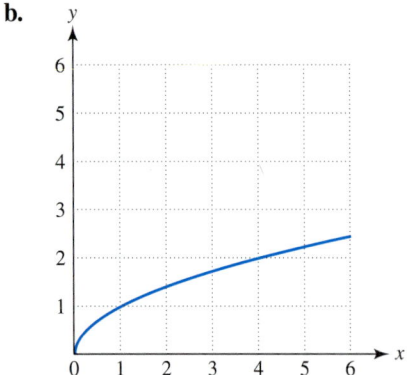

Applications of Power Functions

79. *Trout and Pollution*   Rainbow trout are sensitive to zinc ions in the water. High concentrations are lethal. The average survival times x in minutes for trout in various concentrations of zinc ions y in milligrams per liter (mg/l) are listed in the table. (Source: C. Mason, *Biology of Freshwater Pollution.*)

x (min)	0.5	1	2	3
y (mg/l)	4500	1960	850	525

(a) This data can be modeled by $f(x) = bx^a$, where a and b are constants. Determine b. (*Hint:* Let $f(1) = 1960$.)

(b) Estimate a graphically.

(c) Evaluate $f(4)$ and interpret the result.

80. *Asbestos and Cancer*   Insulation workers who were exposed to asbestos and employed before 1960 experienced an increased likelihood of lung cancer. If a group of insulation workers have a cumulative total of 100,000 years of work experience with their first date of employment x years ago, then the number of lung cancer cases occurring within the group can be modeled by $N(x) = 0.00437x^{3.2}$. (Source: A. Walker, *Observation and Inference.*)

(a) Calculate $N(x)$ when $x = 5, 10,$ and 20. What happens to the likelihood of cancer as x increases?

(b) If x doubles, does the number of cancer cases also double?

81. *Fiddler Crab Size*   **Allometry** is the study of the relative size between different characteristics of an organism. For example, there is frequently a relationship between the height and weight of an animal—taller animals are usually heavier. Allometric relations often can be modeled by $f(x) = bx^a$, where a and b are constants. One study of the male fiddler crab showed a connection between the weight of the large claws and its total body weight. For a crab weighing over 0.75 gram, the weight of its claws can be estimated by $f(x) = 0.445x^{1.25}$. The input x is the weight of the crab in grams and the output $f(x)$ is the weight of the claws in grams. (Sources: J. Huxley, *Problems of Relative Growth; D. Brown and P. Rothery, Models in Biology: Mathematics, Statistics and Computing.*)

(a) Predict the weight of the claws for a two-gram crab.

(b) Approximate graphically the weight of a crab that has 0.5-gram claws.

(c) Solve part (b) symbolically.

82. *Weight and Height*   (Refer to the previous exercise.)   Allometry can be applied to height-weight relationships for humans. The average weight in pounds for men and women can be estimated using a function f of the form $f(x) = bx^{1.7}$, where x represents a person's height in inches and b is a constant that depends on the sex of the individual.

(a) If an average weight for a 68-inch man is 152 pounds, approximate b. Use f to estimate the average weight of a 66-inch man.

(b) If an average weight for a 68-inch woman is 137 pounds, approximate b. Use f to estimate the average weight of a 70-inch woman.

Writing about Mathematics

1. Explain differences and similarities between polynomial and power functions. Consider their formulas, graphs, and domains.

2. Describe ways to determine if a polynomial function is odd, even, or neither. Give examples.

CHECKING BASIC CONCEPTS FOR SECTIONS 3.1 AND 3.2

1. Solve the quadratic equations graphically, numerically, and symbolically.
   (a) $16x^2 = 81$
   (b) $2x^2 + 3x = 2$
   (c) $x^2 = x - 3$

2. Solve each quadratic inequality.
   (a) $2x^2 + 7x - 4 \le 0$
   (b) $2x^2 + 7x - 4 > 0$

3. Graph $f(x) = \frac{1}{4}x^4 - \frac{2}{3}x^3 - \frac{5}{2}x^2 + 6x$.

   (a) Identify any local extrema.
   (b) Find any absolute extrema.
   (c) Approximate the zeros of f.

4. Table $f(x) = \sqrt{25 - x^2}$ starting at $x = -5$, incrementing by 1.
   (a) Is f an odd or even function? Graph f in $[-9, 9, 1]$ by $[-6, 6, 1]$ to support your answer.
   (b) Find the domain and range of f.

3.3 Polynomial Functions and Their Graphs

Graphs of Polynomial Functions • Reflection of Graphs • Piecewise-Defined Polynomial Functions (Optional)

Introduction

The study of higher degree polynomial equations dates back to Old Babylonian civilization in about 1800–1600 B.C. Gottfried Leibnitz (1646–1716) was the first mathematician to generalize polynomial functions of degree n. Many eighteenth-century mathematicians devoted their lives to studying polynomial equations. Today, polynomial functions are used to model a wide variety of real data. We begin by discussing the graphs of polynomial functions. (References: *Historical Topics for the Mathematics Classroom, Thirty-first Yearbook,* NCTM; L. Motz and J. H. Weaver, *The Story of Mathematics.*)

Graphs of Polynomial Functions

In Section 1.4 polynomial functions were defined. Polynomial functions have been used in several applications. Their graphs are continuous with no breaks or sharp corners. The domain of a polynomial function is all real numbers. A polynomial function of degree n can be expressed as

$$f(x) = a_n x^n + \cdots + a_2 x^2 + a_1 x + a_0,$$

where each coefficient a_k is a real number, $a_n \ne 0$, and n is a nonnegative integer. The *leading coefficient* is a_n.

The expression $a_n x^n + \cdots + a_2 x^2 + a_1 x + a_0$ is a **polynomial of degree n** and the equation $a_n x^n + \cdots + a_2 x^2 + a_1 x + a_0 = 0$ is a **polynomial equation of degree n.** Thus, f is a polynomial function, $f(x)$ is a polynomial, and $f(x) = 0$ is a polynomial equation. For example, the function f, given by $f(x) = x^3 - 3x^2 + x - 5$, is a polynomial function, $x^3 - 3x^2 + x - 5$ is a polynomial, and $x^3 - 3x^2 + x - 5 = 0$ is a polynomial equation.

A **turning point** occurs whenever the graph of a polynomial function changes from increasing to decreasing or from decreasing to increasing. Turning points are associated with "hills" or "valleys" on a graph. The y-value at a turning point is either a local maximum or local minimum of the function. In Figure 3.51 the graph has two turning points, $(-2, 8)$ and $(2, -8)$.

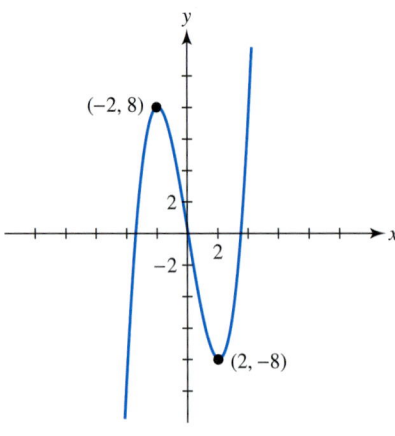

FIGURE 3.51

We will discuss the graphs of polynomial functions starting with degree 0 and continuing to degree 5.

Constant Polynomial Functions. If $f(x) = a$ and $a \neq 0$, then f is both a constant function and a polynomial function of degree 0. (If $a = 0$ then f has an undefined degree.) Its graph is a horizontal line that does not coincide with the x-axis. Graphs of $f_1(x) = 4$ and $f_2(x) = -3$ are shown in Figures 3.52 and 3.53. A graph of a polynomial function of degree 0 has no x-intercepts or turning points.

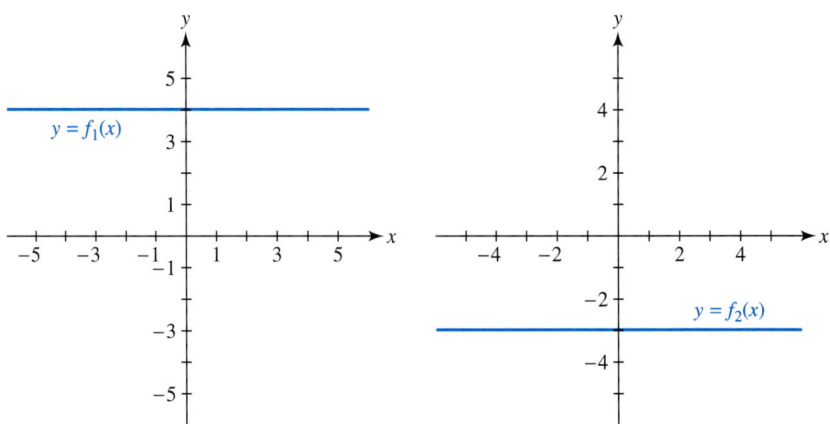

FIGURE 3.52   **FIGURE 3.53**

Linear Polynomial Functions. If $f(x) = ax + b$ and $a \neq 0$, then f is both a linear function and a polynomial function of degree 1. Its graph is a line that is neither horizontal nor vertical. The graphs of $f_1(x) = 2x - 3$ and $f_2(x) = -1.6x + 5$ are shown in Figures 3.54 and 3.55. A polynomial function of degree 1 has one x-intercept and no turning points.

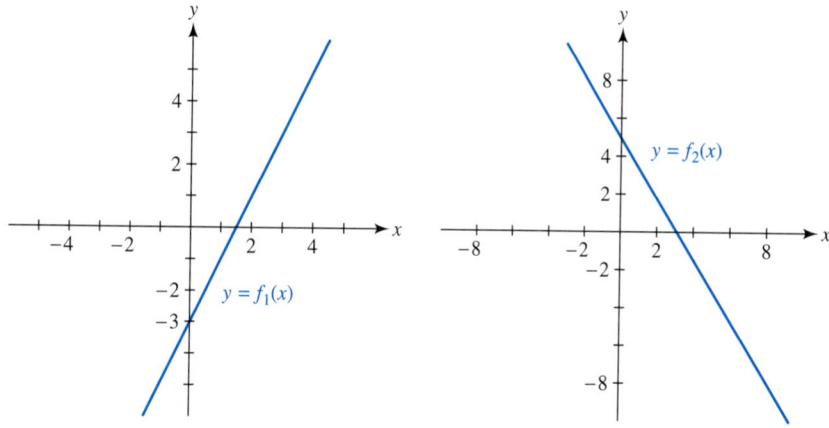

FIGURE 3.54　　　　　　　　　　**FIGURE 3.55**

The graph of $f(x) = ax + b$ with leading coefficient $a > 0$ is a line slop-ing upward. The left side of the graph falls while the right side rises. As one traces from left to right, the y-values become larger without a maximum. We say that the *end behavior* of the graph tends to ∞ on the right and $-\infty$ on the left. (Strictly speaking, the graph of a polynomial has infinite length and does not have an end.)

If the leading coefficient is $a < 0$, then the end behavior is reversed. The line slopes downward from left to right. The y-values on the left side of the graph become large positive values without a maximum and the y-values on the right side become negative without a minimum. The end behavior tends to ∞ on the left and $-\infty$ on the right.

Quadratic Polynomial Functions.　If $f(x) = ax^2 + bx + c$ and $a \neq 0$, then f is both a quadratic function and a polynomial function of degree 2. Its graph is a parabola that either opens upward $(a > 0)$ or downward $(a < 0)$. The graphs of $f_1(x) = 0.5x^2 + 2, f_2(x) = x^2 + 4x + 4$, and $f_3(x) = -x^2 + 3x + 4$ are shown in Figures 3.56–3.58, respectively. Quadratic functions can have zero, one, or two x-intercepts. A parabola has exactly one turning point, which coincides with the vertex.

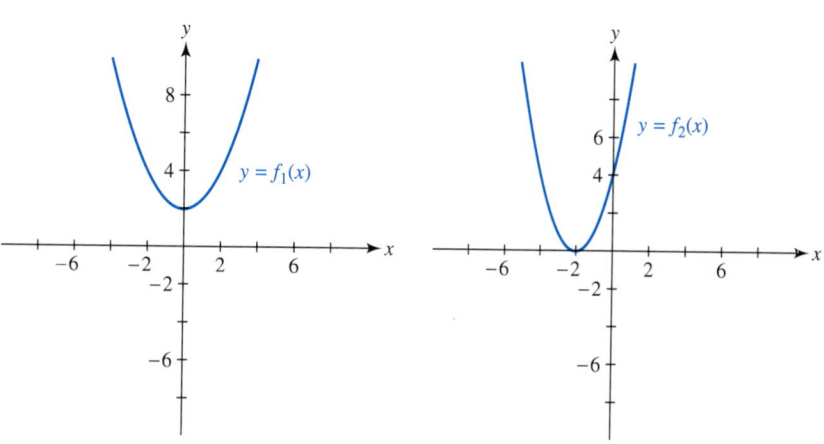

FIGURE 3.56　　　　　　　　　　**FIGURE 3.57**

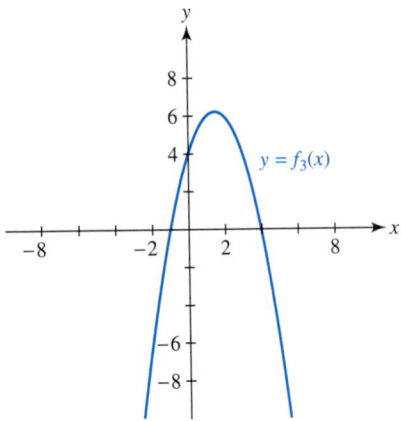

FIGURE 3.58

If $a > 0$, then both sides of the graph go up. The end behavior tends to ∞ on both sides. If $a < 0$ as in Figure 3.58, then the end behavior is reversed and tends to $-\infty$ on both sides.

Cubic Polynomial Functions. If $f(x) = ax^3 + bx^2 + cx + d$ and $a \neq 0$, then f is both a **cubic function** and a polynomial function of degree 3. The graph of a cubic function can have zero or two turning points. The graph of $f_1(x) = -x^3 + 4x + 1$ in Figure 3.59 has two turning points, whereas the graph of $f_2(x) = \frac{1}{4}x^3$ in Figure 3.60 has no turning points.

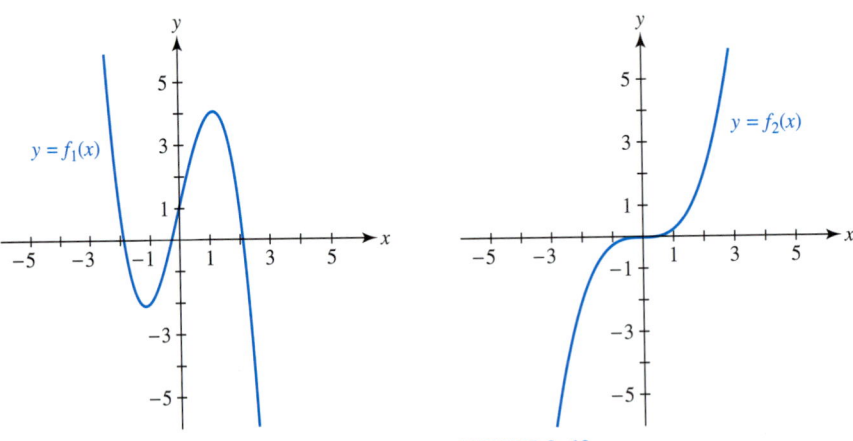

FIGURE 3.59

FIGURE 3.60

If $a > 0$, the graph of a cubic function falls to the left and rises to the right as in Figure 3.60. If $a < 0$, its graph rises to the left and falls to the right as in Figure 3.59. The end behavior of a cubic function is similar to that of a linear function, and tends to ∞ on one side and $-\infty$ on the other. Therefore, its graph must cross the x-axis at least once. A cubic function can have up to three x-intercepts. The graph of $f_3(x) = 0.1x^3 - 0.1x^2 - 2.1x + 4.5$ in Figure 3.61 has two x-intercepts.

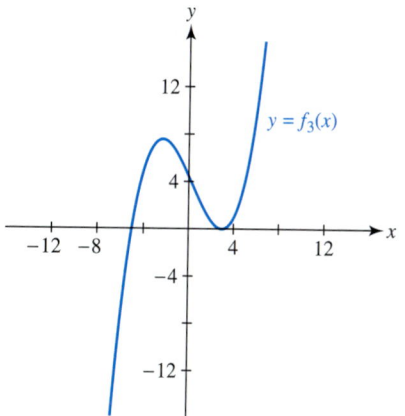

FIGURE 3.61

Quartic Polynomial Functions.   If $f(x) = ax^4 + bx^3 + cx^2 + dx + e$ and $a \neq 0$, then f is both a **quartic function** and a polynomial function of degree 4. The graph of a quartic function can have up to four x-intercepts and three turning points. The graph of $f_1(x) = 0.1x^4 - 0.3x^3 - 1.2x^2 + 1.8x + 2$ in Figure 3.62 is an example. The graph of $f_2(x) = -x^4 + 2x^3$ in Figure 3.63 has one turning point and two x-intercepts, while the graph of $f_3(x) = -x^4 + x^3 + 3x^2 - x - 2$ in Figure 3.64 has three turning points and three x-intercepts.

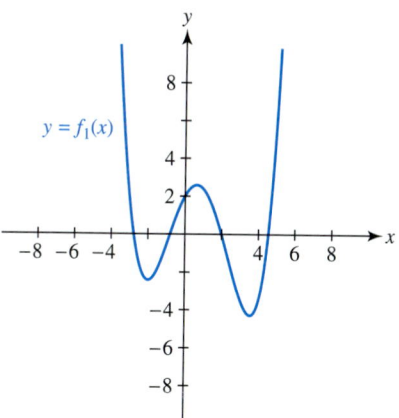

FIGURE 3.62

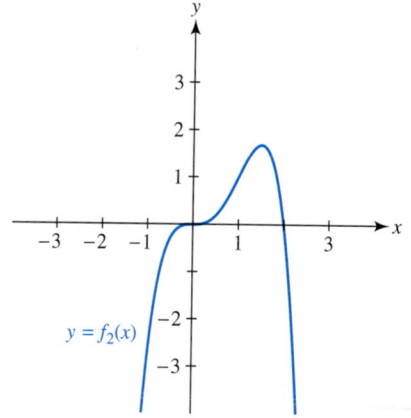

FIGURE 3.63

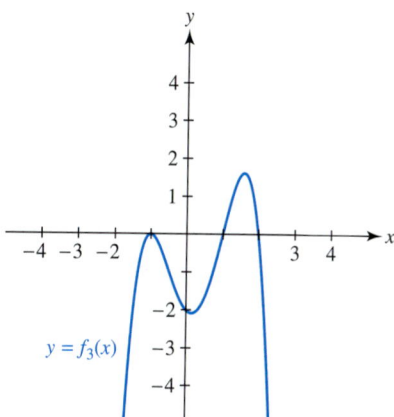

FIGURE 3.64

If $a > 0$, then both sides of the graph of a quartic function go up as in Figure 3.62. If $a < 0$, then both sides of its graph go down as in Figures 3.63 and 3.64. The end behaviors of quartic and quadratic functions are similar.

Critical Thinking

Can a quartic function have both an absolute maximum and an absolute minimum? Explain.

Quintic Polynomial Functions.   If $f(x) = ax^5 + bx^4 + cx^3 + dx^2 + ex + k$ and $a \neq 0$, then f is both a **quintic function** and a polynomial function of degree 5. The graph of a quintic function may have up to five x-intercepts and four turning points. An example is shown in Figure 3.65, given by

$$f_1(x) = 0.01x^5 + 0.03x^4 - 0.63x^3 - 0.67x^2 + 8.46x - 7.2.$$

Other quintic functions are shown in Figures 3.66 and 3.67. They are defined by

$$f_2(x) = \frac{1}{5}x^5 - 3 \quad \text{and}$$

$$f_3(x) = -0.02x^5 - 0.14x^4 + 0.04x^3 + 0.92x^2 - 1.3x + 0.5.$$

The function f_2 has one x-intercept and no turning points. The graph of f_3 appears to have two x-intercepts and two turning points. Notice that the end behavior of a quintic function is similar to linear and cubic functions.

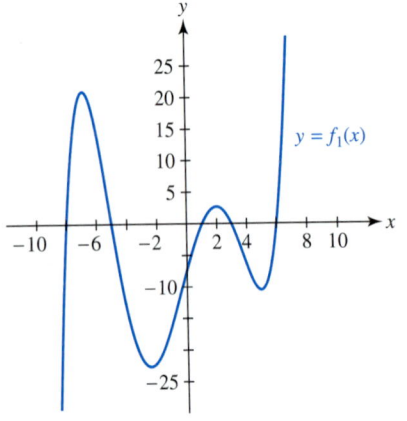

FIGURE 3.65

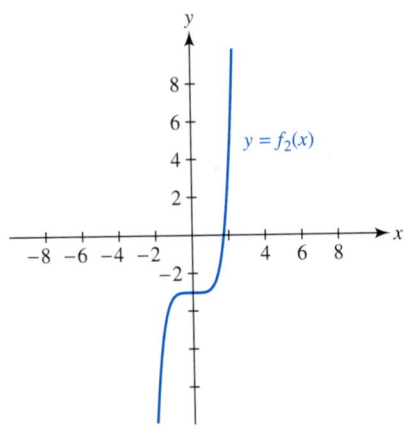

FIGURE 3.66

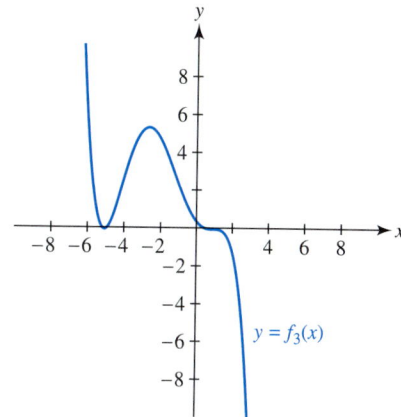

FIGURE 3.67

End behavior of polynomial functions

Let f be a polynomial function with leading coefficient a and degree n.

1. $n \geq 2$ is even.
   $a > 0$ implies the graph of f rises both to the left and to the right.
   $a < 0$ implies the graph of f falls both to the left and to the right.
2. $n \geq 1$ is odd.
   $a > 0$ implies the graph of f falls to the left and rises to the right.
   $a < 0$ implies the graph of f rises to the left and falls to the right.

The maximum numbers of x-intercepts and turning points on the graph of a polynomial function are summarized in the following.

Degree, x-intercepts, and turning points

The graph of a polynomial function of degree $n \geq 1$ has at most n x-intercepts and at most $n - 1$ turning points.

EXAMPLE 1   *Analyzing the graph of a polynomial function*

Figure 3.68 on the following page contains the graph of a polynomial function f.
(a) How many turning points and x-intercepts are there?
(b) Is the leading coefficient a positive or negative? Is the degree odd or even?
(c) Determine the minimum possible degree of f.

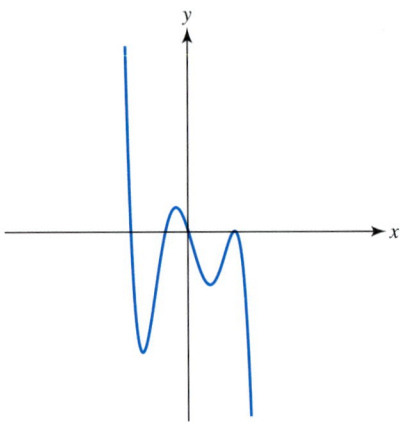

FIGURE 3.68

Solution

(a) There are four turning points corresponding to the two "hills" and two "valleys." There appears to be four x-intercepts.

(b) The left side of the graph rises and the right side falls. Therefore, $a < 0$ and the polynomial function has odd degree.

(c) The graph has four turning points. A polynomial of degree n can have at most $n - 1$ turning points. Therefore, f must be at least degree 5. ∎

Critical Thinking ───────────────────────

Can you sketch the graph of a quadratic function with no turning points, a cubic function with one turning point, or a quartic function with two turning points? Explain.

Reflection of Graphs

In Section 1.6 we discussed translations of graphs. Another type of translation is called a *reflection*. For example, the reflection of the graph of $f(x) = x^2 - x + 2$ across the x-axis is shown in Figure 3.69 as a dashed curve.

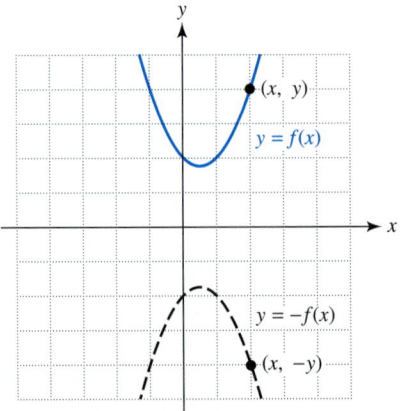

FIGURE 3.69

If (x, y) is a point on the graph of f, then $(x, -y)$ lies on the graph of its reflection across the x-axis as shown in Figure 3.69. Thus, a reflection of $y = f(x)$ is given by $-y = f(x)$, or equivalently, $y = -f(x)$. For example, a reflection of the graph of $f(x) = x^2 - x + 2$ is obtained by graphing $y = -f(x) = -(x^2 - x + 2)$.

If a point (x, y) lies on the graph of a function f, then the point $(-x, y)$ lies on the graph of its reflection across the y-axis as shown in Figure 3.70. Thus, a reflection of $y = f(x)$ is given by $y = f(-x)$. For example, a reflection of the graph of $f(x) = x^3 + x - 2$ across the y-axis is shown in Figure 3.71 as a dashed curve. Since $f(x) = x^3 + x - 2$, its reflection is given by

$$f(-x) = (-x)^3 + (-x) - 2 = -x^3 - x - 2.$$

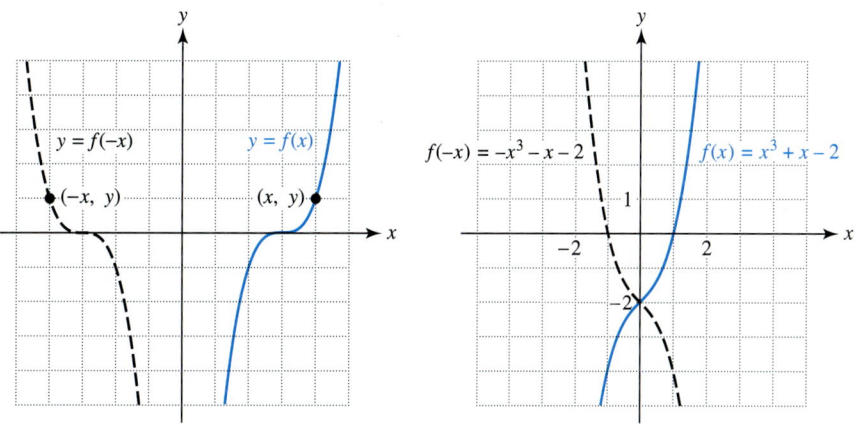

FIGURE 3.70   **FIGURE 3.71**

These results are summarized in the following.

Reflections of graphs across the *x*- and *y*-axes

1. The graph of $y = -f(x)$ is a reflection of the graph of $y = f(x)$ across the x-axis.
2. The graph of $y = f(-x)$ is a reflection of the graph of $y = f(x)$ across the y-axis.

If a graphing calculator is capable of using function notation, equations for reflections of a function f can be entered easily. For example, if $f(x) = (x - 4)^2$, then let $Y_1 = (X - 4)^2$, $Y_2 = -Y_1$, and $Y_3 = Y_1(-X)$. The graph of Y_2 is the reflection of f in the x-axis, and Y_3 is the reflection of Y_1 in the y-axis. See Figures 3.72 and 3.73 on the following page. However, it is not necessary to have this feature to plot reflections.

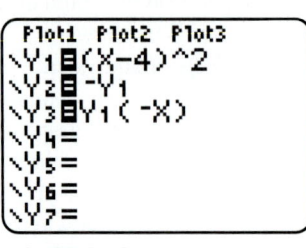

FIGURE 3.72

$[-10, 10, 1]$ by $[-10, 10, 1]$

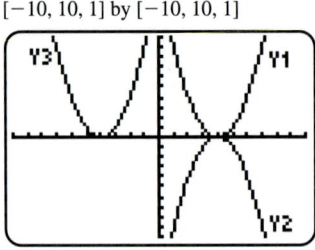

FIGURE 3.73

EXAMPLE 2   *Reflecting a function in the y-axis*

Figure 3.74 shows a partial graph of the number of daylight hours at latitude 60°N. The variable x is measured in months where $x = 0$ corresponds to December 21, the day with the least amount of daylight. The domain is $-6 \le x \le 6$. For example, $x = 2$ represents February 21 and $x = -3$ represents September 21. (Source: J. Williams, *The Weather Almanac 1995*.)

(a) Estimate the number of daylight hours on February 21.
(b) Conjecture the number of daylight hours on October 21 when $x = -2$.
(c) Sketch the left side of the graph for $-6 \le x \le 0$.

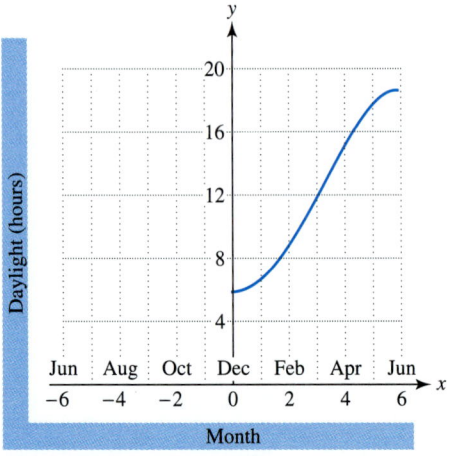

FIGURE 3.74 Daylight Hours at 60°N Latitude

Solution

(a) When $x = 2$ the y-value is about 9. There are nine hours of daylight on February 21 at 60°N latitude.

(b) Since February 21 is two months after the shortest day and October 21 is two months before, they both have approximately the same number of daylight hours. A reasonable conjecture would be nine hours.

(c) The amount of daylight would increase left or right of $x = 0$. There are approximately the same number of daylight hours either x months before or after December 21. The graph should be symmetric about the y-axis as shown in Figure 3.75.

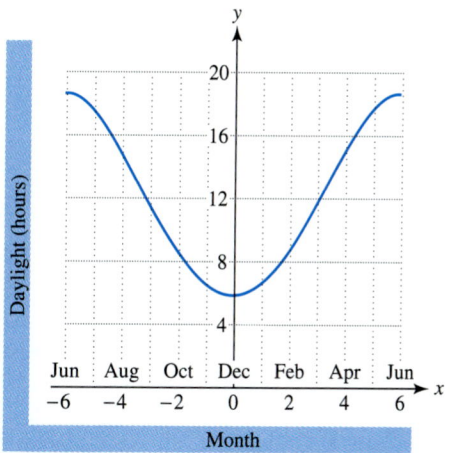

FIGURE 3.75

■

Piecewise-Defined Polynomial Functions (Optional)

In Section 2.4 piecewise-defined functions were discussed. If each piece is a polynomial, then the function is a **piecewise-defined polynomial function** or **piecewise-polynomial function.** An example is given by $f(x)$.

$$f(x) = \begin{cases} x^3 & \text{if } x < 1 \\ x^2 - 1 & \text{if } x \geq 1 \end{cases}$$

The graph of f can be determined by graphing $y = x^3$ when $x < 1$ and graphing $y = x^2 - 1$ when $x \geq 1$. At $x = 1$ there is a break in the graph, where the graph of f is discontinuous. See Figure 3.76.

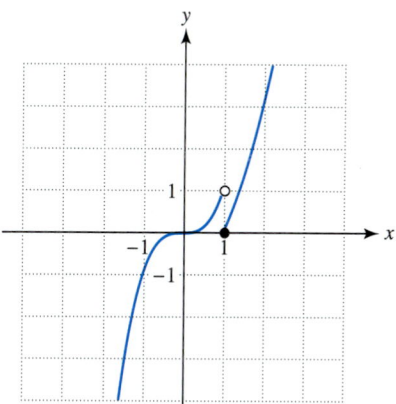

FIGURE 3.76

A cylindrical container has a height of 16 centimeters. Water entered the container at a constant rate until it was completely filled. Then, water was allowed to leak out through a small hole in the bottom. See Figure 3.77. The height of the water in the container was recorded every half minute in Table 3.4 over a five-minute period.

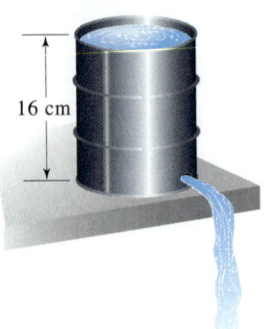

FIGURE 3.77

TABLE 3.4

Time (min)	0	0.5	1.0	1.5	2.0	2.5	3.0	3.5	4.0	4.5	5.0
Height (cm)	0	4	8	12	16	11.6	8.1	5.3	3.1	1.4	0.5

During the first two minutes, the water level rose at a constant rate of four centimeters every half minute or eight centimeters per minute. After two minutes, water was drained at a nonconstant rate. A scatterplot of the data in Table 3.4 is shown in Figure 3.78.

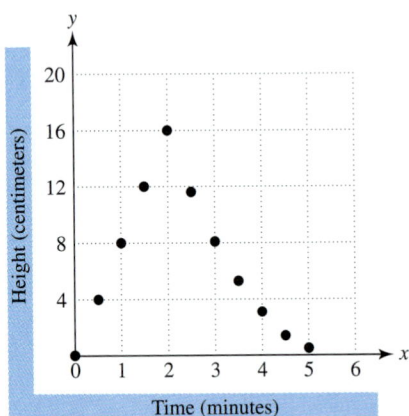

FIGURE 3.78

Water entering the container can be modeled with a linear formula, while the water leaking out can be modeled with a nonlinear formula. The function f models the water level in this example.

$$f(x) = \begin{cases} 8x & \text{if } 0 \le x \le 2 \\ 1.32x^2 - 14.4x + 39.52 & \text{if } 2 < x \le 5 \end{cases}$$

The graph of f together with a scatterplot of the data are shown in Figure 3.79. By defining f in pieces, f is capable of modeling water flowing into and out of the container.

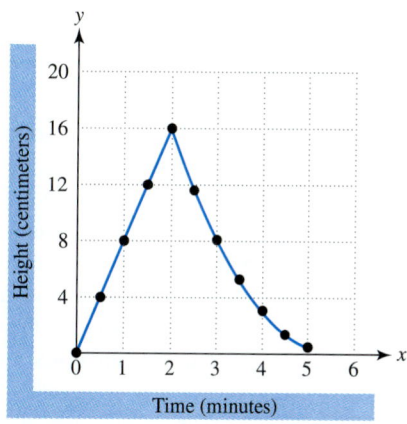

FIGURE 3.79

EXAMPLE 3   *Analyzing a piecewise-defined nonlinear function*

Use the previous function f to complete the following.
(a) Approximate the water level in the container after 1.25 and after 3.2 minutes.
(b) Estimate the time when the water level was five centimeters.

Solution

(a) When $x = 1.25$, water is flowing into the tank and $f(x) = 8x$.

$$f(1.25) = 8(1.25) = 10 \text{ cm}$$

When $x = 3.2$ minutes, water is flowing out of the tank and $f(x) = 1.32x^2 - 14.4x + 39.52$.

$$f(3.2) = 1.32(3.2)^2 - 14.4(3.2) + 39.52 = 6.9568 \approx 7 \text{ cm}$$

(b) The water level in Figure 3.79 equals five centimeters twice—once when water is entering the container and once when it is leaking out. To find these times solve the equations $8x = 5$ and $1.32x^2 - 14.4x + 39.52 = 5$.

$$8x = 5 \quad \text{or} \quad x = \frac{5}{8} = 0.625$$

Thus, after 0.625 minute or 37.5 seconds, the water level was five centimeters. The equation $1.32x^2 - 14.4x + 39.52 = 5$ is solved graphically in Figure 3.80. The solution satisfying $2 < x \leq 5$ is $x \approx 3.56$ minutes.

[2, 5, 1] by [0, 20, 5]

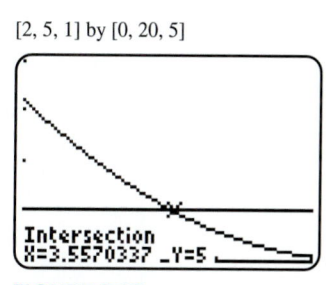

FIGURE 3.80

3.3 PUTTING IT ALL TOGETHER

Higher degree polynomials generally have more complicated graphs. Each additional degree allows the graph to have possibly one more turning point and one more x-intercept. The graph of a polynomial function is continuous and smooth with no breaks or sharp corners. Its domain includes all real numbers. The end behavior of a polynomial always tends to either ∞ or $-\infty$. End behavior describes what happens to the y-values as $|x|$ becomes large.

The following table summarizes some important concepts regarding graphs of polynomial functions.

Function Type	Characteristics	Example Graphs
Constant (degree 0)	No x-intercepts and no turning points	
Linear (degree 1)	One x-intercept and no turning points	
Quadratic (degree 2)	At most two x-intercepts and exactly one turning point	
Cubic (degree 3)	At most three x-intercepts and up to two turning points	
Quartic (degree 4)	At most four x-intercepts and up to three turning points	

Continued

Function Type	Characteristics	Example Graphs
Quintic (degree 5)	At most five x-intercepts and up to four turning points	

Piecewise-defined functions occur when a function is defined by using two or more formulas for different intervals of the domain. If each formula is a polynomial it is called a piecewise-polynomial function. Piecewise-polynomial functions are valuable when describing data that changes its character over time.

The following table lists some results involving reflections.

Type of Reflection	Example	Graphs
To reflect $y = f(x)$ across the x-axis, graph $y = -f(x)$.	If $f(x) = x^2 + 1$, graph $y = -(x^2 + 1)$ $= -x^2 - 1$.	
To reflect $y = f(x)$ across the y-axis, graph $y = f(-x)$.	If $f(x) = x^2 + 3x$, graph $y = (-x)^2 + 3(-x)$ $= x^2 - 3x$.	

3.3 EXERCISES

   Tape 3

Graphs of Polynomial Functions

1. A runner is working out on a straight track. The graph shows the runner's distance y in hundreds of feet from the starting line after x minutes.
   (a) Estimate the turning points.
   (b) Interpret each turning point.

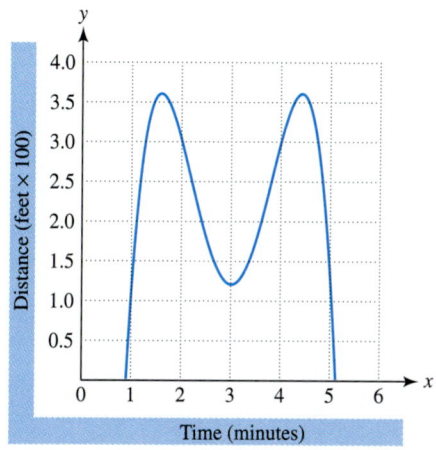

2. A stone is thrown into the air. Its height y in feet after x seconds is shown in the graph.
   (a) Estimate the turning point.
   (b) Interpret this point.

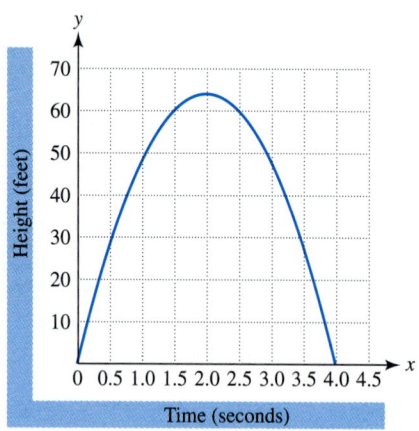

Exercises 3–8: Use the graph of the polynomial function f to complete the following.
   (a) *Estimate the x-intercept(s).*
   (b) *State whether the leading coefficient is positive or negative.*
   (c) *Determine the minimum degree of f.*

3.

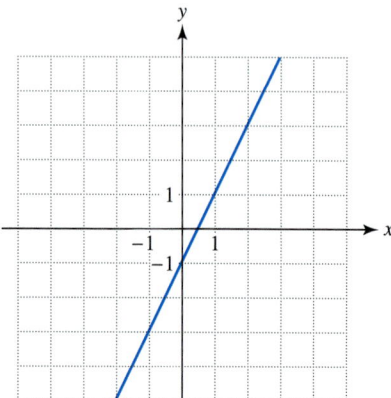

4.

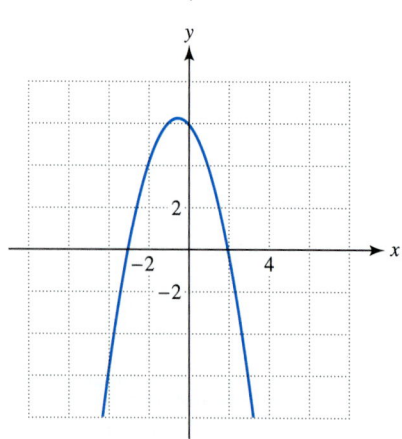

5.

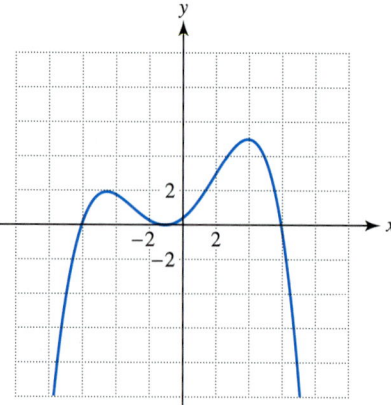

6.

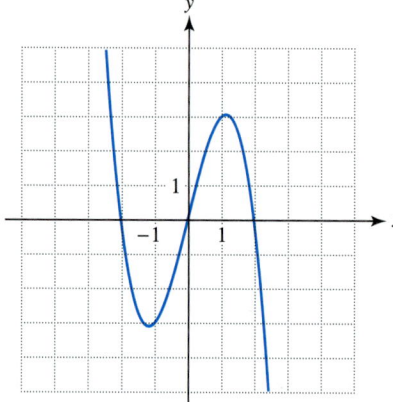

7.

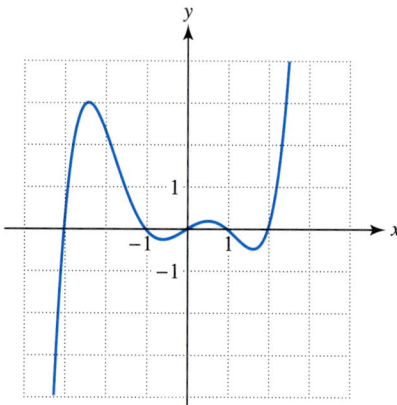

8.

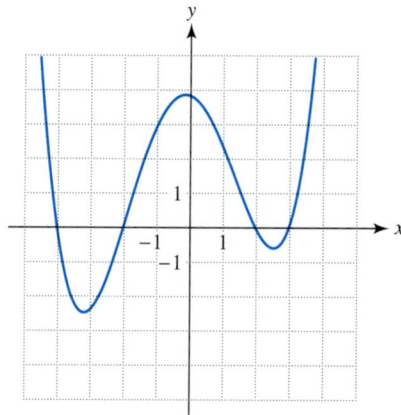

Exercises 9–12: For each polynomial f(x) complete the following.
(a) Graph f(x) in the standard viewing rectangle.
(b) Approximate the coordinates of each turning point.
(c) Use the turning points to find any local extrema.

9. $f(x) = \frac{1}{9}x^3 - 3x$

10. $f(x) = x^2 - 4x - 3$

11. $f(x) = 0.025x^4 - 0.45x^2 - 5$

12. $f(x) = -\frac{1}{8}x^4 + \frac{1}{3}x^3 + \frac{5}{4}x^2 - 3x + 3$

Exercises 13–18: For each polynomial function f complete the following.
(a) State the degree and leading coefficient.
(b) Predict the end behavior of the graph of f.
(c) Support your answer in part (b) by graphing f.

13. $f(x) = x^2 - x^3 - 4$

14. $f(x) = x^4 - 4x^3 + 3x^2 - 3$

15. $f(x) = 0.1x^5 - 2x^2 - 3x + 4$

16. $f(x) = 3x^3 - 2 - x^4$

17. $f(x) = 4 + 2x - \frac{1}{2}x^2$

18. $f(x) = -0.2x^5 + 4x^2 - 3$

Exercises 19 and 20: Graph the functions f, g, and h in the same viewing rectangle. What happens to their graphs as the size of the viewing rectangle increases? Conjecture an explanation.

19. $f(x) = 2x^4$, $g(x) = 2x^4 - 5x^2 + 1$, and $h(x) = 2x^4 + 3x^2 - x - 2$
   i. $[-4, 4, 1]$ by $[-4, 4, 1]$
   ii. $[-10, 10, 1]$ by $[-100, 100, 10]$
   iii. $[-100, 100, 10]$ by $[-10^6, 10^6, 10^5]$

20. $f(x) = -x^3$, $g(x) = -x^3 + x^2 + 2$, and $h(x) = -x^3 - 2x^2 + x - 1$
   i. $[-4, 4, 1]$ by $[-4, 4, 1]$
   ii. $[-10, 10, 1]$ by $[-100, 100, 10]$
   iii. $[-100, 100, 10]$ by $[-10^5, 10^5, 10^4]$

Exercises 21–24: The data table has been generated by a linear, quadratic, or cubic function f. All zeros of f are real numbers located in the interval $[-3, 3]$.
(a) Make a line graph of the data.
(b) Conjecture the degree of f.

21.

x	-3	-2	-1	0	1	2	3
$f(x)$	11	9	7	5	3	1	-1

22.

x	-3	-2	-1	0	1	2	3
$f(x)$	3	-8	-7	0	7	8	-3

23.

x	-3	-2	-1	0	1	2	3
$f(x)$	14	7	2	-1	-2	-1	2

24.

x	-3	-2	-1	0	1	2	3
$f(x)$	-13	-6	-1	2	3	2	-1

Sketching Graphs of Polynomials

Exercises 25–32: Sketch a graph of a polynomial that satisfies the conditions.

25. Degree 3 with three real zeros and a positive leading coefficient

26. Degree 4 with four real zeros and a negative leading coefficient

27. Linear with a negative leading coefficient

28. Cubic with one real zero and a positive leading coefficient

29. Degree 3 with turning points $(-1, 2)$ and $\left(1, \frac{2}{3}\right)$

30. Quartic with turning points $(-1, -1)$, $(0, 0)$, and $(1, -1)$

31. Degree 2 with turning point $(-1, 2)$, passing through $(-3, 4)$ and $(1, 4)$

32. Degree 5 and odd with five x-intercepts and a negative leading coefficient

Applications

33. *Marijuana Use* The table lists the percentage y of high school seniors that had used marijuana within the previous month in the United States for various

years x. In this table $x = 0$ corresponds to 1975 and $x = 20$ to 1995. (Source: Health and Human Services Department.)

x (yr)	0	3	5	10	15	20
y (%)	27	37	33	25	14	21

(a) Make a line graph of the data in $[-2, 22, 1]$ by $[10, 40, 10]$. Discuss any trends in the data.
(b) Conjecture which polynomial models the data best.
  i. $f(x) = -0.8x + 33.2$
  ii. $g(x) = -0.0036x^2 - 0.72x + 33$
  iii. $h(x) = 0.024x^3 - 0.71x^2 + 4.4x + 27$
(c) Test your conjecture by graphing each function with the data.

34. *Aging of America* The table lists the numbers in thousands of Americans expected to be over 100 years old for selected years. (Source: Bureau of the Census.)

Year	Number (thousands)
1994	50
1996	56
1998	65
2000	75
2002	94
2004	110

(a) Graph the data and the functions f, g, and h in $[1993, 2005, 1]$ by $[40, 120, 10]$. Which function appears to model the data best?
  i. $f(x) = 6(x - 1994) + 45$
  ii. $g(x) = 0.4(x - 1994)^2 + 2(x - 1994) + 50$
  iii. $h(x) = -0.06(x - 1994)^3 + 0.5(x - 1994)^2 + 1.7(x - 1994) + 50$
(b) Use your choice from part (a) to estimate the number of Americans over 100 in 2008.

35. *Highway Design* In order to allow enough distance for cars to pass on two-lane highways, engineers calculate minimum sight distances between curves and hills. See the accompanying figure. The table shows the minimum sight distance y in feet for a car traveling at x miles per hour. (Source: L. Haefner, *Introduction to Transportation Systems.*)

x (mph)	20	30	40	50	60	65	70
y (ft)	810	1090	1480	1840	2140	2310	2490

(a) Make a scatterplot of the data. As x increases, describe how y changes.
(b) What minimum degree polynomial might model this data? Find such a polynomial, $D(x)$.
(c) If a car is traveling at 43 miles per hour, estimate the minimum sight distance.

36. *Sunspots* The table shows the number of sunspots during the month of April from 1950 to 1969. (Source: R. Wolff and L. Yaeger, *Visualization of Natural Phenomena.*)

Year	1950	1951	1952	1953	1954
Sunspots	106	109	23	13	1

Year	1955	1956	1957	1958	1959
Sunspots	29	137	165	175	172

Year	1960	1961	1962	1963	1964
Sunspots	120	51	44	43	10

Year	1965	1966	1967	1968	1969
Sunspots	24	45	87	127	120

(a) Make a line graph of the data.
(b) If you were asked to find a polynomial whose graph models this data, what degree polynomial might you try first? Explain your reasoning.

37. *Modeling Temperature* In the accompanying figure the monthly average temperature in degrees Fahrenheit from January to December in Minneapolis is modeled by a polynomial function f, where $x = 1$ corresponds to January and $x = 12$

to December. (Source: A. Miller and J. Thompson, *Elements of Meteorology.*)

(a) Estimate the turning points.

(b) Interpret each turning point.

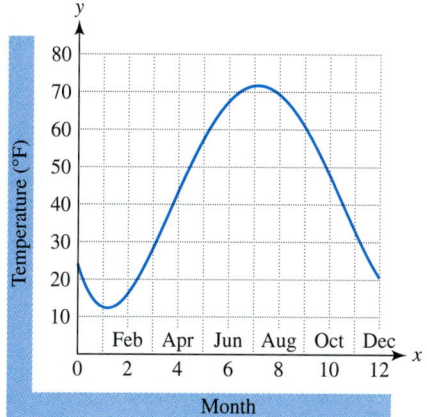

38. *Endangered Species* The total numbers y of endangered and threatened species in the United States are given in the table for various years x. (Source: Fish and Wildlife Services.)

x (yr)	1980	1985	1990	1995
y (total)	786	941	1181	1599

(a) This data may be modeled by $f(x) = a(x - 1980)^2 + 14(x - 1980) + 790$. Plot the data in [1979, 1996, 1] by [700, 1700, 100]. Determine the sign of the leading coefficient a.

(b) Estimate a by plotting the data and f. Adjust a until a satisfactory fit is found.

(c) If current trends continue, predict the number of endangered and threatened species in 2000.

Reflection

Exercises 39–42: Graph f and g if g(x) = −f(x). How does the graph of g compare to the graph of f?

39. $f(x) = x^3 - 1$

40. $f(x) = 0.2x^4 - 2x^2 + 3$

41. $f(x) = -x^2 + 2x + 2$

42. $f(x) = 0.5x^3 - 2x^2 - 3x + 7$

Exercises 43–46: Graph f and g if g(x) = f(−x). How does the graph of g compare to the graph of f?

43. $f(x) = 2x - 6$

44. $f(x) = x^2 - 4x + 2$

45. $f(x) = -x^2 + 7x - 4$

46. $f(x) = x^3 + 3x - 4$

Exercises 47–50: For the given representation of f, graph its reflection across the y-axis and across the x-axis.

47.

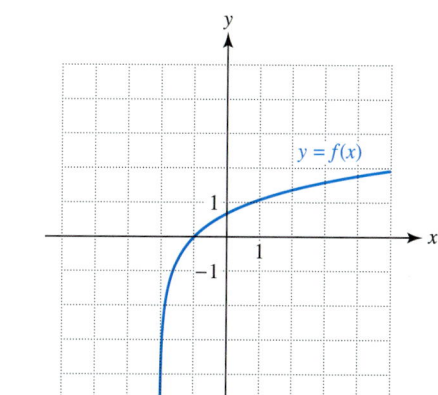

48.

x	-2	-1	0	1	2	3
$f(x)$	5	3	-2	-3	-5	-1

49. $f(x) = x^2 - 4x - 2$

50. f is the square root function.

Piecewise-Defined Functions

Exercises 51–54: Evaluate f(x) at the given values of x.

51. $x = -2$ and 1

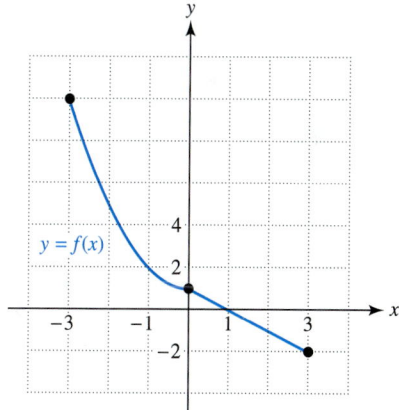

52. $x = -1, 0,$ and 3

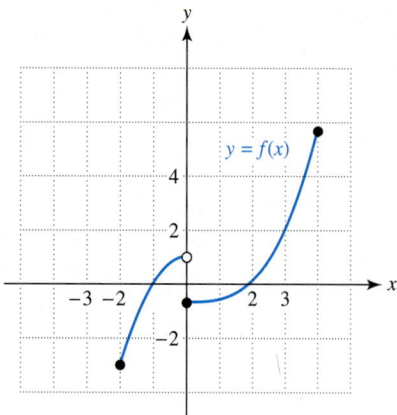

53. $x = -2, 1,$ and 2

$$f(x) = \begin{cases} x^2 + 2x + 6 & \text{if } -5 \leq x < 0 \\ x + 6 & \text{if } 0 \leq x < 2 \\ x^3 + 1 & \text{if } 2 \leq x \leq 5 \end{cases}$$

54. $x = 1975, 1980,$ and 1998

$$f(x) = \begin{cases} 0.2(x - 1970)^3 + 60 & \text{if } 1970 \leq x < 1980 \\ 190 - (x - 1980)^2 & \text{if } 1980 \leq x < 1990 \\ 2(x - 1990) + 100 & \text{if } 1990 \leq x \leq 2000 \end{cases}$$

Exercises 55–58: Complete the following.
  (a) *Sketch a graph of f.*
  (b) *Determine if f is continuous.*

55. $f(x) = \begin{cases} 4 - x^2 & \text{if } -3 \leq x \leq 0 \\ x^2 - 4 & \text{if } 0 < x \leq 3 \end{cases}$

56. $f(x) = \begin{cases} x^2 & \text{if } -2 \leq x < 0 \\ x + 1 & \text{if } 0 \leq x \leq 2 \end{cases}$

57. $f(x) = \begin{cases} 2x & \text{if } -5 \leq x < -1 \\ -2 & \text{if } -1 \leq x < 0 \\ x^2 - 2 & \text{if } 0 \leq x \leq 2 \end{cases}$

58. $f(x) = \begin{cases} 0.5x^2 & \text{if } -4 \leq x \leq -2 \\ x & \text{if } -2 < x < 2 \\ x^2 - 4 & \text{if } 2 \leq x \leq 4 \end{cases}$

59. *Modeling*   An object is lifted rapidly into the air at a constant speed and then dropped. Its height after x seconds is listed in the table.

Time (sec)	0	1	2	3	4	5	6	7
Height (ft)	0	36	72	108	144	128	80	0

(a) Make a line graph of the data. At what time does it appear that the object was dropped?
(b) Identify the time interval when the height could be modeled by a linear function. When could it be modeled by a nonlinear function?
(c) Determine values for the constants m, a, and b so that f models the data.

$$f(x) = \begin{cases} mx & \text{if } 0 \leq x \leq 4 \\ a(x - 4)^2 + b & \text{if } 4 < x \leq 7 \end{cases}$$

60. *Personal Savings*   Personal savings as a percentage of disposable income has varied between 1960 and 1995 as shown in the table. (Source: Bureau of Economic Analysis, Department of Commerce.)

Year	1960	1965	1970	1975
Savings (%)	5.6	7.1	8.0	8.6

Year	1980	1985	1990	1995
Savings (%)	6.0	4.4	5.1	3.9

(a) Make a scatterplot of the data in $[0, 35, 5]$ by $[0, 10, 1]$. Let $x = 0$ correspond to 1960, $x = 5$ to 1965, and so on, until $x = 35$ corresponds to 1995. Identify any trends in the data.
(b) The piecewise-polynomial function f models the data.

$$f(x) = \begin{cases} -0.009x^2 + 0.333x + 5.6 & \text{if } 0 \leq x < 15 \\ -0.00036667x^4 + 0.034733x^3 - \\ \quad 1.1748x^2 + 16.492x - 73.1 & \text{if } 15 \leq x \leq 35 \end{cases}$$

Estimate the percentage of disposable income saved in the years 1972 and 1988.

Writing about Mathematics

Exercises 1–4: Discuss possible local or absolute extrema on the graph of f. Assume that $a > 0$.
1. $f(x) = ax + b$
2. $f(x) = ax^2 + bx + c$
3. $f(x) = ax^3 + bx^2 + cx + d$
4. $f(x) = x^a$

3.4 Real Zeros of Polynomial Functions

Division of Polynomials • Factoring Polynomials • Graphs and
Multiple Zeros • Rational Zeros • Polynomial Equations

Introduction

In Section 3.1 the quadratic formula was used to solve $ax^2 + bx + c = 0$. Are there
similar formulas for higher degree polynomial equations? One of the most spec-
tacular mathematical achievements during the sixteenth century was the discovery
of formulas for solving cubic and quartic equations. This was accomplished by the
Italian mathematicians Tartaglia, Cardano, Fior, del Ferro, and Ferrari between
1515 and 1545. These formulas are quite complicated and typically used only in
computer software packages. Between 1750 and 1780 both Euler and Lagrange
failed at finding symbolic solutions to the quintic equation $ax^5 + bx^4 + cx^3 +
dx^2 + ex + k = 0$. Later, in about 1805, the Italian physician Ruffini proved that
formulas for quintic or higher degree equations were impossible. His results make
it necessary for us to rely on numerical and graphical methods. (Source: H. Eves,
An Introduction to the History of Mathematics.)

Division of Polynomials

First, we briefly review division of natural numbers.

$$\begin{array}{r} \text{quotient} \rightarrow 58 \text{ R } 1 \quad \leftarrow \text{remainder} \\ \text{divisor} \rightarrow 3\,)\overline{175} \leftarrow \text{dividend} \\ \underline{15} \\ 25 \\ \underline{24} \\ 1 \end{array}$$

This result is checked as follows: $3 \cdot 58 + 1 = 175$. The quotient and remainder
also can be expressed as $58\frac{1}{3}$. Since 3 does not divide into 175 evenly, 3 is not a
factor of 175. When the remainder is 0, the divisor is a factor of the dividend.
Division of polynomials is similar to division of natural numbers.

EXAMPLE 1

Dividing polynomials

Divide $2x^3 - 3x^2 - 11x + 7$ by $x - 3$. Check the result.

Solution

Begin by dividing x into $2x^3$.

$$\begin{array}{r} 2x^2 \\ x - 3\,)\overline{2x^3 - 3x^2 - 11x + 7} \\ \underline{2x^3 - 6x^2} \\ 3x^2 - 11x \end{array}$$

$\dfrac{2x^3}{x} = 2x^2$

$2x^2(x - 3) = 2x^3 - 6x^2$

Subtract. Bring down $-11x$.

In the next step, divide x into $3x^2$.

$$
\begin{array}{r}
2x^2 + 3x \\
x - 3 \overline{)\,2x^3 - 3x^2 - 11x + 7} \\
\underline{2x^3 - 6x^2} \\
3x^2 - 11x \\
\underline{3x^2 - 9x} \\
-2x + 7
\end{array}
$$

$\dfrac{3x^2}{x} = 3x$

$3x(x - 3) = 3x^2 - 9x$

Subtract. Bring down 7.

Now divide x into $-2x$.

$$
\begin{array}{r}
2x^2 + 3x - 2 \\
x - 3 \overline{)\,2x^3 - 3x^2 - 11x + 7} \\
\underline{2x^3 - 6x^2} \\
3x^2 - 11x \\
\underline{3x^2 - 9x} \\
-2x + 7 \\
\underline{-2x + 6} \\
1
\end{array}
$$

$\dfrac{-2x}{x} = -2$

$-2(x - 3) = -2x + 6$

Subtract. Remainder is 1.

The quotient is $2x^2 + 3x - 2$ and the remainder is 1. This can be written as $2x^2 + 3x - 2 + \dfrac{1}{x - 3}$ in the same manner as 58 R 1 was expressed as $58\frac{1}{3}$. Polynomial division is checked by multiplying the divisor with the quotient and then adding the remainder.

$$
\begin{aligned}
(x - 3)(2x^2 + 3x - 2) + 1 &= x(2x^2 + 3x - 2) - 3(2x^2 + 3x - 2) + 1 \\
&= 2x^3 + 3x^2 - 2x - 6x^2 - 9x + 6 + 1 \\
&= 2x^3 - 3x^2 - 11x + 7 \checkmark
\end{aligned}
$$

This process illustrates an algorithm for division of polynomials by $x - k$.

Division algorithm for polynomials

For any polynomial $f(x)$ with degree $n \geq 1$ and any number k, there exists a unique polynomial $q(x)$ and a number r such that

$$
f(x) = (x - k)\,q(x) + r.
$$

The degree of $q(x)$ is one less than the degree of $f(x)$ and r is called the *remainder*.

Synthetic Division. There is a shortcut called *synthetic division* that can be used to divide $x - k$ into a polynomial. For example, to divide $x - 2$ into $f(x) =$

$3x^4 - 7x^3 - 4x + 5$, we perform the following steps. The equivalent steps involving long division are shown to the right.

$$
\begin{array}{r|rrrrr}
2 & 3 & -7 & 0 & -4 & 5 \\
  &   & 6 & -2 & -4 & -16 \\
\hline
  & 3 & -1 & -2 & -8 & -11
\end{array}
$$

$$
\begin{array}{r}
3x^3 - x^2 - 2x - 8 \\
x - 2 \overline{)\,3x^4 - 7x^3 + 0x^2 - 4x + 5} \\
\underline{3x^4 - 6x^3} \\
-1x^3 + 0x^2 \\
\underline{-1x^3 + 2x^2} \\
-2x^2 - 4x \\
\underline{-2x^2 + 4x} \\
-8x + 5 \\
\underline{-8x + 16} \\
-11
\end{array}
$$

Notice how the highlighted numbers in the expression for long division correspond to the third row in synthetic division. The remainder is -11, which is the last number in the third row. The quotient, $3x^3 - x^2 - 2x - 8$, is one degree less than $f(x)$. Its coefficients are 3, -1, -2, and -8 and are found in the third row. The steps to divide $f(x)$ by $x - k$ using synthetic division are summarized.

1. Write k to the left and the coefficients of $f(x)$ to the right in the top row. If any power of x does not appear in $f(x)$, include a zero for that term. In this example an x^2-term did not appear, so a 0 is included in the first row.
2. Copy the leading coefficient of $f(x)$ into the third row and multiply it by k. Write the result below the next coefficient of $f(x)$ in the second row. Add the numbers in the second column and place the result in the third row. Repeat the process. In this example, the leading coefficient is 3 and $k = 2$. Since $3 \cdot 2 = 6$, 6 is placed below -7. Then add to obtain $-7 + 6 = -1$. Multiply -1 by 2 and repeat.
3. The last number in the third row is the remainder. If the remainder is 0, then $x - k$ is a factor of $f(x)$. The other numbers in the third row are the coefficients of the quotient in descending powers.

EXAMPLE 2   *Performing synthetic division*

Use synthetic division to divide $2x^3 + 4x^2 - x + 5$ by $x + 2$.

Solution
Let $k = -2$ and perform the following.

$$
\begin{array}{r|rrrr}
-2 & 2 & 4 & -1 & 5 \\
   &   & -4 & 0 & 2 \\
\hline
   & 2 & 0 & -1 & 7
\end{array}
$$

The remainder is 7 and the quotient is $2x^2 - 1$. This result is expressed by the equation

$$
\frac{2x^3 + 4x^2 - x + 5}{x + 2} = 2x^2 - 1 + \frac{7}{x + 2}.
$$ ∎

If we let $x = k$ in the division algorithm, $f(x) = (x - k)q(x) + r$, then

$$
f(k) = (k - k)q(k) + r = r.
$$

Thus, $f(k)$ is equal to the remainder obtained in synthetic division. In Example 2 when $f(x) = 2x^3 + 4x^2 - x + 5$ was divided by $x + 2$ the remainder was 7. It follows that $f(-2) = 7$. This result is summarized by the following theorem.

Remainder theorem

If a polynomial $f(x)$ is divided by $x - k$, the remainder is $f(k)$.

Factoring Polynomials

The polynomial $f(x) = x^2 - 3x + 2$ can be factored as $f(x) = (x - 1)(x - 2)$. Notice that $f(1) = 0$ and $(x - 1)$ is a factor of $f(x)$. Similarly, $f(2) = 0$ and $(x - 2)$ is a factor. This concept is known as the factor theorem and is a consequence of the remainder theorem.

Factor theorem

A polynomial $f(x)$ has a factor $x - k$ if and only if $f(k) = 0$.

The polynomial $f(x) = x^2 + 4x + 4$ can be written as $f(x) = (x + 2)^2$. Since the factor $(x + 2)$ occurs twice in $f(x)$, the zero -2 is called a **zero of multiplicity two.** The polynomial $g(x) = (x + 1)^3(x - 2)$ has zeros -1 and 2 with *multiplicities* 3 and 1, respectively. A graph of g is shown in Figure 3.81, where the x-intercepts coincide with the zeros of g. *Counting multiplicities,* a polynomial of degree n has at most n real zeros. For $g(x)$, the sum of the multiplicities is $3 + 1 = 4$, which equals its degree.

These concepts can be used to find the *complete factored form* of a polynomial.

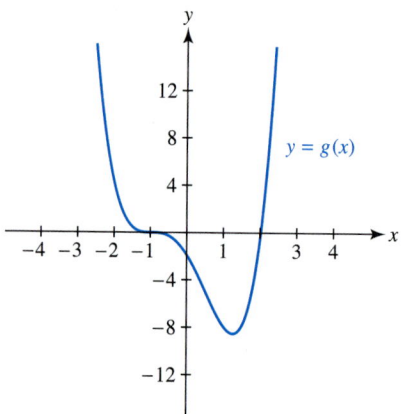

FIGURE 3.81

> ### Complete factored form
>
> Suppose a polynomial
>
> $$f(x) = a_n x^n + \cdots + a_2 x^2 + a_1 x + a_0$$
>
> has n real zeros $c_1, c_2, c_3, \ldots, c_n$, where distinct zeros are listed as many times as their multiplicity. Then $f(x)$ can be written in **complete factored form** as
>
> $$f(x) = a_n(x - c_1)(x - c_2)(x - c_3) \cdots (x - c_n).$$

EXAMPLE 3   *Factoring a polynomial with known zeros*

The polynomial $f(x) = 3x^6 - 9x^5 - 45x^4 + 261x^3 - 486x^2 + 396x - 120$ has real zeros of -5, 1, and 2 with multiplicities 1, 2, and 3, respectively. Write $f(x)$ in complete factored form.

Solution
Start by listing each zero the same number of times as its multiplicity.

$$c_1 = -5, \qquad c_2 = 1, \qquad c_3 = 1, \qquad c_4 = 2, \qquad c_5 = 2, \qquad c_6 = 2$$

The leading coefficient of $f(x)$ is 3. Thus, the complete factored form of $f(x)$ is

$$f(x) = 3(x -(-5))(x - 1)(x - 1)(x - 2)(x - 2)(x - 2) \qquad \text{or}$$
$$f(x) = 3(x + 5)(x - 1)^2(x - 2)^3. \qquad\qquad \blacksquare$$

Sometimes a polynomial may be factored using graphical methods. The next example illustrates this technique.

EXAMPLE 4   *Factoring a cubic polynomial graphically*

Factor $f(x) = 2x^3 - 8x^2 + 3x + 4$.

Solution
Graph $Y_1 = 2X^3 - 8X^2 + 3X + 4$ as in Figure 3.82, where one zero of f is $c_1 \approx -0.51966$. The other two zeros are $c_2 \approx 1.1382$ and $c_3 \approx 3.3815$.

The leading coefficient of $f(x)$ is $a_n = 2$. The complete factored form of f is

$$f(x) = 2(x - c_1)(x - c_2)(x - c_3) \qquad \text{or}$$
$$f(x) \approx 2(x + 0.51966)(x - 1.1382)(x - 3.3815).$$

$[-8, 8, 1]$ by $[-8, 8, 1]$

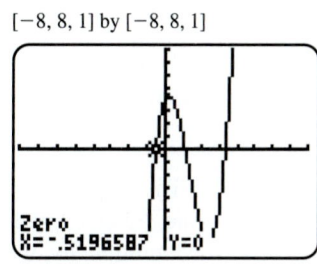

FIGURE 3.82 $\blacksquare$

EXAMPLE 5   *Factoring a polynomial with real zeros*

The polynomial $f(x) = 2x^3 - 2x^2 - 34x - 30$ has a zero of -1. Express $f(x)$ in complete factored form.

Solution
If -1 is a zero, then by the factor theorem $(x + 1)$ is a factor of $2x^3 - 2x^2 - 34x - 30$. To factor $f(x)$, divide $x + 1$ into $2x^3 - 2x^2 - 34x - 30$ using synthetic division.

$$
\begin{array}{r|rrrr}
-1 & 2 & -2 & -34 & -30 \\
   &   & -2 & 4   & 30 \\
\hline
   & 2 & -4 & -30 & 0
\end{array}
$$

The remainder is 0, so $x + 1$ divides evenly into the dividend.

$$2x^3 - 2x^2 - 34x - 30 = (x + 1)(2x^2 - 4x - 30)$$

The quotient $2x^2 - 4x - 30$ can be factored further.

$$
\begin{aligned}
2x^2 - 4x - 30 &= 2(x^2 - 2x - 15) \\
&= 2(x + 3)(x - 5)
\end{aligned}
$$

Thus, the complete factored form is $f(x) = 2(x + 1)(x + 3)(x - 5)$.   ■

Graphs and Multiple Zeros

The polynomial $f(x) = 0.02(x + 3)^3(x - 3)^2$ has zeros -3 and 3 with multiplicities 3 and 2, respectively. At the zero of even multiplicity the graph does not cross the x-axis, whereas the graph does cross the x-axis at the zero of odd multiplicity. See Figure 3.83.

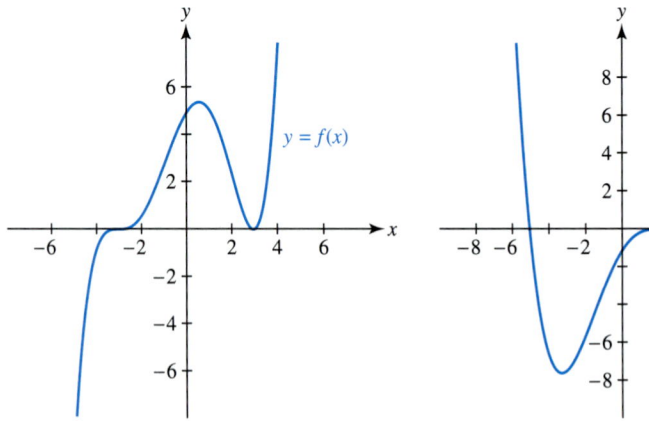

FIGURE 3.83   **FIGURE 3.84**

The graph of $g(x) = 0.03(x + 5)(x - 2)^3$ is shown in Figure 3.84. The zeros of g are -5 and 2 with multiplicities of 1 and 3, respectively. Since both zeros have odd multiplicity, the graph crosses the x-axis at -5 and 2. Notice that the zero 2 has a higher multiplicity and the graph levels off more near $x = 2$ than it does near

$x = -5$. The higher the multiplicity of a zero, the more the graph of a polynomial levels off near the zero.

EXAMPLE 6   *Finding the multiplicity of a zero graphically*

Figure 3.85 shows the graph of a sixth-degree polynomial $f(x)$ with leading coefficient 1. All zeros are integers. Write $f(x)$ in complete factored form.

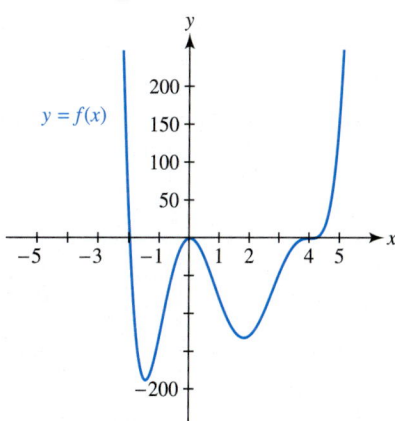

FIGURE 3.85

Solution

The x-intercepts or zeros of f are -2, 0, and 4. Since the graph crosses the x-axis at -2 and 4, these zeros have odd multiplicity. The graph of f levels off more at $x = 4$ than at $x = -2$, so 4 has a higher multiplicity than -2. At $x = 0$ the graph of f does not cross the x-axis. Thus, 0 has even multiplicity. If -2 has multiplicity 1, 0 has multiplicity 2, and 4 has multiplicity 3, then the sum of the multiplicities is $1 + 2 + 3 = 6$, which equals the degree of $f(x)$. Let

$$c_1 = -2, \quad c_2 = 0, \quad c_3 = 0, \quad c_4 = 4, \quad c_5 = 4, \quad \text{and} \quad c_6 = 4.$$

The leading coefficient is 1, so the complete factorization of $f(x)$ is

$$f(x) = 1(x + 2)(x - 0)(x - 0)(x - 4)(x - 4)(x - 4) \quad \text{or}$$
$$f(x) = x^2(x + 2)(x - 4)^3.$$

    Multiple zeros sometimes have physical significance in an application. The next example shows how a multiple zero represents the boundary between an object floating or sinking.

EXAMPLE 7   *Interpreting a multiple zero*

The polynomial $f(x) = 1.0472x^3 - 15.708x^2 + 523.599d$ can be used to find the depth that a ball, 10 centimeters in diameter, sinks in water. See Figure 3.86 on the following page. The constant d is the density of the ball, where the density of water is 1. The smallest *positive* zero of $f(x)$ equals the depth that the sphere sinks. Approximate this depth for each material and interpret the results.

(a) A wood ball with $d = 0.8$
(b) A solid aluminum sphere with $d = 2.7$
(c) A water balloon with $d = 1$

FIGURE 3.86

Solution

(a) Let $d = 0.8$ and graph $Y_1 = 1.0472X^3 - 15.708X^2 + 523.599(0.8)$. In Figure 3.87 the smallest positive zero is near 7.13. This means that the 10-centimeter wood ball sinks about 7.13 centimeters into the water.

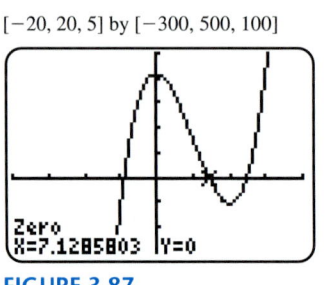

$[-20, 20, 5]$ by $[-300, 500, 100]$

Zero
X=7.1285803   Y=0

FIGURE 3.87

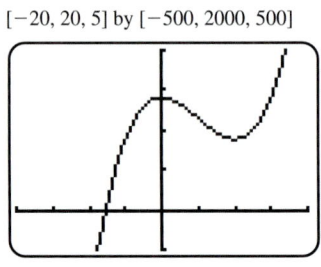

$[-20, 20, 5]$ by $[-500, 2000, 500]$

FIGURE 3.88

(b) Let $d = 2.7$ and graph $Y_1 = 1.0472X^3 - 15.708X^2 + 523.599(2.7)$. In Figure 3.88 there is no smallest positive zero. The aluminum sphere is more dense than water and sinks.

(c) Let $d = 1$ and graph $Y_1 = 1.0472X^3 - 15.708X^2 + 523.599$. The graph in Figure 3.89 appears to have one positive zero near 10 with multiplicity 2. The water balloon has the same density as water and "floats" even with the surface. The value of $d = 1$ represents the boundary between sinking and floating. If the ball floats, $f(x)$ has two positive zeros—if it sinks, $f(x)$ has no positive zeros. With the water balloon there is one positive zero with multiplicity 2 that represents a transition point between floating and sinking. (*Note:* If one zooms in sufficiently, f actually has two zeros close to 10. This is because the coefficients of f are rounded.)

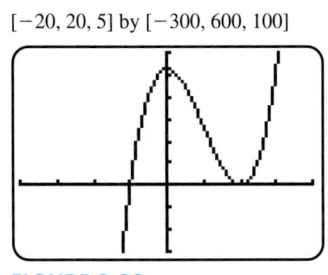

$[-20, 20, 5]$ by $[-300, 600, 100]$

FIGURE 3.89

■

Critical Thinking

Conjecture the depth that a ball with a 10-centimeter diameter will sink in water if $d = 0.5$. Test your conjecture graphically.

Rational Zeros

If a polynomial has a zero that is rational, it can be found using the rational zero test.

> ### Rational zero test
>
> Let $f(x) = a_n x^n + \cdots + a_2 x^2 + a_1 x + a_0$, where $a_n \neq 0$, represent a polynomial function f with integer coefficients. If $\dfrac{p}{q}$ is a rational number written in lowest terms and if $\dfrac{p}{q}$ is a zero of f, then p is a factor of the constant term a_0 and q is a factor of the leading coefficient a_n.

The following example illustrates how to find rational zeros using this test.

EXAMPLE 8   *Finding rational zeros of a polynomial*

Find all rational zeros of $f(x) = 6x^3 - 5x^2 - 7x + 4$. Write $f(x)$ in complete factored form.

Solution

If $\dfrac{p}{q}$ is a rational zero in lowest terms, then p is a factor of the constant term 4 and q is a factor of the leading coefficient 6. The possible values for p and q are as follows.

$$p: \quad \pm 1, \quad \pm 2, \quad \pm 4$$
$$q: \quad \pm 1, \quad \pm 2, \quad \pm 3, \quad \pm 6$$

As a result, any rational zero of $f(x)$ in the form $\dfrac{p}{q}$ must occur in the list

$$\pm \frac{1}{6}, \quad \pm \frac{1}{3}, \quad \pm \frac{1}{2}, \quad \pm \frac{2}{3}, \quad \pm 1, \quad \pm \frac{4}{3}, \quad \pm 2, \quad \text{or} \quad \pm 4.$$

Evaluate $f(x)$ at each value in the list. See Table 3.5.

TABLE 3.5

x	$f(x)$	x	$f(x)$	x	$f(x)$	x	$f(x)$
$\frac{1}{6}$	$\frac{49}{18}$	$\frac{1}{2}$	0	$\frac{2}{3}$	$-\frac{10}{9}$	2	18
$-\frac{1}{6}$	5	$-\frac{1}{2}$	$\frac{11}{2}$	$-\frac{2}{3}$	$\frac{14}{3}$	-2	-50
$\frac{1}{3}$	$\frac{4}{3}$	1	-2	$\frac{4}{3}$	0	4	280
$-\frac{1}{3}$	$\frac{50}{9}$	-1	0	$-\frac{4}{3}$	$-\frac{88}{9}$	-4	-432

From Table 3.5 there are three rational zeros of -1, $\frac{1}{2}$, and $\frac{4}{3}$. Since a third degree polynomial has at most three zeros, the complete factored form of $f(x)$ is

$$f(x) = 6(x + 1)\left(x - \frac{1}{2}\right)\left(x - \frac{4}{3}\right).$$ ∎

Although $f(x)$ in Example 8 had only rational zeros, it is important to realize that many polynomials have irrational zeros. Irrational zeros cannot be found using the rational zero test.

Polynomial Equations

In Section 3.1, factoring was used to solve quadratic equations. Factoring also can be used to solve polynomial equations with degrees greater than two. This is illustrated in the next example.

EXAMPLE 9   *Solving a cubic equation*

Solve $x^3 + 3x^2 - 4x = 0$ symbolically. Support your answer graphically and numerically.

Solution

Symbolic Solution

$x^3 + 3x^2 - 4x = 0$	Given equation
$x(x^2 + 3x - 4) = 0$	Factor out x.
$x(x + 4)(x - 1) = 0$	Factor the quadratic expression.
$x = 0$, $x + 4 = 0$, or $x - 1 = 0$	Zero-product property
$x = 0, -4,$ or 1	Solve.

Graphical Solution   Graph $Y_1 = X^3 + 3X^2 - 4X$ as in Figure 3.90. The x-intercepts are $-4, 0,$ and 1, which correspond to the solutions.

Numerical Solution   Table $Y_1 = X^3 + 3X^2 - 4X$ as in Figure 3.91. The zeros of Y_1 occur at $x = -4, 0,$ and 1.

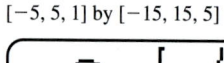

$[-5, 5, 1]$ by $[-15, 15, 5]$

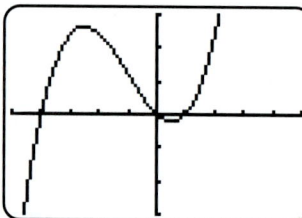

FIGURE 3.90   **FIGURE 3.91**   ∎

MAKING CONNECTIONS
Functions and Equations

Each time a new type of function is defined, we can define a new type of equation. For example, cubic functions given by $f(x) = ax^3 + bx^2 + cx + d$ can be used to define cubic equations of the form $ax^3 + bx^2 + cx + d = 0$. This concept will be used to define other types of equations.

EXAMPLE 10   *Solving a polynomial equation*

Solve $x^4 + 4x^2 - 12 = 0$ symbolically. Support your answer graphically and numerically.

Solution

Symbolic Solution   The equation $x^4 + 4x^2 - 12 = 0$ can be solved in a manner similar to how quadratic equations are solved.

$x^4 + 4x^2 - 12 = 0$	Given equation
$(x^2 + 6)(x^2 - 2) = 0$	Factor.
$x^2 + 6 = 0$ or $x^2 - 2 = 0$	Zero-product property
$x^2 = -6$ or $x^2 = 2$	Zero-product property
$x = \pm\sqrt{2}$	Square root property

Notice that there are no real solutions to $x^2 = -6$, since $x^2 \geq 0$ for every real number x.

Graphical Solution   Graph $Y_1 = X^4 + 4X^2 - 12$. The x-intercepts are approximately $\pm 1.414214 \approx \pm\sqrt{2}$. See Figures 3.92 and 3.93.

$[-4, 4, 1]$ by $[-15, 15, 5]$ 　　　　　 $[-4, 4, 1]$ by $[-15, 15, 5]$

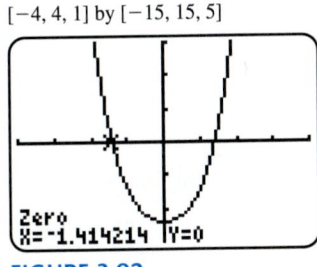

FIGURE 3.92

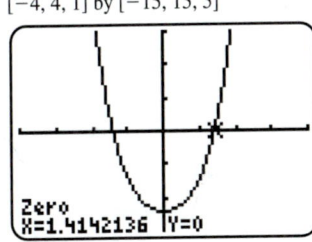

FIGURE 3.93

Numerical Solution   Table $Y_1 = X^4 + 4X^2 - 12$ as in Figure 3.94. From $x = -1.5$ to $x = -1.4$, Y_1 changes from positive to negative. Since the graph of Y_1 is continuous, there is an x-intercept between these two values. In a similar manner there is an x-intercept between $x = 1.4$ and $x = 1.5$. See Figure 3.95.

X	Y1
-1.7	7.9121
-1.6	4.7936
-1.5	2.0625
-1.4	-.3184
-1.3	-2.384
-1.2	-4.166
-1.1	-5.696

Y1◻X^4+4X^2-12

FIGURE 3.94

X	Y1
1.1	-5.696
1.2	-4.166
1.3	-2.384
1.4	-.3184
1.5	2.0625
1.6	4.7936
1.7	7.9121

Y1◻X^4+4X^2-12

FIGURE 3.95

3.4 PUTTING IT ALL TOGETHER

The following table lists some important concepts related to a polynomial $f(x)$.

Concept	Example
Factor theorem $(x - k)$ is a factor of $f(x)$ if and only if $f(k) = 0$.	$f(x) = x^2 + 3x - 4$ and $f(1) = 0$ implies that $(x - 1)$ is a factor of $f(x)$. That is, $f(x) = (x - 1)(x + 4)$.
Complete factored form $f(x) = a_n(x - c_1)(x - c_2)(x - c_3) \cdots (x - c_n)$, where the c_k are zeros of f, listed as many times as their multiplicities.	$f(x) = 3(x - 5)(x + 3)(x + 3)$ $= 3(x - 5)(x + 3)^2$ $c_1 = 5, \quad c_2 = -3, \quad c_3 = -3$
Division algorithm $f(x) = (x - k)\, q(x) + r$, where r is the remainder.	$f(x) = (x - 1)(x^2 - x - 2) + 3$ $q(x) = x^2 - x - 2$ and $r = 3$
Zero with odd multiplicity The graph of $y = f(x)$ crosses the x-axis at a zero of odd multiplicity.	$f(x) = (x + 1)^3(x - 3)$ Both zeros of -1 and 3 have odd multiplicity.
Zero with even multiplicity The graph of $y = f(x)$ intersects but does not cross the x-axis at a zero of even multiplicity.	$f(x) = (x + 1)^2(x - 3)^4$ Both zeros of -1 and 3 have even multiplicity.

3.4 EXERCISES

   Tape 4

Division of Polynomials

Exercises 1–8: Divide the first polynomial by the second. State the quotient and remainder.

1. $x^3 - 2x^2 - 5x + 6, \quad x - 3$
2. $3x^3 - 10x^2 - 27x + 10, \quad x + 2$
3. $2x^4 - 7x^3 - 5x^2 - 19x + 17, \quad x + 1$
4. $x^4 - x^3 - 4x + 1, \quad x - 2$
5. $3x^3 - 7x + 10, \quad x - 1$
6. $x^4 - 16x^2 + 1, \quad x + 4$
7. $x^3 - 2.5x^2 + 0.5x + 1, \quad x + 0.5$
8. $x^4 - 3.5x^3 + 6.5x - 3, \quad x - 1.5$
9. $\dfrac{x^3 - 8x^2 + 15x - 6}{x - 2} = x^2 - 6x + 3$ implies $(x - 2)(x^2 - 6x + 3) = \underline{\quad ? \quad}$.
10. $\dfrac{x^4 - 15}{x + 2} = x^3 - 2x^2 + 4x - 8 + \dfrac{1}{x + 2}$ implies $x^4 - 15 = (x + 2) \times \underline{\quad ? \quad} + \underline{\quad ? \quad}$.

Factoring Polynomials

Exercises 11–18: Write the complete factored form of $f(x)$.

11. $f(x) = 2x^2 - 25x + 77$; zeros: $\dfrac{11}{2}$ and 7
12. $f(x) = 6x^2 + 21x - 90$; zeros: -6 and $\dfrac{5}{2}$
13. $f(x) = x^3 - 2x^2 - 5x + 6$; zeros: $-2, 1,$ and 3
14. $f(x) = x^3 + 6x^2 + 11x + 6$; zeros: $-3, -2,$ and -1
15. $f(x) = -2x^3 + 3x^2 + 59x - 30$; zeros: $-5, \dfrac{1}{2},$ and 6
16. $f(x) = 3x^4 - 8x^3 - 67x^2 + 112x + 240$; zeros: $-4, -\dfrac{4}{3}, 3,$ and 5

17. $f(x) = 2x^4 - 19x^3 + 64x^2 - 89x + 42$;
zeros: 1, 2, 3, and $\dfrac{7}{2}$

18. $f(x) = -2x^4 - 5x^3 + 10x^2 + 15x - 18$;
zeros: $-3, -2, 1$, and $\dfrac{3}{2}$

Exercises 19–22: Write a polynomial in complete factored form that satisfies the stated conditions.

19. Degree 3; leading coefficient 2;
zeros $-3, 2,$ and 10

20. Degree 4; leading coefficient -3;
zeros $-1, 2, 3$, and $\dfrac{17}{4}$

21. Degree 4; leading coefficient -7;
zeros $-10, -\dfrac{7}{3}, -\sqrt{3}$, and $\sqrt{3}$

22. Degree 5; leading coefficient 4;
zeros $-1.7, -2.5, 3, 6,$ and 9.77

Exercises 23–26: The graph of a polynomial $f(x)$ with leading coefficient ± 1 and integer zeros is shown in the figure. Write its complete factored form.

23.

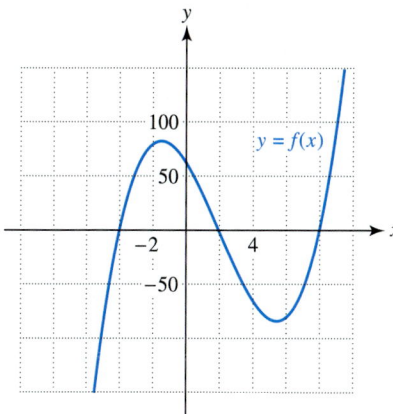

24.

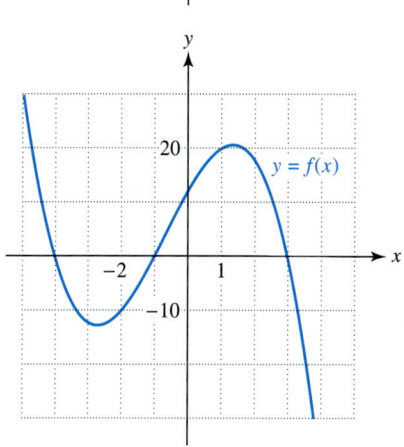

25.

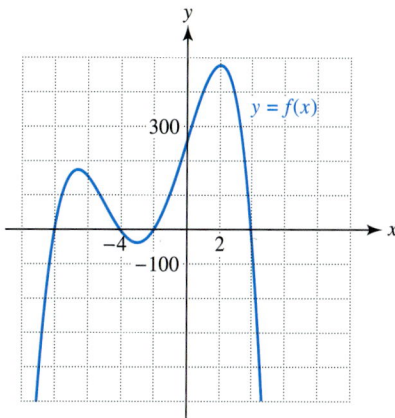

26.

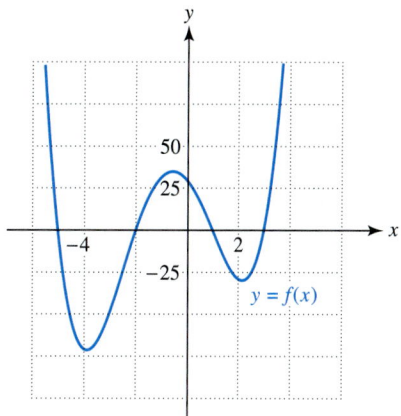

Exercises 27–32: (Refer to Example 4.) Approximate graphically the complete factored form of $f(x)$.

27. $f(x) = x^3 - 1.1x^2 - 5.9x + 0.7$

28. $f(x) = x^3 + x^2 - 18x + 13$

29. $f(x) = -0.7x^3 - 2x^2 + 4x + 2.5$

30. $f(x) = 3x^3 - 46x^2 + 180x - 99$

31. $f(x) = 2x^4 - 1.5x^3 - 24x^2 - 10x + 13$

32. $f(x) = -x^4 + 2x^3 + 20x^2 - 22x - 41$

Exercises 33–36: The graph of a polynomial f(x) is shown in the figure. Estimate the zeros and state whether their multiplicities are odd or even.

33.

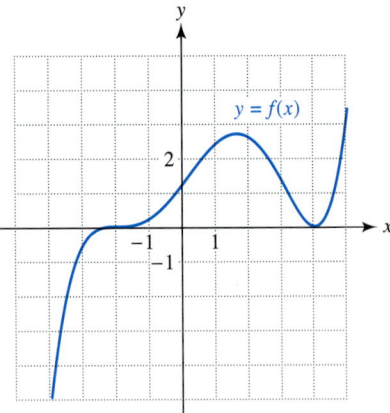

34.

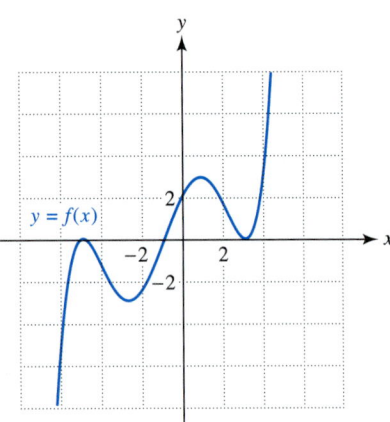

35.

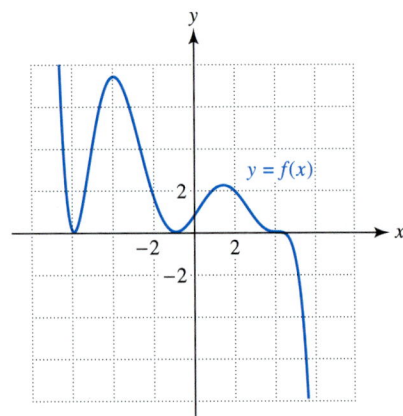

36.

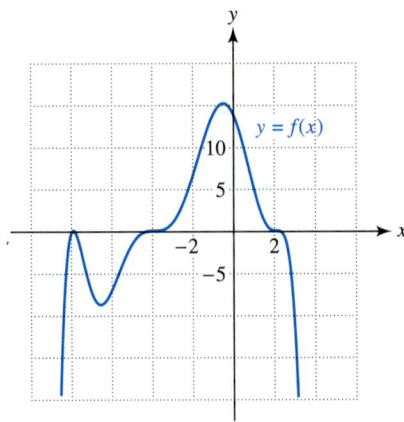

Exercises 37–40: Write a polynomial f(x) in complete factored form that satisfies the conditions. Support your answer graphically.

37. Degree 3; zeros: -1 with multiplicity 2, and 6 with multiplicity 1

38. Degree 4; zeros: 5 and 7, both with multiplicity 2

39. Degree 4; zeros: 2 with multiplicity 3, and 6 with multiplicity 1

40. Degree 5; zeros: -2 with multiplicity 2, and 4 with multiplicity 3

Exercises 41–44: The graph of either a cubic or quartic polynomial f(x) with a leading coefficient of ± 1 and integer zeros is shown. Write the complete factored form of f(x).

41.

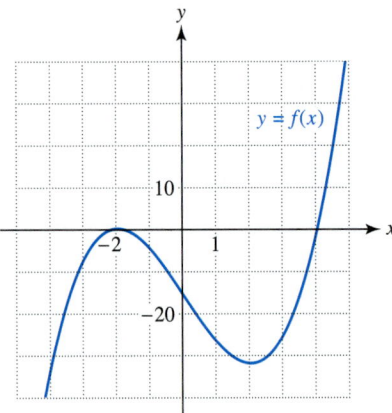

42.

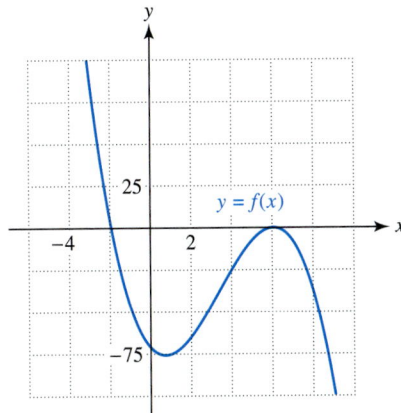

43.

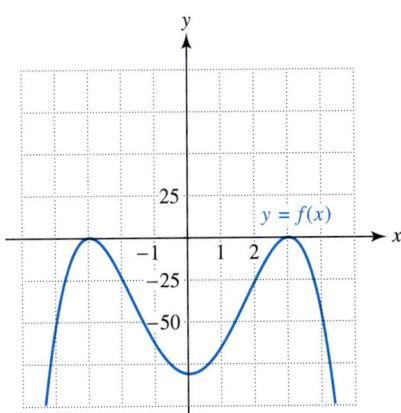

44.

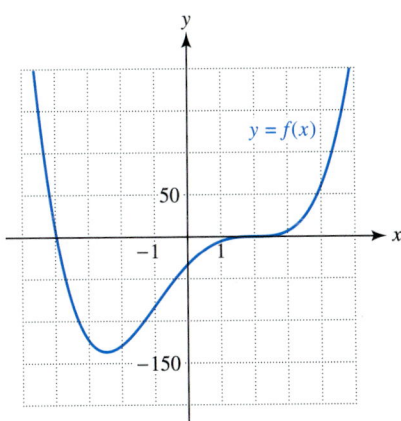

Exercises 45–48: (Refer to Example 5.) Write the complete factored form of the cubic polynomial $f(x)$, given that c is a zero. Support your result by graphing $f(x)$ and its complete factored form in the same viewing rectangle.

45. $f(x) = x^3 - 9x^2 + 23x - 15, \qquad c = 1$

46. $f(x) = 2x^3 + x^2 - 11x - 10, \qquad c = -2$

47. $f(x) = -4x^3 - x^2 + 51x - 36, \qquad c = -4$

48. $f(x) = 3x^3 - 11x^2 - 35x + 75, \qquad c = 5$

You Decide the Method

Exercises 49–52: Solve the equation symbolically, graphically, or numerically. Give reasons for the method you choose.

49. $x + 6 = 0$

50. $x^2 = 4$

51. $x^3 + x + 1 = 0$

52. $2x^5 + x^4 - 3x^2 + 6 = 3$

Rational Zeros

Exercises 53–60: (Refer to Example 8.)

  (a) Use the rational zero test to find any rational zeros of the polynomial $f(x)$.

  (b) Write the complete factored form of $f(x)$.

53. $f(x) = 2x^3 + 3x^2 - 8x + 3$

54. $f(x) = x^3 - 7x + 6$

55. $f(x) = 2x^4 + x^3 - 8x^2 - x + 6$

56. $f(x) = 2x^4 + x^3 - 19x^2 - 9x + 9$

57. $f(x) = 3x^3 - 16x^2 + 17x - 4$

58. $f(x) = x^3 + 2x^2 - 3x - 6$

59. $f(x) = x^3 - x^2 - 7x + 7$

60. $f(x) = 2x^3 - 5x^2 - 4x + 10$

Polynomial Equations

Exercises 61–66: Find the zeros of $f(x)$

  (a) symbolically,

  (b) graphically, and

  (c) numerically.

61. $f(x) = x^3 + x^2 - 6x$

62. $f(x) = 2x^2 - 8x + 6$

63. $f(x) = x^4 - 1$

64. $f(x) = x^4 - 5x^2 + 4$

65. $f(x) = -x^3 + 4x$

66. $f(x) = 6 - 4x - 2x^2$

Exercises 67–74: Solve each polynomial equation symbolically. Support your results graphically and numerically.

67. $x^3 - 25x = 0$

68. $x^4 - x^3 - 6x^2 = 0$

69. $x^4 - x^2 = 2x^2 + 4$

70. $x^4 + 5 = 6x^2$

71. $x^3 - 3x^2 - 18x = 0$

72. $x^4 - x^2 = 0$

73. $2x^3 = 4x^2 - 2x$

74. $x^3 = x$

Applications

75. *Water Pollution*   In one study, freshwater mussels were used to monitor copper discharge into a river from an electroplating works. Copper in high doses can be lethal to aquatic life. The table lists copper concentrations in mussels after 45 days at various distances downstream from the plant. The concentration C is measured in micrograms of copper per gram of mussel x kilometers downstream. (Sources: R. Foster and J. Bates, "Use of mussels to monitor point source industrial discharges"; C. Mason, *Biology of Freshwater Pollution.*)

x	5	21	37	53	59
C	20	13	9	6	5

(a) Describe the relationship between x and C.
(b) This data is modeled by $C(x) = -0.000068x^3 + 0.0099x^2 - 0.653x + 23$. Graph C and the data.
(c) Concentrations above 10 are lethal to mussels. Locate this region in the river.

76. *Floating Ball*   (Refer to Example 7.) Determine the depth that a pine ball with a 10-centimeter diameter sinks in water, if $d = 0.55$.

77. *Modeling Temperature*   Complete the following.
(a) Approximate the complete factored form of $f(x) = -0.184x^3 + 1.45x^2 + 10.7x - 27.9$.
(b) The cubic polynomial $f(x)$ models monthly average temperature at Trout Lake, Canada, in degrees Fahrenheit, where $x = 1$ corresponds to January and $x = 12$ represents December. Interpret the zeros of f.

78. *Average High Temperatures*   The monthly average high temperatures in degrees Fahrenheit at Daytona Beach can be modeled by $f(x) = 0.0151x^4 - 0.438x^3 + 3.60x^2 - 6.49x + 72.5$, where $x = 1$ corresponds to January and $x = 12$ represents December.
(a) Find the average high temperature during March and July.
(b) Graph f in [0.5, 12.5, 1] by [60, 100, 10]. Interpret the graph.
(c) Estimate graphically and numerically when the average high temperature is 80°F.

79. *Computers in the Classroom*   The ratio of students to computers in public schools from 1984 to 1995 is shown in the line graph. (Source: Quality Education Data.)
(a) Use the graph to estimate when this ratio was 45 students per computer.
(b) This data can be modeled by $f(x) = -0.273x^3 + 9.21x^2 - 102.9x + 400.2$, where $x = 4$ represents 1984 and $x = 15$ represents 1995.

Use $f(x)$ to solve part (a) graphically and numerically.

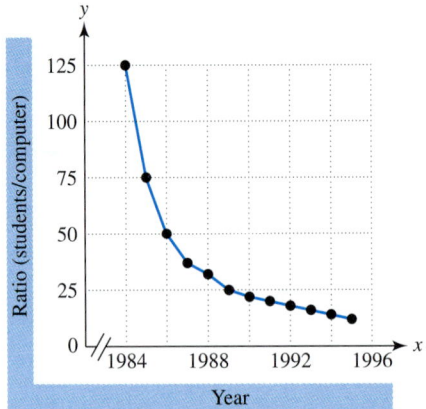

80. *Dog Years*   There is an old saying that every year of a dog's life is equal to seven years for a human. A more accurate approximation is given by the graph of f. Given a dog's age x, where $x \geq 1$, $f(x)$ models the equivalent age in human years. According to the Bureau of the Census, middle age for people begins at age 45. (Source: J. Brearley and A. Nicholas, *This is the Bichon Frise.*)
(a) Use the graph of f to estimate the equivalent age for dogs.
(b) Use $f(x) = -0.001183x^4 + 0.05495x^3 - 0.8523x^2 + 9.054x + 6.748$ to solve part (a) graphically and numerically.

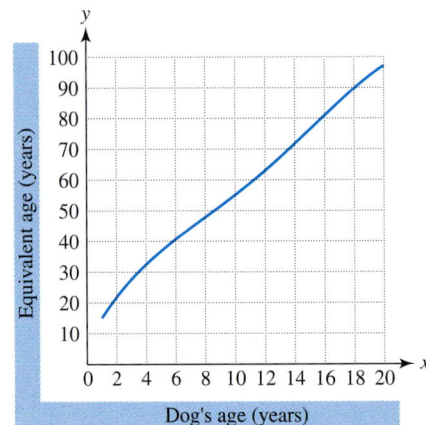

Writing about Mathematics

1. Suppose that $f(x)$ is a quintic polynomial with distinct real zeros. Assuming you have access to technology, explain how to factor $f(x)$ approximately. Have you used the factor theorem? Explain.

2. Explain how to determine graphically whether a zero of a polynomial is a multiple zero. Sketch examples.

CHECKING BASIC CONCEPTS FOR SECTIONS 3.3 AND 3.4

1. Sketch graphs of cubic polynomials with negative leading coefficients and 1, 2, and 3 x-intercepts.

2. Graph $f(x) = -x^2 + 3x + 2$ in $[-10, 10, 1]$ by $[-10, 10, 1]$.
   (a) Graph its reflection across the y-axis.
   (b) Graph its reflection across the x-axis.

3. Solve $x^3 - 2x^2 - 15x = 0$ symbolically, graphically, and numerically.

4. Determine graphically the zeros of $f(x) = x^4 - x^3 - 18x^2 + 16x + 32$. Write $f(x)$ in complete factored form.

5. Divide $x^3 - x^2 + 4x - 4$ by $x - 1$.

3.5 The Fundamental Theorem of Algebra

Complex Numbers • Fundamental Theorem of Algebra • Polynomial Equations with Complex Solutions

Introduction

Mathematics can be both abstract and applied. Abstract mathematics is focused on axioms, theorems, and proofs. It can be derived independently of empirical evidence. Theorems that were proved centuries ago are still valid today. In this sense, abstract mathematics transcends time. Yet, even though mathematics can be developed in an abstract setting—separate from science and all measured data—it also has countless applications. Mathematics is the fuel that drives society's ability to create new products.

There is a common misconception that theoretical mathematics is unimportant, yet many of the new ideas that eventually had great practical importance were first born in the abstract. For example, in 1854 George Boole published *Laws of Thought,* which outlined the basis for Boolean algebra. This was 85 years before the invention of the first digital computer. However, Boolean algebra became the basis by which modern computer hardware operates.

In this section we discuss some important abstract topics in algebra that have had an impact on society. We are privileged to read in a few hours what took people centuries to discover. To ignore either the abstract beauty or the profound applicability of mathematics is like seeing a rose, but never smelling one.

Complex Numbers

Throughout history, people have invented new numbers to solve equations and describe data. Often these new numbers were met with resistance and were regarded as being imaginary or unreal. The number 0 was not invented at the same time as the natural numbers. There was no Roman numeral for 0. No doubt there were skeptics who wondered why one needed a number to represent nothing. Negative numbers also met strong resistance. After all, how could one possibly have -6 apples? The same was true for complex numbers. However, complex numbers are no more imaginary than any other number created by mathematics. Much like Boolean algebra, the purpose of complex numbers was at first theoretical. However, today complex numbers are used in the design of electrical circuits, ships, and airplanes. Basic quantum mechanics and certain features of the theory of relativity would not have been developed without complex numbers. Even the *fractal image* shown in Figure 3.96 would not have been discovered without complex numbers.

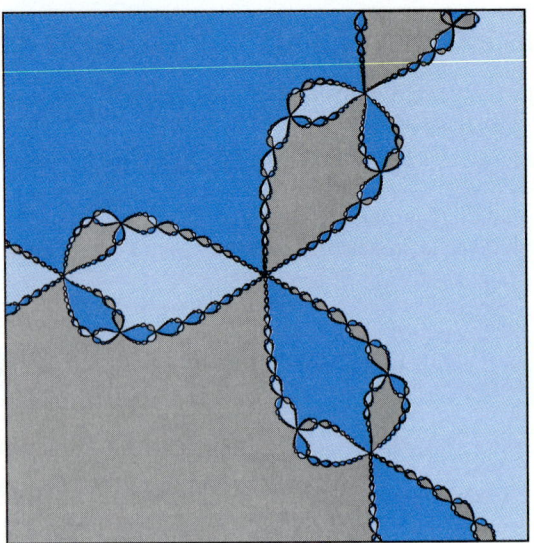

FIGURE 3.96 The Cube Roots of Unity
Source: D. Kincaid and W. Cheney, *Numerical Analysis,* © 1991 by Brooks/Cole Publishing Co. Used with permission.

If we graph $y = x^2 + 1$, there are no x-intercepts. See Figure 3.97. Therefore, the equation $x^2 + 1 = 0$ has no real solutions.

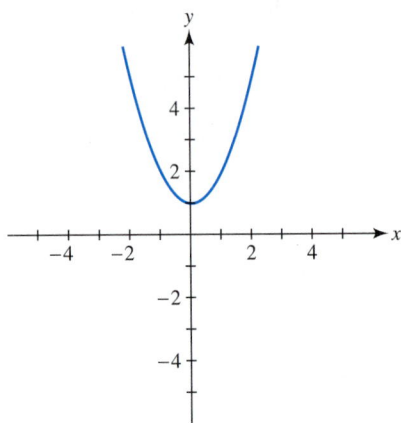

FIGURE 3.97

If we attempt to solve $x^2 + 1 = 0$ symbolically, it results in $x^2 = -1$. Since $x^2 \geq 0$ for any real number x, there is no real number solution. However, we can invent a solution.

$$x^2 = -1$$
$$x = \pm\sqrt{-1}$$

We now define a new number called the *imaginary unit*, denoted by i.

Properties of the imaginary unit i

$$i = \sqrt{-1}, \qquad i^2 = -1$$

By inventing the number i, the solutions to the equation $x^2 + 1 = 0$ are i and $-i$. Using the real numbers and the imaginary unit i, complex numbers can be defined. A **complex number** can be written in *standard form* as $a + bi$, where a and b are real numbers. The **real part** is a and the **imaginary part** is b. Every real number a is also a complex number because it can be written as $a + 0i$. A complex number $a + bi$ with $b \neq 0$ is an **imaginary number.** Table 3.6 lists several complex numbers with their real and imaginary parts.

TABLE 3.6

$a + bi$	$-3 + 2i$	5	$-3i$	$-1 + 7i$	$-5 - 2i$	$4 + 6i$
a	-3	5	0	-1	-5	4
b	2	0	-3	7	-2	6

Using the imaginary unit i, square roots of negative numbers may be written as a complex number. For example, $\sqrt{-3} = i\sqrt{3}$, and $\sqrt{-16} = i\sqrt{16} = 4i$. This can be summarized as follows.

The expression $\sqrt{-a}$

If $a > 0$, then $\sqrt{-a} = i\sqrt{a}$.

Arithmetic operations are also defined for complex numbers.

Addition and Subtraction.   To add the complex numbers $(-2 + 3i)$ and $(4 - 6i)$ simply combine the real and imaginary parts.

$$(-2 + 3i) + (4 - 6i) = -2 + 4 + 3i - 6i$$
$$= 2 - 3i$$

This same process works for subtraction.

$$(5 - 7i) - (8 + 3i) = 5 - 8 - 7i - 3i$$
$$= -3 - 10i$$

Multiplication.   Two complex numbers can be multiplied. The property $i^2 = -1$ is applied when appropriate.

$$(-5 + i)(7 - 9i) = -5(7) + -5(-9i) + (i)(7) + (i)(-9i)$$
$$= -35 + 45i + 7i - 9i^2$$
$$= -35 + 52i - 9(-1)$$
$$= -26 + 52i$$

When performing arithmetic with complex numbers, express the result in standard form $a + bi$.

Division.   The **conjugate** of $a + bi$ is $a - bi$. To find the conjugate, change the sign of the imaginary part b. Table 3.7 lists examples of complex numbers and their conjugates.

TABLE 3.7

$a + bi$	$2 + 5i$	$6 - 3i$	$-2 + 7i$	$-1 - i$	5	$-4i$
$a - bi$	$2 - 5i$	$6 + 3i$	$-2 - 7i$	$-1 + i$	5	$4i$

To simplify the quotient $\dfrac{3 + 2i}{5 - i}$, first multiply both the numerator and the denominator by the conjugate of the denominator.

$$\frac{3 + 2i}{5 - i} = \frac{(3 + 2i)(5 + i)}{(5 - i)(5 + i)} \qquad \text{Multiply by conjugate.}$$

$$= \frac{3(5) + (3)(i) + (2i)(5) + (2i)(i)}{(5)(5) + (5)(i) + (-i)(5) + (-i)(i)} \qquad \text{Expand.}$$

$$= \frac{15 + 3i + 10i + 2i^2}{25 + 5i - 5i - i^2} \qquad \text{Simplify.}$$

$$= \frac{15 + 13i + 2(-1)}{25 - (-1)} \qquad i^2 = -1$$

$$= \frac{13 + 13i}{26} \qquad \text{Simplify.}$$

$$= \frac{1}{2} + \frac{1}{2}i \qquad \frac{a + bi}{c} = \frac{a}{c} + \frac{b}{c}i$$

The last step expresses the quotient as a complex number in standard form.

Evaluating Complex Arithmetic with Technology. Some graphing calculators perform complex arithmetic. The evaluation of the previous examples are shown in Figures 3.98 and 3.99.

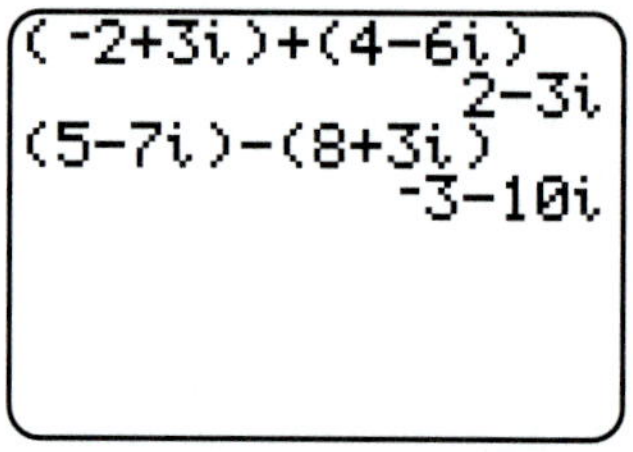

```
( -2+3i )+(4-6i )
              2-3i
(5-7i )-(8+3i )
            -3-10i
```

```
( -5+i )(7-9i )
          -26+52i
(3+2i )/(5-i )
          .5+.5i
```

FIGURE 3.98 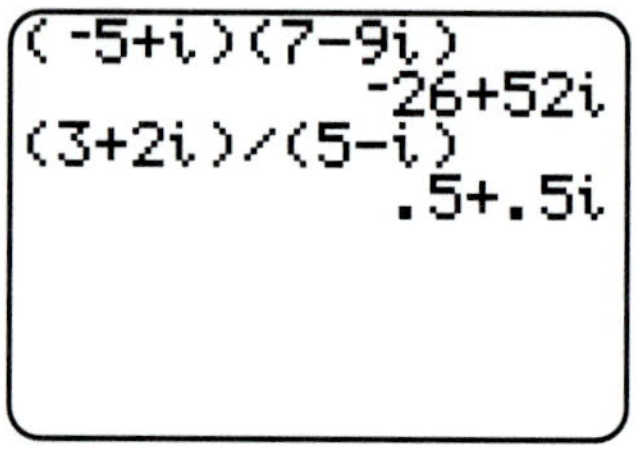   **FIGURE 3.99**

EXAMPLE 1   *Performing complex arithmetic*

Write each expression in standard form. Support your results using a calculator.
(a) $(-3 + 4i) + (5 - i)$
(b) $(-7i) - (6 - 5i)$
(c) $(-3 + 2i)^2$
(d) $\dfrac{17}{4 + i}$

Solution
(a) $(-3 + 4i) + (5 - i) = -3 + 5 + 4i - i = 2 + 3i$
(b) $(-7i) - (6 - 5i) = -6 - 7i + 5i = -6 - 2i$

(c) $(-3 + 2i)^2 = (-3 + 2i)(-3 + 2i)$

$$\begin{aligned} &= 9 - 6i - 6i + 4i^2 \\ &= 9 - 12i + 4(-1) \\ &= 5 - 12i \end{aligned}$$

(d) $\dfrac{17}{4+i} = \dfrac{17}{4+i} \cdot \dfrac{4-i}{4-i}$

$$\begin{aligned} &= \frac{68 - 17i}{16 - i^2} \\ &= \frac{68 - 17i}{17} \\ &= 4 - i \end{aligned}$$

Standard forms can be found using a calculator. See Figures 3.100 and 3.101

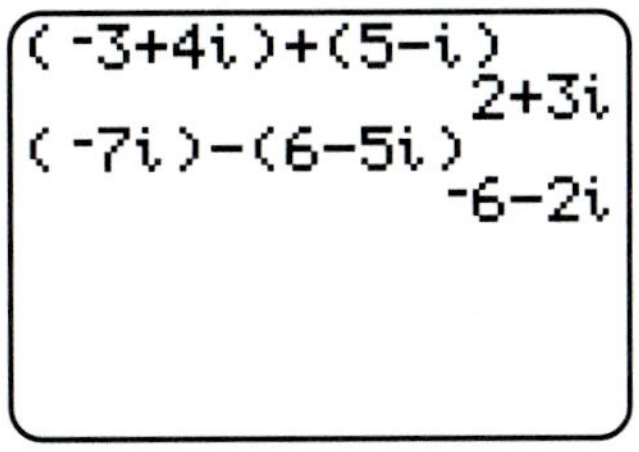
(-3+4i)+(5-i)
2+3i
(-7i)-(6-5i)
-6-2i
FIGURE 3.100

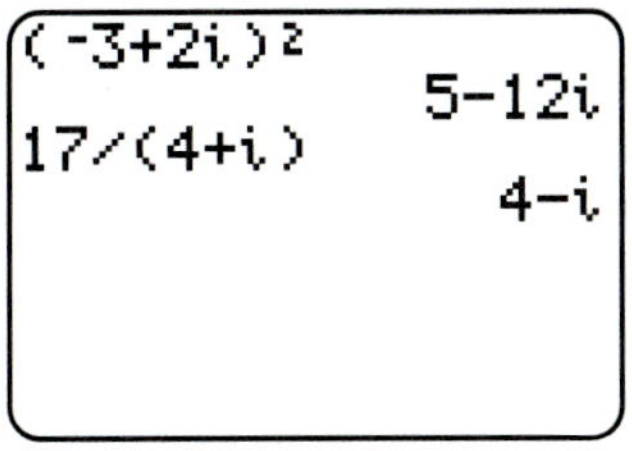
(-3+2i)²
5-12i
17/(4+i)
4-i
FIGURE 3.101

Fundamental Theorem of Algebra

One of the most brilliant mathematicians of all time, Carl Friedrich Gauss, at age 20 proved the fundamental theorem of algebra as part of his doctoral thesis. Although his theorem and proof were completed in 1797, they are still valid today.

> **Fundamental theorem of algebra**
>
> A polynomial $f(x)$ of degree $n \geq 1$ has at least one complex zero.

If $f(x)$ is a polynomial of degree 1 or higher, then by the fundamental theorem of algebra there is a zero c_1 such that $f(c_1) = 0$. By the factor theorem $(x - c_1)$ is a factor of $f(x)$ and

$$f(x) = (x - c_1)q_1(x)$$

for some polynomial $q_1(x)$. If $q_1(x)$ has positive degree, then by the fundamental theorem of algebra there exists a zero c_2 of $q_1(x)$. By the factor theorem $q_1(x)$ can be written as

$$q_1(x) = (x - c_2)q_2(x).$$

Then,

$$\begin{aligned} f(x) &= (x - c_1)q_1(x) \\ &= (x - c_1)(x - c_2)q_2(x). \end{aligned}$$

If $f(x)$ has degree n this process can be continued until $f(x)$ is written in the complete factored form

$$f(x) = a_n(x - c_1)(x - c_2) \cdots (x - c_n),$$

where a_n is the leading coefficient and the c_k are complex (or real) zeros. If each c_k is distinct, then $f(x)$ has n zeros. However, in general the c_k may not be distinct since multiple zeros are possible.

Number of zeros theorem

A polynomial of degree n has at most n distinct zeros.

EXAMPLE 2   *Classifying zeros*

All zeros for the given polynomials are distinct. Determine graphically the number of real zeros and the number of imaginary zeros.
(a) $f(x) = 3x^3 - 3x^2 - 3x - 5$
(b) $g(x) = 2x^2 + x + 1$
(c) $h(x) = -x^4 + 4x^2 + 4$

Solution

(a) The graph of $f(x)$ in Figure 3.102 crosses the x-axis once so there is one real zero. Since f is degree 3 and all zeros are distinct, there are two imaginary zeros.
(b) The graph of $g(x)$ in Figure 3.103 never crosses the x-axis. Since g is degree 2, there are no real zeros and two imaginary zeros.
(c) The graph of $h(x)$ is shown in Figure 3.104. Since h is degree 4, there are two real zeros and the remaining two zeros are imaginary.

$[-10, 10, 1]$ by $[-10, 10, 1]$

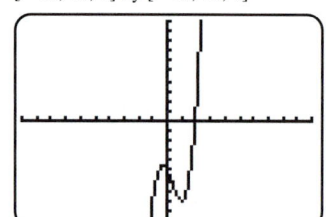

FIGURE 3.102

$[-10, 10, 1]$ by $[-10, 10, 1]$

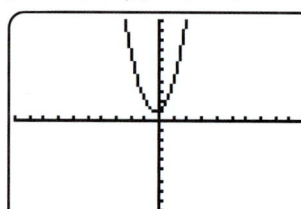

FIGURE 3.103

$[-10, 10, 1]$ by $[-10, 10, 1]$

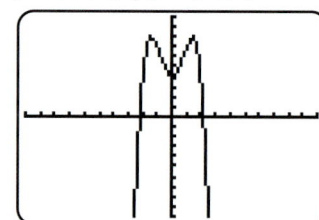

FIGURE 3.104 ∎

EXAMPLE 3   *Constructing a polynomial with prescribed zeros*

Represent a polynomial $f(x)$ of degree 4 with a leading coefficient of 2 and zeros of $-3, 5, i,$ and $-i$ in
(a) complete factored form, and
(b) expanded form.

Solution

(a) Let $a_n = 2, c_1 = -3, c_2 = 5, c_3 = i,$ and $c_4 = -i.$ Then,

$$f(x) = 2(x + 3)(x - 5)(x - i)(x + i).$$

(b) To expand this expression for $f(x)$, perform the following.

$$2(x + 3)(x - 5)(x - i)(x + i) = 2(x + 3)(x - 5)(x^2 + 1)$$
$$= 2(x + 3)(x^3 - 5x^2 + x - 5)$$
$$= 2(x^4 - 5x^3 + x^2 - 5x + 3x^3$$
$$- 15x^2 + 3x - 15)$$
$$= 2(x^4 - 2x^3 - 14x^2 - 2x - 15)$$
$$= 2x^4 - 4x^3 - 28x^2 - 4x - 30$$

Thus, $f(x) = 2x^4 - 4x^3 - 28x^2 - 4x - 30.$ ∎

EXAMPLE 4

Factoring a cubic polynomial with imaginary zeros

Determine the complete factored form for $f(x) = x^3 + 2x^2 + 4x + 8$.

Solution

One zero of the polynomial $f(x)$ is -2. The other two zeros are imaginary. See Figure 3.105.

$[-10, 10, 1]$ by $[-10, 10, 1]$

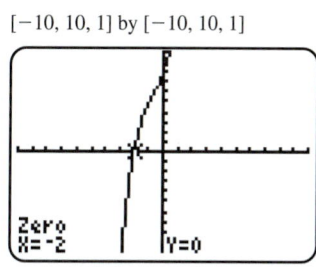

FIGURE 3.105

By the factor theorem $x + 2$ is a factor of $f(x)$. Divide $x + 2$ into $x^3 + 2x^2 + 4x + 8$.

$$
\begin{array}{r}
x^2 + 0x + 4 \\
x + 2 \overline{)x^3 + 2x^2 + 4x + 8} \\
\underline{x^3 + 2x^2} \\
0x^2 + 4x \\
\underline{0x^2 + 0x} \\
4x + 8 \\
\underline{4x + 8} \\
0
\end{array}
$$

By the division algorithm, $x^3 + 2x^2 + 4x + 8 = (x + 2)(x^2 + 4)$. To factor $x^2 + 4$, first find its zeros.

$$x^2 + 4 = 0$$
$$x^2 = -4$$
$$x = \pm\sqrt{-4}$$
$$x = \pm 2i$$

The zeros of $f(x)$ are -2, $2i$, and $-2i$. Its complete factored form is

$$f(x) = (x + 2)(x - 2i)(x + 2i).$$ ∎

Notice that in Example 4 both $2i$ and $-2i$ were zeros of $f(x)$. The numbers $2i$ and $-2i$ are conjugates. This result can be generalized for any polynomial with real coefficients.

> ## Conjugate zeros theorem
>
> If a polynomial $f(x)$ has only real coefficients and if $a + bi$ is a zero of $f(x)$, then the conjugate $a - bi$ is also a zero of $f(x)$.

EXAMPLE 5 Constructing a polynomial with prescribed zeros

Find a cubic polynomial $f(x)$ with real coefficients, a leading coefficient of 2, and zeros of 3 and $5i$. Express f in
(a) complete factored form, and
(b) expanded form.

Solution

(a) Since $f(x)$ has real coefficients, it must also have a third zero of $-5i$, the conjugate of $5i$. Let $c_1 = 3$, $c_2 = 5i$, $c_3 = -5i$, and $a_n = 2$. The complete factored form is

$$f(x) = 2(x - 3)(x - 5i)(x + 5i).$$

(b) To expand $f(x)$ perform the following steps.

$$\begin{aligned} 2(x - 3)(x - 5i)(x + 5i) &= 2(x - 3)(x^2 + 25) \\ &= 2(x^3 - 3x^2 + 25x - 75) \\ &= 2x^3 - 6x^2 + 50x - 150 \end{aligned}$$ ∎

Polynomial Equations with Complex Solutions

The quadratic formula can be used to solve quadratic equations with complex solutions.

EXAMPLE 6 Solving a quadratic equation with complex solutions

Solve $x^2 + 3x + 5 = 0$.

Solution

Let $a = 1$, $b = 3$, and $c = 5$ and apply the quadratic formula.

$$\begin{aligned} x &= \frac{-b \pm \sqrt{b^2 - 4ac}}{2a} \\ &= \frac{-3 \pm \sqrt{3^2 - 4(1)(5)}}{2(1)} \\ &= \frac{-3 \pm \sqrt{-11}}{2} \\ &= \frac{-3 \pm i\sqrt{11}}{2} \\ &= -\frac{3}{2} \pm \frac{\sqrt{11}}{2}i \end{aligned}$$ ∎

Critical Thinking

What is the result if each expression is evaluated? (See Example 6.)

$$\left(-\frac{3}{2} + \frac{\sqrt{11}}{2}i\right)^2 + 3\left(-\frac{3}{2} + \frac{\sqrt{11}}{2}i\right) + 5$$

$$\left(-\frac{3}{2} - \frac{\sqrt{11}}{2}i\right)^2 + 3\left(-\frac{3}{2} - \frac{\sqrt{11}}{2}i\right) + 5$$

Every polynomial equation of degree n can be written in the form

$$a_n x^n + \cdots + a_2 x^2 + a_1 x + a_0 = 0.$$

If we let $f(x) = a_n x^n + \cdots + a_2 x^2 + a_1 x + a_0$ and write $f(x)$ in complete factored form as

$$f(x) = a_n(x - c_1)(x - c_2) \cdots (x - c_n),$$

then the solutions to the polynomial equation are $c_1, c_2, \ldots, c_n$. Solving a polynomial equation with this technique is illustrated in the next two examples.

EXAMPLE 7   *Solving a polynomial equation*

Solve $x^3 = 3x^2 - 7x + 21$.

Solution

Write the equation as $f(x) = 0$, where $f(x) = x^3 - 3x^2 + 7x - 21$. Then graph $f(x)$. Figure 3.106 shows that 3 is a zero of $f(x)$. By the factor theorem, $x - 3$ is a factor of $f(x)$. We divide $x - 3$ into $f(x)$ using synthetic division.

$$
\begin{array}{r|rrrr}
3 & 1 & -3 & 7 & -21 \\
  &   & 3 & 0 & 21 \\
\hline
  & 1 & 0 & 7 & 0
\end{array}
$$

Thus, $x^3 - 3x^2 + 7x - 21 = (x - 3)(x^2 + 7)$.

$x^3 - 3x^2 + 7x - 21 = 0$		$f(x) = 0$
$(x - 3)(x^2 + 7) = 0$		Factor.
$x - 3 = 0$ or $x^2 + 7 = 0$		Zero-product property
$x = 3$ or $x^2 = -7$		Solve.
$x = 3$ or $x = \pm i\sqrt{7}$		Property of i

The solutions are 3 and $\pm i\sqrt{7}$. ∎

[−5, 5, 1] by [−30, 30, 10]

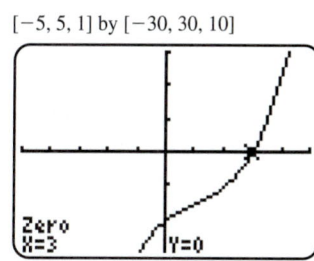

FIGURE 3.106

EXAMPLE 8 *Solving a polynomial equation*

Solve $x^4 + x^2 = x^3$.

Solution

Write the equation as $f(x) = 0$, where $f(x) = x^4 - x^3 + x^2$.

$$x^4 - x^3 + x^2 = 0 \qquad\qquad f(x) = 0$$

$$x^2(x^2 - x + 1) = 0 \qquad\qquad \text{Factor out } x^2.$$

$$x^2 = 0 \quad \text{or} \quad x^2 - x + 1 = 0 \qquad \text{Zero-product property}$$

The only solution to $x^2 = 0$ is 0. To solve $x^2 - x + 1 = 0$ use the quadratic formula.

$$x = \frac{-b \pm \sqrt{b^2 - 4ac}}{2a}$$

$$= \frac{1 \pm \sqrt{(-1)^2 - 4(1)(1)}}{2(1)}$$

$$= \frac{1 \pm \sqrt{-3}}{2}$$

$$= \frac{1 \pm i\sqrt{3}}{2}$$

$$= \frac{1}{2} \pm \frac{\sqrt{3}}{2}i$$

The solutions are 0 and $\dfrac{1}{2} \pm \dfrac{\sqrt{3}}{2}i$. ∎

3.5 PUTTING IT ALL TOGETHER

A complex number has the standard form $a + bi$, where a and b are real numbers. Arithmetic can be performed using complex numbers. The set of complex numbers contain the set of real numbers.

The fundamental theorem of algebra guarantees that every polynomial $f(x)$ of degree n can be written in complete factored form as $f(x) = a_n(x - c_1)(x - c_2) \cdots (x - c_n)$, where a_n is the leading coefficient and the c_k are complex (or real) numbers. The zeros of $f(x)$ are $c_1, c_2, \ldots, c_n$. A polynomial equation of degree n, written as $a_n x^n + \cdots + a_2 x^2 + a_1 x + a_0 = 0$, has at most n solutions.

3.5 EXERCISES

Complex Numbers

Exercises 1–8: Simplify the expression using the imaginary unit i.

1. $\sqrt{-4}$

2. $\sqrt{-16}$

3. $\sqrt{-100}$

4. $\sqrt{-49}$

5. $\sqrt{-23}$

6. $\sqrt{-11}$

7. $\sqrt{-12}$

8. $\sqrt{-32}$

Exercises 9–28: Write the expression in standard form.

9. $3i + 5i$

10. $-7i + 5i$

11. $(3 + i) + (-5 - 2i)$

12. $(-4 + 2i) + (7 + 35i)$

13. $(12 - 7i) - (-1 + 9i)$

14. $(2i) - (-5 + 23i)$

15. $(3) - (4 - 6i)$

16. $(7 + i) - (-8 + 5i)$

17. $(2)(2 + 4i)$

18. $(-5)(-7 + 3i)$

19. $(1 + i)(2 - 3i)$

20. $(-2 + i)(1 - 2i)$

21. $(-3 + 2i)(-2 + i)$

22. $(2 - 3i)(1 + 4i)$

23. $\dfrac{1}{1 + i}$

24. $\dfrac{1 - i}{2 + 3i}$

25. $\dfrac{4 + i}{5 - i}$

26. $\dfrac{10}{1 - 4i}$

27. $\dfrac{2i}{10 - 5i}$

28. $\dfrac{3 - 2i}{1 + 2i}$

Exercises 29–34: Evaluate the expression with a calculator.

29. $(23 - 5.6i) + (-41.5 + 93i)$

30. $(-8.05 - 4.67i) + (3.5 + 5.37i)$

31. $(17.1 - 6i) - (8.4 + 0.7i)$

32. $\left(\dfrac{3}{4} - \dfrac{1}{10}i\right) - \left(-\dfrac{1}{8} + \dfrac{4}{25}i\right)$

33. $(-12.6 - 5.7i)(5.1 - 9.3i)$

34. $(7.8 + 23i)(-1.04 + 2.09i)$

Exercises 35 and 36: Perform the complex division using a calculator. Express your answer as a + bi, where a and b are rounded to four significant digits.

35. $\dfrac{17 - 135i}{18 + 142i}$

36. $\dfrac{141 + 52i}{102 - 31i}$

Exercises 37 and 38: Corrosion in Airplanes   While corrosion in the metal surface of an airplane can be difficult to detect visually, one test used to locate it involves passing an alternating current through a small area on the plane's surface. If the current varies from one region to another, it may indicate that corrosion is occurring. The impedance Z (or opposition to the flow of electricity) of the metal is related to the voltage V and current I by the equation $Z = \dfrac{V}{I}$, where Z, V, and I are complex numbers. Calculate Z for the given values of V and I. (Source: Society for Industrial and Applied Mathematics (SIAM).)

37. $V = 50 + 98i$, $I = 8 + 5i$

38. $V = 30 + 60i$, $I = 8 + 6i$

Zeros of Polynomials

Exercises 39–44: The graph and the formula for a quadratic polynomial f(x) are given.

(a)   *Use the graph to predict the number of real zeros and the number of imaginary zeros.*

(b)   *Find these zeros using the quadratic formula.*

39. $f(x) = 2x^2 - x - 3$

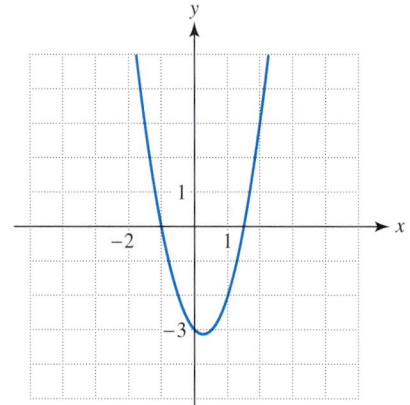

40. $f(x) = -x^2 + 4.6x - 5.29$

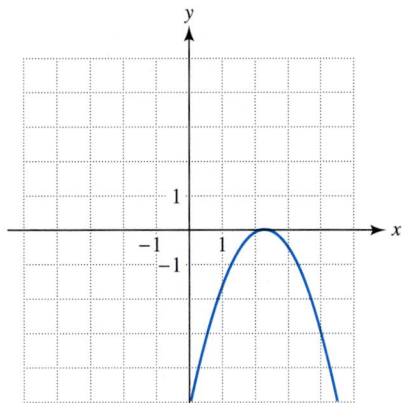

41. $f(x) = x^2 + x + 2$

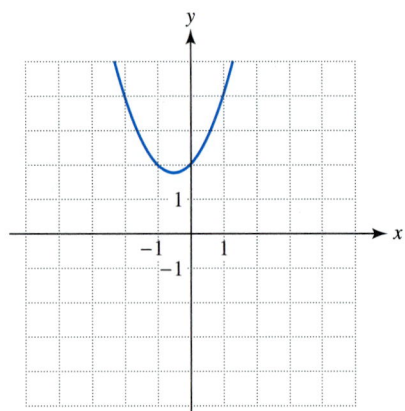

42. $f(x) = -2x^2 + 2x - 3$

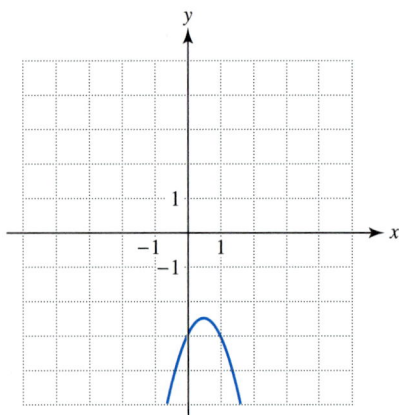

43. $f(x) = 5x^2 + 4x + 1$

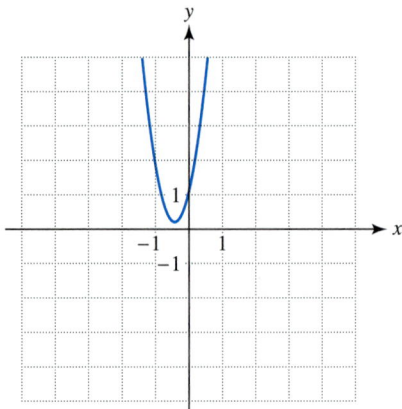

44. $f(x) = 9x^2 - 12x + 4$

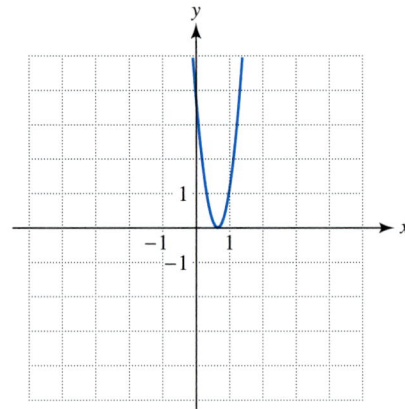

Exercises 45–50: The graph of a polynomial $f(x)$ is given. Determine the number of real zeros and the number of imaginary zeros. Assume that all zeros of $f(x)$ are distinct.

45. $f(x) = x^3 - 5x^2 + 5x - 15$

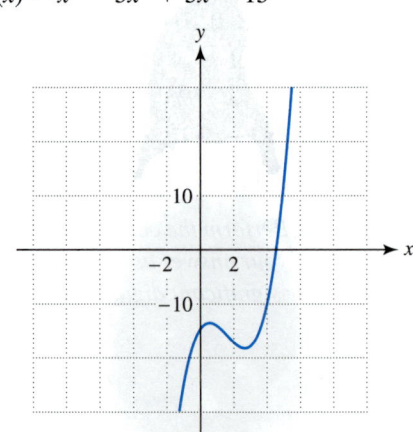

46. $f(x) = -x^3 + 5x^2 - x - 5$

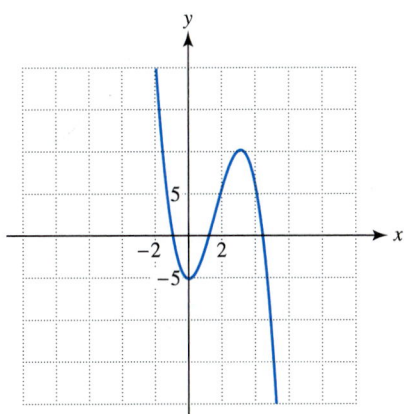

47. $f(x) = x^4 + 2x^3 - 3x^2 + 2x - 3$

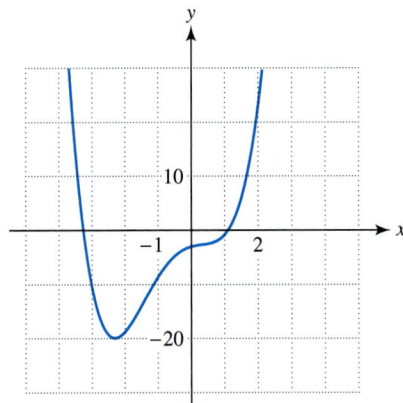

48. $f(x) = x^4 - 5x^2 + 10$

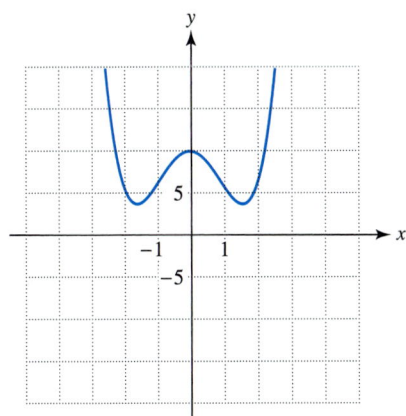

49. $f(x) = x^5 - 6x^4 + 6x^3 - 8x^2 + 9x + 70$

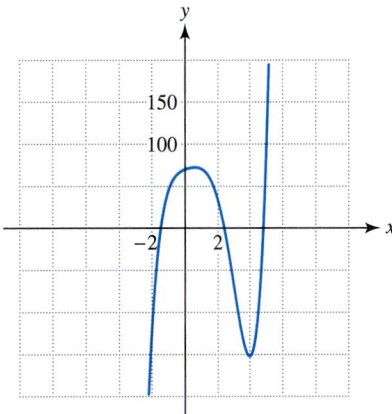

50. $f(x) = 4x^5 + 8x^4 - 31x^3 - 67x^2 - 33x - 100$

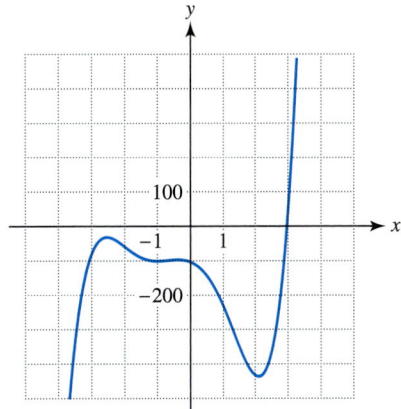

Exercises 51–56: Let a_n denote the leading coefficient of a polynomial $f(x)$.

   (**a**)   *Find the complete factored form of $f(x)$ that satisfies the given conditions.*

   (**b**)   *Express $f(x)$ in expanded form.*

51. Degree 2; $a_n = 1$;   zeros $6i$ and $-6i$

52. Degree 3; $a_n = 5$;   zeros 2, i, and $-i$

53. Degree 3; $a_n = -1$; zeros -1, $2i$, and $-2i$

54. Degree 4; $a_n = 3$;   zeros -2, 4, i, and $-i$

55. Degree 4; $a_n = 10$; zeros 1, -1, $3i$, and $-3i$

56. Degree 2; $a_n = -5$; zeros $1 + i$ and $1 - i$

Exercises 57–64: Express $f(x)$ in complete factored form.

57. $f(x) = x^2 + 25$ **58.** $f(x) = x^2 + 11$

59. $f(x) = 3x^3 + 3x$ **60.** $f(x) = 2x^3 + 10x$

61. $f(x) = x^4 + 5x^2 + 4$ **62.** $f(x) = x^4 + 4x^2$

63. $f(x) = x^4 + 2x^3 + x^2 + 8x - 12$

64. $f(x) = x^3 + 2x^2 + 16x + 32$

Polynomial Equations

Exercises 65–70: Solve the quadratic equation.

65. $x^2 - 4x + 5 = 0$ **66.** $2x^2 + x + 1 = 0$

67. $x^2 = 3x - 5$ **68.** $3x - x^2 = 5$

69. $6 = x^2 + 2x + 10$ **70.** $x(x - 4) = -8$

Exercises 71–78: Solve the polynomial equation.

71. $x^3 + x = 0$

72. $2x^3 - x + 1 = 0$

73. $x^3 = 2x^2 - 7x + 14$

74. $x^2 + x + 2 = x^3$

75. $x^4 + 5x^2 = 0$

76. $x^4 - 2x^3 + x^2 - 2x = 0$

77. $x^4 = x^3 - 4x^2$

78. $x^5 + 9x^3 = x^4 + 9x^2$

Writing about Mathematics

1. The properties of the imaginary unit are $i = \sqrt{-1}$ and $i^2 = -1$.

   (a) Begin simplifying the expressions i, i^2, i^3, i^4, i^5, ..., until a pattern is discovered. For example, $i^3 = i \cdot i^2 = i \cdot (-1) = -i$.

   (b) Summarize your findings by describing how to simplify i^n for any natural number n.

2. Could a cubic function have only imaginary zeros? Explain.

3. Give an example of a polynomial function that has only imaginary zeros and a polynomial function that has only real zeros. Explain how to determine graphically if a function only has imaginary zeros.

4. Explain what the fundamental theorem of algebra implies about factoring polynomials.

3.6 Rational Functions

Rational Functions • Vertical Asymptotes • Horizontal Asymptotes • Identifying Asymptotes • Rational Equations • Variation (Optional)

Introduction

Rational functions are nonlinear functions that frequently occur in applications. For example, rational functions are used to design curves for railroad tracks, predict mortality rates, determine stopping distances on hills, and calculate the average number of people waiting in a line.

Rational Functions

A rational number can be expressed as a ratio $\frac{p}{q}$, where p and q are integers and $q \neq 0$. A rational function is defined in a similar manner, using the concept of a polynomial.

> ### Rational function
>
> A function f represented by $f(x) = \frac{p(x)}{q(x)}$, where $p(x)$ and $q(x)$ are polynomials and $q(x) \neq 0$, is a **rational function.**

The domain of a rational function includes all real numbers *except* the zeros of the denominator $q(x)$. The graph of a rational function is discontinuous at x-values where $q(x) = 0$. Examples of rational functions include the following.

i. $f(x) = \dfrac{2x - 1}{x^2 + 1}$

ii. $g(x) = \dfrac{1}{x}$

iii. $h(x) = \dfrac{x^3 - 2x^2 + 1}{x^2 - 3x + 2}$

Each representation consists of a ratio of two polynomials. The domain of f includes all real numbers since $x^2 + 1 \neq 0$ for any real number x. The domain of g is $\{x \mid x \neq 0\}$. Because $x^2 - 3x + 2 = (x - 1)(x - 2) = 0$ when $x = 1$ or $x = 2$, the domain of h is $\{x \mid x \neq 1 \text{ or } x \neq 2\}$.

Critical Thinking

Is an integer a rational number? Is a polynomial function a rational function?

Vertical Asymptotes

If cars arrive randomly at the exit of a parking ramp, then the average length of the line depends on two factors: the average traffic volume exiting the ramp and the average rate that a parking attendant can wait on cars. For instance, if the average traffic volume is three cars per minute and the parking attendant is serving four cars per minute, then at times a line may form if cars arrive in a random manner. The **traffic intensity** x is the ratio of the average traffic volume to the average working rate of the attendant. In this example, $x = \dfrac{3}{4}$. (Source: F. Mannering and W. Kilareski, *Principles of Highway Engineering and Traffic Control.*)

EXAMPLE 1   *Estimating the length of parking ramp lines*

If the traffic intensity is x, then the average number of cars waiting to exit a parking ramp can be computed by $f(x) = \dfrac{x^2}{2 - 2x}$, where $0 \leq x < 1$.

(a) Evaluate $f(0.5)$ and $f(0.9)$. Interpret the results.
(b) Graph f in $[0, 1, 0.1]$ by $[0, 10, 1]$.
(c) Explain what happens to the length of the line as the traffic intensity approaches 1.

Solution

(a) $f(0.5) = \dfrac{0.5^2}{2 - 2(0.5)} = 0.25$ and $f(0.9) = \dfrac{0.9^2}{2 - 2(0.9)} = 4.05$. This means that if the traffic intensity is 0.5, there is little waiting in line. As the traffic intensity increases to 0.9, the average line has over four cars.

(b) Graph $Y_1 = X^2/(2 - 2X)$ as in Figure 3.107.

[0, 1, 0.1] by [0, 10, 1]

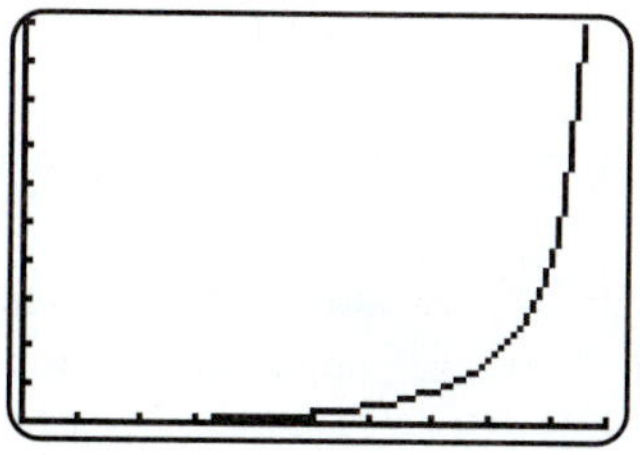

X	Y1
.94	7.3633
.95	9.025
.96	11.52
.97	15.682
.98	24.01
.99	49.005
1	ERROR

Y1 ▉X^2/(2-2X)

FIGURE 3.107 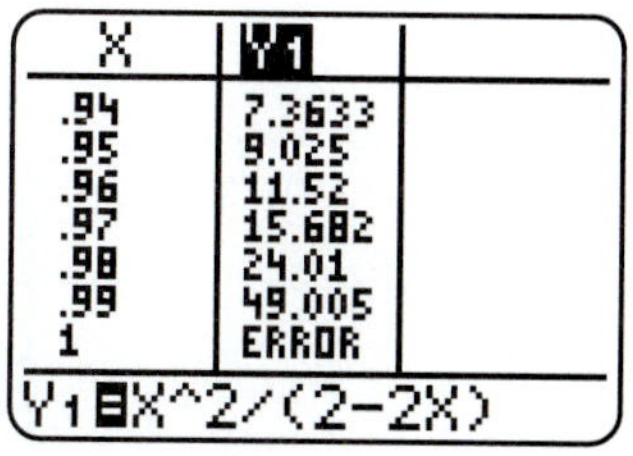   **FIGURE 3.108**

(c) As the traffic intensity x approaches 1 from the left, the graph of f increases rapidly without bound. Numerical support is given in Figure 3.108 on the previous page. With a traffic intensity slightly less than 1, the attendant has difficulty keeping up. If cars occasionally arrive in groups, long lines will form. At $x = 1$ the denominator, $2 - 2x$, equals 0 and $f(x)$ is undefined. See Figure 3.109.

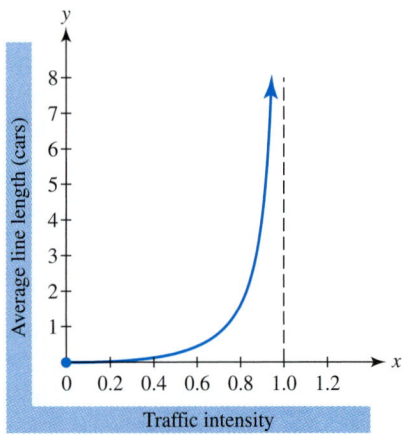

FIGURE 3.109 Parking Ramp Lines

In Figure 3.109, the vertical line $x = 1$ is a *vertical asymptote* of the graph of f. A graph of a different rational function f is shown in Figure 3.110. The $f(x)$-values *decrease without bound* as x approaches 2 from the left. This is denoted by $f(x) \to -\infty$ as $x \to 2^-$. Similarly, the $f(x)$-values *increase without bound* as x approaches 2 from the right. This is expressed as $f(x) \to \infty$ as $x \to 2^+$. The line $x = 2$ is a vertical asymptote of the graph of f.

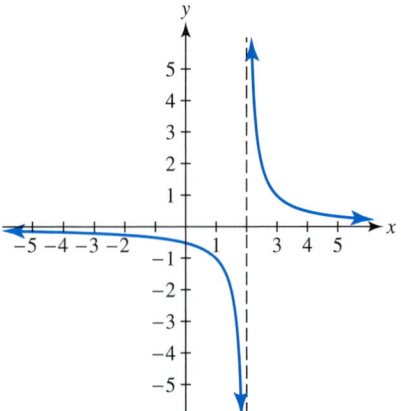

FIGURE 3.110

Vertical asymptote

The line $x = k$ is a **vertical asymptote** of the graph of f if $f(x) \to \infty$ or $f(x) \to -\infty$, as x approaches k from either the left or the right.

If $x = k$ is a vertical asymptote of the graph of f, then k is not in the domain of f. Furthermore, the graph of a rational function f does not cross a vertical asymptote.

Horizontal Asymptotes

Probability sometimes can be described with rational functions. Suppose that a container holds ten identical balls numbered 1 through 10, where one ball has the winning number. Then the probability or likelihood of drawing the winning ball is 1 in 10. This probability can be expressed as the ratio $\frac{1}{10}$. Probabilities vary between 0 and 1, where 1 represents an event that is certain to occur.

If there are x balls, the probability of drawing the winning ball is given by $P(x) = \frac{1}{x}$, where x is a natural number. A graph of P is shown in Figure 3.111. As the number of balls increases, the probability of drawing the winning ball decreases toward zero, without becoming zero. Even if there were one million balls in the container, there still would be a slight chance of drawing the winning ball. As a result, the graph of P comes closer and closer to the x-axis ($y = 0$) without ever actually touching it. The x-axis is a *horizontal asymptote* of the graph of P.

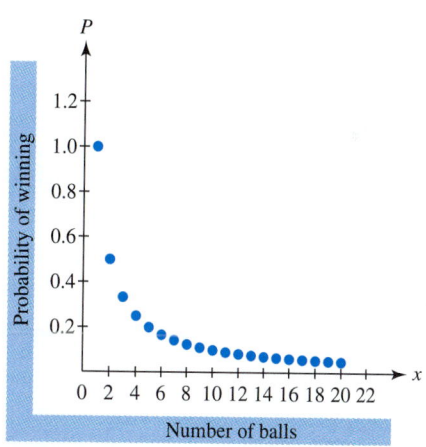

FIGURE 3.111

Horizontal asymptote

The line $y = b$ is a **horizontal asymptote** of the graph of f, if $f(x) \to b$, as x approaches either ∞ or $-\infty$.

The graph of f in Figure 3.112 on the next page is an example of a von Bertalanffy growth curve. It models the length in millimeters of a small fish after x weeks. After several weeks the length of the fish begins to level off near 25 millimeters. Thus, $y = 25$ is a horizontal asymptote of the graph of f. This is denoted by $f(x) \to 25$ as $x \to \infty$. (Source: D. Brown and P. Rothery, *Models in Biology: Mathematics, Statistics and Computing.*)

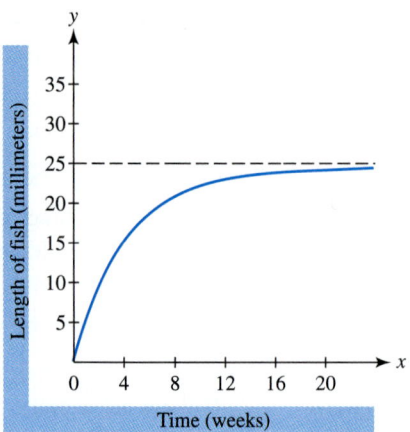

FIGURE 3.112 Size of a Small Fish

Identifying Asymptotes

The graph in Figure 3.111 limits the domain of P to natural numbers because a fraction of a ball is not allowed. However, the implied domain of $f(x) = \dfrac{1}{x}$ is the set of nonzero real numbers.

EXAMPLE 2 *Analyzing the graph of* $f(x) = \dfrac{1}{x}$

The graph of $f(x) = \dfrac{1}{x}$ is shown in Figure 3.113. Numerical values of $f(x)$ are listed in Tables 3.8 and 3.9. Relate these values to the graph of f.

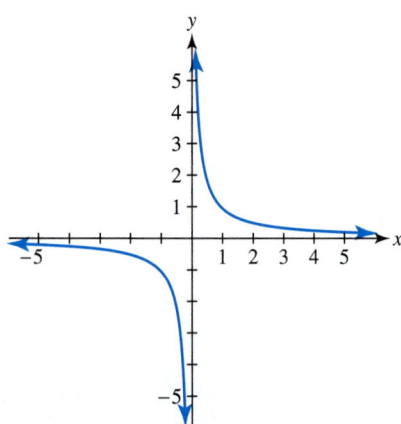

FIGURE 3.113

TABLE 3.8

x	0	0.001	0.01	0.1	1	10	100	1000
$f(x) = \dfrac{1}{x}$	—	1000	100	10	1	0.1	0.01	0.001

TABLE 3.9

x	-1000	-100	-10	-1	-0.1	-0.01	-0.001	0
$f(x) = \dfrac{1}{x}$	-0.001	-0.01	-0.1	-1	-10	-100	-1000	—

Solution

Table 3.8 shows positive values of x. As x approaches 0 from the right, the $f(x)$-values increase without bound. This is denoted by $f(x) \to \infty$ as $x \to 0^+$. Slightly right of the y-axis there are large y-values on the graph of $f(x) = \dfrac{1}{x}$. The y-axis is a vertical asymptote.

As x assumes large positive values in Table 3.8, the $f(x)$-values decrease and tend toward 0. This is expressed as $f(x) \to 0$ as $x \to \infty$. As a result, the graph of f levels off above the x-axis. The x-axis is a horizontal asymptote.

Table 3.9 shows negative values of x. As x approaches 0 from the left, the $f(x)$-values decrease without bound. This is denoted by $f(x) \to -\infty$ as $x \to 0^-$. The graph of $f(x) = \dfrac{1}{x}$ falls rapidly just left of the y-axis, indicating that the y-axis is vertical asymptote.

When x decreases in Table 3.9, the $f(x)$-values are negative and tend toward 0. This is denoted by $f(x) \to 0$ as $x \to -\infty$. The graph of f levels off below the x-axis, which is a horizontal asymptote. ∎

EXAMPLE 3   *Determining horizontal and vertical asymptotes visually*

Use the graph of each rational function to determine any vertical or horizontal asymptotes.

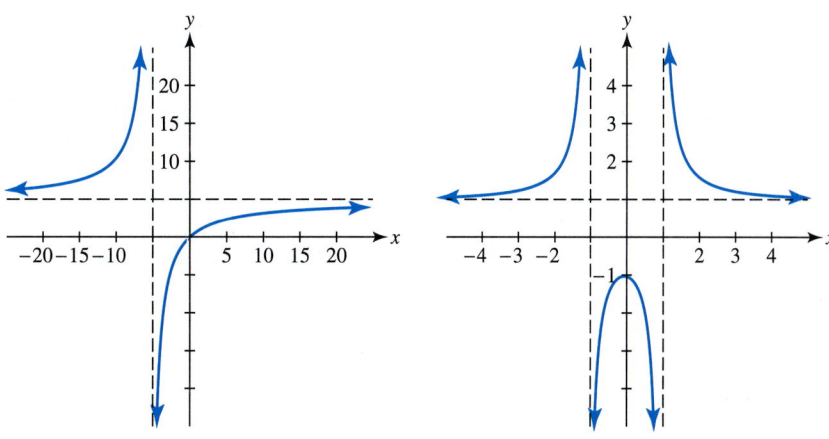

FIGURE 3.114   **FIGURE 3.115**

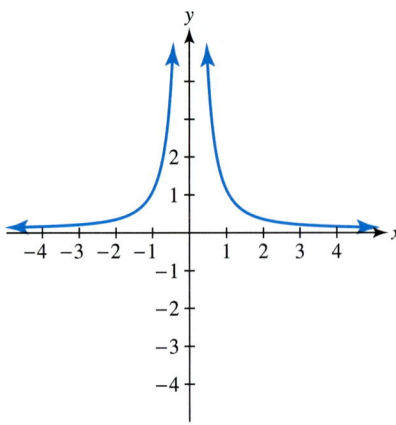

FIGURE 3.116

Solution

In Figure 3.114, $y = 5$ is a horizontal asymptote and $x = -5$ is a vertical asymptote. In Figure 3.115, $x = \pm 1$ are vertical asymptotes and $y = 1$ is a horizontal asymptote. As x approaches 0 from either side in Figure 3.116, the y-values become large without bound. The y-axis ($x = 0$) is a vertical asymptote. As x approaches ∞ or $-\infty$, the graph of f approaches the x-axis from above. The x-axis ($y = 0$) is a horizontal asymptote. ∎

The following can be used to find vertical and horizontal asymptotes.

Identifying vertical and horizontal asymptotes

Let the rational function f be given by

$$f(x) = \frac{p(x)}{q(x)} = \frac{a_n x^n + \cdots + a_2 x^2 + a_1 x + a_0}{b_m x^m + \cdots + b_2 x^2 + b_1 x + b_0},$$

where $a_n \neq 0$ and $b_m \neq 0$.

1. The line $x = k$ is a vertical asymptote if $q(k) = 0$ and $p(k) \neq 0$.
2. (a) If $n < m$, then the x-axis is a horizontal asymptote.

   (b) If $n = m$, then $y = \dfrac{a_n}{b_m}$ is a horizontal asymptote.

   (c) If $n > m$, there are no horizontal asymptotes.

EXAMPLE 4   *Finding asymptotes*

For each rational function, determine any vertical or horizontal asymptotes.

(a) $f(x) = \dfrac{6x - 1}{3x + 3}$

(b) $g(x) = \dfrac{x + 1}{x^2 - 4}$

(c) $h(x) = \dfrac{x^2 - 1}{x + 1}$

Solution

(a) The degrees of the numerator and the denominator are both 1. Since the ratio of the leading coefficients is $\dfrac{6}{3} = 2$, the graph of f has a horizontal asymptote of $y = 2$. This is supported numerically in Figures 3.117 and 3.118, where the y-values approach 2 as the x-values increase or decrease. When $x = -1$ the denominator, $3x + 3$, equals 0 while the numerator, $6x + 1$, does not equal 0. Thus, $x = -1$ is a vertical asymptote. A graph of f is shown in Figure 3.119.

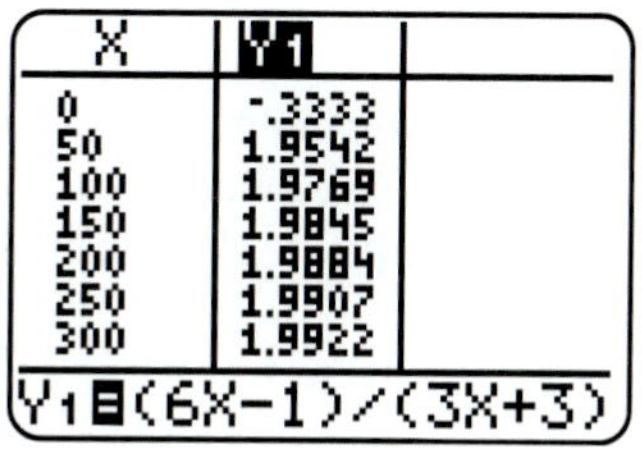

X	Y1
0	-.3333
50	1.9542
100	1.9769
150	1.9845
200	1.9884
250	1.9907
300	1.9922

Y1◘(6X-1)/(3X+3)

FIGURE 3.117

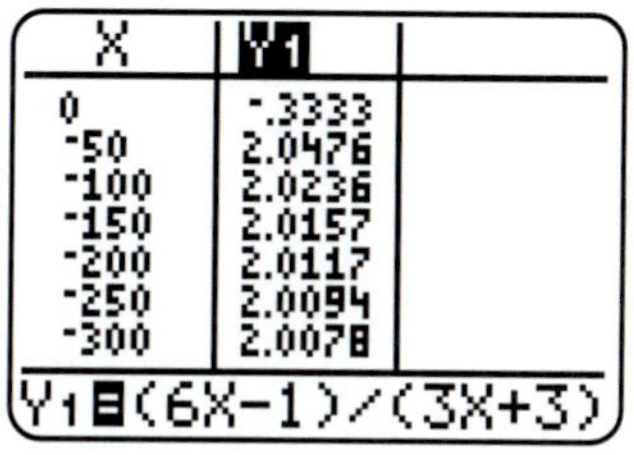

X	Y1
0	-.3333
-50	2.0476
-100	2.0236
-150	2.0157
-200	2.0117
-250	2.0094
-300	2.0078

Y1◘(6X-1)/(3X+3)

FIGURE 3.118

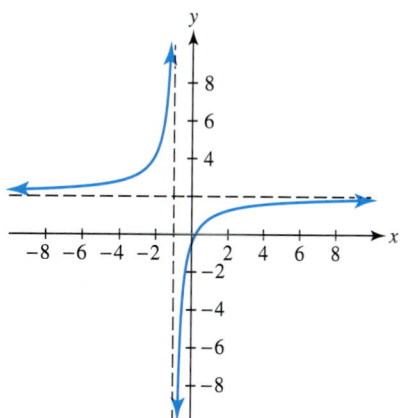

FIGURE 3.119

(b) The degree of the numerator is one less than the degree of the denominator, so the x-axis is a horizontal asymptote. The denominator, $x^2 - 4$, equals 0 and the numerator, $x + 1$, does not equal 0 when $x = \pm 2$. Thus, $x = \pm 2$ are vertical asymptotes. See Figure 3.120 on the following page.

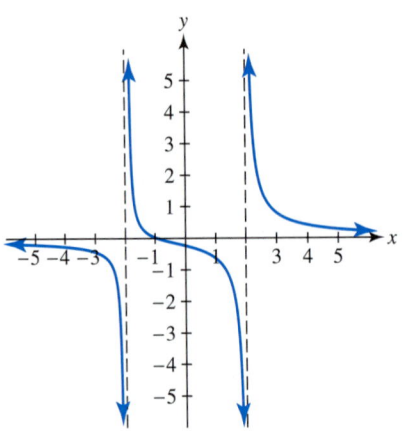

FIGURE 3.120

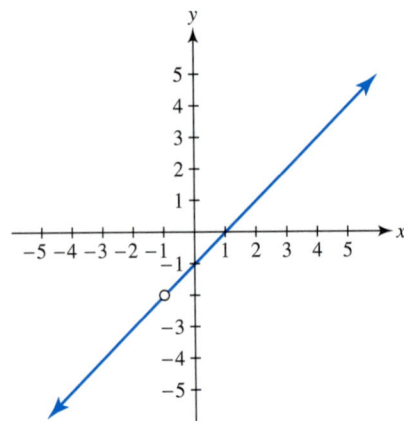

FIGURE 3.121

(c) The degree of the numerator is greater than the degree of the denominator so there are no horizontal asymptotes. When $x = -1$, both numerator and denominator equal 0. We can simplify $h(x)$ as follows.

$$h(x) = \frac{x^2 - 1}{x + 1} = \frac{(x + 1)(x - 1)}{x + 1} = x - 1, \qquad x \neq -1$$

The graph of $h(x)$ is the line $y = x - 1$ with the point $(-1, -2)$ missing. There are no vertical asymptotes. See Figure 3.121. ∎

 Technology Note ———————————————————————

Calculators typically graph in connected or dot mode. If connected mode is used when graphing a rational function, it may appear as though the calculator is graphing vertical asymptotes automatically. However, in most instances the calculator is connecting points inappropriately.

Sometimes rational functions can be graphed in connected mode using a *decimal* or *friendly* viewing rectangle. This is done in Examples 6 and 7.

———————————————————————

EXAMPLE 5 *Analyzing a rational function with technology*

Let $f(x) = \dfrac{2x^2 + 1}{x^2 - 4}$.

(a) Graph f. Find the domain of f.
(b) Identify any vertical or horizontal asymptotes.
(c) Sketch a graph of f that includes the asymptotes.

Solution

(a) A graph of f is shown in Figure 3.122 using dot mode. The function is undefined when $x^2 - 4 = 0$, or when $x = \pm 2$. The domain of f is $D = \{x \mid x \neq \pm 2\}$.

[−6, 6, 1] by [−6, 6, 1]

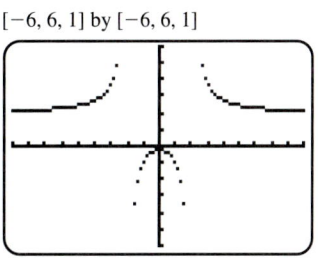

FIGURE 3.122

(b) When $x = \pm 2$ the denominator, $x^2 - 4$, equals 0 and the numerator, $2x^2 + 1$, does not equal 0. Therefore, $x = \pm 2$ are vertical asymptotes. The degree of the numerator equals the degree of the denominator, and the ratio of the leading coefficients is $\frac{2}{1} = 2$. A horizontal asymptote of the graph of f is $y = 2$.

(c) A graph of f and its asymptotes are shown in Figure 3.123.

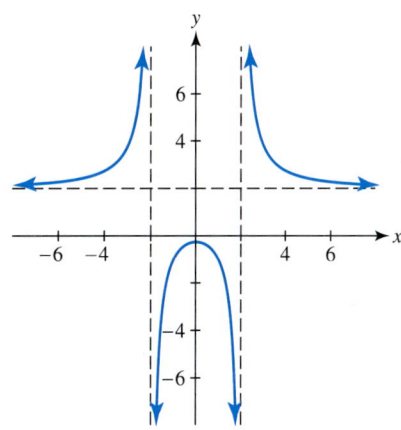

FIGURE 3.123   ■

A third type of asymptote, which is neither vertical nor horizontal, occurs when the numerator of a rational function has degree *one more* than the degree of the denominator. For example, let $f(x) = \dfrac{x^2 + 2}{x - 1}$. If $x - 1$ is divided into $x^2 + 2$ the quotient is $x + 1$ with remainder 3. Thus,

$$f(x) = x + 1 + \frac{3}{x - 1}$$

is an equivalent representation of f. For large values of $|x|$ the ratio $\dfrac{3}{x - 1}$ approaches zero and the graph of f approaches $y = x + 1$. The line $y = x + 1$ is called a **slant asymptote** or **oblique asymptote** of the graph of f. A graph of f with asymptotes $x = 1$ and $y = x + 1$ is shown in Figure 3.124 on the next page.

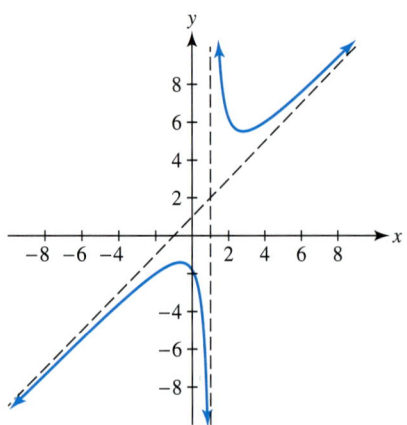

FIGURE 3.124

Rational Equations

A rational equation can be written in the form $f(x) = 0$, where f is a rational function. Rational equations often contain rational expressions. Examples of rational equations include

$$\frac{x^2 - 1}{x^2 + x + 3} = 0, \qquad \frac{3x}{x^3 + x} = 5, \qquad \text{and} \qquad \frac{1}{x - 1} + \frac{1}{x} = 5.$$

Rational equations can be solved symbolically, graphically, and numerically.

EXAMPLE 6 *Solving a rational equation*

Solve $\dfrac{4x}{x - 1} = 6$ symbolically, graphically, and numerically.

Solution

Symbolic Solution

$\dfrac{4x}{x - 1} = 6$	Original equation
$4x = 6(x - 1)$	Cross multiply.
$4x = 6x - 6$	Distributive property
$-2x = -6$	Subtract $6x$.
$x = 3$	Divide by -2.

Graphical Solution Graph $Y_1 = 4X/(X - 1)$ and $Y_2 = 6$. Their graphs intersect at $(3, 6)$, so the solution is 3. See Figure 3.125.

Numerical Solution In Figure 3.126, $Y_1 = Y_2$ when $x = 3$.

$[-9.4, 9.4, 1]$ by $[-9.4, 9.4, 1]$

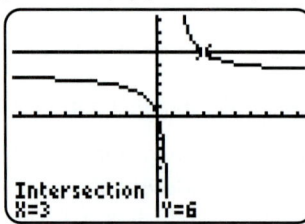

FIGURE 3.125

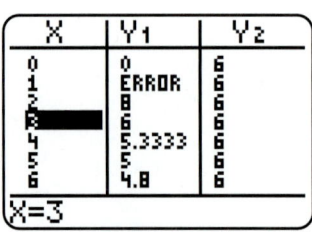

FIGURE 3.126

EXAMPLE 7   *Solving a rational equation*

Solve $\dfrac{6}{x^2} - \dfrac{5}{x} = 1$ symbolically, graphically, and numerically.

Solution
Symbolic Solution

$$\dfrac{6}{x^2} - \dfrac{5}{x} = 1 \qquad \text{Original equation}$$

$$6 - 5x = x^2 \qquad \text{Multiply by } x^2.$$

$$0 = x^2 + 5x - 6 \qquad \text{Add } 5x \text{ and subtract 6.}$$

$$0 = (x + 6)(x - 1) \qquad \text{Factor.}$$

$$x + 6 = 0 \qquad \text{or} \qquad x - 1 = 0 \qquad \text{Zero-product property}$$

$$x = -6 \qquad \text{or} \qquad x = 1 \qquad \text{Solve.}$$

Graphical Solution   Graph $Y_1 = 6/X^2 - 5/X$ and $Y_2 = 1$. The graphs intersect at $(-6, 1)$ and $(1, 1)$. See Figures 3.127 and 3.128. The solutions are -6 and 1.

$[-9.4, 9.4, 1]$ by $[-6.2, 6.2, 1]$ $[-9.4, 9.4, 1]$ by $[-6.2, 6.2, 1]$

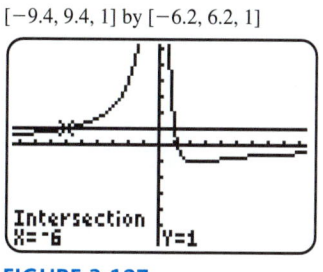

FIGURE 3.127   **FIGURE 3.128**

Numerical Solution   In Figures 3.129 and 3.130, $Y_1 = Y_2$ when $x = -6$ or 1.

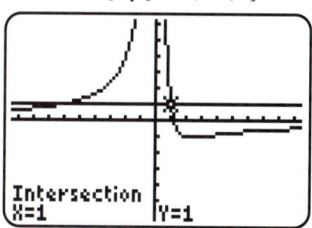

FIGURE 3.129   **FIGURE 3.130**

The next example illustrates the importance of checking possible solutions when solving rational equations.

EXAMPLE 8   *Solving a rational equation*

Solve $\dfrac{1}{x + 3} + \dfrac{1}{x - 3} = \dfrac{6}{x^2 - 9}$ symbolically. Check the result.

Solution

$$\frac{1}{x+3} + \frac{1}{x-3} = \frac{6}{x^2-9} \qquad \text{Given equation}$$

$$(x-3) + (x+3) = 6 \qquad \text{Multiply by } (x+3)(x-3).$$

$$2x = 6 \qquad \text{Simplify.}$$

$$x = 3 \qquad \text{Divide by 2.}$$

When 3 is substituted for x, two expressions in the given equation are undefined. There are no solutions. ∎

Variation (Optional)

In Section 2.3 direct variation was discussed. Sometimes a quantity y varies directly as a power of x. For example, the area A of a circle varies directly as the second power of the radius r. That is, $A = \pi r^2$.

Direct variation as the nth power

Let x and y denote two quantities and n be a positive number. Then y is **directly proportional to the nth power** of x, or y **varies directly as the nth power** of x, if there exists a nonzero number k such that

$$y = kx^n.$$

The number k is called the *constant of variation* or the *constant of proportionality*.

EXAMPLE 9   *Modeling a pendulum*

The time T required for a pendulum to swing back and forth once is called its *period*. See Figure 3.131. The length L of a pendulum is directly proportional to the nth power of T, where n is a positive integer. Table 3.10 lists the period T for various lengths L.

(a) Find the constant of proportionality k and the value of n.
(b) Predict T for a pendulum having a length of five feet.

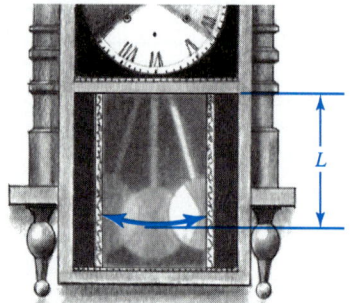

FIGURE 3.131

TABLE 3.10

L (ft)	1.0	1.5	2.0	2.5	3.0	3.5	4.0
T (sec)	1.11	1.36	1.57	1.76	1.92	2.08	2.22

Solution

(a) The equation $L = kT^n$ models the pendulum. It follows that $k = \dfrac{L}{T^n}$. The ratio $\dfrac{L}{T^n}$ equals the constant k for some n. To find k and n see Table 3.11. The ratio is constant when $n = 2$. In this case $k \approx 0.81$ and $L = 0.81T^2$.

TABLE 3.11

L	T	L/T	L/T^2	L/T^3
1.0	1.11	0.90	0.81	0.73
1.5	1.36	1.10	0.81	0.60
2.0	1.57	1.27	0.81	0.52
2.5	1.76	1.42	0.81	0.46
3.0	1.92	1.56	0.81	0.42
3.5	2.08	1.68	0.81	0.39
4.0	2.22	1.80	0.81	0.37

(b) If $L = 5$, then $5 = 0.81T^2$. It follows that $T = \sqrt{5/0.81} \approx 2.48$ seconds. ∎

When two quantities vary inversely, an increase in one quantity results in a decrease in the second quantity. For example, it takes four hours to travel 100 miles at 25 miles per hour and two hours to travel 100 miles at 50 miles per hour. Greater speed results in less travel time. If s represents the average speed of a car and t is the time to travel 100 miles, then $s \cdot t = 100$ or $t = \dfrac{100}{s}$. Doubling the speed cuts the time in half, tripling the speed reduces the time by one-third. The quantities t and s are said to *vary inversely*. The constant of variation is 100.

Inverse variation as the *n*th power

Let x and y denote two quantities and n be a positive number. Then y is **inversely proportional to the *n*th power** of x, or y **varies inversely as the *n*th power** of x, if there exists a nonzero number k such that

$$y = \frac{k}{x^n}.$$

If $y = \dfrac{k}{x}$, then y is **inversely proportional** to x or y **varies inversely as** x.

Inverse variation occurs when measuring the intensity of light. If we increase our distance from a light bulb, the intensity of the light decreases. Intensity I is inversely proportional to the second power of the distance d. The equation $I = \dfrac{k}{d^2}$, models this phenomenon.

EXAMPLE 10

Modeling the intensity of light

At a distance of three meters, a 100-watt bulb produces an intensity of 0.88 watt per square meter. (Source: R. Weidner and R. Sells, *Elementary Classical Physics, Volume 2.*)

(a) Find the constant of proportionality k.

(b) Determine the intensity at a distance of two meters.

Solution

(a) Substitute $d = 3$ and $I = 0.88$ into the equation $I = \dfrac{k}{d^2}$. Solve for k.

$$0.88 = \frac{k}{3^2} \quad \text{or} \quad k = 7.92$$

(b) Since $I = \dfrac{7.92}{d^2}$, let $d = 2$ and find I.

$$I = \frac{7.92}{2^2} = 1.98$$

The intensity at two meters is 1.98 watts per square meter. ■

3.6 PUTTING IT ALL TOGETHER

A rational function is a function that can be represented as a ratio of two polynomials. Unlike polynomials, the graphs of rational functions may have horizontal, vertical, or slant asymptotes. Locating these asymptotes is an important part of the graphing process. A rational function is discontinuous at a vertical asymptote. Its graph never crosses a vertical asymptote.

The following table summarizes some concepts about variation.

Concept	Equation	Example
Varies directly as the nth power of x	$y = kx^n$	Let y vary directly as the third power of x. If the constant of variation is $k = 5$, then $y = 5x^3$.
Varies inversely as the nth power of x	$y = \dfrac{k}{x^n}$	Let y vary inversely as the square of x. If the constant of variation is $k = 3$, then $y = \dfrac{3}{x^2}$.

3.6 EXERCISES

   Tape 4

Rational Functions

Exercises 1–8: Determine whether f is a rational function.

1. $f(x) = \dfrac{x^3 - 5x + 1}{4x - 5}$

2. $f(x) = \dfrac{6}{x^2}$

3. $f(x) = 5x^3 + 6x - 9$

4. $f(x) = \dfrac{x^2 + 1}{\sqrt{x - 8}}$

5. $f(x) = \sqrt[3]{x^2 + 1}$

6. $f(x) = \dfrac{x^4 + 3x^3 + x^{1.2}}{5x^4 - 7x^{3/2} + 2x^2 + x - 17}$

7. $f(x) = \dfrac{4}{x} + 1$

8. $f(x) = \dfrac{1}{x + 1} + \dfrac{1}{x - 1}$

Asymptotes

Exercises 9–12: Identify any horizontal or vertical asymptotes in the graph.

9.

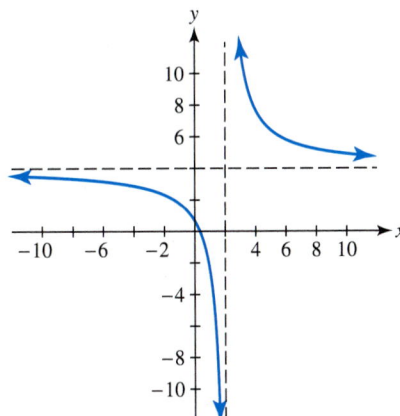

10.

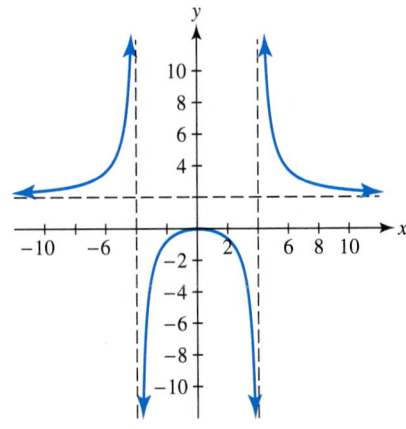

11.

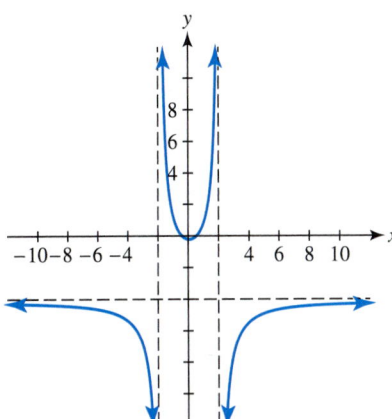

12.

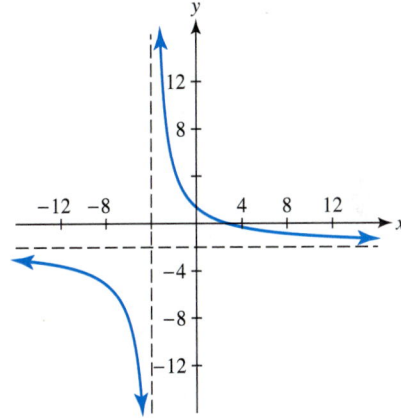

Exercises 13–20: Find any vertical or horizontal asymptotes.

13. $f(x) = \dfrac{4x + 1}{2x - 6}$

14. $f(x) = \dfrac{x + 6}{5 - 2x}$

15. $f(x) = \dfrac{3}{x^2 - 5}$

16. $f(x) = \dfrac{3x^2}{x^2 - 9}$

17. $f(x) = \dfrac{x^4 + 1}{x^2 + 3x - 10}$

18. $f(x) = \dfrac{6x^2 - x - 2}{2x^2 + x - 6}$

19. $f(x) = \dfrac{x^2 - 9}{x + 3}$

20. $f(x) = \dfrac{2x^2 - 3x + 1}{2x - 1}$

Exercises 21–24: Let a be a positive constant. Match f(x) with its graph without using a calculator.

21. $f(x) = \dfrac{a}{x - 1}$

22. $f(x) = \dfrac{2x + a}{x - 1}$

23. $f(x) = \dfrac{x - a}{x + 2}$

24. $f(x) = \dfrac{-2x}{x^2 - a}$

a.

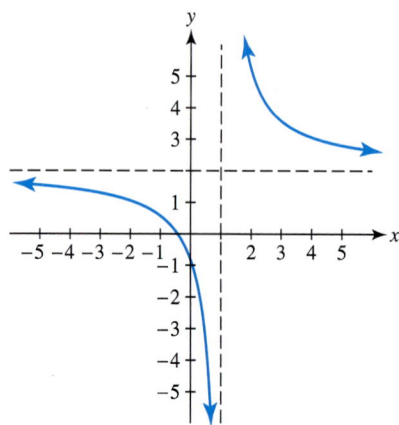

b.

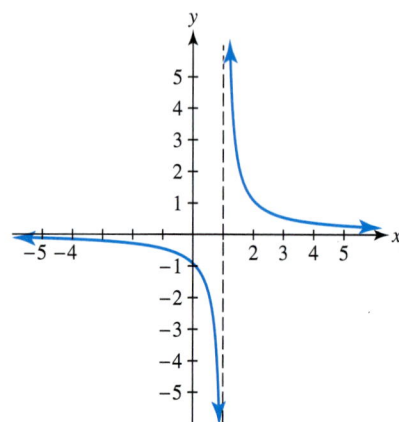

c.

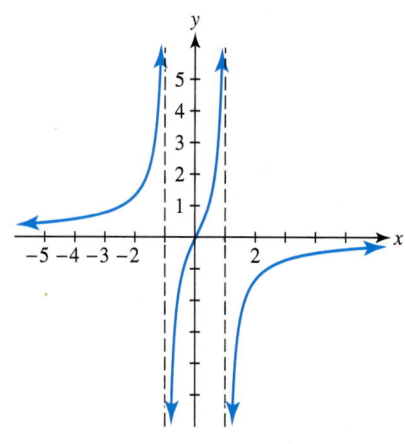

d.

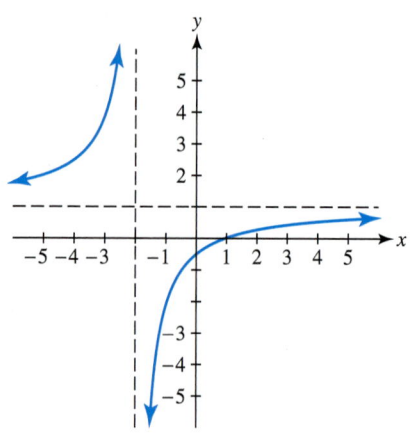

Exercises 25–28: The graph of $f(x) = \dfrac{1}{x}$ was discussed in Example 2. Use translations to sketch a graph of the equation. Include any asymptotes.

25. $y = \dfrac{1}{x} + 2$

26. $y = \dfrac{1}{x - 3}$

27. $y = \dfrac{1}{x + 3} - 1$

28. $y = \dfrac{1}{x - 1} + 3$

Exercises 29–36: Complete the following.
- (a) Find the domain of f.
- (b) Graph f in an appropriate viewing rectangle.
- (c) Find any vertical or horizontal asymptotes.
- (d) Sketch a graph of f that includes any asymptotes.

29. $f(x) = \dfrac{x + 3}{x - 2}$

30. $f(x) = \dfrac{6 - 2x}{x + 3}$

31. $f(x) = \dfrac{4x + 1}{x^2 - 4}$

32. $f(x) = \dfrac{0.5x^2 + 1}{x^2 - 9}$

33. $f(x) = \dfrac{4}{1 - 0.25x^2}$

34. $f(x) = \dfrac{x^2}{1 + 0.25x^2}$

35. $f(x) = \dfrac{x^2 - 4}{x - 2}$

36. $f(x) = \dfrac{4(x - 1)}{x^2 - x - 6}$

Exercises 37 and 38: In the accompanying table, Y_1 represents a rational function. Conjecture the equation for a horizontal asymptote of the graph of Y_1.

37.

X	Y1
50	2.8654
100	2.9314
150	2.9539
200	2.9653
250	2.9722
300	2.9768
350	2.9801

X=50

38.

X	Y1
-10	4.8922
-20	4.9726
-30	4.9878
-40	4.9931
-50	4.9956
-60	4.9969
-70	4.9978

X=-10

Exercises 39–42: Sketch a graph of a function that satisfies the conditions. (Hint: *Graph the asymptotes first.*)

39. Vertical asymptote $x = 3$,
horizontal asymptote $y = 1$

40. Vertical asymptote $x = -4$,
horizontal asymptote $y = -2$

41. Vertical asymptotes $x = \pm 1$,
horizontal asymptote $y = -3$

42. Vertical asymptotes $x = -4$ and $x = 2$,
horizontal asymptote $y = 1$

Exercises 43–48: Write a symbolic representation of a rational function f that satisfies the conditions.

43. Vertical asymptote $x = 5$

44. Horizontal asymptote $y = 7$

45. Vertical asymptote $x = -3$,
horizontal asymptote $y = 1$

46. Vertical asymptote $x = 4$,
horizontal asymptote $y = -3$

47. Vertical asymptotes $x = \pm 3$,
horizontal asymptote $y = 0$

48. Vertical asymptotes $x = -2$ and $x = 4$,
horizontal asymptote $y = 5$

Exercises 49–52: Find any vertical or slant asymptotes of the graph of f. Use a graphing calculator to help sketch a graph of f. Show all asymptotes.

49. $f(x) = \dfrac{x^2 + 1}{x + 1}$

50. $f(x) = \dfrac{2x^2 - 5x - 2}{x - 2}$

51. $f(x) = \dfrac{0.5x^2 - 2x + 2}{x + 2}$

52. $f(x) = \dfrac{0.5x^2 - 5}{x - 3}$

53. *Interpreting an Asymptote* Suppose that an insect population in millions is modeled by $f(x) = \dfrac{10x + 1}{x + 1}$, where $x \geq 0$ is in months.
(a) Graph f in [0, 14, 1] by [0, 14, 1]. Find the equation of the horizontal asymptote.
(b) Determine the initial insect population.

(c) What happens to the population after several months?
(d) Interpret the horizontal asymptote.

54. *Interpreting an Asymptote* Suppose that the population of a species of fish in thousands is modeled by $f(x) = \dfrac{x + 10}{0.5x^2 + 1}$, where $x \geq 0$ is in years.
(a) Graph f in [0, 12, 1] by [0, 12, 1]. What is the horizontal asymptote?
(b) Determine the initial population.
(c) What happens to the population of this fish after many years?
(d) Interpret the horizontal asymptote.

Rational Equations

Exercises 55–60: Solve the rational equation
(a) *symbolically,*
(b) *graphically, and*
(c) *numerically.*

55. $\dfrac{2x}{x + 2} = 6$

56. $\dfrac{3x}{2x - 1} = 3$

57. $2 - \dfrac{5}{x} + \dfrac{2}{x^2} = 0$

58. $\dfrac{1}{x^2} + \dfrac{1}{x} = 2$

59. $\dfrac{1}{x + 1} + \dfrac{1}{x - 1} = \dfrac{1}{x^2 - 1}$

60. $\dfrac{4}{x - 2} = \dfrac{3}{x - 1}$

Exercises 61–64: Find all real solutions to the equation. Check your results.

61. $\dfrac{1}{x + 2} + \dfrac{1}{x} = 1$

62. $\dfrac{2x}{x - 1} = 5 + \dfrac{2}{x - 1}$

63. $\dfrac{1}{x} - \dfrac{2}{x^2} = 5$

64. $\dfrac{1}{x^2 - 2} = \dfrac{1}{x}$

Applications of Rational Functions

65. *Time Spent in Line* Suppose the average number of vehicles arriving at the main gate of an amusement park is equal to 10 per minute, while the average number of vehicles being admitted through the gate per minute is equal to x. Then the average waiting time in minutes for each vehicle at the gate can be computed by $f(x) = \dfrac{x - 5}{x^2 - 10x}$, where $x > 10$. (Source: F. Mannering.)

(a) Estimate graphically the admittance rate x that results in an average wait of 15 seconds.

(b) If one attendant can serve five vehicles per minute, how many attendants are needed to keep the average wait to 15 seconds or less?

66. *Length of Lines*   (Refer to Example 1.) Determine graphically and symbolically the traffic intensity x when the average number of vehicles in line equals three.

67. *Train Curves*   When curves are designed for trains, sometimes the outer rail is elevated or banked, so that a locomotive can safely negotiate the curve at a higher speed. See the accompanying figure. Suppose a circular curve is being designed for 60 miles per hour. The rational function given by $f(x) = \dfrac{2540}{x}$ computes the elevation y in inches of the outer track for a curve with a radius of x feet, where $y = f(x)$. (Source: L. Haefner, *Introduction to Transportation Systems.*)

(a) Evaluate $f(400)$ and interpret its meaning.

(b) Graph f in [0, 600, 100] by [0, 50, 5]. Discuss how the elevation of the outer rail changes with the radius x.

(c) Interpret the horizontal asymptote.

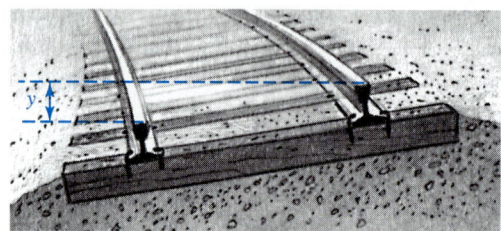

68. *Cost-Benefit*   A cost-benefit function C computes the cost in millions of dollars of implementing a city recycling project having x percent of the citizens participate, where $C(x) = \dfrac{1.2x}{100 - x}$.

(a) Graph C in [0, 100, 10] by [0, 10, 1]. Interpret the graph as x approaches 100.

(b) If 75% participation is expected, determine the cost for the city.

(c) The city plans to spend $5 million on this recycling project. Estimate graphically the percentage of participation that can be expected.

(d) Solve part (c) symbolically.

69. *Mortality Rates*   The table contains incidence ratios by age for deaths from coronary heart disease

(CHD) and lung cancer (LC) when comparing smokers (21–39 cigarettes per day) to nonsmokers. For example, the incidence ratio of 10 means that smokers are 10 times more likely than nonsmokers to die from lung cancer between the ages of 55 and 64. If the incidence ratio is x, then the percentage P of deaths caused by smoking can be calculated using the rational function $P(x) = \dfrac{x - 1}{x}$, where the output is in decimal form. (Source: A. Walker, *Observation and Inference.*)

Age	CHD	LC
55–64	1.9	10
65–74	1.7	9

(a) Calculate the percentage of lung cancer deaths in the age group 65–74 that can be attributed to smoking.

(b) Determine the percentage of deaths from coronary heart disease in the age group from 55–64 that can be attributed to smoking.

70. *Probability*   A container holds x balls numbered 1 through x. Only one ball has the winning number.

(a) Determine a rational function f that computes the probability or likelihood of *not* drawing the winning ball.

(b) What is the domain of f?

(c) Table f starting at $x = 1$, incrementing by 10. What happens to the probability of *not* drawing the winning ball as the number of balls increases?

(d) Interpret the horizontal asymptote of the graph of f.

71. *Stopping Distance*   The **grade** x of a hill is a measure of its steepness. For example, if a road rises 10 feet for every 100 feet of horizontal distance, then it has an uphill grade of $x = \dfrac{10}{100}$ or 10%. See the accompanying figure. Grades are typically kept quite small—usually less than 10%. The stopping distance D for a car traveling at 50 miles per hour on a wet, uphill grade is given by $D(x) = \dfrac{2500}{30(0.3 + x)}$. (Source: L. Haefner.)

(a) Table D starting at $x = 0$, incrementing by 0.05.

(b) Describe what happens to the stopping distance as the hill becomes steeper. Does this agree with your driving experience?

72. (Refer to the previous exercise.) If a car is traveling 50 miles per hour downhill, then its stopping distance on a wet pavement is given by $D(x) = \dfrac{2500}{30(0.3 + x)}$, where $x < 0$ for a downhill grade.
 (a) Evaluate D for $x = 0, -0.05, -0.10, -0.15,$ ..., using a table.
 (b) What happens to the stopping distance as the downhill grade becomes steeper? Does this agree with your intuition?
 (c) The graph of D has a vertical asymptote at $x = -0.3$. Conjecture the physical significance of this asymptote.

73. *Slippery Roads* If a car is moving at 50 miles per hour on a level highway, then its stopping distance is dependent on the road conditions. This distance in feet can be computed by $D(x) = \dfrac{2500}{30x}$, where x is the coefficient of friction between the tires and the road and $0 < x \leq 1$. A smaller value of x indicates that the road is more slippery.
 (a) Graph D in [0, 1, 0.1] by [0, 2000, 500].
 (b) Identify and interpret the vertical asymptote.

74. *Rational Approximation* Computers frequently use rational functions to approximate other types of functions. Use graphing to match the function with its rational approximation for $1 \leq x \leq 15$.
 (a) $f_1(x) = \sqrt{x}$
 (b) $f_2(x) = \sqrt{4x + 1}$
 (c) $f_3(x) = \sqrt[3]{x}$
 (d) $f_4(x) = \dfrac{1 - \sqrt{x}}{1 + \sqrt{x}}$
 i. $r_1(x) = \dfrac{2 - 2x^2}{3x^2 + 10x + 3}$
 ii. $r_2(x) = \dfrac{15x^2 + 75x + 33}{x^2 + 23x + 31}$
 iii. $r_3(x) = \dfrac{10x^2 + 80x + 32}{x^2 + 40x + 80}$
 iv. $r_4(x) = \dfrac{7x^3 + 42x^2 + 30x + 2}{2x^3 + 30x^2 + 42x + 7}$

Variation

Exercises 75–78: Find the constant of proportionality k for the given conditions.

75. $y = \dfrac{k}{x}$, and $y = 2$ when $x = 3$

76. $y = \dfrac{k}{x^2}$, and $y = \dfrac{1}{4}$ when $x = 8$

77. $y = kx^3$, and $y = 64$ when $x = 2$

78. $y = kx^{3/2}$, and $y = 96$ when $x = 16$

Exercises 79–82: Solve the variation problem.

79. Suppose T varies directly as the $\dfrac{3}{2}$ power of x. When $x = 4$, $T = 20$. Find T when $x = 16$.

80. Suppose y varies directly as the second power of x. When $x = 3$, $y = 10.8$. Find y when $x = 1.5$.

81. Let y be inversely proportional to x. When $x = 6$, $y = 5$. Find y when $x = 15$.

82. Let z be inversely proportional to the third power of t. When $t = 5$, $z = 0.08$. Find z when $t = 2$.

Exercises 83–86: Assume that the constant of proportionality is positive.

83. Let y be inversely proportional to x. If x doubles, what happens to y?

84. Let y vary inversely as the second power of x. If x doubles, what happens to y?

85. Suppose y varies directly as the third power of x. If x triples, what happens to y?

86. Suppose y is directly proportional to the second power of x. If x is halved, what happens to y?

Exercises 87 and 88: The data in the table satisfies the equation $y = kx^n$, where n is a positive integer. Determine k and n.

87.

x	2	3	4	5
y	2	4.5	8	12.5

88.

x	3	5	7	9
y	32.4	150	411.6	874.8

Exercises 89 and 90: The data in the table satisfies the equation $y = \dfrac{k}{x^n}$, where n is a positive integer. Determine k and n.

89.

x	2	3	4	5
y	1.5	1	0.75	0.6

90.

x	2	4	6	8
y	9	2.25	1	0.5625

91. *Allometric Growth*   The weight y of a fiddler crab is directly proportional to the 1.25 power of the weight x of its claws. A crab with a body weight of 1.9 grams has claws weighing 1.1 grams. Estimate the weight of a fiddler crab with claws weighing 0.75 gram. (Source: D. Brown.)

92. *Gravity*   The weight of an object varies inversely as the second power of the distance from the center of Earth. The radius of Earth is approximately 4000 miles. If a person weighs 160 pounds on Earth's surface, what would this individual weigh 8000 miles above the surface of Earth?

93. *Hubble Telescope*   The brightness or intensity of starlight varies inversely as the square of its distance from Earth. The Hubble Telescope can see stars whose intensities are $\frac{1}{50}$ of the faintest star now seen by ground-based telescopes. Determine how much farther the Hubble Telescope can see into space than ground-based telescopes. (Source: National Aeronautics and Space Administration.)

94. *Volume*   The volume V of a cylinder with a fixed height is directly proportional to the square of its radius r. If a cylinder with a radius of 10 inches has a volume of 200 cubic inches, what is the volume of a cylinder with the same height and a radius of five inches?

95. *Electrical Resistance*   The electrical resistance R of a wire varies inversely as the square of its diameter d. If a 25-foot wire with a diameter of two millimeters has a resistance of 0.5 ohm, find the resistance of a wire having the same length and a diameter of three millimeters.

96. *Strength of a Beam*   The strength of a rectangular wood beam varies directly as the square of the depth of its cross section. If a beam with a depth of 3.5 inches can support 1000 pounds, how much weight can the same type of beam hold if its depth is 10.5 inches?

Writing about Mathematics

1. Suppose that the symbolic representation of a rational function f is given.
  (a) Explain how to find any vertical or horizontal asymptotes of the graph of f.
  (b) Discuss what a horizontal asymptote represents.

2. Discuss how to find the domain of a rational function symbolically and graphically.

CHECKING BASIC CONCEPTS FOR SECTIONS 3.5 AND 3.6

1. Find a quadratic polynomial $f(x)$ with zeros $\pm 4i$ and leading coefficient 3. Write $f(x)$ in complete factored form and expanded form.

2. Sketch a graph of a quartic function with a negative leading coefficient, two real zeros, and two imaginary zeros.

3. Write $f(x) = x^3 - x^2 + 4x - 4$ in complete factored form.

4. Graph $f(x) = \dfrac{2x - 1}{x + 2}$ using a graphing calculator.
  (a) Find the domain of f.
  (b) Identify any vertical or horizontal asymptotes.
  (c) Sketch a graph of f that includes all asymptotes.

5. Solve $\dfrac{3x - 1}{1 - x} = 1$ graphically, numerically, and symbolically.

CHAPTER **3** Summary

In Chapter 1 the concept of a function was introduced. Since then, several different types of functions have been discussed. Figure 3.132 illustrates how these functions are related.

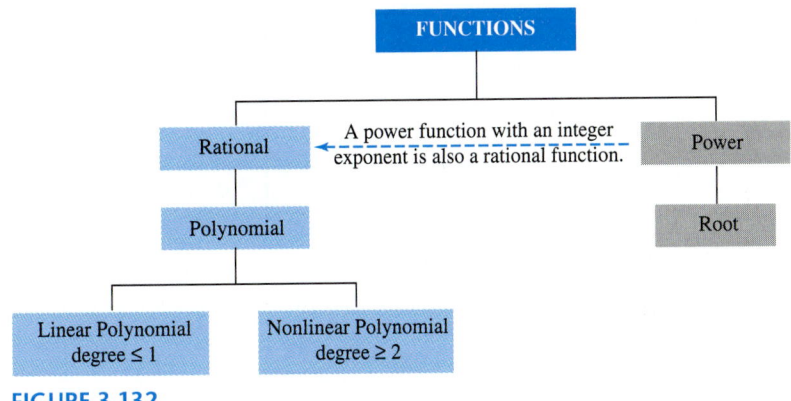

FIGURE 3.132

Two types of functions are rational and power. Although there are exceptions, rational functions and power functions are typically nonlinear functions. Every polynomial function is also a rational function. Polynomial functions of degree 2 or higher are nonlinear functions. Power functions include root functions. The graphs of nonlinear functions are not lines.

The following table includes examples of functions and their graphs.

Type of Function	Examples	Graphs
Linear function $f(x) = ax + b$	$f(x) = 0.5x - 1$ $g(x) = -3x + 2$	
Polynomial function $f(x) = a_n x^n + \cdots +$ $a_2 x^2 + a_1 x + a_0$	$f(x) = x^2 - 1$ $g(x) = x^3 - 4x - 1$	

Continued

Type of Function	Examples	Graphs
Rational function $f(x) = \dfrac{p(x)}{q(x)},$ where $p(x)$ and $q(x)$ are polynomials.	$f(x) = \dfrac{1}{x}$ $g(x) = \dfrac{2x-1}{x+2}$	
Root function $f(x) = x^{1/n},$ where $n \geq 2$ is an integer.	$f(x) = x^{1/2} = \sqrt{x}$ $g(x) = x^{1/3} = \sqrt[3]{x}$	
Power function $f(x) = x^a,$ where a is a constant.	$f(x) = x^{2/3}$ $g(x) = x^{\sqrt{2}}$	

Review Exercises

Exercises 1 and 2: The graph of $f(x) = ax^2 + bx + c$ is given.

   (a) *State whether $a > 0$ or $a < 0$.*

   (b) *Estimate the real solutions to $ax^2 + bx + c = 0$.*

   (c) *Determine if the discriminant is positive, negative, or zero.*

1.

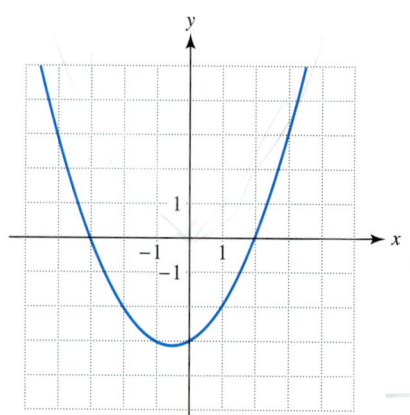

2.

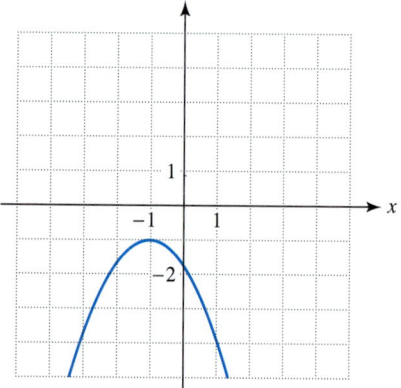

3. Solve $x^2 - x - 20 = 0$ graphically, numerically, and symbolically.

4. Solve $3x^2 + 4x = 1$ symbolically.

5. Solve the quadratic inequality $x^2 - 3x + 2 > 0$ symbolically. Support your results graphically and numerically.

6. Solve $2x^2 + 1.3x \le 0.4$ graphically.

Exercises 7 and 8: The table is a complete representation of f. Decide if f is even, odd, or neither.

7.

x	-4	-2	0	2	4
$f(x)$	13	7	0	-13	-13

8.

x	-5	-3	-1	1	3	5
$f(x)$	-6	2	7	7	2	-6

Exercises 9–12: Determine if f is even, odd, or neither.

9. $f(x) = 2x^6 - 5x^4 - x^2$

10. $f(x) = -5x^3 - 18$

11. $f(x) = 7x^5 + 3x^3 - x$

12. $f(x) = \dfrac{1}{1 + x^2}$

13. The polynomial $f(x) = x^3 - 6x^2 + 11x - 6$ has zeros 1, 2, and 3. Write its complete factored form.

14. Find the complete factored form of $f(x) = x^4 + 2x^3 - 13x^2 - 14x + 24$.

15. Find the complete factored form of $f(x) = x^3 + 2x^2 - 5x - 5$ graphically.

16. Let $f(x)$ be given by

$$f(x) = \begin{cases} 2x & \text{if } 0 \le x < 2 \\ 8 - x^2 & \text{if } 2 \le x \le 4. \end{cases}$$

   (a) Sketch a graph of f. Is f continuous?
   (b) Evaluate $f(1)$ and $f(3)$.
   (c) Solve the equation $f(x) = 2$.

Exercises 17 and 18: Use the graph of f to estimate any
   (a) *local extrema, and*
   (b) *absolute extrema.*

17.

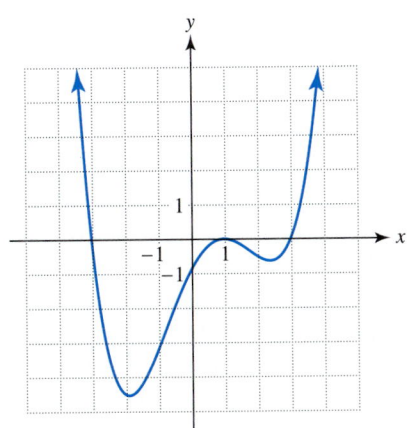

18.

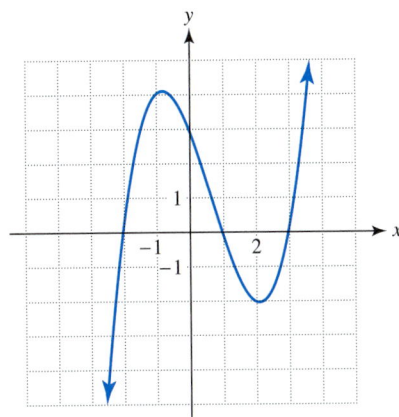

19. Graph $f(x) = -0.25x^4 + 0.67x^3 + 9.5x^2 - 20x - 50$.

   (a) Approximate any local extrema.
   (b) Approximate any absolute extrema.
   (c) Determine x-intervals where f is increasing or decreasing.

20. Let $f(x) = \dfrac{2x^2}{x^2 - 4}$.

   (a) Find the domain of f.
   (b) Identify any vertical or horizontal asymptotes.
   (c) Graph f with a graphing calculator.
   (d) Sketch a graph of f that includes all asymptotes.

21. Graph each quartic polynomial. Count the local maxima, local minima, and x-intercepts.
   (a) $f(x) = x^4 + 2x^3 - 9x^2 - 2x + 20$
   (b) $g(x) = -0.4x^4 + 3.6x^2$
   (c) $h(x) = x^4 + 2x^3 - 21x^2 - 22x + 40$

22. Graph each quintic polynomial. Estimate any turning points.
   (a) $f(x) = 0.03x^5 - 0.21x^4 + 0.21x^3 + 0.57x^2 - 0.48x - 0.6$
   (b) $g(x) = -0.02x^5 + 1$
   (c) $h(x) = 0.01x^5 - 0.09x^4 - 0.22x^3 + 2.72x^2 - 0.96x - 12.8$

23. Sketch a graph of a function f with vertical asymptote $x = -2$ and horizontal asymptote $y = 2$.

24. Find the quotient and remainder when $2x^3 - x^2 - 4x + 1$ is divided by $x - 2$.

25. Solve $9x = 3x^3$.

26. Solve $x^4 - 3x^2 + 2 = 0$.

27. Write the complete factored form of a quintic polynomial $f(x)$ that has zeros -2 and 2 with multiplicities 2 and 3, respectively.

28. Find the complete factored form of $f(x) = x^3 - 2x^2 - 5x + 6$, given $c = 3$ is a zero. Do not use a calculator.

Exercises 29 and 30: Sketch a graph of a polynomial that satisfies the given conditions.

29. Cubic, two x-intercepts, and a positive leading coefficient

30. Quartic with no x-intercepts

31. What is the maximum number of times that a horizontal line can intersect the graph of each type of polynomial?
  (a) linear (degree 1)
  (b) quadratic
  (c) cubic

32. **(a)** Conjecture the maximum number of times that a horizontal line can intersect the graph of a polynomial of degree $n \geq 1$.
  (b) Discuss the maximum number of real solutions to the polynomial equation $a_n x^n + \cdots + a_2 x^2 + a_1 x + a_0 = k$, where k is a constant.

Exercises 33–36: Write the following in standard form.

33. $(2 - 3i) + (-3 + 3i)$

34. $(-5 + 3i) - (-3 - 5i)$

35. $(3 + 2i)(-4 - i)$

36. $\dfrac{3 + 2i}{2 - i}$.

37. Solve $4x^2 + 9 = 0$.

38. Solve $2x^2 + 3 = 2x$.

39. Determine graphically the number of real zeros and the number of imaginary zeros of $f(x) = x^3 - 3x^2 + 3x - 9$.

40. Write a polynomial $f(x)$ in complete factored form that has degree 3, leading coefficient 4, and zeros 1, $3i$, and $-3i$. Then write $f(x)$ in expanded form.

41. Find the complete factorization of $f(x) = 2x^2 + 4$.

42. The graph of a semicircle with a radius of 5 and center $(0, 0)$ is given by $y = \sqrt{25 - x^2}$. Use reflections to graph the entire circle in $[-9, 9, 1]$ by $[-6, 6, 1]$.

Exercises 43 and 44: Use the rational zero test to determine any rational zeros of $f(x)$. Then, find the complete factored form of $f(x)$ without a calculator.

43. $f(x) = 2x^3 + x^2 - 13x + 6$

44. $f(x) = x^3 + x^2 - 11x - 11$

Exercises 45–48: Solve the equation. Check your results.

45. $x^5 = 1024$

46. $x^{1/3} = 4$

47. $\sqrt{x - 2} = x - 4$

48. $x^{3/2} = 27$

Exercises 49 and 50: Solve the rational equation. Support your results graphically and numerically.

49. $\dfrac{5x + 1}{x + 3} = 3$

50. $\dfrac{1}{x} - \dfrac{1}{x^2} + 2 = 0$

Applications

51. *Allometry* (Refer to Exercise 81 in Section 3.2.) One of the earliest studies in allometry was performed by Bryan Robinson during the eighteenth century. He found that the pulse rate of an animal could be approximated by $f(x) = 1607x^{-0.75}$. The input x is the length of the animal in inches, and the output $f(x)$ is the approximate number of heart beats per minute. (Source: H. Lancaster, *Quantitative Methods in Biology and Medical Sciences.*)
  (a) Use f to estimate the pulse rates of a 2-foot dog and a 5.5-foot person.
  (b) What length corresponds to a pulse rate of 400 beats per minute?

52. *Modeling Water Flow* Water is leaking out of a hole in the bottom of a 100-gallon tank. The table lists the volume V of the water after t minutes. Decide whether $f(t) = 0.4t^2 - 9.9t + 100$ or $g(t) = 100 - 9.5t$ models the data better.

t (min)	0	1	2	3	4
V (gal)	100	90.5	81.9	74.1	67.0

53. *Credit Card Debt* The table lists the outstanding balances on Visa and MasterCard credit cards in billions of dollars. (Sources: Bankcard Holders of America, Federal Reserve Board.)

Year	1980	1984	1988	1992	1996
Debt	82	108	172	254	444

  (a) Explain why a linear function does not model this data.
  (b) The data can be modeled by $f(x) = 1.3(x - 1980)^2 + 82$, where x is the year. Solve the equation $f(x) = 212$ graphically, numerically, and symbolically. Interpret the results.

54. *Corporate Mergers* The number of corporate mergers has changed during the past decade. The table lists the approximate numbers of mergers from 1987 to 1995. (Source: Securities Data Company.)

Year	1987	1988	1989	1990	1991
Mergers	3100	3800	5300	5600	5100

Year	1992	1993	1994	1995
Mergers	5300	6200	7500	8800

(a) Make a line graph of the data in [1986, 1996, 1] by [2000, 10,000, 1000].

(b) Conjecture a possible degree of a polynomial that models this data. Explain your reasoning.

55. *Daylight Hours* The accompanying graph shows the deviation in the number of daylight hours from the average of 12 hours at a latitude of 40°N. In this graph x represents the month, where $x = 0$ corresponds to the spring equinox on March 21. For example, $x = 2$ represents May 21 and $x = -3$ represents December 21. At the spring equinox, day and night are of equal length so the deviation is 0. (Source: J. Williams, *The Weather Almanac 1995*.)

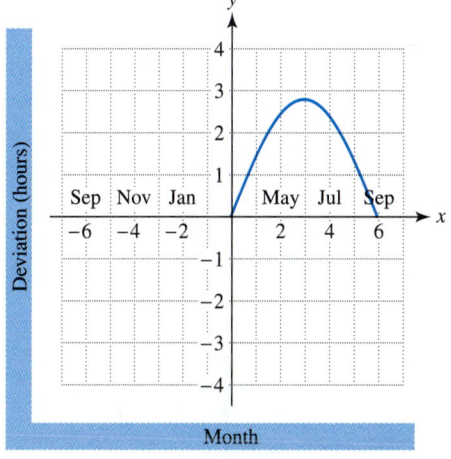

(a) Let f model this deviation in daylight. Is f an odd or even function? Explain.

(b) Sketch the left side of the graph for $-6 \le x \le 0$.

56. *Irrigation and Yield* The table shows how irrigation of rice crops affects yield, where x represents the percent of total area that is irrigated and y is the rice yield in tons per hectare. (1 hectare ≈ 2.47 acres.) (Source: D. Grigg, *The World Food Problem*.)

x	0	20	40	60	80	100
y	1.6	1.8	2.2	3.0	4.5	6.1

(a) The function $f(x) = 0.00051x^2 - 0.00604x + 1.6$ models the data. Graph f and the data in the same viewing rectangle.

(b) Solve the equation $f(x) = 3.7$ graphically, numerically, and symbolically. Interpret the results.

57. *Internet* Downloading files on the Internet can be both slow and unreliable when standard phone lines are being used. Phone modems can transmit 56,000 bits per second, whereas cable TV modems are capable of transmitting 2–10 million bits per second. Demand for this new technology is expected to increase from 1996 to 2000. The table shows this expected increase in millions of accounts. (Source: *New York Times Service, Pioneer Press Research*.)

Year	1996	1997	1998	1999	2000
Accounts (millions)	0.5	1	2	4	7

(a) Let $f(x) = a(x - 1996)^2 + 0.5$. Approximate a so that f models the data.

(b) Solve the equation $f(x) = 10$ and interpret the result.

58. *Falling Object* If an object is dropped from a height h, then the time t for the object to strike the ground is directly proportional to the square root of h. If it requires one second for an object to fall 16 feet, how long does it take for an object to fall 256 feet?

Extended and Discovery Exercises

Exercises 1–4: Reflecting Functions   Computer graphics frequently use reflections. Reflections can speed up the generation of a picture or create a figure that appears perfectly symmetrical. (Source: S. Hoggar, *Mathematics for Computer Graphics*.)

(a) *For the given $f(x)$, constant k, and viewing rectangle, graph*

$$x = k, \qquad y = f(x), \qquad \text{and} \qquad y = f(2k - x).$$

(b) *Generalize how the graph of $y = f(2k - x)$ compares to the graph of $y = f(x)$.*

1. $f(x) = \sqrt{x}$, $k = 2$, $[-1, 8, 1]$ by $[-4, 4, 1]$
2. $f(x) = x^2$, $k = -3$, $[-12, 6, 1]$ by $[-6, 6, 1]$
3. $f(x) = x^4 - 2x^2 + 1$, $k = -6$, $[-15, 3, 1]$ by $[-3, 9, 1]$
4. $f(x) = 4x - x^3$, $k = 5$, $[-6, 18, 1]$ by $[-8, 8, 1]$

Exercises 5–8: Average Rates of Change   *These exercises investigate the relationship between polynomial functions and their average rates of change. For example, the average rate of change of $f(x) = x^2$ from x to $x + 0.001$ for any x can be calculated and graphed as shown in the accompanying figures. The graph of f is a parabola, while the graph of its average rate of change is a line. There is an interesting relationship between a graph of a polynomial function and a graph of its average rate of change. Try to discover what it is by completing the following.*

(a) *Graph each function and its average rate of change from x to $x + 0.001$.*

(b) *Compare the graph of each function to the graph of its average rate of change. How are turning points on the graph of a function related to its average rate of change?*

(c) *Conjecture a generalization of these results. Test your conjecture.*

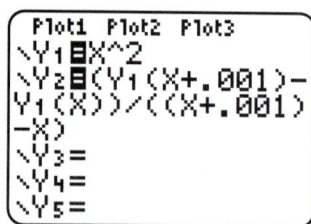

[−10, 10, 1] by [−10, 10, 1]

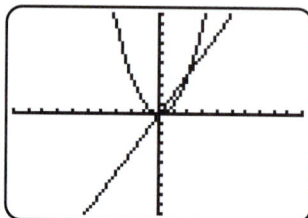

5. *Linear Functions*
$f_1(x) = 3x + 1$ $f_2(x) = -2x + 6$
$f_3(x) = 1.5x - 5$ $f_4(x) = -4x - 2.5$

6. *Quadratic Functions*
$f_1(x) = 2x^2 - 3x + 1$ $f_2(x) = -0.5x^2 + 2x + 2$
$f_3(x) = x^2 + x - 2$ $f_4(x) = -1.5x^2 - 4x + 6$

7. *Cubic Functions*
$f_1(x) = 0.5x^3 - x^2 - 2x + 1$
$f_2(x) = -x^3 + x^2 + 3x - 5$
$f_3(x) = 2x^3 - 5x^2 + x - 3$
$f_4(x) = -x^3 + 3x - 4$

8. *Quartic Functions*
$f_1(x) = 0.05x^4 + 0.2x^3 - x^2 - 2.4x$
$f_2(x) = -0.1x^4 + 0.1x^3 + 1.3x^2 - 0.1x - 1.2$
$f_3(x) = 0.1x^4 + 0.4x^3 - 0.2x^2 - 2.4x - 2.4$

Exercises 9–14: Polynomial Inequalities   *The solution set for a polynomial inequality can be found by first determining the boundary numbers. For example, to solve $f(x) = x^3 - 4x > 0$ begin by solving $x^3 - 4x = 0$. The solutions (boundary numbers) are -2, 0, and 2. Since $f(x) = x^3 - 4x$ is continuous, it is either only positive or only negative on intervals between consecutive zeros. To determine the solution set, we can evaluate test values for each interval as shown in the following.*

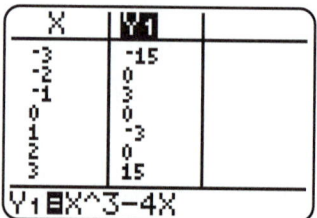

Interval	Test Value	$f(x) = x^3 - 4x$
$(-\infty, -2)$	$x = -3$	$f(-3) = -15 < 0$
$(-2, 0)$	$x = -1$	$f(-1) = 3 > 0$
$(0, 2)$	$x = 1$	$f(1) = -3 < 0$
$(2, \infty)$	$x = 3$	$f(3) = 15 > 0$

We can see that $f(x) > 0$ for $(-2, 0) \cup (2, \infty)$. These results also are supported graphically in the accompanying figure, where the graph of f is above the x-axis when $-2 < x < 0$ or when $x > 2$.

[−6, 6, 1] by [−4, 4, 1]

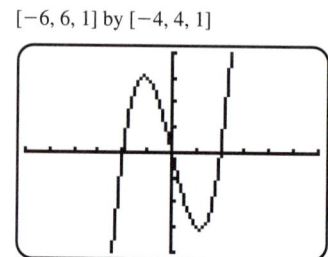

Solve the polynomial inequality.

9. $x^3 - x^2 - 6x > 0$

10. $x^3 - 3x^2 + 2x < 0$

11. $x^3 - 7x^2 + 14x \leq 8$

12. $9x - x^3 \geq 0$

13. $x^4 - 5x^2 + 4 > 0$

14. $1 < x^4$

Exercises 15–20: Rational Inequalities   *Rational inequalities can be solved using many of the same techniques that are used to solve other types of inequalities. However, there is one important difference. For a rational inequality, the boundary between greater than and less* than *can either be an x-value where equality occurs or an x-value where a rational expression is undefined. For example, consider the inequality* $f(x) = \dfrac{2-x}{2x} > 0$. *The solution to the equation* $\dfrac{2-x}{2x} = 0$ *is 2. The rational expression* $\dfrac{2-x}{2x}$ *is undefined when* $x = 0$. *Therefore, we select test values on the intervals* $(-\infty, 0)$, $(0, 2)$, *and* $(2, \infty)$. *From the following we see that* $f(x) > 0$ *for* $(0, 2)$.

X	Y1
-.5	-2.5
0	ERROR
.5	1.5
1	.5
1.5	.16667
2	0
2.5	-.1

Y1 ⊟ (2−X)/(2X)

Interval	Test Value	$f(x) = (2 - x)/2x$
$(-\infty, 0)$	$x = -0.5$	$f(-0.5) = -2.5 < 0$
$(0, 2)$	$x = 1$	$f(1) = 0.5 > 0$
$(2, \infty)$	$x = 2.5$	$f(2.5) = -0.1 < 0$

Notice that Y_1 *changes from negative to positive at* $x = 0$, *where* $f(x)$ *is undefined. These results are supported graphically in the accompanying figure.*

$[-4.7, 4.7, 1]$ by $[-3.1, 3.1, 1]$

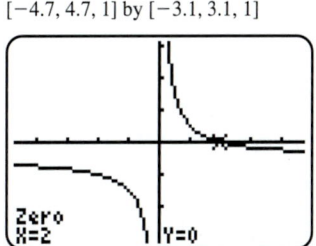

Solve the rational inequality.

15. $\dfrac{3-x}{3x} \geq 0$

16. $\dfrac{x-2}{x+2} > 0$

17. $\dfrac{3-2x}{1+x} < 3$

18. $\dfrac{x+1}{4-2x} \geq 1$

19. $\dfrac{5}{x^2-4} < 0$

20. $\dfrac{x}{x^2-1} \geq 0$

CHAPTER **4** *Inverse Functions*

*I*n 1900 the population of the world was approximately 1.6 billion. At that time the Swedish scientist Svante Arrhenius first predicted a greenhouse effect resulting from emissions of carbon dioxide by the industrialized countries. His classic calculation made use of logarithms and predicted that a doubling of the carbon dioxide concentration in the atmosphere would raise the average global temperature by 7°F to 11°F. Since then, the world population has increased dramatically to approximately 6 billion. With this increase in population, there has been a corresponding increase in emissions of greenhouse gases such as carbon dioxide, methane, and chlorofluorocarbons. These emissions have created the potential to alter the earth's climate and destroy portions of the ozone layer.

When quantities such as population and pollution increase rapidly, nonlinear functions and equations are used to model their growth. In this chapter, exponential and logarithmic functions are introduced. They occur in a wide variety of applications. Examples include acid rain, the decline of the bluefin tuna, air pollution, global warming, salinity of the ocean, demand for liver transplants, diversity of bird species, the relationship between caloric intake and land ownership in develop-

> *T*he important thing is not to stop questioning.
>
> — **Albert Einstein**

ing countries, hurricanes, earthquakes, and increased skin cancer due to a decrease in the ozone layer. Understanding these issues and discovering solutions will require creativity, innovation, and mathematics. This chapter presents some of the mathematical tools necessary for modeling these trends.

Reference: M. Kraljic, *The Greenhouse Effect.*

4.1 Combining Functions

Arithmetic Operations on Functions • Composition of Functions

Introduction

Addition, subtraction, multiplication, and division can be performed on numbers and variables. Arithmetic operations also can be used to combine functions. For example, to model the stopping distance of a car, we compute two quantities. The first quantity is the *reaction distance,* which is the distance that a car travels between the time when a driver first recognizes a hazard and when the brakes are applied. The second quantity is *braking distance,* which is the distance that a car travels after the brakes have been applied. *Stopping distance* is equal to the sum of the reaction distance and the braking distance. One method to determine stopping distance is to find one function that models the reaction distance and a second function that calculates the braking distance. The stopping distance is found by adding the two functions together.

Arithmetic Operations on Functions

Highway engineers frequently assume that drivers have a reaction time of 2.5 seconds or less. During this time a car continues to travel at a constant speed, until a driver is able to move his or her foot from the accelerator to the brake pedal. If a car travels at x miles per hour, then $f(x) = (11/3)x$ computes the distance in feet that a car travels in 2.5 seconds. For example, a driver traveling at 55 miles per hour might have a reaction distance of $f(55) = (11/3)(55) \approx 201.7$ feet.

The braking distance in feet for a car traveling on wet, level pavement at x miles per hour can be approximated by $g(x) = x^2/9$. For instance, a car traveling at 55 miles per hour would need $g(55) = 55^2/9 \approx 336.1$ feet to stop after the brakes have been applied.

The estimated stopping distance at 55 miles per hour is $201.7 + 336.1 = 537.8$ feet. The process of finding stopping distances is shown in Figure 4.1. Algorithm 4.1 computes the reaction distance $f(x)$, while Algorithm 4.2 computes the braking distance $g(x)$. Once $f(x)$ and $g(x)$ have been computed, they are added together. (Source: L. Haefner, *Introduction to Transportation Systems.*)

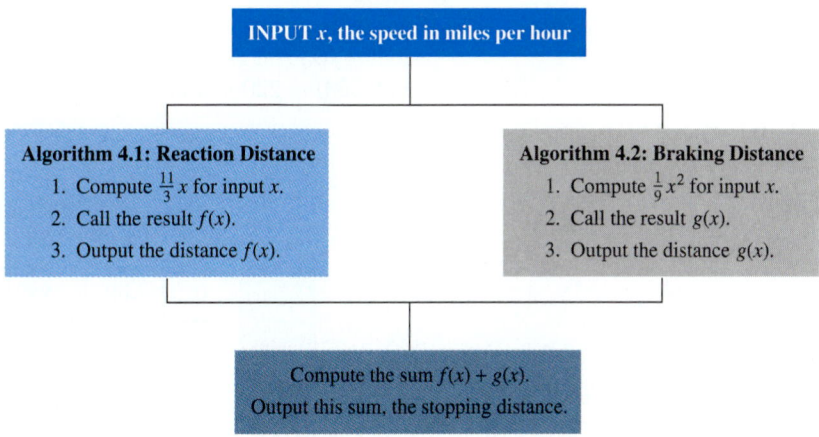

FIGURE 4.1

The concept of finding the sum of two functions also can be represented symbolically, graphically, and numerically as illustrated in the next example.

EXAMPLE 1   *Representing stopping distance symbolically, graphically, and numerically*

A driver traveling at 60 miles per hour with a reaction time of 2.5 seconds attempts to stop on wet, level pavement in order to avoid an accident.
(a) Write a symbolic representation for a function d using f and g that computes the stopping distance for a car traveling at x miles per hour. Evaluate $d(60)$.
(b) Graph f, g, and d in [45, 70, 5] by [0, 800, 100]. Interpret the graph.
(c) Illustrate the relationship between f, g, and d numerically.

Solution
(a) *Symbolic Representation*   Let $d(x)$ be the sum of the two formulas for f and g.

$$d(x) = f(x) + g(x) = \frac{11}{3}x + \frac{1}{9}x^2$$

The stopping distance for a car traveling at 60 miles per hour is

$$d(60) = \frac{11}{3}(60) + \frac{1}{9}(60)^2 = 620 \text{ feet.}$$

(b) *Graphical Representation*   Let $Y_1 = (11/3)X$, $Y_2 = (X^2)/9$, and $Y_3 = Y_1 + Y_2$ as shown in Figure 4.2. For any x-value in the graph, the sum of Y_1 and Y_2 equals Y_3.

[45, 70, 5] by [0, 800, 100]

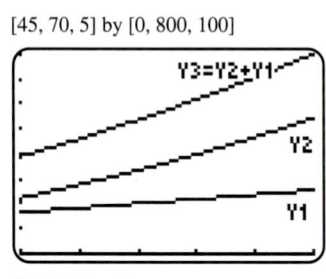

FIGURE 4.2

(c) *Numerical Representation*   Table 4.1 shows numerical representations of $f(x)$, $g(x)$, and $d(x) = f(x) + g(x)$. Notice that values for $d(x)$ can be found by adding $f(x)$ and $g(x)$. For example, when $x = 60$, $f(60) = 220$, $g(60) = 400$, and $d(60) = 220 + 400 = 620$.

TABLE 4.1

x	0	12	24	36	48	60
$f(x)$	0	44	88	132	176	220
$g(x)$	0	16	64	144	256	400
$d(x)$	0	60	152	276	432	620

We now formally define arithmetic operations on functions.

Operations on functions

If $f(x)$ and $g(x)$ both exist, the sum, difference, product, and quotient of two functions f and g are defined by

$$(f + g)(x) = f(x) + g(x),$$
$$(f - g)(x) = f(x) - g(x),$$
$$(fg)(x) = f(x) \cdot g(x), \text{ and}$$
$$\left(\frac{f}{g}\right)(x) = \frac{f(x)}{g(x)}, \text{ where } g(x) \neq 0.$$

The domains of the sum, difference, and product of f and g include x-values that are in both the domain of f and the domain of g. The domain of the quotient f/g includes all x-values in both the domain of f and the domain of g, where $g(x) \neq 0$.

EXAMPLE 2 *Performing arithmetic operations on functions symbolically*

Let $f(x) = 2 + \sqrt{x - 1}$ and $g(x) = x^2 - 4$.
(a) Find the domain of $f(x)$ and $g(x)$. Then find the domains of $(f + g)(x)$, $(f - g)(x)$, $(fg)(x)$, and $(f/g)(x)$.
(b) If possible, evaluate $(f + g)(5)$, $(f - g)(1)$, $(fg)(0)$, and $(f/g)(3)$.
(c) Write expressions for $(f + g)(x)$, $(f - g)(x)$, $(fg)(x)$, and $(f/g)(x)$.

Solution
(a) Whenever $x \geq 1$, $f(x) = 2 + \sqrt{x - 1}$ is defined. Therefore, the domain of f is $\{x | x \geq 1\}$. The domain of $g(x) = x^2 - 4$ is all real numbers. The domains of $f + g$, $f - g$, and fg include all x-values in *both* the domain of f and the domain of g. Thus, their domains are $\{x | x \geq 1\}$.

To determine the domain of f/g, we must also exclude x-values where $g(x) = x^2 - 4 = 0$. This occurs when $x = \pm 2$. Thus, the domain of f/g is $\{x | x \geq 1, x \neq 2\}$. (Note that $x \neq -2$ is satisfied if $x \geq 1$.)

(b) The expressions can be evaluated as follows.

$$(f + g)(5) = f(5) + g(5) = (2 + \sqrt{5 - 1}) + (5^2 - 4) = 4 + 21 = 25$$
$$(f - g)(1) = f(1) - g(1) = (2 + \sqrt{1 - 1}) - (1^2 - 4) = 2 - (-3) = 5$$

$(fg)(0)$ is undefined since 0 is not in the domain of $f(x)$.

$$\left(\frac{f}{g}\right)(3) = \frac{f(3)}{g(3)} = \frac{2 + \sqrt{3 - 1}}{3^2 - 4} = \frac{2 + \sqrt{2}}{5} = \frac{2}{5} + \frac{1}{5}\sqrt{2}$$

(c) The sum, difference, product, and quotient of f and g are calculated as follows.

$$(f + g)(x) = f(x) + g(x) = (2 + \sqrt{x - 1}) + (x^2 - 4) = \sqrt{x - 1} + x^2 - 2$$
$$(f - g)(x) = f(x) - g(x) = (2 + \sqrt{x - 1}) - (x^2 - 4) = \sqrt{x - 1} - x^2 + 6$$
$$(fg)(x) = f(x) \cdot g(x) = (2 + \sqrt{x - 1})(x^2 - 4)$$
$$\left(\frac{f}{g}\right)(x) = \frac{f(x)}{g(x)} = \frac{2 + \sqrt{x - 1}}{x^2 - 4}$$

EXAMPLE 3   *Evaluating combinations of functions graphically, numerically, and symbolically*

If possible, use the given representations of the functions f and g to evaluate $(f + g)(4)$, $(f - g)(-2)$, $(fg)(1)$, and $(f/g)(0)$.

(a)

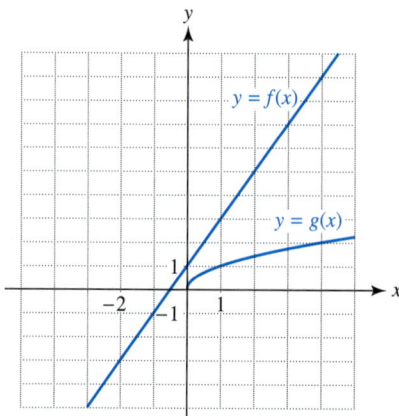

FIGURE 4.3

(b)

TABLE 4.2

x	-2	0	1	4
$f(x)$	-3	1	3	9
$g(x)$	Error	0	1	2

(c) $f(x) = 2x + 1$, $g(x) = \sqrt{x}$

Solution

(a) *Graphical Evaluation*   In Figure 4.3, $f(4) = 9$ and $g(4) = 2$. Thus,

$$(f + g)(4) = f(4) + g(4) = 9 + 2 = 11.$$

Although $f(-2) = -3$, $g(-2)$ is undefined since -2 is not in the domain of g. Thus, $(f - g)(-2)$ is undefined. The domains of f and g include 1, and $(fg)(1) = f(1)g(1) = 3(1) = 3$. The graph of g intersects the origin, so $g(0) = 0$. Thus, $(f/g)(0)$ is undefined.

(b) *Numerical Evaluation*   Numerical evaluation of f and g is performed using Table 4.2. From the table $f(4) = 9$ and $g(4) = 2$. As in part (a),

$$(f + g)(4) = f(4) + g(4) = 9 + 2 = 11.$$

In the table $g(-2)$ is undefined (error), so $(f - g)(-2)$ is also undefined. The calculations of $(fg)(1)$ and $(f/g)(0)$ are done in a similar manner.

(c) *Symbolic Evaluation*   Use the formulas $f(x) = 2x + 1$ and $g(x) = \sqrt{x}$.

$$(f + g)(4) = f(4) + g(4) = (2 \cdot 4 + 1) + \sqrt{4} = 9 + 2 = 11$$
$$(f - g)(-2) = f(-2) + g(-2) = (2 \cdot (-2) + 1) + \sqrt{-2} \text{ is undefined.}$$

$$(fg)(1) = f(1)g(1) = (2 \cdot 1 + 1)\sqrt{1} = 3(1) = 3$$

$$\left(\frac{f}{g}\right)(0) = \frac{f(0)}{g(0)} \text{ is undefined, since } g(0) = 0. \qquad \blacksquare$$

The next example involves an application from business where the difference between two functions occurs.

EXAMPLE 4 *Finding the difference of two functions*

Once the sound track for a compact disc (CD) has been recorded, the cost of producing the master disc can be significant. After this disc has been made, additional discs can be produced inexpensively. In 1997 a typical cost for producing the master disc was $2000, while additional discs cost approximately $2 each. (Source: Windcrest Productions.)

(a) Assuming no other expenses, find a function C that outputs the cost of producing the master disc plus x additional compact discs. Find the cost of making the master disc and 1500 additional compact discs.

(b) Suppose that each CD is sold for $12. Find a function R that computes the revenue received from selling x compact discs. Find the revenue from selling 1500 compact discs.

(c) Assuming that the master CD is not sold, determine a function P that outputs the profit from selling x compact discs. How much profit will there be from selling 1500 discs?

Solution

(a) The cost of producing the master disc for $2000 plus x additional discs at $2 each is given by $C(x) = 2x + 2000$. The $2000 cost is sometimes called a *fixed cost*. The cost of manufacturing the master disc and 1500 additional compact discs is

$$C(1500) = 2(1500) + 2000 = \$5000.$$

(b) The revenue from x discs at $12 each is computed by $R(x) = 12x$. The revenue from selling 1500 compact discs is $R(1500) = 12(1500) = \$18,000$.

(c) Profit P is equal to revenue minus cost. This can be written using function notation.

$$
\begin{aligned}
P(x) = R(x) &- C(x) \\
&= 12x - (2x + 2000) \\
&= 12x - 2x - 2000 \\
&= 10x - 2000
\end{aligned}
$$

If $x = 1500$, $P(1500) = 10(1500) - 2000 = \$13,000$. $\qquad \blacksquare$

Example 5 demonstrates how the quotient of two functions can model AIDS data.

EXAMPLE 5 *Modeling AIDS data using the quotient of two functions*

Table 4.3 lists both the cumulative numbers of AIDS cases and cumulative deaths reported for selected years. The numbers of AIDS cases are modeled by $f(x) = 3200(x - 1982)^2 + 1586$ and the numbers of deaths are modeled by $g(x) = 1900(x - 1982)^2 + 619$. (Source: Department of Health and Human Services.)

TABLE 4.3

Year	Cases	Deaths
1982	1586	619
1984	10,927	5605
1986	41,910	24,593
1988	106,304	61,911
1990	196,576	120,811
1992	329,205	196,283
1994	441,528	270,533

(a) Graph $h(x) = \dfrac{g(x)}{f(x)}$ in [1982, 1994, 2] by [0, 1, 0.1]. Interpret the graph.

(b) Compute the ratio $\dfrac{\text{Deaths}}{\text{Cases}}$ for the data in Table 4.3. Compare these ratios with the results from part (a).

Solution

(a) Let $h(x) = \dfrac{1900(x - 1982)^2 + 619}{3200(x - 1982)^2 + 1586}$. A graph of h is shown in Figure 4.4. As x increases, the graph of f becomes nearly horizontal. By tracing this graph we can see that the ratio of AIDS deaths to cases remains near 0.6 from 1986 to 1994. This means that approximately 60% of the people who contracted AIDS have died.

[1982, 1994, 2] by [0, 1, 0.1]

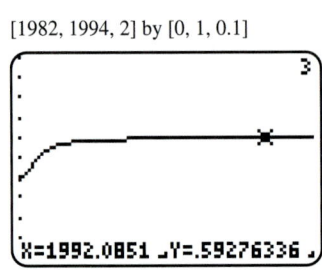

X=1992.0851 ⌐Y=.59276336

FIGURE 4.4

(b) In 1982 the ratio of AIDS deaths to cases was $\dfrac{619}{1586} \approx 0.39$. Other ratios in Table 4.4 are calculated similarly. In later years the ratios are near 0.6, which agrees with part (a).

TABLE 4.4

Year	1982	1984	1986	1988	1990	1992	1994
Ratio	0.39	0.51	0.59	0.58	0.61	0.60	0.61

Composition of Functions

Many tasks in life are performed in sequence. For example, to go to a movie we might get into a car, drive to the movie theater, and get out of the car. The order in which these tasks are performed is important.

A similar situation occurs with functions. For example, to convert miles to inches we might first convert miles to feet and then feet to inches. Since there are 5280 feet in a mile, $f(x) = 5280x$ converts x miles to an equivalent number of feet. Then $g(x) = 12x$ changes feet to inches. To convert x miles to inches, we combine the functions f and g in sequence. Figure 4.5 illustrates how to convert 5 miles to inches. First, $f(5) = 5280(5) = 26,400$. Then the output of 26,400 feet from f is used as input for g. The number of inches in 26,400 feet is $g(26,400) = 12(26,400) = 316,800$. This computation is called the *composition* of g and f.

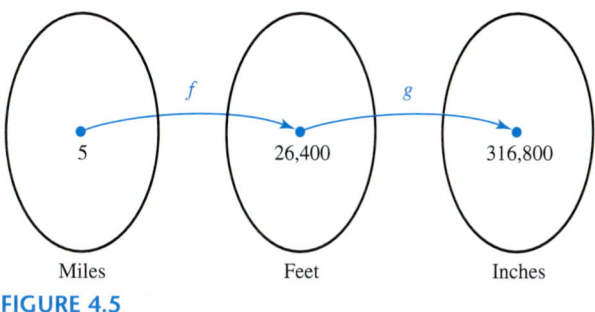

FIGURE 4.5

Algorithms can be used to represent composition of functions as shown in Figure 4.6. Algorithm 4.3 computes $f(x) = 5280x$ and Algorithm 4.4 computes $g(x) = 12x$. Computing a composition of two functions is fundamentally different than the addition of two functions. In Figure 4.1, both Algorithm 4.1 and Algorithm 4.2 receive the *same* input x. Then, their outputs are added. However, in Figure 4.6 the *output* $f(x)$ from Algorithm 4.3 provides the *input* for Algorithm 4.4.

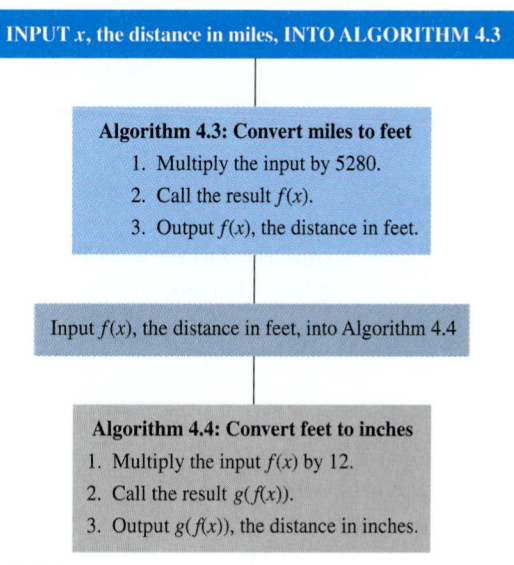

INPUT x, the distance in miles, INTO ALGORITHM 4.3

Algorithm 4.3: Convert miles to feet
1. Multiply the input by 5280.
2. Call the result $f(x)$.
3. Output $f(x)$, the distance in feet.

Input $f(x)$, the distance in feet, into Algorithm 4.4

Algorithm 4.4: Convert feet to inches
1. Multiply the input $f(x)$ by 12.
2. Call the result $g(f(x))$.
3. Output $g(f(x))$, the distance in inches.

FIGURE 4.6

The composition of g and f shown in Figures 4.5 and 4.6 can be expressed symbolically. The symbol ∘ is used to denote composition of two functions.

$$(g \circ f)(5) = g(f(5)) \qquad \text{First compute } f(5).$$
$$= g(5280 \cdot 5) \qquad f(x) = 5280x$$
$$= g(26{,}400) \qquad \text{Simplify.}$$
$$= 12(26{,}400) \qquad g(x) = 12x$$
$$= 316{,}800 \qquad \text{Simplify.}$$

A distance of 5 miles is equivalent to 316,800 inches. The concept of composition of two functions is now defined formally.

Composition of functions

If f and g are functions, then the **composite function** $g \circ f$, or **composition** of g and f is defined by

$$(g \circ f)(x) = g(f(x)).$$

The domain of $g \circ f$ is all x in the domain of f such that $f(x)$ is in the domain of g.

EXAMPLE 6 *Finding a symbolic representation of a composite function*

Find a symbolic representation for the composite function $g \circ f$ that converts x miles into inches. Show that the composition $(g \circ f)(x)$ is not equivalent to the product $(gf)(x)$.

Solution
Let $f(x) = 5280x$ and $g(x) = 12x$.

$$(g \circ f)(x) = g(f(x)) \qquad \text{Definition of composition.}$$
$$= g(5280x) \qquad f(x) = 5280x \text{ is the input for } g.$$
$$= 12(5280x) \qquad g \text{ multiplies the input by 12.}$$
$$= 63{,}360x \qquad \text{Simplify.}$$

Thus, $(g \circ f)(x) = 63{,}360x$ converts x miles to inches. The product $(gf)(x)$ is computed as follows.

$$(gf)(x) = g(x) \cdot f(x) \qquad \text{Product of two functions}$$
$$= (12x)(5280x) \qquad f(x) = 5280x,\ g(x) = 12x$$
$$= 63{,}360x^2 \qquad \text{Simplify.}$$

Notice that $(g \circ f)(x) = 63{,}360x \neq 63{,}360x^2 = (gf)(x)$. ∎

Composition of functions also can be performed using numerical and graphical representations as illustrated in the next two examples.

EXAMPLE 7 *Evaluating a composite function numerically*

Ozone in the stratosphere filters out approximately 90% of the harmful ultraviolet (UV) rays from the sun. Depletion of the ozone layer has caused an increase in the amount of UV radiation reaching the surface of the earth. An increase in UV radiation is associated with skin cancer. In Table 4.5 the function f computes the approximate percent increase in UV radiation for an x percent decrease in the thickness of the ozone layer. The function g shown in Table 4.6 computes the expected percent increase in cases of skin cancer for an x percent increase in UV radiation. (Source: R. Turner, D. Pearce, and I. Bateman, *Environmental Economics*.)

TABLE 4.5 Percent Increase in UV Radiation

x	0	1	2	3	4	5	6
$f(x)$	0	1.5	3.0	4.5	6.0	7.5	9.0

TABLE 4.6 Percent Increase in Skin Cancer

x	0	1.5	3.0	4.5	6.0	7.5	9.0
$g(x)$	0	5.25	10.5	15.75	21.0	26.25	31.5

(a) Find $(g \circ f)(2)$ and interpret this calculation.

(b) Determine a tabular representation for $g \circ f$. Describe what $(g \circ f)(x)$ computes.

Solution

(a) $(g \circ f)(2) = g(f(2)) = g(3.0) = 10.5$. This means that a 2% decrease in the thickness of the ozone layer results in a 3% increase in UV radiation. This could cause a 10.5% increase in skin cancer.

(b) The values for $(g \circ f)(x)$ can be found in a manner similar to part (a). See Table 4.7.

TABLE 4.7

x	0	1	2	3	4	5	6
$(g \circ f)(x)$	0	5.25	10.5	15.75	21.0	26.25	31.5

The composition $(g \circ f)(x)$ computes the percent increase in cases of skin cancer resulting from an x percent decrease in the ozone layer. ∎

EXAMPLE 8 *Evaluating a composite function graphically*

Cities are made up of large amounts of concrete and asphalt that heat up in the daytime from sunlight but do not cool off completely at night. As a result, urban areas tend to be warmer than the surrounding rural areas. This effect is called the **urban heat island** and has been documented in cities throughout the world. In Figure 4.7

the function f computes the average increase in nighttime summer temperatures in degrees Celsius at Sky Harbor Airport in Phoenix from 1948 to 1990. In this graph 1948 is the base year with a zero temperature increase. This rise in urban temperature increased peak demand for electricity. In Figure 4.8 the function g computes the percent increase in electrical demand for an average nighttime temperature increase of x degrees Celsius. (Source: W. Cotton and R. Pielke, *Human Impacts on Weather and Climate.*)

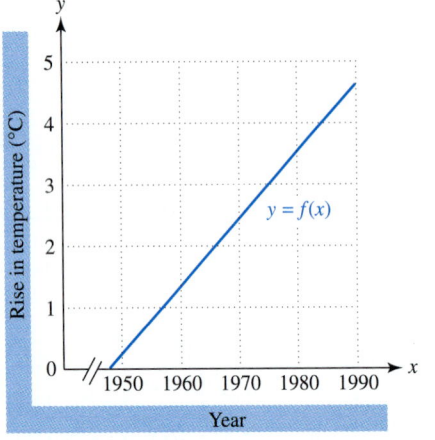

FIGURE 4.7 Nighttime Temperature Increase

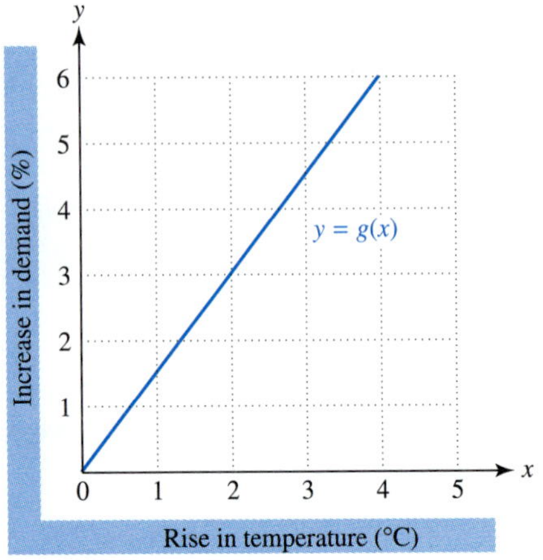

FIGURE 4.8 Percent Increase in Electrical Demand

(a) Evaluate $(g \circ f)(1975)$ graphically.
(b) Interpret $(g \circ f)(x)$.

Solution

(a) To evaluate $(g \circ f)(1975) = g(f(1975))$ graphically, first find $f(1975)$. In Figure 4.9, $f(1975) \approx 3$. Next let 3 be input for $g(x)$. In Figure 4.10, $g(3) \approx 4.5$. Thus, $(g \circ f)(1975) = g(f(1975)) \approx g(3) \approx 4.5$. In 1975 the average nighttime temperature had risen about 3°C since 1948. This resulted in a 4.5% increase in peak demand for electricity.

(b) $(g \circ f)(x)$ computes the percent increase in peak electrical demand during year x.

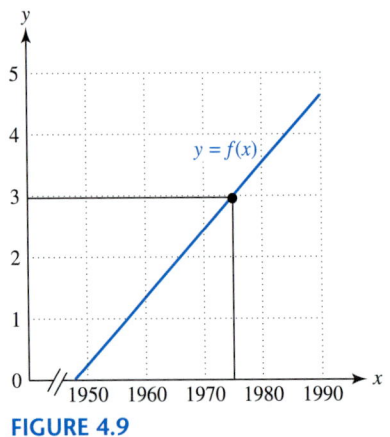

FIGURE 4.9

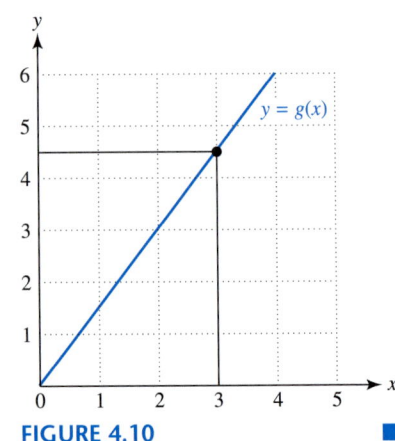

FIGURE 4.10

EXAMPLE 9

Evaluating a composite function symbolically

Let $f(x) = x^2 + 3x + 2$ and $g(x) = \dfrac{1}{x}$.

(a) Evaluate $(f \circ g)(2)$ and $(g \circ f)(2)$. How do they compare?
(b) Find symbolic expressions for $(f \circ g)(x)$ and $(g \circ f)(x)$. Are they equivalent expressions?
(c) Find the domains of $(f \circ g)(x)$ and $(g \circ f)(x)$.

Solution

(a) $(f \circ g)(2) = f(g(2)) = f(0.5) = (0.5)^2 + 3(0.5) + 2 = 3.75.$

$(g \circ f)(2) = g(f(2)) = g(12) = \dfrac{1}{12} \approx 0.0833.$

The results are not equal.

(b) Symbolic representations for $(f \circ g)(x)$ and $(g \circ f)(x)$ also can be found.

$$(f \circ g)(x) = f(g(x)) = f\left(\frac{1}{x}\right) = \left(\frac{1}{x}\right)^2 + 3\left(\frac{1}{x}\right) + 2 = \frac{1}{x^2} + \frac{3}{x} + 2$$

$$(g \circ f)(x) = g(f(x)) = g(x^2 + 3x + 2) = \frac{1}{x^2 + 3x + 2}$$

The expressions for $(f \circ g)(x)$ and $(g \circ f)(x)$ are not equivalent.

(c) The domain of f is all real numbers and the domain of g is $\{x \mid x \neq 0\}$. The domain of $(f \circ g)(x) = f(g(x))$ consists of all x in the domain of g such that $g(x)$ is in the domain of f. Thus, the domain of $(f \circ g)(x) = \dfrac{1}{x^2} + \dfrac{3}{x} + 2$ is $\{x \mid x \neq 0\}$.

The domain of $(g \circ f)(x) = g(f(x))$ consists of all x in the domain of f such that $f(x)$ is in the domain of g. Since $x^2 + 3x + 2 = 0$ when $x = -1$ or -2, the domain of $(g \circ f)(x) = \dfrac{1}{x^2 + 3x + 2}$ is $\{x \mid x \neq -1 \text{ or } -2\}$. ■

EXAMPLE 10

Writing a function as a composition of two functions

Find functions f and g so that $h(x) = (g \circ f)(x)$.

(a) $h(x) = (x + 3)^2$
(b) $h(x) = \sqrt{2x - 7}$
(c) $h(x) = \dfrac{1}{x^2 + 2x}$

Solution

(a) Let $f(x) = x + 3$ and $g(x) = x^2$. Then,

$$(g \circ f)(x) = g(f(x)) = g(x + 3) = (x + 3)^2.$$

(b) Let $f(x) = 2x - 7$ and $g(x) = \sqrt{x}$. Then,

$$(g \circ f)(x) = g(f(x)) = g(2x - 7) = \sqrt{2x - 7}.$$

(c) Let $f(x) = x^2 + 2x$ and $g(x) = \dfrac{1}{x}$. Then,

$$(g \circ f)(x) = g(f(x)) = g(x^2 + 2x) = \frac{1}{x^2 + 2x}. \qquad ■$$

4.1 PUTTING IT ALL TOGETHER

Addition, subtraction, multiplication, and division can be used to combine functions. Composition of functions is a method of combining functions, and is fundamentally different from arithmetic operations on functions. In the composition $(g \circ f)(x)$, the *output $f(x)$* is used as *input* for g.

The following table summarizes some concepts involved with combining functions.

Concept	Notation	Examples
Sum of two functions	$(f + g)(x) = f(x) + g(x)$	$f(x) = x^2,\ g(x) = 2x + 1$ $(f + g)(3) = f(3) + g(3)$ $\qquad = 9 + 7 = 16$ $(f + g)(x) = f(x) + g(x)$ $\qquad = x^2 + 2x + 1$
Difference of two functions	$(f - g)(x) = f(x) - g(x)$	$f(x) = 3x,\ g(x) = 2x + 1$ $(f - g)(1) = f(1) - g(1)$ $\qquad = 3 - 3 = 0$ $(f - g)(x) = f(x) - g(x)$ $\qquad = 3x - (2x + 1)$ $\qquad = x - 1$
Product of two functions	$(fg)(x) = f(x) \cdot g(x)$	$f(x) = x^3,\ g(x) = 1 - 3x$ $(fg)(-2) = f(-2) \cdot g(-2)$ $\qquad = (-8)(7) = -56$ $(fg)(x) = f(x) \cdot g(x)$ $\qquad = x^3(1 - 3x)$ $\qquad = x^3 - 3x^4$
Quotient of two functions	$\left(\dfrac{f}{g}\right)(x) = \dfrac{f(x)}{g(x)},\ g(x) \ne 0$	$f(x) = x^2 - 1,\ g(x) = x + 2$ $\left(\dfrac{f}{g}\right)(2) = \dfrac{f(2)}{g(2)}$ $\qquad = \dfrac{3}{4}$ $\left(\dfrac{f}{g}\right)(x) = \dfrac{f(x)}{g(x)}$ $\qquad = \dfrac{x^2 - 1}{x + 2},\ x \ne -2$
Composition of two functions	$(g \circ f)(x) = g(f(x))$	$f(x) = x^3,\ g(x) = x^2 - 2x + 1$ $(g \circ f)(2) = g(f(2)) = g(8)$ $\qquad = 64 - 16 + 1 = 49$ $(g \circ f)(x) = g(f(x))$ $\qquad = g(x^3)$ $\qquad = (x^3)^2 - 2(x^3) + 1$ $\qquad = x^6 - 2x^3 + 1$

4.1 EXERCISES

   Tape 5

Arithmetic Operations on Functions

Exercises 1 and 2: Use the table to evaluate each expression, if possible.

 (a) $(f + g)(2)$ (b) $(f - g)(4)$
 (c) $(fg)(-2)$ (d) $(f/g)(0)$

1.

x	-2	0	2	4
$f(x)$	0	5	7	10
$g(x)$	6	0	-2	5

2.

x	-2	0	2	4
$f(x)$	-4	8	5	0
$g(x)$	2	-1	4	0

3. Use the table in Exercise 1 to complete the following table.

x	-2	0	2	4
$(f + g)(x)$				
$(f - g)(x)$				
$(fg)(x)$				
$(f/g)(x)$				

4. Use the table in Exercise 2 to complete the table in Exercise 3.

Exercises 5 and 6: Use the graph to evaluate each expression.

5. (a) $(f + g)(2)$ (b) $(f - g)(1)$
  (c) $(fg)(0)$ (d) $(f/g)(1)$

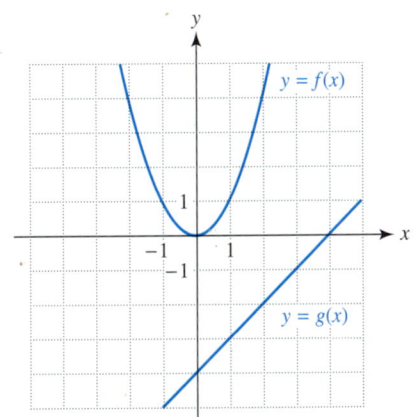

6. (a) $(f + g)(1)$ (b) $(f - g)(0)$
  (c) $(fg)(-1)$ (d) $(f/g)(1)$

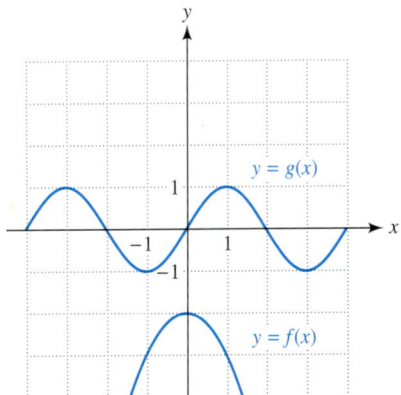

Exercises 7 and 8: Use f(x) and g(x) to evaluate each expression.

7. $f(x) = 2x - 3, g(x) = 1 - x^2$
  (a) $(f + g)(3)$ (b) $(f - g)(-1)$
  (c) $(fg)(0)$ (d) $(f/g)(2)$

8. $f(x) = 4x - x^3, g(x) = x + 3$
  (a) $(g + g)(-2)$ (b) $(f - g)(0)$
  (c) $(gf)(1)$ (d) $(g/f)(-3)$

Exercises 9–14: Use f(x) and g(x) to find each expression. Identify its domain.

 (a) $(f + g)(x)$
 (b) $(f - g)(x)$
 (c) $(fg)(x)$
 (d) $(f/g)(x)$

9. $f(x) = 2x, g(x) = x^2$

10. $f(x) = 1 - 4x, g(x) = 3x + 1$

11. $f(x) = x - \sqrt{x - 1}, g(x) = x + \sqrt{x - 1}$

12. $f(x) = \sqrt{1 - x}, g(x) = x^3$

13. $f(x) = \dfrac{1}{x + 1}, g(x) = \dfrac{3}{x + 1}$

14. $f(x) = 1 - x^2, g(x) = 3x^2 + 5$

15. Let $f(x) = \sqrt{x}$ and $g(x) = x + 1$.
  (a) Graph $f(x)$, $g(x)$, and $(f + g)(x)$ in $[0, 9, 1]$ by $[0, 15, 1]$.
  (b) Explain how the graph of $(f + g)(x)$ can be found using the graphs of $f(x)$ and $g(x)$.

16. Let $f(x) = 0.2x + 5$ and $g(x) = 0.1x + 1$.
  (a) Graph $f(x)$, $g(x)$, and $(f - g)(x)$ in $[0, 10, 1]$ by $[0, 10, 1]$.
  (b) Explain how the graph of $(f - g)(x)$ can be found using the graphs of $f(x)$ and $g(x)$.

Composition of Functions

Exercises 17 and 18: Complete numerical representations for the functions f and g are given. Evaluate the expression, if possible.

   (a) $(g \circ f)(1)$
   (b) $(f \circ g)(4)$
   (c) $(f \circ f)(3)$

17.

x	1	2	3	4
$f(x)$	4	3	1	2

x	1	2	3	4
$g(x)$	2	3	4	5

18.

x	1	3	4	6
$f(x)$	2	6	5	7

x	2	3	5	7
$g(x)$	4	2	6	0

Exercises 19 and 20: Use the graph to evaluate each expression.

19. (a) $(f \circ g)(4)$
      (b) $(g \circ f)(3)$

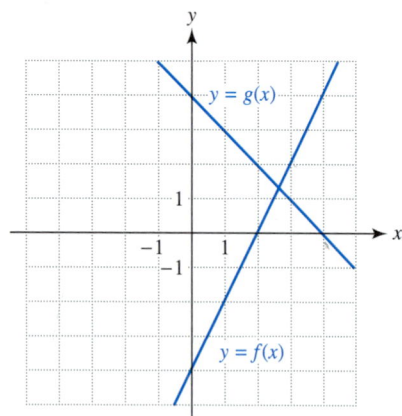

20. (a) $(f \circ g)(2)$
      (b) $(g \circ g)(0)$

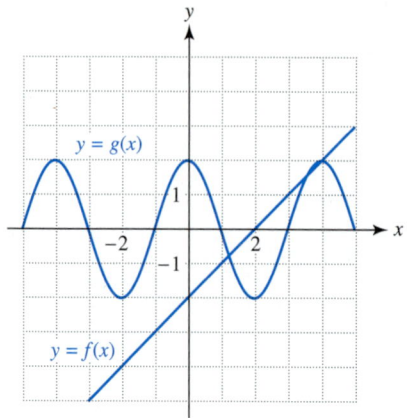

Exercises 21 and 22: Use f(x) and g(x) to evaluate each expression.

21. $f(x) = \sqrt{x + 5}$, $g(x) = x^2$
     (a) $(f \circ g)(2)$
     (b) $(g \circ f)(-1)$

22. $f(x) = |x^2 - 4|$, $g(x) = 2x^2 + x + 1$
     (a) $(f \circ g)(1)$
     (b) $(g \circ f)(-3)$

Exercises 23–28: Use f(x) and g(x) to find each of the following. Identify its domain.

   (a) $(f \circ g)(x)$
   (b $(g \circ f)(x)$
   (c) $(f \circ f)(x)$

23. $f(x) = x^3$, $g(x) = x^2 + 3x - 1$

24. $f(x) = 2 - x$, $g(x) = \dfrac{1}{x^2}$

25. $f(x) = x^2$, $g(x) = \sqrt{1 - x}$

26. $f(x) = x + 2$, $g(x) = x^4 + x^2 - 3x - 4$

27. $f(x) = 2 - 3x$, $g(x) = x^3$

28. $f(x) = \sqrt{x}$, $g(x) = 1 - x^2$

Exercises 29–36: Find functions f and g so that h(x) = (g \circ f)(x).

29. $h(x) = \sqrt{x - 2}$

30. $h(x) = (x + 2)^4$

31. $h(x) = 5(x + 2)^2 - 4$

32. $h(x) = \dfrac{1}{x + 2}$

33. $h(x) = 4(2x + 1)^3$

34. $h(x) = \sqrt[3]{x^2 + 1}$

35. $h(x) = (x^3 - 1)^2$

36. $h(x) = 4(x - 5)^{-2}$

Applications

37. *Hollywood Movies* From 1991 to 1996 the cost to produce and market Hollywood movies increased. The function f computes the average amount spent to produce a movie, while g computes the average cost to market a movie. Amounts in the table are given in millions of dollars. (Source: Motion Picture Association of America.)

x (yr)	1991	1992	1993	1994	1995	1996
$f(x)$ ($ millions)	26	28	30	33	37	41
$g(x)$ ($ millions)	12	14	14	17	17	20

(a) Make a table for a function h that computes the average cost to produce *and* market a movie in the year x.
(b) Write an equation that relates $f(x)$, $g(x)$, and $h(x)$.

38. *Petroleum Spillage* Large amounts of petroleum products enter the oceans each year. In the table, f computes the total amount of petroleum that enters the world's oceans. The function g computes petroleum spillage into the oceans caused by oil tankers. Amounts are given in thousands of tons. (Source: B. Freedman, *Environmental Ecology.*)

Year	1973	1979	1981	1983	1989
$f(x)$	6110	4670	3570	3200	570
$g(x)$	1380	900	1050	1100	—

(a) Define a function h by $h(x) = f(x) - g(x)$. Make a table for $h(x)$. What is the domain of h?
(b) Interpret what h computes.

39. *Acid Rain* A common air pollutant responsible for acid rain is sulfur dioxide (SO_2). Emissions of SO_2 from burning coal during year x are computed by $f(x)$ in the table. Emissions of SO_2 from burning oil are computed by $g(x)$. Amounts are given in millions of tons. (Source: B. Freedman.)

x	1860	1900	1940	1970	2000
$f(x)$	2.4	12.6	24.2	32.4	55.0
$g(x)$	0.0	0.2	2.3	17.6	23.0

(a) Evaluate $(f + g)(1970)$.
(b) Interpret $(f + g)(x)$.
(c) Make a table for $(f + g)(x)$.

40. (Refer to the previous exercise.) Make a table for a function h defined by $h(x) = (f/g)(x)$. Round values for $h(x)$ to the nearest hundredth. What information does h give regarding the use of coal and oil in the United States?

41. *Methane Emissions* Methane is a greenhouse gas. It lets sunlight into the atmosphere but blocks heat from escaping the earth's atmosphere. Methane is a by-product of burning fossil fuels. In the table, f models the predicted methane emissions in millions of tons produced by developed countries. The function g models the same emissions for developing countries. (Source: A. Nilsson, *Greenhouse Earth.*)

x	1990	2000	2010	2020	2030
$f(x)$	27	28	29	30	31
$g(x)$	5	7.5	10	12.5	15

(a) Make a table for a function h that models the total predicted methane emissions for developed *and* developing countries.
(b) Write an equation that relates $f(x)$, $g(x)$, and $h(x)$.

42. (Refer to the previous exercise.) The accompanying figure shows graphical representations of the functions f and g that model methane emissions. Use their graphs to sketch a graph of the function h.

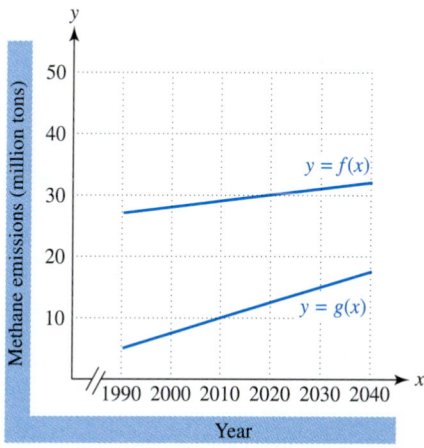

43. (Refer to the two previous exercises.) Symbolic representations for f and g are $f(x) = 0.1x - 172$ and $g(x) = 0.25x - 492.5$, where x is the year. Find a symbolic representation for h.

44. *Stopping Distance* (Refer to Example 1.) A car is traveling at 60 miles per hour. The driver has a reaction time of 1.25 seconds.

(a) Determine a function f that computes the reaction distance for this driver.

(b) Find a symbolic representation for a function d that computes the stopping distance for this driver traveling at x miles per hour.

(c) Evaluate $d(60)$ and interpret the result.

45. *China's Energy Production* Future energy production of China is shown in the figure. The function f computes total coal production, and the function g computes total coal *and* oil production. Energy units are in million metric tons of oil equivalent (Mtoe). Let the function h compute China's future oil production. (Source: A. Pascal, *Global Energy: The Changing Outlook.*)

(a) Write an equation that relates $f(x)$, $g(x)$, and $h(x)$.

(b) Evaluate $h(1995)$ and $h(2000)$.

(c) Determine a symbolic representation for h. (*Hint: h is linear.*)

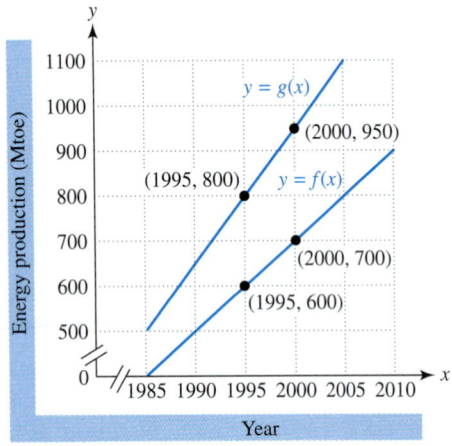

46. (Refer to the previous exercise.) Use the table for f and g to make a table for h.

x	1990	1995	2000	2005
$f(x)$	500	600	700	800
$g(x)$	650	800	950	1100

47. *Freshwater Nutrients* Essential elements, such as nitrogen and phosphorus, are found in aquatic plants and freshwater. In the table, f computes the concentration of these elements in aquatic plants and g computes this concentration in freshwater. Concentrations are expressed as percentages by weight. The following abbreviations are used for the elements: C-carbon, Si-silicon, N-nitrogen, K-potassium, and P-phosphorus. For example,

$f(N) = 0.7$ means that by weight, 0.7% of aquatic freshwater plants is nitrogen, whereas $g(N) = 0.000023$ implies that 0.000023% of the freshwater is made up of nitrogen. (Source: B. Freedman.)

x	C	Si	N	K	P
$f(x)$	6.5	1.3	0.7	0.3	0.08
$g(x)$	0.0012	0.00065	0.000023	0.00023	0.000001

(a) Make a table of a function defined by $h(x) = (f/g)(x)$. Round values to the nearest hundred.

(b) Interpret what h computes.

48. *Profit* (Refer to Example 4.) Determine a profit function P that results if the compact discs are sold for $15 each. Find the profit from selling 3000 compact discs.

49. *Epidemics* William Farr was one of the first to understand how epidemics spread and eventually die out. In 1865 there had been a cattle plague in England. In the table, $f(x)$ computes the new cases of the disease reported at four-week intervals after x weeks. (Source: H. Lancaster, *Quantitative Methods in Biological and Medical Sciences.*)

x	4	8	12	16
$f(x)$	9597	18,817	33,835	47,191

(a) Define $g(x) = \dfrac{f(x+4)}{f(x)}$ for $x = 4, 8, 12$. Make a table for $g(x)$.

(b) Interpret $g(x)$. Farr predicted that the epidemic would be over in five more four-week periods. What do you think led Farr to believe that the plague would subside?

50. *Skin Cancer* In Example 7 the functions f and g are both linear.

(a) Find symbolic representations for f and g.

(b) Determine $(g \circ f)(x)$.

(c) Evaluate $(g \circ f)(3.5)$ and interpret the result.

51. *Swimming Pools* In the accompanying figures on the next page, f computes the cubic feet of water contained in a pool after x days, and g converts cubic feet to gallons.

(a) Estimate the gallons of water contained in the pool when $x = 2$.

(b) Interpret $(g \circ f)(x)$.

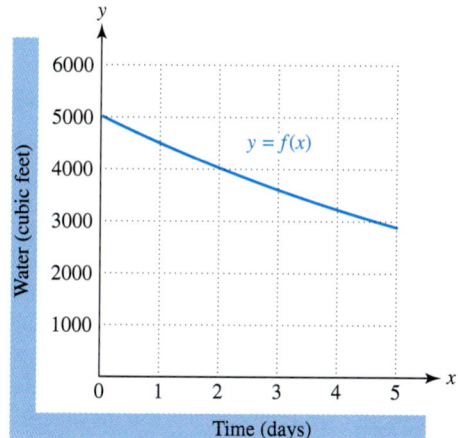

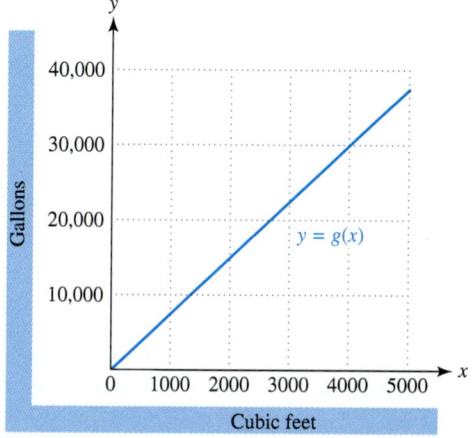

52. *Temperature* The figures show graphs of f and g. The function f computes the temperature on a summer day after x hours, while g converts Fahrenheit temperature to Celsius temperature.
(a) Evaluate $(g \circ f)(2)$.
(b) Interpret $(g \circ f)(x)$.

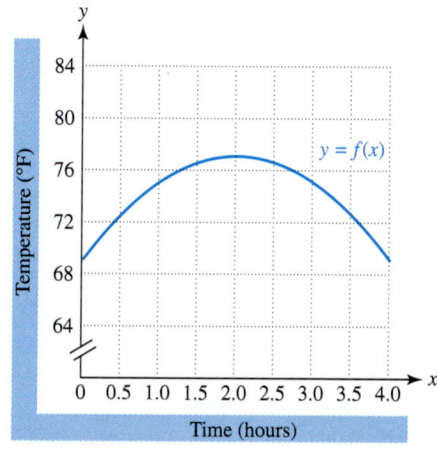

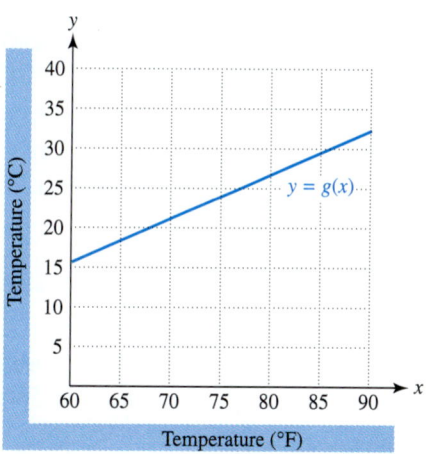

53. *Urban Heat Island* (Refer to Example 8.) The functions f and g are given by

$$f(x) = 0.11(x - 1948) \text{ and } g(x) = 1.5x.$$

(a) Evaluate $(g \circ f)(1960)$.
(b) Find $(g \circ f)(x)$.
(c) What type of functions are f, g, and $g \circ f$?

54. Show that if $f(x) = ax + b$ and $g(x) = cx + d$, then $(g \circ f)(x)$ also represents a linear function. Find the slope of the graph of $(g \circ f)(x)$.

Writing about Mathematics

1. Describe differences between $(fg)(x)$ and $(f \circ g)(x)$. Give examples.

2. Describe differences between $(f \circ g)(x)$ and $(g \circ f)(x)$. Give examples.

4.2 Inverse Functions and Their Representations

Inverse Operations • One-to-One Functions • Definition of Inverse Function • Representations of Inverse Functions • Inverse Functions and Equations

Introduction

Many actions are reversible. A closed door can be opened—an open door can be closed. One hundred dollars can be withdrawn from or deposited into a savings account. These actions undo or cancel each other. But not all actions are reversible. Explosions and weather are two examples. In mathematics this basic concept of reversing a calculation and arriving at an original result is associated with an inverse.

Many times discovering an inverse function is essential to the success of a new product. For example, in electronic communication a function is used to code voices and pictures into a form that can be transmitted. After they arrive at their destination, an inverse function must decode the transmission back into sounds and pictures that people understand. It is essential that an inverse function output the correct picture and sound without any static. This is one reason why mathematics is important in the development of clear, electronic communication.

Inverse Operations

Throughout time, codes have been used for secrecy. Every code requires a consistent method to decode it. A simple code that was used during the time of the Roman Empire is shown in Table 4.8.

TABLE 4.8

Letter	A	B	C	D	E	F	G	H	I	J	K	L	M
Code	C	D	E	F	G	H	I	J	K	L	M	N	O

Letter	N	O	P	Q	R	S	T	U	V	W	X	Y	Z
Code	P	Q	R	S	T	U	V	W	X	Y	Z	A	B

Source: A. Sinkov, *Elementary Cryptanalysis: A Mathematical Approach.*

Table 4.8 can be used to code the word HELP as JGNR and decode it back to HELP. Coding and decoding are inverse actions or operations. It is essential that both the coding and decoding procedures give exactly one output for each input.

Actions and their inverses occur in everyday life. Suppose a person opens a car door, gets in, and starts the engine. What are the inverse actions? They are turn off the engine, get out, and close the car door. Notice that the order must be reversed in addition to applying the inverse operation at each step.

In mathematics there are basic operations that can be considered inverse operations of each other. For example, if we begin with 10 and add 5, the result is 15.

To undo this operation, subtract 5 from 15 to obtain 10. Addition and subtraction are inverse operations. The same is true for multiplication and division. If we multiply a number by 2 and then divide by 2, the original number is obtained. Division and multiplication are inverse operations.

EXAMPLE 1   *Finding inverse actions and operations*

For each of the following, state the inverse actions or operations.
(a) Put on a coat and go outside.
(b) Subtract 7 from x and divide the result by 2.

Solution
(a) To find the inverse actions reverse the order and apply the inverse action at each step. The inverse actions would be to come inside and take off the coat.
(b) We must reverse the order and apply the inverse operation at each step. The inverse operations would be to multiply x by 2 and add 7. The original operations could be expressed as $\dfrac{x-7}{2}$, while the inverse operations could be written as $2x + 7$.   ∎

Inverse operations can be described by functions. If $f(x) = x + 5$, then the *inverse function* of f is given by $g(x) = x - 5$. For example, $f(5) = 10$, and $g(10) = 5$. If input x produces output y with function f, input y produces output x with function g. This can be seen numerically in Table 4.9.

TABLE 4.9

x	0	5	10	15
$f(x)$	5	10	15	20

x	5	10	15	20
$g(x)$	0	5	10	15

When functions f and g are applied in sequence, the output of f is used as input for g. This is composition of functions.

$$(g \circ f)(x) = g(f(x)) \qquad \text{Definition of composition}$$
$$= g(x + 5) \qquad f(x) = x + 5$$
$$= (x + 5) - 5 \qquad g \text{ subtracts 5 from its input.}$$
$$= x \qquad \text{Simplify.}$$

The composition $g \circ f$ with input x produces output x. The same action occurs for $f \circ g$.

$$(f \circ g)(x) = f(g(x)) \qquad \text{Definition of composition}$$
$$= f(x - 5) \qquad g(x) = x - 5$$
$$= (x - 5) + 5 \qquad f \text{ adds 5 to its input.}$$
$$= x \qquad \text{Simplify.}$$

Another pair of inverse functions are given by $f(x) = x^3$ and $g(x) = \sqrt[3]{x}$. If g is the inverse function of f, then the notation $g = f^{-1}$ is used. Before a formal definition of inverse functions is given, we will discuss the concept of one-to-one functions.

One-to-One Functions

Does every function have an inverse function? The next example answers this question.

EXAMPLE 2   *Determining if a function has an inverse function*

Table 4.10 represents a function f that computes the percentage of the time that the sky is cloudy in Augusta, Georgia, where x corresponds to the standard numbers for the months. Determine if f has an inverse function.

TABLE 4.10   Cloudy Skies in Augusta

x (month)	1	2	3	4	5	6	7	8	9	10	11	12
$f(x)$ (%)	43	40	39	29	28	26	27	25	30	26	31	39

Source: J. Williams, *The Weather Almanac 1995.*

Solution

For each input, f computes exactly one output. For example, $f(3) = 39$ means that during March the sky is cloudy 39% of the time. If f has an inverse function, the inverse must receive 39 as input and produce exactly one output. Both March and December have cloudy skies 39% of the time. Given an input of 39, it is impossible for an inverse *function* to output both 3 and 12. Therefore, f does not have an inverse function.   ■

If two different inputs of f produce the same output, then an inverse function of f does not exist. On the other hand, if different inputs always produce different outputs, f is a *one-to-one function*. Every one-to-one function has an inverse function.

One-to-one function

A function f is a **one-to-one function** if, for elements c and d in the domain of f,

$$c \neq d \quad \text{implies} \quad f(c) \neq f(d).$$

EXAMPLE 3   *Determining if a function is one-to-one graphically*

Use the graph of f to determine if f is a one-to-one function in Figures 4.11 and 4.12 on the next page.

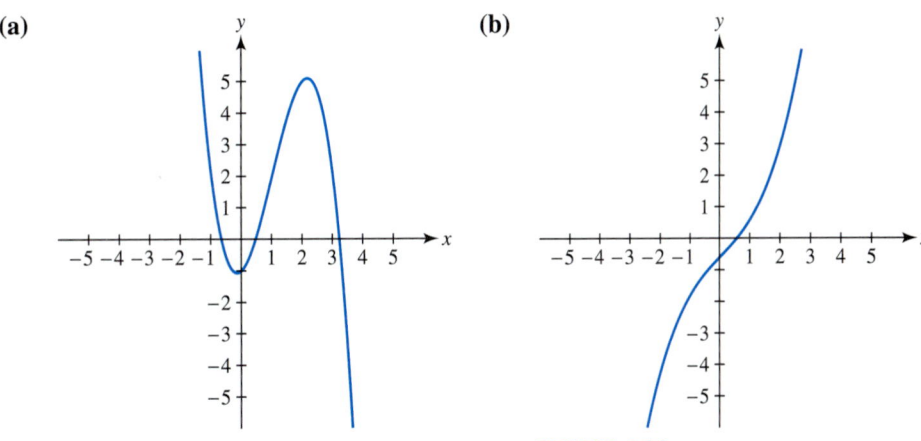

FIGURE 4.11

FIGURE 4.12

Solution

(a) To decide if the graph represents a one-to-one function, determine if different inputs (x-values) always correspond to different outputs (y-values). In Figure 4.13 the horizontal line $y = 2$ intersects the graph of f at $(-1, 2)$, $(1, 2)$, and $(3, 2)$. This means that $f(-1) = f(1) = f(3) = 2$. Three distinct inputs, -1, 1, and 3, produce the same output, 2. Therefore, f is not one-to-one and does not have an inverse function.

(b) Any horizontal line will intersect the graph of f at most once. For example, if the line $y = 3$ is graphed, it intersects the graph of f at one point, $(2, 3)$. See Figure 4.14. Only input 2 results in output 3. This is true in general. Therefore, f is one-to-one and has an inverse function.

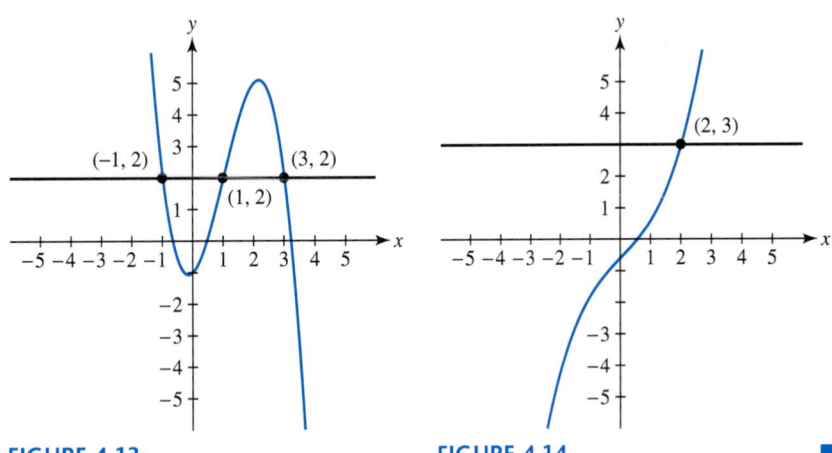

FIGURE 4.13

FIGURE 4.14

■

Note: To show that f is not one-to-one, it was not necessary to find the actual points of intersection—we only have to show that a horizontal line can intersect the graph of f more than once.

The technique of visualizing horizontal lines to determine if a graph represents a one-to-one function is called the *horizontal line test*.

> ### Horizontal line test
>
> If every horizontal line intersects the graph of a function f at most once, then f is a one-to-one function.

Critical Thinking

Use the horizontal line test to explain why a nonconstant, linear function has an inverse function, whereas a quadratic function does not.

Definition of Inverse Function

We now define an inverse function.

> ### Inverse function
>
> Let f be a one-to-one function. Then g is the **inverse function** of f, if
>
> $$(f \circ g)(x) = x \quad \text{for every } x \text{ in the domain of } g, \text{ and}$$
> $$(g \circ f)(x) = x \quad \text{for every } x \text{ in the domain of } f.$$
>
> Under these conditions g is denoted by f^{-1}.

EXAMPLE 4   *Finding and verifying an inverse function*

The function given by $f(x) = \frac{18}{25}x + 2$ computes the proper crutch length for a person x inches tall. (See Example 6, Section 1.3.)

(a) Explain why f is a one-to-one function.
(b) Find a symbolic representation for f^{-1}.
(c) Verify that this is the inverse function of f.
(d) Interpret the meaning of $f^{-1}(56)$.

Solution

(a) Since f is a linear function, its graph is a line with a slope of $\frac{18}{25}$. Every horizontal line will intersect it at most once. See Figure 4.15. By the horizontal line test f is one-to-one.

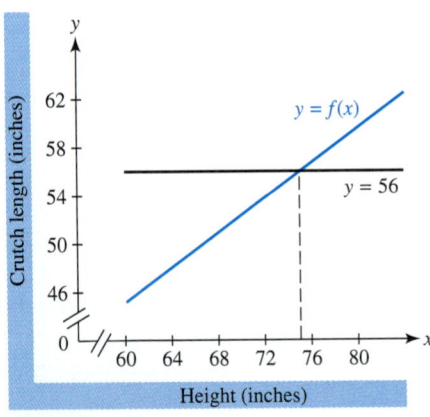

FIGURE 4.15

(b) The function f multiplies input x by $\frac{18}{25}$ and adds 2. Therefore, f^{-1} subtracts 2 from x and divides by $\frac{18}{25}$, or equivalently, multiplies by $\frac{25}{18}$. Thus,

$$f^{-1}(x) = \frac{25}{18}(x - 2).$$

(c) We must show that $(f^{-1} \circ f)(x) = x$ and $(f \circ f^{-1})(x) = x$.

$$(f^{-1} \circ f)(x) = f^{-1}(f(x)) \qquad \text{Definition of composition}$$

$$= f^{-1}\left(\frac{18}{25}x + 2\right) \qquad f(x) = \frac{18}{25}x + 2$$

$$= \frac{25}{18}\left(\left(\frac{18}{25}x + 2\right) - 2\right) \qquad f^{-1}(x) = \frac{25}{18}(x - 2)$$

$$= \frac{25}{18}\left(\frac{18}{25}x\right) \qquad \text{Simplify.}$$

$$= x \;\checkmark \qquad \text{It checks.}$$

Similarly,

$$(f \circ f^{-1})(x) = f(f^{-1}(x)) \qquad \text{Definition of composition}$$

$$= f\left(\frac{25}{18}(x - 2)\right) \qquad f^{-1}(x) = \frac{25}{18}(x - 2)$$

$$= \frac{18}{25}\left(\frac{25}{18}(x - 2)\right) + 2 \qquad f(x) = \frac{18}{25}x + 2$$

$$= (x - 2) + 2 \qquad \text{Simplify.}$$

$$= x \;\checkmark. \qquad \text{It checks.}$$

(d) The expression $f(x)$ calculates the proper crutch length for a person x inches tall, and $f^{-1}(x)$ computes the height of a person requiring a crutch x inches long. Therefore, $f^{-1}(56) = \frac{25}{18}(56 - 2) = 75$ means that a 56-inch crutch would be appropriate for a person 75 inches tall. See Figure 4.15. ∎

MAKING CONNECTIONS
The Notation f^{-1} and Negative Exponents

If a represents a real number, then $a^{-1} = \frac{1}{a}$. For example, $4^{-1} = \frac{1}{4}$. On the other hand, if f represents a function, then $f^{-1}(x) \neq \frac{1}{f(x)}$. Instead, $f^{-1}(x)$ represents the inverse function of f. For example, if $f(x) = 5x$, then $f^{-1}(x) = \frac{x}{5}$.

Representations of Inverse Functions

Given a numerical, graphical, or symbolic representation of a one-to-one function, it is often possible to determine the same type of representation for its inverse function. We begin by discussing numerical representations.

Numerical Representations. In Table 4.11 on the next page, f has domain $D = \{1940, 1970, 1995\}$. It computes the percentage of the U.S. population with four or more years of college in year x.

TABLE 4.11

x	1940	1970	1995
$f(x)$	5	11	23

Source: Bureau of the Census.

Function f is one-to-one because different inputs produce different outputs. Therefore, f^{-1} exists. Since $f(1940) = 5$, it follows that $f^{-1}(5) = 1940$. In a similar manner, $f^{-1}(11) = 1970$ and $f^{-1}(23) = 1995$. Table 4.12 shows a numerical representation of f^{-1}.

TABLE 4.12

x	5	11	23
$f^{-1}(x)$	1940	1970	1995

The domain of f is {1940, 1970, 1995} and the range is {5, 11, 23}. The domain of f^{-1} is {5, 11, 23} and the range of f^{-1} is {1940, 1970, 1995}. The functions f and f^{-1} interchange domains and ranges. This result is true in general for inverse functions.

The diagrams shown in Figures 4.16 and 4.17 demonstrate this property. To obtain f^{-1} from f, the arrows for f are simply reversed. This reversal causes the domains and ranges to be interchanged.

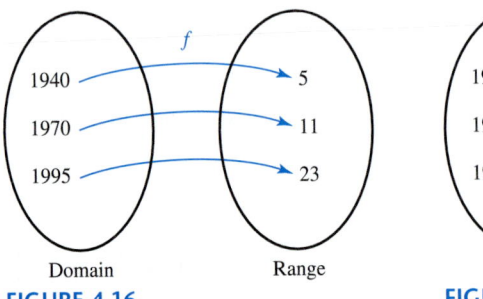

FIGURE 4.16

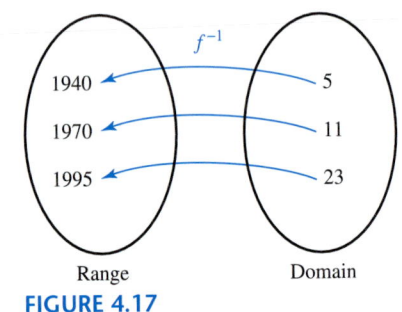

FIGURE 4.17

Figure 4.18 shows a function f that is not one-to-one. In Figure 4.19 the arrows defining f have been reversed. This is a relation. However, this relation does not represent the inverse *function*, since input 4 produces outputs 1 and 2.

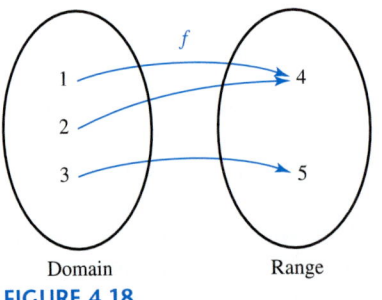

FIGURE 4.18

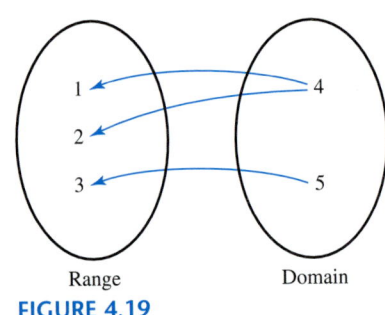

FIGURE 4.19

Graphical Representations. If the point (2, 5) lies on the graph of f, then $f(2) = 5$ and $f^{-1}(5) = 2$. Therefore, the point (5, 2) must lie on the graph of f^{-1}. In general, if the point (a, b) lies on the graph of f, then the point (b, a) lies on the graph of f^{-1}. Refer to Figure 4.20. If a line segment is drawn between the points (a, b) and (b, a), the line $y = x$ is a perpendicular bisector of this line segment. Figure 4.21 shows pairs of points in the form (a, b) and (b, a). Figure 4.22 contains continuous graphs of f and f^{-1} passing through these points. The graph of f^{-1} is a *reflection* of the graph of f across the line $y = x$.

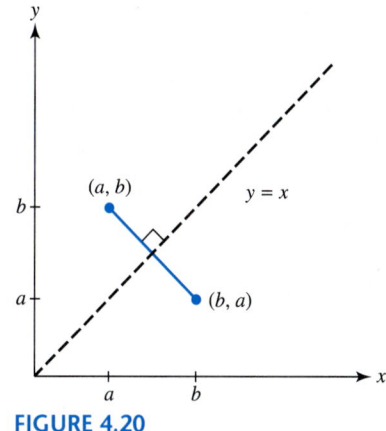

FIGURE 4.20

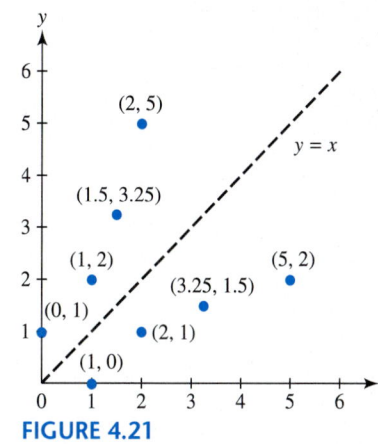

FIGURE 4.21

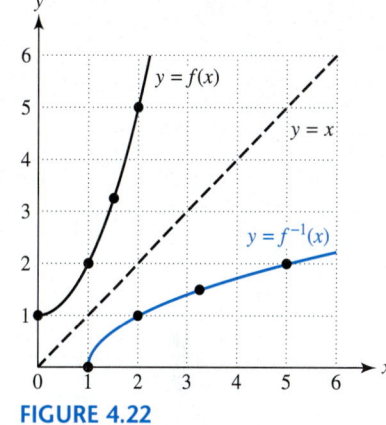

FIGURE 4.22

Many graphing calculators use this relationship to generate a graph of f^{-1}. Given a formula for $f(x)$, the calculator computes $y = f(x)$ for different inputs x. Rather than plotting the point (x, y) at each step, it simply plots (y, x). A graphing calculator can perform this task even if f is *not* one-to-one. However, in this case the resulting graph does not represent an inverse *function*.

EXAMPLE 5   *Representing an inverse function graphically*

Let $f(x) = x^3 + 2$. Graph f. Then sketch a graph of f^{-1}.

Solution
Figure 4.23 shows a graph of f. To sketch a graph of f^{-1} reflect the graph of f across the line $y = x$. The graph of f^{-1} appears as though it were the image of the graph of f in a mirror located along $y = x$. See Figure 4.24.

$[-5, 5, 1]$ by $[-5, 5, 1]$

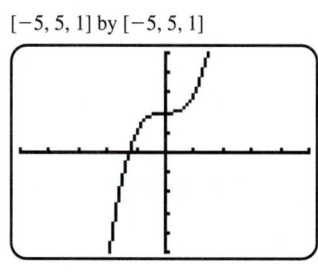

FIGURE 4.23

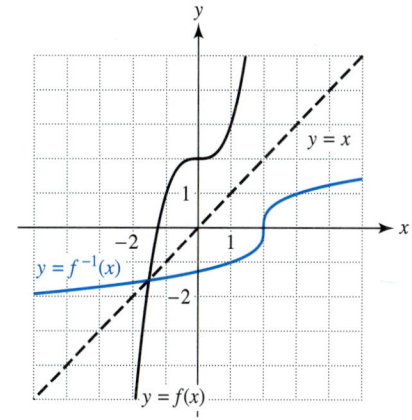

FIGURE 4.24

Symbolic Representations. Let f be one-to-one. To find f^{-1}, we apply the inverse operations in reverse order. A symbolic way to do this is to solve the right side of the equation $y = f(x)$ for x. In the process we obtain the equivalent equation $f^{-1}(y) = x$. The expression opposite x is a symbolic representation for f^{-1} in terms of the variable y. This technique is shown in the next example.

EXAMPLE 6 *Finding an inverse function symbolically*

The function given by $f(x) = \dfrac{9}{5}x + 32$ converts a Celsius temperature to an equivalent Fahrenheit temperature.
(a) Find a symbolic representation for f^{-1} and interpret the result.
(b) What Celsius temperature is equivalent to 68°F?

Solution
(a) The inverse can be found symbolically by solving the equation $y = f(x)$ for x.

$$y = \frac{9}{5}x + 32 \qquad \text{\small $y = f(x)$}$$

$$y - 32 = \frac{9}{5}x \qquad \text{\small Subtract 32.}$$

$$\frac{5}{9}(y - 32) = x \qquad \text{\small Divide by $\frac{9}{5}$ or multiply by $\frac{5}{9}$.}$$

To write f^{-1} in terms of x, interchange x and y.

$$\frac{5}{9}(x - 32) = y \qquad \text{\small Interchange x and y.}$$

$$f^{-1}(x) = \frac{5}{9}(x - 32) \qquad \text{\small Write f^{-1} using the variable x.}$$

The function f converts Celsius temperature to Fahrenheit temperature. Therefore, f^{-1} converts Fahrenheit temperature to Celsius temperature.

(b) We can use f^{-1} to find the equivalent Celsius temperature for 68°F.

$$f^{-1}(68) = \frac{5}{9}(68 - 32) = \frac{5}{9}(36) = 20$$

The temperature 20°C is equivalent to 68°F. ∎

This symbolic technique for finding an inverse function is now summarized.

> **Finding a symbolic representation for f^{-1}**
>
> To find a symbolic representation for f^{-1}, perform the following steps.
>
> 1. Verify that f is a one-to-one function.
> 2. Solve the equation $y = f(x)$ for x, resulting in the equation $x = f^{-1}(y)$.
> 3. Interchange x and y to obtain $y = f^{-1}(x)$.
>
> To verify $f^{-1}(x)$ is correct, show $(f^{-1} \circ f)(x) = x$ and $(f \circ f^{-1})(x) = x$.

The graph of $f(x) = x^5 + 3x^3 - x^2 + 5x - 3$ is shown in Figure 4.25. By the horizontal line test, f is a one-to-one function and f^{-1} exists. However, it would be difficult to obtain a symbolic representation for f because we would need to solve the equation $y = x^5 + 3x^3 - x^2 + 5x - 3$ for x. It is important to realize that there are many functions that have inverses, but their inverses can either be difficult or impossible to find symbolically.

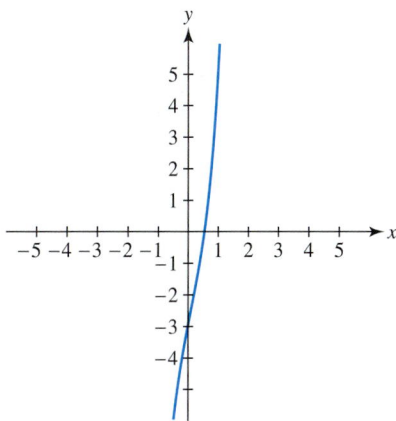

FIGURE 4.25

EXAMPLE 7   *Finding symbolic, graphical, and numerical representations*

Let $f(x) = (x - 1)^2$ with domain $x \geq 1$. Find $f^{-1}(x)$ and identify its domain and range. Support your results graphically and numerically.

Solution

Symbolic Representation   Since the domain of f is restricted to $x \geq 1$, the graph of f is the right half of a parabola. Thus, f is one-to-one. To find $f^{-1}(x)$, solve the equation $y = f(x)$ for x.

$$y = (x - 1)^2 \qquad \text{\textcolor{blue}{$y = f(x)$}}$$
$$\sqrt{y} = x - 1 \qquad \text{\textcolor{blue}{Take the positive square root.}}$$
$$\sqrt{y} + 1 = x \qquad \text{\textcolor{blue}{Add 1 to both sides.}}$$
$$\sqrt{x} + 1 = y \qquad \text{\textcolor{blue}{Interchange x and y.}}$$

Thus, $f^{-1}(x) = \sqrt{x} + 1$. Since the domain $x \geq 1$ implies $x - 1 \geq 0$, the positive square root is selected in the second step.

The domain of f^{-1} is the range of f. Since $f(x) = (x - 1)^2 \geq 0$, the domain of f^{-1} is $D = \{x \mid x \geq 0\}$. The range R of f^{-1} is the domain of f, so $R = \{y \mid y \geq 1\}$.

Graphical Representation   Graph $Y_1 = (X - 1)^2/(X \geq 1)$, $Y_2 = \sqrt{(X)} + 1$, and $Y_3 = X$. The graph of Y_2 is a reflection of Y_1 in the line $y = x$. See Figure 4.26. (*Note:* Division by $(X \geq 1)$ in the expression for Y_1 causes the graph of $y = (x - 1)^2$ to appear only when $x \geq 1$. This technique is called *logical division*.)

Numerical Representation   Table $Y_1 = (X - 1)^2$ and $Y_2 = \sqrt{(X)} + 1$. Notice that the ordered pairs for Y_1 and Y_2 are interchanged because they represent inverse functions. See Figures 4.27 and 4.28 on the next page.

[0, 9, 1] by [0, 6, 1]

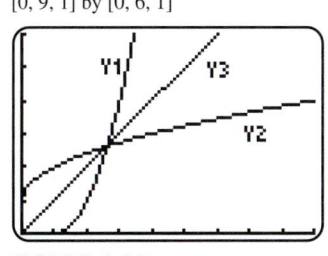

FIGURE 4.26

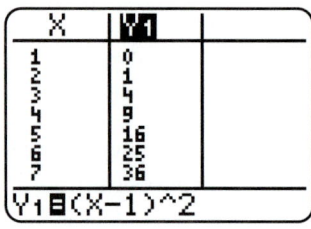

FIGURE 4.27   **FIGURE 4.28**   ■

Inverse Functions and Equations

In Example 4, $f(x) = \dfrac{18}{25}x + 2$ computes the correct crutch length for a person x inches tall. To determine the height of a person who requires a 51-inch crutch, solve the equation $f(x) = 51$. We can use $f^{-1}(x) = \dfrac{25}{18}(x - 2)$ to find the solution.

$$f(x) = 51 \qquad \text{Original equation}$$
$$f^{-1}(f(x)) = f^{-1}(51) \qquad \text{Apply } f^{-1} \text{ to both sides.}$$
$$x = f^{-1}(51) \qquad f^{-1}(f(x)) = x$$

The solution to $f(x) = 51$ is

$$x = f^{-1}(51) = \frac{25}{18}(51 - 2) \approx 68.06.$$

Thus, a 51-inch crutch would be appropriate for someone 68 inches tall. These results are generalized by the following statement.

> ### Solving an equation with the inverse function
>
> Let f be a one-to-one function and k be a constant in the range of f. Then,
> $$f(x) = k \qquad \text{is equivalent to} \qquad x = f^{-1}(k).$$

4.2 PUTTING IT ALL TOGETHER   ■ ■ ■

An inverse function f^{-1} will undo the computation performed by f. Inverse functions interchange inputs (domains) and outputs (ranges). That is, if $f(a) = b$ then $f^{-1}(b) = a$ for every a in the domain of f. A function f has an inverse function if and only if it is a one-to-one function. If a function is one-to-one, different inputs cannot result in the same output.

Both functions and their inverses have verbal, numerical, graphical, and symbolic representations. The following table summarizes these representations.

Verbal Representations

Given a verbal representation of f, apply the inverse operations in reverse order.

Example: The function f multiplies 2 times x and then adds 25.
The function f^{-1} subtracts 25 from x and divides the result by 2.

Numerical Representations

x	1	2	3
$f(x)$	0	5	7

x	0	5	7
$f^{-1}(x)$	1	2	3

Graphical Representations

The graph of f^{-1} can be obtained by reflecting the graph of f in the line $y = x$.

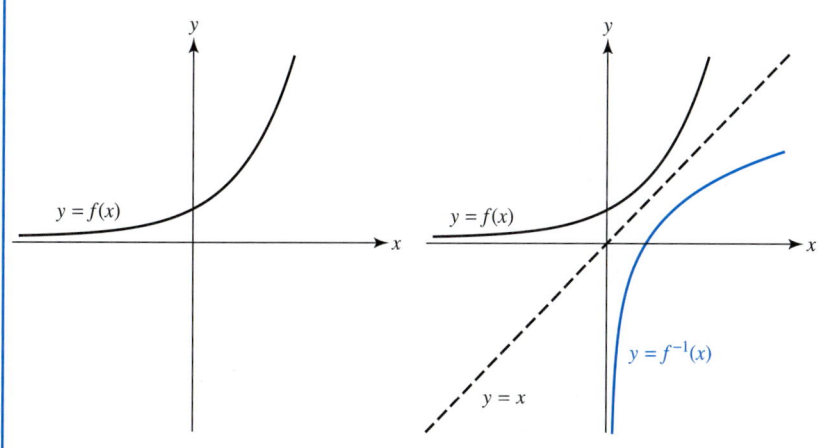

Symbolic Representations

If possible, solve the equation $y = f(x)$ for x. For example, let $f(x) = 3x - 5$.

$$y = 3x - 5 \quad \text{is equivalent to} \quad \frac{y + 5}{3} = x.$$

Interchange x and y to obtain $f^{-1}(x) = \dfrac{x + 5}{3}$.

4.2 EXERCISES   Tape 5

Inverse Operations

Exercises 1–4: State the inverse action or actions.

1. Opening a window
2. Climbing up a ladder
3. Walking into a classroom, sitting down, and opening a book
4. Opening the door and turning on the lights

Exercises 5–12: Describe verbally the inverse of the statement. Then, express both the statement and its inverse symbolically.

5. Add 2 to x.
6. Multiply x by 5.
7. Subtract 2 from x and multiply the result by 3.
8. Divide x by 20 and add 10.

9. Take the cube root of x and add 1.

10. Multiply x by -2 and add 3.

11. Take the reciprocal of a nonzero number x.

12. Take the square root of a positive number x.

One-to-One Functions

Exercises 13–16: Use the graph of f to determine if f is one-to-one.

13.

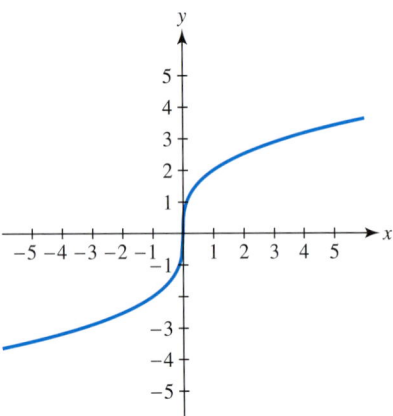

14.

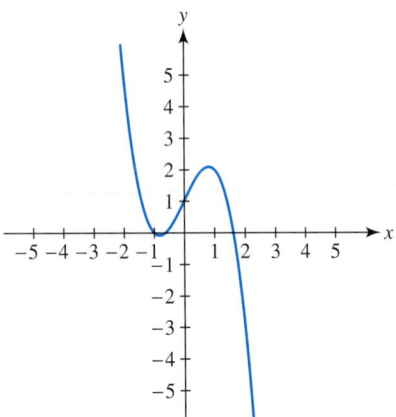

15.

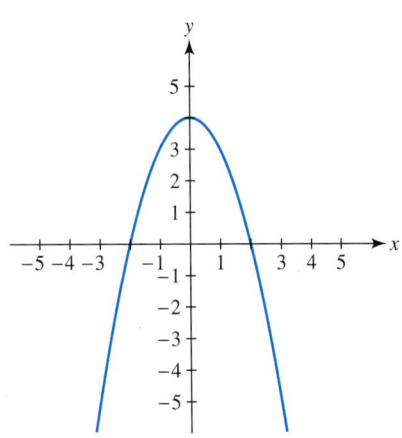

16.

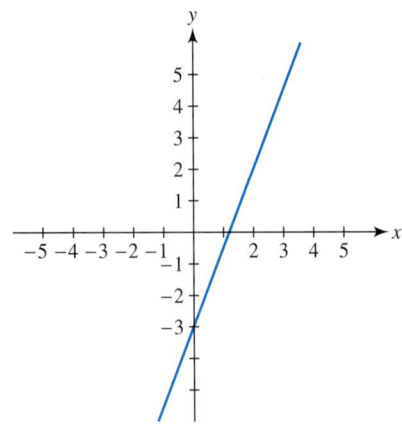

Exercises 17–20: Use the table to determine if f is one-to-one.

17.

x	1	2	3	4
$f(x)$	4	3	3	5

18.

x	-2	0	2	4
$f(x)$	4	2	0	-2

19.

x	0	2	4	6	8
$f(x)$	-1	0	4	1	-3

20.

x	-2	-1	0	1	2
$f(x)$	4	1	0	1	4

Exercises 21–28: Use f(x) to determine if f is one-to-one. Justify your answer graphically.

21. $f(x) = 2x - 7$

22. $f(x) = x^2 - 1$

23. $f(x) = |x - 1|$

24. $f(x) = x^3$

25. $f(x) = \dfrac{1}{1 + x^2}$

26. $f(x) = \dfrac{1}{x}$

27. $f(x) = 3x - x^3$

28. $f(x) = x^{2/3}$

Exercises 29–32: Decide if the situation could be modeled by a one-to-one function.

29. The distance between the ground and a person who is riding a Ferris wheel after x seconds

30. The cumulative numbers of AIDS cases from 1980 to present

31. The total federal debt from 1980 to 1997

32. The height y of a stone thrown upward after x seconds

Numerical Representations of Inverse Functions

Exercises 33–36: Use the table for $f(x)$ to find a table for $f^{-1}(x)$. Identify the domains and ranges of f and f^{-1}.

33.

x	1	2	3
$f(x)$	5	7	9

34.

x	1	10	100
$f(x)$	0	1	2

35.

x	0	1	2	3	4
$f(x)$	0	1	4	9	16

36.

x	-2	-1	0	1	2
$f(x)$	$\dfrac{1}{4}$	$\dfrac{1}{2}$	1	2	4

Exercises 37–40: Use $f(x)$ to complete the table for $f^{-1}(x)$.

37. $f(x) = x + 5$

x	-3	0	3	6
$f^{-1}(x)$				

38. $f(x) = 4x$

x	0	2	4	6
$f^{-1}(x)$				

39. $f(x) = x^3$

x	-8	-1	8	27
$f^{-1}(x)$				

40. $f(x) = \dfrac{1}{x}$

x	-3	-1	4	5
$f^{-1}(x)$				

Exercises 41–44: Use the tables to evaluate the following.

x	0	1	2	3	4
$f(x)$	1	3	5	4	2

x	-1	1	2	3	4
$g(x)$	0	2	1	4	5

41. $(f \circ g^{-1})(1)$ **42.** $(g^{-1} \circ g^{-1})(2)$

43. $(g \circ f^{-1})(5)$ **44.** $(f^{-1} \circ g)(4)$

Graphical Representations of Inverse Functions

45. *Interpreting an Inverse* The graph of f computes the number of dollars in a savings account after x years. Estimate each of the following. Interpret what $f^{-1}(x)$ computes.
  (a) $f(1)$
  (b) $f^{-1}(110)$
  (c) $f^{-1}(160)$

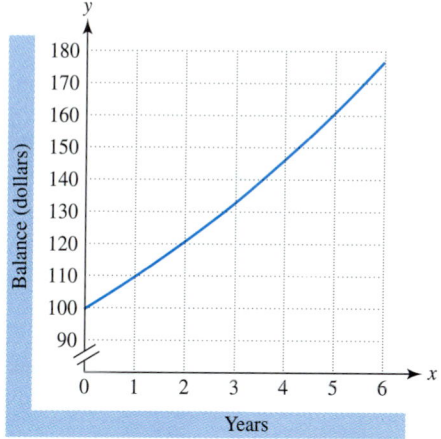

46. *Interpreting an Inverse* The graph of f on the next page computes the Celsius temperature of a pan of water after x minutes. Estimate each of the following. Interpret what $f^{-1}(x)$ computes.
  (a) $f(4)$
  (b) $f^{-1}(90)$
  (c) $f^{-1}(80)$

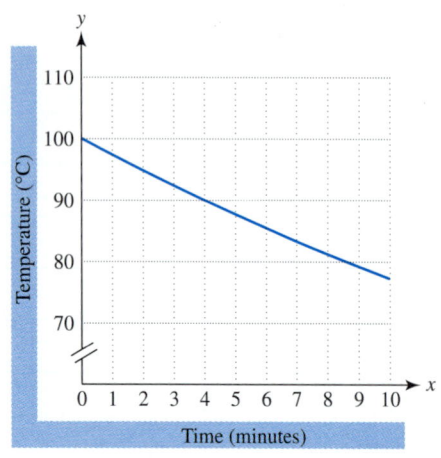

Exercises 47–50: Use the graph of f to sketch a graph of
f^{-1}.

47.

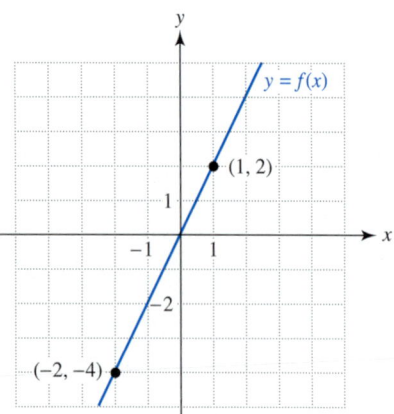

48.

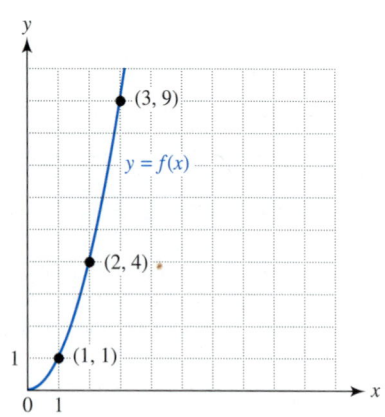

49.

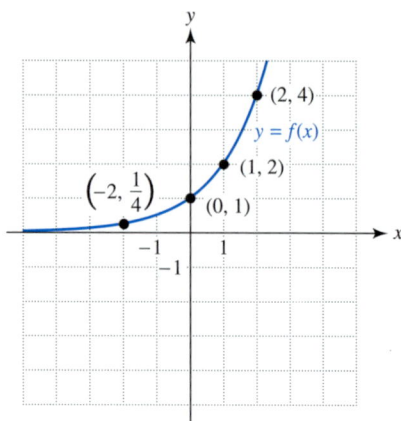

50.

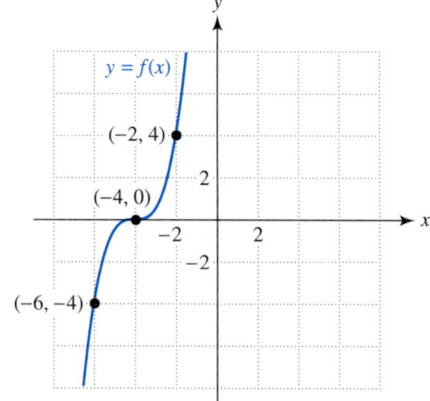

Symbolic Representations of Inverse Functions

Exercises 51–56: Find a symbolic representation for
$f^{-1}(x)$. *Support your results graphically and numerically.*

51. $f(x) = \sqrt[3]{x}$ **52.** $f(x) = 2x$

53. $f(x) = -2x + 10$ **54.** $f(x) = x^3 + 2$

55. $f(x) = 3x - 1$ **56.** $f(x) = \dfrac{x-1}{2}$

Exercises 57–64: Find a symbolic representation for
$f^{-1}(x)$. *Identify the domain and range of* f^{-1}.

57. $f(x) = 5x - 15$

58. $f(x) = (x + 3)^2, x \geq -3$

59. $f(x) = \dfrac{1}{x+3}$

60. $f(x) = \dfrac{2}{x-1}$

61. $f(x) = 2x^3$

62. $f(x) = 1 - 4x^3$

63. $f(x) = x^2, x \geq 0$

64. $f(x) = \sqrt[3]{1-x}$

Exercises 65–70: Verify that f and g are inverse functions.

65. $f(x) = 5x$, $g(x) = \dfrac{x}{5}$

66. $f(x) = x^3 + 10$, $g(x) = \sqrt[3]{x - 10}$

67. $f(x) = 4x + 1$, $g(x) = \dfrac{x - 1}{4}$

68. $f(x) = \dfrac{1}{2}x - 5$, $g(x) = 2x + 10$

69. $f(x) = \sqrt[3]{x + 7}$, $g(x) = x^3 - 7$

70. $f(x) = x^2 + 5$, $x \geq 0$, $g(x) = \sqrt{x - 5}$

Solving Equations with Inverse Functions

71. *Interpreting an Inverse* Let $f(x)$ compute the distance traveled in miles after x hours by a car with a velocity of 60 miles per hour.
 (a) Explain what $f^{-1}(x)$ computes.
 (b) Interpret the solution to the equation $f(x) = 200$.
 (c) Explain how to solve the equation in part (b) using $f^{-1}(x)$.

72. *Interpreting an Inverse* Let $f(x)$ compute the height in feet of a rocket after x seconds of flight.
 (a) Explain what $f^{-1}(x)$ computes.
 (b) Interpret the solution to the equation $f(x) = 5000$.
 (c) Explain how to solve the equation in part (b) using $f^{-1}(x)$.

Exercises 73–76: Use f^{-1} to solve the equation $f(x) = k$, where k is a constant. Check the solution using $f(x)$.

73. Solve $f(x) = 12$, where $f^{-1}(x) = \dfrac{x}{4}$ and $f(x) = 4x$.

74. Solve $f(x) = 9$, where $f^{-1}(x) = \sqrt[3]{x - 1}$ and $f(x) = x^3 + 1$.

75. Solve $f(x) = \dfrac{1}{10}$, where $f^{-1}(x) = \dfrac{1}{x + 1}$ and $f(x) = \dfrac{1 - x}{x}$.

76. Solve $f(x) = -5$, where $f^{-1}(x) = 2x + 1$ and $f(x) = \dfrac{x - 1}{2}$.

77. *Modeling AIDS* Let $f(x)$ approximate the cumulative numbers of AIDS cases in the United States during the year x, where $1982 \leq x \leq 1994$. Then,
$$f^{-1}(x) = 1982 + \sqrt{\dfrac{x - 1586}{3200}}.$$
 (a) Explain what f^{-1} computes.
 (b) Interpret the solution to the equation $f(x) = 260,000$.
 (c) Use $f^{-1}(x)$ to solve the equation in part (b).

78. *Crutch Length* (Refer to Example 4.) Solve the equation $f(x) = 47$ and interpret the solution.

79. *Advertising Costs* The line graph represents a function f that computes the cost in millions of dollars of a 30-second commercial during a Super Bowl telecast. Perform each calculation and interpret the results. (Source: *USA Today.*)
 (a) Evaluate $f(1995)$.
 (b) Solve $f(x) = 1$.
 (c) Evaluate $f^{-1}(1)$.

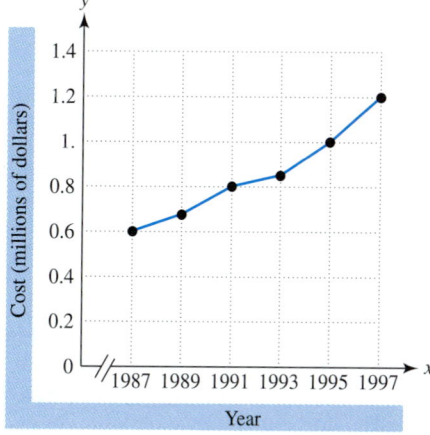

80. *Social Security* The line graph represents a function f that models the numbers in millions of Social Security recipients from 1990 to 2030. (Source: Social Security Administration.)

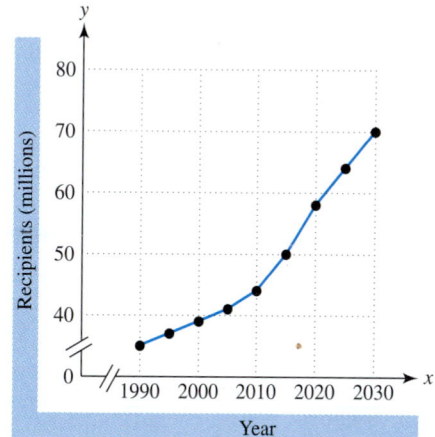

 (a) Solve the equation $f(x) = 50$.
 (b) Evaluate $f^{-1}(50)$.

Applications

81. *Acid Rain and Salmon* Acid rain has caused a decline in the overall Atlantic salmon yield in southern Norway. The accompanying table shows this yield in metric tons for seven rivers with acidic water. (Source: B. Freedman, *Environmental Ecology.*)

Year	Metric Tons
1900	30
1918	9
1923	12
1940	4
1948	10
1958	5
1970	0

(a) Let a line graph of the data represent a function f. Graph f in [1885, 1975, 5] by [−5, 35, 5].

(b) Determine if f is a one-to-one function. Does f have an inverse function?

82. *Juvenile Homicides*   The accompanying table shows the number of juvenile homicides in Boston for the years 1990 through 1996. (Source: Boston Police Department.)

Year	1990	1991	1992	1993	1994	1995	1996
Homicides	12	10	8	16	6	4	0

(a) Make a scatterplot of the data in the window [1989, 1997, 1] by [0, 18, 2]. Does this graph represent a one-to-one function?

(b) Make a line graph of this data. Could the line graph represent a one-to-one function?

(c) Suppose that a continuous function f is both increasing and decreasing. Could f be a one-to-one function? Explain.

83. *Converting Units*   The accompanying tables represent a function f that converts yards to feet and a function g that converts miles to yards. Evaluate each expression and interpret the results.

x (yd)	1760	3520	5280	7040	8800
$f(x)$ (ft)	5280	10,560	15,840	21,120	26,400

x (mi)	1	2	3	4	5
$g(x)$ (yd)	1760	3520	5280	7040	8800

(a) $(f \circ g)(2)$
(b) $f^{-1}(26{,}400)$
(c) $(g^{-1} \circ f^{-1})(21{,}120)$

84. (Refer to previous exercise.)
(a) Find symbolic representations for $f(x)$, $g(x)$, and $(f \circ g)(x)$.
(b) Express $(g^{-1} \circ f^{-1})(x)$ symbolically. What does this function compute?

85. *Converting Units*   The accompanying tables represent a function f that converts tablespoons to cups and a function g that converts cups to quarts. Evaluate each expression and interpret the results.

x (tbsp)	32	64	96	128
$f(x)$ (c)	2	4	6	8

x (c)	2	4	6	8
$g(x)$ (qt)	0.5	1	1.5	2

(a) $(g \circ f)(96)$
(b) $g^{-1}(2)$
(c) $(f^{-1} \circ g^{-1})(1.5)$

86. (Refer to previous exercise.)
(a) Find symbolic representations for $f(x)$, $g(x)$, and $(g \circ f)(x)$.
(b) Express $(f^{-1} \circ g^{-1})(x)$ symbolically. What does this function compute?

87. *Air Pollution*   Tiny particles suspended in the air are necessary for clouds to form. Because of this fact, experts believe that air pollutants may cause an increase in cloud cover. From 1930 to 1980 the percentage of cloud cover over the world's oceans was monitored. The linear function given by $f(x) = 0.06(x − 1930) + 62.5$, where $1930 \leq x \leq 1980$, approximates this percentage. (Source: W. Cotton and R. Pielke, *Human Impacts on Weather and Climate*.)

(a) Evaluate $f(1930)$ and $f(1980)$. How did the amount of cloud cover over the oceans change during this 50-year period?

(b) What does $f^{-1}(x)$ compute?

(c) Use your results from part (a) to evaluate $f^{-1}(62.5)$ and $f^{-1}(65.5)$.

(d) Find $f^{-1}(x)$ symbolically. Support your result graphically and numerically.

88. *Rise in Sea Level*   Due to the greenhouse effect, the global sea level could rise through thermal expansion and partial melting of the polar ice caps. The table represents a function f that models this expected rise in sea level in centimeters for the year x. (This model assumes no changes in current trends.) (Source: A. Nilsson, *Greenhouse Earth*.)

x (yr)	1990	2000	2030	2070	2100
$f(x)$ (cm)	0	1	18	44	66

(a) Is f a one-to-one function? Explain.
(b) Use $f(x)$ to find a table for $f^{-1}(x)$. Interpret f^{-1}.

Drawing Inverses with Technology

Exercises 89 and 90: Use your calculator to graph f, f^{-1}, and the line $y = x$ in $[-9, 9, 1]$ by $[-6, 6, 1]$.

89. $f(x) = x^3 - x^2 + x + 1$

90. $f(x) = 0.1x^5 + 0.1x^3 + 0.2x + 1$

Writing about Mathematics

1. Explain how to find verbal, numerical, graphical, and symbolic representations of an inverse function. Give examples.

2. Can a one-to-one function have more than one x-intercept or more than one y-intercept? Explain.

3. If f is one-to-one with domain D and range R, find the domain and range of f^{-1}. Give examples.

4. If $f(x) = ax^2 + bx + c$ with $a \neq 0$, does $f^{-1}(x)$ exist? Explain.

CHECKING BASIC CONCEPTS FOR SECTIONS 4.1 AND 4.2

1. Use the table to evaluate each expression, if possible.

x	-2	-1	0	1	2
$f(x)$	0	1	-2	-1	2
$g(x)$	1	-2	-1	2	0

  (a) $(f + g)(1)$ **(b)** $(f - g)(-1)$
  (c) $(fg)(0)$ **(d)** $(f/g)(2)$
  (e) $(f \circ g)(2)$ **(f)** $(g \circ f)(-2)$

2. Let $f(x) = x^2 + 3x - 2$ and $g(x) = 3x - 1$. Find each expression.
  (a) $(f + g)(x)$
  (b) $(f/g)(x)$
  (c) $(f \circ g)(x)$

3. If $f(x) = 5 - 2x$, find $f^{-1}(x)$ symbolically. Support your results graphically and numerically.

4.3 Exponential Functions

Representations of Exponential Functions • Compound Interest • The Natural Exponential Function • Exponential Models

Introduction

Even though modern exponential notation was not developed until 1637 by the great French mathematician, René Descartes, some of the earliest applications involving exponential functions occurred in the calculation of interest. The custom of charging interest dates back to at least 2000 B.C. in ancient Babylon, where interest rates ran as high as 33%. Today, exponential functions are not only used to calculate interest, but are also used to model a wide variety of phenomena found in business, biology, medicine, engineering, and education. This section discusses exponential functions and their representations. (Source: *Historical Topics for the Mathematics Classroom, Thirty-first Yearbook,* NCTM.)

Representations of Exponential Functions

If a small amount of money is deposited into a bank account, it may not earn much interest. However, if the interest is reinvested, the account balance can grow substantially over time. A similar situation has occurred with the world population. A larger population usually results in more births, which then results in an even larger population.

If the growth of a quantity is directly proportional to the amount or number present, then it often can be modeled by an *exponential function.*

Exponential function

A function f represented by

$$f(x) = a^x, \qquad a > 0 \text{ and } a \neq 1,$$

is an **exponential function with base a.**

Examples of exponential functions include

$$f(x) = 2^x, \qquad g(x) = 10^x, \qquad \text{and} \qquad h(x) = 0.5^x.$$

In each representation the variable x occurs as an exponent. The bases of f, g, and h are 2, 10, and 0.5, respectively. Their graphs are shown in Figures 4.29–4.31. The domain of an exponential function is all real numbers, and the range is all positive real numbers.

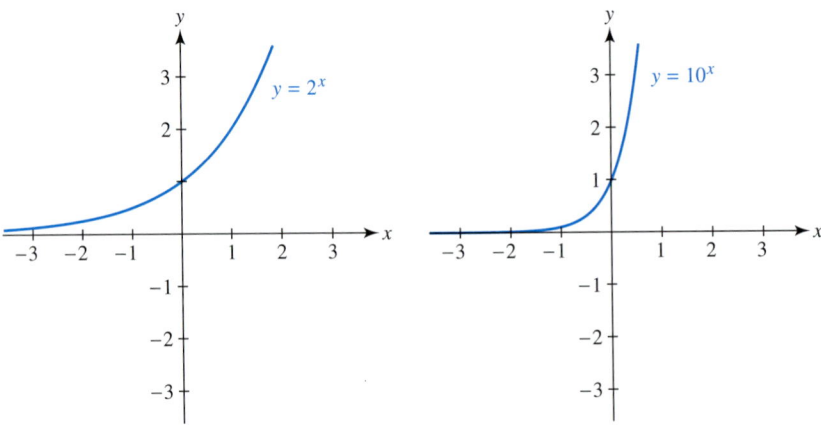

FIGURE 4.29   **FIGURE 4.30**

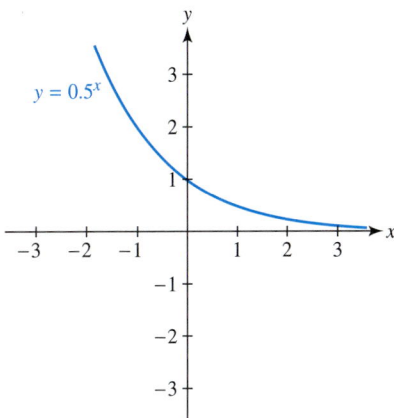

FIGURE 4.31

To investigate the graphs of exponential functions, graph $y = 1^x$, $y = 1.3^x$, $y = 1.7^x$, and $y = 2.5^x$ as shown in Figure 4.32. (Note that $f(x) = 1^x$ does *not* represent an exponential function.) As a increases, the graph of $y = a^x$ increases at a faster rate. On the other hand, the graphs of $y = 0.7^x$, $y = 0.5^x$, and $y = 0.15^x$ decrease faster as a decreases. See Figure 4.33. The graph of $y = a^x$ is increasing when $a > 1$ and decreasing when $0 < a < 1$.

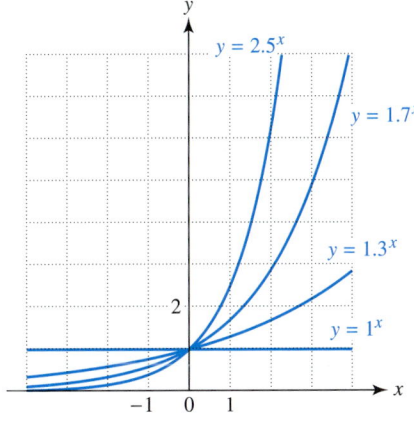

FIGURE 4.32   **FIGURE 4.33**

MAKING CONNECTIONS
Exponential and Polynomial Functions

The function represented by $f(x) = 2^x$ is an exponential function. The exponent x is a variable. The function given by $g(x) = x^2$ is a polynomial function. The exponent 2 is a constant.

EXAMPLE 1   *Comparing exponential and polynomial functions graphically and numerically*

Compare $f(x) = 3^x$ and $g(x) = x^3$ graphically and numerically for $x \geq 0$.

Solution

Graphical Comparison The graphs $Y_1 = 3^X$ and $Y_2 = X^3$ are shown in Figure 4.34. For $x \geq 6$, the graph of the exponential function Y_1 increases significantly faster than the graph of the polynomial function Y_2.

Numerical Comparison Table $Y_1 = 3^X$ and $Y_2 = X^3$ as shown in Figure 4.35. The values for Y_1 increase faster than the values for Y_2.

[0, 12, 1] by [0, 10,000, 1000]

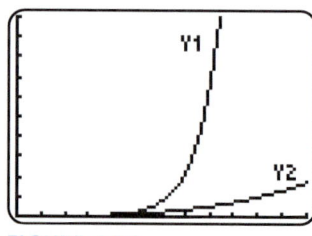

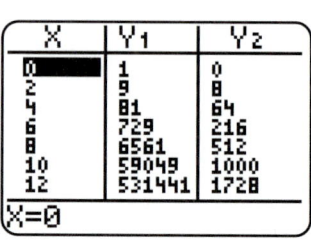

FIGURE 4.34   **FIGURE 4.35**   ■

The results of Example 1 are true in general. For large enough inputs, exponential functions with $a > 1$ eventually become greater than any polynomial function. This is an example of *exponential growth*. When $0 < a < 1$, *exponential decay* occurs.

Review Note *Properties of Exponents*

The following are important properties of exponents. It is assumed that $a > 0$ and that r and s are real numbers.

Property	*Example*
1. $a^0 = 1$	5. $1^0 = 1$
2. $a^r a^s = a^{r+s}$	$2^4 2^3 = 2^{4+3} = 2^7 = 128$
3. $a^{-r} = \dfrac{1}{a^r}$	$4^{-1/2} = \dfrac{1}{4^{1/2}} = \dfrac{1}{\sqrt{4}} = \dfrac{1}{2}$
4. $\dfrac{a^r}{a^s} = a^{r-s}$	$\dfrac{3^2}{3^5} = 3^{2-5} = 3^{-3} = \dfrac{1}{3^3} = \dfrac{1}{27}$
5. $(a^r)^s = a^{rs}$	$(4^3)^{1/3} = 4^1 = 4$

In Section 3.3 we learned that the graph of $y = f(-x)$ is a reflection of the graph of $y = f(x)$ across the y-axis. As a result, the graph of $y = a^{-x}$ is a reflection of $y = a^x$ in the y-axis. For example, let $f(x) = 2^x$ and $f(-x) = 2^{-x}$. Then the graphs of $y = 2^x$ and $y = 2^{-x}$ are reflections of each other across the y-axis. See Figure 4.36. Note that by properties of exponents $2^{-x} = \left(\dfrac{1}{2}\right)^x$.

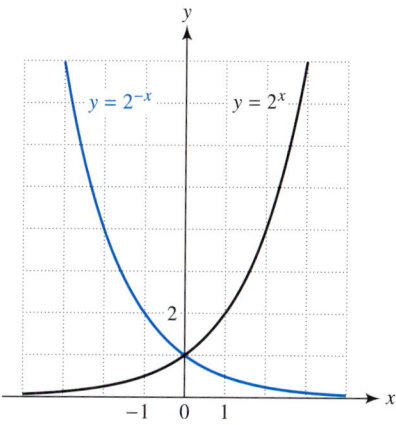

FIGURE 4.36

Compound Interest

Suppose $1000 is deposited in an account paying 10% annual interest. At the end of one year, the account will contain $1000 plus 10% of $1000, or $100, in interest. Let A_0 represent the **principal** or initial amount deposited. Then, the amount A_1 in the account after one year can be computed as follows.

$$
\begin{aligned}
A_1 &= A_0 + (0.10)A_0 && \text{Principal plus interest} \\
&= A_0(1 + 0.10) && \text{Factor.} \\
&= 1000(1.10) && A_0 = \$1000 \\
&= 1100 && \text{Simplify.}
\end{aligned}
$$

The sum of the principal and interest is $1100.

During the second year, the account earns interest on $1100. The amount A_2 after the second year will be equal to A_1 plus 10% of A_1.

$$
\begin{aligned}
A_2 &= A_1 + (0.10)A_1 \\
&= A_1(1 + 0.10) && \text{Factor.} \\
&= A_0(1 + 0.10)(1 + 0.10) && A_1 = A_0(1 + 0.10) \\
&= A_0(1 + 0.10)^2 \\
&= 1000(1.10)^2 \\
&= 1210
\end{aligned}
$$

After two years there will be $1210.

We would like to determine a general formula for A_n, the amount in the account after n years. To do this, compute A_3 and observe a pattern.

$$
\begin{aligned}
A_3 &= A_2 + (0.10)A_2 \\
&= A_2(1 + 0.10) \\
&= A_0(1 + 0.10)^2(1 + 0.10) && A_2 = A_0(1 + 0.10)^2 \\
&= A_0(1 + 0.10)^3 \\
&= 1000(1.10)^3 \\
&= 1331
\end{aligned}
$$

The amount is \$1331. We can see that in general, $A_n = A_0(1 + 0.10)^n$. This type of interest is said to be *compounded annually,* since it is paid once a year.

If the interest rate had been r, expressed in decimal form, then $A_n = A_0(1 + r)^n$. Notice that in the expression $A_0(1 + r)^n$, the variable n occurs as an exponent.

EXAMPLE 2   *Calculating an account balance*

If the principal is \$2000 and the interest rate is 8% compounded annually, calculate the account balance after 4 years.

Solution

The initial amount is $A_0 = 2000$, the interest rate is $r = 0.08$, and the number of years is $n = 4$.

$$A_4 = 2000(1 + 0.08)^4 \approx 2720.98$$

After 4 years the account contains \$2720.98. ∎

In most savings accounts, interest is paid more often than once a year. In this case a smaller amount of interest is paid more frequently. For example, suppose \$1000 is deposited in an account paying 10% annual interest, compounded quarterly. After three months the interest would amount to one-fourth of 10% or 2.5% of \$1000. The account balance would be $1000(1 + 0.025) = \$1025$. During the next three-month period, interest would be paid on the \$1025. In a manner similar to annual compounding, the balance would be $\$1000(1 + 0.025)^2 \approx \$1050.63$ after six months, $\$1000(1 + 0.025)^3 \approx \$1076.89$ after nine months, and $\$1000(1 + 0.025)^4 \approx \$1103.81$ after a year. With annual compounding the amount is \$1100. The difference of \$3.81 is due to compounding quarterly. Although this amount is small after one year, compounding more frequently can have a dramatic effect over a long period of time.

Compound interest

If A_0 dollars is deposited in an account paying an annual rate of interest r, compounded (paid) m times per year, then after n years the account will contain A_n dollars, where

$$A_n = A_0\left(1 + \frac{r}{m}\right)^{mn}.$$

EXAMPLE 3   *Comparing compound interest*

Suppose \$1000 is deposited by a 20-year-old worker in an Individual Retirement Account (IRA) that pays an annual interest rate of 12%. Describe the effect on the balance at age 65, if interest were compounded annually and quarterly.

Solution

Compounded Annually   Let $A_0 = 1000$, $r = 0.12$, $m = 1$, and $n = 65 - 20 = 45$.

$$A_{45} = 1000(1 + 0.12)^{45} \approx \$163{,}987.60$$

Compounded Quarterly   Let $A_0 = 1000$, $r = 0.12$, $m = 4$, and $n = 45$.

$$A_{45} = 1000\left(1 + \frac{0.12}{4}\right)^{4(45)} = 1000(1 + 0.03)^{180} \approx \$204{,}503.36$$

Quarterly compounding results in an increase of $40,515.76! ∎

Critical Thinking

In Example 3, conjecture the effect on the IRA balance after 45 years if the interest rate were 6% instead of 12%. Test your conjecture.

The Natural Exponential Function

In Example 3, compounding interest quarterly rather than annually made a significant difference in the balance after 45 years. What would happen if interest were compounded daily or even hourly? Would there be a limit to the amount of interest that could be received? To answer these questions, suppose $1 was deposited in an account at the very high interest rate of 100%. Table 4.13 shows the amount of money after one year, compounding m times during the year. The first column represents the time interval between compound interest payments. In Table 4.13, the formula $A_n = A_0\left(1 + \dfrac{r}{m}\right)^{mn}$ reduces to $A_1 = \left(1 + \dfrac{1}{m}\right)^m$, since $A_0 = n = r = 1$.

TABLE 4.13

Time	m	A_1
Year	1	2.000000
Month	12	2.613035
Day	365	2.714567
Hour	8760	2.718127
Minute	525,600	2.718279
Second	31,536,000	2.718282

Notice that A_1 levels off near $2.72. If compounding were done more frequently, by letting m become large without bound, it would be called *continuous compounding*. The expression $\left(1 + \dfrac{1}{m}\right)^m$ reaches a limit of approximately 2.718281828. This numeric value is so important in mathematics that it has been given its own symbol, e, sometimes called **Euler's number.** The number e has many of the same characteristics as π. Its decimal expansion never terminates or repeats in a pattern. It is an irrational number.

Value of e

To ten significant digits, $e \approx 2.718281828$.

Continuous compounding can be applied to population growth. Compounding annually would mean that all births and deaths occur on December 31. Similarly, compounding quarterly would mean that births and deaths occur at the end of March, June, September, and December. In large populations, births and deaths occur *continuously* throughout the year. Compounding continuously is a *natural* way to model large populations.

The natural exponential function

The function f, represented by $f(x) = e^x$, is the **natural exponential function.**

EXAMPLE 4   *Evaluating the natural exponential function*

Use a calculator to approximate to five significant digits $f(x) = e^x$ when $x = 1, 0.5$, and -2.56.

Solution

Figure 4.37 shows that $f(1) = e^1 \approx 2.7183$, $f(0.5) = e^{0.5} \approx 1.6487$, and $f(-2.56) = e^{-2.56} \approx 0.077305$.

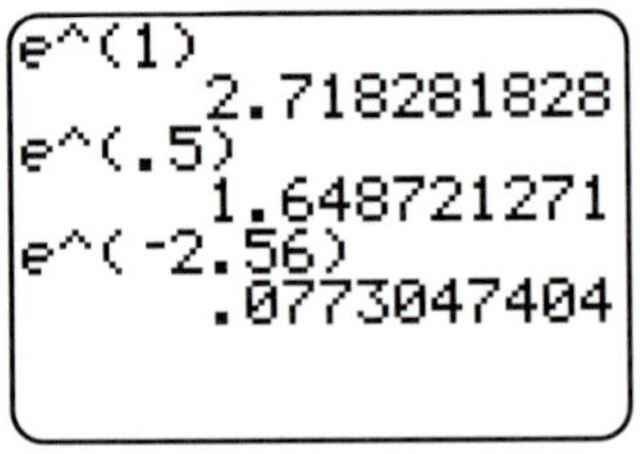

```
e^(1)
          2.718281828
e^(.5)
          1.648721271
e^(-2.56)
          .0773047404
```

FIGURE 4.37   ∎

Critical Thinking

Graph $y = 2^x$ and $y = 3^x$ on the same coordinate axes, using the viewing rectangle $[-3, 3, 1]$ by $[0, 4, 1]$. Conjecture how the graph of $y = e^x$ will appear. Test your conjecture.

When interest is compounded continuously, the following can be used to compute the amount of money in an account.

Continuously compounded interest

If A_0 dollars is deposited in an account paying an annual rate of interest r, compounded continuously, then after n years the account will contain A_n dollars, where

$$A_n = A_0 e^{rn}.$$

EXAMPLE 5   *Calculating continuously compounded interest*

Suppose $1000 is deposited into an IRA with an interest rate of 12%, compounded continuously.
(a) How much money will there be after 45 years?
(b) Determine graphically and numerically the ten-year period when the account balance increases the most.

Solution

(a) Let $A_0 = 1000$, $r = 0.12$, and $n = 45$. Then, $A_{45} = 1000e^{(0.12)45} \approx$ $221,406.42$. This is more than the $204,503.36$ that resulted from compounding quarterly in Example 3.

(b) Graph $Y_1 = 1000e^{\wedge}(.12X)$ as shown in Figure 4.38. From the graph we can see that the account balance increases the most during the last ten years. Numerical support for this conclusion is shown in Figure 4.39.

[0, 45, 5] by [0, 250,000, 100,000]

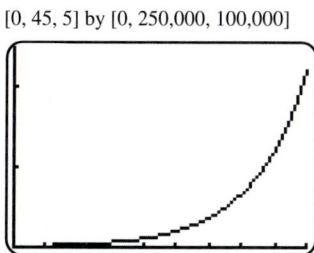

X	Y1
15	6049.6
20	11023
25	20086
30	36598
35	66686
40	121510
45	221406

$Y_1 = 1000e^{\wedge}(.12X)$

FIGURE 4.38   **FIGURE 4.39**   ■

Exponential Models

Traffic flow on highways is modeled using exponential functions whenever the traffic patterns are random.

EXAMPLE 6   *Modeling traffic flow*

An intersection has an hourly average volume of 360 vehicles, where vehicles arrive at the intersection randomly. Highway engineers estimate the likelihood or probability that no vehicle will enter the intersection within a period of x seconds using $f(x) = (0.905)^x$. (Source: F. Mannering and W. Kilareski, *Principles of Highway Engineering and Traffic Analysis.*)

(a) Compute $f(5)$ and $f(60)$. Interpret the results.

(b) Solve the equation $f(x) = 0.5$ graphically. Interpret the solution.

Solution

(a) $f(5) = (0.905)^5 \approx 0.61$ and $f(60) = (0.905)^{60} \approx 0.0025$. There is a 61% chance that no vehicle will enter the intersection during any five-second interval, and a 0.25% chance that no vehicle will enter the intersection within a 60-second interval.

(b) Let $Y_1 = 0.905^{\wedge}X$ and $Y_2 = 0.5$. Their graphs intersect near (6.94, 0.5) as shown in Figure 4.40. There is a 50% chance that no car will enter the intersection in a seven-second interval.

[0, 10, 1] by [0, 1.5, 0.5]

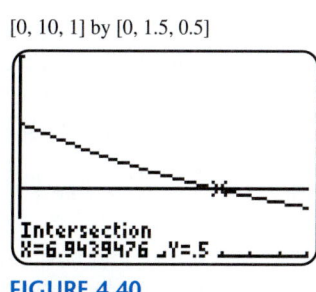

Intersection
X=6.9439476 Y=.5

FIGURE 4.40   ■

EXAMPLE 7   *Modeling the growth of* E. coli *bacteria*

A type of bacteria that inhabits the intestines of animals is named *E. coli* (*Escherichia coli*). This bacteria is capable of rapid growth and can be dangerous to humans—especially children. In one study, *E. coli* bacteria were capable of doubling in number every 49.5 minutes. Their number N after x minutes could be modeled by $N(x) = N_0 e^{0.014x}$. Suppose $N_0 = 500{,}000$ is the initial number of bacteria per milliliter. (Source: G. S. Stent, *Molecular Biology of Bacterial Viruses.*)

(a) Conjecture the number of bacteria per milliliter after 99 minutes. Verify your conjecture.

(b) Determine graphically and numerically the elapsed time when there were 25 million bacteria per milliliter.

Solution

(a) Since the bacteria double every 49.5 minutes, there would be 1,000,000 per milliliter after 49.5 minutes and 2,000,000 after 99 minutes. This is verified by evaluating

$$N(99) = 500{,}000 e^{0.014(99)} \approx 2{,}000{,}000.$$

(b) *Graphical Solution*   Solve $N(x) = 25{,}000{,}000$ by graphing $Y_1 = 500000e^{\wedge}(0.014X)$ and $Y_2 = 25000000$. Their graphs intersect near $(279.4, 25{,}000{,}000)$ as shown in Figure 4.41. Thus, in a one milliliter sample, half a million *E. coli* bacteria could increase to 25 million in approximately 279 minutes or 4 hours and 39 minutes.

Numerical Solution   Numerical support is shown in Figure 4.42, where Y_1 equals 25 million between $x = 279$ and 280.

$[0, 400, 100]$ by $[0, 3 \times 10^7, 1 \times 10^7]$

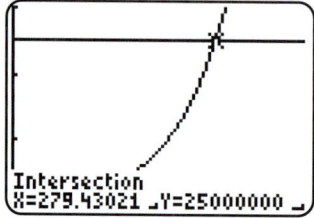

FIGURE 4.41

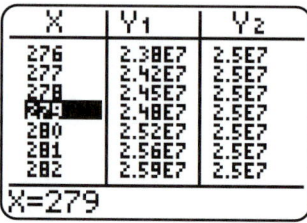

FIGURE 4.42

Logistic Functions.   In real life, the populations of bacteria, insects, and animals do not continue to grow indefinitely. Initially, population growth may be slow. Then, as their numbers increase, so does the rate of growth. After a region has become heavily populated or saturated, the population usually levels off because of limited resources. This phenomenon also occurs with the height of a tree. When a tree is small its growth is correspondingly small. Gradually a tree starts to grow more rapidly. After many years, it becomes a mature tree and its height levels off.

This type of growth may be modeled by a **logistic function** represented by

$$f(x) = \frac{c}{(1 + ba^x)},$$ where a, b, and c are constants. A typical graph of a logistic function f is shown in Figure 4.43. The graph of f is referred to as a *sigmoidal curve*. The next example demonstrates how a logistic function can be used to describe the growth of a yeast culture.

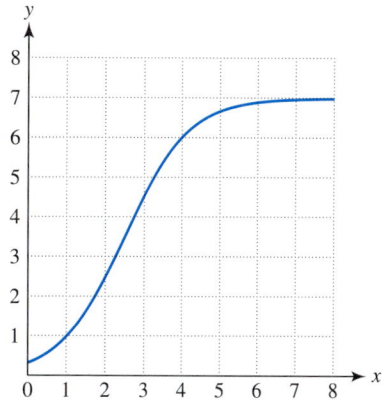

FIGURE 4.43 A Sigmoidal Curve

EXAMPLE 8   *Modeling logistic growth*

One of the earliest studies about population growth was done using yeast plants in 1913. (See Exercise 25, Section 1.5.) A small amount of yeast was placed in a container with a fixed amount of nourishment. The units of yeast were recorded every two hours. The data is listed in Table 4.14.

TABLE 4.14

Time	0	2	4	6	8	10	12	14	16	18
Yeast	9.6	29.0	71.1	174.6	350.7	513.3	594.8	640.8	655.9	661.8

(a) Make a scatterplot of the data. Describe the growth.

(b) This data is modeled by $f(x) = \dfrac{663}{1 + 71.6(0.579)^x}$. Graph f and the data.

(c) Approximate graphically the time when the amount of yeast was 200 units.

Solution

(a) A scatterplot of the data is shown in Figure 4.44. The yeast increase slowly at first. Then they grow more rapidly, until the amount of yeast gradually levels off. The limited amount of nourishment causes this leveling off.

(b) The graph of $Y_1 = 663/(1 + 71.6*0.579^X)$ together with the data is shown in Figure 4.45. The function f models the real data remarkably well.

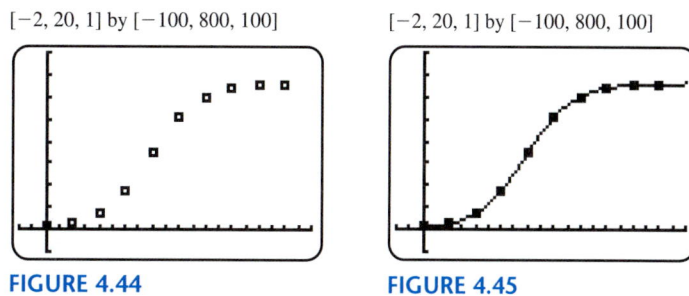

$[-2, 20, 1]$ by $[-100, 800, 100]$ $[-2, 20, 1]$ by $[-100, 800, 100]$

FIGURE 4.44   **FIGURE 4.45**

(c) The graphs of Y_1 and $Y_2 = 200$ intersect near $(6.28, 200)$ as shown in Figure 4.46 on the next page. The amount of yeast reached 200 after about 6.28 hours.

$[-2, 20, 1]$ by $[-100, 800, 100]$

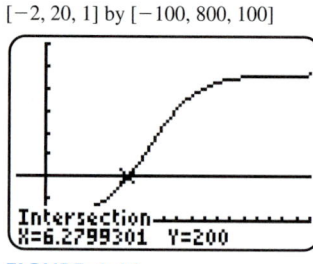

FIGURE 4.46

Critical Thinking

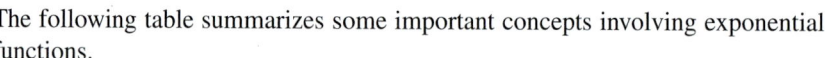

In Example 8, suppose that after 18 hours the experiment had been extended and more nourishment was provided for the yeast plants. Sketch a possible graph of the amount of yeast.

4.3 PUTTING IT ALL TOGETHER

The following table summarizes some important concepts involving exponential functions.

Concept	Examples	Comments
Exponential function $f(x) = a^x$ $a > 0, \quad a \neq 1$	$f(x) = 2^x$ $f(x) = 1.2^x$ $f(x) = 0.5^x$	If $a > 1$ then exponential growth occurs, whereas if $0 < a < 1$ exponential decay occurs.
Interest compounded m times per year $A_n = A_0\left(1 + \dfrac{r}{m}\right)^{mn}$	$500 at 8%, compounded monthly for three years yields: $500\left(1 + \dfrac{0.08}{12}\right)^{12(3)}$ $\approx \$635.12.$	A_0 is the principal, r is the interest rate, m is the number of times interest is paid each year, n is the number of years, and A_n is the amount after n years.
The number e	$e \approx 2.718282$	The number e is an important number in mathematics, much like π.
Natural exponential function	$f(x) = e^x$	This function is an exponential function with base e.
Interest compounded continuously $A_n = A_0 e^{rn}$	$500 at 8% compounded continuously for three years yields: $500e^{0.08(3)} \approx \$635.62.$	A_0 is the principal, r is the interest rate, n is the number of years, and A_n is the amount after n years.

4.3 EXERCISES

   Tape 5

Exponents

Exercises 1–12: Simplify the expression without a calculator.

1. 2^{-3}

2. $(-3)^{-2}$

3. $4^{1/2}$

4. $\left(\dfrac{1}{2}\right)^{-3}$

5. $27^{2/3}$

6. $8^{-2/3}$

7. $\left(\dfrac{1}{8}\right)^{-1}$

8. $a^{0.5}a^{1.5}$

9. $4^{1/6}4^{1/3}$

10. $\dfrac{9^{5/6}}{9^{1/3}}$

11. $e^x e^x$

12. $16^{-1/2}27^{1/3}$

Compound Interest

Exercises 13–18: Use the compound interest formula to determine the final value of each amount.

13. $600 at 7% compounded annually for 5 years

14. $2300 at 11% compounded semiannually for 10 years

15. $950 at 3% compounded daily for 20 years (*Hint:* Let $m = 365$.)

16. $3300 at 8% compounded quarterly for 2 years

17. $2000 at 10% compounded continuously for 8 years

18. $100 at 19% compounded continuously for 50 years

19. *Continuous Compounding* A principal of $500 is deposited in a savings account with an annual interest rate of 7.2% compounded continuously. Make a table of the amount of money in the account over the next 7 years.

20. Complete the previous exercise if the interest rate were 12% compounded continuously.

21. *Interest* A principal of $100 is deposited into an account paying 5% interest compounded monthly. In a similar account, $200 is deposited. Conjecture how the amounts in each account will compare after 10 years. Explain your reasoning.

22. *Interest* A principal of $2000 is deposited into two different accounts. One account pays 10% interest compounded annually, while another pays 5% interest compounded annually.
  (a) Conjecture if the 10% account will accrue twice the interest as the 5% account after 5 years.
  (b) Verify your conjecture by computing the actual amounts.

23. *Federal Debt* In fiscal year 1996 the federal budget deficit was $107.3 billion. At the same time, 30-year treasury bonds were paying 6.83% interest. Suppose the American taxpayer loaned $107.3 billion to the federal government at 6.83% compounded annually. If the federal government waited 30 years to pay the entire amount back, including the interest, how much would this be? (Source: Department of the Treasury.)

24. In the previous exercise suppose that interest rates were 2% higher. How much would the federal government owe after 30 years? Is the national debt sensitive to interest rates?

25. *Annuity* If x dollars is deposited every two weeks (26 times per year) into an account paying an annual interest rate r, expressed in decimal form, then the amount A_n in the account after n years can be approximated by the formula

$$A_n = x\left[\frac{(1 + r/26)^{26n} - 1}{(r/26)}\right].$$

If $50 is deposited every two weeks into an account paying 8% interest, approximate the amount in the account after 10 years.

26. (Refer to the previous exercise.) Suppose a retirement account pays 10% annual interest. Determine how much a 20-year-old worker should deposit in this account every two weeks, in order to have one million dollars at age 65.

Exponential Functions

27. Match the symbolic representation of f with its graphical representation (a–d). Do not use a calculator.
  i. $f(x) = e^x$
  ii. $f(x) = 3^{-x}$
  iii. $f(x) = 1.5^x$
  iv. $f(x) = 0.99^x$

  (a)

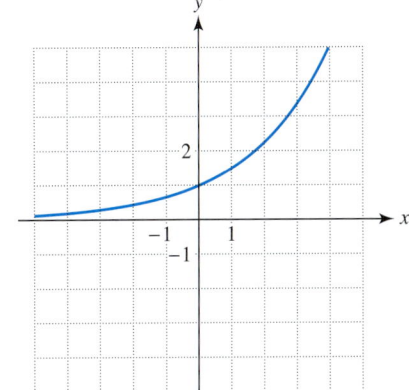

(b)

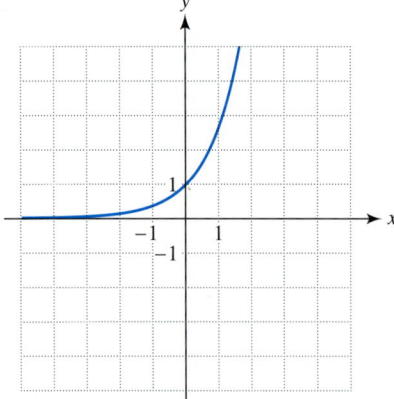

(c)

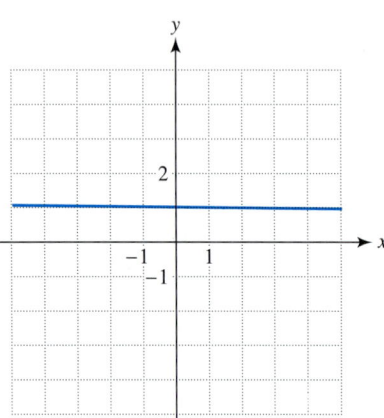

(d)

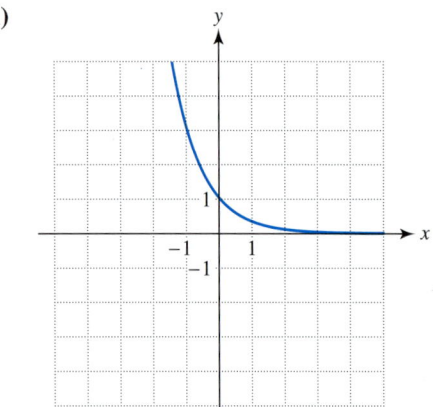

(a)

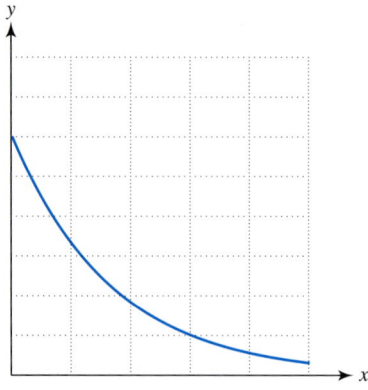

(b)

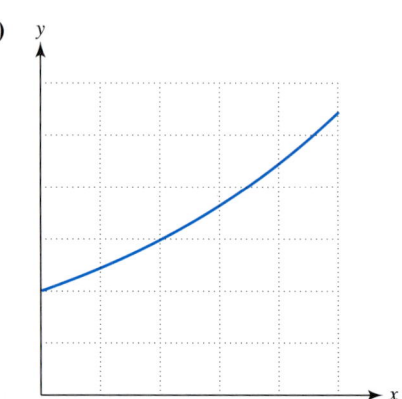

(c)

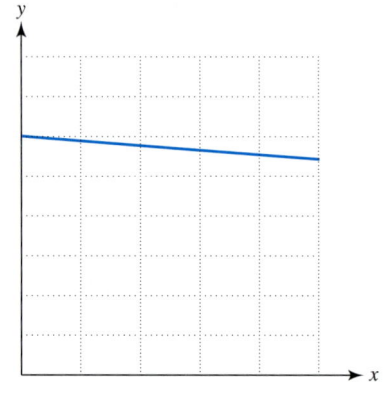

(d)

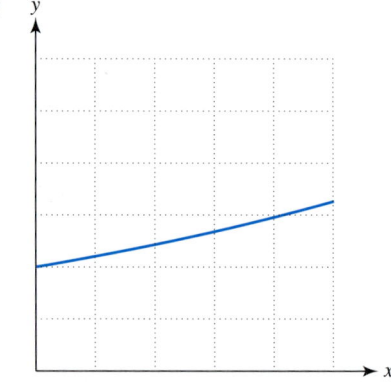

28. *Modeling Phenomena*   Match the situation with the graph that models it best.
    i.   Balance of an account after x years earning 10% interest compounded continuously
    ii.   Balance of an account after x years earning 5% interest compounded annually
    iii.   Air pressure in a car tire with a large hole in it after x minutes
    iv.   Air pressure in a car tire with a pinhole in it after x minutes

Exercises 29 and 30: The graph of $y = f(x)$ is shown in the accompanying figure. Sketch a graph of each equation using translations of graphs and reflections. Do not use a graphing calculator.

29. $f(x) = 2^x$
  (a) $y = 2^x - 2$
  (b) $y = 2^{x-1}$
  (c) $y = 2^{-x}$
  (d) $y = -2^x$

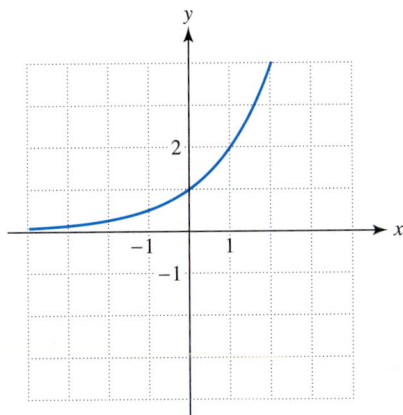

30. $f(x) = e^{-0.5x}$
  (a) $y = -e^{-0.5x}$
  (b) $y = e^{-0.5x} - 3$
  (c) $y = e^{-0.5(x-2)}$
  (d) $y = e^{0.5x}$

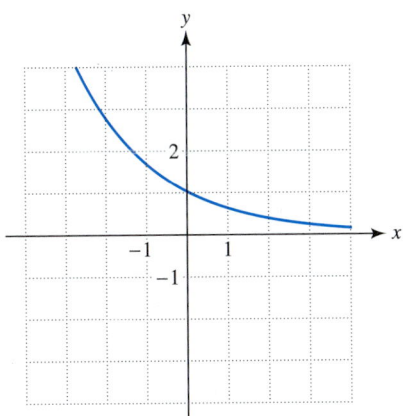

Exercises 31–36: Approximate $f(a)$ to four significant digits.

31. $f(x) = e^x, a = 3.1$

32. $f(x) = e^{2x}, a = -0.43$

33. $f(x) = 4e^{-1.2x}, a = -2.4$

34. $f(x) = -2.1e^{-0.71x}, a = 1.9$

35. $f(x) = \dfrac{e^x - e^{-x}}{2}, a = -0.7$

36. $f(x) = 4(e^{-0.3x} - e^{-0.6x}), a = 1.6$

Applications

37. *Population*   In July 1994 the population of New York state in millions was modeled by $f(x) = 18.2e^{0.001x}$, whereas the population of Florida in millions was modeled by $g(x) = 14e^{0.0168x}$. In both formulas x is the year, where $x = 0$ corresponds to July 1994. (Source: Bureau of the Census.)
  (a) Find the population of New York and Florida in July 1994.
  (b) Assuming these trends continue, estimate graphically the year when the population of Florida will equal the population of New York. What will their populations be at this time?

38. *Seal Counts*   In order to have their pups, female seals remain on shore for about 30 days. The table lists the actual seal counts every four days during a breeding season. (Source: D. Brown and P. Rothery, *Models in Biology: Mathematics, Statistics and Computing.*)

Day	0	4	8	12	16	20	24	28
Count	24	51	147	342	594	802	891	931

Day	32	36	40	44	48	52	56
Count	1001	955	785	544	362	196	112

Determine if $f(x) = -1.22x^2 + 73.3x - 213$ or $g(x) = 1001e^{-0.5(x-30)^2/159}$ models the data better using graphing.

39. *Swimming Pool Maintenance*   Chlorine is frequently used to disinfect swimming pools. The chlorine concentration should remain between 1.5 and 2.5 parts per million. On warm, sunny days with many swimmers agitating the water, 30% of the chlorine can dissipate into the air or combine with other chemicals. (Source: D. Thomas, *Swimming Pool Operators Handbook.*)
  (a) Let $f(x) = 2.5(0.7)^x$ model the amount of chlorine in a pool after x days. What is the initial concentration of chlorine in the pool?
  (b) If no chlorine is added, estimate graphically and numerically the number of days before chlorine should be added.

40. *Filters*   Impurities from water are frequently removed using filters. Suppose that a one-inch filter allows 10% of the impurities to pass through it. The other 90% is trapped in the filter.

(a) Make a table of the percentage of the impurities passing through x filters. Let $x = 1, 2, 3, 4,$ and 5.

(b) If only 0.001% of the contaminants are allowed to pass through, how many filters should be used?

41. *Computer Growth*   Between 1971 and 1995 the numbers of transistors that could be placed on a single chip is shown in the table and modeled by $f(x) = 0.00168(1.39)^{x-1970}$. The input x is the year and the output $f(x)$ is the number of transistors in millions. (Source: Intel.)

Year	Chip	Transistors (millions)
1971	4004	0.0023
1986	386DX	0.275
1989	486DX	1.2
1993	Pentium	3.3
1995	P6	5.5

(a) Graph f and the data pairs (year, transistors) in $[1968, 1998, 2]$ by $[-1, 6, 1]$.

(b) Use $f(x)$ to estimate an average annual percentage increase in the number of transistors between 1971 and 1995.

42. *Incubation Time of AIDS*   With some diseases such as AIDS, there is a delay time between the initial time of infection and the time when symptoms begin to occur. This delay time can vary greatly from one individual to the next, but it appears to have an overall pattern. In one study $N(x) = N_0 e^{-0.1x^2/2}$ models the delay time for AIDS patients receiving infected blood transfusions between 1978 and 1986. The formula $N(x)$ computes the number of people who have not developed AIDS x years after their initial infection. (Source: G. Medley, "Incubation period of AIDS in patients infected via blood transfusions.")

(a) Evaluate $N(0)$ and interpret N_0.

(b) Let $N_0 = 100$. Graphically approximate the number of years before 50% of the patients have developed symptoms of AIDS.

43. *Tree Density*   Ecologists studied the spacing between individual trees in a forest found in British Columbia. This lodgepole pine forest was 40–50 years old and had approximately 1600 trees per acre that were randomly spaced. The probability or likelihood that there is at least one tree located in a circle with a radius of x feet could be estimated by $P(x) = 1 - e^{-0.1144x}$. For example, $P(4) \approx 0.37$ means that if a person picks a point at random in the forest, there is a 37% chance that at least one tree

will be located within 4 feet. See the accompanying figure. (Source: E. Pielou, *Populations and Community Ecology.*)

(a) Evaluate $P(2)$ and $P(20)$, and interpret the results.

(b) Graph P. Explain verbally why it is logical for P to be an increasing function.

(c) Solve $P(x) = 0.5$ and interpret the result.

44. *E. Coli Growth*   (Refer to Example 7.)

(a) Approximate the number of E. coli after 3 hours.

(b) Estimate graphically the elapsed time when there are 10 million bacteria per milliliter.

45. *Half-Life*   When a material undergoes radioactive decay, each atom changes its atomic structure by emitting tiny particles and X-rays. Each type of radioactive material has its own unique **half-life,** the time required for half of the atoms in the initial sample to decay. Radioactive cesium-137 was emitted in large amounts in the Chernobyl nuclear power station accident in Russia on April 26, 1986. The amount of cesium remaining after x years in an initial sample of 100 milligrams can be described by $A(x) = 100e^{-0.02295x}$. (Source: C. Mason, *Biology of Freshwater Pollution.*)

(a) How much is remaining after 50 years? Is the half-life of cesium more or less than 50 years?

(b) Estimate graphically and numerically the half-life of cesium-137.

46. *Radioactive Carbon-14*   (Refer to the previous exercise.) Radioactive carbon-14 is found in all living things and is used to determine the age of plants and animals that have died in the past. The amount of carbon-14 remaining after x years is given by $A(x) = A_0 e^{-0.000121x}$, where A_0 is the initial amount of carbon-14. Determine the half-life of carbon-14. (Source: C. Mason.)

47. *Thickness of Runways*   Heavier aircraft require runways with thicker pavement for landings and takeoffs. A pavement 6 inches thick can accommodate an aircraft weighing 80,000 pounds, whereas a 12-inch pavement is necessary for a 350,000-pound plane. The relation between pavement thickness x in inches and gross weight y in thousands of pounds can be modeled by $y = 18.29(1.279)^x$. Complete the table. Round values to the nearest thousand pounds. (Source: Federal Aviation Administration.)

x (inches)	6	7.5	9	10.5	12
y (lb × 1000)	80	?	?	?	350

48. *Survival of Reindeer*   For all types of animals, the percentage that survive into the next year decreases. In one study, the survival rate of a sample of reindeer was modeled by $S(t) = 100(0.999993)^{t^5}$. The function S outputs the percentage of reindeer that survive t years. (Source: D. Brown.)

(a) Evaluate $S(4)$ and $S(15)$. Interpret the results.
(b) Graph S in [0, 15, 5] by [0, 110, 10]. Interpret the graph.

49. *Trains and Horsepower*   The faster a locomotive travels, the more horsepower is needed. The function given by $H(x) = 0.157(1.033)^x$ calculates this horsepower for a level track. The input x is in miles per hour and the output $H(x)$ is the horsepower required per ton of cargo. (Source: L. Haefner, *Introduction to Transportation Systems.*)
(a) Graph H in [0, 100, 10] by [0, 5, 1]. Is H increasing or decreasing?
(b) Evaluate $H(30)$ and interpret the result.
(c) Determine the horsepower needed to move a 5000-ton train 30 miles per hour.
(d) Some types of locomotives are rated for 1350 horsepower. How many locomotives of this type would be needed in part (c)?

50. *Drive or Fly?*   Is there a relation between when a person decides to travel by plane rather than by car? In a study of one city, it was found $f(x) = \dfrac{100.2}{1 + 0.00183(853)^x}$ modeled this decision process. The input x is the ratio of driving time to flying time. The output $f(x)$ is the percentage of people who would drive. For example, $f(0.6) \approx 91$ means that if driving a car requires 0.6 of the time to fly, 91% of people would drive. (Source: Department of Commerce; L. Haefner.)
(a) Evaluate $f(1.3)$ and interpret the result.
(b) Estimate graphically the ratio x when 50% of the people would drive.

51. *Heart Disease*   As age increases, so does the likelihood of coronary heart disease (CHD). The fraction of people x years old with some CHD is modeled by $f(x) = \dfrac{0.9}{1 + 271(0.885)^x}$. (Source: D. Hosmer and S. Lemeshow, *Applied Logistic Regression.*)
(a) Evaluate $f(25)$ and $f(65)$. Interpret the results.
(b) At what age does this likelihood equal 50%?

52. *Modeling Tree Growth*   The height of a tree in feet after x years is modeled by $f(x) = \dfrac{50}{1 + 47.5e^{-0.22x}}$.
(a) Table f starting at $x = 10$, incrementing by 10. What appears to be the maximum height of the tree?
(b) Graph f and identify the horizontal asymptote. Explain its significance.

Writing about Mathematics

1. Give one example of an exponential function f. Find verbal, graphical, numerical, and symbolic representations of f.

2. Discuss the domain and range of an exponential function f. Is f one-to-one? Explain.

4.4 Logarithmic Functions

Representations of the Common Logarithmic Function • Solving Equations • Logarithms with Other Bases • General Logarithmic Functions and Equations • Properties of Logarithms • Graphs of Logarithmic Functions

Introduction

In Chapter 1, numbers were introduced as a means to measure and quantify data. Then, functions were developed to model and describe data. In Chapters 2 and 3, applications involving functions frequently resulted in equations. Although we have solved many equations, one equation that we have not solved *symbolically* is $a^x = k$. This exponential equation occurs frequently in applications. This section discusses logarithmic functions and how they can be used to solve exponential equations.

Representations of the Common Logarithmic Function

In applications measurements can vary greatly in size. Table 4.15 lists some examples with the approximate distances across each.

TABLE 4.15

Object	Distance (m)
Atom	10^{-9}
Protozoa	10^{-4}
Small Asteroid	10^2
Earth	10^7
Universe	10^{26}

Source: C. Ronan, *The Natural History of the Universe.*

Each distance is listed in the form 10^k for some k. The value of k distinguishes one measurement from another. The *common logarithmic function* or *base-10 logarithmic function* denoted by *log* outputs k if the input x can be expressed as 10^k for some real number k. For example, $\log 10^{-9} = -9$, $\log 10^7 = 7$, and $\log 10^\pi = \pi$. A partial numerical representation of $f(x) = \log x$ is given in Table 4.16.

TABLE 4.16

x	10^{-4}	$10^{-3.2}$	10^{-2}	10^{-1}	$10^{-0.4}$	10^0	$10^{0.5}$	10^1	10^2	$10^{4.1}$
$\log x$	-4	-3.2	-2	-1	-0.4	0	0.5	1	2	4.1

The points $(10^{-1}, -1)$, $(10^0, 0)$, $(10^{0.5}, 0.5)$, and $(10^1, 1)$ are located on the graph of $y = \log x$. They are plotted in Figure 4.47. Any *positive* real number x can be expressed as $x = 10^k$ for some real number k. In this case $\log x = \log 10^k = k$. The graph of $y = \log x$ is a continuous curve as shown in Figure 4.48. The y-axis is a vertical asymptote. The common logarithmic function is one-to-one and always increasing. Its domain is all positive real numbers, and its range is all real numbers.

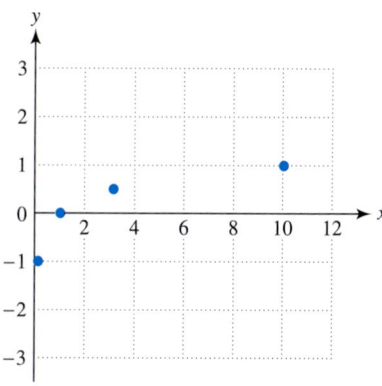

FIGURE 4.47

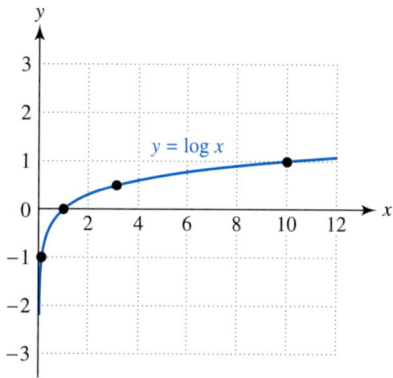

FIGURE 4.48 The Common Logarithmic Function

We have shown *verbal, numerical, graphical,* and *symbolic* representations of the common logarithmic function. A formal definition of the common logarithm is now given.

Common logarithm

The **common logarithm of a positive number** x, denoted by log x, is defined by

$$\log x = k \text{ if and only if } x = 10^k,$$

where k is a real number. The function given by

$$f(x) = \log x$$

is called the **common logarithmic function.**

The common logarithmic function outputs an *exponent k,* which may be positive, negative, or zero. However, a valid input must be positive.

EXAMPLE 1   *Evaluating common logarithms*

Simplify each logarithm.
(a) log 1
(b) $\log \dfrac{1}{1000}$
(c) $\log \sqrt{10}$
(d) log 12

Solution
(a) Since $1 = 10^0$, $\log 1 = \log 10^0 = 0$, as supported by Table 4.16.
(b) $\log \dfrac{1}{1000} = \log 10^{-3} = -3$.
(c) $\log \sqrt{10} = \log\left(10^{1/2}\right) = \dfrac{1}{2}$.
(d) To evaluate log 12, find k so that $10^k = 12$. Since $10^1 = 10$ and $10^2 = 100$, log 12 lies between 1 and 2. Graph $Y_1 = 10^{\wedge}X$ and $Y_2 = 12$ as shown in Figure 4.49 on the next page. The point of intersection is near (1.0791812, 12). Thus, log 12 ≈ 1.0791812.

$[-3, 3, 1]$ by $[0, 15, 1]$

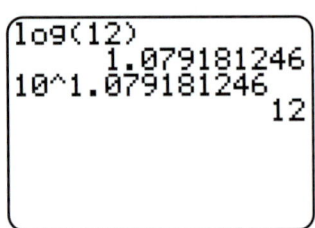

FIGURE 4.49 ∎

Much like the square root function, the common logarithmic function does not have an easy-to-evaluate formula. For example, $\sqrt{4} = 2$ and $\sqrt{100} = 10$ can be calculated mentally, but for $\sqrt{2}$ we usually rely on a calculator. Similarly, $\log 100 = 2$ can be found mentally, whereas $\log 12$ can be approximated using a calculator. See Figure 4.50. To check that $\sqrt{2} \approx 1.414$, multiply $1.414 \times 1.414 \approx 2$. To check that $\log 12 \approx 1.0791812$, evaluate $10^{1.0791812} \approx 12$. Another similarity between the square root function and the common logarithmic function is that their domains do not include negative numbers. If outputs are restricted to real numbers, both $\sqrt{-3}$ and $\log(-3)$ are undefined expressions.

```
log(12)
            1.079181246
10^1.079181246
                     12
```

FIGURE 4.50

The human ear is extremely sensitive and is able to detect pressures on the eardrum ranging from 10^{-16} w/cm² (watts per square centimeter) to 10^{-4} w/cm². Sound with an intensity of 10^{-4} w/cm² is painful to the human eardrum. Because of this wide range of intensities, scientists use logarithms to measure sound in decibels.

EXAMPLE 2 *Applying common logarithms to sound*

Sound levels in decibels (db) can be computed by $f(x) = 10 \log(10^{16}x)$, where x is the intensity of the sound in watts per square centimeter. (Source: R. Weidner and R. Sells, *Elementary Classical Physics.*)

(a) At what decibel level does the threshold for pain occur?
(b) Table f for the intensities $x = 10^{-13}, 10^{-12}, 10^{-11}, \ldots, 10^{-4}$.
(c) If the intensity x increases by a factor of 10, what is the corresponding increase in decibels?

Solution

(a) The threshold for pain occurs when $x = 10^{-4}$.

$$f(10^{-4}) = 10 \log(10^{16} \cdot 10^{-4}) \qquad \text{Substitute } x = 10^{-4}.$$
$$= 10 \log(10^{12}) \qquad \text{Add exponents when multiplying like bases.}$$
$$= 10(12) \qquad \text{Evaluate the logarithm.}$$
$$= 120 \qquad \text{Multiply.}$$

The human eardrum begins to hurt at 120 decibels.

(b) Other values in Table 4.17 are computed in a similar manner.

TABLE 4.17

x (w/cm²)	10^{-13}	10^{-12}	10^{-11}	10^{-10}	10^{-9}	10^{-8}	10^{-7}	10^{-6}	10^{-5}	10^{-4}
$f(x)$ (db)	30	40	50	60	70	80	90	100	110	120

(c) From Table 4.17, a tenfold increase in x results in an increase of 10 decibels. ■

Critical Thinking

If the sound level increases by 60 db, by what factor does the intensity increase?

Solving Equations

The graphs of $y = 10^x$ and $y = \log x$ are shown in Figure 4.51. Notice that they are reflections of each other across the line $y = x$. Both functions are one-to-one.

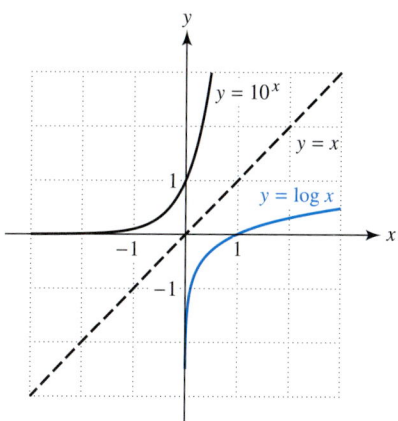

FIGURE 4.51

If $f(x) = 10^x$, then its inverse function is given by $f^{-1}(x) = \log x$. To see this numerically, consider Table 4.18. Notice that $\log 10^x = x$, or equivalently, $f^{-1}(f(x)) = x$ for each input x. In a similar manner, $10^{\log x} = x$ for any positive x.

TABLE 4.18

x	-2	-1	0	1	2
10^x	10^{-2}	10^{-1}	10^0	10^1	10^2

x	10^{-2}	10^{-1}	10^0	10^1	10^2
$\log x$	-2	-1	0	1	2

> ### Inverse properties of the common logarithm
>
> The following inverse properties hold for the common logarithm.
>
> $$\log 10^x = x \qquad \text{for any real number } x, \text{ and}$$
> $$10^{\log x} = x \qquad \text{for any positive number } x.$$

Section 4.2 demonstrated how the inverse function f^{-1} can be used to solve an equation of the form $f(x) = k$, where f is a one-to-one function and k is a constant in the range of f.

$$f(x) = k$$

$$f^{-1}(f(x)) = f^{-1}(k) \qquad \text{Apply } f^{-1} \text{ to both sides.}$$

$$x = f^{-1}(k) \qquad \text{Property of inverse functions}$$

For example, let $f(x) = 10^x$ and $f^{-1}(x) = \log x$.

$$10^x = 5 \qquad f(x) = k$$

$$\log 10^x = \log 5 \qquad f^{-1}(f(x)) = f^{-1}(k)$$

$$x = \log 5 \qquad x = f^{-1}(k)$$

Converting $10^x = 5$ to the equivalent equation $x = \log 5$ is sometimes referred to as changing *exponential form* to *logarithmic form*. In the second step, the common logarithm is taken of both sides of the equation. We include this second step to emphasize the fact that properties of inverse functions are being used to solve these equations.

EXAMPLE 3   *Solving equations of the form $10^x = k$*

Solve each equation, if possible.
(a) $10^x = 0.001$
(b) $10^x = 55$
(c) $10^x = -5$

Solution

(a) Take the common logarithm of both sides of the equation $10^x = 0.001$. Then,

$$\log 10^x = \log 0.001 \qquad \text{or} \qquad x = \log 10^{-3} = -3.$$

(b) In a similar manner, $10^x = 55$ is equivalent to $x = \log 55 \approx 1.7404$.
(c) The equation $10^x = -5$ has no real solution, because -5 is not in the range of 10^x. Figure 4.52 shows the graphs of $Y_1 = 10^{\wedge}X$ and $Y_2 = -5$.

$$[-6, 6, 1] \text{ by } [-6, 6, 1]$$

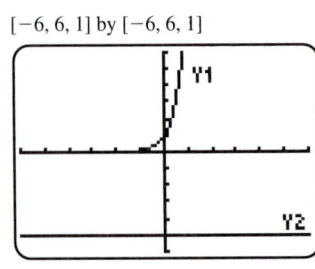

FIGURE 4.52

We can use a similar technique to solve the equation $\log x = k$. If $f(x) = \log x$, then $f^{-1}(x) = 10^x$ and

$$f^{-1}(f(x)) = f^{-1}(\log x) = 10^{\log x} = x.$$

To solve $\log x = k$, apply $f^{-1}(x) = 10^x$ to both sides of the equation. This is equivalent to having each side of the equation become an exponent with a base of 10. When an exponential function is applied to both sides of an equation, the phrase "exponentiate both sides" may be used.

$$\log x = k$$
$$10^{\log x} = 10^k \qquad \text{Exponentiate both sides.}$$
$$x = 10^k \qquad \text{Inverse property}$$

Converting the equation $\log x = k$ to the equivalent equation $x = 10^k$ is called changing *logarithmic form* to *exponential form*.

EXAMPLE 4   *Solving equations of the form log x = k*

Solve each equation.
(a) $\log x = 3$
(b) $\log x = -2$
(c) $\log x = 2.7$

Solution

(a)
$$\log x = 3 \qquad \text{Given equation}$$
$$10^{\log x} = 10^3 \qquad \text{Exponentiate both sides.}$$
$$x = 10^3 \qquad \text{Inverse property}$$
$$x = 1000 \qquad \text{Simplify.}$$

(b) Similarly, $\log x = -2$ is equivalent to $x = 10^{-2} = 0.01$.
(c) $\log x = 2.7$ is equivalent to $x = 10^{2.7} \approx 501.2$. Graphical support is shown in Figure 4.53, where the graphs of $Y_1 = \log X$ and $Y_2 = 2.7$ intersect near $(501.2, 2.7)$.

[0, 700, 100] by [0, 4, 1]

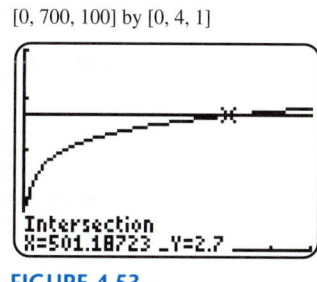

FIGURE 4.53   ∎

Air pollutants frequently cause acid rain. A measure of the acidity is pH, which ranges between 1 and 14. Pure water is neutral and has a pH of 7. Acidic solutions have a pH less than 7, whereas alkaline solutions have a pH greater than 7. A pH value measures the concentration of the hydrogen ions in a solution. It can be computed by $f(x) = -\log x$, where x represents the hydrogen ion concentration in moles per liter. Pure water exposed to normal carbon dioxide in the atmosphere has a pH of 5.6. If the pH of a lake drops below this level, it is indicative of an *acid lake*.

EXAMPLE 5   *Analyzing acid rain*

In rural areas of Europe, rainwater typically has a hydrogen ion concentration of $x = 10^{-4.7}$. (Source: G. Howells, *Acid Rain and Acid Water.*)

(a) Find its pH. What effect might this have on a lake with a pH of 5.6?

(b) Seawater has a pH of 8.2. Compared to seawater, how many times greater is the hydrogen ion concentration in rainwater from rural Europe?

Solution

(a) $f(10^{-4.7}) = -\log 10^{-4.7} = -(-4.7) = 4.7$. The pH is 4.7, which could cause the lake to become more acidic.

(b) To find the hydrogen ion concentration in seawater, solve $f(x) = 8.2$ for x.

$$-\log x = 8.2 \qquad \text{\color{blue}{$f(x) = 8.2$}}$$
$$\log x = -8.2 \qquad \text{\color{blue}{Multiply by -1.}}$$
$$10^{\log x} = 10^{-8.2} \qquad \text{\color{blue}{Exponentiate both sides.}}$$
$$x = 10^{-8.2} \qquad \text{\color{blue}{Inverse property}}$$

Since $\dfrac{10^{-4.7}}{10^{-8.2}} = 10^{3.5}$, the hydrogen ion concentration in the rainwater is $10^{3.5} \approx 3162$ times greater than it is in seawater. ∎

Logarithms with Other Bases

The development of base-a logarithms can be done with any positive base $a \neq 1$. For example, in computer science base-2 logarithms are frequently used. A partial numerical representation of the base-2 logarithmic function, denoted by $f(x) = \log_2 x$, is shown in Table 4.19. If x can be expressed in the form $x = 2^k$ for some k, then $\log_2 x = k$.

TABLE 4.19

x	$2^{-3.1}$	2^{-2}	$2^{-0.5}$	2^0	$2^{0.5}$	2^2	$2^{3.1}$
$\log_2 x$	-3.1	-2	-0.5	0	0.5	2	3.1

In a similar manner, a partial numerical representation for base-e logarithms is shown in Table 4.20. The base-e logarithm is referred to as the **natural logarithm** and denoted by either $\log_e x$ or $\ln x$. Natural logarithms are used in mathematics, science, economics, and technology.

TABLE 4.20

x	$e^{-3.1}$	e^{-2}	$e^{-0.5}$	e^0	$e^{0.5}$	e^2	$e^{3.1}$
$\ln x$	-3.1	-2	-0.5	0	0.5	2	3.1

Critical Thinking

Make a partial numerical representation for a base-4 logarithm. Evaluate $\log_4 16$.

A base-*a* logarithm is now defined.

> ## Logarithm
>
> The **logarithm with base *a* of a positive number *x*,** denoted by $\log_a x$, is defined by
>
> $$\log_a x = k \text{ if and only if } x = a^k,$$
>
> where $a > 0$, $a \neq 1$, and k is a real number. The function, given by
>
> $$f(x) = \log_a x,$$
>
> is called the **logarithmic function with base *a*.**

Remember that *a logarithm is an exponent.* The expression $\log_a x$ is the exponent k such that $a^k = x$. Logarithms with base *a* satisfy inverse properties similar to those for common logarithms.

> ## Inverse properties
>
> The following inverse properties hold for logarithms with base *a*.
>
> $$\log_a a^x = x \qquad \text{for any real number } x, \text{ and}$$
> $$a^{\log_a x} = x \qquad \text{for any positive number } x.$$

EXAMPLE 6   *Evaluating logarithms*

Evaluate each logarithm.
(a) $\log_2 8$

(b) $\log_5 \dfrac{1}{25}$

(c) $\log_7 49$
(d) $\ln e^{-7}$

Solution
(a) To determine $\log_2 8$, express 8 as 2^k for some k. Since $8 = 2^3$,
$\log_2 8 = \log_2 2^3 = 3$.

(b) $\log_5 \dfrac{1}{25} = \log_5 \dfrac{1}{5^2} = \log_5 5^{-2} = -2$

(c) $\log_7 49 = \log_7 7^2 = 2$
(d) $\ln e^{-7} = \log_e e^{-7} = -7$ ■

General Logarithmic Functions and Equations

If $f(x) = a^x$ then f is a one-to-one function. Its inverse is given by $f^{-1}(x) = \log_a x$. To solve the equation $a^x = k$, take the base-*a* logarithm of both sides.

EXAMPLE 7   *Solving equations of the form $a^x = k$*

Solve each equation.

(a) $3^x = \dfrac{1}{27}$

(b) $e^x = 5$

Solution

(a) $3^x = \dfrac{1}{27}$ Given equation

   $\log_3 3^x = \log_3 \dfrac{1}{27}$ Take the base-3 logarithm of both sides.

   $\log_3 3^x = \log_3 3^{-3}$ Properties of exponents

   $x = -3$ Inverse property: $\log_a a^k = k$

(b) Take the natural logarithm of both sides. Then,

$$\ln e^x = \ln 5 \text{ is equivalent to } x = \ln 5 \approx 1.6094.$$

Many calculators are able to compute natural logarithms. The evaluation of $\ln 5$ is shown in Figure 4.54. Notice that $e^{1.609437912} \approx 5$.

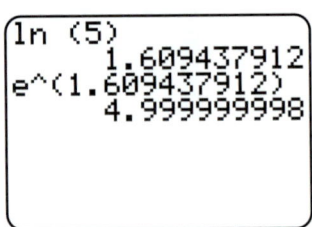

FIGURE 4.54   ■

To solve the equation $\log_a x = k$, exponentiate both sides of the equation using base a. This is illustrated in the next example.

EXAMPLE 8   *Solving equations of the form $\log_a x = k$*

Solve each equation.
(a) $\log_2 x = 5$
(b) $\log_5 x = -2$
(c) $\ln x = 4.3$

Solution

(a) $\log_2 x = 5$ Given equation

   $2^{\log_2 x} = 2^5$ Exponentiate both sides using the base 2.

   $x = 2^5$ Inverse property: $a^{\log_a x} = x$

   $x = 32$ Simplify.

(b) In a similar manner, $\log_5 x = -2$ is equivalent to $x = 5^{-2} = \dfrac{1}{25}$.

(c) $\ln x = 4.3$ is equivalent to $x = e^{4.3} \approx 73.7$.   ■

Properties of Logarithms

The discovery of logarithms by John Napier (1550–1617) played an important role in the history of science. Logarithms were instrumental for Johannes Kepler (1571–1630) to calculate the positions of the planet Mars, which led to his discovery of the laws of planetary motion. Kepler's laws were used by Isaac Newton (1642–1727) to discover the universal laws of gravitation. Although calculators and computers have made tables of logarithms obsolete, applications involving logarithms still play an important role in modern day computation. One reason for this is that logarithms possess several important properties.

Properties of logarithms

For positive numbers m, n, and $a \neq 1$, and any real number r:

1. $\log_a 1 = 0$
2. $\log_a m + \log_a n = \log_a (mn)$
3. $\log_a m - \log_a n = \log_a \left(\dfrac{m}{n} \right)$
4. $\log_a (m^r) = r \log_a m$

The properties of logarithms are a direct result of the properties of exponents and the inverse property $\log_a a^k = k$ as shown in the following.

Property 1:   Since $a^0 = 1$, it follows that $\log_a 1 = \log_a a^0 = 0$.

Examples: $\log 1 = \log 10^0 = 0$ and $\ln 1 = \ln e^0 = 0$.

Property 2:   If m and n are positive numbers, then we can write $m = a^c$ and $n = a^d$ for some real numbers c and d.

$$\log_a m + \log_a n = \log_a a^c + \log_a a^d = c + d$$
$$\log_a (mn) = \log_a (a^c a^d) = \log_a (a^{c+d}) = c + d$$

Thus, $\log_a m + \log_a n = \log_a (mn)$.

Example: Let $m = 100$ and $n = 1000$.

$$\log m + \log n = \log 100 + \log 1000 = \log 10^2 + \log 10^3 = 2 + 3 = 5$$
$$\log (mn) = \log (100 \cdot 1000) = \log 100{,}000 = \log 10^5 = 5$$

Property 3:   Let $m = a^c$ and $n = a^d$ for some real numbers c and d.

$$\log_a m - \log_a n = \log_a a^c - \log_a a^d = c - d$$
$$\log_a \left(\frac{m}{n} \right) = \log_a \left(\frac{a^c}{a^d} \right) = \log_a (a^{c-d}) = c - d$$

Thus, $\log_a m - \log_a n = \log_a \left(\dfrac{m}{n} \right)$.

Example: Let $m = 100$ and $n = 1000$.

$$\log m - \log n = \log 100 - \log 1000 = \log 10^2 - \log 10^3 = 2 - 3 = -1$$
$$\log \left(\frac{m}{n} \right) = \log \left(\frac{100}{1000} \right) = \log \left(\frac{1}{10} \right) = \log (10^{-1}) = -1$$

Property 4: Let $m = a^c$ and r be any real number.

$$\log_a m^r = \log_a (a^c)^r = \log_a (a^{cr}) = cr$$
$$r \log_a m = r \log_a a^c = rc$$

Thus, $\log_a (m^r) = r \log_a m$.

Example: Let $m = 100$ and $r = 3$.

$$\log m^r = \log 100^3 = \log 1{,}000{,}000 = \log 10^6 = 6$$
$$r \log m = 3 \log 100 = 3 \log 10^2 = 3 \cdot 2 = 6$$

Caution: $\log_a (m + n) \neq \log_a m + \log_a n;$ $\log_a(m - n) \neq \log_a m - \log_a n$

EXAMPLE 9   *Analyzing sound with decibels*

In Example 2 sound levels in decibels (db) were computed by $f(x) = 10 \log (10^{16} x)$.
(a) Use the properties of logarithms to simplify the symbolic representation for f.
(b) Ordinary conversation has an intensity of $x = 10^{-10}$ w/cm^2. Find the decibel level.

Solution
(a) To simplify the formula use Property 2.

$$
\begin{aligned}
f(x) &= 10 \log (10^{16} x) \\
&= 10(\log 10^{16} + \log x) &&\text{Property 2} \\
&= 10(16 + \log x) &&\text{Evaluate the logarithm.} \\
&= 160 + 10 \log x &&\text{Distributive property}
\end{aligned}
$$

(b) $f(10^{-10}) = 160 + 10 \log (10^{-10}) = 160 + 10(-10) = 160 - 100 = 60$
Ordinary conversation occurs at about 60 decibels. ∎

The next example demonstrates how the properties of logarithms can be used to simplify logarithmic expressions.

EXAMPLE 10   *Applying properties of logarithms*

Write each expression as one term.

(a) $\ln 2e + \ln \dfrac{1}{e}$

(b) $\log_2 27 + \log_2 x^3$

(c) $\log x^3 - \log x^2$

Solution
(a) By Property 2, $\ln 2e + \ln \dfrac{1}{e} = \ln \left(2e \cdot \dfrac{1}{e} \right) = \ln 2.$

(b) By Properties 2 and 4, $\log_2 27 + \log_2 x^3 = \log_2 (27 x^3) = \log_2 (3x)^3 = 3 \log_2 (3x).$

(c) By Property 3, $\log x^3 - \log x^2 = \log \dfrac{x^3}{x^2} = \log x$. Property 4 could also be used as follows.

$$\log x^3 - \log x^2 = 3 \log x - 2 \log x = \log x \qquad \blacksquare$$

EXAMPLE 11   *Applying properties of logarithms*

Expand each expression.
(a) $\log 2x^4$
(b) $\ln \dfrac{7x^3}{k}$

Solution

(a)
$$\begin{aligned} \log 2x^4 &= \log 2 + \log x^4 && \text{Property 2}\\ &= \log 2 + 4 \log x && \text{Property 4} \end{aligned}$$

(b)
$$\begin{aligned} \ln \frac{7x^3}{k} &= \ln 7x^3 - \ln k && \text{Property 3}\\ &= \ln 7 + \ln x^3 - \ln k && \text{Property 2}\\ &= \ln 7 + 3 \ln x - \ln k && \text{Property 4} \end{aligned} \qquad \blacksquare$$

Occasionally it is necessary to evaluate or graph a logarithmic function with a base other than 10 or e. This can be done using a change of base formula.

Change of base formula

Let x, $a \neq 1$, and $b \neq 1$ be positive real numbers. Then,

$$\log_a x = \frac{\log_b x}{\log_b a}.$$

The change of base formula can be derived as follows.

$$\begin{aligned} y &= \log_a x\\ a^y &= a^{\log_a x} && \text{Exponentiate both sides.}\\ a^y &= x && \text{Inverse property}\\ \log_b a^y &= \log_b x && \text{Take base-}b\text{ logarithm of both sides.}\\ y \log_b a &= \log_b x && \text{Property 4}\\ y &= \frac{\log_b x}{\log_b a} && \text{Divide by } \log_b a.\\ \log_a x &= \frac{\log_b x}{\log_b a} && \text{Substitute } \log_a x \text{ for } y. \end{aligned}$$

To calculate $\log_2 5$, evaluate $\dfrac{\log 5}{\log 2} \approx 2.322$. The change of base formula was used with $x = 5$, $a = 2$, and $b = 10$. We could have also evaluated $\dfrac{\ln 5}{\ln 2} \approx 2.322$. Similarly, $\log_7 81 = \dfrac{\log 81}{\log 7} \approx 2.258$.

Graphs of Logarithmic Functions

Figure 4.55 shows graphs of $f(x) = 2^x$ and $f^{-1}(x) = \log_2 x$. The graph of f^{-1} is a reflection of f in the line $y = x$. The rapid growth in f is called *exponential growth.* On the other hand, the graph of f^{-1} begins to level off. This slow rate of growth is called *logarithmic growth.* Since the range of f is all positive numbers, the domain of f^{-1} is all positive numbers. Notice that when $0 < x < 1$, $f^{-1}(x) = \log_2 x$ outputs negative numbers, and when $x > 1$, $f^{-1}(x)$ outputs positive numbers.

Graphs of $g(x) = e^x$ and $g^{-1}(x) = \ln x$ are shown in Figure 4.56. Since $e \approx 2.7183 > 2$, the graph of g increases faster than f, while the graph of g^{-1} levels off faster than the graph of f^{-1}. Similar results occur when the base increases to 10. The graphs of $h(x) = 10^x$ and $h^{-1}(x) = \log x$ are shown in Figure 4.51.

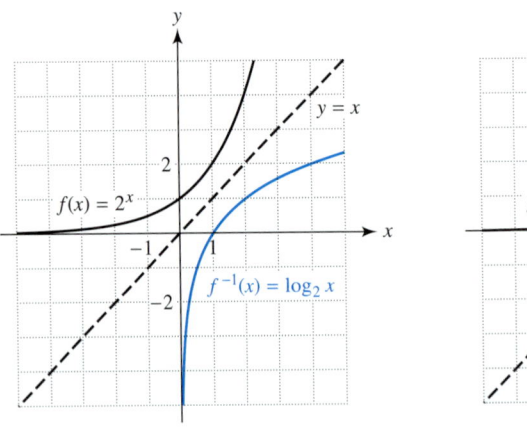

FIGURE 4.55   **FIGURE 4.56**

The change of base formula can be used to graph logarithmic functions with bases other than 10 or e.

EXAMPLE 12

Using the change of base formula

Estimate graphically the solution to the equation $\log_2 (x^3 + x - 1) = 5$.

Solution

Graph $Y_1 = \log (X^3 + X - 1)/\log (2)$ and $Y_2 = 5$. See Figure 4.57. Their graphs intersect near the point (3.104, 5). The solution is $x \approx 3.104$.

$[-10, 10, 1]$ by $[-10, 10, 1]$

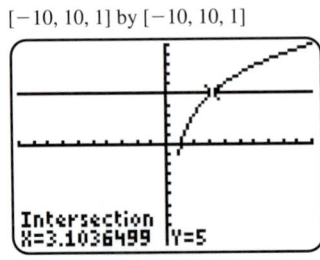

Intersection
X=3.1036499 Y=5

FIGURE 4.57

4.4 PUTTING IT ALL TOGETHER

The following table summarizes some important concepts about base-*a* logarithms. Common and natural logarithms satisfy the same properties.

Logarithms with Base *a* of a Positive Number *x*
Notation: $\log_a x$ ($a > 0$ and $a \neq 1$)
Meaning: $\log_a x = k$ if and only if $a^k = x$.
Example: $\log_2 16 = 4$, since $2^4 = 16$.
Inverse: $f(x) = \log_a x$ and $g(x) = a^x$ are inverse functions.
Inverse properties: $\log_a a^x = x$ and $a^{\log_a x} = x$

Change of Base Formula
Let x, $a \neq 1$, and $b \neq 1$ be positive real numbers. Then,
$$\log_a x = \frac{\log_b x}{\log_b a}.$$
Example: The expression $\log_3 6$ is equivalent to either
$$\frac{\ln 6}{\ln 3} \quad \text{or} \quad \frac{\log 6}{\log 3}.$$

Properties of Logarithms
For positive numbers m, n, and $a \neq 1$, and any real number r:
1. $\log_a 1 = 0$
2. $\log_a m + \log_a n = \log_a (mn)$
3. $\log_a m - \log_a n = \log_a \left(\dfrac{m}{n}\right)$
4. $\log_a (m^r) = r \log_a m$

4.4 EXERCISES

  Tape 5

Common Logarithms

Exercises 1 and 2: Evaluate the following expressions mentally. Check your results with a calculator.

1. **(a)** $\log 10$ **(b)** $\log 10{,}000$
    (c) $20 \log 0.1$ **(d)** $\log 10 + \log 0.001$

2. **(a)** $\log 100$ **(b)** $\log 1{,}000{,}000$
    (c) $5 \log 0.01$ **(d)** $\log 0.1 - \log 1000$

Exercises 3 and 4: Determine mentally an integer n so that the logarithm is between n and n + 1. Check your result with a calculator.

3. **(a)** $\log 79$ **(b)** $\log 500$
    (c) $\log 5$ **(d)** $\log 0.5$

4. **(a)** $\log 63$ **(b)** $\log 5000$
    (c) $\log 9$ **(d)** $\log 0.04$

5. Solve each equation.
    (a) $10^x = 1000$
    (b) $10^x = 1{,}000{,}000$
    (c) $10^x = 0.01$
    (d) $10^x = 0.0001$

6. Evaluate each expression. Explain how the results are related to the previous exercise.
    (a) $\log 1000$
    (b) $\log 1{,}000{,}000$
    (c) $\log 0.01$
    (d) $\log 0.0001$

7. Solve each equation symbolically. Support your results graphically and numerically.
    (a) $10^x = 250$
    (b) $10^x = 4$
    (c) $10^x = 0.5$

8. Find the exact value of each expression. Use a calculator to support your results.
   (a) $\log \sqrt{1000}$
   (b) $\log \sqrt[3]{10}$
   (c) $\log \sqrt{10} + \log \sqrt[5]{0.1}$
   (d) $\log \sqrt{0.02} - \log \sqrt{2}$

General Logarithms

Exercises 9–16: Evaluate the logarithm.

9. $\log_2 64$

10. $\log_2 \dfrac{1}{4}$

11. $\log_4 2$

12. $\log_3 9$

13. $\ln 1$

14. $\ln e$

15. $\log_a \dfrac{1}{a}$

16. $\log_a (a^2 \cdot a^3)$

Exercises 17–20: Complete the table without the aid of a calculator.

17. $f(x) = \log_2 x$

x	$\dfrac{1}{16}$	1	8	32
$f(x)$				

18. $f(x) = \log_4 x$

x	$\dfrac{1}{4}$	4	16	64
$f(x)$				

19. $f(x) = 2 \log_2 (x - 5)$

x	5	6	7	21
$f(x)$				

20. $f(x) = 2 \log_3 (2x)$

x	$\dfrac{1}{18}$	$\dfrac{3}{2}$	$\dfrac{9}{2}$
$f(x)$			

Exercises 21–24: Use the change of base formula to approximate each logarithm to four significant digits.

21. $\log_2 25$

22. $\log_3 67$

23. $\log_5 130$

24. $\log_6 0.77$

Solving Equations

Exercises 25–32: Solve the equation symbolically. Support your result graphically and numerically.

25. $2^x = 72$

26. $3^x = 101$

27. $5^x = 0.25$

28. $4^x = 11$

29. $e^x = 25$

30. $e^x = 0.23$

31. $e^{-x} = 3$

32. $e^{-x} = \dfrac{1}{2}$

Exercises 33–38: Solve the equation symbolically. Support your result graphically and numerically.

33. $\log x = 2.3$

34. $\log_2 x = -3$

35. $\log_2 x = 1.2$

36. $\log_4 x = 3.7$

37. $\ln x = -2$

38. $\ln x = 2$

Properties of Logarithms

Exercises 39–46: Write each expression as one term.

39. $\log 2 + \log 3$

40. $\log \sqrt{2} + \log \sqrt[3]{2}$

41. $\ln \sqrt{5} - \ln 25$

42. $\ln 33 - \ln 11$

43. $\log \sqrt{x} + \log x^2 - \log x$

44. $\log \sqrt[4]{x} + \log x^4 - \log x^2$

45. $\ln \dfrac{1}{e^2} + \ln 2e$

46. $\ln 4e^3 - \ln 2e^2$

Exercises 47–52: (Refer to Example 11.) Expand each expression.

47. $\log \dfrac{x^2}{3}$

48. $\log 3x^6$

49. $\ln \dfrac{2x^7}{3k}$

50. $\ln \dfrac{kx^3}{5}$

51. $\log_2 4k^2x^3$

52. $\log \dfrac{5kx^2}{11}$

Exercises 53–60: Complete the following.
   (a) *Table $f(x)$ and $g(x)$ starting at $x = 1$, increment-ing by 1. Conjecture whether $f(x) = g(x)$.*
   (b) *If possible, show that $f(x) = g(x)$.*

53. $f(x) = \log 3x + \log 2x$, $g(x) = \log 6x^2$

54. $f(x) = \log 2 + \log x$, $g(x) = \log 2x$

55. $f(x) = \ln 3x - \ln 2x$, $g(x) = \ln x$

56. $f(x) = \ln 2x^2 - \ln x$, $g(x) = \ln 2x$

57. $f(x) = \log x^4$, $g(x) = 4 \log x$

58. $f(x) = \log x^2 + \log x^3$, $g(x) = 5 \log x$

59. $f(x) = \ln x^4 - \ln x^2$, $g(x) = 2 \ln x$

60. $f(x) = (\ln x)^2$, $g(x) = 2 \ln x$

Graphs of Logarithmic Equations

Exercises 61 and 62: Use the graph of f to sketch a graph of f^{-1}. Give a symbolic representation of f^{-1}.

61. $f(x) = e^x$

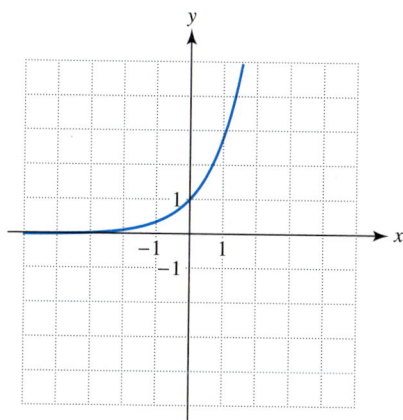

62. $f(x) = \log_4 x$

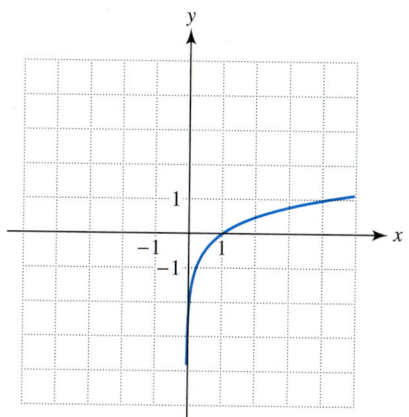

Exercises 63–66: Match the function f with its graph (a–d). Support your answer graphically.

63. $f(x) = \log_2 x$

64. $f(x) = \log x$

65. $f(x) = \log (x + 2)$

66. $f(x) = 2 + \log_2 x$

(a)

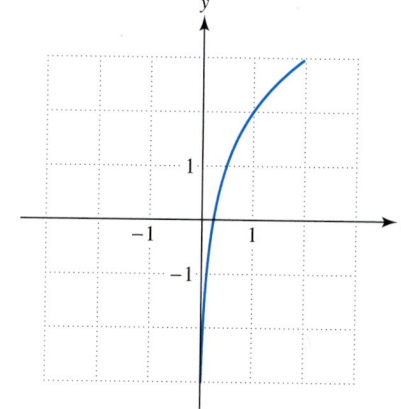

(b)

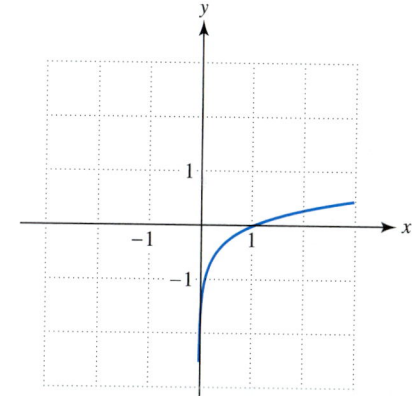

(c)

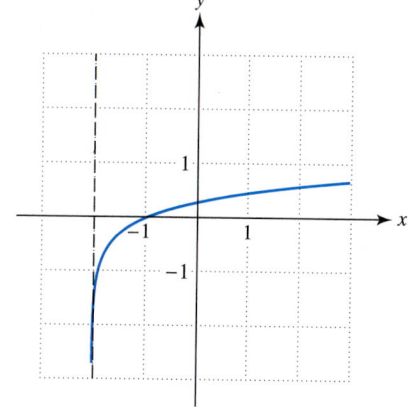

(d)

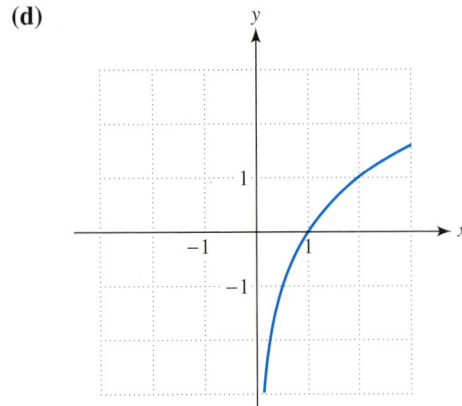

Exercises 67–70: Solve the equation graphically. Express any solutions using four significant digits.

67. $\log_2(x^3 + x^2 + 1) = 7$

68. $\log_3(1 + x^2 + 2x^4) = 4$

69. $\log_2(x^2 + 1) = 5 - \log_3(x^4 + 1)$

70. $\ln(x^2 + 2) = \log_2(10 - x^2)$

Applications

71. *Runway Length* There is a mathematical relation between an airplane's weight x and the runway length required at takeoff. For some airplanes the minimum runway length in thousands of feet may be modeled by $L(x) = 3 \log x$, where x is measured in thousands of pounds. (Source: L. Haefner, *Introduction to Transportation Systems.*)

(a) Graph L in [0, 50, 10] by [0, 6, 1]. Interpret the graph.

(b) If the weight of an airplane increases tenfold from 10,000 to 100,000 pounds, does the length of the required runway also increase by a factor of 10? Explain.

(c) Generalize your answer from part (b).

72. *Modeling Algae Growth* When sewage was accidentally dumped into Lake Tahoe, the concentration of the algae *Selenastrum* increased from 1000 cells per milliliter to approximately 1,000,000 cells

per milliliter within a six-day period. Let f model the algae concentration after x days have elapsed. (Source: A. Payne, "Responses of the three test algae of the algal assay procedure: bottle test.")

(a) Compute the average rate of change in f from 0 to 6 days.

(b) The specific growth rate r is defined by $r = \dfrac{\log N_2 - \log N_1}{x_2 - x_1}$, where N_1 is the algae concentration at time x_1 and N_2 is the algae concentration at time x_2. Compute r in this example.

(c) Conjecture why environmental scientists might use the specific growth rate r, rather than the average rate of change, to describe algae growth.

73. *Acid Rain* (Refer to Example 5.) Find the hydrogen ion concentration for the following pH levels of acid rain. (Source: G. Howells.)

(a) 4.92 (pH of rain at Amsterdam Islands in the Indian Ocean)

(b) 3.9 (pH of rain at some locations in the eastern United States)

74. *Allometry* (See Exercises 81 and 82, Section 3.2.) The equation $y = bx^a$ is used in applications involving allometry. Another form of this equation is $\log y = \log b + a \log x$. Use properties of logarithms to obtain this second equation from the first. (Source: H. Lancaster, *Quantitative Methods in Biological and Medical Sciences.*)

75. *Telecommuting* Some workers use technology such as fax machines, E-mail, computers, and multiple phone lines to work at home, rather than in the office. However, because of the need for teamwork and collaboration in the workplace, fewer employees are telecommuting than expected. The table lists the expected telecommuters in millions from 1997 to 2006. (Source: *USA Today.*)

Year	1997	1998	1999	2000	2001
Telecommuters	9.2	9.6	10.0	10.4	10.6

Year	2002	2003	2004	2005	2006
Telecommuters	11.0	11.1	11.2	11.3	11.4

(a) Conjecture which of the following models the data better,

$$f_1(x) = 9.2(1.03)^{x-1997} \quad \text{or}$$
$$f_2(x) = 9.2 + \ln(x - 1996).$$

(b) Support your conjecture by graphing f_1, f_2, and the data in [1996, 2007, 1] by [8, 12, 1].

76. *Interest* The following table lists the interest rates for certificates of deposit during January 1997. (Source: *USA Today*.)

Time	Yield
6-month	4.75%
1-year	5.03%
$2\frac{1}{2}$-year	5.25%
5-year	5.54%

(a) Let x represent time in years. Plot the data in [0, 6, 1] by [4.7, 5.7, 0.1].

(b) Conjecture which of the following models the yield better,

$$f_1(x) = 5 + 0.33 \ln x \quad \text{or} \quad f_2(x) = 4.8(1.03)^x.$$

(c) Use graphing to support your answer.

77. *Hurricanes* Hurricanes are some of the largest storms on earth. They are very low pressure areas with diameters of over 500 miles. The barometric air pressure in inches of mercury at a distance of x miles from the eye of a typical hurricane is modeled by $f(x) = 0.48 \ln(x + 1) + 27$. (Source: A. Miller and R. Anthes, *Meteorology*.)

(a) Evaluate $f(0)$ and $f(100)$. Interpret the results.

(b) Graph f in [0, 250, 50] by [25, 30, 1]. Describe how air pressure changes as one moves away from the eye of the hurricane.

78. *Predicting Wind Speed* Wind speed typically varies in the first 20 meters above the ground. Close to the ground wind speed is often less than it is at 20 meters above the ground. For this reason, the National Weather Service usually measures wind speeds at heights between 5 and 10 meters. For a particular day, let $f(x) = 1.2 \ln x + 2.3$ compute the wind speed in meters per second at a height x meters above the ground. (Source: A. Miller.)

(a) Find the wind speed at a height of 10 meters.

(b) Graph f in [0, 20, 5] by [0, 7, 1]. Interpret the graph.

(c) Estimate graphically the height where the wind speed is 5 meters per second.

79. *Greenhouse Effect* According to one model, the future increases in average global temperatures can be estimated using $T = 6.5 \ln (C/280)$, where C is the concentration of atmospheric carbon dioxide in parts per million (ppm) and T is in degrees Fahrenheit. Let future amounts of carbon dioxide

be modeled by $C(x) = 364(1.005)^x$, where $x = 0$ corresponds to the year 2000 and $x = 100$ to 2100. (Source: W. Clime, *The Economics of Global Warming*.)

(a) Use composition of functions to write T as a function of x. Evaluate T when $x = 100$ and interpret the result.

(b) Graph $C(x)$ in [0, 200, 50] by [0, 1000, 100] and $T(x)$ in [0, 200, 50] by [0, 10, 1]. Describe each graph.

(c) How does an exponential growth in carbon dioxide concentrations affect the rise in global temperature?

80. *Decibels* If the intensity x of a sound increases by a factor of 10, then the decibel level increases by how much? (*Hint:* Let $f(x) = 160 + 10 \log x$, and evaluate $f(10x)$.)

81. *Earthquakes* The Richter scale is used to measure the intensity of earthquakes. Intensity corresponds to the amount of energy released by an earthquake. If an earthquake has an intensity of x then its *magnitude*, as computed by the Richter scale, is given by $R(x) = \log \dfrac{x}{I_0}$, where I_0 is the intensity of a small, measurable earthquake.

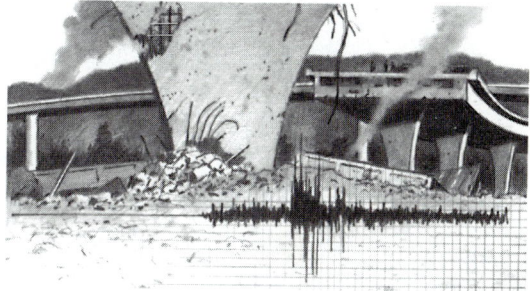

(a) On July 26, 1963, an earthquake in Yugoslavia had a magnitude of 6.0 on the Richter scale, and on August 19, 1977, an earthquake in Indonesia measured 8.0. Find the intensity x for each of these earthquakes if $I_0 = 1$.

(b) How many times more intense was the Indonesian earthquake than the Yugoslavian earthquake?

82. *Growth in Salary* Suppose that a person's salary is initially $30,000 and is modeled by $f(x)$, where x represents the number of years of experience. Use $f(x)$ to numerically approximate the years of experience when the salary exceeds $60,000.

(a) $f(x) = 30,000(1.1)^x$

(b) $f(x) = 30,000 \log (10 + x)$

Would most people prefer that their salaries increase exponentially or logarithmically?

Writing about Mathematics

1. Describe the relationship between an exponential function and a logarithmic function with the same base. Explain why logarithms are needed to solve exponential equations.

2. Give verbal, numerical, graphical, and symbolic representations of a base-5 logarithmic function.

CHECKING BASIC CONCEPTS FOR SECTIONS 4.3 AND 4.4

1. If the principal is $1200 and the interest rate is 9.5% compounded monthly, calculate the account balance after 4 years. Determine the balance if the interest is compounded continuously.

2. Evaluate the following to the nearest hundredth.
   (a) $f(1.2)$ if $f(x) = e^{2x}$
   (b) $f(3.4)$ if $f(x) = 2.1(2^x + 2^{-x})$

3. Explain verbally what $\log_2 15$ represents. Is it equal to an integer? (Do not use a calculator.)

4. Evaluate each of the following logarithms.
   (a) $\log_6 36$
   (b) $\log \sqrt{10} + \log 0.01$
   (c) $\ln \dfrac{1}{e^2}$

4.5 Exponential and Logarithmic Equations

Exponential Equations • Logarithmic Equations • Nonlinear Regression (Optional)

Introduction

The population of the world has grown from 1 billion in 1850 to 5.8 billion in 1996. As a result of this rapid growth, heavy demands have been made on the environment. Exponential functions and equations are often used to model this rapid growth. In order to solve exponential equations, logarithms frequently are used. In this section we discuss applications involving exponential and logarithmic equations.

Exponential Equations

We will refer to an equation where a variable occurs in the exponent of an expression as an *exponential equation*. In Section 4.4 exponential equations in the form $a^x = k$ were solved by taking a logarithm of both sides. This technique is fundamental to solving exponential equations as illustrated in the next example.

EXAMPLE 1 *Solving an exponential equation*

Revenues in the U.S. from all forms of legal gambling increased between 1991 and 1995. The function represented by $f(x) = 26.6e^{0.131x}$ models these revenues in billions of dollars. In this formula x represents the year, where $x = 0$ corresponds to 1991. (Source: *International Gaming & Wagering Business.*)

(a) Estimate gambling revenues in 1995 to the nearest billion dollars.
(b) Determine symbolically the year when these revenues reached $30 billion.
(c) Support your result in part (b) graphically and numerically.

Solution

(a) Since $x = 4$ corresponds to 1995, evaluate $f(4)$.

$$f(4) = 26.6e^{0.131(4)} \approx 44.9$$

In 1995 gambling revenues were approximately $45 billion.

(b) *Symbolic Solution* Solve the exponential equation $f(x) = 30$.

$26.6e^{0.131x} = 30$	$f(x) = 30$
$e^{0.131x} = \dfrac{30}{26.6}$	Divide by 26.6.
$\ln e^{0.131x} = \ln \dfrac{30}{26.6}$	Take the natural logarithm of both sides.
$0.131x = \ln \dfrac{30}{26.6}$	Inverse property: $\ln e^k = k$
$x = \dfrac{\ln (30/26.6)}{0.131}$	Divide by 0.131.
$x \approx 0.92$	Approximate x.

Gambling revenues reached $30 billion when $x \approx 1$, or in 1992.

(c) *Graphical Solution* Graph $Y_1 = 26.6e^{\wedge}(.131X)$ and $Y_2 = 30$. In Figure 4.58 their graphs intersect near $(0.92, 30)$.

Numerical Solution Numerical support is shown in Figure 4.59, where $Y_1 \approx Y_2 = 30$ when $x = 1$.

[0, 5, 1] by [0, 50, 10]

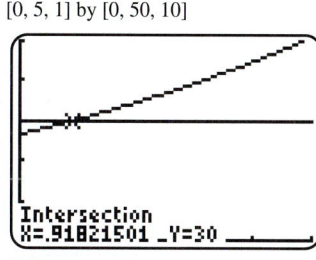

FIGURE 4.58   **FIGURE 4.59**   ■

To solve an exponential equation with a base other than 10 or e, Property 4 of logarithms may be used. This technique is demonstrated in the next example.

EXAMPLE 2

Modeling the decline of bluefin tuna

Bluefin tuna are large fish that can weigh 1500 pounds and swim at speeds of 55 miles per hour. Because they are used for sushi, a prime fish can be worth over $30,000. As a result, the western Atlantic bluefin tuna have been exploited, and their numbers have declined exponentially. Their numbers in thousands from 1974 to 1991 can be modeled by $f(x) = 230(0.881)^x$, where x is the year and $x = 0$ corresponds to 1974. (Source: B. Freedman, *Environmental Ecology*.)

(a) Estimate the number of bluefin tuna in 1974 and 1991.
(b) Determine symbolically the year when their numbers were 50 thousand.
(c) Support your result in part (b) graphically and numerically.

Solution

(a) To determine their numbers in 1974 and 1991, evaluate $f(0)$ and $f(17)$.

$$f(0) = 230(0.881)^0 = 230(1) = 230$$

$$f(17) = 230(0.881)^{17} \approx 26.7$$

Bluefin tuna decreased from 230 thousand in 1974 to less than 27 thousand in 1991.

(b) *Symbolic Solution*   Solve the equation $f(x) = 50$.

$230(0.881^x) = 50$	$f(x) = 50$
$0.881^x = \dfrac{5}{23}$	Divide by 230.
$\ln 0.881^x = \ln \dfrac{5}{23}$	Take the natural logarithm of both sides.
$x \ln 0.881 = \ln \dfrac{5}{23}$	Property 4: $\ln m^r = r \ln m$
$x = \dfrac{\ln (5/23)}{\ln 0.881}$	Divide by $\ln 0.881$.
$x \approx 12.04$	Approximate x.

The population of the bluefin tuna was about 50 thousand in 1974 + 12.04 $\approx$ 1986.

(c) *Graphical Solution*   Graph $Y_1 = 230*.881\wedge X$ and $Y_2 = 50$. In Figure 4.60 their graphs intersect near (12.04, 50).

Numerical Solution   Numerical support is shown in Figure 4.61, where $Y_1 \approx Y_2 = 50$ when $x = 12$.

[0, 17, 1] by [0, 250, 50]

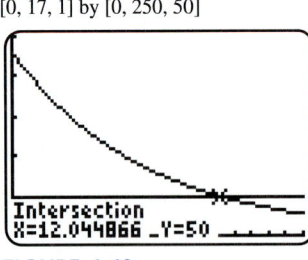

FIGURE 4.60   **FIGURE 4.61**   ■

In real applications, a modeling function is seldom provided with the data. However, it is not uncommon to be given the general form of a function that describes a data set.

EXAMPLE 3 *Finding a modeling function*

The gap between available organs for liver transplants and people who need them has widened. The number of individuals waiting for liver transplants in 1988 was 2329, and in 1995 it increased to 10,532. These numbers can be modeled by $f(x) = Ca^{(x-1988)}$, where C and a are constants. (Source: United Network for Organ Sharing.)

(a) Approximate C and a.

(b) Use f to estimate the number of individuals who could be waiting to receive a liver transplant in the year 1998.

Solution

(a) In 1988 there were 2329 waiting, so $f(1988) = 2329$. This equation can be used to find C.

$$f(1988) = Ca^{(1988-1988)} = Ca^0 = C(1) = 2329$$

Thus, $f(x) = 2329a^{(x-1988)}$. In 1995 the number waiting was 10,532.

$$f(1995) = 10{,}532$$
$$2329a^{(1995-1988)} = 10{,}532 \qquad \text{Substitute.}$$
$$2329a^7 = 10{,}532 \qquad \text{Simplify.}$$
$$a^7 = \frac{10{,}532}{2329} \qquad \text{Divide by 2329.}$$
$$a = \left(\frac{10{,}532}{2329}\right)^{1/7} \qquad \text{Take the seventh root of both sides.}$$
$$a \approx 1.2406 \qquad \text{Approximate } a.$$

Thus, let $f(x) = 2329(1.2406)^{(x-1988)}$.

(b) To estimate this number in the year 1998, evaluate $f(1998)$.

$$f(1998) = 2329(1.2406)^{(1998-1988)} \approx 20{,}113$$

In 1998 over 20,000 people may need liver transplants. ∎

Exponential equations can occur in many different forms. Although some types of exponential equations cannot be solved symbolically, Example 4 shows three equations that can.

EXAMPLE 4 *Solving exponential equations symbolically*

Solve each equation.

(a) $10^{x+2} = 10^{3x}$

(b) $4e^{-2x} - 2 = 1$

(c) $\left(\dfrac{1}{4}\right)^{x-1} = \dfrac{1}{10}$

Solution

(a) Start by taking the common logarithm of both sides.

$$10^{x+2} = 10^{3x}$$

$$\log 10^{x+2} = \log 10^{3x} \qquad \text{Take the common logarithm.}$$

$$x + 2 = 3x \qquad \text{Inverse property: } \log 10^k = k$$

$$2 = 2x \qquad \text{Subtract } x.$$

$$x = 1 \qquad \text{Solve.}$$

(b) Begin by solving for e^{-2x}.

$$4e^{-2x} - 2 = 1$$

$$4e^{-2x} = 3 \qquad \text{Add 2.}$$

$$e^{-2x} = \frac{3}{4} \qquad \text{Divide by 4.}$$

$$\ln e^{-2x} = \ln\frac{3}{4} \qquad \text{Take the natural logarithm.}$$

$$-2x = \ln\frac{3}{4} \qquad \text{Inverse property: } \ln e^k = k$$

$$x = -\frac{1}{2}\ln\frac{3}{4} \approx 0.144 \qquad \text{Divide by } -2 \text{ and approximate.}$$

(c) Begin by taking the common logarithm of both sides.

$$\left(\frac{1}{4}\right)^{x-1} = \frac{1}{10}$$

$$\log\left(\frac{1}{4}\right)^{x-1} = \log\frac{1}{10} \qquad \text{Take the common logarithm.}$$

$$(x-1)\log\left(\frac{1}{4}\right) = -1 \qquad \text{Property 4: } \log m^r = r\log m$$

$$(x-1) = \frac{-1}{\log(1/4)} \qquad \text{Divide by } \log\frac{1}{4}.$$

$$x = 1 - \frac{1}{\log(1/4)} \approx 2.661 \qquad \text{Add 1 and approximate.}$$

This also could have been solved by taking the natural logarithm of both sides. ∎

The following real-data example can be solved graphically with ease. However, a symbolic solution is more difficult and will not be attempted.

For medical reasons, dyes may be injected into the bloodstream to determine the health of internal organs. In one study involving animals, the dye BSP was injected to assess the blood flow in the liver. The results are listed in Table 4.21 on the next page, where x represents the elapsed time in minutes and y is the concentration of the dye in the bloodstream in milligrams per milliliter (mg/ml). (Sources: F. Harrison, "The measurement of liver blood flow in conscious calves"; D. Brown and P. Rothery, *Models in Biology: Mathematics, Statistics and Computing.*)

TABLE 4.21

x (min)	1	2	3	4	5	7	9	13	16	19	22
y (mg/ml)	0.102	0.077	0.057	0.045	0.036	0.023	0.015	0.008	0.005	0.004	0.003

EXAMPLE 5   *Solving an exponential equation graphically*

The data in Table 4.21 was modeled by $f(x) = 0.133(0.878(0.73^x) + 0.122(0.92^x))$.
(a) Graph f together with the data. Comment on the fit.
(b) Determine the y-intercept and interpret the result.
(c) Use the intersection-of-graphs method to estimate the elapsed time when the concentration of the dye was 50% of its initial amount.

Solution
(a) Graph $Y_1 = .133(.878*.73^\wedge X + .122*.92^\wedge X)$ and the data as shown in Figure 4.62. The graph of f fits the data remarkably well.
(b) Since $f(0) = 0.133$, the y-intercept is 0.133. The value of 0.133 mg/ml represents the initial concentration of dye in the bloodstream.
(c) Fifty percent of 0.133 is 0.0665 mg/ml. The graphs of Y_1 and $Y_2 = 0.0665$ in Figure 4.63 intersect near (2.49, 0.0665). The concentration of the dye decreased to half its initial level after about 2.5 minutes.

$[-2, 23, 5]$ by $[-0.02, 0.15, 0.01]$ $[-2, 23, 5]$ by $[-0.02, 0.15, 0.01]$

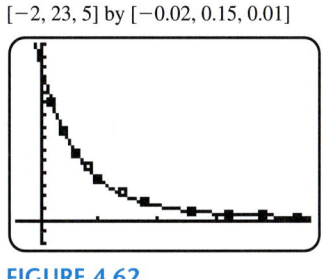

FIGURE 4.62

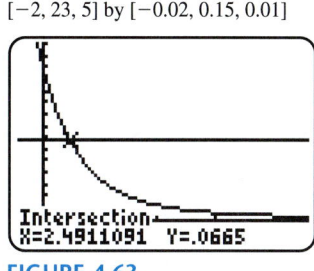

FIGURE 4.63

Logarithmic Equations

Logarithmic equations contain logarithms. Like exponential equations, logarithmic equations also occur in applications. To solve a logarithmic equation, we will use the inverse property $a^{\log_a x} = x$.

EXAMPLE 6   *Solving a logarithmic equation symbolically*

In developing countries there is a relationship between the amount of land a person owns and the average daily calories consumed. This relationship is modeled by $f(x) = 280 \ln (x + 1) + 1925$, where x is the amount of land owned in acres and $0 \le x \le 4$. (Source: D. Grigg, *The World Food Problem.*)
(a) Find the average caloric intake for a person in a developing country who owns no land.
(b) Graph f in [0, 4, 1] by [1800, 2500, 100]. Interpret the graph.
(c) Determine symbolically the number of acres owned for the average intake to be 2000 calories.

Solution

(a) Since $f(0) = 280 \ln (0 + 1) + 1925 = 1925$, a person without any land consumes an average of 1925 calories.

(b) Graph $Y_1 = 280 \ln (X + 1) + 1925$ as shown in Figure 4.64. As the amount of land x increases, the caloric intake y also increases. However, it begins to level off. This would be expected because there is a limit to the number of calories an average person would eat, regardless of his or her economic status.

[0, 4, 1] by [1800, 2500, 100]

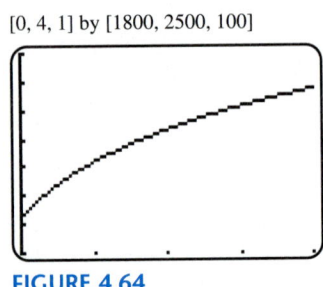

FIGURE 4.64

(c) Solve the equation $f(x) = 2000$.

$$280 \ln (x + 1) + 1925 = 2000$$

$$280 \ln (x + 1) = 75 \qquad \text{Subtract 1925.}$$

$$\ln (x + 1) = \frac{75}{280} \qquad \text{Divide by 280.}$$

$$e^{\ln (x+1)} = e^{75/280} \qquad \text{Exponentiate both sides using base } e.$$

$$x + 1 = e^{75/280} \qquad \text{Inverse property: } e^{\ln k} = k$$

$$x = e^{75/280} - 1 \qquad \text{Subtract 1.}$$

$$x \approx 0.307 \qquad \text{Approximate.}$$

A person owning about 0.3 acre has an average intake of 2000 calories. ■

Like exponential equations, logarithmic equations can occur in many different forms. The next example illustrates two equations that can be solved symbolically.

EXAMPLE 7 *Solving logarithmic equations symbolically*

Solve each equation.
(a) $\log (2x + 1) = 2$
(b) $\log_2 4x = 2 - \log_2 x$

Solution

(a) To solve the equation exponentiate both sides of the equation using base 10.

$$\log (2x + 1) = 2$$

$$10^{\log (2x+1)} = 10^2 \qquad \text{Exponentiate both sides using base 10.}$$

$$2x + 1 = 100 \qquad \text{Inverse property: } 10^{\log k} = k$$

$$x = 49.5 \qquad \text{Solve for } x.$$

(b) To solve this equation we apply properties of logarithms.

$$\log_2 4x = 2 - \log_2 x$$

$$\log_2 4x + \log_2 x = 2 \qquad \text{Add } \log_2 x.$$

$$\log_2 4x^2 = 2 \qquad \text{Property 2: } \ln m + \ln n = \ln (mn)$$

$$2^{\log_2 4x^2} = 2^2 \qquad \text{Exponentiate both sides using base 2.}$$

$$4x^2 = 4 \qquad \text{Inverse property: } 2^{\log_2 k} = k$$

$$x = \pm 1 \qquad \text{Solve for } x.$$

However, $x = -1$ is not a solution since $\log_2 x$ is undefined for negative values of x. Thus, the only solution is $x = 1$. ∎

EXAMPLE 8   *Modeling the life span of a robin*

The life span of 129 robins was monitored over a four-year period in one study. The equation $y = \dfrac{2 - \log(100 - x)}{0.42}$ can be used to calculate the number of years y for x percent of the robin population to die. For example, to find the time when 40% of the robins had died, substitute $x = 40$ into the equation. The result is $y \approx 0.53$, or about half a year. (Source: D. Lack, *The Life of a Robin.*)

(a) Graph y in [0, 100, 10] by [0, 5, 1]. Interpret the graph.

(b) Estimate graphically the percentage of the robins that died after two years.

Solution

(a) The graph of $Y_1 = (2 - \log(100 - X))/.42$ is shown in Figure 4.65. At first, the graph increases slowly between 0% and 60%. Since years are listed on the y-axis and percentage on the x-axis, this slow increase in f indicates that not many years pass by before a large percentage of the robins die. When $x \geq 60$, the graph begins to increase dramatically. This indicates that a small percentage has a comparatively long life span.

(b) To determine the percentage that died after two years, graph Y_1 and $Y_2 = 2$. Their graphs intersect near (85.5, 2) as shown in Figure 4.66. After two years, approximately 85.5% of the robins had died—a surprisingly high percentage.

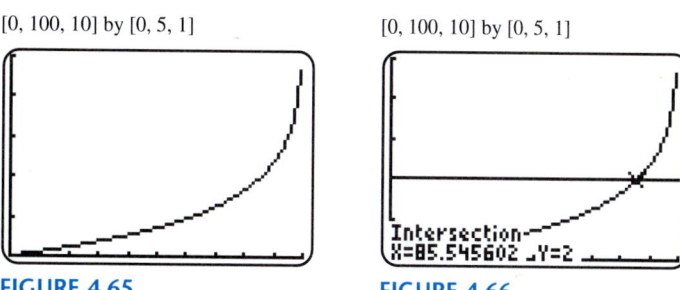

[0, 100, 10] by [0, 5, 1]   [0, 100, 10] by [0, 5, 1]

FIGURE 4.65   **FIGURE 4.66**   ∎

Nonlinear Regression (Optional)

In Section 2.5, linear regression based on least-squares fit was discussed. The least-squares method can also be applied to nonlinear functions.

EXAMPLE 9 *Using nonlinear least-squares regression*

The National Institute for Automotive Service Excellence (ASE) reported that the number of females working in automotive repair is increasing. Table 4.22 shows the net total of female ASE-certified technicians.

TABLE 4.22

Year	1988	1989	1990	1991	1992	1993	1994	1995
Total	556	614	654	737	849	1086	1329	1592

Source: National Institute for Automotive Service Excellence.

(a) Let y be the number of female technicians and x be the year, where $x = 0$ corresponds to 1988 and $x = 7$ to 1995. Use nonlinear least-squares regression to determine a cubic function f that models this data. Graph f together with the data in $[-1, 8, 1]$ by $[400, 1700, 100]$.

(b) Repeat part (a), except this time find a function g defined by $g(x) = Ca^x$.

(c) Does f or g fit the data in Table 4.22 better? If present trends continue, estimate the number of certified female technicians in the year 2000. Round your answer to the nearest hundred.

Solution

(a) Apply cubic regression to the data points. Two resulting screens are shown in Figures 4.67 and 4.68. The data is modeled by $f(x) = 0.8005x^3 + 14.39x^2 + 8.127x + 568.5$. The data and f are graphed in Figure 4.69.

$[-1, 8, 1]$ by $[400, 1700, 100]$

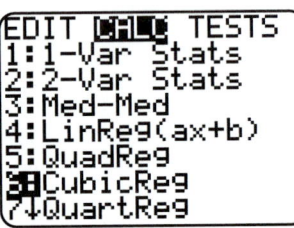

FIGURE 4.67

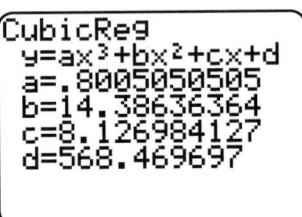

FIGURE 4.68

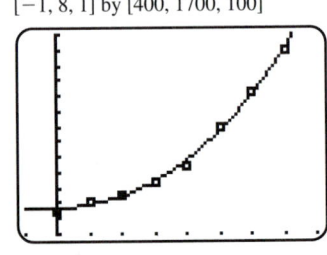

FIGURE 4.69

(b) To find a function g of the form $g(x) = Ca^x$ that models the data, use exponential regression as shown in Figures 4.70 and 4.71. The data is modeled by $g(x) = 507.1(1.166)^x$. A graph of g and the data are shown in Figure 4.72.

$[-1, 8, 1]$ by $[400, 1700, 100]$

FIGURE 4.70

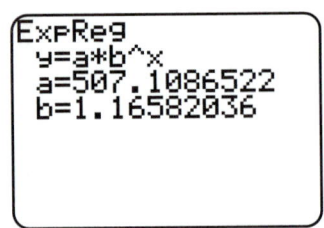

FIGURE 4.71

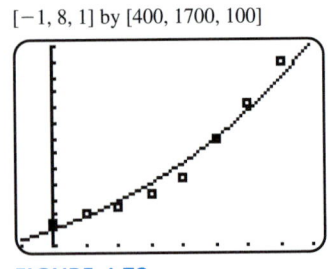

FIGURE 4.72

(c) From the graphs, it appears that f models the data better than g. Since $x = 12$ corresponds to the year 2000, evaluate

$$f(12) = 0.8005(12)^3 + 14.39(12)^2 + 8.127(12) + 568.5 \approx 4121.$$

If present trends continue, the number of certified female automotive technicians could reach approximately 4100 in the year 2000.   ∎

4.5 PUTTING IT ALL TOGETHER

The following table summarizes techniques that can be used to solve certain forms of exponential and logarithmic equations symbolically.

Exponential Equations	
Form of equation:	$Ca^x = k$
Method of solution:	Solve for a^x. Then take a logarithm of both sides to solve for x.

Logarithmic Equations	
Form of equation:	$C \log_a x = k$
Method of solution:	Solve for $\log_a x$. Then exponentiate both sides of the equation using base a.
Form of equation:	$\log_a(bx) \pm \log_a(cx) = k$
Method of solution:	When more than one logarithm with the same base occurs in an equation, first combine logarithms by applying the properties of logarithms. Check any solutions.

4.5 EXERCISES   Tape 5

Solving Exponential and Logarithmic Equations

Exercises 1–4: The graphical and symbolic representations of f and g are given.

(a) *Use the graph to estimate the solution of $f(x) = g(x)$.*

(b) *Solve $f(x) = g(x)$ symbolically.*

1. $f(x) = e^x$, $g(x) = 7.5$

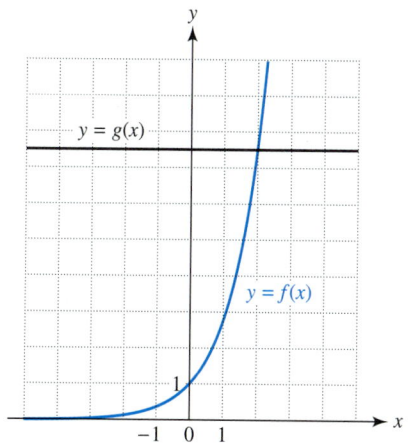

2. $f(x) = 0.1(10^x)$, $g(x) = 0.5$

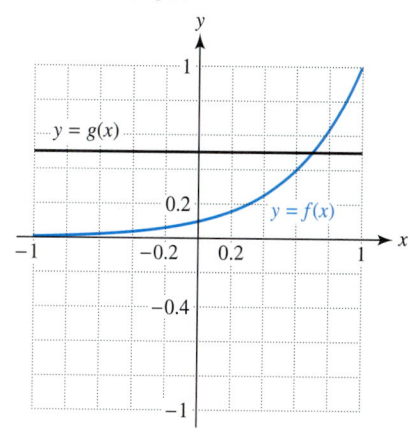

3. $f(x) = 10^{0.2x}$, $g(x) = 2.5$

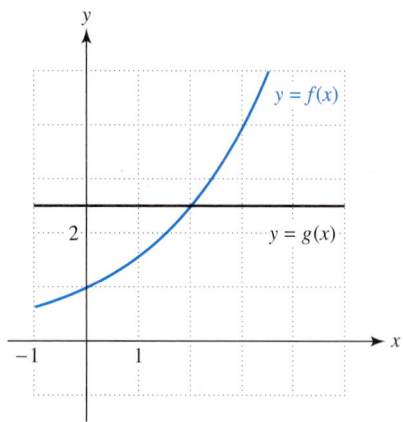

4. $f(x) = e^{-0.7x}$, $g(x) = 2$

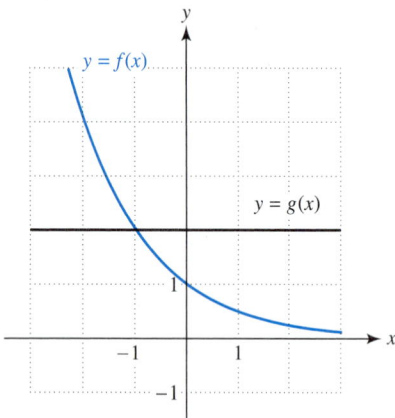

Exercises 5–20: Use common or natural logarithms to solve the exponential equation symbolically. Support your results graphically and numerically.

5. $4e^x = 5$

6. $2e^{-x} = 8$

7. $2(10^x) + 5 = 45$

8. $100 - 5(10^x) = 7$

9. $2.5e^{-1.2x} = 1$

10. $9.5e^{0.005x} = 19$

11. $1.2(0.9^x) = 0.6$

12. $0.05(1.15)^x = 5$

13. $4(1.1^{x-1}) = 16$

14. $3(2^{x-2}) = 99$

15. $10^{(x^2)} = 10^{3x-2}$

16. $e^{2x} = e^{5x-3}$

17. $\left(\dfrac{1}{5}\right)^x = -5$

18. $2^x = -4$

19. $10^{1-x} = 7$

20. $e^{2x}e^x = e^6$

Exercises 21 and 22: Continuous Compounding Suppose that A_0 dollars is deposited in a savings account paying 9% interest compounded continuously. After x years the account will contain $A(x) = A_0e^{0.09x}$ dollars.

    (a) *Solve the equation $A(x) = k$ symbolically for the given values of A_0 and k. Support your results graphically and numerically.*

    (b) *Interpret your results.*

21. $A_0 = 500$ and $k = 750$

22. $A_0 = 1000$ and $k = 2000$

Exercises 23–38: Solve the logarithmic equation symbolically. Support your answer graphically and numerically.

23. $3 \log x = 2$

24. $5 \ln x = 10$

25. $\ln 2x = 5$

26. $\ln 4x = 1.5$

27. $\log 2x^2 = 2$

28. $\log (2 - x) = 0.5$

29. $160 + 10 \log x = 50$

30. $160 + 10 \log x = 120$

31. $\ln x + \ln x^2 = 3$

32. $\log x^5 = 4 + 3 \log x$

33. $2 \log_2 x = 4.2$

34. $3 \log_2 3x = 1$

35. $\log x + \log 2x = 2$

36. $\ln 2x + \ln 3x = \ln 6$

37. $\ln (x - 1) = 1$

38. $\log x + \log (2x + 5) = \log 7$

Applications

39. *Credit Cards* From 1987 to 1996 the number of Visa cards and MasterCards was up 80% to 376 million. The function given by $f(x) = 36.2e^{0.14x}$ models the amount of credit card spending from Thanksgiving to Christmas in billions of dollars. In this formula $x = 0$ corresponds to 1987 and $x = 9$ to 1996. (Source: National Credit Counseling Services.)

    (a) Determine symbolically the year when this amount reached $55 billion.

    (b) Solve part (a) graphically and numerically.

40. *Salinity* The salinity of the oceans changes with latitude and with depth. In the tropics, the salinity increases on the surface of the ocean due to rapid evaporation. In the higher latitudes, there is less evaporation and rainfall causes the salinity to be less on the surface than at lower depths. The function given by $f(x) = 31.5 + 1.1 \log (x + 1)$ models salinity to depths of 1000 meters at a latitude of 57.5°N. The input x is the depth in meters and the output $f(x)$ is in grams of salt per kilogram of seawater. (Source: D. Hartman, *Global Physical Climatology*.)

(a) Table f starting at $x = 0$, incrementing by 100. Discuss any trends.

(b) Numerically approximate the depth where the salinity equals 33.

41. *Life Span of Robins*   Solve Example 8, part (b) symbolically.

42. *Atmospheric Pressure*   As altitude increases, air pressure decreases. The atmospheric pressure in millibars (mb) at a given altitude x in meters is listed in the table. (Source: A. Miller and J. Thompson, *Elements of Meteorology*.)

Altitude (m)	0	5000	10,000	15,000
Pressure (mb)	1013	541	265	121

Altitude (m)	20,000	25,000	30,000
Pressure (mb)	55	26	12

(a) The data can be modeled by $f(x) = 1013e^{kx}$. Approximate the constant k.

(b) Graph f and the data in the same viewing rectangle. Comment on the fit.

43. *Diversity*   In 1995 the U.S. racial mix was 75.3% white, 9.0% Hispanic, 12.0% African American, 2.9% Asian/Pacific Islanders, and 0.8% Native American. Their respective percentages of executives, managers, and administrators in private-industry communication firms were 84.0%, 9.5%, 3.4%, 2.0%, and 0.0%. If the percentage P of African American executives, managers, and administrators increases at an annual rate of 7% per year, determine symbolically the year when their representation will reach 12.0% of the total number of executives, managers, and administrators in communications. (*Hint:* Let $P(x) = 3.4(1.07)^x$, where $x = 0$ corresponds to 1995, and solve $P(x) = 12$.) (Source: Labor Department's Glass Ceiling Commission.)

44. *Midair Near Collisions*   The table shows the number of airliner near collisions y in year x. (Source: Federal Aviation Administration.)

x	1989	1991	1993	1995
y	131	78	44	34

(a) Approximate constants C and a so that $f(x) = Ca^{(x - 1989)}$ models the data.

(b) Support your answer graphically.

45. *Liver Transplants*   (Refer to Example 3.) Determine symbolically when the number of individuals waiting for liver transplants might reach 20,000. Support your answer numerically.

46. *Wind Speed*   Solve part (c) of Exercise 78 in Section 4.4 symbolically.

Exercises 47 and 48: Radioactive Carbon Dating While animals are alive they breathe both carbon dioxide and oxygen. Because a small portion of normal atmospheric carbon dioxide is made up of radioactive carbon-14, a fixed percentage of their bodies is composed of carbon-14. When an animal dies, it quits breathing and the carbon-14 disintegrates without being replaced. One method to determine the time since an animal died is to estimate the percentage of carbon-14 remaining in its bones. The **half-life** *of carbon-14 is 5730 years. This means that half of the original amount of carbon-14 in a fossil will be remaining after 5730 years. The percent P in decimal form of carbon-14 remaining after x years is given by $P(x) = e^{-0.000121x}$. For each percent P, approximate the age of a fossil.*

47. $P = 60\%$

48. $P = 1\%$

49. *Greenhouse Gases*   If the current trends continue, future concentrations of atmospheric carbon dioxide (CO_2) in parts per million (ppm) are shown in the accompanying table. The CO_2 concentration in the year 2000 is greater than it has been at any time in the previous 160,000 years. The increase in concentration of CO_2 has been accelerated by the burning of fossil fuels and by deforestation. (Source: R. Turner, *Environmental Economics*.)

Year	2000	2050	2100	2150	2200
CO_2 (ppm)	364	467	600	769	987

(a) Let $x = 0$ correspond to 2000 and $x = 200$ to 2200. Find values for A_0 and a so that $f(x) = A_0 a^x$ models the data.

(b) Estimate graphically the year when the carbon dioxide concentration could be double the preindustrial level of 280 ppm.

50. *Greenhouse Gases* Chlorofluorocarbons (CFCs) are gases created by people that increase the greenhouse effect. CFC-12 is a type of chlorofluorocarbon used in refrigeration, air conditioning, and insulating foams. Over a 100-year period, the greenhouse effect from CFC-12 is 7000 times stronger than an equal amount of carbon dioxide. Concentrations in parts per billion (ppb) of CFC-12 can be modeled by $f(x) = 0.48e^{0.04x}$, where $x = 0$ corresponds to 1990. (Source: D. Wuebbles and J. Edmonds, *Primer on Greenhouse Gases.*)

(a) Determine symbolically the year when the concentration of CFC-12 could be double its 1990 level.

(b) Solve part (a) graphically.

51. *Reducing Carbon Emissions* When fossil fuels are burned, carbon is released into the atmosphere. Governments could reduce carbon emissions by placing a tax on fossil fuels. For example, a higher tax might lead people to use less gasoline. The **cost-benefit** equation $\ln(1 - P) = -0.0034 - 0.0053x$ estimates the relationship between a tax of x dollars per ton of carbon and the percent P reduction in emissions of carbon, where P is in decimal form. Determine P when $x = 60$. Interpret the result. (Source: W. Clime, *The Economics of Global Warming.*)

52. (Refer to the previous exercise.) The cost-benefit function from the previous exercise can be represented as $P(x) = 1 - e^{-0.0034-0.0053x}$.

(a) Graph P in [0, 1000, 100] by [0, 1, 0.1]. Discuss the benefit of continuing to raise taxes on fossil fuels.

(b) According to this model, what tax would result in a 50% reduction in carbon emissions? Solve this symbolically. Support your result graphically.

Nonlinear Regression

Exercises 53 and 54: Exponential Regression The table contains data that can be modeled by a function of the form $f(x) = ba^x$. Use regression to find the constants a and b to the nearest hundredth. Graph f and the data.

53.

x	2	3	4	5
y	2.56	2.84	3.12	3.45

54.

x	5	10	15	20
y	0.65	0.84	1.09	1.43

Exercises 55 and 56: Power Regression The table contains data that can be modeled by a function of the form $f(x) = bx^a$. Use regression to find the constants a and b to the nearest hundredth. Graph f and the data.

55.

x	2	4	6	8
$f(x)$	3.7	4.2	4.6	4.9

56.

x	3	6	9	12
$f(x)$	23.8	58.5	99.2	144

57. *Atmospheric Density* The table lists the atmospheric density y in kilograms per cubic meter (kg/m^3) at an altitude of x meters. (Source: A. Miller.)

x (m)	0	5000	10,000	15,000
y (kg/m^3)	1.2250	0.7364	0.4140	0.1948

x (m)	20,000	25,000	30,000
y (kg/m^3)	0.0889	0.0401	0.0184

(a) Use regression to estimate the constants a and b so that $f(x) = ba^x$ models the data.

(b) Predict the density at 7000 meters. (The actual value is 0.59 kg/m^3.)

58. *Bird Populations* Near New Guinea there is a relationship between the number of bird species found on an island and the size of the island. The accompanying table lists species of birds y found on an island with an area of x square kilometers. (Source: B. Freedman, *Environmental Ecology.*)

x (km^2)	0.1	1	10	100	1000
y (species)	10	15	20	25	30

    (a) Use regression to estimate the constants a and b so that $f(x) = a + b \ln x$ models the data.

    (b) Predict the number of bird species on an island of 5000 square kilometers.

59. *Fertilizer Usage* Between 1950 and 1980 the use of chemical fertilizers increased. The table lists worldwide average usage y in kilograms per hectare of cropland during year x, where $x = 0$ corresponds to 1950. (*Note:* 1 hectare $\approx$ 2.47 acres.) (Source: D. Grigg.)

x	0	13	22	29
y	12.4	27.9	54.3	77.1

    (a) Graph the data in $[-2, 32, 5]$ by $[0, 80, 10]$. Is the data linear?

    (b) Use regression to estimate the constants a and b so that $f(x) = ba^x$ models the data.

    (c) Predict the chemical fertilizer usage in 1989. The actual value was 98.7 kilograms per hectare. What does this indicate about usage of fertilizer during the 1980s?

60. *Social Security* If major reform occurs in the Social Security system, individuals may be able to invest some of their contributions into individual accounts. Many of these accounts would be managed by financial firms that charge fees. The following table lists the amount in billions of dollars that may be collected if fees are 0.93% of the assets each year. (Source: Social Security Advisory Council.)

Year	1998	2005	2010	2015	2020
Fee ($ billions)	1.4	20	41	80	136

    (a) Use regression to find a quadratic function f that models the data.

    (b) Graph f and the data in $[1995, 2025, 5]$ by $[0, 150, 10]$.

    (c) Repeat parts (a) and (b) using a cubic function g.

    (d) Which function models the data better?

Writing about Mathematics

1. Explain how to solve the equation $Ca^x = k$ symbolically, where C and k are constants. Demonstrate your method.

2. Explain how to solve the equation $b \log_a x = k$ symbolically and graphically, where b and k are constants. Demonstrate your method.

CHECKING BASIC CONCEPTS FOR SECTION 4.5

1. Solve each equation symbolically. Support your results graphically and numerically.
    (a) $e^x = 5$
    (b) $10^x = 25$
    (c) $\log x = 1.5$

2. Solve the equation symbolically. Support your results graphically and numerically.
    (a) $2e^x + 1 = 25$
    (b) $\log 2x = 2.3$
    (c) $\log x^2 = 1$

CHAPTER 4 Summary

The concept of an inverse occurs in everyday life. An inverse action will undo an earlier action. For example, the inverse actions of taking out a calculator and turning it on are turning the calculator off and putting it away. By reversing the order of the actions and performing the inverse action at each step, the original situation is restored.

The same is true for functions in mathematics. When two functions are combined in sequence, it is called *composition of functions.* When one function will undo the calculations performed by the other function, they are called *inverse functions.* If $f(a) = b$, then $f^{-1}(b) = a$. For instance, suppose f multiplies x by 2 and then adds 1. The inverse function of f must reverse the order of the steps and replace each operation with its inverse operation. The inverse subtracts 1 from x and then divides by 2. Thus, $f(x) = 2x + 1$ and $f^{-1}(x) = \dfrac{x-1}{2}$. A function f has an inverse if f is a *one-to-one* function. Different inputs always result in different outputs for a one-to-one function.

Exponential functions can be written as $f(x) = a^x$. The variable x occurs in the exponent of the expression. The constant a is called the *base,* where $a > 0$ and $a \neq 1$. *Logarithmic functions* are inverses of exponential functions. The expression $\log_a x$ equals the *exponent k* that satisfies $a^k = x$. For example, $\log_2 8 = 3$ since $2^3 = 8$.

Inverse functions can be used to solve equations in the form $f(x) = k$, where f is a one-to-one function and k is a constant in the range of f. The solution to $f(x) = k$ is $x = f^{-1}(k)$. For example, if $f(x) = 10^x$, then the solution to $10^x = 50$ is $x = f^{-1}(50) = \log 50$.

Two important properties of logarithms are the *inverse properties:* $\log_a a^x = x$ and $a^{\log_a x} = x$. The calculations of a^x and $\log_a x$ are inverse operations, just like the computation of x^3 and $\sqrt[3]{x}$ are inverse operations. Although logarithms of any positive base $a \neq 1$ are possible, the *common logarithm* and *natural logarithm* are the most frequently used logarithms.

Review Exercises

1. Use the table to evaluate each expression, if possible.

x	-1	0	1	3
$f(x)$	3	5	7	9
$g(x)$	-2	0	1	9

(a) $(f + g)(1)$
(b) $(f - g)(3)$
(c) $(fg)(-1)$
(d) $(f/g)(0)$

2. Use the graph to evaluate each expression.
(a) $(f - g)(2)$
(b) $(fg)(0)$

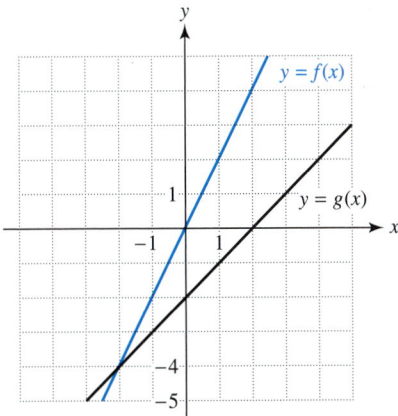

3. Use $f(x) = x^2$ and $g(x) = 1 - x$ to evaluate each expression.
   (a) $(f + g)(3)$ **(b)** $(f - g)(-2)$
   (c) $(fg)(1)$ **(d)** $(f/g)(3)$

4. Use $f(x) = x^2 + 3x$ and $g(x) = x^2 - 1$ to find each expression. Identify its domain.
   (a) $(f + g)(x)$ **(b)** $(f - g)(x)$
   (c) $(fg)(x)$ **(d)** $(f/g)(x)$

5. Numerical representations for f and g are given. Evaluate the expressions.
   (a) $(g \circ f)(-2)$
   (b) $(f \circ g)(3)$

x	−2	0	2	4
$f(x)$	1	4	3	2

x	1	2	3	4
$g(x)$	2	4	−2	0

6. Use the graph to evaluate each expression.
   (a) $(f \circ g)(2)$
   (b) $(g \circ f)(0)$

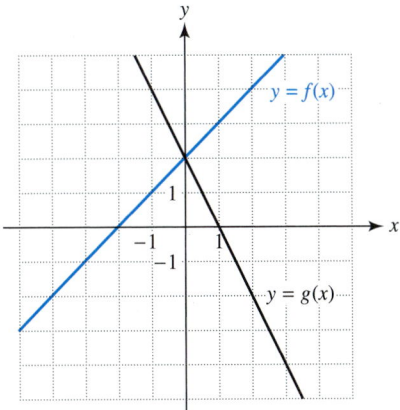

7. Use $f(x) = \sqrt{x}$ and $g(x) = x^2 + x$ to evaluate each expression.
   (a) $(f \circ g)(2)$
   (b) $(g \circ f)(9)$

8. Use $f(x) = x^2 + 1$ and $g(x) = x^3 - x^2 + 2x + 1$ to find the following.
   (a) $(f \circ g)(x)$
   (b) $(g \circ f)(x)$

Exercises 9 and 10: Describe the inverse operations of the given statement. Then, express both the statement and its inverse symbolically.

9. Divide x by 10 and add 6.

10. Subtract 5 from x and take the cube root.

Exercises 11 and 12: Determine if f is one-to-one.

11. $f(x) = 3x - 1$

12. $f(x) = 3x^2 - 2x + 1$

13. Use the table to determine if f is one-to-one.

x	−1	−2	0	4
$f(x)$	−5	4	3	4

14. Use the table of f to determine f^{-1}. Identify the domains and ranges of f and f^{-1}.

x	−1	0	4	6
$f(x)$	6	4	3	1

15. The function f computes the dollars in a savings account after x years. Use the graph to evaluate the following. Interpret f^{-1}.
   (a) $f(1)$
   (b) $f^{-1}(1200)$

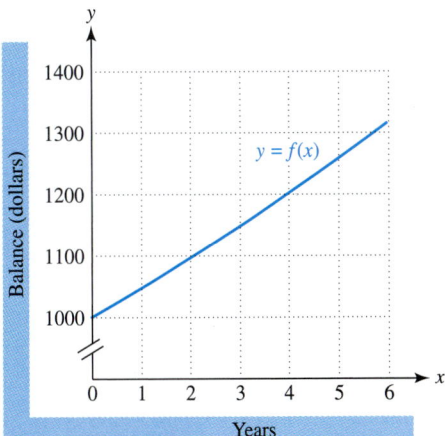

16. Use the graph of f to sketch a graph of f^{-1}.

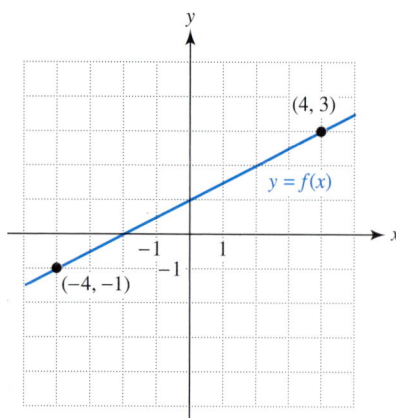

Exercises 17 and 18: Find $f^{-1}(x)$.

17. $f(x) = 3x - 5$

18. $f(x) = \sqrt[3]{x} + 1$

19. If $f(x) = \dfrac{1}{x + 7}$, find $f^{-1}(x)$.

20. Verify that $f(x) = 2x - 1$ and $g(x) = \dfrac{x + 1}{2}$ are inverses.

21. Determine the final value of $1200 invested at 9% compounded semiannually for 3 years.

22. Determine the final value of $500 invested at 6.5% compounded continuously for 8 years.

Exercises 23 and 24: Evaluate $f(a)$ for the given a. Approximate the result to five significant digits.

23. $f(x) = e^{-2x}$, $a = -1.2$

24. $f(x) = 4e^{0.23x}$, $a = 5.7$

25. Solve $e^x = 19$ symbolically. Support your result graphically and numerically.

26. Solve $2^x - x^2 = x$ graphically. Round each solution to four significant digits.

Exercises 27–30: Evaluate the expression without a calculator.

27. $\log 1000$

28. $\log 0.001$

29. $10 \log 0.01 + \log \dfrac{1}{10}$

30. $\log 100 + \log \sqrt[3]{10}$

Exercises 31–34: Evaluate the logarithm without a calculator.

31. $\log_3 9$

32. $\log_5 \dfrac{1}{25}$

33. $\ln e$

34. $\log_2 32$

Exercises 35 and 36: Approximate the logarithm to four significant digits.

35. $\log_3 18$

36. $\log_2 173$

Exercises 37–42: Solve the equation using common or natural logarithms. Support your results graphically and numerically.

37. $10^x = 125$

38. $1.5^x = 55$

39. $e^{0.1x} = 5.2$

40. $4e^{2x} - 5 = 3$

41. $5^{-x} = 10$

42. $3(10^{-x}) = 6$

Exercises 43–46: Solve the equation symbolically.

43. $\log x = 1.5$ **44.** $\log_3 x = 4$

45. $\ln x = 3.4$ **46.** $\log x = 2.2$

Exercises 47 and 48: Use properties of logarithms to write each expression as one term.

47. $\log 6 + \log 5x$

48. $\log \sqrt{3} - \log \sqrt[3]{3}$

49. Expand $\ln \dfrac{4}{x^2}$.

50. Expand $\log \dfrac{4x^3}{k}$.

Exercises 51–54: Solve the logarithmic equation symbolically. Support your results graphically and numerically.

51. $8 \log x = 2$ **52.** $\ln 2x = 2$

53. $2 \log 3x + 5 = 15$ **54.** $5 \log_2 x = 25$

55. Suppose that b is the y-intercept on the graph of a one-to-one function f. What is the x-intercept on the graph of f^{-1}? Explain your reasoning.

56. Let f be a linear function given by $f(x) = ax + b$ with $a \neq 0$.
   (a) Show that f^{-1} is also a linear function by finding $f^{-1}(x)$.
   (b) How is the slope of the graph of f related to the slope of the graph of f^{-1}?

57. *Diversity* In 1995 the U.S. racial mix was 75.3% white, 9.0% Hispanic, 12.0% African American,

2.9% Asian/Pacific Islanders, and 0.8% Native American. Their respective percentages of executives, managers, and administrators in private-industry retail trade were 80.8%, 4.8%, 4.9%, 5.2%, and 0.0%. If the percentage P of Hispanic executives, managers, and administrators increases at a rate of 5% per year, estimate the year when their representation will reach 9.0% of the total number in retail trade. (*Hint:* Let $P(x) = 4.8(1.05)^x$, where $x = 0$ corresponds to 1995, and solve $P(x) = 9$.) (Source: Labor Department's Glass Ceiling Commission.)

58. *Combining Functions* The total number of gallons of water passing through a pipe after x seconds is computed by $f(x) = 10x$. Another pipe delivers $g(x) = 5x$ gallons per second. Find a function h that computes the total volume of water passing through both pipes in x seconds.

59. *Converting Units* The accompanying figures show graphs of a function f that converts fluid ounces to pints and a function g that converts pints to quarts. Evaluate each expression. Interpret the results.
   (a) $(g \circ f)(32)$
   (b) $f^{-1}(1)$
   (c) $(f^{-1} \circ g^{-1})(1)$

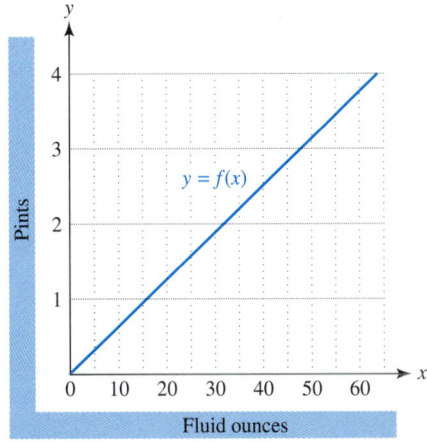

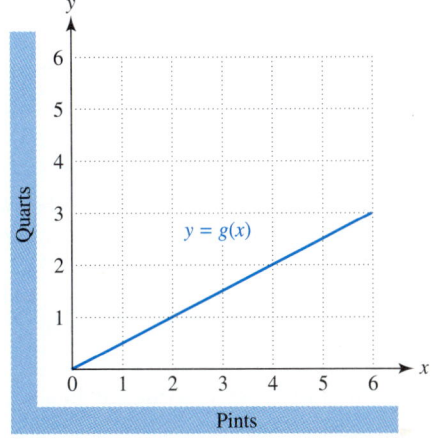

60. *Population and Tax Revenue* The accompanying graph of f shows the population in thousands of a city from 1980 until 2000. The graph of g computes the expected tax revenue in millions of dollars that can be raised from a population of x thousand. Evaluate and interpret $(g \circ f)(1988)$.

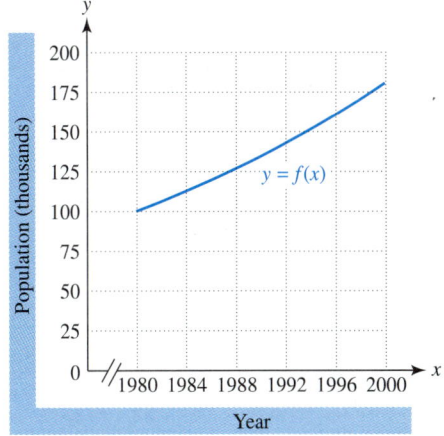

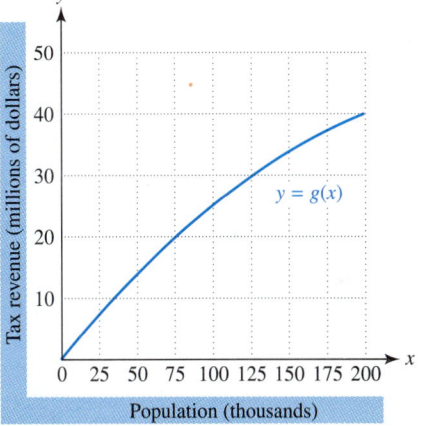

61. *Modeling Growth* The function given by $W(x) = 175.6(1 - 0.66e^{-0.24x})^3$ models the weight in milligrams of a small fish called the *Lebistes reticulatus* after x weeks, where $0 \le x \le 14$. (Source: D. Brown and P. Rothery, *Models in Biology: Mathematics, Statistics and Computing.*)
   (a) Evaluate $W(1)$.
   (b) Solve the equation $W(x) = 50$ graphically. Interpret the result.

62. *Test Scores* Let scores on a standardized test be modeled by $f(x) = 36e^{-(x-20)^2/49}$. The function f computes the number in thousands of people that received score x.
   (a) Graph f in [0, 40, 10] by [0, 40, 10]. What score did the largest number of people receive?
   (b) Solve the equation $f(x) = 30$ symbolically. Interpret your result.
   (c) Solve part (b) graphically.

63. *Modeling Epidemics*   In 1666 the village of Eyam, located in England, experienced an outbreak of the Great Plague. Out of 261 people in the community, only 83 people survived. The accompanying table shows a function f that computes the number of people who had not (yet) been infected after x days. (Source: G. Raggett, "Modeling the Eyam plague.")

x	0	15	30	45	60	75	90	125
$f(x)$	254	240	204	150	125	103	97	83

(a)   Use a table to represent a function g that computes the number of people in Eyam that were infected after x days.

(b)   Write an equation that shows the relationship between $f(x)$ and $g(x)$.

(c)   Use graphing to decide which equation represents $g(x)$ better,

$$y_1 = \frac{171}{1 + 18.6e^{-0.0747x}} \quad \text{or} \quad y_2 = 18.3(1.024)^x.$$

(d)   Use your results from parts (b) and (c) to find a formula for $f(x)$.

64. *Greenhouse Gases*   Methane is a greenhouse gas that is produced when fossil fuels are burned. In 1600 methane had an atmospheric concentration of 700 parts per billion (ppb), whereas in 2000 its concentration is expected to be about 1700 ppb. (Source: D. Wuebbles and J. Edmonds, *Primer on Greenhouse Gases.*)

(a)   Let $x = 0$ correspond to 1600 and $x = 400$ to 2000. Then, $f(x) = 700(10^{0.000964x})$ models methane concentrations. Graph f for $0 \le x \le 400$ and explain why f is one-to-one.

(b)   Given $f^{-1}(x) = \frac{1}{0.000964} \log\left(\frac{x}{700}\right)$, evaluate $f^{-1}(1000)$ and interpret the result.

65. *Exponential Regression*   The data in the table can be modeled by $f(x) = ba^x$. Use regression to estimate the constants a and b. Graph f and the data.

x	1	2	3	4
y	2.59	1.92	1.42	1.05

66. *Power Regression*   The data in the table can be modeled by $f(x) = bx^a$. Use regression to estimate the constants a and b. Graph f and the data.

x	2	3	4	5
y	1.70	2.08	2.40	2.68

Extended and Discovery Exercises

1. *Lunar Orbits*   The table lists the orbital distances and periods of several moons of Jupiter. Let x represent the distance and y the period. Find a function f that models this data. Try more than one type of function such as linear, quadratic, power, and exponential. Graph each function and the data. Discuss which function models the data best. (Source: M. Zeilik, *Introductory Astronomy and Astrophysics.*)

Moons of Jupiter	Distance (10^3km)	Period (days)
Metis	128	0.29
Almathea	181	0.50
Thebe	222	0.67
Io	422	1.77
Europa	671	3.55
Ganymede	1070	7.16
Callisto	1883	16.69

2. *Global Warming*   Greenhouse gases such as carbon dioxide trap heat from the sun. Presently, the net incoming solar radiation reaching the earth's surface is approximately 240 watts per square meter (w/m^2). Any portion of this amount that is due to greenhouse gases is called **radiative forcing.** The accompanying table lists the estimated increase in radiative forcing R over 1750 levels. (Source: A. Nilsson, *Greenhouse Earth.*)

x (year)	1800	1850	1900	1950	2000
$R(x)$ (w/m^2)	0.2	0.4	0.6	1.2	2.4

(a)   Estimate constants C and k so that $R(x) = Ce^{kx}$ models the data. Let $x = 0$ correspond to 1800.

(b)   Determine symbolically when the additional radiative forcing could reach 3 w/m^2. Support your answer graphically.

3. *Global Warming*   (Refer to the previous exercise.) The relationship between radiative forcing R and the increase in average global temperature T in degrees Fahrenheit can be modeled by $T(R) = 1.03R$. For example, $T(2) = 1.03(2) = 2.06$ means that if the earth's atmosphere traps an additional 2 watts per square meter, then the average global temperature may increase by 2.06°F *if* all other factors remained constant. (Source: W. Clime, *The Economics of Global Warming.*)

(a) Use $R(x)$ from the previous exercise to express $(T \circ R)(x)$ symbolically, where x is the year and $x = 0$ corresponds to the year 1800.

(b) Evaluate $(T \circ R)(100)$ and interpret its meaning.

Exercises 4 and 5: Evaluate the expression as accurately as possible. Try to decide if it is precisely an integer. (Hint: Use computer software capable of calculating a large number of significant digits.)

4. $\left(\dfrac{1}{\pi} \ln(640{,}320^3 + 744)\right)^2$ (Source: I. J. Good, "What is the most amazing approximate integer in the universe?")

5. $e^{\pi\sqrt{163}}$ (Source: W. Cheney and D. Kincaid, *Numerical Mathematics and Computing.*)

CHAPTER 5 *Trigonometric Functions*

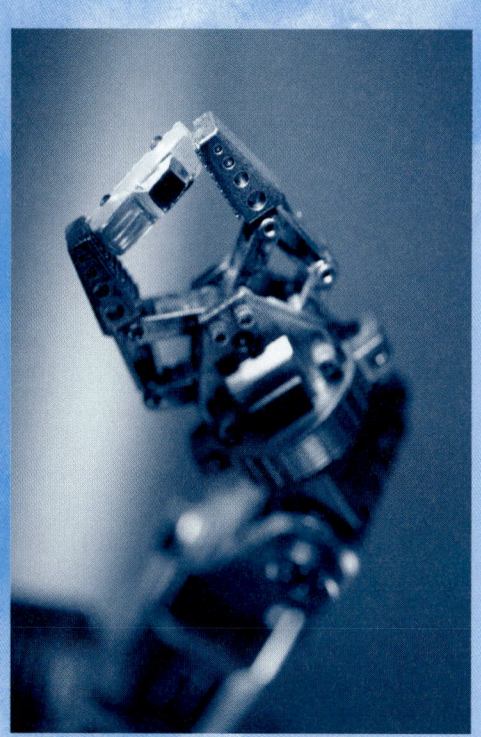

*T*rigonometry has been used for millennia to solve problems involving astronomy, surveying, and construction. The word trigonometry is derived from two Greek words, which mean triangle measurement. In fact, the Greek astronomer Hipparchus is usually given credit for first studying the trigonometric properties of angles. Trigonometry originated when people tried to correlate shadow lengths with the time of day. Prior to the fifteenth century, astronomy had the greatest influence on the development of trigonometry, and it was not until the thirteenth century that astronomy and trigonometry could be regarded as separate entities.

In modern times trigonometric relationships have been viewed more in terms of functions and graphs. One reason for this is the explosion of new technologies in our society. Trigonometric functions often are used to model phenomena that involve periodic motion or rotation. Some examples include average monthly temperature, daylight hours, tides, tidal currents, the Global Positioning System (GPS), movement of robotic arms, electricity, highway design,

> *E*ducation is not the filling of a pail, but the lighting of a fire.
> — **William Butler Yeats**

orbits of satellites, the phases of the moon, and music. Trigonometric functions have been essential to the development of our modern society. They are similar to other types of functions that we have studied and have verbal, graphical, numerical, and symbolic representations.

Sources: H. Freebury, *A History of Mathematics; Historical Topics for the Mathematics Classroom, Thirty-first Yearbook,* NCTM.

5.1 Angles and their Measure

Angles • Degree Measure • Radian Measure
• Arc Length • Area of a Sector

Introduction

The concept of an angle dates back thousands of years. Degree measure began in Babylonia (5000–4000 B.C.), where angles were frequently used in astronomy. Astronomy was important to society because of its connections to the calendar, the seasons, and planting times. Around 1873 a second unit to measure angles, called a *radian,* was developed independently by Thomas Muir and James Thomson, a mathematician and a physicist. Radian measure is used because it often results in simpler formulas. (Source: *Historical Topics for the Mathematics Classroom, Thirty-first Yearbook,* NCTM.)

Angles

An **angle** is formed by rotating a ray about its endpoint. The starting position of the ray is called the **initial side** and the final position of the ray is the **terminal side.** If the rotation of the ray is counterclockwise, the angle has *positive measure,* whereas if the rotation is clockwise, the angle has *negative measure.* For simplicity we will refer to an angle as being **positive** or **negative**. The endpoint of the ray is called the **vertex** of the angle. See Figures 5.1 and 5.2, where the Greek letter θ (theta) has been used to denote an angle.

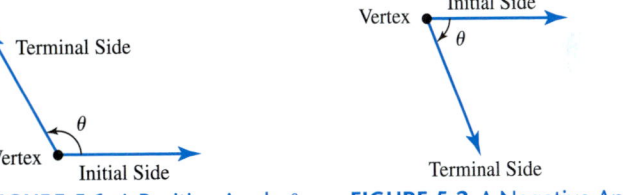

FIGURE 5.1 A Positive Angle θ **FIGURE 5.2** A Negative Angle θ

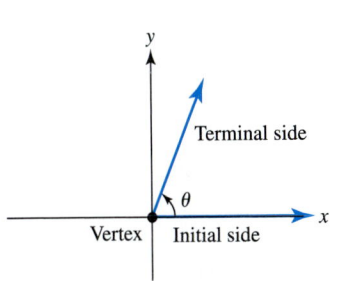

FIGURE 5.3 Standard Position

If the vertex is positioned so it corresponds to the origin in the *xy*-plane and the initial side coincides to the positive *x*-axis, then the angle is in **standard position** as shown in Figure 5.3.

There are two common systems to measure the size of an angle: *degree measure* and *radian measure.*

Degree Measure

In degree measure, one complete rotation of a ray about its endpoint contains 360 degrees. One degree, denoted by 1°, represents $\frac{1}{360}$ of a complete rotation. Figure 5.4 on the following page shows some examples of angles with their degree measure.

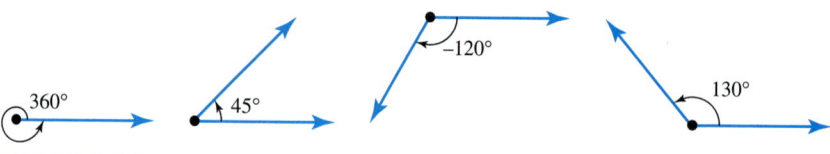

FIGURE 5.4 Degree Measure

A **right angle** has measure 90°, and a **straight angle** has measure 180°. The measure of an **acute angle** is greater than 0° but less than 90°, whereas the measure of an **obtuse angle** is greater than 90° but less than 180°. Examples are shown in Figure 5.5.

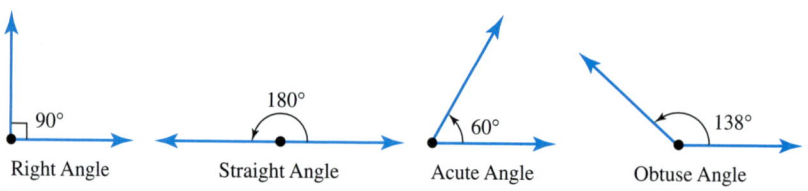

FIGURE 5.5 Types of Angles

We will use the Greek letters α (alpha), β (beta), γ (gamma), and θ (theta) to denote angles. For simplicity, we sometimes refer to an angle θ having measure 45° as a 45° angle, or an angle of 45°. This may be expressed as $\theta = 45°$. Two different angles with the same initial and terminal sides are **coterminal angles.** Examples of coterminal angles are shown in Figure 5.6.

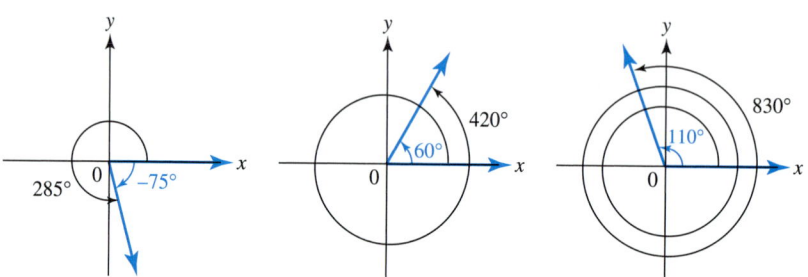

FIGURE 5.6 Coterminal Angles

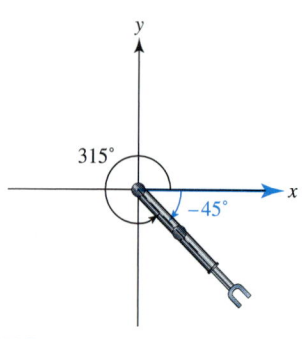

FIGURE 5.7 A Robotic Arm

Use of robotic arms has increased dramatically throughout society. Robots assemble products, paint vehicles, prepare fast foods, and even link space stations together. Coterminal angles have applications in robotics. Suppose a robotic arm, called a *simple polar manipulator,* picks up a bolt at an initial position corresponding to the positive x-axis and inserts the bolt at a final position as illustrated in Figure 5.7. There are different angles that the shoulder of the robotic arm could rotate through to accomplish this task. For example, the arm could either rotate 45° clockwise or 315° counterclockwise. These rotations are represented by coterminal angles of −45° or 315°. Either angle accomplishes the task. However, an angle of −45° is usually preferable, since it involves less movement for the robotic arm and saves time. (Source: J. Craig, *Introduction to Robotics: Mechanics and Control.)*

EXAMPLE 1 *Finding coterminal angles*

Find three angles coterminal to $\theta = 45°$, where θ is in standard position. Sketch these angles in standard position.

Solution

We can find coterminal angles by either adding or subtracting multiples of 360° to θ.

i. $45° + 360° = 405°$ **ii.** $45° + 2(360°) = 765°$ **iii.** $45° - 360° = -315°$

The angles 405°, 765°, and −315° are all coterminal to a 45° angle. These three angles and θ are sketched in Figure 5.8.

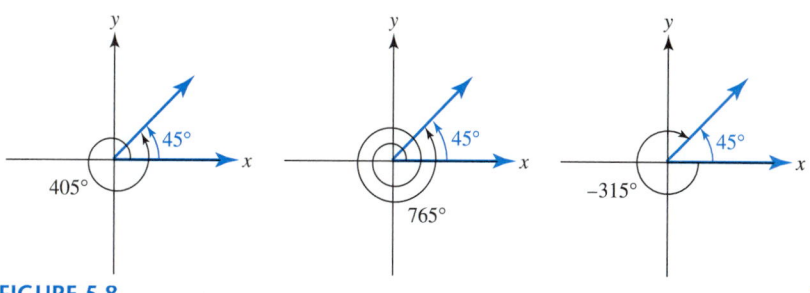

FIGURE 5.8

Two positive angles are **complementary angles** if their sum equals 90° and are **supplementary angles** if their sum is 180°. For example, $\alpha = 35°$ and $\beta = 55°$ are complementary angles, whereas $\alpha = 60°$ and $\beta = 120°$ are supplementary angles. See Figures 5.9 and 5.10.

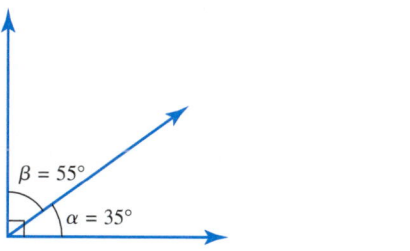

FIGURE 5.9 Complementary Angles   **FIGURE 5.10** Supplementary Angles

Fractions of a degree may be measured using **minutes** and **seconds.** One minute, written $1'$, equals $\dfrac{1}{60}$ of a degree and one second, written $1''$, equals $\dfrac{1}{60}$ of a minute or $\dfrac{1}{3600}$ of a degree. The measurement $25°45'30''$ represents 25 degrees, 45 minutes, 30 seconds. Expressed in decimal degrees, this measurement is

$$24°45'30'' = 24° + \left(\frac{45}{60}\right)^° + \left(\frac{30}{3600}\right)^° = 24.758\overline{3}°.$$

EXAMPLE 2 *Finding complementary and supplementary angles*

Find angles that are complementary and supplementary to $\alpha = 34°19'42''$.

Solution

If angle β is complementary to α, then $\beta = 90° - \alpha$.

$$\beta = 90° - 34°19'42''$$
$$= 89°59'60'' - 34°19'42'' \qquad 90° = 89°59'60''$$
$$= 55°40'18''$$

A supplementary angle to α is given by $\gamma = 180° - \alpha$.

$$\gamma = 180° - 34°19'42''$$
$$= 179°59'60'' - 34°19'42'' \qquad 180° = 179°59'60''$$
$$= 145°40'18''$$ ∎

Technology Note

Some calculators are capable of performing arithmetic using degrees, minutes, and seconds as shown in Figures 5.11 and 5.12.

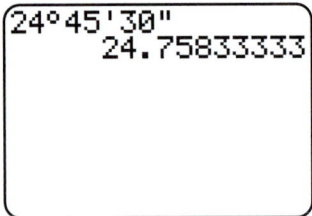

FIGURE 5.11 Degree Mode

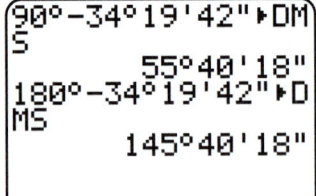

FIGURE 5.12 Degree Mode

A new and exciting technology is the **Global Positioning System** (GPS). This system involves 24 satellites in nearly circular orbits that can be used to determine locations and velocities on Earth with a high degree of accuracy. See Figure 5.13. Private individuals can purchase handheld GPS devices that determine coordinates within 100 meters and velocities within 2 meters per second. Although GPS is capable of locating an object within a few feet, this service is not available to the general public for security reasons. (Source: J. Sickle, *GPS for Land Surveyors.*)

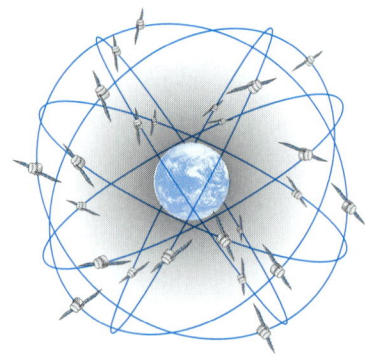

FIGURE 5.13 GPS Satellite Coverage

EXAMPLE 3   *Determining the accuracy of GPS*

Some GPS receivers display the latitude of a location to within $\pm\dfrac{1}{60{,}000}^{\circ}$. Convert this error to seconds. (Source: G. West, *"Differential GPS—how accurate is it?"*)

Solution

Since there are 3600 seconds in one degree it follows that

$$\pm\frac{1}{60{,}000}^{\circ} \times 3600 = \pm\frac{3}{50}'' = \pm 0.06''$$

Thus, this GPS device can display a latitude measurement to within $\pm 0.06''$. ■

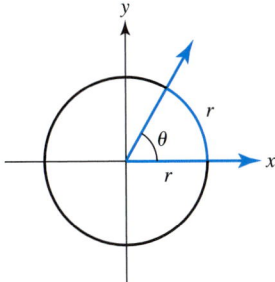

FIGURE 5.14 One Radian

Radian Measure

A second unit of angle measure is *radians.* Radian measure is a common form of measurement in many technical fields, including calculus. Radian measure often results in formulas being simpler and easier to use. Two examples of this are arc length and area of a sector, which are introduced later in this section.

Angle θ in Figure 5.14 has a measure of one *radian.* The vertex of θ is located at the center of the circle, while its initial and terminal sides intercept an arc whose length is equal to the radius of the circle.

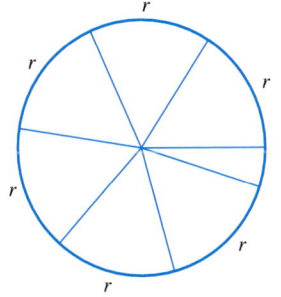

FIGURE 5.15 One Revolution Contains 2π Radians

Radian measure

An angle that has its vertex at the center of a circle and intercepts an arc on the circle equal in length to the radius of the circle has a measure of **one radian.**

The circumference of a circle is $C = 2\pi r$. If we mark off distances of r along the circumference of a circle, it will appear as in Figure 5.15, where $2\pi \approx 6.28$ distances of r are shown. Therefore, one rotation contains $2\pi \approx 6.28$ radians. It follows that 360° is equivalent to 2π radians. Radian measure can be compared to degree measure using proportions. Since 180° is equivalent to π radians, it follows that

$$\frac{\text{radian measure}}{\text{degree measure}} = \frac{\pi}{180°}.$$

Solving for radian measure results in

$$\text{radian measure} = \text{degree measure} \times \frac{\pi}{180°},$$

while solving for degree measure results in

$$\text{degree measure} = \text{radian measure} \times \frac{180°}{\pi}.$$

The following statements verbally summarize the above discussion.

> ### Converting between degrees and radians
>
> To convert *degrees to radians,* multiply a degree measure by $\dfrac{\pi}{180°}$.
>
> To convert *radians to degrees,* multiply a radian measure by $\dfrac{180°}{\pi}$.

EXAMPLE 4   *Converting degrees to radians*

Convert each degree measure to radian measure.

(a) $90°$ **(b)** $225°$

Solution

(a) To convert degrees to radians, multiply by $\dfrac{\pi}{180°}$.

$$90° \times \frac{\pi}{180°} = \frac{\pi}{2} \text{ radians}$$

Thus, $90°$ are equivalent to $\dfrac{\pi}{2}$ radians.

(b) $225° \times \dfrac{\pi}{180°} = \dfrac{5\pi}{4}$ radians.   ∎

Table 5.1 shows some equivalent measures in degrees and radians.

TABLE 5.1

Degrees	0°	30°	45°	60°	90°	180°	360°
Radians	0	$\dfrac{\pi}{6}$	$\dfrac{\pi}{4}$	$\dfrac{\pi}{3}$	$\dfrac{\pi}{2}$	π	2π

EXAMPLE 5   *Converting radians to degrees*

Convert each radian measure to degree measure.

(a) $\dfrac{4\pi}{3}$ **(b)** $\dfrac{5\pi}{6}$

Solution

(a) To convert radians to degrees, multiply by $\dfrac{180°}{\pi}$.

$$\frac{4\pi}{3} \times \frac{180°}{\pi} = 240°$$

Thus, $\dfrac{4\pi}{3}$ radians are equivalent to $240°$.

(b) $\dfrac{5\pi}{6} \times \dfrac{180°}{\pi} = 150°.$ ∎

Technology Note   *Converting Angle Measures*

Some calculators can convert degrees to radians and radians to degrees as shown in Figures 5.16 and 5.17, respectively. When converting from degrees to radians, many calculators will only give decimal approximations rather than exact values. For example, $\dfrac{\pi}{2}$ may be expressed as 1.570796327.

```
90°
          1.570796327
225°
          3.926990817
```

```
(4π/3)ʳ
                    240
(5π/6)ʳ
                    150
```

FIGURE 5.16 Radian Mode   **FIGURE 5.17** Degree Mode

Arc Length

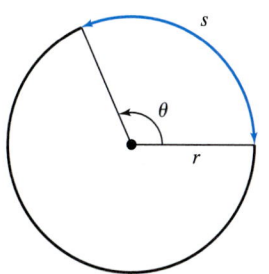

FIGURE 5.18 Arc Length *s*

From geometry we know that the arc length s on a circle is proportional to the measure of the central angle θ. See Figure 5.18. A central angle of 2π radians corresponds to an arc length that equals the circumference $C = 2\pi r$. Using proportions,

$$\frac{s}{\theta} = \frac{2\pi r}{2\pi}$$

which simplifies to $s = r\theta$.

> ### Arc length
>
> The **arc length** s intercepted on a circle of radius r by a central angle of θ *radians* is given by
>
> $$s = r\theta.$$

Note: Angle θ *must* be in radian measure when using the arc length formula $s = r\theta$.

EXAMPLE 6   *Finding arc length*

A circle has a radius of 25 inches. Find the length of an arc intercepted by a central angle of 45°.

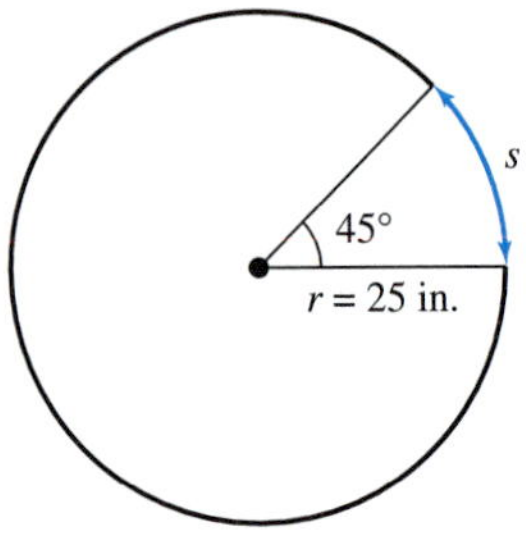

Solution

First convert 45° to radian measure.

$$45° \times \frac{\pi}{180°} = \frac{\pi}{4}$$

The arc length s is given by

$$s = r\theta$$
$$= 25\left(\frac{\pi}{4}\right)$$
$$= 6.25\pi \text{ inches}$$

The arc length shown in Figure 5.19 is $6.25\pi \approx 19.6$ inches. ■

FIGURE 5.19

EXAMPLE 7 *Finding distance between cities*

Albuquerque, New Mexico and Glasgow, Montana have the same longitude of 106°37′ W. The latitude of Albuquerque is 35°03′ N and the latitude of Glasgow is 48°13′ N. If the radius of Earth is approximately 3955 miles, estimate the distance between Albuquerque and Glasgow. See Figure 5.20. *(Source: J. Williams, The Weather Almanac 1995.)*

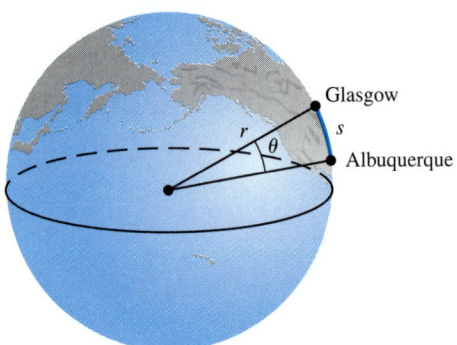

FIGURE 5.20 Distance Between Two Cities (Not to scale)

Solution

This distance can be estimated using the arc length formula. Start by converting $\theta = 48°13′ - 35°03′ = 13°10′$ to radian measure.

$$\left[13° + \left(\frac{10}{60}\right)°\right] \times \frac{\pi}{180°} \approx 0.2298 \text{ radian}$$

The distance between Albuquerque and Glasgow is approximated by

$$s = r\theta$$
$$\approx 3955(0.2298)$$
$$\approx 909 \text{ miles.}$$ ■

The human joint that can be flexed the fastest is the wrist, which can rotate through 90°, or $\frac{\pi}{2}$ radians, in 0.045 second while holding a tennis racket. See

FIGURE 5.21 Flexing the Wrist in Tennis

Figure 5.21. **Angular speed** ω (omega) measures the speed of rotation and is defined by

$$\omega = \frac{\theta}{t},$$

where θ is the angle of rotation and t is time. The angular speed of a human wrist holding a tennis racket is

$$\frac{\pi/2}{0.045} \approx 34.9 \text{ rad/sec,}$$

or about 35 radians per second.

The **linear speed** v at which the tip of the racket travels as a result of flexing the wrist is given by $v = r\omega$, where r is the radius from the end of the racket to the wrist joint. If $r = 2$ feet, then the speed at the tip of the racket is

$$v = r\omega \approx (2)(35) = 70 \text{ ft/sec,}$$

or about 48 miles per hour. Since the arm rotates at the shoulder, the final speed of the racket is considerably faster. (Source: J. Cooper and R. Glassow, *Kinesiology.*)

Critical Thinking

A human shoulder can rotate at about 25 radians per second. Estimate how much this rotation increases the speed of a racket.

EXAMPLE 8

Finding the speed of a GPS satellite

Each of the 24 satellites used in the GPS is located 16,526 miles from the center of Earth and has a nearly circular orbit with a period of 12 hours. (Source: Y. Zhoa, *Vehicle Location and Navigation Systems.*)

(a) Find the angular speed of a satellite.
(b) Estimate the linear speed of a satellite using the formula $v = r\omega$.

Solution

(a) A GPS satellite circles Earth once every 12 hours. Its angular speed is

$$\frac{2\pi}{12} = \frac{\pi}{6} \approx 0.5236 \text{ rad/hr.}$$

(b) Its linear speed is $v = (16,526)(0.5236) \approx 8653$ miles per hour. ∎

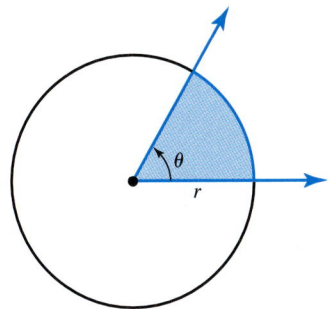

FIGURE 5.22 Sector of a Circle

Area of a Sector

The **sector of a circle** is the portion of the interior of a circle intercepted by a central angle. The shaded region in Figure 5.22 shows a sector of a circle with radius r and central angle θ.

The area of a sector is proportional to the measure of the central angle. If the central angle is 2π radians, then the area of the sector is the entire interior of the circle, which has an area of πr^2. Using proportions,

$$\frac{\text{area of a sector}}{\theta} = \frac{\pi r^2}{2\pi}.$$

Solving the equation for area of the sector results in

$$\text{area of a sector} = \frac{1}{2}r^2\theta.$$

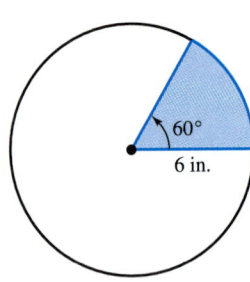

FIGURE 5.23

Area of a sector

The **area of a sector** A of a circle of radius r and central angle θ in *radians* is given by

$$A = \frac{1}{2}r^2\theta.$$

Note: Angle θ *must* be in radian measure when using the area formula $A = \frac{1}{2}r^2\theta$.

EXAMPLE 9 *Finding the area of a sector*

A circle has a radius of 6 inches. Find the area of the sector if its central angle is 60°.

Solution

Since 60° is equivalent to $\dfrac{\pi}{3}$ radians, the area of the sector is given by

$$A = \frac{1}{2}r^2\theta$$

$$= \frac{1}{2}(6)^2\left(\frac{\pi}{3}\right)$$

$$= 6\pi \text{ square inches.}$$

This area is $6\pi \approx 18.8$ square inches as illustrated in Figure 5.23. ∎

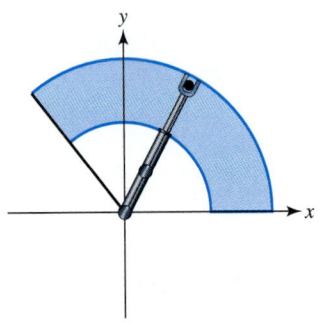

FIGURE 5.24 Work Space of a Robotic Arm

Consider the robotic arm shown in Figure 5.24. The *work space* of the robotic hand is the shaded region and corresponds to the places that the hand can reach either by rotating or changing its length. (Source: W. Stadler, *Analytical Robotics and Mechatronics.*)

EXAMPLE 10 *Finding the area of the work space for a robotic arm*

Suppose that the arm in Figure 5.24 can rotate between $\theta = 10°$ and $\theta = 130°$. If the length of the robotic arm can vary between 5 inches and 20 inches, find the area of its work space.

Solution

The work space can be thought of as a large sector having radius $r_1 = 20$ inches with a small sector of radius $r_2 = 5$ inches removed. See Figure 5.25 on the next page. The arm can rotate through $130° - 10° = 120°$ or $2\pi/3$ radians. The area A of the work space is as follows.

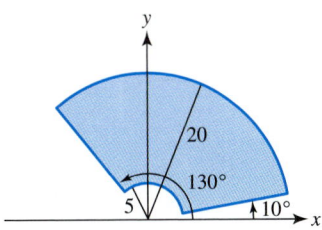

FIG. 5.25

$$A = \frac{1}{2}r_1^2\theta - \frac{1}{2}r_2^2\theta$$

$$= \frac{1}{2}\theta(r_1^2 - r_2^2)$$

$$= \frac{1}{2}\left(\frac{2\pi}{3}\right)(20^2 - 5^2)$$

$$= 125\pi$$

The work space is $125\pi \approx 392.7$ square inches.

5.1 PUTTING IT ALL TOGETHER

Some concepts involving angles are summarized in the following table.

Concept	Explanation	Examples
Degree Measure	One complete rotation contains 360°.	A right angle contains 90°. A straight angle contains 180°. An acute angle α satisfies $0° < \alpha < 90°$. An obtuse angle β satisfies $90° < \beta < 180°$.
Radian Measure	One complete rotation contains 2π radians.	2π radians are equivalent to 360°. π radians are equivalent to 180°. $\frac{\pi}{2}$ radians are equivalent to 90°. $\frac{\pi}{3}$ radians are equivalent to 60°. $\frac{\pi}{4}$ radian is equivalent to 45°. $\frac{\pi}{6}$ radian is equivalent to 30°.
Arc Length	$s = r\theta$, where θ is in radians.	If $r = 12$ feet and $\theta = 90°$, then $s = 12 \times \frac{\pi}{2} = 6\pi \approx 18.8$ feet
Area of a Sector	$A = \frac{1}{2}r^2\theta$, where θ is in radians.	If $r = 6$ inches and $\theta = 45°$, then $A = \frac{1}{2}(6)^2\left(\frac{\pi}{4}\right) = 4.5\pi \approx 14.1$ square inches.
Angular Speed	$\omega = \frac{\theta}{t}$, where θ is the angle of rotation and t is time.	If $\theta = 5$ radians and $t = 0.1$ second then $\omega = \frac{5}{0.1} = 50$ radians per second.
Linear Speed of a Rotating Object	$v = r\omega$, where r is the radius and ω is the angular speed.	If $r = 3$ feet and $\omega = 5$ radians per second, then $v = (3)(5) = 15$ feet per second.

5.1 EXERCISES

   Tape 8

Angles

Exercises 1 and 2: Sketch the following angles in standard position.

1. (a) 45°   (b) −150°

    (c) $\dfrac{\pi}{3}$ (d) $-\dfrac{3\pi}{4}$

2. (a) −90°   (b) 225°

    (c) $-\dfrac{2\pi}{3}$ (d) $\dfrac{\pi}{6}$

Exercises 3–10: Sketch an angle θ in standard position that satisfies the conditions.

3. Acute

4. Obtuse

5. A positive straight angle

6. Complementary to 60°

7. Positive and the terminal side lies in quadrant III

8. Negative and the terminal side lies in quadrant IV

9. Negative and coterminal to $\alpha = 90°$ if α is in standard position

10. Positive and coterminal to $\alpha = -135°$ if α is in standard position

11. What fraction of a complete revolution are each of the following angles?

    (a) 90°   (b) 30°

    (c) $\dfrac{\pi}{3}$ (d) $\dfrac{\pi}{4}$

12. What angle is its own complement? What angle is its own supplement?

Exercises 13–16: Use the figure to determine the radian measure of angle θ. Then approximate the degree measure of θ to the nearest tenth of a degree.

13. 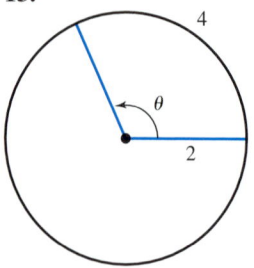   14.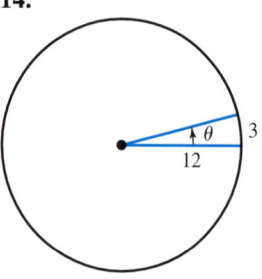

15. 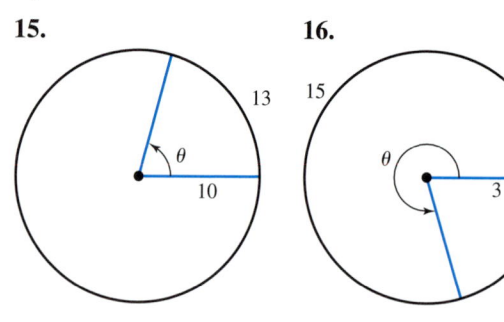   16.

Exercises 17–22: Find the complementary angle α and the supplementary angle β to θ.

17. $\theta = 55.9°$ 18. $\theta = 71.5°$

19. $\theta = 85°23'45''$ 20. $\theta = 5°45'30''$

21. $\theta = 23°40'35''$ 22. $\theta = 67°25'10''$

Exercises 23–26: Express the angle in decimal degrees.

23. 125°15'   24. 15°30'

25. 108°45'36''   26. 256°06'12''

Exercises 27–30: Convert the given angle to degrees, minutes, and seconds.

27. 125.3°   28. 15.25°

29. 51.36°   30. 22.46°

Exercises 31 and 32: Convert each angle from radian measure to degree measure.

31. (a) $\dfrac{\pi}{6}$ (b) $\dfrac{\pi}{15}$

    (c) $-\dfrac{5\pi}{3}$ (d) $-\dfrac{7\pi}{6}$

32. (a) $-\dfrac{\pi}{12}$ (b) $-\dfrac{5\pi}{2}$

    (c) $\dfrac{17\pi}{15}$ (d) $\dfrac{5\pi}{6}$

Exercises 33 and 34: Convert each angle from degree measure to radian measure.

33. (a) 45°   (b) 135°
    (c) −120°   (d) −210°

34. (a) 105°   (b) 245°
    (c) −255°   (d) −80°

Exercises 35–42: Find a positive angle and a negative angle that are coterminal to the given angle.

35. 150°   **36.** 65°

37. −72°   **38.** −330°

39. $\dfrac{\pi}{2}$ **40.** $\dfrac{5\pi}{6}$

41. $-\dfrac{\pi}{5}$ **42.** $-\dfrac{2\pi}{3}$

Arc Length

Exercises 43–48: Use the formula $s = r\theta$ to determine the missing value in the figure.

43.   **44.**

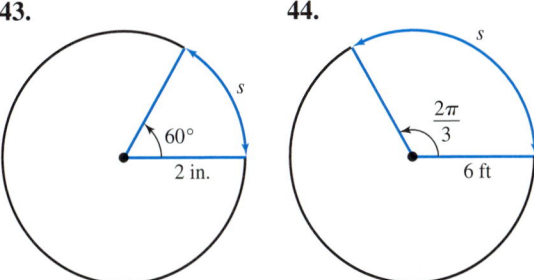

45.   **46.**

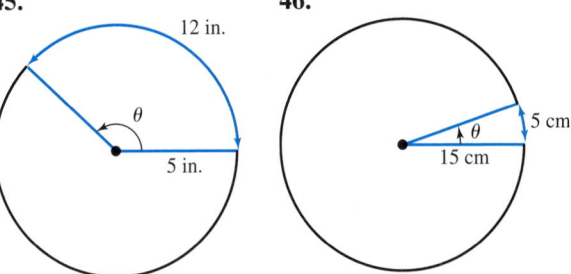

47.   **48.**

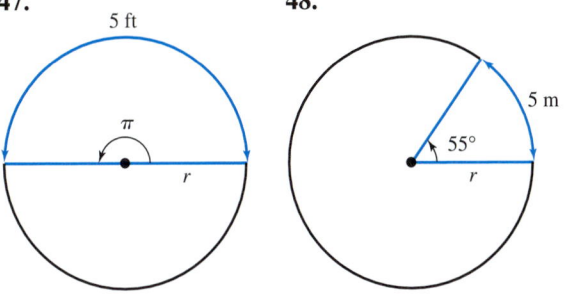

Exercises 49–54: Find the length of the arc intercepted by a central angle θ in a circle of radius r.

49. $r = 3$ m, $\theta = \dfrac{\pi}{12}$ **50.** $r = 7.3$ mm, $\theta = \dfrac{7\pi}{4}$

51. $r = 12$ ft, $\theta = 15°$ **52.** $r = 5$ cm, $\theta = 240°$

53. $r = 2$ mi, $\theta = 1°45'$

54. $r = 3$ mi, $\theta = 4°15'09''$

Exercises 55–58: A minute hand on a clock is 4 inches long. Determine how far the tip of the minute hand travels between the given times. Find the linear speed of the tip.

55. 10:15 A.M., 10:30 A.M.

56. 1:00 P.M., 1:40 P.M.

57. 3:00 P.M., 4:15 P.M.

58. 11:00 A.M., 1:25 P.M.

59. A bicycle has a tire 26 inches in diameter that is rotating at 15 radians per second. Approximate the speed of the bicycle in feet per second and in miles per hour.

60. The wheels on a skate board have a diameter of 2.25 inches. If a skate boarder is traveling downhill at 15 miles per hour, determine the angular velocity of the wheels in radians per second.

Area of a Sector

Exercises 61–64: Find the area of the shaded sector.

61.   **62.**

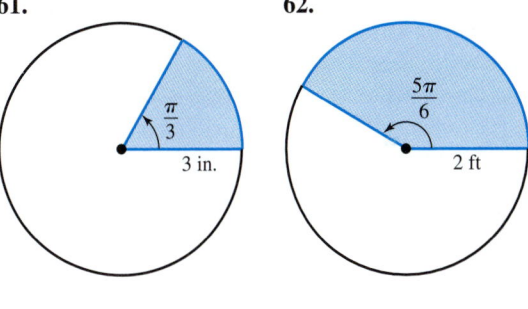

63.   **64.**

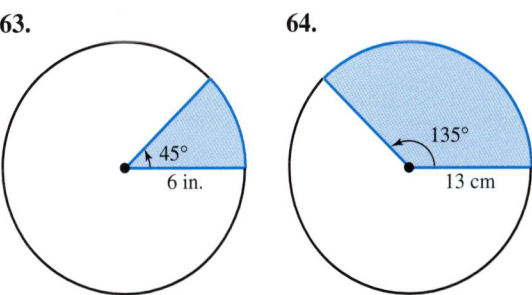

Exercises 65–68: Find the area of the sector of a circle having radius r and central angle θ.

65. $r = 13.1$ cm, $\theta = \dfrac{\pi}{15}$ **66.** $r = 7.3$ m, $\theta = \dfrac{5\pi}{4}$

67. $r = 1.5$ ft, $\theta = 30°$ **68.** $r = 5.5$ in., $\theta = 225°$

Exercises 69–72: Robotics (Refer to Example 10.) Find the area of the work space for a robotic arm that can rotate between angles θ_1 and θ_2 and can change its length from r_1 to r_2. See the accompanying figure.

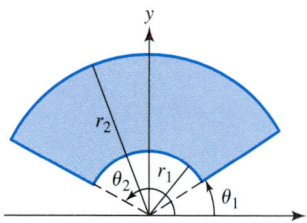

69. $\theta_1 = -45°$, $\theta_2 = 90°$, $r_1 = 6$ in., $r_2 = 26$ in.
70. $\theta_1 = -60°$, $\theta_2 = 60°$, $r_1 = 0.5$ ft, $r_2 = 2.5$ ft
71. $\theta_1 = 15°$, $\theta_2 = 195°$, $r_1 = 21$ cm, $r_2 = 95$ cm
72. $\theta_1 = 43°$, $\theta_2 = 178°$, $r_1 = 0.4$ m, $r_2 = 1.8$ m

Applications

73. *Distance Between Cities* (Refer to Example 7.) Daytona Beach, Florida and Akron, Ohio have nearly the same longitude of 81° W. The latitude of Daytona Beach is 29°11′ and the latitude of Akron is 40°55′. Approximate the distance between these two cities if the average radius of Earth is 3955 miles. (Source: J. Williams.)

74. *Nautical Miles* Nautical miles are used by ships and airplanes. They are different than statute miles, which equal 5280 feet. A nautical mile is defined to be the arc length along the equator intercepted by a central angle *AOB* of 1 minute as illustrated in the figure below at the left. If the equatorial radius of Earth is 3963 miles, use the arc length formula to approximate the number of statute miles in one nautical mile. Round your answer to three significant digits.

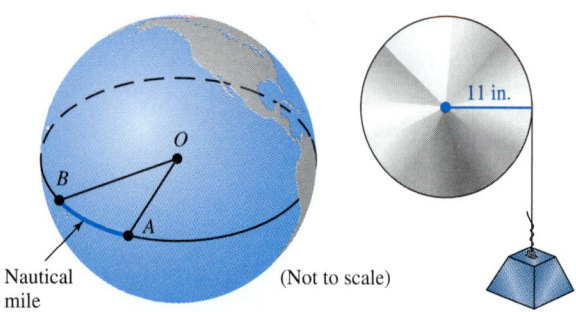

Nautical mile   (Not to scale)

75. *Pulleys* Approximate how many inches the weight in the figure above at the right will rise if $r = 11$ inches and the pulley is rotated through an angle of 75.3°.

76. *Pulleys* Use the figure in the previous exercise to estimate the angle θ that the pulley should be rotated through to raise the weight 5 inches.

77. *Bicycle Chain Drive* The figure shows the chain drive of a bicycle. The diameter of the sprocket wheel that the pedals are attached to is 7.5 inches and the diameter of the other sprocket wheel is 3 inches.

(a) If the pedals are rotated one revolution, determine the number of revolutions that the bicycle tire rotates.

(b) If the bicycle has a tire with a 26-inch diameter, determine how fast the bicycle travels in feet per second if the pedals turn through two revolutions per second.

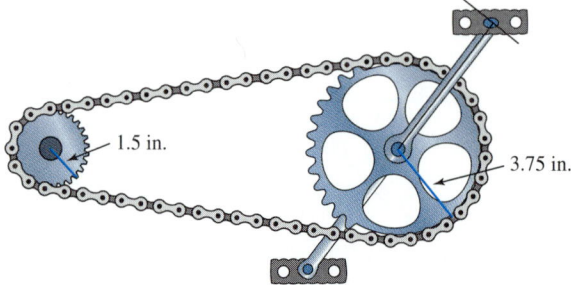

78. *Wind Speed* One of the most common ways to measure wind speed is with a *three-cup anemometer* as shown in the accompanying figure. The cups catch the wind and cause the vertical shaft to rotate. At lower wind speeds the cups move at approximately the same speed as the wind. If the cups are rotating 5 times per second with a radius of 6 inches, estimate the wind speed in miles per hour. (Source: J. Navarra, *Atmosphere, Weather and Climate.*)

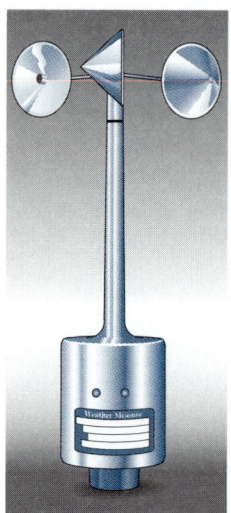

79. *Velocity of Planets* The average distance D in millions of miles from the sun and the orbital period P in years of various planets are given. Assuming that the orbits are circular, approximate the average orbital velocity in miles per hour for each planet. Discuss the effect that average distance from the sun has on orbital velocity. (Source: C. Ronan, *The Natural History of the Universe.*)

(a) Venus: $D = 67.2$, $P = 0.615$

(b) Earth: $D = 92.9$, $P = 1$

(c) Jupiter: $D = 483.6$, $P = 11.86$

(d) Neptune: $D = 2794$, $P = 164.8$

80. *Speed of a Propeller* A 90-horsepower outboard motor at full-throttle will rotate its propeller at 5000 revolutions per minute. Find the angular velocity of the propeller in radians per second. What is the linear speed in inches per second of a point at the tip of the propeller if its diameter is 10 inches?

81. *Surveying* The *subtense bar method* is a technique used in surveying to measure distances. A subtense bar, which is usually 2 meters long, is shown in the accompanying figure connecting points P and Q. If the distance d from the surveyor to the bar is large, then there is little difference between the length of the subtense bar and the arc connecting P and Q. Similarly, there is little difference between d and the radius r of the arc intercepted by the subtense bar. If θ is measured to be $0.835°$, approximate d using the arc length formula. (Source: I. Mueller and K. Ramsayer, *Introduction to Surveying.*)

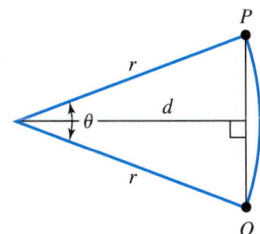

82. *Diameter of the Moon* (Refer to the previous exercise.) The distance to the moon is approximately 238,900 miles. Use the arc length formula to estimate the diameter d of the moon if angle θ in the accompanying figure is measured to be $0.517°$.

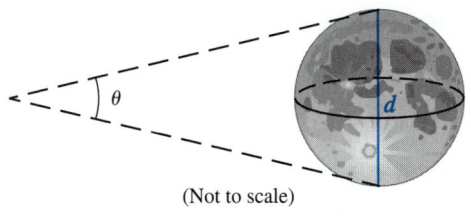

(Not to scale)

83. *Global Positioning System* For civilians, a GPS location device can consistently determine latitude to within $\pm 0.001°$. If the average radius of Earth is 3955 miles, approximate how accurately the north-south position of an object can be found by a civilian. (Source: Y. Zhao.)

84. *Doppler Radar* Radar is used to identify severe weather. If Doppler radar can detect weather within a 240-mile radius and creates a new image every 48 seconds, find the area scanned by the radar in one second.

85. *Arc Length Formula* Modify the arc length formula $s = r\theta$ so that angle θ can be given in degrees rather than radians. Which of the two formulas is simpler?

86. *Area of a Sector Formula* Modify the area formula $A = \frac{1}{2}r^2\theta$ so that angle θ can be given in degrees rather than radians. Which of the two formulas is simpler?

87. *Solar Power Plant* A 150-megawatt solar power plant requires approximately 475,000 square meters of land to collect the required amount of energy from sunlight. (Source: C. Winter, *Solar Power Plants.*)

(a) If this land area is circular, approximate its radius.

(b) If this land area is a sector of a circle with $\theta = 70°$, approximate its radius.

88. *Location of the North Star* Presently the north star, Polaris, is located near the true north pole. However, because Earth is inclined $23.5°$, the moon's gravitational pull on Earth is uneven. As a result, Earth precesses like a spinning top and the direction of the celestial north pole traces out a circular path once every 26,000 years as shown below. For example, in the year 14,000 the star Vega and not Polaris will be located at the celestial north pole. As viewed from the center C of this circular path, calculate the angle in seconds that the celestial north pole moves each year. (Source: M. Zeilik, *Introductory Astronomy and Astrophysics.*)

89. *Measuring the Circumference of Earth*   The first accurate estimate of the distance around Earth was done by the Greek astronomer Eratosthenes (276–195 B.C.), who noted that the noontime position of the sun at the summer solstice differed by 7°12′ from the city of Syene to the city of Alexander. See the accompanying figure. The distance between these two cities is 496 miles. Use the arc length formula to estimate the radius of Earth. Then find the circumference of Earth. (Source: M. Zeilik.)

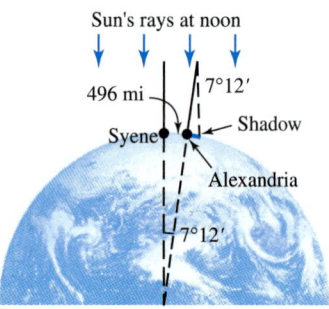

90. *Club Speed in Golf*   The shoulder joint can rotate at about 25 radians per second. Assuming that a golfer's arm is straight and the distance from the shoulder to the club head is 5 feet, estimate the linear speed of the club head from shoulder rotation. (Source: J. Cooper.)

Writing about Mathematics

1.   Give definitions for one degree and one radian. Compare these two units of angle measure. Which unit of measure do you prefer? Explain why.

2.   Suppose a central angle θ of a circle remains fixed. Describe what happens to the arc length intercepted by θ and the area of the corresponding sector as the radius r doubles and triples.

5.2 Right Triangle Trigonometry

Basic Concepts of Trigonometric Functions • Applications of Right Triangle Trigonometry • Complementary Angles and Cofunctions

Introduction

A right triangle is a basic geometric shape that occurs in many applications such as astronomy, surveying, construction, highway design, GPS, and aerial photography. Trigonometric functions are used to *solve* triangles. Solving a triangle involves finding the measure of each side and angle in the triangle. The sine function is one of the earliest trigonometric functions and it dates back to the Greeks, where they invented a similar function called the *chord function*. However, it was not until 1550 that the sine function was formally defined in terms of right triangles by Georg Rhaeticus. The six trigonometric functions are similar to other functions that we have encountered previously. They have verbal, symbolic, graphical, and numerical representations and can be used to model data and a variety of physical phenomena. (Source: *Historical Topics for the Mathematics Classroom, Thirty-first Yearbook*, NCTM.)

Basic Concepts of Trigonometric Functions

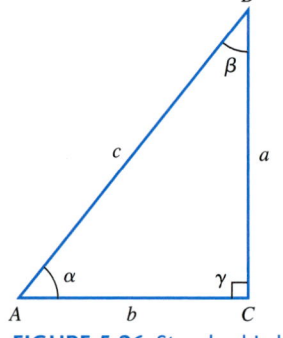

FIGURE 5.26 Standard Labeling

The following **standard labeling** is used to designate vertices, angles, and sides of triangle ABC as shown in Figure 5.26. The vertices are denoted by A, B, and C, the angles by α, β, and γ, and the length of the sides opposite these angles by a, b, and c. If triangle ABC is a right triangle, then we let $\gamma = 90°$.

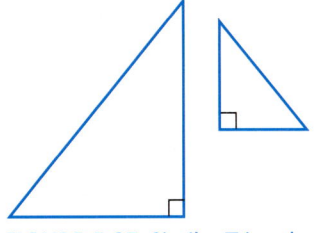

FIGURE 5.27 Similar Triangles

Many important ideas in trigonometry depend on the properties of similar triangles. **Similar triangles** have congruent corresponding angles, but similar triangles are not necessarily the same size. An example of similar triangles is shown in Figure 5.27. Corresponding sides of similar triangles are proportional.

The right triangles ABC and $AB'C'$ shown in Figure 5.28 are similar triangles. Using the properties of similar triangles the following ratios are equal.

$$\frac{BC}{AB} = \frac{B'C'}{AB'}$$

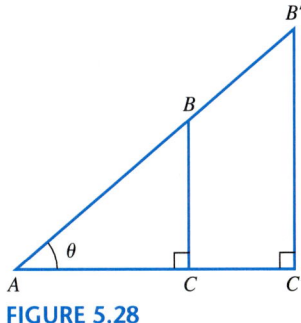

FIGURE 5.28

That is, the ratio of the side opposite angle θ to the hypotenuse is constant for a given angle θ and does not depend on the size of the right triangle. If the measure of θ changes then the ratio of the side opposite to the hypotenuse also changes. This concept can be used to define a new function called the *sine function*. That is, if θ is an acute angle in a right triangle as shown in Figure 5.29, then we define the sine of θ as

$$\sin\theta = \frac{\text{side opposite}}{\text{hypotenuse}},$$

where $\sin \theta$ denotes the sine function with input θ.

Critical Thinking

Is it possible that $\sin \theta > 1$ for some angle θ? Explain your reasoning.

EXAMPLE 1 *Evaluating the sine function*

Find $\sin 30°$. Support your answer using a calculator.

Solution

Since the sine function depends only on the measure of θ, we can choose any size right triangle to evaluate $\sin \theta$. For convenience let the length of the hypotenuse equal 2 as shown in Figure 5.30. From geometry we know that the length of the shortest leg in a 30°–60° right triangle is half the hypotenuse. Thus, the side opposite equals 1 and

$$\sin 30° = \frac{\text{side opposite}}{\text{hypotenuse}} = \frac{1}{2}.$$

This result is supported in Figure 5.31, where the sine function has been evaluated at 30° using a calculator set in degree mode.

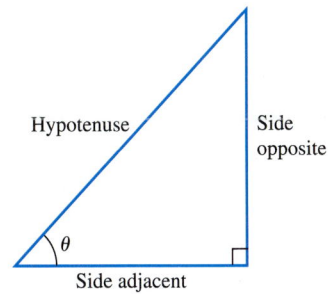

FIGURE 5.29

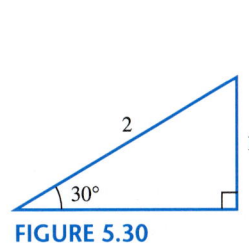

FIGURE 5.30

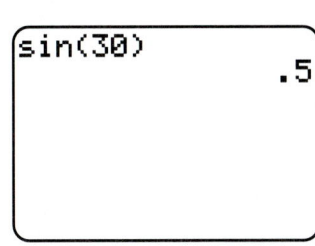

FIGURE 5.31 Degree Mode

Using Figures 5.28 and 5.29, we can define other trigonometric functions. Since,

$$\frac{AC}{AB} = \frac{A'C'}{AB'}$$

the ratio of the side adjacent to the hypotenuse is constant for a fixed angle θ and does not depend on the size of the right triangle. We define the *cosine function* to be

$$\cos\theta = \frac{\text{side adjacent}}{\text{hypotenuse}}.$$

The six trigonometric functions of angle θ are called *sine, cosine, tangent, cosecant, secant,* and *cotangent.* We use the customary abbreviations for each trigonometric function.

Right triangle-based definitions of trigonometric functions

Let θ be an acute angle in a right triangle. Then the six trigonometric functions of θ may be evaluated as follows.

$$\sin\theta = \frac{\text{side opposite}}{\text{hypotenuse}}, \quad \cos\theta = \frac{\text{side adjacent}}{\text{hypotenuse}}, \quad \tan\theta = \frac{\text{side opposite}}{\text{side adjacent}}$$

$$\csc\theta = \frac{\text{hypotenuse}}{\text{side opposite}}, \quad \sec\theta = \frac{\text{hypotenuse}}{\text{side adjacent}}, \quad \cot\theta = \frac{\text{side adjacent}}{\text{side opposite}}$$

The next example illustrates how to evaluate the trigonometric functions.

EXAMPLE 2   *Evaluating trigonometric functions*

Consider the right triangle shown in Figure 5.32. Find the six trigonometric functions of θ.

Solution

In triangle ABC the side opposite angle θ is $a = 8$ and the hypotenuse is $c = 17$. To find the adjacent side b we apply the Pythagorean theorem.

$$c^2 = a^2 + b^2 \qquad \text{Pythagorean theorem}$$
$$b^2 = c^2 - a^2 \qquad \text{Solve for } b^2.$$
$$b^2 = 17^2 - 8^2 \qquad \text{Let } c = 17 \text{ and } a = 8.$$
$$b^2 = 225 \qquad \text{Simplify.}$$
$$b = 15 \qquad \text{Solve for } b, \text{ where } b > 0.$$

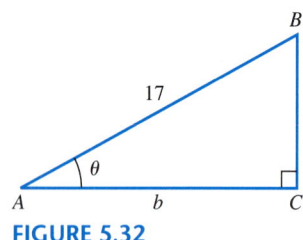

FIGURE 5.32

Thus, the six trigonometric functions of θ are as follows.

$$\sin\theta = \frac{\text{side opposite}}{\text{hypotenuse}} = \frac{8}{17} \qquad \csc\theta = \frac{\text{hypotenuse}}{\text{side opposite}} = \frac{17}{8}$$

$$\cos\theta = \frac{\text{side adjacent}}{\text{hypotenuse}} = \frac{15}{17} \qquad \sec\theta = \frac{\text{hypotenuse}}{\text{side adjacent}} = \frac{17}{15}$$

$$\tan\theta = \frac{\text{side opposite}}{\text{side adjacent}} = \frac{8}{15} \qquad \cot\theta = \frac{\text{side adjacent}}{\text{side opposite}} = \frac{15}{8}$$

■

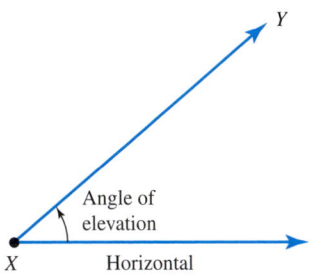

FIGURE 5.33 Angle of Elevation

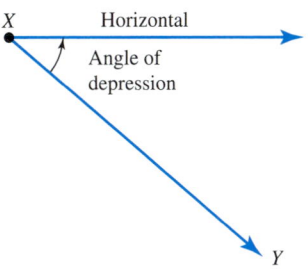

FIGURE 5.34 Angle of Depression

If an object is located above the horizontal, then the angle between the horizontal and the line of sight XY is called the **angle of elevation.** See Figure 5.33. If an object is located below the horizontal, then the angle between the horizontal and the line of sight XY is called the **angle of depression.** See Figure 5.34.

Trigonometry allows people to determine distances and heights without measuring them directly. For example, the altitude of the cloud base is particularly important at airports. (The cloud base is where the lowest layer of clouds begins to form.) Although it is not practical to measure the altitude of the cloud base directly, trigonometry can indirectly determine this height at nighttime. In Figure 5.35 a bright spotlight is directed vertically upward. It creates a bright spot on the cloud base. From a known horizontal distance d from the spotlight, the angle of elevation θ is measured. The side adjacent to θ is d and the side opposite is h, where h represents the height of the cloud base. (Source: F. Cole, *Introduction to Meteorology.*) It follows that

$$\tan \theta = \frac{\text{side opposite}}{\text{side adjacent}} = \frac{h}{d}.$$

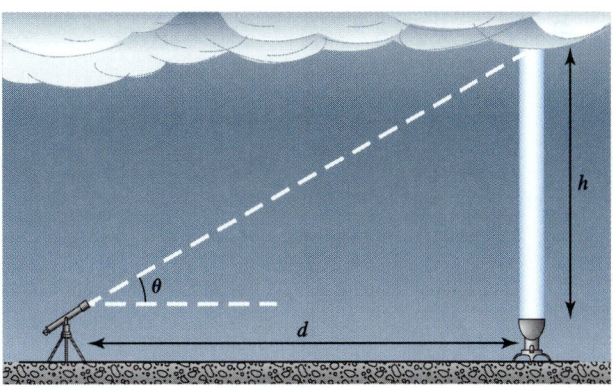

FIGURE 5.35

EXAMPLE 3 *Determining the cloud base*

Suppose that in Figure 5.35 $\theta = 55°$ and $d = 1150$ feet. Estimate the height of the cloud base.

Solution

Solve the equation $\tan \theta = \dfrac{h}{d}$ for h and then substitute values for θ and d.

$$\tan \theta = \frac{h}{d}$$
$$h = d \tan \theta \qquad \text{Cross multiply.}$$
$$= 1150 \tan 55° \qquad \text{Substitute for } d \text{ and } \theta.$$
$$\approx 1150(1.4281) \qquad \text{Approximate } \tan 55°.$$
$$\approx 1642 \text{ feet} \qquad \text{Multiply.}$$

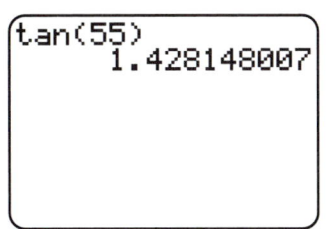

FIGURE 5.36 Degree Mode

Thus, the cloud base is about 1642 feet. A calculator was used to approximate $\tan 55°$ as shown in Figure 5.36. ∎

In most applications, calculators are used to approximate values of the trigonometric functions. However, with the aid of geometry we can determine exact values for the trigonometric functions of some special angles such as 30°, 45°, and 60°.

EXAMPLE 4 *Evaluating trigonometric functions*

Evaluate the six trigonometric functions of $\theta = 45°$. Support your answer using a calculator.

Solution

Begin by drawing a right triangle with a 45° angle as shown in Figure 5.37.

The lengths of the legs in this triangle are equal. Since the size of the right triangle does not affect the values of the trigonometric functions, let the lengths of both legs equal 1. Using the Pythagorean theorem we can find the length of the hypotenuse as follows.

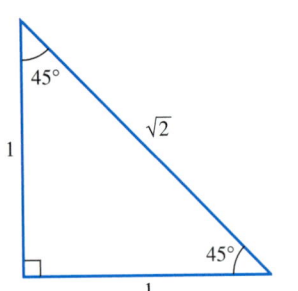

FIGURE 5.37 A 45°–45° Right Triangle

$$c^2 = a^2 + b^2 \qquad \text{Pythagorean theorem}$$
$$c^2 = 1^2 + 1^2 \qquad a = b = 1$$
$$c^2 = 2 \qquad \text{Simplify.}$$
$$c = \sqrt{2} \qquad \text{Solve for } c, \text{ where } c > 0.$$

The hypotenuse has length $\sqrt{2}$. Evaluating the six trigonometric functions gives the following.

$$\sin 45° = \frac{\text{side opposite}}{\text{hypotenuse}} = \frac{1}{\sqrt{2}} \qquad \csc 45° = \frac{\text{hypotenuse}}{\text{side opposite}} = \frac{\sqrt{2}}{1} = \sqrt{2}$$

$$\cos 45° = \frac{\text{side adjacent}}{\text{hypotenuse}} = \frac{1}{\sqrt{2}} \qquad \sec 45° = \frac{\text{hypotenuse}}{\text{side adjacent}} = \frac{\sqrt{2}}{1} = \sqrt{2}$$

$$\tan 45° = \frac{\text{side opposite}}{\text{side adjacent}} = \frac{1}{1} = 1 \qquad \cot 45° = \frac{\text{side adjacent}}{\text{side opposite}} = \frac{1}{1} = 1$$

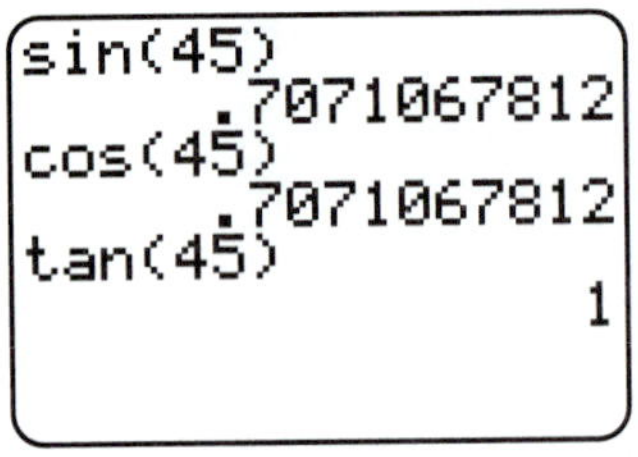

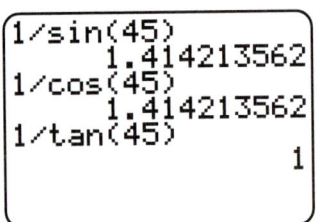

FIGURE 5.38

These results are supported in Figures 5.38 and 5.39 using a calculator. Note that

$$\frac{1}{\sqrt{2}} \approx 0.7071067812 \quad \text{and} \quad \sqrt{2} \approx 1.414213562.$$

In Example 4, $\sin 45° = \frac{1}{\sqrt{2}}$ and $\csc \theta = \frac{\sqrt{2}}{1}$. Since

$$\sin \theta = \frac{\text{side opposite}}{\text{hypotenuse}} \quad \text{and} \quad \csc \theta = \frac{\text{hypotenuse}}{\text{side opposite}}$$

FIGURE 5.39

it follows that $\csc \theta = \frac{1}{\sin \theta}$ in general. That is, the values of $\csc \theta$ and $\sin \theta$ are *reciprocals*. In a similar manner,

$$\sec \theta = \frac{1}{\cos \theta} \quad \text{and} \quad \cot \theta = \frac{1}{\tan \theta}.$$

Technology Note *Evaluating Trigonometric Functions—*

Most calculators have keys to evaluate the sine, cosine, and tangent functions, but do not have keys to evaluate the cosecant, secant, and cotangent functions. These three functions may be evaluated using the following *reciprocal identities.*

$$\csc \theta = \frac{1}{\sin \theta}, \qquad \sec \theta = \frac{1}{\cos \theta}, \qquad \cot \theta = \frac{1}{\tan \theta}$$

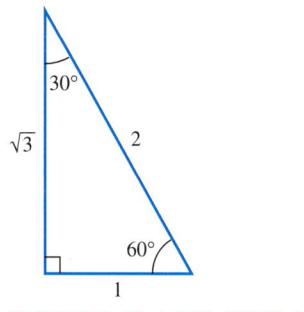

FIGURE 5.40 A 30°–60° Right Triangle

Caution: Do not use the $\sin^{-1}$, $\cos^{-1}$, and $\tan^{-1}$ keys on your calculator to evaluate reciprocals since they represent inverse functions, which will be discussed in Section 5.6. You may use the x^{-1} key to evaluate reciprocals instead.

Using the right triangles in Figures 5.37 and 5.40, the six trigonometric functions can be evaluated at 30°, 45°, and 60° without the aid of a calculator.

Table 5.2 lists the values of the six trigonometric functions for selected angles.

TABLE 5.2

θ	$\sin \theta$	$\cos \theta$	$\tan \theta$	$\csc \theta$	$\sec \theta$	$\cot \theta$
30°	$\dfrac{1}{2}$	$\dfrac{\sqrt{3}}{2}$	$\dfrac{1}{\sqrt{3}}$	2	$\dfrac{2}{\sqrt{3}}$	$\sqrt{3}$
45°	$\dfrac{1}{\sqrt{2}}$	$\dfrac{1}{\sqrt{2}}$	1	$\sqrt{2}$	$\sqrt{2}$	1
60°	$\dfrac{\sqrt{3}}{2}$	$\dfrac{1}{2}$	$\sqrt{3}$	$\dfrac{2}{\sqrt{3}}$	2	$\dfrac{1}{\sqrt{3}}$

Note: The values in Table 5.2 may be written so that denominators are rational. For example,

$$\frac{1}{\sqrt{3}} = \frac{1}{\sqrt{3}} \times \frac{\sqrt{3}}{\sqrt{3}} = \frac{\sqrt{3}}{3} \quad \text{and}$$

$$\frac{2}{\sqrt{3}} = \frac{2}{\sqrt{3}} \times \frac{\sqrt{3}}{\sqrt{3}} = \frac{2\sqrt{3}}{3}.$$

We can use the table feature to evaluate trigonometric functions. Numerical representations of the six trigonometric functions are shown in Figures 5.41–5.46. Notice that for $0° \leq \theta \leq 90°$, outputs from the sine function increase from 0 to 1, while outputs from the cosine function decrease from 1 to 0. The cosecant, secant, and cotangent functions have been evaluated using the reciprocal identities.

X	Y1
0	0
15	.25882
30	.5
45	.70711
60	.86603
75	.96593
90	1

Y1 ∎ sin(X)

FIGURE 5.41 Sine Function

X	Y1
0	1
15	.96593
30	.86603
45	.70711
60	.5
75	.25882
90	0

Y1 ∎ cos(X)

FIGURE 5.42 Cosine Function

X	Y1
0	0
15	.26795
30	.57735
45	1
60	1.7321
75	3.7321
90	ERROR

Y1 ∎ tan(X)

FIGURE 5.43 Tangent Function

X	Y1
0	ERROR
15	3.8637
30	2
45	1.4142
60	1.1547
75	1.0353
90	1

Y1 ∎ 1/sin(X)

FIGURE 5.44 Cosecant Function

X	Y1
0	1
15	1.0353
30	1.1547
45	1.4142
60	2
75	3.8637
90	ERROR

Y1 ∎ 1/cos(X)

FIGURE 5.45 Secant Function

X	Y1
0	ERROR
15	3.7321
30	1.7321
45	1
60	.57735
75	.26795
90	ERROR

Y1 ∎ 1/tan(X)

FIGURE 5.46 Cotangent Function

Applications of Right Triangle Trigonometry

For centuries astronomers wanted to know how far it was to the stars. It was not until 1838 that the astronomer Friedrich Bessel determined the distance to a star called 61 Cygni. He used a *parallax* method that relied on the measurement of very small angles. See Figure 5.47. As Earth revolves around the sun, the observed parallax of 61 Cygni is $\theta \approx 0.0000811°$. Because stars are so distant, parallax angles are very small. (Sources: H. Freebury, *A History of Mathematics;* M. Zeilik, *Introductory Astronomy and Astrophysics.*)

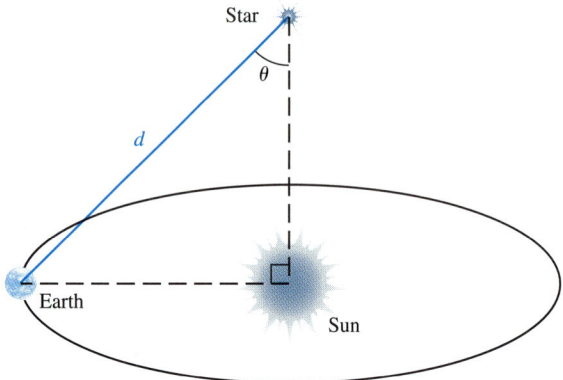

FIGURE 5.47 Parallax of a Star (Not to scale)

EXAMPLE 5   *Calculating the distance to a star*

One of the nearest stars is Alpha Centauri, which has a parallax of $\theta \approx 0.000212°$. (Source: M. Zeilik.)

(a) Calculate the distance to Alpha Centauri if the Earth-Sun distance is 93,000,000 miles.
(b) A light-year is defined to be the distance that light travels in one year and equals about 5.9 trillion miles. Find the distance to Alpha Centauri in light-years.

Solution
(a) Let d represent the distance between Earth and Alpha Centauri. From Figure 5.47, it can be seen that

$$\sin \theta = \frac{93,000,000}{d} \quad \text{or} \quad d = \frac{93,000,000}{\sin \theta}.$$

Substituting for θ gives the following result.

$$d = \frac{93,000,000}{\sin 0.000212°} \approx 2.51 \times 10^{13} \text{ miles}$$

(b) This distance equals $\dfrac{2.51 \times 10^{13}}{5.9 \times 10^{12}} \approx 4.3$ light-years.   ∎

Water is often an obstacle to surveyors in the field when measuring distances between two points. For example, to measure the distance between points P and Q in Figure 5.48 a baseline PR, perpendicular to PQ, is determined. Angle PRQ is then measured. Using right triangle trigonometry, the length of PQ can be determined. (Source: P. Kissam, *Surveying Practice.*)

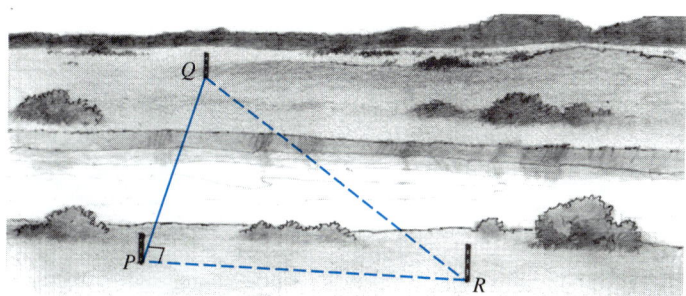

FIGURE 5.48

EXAMPLE 6   *Finding distance*

Suppose in Figure 5.48 the length of PR is 94.75 feet and angle PRQ has measure $41°35'45''$. Estimate the distance between points P and Q.

Solution

Let angle PRQ be θ. Since $\tan\theta = \dfrac{PQ}{PR}$, it follows that

$$PQ = PR\tan\theta = 94.75\tan(41°35'45'') \approx 84.11 \text{ feet.} \quad\blacksquare$$

Trigonometric identities are frequently used to simplify formulas. For example, since

$$\sin\theta = \frac{\text{side opposite}}{\text{hypotenuse}} \quad \text{and} \quad \cos\theta = \frac{\text{side adjacent}}{\text{hypotenuse}},$$

it follows that

$$\frac{\sin\theta}{\cos\theta} = \frac{\text{side opposite}}{\text{side adjacent}} = \tan\theta.$$

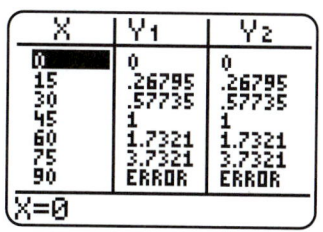

FIGURE 5.49

The expressions $\tan\theta$ and $\dfrac{\sin\theta}{\cos\theta}$ are equivalent and the equation

$$\tan\theta = \frac{\sin\theta}{\cos\theta}$$

is an identity. This equation is true for all values of θ for which $\cos\theta \neq 0$. Numerical support is shown in Figure 5.49, where $Y_1 = \tan(X)$ and $Y_2 = \sin(X)/\cos(X)$.

In the next example we derive a formula that is used in both the design of highways and the calculation of satellite orbits.

EXAMPLE 7   *Deriving a formula for the design of highway curves*

One common type of highway curve is a *simple horizontal curve*. It consists of two straight segments of highway connected by a circular arc with radius r as shown in Figure 5.50. The distance d is called the *external distance*. (Source: F. Mannering, *Principles of Highway Engineering and Traffic Analysis*.)

(a) Derive a formula for d that involves r and θ.
(b) Find d for a curve with a 750-foot radius and $\theta = 36°$.

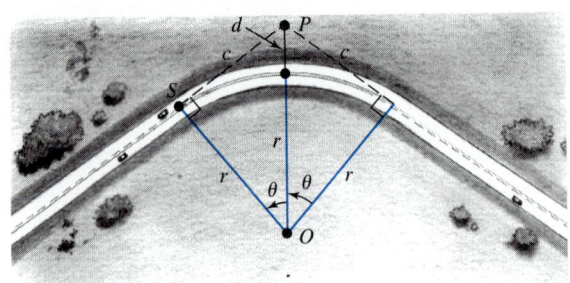

FIGURE 5.50 A Simple Horizontal Curve

Solution

(a) Since $\tan\theta = \dfrac{c}{r}$, it follows that $c = r\tan\theta$. Then,

$$\sin\theta = \frac{c}{r+d} \qquad \text{Use triangle } OSP.$$

$$\sin\theta = \frac{r\tan\theta}{r+d} \qquad \text{Substitute } r\tan\theta \text{ for } c.$$

$$(r+d)\sin\theta = r\tan\theta \qquad \text{Cross multiply.}$$

$$r+d = \frac{r\tan\theta}{\sin\theta} \qquad \text{Divide by } \sin\theta.$$

$$r+d = \frac{r\,(\sin\theta/\cos\theta)}{\sin\theta} \qquad \text{Apply the identity } \tan\theta = \frac{\sin\theta}{\cos\theta}.$$

$$r+d = \frac{r}{\cos\theta} \qquad \text{Simplify.}$$

$$d = \frac{r}{\cos\theta} - r \qquad \text{Subtract } r.$$

$$d = r\left(\frac{1}{\cos\theta} - 1\right) \qquad \text{Factor out } r.$$

(b) $d = 750\left(\dfrac{1}{\cos 36°} - 1\right) \approx 177$ feet.   ■

The Global Positioning System (GPS) provides the technology to accurately determine locations and velocities of objects such as cars. There are 24 satellites in the GPS that orbit Earth every 12 hours. Figure 5.51 illustrates a GPS satellite over the equator. (Source: Y. Zhao, *Vehicle Location and Navigation Systems.*)

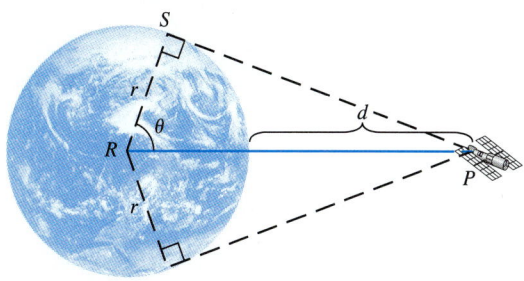

FIGURE 5.51

The formula derived in Example 7 can be used to find the altitude of the orbit of a GPS satellite orbit.

EXAMPLE 8 *Analyzing GPS satellite communication*

The equatorial radius of Earth is 3963 miles. If $\theta = 76.1°$ in Figure 5.51, estimate the altitude of the orbit of the GPS satellite above Earth's surface. (Source: A. Leick, *GPS Satellite Surveying.*)

Solution

Let d represent the altitude of the GPS satellite. Notice that d in Figure 5.51 corresponds to the external distance d in Figure 5.50.

$$d = r\left(\frac{1}{\cos\theta} - 1\right)$$

$$d = 3963\left(\frac{1}{\cos 76.1°} - 1\right)$$

$$d \approx 12{,}534$$

The altitude of a GPS satellite orbit is about 12,534 miles. ∎

Complementary Angles and Cofunctions

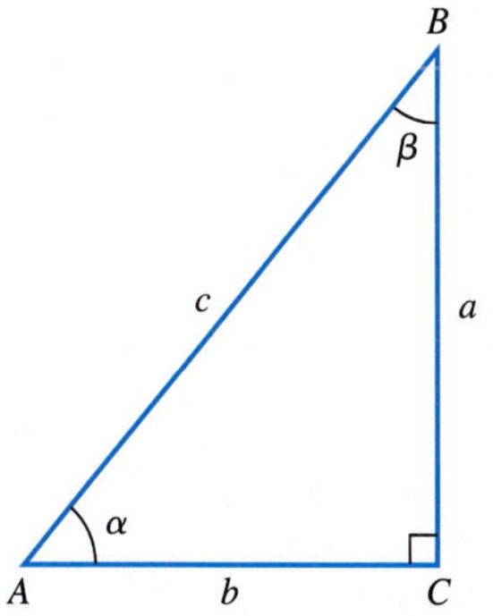

In Figure 5.52, α and β are complementary angles since their measures sum to 90°. The six trigonometric functions for α and β can be expressed as follows.

$$\sin\alpha = \frac{a}{c} = \cos\beta \qquad \cos\alpha = \frac{b}{c} = \sin\beta$$

$$\tan\alpha = \frac{a}{b} = \cot\beta \qquad \cot\alpha = \frac{b}{a} = \tan\beta$$

$$\sec\alpha = \frac{c}{b} = \csc\beta \qquad \csc\alpha = \frac{c}{a} = \sec\beta$$

FIGURE 5.52

Notice that the value of a trigonometric function of α equals the value of the trigonometric cofunction for β. For example, $\sin\alpha = \cos\beta$ and $\sin\beta = \cos\alpha$. This is how cofunctions were named. In 1620 Edmund Gunter combined the words "complement" and "sine" to obtain *co*sine. In a similar manner the cosecant and cotangent functions are the "complementary functions" of the secant and tangent functions, respectively, and their names were shortened to *co*secant and *co*tangent.

Cofunction identities

$$\sin\theta = \cos(90° - \theta) \qquad \cos\theta = \sin(90° - \theta)$$
$$\tan\theta = \cot(90° - \theta) \qquad \cot\theta = \tan(90° - \theta)$$
$$\sec\theta = \csc(90° - \theta) \qquad \csc\theta = \sec(90° - \theta)$$

EXAMPLE 9   *Evaluating trigonometric functions using complementary angles*

Write an equivalent expression using a cofunction. Then evaluate the expression using a calculator.

(a) $\cot 23°$ (b) $\sec 70°$ (c) $\cos 12°$

Solution

(a) The complementary angle of $23°$ is $90° - 23° = 67°$. Thus,

$$\cot 23° = \tan 67° \approx 2.3559.$$

(b) $\sec 70° = \csc 20° = \dfrac{1}{\sin 20°} \approx 2.9238$

(c) $\cos 12° = \sin 78° \approx 0.9781$

5.2 PUTTING IT ALL TOGETHER

The following table summarizes some properties of right triangle trigonometry.

Right Triangle Trigonometry

Let θ be an acute angle in a right triangle ABC.

$$\sin\theta = \frac{\text{side opposite}}{\text{hypotenuse}}, \qquad \cos\theta = \frac{\text{side adjacent}}{\text{hypotenuse}}, \qquad \tan\theta = \frac{\text{side opposite}}{\text{side adjacent}}$$

$$\csc\theta = \frac{\text{hypotenuse}}{\text{side opposite}}, \qquad \sec\theta = \frac{\text{hypotenuse}}{\text{side adjacent}}, \qquad \cot\theta = \frac{\text{side adjacent}}{\text{side opposite}}$$

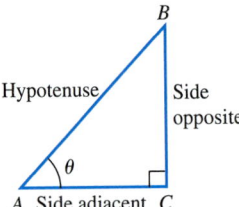

Let α and β be complementary angles.

$$\sin\alpha = \cos(90° - \alpha) = \cos\beta$$
$$\tan\alpha = \cot(90° - \alpha) = \cot\beta$$
$$\sec\alpha = \csc(90° - \alpha) = \csc\beta$$

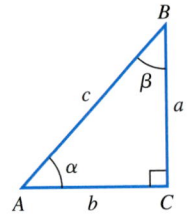

5.2 EXERCISES

  **Tape 8**

Basic Concepts

Exercises 1–4: Sketch a right triangle with the following properties. Label the measure of each angle and side.

1. Acute angles of 30° and 60°, and a hypotenuse with length 2

2. Acute angle of 45° and a leg with length 1

3. Isosceles and a hypotenuse with length 4

4. Acute angle of 60° and the shorter leg with length 3

Exercises 5–8: Find the six trigonometric functions of θ.

5.

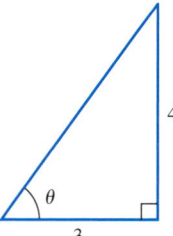

6.

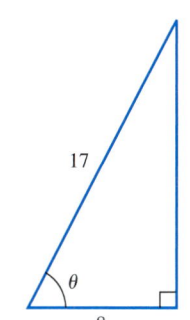

7.

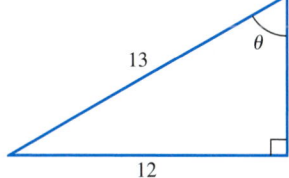

8.

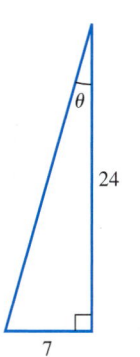

Exercises 9–16: Find the lengths of the unknown sides in the right triangle.

9.

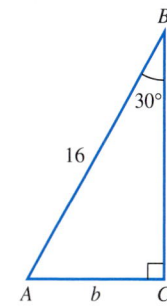

10.

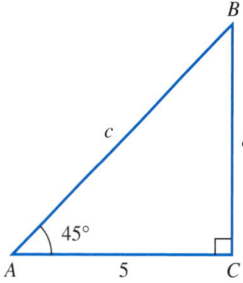

11.

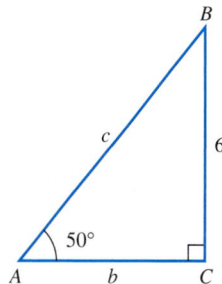

12.

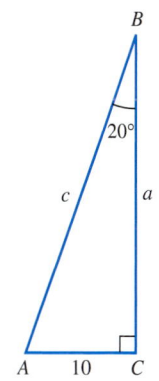

13.

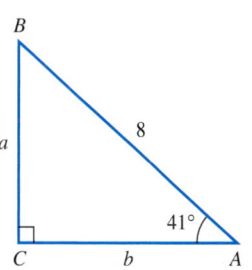

14.

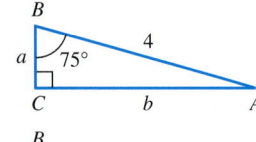

15.

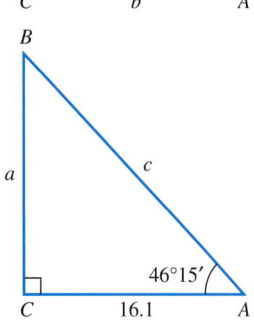

16.

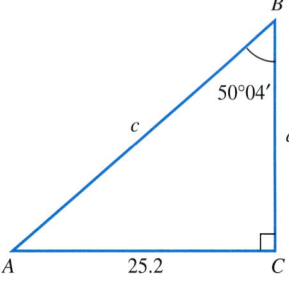

Exercises 17–22: Let θ be an acute angle. Find the unknown trigonometric value using the given information.

17. $\sec\theta$ if $\cos\theta = \dfrac{1}{3}$

18. $\cot\theta$ if $\tan\theta = 5$

19. $\csc\theta$ if $\sin\theta = \dfrac{12}{13}$

20. $\sin\theta$ if $\csc\theta = \dfrac{5}{4}$

21. $\tan\theta$ if $\sin\theta = \dfrac{4}{5}$ and $\cos\theta = \dfrac{3}{5}$

22. $\cos\theta$ if $\sin\theta = \dfrac{3}{5}$ (*Hint:* Sketch a triangle.)

Exercises 23–30: Approximate the six trigonometric functions of θ using four significant digits.

23. $25°$

24. $77°$

25. $5°35'$

26. $85°35'33''$

27. $13°45'30''$

28. $45°44'$

29. $1.05°$

30. $0.000161°$

Exercises 31–34: (Refer to Example 9.) Write an equivalent expression using a cofunction. Approximate the expression to four decimal places using a calculator.

31. **(a)** $\sin 70°$ **(b)** $\cos 40°$

32. **(a)** $\cot 23°$ **(b)** $\tan 48°$

33. **(a)** $\csc 49°$ **(b)** $\sec 63°$

34. **(a)** $\cot 87°$ **(b)** $\sec 72°$

Applications

35. *Height of the Cloud Base* (Refer to Example 3 and Figure 5.35.) From a distance of 1500 feet from the spotlight, the angle of elevation θ equals $37°30'$. Find the height of the cloud base. (Source: F. Cole.)

36. *Weather Tower* A 410-foot weather tower used to measure wind speed has a guy wire attached to it 175 feet above the ground. The angle between the wire and the vertical tower is $57°$ as shown in the figure. Approximate the length of the guy wire. (Source: Brookhaven National Laboratory.)

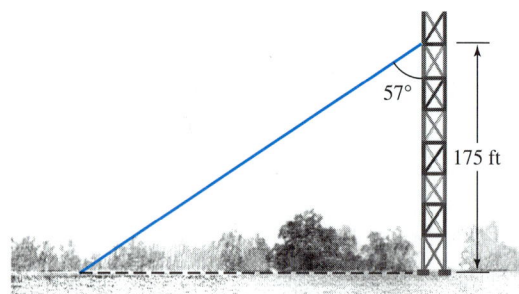

37. *Height of a Tree* One hundred feet from the trunk of a tree on level ground, the angle of elevation of the top of the tree is $35°$. Estimate the height of the tree to the nearest foot.

38. *Height of a Building* From a window 30 feet above the street the angle of elevation to the top of the building across the street is $50°$ and the angle of depression to the base of this building is $20°$. See the accompanying figure. Find the height of the building across the street.

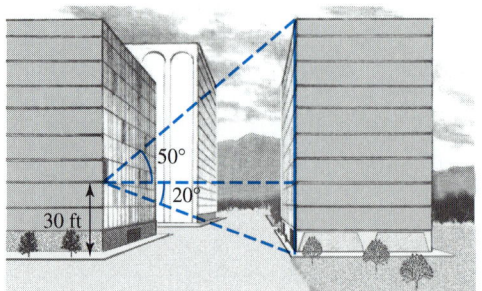

39. *Distance to Nearby Stars* (Refer to Example 5 and Figure 5.47.) The table lists the parallax θ in degrees for some nearby stars. Approximate the distance in miles from Earth to each star. Estimate this distance in light-years.

Star	θ (degrees)
Barnard's Star	1.52×10^{-4}
Sirus	1.05×10^{-4}
61 Cygni	8.11×10^{-5}
Procyon	7.97×10^{-5}

Source: M. Zeilik, *Introductory Astronomy and Astrophysics.*

40. *Parallax and Distance* (Refer to the previous exercise.) When the parallax θ is equal in measure to one second, a star is said to have a distance from Earth of 1 parsec. If the distance between Earth and the sun is 93,000,000 miles, approximate the number of miles in 1 parsec using two significant digits. How many light-years is this? (Source: M. Zeilik.)

41. *Observing Mercury* The planet Mercury is closer to the sun than Earth. For this reason it can only be observed low in the horizon around sunset or sunrise. See the accompanying figure, where angle θ is called the *elongation*. Because Mercury's orbit is not circular, the elongation varies between 18° and 28°. Approximate the minimum and maximum distances between Mercury and the sun using two significant digits. (Source: M. Zeilik.)

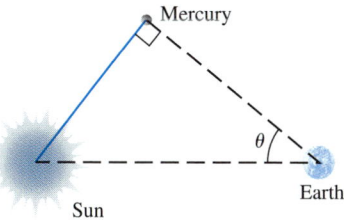

42. *Observing Venus* (Refer to the previous exercise.) The orbit of Venus is nearly circular with an elongation of 48°. Estimate the distance between Venus and the sun.

43. *Orbital Height of a Satellite* (Refer to Example 8 and Figure 5.51.) If $\theta = 67.3°$ for a satellite, find the altitude of its orbit. Discuss how the altitude of a satellite orbit changes as θ increases.

44. *Height of a Mountain* From a point A the angle of elevation of Mount Kilimanjaro in Africa is 13.7° and from a point B directly behind A, the angle of elevation is 10.4°. See the accompanying figure. If the distance between A and B is 5 miles, approximate the height of Mount Kilimanjaro.

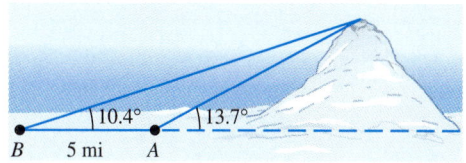

45. *Heights of Lunar Mountains* The lunar mountain peak Huygens has a height of 21,000 feet. The shadow of Huygens on a photograph was 2.8 mm, while the nearby mountain Bradley had a shadow of 1.8 mm on the same photograph. Use similar triangles to calculate the height of Bradley. (Source: T. Webb, *Celestial Objects for Common Telescopes.*)

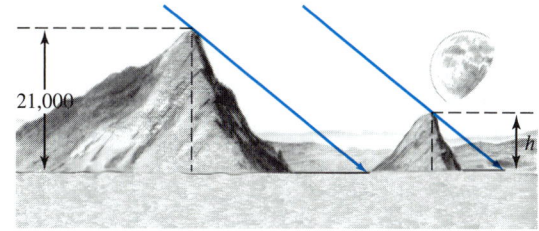

46. *Depths of Moon Craters* (Refer to the previous exercise.) The depths of unknown craters on the moon can be found by comparing their shadows to shadows of nearby craters with known depths. The crater Aristillus is 11,000 feet deep and its shadow was measured as 1.5 mm on a photograph. Its companion crater, Autolycus, had a shadow of 1.3 mm on the same photograph. Estimate the depth of the crater Autolycus. (Source: T. Webb.)

47. *Surveying* (Refer to Example 6 and Figure 5.48.) Find the distance from P to Q if PR is 85.62 feet and angle PQR is 23°45′30″.

48. *Aerial Photography* An aerial photograph is taken directly above a building. The length of the building's shadow is 48 feet when the angle of elevation of the sun is 35.3°. Estimate the height of the building.

49. *Surveying* The *subtense bar method* is used by surveyors to find a distance d between two points P and Q. A subtense bar 2 meters long is centered at Q, perpendicular to the line of sight from P to Q as shown in the figure. If angle θ is measured, then d can be found using trigonometry. (Source: I. Mueller and K. Ramsayer, *Introduction to Surveying*.)

(a) Find a formula for d involving a trigonometric function of θ.

(b) Find d if $\theta = 1°45'15''$.

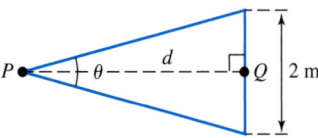

50. (Refer to the previous exercise.) A variation of the subtense bar method that is used to determine larger distances between two points P and Q is shown in the figure. A subtense bar with length b is placed between the points P and Q so that the line of sight connecting P and Q is a perpendicular bisector. If $\alpha = 0.63°$, $\beta = 0.78°$, and $b = 2$ meters, find the distance from P to Q. (Source: I. Mueller.)

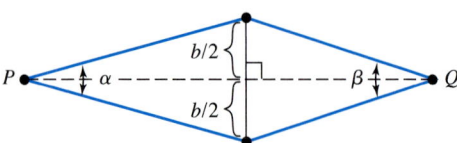

51. *Highway Curve Design* Highway curves are sometimes banked so that the outside of the curve is slightly elevated or inclined above the inside of the curve as shown in the figure. This inclination is called the *superelevation*. It is important that both the curve's radius and superelevation are correct for a given speed limit. The relationship between a car's velocity v in feet per second, the safe radius r of the curve in feet, and the superelevation θ in degrees is given by $r = \dfrac{v^2}{4.5 + 32.2 \tan\theta}$. (Source: F. Mannering.)

(a) A curve has a speed limit of 66 feet per second (45 mph) and a superelevation of $\theta = 3°$. Approximate the safe radius r.

(b) Find r if $\theta = 5°$.

(c) Conjecture how increasing θ affects the safe radius r. Verify your conjecture by tabling r starting at $\theta = 0$ and incrementing by 1. Let $v = 66$.

52. (Refer to the previous exercise.) A highway curve has a radius $r = 1150$ feet and has a superelevation of $\theta = 2.1°$. What should be the speed limit (in miles per hour) for this curve?

53. *Highway Design* (Refer to Example 7 and Figure 5.50.) Find the external distance d for a highway curve with $r = 625$ feet and $\theta = 54°$.

54. *Highway Design* (Refer to Example 7.) A simple horizontal curve is shown in the accompanying figure. The points P and S mark the beginning and end of the curve. Let Q be the point of intersection where the two straight sections of highway leading into the curve would meet if extended. The radius of the curve is r and the angle θ denotes how many degrees the curve turns. If $r = 765$ feet and $\theta = 83°$, find the distance between P and Q. (Source: F. Mannering.)

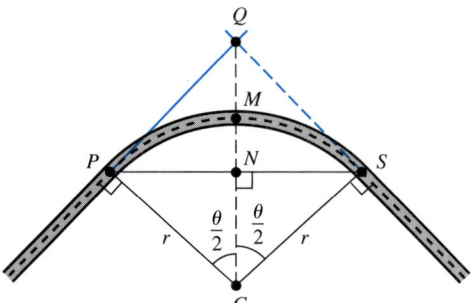

Identities

55. (a) Use the right triangle in the accompanying figure to show symbolically that

$$\sin^2\theta + \cos^2\theta = 1,$$

where θ is an acute angle in a right triangle. Note that $\sin^2\theta = (\sin\theta)^2$ and $\cos^2\theta = (\cos\theta)^2$. (*Hint:* Apply the definitions of the sine and cosine functions and then use the Pythagorean theorem.)

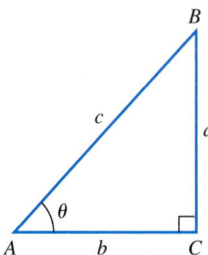

(b) Support your result numerically by tabling $Y_1 = (\sin(X))^2 + (\cos(X))^2$ starting at $x = 0$ and incrementing by 10. Use degree mode.

56. Repeat the previous exercise and show that $\sec^2\theta - \tan^2\theta = 1$.

Writing about Mathematics

1. Most calculators have built-in keys to compute the sine, cosine, and tangent functions, but not the secant, cosecant, and cotangent functions. Is it possible to evaluate all of the trigonometric functions with this type of calculator? Explain and include examples.

2. The sine function is defined in terms of right triangles as $\sin\theta = \dfrac{\text{side opposite}}{\text{hypotenuse}}$. Suppose that a fixed angle θ occurs in two different right triangles. If the length of the hypotenuse in the first triangle has twice the length of the hypotenuse in the second triangle, what can be said about the sides opposite in each triangle? How does the value $\sin\theta$ compare in each triangle? Explain.

CHECKING BASIC CONCEPTS FOR SECTIONS 5.1 AND 5.2

1. Find the radian measure of each angle.
   (a) $45°$
   (b) $75°$

2. Find the degree measure of each angle.
   (a) $\dfrac{\pi}{6}$
   (b) $\dfrac{5\pi}{6}$

3. Find the arc length intercepted by a central angle of $\theta = 30°$ in a circle with radius $r = 12$ inches. Calculate the area of the sector determined by r and θ.

4. Use the right triangle in the accompanying figure to find the six trigonometric functions of θ.

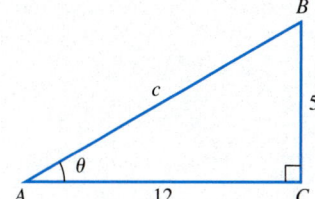

5. Evaluate the six trigonometric functions of $\theta = 60°$ by hand. Support your results using a calculator.

6. If $\alpha = 63°$ and $a = 9$ in right triangle ABC, find the length of the hypotenuse.

5.3 The Sine and Cosine Functions and Their Graphs

Definitions • The Unit Circle • Representations of the Sine and the Cosine Functions • Modeling with the Sine and Cosine Functions

Introduction

The sine and cosine functions are not only used in applications involving right triangles, they are also used to model phenomena involving rotation and periodic motion. To accomplish this, we must extend the domains of the trigonometric functions from acute angles to angles of any measure. This will allow us to model a wide variety of phenomena such as biorhythms, weather, tides, electricity, and robotic arms.

Definitions

Robotics is a rapidly growing field requiring extensive mathematics. One basic problem that occurs when designing a robotic arm is determining the location of the robot's hand. Suppose we have a robotic arm that rotates at the shoulder and is controlled by changing the angle θ and the length of the arm r as illustrated in Figure 5.53. We would like to find a relation between the xy-coordinates of the hand and the settings for r and θ. (Source: W. Stadler, *Analytical Robotics and Mechatronics*.)

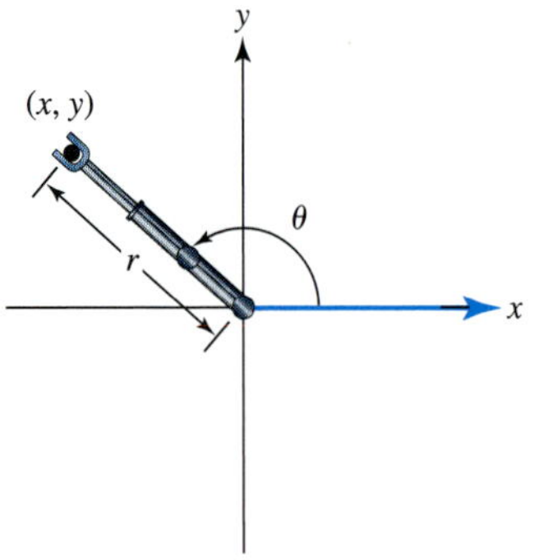

FIGURE 5.53

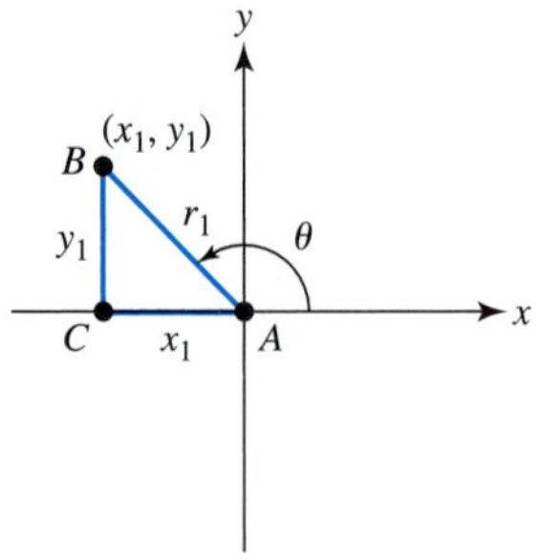

FIGURE 5.54

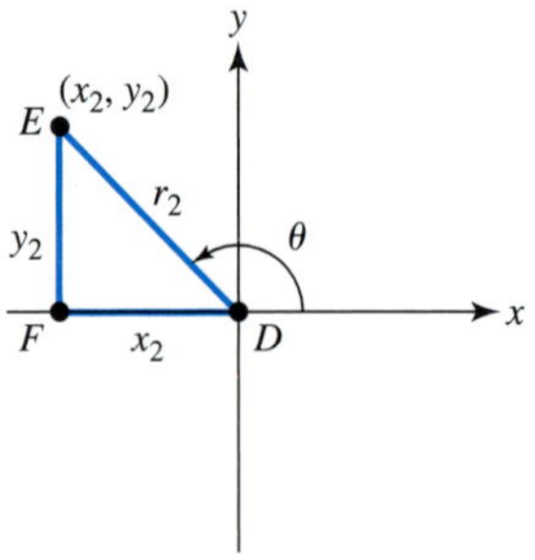

FIGURE 5.55

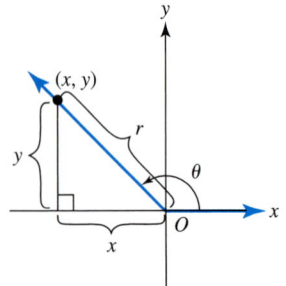

FIGURE 5.56

Notice that for a fixed angle θ, if the length of the arm is changed from r_1 to r_2, triangle ABC in Figure 5.54 and triangle DEF in Figure 5.55 are similar triangles. Thus, the following ratios are equal and depend only on the measure of θ.

$$\frac{x_1}{r_1} = \frac{x_2}{r_2} \quad \text{and} \quad \frac{y_1}{r_1} = \frac{y_2}{r_2}$$

In Figure 5.56 the Pythagorean theorem gives $r^2 = x^2 + y^2$. Since $r > 0$, it follows that $r = \sqrt{x^2 + y^2}$. The ratios $\frac{y}{r}$ and $\frac{x}{r}$ can be used to define the *sine* and *cosine* functions of any angle θ.

The sine and cosine functions of any angle θ

Let angle θ be in standard position with the point (x, y) lying on the angle's terminal side. If $r = \sqrt{x^2 + y^2}$, then

$$\sin\theta = \frac{y}{r} \quad \text{and} \quad \cos\theta = \frac{x}{r} \quad (r \neq 0).$$

Although the terminal side of θ was shown in the second quadrant, these definitions are valid for any angle θ.

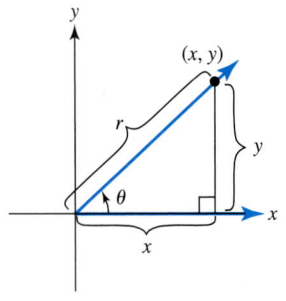

FIGURE 5.57

MAKING CONNECTIONS
Right Triangle Trigonometry

If $0° < \theta < 90°$, then x corresponds to the length of the adjacent side, y corresponds to the length of the opposite side, and r corresponds to the length of the hypotenuse. See Figure 5.57. These new definitions for sine and cosine are equivalent to the definitions presented in Section 5.2 when θ is an acute angle.

EXAMPLE 1 *Evaluating sine and cosine for coterminal angles*

Suppose a robotic hand is located at the point $(15, -8)$, where all units are in inches.
(a) Find the length of the arm.
(b) Let α satisfy $0° \leq \alpha < 360°$ and represent the angle between the positive x-axis and the robotic arm. Find $\sin\alpha$ and $\cos\alpha$.
(c) Let β satisfy $-360° \leq \beta < 0°$ and represent the angle between the positive x-axis and the robotic arm. Find $\sin\beta$ and $\cos\beta$. How do the values for $\sin\beta$ and $\cos\beta$ compare with the values for $\sin\alpha$ and $\cos\alpha$?

Solution

(a) Since $r = \sqrt{15^2 + (-8)^2} = 17$, the length of the arm is 17 inches. See Figure 5.58.

(b) Let $x = 15$, $y = -8$, and $r = 17$. Then,

$$\sin \alpha = \frac{y}{r} = -\frac{8}{17} \quad \text{and} \quad \cos \alpha = \frac{15}{17}.$$

(c) In Figure 5.59, β satisfies $-360° \le \beta < 0°$. Since the values of x, y, and r do not change, the trigonometric values of β are the same as those for α in part (b).

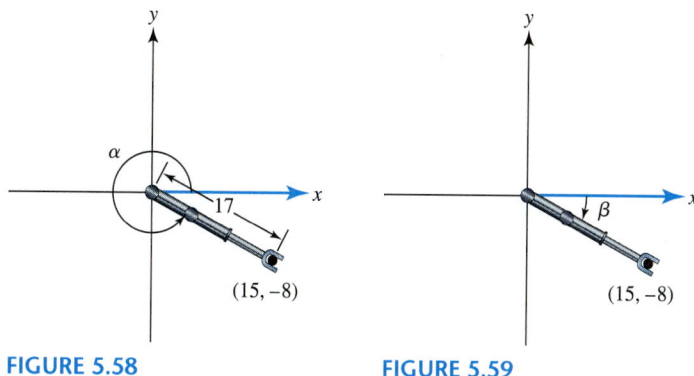

FIGURE 5.58   **FIGURE 5.59**   ∎

The results of Example 1 can be generalized. If α and β are coterminal angles, then

$$\sin \alpha = \sin \beta \quad \text{and} \quad \cos \alpha = \cos \beta.$$

The angles θ and $\theta + 360°$ are coterminal for any θ. Their terminal sides pass through the same point (x, y) as shown in Figure 5.60. Therefore, for all θ $\sin \theta = \sin(\theta + 360°)$ and $\cos \theta = \cos(\theta + 360°)$. In general, if n is any integer, then

$$\sin \theta = \sin(\theta + n \cdot 360°) \quad \text{and} \quad \cos \theta = \cos(\theta + n \cdot 360°).$$

FIGURE 5.60

As a result, we say that the sine and cosine functions are **periodic** with **period** $360°$ (or 2π radians.) For example,

$$\sin 90° = \sin(90° + 360°) = \sin(90° - 2 \cdot 360°).$$

```
sin(90)
                    1
sin(90+360)
                    1
sin(90-2*360)
                    1
```

FIGURE 5.61 Degree Mode

See Figure 5.61.

If an angle is a multiple of 30° or 45°, exact evaluation of the sine function or cosine function is possible by hand. However, in most applications calculators are used to obtain numerical approximations. The next example illustrates an angle where the sine and cosine functions can be evaluated exactly by hand.

EXAMPLE 2 *Finding values of sin θ and cos θ*

Find $\sin 120°$ and $\cos 120°$. Support your answer using a calculator.

Solution

When evaluating the sine or cosine function by hand, we can select any positive value for r and not change the resulting values for $\sin \theta$ and $\cos \theta$. For convenience we let $r = 2$ as shown in Figure 5.62. The length of the shortest leg in a 30°–60°

right triangle is half the hypotenuse. Since the terminal side of θ is in the second quadrant, $x < 0$ and so $x = -1$. Next we find y.

$$x^2 + y^2 = r^2 \qquad \text{Pythagorean theorem}$$
$$y^2 = r^2 - x^2 \qquad \text{Subtract } x^2.$$
$$y^2 = 2^2 - (-1)^2 \qquad \text{Let } r = 2 \text{ and } x = -1.$$
$$y = \pm\sqrt{3} \qquad \text{Square root property}$$

Since the terminal side of θ is in the second quadrant, $y > 0$ and so $y = \sqrt{3}$. Thus,

$$\sin 120° = \frac{y}{r} = \frac{\sqrt{3}}{2} \approx 0.8660, \qquad \text{and}$$

$$\cos 120° = \frac{x}{r} = -\frac{1}{2} = -0.5.$$

These results are supported in Figure 5.63.

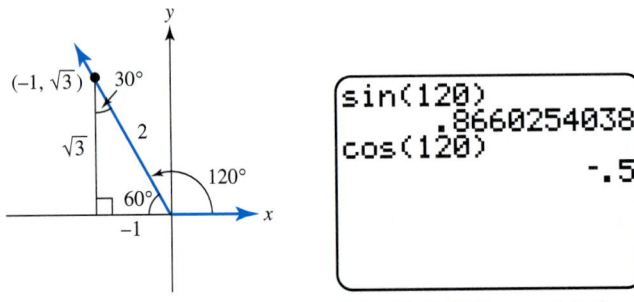

FIGURE 5.62   **FIGURE 5.63** Degree Mode   ∎

The values of the sine and cosine functions for the **quadrantal angles** $0°$, $90°$, $180°$, and $270°$ are shown in Table 5.3. Try to verify these values using a calculator.

TABLE 5.3

θ	$0°$	$90°$	$180°$	$270°$
$\sin \theta$	0	1	0	-1
$\cos \theta$	1	0	-1	0

Trigonometric functions can also be evaluated using radian mode. For example, since $90°$ is equivalent to $\dfrac{\pi}{2}$ radians, it follows that

$$\sin\frac{\pi}{2} = 1 \qquad \text{and} \qquad \cos\frac{\pi}{2} = 0.$$

```
sin(π/2)
                    1
cos(π/2)
                    0
```

FIGURE 5.64 Radian Mode

In Figure 5.64 a calculator has been used to perform these evaluations in *radian mode*.

Grade or *slope* is a measure of steepness and indicates whether a highway is uphill or downhill. A 5% grade indicates that a road is increasing five vertical feet for each 100-foot increase in horizontal distance. *Grade resistance R* is the gravitational force acting on a vehicle given by

$$R = W \sin\theta,$$

where W is the weight of the vehicle and θ is the angle associated with the grade. See Figure 5.65. For an uphill grade $\theta > 0$ and for a downhill grade $\theta < 0$. (Source: F. Mannering, *Principles of Highway Engineering and Traffic Control.*)

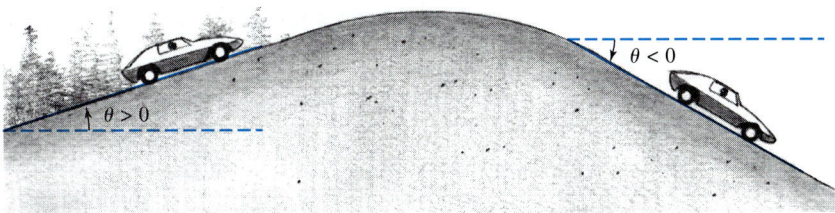

FIGURE 5.65

EXAMPLE 3   *Calculating the grade resistance*

A downhill highway grade is modeled by the line $y = -0.06x$.
(a) Find the grade of the road.
(b) Determine the grade resistance for a 3000-pound car. Interpret the result.

Solution

(a) The slope of the line is -0.06, so when x *increases* by 100 feet, y *decreases* by 6 feet. See Figure 5.66. Thus, this road has a grade of -6%.

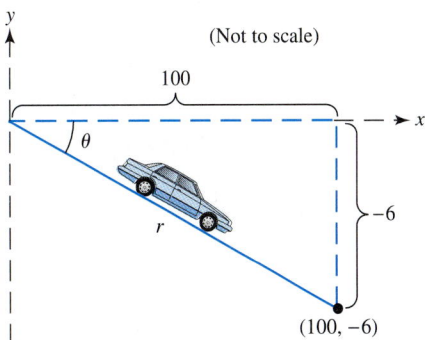

FIGURE 5.66 A Downhill Grade of -6%

(b) First we must find $\sin\theta$. From Figure 5.66 we see that the point $(100, -6)$ lies on the terminal side of θ. Since

$$r = \sqrt{100^2 + (-6)^2} = \sqrt{10{,}036}$$

it follows that

$$\sin\theta = \frac{y}{r} = \frac{-6}{\sqrt{10{,}036}}.$$

The grade resistance is

$$R = W\sin\theta$$
$$= 3000\left(\frac{-6}{\sqrt{10{,}036}}\right)$$
$$\approx -179.7 \text{ lb.}$$

On this stretch of highway, gravity would pull a 3000-pound vehicle *downhill* with a force of about 180 pounds. Note that a downhill grade results in a negative grade resistance. ∎

The Unit Circle

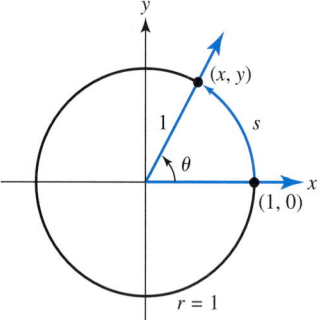

FIGURE 5.67

Trigonometric functions were first associated with angles. As applications became more diverse, real numbers were included in the domains of the six trigonometric functions. Real numbers were needed to represent quantities such as time and distance. To extend the domain of the sine and cosine functions to include all real numbers we will introduce the unit circle.

The **unit circle** has radius 1 and equation $x^2 + y^2 = 1$. Let (x, y) be a point on the unit circle and let s be the arc length along the unit circle from the point $(1, 0)$ to the point (x, y) determined by a counterclockwise rotation. See Figure 5.67. Since $r = 1$, the arc length formula $s = r\theta$ reduces to $s = \theta$. That is, the real number s representing arc length is numerically equal to the radian measure of θ. This discussion gives motivation for the following.

> ### Trigonometric functions of real numbers
>
> The value of a trigonometric function at the real number s is equal to its value at s radians.

Note: To evaluate a trigonometric function of a real number s with a calculator, use *radian* mode.

If (x, y) is a point on the unit circle, lying on the terminal side of an angle θ in standard position, then

$$\sin\theta = \frac{y}{r} = y \qquad \text{and} \qquad \cos\theta = \frac{x}{r} = x$$

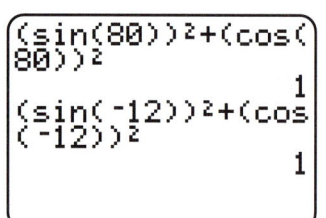

FIGURE 5.68 Degree Mode

since $r = 1$. Points (x, y) on the unit circle can be expressed as $(\cos\theta, \sin\theta)$. Because trigonometric functions can be defined using the unit circle, they are referred to as **circular functions.**

The unit circle is described by $x^2 + y^2 = 1$. Since $x = \cos\theta$ and $y = \sin\theta$, it follows that

$$(\cos\theta)^2 + (\sin\theta)^2 = 1, \qquad \text{or equivalently,}$$

$$\sin^2\theta + \cos^2\theta = 1.$$

This equation is an example of a trigonometric identity. An identity is true for all meaningful values of the variable. In Figures 5.68 and 5.69 this identity is evaluated for different values of θ. In every case the result is 1, regardless of whether θ is measured in degrees or radians.

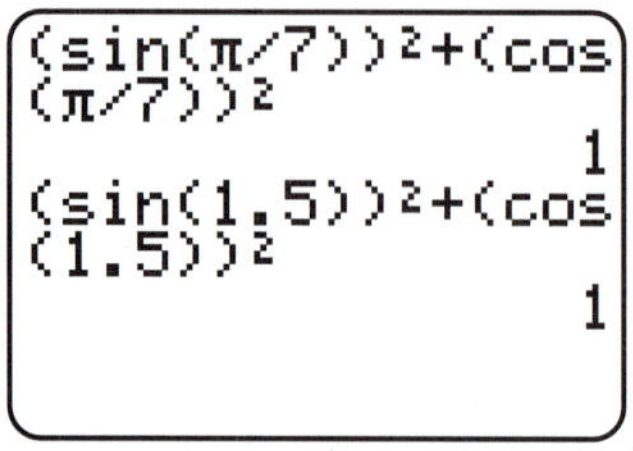

FIGURE 5.69 Radian Mode

Representations of the Sine and Cosine Functions

Like other functions that we have studied, the sine and cosine functions also have symbolic, verbal, numerical, and graphical representations. Both functions are nonlinear.

The Sine Function. A *symbolic representation* of the sine function is $f(t) = \sin t$. The domain of the sine function is all real numbers. There is no simple formula that can be used to evaluate the sine function. Instead we generally rely on other methods for its evaluation. We can use the concept of an algorithm to provide a *verbal representation* of the sine function.

ALGORITHM 5.1 Evaluating the Sine Function at a Real Number t
>*Step 1:* Input a real number t.
>*Step 2:* Let θ be an angle of t radians in standard position.
>*Step 3:* Determine the point (x, y) where the terminal side of θ intersects the unit circle.
>*Step 4:* Output y, the value of $\sin t$.

In Algorithm 5.1, Step 3 is typically the most difficult step to perform. Many times the exact coordinates of the point (x, y) cannot be determined. Other times, geometry must be used to determine its precise coordinates. In most instances we will use a calculator to approximate $\sin t$ for a given value of t. A *partial numerical representation* of $f(t) = \sin t$ is shown in Table 5.4. Since outputs from the sine function correspond to a y-coordinate on the unit circle, the range of the sine function is $-1 \le y \le 1$.

$[-2\pi, 2\pi, \pi/2]$ by $[-3, 3, 1]$

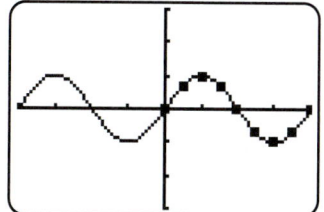

FIGURE 5.70 The Sine Function

TABLE 5.4

t	0	$\dfrac{\pi}{4}$	$\dfrac{\pi}{2}$	$\dfrac{3\pi}{4}$	π	$\dfrac{5\pi}{4}$	$\dfrac{3\pi}{2}$	$\dfrac{7\pi}{4}$	2π
$\sin t$	0	$\dfrac{1}{\sqrt{2}}$	1	$\dfrac{1}{\sqrt{2}}$	0	$-\dfrac{1}{\sqrt{2}}$	-1	$-\dfrac{1}{\sqrt{2}}$	0

Graphical representations of $f(t) = \sin t$ are shown in Figures 5.70 and 5.71, where the points in Table 5.4 are also plotted. The graph of the sine function increases from 0 to 1 for $0 \le t \le \pi/2$, decreases from 1 to -1 for $\pi/2 \le t \le 3\pi/2$, and then increases from -1 to 0 for $3\pi/2 \le t \le 2\pi$. The graph repeats itself both to the left and the right every 2π units on the x-axis since the sine function has period 2π or $360°$.

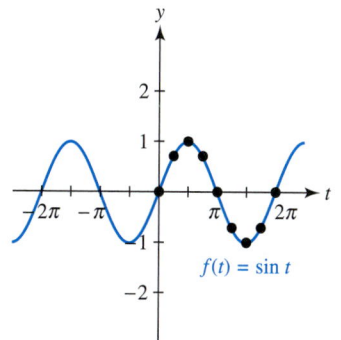

FIGURE 5.71 The Sine Function

MAKING CONNECTIONS
The Unit Circle and the Sine Graph
The graph of the sine function may be better understood using a unit circle. If a point P on the unit circle starts at $(1, 0)$ and rotates counterclockwise, then the y-coordinate of P corresponds to $\sin \theta$. Refer to Figure 5.67. As a result, $\sin \theta$ first increases from 0 to 1, then decreases from 1 to -1, and finally increases from -1 to 0, where P completes one rotation and returns to the point $(1, 0)$. If P continues to rotate, it passes through the same points a second time and so the sine function is periodic with 2π or $360°$.

EXAMPLE 4 *Evaluating the sine function*

Evaluate $f(t) = \sin t$ at $t = \dfrac{3\pi}{2}$ using verbal, numerical, and graphical representations. Support your answer by evaluating $\sin\left(\dfrac{3\pi}{2}\right)$ with a calculator.

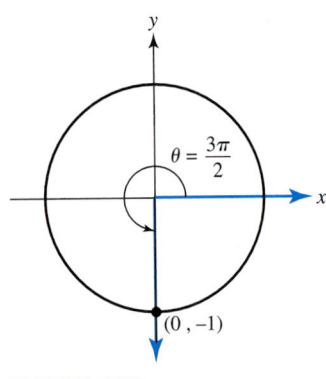

FIGURE 5.72

Solution

Verbal Representation   Apply Algorithm 5.1. An angle θ of $\frac{3\pi}{2}$ radians in standard position has a terminal side that intersects the unit circle at the point $(0, -1)$. See Figure 5.72. Therefore, $\sin\left(\frac{3\pi}{2}\right) = -1$.

Numerical Representation   Using *radian mode*, table $Y_1 = \sin(X)$ starting at $x = 0$, incrementing by $\pi/4$. From Figure 5.73 we see that when $x = \frac{3\pi}{2} \approx 4.7124$, $\sin\left(\frac{3\pi}{2}\right) = -1$.

[0, 2π, π/2] by [−2, 2, 1]

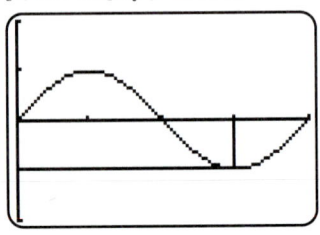

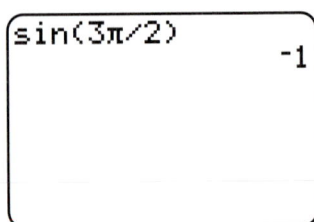

FIGURE 5.73   **FIGURE 5.74**   **FIGURE 5.75**

Graphical Representation   From the graph of $f(t) = \sin t$ in Figure 5.74, it can be seen that $\sin\frac{3\pi}{2} = -1$. Figure 5.75 supports these results. ∎

The Cosine Function.   The cosine function, *represented symbolically* by $f(t) = \cos t$, has verbal, numerical, and graphical representations similar to those of the sine function. Its domain is all real numbers. Like the sine function, the cosine function does not have a simple formula. However, the following verbal representation sometimes can be used to evaluate $\cos t$.

[−2π, 2π, π/2] by [−3, 3, 1]

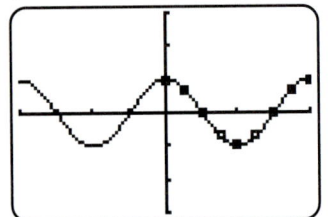

FIGURE 5.76 The Cosine Function

ALGORITHM 5.2   Evaluating the Cosine Function at a Real Number *t*
 Step 1:   Input a real number t.
 Step 2:   Let θ be an angle of t radians in standard position.
 Step 3:   Determine the point (x, y) where the terminal side of θ intersects the unit circle.
 Step 4:   Output x, the value of $\cos t$.

A *partial numerical representation* of $f(t) = \cos t$ is shown in Table 5.5. Since an output from the cosine function corresponds to an x-coordinate on the unit circle, the range of the cosine function is from -1 to 1. Like the sine function, the cosine function is also periodic with 2π or $360°$.

TABLE 5.5

t	0	$\frac{\pi}{4}$	$\frac{\pi}{2}$	$\frac{3\pi}{4}$	π	$\frac{5\pi}{4}$	$\frac{3\pi}{2}$	$\frac{7\pi}{4}$	2π
$\cos t$	1	$\frac{1}{\sqrt{2}}$	0	$-\frac{1}{\sqrt{2}}$	-1	$-\frac{1}{\sqrt{2}}$	0	$\frac{1}{\sqrt{2}}$	1

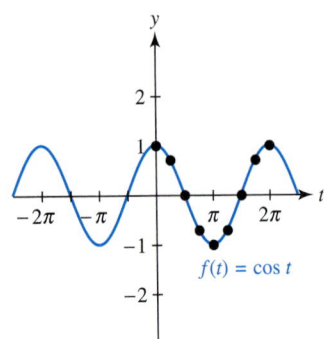

FIGURE 5.77 The Cosine Function

Graphical representations of $f(t) = \cos t$ are shown in Figures 5.76 and 5.77, where the points in Table 5.5 are also plotted. Notice that the overall shape of the graph of the cosine function is similar to the graph of the sine function. However, the cosine function decreases from 1 to -1 for $0 \leq t \leq \pi$ and increases from -1 to 1 for $\pi \leq t \leq 2\pi$.

Critical Thinking

If a point P on the unit circle starts at (1, 0) and rotates counterclockwise, then the x-coordinate of P corresponds to cos θ. Explain how this relates to the graph of the cosine function.

EXAMPLE 5   *Evaluating the cosine function*

Evaluate $f(t) = \cos t$ at $t = \dfrac{5\pi}{6}$ using verbal, numerical, and graphical representations. Support your answer by evaluating $\cos \left(\dfrac{5\pi}{6} \right)$ with a calculator.

Solution

Verbal Representation   Apply Algorithm 5.2. An angle θ of $\dfrac{5\pi}{6}$ radians in standard position has a terminal side that intersects the unit circle in the second quadrant. See Figure 5.78. To find the point of intersection (x, y) notice that the hypotenuse of the 30°–60° right triangle has radius 1 and the shorter leg has length $y = 1/2$. We must determine x *symbolically.*

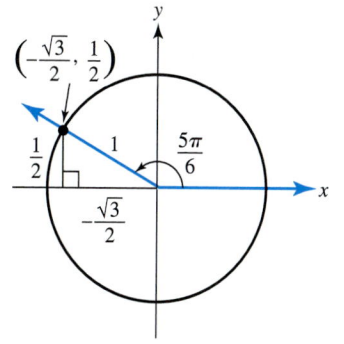

FIGURE 5.78

$$x^2 + y^2 = 1 \qquad \text{Equation of the unit circle}$$

$$x^2 + \left(\frac{1}{2} \right)^2 = 1 \qquad y = \frac{1}{2}$$

$$x^2 = \frac{3}{4} \qquad \text{Solve for } x^2.$$

$$x = \pm \frac{\sqrt{3}}{2} \qquad \text{Square root property}$$

Since the point (x, y) is located in the second quadrant, choose $x = -\dfrac{\sqrt{3}}{2}$. It then follows that $\cos \dfrac{5\pi}{6} = -\dfrac{\sqrt{3}}{2} \approx -0.8660$.

Numerical Representation   Table $Y_1 = \cos(X)$ starting at $x = 0$, incrementing by $\pi/6$. From Figure 5.79 we can see that when $x = 5\pi/6 \approx 2.618$, $\cos(5\pi/6) \approx -0.8660$.

$[0, 2\pi, \pi/2]$ by $[-2, 2, 1]$

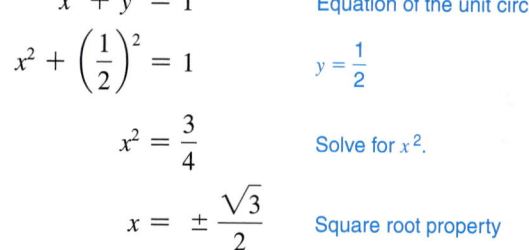

FIGURE 5.79

$[0, 2\pi, \pi/2]$ by $[-2, 2, 1]$

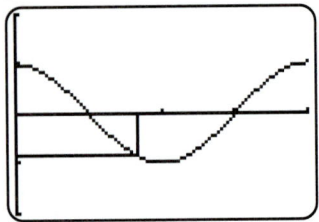

FIGURE 5.80

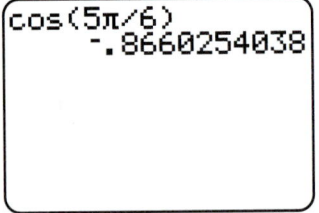

FIGURE 5.81

Graphical Representation From the graph of $Y_1 = \cos(X)$ in Figure 5.80, we can see that $\cos(5\pi/6) \approx -0.9$. Figure 5.81 supports these results. ∎

Modeling with the Sine and Cosine Functions

Periodic graphs that are similar in shape to the graph of the sine and cosine functions are **sinusoidal.** Of all the periodic graphs, sinusoidal graphs are the most important in applications because they occur in nearly every aspect of physical science.

Because the moon orbits Earth, we observe different phases of the moon during the period of a month. In Figure 5.82 angle θ is called the *phase angle*. The *phase F* of the moon is computed by

$$F(\theta) = \frac{1}{2}(1 - \cos\theta),$$

and gives the fraction of the moon's face that is illuminated by the sun. (Source: P. Duffet-Smith, *Practical Astronomy with Your Calculator.*)

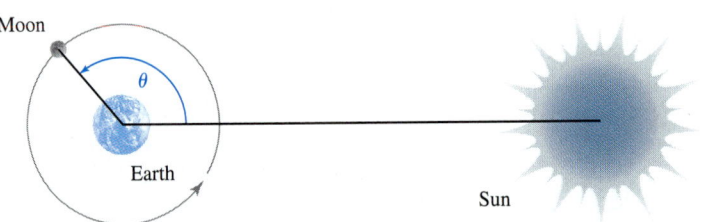

FIGURE 5.82 Phase Angle θ of the Moon

EXAMPLE 6 *Calculating phases of the moon*

Analyze the phases of the moon graphically, numerically, and symbolically.

Solution

Graphical Representation The graph of $Y_1 = 0.5(1 - \cos(X))$ is shown in Figure 5.83. Notice that the phases of the moon are periodic. Each peak represents a full moon and each valley represents a new moon.

$[0, 6\pi, \pi]$ by $[-2, 2, 1]$

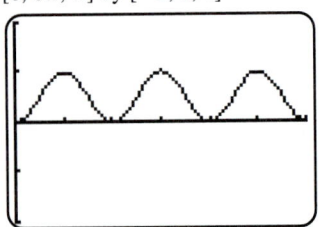

FIGURE 5.83

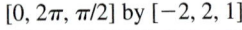

FIGURE 5.84

Numerical Representation Table Y_1 starting at $\theta = 0$, incrementing by $\pi/4$ as shown in Figure 5.84. When $\theta = 0$, $F = 0$ and there is a new moon. At $\theta = \pi/2 \approx 1.5708$, $F = 0.5$, which corresponds to the first quarter. The phase increases to 1 (full moon) at $\theta = \pi$ and then starts to decrease. At $\theta = 3\pi/2 \approx 4.7124$, $F = 0.5$, which represents the last quarter.

Symbolic Representation The cosine function decreases from 1 to -1 for $0 \le \theta \le \pi$, so $F = 0.5(1 - \cos\theta)$ varies between a minimum of 0 (when $\theta = 0$ and $\cos\theta = 1$) and a maximum of 1 (when $\theta = \pi$ and $\cos\theta = -1$). At $\theta = \pi/2$ and $3\pi/2$, $\cos\theta = 0$ and $F = 0.5$. ∎

The study of *biological clocks* is a fascinating field. Many living organisms undergo regular biological rhythms. A simple example is a flower that opens during daylight and closes at nighttime. One amazing result is that flowers often continue to open and close even when they are placed in continual darkness. (Source: F. Brown, *The Biological Clock.*)

EXAMPLE 7   *Modeling biological rhythms*

Some types of water plants are *luminescent*—they radiate a type of "cold" light that is similar to the light emitted from a firefly. A sample of a luminescent plant (*Gonyaulax polyedra*) was put into continual dim light. The luminescence was measured at six-hour intervals and the data in Table 5.6 summarizes the results. Noon corresponds to $t = 0$ and the *y*-units of luminescence are arbitrary. (Source: E. Bunning, *The Physiological Clock.*)

TABLE 5.6

t (hr)	0	6	12	18	24	30	36	42	48
y (luminescence)	1	4	7	4	1	4	7	4	1

(a) Make a scatterplot of the data in $[-6, 54, 6]$ by $[0, 8, 1]$. Interpret the data.

(b) Graph the data and $f(t) = 3 \sin (0.27t - 1.7) + 4$. Comment on how well f models the data.

(c) Estimate the luminescence when $t = 33$.

Solution

$[-6, 54, 6]$ by $[0, 8, 1]$

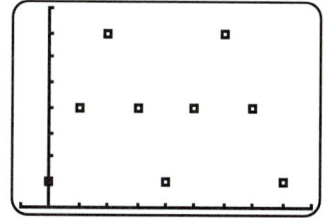

FIGURE 5.85

$[-6, 54\ 6]$ by $[0, 8, 1]$

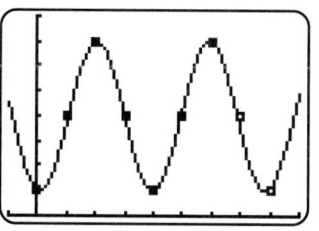

FIGURE 5.86

(a) A scatterplot of the data is shown in Figure 5.85. Luminescence appears to be periodic, increasing and decreasing at regular intervals even though the lighting was always dim. The plant is the most luminescent during times that correspond to midnight: $t = 12$ and 36. It was least luminescent at times corresponding to noon: $t = 0$, 24, and 48.

(b) Graph $Y_1 = 3 \sin (.27X - 1.7) + 4$ and the data as shown in Figure 5.86. The graph of f models the periodic data quite well.

(c) Evaluate $f(t)$ at $t = 33$ using radian mode.

$$f(33) = 3 \sin (0.27(33) - 1.7) + 4 \approx 6.4$$

The luminescence was about 6.4 after 33 hours or at 9 P.M.   ■

Sinusoidal curves occur when modeling the voltage in a common electrical outlet. An electrical circuit can sometimes be compared to water flowing through a hose. *Voltage* in an electrical circuit corresponds to the water pressure. Higher water pressure causes more water to flow through a hose in the same way that higher voltage causes more current to flow through a wire. Electrical current (or *amperage*) corresponds to the amount of water flowing through a hose. More electrons moving through a wire per second result in greater amperage. Electrical *resistance* can be compared to the diameter of a hose. A smaller diameter reduces the flow of water in the same way that increasing electrical resistance reduces current in a wire.

Common household current is called *alternating current* (a. c. current) because both the current and voltage change direction in a wire 120 times per second.

EXAMPLE 8   *Analyzing household current*

The voltage V in a household outlet can be modeled by $V(t) = 160 \sin (120\pi t)$, where t represents time in seconds.

(a) Graph V in $[0, 1/15, 1/60]$ by $[-200, 200, 50]$. Describe the voltage.

(b) Evaluate $V(1/240)$ and interpret the result.

[0, 1/15, 1/60] by [−200, 200, 50]

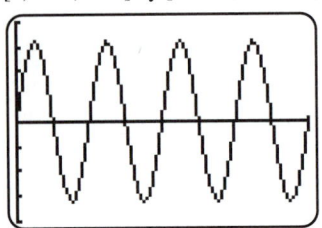

FIGURE 5.87

(c) One way to estimate the "average" voltage in a circuit is to use the *root mean square* voltage, which equals the maximum voltage divided by $\sqrt{2}$. Find this "average" voltage.

Solution

(a) The graph of *V* in Figure 5.87 is sinusoidal and varies between −160 volts and 160 volts. This corresponds to the fact that voltage is changing direction in a household outlet. When the graph is above the *x*-axis the voltage is one direction, and when it is below the *x*-axis voltage is in the opposite direction.

(b) $V(1/240) = 160 \sin(\pi/2) = 160(1) = 160$. After 1/240 second, the voltage is 160 volts.

(c) The maximum voltage is 160 volts. The "average" voltage is $160/\sqrt{2} \approx 113$ volts. Common household electricity is often rated at 110–120 volts. ∎

5.3 PUTTING IT ALL TOGETHER

In this section we extended the domains of the sine and cosine functions to include any angle θ and any real number *t*. Both of these functions are nonlinear functions and their ranges include values satisfying $-1 \leq y \leq 1$. Some concepts about the sine and cosine functions are summarized in the following table.

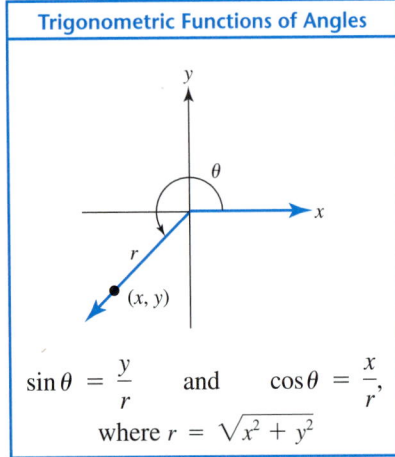

Trigonometric Functions of Angles

$$\sin\theta = \frac{y}{r} \quad \text{and} \quad \cos\theta = \frac{x}{r},$$

$$\text{where } r = \sqrt{x^2 + y^2}$$

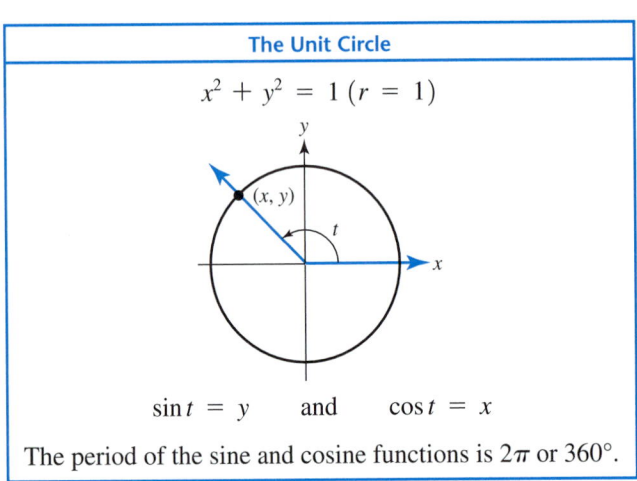

The Unit Circle

$$x^2 + y^2 = 1 \ (r = 1)$$

$$\sin t = y \quad \text{and} \quad \cos t = x$$

The period of the sine and cosine functions is 2π or 360°.

5.3 EXERCISES

   Tape 8

Basic Concepts

Exercises 1–4: The xy-coordinates of the hand for a robotic arm are shown in the figure.

(a) *Find the length of the arm.*

(b) *Find the sine and cosine functions for the angle θ.*

1.

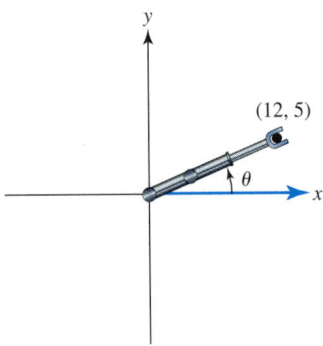

2.

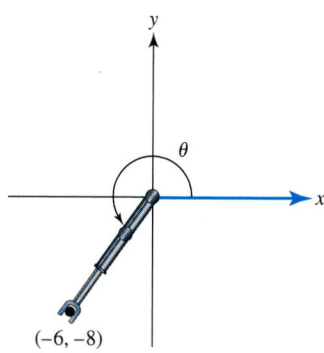

$(-6, -8)$

6.

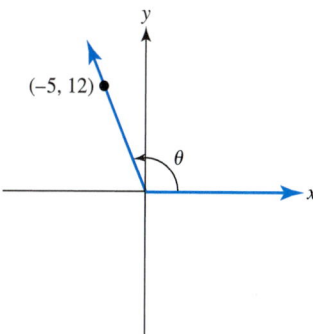

$(-5, 12)$

3.

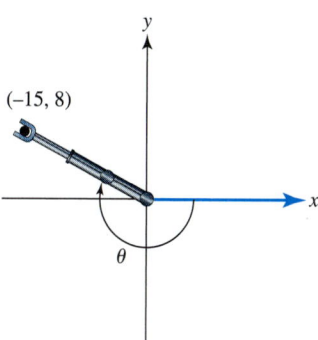

$(-15, 8)$

7.

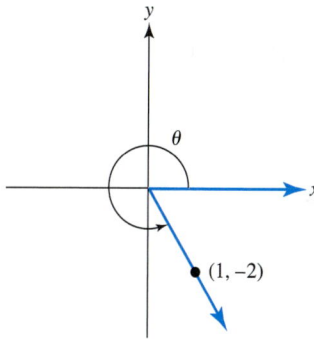

$(1, -2)$

4.

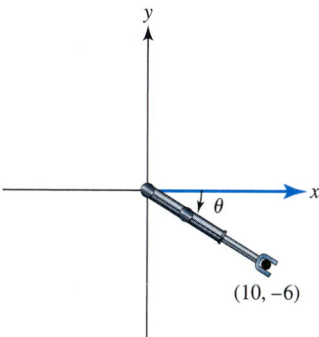

$(10, -6)$

8.

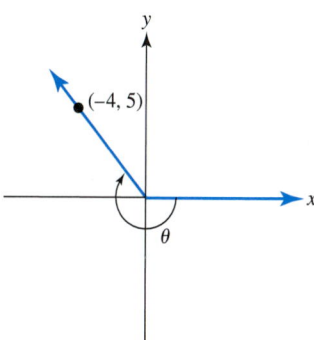

$(-4, 5)$

Exercises 5–10: Find sin θ and cos θ.

5.

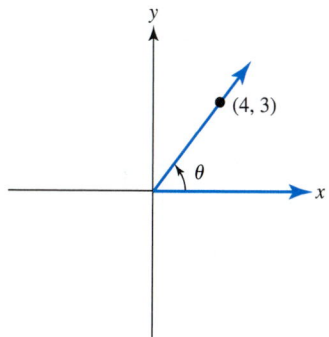

$(4, 3)$

9.

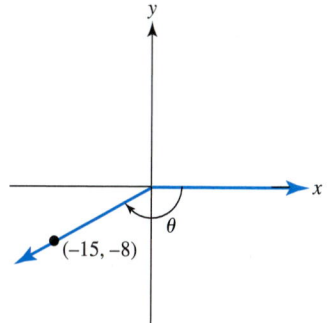

$(-15, -8)$

10.

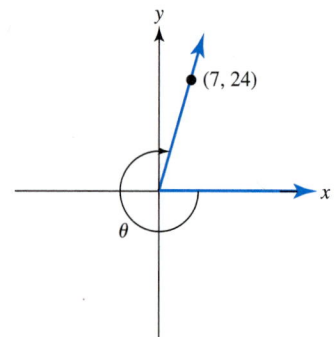

Exercises 11–18: *(Refer to Example 2.) Find the sine and cosine functions by hand for the given angle. Then support your answer using a calculator.*

11. $45°$

12. $150°$

13. $-30°$

14. $-180°$

15. $\dfrac{\pi}{3}$

16. $\dfrac{5\pi}{4}$

17. $-\dfrac{\pi}{2}$

18. -2π

Exercises 19–22: *The terminal side of an angle θ in standard position lies on the line in the given quadrant. Find the sine and cosine functions of θ. (Hint: Find a point (x, y) lying on the terminal side of θ.)*

19. $y = 2x,$ Quadrant I

20. $y = -\dfrac{1}{2}x,$ Quadrant II

21. $y = -3x,$ Quadrant IV

22. $y = \dfrac{3}{4}x,$ Quadrant III

Exercises 23–30: *Approximate the sine and cosine of each angle to four decimal places.*

23. $93.2°$

24. $-43°$

25. $123°50'$

26. $12°40'45''$

27. -4

28. 1.56

29. $\dfrac{11\pi}{7}$

30. $-\dfrac{7\pi}{5}$

Exercises 31–34: *The figure shows angle θ in standard position with its terminal side intersecting the unit circle. Evaluate sin θ and cos θ.*

31.

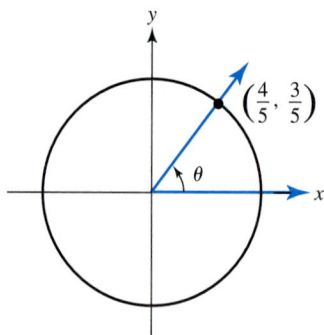

32.

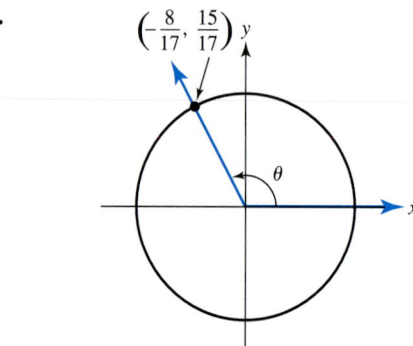

33.

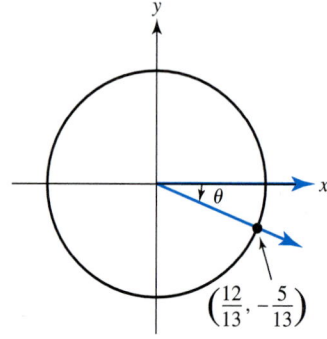

34.

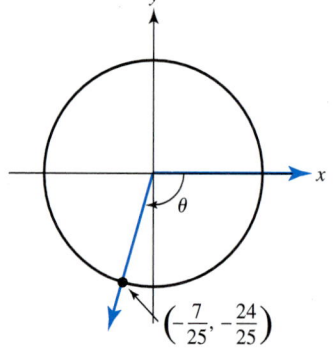

Representations of Functions

Exercises 35–40: Use the accompanying graph of
$f(t) = \sin t$ *to complete the following.*
(a) *Evaluate $f(t)$ graphically for the given t.*
(b) *Support your result by evaluating $f(t)$ with a calculator.*

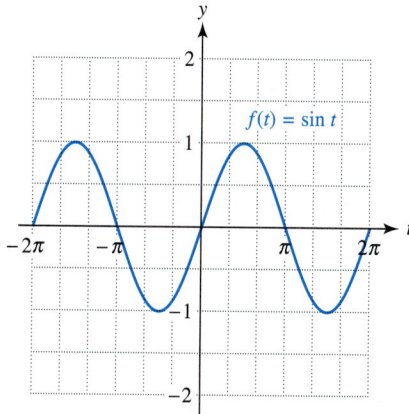

35. $t = 0$ **36.** $t = \pi$

37. $t = \dfrac{\pi}{2}$ **38.** $t = \dfrac{3\pi}{2}$

39. $t = -\dfrac{\pi}{6}$ **40.** $t = -\dfrac{\pi}{2}$

Exercises 41–46: Use the accompanying graph of
$f(t) = \cos t$ *to complete the following.*
(a) *Evaluate $f(t)$ graphically for the given t.*
(b) *Support your result by evaluating $f(t)$ with a calculator.*

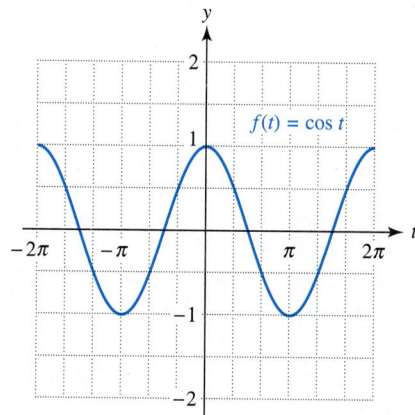

41. $t = 0$ **42.** $t = -\dfrac{\pi}{2}$

43. $t = \dfrac{\pi}{3}$ **44.** $t = \dfrac{\pi}{2}$

45. $t = -\dfrac{3\pi}{2}$ **46.** $t = \pi$

Exercises: 47–52: Evaluate $f(t)$ for the given t
(a) *verbally,*
(b) *numerically, and*
(c) *graphically.*

Support your results by evaluating $f(t)$ with a calculator.

47. $f(t) = \sin t, \quad t = -\pi$

48. $f(t) = \cos t, \quad t = \dfrac{\pi}{2}$

49. $f(t) = \cos t, \quad t = \dfrac{3\pi}{4}$

50. $f(t) = \sin t, \quad t = \dfrac{5\pi}{6}$

51. $f(t) = \sin t, \quad t = -\dfrac{\pi}{2}$

52. $f(t) = \cos t, \quad t = 2\pi$

53. Graph $f(t) = \sin t$ and $g(t) = \cos t$ in the same viewing rectangle. Then translate the graph of f to the left $\pi/2$ units by graphing $f(t + \pi/2)$. How does this translated graph compare to the graph of g?

54. (Continuation of the previous exercise) Translate the graph of $g(t) = \cos t$ to the right $3\pi/2$ units by graphing $g(t - 3\pi/2)$. How does this translated graph compare to the graph of $f(t) = \sin t$?

Exercises 55 and 56: The given algorithm computes a function f. Give
(a) *symbolic,*
(b) *graphical, and*
(c) *numerical*
representations for f. For the numerical representation, table f starting at $t = 0$, incrementing by $\pi/4$.

55. **ALGORITHM** **Computing a Function for a Real Number t**
Step 1: Input a real number t.
Step 2: Let θ be an angle of $2t$ radians in standard position.
Step 3: Determine the point (x, y) where the terminal side of θ intersects the unit circle.
Step 4: Output x.

56. **ALGORITHM** **Computing a Function for a Real Number t**
Step 1: Input a real number t.
Step 2: Let θ be an angle of t radians in standard position.
Step 3: Determine the point (x, y) where the terminal side of θ intersects the unit circle.
Step 4: Output $3y + 1$.

57. Modify Algorithm 5.1 so that it computes the function $f(t) = 2 \sin t$. Compute $f(\pi/2)$ and check your result using a calculator.

58. Modify Algorithm 5.2 so that it computes the function $f(t) = \cos(2t)$. Compute $f(\pi/4)$ and check your result using a calculator.

59. Give graphical and numerical representations of $f(t) = 2 \sin t$ and $g(t) = \sin(2t)$. For the numerical representation start at $t = 0$, incrementing by $\pi/4$. Are f and g the same functions? Explain.

60. Give graphical and numerical representations of $f(t) = 3 \cos t$ and $g(t) = \cos(3t)$. Are f and g the same functions? Explain.

Exercises 61–64: Graph f in $[-2\pi, 2\pi, \pi/2]$ by $[-4, 4, 1]$ and identify the range of f.

61. (a) $f(t) = 3 \sin t$
 (b) $f(t) = \sin(3t)$

62. (a) $f(t) = 2 \cos t$
 (b) $f(t) = \cos(2t)$

63. (a) $f(t) = 2 \cos(t) + 1$
 (b) $f(t) = \cos(2t) - 1$

64. (a) $f(t) = 3 \sin(t) + 1$
 (b) $f(t) = \sin(3t) - 1$

Exercises 65–68: Conjecture how the graph of f will appear in $[-2\pi, 2\pi, \pi/2]$ by $[-4, 4, 1]$. Then, test your conjecture by graphing f.

65. (a) $f(t) = -\sin t$
 (b) $f(t) = -\cos t$

66. (a) $f(t) = \sin(-t)$
 (b) $f(t) = \cos(-t)$

67. (a) $f(t) = \sin(t) + 1$
 (b) $f(t) = \cos(t) - 1$

68. (a) $f(t) = 2 \sin t$
 (b) $f(t) = 2 \cos t$

Applications

69. *Highway Grade* (Refer to Example 3.) Suppose an uphill grade of a highway can be modeled by the line $y = 0.03x$.

 (a) Find the grade of the hill.
 (b) Determine the grade resistance for a gravel truck weighing 25,000 pounds.

70. *Modeling Biological Rhythms* (Refer to Example 7.) The accompanying table shows the time y when a typical flying squirrel becomes active in the afternoon during each month x. (Source: J. Harker, *The Physiology of Diurnal Rhythms*.)

x (month)	1	2	3	4	5	6
y (P. M.)	4:30	5:15	5:45	6:30	7:00	7:30

x (month)	7	8	9	10	11	12
y (P. M.)	7:45	7:30	6:45	6:15	5:00	4:15

 (a) Make a scatterplot of the data in [0, 13, 1] by [3, 8, 1]. (*Hint:* Represent 4:30 P.M. by 4.5.)
 (b) The function given by

$$f(t) = 1.9 \sin(0.42x - 1.2) + 5.7$$

 models the time of sunset, where x represents the month. Graph f and the data together. Interpret the graph.

71. *Voltage* (Refer to Example 8.) Electric ranges and ovens often use a higher voltage than that found in normal household outlets. This voltage can be modeled by

$$V(t) = 310 \sin(120\pi t),$$

where t represents time in seconds.

 (a) In [0, 1/15, 1/60] by [−400, 400, 100] graph V. Describe the voltage in the circuit.
 (b) Evaluate $V(1/120)$ and interpret the result.
 (c) Approximate the root mean square voltage.

72. *Voltage and Phase* If two different voltages can be modeled by $V(t) = V_0 \cos(120\pi t)$ and $U(t) = U_0 \cos(120\pi t + \phi)$ then the *phase difference* equals ϕ. Let $V(t) = 160 \cos(120\pi t)$ and $U(t) = 100 \cos(120\pi t + \pi/2)$, where t is in seconds. (Source: A. Howatson, *Electrical Circuits and Systems*.)

 (a) If the phase difference is $\pm \dfrac{\pi}{2}$, V and U are said to be exactly out of phase or in *quadrature*. If the phase difference is equal to π, they are said to be in opposite phase or in *antiphase*. Decide if V and U are in quadrature or antiphase.
 (b) Graph V and U in [0, 0.04, 0.01] by [−200, 200, 50]. How are the turning points and zeros of V related to the turning points and zeros of U?
 (c) Modify the formula for U so that it is in antiphase with V. Graph V and U. How are the turning points and zeros of V related to the turning points and zeros of U?

73. *Dead Reckoning* Suppose an airplane flies the path shown in the figure, where distance is measured in hundreds of miles. Approximate the final coordinates of the airplane if its initial coordinates are (0, 0).

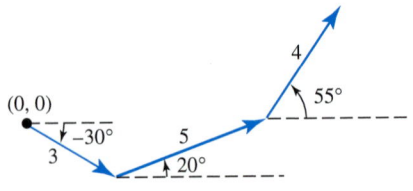

74. *Solar Constant* The solar constant measures the amount of energy reaching Earth from the sun. On average, it is equal to 1367 watts per square meter.

However, because Earth's orbit around the sun is not exactly circular, the solar constant is not *exactly* constant. The percent change in the solar constant can be computed by

$$f(t) = 3.4 \sin\left(\frac{2\pi(82.8 - t)}{365.25}\right),$$

where t is in days and $t = 1$ corresponds to January 1 of a leap year. (Source: C. Winter, *Solar Power Plants.*)

(a) Graph f in [0, 1461, 365.25] by [−4, 4, 1]. This represents the percent change in the solar constant over a four-year period.

(b) Find the maximum percent change. When does the maximum value for the solar constant occur?

(c) Interpret the graph.

75. *Highway Design*   When an automobile travels along a circular curve with radius r, trees and buildings situated on the inside of the curve can obstruct a driver's vision. See the accompanying figure. To ensure a safe stopping distance S, the *minimum* distance d that should be cleared on the inside of a highway curve is given by the equation

$$d = r\left(1 - \cos\frac{\beta}{2}\right),$$ where β in radians is deter-

mined by $\beta = \dfrac{S}{r}$. (Source: F. Mannering.)

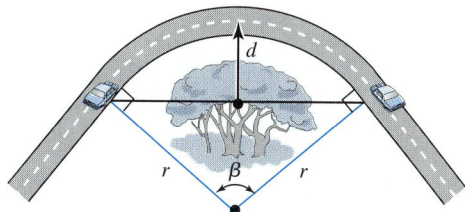

(a) At 45 miles per hour, $S = 390$ feet. If $r = 600$ feet, approximate d.

(b) At 60 miles per hour, $S = 620$ feet. Approximate d for the same curve.

(c) Discuss how the speed limit affects the amount of land that should be cleared on the inside of the curve.

76. *Phases of the Moon*   (Refer to Exercise 6.) Find all phase angles θ that correspond to a full moon and all phase angles that correspond to a new moon. Assume that θ can be any angle.

77. *Stopping Distance and Grade*   The braking distance D in feet for a typical automobile to change its velocity from V_1 to V_2 on dry pavement can be estimated using the equation

$$D = \frac{1.05(V_1{}^2 - V_2{}^2)}{27 + 64.4 \sin\theta}.$$

In this equation θ represents the angle of the grade of the highway and velocity is measured in feet per second. See Figure 5.65 (Source: F. Mannering.)

(a) Compute the number of feet required to slow a car from 88 feet per second (60 mph) to 44 feet per second (30 mph), while traveling uphill with $\theta = 3°$.

(b) Repeat part (a) with $\theta = -3°$.

(c) How is braking distance affected by θ? Does this agree with your driving experience?

78. *Stopping Distance*   (Refer to the previous exercise.) An automobile is traveling at 88 feet per second (60 mph) on a highway with $\theta = -4°$. The driver sees a stalled truck in the road and applies the brakes 250 feet from the vehicle. Assuming that a collision cannot be avoided, how fast is the car traveling when it collides with the truck?   (*Hint:* Let $D = 250$ and find V_2.)

79. *Music and the Sine Function*   A *pure tone* can be modeled by a sine wave. Pure tones typically sound dull and uninteresting. An example of a pure tone is the sound heard when a tuning fork is lightly struck. The pure tone of the first A-note above middle C can be modeled by $f(t) = \sin(880\pi t)$. (Source: J. Pierce, *The Science of Musical Sound.*)

(a) In [0, 1/100, 1/880] by [−1.5, 1.5, 0.5] graph f.

(b) Find the period P of this tone.

(c) Frequency gives the number of vibrations or cycles per second in a sinusoidal graph. The human ear can hear frequencies from 16.4 to 16,000 cycles per second. Frequency F may be determined using the equation $F = \dfrac{1}{P}$, where P is the period. Find the frequency of this A-note.

80. *Music*   (Continuation of the previous exercise) Middle C has a frequency of 261.6 cycles per second and can be modeled by $g(t) = \sin(523.2\pi t)$. (Source: J. Pierce.)

(a) Estimate graphically the period of middle C.

(b) Graph f from the previous exercise and g in [0, 1/100, 1/880] by [−1.5, 1.5, 0.5]. Compare their graphs.

Writing about Mathematics

1. Discuss whether the sine and cosine functions are linear or nonlinear functions. Use graphical and numerical representations to justify your reasoning.

2. Describe two different ways to define the sine and cosine functions. Compare and contrast these definitions. Give examples.

5.4 Other Trigonometric Functions and Their Graphs

Definitions and Basic Identities • Representations of Other
Trigonometric Functions • Modeling with Trigonometric Functions

Introduction

Like the sine and cosine functions, we can extend the domains of the tangent, cotangent, secant, and cosecant functions to include any angle θ or any real number t. This will be accomplished using techniques similar to those presented in the previous section.

The origins of the tangent and cotangent functions were not with astronomy, as was the case with the sine function. Rather, they were developed as part of surveying land, finding heights of objects, and determining time. The cotangent function was computed as early as 1500 B.C. in Egypt, where sundials were used to determine time. Depending on the position of the sun in the sky, a vertical stick casts shadows of different lengths. See Figure 5.88. If θ represents the angle of elevation of the sun and the stick is one unit long, then the length of the shadow equals $\cot \theta$. This is a simple device for evaluating the cotangent function. (Source: *Historical Topics for the Mathematics Classroom, Thirty-first Yearbook.* NCTM.)

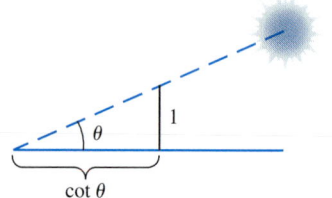

FIGURE 5.88 Modeling Shadow Length

Definitions and Basic Identities

Using Figure 5.89, the other trigonometric functions may be defined for any angle θ in a manner similar to how the sine and cosine functions were defined.

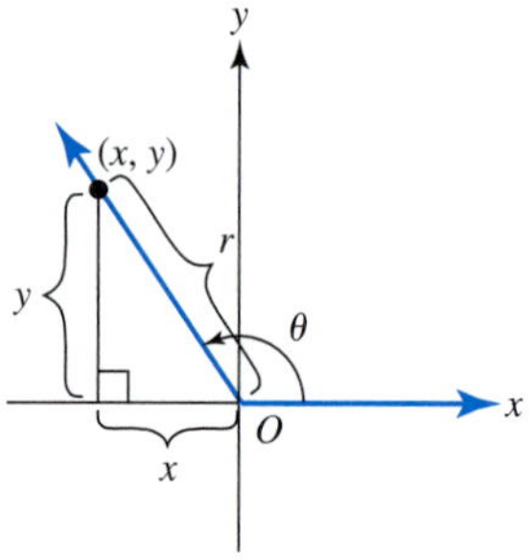

FIGURE 5.89

Trigonometric functions of any angle θ

Let (x, y) be a point other than the origin on the terminal side of an angle θ in standard position. If $r = \sqrt{x^2 + y^2}$, then the six trigonometric functions are as follows.

$$\sin \theta = \frac{y}{r} \qquad\qquad \csc \theta = \frac{r}{y} \, (y \neq 0)$$

$$\cos \theta = \frac{x}{r} \qquad\qquad \sec \theta = \frac{r}{x} \, (x \neq 0)$$

$$\tan \theta = \frac{y}{x} \, (x \neq 0) \qquad \cot \theta = \frac{x}{y} \, (y \neq 0)$$

The domains of both the sine and cosine functions include all angles. However, the cosecant and cotangent functions are undefined when $y = 0$, which corresponds to angles whose terminal sides lie on the x-axis. Examples include $0°$, $\pm 180°$, and $\pm 360°$. The domains of the cosecant and cotangent functions are as follows.

$$D = \{\theta \mid \theta \neq 180° \cdot n\}, \quad n \text{ an integer}$$

Similarly, the tangent and secant functions are undefined whenever $x = 0$. This corresponds to angles whose terminal sides lie on the y-axis. Examples include

$\pm 90°$, $\pm 270°$, and $\pm 450°$. The domains of the tangent and secant functions are as follows.

$$D = \{\theta \mid \theta \neq 90° + 180°{\cdot}n\}, \quad n \text{ an integer}$$

In some situations we can evaluate the trigonometric functions without a calculator. This is illustrated in the next example.

EXAMPLE 1   *Finding values of trigonometric functions*

The point $(5, -12)$ is located on the terminal side of an angle θ in standard position. Find the six trigonometric functions of θ.

Solution

There are many coterminal angles that have the point $(5, -12)$ located on their terminal sides. However, the trigonometric values for coterminal angles are equal. In Figure 5.90 one possibility for θ is shown. Begin by calculating r.

$$\begin{aligned} r &= \sqrt{x^2 + y^2} \\ &= \sqrt{5^2 + (-12)^2} \\ &= 13 \end{aligned}$$

Since $x = 5$, $y = -12$, and $r = 13$, the values of the six trigonometric functions are as follows.

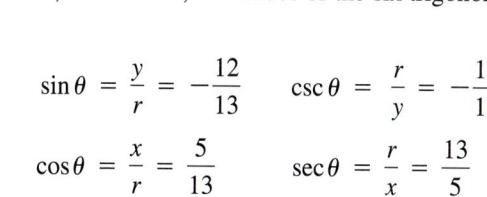

$$\sin \theta = \frac{y}{r} = -\frac{12}{13} \qquad \csc \theta = \frac{r}{y} = -\frac{13}{12}$$

$$\cos \theta = \frac{x}{r} = \frac{5}{13} \qquad \sec \theta = \frac{r}{x} = \frac{13}{5}$$

$$\tan \theta = \frac{y}{x} = -\frac{12}{5} \qquad \cot \theta = \frac{x}{y} = -\frac{5}{12}$$

Since $\sin \theta = \frac{y}{r}$ and $\csc \theta = \frac{r}{y}$, it follows that $\sin \theta$ and $\csc \theta$ are reciprocals.
This may be expressed using either of the *reciprocal identities,*

$$\sin \theta = \frac{1}{\csc \theta} \qquad \text{or} \qquad \csc \theta = \frac{1}{\sin \theta}.$$

In Example 1 it was shown that $\sin \theta = -\frac{12}{13}$. It follows that

$$\csc \theta = \frac{1}{-12/13} = -\frac{13}{12}.$$

Identities are equations that are true for all meaningful values of a variable. Reciprocal identities also hold for $\tan \theta$ and $\cot \theta$ as well as $\cos \theta$ and $\sec \theta$.

Reciprocal identities

$$\sin \theta = \frac{1}{\csc \theta} \qquad \cos \theta = \frac{1}{\sec \theta} \qquad \tan \theta = \frac{1}{\cot \theta}$$

$$\csc \theta = \frac{1}{\sin \theta} \qquad \sec \theta = \frac{1}{\cos \theta} \qquad \cot \theta = \frac{1}{\tan \theta}$$

FIGURE 5.90

(5, −12)

The expressions $\tan\theta$, $\cot\theta$, $\sec\theta$, and $\csc\theta$ can all be written in terms of the $\sin\theta$ and $\cos\theta$. For example,

$$\cot\theta = \frac{x}{y} = \frac{x/r}{y/r} = \frac{\cos\theta}{\sin\theta}.$$

Other *quotient identities* can be obtained similarly.

Quotient identities

$$\tan\theta = \frac{\sin\theta}{\cos\theta} \qquad \cot\theta = \frac{\cos\theta}{\sin\theta} \qquad \sec\theta = \frac{1}{\cos\theta} \qquad \csc\theta = \frac{1}{\sin\theta}$$

EXAMPLE 2   Using the quotient identities

If $\sin\theta = \dfrac{3}{5}$ and $\cos\theta = -\dfrac{4}{5}$, find the other four trigonometric functions of θ.

Solution
To find the other four trigonometric functions apply the quotient identities.

$$\tan\theta = \frac{\sin\theta}{\cos\theta} = \frac{3/5}{-4/5} = -\frac{3}{4}$$

$$\cot\theta = \frac{\cos\theta}{\sin\theta} = \frac{-4/5}{3/5} = -\frac{4}{3}$$

$$\sec\theta = \frac{1}{\cos\theta} = \frac{1}{-4/5} = -\frac{5}{4}$$

$$\csc\theta = \frac{1}{\sin\theta} = \frac{1}{3/5} = \frac{5}{3}$$ ∎

Suppose the terminal side of an angle θ in standard position intersects the unit circle at the point (x, y). In this case $r = 1$ and the definitions of the six trigonometric functions become as follows.

$$\sin\theta = y \qquad \cos\theta = x \qquad \tan\theta = \frac{y}{x}$$

$$\csc\theta = \frac{1}{y} \qquad \sec\theta = \frac{1}{x} \qquad \cot\theta = \frac{x}{y}$$

Like the sine and cosine functions, evaluating the other trigonometric functions at a real number t is equivalent to evaluating these functions at t radians.

Representations of Other Trigonometric Functions

Symbolic representations for the tangent, cotangent, secant, and cosecant functions are given by

$$f(t) = \tan t, \quad g(t) = \cot t, \quad h(t) = \sec t, \quad \text{and} \quad k(t) = \csc t.$$

Like sin t and cos t, there are no simple formulas to evaluate these four trigonometric functions. However, these functions do have verbal, numerical, and graphical representations. We begin by discussing representations of the tangent function.

The Tangent Function.   Figure 5.91 shows angle θ in standard position with its terminal side passing through the point (x, y). The slope of this terminal side is

$$ m = \frac{y - 0}{x - 0} = \frac{y}{x} = \tan \theta. $$

It can be shown that the slope of the terminal side of θ equals $\tan \theta$, regardless of the quadrant containing θ. Using this fact, we will analyze the tangent function by letting θ vary from $-\pi/2$ to $\pi/2$, or equivalently, $-90°$ to $90°$.

When $\theta = -\pi/2$ the slope of the terminal side of θ is undefined, and therefore, $\tan(-\pi/2)$ is undefined. Refer to Figure 5.92. For values of θ slightly greater than $-\pi/2$, the slope of the terminal side of θ is both negative and large in absolute value. At $\theta = -\pi/4$, the slope of the terminal side is $m = -1$ and $\tan(-\pi/4) = -1$. The slope is negative and increasing for $-\pi/2 < \theta < 0$. When $\theta = 0$ then $m = 0$ and $\tan 0 = 0$.

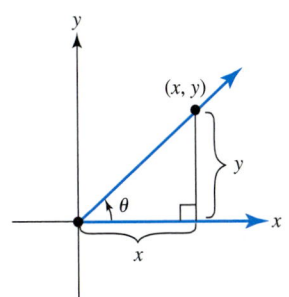

FIGURE 5.91

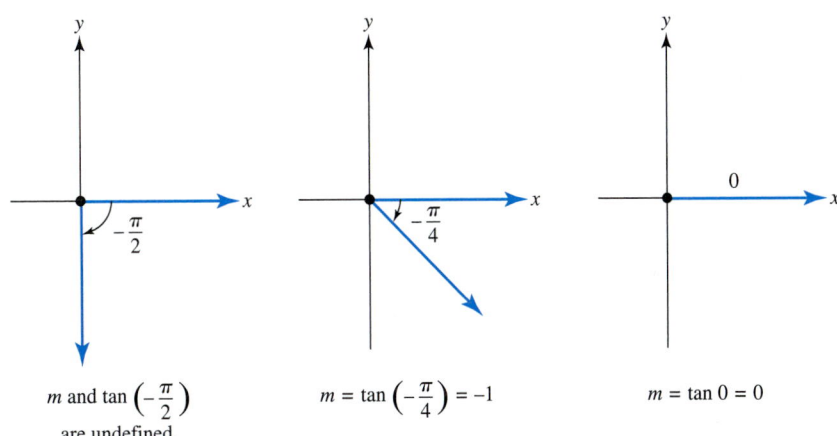

m and $\tan\left(-\dfrac{\pi}{2}\right)$ are undefined. $m = \tan\left(-\dfrac{\pi}{4}\right) = -1$ $m = \tan 0 = 0$

FIGURE 5.92

As θ increases from 0 to $\pi/2$, the slope m is increasing and positive. Refer to Figure 5.93. At $\theta = \pi/4$ the slope is $m = 1$ and $\tan(\pi/4) = 1$. As θ increases toward $\pi/2$ the slope of the terminal side increases without bound and becomes undefined when $\theta = \pi/2$. Thus, $\tan(\pi/2)$ is undefined.

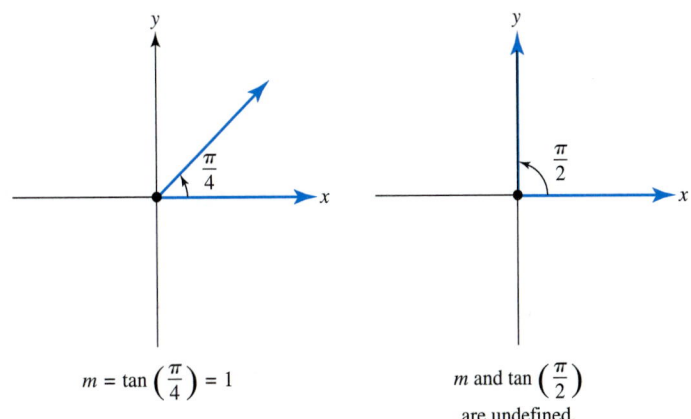

$m = \tan\left(\dfrac{\pi}{4}\right) = 1$ m and $\tan\left(\dfrac{\pi}{2}\right)$ are undefined.

FIGURE 5.93

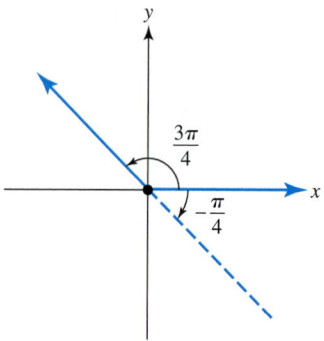

FIGURE 5.94

As θ becomes greater than $\pi/2$ the values of $\tan \theta$ repeat. For example, when $\theta = 3\pi/4$, the slope of the terminal side of θ is equal to the slope of the terminal side of $\theta = -\pi/4$. See Figure 5.94. As a result, the tangent function is periodic with π or $180°$. We now represent the tangent function numerically, graphically, and verbally.

A *partial numerical representation* of $f(t) = \tan t$ for selected values of t satisfying $-\pi/2 \le t \le \pi/2$ is shown in Table 5.7. (A dash indicates that $\tan t$ is undefined.)

TABLE 5.7

t	$-\dfrac{\pi}{2}$	$-\dfrac{\pi}{3}$	$-\dfrac{\pi}{4}$	$-\dfrac{\pi}{6}$	0	$\dfrac{\pi}{6}$	$\dfrac{\pi}{4}$	$\dfrac{\pi}{3}$	$\dfrac{\pi}{2}$
$\tan t$	—	$-\sqrt{3}$	-1	$-\dfrac{1}{\sqrt{3}}$	0	$\dfrac{1}{\sqrt{3}}$	1	$\sqrt{3}$	—

A *graphical representation* of $y = \tan t$ and the data in Table 5.7 appear in Figure 5.95, where t is measured in radians. Vertical asymptotes occur when the $\tan t$ is undefined at $t = \pm\, \pi/2$. Since the tangent function is periodic with π (or $180°$), we can extend the graph of $y = \tan t$ as in Figure 5.96. Notice that vertical asymptotes occur at

$$t = \pm\frac{\pi}{2},\ \pm\frac{3\pi}{2},\ \pm\frac{5\pi}{2},\ \ldots$$

where $\tan t$ is undefined.

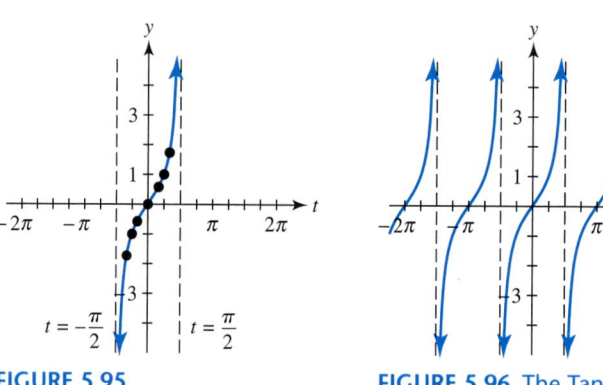

FIGURE 5.95   **FIGURE 5.96** The Tangent Function

A *verbal representation* of the tangent function is given by Algorithm 5.3.

ALGORITHM 5.3 Evaluating the Tangent Function at a Real Number t
   Step 1: Input a real number t.
   Step 2: Let θ be an angle of t radians in standard position.
   Step 3: Determine the point (x, y) where the terminal side of θ intersects the unit circle.
   Step 4: If $x \ne 0$, then output $\dfrac{y}{x}$, the value of $\tan t$.

EXAMPLE 3   *Evaluating the tangent function verbally, numerically, and graphically*

Evaluate $f(t) = \tan t$ at $t = -\pi/4$ using verbal, numerical, and graphical representations. Support your answer by evaluating $\tan(-\pi/4)$ with a calculator.

Solution

Verbal Representation   Apply Algorithm 5.3. If $\theta = -\pi/4$ is in standard position, the terminal side intersects the unit circle at the point $(1/\sqrt{2}, -1/\sqrt{2})$. See Figure 5.97. Therefore,

$$\tan\left(-\frac{\pi}{4}\right) = \frac{y}{x} = \frac{-1/\sqrt{2}}{1/\sqrt{2}} = -1.$$

Numerical Representation   Table $f(t) = \tan t$ starting at $x = -\pi/2$, incrementing by $\pi/4$. From Figure 5.98 we can see that at $x = -\pi/4 \approx -0.7854$, $\tan(-\pi/4) = -1$.

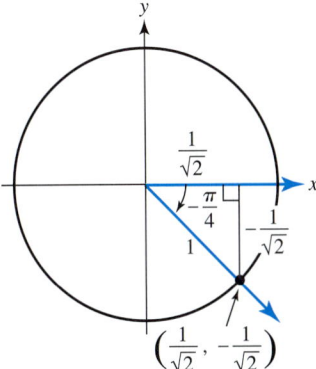

FIGURE 5.97

$[-\pi, \pi, \pi/4]$ by $[-4, 4, 1]$

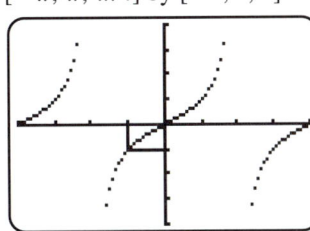

FIGURE 5.98   **FIGURE 5.99** Dot Mode   **FIGURE 5.100**

Graphical Representation   The graph of $Y_1 = \tan(X)$ in dot mode is shown in Figure 5.99. We see that $\tan(-\pi/4) = -1$. Figure 5.100 supports this result. (*Note:* If dot mode is not used, a graphing calculator may connect points inappropriately, resulting in lines that resemble but are not vertical asymptotes.)   ■

The Cosecant Function.   A *verbal representation* of the cosecant function is given in Algorithm 5.4.

ALGORITHM 5.4 Evaluating the Cosecant Function at a Real Number t
   Step 1: Input a real number t.
   Step 2: Let θ be an angle of t radians in standard position.
   Step 3: Determine the point (x, y) where the terminal side of θ intersects the unit circle.
   Step 4: If $y \neq 0$, then output $\dfrac{1}{y}$, the value of csc t.

For example, if $t = \dfrac{7\pi}{6}$, then the terminal side of $\theta = \dfrac{7\pi}{6}$ intersects the unit circle

at the point $(-\sqrt{3}/2, -1/2)$. See Figure 5.101. It follows that

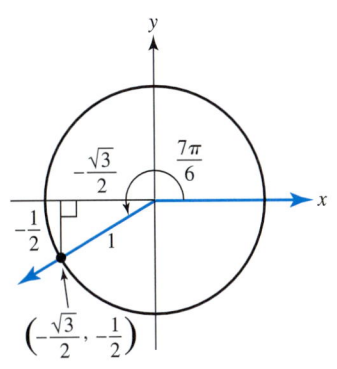

FIGURE 5.101

$$\csc\frac{7\pi}{6} = \frac{1}{y} = \frac{1}{-1/2} = -2.$$

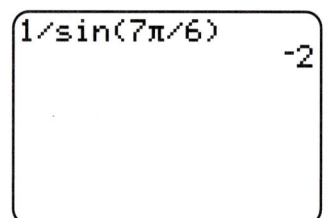

FIGURE 5.102

This result is supported in Figure 5.102 by evaluating $\dfrac{1}{\sin (7\pi/6)}$.

A *partial numerical representation* of $f(t) = \csc t$ is shown in Table 5.8 for selected values of t satisfying $-\pi \le t \le \pi$. Note that $\csc t = 1/\sin t$. (A dash indicates that $\csc t$ is undefined.)

TABLE 5.8

t	$-\pi$	$-\dfrac{3\pi}{4}$	$-\dfrac{\pi}{2}$	$-\dfrac{\pi}{4}$	0	$\dfrac{\pi}{4}$	$\dfrac{\pi}{2}$	$\dfrac{3\pi}{4}$	π
$\csc t$	—	$-\sqrt{2}$	-1	$-\sqrt{2}$	—	$\sqrt{2}$	1	$\sqrt{2}$	—

A *graphical representation* of $\csc t$ is given in Figure 5.103. The domain of the cosecant function is $D = \{t \mid t \ne \pi n\}$, where n is an integer and vertical asymptotes occur whenever $t = \pi n$. Its period equals 2π.

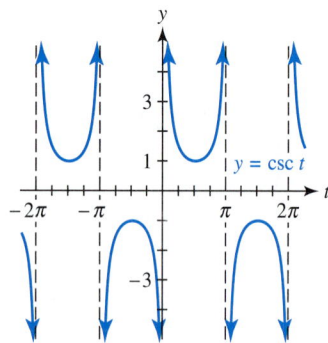

FIGURE 5.103 The Cosecant Function

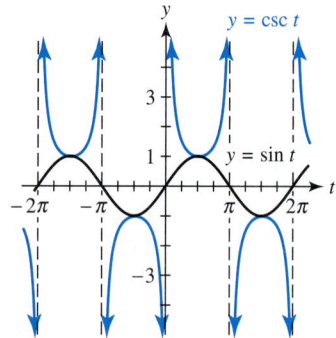

FIGURE 5.104 The Sine and Cosecant Functions

Since $\csc t = 1/\sin t$, we can sketch the graph of $y = \csc t$ by using the graph of $y = \sin t$ as an aid. See Figure 5.104. When $\sin t = 0$, $\csc t$ is undefined and a vertical asymptote occurs on the graph of $\csc t$. When $\sin t = \pm 1$, $\csc t = \pm 1$. Whenever $\sin t$ increases, $\csc t$ decreases and whenever $\sin t$ decreases, $\csc t$ increases. Since $|\sin t| \le 1$ for all t, it follows that $|\csc t| \ge 1$ for all $t \ne \pi n$.

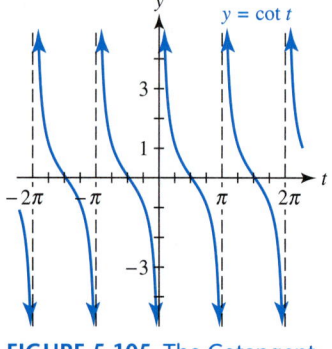

FIGURE 5.105 The Cotangent Function

The Cotangent Function. Verbal and numerical representations are possible for the cotangent function. A graphical representation of $y = \cot t$ is shown in Figure 5.105. Its period is π and its domain is $D = \{t \mid t \ne \pi n\}$, where n is an integer. Vertical asymptotes occur at $t = \pi n$.

Technology Note *Graphing in Degrees*

Trigonometric functions may be represented graphically and numerically in either radian or degree mode. If a friendly window is selected using degree mode, the graph of a trigonometric function may be accurately graphed in connected mode. Graphical and numerical representations of the cosecant and cotangent functions are shown in Figures 5.106–5.109 on the next page.

$[-352.5°, 352.5°, 90°]$ by $[-4, 4, 1]$

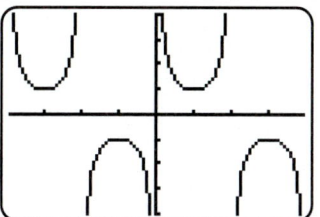

FIGURE 5.106 The Cosecant Function

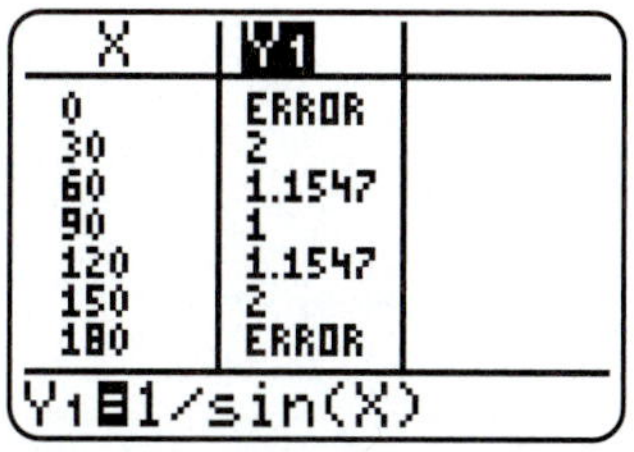

FIGURE 5.107 The Cosecant Function

$[-352.5°, 352.5°, 90°]$ by $[-4, 4, 1]$

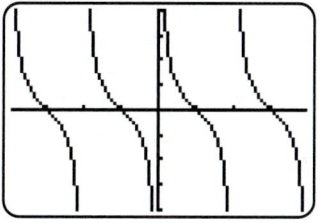

FIGURE 5.108 The Cotangent Function

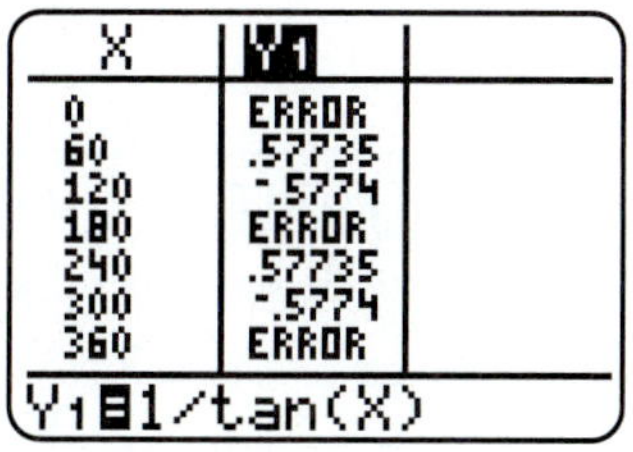

FIGURE 5.109 The Cotangent Function

The Secant Function. Verbal and numerical representations are also possible for the secant function. A graphical representation of $y = \sec t$ is shown in Figure 5.110. Its period is 2π and its domain is $D = \{t \mid t \neq \pi/2 + \pi n\}$, where n is an integer. Vertical asymptotes occur at $t = \dfrac{\pi}{2} + \pi n$. The graph of the cosine function may be used as an aid to graph the secant function, since the zeros of $y = \cos t$ correspond to the vertical asymptotes on the graph of $y = \sec t$. See Figure 5.111.

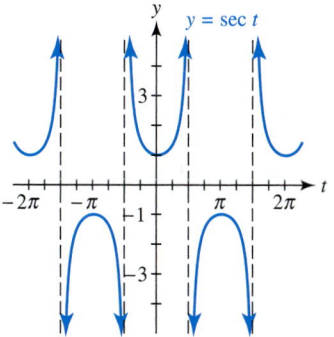

FIGURE 5.110 The Secant Function

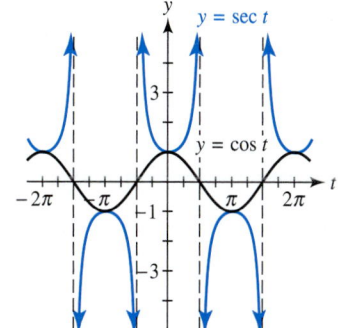

FIGURE 5.111 The Cosine and Secant Functions

Critical Thinking

Give verbal and numerical representations of the secant function.

The next two examples illustrate how to evaluate the six trigonometric functions by hand and with a calculator.

EXAMPLE 4   *Evaluating the trigonometric functions*

If $\theta = \dfrac{3\pi}{2}$, find the six trigonometric functions of θ.

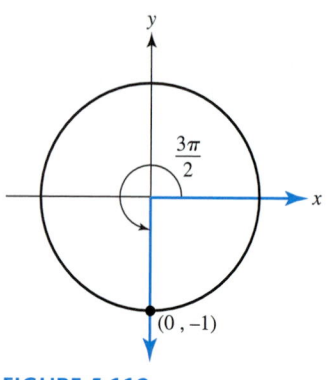

FIGURE 5.112

Solution

If θ is in standard position, then its terminal side intersects the unit circle at the point $(0, -1)$. See Figure 5.112. Thus, $x = 0$ and $y = -1$. The values of the six trigonometric functions are as follows.

$$\sin \frac{3\pi}{2} = y = -1 \qquad\qquad \csc \frac{3\pi}{2} = \frac{1}{y} = \frac{1}{-1} = -1$$

$$\cos \frac{3\pi}{2} = x = 0 \qquad\qquad \sec \frac{3\pi}{2} = \frac{1}{x} \text{ is undefined, since } x = 0$$

$$\tan \frac{3\pi}{2} = \frac{y}{x} \text{ is undefined, since } x = 0 \qquad \cot \frac{3\pi}{2} = \frac{x}{y} = \frac{0}{-1} = 0 \quad ∎$$

EXAMPLE 5 *Using a calculator to evaluate trigonometric functions*

Find values of the six trigonometric functions of θ.
(a) $\theta = 102.6°$
(b) $\theta = 2.56$

Solution

(a) Most calculators do not have special keys for secant, cosecant, and cotangent. To evaluate these functions, we use the reciprocal identities. For example, to find sec (102.6°), evaluate 1/cos (102.6°). The values of the trigonometric functions are shown in Figures 5.113 and 5.114.

```
sin(102.6)
          .9759167619
cos(102.6)
         -.2181432414
tan(102.6)
         -4.473742829
```

```
1/sin(102.6)
          1.024677553
1/cos(102.6)
         -4.584143857
1/tan(102.6)
         -.2235264829
```

FIGURE 5.113 Degree Mode   **FIGURE 5.114** Degree Mode

(b) Since there is no degree symbol, use radian mode. The values of the six trigonometric functions are shown in Figures 5.115 and 5.116.

```
sin(2.56)
          .5493554364
cos(2.56)
         -.8355887771
tan(2.56)
         -.6574471217
```

```
1/sin(2.56)
          1.820315107
1/cos(2.56)
         -1.196760928
1/tan(2.56)
         -1.521034874
```

FIGURE 5.115 Radian Mode   **FIGURE 5.116** Radian Mode   ∎

Modeling with Trigonometric Functions

When a stick is partially put into a glass of water, it appears to bend at the surface of the water. The same phenomenon occurs when starlight enters Earth's atmosphere. To an observer on the ground, a star's apparent position α is different from its true position. This is referred to as *atmospheric refraction.* See Figure 5.117. The amount that starlight is bent is given by angle θ measured in seconds, where $\theta = 57.3 \tan \alpha$ with $0° \le \alpha \le 45°$. (Source: W. Schlosser, *Challenges of Astronomy.*)

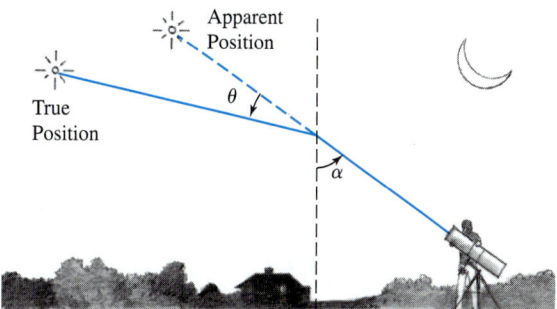

FIGURE 5.117 Atmospheric Refraction

EXAMPLE 6   *Calculating refraction of starlight*

Refer to Figure 5.117.
(a) Calculate θ when $\alpha = 32°35'21''$.
(b) Is starlight bent more or less for stars that appear lower in the horizon? Explain.

Solution
(a) $\theta = 57.3 \tan (32°35'21'') \approx 36.6''$. The star is actually $36.6''$ lower in the sky than it appears.
(b) As α increases from $0°$ to $45°$ the star appears lower in the horizon, while the graph of $y = \tan \alpha$ increases. Thus, $\theta = 57.3 \tan \alpha$ also increases and starlight is bent more for stars that appear lower in the horizon. This is supported numerically in Figure 5.118, where $Y_1 = 57.3 \tan (X)$ increases as x increases. ∎

X	Y1
0	0
9	9.0754
18	18.618
27	29.196
36	41.631
45	57.3
54	78.867

$Y_1 = 57.3\tan(X)$

FIGURE 5.118

The shortest path through Earth's atmosphere for the sun's rays occurs when the sun is directly overhead. As the sun moves lower in the horizon, sunlight travels through more atmosphere. See Figure 5.119, where $d = h \csc \theta$. If the angle of elevation of the sun is θ, then the path length d of sunlight through the atmosphere increases by a factor of $\csc \theta$ from its noontime value of h. (Due to the curvature of Earth, this model is not accurate when $\theta < 20°$ and the sun is positioned near the horizon. See Exercise 92.) When sunlight passes through more atmosphere, less ultraviolet light reaches Earth's surface. This is one reason why some experts recommend sun tanning either earlier or later in the day. (Source: C. Winter, *Solar Power Plants.*)

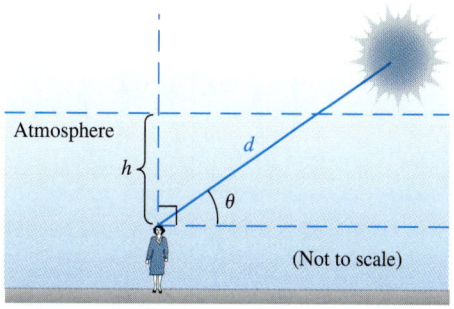

FIGURE 5.119

EXAMPLE 7 *Measuring intensity of the sun*

Assuming that the sun is directly overhead at noon, calculate the percent increase in atmospheric distance that sunlight must pass through at 10 A.M. compared to noon.

Solution

From Figure 5.119, $\theta = 90°$ at noon. The sun moves 360° in the sky in 24 hours or 15° per hour. Thus, two hours earlier $\theta = 60°$. Since csc 90° = 1 and csc 60° = 1/sin 60° ≈ 1.15, this means that sunlight travels through about 15% more atmosphere at 10 A.M. than at noon. ∎

5.4 PUTTING IT ALL TOGETHER

There are six trigonometric functions. The following table summarizes the domain, range, and period of each trigonometric function. In this table n is an integer and t is a real number.

Function	Domain	Range	Period
sin t	$-\infty < t < \infty$	$-1 \le \sin t \le 1$	2π
cos t	$-\infty < t < \infty$	$-1 \le \cos t \le 1$	2π
tan t	$t \ne \pi/2 + \pi n$	$-\infty < \tan t < \infty$	π
cot t	$t \ne \pi n$	$-\infty < \cot t < \infty$	π
sec t	$t \ne \pi/2 + \pi n$	$|\sec t| \ge 1$	2π
csc t	$t \ne \pi n$	$|\csc t| \ge 1$	2π

Graphs of the six trigonometric functions, including any asymptotes, are as follows.

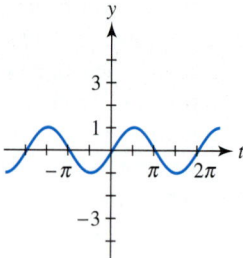

The Sine Function

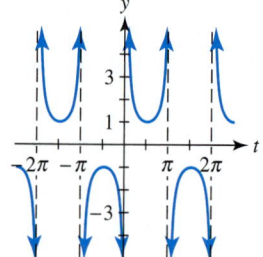

The Cosecant Function

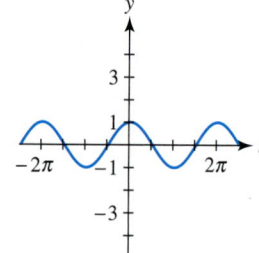

The Cosine Function

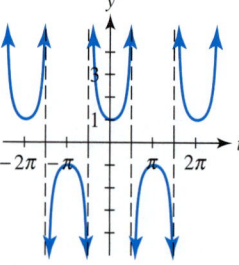

The Secant Function

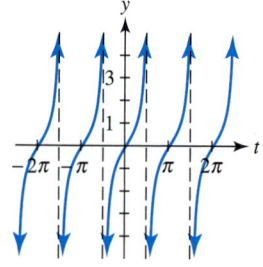

The Tangent Function

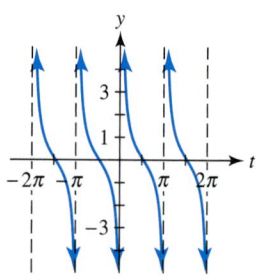

The Cotangent Function

The following figure can be used to evaluate the trigonometric functions for certain angles. For example, if $\theta = 120°$ or $2\pi/3$ radians is in standard position, then its terminal side intersects the unit circle at the point $(-1/2, \sqrt{3}/2)$. It follows that $\cos 120° = -1/2$ and $\sin 120° = \sqrt{3}/2$. Other trigonometric values can be found similarly.

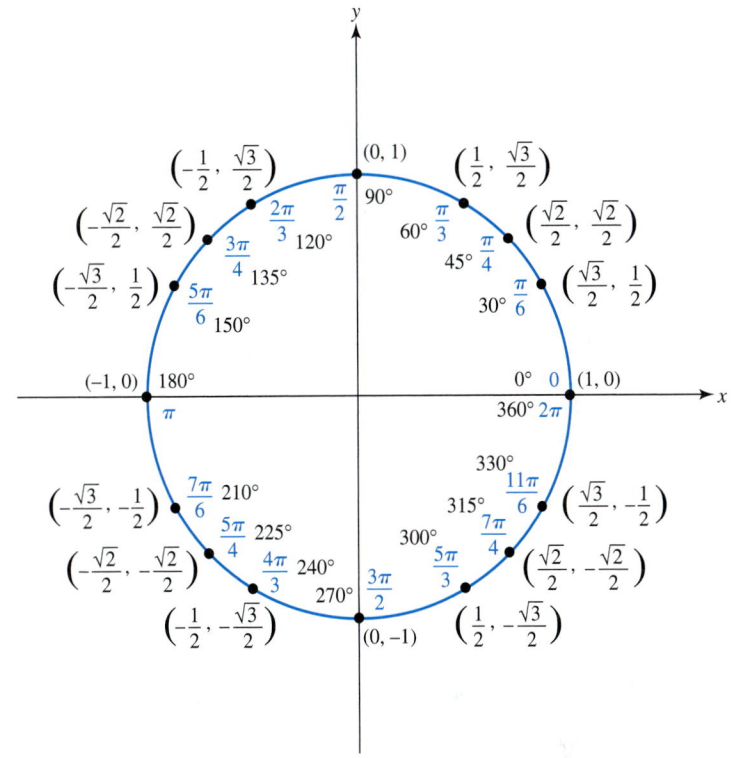

The Unit Circle and Special Angles

5.4 EXERCISES   Tape 9

Basic Concepts

Exercises 1 and 2: The xy-coordinates of the hand for a robotic arm are shown in the figure.
   (a) *Find the length of the arm.*
   (b) *Find the six trigonometric functions of θ.*

1.

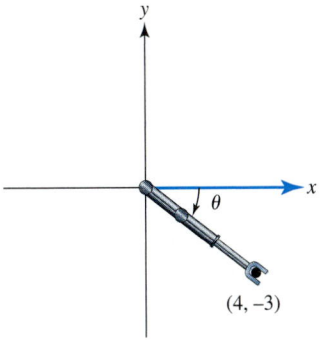

(4, −3)

2.

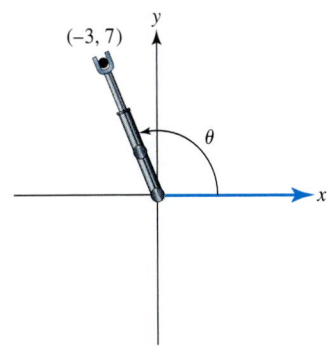

(−3, 7)

Exercises 3–8: Find the six trigonometric functions of θ.

3.

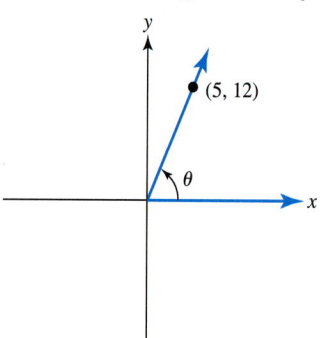

4.

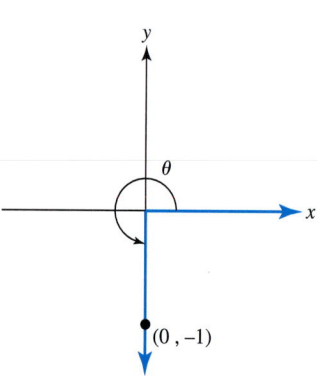

5.

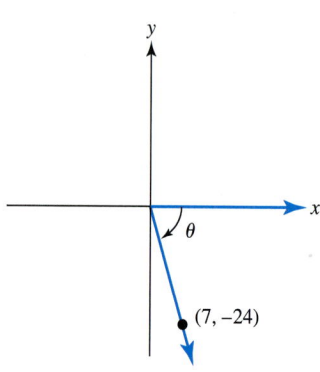

6.

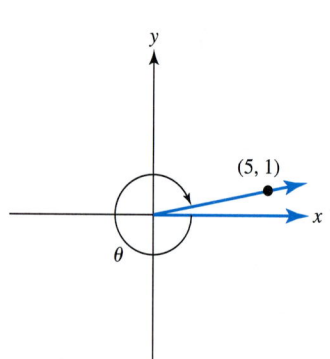

7.

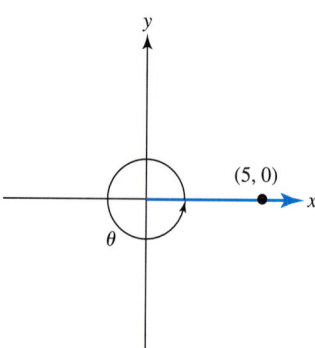

8.

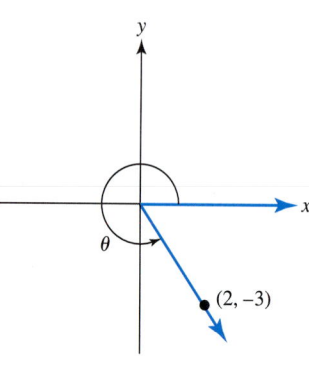

Exercises 9–14: Graph f. Discuss the symmetry of the graph.

9. $f(t) = \sin t$

10. $f(t) = \cos t$

11. $f(t) = \tan t$

12. $f(t) = \cot t$

13. $f(t) = \sec t$

14. $f(t) = \csc t$

Exercises 15–22: Find the six trigonometric functions of the given angle by hand. Support your answer using a calculator.

15. $90°$

16. $135°$

17. $-45°$

18. $-180°$

19. π

20. $\dfrac{3\pi}{4}$

21. $-\dfrac{\pi}{3}$

22. $-\dfrac{5\pi}{6}$

Exercises 23–26: The terminal side of an angle θ lies on the line in the given quadrant. Find the six trigonometric functions of θ. How does the slope of the line compare to tan θ?

23. $y = -4x$, Quadrant II

24. $y = \frac{1}{2}x$, Quadrant I

25. $y = 6x$, Quadrant III

26. $y = -\frac{2}{3}x$, Quadrant IV

Exercises 27–34: Approximate the following to four decimal places.

27. **(a)** $\sin 93.2°$ **(b)** $\csc 93.2°$

28. **(a)** $\cos(-43°)$ **(b)** $\sec(-43°)$

29. **(a)** $\tan 234°33'$ **(b)** $\cot 234°33'$

30. **(a)** $\sec 123°44'25''$ **(b)** $\cos 123°44'25''$

31. **(a)** $\cot(-4)$ **(b)** $\tan(-4)$

32. **(a)** $\csc 1.56$ **(b)** $\sin 1.56$

33. **(a)** $\cos\left(\frac{11\pi}{7}\right)$ **(b)** $\sec\left(\frac{11\pi}{7}\right)$

34. **(a)** $\tan\left(\frac{7\pi}{5}\right)$ **(b)** $\cot\left(\frac{7\pi}{5}\right)$

Exercises 35–38: The accompanying figure shows angle θ in standard position with its terminal side intersecting the unit circle. Evaluate the six trigonometric functions of θ.

35.

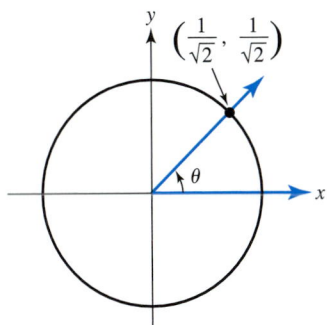

36.

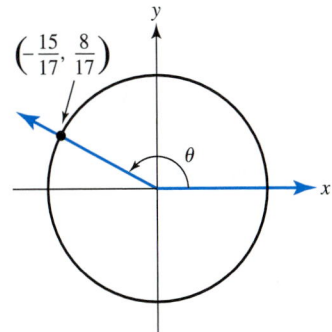

37.

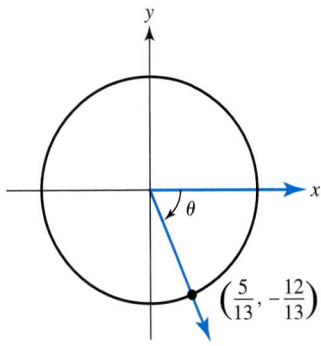

38.

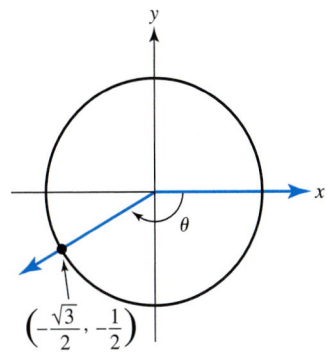

Exercises 39–42: Graph f using degree mode in a friendly window such as $[-352.5°, 352.5°, 90°]$ by $[-4, 4, 1]$. Then use the trace feature to evaluate $f(t)$ at $t = -90°, -45°, 0°, 45°, 90°$, if possible.

39. $f(t) = \tan t$ **40.** $f(t) = \cot t$

41. $f(t) = \csc t$ **42.** $f(t) = \sec t$

Exercises 43–48: Graph f in $[-2\pi, 2\pi, \pi/2]$ by $[-4, 4, 1]$ using dot mode. Use the graph to identify the domain, range, and period of f.

43. $f(t) = \sin t$ **44.** $f(t) = \cos t$

45. $f(t) = \tan t$ **46.** $f(t) = \cot t$

47. $f(t) = \sec t$ **48.** $f(t) = \csc t$

Exercises 49 and 50: Graph f and g in $[-2\pi, 2\pi, \pi/2]$ by $[-4, 4, 1]$ using dot mode. Explain how the graph of f is related to the graph of g.

49. $f(t) = \sin t$, $g(t) = \csc t$

50. $f(t) = \cos t$, $g(t) = \sec t$

51. Determine graphically or numerically where $f(t) = 0$ for $-2\pi < t \le 2\pi$.
(a) $f(t) = \sin t$ **(b)** $f(t) = \cos t$ **(c)** $f(t) = \tan t$
(d) $f(t) = \csc t$ **(e)** $f(t) = \sec t$ **(f)** $f(t) = \cot t$

52. Determine graphically or numerically where $f(t)$ is undefined for $-2\pi < t \le 2\pi$.
(a) $f(t) = \sin t$ **(b)** $f(t) = \cos t$ **(c)** $f(t) = \tan t$
(d) $f(t) = \csc t$ **(e)** $f(t) = \sec t$ **(f)** $f(t) = \cot t$

Exercises 53–58: Determine the values of the trigono-metric functions of θ using the given information.

53. $\sin\theta = \dfrac{3}{5}$ and $\cos\theta = \dfrac{4}{5}$

54. $\sin\theta = -\dfrac{7}{25}$ and $\cos\theta = \dfrac{24}{25}$

55. $\csc\theta = -\dfrac{17}{15}$ and $\sec\theta = -\dfrac{17}{8}$

56. $\csc\theta = 2$ and $\sec\theta = -\dfrac{2}{\sqrt{3}}$

57. $\tan\theta = \dfrac{5}{12}$ and $\cos\theta = \dfrac{12}{13}$
 (*Hint:* $\sin\theta = \tan\theta\cos\theta$.)

58. $\sin\theta = \dfrac{3}{5}$ and $\cot\theta = -\dfrac{4}{3}$

59. *Horizontal Compression* Compare the graphs of f and g.
 (a) $f(x)=\csc x,\quad g(x)=\csc(2x)$
 (b) $f(x)=\cot x,\quad g(x)=\cot(3x)$
 (c) $f(x)=\tan x,\quad g(x)=\tan(4x)$
 (d) $f(x)=\sec x,\quad g(x)=\sec(2x)$

60. *Horizontal Stretching* Compare the graphs of f and g.
 (a) $f(x)=\cot x,\quad g(x)=\cot\left(\dfrac{1}{2}x\right)$
 (b) $f(x)=\csc x,\quad g(x)=\csc\left(\dfrac{1}{3}x\right)$
 (c) $f(x)=\tan x,\quad g(x)=\tan\left(\dfrac{1}{4}x\right)$
 (d) $f(x)=\sec x,\quad g(x)=\sec\left(\dfrac{1}{2}x\right)$

Exercises 61–64: For each pair of trigonometric functions conjecture how many times their graphs will intersect on the interval $[0, 2\pi]$. Support your conjecture by graphing.

61. $y=\sin x,\quad y=\cos x$

62. $y=\cos x,\quad y=\tan x$

63. $y=\sin x,\quad y=\sec x$

64. $y=\cot x,\quad y=\tan x$

Representations of Functions

Exercises 65–68: If possible, evaluate $f(t)=\tan t$ and $g(t)=\cot t$ at the given value of t
 (a) *verbally,*
 (b) *numerically, and*
 (c) *graphically.*

65. $t=0$

66. $t=\dfrac{\pi}{3}$

67. $t=-\dfrac{\pi}{4}$

68. $t=-\dfrac{\pi}{2}$

Exercises 69–72: If possible, evaluate the $f(t)=\sec t$ and $g(t)=\csc t$ at the given value of t
 (a) *verbally,*
 (b) *numerically, and*
 (c) *graphically.*

69. $t=\pi$

70. $t=\dfrac{\pi}{2}$

71. $t=-\dfrac{\pi}{2}$

72. $t=-\dfrac{3\pi}{4}$

Exercises 73 and 74: The given algorithm computes a function f. Give
 (a) *symbolic,*
 (b) *graphical, and*
 (c) *numerical*
representations for f. For the numerical representation, table f starting at $t=0$, incrementing by $\pi/4$.

73. **ALGORITHM Computing a Function at a Real Number t**
 Step 1: Input a real number t.
 Step 2: Let θ be an angle of $2t$ radians in standard position.
 Step 3: Determine the point (x, y) where the terminal side of θ intersects the unit circle.
 Step 4: If $y\neq 0$, output $\dfrac{x}{y}$.

74. **ALGORITHM Computing a Function at a Real Number t**
 Step 1: Input a real number t.
 Step 2: Let θ be an angle of t radians in standard position.
 Step 3: Determine the point (x, y) where the terminal side of θ intersects the unit circle.
 Step 4: If $x\neq 0$, output $\dfrac{2}{x}$.

75. Modify Algorithm 5.3 so that it computes the function $f(t)=\tan t-5$. Compute $f(0)$ and check your result using a calculator.

76. Modify Algorithm 5.4 so that it computes the function $f(t)=\csc(4t)$. Compute $f(\pi/8)$ and check your result using a calculator.

Exercises 77–82: Determine which of the six trigonometric functions is represented by the table without using a calculator.

77.

x	0	$\dfrac{\pi}{2}$	π	$\dfrac{3\pi}{2}$	2π
y	0	1	0	-1	0

78.

x	0	$\frac{\pi}{4}$	$\frac{\pi}{2}$	$\frac{3\pi}{4}$	π
y	0	1	Error	-1	0

79.

x	0	$\frac{\pi}{4}$	$\frac{\pi}{2}$	$\frac{3\pi}{4}$	π
y	Error	1	0	-1	Error

80.

x	0	$\frac{\pi}{2}$	π	$\frac{3\pi}{2}$	2π
y	1	0	-1	0	1

81.

x	0	$\frac{\pi}{2}$	π	$\frac{3\pi}{2}$	2π
y	1	Error	-1	Error	1

82.

x	0	$\frac{\pi}{2}$	π	$\frac{3\pi}{2}$	2π
y	Error	1	Error	-1	Error

Exercises 83 and 84: The function f can be represented by either $f(t) = a\cos t$ or $f(t) = a\sin t$. Use the table to determine the symbolic representation of f. Note that values may be approximate.

83.

t	0	$\frac{\pi}{4}$	$\frac{\pi}{2}$	$\frac{3\pi}{4}$	π
$f(t)$	3	2.121	0	-2.121	-3

84.

t	0	$\frac{\pi}{4}$	$\frac{\pi}{2}$	$\frac{3\pi}{4}$	π
$f(t)$	0	-1.414	-2	-1.414	0

Exercises 85 and 86: The function f can be represented by either $f(t) = a\tan t$ or $f(t) = a\cot t$. Use the table to determine the symbolic representation for f.

85.

t	0	$\frac{\pi}{4}$	$\frac{\pi}{2}$	$\frac{3\pi}{4}$	π
$f(t)$	0	-3	Error	3	0

86.

t	0	$\frac{\pi}{4}$	$\frac{\pi}{2}$	$\frac{3\pi}{4}$	π
$f(t)$	Error	2	0	-2	Error

Exercises 87 and 88: The function f can be represented by either $f(t) = a\sec t$ or $f(t) = a\csc t$. Use the table to determine the symbolic representation for f. Note that values may be approximate.

87.

t	0	$\frac{\pi}{6}$	$\frac{\pi}{3}$	$\frac{\pi}{2}$	π
$f(t)$	Error	8	4.618	4	Error

88.

t	0	$\frac{\pi}{6}$	$\frac{\pi}{3}$	$\frac{\pi}{2}$	π
$f(t)$	-2	-2.309	-4	Error	2

Applications

89. *Shadow Length* In the introduction it was discussed how shadows can be used to compute the cotangent function.
   (a) Graph $f(t) = \cot\theta$ in $[0, \pi, \pi/4]$ by $[-4, 4, 1]$.
   (b) Let $\theta = 0$ correspond to sunrise, $\theta = \pi/2$ to noon, and $\theta = \pi$ to sunset. Explain how the graph of f models the length of a shadow cast by a vertical stick with length one.

90. *GPS Satellite Communication* Artificial satellites that orbit Earth often use VHF signals to communicate with the ground. Because VHF signals travel in straight lines, a satellite orbiting Earth can only communicate with a fixed location on the ground during certain times. The height h in miles of an orbit with communication time T is given by

$$h = 3955\left(\sec\left(\pi T/P\right) - 1\right),$$

where P is the period for the satellite to orbit Earth. Suppose a GPS satellite orbit has a period of $P = 12$ hours and can communicate with a person at the north pole for $T = 5.08$ hours each orbit. Approximate the height h of its orbit. (Sources: W. Schlosser, *Challenges of Astronomy*; Y. Zhao.)

91. *Intensity of the Sun* (Refer to Example 7) If the sun is directly overhead at noon, calculate the percent increase in atmospheric distance that sunlight must pass through at 3:00 P.M.

92. *Intensity of the Sun* (Refer to the previous exercise.) The formula

$$y_1 = \csc\theta = \frac{1}{\sin\theta}$$

presented in Example 7 to calculate the path length of sunlight through the atmosphere relative to noon is not accurate when the sun's elevation is less than

20°. A more accurate formula for small values of θ is given by

$$y_2 = \frac{1}{\sin\theta + 0.5(6° + \theta)^{-1.64}},$$

where θ is measured in degrees. Table y_1 and y_2 starting at $\theta = 2°$ and incrementing by 1°. How do the values of y_1 and y_2 compare as θ increases? (Source: C. Winter.)

Exercises 93 and 94: Projectile Flight If a projectile is fired with an initial velocity of v feet per second at an angle θ with the horizontal, it will follow a parabolic path described by $y = \dfrac{-16x^2}{v^2\cos^2\theta} + x\tan\theta$. See the accompanying figure.

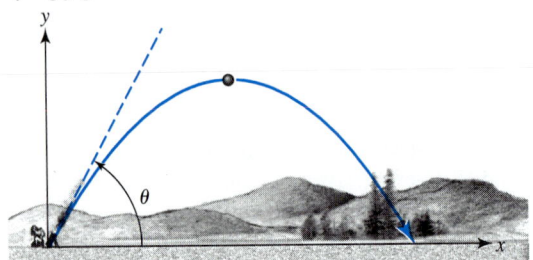

93. If $v = 750$ and $\theta = 30°$, graph the path of the projectile.
  (a) Determine the coordinates when the maximum height occurs.
  (b) Assuming the ground is flat, find the total distance traveled by the projectile graphically.

94. If $v = 500$ and $\theta = 45°$, graph the path of the projectile.
  (a) Determine the maximum height that occurs graphically.
  (b) Assume that the projectile is fired toward a hill that rises with a constant slope of $\frac{1}{4}$. Approximate graphically the total horizontal distance traveled by the projectile.

Writing about Mathematics

1. Discuss whether any of the six trigonometric functions are linear functions. Use graphical or numerical representations to justify your reasoning.

2. If α and β are coterminal angles, what can be said about the six trigonometric functions of α and β? Explain your answer using an example.

CHECKING BASIC CONCEPTS FOR SECTIONS 5.3 AND 5.4

1. Find the six trigonometric functions of θ, if θ is in standard position and its terminal side passes through the point $(-7, 6)$.

2. Evaluate $\sin\dfrac{\pi}{4}$ verbally, graphically, and numerically.

3. Sketch a graph of the sine, cosine, and tangent functions in $[-2\pi, 2\pi, \pi/2]$ by $[-4, 4, 1]$. Support your results using a graphing calculator.

4. Evaluate the six trigonometric functions by hand at the given real number. Support your results using a calculator.
  (a) $-\pi$ **(b)** $\dfrac{\pi}{2}$

5.5 Modeling with Trigonometric Functions

Transformations of Trigonometric Graphs • Models Involving Trigonometric Functions • Simple Harmonic Motion (Optional)

Introduction

In Section 1.6 we used transformations of graphs to model data. Trigonometric graphs can be used to model periodic data. For example, monthly average temperatures are usually periodic. They can vary dramatically between January and

December, but tend to be periodic from one year to the next. Tides are also periodic. By performing basic transformations of graphs we can model a variety of phenomena.

Transformations of Trigonometric Graphs

Before modeling real data, we will discuss how the constants a and b affect the graphs of $y = a \sin bx$ and $y = a \cos bx$. The constant a controls the *amplitude* of a sinusoidal wave. If $y = 3 \sin x$ then the amplitude is $|a| = 3$. The graph of $Y_1 = \sin(X)$ oscillates between -1 and 1, whereas the graph of $Y_2 = 3 \sin(X)$ oscillates between -3 and 3. See Figure 5.120.

The constant b controls the number of oscillations in each interval of length 2π. For example, if $b = 2$ then there are two complete oscillations in every interval of length 2π. As a result, the graph repeats every π units and the period of both $y = \sin 2x$ and $y = \cos 2x$ is π. The *period P* of a sinusoidal graph can be computed using the formula $P = \dfrac{2\pi}{b}$, where $b > 0$.

When $b > 1$ the graph of $y = \sin bx$ is horizontally *compressed* compared to the graph of $y = \sin x$. When $0 < b < 1$ the graph of $y = \sin bx$ is horizontally *stretched* compared to the graph of $y = \sin x$. This is illustrated in the following example.

[−2π, 2π, π/2] by [−4, 4, 1]

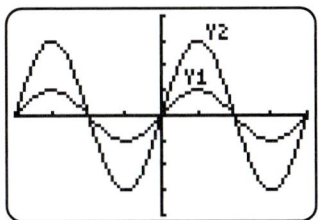

FIGURE 5.120

EXAMPLE 1   *Stretching and compressing the sine graph*

Graph $y = \sin bx$ with $b = 0.5, 1, 2,$ and 4 for $-2\pi \le x \le 2\pi$. Discuss the effect that b has on each graph.

Solution
The graphs of $Y_1 = \sin(0.5X)$, $Y_2 = \sin(X)$, $Y_3 = \sin(2X)$, and $Y_4 = \sin(4X)$ are shown in Figures 5.121–5.123. As the value of b increases, the number of oscillations increases proportionally. When $b = 0.5$, the period of Y_1 is $\dfrac{2\pi}{0.5} = 4\pi$. There is one complete oscillation or cycle every 4π units along the x-axis. Similarly, the period of Y_2 is $\dfrac{2\pi}{1} = 2\pi$, the period of Y_3 is $\dfrac{2\pi}{2} = \pi$, and the period of Y_4 is $\dfrac{2\pi}{4} = \dfrac{\pi}{2}$.

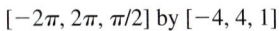

[−2π, 2π, π/2] by [−4, 4, 1]

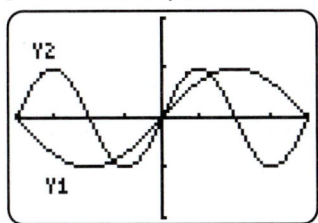

FIGURE 5.121

[−2π, 2π, π/2] by [−4, 4, 1]

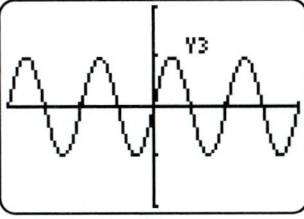

FIGURE 5.122

[−2π, 2π, π/2] by [−4, 4, 1]

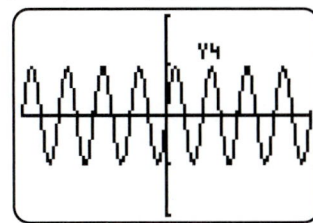

FIGURE 5.123 ∎

EXAMPLE 2   *Transforming sinusoidal graphs*

Conjecture the form of the graph of f. Then graph f in $[-2\pi, 2\pi, \pi/2]$ by $[-4, 4, 1]$.
(a) $f(x) = 3 \cos(2x) + 1$
(b) $f(x) = -2 \sin(x - \pi/2)$

Solution

(a) The graph of f can be obtained from the graph of the cosine function by performing the following steps. Shorten the period to π, increase the amplitude to 3, and shift the graph upward 1 unit. These three steps are shown in Figures 5.124–5.126, where $Y_1 = \cos(2X)$, $Y_2 = 3\cos(2X)$, and $Y_3 = 3\cos(2X) + 1$ are graphed, respectively.

$[-2\pi, 2\pi, \pi/2]$ by $[-4, 4, 1]$

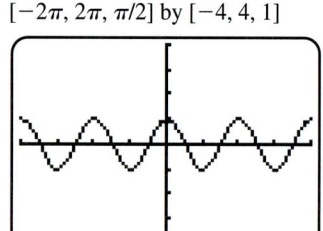

FIGURE 5.124 Reduce Period

$[-2\pi, 2\pi, \pi/2]$ by $[-4, 4, 1]$

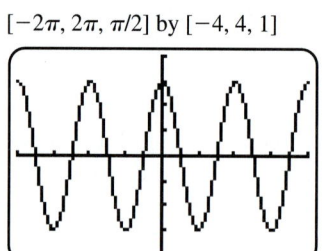

FIGURE 5.125 Increase Amplitude

$[-2\pi, 2\pi, \pi/2]$ by $[-4, 4, 1]$

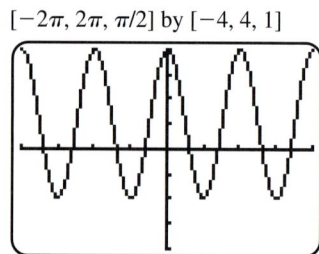

FIGURE 5.126 Shift Upward

(b) The graph of f can be obtained from the graph of the sine function by shifting the graph right $\pi/2$ units and increasing the amplitude to 2. The negative sign will cause the graph to be reflected across the x-axis. These steps are shown in Figures 5.127–5.129, where $Y_1 = \sin(X - \pi/2)$, $Y_2 = 2\sin(X - \pi/2)$, and $Y_3 = -2\sin(X - \pi/2)$ are graphed, respectively.

$[-2\pi, 2\pi, \pi/2]$ by $[-4, 4, 1]$

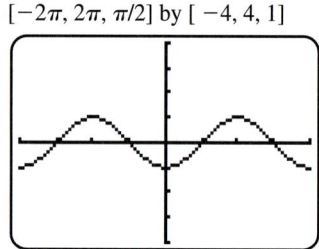

FIGURE 5.127 Shift Right

$[-2\pi, 2\pi, \pi/2]$ by $[-4, 4, 1]$

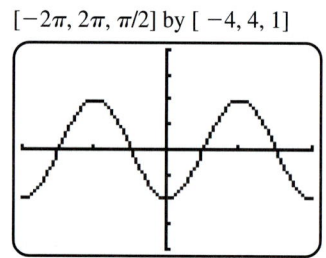

FIGURE 5.128 Increase Amplitude

$[-2\pi, 2\pi, \pi/2]$ by $[-4, 4, 1]$

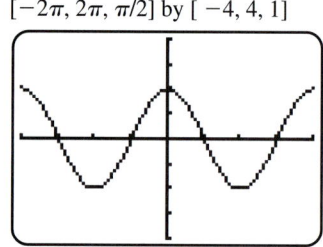

FIGURE 5.129 Reflect Graph ■

The graph of f was translated horizontally in Example 2(b). A horizontal shift of a trigonometric graph is called a *phase shift*. The phase shift for $f(x) = -2\cos(x - \pi/2)$ is $c = \pi/2$, since its graph is translated right $\pi/2$ units compared to the graph of the cosine function. Our discussion is summarized in the following.

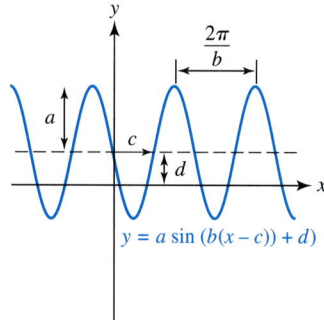

FIGURE 5.130
Transforming the Sine Graph

Amplitude, period, phase shift and vertical shift

The **amplitude, period,** and **phase shift** for the graphs of

$$y = a\sin(b(x - c)) + d \quad \text{and} \quad y = a\cos(b(x - c)) + d$$

with $b > 0$ may be determined as follows.

$$\text{Amplitude} = |a|, \qquad \text{Period} = \frac{2\pi}{b}, \qquad \text{Phase shift} = c$$

A **vertical shift** of $|d|$ units upward occurs when $d > 0$ and $|d|$ units downward when $d < 0$. The graph of $y = a\sin(b(x - c)) + d$ in Figure 5.130 illustrates the role of each constant, assuming all constants are positive.

Sketching Graphs by Hand. If desired, graphs of trigonometric functions can be sketched by hand. To graph $y = \sin x$ by hand it is helpful to first locate five *key points*. They are labeled in Figure 5.131. Notice that they are equally spaced along the interval $[0, 2\pi]$ on the *x*-axis.

$(0, 0)$	$\left(\dfrac{\pi}{2}, 1\right)$	$(\pi, 0)$	$\left(\dfrac{3\pi}{2}, -1\right)$	$(2\pi, 0)$
x-intercept	Maximum	x-intercept	Minimum	x-intercept

Five key points on the graph of $y = \cos x$ are shown in Figure 5.132.

$(0, 1)$	$\left(\dfrac{\pi}{2}, 0\right)$	$(\pi, -1)$	$\left(\dfrac{3\pi}{2}, 0\right)$	$(2\pi, 1)$
Maximum	x-intercept	Minimum	x-intercept	Maximum

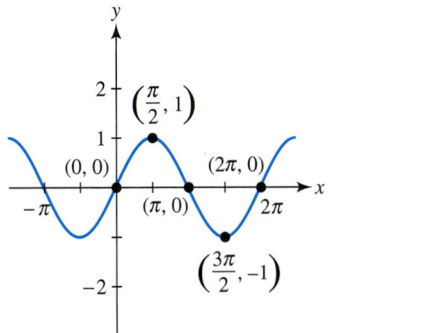

FIGURE 5.131 Key Points on the Sine Graph   **FIGURE 5.132** Key Points on the Cosine Graph

One way to sketch a sinusoidal graph by hand is to find the transformed key points. Using these points and the fact that the graph is periodic, a sketch can be obtained. This is illustrated for the sine function in the next example.

EXAMPLE 3 *Sketching sinusoidal graphs by hand*

Find the amplitude, period, and phase shift of the graph of $f(x) = 3 \sin 2x$. Sketch a graph of *f* by hand on the interval $[-2\pi, 2\pi]$.

Solution

If $f(x) = 3 \sin 2x$, then the amplitude is 3 and the period is π. There is no horizontal or vertical shifting. The graph of *f* passes through the point $(0, 0)$ and completes one oscillation in π units. Thus, the graph passes through the point $(\pi, 0)$. An amplitude of 3 will change the maximum and minimum *y*-values to 3 and -3, respectively. The transformed key points are equally spaced on the graph of *f*.

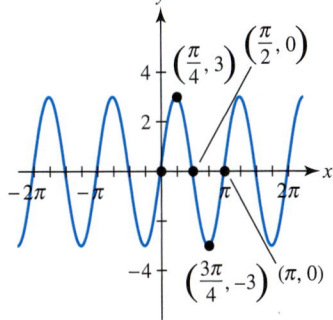

FIGURE 5.133

$(0, 0)$	$\left(\dfrac{\pi}{4}, 3\right)$	$\left(\dfrac{\pi}{2}, 0\right)$	$\left(\dfrac{3\pi}{4}, -3\right)$	$(\pi, 0)$
x-intercept	Maximum	x-intercept	Minimum	x-intercept

These points and the graph of *f* are plotted in Figure 5.133. ∎

Transformations of Other Graphs. The graphs of the other four trigonometric functions can also be transformed. The period of both the tangent function and the cotangent function is π. As a result, the graphs of $y = \tan(b(x - c))$ and $y = \cot(b(x - c))$ have a period of $P = \dfrac{\pi}{b}$ and a phase shift of *c*. Since the ranges of the tangent and cotangent functions include all real numbers, their graphs do not have an amplitude. The next example illustrates tangent and cotangent graphs. Graphs of the secant and cosecant functions are done in Exercises 43–48.

EXAMPLE 4   *Graphing other trigonometric functions*

Find the period and phase shift for the graph of *f*. Graph *f* on the interval $[-2\pi, 2\pi]$ using dot mode. Identify where asymptotes occur in the graph.

(a) $f(x) = \tan \frac{1}{2}x$

(b) $f(x) = \cot \left(x + \frac{\pi}{2} \right) + 1$

$[-2\pi, 2\pi, \pi/2]$ by $[-4, 4, 1]$

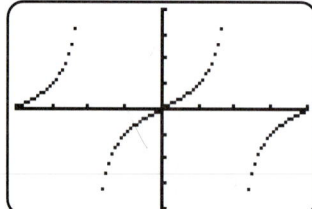

FIGURE 5.134

$[-2\pi, 2\pi, \pi/2]$ by $[-4, 4, 1]$

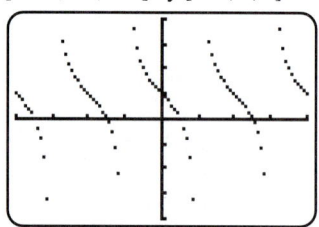

FIGURE 5.135

Solution

(a) If $f(x) = \tan \frac{1}{2}x$, then $b = \frac{1}{2}$ and $c = 0$. The period is $P = \frac{\pi}{b} = 2\pi$ and there is no phase shift. Graph $Y_1 = \tan(.5X)$ as shown in Figure 5.134. On the interval $[-2\pi, 2\pi]$, vertical asymptotes occur at $x = \pm \pi$. The graph of *f* is horizontally stretched by a factor of 2 compared to the graph of the tangent function.

(b) If $f(x) = \cot \left(x + \frac{\pi}{2} \right) + 1$ then $b = 1$ and $c = -\frac{\pi}{2}$. The period is π, the phase shift is $-\pi/2$, and the graph is shifted upward 1 unit. Vertical asymptotes occur at $x = \pm \frac{\pi}{2}$ and $x = \pm \frac{3\pi}{2}$. The graph of $Y_1 = 1/\tan(X + \pi/2) + 1$ is shown in Figure 5.135. ∎

Models Involving Trigonometric Functions

The monthly average temperatures for Prince George, Canada are shown in Table 5.9, where the months have been assigned the standard numbers. (Source: A. Miller, *Elements of Meteorology*.)

TABLE 5.9

Month	1	2	3	4	5	6	7	8	9	10	11	12
Temperature (°F)	15	19	28	41	50	55	59	57	50	41	28	19

$[0, 25, 2]$ by $[-5, 70, 10]$

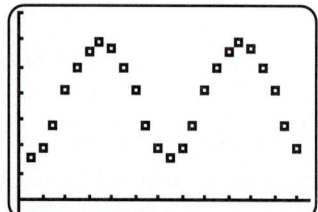

FIGURE 5.136

Since the data is periodic, it is plotted over a two-year period in Figure 5.136. For example, both $x = 1$ and $x = 13$ correspond to January. Notice that the scatterplot suggests that a sinusoidal graph might model this data.

Data of the type shown in Figure 5.136 sometimes may be modeled by

$$f(x) = a \sin(b(x - c)) + d,$$

where *a*, *b*, *c*, and *d* are constants. To estimate values for these constants, we will perform transformations on the graph of $y = \sin x$.

The addition of a constant *d* causes a vertical shift to the graph of *f*. The monthly average temperatures at Prince George vary from 15°F to 59°F. The average of these two temperatures is 37°F. If we let $d = 37$ and graph $y = \sin(x) + 37$,

[0, 25, 2] by [−5, 70, 10]

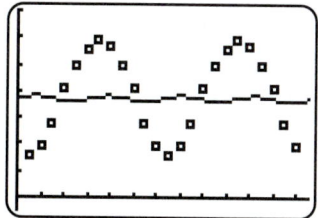

FIGURE 5.137

[0, 25, 2] by [−5, 70, 10]

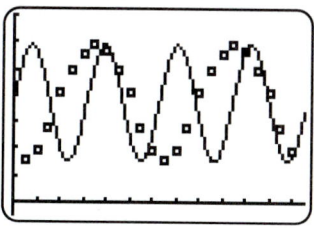

FIGURE 5.138

the resulting graph is shown in Figure 5.137. Compared to the graph of the sine function, this graph is shifted upward 37 units.

From Figure 5.137 it is apparent that the amplitude of the oscillations in the graph of $y = \sin(x) + 37$ are not large enough to model the data. The monthly average temperatures have a range of 59°F to 15°F = 44°F. If we let $a = \dfrac{44}{2} = 22$, the peaks and valleys on the graph of $y = 22\sin(x) + 37$ will model the data better. See Figure 5.138.

In Figure 5.138 the oscillations are too frequent, which indicate that the period is too small. Since the temperature cycles every 12 months, the period is $P = \dfrac{2\pi}{b} = 12$. Thus,

$$b = \frac{2\pi}{12} = \frac{\pi}{6} \approx 0.524.$$

The graph of $y = 22\sin\left(\dfrac{\pi}{6}x\right) + 37$ is shown in Figure 5.139.

We can obtain a reasonable fit by shifting the graph horizontally to the right. The maximum of $y = 22\sin\left(\dfrac{\pi}{6}x\right) + 37$ is at $x = 3$, which corresponds to March. We would like this maximum to occur in July ($x = 7$), so we translate the graph 4 units to the right by replacing x with $x - 4$ to obtain

$$y = 22\sin\left(\frac{\pi}{6}(x - 4)\right) + 37.$$

This equation is graphed in Figure 5.140. The phase shift is $c = 4$.

[0, 25, 2] by [−5, 70, 10]

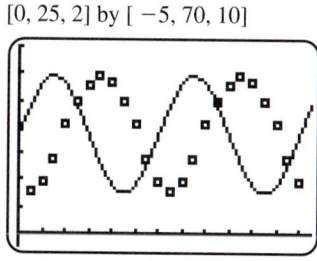

FIGURE 5.139

[0, 25, 2] by [−5, 70, 10]

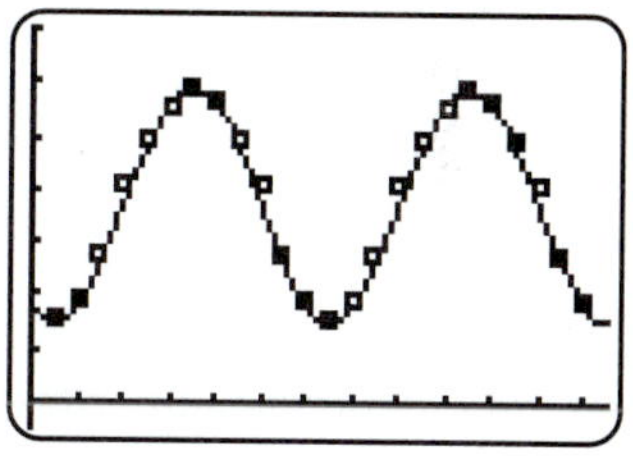

FIGURE 5.140

Critical Thinking

Is the value of $c = 4$ the only value that can be used to model the temperature data? Explain your reasoning.

Technology Note   *Sine Regression*

Linear and nonlinear regression were introduced in Sections 2.5 and 4.5, respectively. Regression can also be performed using a sinusoidal function of the form

$$f(x) = a\sin(bx + c) + d.$$

Sine regression may be performed on a two-year interval of the temperature data in Table 5.9. See Figures 5.141 and 5.142 on the next page. This function is similar to the one that we found except that the variable c in the regression equation does *not* represent the phase shift.

```
EDIT CALC TESTS
7↑QuartReg
8:LinReg(a+bx)
9:LnReg
0:ExpReg
A:PwrReg
B:Logistic
C:SinReg
```

FIGURE 5.141

```
SinReg
y=a*sin(bx+c)+d
a=21.7399239
b=.5207209998
c=-2.06480837
d=38.38287503
```

FIGURE 5.142

Tides represent the largest collective motion of water on Earth and cause water levels in oceans to vary during the course of a day. Tides usually occur once or twice a day and change with the phases of the moon. In deep oceans water levels may vary little, whereas near coasts it can vary as much as 24 feet. The largest tides occur when the moon is either full or new.

EXAMPLE 5 *Modeling tides*

Figure 5.143 shows a function f that models the tides in feet at Clearwater Beach, Florida x hours after midnight starting on August 26, 1998. (*Source: D. Pentcheff, WWW Tide and Current Predictor.*)

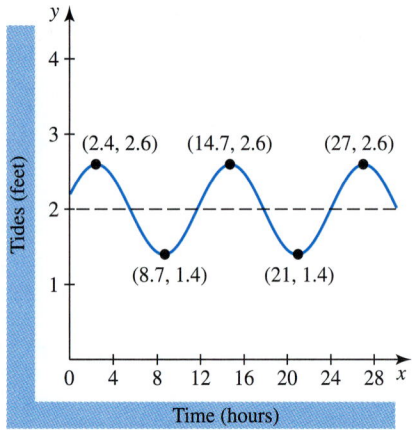

FIGURE 5.143 Tides at Clearwater Beach

(a) Find the time between high tides.
(b) What is the difference in water levels between high tide and low tide?
(c) Determine a, b, c, and d so that $f(x) = a \cos(b(x - c)) + d$ models the data. Graph f and the data in the same viewing rectangle.

Solution

(a) A high tide level corresponds to a peak on the graph. The time between peaks is 12.3 hours, since $14.7 - 2.4 = 12.3$ and $27 - 14.7 = 12.3$.
(b) High tide levels were 2.6 feet and low tide levels were 1.4 feet. Their difference is 1.2 feet.
(c) The amplitude of f is given by

$$a = \frac{2.6 - 1.4}{2} = 0.6.$$

Since the period is 12.3 hours, the value of b is

$$b = \frac{2\pi}{12.3} \approx 0.511.$$

[0, 28, 2] by [0, 4, 1]

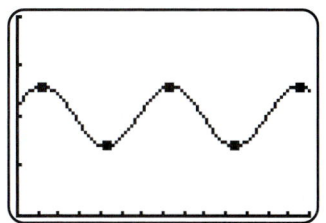

FIGURE 5.144

The average of high tide and low tide is $\frac{2.6 + 1.4}{2} = 2$, so we let $d = 2$.
Finally, a peak occurs at about 2.4 hours after midnight. Since midnight corresponds to $x = 0$ and the cosine function has a peak at $x = 0$, we translate the graph of f right 2.4 units by letting $c = 2.4$. Thus,

$$f(x) = 0.6 \cos{(0.511(x - 2.4))} + 2.$$

A graph of f and the data is shown in Figure 5.144. ∎

In the next example we use the cosine function to model daylight hours at San Antonio, Texas.

EXAMPLE 6   *Modeling daylight hours*

San Antonio, Texas has a latitude of 29.5°N. Table 5.10 lists the average number of daylight hours on the first day of each month at San Antonio. (Source: J. Williams, *The Weather Almanac 1995.*)

TABLE 5.10

Month	1	2	3	4	5	6	7	8	9	10	11	12
Daylight (hr)	10.2	10.7	11.5	12.5	13.3	13.9	14.1	13.6	12.7	11.9	11.0	10.4

(a) Plot the data over a two-year period.
(b) Model this data using $f(x) = a \cos{(b(x - c))} + d$.
(c) Estimate the daylight hours on February 15.

Solution
(a) A scatterplot of the data is shown in Figure 5.145 on the next page. The graph suggests that a sinusoidal graph might model this data.
(b) The maximum number of daylight hours is 14.1 and the minimum is 10.2. The average of these values is $\frac{14.1 + 10.2}{2} = 12.15$ and their difference is $14.1 - 10.2 = 3.9$. Let $d = 12.15$ and $a = \frac{3.9}{2} = 1.95$. Since daylight hours cycle every 12 months, let $b = \frac{2\pi}{12} = \frac{\pi}{6}$. The maximum of the cosine graph occurs at $x = 0$. The maximum in the table occurs on July 1 ($x = 7$) so let $c = 7$. The graph of $f(x) = 1.95 \cos{\left(\frac{\pi}{6}(x - 7)\right)} + 12.15$ together with the data is shown in Figure 5.146 on the following page. A slightly better fit is obtained if we let $c = 6.6$ as shown in Figure 5.147.
(c) February 15 corresponds to $x \approx 2.5$.

$$f(2.5) = 1.95 \cos{\left(\frac{\pi}{6}(2.5 - 6.6)\right)} + 12.15 \approx 11.1 \text{ hours.}$$

[0, 25, 4] by [8, 16, 1]   [0, 25, 4] by [8, 16, 1]   [0, 25, 4] by [8, 16, 1]

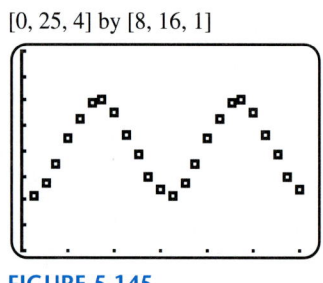 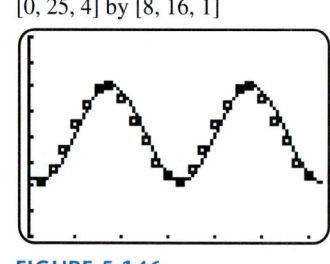 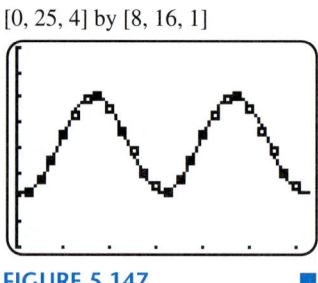

FIGURE 5.145   **FIGURE 5.146**   **FIGURE 5.147**   ■

Critical Thinking

A phase shift of $c = 7$ corresponds to the maximum number of daylight hours occurring on July 1. Conjecture why $c = 6.6$ fits the data slightly better.

A *sag curve* occurs when a highway goes downhill and then uphill. Improperly designed sag curves can be dangerous at night because a vehicle's headlights point downward and may not illuminate the uphill portion of the highway as illustrated in Figure 5.148. The minimum safe length L for a typical sag curve with a 40-mile-per-hour speed limit is computed by the equation

$$L = \frac{2700}{h + 3 \tan \alpha}.$$

The variable h represents the height of the headlights above the road surface and α represents a small angle associated with the vertical alignment of the headlight shown in Figure 5.148. (Source: F. Mannering, *Principles of Highway Engineering and Traffic Analysis.*)

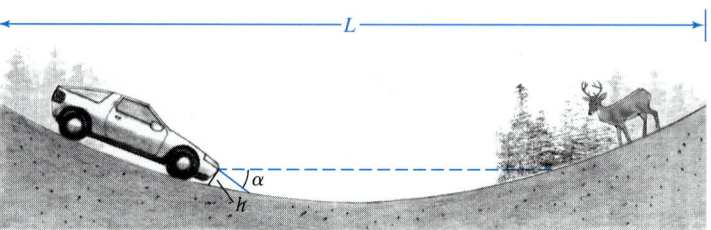

FIGURE 5.148 Designing a Safe Sag Curve

EXAMPLE 7   *Designing a sag curve*

Calculate L for a car with headlights 2.5 feet above the ground and $\alpha = 1°$. Repeat the calculations for a truck with headlights 4 feet above the ground and $\alpha = 2°$.

Solution

For the car, let $h = 2.5$ and $\alpha = 1°$.

$$L = \frac{2700}{2.5 + 3 \tan 1°}$$

$$\approx 1058 \text{ feet}$$

For the truck, let $h = 4$ and $\alpha = 2°$.

$$L = \frac{2700}{4 + 3 \tan 2°}$$

$$\approx 658 \text{ feet}$$

Since both cars and trucks use the same highways, engineers typically use the larger of the two distances to design a safe sag curve. ■

Simple Harmonic Motion (Optional)

Harmonic motion occurs frequently in nature and physical science. A particle or object undergoing small oscillations about a point of stable equilibrium executes simple harmonic motion. For example, if a string on a guitar is plucked gently, any point on the string will vibrate back and forth about its natural position of equilibrium. If a pendulum on a clock is pulled to one side, the pendulum will swing back and forth about its stable, vertical position. A small weight on a spring will bounce up and down when displaced from its natural length. All of these situations are examples of simple harmonic motion.

When an object undergoes simple harmonic motion, its distance or displacement from its stable or natural position can be modeled by either

$$s(t) = a \sin bt \quad \text{or} \quad s(t) = a \cos bt,$$

where $s(t)$ represents the displacement being experienced at time t. To better understand this, consider the spring and weight shown in Figure 5.149. If $s = 0$ corresponds to the spring's natural length, then $s = -2$ inches indicates that the spring is stretched 2 units beyond its natural length and $s = 2$ indicates that the spring is compressed 2 units. When a spring is stretched and let go, it will oscillate up and down. If the displacement s is plotted after t seconds, a sinusoidal graph results as shown in Figure 5.150.

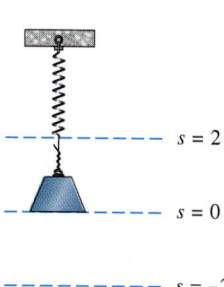

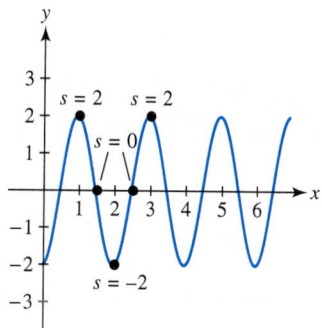

FIGURE 5.149 Displacement of a Spring   FIGURE 5.150 Modeling Oscillations of a Spring

The amplitude of each oscillation equals $|a|$ and the period P is given by $\frac{2\pi}{b}$. The number of oscillations per unit time is the **frequency.** The frequency F equals the reciprocal of the period P. That is, $F = \frac{1}{P}$. Since $P = \frac{2\pi}{b}$ it follows that $F = \frac{b}{2\pi}$, or equivalently, $b = 2\pi F$. Substituting for b in the formulas for $s(t)$ results in

$$s(t) = a \sin (2\pi F t) \quad \text{or} \quad s(t) = a \cos (2\pi F t).$$

EXAMPLE 8   *Modeling simple harmonic motion*

Suppose that the weight and spring in Figure 5.149 have a period of 0.4 second. Initially the weight is lifted 3 inches above its natural length and then let go.
(a) Find an equation that models the displacement s of the weight.

(b) Estimate s after 0.92 second. Determine if the weight is moving upward or downward at this time. Support your results graphically and numerically.

Solution

(a) Since the displacement is nonzero at $t = 0$, let $s(t) = a \cos(2\pi F t)$. The spring is initially displaced 3 inches, so the amplitude is 3. Let $a = 3$. The period is $P = 0.4$, so the frequency is

$$F = \frac{1}{0.4} = 2.5.$$

This indicates that the weight and spring oscillate up and down 2.5 times per second. It follows that $b = 2\pi F = 5\pi$ and $s(t) = 3 \cos(5\pi t)$. Note that the initial position is $s(0) = 3$ inches.

(b) After 0.92 second, the displacement is

$$s(0.92) = 3 \cos(5\pi(0.92)) \approx -0.927.$$

The weight is about 0.93 inch *below* its natural position after 0.92 second.

The weight initially moves downward and oscillates with a period of 0.4 second. During the time intervals $(0, 0.2)$, $(0.4, 0.6)$, and $(0.8, 1.0)$ the weight is moving downward, and during the intervals $(0.2, 0.4)$, $(0.6, 0.8)$, and $(1.0, 1.2)$ the weight is moving upward. Thus, the weight is moving downward when $t = 0.92$.

By tracing the graph of $s(t) = 3 \cos(5\pi t)$, we see that $s(0.92) \approx -0.927$. See Figure 5.151. Since the graph is decreasing in this region the weight is moving downward. Numerical support is also given in Figure 5.152. Notice that the Y_1 values are decreasing so the weight is moving downward at $t = 0.92$. ∎

$[0, 1.6, 0.2]$ by $[-4, 4, 1]$

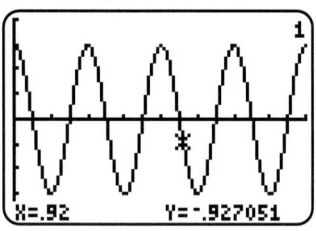

FIGURE 5.151

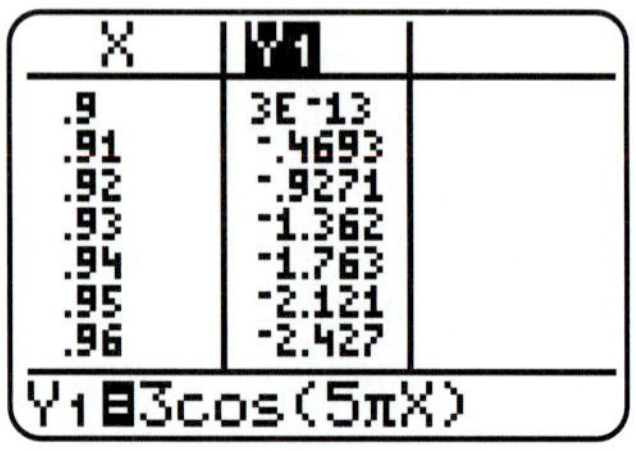

FIGURE 5.152

5.5 PUTTING IT ALL TOGETHER

Some concepts about trigonometric models are summarized in the following table.

Concept	Explanation	Examples
Sinusoidal Model $s(t) = a \sin(b(x - c)) + d$ or $s(t) = a \cos(b(x - c)) + d$ with $b > 0$	Amplitude $= \|a\|$ Period $= \dfrac{2\pi}{b}$ Phase shift $= c$ Vertical shift $= \|d\|$ upward if $d > 0$ downward if $d < 0$ If $b = 2\pi F$ then F represents the frequency.	Let $s(t) = 3 \sin(2(x - \pi)) - 5$. Amplitude $= 3$. Period $= \dfrac{2\pi}{2} = \pi$ Phase shift $= \pi$. Vertical shift downward 5 units. The frequency is $F = \dfrac{b}{2\pi} = \dfrac{1}{\pi}$, which is approximately 0.32 oscillation per unit time.
Simple Harmonic Motion $s(t) = a \sin bt$ or $s(t) = a \cos bt$	An object that oscillates about a stable equilibrium point undergoes simple harmonic motion.	A pendulum on a clock A weight on a spring

5.5 EXERCISES

  Tape 9

Graphs of Trigonometric Functions

Exercises 1–4: Match the physical situation with the graph (a–d) that models it best.

1. The height y in feet that a person is above the ground while riding a Ferris wheel after t seconds, where $t = 0$ corresponds to when the person began the ride

2. The number of hours y of darkness at 30°N latitude during month t, where $t = 1$ corresponds to January

3. The length y of a shadow cast by a horizontal stick of length 1 on a vertical wall between sunrise and noon, where angle t is shown in the figure

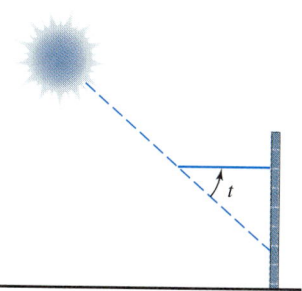

4. The length y of a shadow cast by a vertical stick of length 1 between sunrise and noon, where t is the angle of elevation of the sun

a.

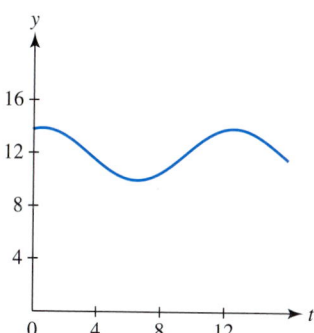

b.

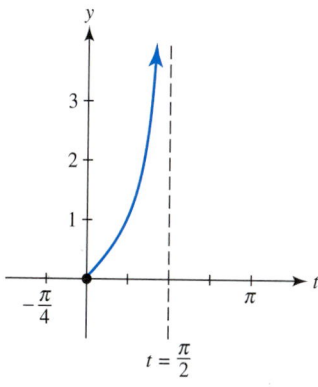

c.

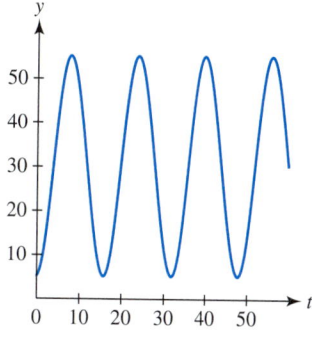

d.

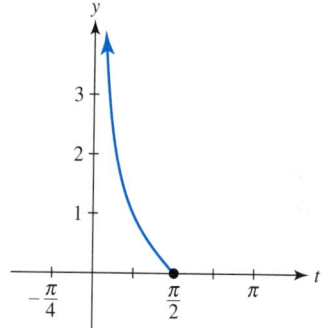

Exercises 5–8: The graph of a trigonometric function f represented by $f(t) = a \sin(b(x - c))$ is shown, where a, b, and c are nonnegative. State the amplitude, period, and phase shift.

5.

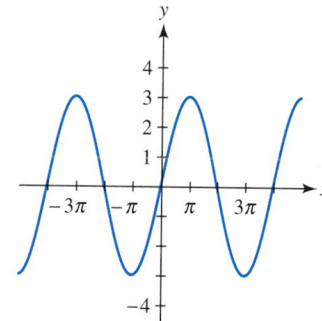

6.

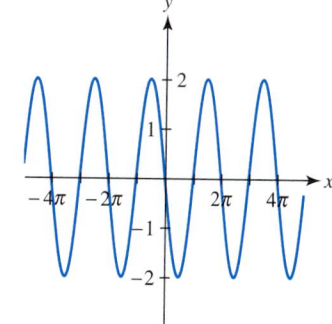

7.

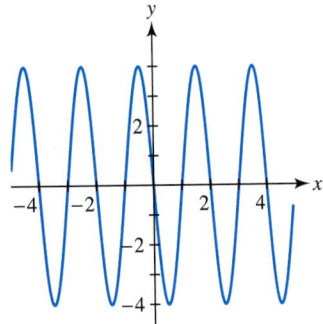

b.

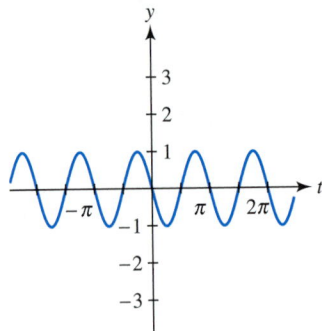

8.

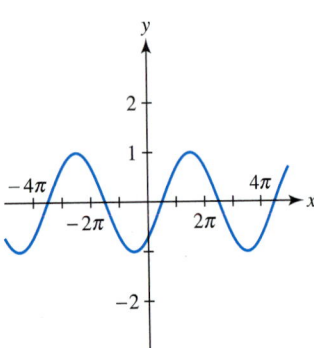

c.

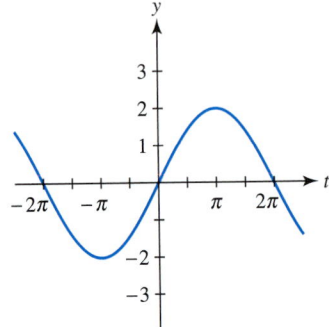

Exercises 9–14: Match the function f with its graph (a–f). Do not use a calculator.

9. $f(t) = 2 \sin\left(\frac{1}{2}t\right)$

10. $f(t) = -\sin(2t)$

11. $f(t) = 3 \cos(\pi t)$

12. $f(t) = 2 \sin\left(t - \frac{\pi}{4}\right)$

13. $f(t) = \cos\left(t + \frac{\pi}{2}\right)$

14. $f(t) = -3 \cos t$

d.

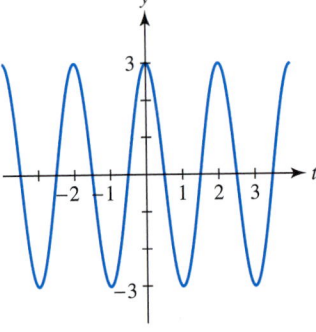

a.

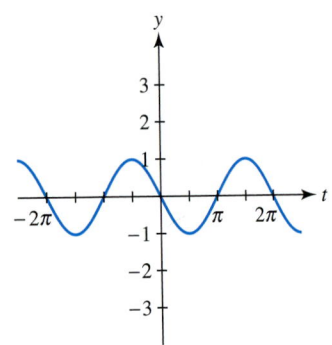

e.

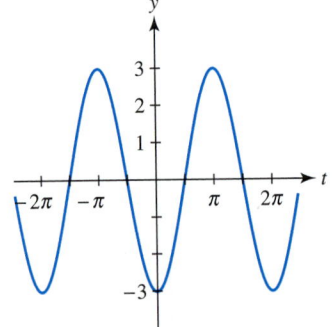

f.

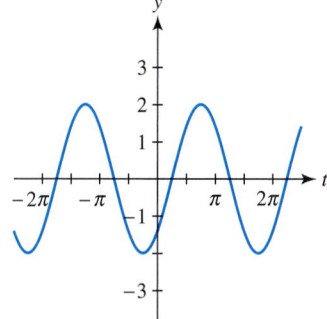

30.

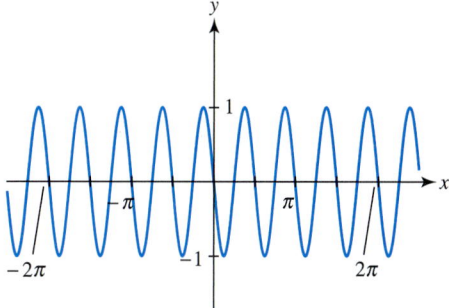

Exercises 15–20: Sketch the graph of f by hand on the interval $[-2\pi, 2\pi]$.

15. $f(t) = 2\sin t$

16. $f(t) = \cos 2t$

17. $f(t) = \cos(\pi t) + 2$

18. $f(t) = 2\sin\left(t - \dfrac{\pi}{2}\right)$

19. $f(t) = -\sin(2(t + \pi))$

20. $f(t) = -3\cos\dfrac{1}{2}t$

31.

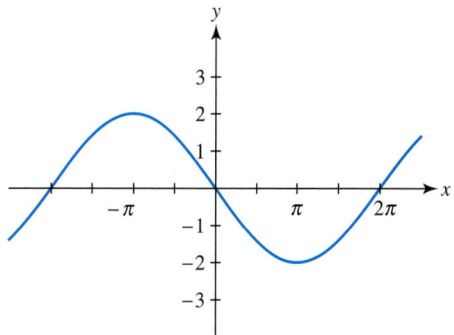

Exercises 21–28: Graph f in $[-2\pi, 2\pi, \pi/2]$ *by* $[-4, 4, 1]$. *State the amplitude, period, and phase shift.*

21. $f(t) = 2\sin(2t)$

22. $f(t) = -3\sin(t - \pi)$

23. $f(t) = \dfrac{1}{2}\cos\left(3\left(t + \dfrac{\pi}{3}\right)\right)$

24. $f(t) = 1.5\cos\left(\dfrac{1}{2}\left(t + \dfrac{\pi}{2}\right)\right)$

25. $f(t) = -\sin(4t)$

26. $f(t) = -2.5\cos\left(2t + \dfrac{\pi}{2}\right)$

27. $f(t) = -\cos(2\pi t) + 1$

28. $f(t) = -2.5\sin\left(\pi t + \dfrac{\pi}{2}\right) - 1$

32.

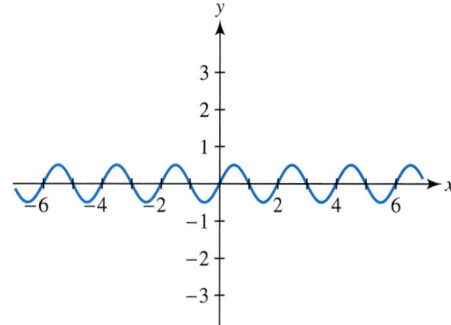

Exercises 29–32: A graph of the equation $y = a\sin bx$ *is shown, where b is a positive constant. Estimate the values of a and b.*

29.

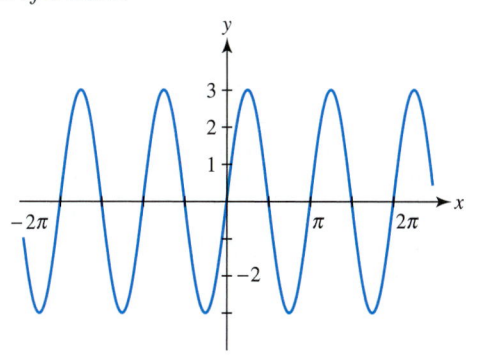

Exercises 33–36: Determine an equation in the form $y = a\sin(b(x - c))$ *for the graph shown in the exercise. Assume that a, b, and c are nonnegative.*

33. Exercise 5   **34.** Exercise 6

35. Exercise 7   **36.** Exercise 8

Exercises 37–42: (Refer to Example 4.) Find the period and phase shift for the graph of f. Graph f on the interval $[-2\pi, 2\pi]$ *using dot mode. Identify where asymptotes occur in the graph.*

37. $f(t) = \tan 2t$

38. $f(t) = \tan\dfrac{1}{2}t$

39. $f(t) = \tan\left(t - \dfrac{\pi}{2}\right)$

40. $f(t) = \cot\left(\dfrac{1}{3}\left(t - \dfrac{\pi}{2}\right)\right)$

41. $f(t) = -\cot 2t$

42. $f(t) = -\cot\left(t + \dfrac{\pi}{2}\right)$

Exercises 43–48: Use the directions for Exercises 37–42 to graph f.

43. $f(t) = \sec \dfrac{1}{2} t$

44. $f(t) = \sec\left(2\left(t - \dfrac{\pi}{2}\right)\right)$

45. $f(t) = \csc(t - \pi)$

46. $f(t) = -\csc 2t$

47. $f(t) = \sec\left(\dfrac{1}{3}\left(t - \dfrac{\pi}{6}\right)\right)$

48. $f(t) = \csc(\pi(t - 1))$

Applications

49. *Average Temperatures*   The graph models the monthly average temperature y in degrees Fahrenheit for a city in Canada, where x is the month.
  (a) Find the maximum and minimum monthly average temperatures.
  (b) Find the amplitude and period. Interpret the results.
  (c) Explain what the x-intercepts represent.

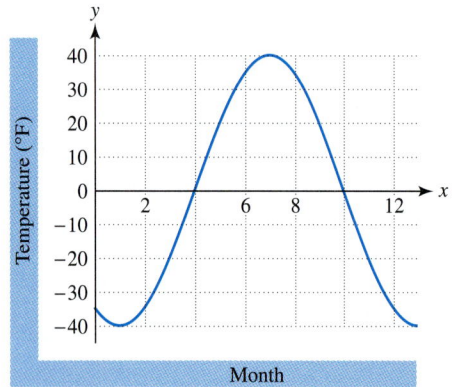

50. *Average Temperatures*   The graph in the previous exercise is given by $y = 40 \cos\left(\dfrac{\pi}{6}(x - 7)\right)$. Modify this equation to model the following situations.
  (a) The maximum monthly average temperature is 50°F and the minimum is −50°F.
  (b) The maximum monthly average temperature is 60°F and the minimum is −20°F.

(c) The maximum monthly average temperature occurs in August and the minimum occurs in February.

51. *Average Temperatures*   The monthly average temperatures in degrees Fahrenheit at Mould Bay, Canada may be modeled by the equation

$$f(x) = 34 \sin\left(\dfrac{\pi}{6}(x - 4.3)\right),$$

where x is the month and $x = 1$ corresponds to January. (Source: A. Miller.)
  (a) Graph f over the two-year interval $1 \le x \le 25$. Find the amplitude, period and phase shift.
  (b) Approximate the average temperature during May and December.
  (c) Conjecture an approximation for the *yearly* average temperature at Mould Bay.

52. *Average Temperatures*   The monthly average temperatures in degrees Fahrenheit at Austin, Texas are given by $f(x) = 17.5 \sin\left(\dfrac{\pi}{6}(x - 4)\right) + 67.5$,

where x is the month and $x = 1$ corresponds to January. (Source: A. Miller.)
  (a) Graph f over the two-year interval $1 \le x \le 25$. Determine the amplitude, period, phase shift, and vertical shift.
  (b) Determine the maximum and minimum monthly average temperature and the months when they occur.
  (c) Conjecture how the *yearly* average temperature might be related to $f(x)$.

53. *Modeling Temperatures*   The monthly average temperatures in Vancouver, Canada are shown in the table.

Month	1	2	3	4	5	6
Temperature (°F)	36	39	43	48	55	59

Month	7	8	9	10	11	12
Temperature (°F)	64	63	57	50	43	39

Source: A. Miller.

  (a) Plot the average monthly temperature over a 24-month period by letting $x = 1$ and $x = 13$ correspond to January.
  (b) Find the constants a, b, c, and d so that $f(x) = a \sin(b(x - c)) + d$ models the data.
  (c) Graph f together with the data.

54. *Modeling Temperatures*   The monthly average temperatures in Chicago, Illinois are shown in the table.

Month	1	2	3	4	5	6
Temperature (°F)	25	28	36	48	61	72

Month	7	8	9	10	11	12
Temperature (°F)	74	75	66	55	39	28

Source: A. Miller.

(a) Plot the monthly average temperature over a 24-month period by letting $x = 1$ and $x = 13$ correspond to January.

(b) Determine a function of the form $f(x) = a \sin(b(x - c)) + d$ that models the data.

(c) Graph f and the data together.

55. *Modeling Temperatures* The monthly average high temperatures in Augusta, Georgia are shown in the table.

Month	1	2	3	4	5	6
Temperature (°F)	58	60	68	77	82	90

Month	7	8	9	10	11	12
Temperature (°F)	92	91	83	77	68	60

Source: J. Williams.

(a) Model this data using a function of the form

$$f(x) = a \cos(b(x - c)) + d.$$

(b) Are different values for c possible? Explain.

56. *Modeling Temperatures* The maximum monthly average temperature in Anchorage, Alaska is 57°F and the minimum is 12°F.

(a) Using only these two temperatures, determine $f(x) = a \cos(b(x - c)) + d$ so that $f(x)$ models the monthly average temperatures in Anchorage.

(b) Graph f and the actual data in the table over a two-year period.

Month	1	2	3	4	5	6
Temperature (°F)	12	18	23	36	46	55

Month	7	8	9	10	11	12
Temperature (°F)	57	55	48	36	23	16

Source: A. Miller.

57. *Highway Design* (Refer to Example 7.) Calculate the minimum length L of a typical sag curve on a highway with a 40-mile-per-hour speed limit for a car with headlights 2 feet above the ground and alignment set at $\alpha = 1.5°$.

58. *Highway Design* Repeat the previous exercise for a truck with headlights 3.5 feet above the ground and alignment set at $\alpha = 2.5°$.

59. *Daylight Hours* The graph models the daylight hours at 60°N latitude, where $x = 1$ corresponds to January 1, $x = 2$ to February 1, and so on.

(a) Estimate the maximum number of daylight hours. When does this occur?

(b) Estimate the minimum number of daylight hours. When does this occur?

(c) Interpret the amplitude and period.

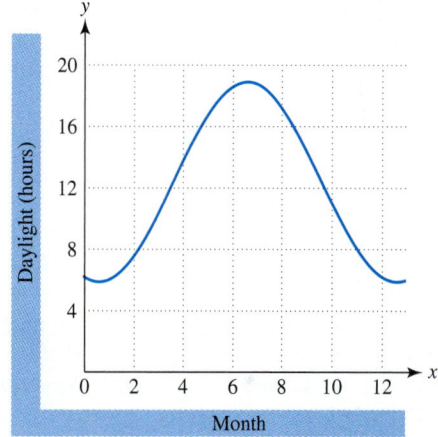

60. *Daylight Hours* The graph in the previous exercise is given by $y = 6.5 \sin\left(\dfrac{\pi}{6}(x - 3.65)\right) + 12.4.$ Modify this equation to model the following situations.

(a) At 50°N latitude the maximum daylight is about 16.3 hours and the minimum is about 8.3 hours.

(b) The daylight hours at 60°S latitude

(c) The daylight hours at the equator

61. *Average Precipitation* The graph on the next page models the monthly average precipitation in inches at Mount Adams, Washington over a three-year period, where x is the month.

(a) Find the maximum and minimum monthly average precipitation.

(b) Find the amplitude and interpret the result.

(c) Could a graph that models precipitation have an x-intercept? If it could, what would this indicate?

(d) Estimate the yearly average precipitation.

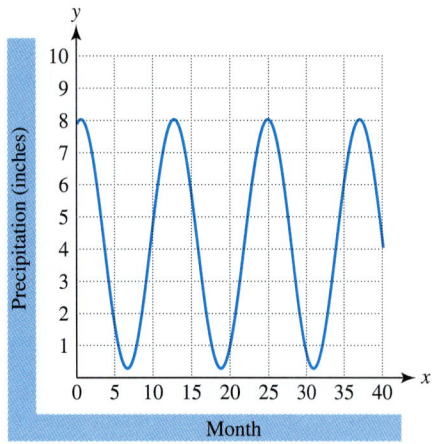

62. *Average Precipitation* Suppose that the monthly average precipitation at a particular location varies sinusoidally between a maximum of 6 inches in January to a minimum of 2 inches in July. Let $t = 1$ correspond to January and $t = 12$ correspond to December.

(a) Find values for a, b, c, and d so that

$$f(t) = a \cos (b(t - c)) + d$$

models these conditions.

(b) Support your result in part (a) by evaluating $f(1)$ and $f(7)$.

63. *Tides and Periodic Functions* The accompanying figure shows the tides at Santa Monica Bay, California on August 28, 1997. This type of tide is *semi-diurnal*, since high tides occur twice a day. (Source: Zihua Software.)

(a) Estimate the time between the two high tides shown in the graph.

(b) Explain why $f(x) = a \sin (b(x - c)) + d$ cannot model these tide levels accurately.

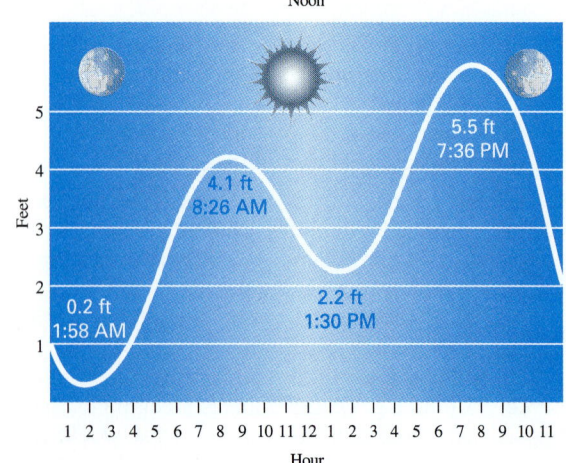

64. *Modeling Tidal Currents* Tides cause ocean currents to flow into and out of harbors and canals. The table shows the speed of the ocean current at Cape Cod Canal in bogo-knots (bk) x hours after midnight on August 26, 1998. (Source: D. Pentcheff, *WWW Tide and Current Predictor.*)

Time (hr)	3.7	6.75	9.8	13.0	16.1	22.2
Current (bk)	−18	0	18	0	−18	18

(a) Find constants a, b, c, and d so that

$$f(x) = a \cos(b(x - c)) + d$$

models the data in the table.

(b) Graph f and the data in $[0, 24, 4]$ by $[-20, 20, 5]$. Interpret the graph.

65. *Ocean Temperatures* The graph models the Gulf of Mexico water temperatures in degrees Fahrenheit at St. Petersburg, Florida. (Source: J. Williams.)

(a) Estimate the maximum and minimum water temperatures. When do they occur?

(b) What would happen to the amplitude of the graph if the minimum water temperature decreased to 50°F? Make a sketch of this situation.

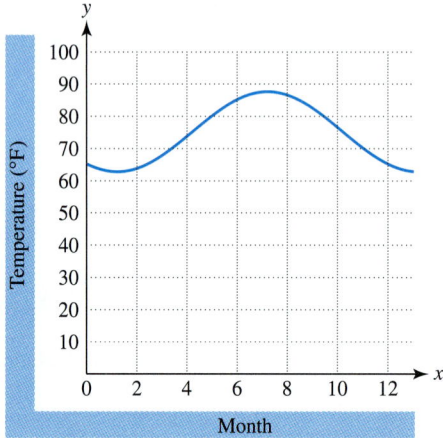

66. *Ocean Temperatures* The graph in the previous exercise is described by the equation

$$y = 12.4 \sin \left(\frac{\pi}{6}(x - 4.2) \right) + 75.$$

Modify this equation to model the following situations.

(a) The monthly average water temperatures vary between 60°F and 90°F.

(b) The monthly average water temperatures vary between 50°F and 70°F.

67. *Modeling Ocean Temperature* The following table lists the monthly average ocean temperatures in degrees Fahrenheit at Veracruz, Mexico.

Month	1	2	3	4	5	6
Temperature (°F)	72	73	74	78	81	83

Month	7	8	9	10	11	12
Temperature (°F)	84	85	84	82	78	74

Source: J. Williams.

   (a) Make a scatterplot of the data over a two-year period.
   (b) Find the constants *a, b, c,* and *d* so that

$$f(x) = a \sin(b(x - c)) + d$$

     models the data.

68. *Interpreting a Model* Graph $y =$

$$20 + 15 \sin \frac{\pi t}{12}$$

in [0, 12, 3] by [0, 50, 10]. Let *y* represent the outdoor temperature in degrees Celsius at time *t* in hours, where $t = 0$ corresponds to 9 a.m. Interpret the graph.

69. *Carbon Dioxide Levels in Hawaii* At Mauna Loa, Hawaii atmospheric carbon dioxide levels in parts per million (ppm) have been measured regularly since 1958. The equation

$$L(x) = 0.022x^2 + 0.55x + 316 + 3.5 \sin(2\pi x)$$

may be used to model these levels, where *x* is the year and $x = 0$ corresponds to 1960.
   (a) Graph *L* in [20, 35, 5] by [320, 370, 10] and interpret the graph.
   (b) The function *L* is represented by the sum of a quadratic function and a sine function. How does each function affect the shape of the graph? Discuss reasons for each function. (Source: A. Nilsson, *Greenhouse Earth*.)

70. *Carbon Dioxide Levels in Alaska* (Refer to the previous exercise.) The carbon dioxide content in the atmosphere at Barrow, Alaska in parts per million (ppm) can be modeled using the equation

$$C(x) = 0.04x^2 + 0.6x + 330 + 7.5 \sin(2\pi x),$$

where $x = 0$ corresponds to 1970. (Source: M. Zeilik, *Introductory Astronomy and Astrophysics*.)
   (a) Graph *C* in [10, 25, 5] by [320, 380, 10]. Compare it with the graph for *L* in the previous exercise.

   (b) Discuss possible reasons for differences between the two graphs.

Simple Harmonic Motion

Exercises 71–74: Springs (Refer to Example 8.) Suppose that a weight on a spring has an initial position of $s(0)$ and a period of P.
   (a) *Find a function s given by $s(t) = a \cos(2\pi Ft)$ that models the displacement of the weight.*
   (b) *Evaluate $s(1)$. Is the weight moving upward, downward, or neither when $t = 1$? Support your results graphically or numerically.*

71. $s(0) = 2$ inches, $P = 0.5$ second
72. $s(0) = 5$ inches, $P = 1.5$ seconds
73. $s(0) = -3$ inches, $P = 0.8$ second
74. $s(0) = -4$ inches, $P = 1.2$ seconds

Exercises 75–78: Music A note on the piano has the given frequency F. Suppose the maximum displacement at the center of the piano wire is given by $s(0)$. Find constants a and b so that $s(t) = a \cos bt$ models this displacement. Graph f in [0, 0.05, 0.01] by [−0.3, 0.3, 0.1].

75. $F = 27.5, s(0) = 0.21$
76. $F = 110, \ s(0) = 0.11$
77. $F = 55, \quad s(0) = 0.14$
78. $F = 220, \ s(0) = 0.06$

Exercises 79–82: Sine Regression Use regression to find constants a, b, c, and d so that

$$f(x) = a \sin(bx + c) + d,$$

models the real data given in the exercise. Graph the data and f together.

79. Exercise 53
80. Exercise 54
81. Exercise 55
82. Exercise 56

Writing about Mathematics

1. Discuss how the constants *a, b, c,* and *d* affect the graph of $y = a \sin(b(x - c)) + d$. Give an example.

2. Discuss some types of real data that could be modeled by $y = a \cos(b(x - c)) + d$. Give an example.

5.6 Inverse Trigonometric Functions

Review of Inverses • The Inverse Sine Function • The Inverse Cosine Function • The Inverse Tangent Function • Solving Triangles and Equations

Introduction

In construction it is sometimes necessary to determine angles. For example, the pitch or slope of a roof frequently is expressed as the ratio $\frac{k}{12}$, where k represents a k-foot rise for every 12 feet of run in horizontal distance. See Figure 5.153. A typical roof pitch for homes is $\frac{6}{12}$. To correctly cut the rafters, a carpenter needs to know the measure of angle θ. This problem can be solved easily using inverse trigonometric functions. Before introducing inverse trigonometric functions we briefly review inverse functions.

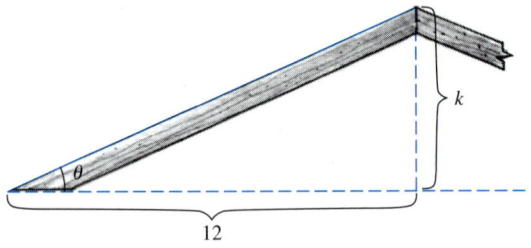

FIGURE 5.153

Review of Inverses

The inverse function f^{-1} will "undo" the computation performed by f. If $f(a) = b$, then $f^{-1}(b) = a$, and as a result, f and f^{-1} interchange domains and ranges. The inverse function of $f(x) = 3x$ is $f^{-1}(x) = \frac{x}{3}$, since dividing by 3 is the inverse operation of multiplying by 3. A function f must be one-to-one for f^{-1} to exist. A function is one-to-one if different inputs *always* produce different outputs. The horizontal line test can be used to determine if a function is one-to-one. Inverse functions can be represented verbally, numerically, graphically, and symbolically as illustrated in the next example.

EXAMPLE 1 *Finding representations of an inverse function*

The function given by $f(x) = -5.5x + 80$ computes the Fahrenheit temperature at an altitude of x-thousand feet when the ground level temperature is 80°F. Find verbal, symbolic, numerical, and graphical representations of f and f^{-1}. Interpret f^{-1}. (Source: A. Miller and R. Anthes, *Meteorology*.)

Solution

Verbal Representation The function f multiplies the input x by -5.5 and then adds 80 to the result. To find the inverse function apply the inverse operations in reverse order. That is, f^{-1} subtracts 80 from the input x and divides the result by -5.5.

Symbolic Representation   Symbolic representations of f^{-1} include either

$$f^{-1}(x) = \frac{x - 80}{-5.5} \quad \text{or} \quad f^{-1}(x) = \frac{80 - x}{5.5}.$$

Numerical Representation   A partial numerical representation of f is shown in Table 5.11.

TABLE 5.11

x	0	2	4	6	8
$f(x)$	80	69	58	47	36

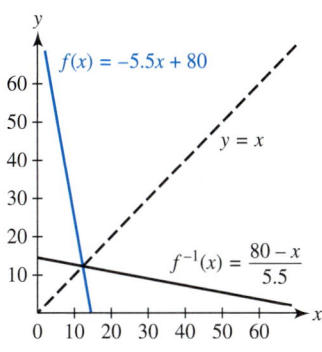

$f(x) = -5.5x + 80$

$y = x$

$f^{-1}(x) = \dfrac{80 - x}{5.5}$

FIGURE 5.154

Since $f(a) = b$ implies $f^{-1}(b) = a$, it follows that a partial numerical representation of $f^{-1}(x)$ is shown in Table 5.12.

TABLE 5.12

x	36	47	58	69	80
$f^{-1}(x)$	8	6	4	2	0

Graphical Representation   The graph of f^{-1} can be found by reflecting the graph of f across the line $y = x$. See Figure 5.154.

Since $f(x)$ computes the Fahrenheit temperature at x-thousand feet, $f^{-1}(x)$ computes the altitude in thousands of feet at which the temperature is x degrees Fahrenheit. ∎

The Inverse Sine Function

A partial numerical representation of the sine function is shown in Table 5.13.

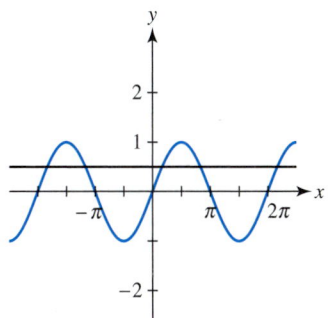

FIGURE 5.155 The Sine Function

TABLE 5.13

x	0	$\dfrac{\pi}{6}$	$\dfrac{\pi}{4}$	$\dfrac{\pi}{3}$	$\dfrac{\pi}{2}$	$\dfrac{2\pi}{3}$	$\dfrac{3\pi}{4}$	$\dfrac{5\pi}{6}$	π
$f(x)$	0	$\dfrac{1}{2}$	$\dfrac{1}{\sqrt{2}}$	$\dfrac{\sqrt{3}}{2}$	1	$\dfrac{\sqrt{3}}{2}$	$\dfrac{1}{\sqrt{2}}$	$\dfrac{1}{2}$	0

Notice that different inputs do not always result in different outputs. Therefore, the sine function is not one-to-one and so an inverse function does not exist. For example, $f(0) = 0$ and $f(\pi) = 0$, so $f^{-1}(0)$ cannot be defined as a single output, which is necessary for an inverse *function* to exist. By the horizontal line test we also can see that the sine function is not one-to-one. See Figure 5.155. It is possible for a horizontal line to intersect the sine graph an infinite number of times.

If we restrict the domain of $f(x) = \sin x$ to $-\pi/2 \leq x \leq \pi/2$, as shown in Figure 5.156, the graph of $y = \sin x$ is one-to-one since a horizontal line intersects this graph at most once. On this restricted domain the sine function has a unique inverse called the *inverse sine function*.

$\left(\dfrac{\pi}{2}, 1\right)$

$\left(-\dfrac{\pi}{2}, -1\right)$

FIGURE 5.156 Restricting the Domain

> ## Inverse sine function
>
> The **inverse sine function,** denoted by $\sin^{-1} x$ or $\arcsin x$, is defined by the following: $y = \sin^{-1} x$ or $y = \arcsin x$ means $x = \sin y$ for $-1 \le x \le 1$ and y in the interval $[-\pi/2, \pi/2]$.

Note: When evaluating the inverse sign function, it may be helpful to think of $\sin^{-1} x$ as an *angle* θ, where $\sin \theta = x$ and θ satisfies $-\pi/2 \le \theta \le \pi/2$.

The next example illustrates how to evaluate the inverse sine function.

EXAMPLE 2 *Evaluating the inverse sine function*

Evaluate each of the following by hand and then support your results with a calculator.

(a) $\sin^{-1} 1$

(b) $\arcsin\left(-\dfrac{1}{2}\right)$

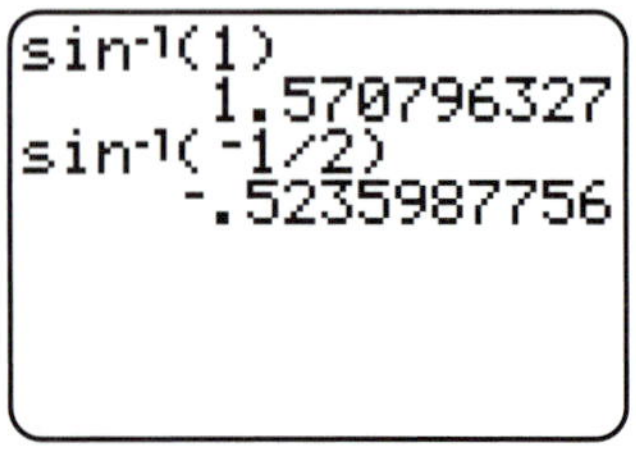

```
sin-1(1)
         1.570796327
sin-1(-1/2)
         -.5235987756
```

FIGURE 5.157 Radian Mode

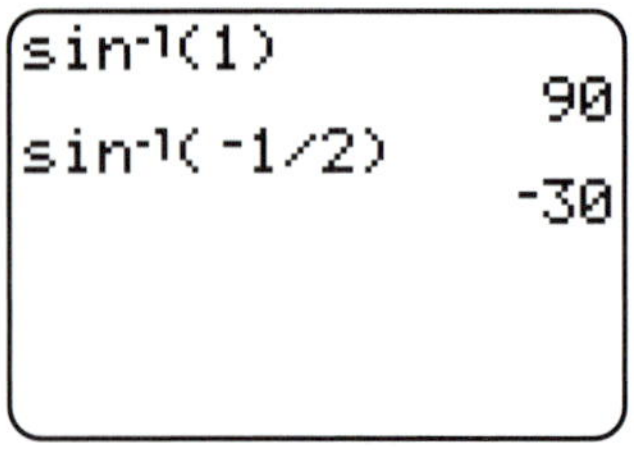

```
sin-1(1)
              90
sin-1(-1/2)
             -30
```

FIGURE 5.158 Degree Mode

Solution

(a) The expression $\sin^{-1} 1$ represents the angle (or real number) θ whose sine equals 1 and satisfies $-\pi/2 \le \theta \le \pi/2$. Thus, $\theta = \sin^{-1}(1) = \pi/2 \approx 1.57$.

(b) The expression $\arcsin\left(-\dfrac{1}{2}\right)$ represents the angle (or real number) θ whose sine equals $-\dfrac{1}{2}$ and satisfies $-\pi/2 \le \theta \le \pi/2$. Thus, $\theta = \arcsin\left(-\dfrac{1}{2}\right) = -\dfrac{\pi}{6} \approx -0.52$. Figures 5.157 and 5.158 support these results both in radian mode and degree mode. ∎

Functions and their inverses interchange domains and ranges. The range of the sine function is $-1 \le y \le 1$. Therefore, the domain of the inverse sine function is $-1 \le x \le 1$. Since the domain of the sine function has been restricted to $-\pi/2 \le x \le \pi/2$, it follows that the range of the inverse sine function is $-\pi/2 \le y \le \pi/2$. That is, $\sin^{-1} x$ outputs angles only in the interval $[-\pi/2, \pi/2]$. The following are properties of the inverse sine function.

$$\sin^{-1}(\sin x) = x \quad \text{for} \quad -\frac{\pi}{2} \le x \le \frac{\pi}{2}$$

$$\sin(\sin^{-1} x) = x \quad \text{for} \quad -1 \le x \le 1$$

MAKING CONNECTIONS
Notation and Inverse Functions

When inverse functions were introduced in Section 4.2, it was discussed that $f^{-1}(x) \ne \dfrac{1}{f(x)}$. The same is true for the inverse sine function: $\sin^{-1} x \ne \dfrac{1}{\sin x}$.

For example, $\sin^{-1} 1 = \dfrac{\pi}{2} \approx 1.57$ and $\dfrac{1}{\sin 1} \approx 1.19$.

EXAMPLE 3 *Finding representations of the inverse sine function*

Represent the inverse sine function verbally, numerically, graphically, and symbolically.

Solution

Verbal Representation   To compute $\sin^{-1}x$ for $-1 \le x \le 1$, determine the angle (or real number) θ such that $\sin\theta = x$ and $-\pi/2 \le \theta \le \pi/2$.

Numerical Representation   Table 5.14 shows a partial numerical representation of $\sin x$ on the interval $[-\pi/2, \pi/2]$. It follows that a partial numerical representation of $\sin^{-1}x$ is shown in Table 5.15. Notice that if $\sin a = b$ then $\sin^{-1}b = a$, provided $-\pi/2 \le a \le \pi/2$.

TABLE 5.14

x	$-\dfrac{\pi}{2}$	$-\dfrac{\pi}{3}$	$-\dfrac{\pi}{4}$	$-\dfrac{\pi}{6}$	0	$\dfrac{\pi}{6}$	$\dfrac{\pi}{4}$	$\dfrac{\pi}{3}$	$\dfrac{\pi}{2}$
$\sin x$	-1	$-\dfrac{\sqrt{3}}{2}$	$-\dfrac{1}{\sqrt{2}}$	$-\dfrac{1}{2}$	0	$\dfrac{1}{2}$	$\dfrac{1}{\sqrt{2}}$	$\dfrac{\sqrt{3}}{2}$	1

TABLE 5.15

x	-1	$-\dfrac{\sqrt{3}}{2}$	$-\dfrac{1}{\sqrt{2}}$	$-\dfrac{1}{2}$	0	$\dfrac{1}{2}$	$\dfrac{1}{\sqrt{2}}$	$\dfrac{\sqrt{3}}{2}$	1
$\sin^{-1}x$	$-\dfrac{\pi}{2}$	$-\dfrac{\pi}{3}$	$-\dfrac{\pi}{4}$	$-\dfrac{\pi}{6}$	0	$\dfrac{\pi}{6}$	$\dfrac{\pi}{4}$	$\dfrac{\pi}{3}$	$\dfrac{\pi}{2}$

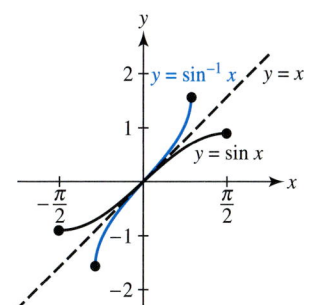

FIGURE 5.159   The Inverse Sine Function

Graphical Representation   The graph of $y = \sin^{-1}x$ can be found by reflecting the graph of $y = \sin x$ for $-\pi/2 \le x \le \pi/2$ across the line $y = x$ as shown in Figure 5.159.

Symbolic Representation   A symbolic representation of the inverse sine function can be written as either $f(x) = \sin^{-1}x$ or $f(x) = \arcsin x$. There is no simple formula to evaluate $\sin^{-1}x$. ■

EXAMPLE 4   *Evaluating the inverse sine function numerically and graphically*

Numerically approximate each of the following with a calculator. Express your result in both radians and degrees. Support your answer graphically.
(a) $\sin^{-1} 0.7$
(b) $\arcsin(2.1)$

Solution

(a) In Figure 5.160, $\sin^{-1} 0.7$ is evaluated in radian mode. We see that $\sin^{-1} 0.7 \approx 0.775$ radians, or equivalently, $44.4°$.

To determine $\sin^{-1} 0.7$ graphically let $Y_1 = \sin X$ and $Y_2 = 0.7$ on the interval $[-\pi/2, \pi/2]$. Their point of intersection provides the angle θ that satisfies $\sin\theta = 0.7$ and $-\pi/2 \le \theta \le \pi/2$. From Figure 5.161 we see that $\theta \approx 0.775$.

$[-\pi/2, \pi/2, \pi/4]$ by $[-2, 2, 1]$

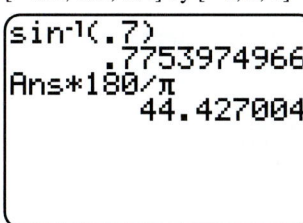

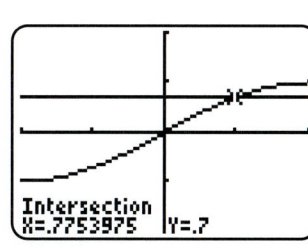

FIGURE 5.160 Radian Mode   **FIGURE 5.161**

$[-\pi/2, \pi/2, \pi/4]$ by $[-3, 3, 1]$

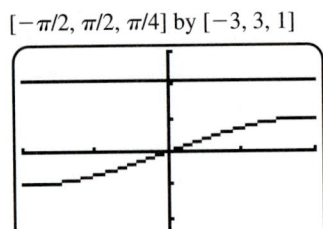

FIGURE 5.162

(b) To evaluate arcsin 2.1 we must find an angle θ such that $\sin \theta = 2.1$. Since $-1 \le \sin \theta \le 1$, $\theta = $ arcsin 2.1 is undefined. This is supported in Figure 5.162, where the graph of $y = \sin x$ never intersects the line $y = 2.1$. ∎

In track and field when an athlete throws the shot, the distance that the shot travels depends on the angle θ that the initial direction of the shot makes with the horizontal. Angle θ in Figure 5.163 is called the *projection angle*. The optimum projection angle θ that results in maximum distance for the shot may be calculated by

$$\theta = \sin^{-1}\sqrt{\frac{v^2}{2v^2 + 64.4h}},$$

where v is the initial speed in feet per second of the shot and h is the height of the shot when it is released. (Source: J. Cooper and R. Glassow, *Kinesiology.*)

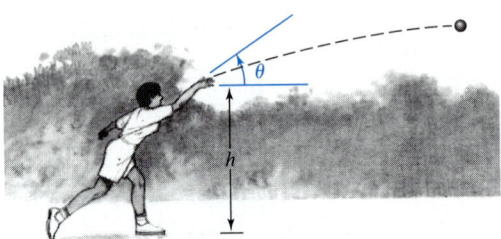

FIGURE 5.163 Projection Angle θ

EXAMPLE 5   *Finding the optimum projection angle for a shot-putter*

Suppose that an athlete releases a shot 8 feet above the ground with velocity v. Give graphical and numerical representations of the optimum projection angle θ. Interpret the results.

Solution

Graph and table $Y_1 = \sin^{-1}(\sqrt{(X^2/(2X^2 + 64.4*8))})$ as shown in Figures 5.164 and 5.165. We can see that the faster a person throws the shot, the greater the optimal projection angle θ becomes. For example, if a shot is thrown at 25 feet per second then $\theta \approx 36.5°$, whereas if the shot is thrown at 50 feet per second then $\theta \approx 42.3°$.

$[0, 60, 10]$ by $[0, 50, 10]$

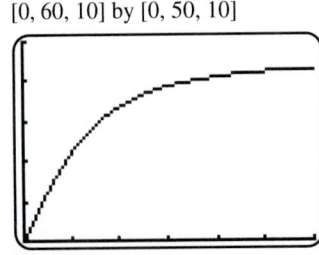

X	Y1
20	33.469
25	36.515
30	38.571
35	39.997
40	41.014
45	41.76
50	42.32

X=25

FIGURE 5.164 Degree Mode   **FIGURE 5.165** Degree Mode   ∎

Critical Thinking

If a cannon ball is shot from ground level, what is the optimum projection angle?

The Inverse Cosine Function

By the horizontal line test, the cosine function is not one-to-one as illustrated in Figure 5.166. If we restrict the domain of $f(x) = \cos x$ to $0 \leq x \leq \pi$, then the resulting function is one-to-one and has an inverse function. See Figure 5.167. This inverse function is called the *inverse cosine function.*

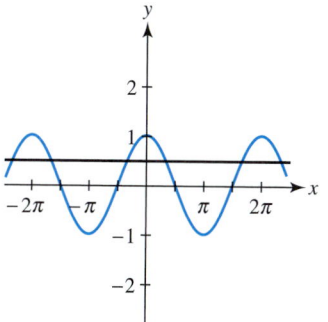

FIGURE 5.166 The Cosine Function   FIGURE 5.167 Restricting the Domain

---

Inverse cosine function

The **inverse cosine function,** denoted by $\cos^{-1} x$ or $\arccos x$, is defined by the following: $y = \cos^{-1} x$ or $y = \arccos x$ means $x = \cos y$ for $-1 \leq x \leq 1$ and y in the interval $[0, \pi]$.

EXAMPLE 6   *Evaluating the inverse cosine function*

Evaluate each of the following.
(a) $\cos^{-1} 1$
(b) $\arccos(-0.75)$

Solution

(a) The expression $\cos^{-1} 1$ represents the angle (or real number) θ whose cosine equals 1 and satisfies $0 \leq \theta \leq \pi$. Thus, $\theta = \cos^{-1}(1) = 0$.

(b) A calculator is often necessary to evaluate inverse trigonometric functions. In Figure 5.168, $\cos^{-1}(-0.75) \approx 2.42$ radians or $138.6°$. For graphical support let $Y_1 = \cos X$ and $Y_2 = -0.75$ on the interval $[0, \pi]$. Their graphs intersect at $\theta \approx 2.42$ or $138.6°$ as shown in Figure 5.169.

$[0, \pi, \pi/4]$ by $[-2, 2, 1]$

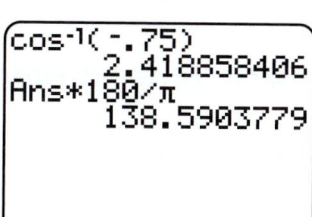

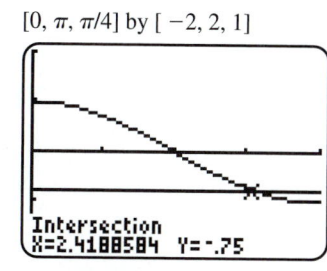

FIGURE 5.168 Radian Mode   FIGURE 5.169

The following are properties of the inverse cosine function.

$$\cos^{-1}(\cos x) = x \quad \text{for } 0 \le x \le \pi$$
$$\cos(\cos^{-1} x) = x \quad \text{for } -1 \le x \le 1$$

EXAMPLE 7 *Finding representations of the inverse cosine function*

Represent the inverse cosine function verbally, numerically, graphically, and symbolically.

Solution

Verbal Representation To compute $\cos^{-1} x$ for $-1 \le x \le 1$ determine the angle θ (or real number) such that $\cos \theta = x$ and $0 \le \theta \le \pi$.

Numerical Representation A partial numerical representation of $\cos x$ on the interval $[0, \pi]$ is shown in Table 5.16.

TABLE 5.16

x	0	$\dfrac{\pi}{6}$	$\dfrac{\pi}{4}$	$\dfrac{\pi}{3}$	$\dfrac{\pi}{2}$	$\dfrac{2\pi}{3}$	$\dfrac{3\pi}{4}$	$\dfrac{5\pi}{6}$	π
$\cos x$	1	$\dfrac{\sqrt{3}}{2}$	$\dfrac{1}{\sqrt{2}}$	$\dfrac{1}{2}$	0	$-\dfrac{1}{2}$	$-\dfrac{1}{\sqrt{2}}$	$-\dfrac{\sqrt{3}}{2}$	-1

A partial numerical representation of $\cos^{-1} x$ is shown in Table 5.17. Notice that if $\cos a = b$, then $\cos^{-1} b = a$, provided that $0 \le a \le \pi$.

TABLE 5.17

x	-1	$-\dfrac{\sqrt{3}}{2}$	$-\dfrac{1}{\sqrt{2}}$	$-\dfrac{1}{2}$	0	$\dfrac{1}{2}$	$\dfrac{1}{\sqrt{2}}$	$\dfrac{\sqrt{3}}{2}$	1
$\cos^{-1} x$	π	$\dfrac{5\pi}{6}$	$\dfrac{3\pi}{4}$	$\dfrac{2\pi}{3}$	$\dfrac{\pi}{2}$	$\dfrac{\pi}{3}$	$\dfrac{\pi}{4}$	$\dfrac{\pi}{6}$	0

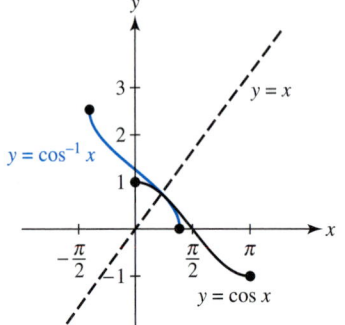

FIGURE 5.170 The Inverse Cosine Function

Graphical Representation The graph of $y = \cos^{-1} x$ can be found by reflecting the graph of $y = \cos x$ for $0 \le x \le \pi$ across the line $y = x$ as shown in Figure 5.170.

Symbolic Representation A symbolic representation of the inverse cosine function can be written as either $f(x) = \cos^{-1} x$ or $f(x) = \arccos x$. There is no simple formula to evaluate $\cos^{-1} x$. ∎

```
cos-1(cos(90))
            90
cos-1(cos(-270))
            90
```

FIGURE 5.171 Degree Mode

Critical Thinking

Explain the results in Figure 5.171. The calculator was in degree mode.

The Inverse Tangent Function

By the horizontal line test, the tangent function is not one-to-one as shown in Figure 5.172. If we restrict the domain of $f(x) = \tan x$ to $-\pi/2 < x < \pi/2$, then the resulting function is one-to-one and has an inverse function. See Figure 5.173. This inverse function is called the *inverse tangent function*.

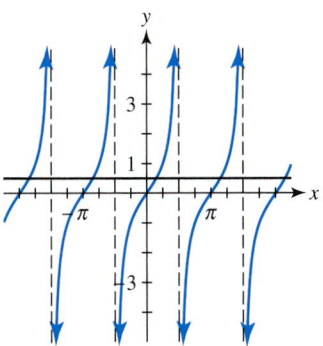

FIGURE 5.172 The Tangent Function

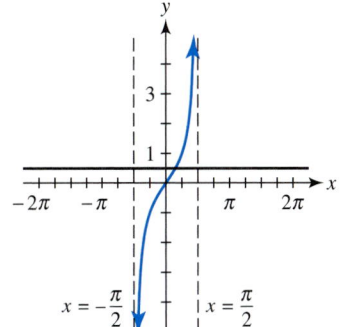

FIGURE 5.173 Restricting the Domain

Inverse tangent function

The **inverse tangent function,** denoted by $\tan^{-1} x$ or arctan x, is defined by the following: $y = \tan^{-1} x$ or $y = \arctan x$ means $x = \tan y$ for y in the interval $(-\pi/2, \pi/2)$.

The following are properties of the inverse tangent function.

$$\tan^{-1}(\tan x) = x \quad \text{for} \quad -\frac{\pi}{2} < x < \frac{\pi}{2}$$

$$\tan(\tan^{-1} x) = x \quad \text{for all real numbers } x$$

EXAMPLE 8   *Evaluating the inverse tangent function*

Evaluate each of the following. Support your answer using a calculator.
(a) $\tan^{-1} 1$

(b) $\arctan\left(-\sqrt{3}\right)$

Solution

(a) The expression $\tan^{-1} 1$ represents the angle (or real number) θ whose tangent equals 1 and satisfies $-\pi/2 < \theta < \pi/2$. Thus, $\theta = \tan^{-1}(1) = \pi/4 \approx 0.7854$.

(b) The expression $\arctan\left(-\sqrt{3}\right)$ represents the angle (or real number) θ whose tangent equals $-\sqrt{3}$ and satisfies $-\pi/2 < \theta < \pi/2$. It follows that $\theta = \arctan\left(-\sqrt{3}\right) = -\pi/3 \approx -1.047$. Numerical support is shown in Figure 5.174. ∎

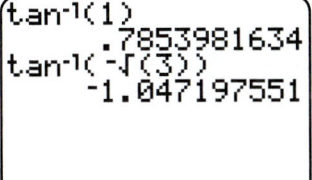

FIGURE 5.174 Radian Mode

Critical Thinking

Give verbal, numerical, and graphical representations of $y = \tan^{-1} x$.

The next example illustrates the use of the inverse tangent function in robotics.

EXAMPLE 9 *Using robots to spray paint*

In industry it is common to use robots to spray paint. The robotic arm in the accompanying figure is being used to paint a flat surface. Because the spray gun must move at a constant speed v, parallel to the surface being painted, the angle of the arm θ_1 and the angle of the spray gun θ_2 must be continually adjusted. Using Figure 5.175, it can be shown that

$$\theta_1 = \arctan \frac{h}{vt} \quad \text{and}$$

$$\theta_2 = 90° - \theta_1,$$

where $t > 0$ is time in seconds. (Source: W. Stadler, *Analytical Robotics and Mechatronics.*)

(a) Let $v = 3$ inches per second and $h = 24$ inches. Graph θ_1 in [0, 25, 5] by [0, 100, 10] using *degree mode*. Describe how θ_1 changes over this 25-second interval.

(b) Determine the degree measure of θ_1 and θ_2 after 10 seconds.

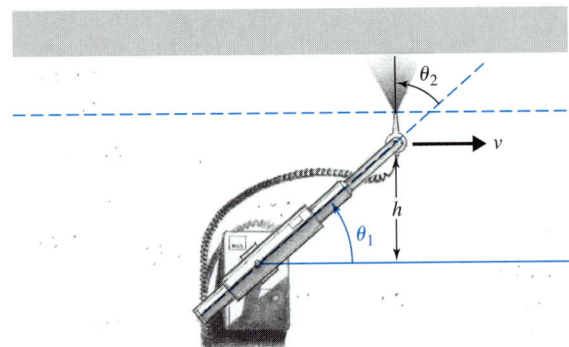

FIGURE 5.175

[0, 25, 5] by [0, 100, 10]

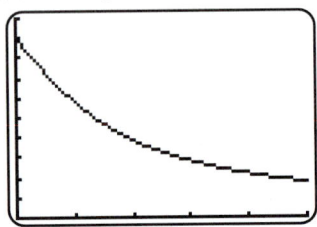

FIGURE 5.176 Degree Mode

Solution

(a) Substitute $h = 24$ and $v = 3$. Graph $Y_1 = \tan^{-1}(24/(3X))$ as shown in Figure 5.176. Initially, the robotic arm is vertical and $\theta_1 = 90°$. As the spray gun moves to the right, angle θ_1 decreases—faster at first and then more slowly.

(b) When $t = 10$, $\theta_1 = \arctan\left(\dfrac{24}{3(10)}\right) \approx 38.7°$ and $\theta_2 = 90° - \theta_1 \approx 51.3°$. ■

Critical Thinking

In Example 9 it was given that $\theta_1 = \arctan\dfrac{h}{vt}$ and $\theta_2 = 90° - \theta_1$. Use Figure 5.175 to verify this.

Solving Triangles and Equations

In Figure 5.177 *standard labeling* is used to denote the vertices, sides, and angles of a right triangle. Finding the measures of the angles and sides in a triangle is called *solving a triangle*. The next example illustrates this process.

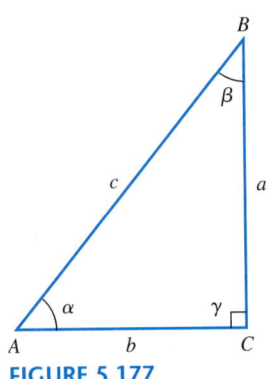

FIGURE 5.177

EXAMPLE 10 *Solving a right triangle*

Solve triangle ABC if $a = 5$ and $c = 13$. See Figure 5.178.

Solution

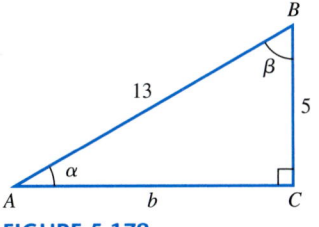

FIGURE 5.178

We are given $a = 5$, $c = 13$, and $\gamma = 90°$. We must find b, α, and β. We begin by finding b using the Pythagorean theorem.

$$a^2 + b^2 = c^2$$
$$b^2 = c^2 - a^2$$
$$= 13^2 - 5^2$$
$$= 144$$

Thus, $b = 12$. We can find angle α as follows.

$$\sin \alpha = \frac{5}{13}$$

$$\alpha = \sin^{-1} \frac{5}{13}$$

$$\approx 22.6°$$

Since β is complementary to α, $\beta = 90° - 22.6° = 67.4°$. ■

Note: There is more than one way to solve the triangle in Example 10. We could have let $\alpha = \tan^{-1} \frac{5}{12} \approx 22.6°$ and $\beta = \cos^{-1} \frac{5}{13} \approx 67.4°$. There are other possibilities.

Grade resistance is the force F that causes a car to roll down a hill. It can be calculated by $F = W \sin \theta$, where θ represents the angle of the grade. In this equation W represents the weight of the the vehicle. (Refer to Figure 5.65 in Section 5.3.)

EXAMPLE 11 *Calculating highway grade*

Find the angle θ for which a 3000-pound car has grade resistance of 500 pounds.

Solution

Solve the equation for θ.

$F = W \sin \theta$	Given equation
$500 = 3000 \sin \theta$	Let $W = 3000$ and $F = 500$.
$\sin \theta = \dfrac{1}{6}$	Solve for $\sin \theta$.
$\theta = \sin^{-1} \dfrac{1}{6}$	Property of inverse sine
$\approx 9.6°$	Approximate.

Thus, if a road is inclined at approximately 9.6°, a 3000-pound car would experience a force of 500 pounds pulling downhill. ■

During the course of a month, the moon goes through different phases as illustrated in Figure 5.179. (See also Example 6 in Section 5.3.) Angle θ is called the *phase angle* of the moon and varies from $0°$ to $360°$ as the moon completes one orbit of Earth. The *phase F* of the moon may be determined by the equation

$$F = \frac{1}{2}(1 - \cos \theta).$$

For example, if $\theta = 180°$ then $F = \frac{1}{2}(1 - \cos 180°) = 1$ and 100% of the moon is illuminated. This corresponds to a full moon. (Source: P. Duffett-Smith, *Practical Astronomy with your Calculator.*)

FIGURE 5.179 Phase Angle θ

EXAMPLE 12 *Determining the phase angle*

Find the phase angle θ associated with $F = 0.5$, assuming that $0° \leq \theta \leq 180°$. Interpret your result.

Solution

$F = \frac{1}{2}(1 - \cos \theta)$	Given equation
$0.5 = \frac{1}{2}(1 - \cos \theta)$	$F = 0.5$
$1 = 1 - \cos \theta$	Multiply by 2
$\cos \theta = 0$	Solve for $\cos \theta$.
$\theta = \cos^{-1} 0$	Property of inverses
$\theta = 90°$	Evaluate.

When 50% of the face of the moon is illuminated (first quarter), the phase angle is 90°. ∎

Critical Thinking

Refer to the introduction of this section. If the roof pitch is $\frac{6}{12}$, at what angle θ should the rafters be cut?

5.6 PUTTING IT ALL TOGETHER

The six trigonometric functions are not one-to-one functions. However, by restricting their domains to appropriate intervals, inverse trigonometric functions can be defined. In this section we defined inverse functions for the sine, cosine, and tangent functions. Inverse functions can be used to solve triangles and trigonometric equations.

The following table summarizes some properties of inverse trigonometric functions.

Inverse Sine Function	Inverse Cosine Function	Inverse Tangent Function
Description $f(x) = \sin^{-1} x$ computes the angle or number in $[-\pi/2, \pi/2]$ whose sine equals x, where $-1 \le x \le 1$.	*Description* $f(x) = \cos^{-1} x$ computes the angle or number in $[0, \pi]$ whose cosine equals x, where $-1 \le x \le 1$.	*Description* $f(x) = \tan^{-1} x$ computes the angle or number in $(-\pi/2, \pi/2)$ whose tangent equals x, where x is any real number.
Domain $\{x \mid -1 \le x \le 1\}$	*Domain* $\{x \mid -1 \le x \le 1\}$	*Domain* $\{x \mid -\infty < x < \infty\}$ (all real numbers)
Range $\left\{ y \mid -\dfrac{\pi}{2} \le y \le \dfrac{\pi}{2} \right\}$	*Range* $\{y \mid 0 \le y \le \pi\}$	*Range* $\left\{ y \mid -\dfrac{\pi}{2} < y < \dfrac{\pi}{2} \right\}$
Inverse Properties $\sin^{-1}(\sin x) = x$ for $-\dfrac{\pi}{2} \le x \le \dfrac{\pi}{2}$ $\sin(\sin^{-1} x) = x$ for $-1 \le x \le 1$	*Inverse Properties* $\cos^{-1}(\cos x) = x$ for $0 \le x \le \pi$ $\cos(\cos^{-1} x) = x$ for $-1 \le x \le 1$	*Inverse Properties* $\tan^{-1}(\tan x) = x$ for $-\dfrac{\pi}{2} < x < \dfrac{\pi}{2}$ $\tan(\tan^{-1} x) = x$ for all real numbers x

5.6 EXERCISES

   Tape 9

Review of Inverses

1. In order for a function f to have an inverse, f must be _____.

2. A function is one-to-one if different inputs always result in _____ outputs.

3. If $f(\pi) = -1$, then $f^{-1}(-1) =$ ____.

4. If $f(c) = d$, then $f^{-1}(d) =$ ____.

5. If $f^{-1}(0) = 1$, then $f(1) =$ ____.

6. If $f^{-1}(b) = a$, then $f(a) =$ ____.

Exercises 7–12: Find a symbolic representation for $f^{-1}(x)$.

7. $f(x) = 3x$

8. $f(x) = 5x - 4$

9. $f(x) = \sqrt[3]{x}$

10. $f(x) = x^2, x \ge 0$

11. $f(x) = (x - 1)^2, x \ge 1$

12. $f(x) = x^3 - 1$

Exercises 13–16: Give verbal, numerical, graphical, and symbolic representations of $f^{-1}(x)$.

13. $f(x) = x + 6$

14. $f(x) = \dfrac{x}{3}$

15. $f(x) = 2x - 1$

16. $f(x) = \sqrt[3]{x - 1}$

Exercises 17 and 18: A graph of a function f that is not one-to-one is given. Estimate the largest interval $[a, b]$ where f is one-to-one, assuming $a < 0 < b$.

17.

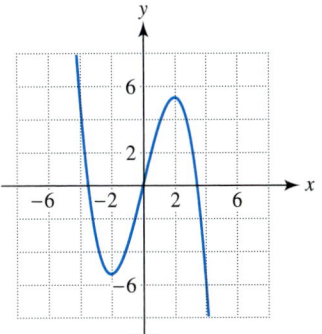

18.

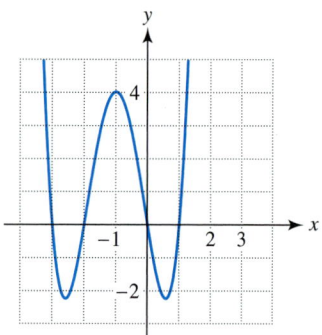

Inverse Trigonometric Functions

19. Since $\sin \dfrac{\pi}{2} = 1$ and $\dfrac{\pi}{2}$ is in the interval $[-\pi/2, \pi/2]$, $\sin^{-1} 1 =$ _____.

20. Since $\cos \dfrac{\pi}{3} = \dfrac{1}{2}$ and $\dfrac{\pi}{3}$ is in the interval $[0, \pi]$, $\cos^{-1} \dfrac{1}{2} =$ _____.

21. Since $\tan\left(-\dfrac{\pi}{4}\right) = -1$ and $-\dfrac{\pi}{4}$ is in the interval $(-\pi/2, \pi/2)$, $\tan^{-1}(-1) =$ _____.

22. Since $\sin\left(-\dfrac{\pi}{6}\right) = -\dfrac{1}{2}$ and $-\dfrac{\pi}{6}$ is in the interval $[-\pi/2, \pi/2]$, $\sin^{-1}\left(-\dfrac{1}{2}\right) =$ _____.

23. Since $\cos\left(\dfrac{2\pi}{3}\right) = -\dfrac{1}{2}$ and $\dfrac{2\pi}{3}$ is in the interval $[0, \pi]$, $\cos^{-1}\left(-\dfrac{1}{2}\right) =$ _____.

24. Since $\tan\left(\dfrac{\pi}{3}\right) = \sqrt{3}$ and $\dfrac{\pi}{3}$ is in the interval $(-\pi/2, \pi/2)$, $\tan^{-1}\sqrt{3} =$ _____.

Exercises 25–30: Evaluate each of the following. Give results in both radians and degrees.

25. **(a)** $\sin^{-1} 1$
    (b) $\arcsin 0$
    (c) $\arcsin\left(-\dfrac{\sqrt{3}}{2}\right)$

26. **(a)** $\arcsin \dfrac{1}{2}$
    (b) $\sin^{-1}(-2)$
    (c) $\sin^{-1}(-1)$

27. **(a)** $\cos^{-1} 0$
    (b) $\arccos(-1)$
    (c) $\cos^{-1}\left(\dfrac{1}{2}\right)$

28. **(a)** $\arccos \dfrac{\sqrt{3}}{2}$
    (b) $\cos^{-1}\left(-\dfrac{1}{2}\right)$
    (c) $\arccos 1$

29. **(a)** $\tan^{-1} 1$
    (b) $\arctan(-1)$
    (c) $\tan^{-1}\sqrt{3}$

30. **(a)** $\arctan(-\sqrt{3})$
    (b) $\tan^{-1} 0$
    (c) $\tan^{-1}\left(-\dfrac{1}{\sqrt{3}}\right)$

Exercises 31 and 32: Approximate the following to a hundredth of a radian and a tenth of a degree. Support your answer graphically.

31. **(a)** $\sin^{-1} 1.5$
    (b) $\tan^{-1} 10$
    (c) $\arccos(-0.75)$

32. **(a)** $\cos^{-1}\left(-\dfrac{1}{3}\right)$
    (b) $\arcsin(-0.54)$
    (c) $\arctan(-2.5)$

Exercises 33 and 34: Evaluate each expression using the figure to obtain either α or β.

33. **(a)** $\tan^{-1} \dfrac{4}{3}$
    (b) $\sin^{-1} \dfrac{3}{5}$
    (c) $\arccos \dfrac{3}{5}$

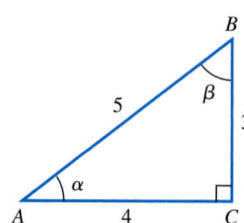

34. **(a)** $\arcsin \dfrac{12}{13}$

    (b) $\cos^{-1} \dfrac{5}{13}$

    (c) $\tan^{-1} \dfrac{5}{12}$

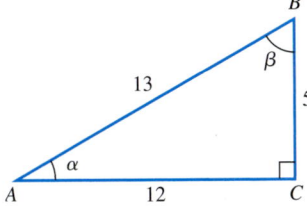

Exercises 35–40: Evaluate each expression.

35. $\sin(\sin^{-1} 1)$

36. $\sin^{-1} \left(\sin \dfrac{\pi}{4} \right)$

37. $\cos^{-1} \left(\cos \dfrac{5\pi}{4} \right)$

38. $\cos(\cos^{-1}(-3))$

39. $\tan(\tan^{-1}(-3))$

40. $\tan^{-1} \left(\tan \dfrac{\pi}{5} \right)$

Exercises 41 and 42: Support each identity by graphing the left side of the equation in $[-4.7, 4.7, 1]$ by $[-3.1, 3.1, 1]$ and comparing it to the graph of $y = x$.

41. **(a)** $\sin(\sin^{-1} x) = x$, if $-1 \le x \le 1$

    (b) $\cos(\cos^{-1} x) = x$, if $-1 \le x \le 1$

    (c) $\tan(\tan^{-1} x) = x$

42. **(a)** $\sin^{-1}(\sin x) = x$, if $-\pi/2 \le x \le \pi/2$

    (b) $\cos^{-1}(\cos x) = x$, if $0 \le x \le \pi$

    (c) $\tan^{-1}(\tan x) = x$, if $-\pi/2 < x < \pi/2$

Exercises 43 and 44: Evaluate the expression in degree mode. Conjecture a generalization of the result and then test your conjecture.

43. **(a)** $\sin^{-1} \dfrac{3}{5} + \cos^{-1} \dfrac{3}{5}$

    (b) $\sin^{-1} \dfrac{1}{3} + \cos^{-1} \dfrac{1}{3}$

    (c) $\sin^{-1} \dfrac{2}{7} + \cos^{-1} \dfrac{2}{7}$

44. **(a)** $\tan^{-1} \dfrac{3}{4} + \tan^{-1} \dfrac{4}{3}$

    (b) $\tan^{-1} \dfrac{5}{12} + \tan^{-1} \dfrac{12}{5}$

    (c) $\tan^{-1} \dfrac{1}{4} + \tan^{-1} 4$

Exercises 45–50: (Refer to Example 10.) Solve the right triangle.

45.

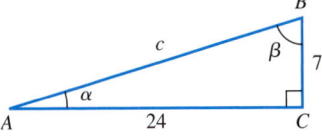

46.

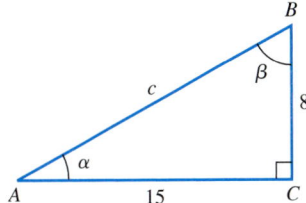

47.

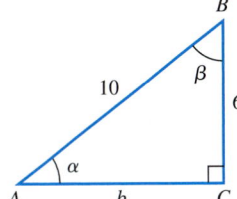

48.

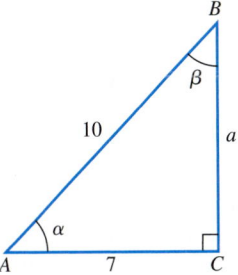

49.

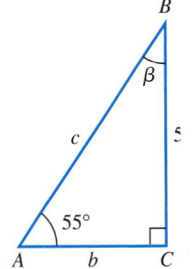

50.

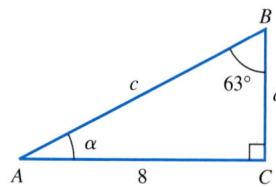

Exercises 51–54: Let θ be an acute angle. Evaluate the indicated trigonometric function of θ, where θ is an acute angle determined by an inverse trigonometric function. (Hint: Make a sketch of a right triangle containing angle θ.)

51. $\tan \theta$, if $\theta = \sin^{-1} x$

52. $\sin \theta$, if $\theta = \tan^{-1} \dfrac{x}{\sqrt{1 - x^2}}$

53. $\cos \theta$, if $\theta = \sin^{-1} \dfrac{x}{\sqrt{1 + x^2}}$

54. $\tan \theta$, if $\theta = \cos^{-1} \dfrac{1}{x}$

Solving Trigonometric Equations

Exercises 55–60: Solve the equation for θ, where $0° \le \theta \le 90°$.

55. $\sin \theta = 1$ **56.** $\cos \theta = \dfrac{1}{2}$

57. $\tan \theta = 1$ **58.** $\sin \theta = \dfrac{\sqrt{3}}{2}$

59. $\cos \theta = 0$ **60.** $\tan \theta = \dfrac{1}{\sqrt{3}}$

Exercises 61–66: Solve the equation for θ, where θ is an acute angle. Approximate θ to the nearest tenth of a degree.

61. $2 \cos \theta = \dfrac{1}{4}$ **62.** $3 \sin \theta = \dfrac{4}{5}$

63. $\tan \theta - 1 = 5$ **64.** $4 \cos \theta + 1 = 6$

65. $\sin^2 \theta = 0.87$ **66.** $\tan^3 \theta - 2 = 1.65$

Exercises 67–74: Solve the equation for t where t is a real number in the given interval. Approximate t to three significant digits. Support your result graphically.

67. $\tan t = -\dfrac{1}{5}, \left(-\dfrac{\pi}{2}, \dfrac{\pi}{2} \right)$

68. $\sin t = -\dfrac{1}{3}, \left[-\dfrac{\pi}{2}, \dfrac{\pi}{2} \right]$

69. $\cos t = 0.452, [0, \pi]$

70. $\tan t = 5.67, \left(-\dfrac{\pi}{2}, \dfrac{\pi}{2} \right)$

71. $2 \sin t = -0.557, \left[-\dfrac{\pi}{2}, \dfrac{\pi}{2} \right]$

72. $3 \cos t + 1 = 0.333, [0, \pi]$

73. $\cos^2 t = \dfrac{1}{25}, [0, \pi]$

74. $\sin^2 t = \dfrac{1}{16}, \left[-\dfrac{\pi}{2}, \dfrac{\pi}{2} \right]$

Applications

75. *Angle of Elevation*   Find the angle of elevation θ of the top of a 50-foot tree at a distance of 85 feet. See the accompanying figure.

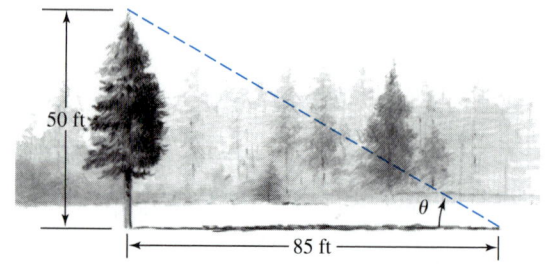

76. *Angle of Elevation*   A 28-foot building casts a 40-foot shadow on level ground. Estimate the angle of elevation θ of the sun to the nearest tenth of a degree. See the accompanying figure.

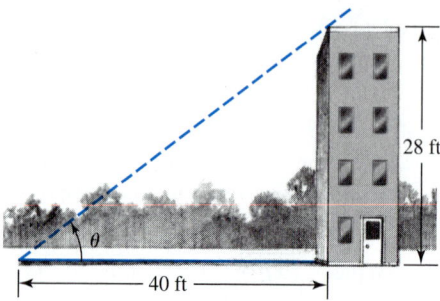

77. *Robotics*   Approximate the angle θ if the robotic hand is located at the following points, where $-90° < \theta < 90°$. See the accompanying figure.

  (a) $(5, 11)$ **(b)** $(1, -3)$

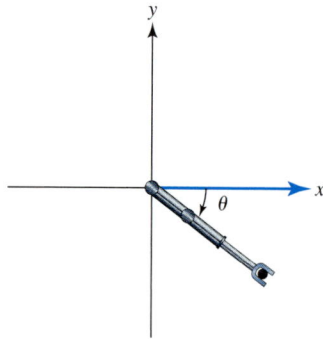

78. *Grade Resistance* (Refer to Example 11.) Approximate θ to the nearest tenth of a degree for the given grade resistance F and vehicle weight W.

(a) $F = 400$ lb, $\quad W = 5000$ lb

(b) $F = 130$ lb, $\quad W = 3500$ lb

(c) $F = -200$ lb, $\quad W = 4000$ lb

79. *Designing Steps* Steps are being attached to a deck as shown in the figure. The bottom of the steps should land 10 feet from the deck. If the deck is 4 feet above the ground, estimate the angle θ that the side boards of the step should be cut so that the ends lie flat on level ground.

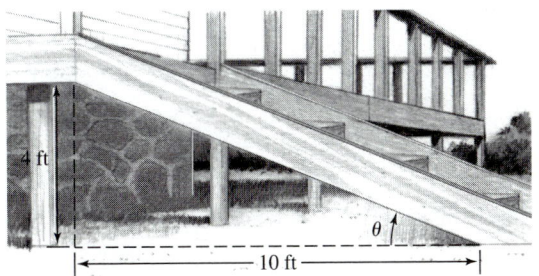

4 ft

θ

10 ft

80. *Designing Steps* (Refer to the previous exercise.) If the length of the side boards for the steps is 4 feet and the deck is 2 feet above the ground, estimate angle θ.

81. *Phases of the Moon* (Refer to Example 12.) Find the phase angle θ for the given phase F. Assume that $0° \le \theta \le 180°$.

(a) $F = \dfrac{1}{4}$ $\qquad$ (b) $F = \dfrac{3}{4}$

82. *Roof Pitch* The pitch or slope of a roof may be expressed in the form $k/12$, where k represents a k-foot rise for every 12 feet of run in horizontal distance. Determine angle θ in Figure 5.153 for each pitch.

(a) $\dfrac{3}{12}$ $\qquad$ (b) $\dfrac{4}{12}$

(c) $\dfrac{6}{12}$ $\qquad$ (d) $\dfrac{12}{12}$

83. *Shot Put* (Refer to Example 5.) Suppose that a shot is released 7 feet above the ground with a velocity of 43 feet per second. Find the optimum projection angle.

84. *Shot-Putting on the Moon* Repeat Example 5 if the shot is thrown on the moon and the optimal projection angle is given by

$$\theta = \sin^{-1} \sqrt{\frac{v^2}{2v^2 + 10.2h}}.$$

85. *Calculating Daylight Hours* The ability to calculate the number of daylight hours H at any location is important for estimating the potential solar energy production. The value of H on the longest day can be calculated using the formula

$$\cos(0.1309\,H) = -0.4336 \tan L,$$

where L is the latitude. Using *radian* mode, calculate the greatest number of daylight hours H during the year for the various cities and their latitudes. (Source: C. Winter, *Solar Power Plants.*)

(a) Akron, Ohio; $L = 40°55'$

(b) Corpus Christi, Texas; $L = 27°46'$

(c) Richmond, Virginia; $L = 37° 30'$

86. *Shortest Day* (Refer to the previous exercise.) The minimum value of H can be calculated using the formula

$$\cos(0.1309\,H) = 0.4336 \tan L.$$

Find the least number of daylight hours at the following locations.

(a) Anchorage, Alaska; $L = 61° 10'$

(b) Atlantic City, New Jersey; $L = 39°27'$

(c) Honolulu, Hawaii; $L = 21°20'$

87. *Snell's Law* When a ray of light enters water, it is bent. This is because light travels slower in water than in air. This change in direction can be calculated using Snell's law. See the accompanying figure. The angles θ_1 and θ_2 are related by the equation

$$n_1 \sin \theta_1 = n_2 \sin \theta_2,$$

where n_1 and n_2 are constants called *indexes of refraction*. For air $n_1 = 1$ and for water $n_2 = 1.33$. If a ray of light enters the water with $\theta_1 = 40°$, estimate θ_2. (Source: R. Weidner and R. Sells, *Elementary Classical Physics, Vol. 2.*)

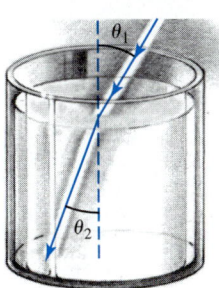

θ_1

θ_2

88. (Refer to the previous exercise.) Calculate θ_2 if a ray of light enters alcohol with $\theta_1 = 20°$ and $n_2 = 1.36$.

89. *Movie Screen* A 10-foot-high movie screen is mounted on a vertical wall so that the bottom of the screen is 6 feet above a horizontal floor. A person sits on a level floor x feet from the screen. If eye level is 3 feet above the floor, then angle θ in the accompanying figure can be expressed as

$$\theta = \tan^{-1}\left(\frac{10x}{x^2 + 39}\right).$$

Graph θ in [0, 50, 10] by [0, 50, 10] using degree mode. Determine where a person should sit to maximize θ.

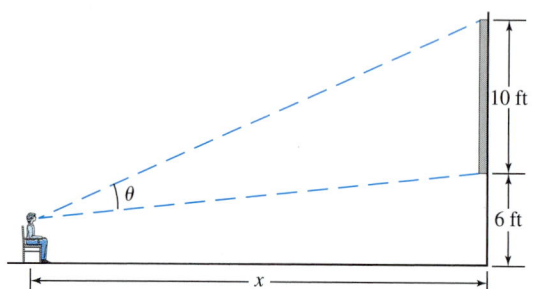

90. *Robotics* (Refer to Example 9.)
  (a) Let $v = 5$ inches per second and $h = 18$. Graph $\theta_1(t) = \tan^{-1}\left(\frac{h}{vt}\right)$ in [0, 15, 5] by [0, 100, 10] using *degree mode*. Describe how θ_1 changes over this 15-second interval.
  (b) Determine the degree measure of θ_1 and θ_2 after 5 seconds.

Writing about Mathematics

1. Explain verbally what each expression computes. Give examples.
  (a) $\sin^{-1} x$ **(b)** $\cos^{-1} x$
  (c) $\tan^{-1} x$

2. Explain verbally why $\sin^{-1}\left(\sin \frac{\pi}{2}\right) = \frac{\pi}{2}$, but

$$\sin^{-1}\left(\sin \frac{5\pi}{2}\right) \neq \frac{5\pi}{2}.$$

Give a similar example using $\cos x$ and $\cos^{-1} x$.

CHECKING BASIC CONCEPTS FOR SECTIONS 5.5 AND 5.6

1. Graph $f(t) = 3 \sin(2(t - \pi/4))$ in $[-\pi, \pi, \pi/4]$ by $[-4, 4, 1]$. State the amplitude, period, and phase shift.

2. The accompanying table contains data that can be modeled by $f(t) = a \cos(bt)$. Find values for a and b. Support your results numerically by making a table of f.

t	0	0.5	1.0	1.5	2.0	2.5	3.0
$f(t)$	2	0	−2	0	2	0	−2

3. Evaluate each of the following by hand, expressing your answer in degrees. Support your results using a calculator.

  (a) $\sin^{-1} 0$
  (b) $\cos^{-1}(-1)$
  (c) $\tan^{-1}(-1)$
  (d) $\sin^{-1} \frac{1}{2}$
  (e) $\tan^{-1} \sqrt{3}$

4. Solve the right triangle if $a = 30$ and $b = 40$.

5. Use a calculator to solve the equation, where t is in the indicated interval.
  (a) $\sin t = 0.55$, $[-\pi/2, \pi/2]$
  (b) $\cos t = -0.35$, $[0, \pi]$
  (c) $\tan t = -2.9$, $(-\pi/2, \pi/2)$

CHAPTER 5 Summary

Angles can be measured in either degrees or radians. Radian measure is typically used in technical fields and advanced mathematics. There are $360°$ or 2π radians in one complete revolution. As a result, $1°$ equals $\dfrac{\pi}{180} \approx 0.0175$ radian and 1 radian equals $\dfrac{180°}{\pi} \approx 57.3°$. The length s of an arc intercepted on a circle of radius r by a central angle θ is given by $s = r\theta$. Similarly, the area A of a sector of a circle is given by $A = \dfrac{1}{2}r^2\theta$. In both formulas, θ must be in *radians*.

The six trigonometric functions are sine, cosine, tangent, cosecant, secant, and cotangent. If θ represents an angle in standard position with a point (x, y) on its terminal side and $r = \sqrt{x^2 + y^2}$, then

$$\sin\theta = \frac{y}{r}, \qquad \cos\theta = \frac{x}{r}, \qquad \tan\theta = \frac{y}{x},$$

$$\csc\theta = \frac{r}{y}, \qquad \sec\theta = \frac{r}{x}, \qquad \cot\theta = \frac{x}{y}.$$

The trigonometric functions can also be defined using right triangles or the unit circle. Right triangle trigonometry assumes that θ is an acute angle, whereas unit circle trigonometry often assumes that θ is a real number. When evaluating a trigonometric function at a real number t, evaluate the function at t radians.

Each trigonometric function can be represented verbally, graphically, numerically, and symbolically. Trigonometric functions do not have formulas that are easy to evaluate. The domains of the sine and cosine functions include all real numbers and their graphs are continuous. The graphs of the other trigonometric functions contain discontinuities, where vertical asymptotes occur.

Trigonometric functions frequently are used to model phenomena that involve rotation or periodic data. Two common representations of functions used to model periodic data are

$$f(t) = a\sin\left(b(x - c)\right) + d \quad \text{and} \quad g(t) = a\cos\left(b(x - c)\right) + d.$$

In these formulas, $|a|$ is the amplitude and c is the phase shift. The period is given by $2\pi/b$ and their graphs are centered vertically on the horizontal line $y = d$.

None of the six trigonometric functions are one-to-one. Therefore, general inverse functions do not exist. However, if the domain of a trigonometric function is restricted to an interval where it is one-to-one, then an inverse trigonometric function may be defined. In this chapter we defined three inverse trigonometric functions represented by $\sin^{-1} x$, $\cos^{-1} x$, and $\tan^{-1} x$. The expression $\sin^{-1} x$ represents the angle θ in the interval $[-\pi/2, \pi/2]$ whose sine equals x. Similar statements can be made for $\cos^{-1} x$ and $\tan^{-1} x$. Inverse trigonometric functions are often used to solve trigonometric equations and right triangles.

Review Exercises

1. Sketch the following angles in standard position.

    (a) 60°

    (b) −120°

    (c) $\dfrac{3\pi}{2}$

    (d) $-\dfrac{5\pi}{6}$

2. Find the complementary angle and the supplementary angle to $\theta = 61°40'$.

3. Convert each angle from radian measure to degree measure.

    (a) $\dfrac{\pi}{3}$

    (b) $\dfrac{\pi}{36}$

    (c) $-\dfrac{5\pi}{6}$

    (d) $-\dfrac{7\pi}{4}$

4. Convert each angle from degree measure to radian measure.

    (a) 30°

    (b) 165°

    (c) −90°

    (d) −105°

5. Find the length of the arc intercepted by a central angle $\theta = 60°$ and a radius $r = 6$ feet.

6. Find the area of the sector of a circle having a radius $r = 5$ inches and a central angle $\theta = 150°$.

7. Find the six trigonometric functions of θ.

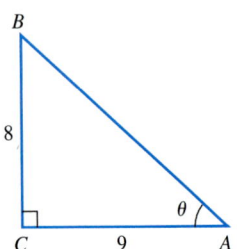

8. Solve triangle ABC.

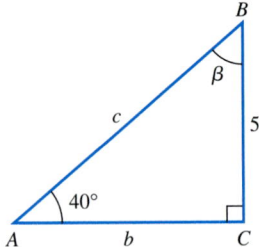

9. Find $\csc \theta$ if $\sin \theta = \dfrac{1}{3}$.

10. Find $\cot \theta$ if $\sin \theta = \dfrac{5}{13}$ and $\cos \theta = -\dfrac{12}{13}$.

Exercises 11 and 12: Approximate the six trigonometric functions of θ using four significant digits.

11. $\theta = 25°$

12. $\theta = -\dfrac{6\pi}{7}$

Exercises 13 and 14: Find the six trigonometric functions of angle θ.

13.

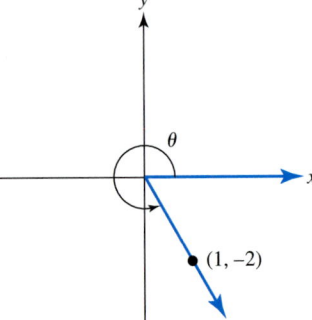

14.

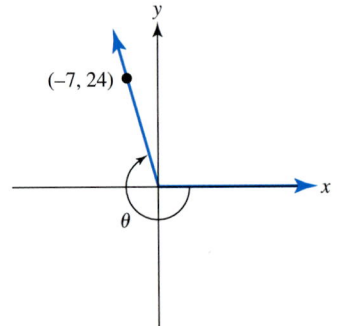

Exercises 15 and 16: Find the six trigonometric functions of the given angle by hand. Support your answer using a calculator.

15. $-45°$

16. $\dfrac{5\pi}{6}$

Exercises 17 and 18: The accompanying figure shows angle θ in standard position with its terminal side intersecting the unit circle. Find the six trigonometric functions of θ.

17.

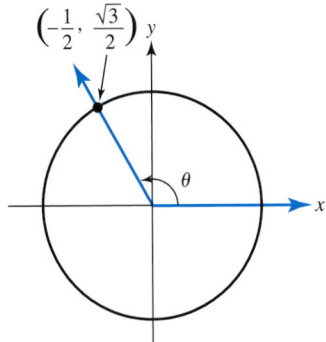

$\left(-\dfrac{1}{2}, \dfrac{\sqrt{3}}{2}\right)$

18.

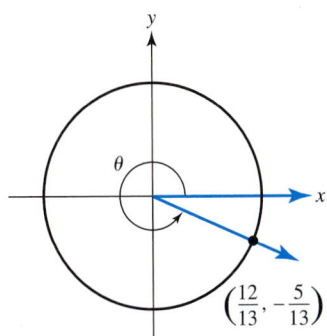

$\left(\dfrac{12}{13}, -\dfrac{5}{13}\right)$

Exercises 19–22: Evaluate the function f at the given value of t

  (a) *verbally,*

  (b) *numerically, and*

  (c) *graphically.*

Support your results by evaluating f(t) directly with a calculator.

19. $f(t) = \sin t, \ t = -\dfrac{\pi}{2}$

20. $f(t) = \cos t, \ t = \pi$

21. $f(t) = \tan t, \ t = -3\pi$

22. $f(t) = \csc t, \ t = \dfrac{\pi}{2}$

23. Find the other trigonometric functions of θ if $\sin\theta = -\dfrac{4}{5}$ and $\cos\theta = \dfrac{3}{5}$.

24. Convert $65°45'36''$ to decimal degrees.

Exercises 25 and 26: Determine which of the six trigonometric functions is represented by the table. Check your results with a calculator.

25.

x	$-\dfrac{\pi}{2}$	$-\dfrac{\pi}{6}$	0	$\dfrac{\pi}{6}$	$\dfrac{\pi}{2}$
y	-1	$-\dfrac{1}{2}$	0	$\dfrac{1}{2}$	1

26.

x	$-\dfrac{\pi}{3}$	$-\dfrac{\pi}{4}$	0	$\dfrac{\pi}{4}$	$\dfrac{\pi}{3}$
y	$-\sqrt{3}$	-1	0	1	$\sqrt{3}$

Exercises 27–30: Graph f in $[-2\pi, 2\pi, \pi/2]$ by $[-4, 4, 1]$. State the amplitude, period, and phase shift.

27. $f(t) = 3\cos(2t)$

28. $f(t) = -2\sin(2t + \pi)$

29. $f(t) = \sin\left(3\left(t - \dfrac{\pi}{3}\right)\right)$

30. $f(t) = 1.5\cos\left(\dfrac{1}{2}\left(t - \dfrac{\pi}{2}\right)\right)$

Exercises 31 and 32: A graph of $y = a\cos bx$ is shown, where b is a positive constant. Estimate the values of a and b.

31.

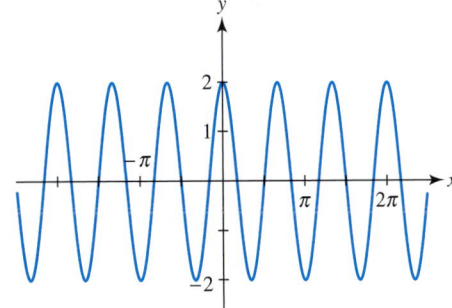

32.

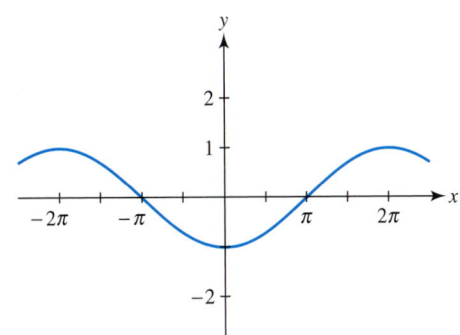

Exercises 33–36: Graph f in $[-2\pi, 2\pi, \pi/2]$ *by* $[-4, 4, 1]$ *using dot mode. State the period and phase shift.*

33. $f(t) = \cot 2t$

34. $f(t) = \tan\left(t - \dfrac{\pi}{2}\right)$

35. $f(t) = \csc(t + \pi)$

36. $f(t) = \sec 2t$

Exercises 37 and 38: If possible, evaluate each of the following in both radians and degrees.

37. **(a)** $\sin^{-1}(-1)$

(b) $\arccos \dfrac{1}{2}$

(c) $\tan^{-1} 1$

38. **(a)** $\arcsin 3$

(b) $\cos^{-1} 0$

(c) $\arctan\left(-\sqrt{3}\right)$

39. Approximate the following to a hundredth of a radian and a tenth of a degree.

(a) $\sin^{-1}(-0.6)$

(b) $\tan^{-1} 5$

(c) $\arccos(0.12)$

40. Evaluate each expression.

(a) $\sin(\sin^{-1} 0.5)$

(b) $\tan^{-1}(\tan 45°)$

(c) $\cos^{-1}\left(\cos \dfrac{3\pi}{2}\right)$

Exercises 41 and 42: Solve the right triangle ABC.

41.

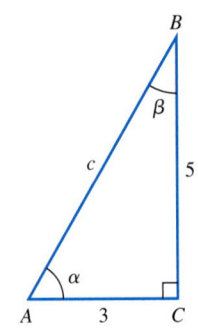

42.

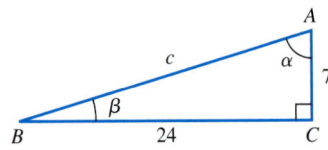

Exercises 43–46: Solve the equation for θ, *where* $0° \le \theta \le 90°$.

43. $\sin \theta = \dfrac{1}{2}$

44. $\cos \theta = 1$

45. $\tan \theta = \dfrac{1}{\sqrt{3}}$

46. $\sin \theta = 0$

Exercises 47–50: Solve the equation for θ, *where* θ *is an acute angle. Approximate* θ *to the nearest tenth of a degree.*

47. $\cos \theta = \dfrac{1}{5}$

48. $3 \sin \theta = \dfrac{15}{13}$

49. $2 \tan \theta - 1 = 5$

50. $2 \cos \theta = \dfrac{4}{7}$

Exercises 51–54: Solve the equation for t, where t is a real number located in the indicated interval. Approximate t to four significant digits. Support your result graphically.

51. $\tan t = -\dfrac{3}{4}, \left(-\dfrac{\pi}{2}, \dfrac{\pi}{2}\right)$

52. $\sin t = -\dfrac{3}{5}, \left[-\dfrac{\pi}{2}, \dfrac{\pi}{2}\right]$

53. $2 \cos t = 1.8, [0, \pi]$

54. $3 \tan t + 2 = 4.7, \left(-\dfrac{\pi}{2}, \dfrac{\pi}{2}\right)$

55. *Height of a Tree* Eighty feet from the trunk of a tree on level ground the angle of elevation of the top of the tree is 48°. Estimate the height of the tree to the nearest foot.

56. *Angle of Elevation* Find the angle of elevation of the top of a 35-foot building at a horizontal distance of 52 feet.

57. *Grade Resistance* Approximate θ to the nearest tenth of a degree for the given grade resistance F and vehicle weight W, where $F = W \sin \theta$.

(a) $F = 350$ lb, $W = 6000$ lb

(b) $F = 160$ lb, $W = 4500$ lb

58. *Highway Grade* (Refer to the previous exercise.) Suppose an uphill grade of a highway can be modeled by the line $y = 0.05x$.

(a) Find the grade of the hill.

(b) Determine the grade resistance for a gravel truck weighing 30,000 pounds.

59. *Distance Between Cities* Cheyenne, Wyoming and Colorado Springs, Colorado have nearly the same longitude of 104°45′ W. The latitude of Cheyenne is 41°09′ and the latitude of Colorado Springs is 38°49′. Approximate the distance between these two cities if the average radius of Earth is 3955 miles. (Source: J. Williams, *The Weather Almanac 1995.*)

60. *Safe Distance for a Tree* From a distance of 45 feet from the base of a tree, the angle of elevation to the top of a tree is 57° as shown in the figure. A building is located 52 feet from the base of the tree. Determine if the tree could fall in a storm and damage the building.

61. *Modeling Temperatures* The monthly average low temperatures in Green Bay, Wisconsin are shown in the table.

Month	1	2	3	4	5	6
Temperature (°F)	6	10	22	35	45	52

Month	7	8	9	10	11	12
Temperature (°F)	58	56	48	38	26	11

Source: A. Miller and J. Thompson, *Elements of Meteorology.*

(a) Plot the average monthly temperature over a 24-month period by letting $x = 1$ and $x = 13$ correspond to January.

(b) Find the constants a, b, c, and d so that

$$f(x) = a \cos (b(x - c)) + d$$

models the data.

(c) Graph f together with the data.

62. *Light and Sinusoidal Waves* Light frequently is modeled using a sinusoidal wave. Light waves have very high frequencies compared to sound waves.

Each color of light has a different frequency. The function given by $f(t) = \sin(2\pi F t)$ can be used to model various colors of light, where F is the frequency in cycles per second. Graph f in $[0, 10^{-14}, 10^{-15}]$ by $[-1.5, 1.5, 0.5]$. Find the period P for each color of light, where $P = \dfrac{1}{F}$. Interpret the results. (Source: R. Weidner and R. Sells, *Elementary Classical Physics, Vol. 2.*)

(a) Violet: $F = 7.5 \times 10^{14}$

(b) Green: $F = 6 \times 10^{14}$

(a) Red: $F = 4 \times 10^{14}$

63. *Vehicle Navigation Systems* Vehicle location and navigation systems for cars are becoming increasingly popular. Devices that depend on Earth's magnetic field are not sufficiently accurate because they can be affected by going through a car wash, having the rear defroster turned on, or even traveling near a large metal truck. The accompanying graph approximates typical errors when using these devices. In this graph 0° corresponds to north, 90° to east, 180° to south, and 270° to west. For example, if a car is traveling in a direction corresponding to 180°, the error in magnetic compass may be −4°. Thus, the car is actually traveling in a direction of 184°. Find the equation of the graph using $y = a \sin(x - c)$, where $0° \le x \le 360°$. (Source: Y. Zhao, *Vehicle Location and Navigation Systems.*)

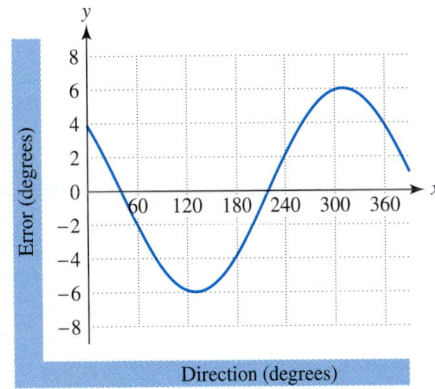

Extended and Discovery Exercises

1. *Surveying* The first fundamental problem of surveying is to determine the coordinates of a point Q given the coordinates of a point P, the distance between P and Q, and the bearing θ from P to Q.

See the accompanying figure. (Source: I. Mueller and K. Ramsayer, *Introduction to Surveying.*)

(a) Find a formula for the coordinates (x_Q, y_Q) of the point Q given θ, the coordinates (x_P, y_P) of P, and the distance d between P and Q.

(b) Use your formula to determine (x_Q, y_Q) if $(x_P, y_P) = (152, 186)$, $\theta = 23.2°$, and $d = 208$ feet.

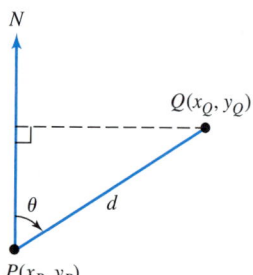

2. *Highway Grade* (Refer to Example 3, Section 5.3.) Complete the table for the trigonometric values of θ to four significant digits.

θ	$\sin \theta$	$\tan \theta$
0°		
1°		
2°		
3°		
4°		

(a) How do $\sin \theta$ and $\tan \theta$ compare for small values of θ?

(b) Highway grades are usually small. Give an approximation to the grade resistance given by $F = W \sin \theta$ that uses the tangent function instead of the sine function.

(c) A stretch of highway has a 4-foot vertical rise for every 100 feet of horizontal run. Use your approximation to estimate the grade resistance for a 3000-pound car.

(d) Compare your result to the exact answer using $F = W \sin \theta$.
(Source: F. Mannering and W. Kilareski, *Principals of Highway Engineering and Traffic Control.*)

3. *Average Temperature* The maximum average monthly temperature in Buenos Aires, Argentina is 74°F and the minimum average monthly temperature is 49°F.

(a) Using these two temperatures conjecture values for a, b, c, and d, so that
$$f(x) = a \cos (b(x - c)) + d$$
models the monthly average temperature.

(b) On the same coordinate axes, graph f for a two year period together with the actual data values found in the accompanying table. Are your results as good as you expected? Explain.

(c) Buenos Aires is located in the Southern Hemisphere. Discuss the effect that this has on the graph of f compared to a city in the Northern Hemisphere.

Month	1	2	3	4	5	6
Temperature (°F)	74	73	69	61	55	50

Month	7	8	9	10	11	12
Temperature (°F)	49	51	55	60	66	71

Source: A. Miller and J. Thompson, *Elements of Meteorology.*

4. (Refer to the previous exercises.) The maximum average monthly temperature in Melbourne, Australia is 68°F and the minimum is 49°F.

(a) Determine $f(x) = a \cos (b(x - c)) + d$ so that f models the monthly average temperature in Melbourne.

(b) Graph f together with the actual data shown in the accompanying table over a two-year period.

Month	1	2	3	4	5	6
Temperature (°F)	68	68	65	60	54	51

Month	7	8	9	10	11	12
Temperature (°F)	49	51	54	58	61	65

Source: A. Miller.

5. Suppose two cities, one in the Northern Hemisphere and one in the Southern Hemisphere, both have average monthly temperatures that can be modeled by
$$f(x) = a \cos (b(x - c)) + d,$$
where x is the month. Suppose also that for both cities the warmest monthly average temperature is 80°F and the coldest is 40°F.

(a) Find $f(x)$ for each city. (Assume that the warmest month for the northern city is July and the coldest month is January, while the opposite is true for the southern city.)

(b) Explain how $f(x)$ for one city could be used to obtain $f(x)$ for the other city.

CHAPTER 6 *Trigonometric Identities and Equations*

*M*usic is both art and science. During the Greek and Roman eras, music played an important role in philosophy and science. Although Pythagoras is usually associated with the Pythagorean theorem, in 500 B.C. he also discovered the mathematical relationships between lengths of strings and musical intervals. This discovery of the mathematical ratios that governed pitch and motion was the beginning of the science of musical sound. In the Middle Ages, music was studied with arithmetic, geometry, and astronomy as part of the liberal arts curriculum. Later in 1862 the psychologist and scientist Hermann von Helmholtz published his classic work that opened a new direction for music using mathematics and technology. Then in 1957, Max Mathews created complex musical sounds with a computer. Mathematics has had an essential role in the development and reproduction of music.

The communications industry also uses mathematics. For example, each number on a touch-tone phone has a unique sound that is a combination of two different tones. These unique tones are easily distinguished when transmitted over phone lines. As a result, consumers can communicate with banks and other businesses using touch-tone phones. Trigonometric functions play an important role in clear, reliable communication systems.

*T*he struggle is what teaches us.
— **Sue Grafton**

Many applications of mathematics began as theoretical mathematics. This is why both pure and applied mathematics are important to society. In this chapter we introduce the concept of verifying an identity. Identities are important because they allow us to write trigonometric expressions in simpler and more convenient forms. Verifying identities requires both effort and concentration. However, by learning to manipulate trigonometric expressions, we will be able to solve complex problems and equations.

Source: J. Pierce, *The Science of Musical Sound.*

6.1 Fundamental Identities

Reciprocal and Quotient Identities • Pythagorean Identities • Negative-Angle Identities

Introduction

Trigonometric expressions can often be written in more than one way. For example, $\cot \theta$ is equivalent to $\dfrac{\cos \theta}{\sin \theta}$. The equation

$$\cot \theta = \frac{\cos \theta}{\sin \theta}$$

is a trigonometric identity. This identity is true for every value of θ, provided $\sin \theta \neq 0$. Trigonometric identities are used to help solve equations and model physical phenomena in music, science, and electricity. They are also used in calculus. We begin our discussion with the reciprocal and quotient identities.

Reciprocal and Quotient Identities

In Section 5.4 the following definitions were presented for the trigonometric functions of any angle θ.

Trigonometric functions of any angle θ

Let (x, y) be a point other than the origin on the terminal side of an angle θ in standard position. If $r = \sqrt{x^2 + y^2}$, then the six trigonometric functions are as follows.

$$\sin \theta = \frac{y}{r} \qquad\qquad \csc \theta = \frac{r}{y} \; (y \neq 0)$$

$$\cos \theta = \frac{x}{r} \qquad\qquad \sec \theta = \frac{r}{x} \; (x \neq 0)$$

$$\tan \theta = \frac{y}{x} \; (x \neq 0) \qquad \cot \theta = \frac{x}{y} \; (y \neq 0)$$

These definitions allow us to write several identities. For example, since

$$\cos \theta = \frac{x}{r} \qquad \text{and} \qquad \sec \theta = \frac{r}{x},$$

it follows that

$$\sec \theta = \frac{1}{x/r} = \frac{1}{\cos \theta} \qquad \text{and} \qquad \cos \theta = \frac{1}{r/x} = \frac{1}{\sec \theta}.$$

These identities are examples of *reciprocal identities*.

Reciprocal identities

$$\sin \theta = \frac{1}{\csc \theta} \qquad \cos \theta = \frac{1}{\sec \theta} \qquad \tan \theta = \frac{1}{\cot \theta}$$

$$\csc \theta = \frac{1}{\sin \theta} \qquad \sec \theta = \frac{1}{\cos \theta} \qquad \cot \theta = \frac{1}{\tan \theta}$$

EXAMPLE 1   *Applying a reciprocal identity*

In Example 8, Section 5.2 the equation

$$d = r\left(\frac{1}{\cos \theta} - 1 \right)$$

was used to calculate the height of an orbit for a GPS satellite. Use a reciprocal identity to rewrite this equation. (Source: A. Leick, *GPS Satellite Surveying.*)

Solution

Since $\sec \theta = \dfrac{1}{\cos \theta}$, we can express the equation as

$$d = r(\sec \theta - 1). \qquad\qquad\blacksquare$$

Because $\sin \theta = \dfrac{y}{r}$ and $\cos \theta = \dfrac{x}{r}$, it is possible to write the other four trigonometric functions in terms of $\sin \theta$ and $\cos \theta$. For example,

$$\tan \theta = \frac{y}{x} = \frac{y/r}{x/r} = \frac{\sin \theta}{\cos \theta}.$$

This identity is an example of a *quotient identity* that can be supported graphically and numerically by letting $Y_1 = \tan(X)$ and $Y_2 = \sin(X)/\cos(X)$. The graph of Y_1 is shown in Figure 6.1 and the graph of Y_2 is shown in Figure 6.2. (Degree mode has been used.) Their graphs appear to be identical. Numerical support for this quotient identity is shown in Figure 6.3. Notice that $Y_1 = Y_2$ for each value of x whenever $\cos x \neq 0$.

$[-352.5°, 352.5°, 90°]$ by $[24, 4, 1]$ $[-352.5°, 352.5°, 90°]$ by $[-4, 4, 1]$

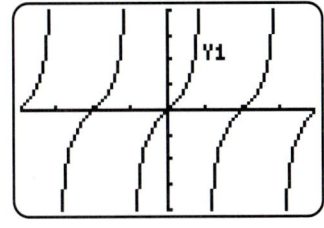

FIGURE 6.1

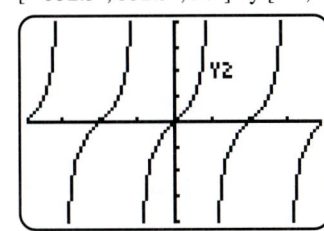

FIGURE 6.2

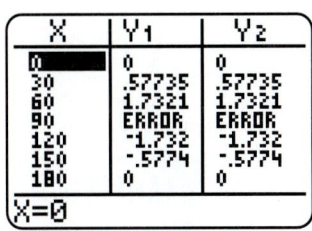

FIGURE 6.3 Degree Mode

The quotient identities are given in the following. The third and fourth identities are also reciprocal identities.

Quotient identities

$$\tan \theta = \frac{\sin \theta}{\cos \theta} \qquad \cot \theta = \frac{\cos \theta}{\sin \theta} \qquad \sec \theta = \frac{1}{\cos \theta} \qquad \csc \theta = \frac{1}{\sin \theta}$$

Technology Note

Figures 6.1 and 6.2 are graphed in degrees using a *friendly window*. If a different window is used, a graphing calculator may graph solid, vertical lines that resemble asymptotes. These lines result from a calculator connecting points inappropriately and are sometimes called *pseudo-asymptotes*. Dot mode can also be used to avoid pseudo-asymptotes. See Figure 6.4 and 6.5.

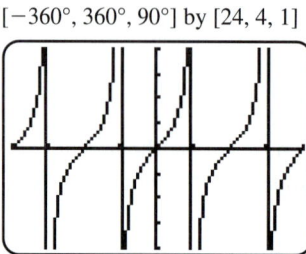

$[-360°, 360°, 90°]$ by $[24, 4, 1]$ $[-360°, 360°, 90°]$ by $[-4, 4, 1]$

FIGURE 6.4 Pseudo-Asymptotes   **FIGURE 6.5** Dot Mode

If $\sin \theta$ and $\cos \theta$ are known, the quotient identities can be used to find the other four trigonometric functions of θ as illustrated in the next example.

EXAMPLE 2   *Using the quotient identities*

If $\sin \theta = \dfrac{7}{25}$ and $\cos \theta = -\dfrac{24}{25}$, find the other four trigonometric functions of θ.

Solution

We can use the quotient identities to find $\tan \theta$, $\cot \theta$, $\sec \theta$, and $\csc \theta$.

$$\tan \theta = \frac{\sin \theta}{\cos \theta} = \frac{7/25}{-24/25} = -\frac{7}{24}$$

$$\cot \theta = \frac{\cos \theta}{\sin \theta} = \frac{-24/25}{7/25} = -\frac{24}{7}$$

$$\sec \theta = \frac{1}{\cos \theta} = \frac{1}{-24/25} = -\frac{25}{24}$$

$$\csc \theta = \frac{1}{\sin \theta} = \frac{1}{7/25} = \frac{25}{7}$$

■

EXAMPLE 3   *Using identities to find trigonometric values*

If $\tan \theta = -\dfrac{8}{15}$ and $\cos \theta = -\dfrac{15}{17}$, find the other four trigonometric functions of θ.

Solution

Using the reciprocal identities we can find $\cot\theta$ and $\sec\theta$.

$$\cot\theta = \frac{1}{\tan\theta} = \frac{1}{-8/15} = -\frac{15}{8}$$

$$\sec\theta = \frac{1}{\cos\theta} = \frac{1}{-15/17} = -\frac{17}{15}$$

To find $\sin\theta$, consider the following.

$$\tan\theta\cos\theta = \frac{\sin\theta}{\cos\theta}\cdot\cos\theta = \sin\theta$$

Thus,

$$\sin\theta = \tan\theta\cos\theta = \left(-\frac{8}{15}\right)\left(-\frac{15}{17}\right) = \frac{8}{17}$$

and using a reciprocal identity

$$\csc\theta = \frac{1}{\sin\theta} = \frac{1}{8/17} = \frac{17}{8}.$$ ■

Pythagorean Identities

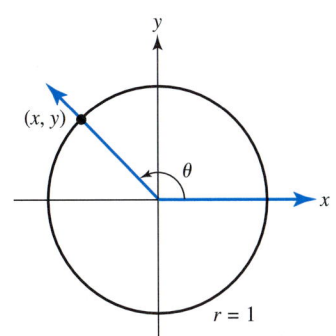

FIGURE 6.6

In Section 5.3 the unit circle, shown in Figure 6.6, was used to define the sine and cosine functions.

$$\sin\theta = y \qquad \text{and} \qquad \cos\theta = x$$

An equation for the unit circle is given by $x^2 + y^2 = 1$. By substitution it follows that

$$(\cos\theta)^2 + (\sin\theta)^2 = 1$$

or equivalently,

$$\cos^2\theta + \sin^2\theta = 1.$$

This identity can be supported graphically and numerically by letting

$$Y_1 = (\cos(X))^2 + (\sin(X))^2$$

as shown in Figures 6.7 and 6.8. Either radian or degree mode may be used.

$[-352.5°, 352.5°, 90°]$ by $[-2, 2, 1]$

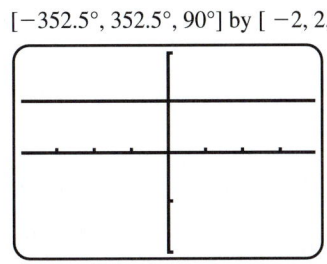

FIGURE 6.7

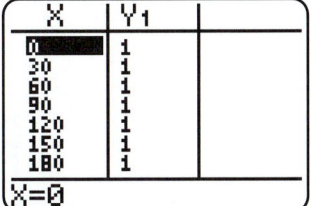

FIGURE 6.8 Degree Mode

```
cos (30)^2
                    -1
(cos (30))^2
                   .75
```

FIGURE 6.9 Degree Mode

Technology Note

On some graphing calculators $\cos^2 x$ must be entered as $(\cos(X))^\wedge 2$, rather than $\cos(X)^\wedge 2$, since $\cos(X)^\wedge 2$ may be interpreted as $\cos(x^2)$. See Figure 6.9.

We also can verify the identity $\cos^2\theta + \sin^2\theta = 1$ geometrically when θ is acute using the right triangle shown in Figure 6.10.

$$x^2 + y^2 = r^2 \qquad \text{Pythagorean theorem}$$

$$\frac{x^2}{r^2} + \frac{y^2}{r^2} = \frac{r^2}{r^2} \qquad \text{Divide both sides by } r^2$$

$$\left(\frac{x}{r}\right)^2 + \left(\frac{y}{r}\right)^2 = 1 \qquad \text{Properties of exponents}$$

$$\cos^2\theta + \sin^2\theta = 1 \qquad \cos\theta = \frac{x}{r}, \sin\theta = \frac{y}{r}$$

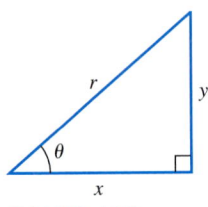

FIGURE 6.10

For this reason, $\sin^2\theta + \cos^2\theta = 1$ is an example of a Pythagorean identity. There are two other Pythagorean identities that can be derived as follows.

$$\sin^2\theta + \cos^2\theta = 1$$

$$\frac{\sin^2\theta}{\cos^2\theta} + \frac{\cos^2\theta}{\cos^2\theta} = \frac{1}{\cos^2\theta} \qquad \text{Divide by } \cos^2\theta.$$

$$\tan^2\theta + 1 = \sec^2\theta \qquad \tan\theta = \frac{\sin\theta}{\cos\theta}, \sec\theta = \frac{1}{\cos\theta}$$

In a similar manner, a third Pythagorean identity can be found.

$$\sin^2\theta + \cos^2\theta = 1$$

$$\frac{\sin^2\theta}{\sin^2\theta} + \frac{\cos^2\theta}{\sin^2\theta} = \frac{1}{\sin^2\theta} \qquad \text{Divide by } \sin^2\theta.$$

$$1 + \cot^2\theta = \csc^2\theta \qquad \cot\theta = \frac{\cos\theta}{\sin\theta}, \csc\theta = \frac{1}{\sin\theta}$$

Critical Thinking

Use the triangle in Figure 6.10 to justify the Pythagorean identities

$$1 + \tan^2\theta = \sec^2\theta \qquad \text{and} \qquad 1 + \cot^2\theta = \csc^2\theta,$$

when θ is acute.

This discussion is summarized in the following.

Pythagorean identities

$$\sin^2\theta + \cos^2\theta = 1 \qquad 1 + \tan^2\theta = \sec^2\theta \qquad 1 + \cot^2\theta = \csc^2\theta$$

The next example illustrates how identities are used in electronic technology.

EXAMPLE 4   *Applying a Pythagorean identity to radios*

Tuners in radios are used to select the radio station played by adjusting the frequency. These tuners may contain an inductor L and a capacitor C as illustrated in Figure 6.11. The energy stored in the inductor is given by $L(t) = k \sin^2(2\pi Ft)$ and the energy stored in the capacitor is given by $C(t) = k \cos^2(2\pi Ft)$, where F is the frequency of the radio station and k is a constant. The total energy E in the circuit is given by $E(t) = L(t) + C(t)$. Show that E is a constant function. (Source: R. Weidner, and R. Sells, *Elementary Classical Physics,* Vol. 2.)

FIGURE 6.11 An Inductor and a Capacitor

Solution

$$E(t) = L(t) + C(t) \qquad \text{Given equation}$$

$$= k \sin^2(2\pi Ft) + k \cos^2(2\pi Ft) \qquad \text{Substitute.}$$

$$= k \left(\sin^2(2\pi Ft) + \cos^2(2\pi Ft) \right) \qquad \text{Factor.}$$

$$= k\,(1) \qquad \sin^2\theta + \cos^2\theta = 1 \ (\theta = 2\pi Ft)$$

$$= k \qquad k \text{ is constant.} \qquad \blacksquare$$

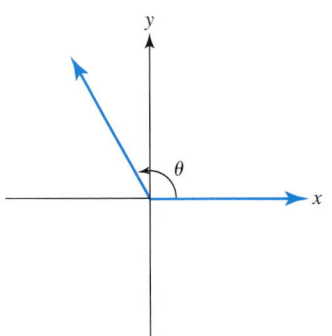

FIGURE 6.12 Angle θ in Quadrant II

If an angle θ is in standard position and its terminal side lies in quadrant II, as shown in Figure 6.12, then θ is *contained in quadrant II* or θ is a *second quadrant angle*. Similar statements can be made for angles whose terminal sides lie in other quadrants.

Let the point (x, y) lie on the terminal side of θ with $r = \sqrt{x^2 + y^2}$. Then $\sin \theta = \dfrac{y}{r}$ is positive when $y > 0$ and negative when $y < 0$. As a result, $\sin \theta$ is positive for first and second quadrant angles and negative for third and fourth quadrant angles. In Figure 6.13 the signs of the six trigonometric functions in each quadrant are listed. Notice that the six trigonometric functions are all positive in the first quadrant.

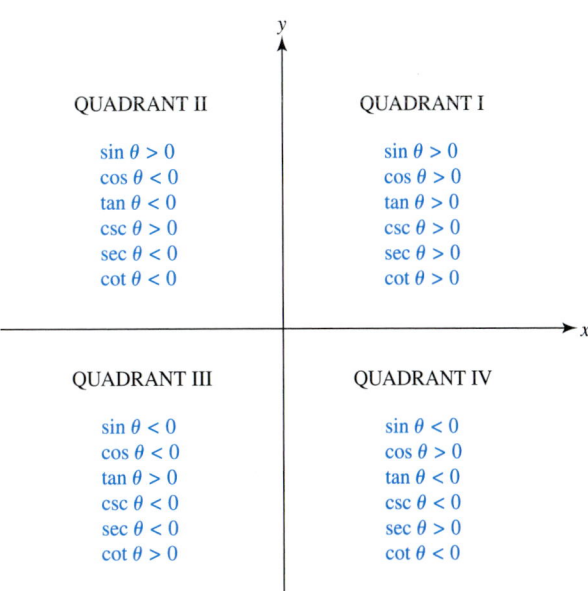

QUADRANT II	QUADRANT I
$\sin \theta > 0$	$\sin \theta > 0$
$\cos \theta < 0$	$\cos \theta > 0$
$\tan \theta < 0$	$\tan \theta > 0$
$\csc \theta > 0$	$\csc \theta > 0$
$\sec \theta < 0$	$\sec \theta > 0$
$\cot \theta < 0$	$\cot \theta > 0$
QUADRANT III	QUADRANT IV
$\sin \theta < 0$	$\sin \theta < 0$
$\cos \theta < 0$	$\cos \theta > 0$
$\tan \theta > 0$	$\tan \theta < 0$
$\csc \theta < 0$	$\csc \theta < 0$
$\sec \theta < 0$	$\sec \theta > 0$
$\cot \theta > 0$	$\cot \theta < 0$

FIGURE 6.13

EXAMPLE 5   *Finding the quadrant containing an angle*

If $\sin \theta > 0$ and $\cos \theta < 0$, find the quadrant containing θ. Support your results graphically and numerically.

Solution

If $\sin \theta > 0$, $\cos \theta < 0$ then any point (x, y) on the terminal side of θ must satisfy $y > 0$ and $x < 0$. Thus, θ is contained in quadrant II.

Graphical Support   In Figure 6.14 the graphs of $Y_1 = \sin(X)$ and $Y_2 = \cos(X)$ are shown. Notice that when $\theta = 135°$ (a second quadrant angle) the graph of $\sin \theta$ is above the x-axis and the graph of $\cos \theta$ is below the x-axis.

Numerical Support   In Figure 6.15 numerical support is given, where angles in quadrant II have positive sine values and negative cosine values.

[0, 352.5°, 90°] by [−2, 2, 1]

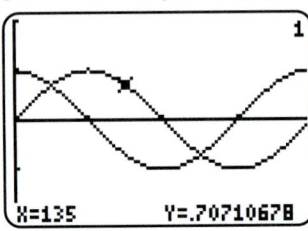

FIGURE 6.14

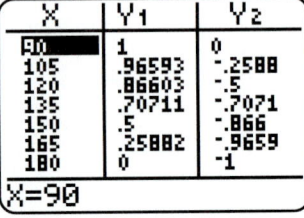

FIGURE 6.15 Degree Mode   ∎

If a trigonometric function of θ is known and the quadrant containing θ is also known, then we can find the other five trigonometric functions of θ as illustrated in the next two examples.

EXAMPLE 6   *Using identities to find trigonometric values*

If $\sin \theta = -\dfrac{3}{5}$ and θ is a third quadrant angle, find the values of the other trigonometric functions.

Solution

Start by finding $\cos \theta$.

$$\sin^2 \theta + \cos^2 \theta = 1 \qquad\qquad \text{Pythagorean identity}$$

$$\cos^2 \theta = 1 - \sin^2 \theta \qquad\qquad \text{Subtract } \sin^2 \theta.$$

$$\cos \theta = \pm\sqrt{1 - \sin^2 \theta} \qquad\qquad \text{Square root property}$$

$$\cos \theta = \pm\sqrt{1 - (-3/5)^2} \qquad\qquad \text{Substitute for } \sin \theta.$$

$$\cos \theta = \pm\frac{4}{5} \qquad\qquad \text{Simplify.}$$

In quadrant III, $x < 0$ and so $\cos \theta = -\dfrac{4}{5}$. The other four trigonometric functions of θ can be found using the quotient identities.

$$\tan \theta = \frac{\sin \theta}{\cos \theta} = \frac{-3/5}{-4/5} = \frac{3}{4}$$

$$\cot \theta = \frac{\cos \theta}{\sin \theta} = \frac{-4/5}{-3/5} = \frac{4}{3}$$

$$\sec\theta = \frac{1}{\cos\theta} = \frac{1}{-4/5} = -\frac{5}{4}$$

$$\csc\theta = \frac{1}{\sin\theta} = \frac{1}{-3/5} = -\frac{5}{3}$$ ∎

EXAMPLE 7   *Using identities to find trigonometric values*

If $\tan\theta = -\dfrac{7}{3}$ and $\sin\theta > 0$, find the values of the other trigonometric functions.

Solution

Since $\tan\theta = -\dfrac{7}{3}$, it follows that $\cot\theta = -\dfrac{3}{7}$. Next we find $\sec\theta$.

$$\sec^2\theta = 1 + \tan^2\theta \qquad \text{Pythagorean identity}$$

$$\sec\theta = \pm\sqrt{1 + \tan^2\theta} \qquad \text{Square root property}$$

$$= \pm\sqrt{1 + (-7/3)^2} \qquad \text{Substitute for } \tan\theta$$

$$= \pm\frac{\sqrt{58}}{3} \qquad \text{Simplify.}$$

Since $\tan\theta < 0$ and $\sin\theta > 0$, θ is a second quadrant angle. It follows that

$$\sec\theta = -\frac{\sqrt{58}}{3} \quad \text{and} \quad \cos\theta = -\frac{3}{\sqrt{58}}.$$

Since $\sin\theta = \tan\theta \cos\theta$,

$$\sin\theta = \left(-\frac{7}{3}\right)\left(-\frac{3}{\sqrt{58}}\right) = \frac{7}{\sqrt{58}}.$$

Using a reciprocal identity

$$\csc\theta = \frac{1}{\sin\theta} = \frac{\sqrt{58}}{7}.$$ ∎

In the next example we write one trigonometric function in terms of another.

EXAMPLE 8   *Using fundamental identities to find trigonometric expressions*

If $\sin\theta = x$ and θ is a quadrant IV angle, find an expression for $\sec\theta$. Approximate $\sec\theta$ if $\sin\theta = -0.7813$.

Solution

We begin by writing $\sec\theta$ in terms of $\sin\theta$.

$$\sin^2\theta + \cos^2\theta = 1 \qquad \text{Pythagorean identity}$$

$$\cos^2\theta = 1 - \sin^2\theta \qquad \text{Subtract } \sin^2\theta.$$

$$\cos\theta = \pm\sqrt{1 - \sin^2\theta} \qquad \text{Square root property}$$

$$\sec\theta = \pm\frac{1}{\sqrt{1 - \sin^2\theta}} \qquad \sec\theta = \frac{1}{\cos\theta}$$

$$\sec\theta = \pm\frac{1}{\sqrt{1 - x^2}} \qquad \sin\theta = x$$

Angle θ is in quadrant IV. Thus, $\sec\theta > 0$ and

$$\sec\theta = \frac{1}{\sqrt{1-x^2}}.$$

Since $\sin\theta = x$, let $x = -0.7813$. Then,

$$\sec\theta = \frac{1}{\sqrt{1-(-0.7813)^2}} \approx 1.602. \qquad \blacksquare$$

Negative-Angle Identities

In Section 3.2 we discussed odd and even functions. The graphs of odd functions are symmetric with respect to the origin and the graphs of even functions are symmetric to the *y*-axis. Symbolically, we say that an odd function *f* satisfies

$$f(-x) = -f(x)$$

for all *x* in the domain of *f*. Similarly, an even function *f* satisfies

$$f(-x) = f(x)$$

for all *x* in its domain. The graphs of all six trigonometric functions have symmetry as shown in Figures 6.16–6.21.

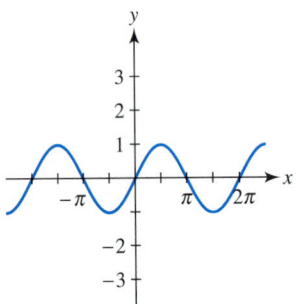

FIGURE 6.16 The Sine Function (An Odd Function)

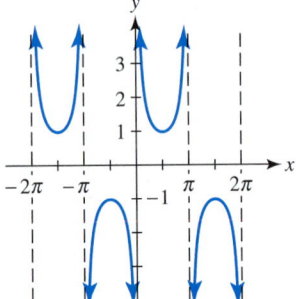

FIGURE 6.17 The Cosecant Function (An Odd Function)

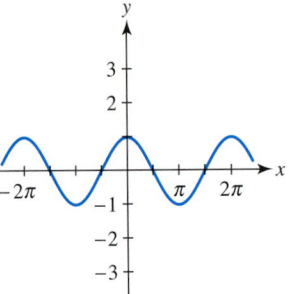

FIGURE 6.18 The Cosine Function (An Even Function)

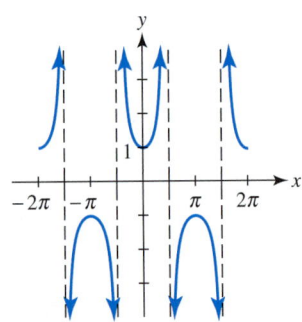

FIGURE 6.19 The Secant Function (An Even Function)

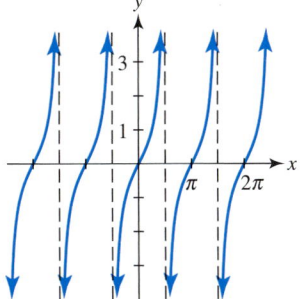

FIGURE 6.20 The Tangent Function (An Odd Function)

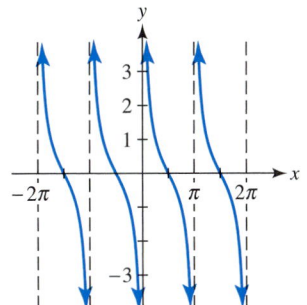

FIGURE 6.21 The Cotangent Function (An Odd Function)

The cosine and secant functions are even functions, while the other four trigonometric functions are odd functions. Numerical support is shown in Figures 6.22–6.27, where degree mode has been used.

X	Y1
⁻60	⁻.866
⁻40	⁻.6428
⁻20	⁻.342
0	0
20	.34202
40	.64279
60	.86603

Y₁**∎**sin(X)

FIGURE 6.22 Sine Function

X	Y1
⁻60	.5
⁻40	.76604
⁻20	.93969
0	1
20	.93969
40	.76604
60	.5

Y₁**∎**cos(X)

FIGURE 6.23 Cosine Function

X	Y1
⁻60	⁻1.732
⁻40	⁻.8391
⁻20	⁻.364
0	0
20	.36397
40	.8391
60	1.7321

Y₁**∎**tan(X)

FIGURE 6.24 Tangent Function

X	Y1
⁻60	⁻1.155
⁻40	⁻1.556
⁻20	⁻2.924
0	ERROR
20	2.9238
40	1.5557
60	1.1547

Y₁**∎**1/sin(X)

FIGURE 6.25 Cosecant Function

X	Y1
⁻60	2
⁻40	1.3054
⁻20	1.0642
0	1
20	1.0642
40	1.3054
60	2

Y₁**∎**1/cos(X)

FIGURE 6.26 Secant Function

X	Y1
⁻60	⁻.5774
⁻40	⁻1.192
⁻20	⁻2.747
0	ERROR
20	2.7475
40	1.1918
60	.57735

Y₁**∎**1/tan(X)

FIGURE 6.27 Cotangent Function

Notice that for even functions the sign of the input does not affect the output, whereas for an odd function changing the sign of the input only changes the sign of the output. (Slight differences are due to rounding by the calculator.) These results can be expressed symbolically using the negative-angle identities.

Negative-angle identities

$$\sin(-\theta) = -\sin\theta \qquad \cos(-\theta) = \cos\theta \qquad \tan(-\theta) = -\tan\theta$$

$$\csc(-\theta) = -\csc\theta \qquad \sec(-\theta) = \sec\theta \qquad \cot(-\theta) = -\cot\theta$$

EXAMPLE 9   *Modeling temperature*

The monthly average high temperatures in degrees Fahrenheit at Chattanooga, Tennessee can be modeled by

$$f(x) = 21 \cos\left(\frac{\pi x}{6}\right) + 70,$$

where x is the month with $x = -6$ corresponding to January, $x = 0$ to July, and $x = 5$ to December. (Source: J. Williams, *The Weather Almanac*.)

(a) Graph f in $[-6, 6, 1]$ by $[40, 100, 10]$. Interpret symmetry in the graph.
(b) Table f and discuss whether f is an even or odd function.
(c) Express this symmetry symbolically.

Solution

(a) Graph $Y_1 = 21 \cos(\pi X/6) + 70$ as shown in Figure 6.28 in radian mode. The graph is symmetric with respect to the y-axis. This type of symmetry implies that the monthly average high temperatures x months before July or x months after July are equal. For example, April ($x = -3$), which is three months before July, and October ($x = 3$), which is three months after July, both have the same average high temperature of 70°.

$[-6, 6, 1]$ by $[40, 100, 10]$

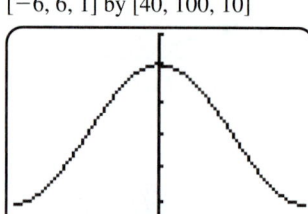

FIGURE 6.28

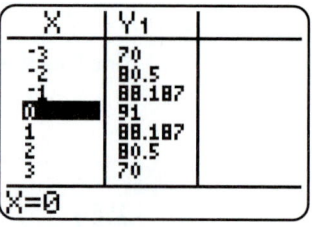

FIGURE 6.29 Radian Mode

(b) Table Y_1 as shown in Figure 6.29. Notice that f is an even function since the sign of the input does not affect the output.
(c) Symbolically this can be expressed as $f(-x) = f(x)$. ∎

6.1 PUTTING IT ALL TOGETHER

In this section we discussed the reciprocal identities, the quotient identities, the Pythagorean identities, and the negative-angle identities. Collectively, they are called fundamental identities. If $\sin\theta$ and $\cos\theta$ are known, then the quotient identities may be used to find the remaining four trigonometric functions of θ. The Pythagorean identities may be used to find trigonometric values when only one trigonometric value is known. The negative-angle identities may be used to determine if a trigonometric function is an odd function or an even function. The cosine and secant are even functions, while the sine, cosecant, tangent, and cotangent are odd functions.

Reciprocal Identities		
$\sin\theta = \dfrac{1}{\csc\theta}$	$\cos\theta = \dfrac{1}{\sec\theta}$	$\tan\theta = \dfrac{1}{\cot\theta}$
$\csc\theta = \dfrac{1}{\sin\theta}$	$\sec\theta = \dfrac{1}{\cos\theta}$	$\cot\theta = \dfrac{1}{\tan\theta}$

Quotient Identities			
$\tan\theta = \dfrac{\sin\theta}{\cos\theta}$	$\cot\theta = \dfrac{\cos\theta}{\sin\theta}$	$\sec\theta = \dfrac{1}{\cos\theta}$	$\csc\theta = \dfrac{1}{\sin\theta}$

Pythagorean Identities		
$\sin^2\theta + \cos^2\theta = 1$	$1 + \tan^2\theta = \sec^2\theta$	$1 + \cot^2\theta = \csc^2\theta$

Negative-Angle Identities		
$\sin(-\theta) = -\sin\theta$	$\cos(-\theta) = \cos\theta$	$\tan(-\theta) = -\tan\theta$
$\csc(-\theta) = -\csc\theta$	$\sec(-\theta) = \sec\theta$	$\cot(-\theta) = -\cot\theta$

6.1 EXERCISES   Tape 10

Fundamental Identities

Exercises 1–6: Use a reciprocal identity to find the indicated trigonometric function of θ.

1. $\cot\theta$ if $\tan\theta = \dfrac{1}{2}$

2. $\csc\theta$ if $\sin\theta = -\dfrac{5}{6}$

3. $\sec\theta$ if $\cos\theta = \dfrac{2}{7}$

4. $\tan\theta$ if $\cot\theta = -\dfrac{3}{7}$

5. $\cos\theta$ if $\sec\theta = -4$

6. $\sin\theta$ if $\csc\theta = 7$

Exercises 7–12: (Refer to Example 5.) Determine the quadrant containing θ. Support your result graphically or numerically.

7. $\sin\theta < 0$ and $\cos\theta > 0$

8. $\tan\theta > 0$ and $\cos\theta < 0$

9. $\sec\theta < 0$ and $\sin\theta < 0$

10. $\csc\theta > 0$ and $\tan\theta > 0$

11. $\cot\theta < 0$ and $\sin\theta > 0$

12. $\cos\theta > 0$ and $\cot\theta < 0$

Exercises 13–22: Find the other trigonometric functions of θ.

13. $\sin\theta = \dfrac{3}{5}$ and $\cos\theta = -\dfrac{4}{5}$

14. $\tan\theta = -\dfrac{12}{5}$ and $\cos\theta = \dfrac{5}{13}$

15. $\cot\theta = \dfrac{7}{24}$ and $\sin\theta = -\dfrac{24}{25}$

16. $\sin\theta = \dfrac{12}{13}$ and $\cos\theta = \dfrac{5}{13}$

17. $\cos\theta = \dfrac{1}{2}$ and $\sin\theta < 0$

18. $\csc\theta = \sqrt{3}$ and $\cos\theta < 0$

19. $\sec\theta = 3$ and θ in Quadrant IV

20. $\tan\theta = \dfrac{3}{4}$ and θ in Quadrant I

21. $\csc\theta = \dfrac{7}{3}$ and θ in Quadrant II

22. $\cos\theta = -\dfrac{3}{5}$ and θ in Quadrant III

Exercises 23–28: Use a negative-angle identity to write an equivalent trigonometric expression involving a positive angle.

23. $\sin(-13°)$

24. $\cos\left(-\dfrac{\pi}{7}\right)$

25. $\tan\left(-\dfrac{\pi}{11}\right)$

26. $\cot(-75°)$

27. $\sec\left(-\dfrac{2\pi}{5}\right)$

28. $\csc(-160°)$

Exercises 29–32: Conjecture the result when the final expression is evaluated. Explain your reasoning.

29.
```
cos(X)
cos(-X)
              .5
```

30.
```
-tan(X)
tan(-X)
              -1
```

31.
```
(tan(X))^2
1/(cos(X))^2
              3
```

32.
```
(sin(X))^2
1-(cos(X))^2
              .75
```

Exercises 33–36: Conjecture if the equation represents an identity. Support your answer by tabling the left and right sides of the equation.

33. $\sin^2\theta + \cos^2\theta + \tan^2\theta = \sec^2\theta$

34. $\sec\theta\cot\theta = \csc\theta$

35. $(\sin\theta + \cos\theta)^2 = 1$

36. $\cot^2\theta - \csc^2\theta = 1$

Exercises 37–40: The partial numerical representation suggests a fundamental identity. Conjecture this identity and explain your reasoning.

37.

x	0°	30°	60°	90°	120°	150°
$\tan^2 x$	0	1/3	3	Error	3	1/3
$\sec^2 x$	1	4/3	4	Error	4	4/3

38.

x	0°	30°	60°	90°	120°	150°
$\sin^2 x$	0	1/4	3/4	1	3/4	1/4
$\cos^2 x$	1	3/4	1/4	0	1/4	3/4

39.

x	0°	60°	120°	180°	240°	300°
$\cos x$	1	1/2	$-1/2$	-1	$-1/2$	1/2
$\sec x$	1	2	-2	-1	-2	2

40.

x	30°	90°	150°	210°	270°	330°
$\sin x$	1/2	1	1/2	$-1/2$	-1	$-1/2$
$\sin(-x)$	$-1/2$	-1	$-1/2$	1/2	1	1/2

41. Consider the table of values for $\tan\theta$ and $\cot\theta$. Does the equation $\tan\theta = \cot\theta$ represent an identity? Explain your answer.

x	45°	135°	225°	315°	405°
$\tan x$	1	-1	1	-1	1
$\cot x$	1	-1	1	-1	1

42. Consider the table of values for $\sec^2 \theta$ and $\csc^2 \theta$. Does the equation $\sec^2 \theta = \csc^2 \theta$ represent an identity? Explain your answer.

x	45°	135°	225°	315°	405°
$\sec^2 x$	2	2	2	2	2
$\csc^2 x$	2	2	2	2	2

43. Table $Y_1 = \sin(X)$ and $Y_2 = -\sin(X)$ in degree mode, starting at $x = 0°$, incrementing by 180°. Does the equation $\sin x = -\sin x$ represent an identity? Explain your reasoning.

44. Table $Y_1 = \sin(X)$ and $Y_2 = \cos(X)$ in degree mode, starting at $x = 45°$, incrementing by 180°. Does the equation $\sin x = \cos x$ represent an identity? Explain your reasoning.

45. A student writes "$\cos^2 + \sin^2 = 1$." Comment on the correctness of this expression.

46. Since $\sec^2 \theta = 1 + \tan^2 \theta$, does it follow that $\sec \theta = 1 + \tan \theta$? Explain your reasoning.

Exercises 47–52: Simplify each expression using fundamental identities. Give numerical support to your result by letting Y_1 be the given expression and Y_2 be the simplified expression.

47. $\sec \theta \cos \theta$

48. $\tan \theta \cot \theta$

49. $\sin \theta \csc \theta$

50. $\tan \theta \cos \theta$

51. $(\sin^2 \theta + \cos^2 \theta)^3$

52. $(1 + \tan^2 \theta) \cos^2 \theta$

Exercises 53–56: (Refer to Example 8.) Write the trigonometric function in terms of x. Then evaluate this trigonometric function if $x = 0.5126$.

53. $\cos \theta$, if $\sin \theta = x$ and θ is acute

54. $\sec \theta$, if $\cos \theta = x$

55. $\sin \theta$, if $\cot \theta = x$ and θ is in Quadrant III

56. $\tan \theta$, if $\cos \theta = x$ and θ is in Quadrant IV

Discovery

Exercises 57–62: Products of Odd and Even Functions Complete the following.

   (a) *Using radian mode, graph the product $h(x) = f(x)g(x)$.*

   (b) *Use the graph to decide if h is an odd or even function.*

   (c) *Conjecture a generalization for products of odd and even functions. Compare this to products of odd and even integers.*

57. $f(x) = \cos x$ (even), $g(x) = \sin x$ (odd)

58. $f(x) = x$ (odd), $g(x) = \cos x$ (even)

59. $f(x) = \sqrt[3]{x}$ (odd), $g(x) = \sin x$ (odd)

60. $f(x) = \tan x$ (odd), $g(x) = \sin x$ (odd)

61. $f(x) = x^2$ (even), $g(x) = \cos x$ (even)

62. $f(x) = \cos x$ (even), $g(x) = \sec x$ (even)

Applications

63. *Distance to the Stars* In Example 5, Section 5.2 the distance d to a star was found using

$$d = \frac{93{,}000{,}000}{\sin \theta},$$

where θ is the parallax of the star. Use a reciprocal identity to rewrite this formula.

64. *Height of a Building* If the angle of elevation of the sun is θ, then a building 40 feet high will cast a shadow x feet long, where $x = \dfrac{40}{\tan \theta}$. Use a reciprocal identity to rewrite this formula.

65. *Oscillating Spring* The distance or displacement y of a weight attached to an oscillating spring from its natural position is modeled by $y = 4\cos(2\pi t)$, where t is in seconds. See the accompanying figure. Potential energy is the energy of position and is given by $P = ky^2$, where k is a constant. The weight has the greatest potential energy when the spring is stretched the most. (Source: R. Weidner and R. Sells, *Elementary Classical Physics,* Vol. 1.)

   (a) Write an expression for P involving the cosine function.

   (b) Let $k = 2$ and graph P in [0, 2, 0.5] by [−1, 40, 8]. For $0 \le t \le 2$, at what times is P maximum and at what times is P minimum? Interpret your result.

   (c) Use a fundamental identity to write P in terms of the sine function.

66. *Energy in an Oscillating Spring* (Refer to the previous exercise.) Two types of mechanical energy are kinetic energy and potential energy. Kinetic energy is the energy of motion and potential energy is the energy of position. A stretched spring has potential energy, which is converted to kinetic energy when it is released. If the potential energy of a weight attached to a spring is $P(t) = k \cos^2 (4\pi t)$, where k is a constant and t is in seconds, then its kinetic energy is given by $K(t) = k \sin^2 (4\pi t)$. The total mechanical energy E is given by

$$E(t) = P(t) + K(t).$$

(a) If $k = 2$ graph P, K, and E in $[0, 0.5, 0.25]$ by $[-1, 3, 1]$. Interpret the graph.
(b) Table K, P, and E starting at $t = 0$, incrementing by 0.05. Interpret the results.
(c) Use a fundamental identity to derive a simplified expression for $E(t)$.

67. *Radio Tuners* (Refer to Example 4.) Let the energy stored in the inductor be given by $L(t) = 3 \cos^2 (6,000,000t)$ and the energy in the capacitor be given by $C(t) = 3 \sin^2 (6,000,000t)$, where t is time in seconds. The total energy E in the circuit is given by $E(t) = L(t) + C(t)$.
(a) Graph L, C, and E in $[0, 10^{-6}, 10^{-7}]$ by $[-1, 4, 1]$. Interpret the graph.
(b) Table L, C, and E starting at $t = 0$, incrementing by 10^{-7}. Interpret your results.
(c) Use a fundamental identity to derive a simplified expression for $E(t)$.

68. *Intensity of a Lamp* According to Lambert's law, the intensity of light from a single source on a flat surface at point P is given by $I = k \cos^2 \theta$, where k is a constant. See the accompanying figure. (Source: C. Winter, *Solar Power Plants*.)
(a) Let $k = 1$ and graph I in $[-90°, 90°, 45°]$ by $[-1, 2, 1]$. For what value of θ is I maximum?
(b) Write I in terms of the sine function.

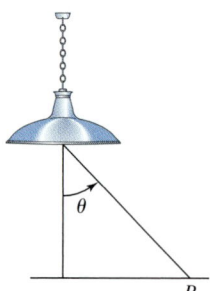

Writing about Mathematics

1. Explain in your own words what a trigonometric identity is. Give two examples with numerical and graphical support.

2. Answer each of the following.
(a) Give two characteristics of an even function. Which of the trigonometric functions are even?
(b) Give two characteristics of an odd function. Which of the trigonometric functions are odd?

6.2 Verifying Identities

Simplifying Trigonometric Expressions • Symbolic Verification with Graphical and Numerical Support • Using Right Triangles to Justify Trigonometric Identities (Optional)

Introduction

In Example 4 of the previous section we used the identity $\sin^2 \theta + \cos^2 \theta = 1$ to show that the energy stored in a particular type of electrical circuit is constant. In Example 7, Section 5.2 the identity

$$\tan \theta = \frac{\sin \theta}{\cos \theta}$$

was used to derive the external distance d for a highway curve.

Trigonometric identities are used both in applications and calculus. Before we can use an identity, we must verify that it is correct. Although the equality of two trigonometric expressions can be *supported* graphically and numerically, symbolic *verification* is necessary to be certain that an equation is indeed an identity.

Simplifying Trigonometric Expressions

Many of the algebraic skills that you have already learned can be used to simplify trigonometric expressions. For example, suppose we would like to multiply the following expression.

$$(1 - \cos\theta)(1 + \cos\theta)$$

In algebra we learned that

$$(1 - x)(1 + x) = 1 + x - x - x^2$$
$$= 1 - x^2.$$

If we substitute $\cos\theta$ for x then

$$(1 - \cos\theta)(1 + \cos\theta) = 1 + \cos\theta - \cos\theta - \cos^2\theta$$
$$= 1 - \cos^2\theta.$$

In algebra we do not simplify $1 - x^2$ further. However, since $\sin^2\theta + \cos^2\theta = 1$, it follows that $\sin^2\theta = 1 - \cos^2\theta$. As a result,

$$(1 - \cos\theta)(1 + \cos\theta) = \sin^2\theta.$$

If we let $Y_1 = (1 - \cos(X))(1 + \cos(X))$ and $Y_2 = (\sin(X))^2$, we can support this result graphically as shown in Figures 6.30 and 6.31. Either radian or degree mode may be used. Notice that the graphs of Y_1 and Y_2 appear to be identical. Numerical support is shown in Figure 6.32, where $Y_1 = Y_2$ for each value of x listed.

$[-2\pi, 2\pi, \pi/2]$ by $[-2, 2, 1]$ $[-2\pi, 2\pi, \pi/2]$ by $[-2, 2, 1]$

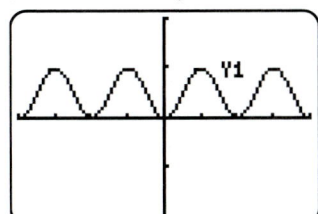 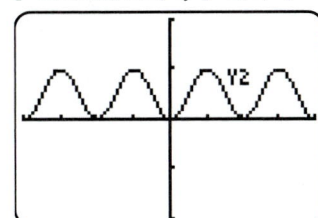

FIGURE 6.30   **FIGURE 6.31**   **FIGURE 6.32** Radian Mode

MAKING CONNECTIONS
Algebraic and Trigonometric Expressions

Many of the techniques we use to simplify algebraic expressions can also be used to simplify trigonometric expressions. Here are some examples.

$\tan^2\theta - 4$

     $= (\tan\theta - 2)(\tan\theta + 2)$ is similar to $x^2 - 4 = (x - 2)(x + 2)$.

$\cos\theta(\sin\theta + \cos\theta)$

     $= \cos\theta\sin\theta + \cos^2\theta$ is similar to $x(y + x) = xy + x^2$.

$\dfrac{\sin\theta}{\cos\theta} + \dfrac{1}{\cos\theta} = \dfrac{\sin\theta + 1}{\cos\theta}$ is similar to $\dfrac{y}{x} + \dfrac{1}{x} = \dfrac{y + 1}{x}$.

The next example illustrates the addition of two trigonometric expressions.

EXAMPLE 1   *Adding two trigonometric expressions*

Write $\tan t + \cot t$ as a product of two trigonometric functions.

Solution

We begin by writing $\tan t$ and $\cot t$ as ratios involving $\sin t$ and $\cos t$.

$$\tan t + \cot t = \frac{\sin t}{\cos t} + \frac{\cos t}{\sin t}$$

In algebra we combine $\frac{y}{x} + \frac{x}{y}$ using the common denominator xy as follows.

$$\frac{y}{x} + \frac{x}{y} = \frac{y}{x} \cdot \frac{y}{y} + \frac{x}{y} \cdot \frac{x}{x} \qquad \text{Multiply each ratio by 1.}$$

$$= \frac{y^2}{xy} + \frac{x^2}{xy} \qquad \text{Simplify.}$$

$$= \frac{y^2 + x^2}{xy} \qquad \text{Add.}$$

Now substitute $\cos t$ for x and $\sin t$ for y.

$$\tan t + \cot t = \frac{\sin t}{\cos t} + \frac{\cos t}{\sin t} \qquad \text{Quotient identities}$$

$$= \frac{\sin t}{\cos t} \cdot \frac{\sin t}{\sin t} + \frac{\cos t}{\sin t} \cdot \frac{\cos t}{\cos t} \qquad \text{Multiply each ratio by 1.}$$

$$= \frac{\sin^2 t}{\cos t \sin t} + \frac{\cos^2 t}{\cos t \sin t} \qquad \text{Simplify.}$$

$$= \frac{\sin^2 t + \cos^2 t}{\cos t \sin t} \qquad \text{Add.}$$

$$= \frac{1}{\cos t \sin t} \qquad \sin^2 t + \cos^2 t = 1$$

$$= \sec t \csc t \qquad \sec t = \frac{1}{\cos t}; \csc t = \frac{1}{\sin t}$$

Thus, $\tan t + \cot t$ is equivalent to $\sec t \csc t$. ∎

Note: To simplify an expression, it is *not* necessary to first write the equation using x and y as was done in Example 1. However, sometimes you may find this technique helpful.

EXAMPLE 2   *Factoring a trigonometric expression*

Factor each expression.
(a) $\sec^2 \theta - 1$
(b) $2 \sin^2 t + \sin t - 1$

Solution

(a) In algebra we factor $x^2 - 1$ as $(x - 1)(x + 1)$. This can be applied to the given expression.

$$\sec^2 \theta - 1 = (\sec \theta - 1)(\sec \theta + 1)$$

(b) Since $2y^2 + y - 1$ can be factored as $(2y - 1)(y + 1)$, it follows that

$$2 \sin^2 t + \sin t - 1 = (2 \sin t - 1)(\sin t + 1). \quad ■$$

Trigonometric expressions are used in applications involving electricity. This is illustrated in the next example.

EXAMPLE 3   *Analyzing electromagnets*

Electromagnets are used in a variety of situations, such as lifting scrap metal, ringing door bells, and opening door locks in apartments. Let the wattage W consumed by an electromagnet at t seconds be

$$W(t) = 100 \sin^2 (120 \pi t),$$

and the voltage V in the circuit be

$$V(t) = 160 \cos (120 \pi t).$$

(Source: A. Howatson, *Electrical Circuits and Systems.*)
(a) Express $W(t)$ in terms of the cosine function. When V is maximum or minimum, what is the value of W? Explain.
(b) Support your answer in part (a) by graphing W and V in [0, 1/15, 1/60] by [-180, 180, 20].

[0, 1/30, 1/60] by [-180, 180, 20]

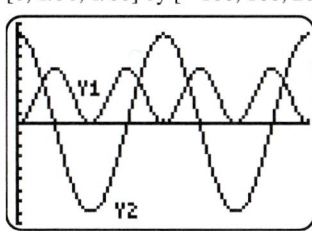

FIGURE 6.33

Solution

(a) Let $\theta = 120 \pi t$. Since $\sin^2 \theta = 1 - \cos^2 \theta$, W can be expressed as

$$W(t) = 100(1 - \cos^2 (120 \pi t)).$$

Since $V(t) = 160 \cos (120 \pi t)$, the voltage is maximum (160) or minimum (-160), whenever $\cos (120 \pi t) = \pm 1$. When $\cos (120 \pi t) = \pm 1$, it follows that $W(t) = 100(1 - (\pm 1)^2) = 0$.
(b) Graph $Y_1 = 100(\sin (120 \pi X))\text{\textasciicircum}2$ and $Y_2 = 160 \cos (120 \pi X)$ as shown in Figure 6.33. The wattage Y_1 is 0 whenever the voltage Y_2 is ± 160.   ∎

Symbolic Verification with Graphical and Numerical Support

When verifying that an equation is an identity, we usually begin with one side of the equation and write a sequence of equivalent expressions until it is transformed into the other side. This is illustrated in the next example.

EXAMPLE 4   *Verifying an identity*

Verify that $\dfrac{\tan \theta}{\sec \theta} = \sin \theta$ symbolically. Give graphical and numerical support.

Solution

Symbolic Verification   We will start with the more complicated expression $\dfrac{\tan \theta}{\sec \theta}$ and simplify it to $\sin \theta$. Begin by writing $\tan \theta$ and $\sec \theta$ in terms of $\sin \theta$ and $\cos \theta$.

$$\frac{\tan \theta}{\sec \theta} = \frac{\sin \theta / \cos \theta}{1 / \cos \theta} \qquad \text{Quotient identities}$$

$$= \frac{\sin \theta}{\cos \theta} \cdot \frac{\cos \theta}{1} \qquad \text{Invert and multiply.}$$

$$= \sin \theta \qquad \text{Cancel and simplify.}$$

These steps verify that $\dfrac{\tan\theta}{\sec\theta} = \sin\theta$ is an identity.

Graphical Support Graph $Y_1 = \tan(X)/(1/\cos(X))$ and $Y_2 = \sin(X)$ as shown in Figures 6.34 and 6.35. Their graphs appear to be identical.

Numerical Support See Figure 6.36. Notice that when $\theta = \pi/2 \approx 1.5708$, the ratio $\tan\theta/\sec\theta$ is undefined, whereas $\sin\theta = 1$. However, the equation $\tan\theta/\sec\theta = \sin\theta$ is nonetheless an identity because this equation is true whenever both expressions are defined.

$[-2\pi, 2\pi, \pi/2]$ by $[-2, 2, 1]$

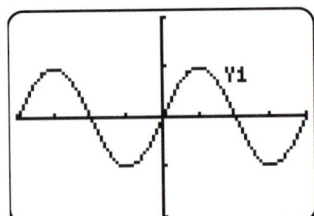

FIGURE 6.34

$[-2\pi, 2\pi, \pi/2]$ by $[-2, 2, 1]$

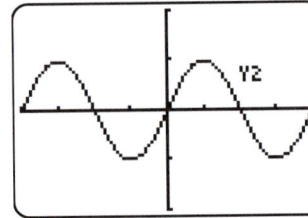

FIGURE 6.35

X	Y1	Y2
0	0	0
.5236	.5	.5
1.0472	.86603	.86603
1.5708	ERROR	1
2.0944	.86603	.86603
2.618	.5	.5
3.1416	0	0

X=0

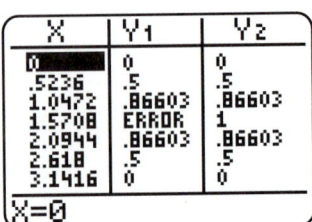

FIGURE 6.36 Radian Mode ■

Sometimes it is possible to use graphical and numerical methods to show that an equation is *not* an identity. However, we cannot use graphical and numerical methods to *verify* or *prove* that an equation is an identity. This is illustrated in the next example.

EXAMPLE 5 *Using graphical and numerical support for identities*

Use graphical and numerical methods to conjecture if an equation might be an identity. If possible, verify symbolically that the equation is an identity.

(a) $\csc\theta - \cos\theta = \sin\theta\tan\theta$

(b) $\dfrac{\sin t}{1 - \cos t} = \dfrac{1 + \cos t}{\sin t}$

Solution

(a) Graph $Y_1 = 1/\sin(X) - \cos(X)$ and $Y_2 = \sin(X)\tan(X)$ as shown in Figures 6.37 and 6.38, where degree mode has been used with a friendly window. Their graphs are different. A table of Y_1 and Y_2 is shown in Figure 6.39, where there exists values of x where $Y_1 \neq Y_2$. The equation $\csc\theta - \cos\theta = \sin\theta\tan\theta$ is *not* an identity, and therefore, *cannot* be verified symbolically.

$[-352.5°, 352.5°, 90°]$ by $[-4, 4, 1]$

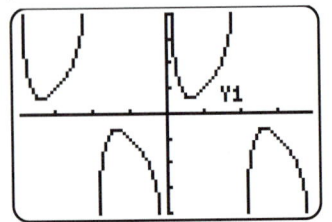

FIGURE 6.37

$[-352.5°, 352.5°, 90°]$ by $[-4, 4, 1]$

FIGURE 6.38

X	Y1	Y2
0	ERROR	0
15	2.8978	.06935
30	1.134	.28868
45	.70711	.70711
60	.6547	1.5
75	.77646	3.6049
90	1	ERROR

X=0

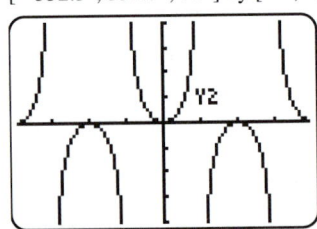

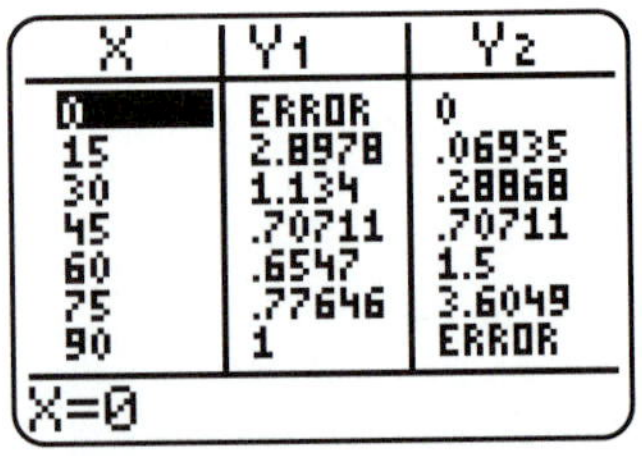

FIGURE 6.39 Degree Mode

(b) Graph $Y_1 = \sin(X)/(1 - \cos(X))$ and $Y_2 = (1 + \cos(X)/\sin(X)$ as shown in Figures 6.40 and 6.41. Their graphs appear to be identical. In Figure 6.42 we see that $Y_1 = Y_2$ for each x-value in the table. Since it is possible that there exists x-values where $Y_1 \neq Y_2$, we can only conjecture that the equation *may* be an identity. We now verify this symbolically.

$$\frac{\sin t}{1 - \cos t} = \frac{\sin t}{1 - \cos t} \cdot \frac{1 + \cos t}{1 + \cos t} \qquad \text{Multiply the ratio by 1.}$$

$$= \frac{\sin t(1 + \cos t)}{1 - \cos^2 t} \qquad \text{Simplify.}$$

$$= \frac{\sin t(1 + \cos t)}{\sin^2 t} \qquad \sin^2 t = 1 - \cos^2 t$$

$$= \frac{1 + \cos t}{\sin t} \qquad \text{Cancel } \sin t.$$

The given equation is an identity.

$[-352.5°, 352.5°, 90°]$ by $[-4, 4, 1]$

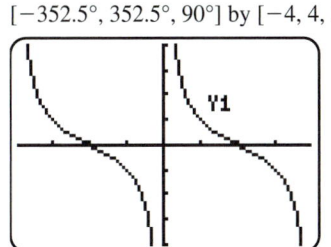

FIGURE 6.40

$[-352.5°, 352.5°, 90°]$ by $[-4, 4, 1]$

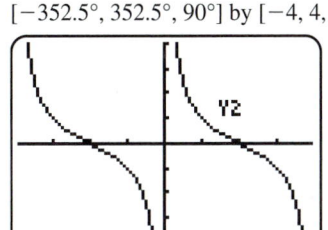

FIGURE 6.41

X	Y1	Y2
15	7.5958	7.5958
30	3.7321	3.7321
45	2.4142	2.4142
60	1.7321	1.7321
75	1.3032	1.3032
90	1	1
105	.76733	.76733

X=15

FIGURE 6.42 Degree Mode ∎

EXAMPLE 6 *Verifying an identity*

Verify that $(\cos\theta + \sin\theta)^2 = 1 + 2\sin\theta\cos\theta$ symbolically. Give graphical and numerical support.

Solution

Symbolic Verification In algebra the expression $(x + y)^2$ can be expanded as follows.

$$(x + y)^2 = (x + y)(x + y)$$
$$= x^2 + xy + yx + y^2$$
$$= x^2 + 2xy + y^2$$

We can perform similar steps with the left side of the trigonometric equation.

$$(\cos\theta + \sin\theta)^2 = (\cos\theta + \sin\theta)(\cos\theta + \sin\theta)$$
$$= \cos^2\theta + \cos\theta\sin\theta + \sin\theta\cos\theta + \sin^2\theta$$
$$= \cos^2\theta + 2\cos\theta\sin\theta + \sin^2\theta$$
$$= 1 + 2\cos\theta\sin\theta$$

The last step is true since $\sin^2\theta + \cos^2\theta = 1$.

Graphical Support For graphical support let

$$Y_1 = (\cos(X) + \sin(X))^2 \text{ and } Y_2 = 1 + 2\cos(X)\sin(X).$$

The graphs of Y_1 and Y_2 are shown in Figures 6.43 and 6.44 on the following page and appear to be identical.

Numerical Support In Figure 6.45, $Y_1 = Y_2$ for each *x*-value in the table.

$[-2\pi, 2\pi \ \pi/2]$ by $[-2, 4, 1]$

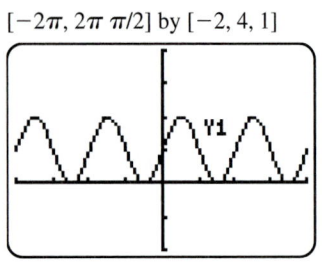

FIGURE 6.43

$[-2\pi, 2\pi \ \pi/2]$ by $[-2, 4, 1]$

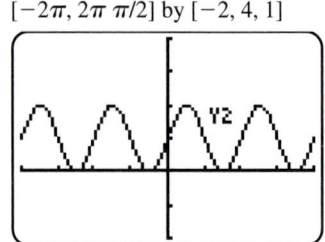

FIGURE 6.44

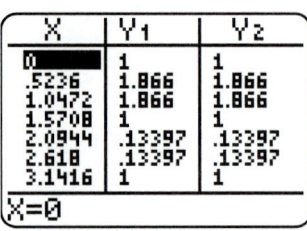

FIGURE 6.45 Radian Mode ∎

$[-2\pi, 2\pi \ \pi/2]$ by $[-4, 4, 1]$

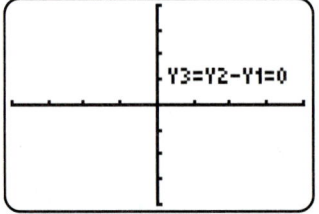

FIGURE 6.46

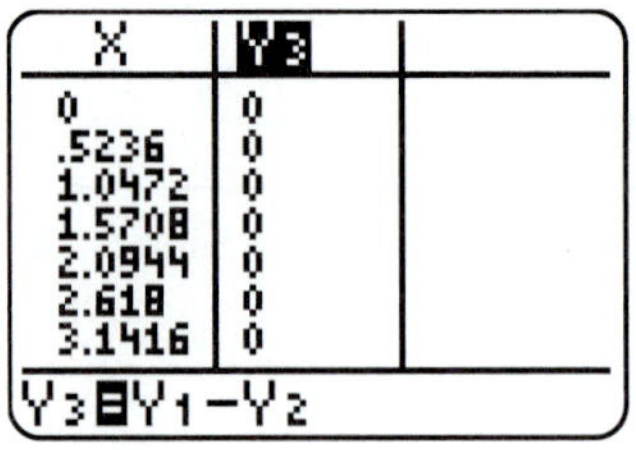

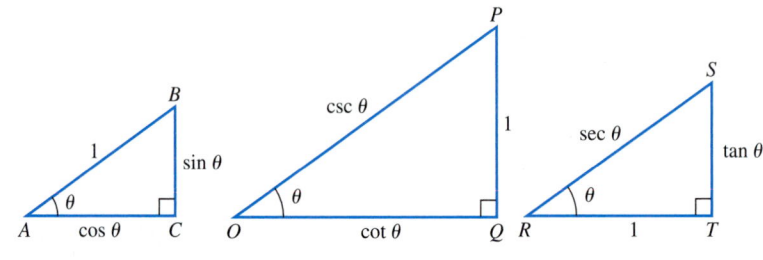

FIGURE 6.47

Technology Note

For graphical and numerical support of identities, we could let Y_1 equal the left side of the equation and Y_2 equal the right side. Since $Y_1 = Y_2$ it follows that $Y_3 = Y_1 - Y_2 = 0$ and the graph of Y_3 will coincide with the *x*-axis $(y = 0)$, while all values for Y_3 in a table will equal 0 to within the accuracy of the calculator. This technique is shown in Figures 6.46 and 6.47 for Y_1 and Y_2 from Example 6.

Using Right Triangles to Justify Trigonometric Identities (Optional)

Right triangles can be used to justify identities involving acute angles. This can be accomplished using the three triangles in Figures 6.48.

FIGURE 6.48

For example, in triangle *RST* the side adjacent θ equals 1. Since

$$\sec\theta = \frac{\text{hypotenuse}}{\text{side adjacent}} \quad \text{and} \quad \tan\theta = \frac{\text{side opposite}}{\text{side adjacent}},$$

it follows that the hypotenuse equals $\sec\theta$ and the side opposite equals $\tan\theta$. The sides for triangles *ABC* and *OPQ* can be verified in a similar manner. Notice that the triangles *ABC*, *OPQ*, and *RST* are similar because corresponding angles are equal.

Applying the Pythagorean theorem to each triangle, the following Pythagorean identities are immediate.

$$\sin^2\theta + \cos^2\theta = 1 \qquad 1 + \cot^2\theta = \csc^2\theta \qquad 1 + \tan^2\theta = \sec^2\theta$$

Now consider triangles *RST* and *ABC*. By similar triangles,

$$\frac{ST}{RS} = \frac{BC}{AB} \quad \text{or equivalently,} \quad \frac{\tan\theta}{\sec\theta} = \frac{\sin\theta}{1}.$$

The identity $\dfrac{\tan \theta}{\sec \theta} = \sin \theta$ was verified in Example 4 for any angle θ.

EXAMPLE 7   *Using right triangles to justify an identity*

Use the triangles in Figure 6.48 to justify the identity

$$\csc \theta + \frac{\sec \theta}{\tan \theta} = \frac{2}{\sin \theta},$$

when θ is acute.

Solution
Using similar triangles,

$$\frac{\csc \theta}{1} = \frac{1}{\sin \theta} \qquad \frac{OP}{PQ} = \frac{AB}{BC}$$

$$\frac{\sec \theta}{\tan \theta} = \frac{1}{\sin \theta} \qquad \frac{RS}{ST} = \frac{AB}{BC}$$

Adding these two equations results in the required identity.

$$\csc \theta + \frac{\sec \theta}{\tan \theta} = \frac{2}{\sin \theta} \qquad\blacksquare$$

6.2 PUTTING IT ALL TOGETHER

Becoming proficient at verifying identities requires practice. Many of the skills learned in algebra can be used to help verify identities. The following table lists some suggestions that may be helpful.

Suggestions for Verifying Identities

1. Become familiar with the fundamental identities found in Section 6.1.
2. Use your knowledge of how to simplify algebraic expressions to help guide you. This is particularly true when factoring or combining ratios.
3. When verifying an identity, start by simplifying the more complicated side of the equation. Otherwise, choose a side of the equation that you can transform into a different expression.
4. If you are simplifying the left side of the equation, then work toward making the left side appear more like the right. For example, if the left side contains an addition sign but the right side does not, then add the terms on the left side.
5. If you are uncertain how to proceed, one strategy is to write each trigonometric function in terms of sine and cosine and then simplify.
6. If a ratio contains $1 + \sin \theta$, it is sometimes helpful to multiply the numerator and denominator by $1 - \sin \theta$. Then,

$$(1 + \sin \theta)(1 - \sin \theta) = 1 - \sin^2 \theta = \cos^2 \theta.$$

Similar statements can be made for $1 - \sin \theta$, $1 + \cos \theta$, and $1 - \cos \theta$. See Example 5.

6.2 EXERCISES

Tape 10

Algebraic and Trigonometric Expressions

Exercises 1–4: Multiply the algebraic expression. Then multiply the corresponding trigonometric expression. If possible simplify the resulting trigonometric expression.

1. **(a)** $(1 + x)(1 - x)$
   (b) $(1 + \sin \theta)(1 - \sin \theta)$
2. **(a)** $(x - 1)(x + 1)$
   (b) $(\csc \theta - 1)(\csc \theta + 1)$
3. **(a)** $x(x - 1)$
   (b) $\sec \theta(\sec \theta - 1)$
4. **(a)** $(x + 1)(2x - 1)$
   (b) $(\tan \theta + 1)(2 \tan \theta - 1)$

Exercises 5–8: Factor the algebraic expression. Then factor the corresponding trigonometric expression.

5. **(a)** $x^2 + 2x + 1$
   (b) $\cos^2 \theta + 2 \cos \theta + 1$
6. **(a)** $2x^2 - 3x + 1$
   (b) $2 \sin^2 t - 3 \sin t + 1$
7. **(a)** $x^2 - 2x$
   (b) $\sec^2 t - 2 \sec t$
8. **(a)** $3x - 9x^2$
   (b) $3 \tan \theta - 9 \tan^2 \theta$

Exercises 9–16: Simplify the algebraic expression. Then simplify the corresponding trigonometric expression. If possible simplify the resulting trigonometric expression.

9. **(a)** $\dfrac{1}{1 - x} + \dfrac{1}{1 + x}$

   (b) $\dfrac{1}{1 - \cos \theta} + \dfrac{1}{1 + \cos \theta}$

10. **(a)** $x + \dfrac{1}{x}$

    (b) $\tan t + \dfrac{1}{\tan t}$

11. **(a)** $\dfrac{x}{y} + \dfrac{y}{x}$

    (b) $\dfrac{\cos t}{\sin t} + \dfrac{\sin t}{\cos t}$

12. **(a)** $\dfrac{1}{y} - \dfrac{x^2}{y}$

    (b) $\dfrac{1}{\sin \theta} - \dfrac{\cos^2 \theta}{\sin \theta}$

13. **(a)** $\dfrac{1}{1/y^2} + \dfrac{1}{1/x^2}$

    (b) $\dfrac{1}{\csc^2 t} + \dfrac{1}{\sec^2 t}$

14. **(a)** $\left(\dfrac{1}{x} + x\right)^2$

    (b) $(\cot \theta + \tan \theta)^2$

15. **(a)** $\dfrac{x/y}{1/y}$

    (b) $\dfrac{\cot \theta}{\csc \theta}$

16. **(a)** $\dfrac{1 - x^2}{1 + x}$

    (b) $\dfrac{1 - \cos^2 \theta}{1 + \cos \theta}$

Simplifying Expressions

Exercises 17–24: Perform the indicated operations and simplify. Support your result graphically and numerically.

17. $\cos \theta \tan \theta$
18. $\sin^2 \theta \csc \theta$
19. $\tan \theta (\cos \theta - \csc \theta)$
20. $(\sin \theta - \cos \theta)^2$
21. $(1 + \tan t)^2$
22. $(\sin t - 1)(\sin t + 1)$
23. $\dfrac{\sec^2 \theta - 1}{\sec^2 \theta}$
24. $\sin^2 t (1 + \cot^2 t)$

Exercises 25–30: Factor the trigonometric expression and simplify if possible.

25. $1 - \tan^2 \theta$
26. $\sin^2 t - \cos^2 t$
27. $\sec^2 t - \sec t - 6$
28. $\cos \theta \sin^2 \theta + \cos^3 \theta$
29. $\tan^4 \theta + 3 \tan^2 \theta + 2$
30. $\sin^4 t - \cos^4 t$

Verifying Identities

Exercises 31–38: Verify the identity. Give graphical and numerical support.

31. $\cot \theta \sin \theta = \cos \theta$
32. $\tan \theta \cos \theta = \sin \theta$
33. $(1 - \cos^2 \theta)(1 + \tan^2 \theta) = \tan^2 \theta$
34. $\cos^2 \theta(1 + \cot^2 \theta) = \cot^2 \theta$
35. $\cos t (\tan t - \sec t) = \sin t - 1$
36. $\dfrac{\cos \theta}{1 - \sin \theta} = \sec \theta + \tan \theta$
37. $\dfrac{\tan (-\theta)}{\sin (-\theta)} = \sec \theta$
38. $\tan^2 t - \sin^2 t = \tan^2 t \sin^2 t$

Exercises 39–44: (Refer to Example 5.) Use graphical and numerical methods to conjecture if the equation is an identity. If possible, verify that the equation is an identity.

39. $\sin^2 \theta + \cos^2 \theta = (\sin \theta + \cos \theta)^2$

40. $\dfrac{\tan^2 \theta + 1}{\sec \theta} = \sec \theta$

41. $\dfrac{\sin^2 t}{\cos t} = \sec t - \cos t$

42. $(1 - \sin t)^2 = 1 - \sin^2 t$

43. $\dfrac{\sin t + \cos t}{\sin t} = 1 + \cos t$

44. $\sec^4 \theta - \sec^2 \theta = \tan^4 \theta + \tan^2 \theta$

Exercises 45–64: Verify the identity.

45. $\sec^2 \theta - 1 = \tan^2 \theta$

46. $\dfrac{\csc^2 \theta}{\cot \theta} = \csc \theta \sec \theta$

47. $\dfrac{\tan^2 t}{\sec t} = \sec t - \cos t$

48. $\dfrac{\sec^2 \theta - 1}{\sec^2 \theta} = \sin^2 \theta$

49. $\sec^2 t + \csc^2 t = \sec^2 t \csc^2 t$

50. $\dfrac{\sec t}{1 + \sec t} = \dfrac{1}{\cos t + 1}$

51. $(\sec t - 1)(\sec t + 1) = \tan^2 t$

52. $\csc^4 \theta - \cot^4 \theta = \csc^2 \theta + \cot^2 \theta$

53. $\dfrac{1 - \sin^2 \theta}{\cos \theta} = \cos \theta$

54. $\dfrac{\tan^2 t - 1}{1 + \tan^2 t} = 1 - 2\cos^2 t$

55. $\dfrac{\sin^4 t - \cos^4 t}{\sin^2 t - \cos^2 t} = 1$

56. $\dfrac{\sec t}{\tan t} - \dfrac{\tan t}{\sec t} = \cos t \cot t$

57. $\dfrac{\cot^2 t}{\csc t + 1} = \csc t - 1$

58. $\sec \theta - \cos \theta = \tan \theta \sin \theta$

59. $\dfrac{\cot t}{\cot t + 1} = \dfrac{1}{1 + \tan t}$

60. $\cos^4 t - \sin^4 t = 2\cos^2 t - 1$

61. $\dfrac{1}{1 - \sin t} + \dfrac{1}{1 + \sin t} = 2\sec^2 t$

62. $\cot \theta + \tan \theta = \csc \theta \sec \theta$

63. $\dfrac{\csc t + \cot t}{\csc t - \cot t} = (\csc t + \cot t)^2$

64. $\dfrac{\csc t}{1 + \csc t} - \dfrac{\csc t}{1 - \csc t} = 2\sec^2 t$

Right Triangles and Identities

Exercises 65–70: (Refer to Example 7.) Use right triangles in the accompanying figure to help justify the identity.

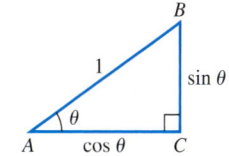

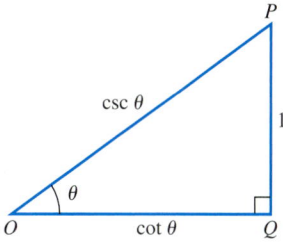

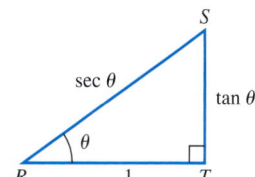

65. $\sec^2 \theta - \tan^2 \theta = 1$

66. $\dfrac{1}{\cot \theta} = \dfrac{\sin \theta}{\cos \theta}$

67. $\dfrac{\cot \theta}{\csc \theta} = \cos \theta$

68. $\dfrac{\csc \theta}{\sec \theta} = \dfrac{1}{\tan \theta}$

69. $\sin^2 \theta + \cos^2 \theta = \csc^2 \theta - \cot^2 \theta$

70. $\dfrac{1}{\cos \theta} - \dfrac{1}{\sin \theta} = \sec \theta - \dfrac{\sec \theta}{\tan \theta}$

71. *Electromagnets* (Refer to Example 3.) Let the wattage consumed by an electromagnet be given by $W(t) = 5\cos^2(120\pi t)$ and the voltage be given by $V(t) = 25\sin(120\pi t)$, where t is in seconds.
  (a) Express $W(t)$ in terms of the sine function. When V is maximum or minimum, what is the value of W? Explain.
  (b) Support your answer in part (a) by graphing W and V in $[0, 1/15, 1/60]$ by $[-30, 30, 10]$.

Writing about Mathematics

1. Create two trigonometric identities of your own. Verify each identity symbolically and then give graphical and numerical support.

2. Explain how to show that an equation is not an identity. Give an example.

CHECKING BASIC CONCEPTS FOR SECTIONS 6.1 AND 6.2

1. Determine the quadrant containing θ if $\cot \theta > 0$ and $\sin \theta < 0$.

2. Find the other trigonometric functions of θ using the given information.
   - (a) $\sin \theta = \dfrac{5}{13}$ and $\cos \theta = -\dfrac{12}{13}$
   - (b) $\sec \theta = \dfrac{5}{4}$ and $\sin \theta < 0$
   - (c) $\tan \theta = -\dfrac{1}{2}$ and $\cos \theta = \dfrac{2}{\sqrt{5}}$

3. Simplify each expression. Give graphical and numerical support for your result.
   - (a) $(1 - \sin \theta)(1 + \sin \theta)$
   - (b) $\tan^2 t \csc^2 t - 1$

4. Factor the trigonometric expression.
   - (a) $\tan^2 t - 1$
   - (b) $3 \sin^2 t + \sin t - 2$

5. Verify each identity. Give graphical and numerical support.
   - (a) $(1 - \sin^2 \theta)(1 + \cot^2 \theta) = \cot^2 \theta$
   - (b) $\dfrac{\cot^2 t}{\csc t} = \csc t - \sin t$

6.3 Trigonometric Equations

Reference Angles • Solving Trigonometric Equations

Introduction

In previous chapters we learned how applications involving functions resulted in the need to solve equations. In a similar manner, applications involving trigonometric functions result in the need to solve trigonometric equations. In Example 6, Section 5.5 we learned that the number of daylight hours near 30°N latitude can be modeled by

$$f(x) = 1.95 \cos\left(\frac{\pi}{6}(x - 6.6)\right) + 12.15,$$

where $x = 1$ corresponds to January 1, $x = 2$ to February 1, and so on. To estimate when there are 11 hours of daylight we can solve the trigonometric equation $f(x) = 11$ or

$$1.95 \cos\left(\frac{\pi}{6}(x - 6.6)\right) + 12.15 = 11.$$

Like other types of equations, trigonometric equations can be solved graphically, numerically, and symbolically. We begin by discussing reference angles, which will be used to solve trigonometric equations.

Reference Angles

A **reference angle** for an angle θ, written θ_R, is the acute angle made by the terminal side of θ and the x-axis. It is assumed that θ is in standard position and its terminal side does not lie on either the x- or y-axis. Examples of reference angles are shown in Figures 6.49–6.52.

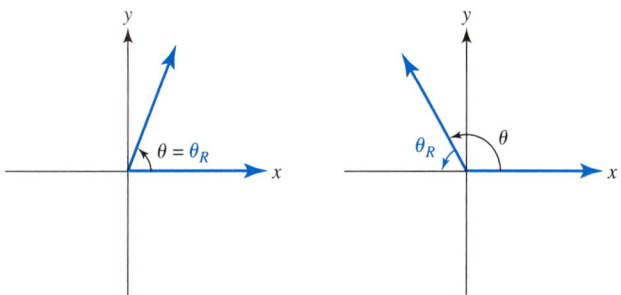

FIGURE 6.49 θ in Quadrant I   **FIGURE 6.50** θ in Quadrant II

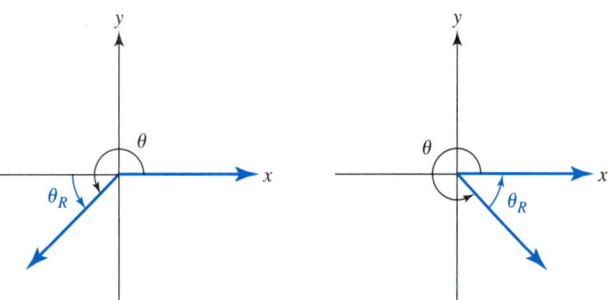

FIGURE 6.51 θ in Quadrant III   **FIGURE 6.52** θ in Quadrant IV

EXAMPLE 1   *Finding reference angles*

Find the reference angle for θ.
(a) $\theta = 43°$

(b) $\theta = \dfrac{2\pi}{3}$

(c) $\theta = -55°$

(d) $\theta = -\dfrac{3\pi}{4}$

Solution

(a) Since θ is in quadrant I, θ and θ_R are equal. Thus, $\theta_R = 43°$.

(b) The terminal side of $\theta = \dfrac{2\pi}{3}$ lies in quadrant II. This is similar to Figure 6.50.

In this case, the acute angle between the terminal side of θ and the x-axis is given by

$$\theta_R = \pi - \theta = \pi - \frac{2\pi}{3} = \frac{\pi}{3}.$$

(c) The terminal side of $\theta = -55°$ lies in quadrant IV and the acute angle between it and the *x*-axis is $\theta_R = 55°$. See Figure 6.53.

(d) The terminal side of $\theta = -\dfrac{3\pi}{4}$ (or $-135°$) lies in quadrant III and the acute angle between it and the *x*-axis is $\theta_R = \dfrac{\pi}{4}$. See Figure 6.54.

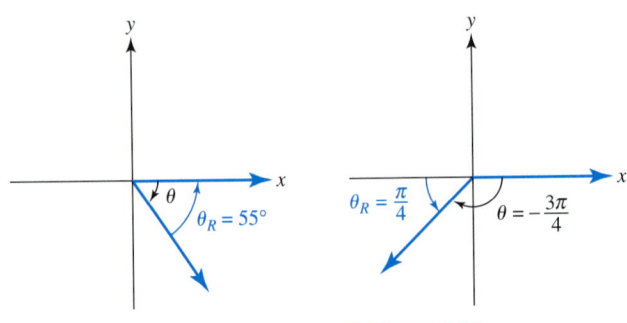

FIGURE 6.53   FIGURE 6.54

The reference angle is important because it can help determine trigonometric values.

> ### Reference angles and trigonometric functions
>
> Let θ be an angle in standard position with reference angle θ_R. Then
>
> $$|\sin\theta| = \sin\theta_R \qquad |\cos\theta| = \cos\theta_R \qquad |\tan\theta| = \tan\theta_R$$
> $$|\csc\theta| = \csc\theta_R \qquad |\sec\theta| = \sec\theta_R \qquad |\cot\theta| = \cot\theta_R.$$
>
> The signs of the trigonometric functions of θ are determined by the quadrant containing θ.

EXAMPLE 2   *Solving trigonometric equations using reference angles*

Solve the following equations for θ in $[0°, 360°)$.

(a) $\sin\theta = -\dfrac{1}{2}$

(b) $\tan\theta = 1$

Solution

(a) We start by solving the equation $\sin\theta_R = \dfrac{1}{2}$. The solution to this equation is

$\theta_R = \sin^{-1}\dfrac{1}{2} = 30°$. The sine function is negative in quadrants III and IV.

Therefore, the solution to $\sin\theta = -\dfrac{1}{2}$ is an angle θ located in either quadrant III or quadrant IV that has a reference angle of $30°$. There are two such angles: $210°$ and $330°$. See Figures 6.55 and 6.56.

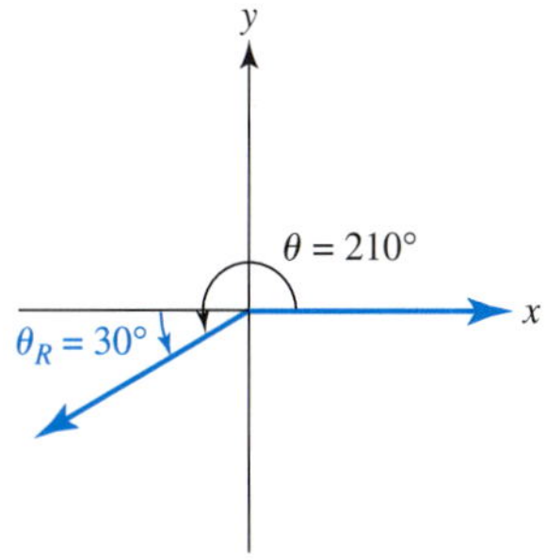

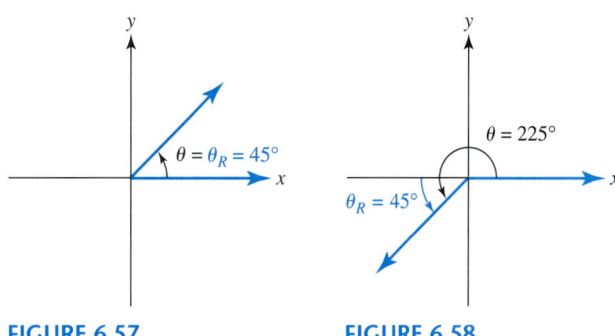

FIGURE 6.55 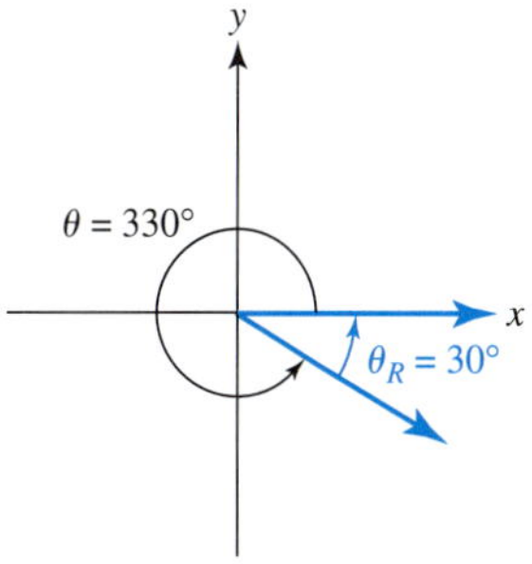   FIGURE 6.56

(b) The solution to $\tan \theta_R = 1$ is $\theta_R = \tan^{-1} 1 = 45°$. The tangent function is positive in quadrants I and III. Thus, $\theta = 45°$ and $\theta = 225°$ are solutions. See Figures 6.57 and 6.58.

FIGURE 6.57   FIGURE 6.58   ■

Critical Thinking

Let $0 < \theta < 2\pi$. Find expressions for θ_R given the quadrant containing θ.

Solving Trigonometric Equations

In the previous section we verified trigonometric *identities*. Identities are equations that are true for all meaningful values of the variable. In this section we discuss trigonometric equations that are *conditional*. Conditional equations are satisfied by some but not all values of the variable. For example,

$$\cos \theta = 1$$

is a conditional trigonometric equation since the cosine function equals 1 only for certain values of θ, such as $\theta = 0°$ or $\theta = 360°$.

EXAMPLE 3   *Solving trigonometric equations*

Solve $2 \sin \theta - 1 = 1$ on the interval $[0°, 360°)$ symbolically, graphically, and numerically.

Solution

Symbolic Solution Begin by solving the given equation for $\sin \theta$.

$$2 \sin \theta - 1 = 1 \qquad \text{Given equation}$$
$$2 \sin \theta = 2 \qquad \text{Add 1.}$$
$$\sin \theta = 1 \qquad \text{Divide by 2.}$$

The only solution to $\sin \theta = 1$ on the interval $[0°, 360°)$ is $\theta = \sin^{-1} 1 = 90°$.

Graphical Solution Graph $Y_1 = 2 \sin(X) - 1$ and $Y_2 = 1$ as in Figure 6.59. Their graphs intersect at $x = 90°$.

Numerical Solution Table Y_1 and Y_2 starting at $x = 0°$, incrementing by $30°$, as in Figure 6.60. From the table we see that $Y_1 = Y_2$ when $x = 90°$.

$[0°, 360°, 90°]$ by $[-4, 4, 1]$

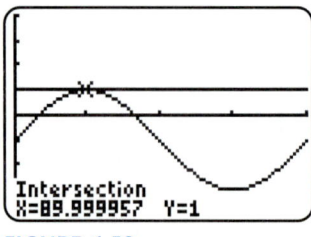

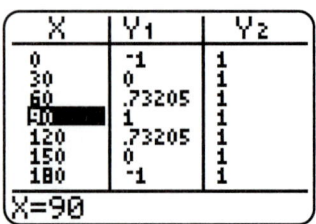

FIGURE 6.59   **FIGURE 6.60**   ■

In the next example we find all the phase angles associated with a particular phase F of the moon. The fraction of the moon that appears illuminated is called the *phase F*. The *phase angle θ* is shown in Figure 6.61. Refer to Example 12, Section 5.6. (Source: M. Zeilik, *Introduction to Astronomy and Astrophysics.*)

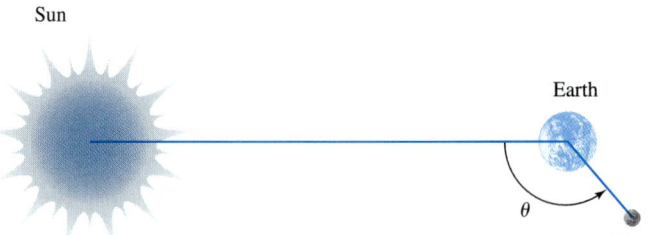

FIGURE 6.61 Phase Angle θ

EXAMPLE 4   *Finding phase angles for the moon*

The phase F associated with a phase angle θ is given by

$$F = \frac{1}{2}(1 - \cos \theta).$$

Find all phase angles θ in degrees when $F = 0.75$. Solve the equation symbolically and then give both graphical and numerical support. (Note that $F = 0.5$ corresponds to a quarter moon and $F = 1$ corresponds to a full moon.)

Solution

Symbolic Solution Let $F = 0.75$ and solve the given equation.

$$0.75 = \frac{1}{2}(1 - \cos \theta) \qquad \text{Let } F = 0.75.$$

$$1.5 = 1 - \cos \theta \qquad \text{Multiply by 2.}$$

$$\cos \theta = -0.5 \qquad \text{Solve for } \cos \theta.$$

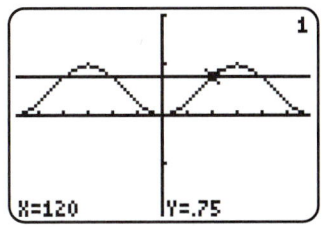

$[-352.5°, 352.5°, 60°]$ by $[-2, 2, 1]$

FIGURE 6.62

X	Y1	Y2
-360	0	.75
-240	.75	.75
-120	.75	.75
0	0	.75
120	.75	.75
240	.75	.75
360	0	.75

$Y_1 \blacksquare .5(1-\cos(X))$

FIGURE 6.63

Start by solving the equation $\cos\theta_R = 0.5$. The solution is $\theta_R = \cos^{-1} 0.5 = 60°$. The cosine function is negative in quadrants II and III. Angles in these quadrants with a 60° reference angle are 120° and 240°. Since the cosine function is periodic with 360°, all solutions can be written in the form

$$\theta = 120° + 360° \cdot n \qquad \text{or} \qquad \theta = 240° + 360° \cdot n,$$

where n is an integer. For example, 120°, 120° ± 360°, and 120° ± 720° are solutions, as well as 240°, 240° ± 360°, and 240° ± 720°.

Graphical Solution Graph $Y_1 = 0.5(1 - \cos(X))$ and $Y_2 = 0.75$ in degree mode as shown in Figure 6.62. By tracing the graph we can see that the graphs intersect at ± 120°, ± 240°, which supports the symbolic results.

Numerical Solution Table Y_1 and Y_2 starting at $x = -360°$, incrementing by 120°. Figure 6.63 supports the symbolic results, where $Y_1 = Y_2 = 0.75$ when $x = \pm 120°$ and $x = \pm 240°$. ∎

Many of the techniques used to solve polynomial equations can be applied to trigonometric equations as illustrated in the next example.

EXAMPLE 5   *Solving trigonometric equations*

Solve each equation for $0 \le t < 2\pi$ symbolically, graphically, and numerically.
(a) $2 \cot t + 1 = -1$
(b) $2 \sin^2 t - 5 \sin t + 2 = 0$

Solution
(a) *Symbolic Solution* In algebra, the equation $2x + 1 = -1$ implies $x = -1$. In a similar manner,

$$2 \cot t + 1 = -1 \quad \text{implies} \quad \cot t = -1.$$

If $\cot t = -1$, then $\tan t = -1$ and t has a reference angle of $\tan^{-1} 1 = \dfrac{\pi}{4}$. The cotangent is negative in quadrants II and IV so the solutions to $\cot t = -1$ in $[0, 2\pi)$ are $t = \dfrac{3\pi}{4} \approx 2.3562$ and $t = \dfrac{7\pi}{4} \approx 5.4978$. Since cotangent has a period of π, all solutions can be expressed in the form

$$t = \frac{3\pi}{4} + \pi n \qquad \text{or} \qquad t = \frac{7\pi}{4} + \pi n,$$

where n is an integer. These solutions are equivalent to $t = \dfrac{3\pi}{4} + \pi n$ since the difference between $\dfrac{7\pi}{4}$ and $\dfrac{3\pi}{4}$ is π.

Graphical Solution Graph $Y_1 = 2/\tan(X) + 1$ and $Y_2 = -1$ as in Figures 6.64–6.65 on the next page. The graphs intersect when $x \approx 2.3562$ and $x \approx 5.4978$.

Numerical Solution Table Y_1 and Y_2 starting at $\pi/2$, incrementing by $\pi/4$ as in Figure 6.66 on the next page. We see that $Y_1 = Y_2 = -1$ when $x \approx 2.3562$ and $x \approx 5.4978$.

$[0, 2\pi, \pi/4]$ by $[-4, 4, 1]$

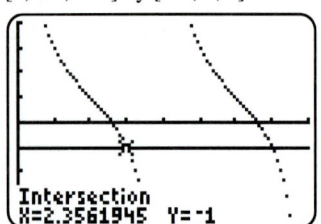

FIGURE 6.64 Dot Mode

$[0, 2\pi, \pi/4]$ by $[-4, 4, 1]$

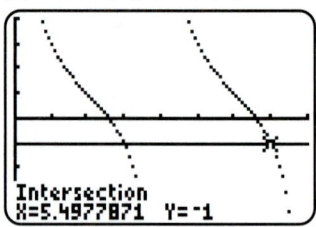

FIGURE 6.65 Dot Mode

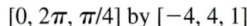

FIGURE 6.66

(b) *Symbolic Solution*   In algebra, the equation $2x^2 - 5x + 2 = 0$ may be solved by factoring.

$$2x^2 - 5x + 2 = (2x - 1)(x - 2) = 0$$

The solutions are $x = \dfrac{1}{2}$ and 2. In a similar manner, we can factor a trigonometric expression.

$$2\sin^2 t - 5\sin t + 2 = (2\sin t - 1)(\sin t - 2) = 0$$

We must solve $\sin t = \dfrac{1}{2}$ and $\sin t = 2$. If $\sin t = \dfrac{1}{2}$, the reference angle is $\sin^{-1}\dfrac{1}{2} = \dfrac{\pi}{6}$. The sine function is positive in quadrants I and II so the solutions in $[0, 2\pi)$ are $t = \dfrac{\pi}{6} \approx 0.5236$ and $t = \dfrac{5\pi}{6} \approx 2.6180$. Since $-1 \le \sin t \le 1$ for all t, the equation $\sin t = 2$ has no solutions. The sine function is periodic with 2π and all solutions to the given equation can be expressed as

$$t = \frac{\pi}{6} + 2\pi n \qquad \text{or} \qquad t = \frac{5\pi}{6} + 2\pi n.$$

Graphical Solution   Graph $Y_1 = 2(\sin(X))^2 - 5\sin(X) + 2$ as in Figure 6.67 and Figure 6.68. The zeros of this graph support the symbolic solution.

Numerical Solution   Table Y_1 starting at 0, incrementing by $\pi/6$ as in Figure 6.69. This table supports the symbolic results.

$[0, 2\pi, \pi/3]$ by $[-4, 4, 1]$

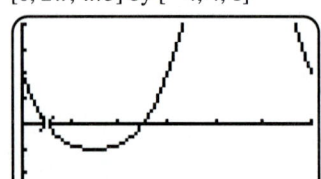

FIGURE 6.67

$[0, 2\pi, \pi/3]$ by $[-4, 4, 1]$

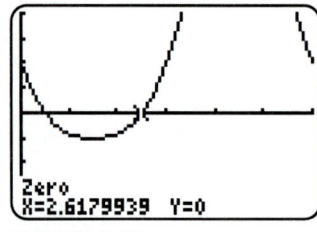

FIGURE 6.68

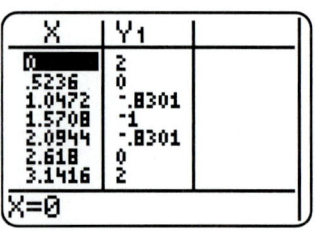

FIGURE 6.69   ∎

MAKING CONNECTIONS
Polynomial and Trigonometric Equations

A polynomial equation of degree *n* has at most *n* solutions. However, a trigonometric equation usually has an infinite number of solutions. This was illustrated in Examples 4 and 5.

Solar power plants are interested in the number of daylight hours during different times of the year and at different latitudes. In the next example, we solve the trigonometric equation presented in the introduction to determine when there are 11 hours of daylight at 30°N latitude.

EXAMPLE 6   *Analyzing daylight hours*

The number of daylight hours at 30°N latitude can be modeled by

$$f(x) = 1.95 \cos\left(\frac{\pi}{6}(x - 6.6)\right) + 12.15,$$

where $x = 1$ corresponds to January 1, $x = 2$ to February 1, and so on. Estimate graphically and numerically when there are 11 hours of daylight.

Solution

Graphical Solution   Graph $Y_1 = 1.95 \cos(\pi/6(x - 6.6)) + 12.15$ and $Y_2 = 11$ in radian mode. Their graphs intersect near $x = 2.4$ and $x = 10.8$. See Figures 6.70 and 6.71. These values correspond to about February 11 and October 25. (Note that four tenths of February is $0.4 \times 28 \approx 11$ days and eight tenths of October is $0.8 \times 31 \approx 25$ days.)

Numerical Solution   Table Y_1 starting at $x = 2$, incrementing by 0.1. The table in Figure 6.72 shows that $Y_1 \approx 11$ when $x = 2.4$. By scrolling down the table, $Y_1 \approx 11$ when $x = 10.8$.

[0, 13, 1] by [8, 16, 1]   [0, 13, 1] by [8, 16, 1]

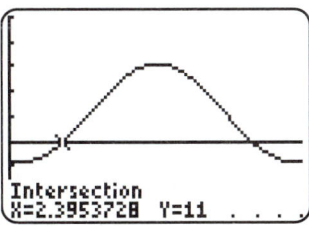

FIGURE 6.70

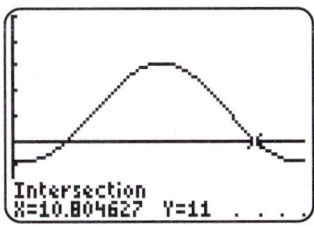

FIGURE 6.71

X	Y₁	Y₂
2	10.701	11
2.1	10.771	11
2.2	10.845	11
2.3	10.923	11
2.4	11.004	11
2.5	11.088	11
2.6	11.175	11

X=2.4

FIGURE 6.72 Radian Mode ■

Highway curves are sometimes banked so that the outside of the curve is slightly elevated or inclined as shown in Figure 6.73 on the next page. This inclination is called the *superelevation*. The relationship between a car's velocity *v* in feet per second, the safe radius *r* of the curve in feet, and the superelevation θ in degrees is given by $r = \dfrac{v^2}{4.5 + 32.2 \tan \theta}$. (Source: F. Mannering and W. Kilareski, *Principles of Highway Engineering and Traffic Analysis.*)

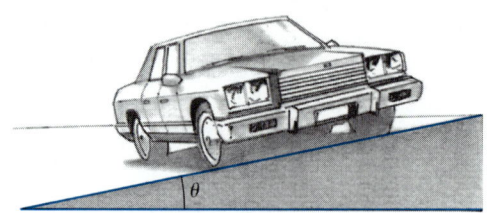

FIGURE 6.73

EXAMPLE 7   *Determining superelevation for a highway curve*

A highway curve with a radius of 700 feet and a speed limit of 88 feet per second (60 mph) is being designed. Find the appropriate superelevation for the curve.

Solution

Let $r = 700$ and $v = 88$ and then solve the equation for θ.

$$700 = \frac{88^2}{4.5 + 32.2 \tan \theta} \qquad \text{Let } r = 700 \text{ and } v = 88.$$

$$4.5 + 32.2 \tan \theta = \frac{88^2}{700} \qquad \text{Properties of ratios.}$$

$$32.2 \tan \theta = \frac{88^2}{700} - 4.5 \qquad \text{Subtract 4.5.}$$

$$\tan \theta = \frac{88^2/700 - 4.5}{32.2} \qquad \text{Divide by 32.2.}$$

$$\tan \theta \approx 0.2038 \qquad \text{Approximate.}$$

$$\theta \approx \tan^{-1} 0.2038 \approx 11.5° \qquad \text{Apply the inverse tangent.}$$

The superelevation should be about $11.5°$. ∎

Some trigonometric equations contain more than one type of trigonometric function. In these situations it is sometimes helpful to use trigonometric identities to rewrite the equation in terms of one trigonometric function. This is illustrated in the next example.

EXAMPLE 8   *Solving a trigonometric equation*

Solve $2 \tan \theta = \sec^2 \theta$ symbolically on the interval $[0°, 360°)$.

Solution

This equation contains two different trigonometric functions. We begin by applying the identity $1 + \tan^2 \theta = \sec^2 \theta$ to rewrite the equation only in terms of $\tan \theta$.

$$2 \tan \theta = \sec^2 \theta \qquad \text{Given equation}$$

$$2 \tan \theta = 1 + \tan^2 \theta \qquad \sec^2 \theta = 1 + \tan^2 \theta$$

$$\tan^2 \theta - 2 \tan \theta + 1 = 0 \qquad \text{Rewrite the equation.}$$

$$(\tan \theta - 1)(\tan \theta - 1) = 0 \qquad \text{Factor.}$$

$$\tan \theta = 1 \qquad \text{Solve for } \tan \theta.$$

The solutions are $\theta = 45°$ and $225°$. See Example 2(b). ∎

One step in the process used by astronomers to calculate the position of a planet as it orbits the sun involves finding the solution of Kepler's equation. Kepler's equation is a trigonometric equation that *cannot* be solved symbolically. It must be solved either graphically or numerically. In real applications, it is not uncommon to encounter equations that cannot be solved symbolically.

EXAMPLE 9   *Solving the equation of Kepler*

One example of Kepler's equation is $\theta = 0.087 + 0.093 \sin \theta$. Solve this equation graphically and numerically. The following source may be used to learn more about Kepler's equation. (Source: J. Meeus, *Astronomical Algorithms.*)

Solution

The given equation is equivalent to

$$\theta - 0.087 - 0.093 \sin \theta = 0.$$

To solve this equation graphically, let $Y_1 = X - 0.087 - 0.093 \sin X$ and apply the x-intercept method. (Be sure to use radian mode.) The x-intercept is near 0.096 as shown in Figure 6.74. Numerical support is shown in Figure 6.75, where $Y_1 \approx 0$ when $x = 0.096$.

[0, 0.2, 0.05] by [−0.2, 0.2, 0.1]

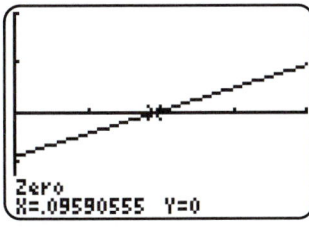

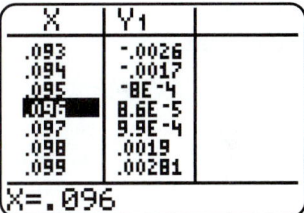

FIGURE 6.74

FIGURE 6.75 Radian Mode

6.3 PUTTING IT ALL TOGETHER

A reference angle θ_R for an angle θ in standard position is the acute angle made by the terminal side of θ and the x-axis. A reference angle can be used to help solve trigonometric equations.

Algebraic skills such as factoring and solving equations can also be used to solve trigonometric equations. Unlike polynomial equations, which always have a finite number of solutions, trigonometric equations frequently have an infinite number of solutions.

The following table gives examples of the concepts presented in this section.

Reference Angles

If $\theta = 155°$ then $\theta_R = 25°$.

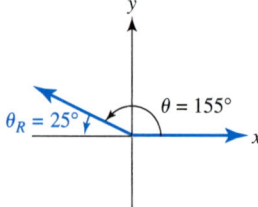

If $\theta = \dfrac{11\pi}{6}$ then $\theta_R = \dfrac{\pi}{6}$.

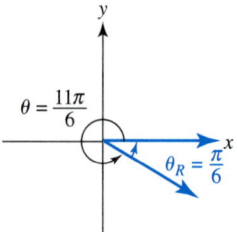

Trigonometric Equations

$\sqrt{3}\,\tan\theta = -1$

$$\tan\theta = -\frac{1}{\sqrt{3}}$$

The reference angle is $\theta_R = \tan^{-1}\left(\dfrac{1}{\sqrt{3}}\right) =$ 30°. Since $\tan\theta$ is negative in quadrants II and IV, the solutions in [0°, 360°) are 150° and 330°.

$\cos^2 t - 3\cos t + 2 = 0$
$(\cos t - 2)(\cos t - 1) = 0$
$\cos t = 2$ or $\cos t = 1$

$\cos t = 2$ has no solutions. The only solution to $\cos t = 1$ on $[0, 2\pi)$ is $t = 0$. All solutions can be written as $t = 2\pi n$, where n is an integer.

6.3 EXERCISES

Tape 10

Reference Angles

Exercises 1–10: Find the reference angle for θ.

1. $\theta = 120°$
2. $\theta = 230°$
3. $\theta = 85°$
4. $\theta = -130°$
5. $\theta = -65°$
6. $\theta = 340°$
7. $\theta = \dfrac{5\pi}{6}$
8. $\theta = \dfrac{7\pi}{4}$
9. $\theta = -\dfrac{2\pi}{3}$
10. $\theta = -\dfrac{5\pi}{4}$

Solving Equations

Exercises 11–14: Use the graph to estimate any solutions to the given equation for $0 \le t < 2\pi$.

11. $\sin t = \cos t$

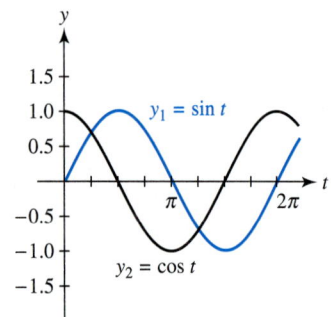

12. $\csc t = \sec t$

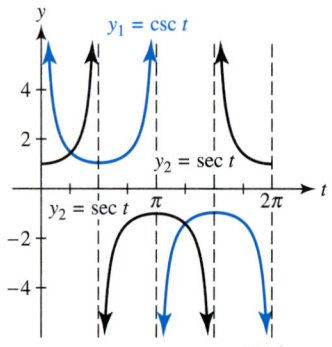

13. $3\cot t = 2\sin t$

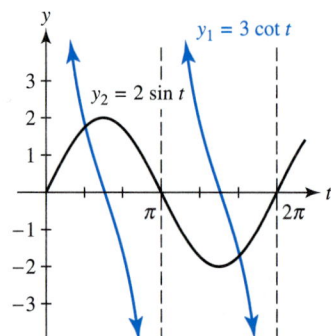

14. $2\cos^2 t = 1 - \cos t$

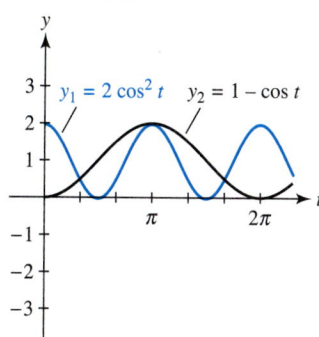

$y_1 = 2\cos^2 t \qquad y_2 = 1 - \cos t$

Exercises 15–18: Use the accompanying table to find the solutions to the given equation on $[0°, 360°)$. Then, write all solutions to the equation.

15. $1 - 4\cos^2 \theta = 0$

X	Y1
0	-3
60	0
120	0
180	-3
240	0
300	0
360	-3

$Y_1 \boxminus 1 - 4(\cos(X))^2$

16. $\tan \theta - \sin \theta = 0$

X	Y1
0	0
60	.86603
120	-2.598
180	0
240	2.5981
300	-.866
360	0

$Y_1 \boxminus \tan(X) - \sin(X)$

17. $\tan \theta - \dfrac{1}{\sqrt{3}} = 0$

X	Y1
30	0
90	ERROR
150	-1.155
210	0
270	ERROR
330	-1.155
390	0

$Y_1 \boxminus \tan(X) - 1/\sqrt{(3)}$

18. $2\sin^2 \theta - 1 = 0$

X	Y1
45	0
90	1
135	0
180	-1
225	0
270	1
315	0

$Y_1 \boxminus 2(\sin(X))^2 - 1$

Exercises 19–26: Solve the equation for θ, where $0° \le \theta < 360°$.

19. (a) $\sin \theta = 1$
 (b) $\sin \theta = -1$

20. (a) $\cos \theta = \dfrac{1}{2}$

 (b) $\cos \theta = -\dfrac{1}{2}$

21. (a) $\tan \theta = \sqrt{3}$
 (b) $\tan \theta = -\sqrt{3}$

22. (a) $\cot \theta = 1$
 (b) $\cot \theta = -1$

23. (a) $\sec \theta = 2$
 (b) $\sec \theta = -2$

24. (a) $\csc \theta = \sqrt{2}$
 (b) $\csc \theta = -\sqrt{2}$

25. (a) $\sin \theta = 3$
 (b) $\sin \theta = -3$

26. (a) $\cos \theta = \dfrac{\sqrt{3}}{2}$

 (b) $\cos \theta = -\dfrac{\sqrt{3}}{2}$

Exercises 27–34: Solve the algebraic equation for x. Then solve the trigonometric equation symbolically, graphically, and numerically for $0° \le \theta < 360°$.

27. (a) $2x - 1 = 0$
 (b) $2\sin \theta - 1 = 0$

28. (a) $x - 1 = 0$
 (b) $\cot \theta - 1 = 0$

29. (a) $x^2 = x$
 (b) $\sin^2 \theta = \sin \theta$

30. (a) $x^2 - x = 0$
 (b) $\cos^2 \theta - \cos \theta = 0$

31. (a) $x^2 + 1 = 2$
 (b) $\tan^2 \theta + 1 = 2$

32. (a) $(x - 1)(x + 1) = 0$
 (b) $(\sin \theta - 1)(\sin \theta + 1) = 0$

33. (a) $x^2 + x = 2$
 (b) $\cos^2 \theta + \cos \theta = 2$

34. (a) $2x^2 + 3x = -1$
 (b) $2\sin^2 \theta + 3\sin \theta = -1$

Exercises 35–42: Solve the equation for t on the interval $[0, 2\pi)$ symbolically. Support your answer graphically and numerically.

35. $\tan^2 t - 3 = 0$

36. $2\sin t = \sqrt{3}$

37. $3 \cos t + 4 = 0$

38. $\cos^2 t + \cos t - 6 = 0$

39. $\sin t \cos t = \cos t$

40. $\cos^2 t - \sin^2 t = 0$

41. $\sec^2 t = 2 \tan t$

42. $2 \sin^2 t - 3 \cos t = 3$

Exercises 43–54: Find all solutions to the equation. Express your results in both degrees and radians.

43. $\tan^2 t - 1 = 0$

44. $2 \cos t = -1$

45. $3 \cos t - 5 = 0$

46. $\sin^2 t + \sin t - 20 = 0$

47. $\cos t \sin t = \sin t$

48. $2 \cos^2 t - 1 = 0$

49. $\csc^2 t = 2 \cot t$

50. $\cos^2 t - 2 \sin t - 1 = 0$

51. $\sin^2 t \cos^2 t = 0$

52. $2 \cot^2 t \sin t - \cot^2 t = 0$

53. $\sin t + \cos t = 1$ (*Hint:* Square both sides.)

54. $\sin t - \cos t = 1$

Exercises 55–60: The following equations cannot be solved symbolically. Approximate to two significant digits any solutions on $[0, 2\pi)$ graphically and numerically.

55. $\tan x = x$

56. $x - \cos x = 0$

57. $\sin x = (x - 1)^2$

58. $\sin^2 x - \ln x = 0$

59. $2x \cos (x + 1) = \sin (\cos x)$

60. $e^{-0.1x} \cos x = x \sin x$

Applications

61. *Phases of the Moon* (Refer to Example 4.) Find all phase angles in degrees where $F = 1$ or $F = 0.25$
   (a) graphically,
   (b) numerically, and
   (c) symbolically.

62. *Designing Highway Curves* (Refer to Example 7.) A highway curve with a radius of 800 feet and a speed limit of 66 feet per second (45 mph) is being designed. Find the appropriate superelevation for the curve.

63. *Daylight Hours* (Refer to Example 6.) The number of daylight hours y at 60°N latitude can be modeled by

$$y = 6.5 \sin \left(\frac{\pi}{6}(x - 3.65) \right) + 12.4,$$

where $x = 1$ corresponds to January 1, $x = 2$ to February 1, and so on. Estimate graphically and numerically when there are 9 hours of daylight. (Source: J. Williams.)

64. *Average Temperatures* The monthly average high temperature y in degrees Fahrenheit at Phoenix, Arizona can be modeled by

$$y = 20.3 \sin (0.53x - 2.18) + 83.8,$$

where $x = 1$ corresponds to January, $x = 2$ to February, and so on. Estimate graphically and numerically when the average monthly high temperature is 93°F. (Source: J. Williams.)

65. *Daylight Hours* Solve Exercise 63 symbolically.

66. *Average Temperatures* Solve Exercise 64 symbolically.

67. *Maximum Monthly Sunshine* The maximum number of hours of sunshine each month are listed in the table for 50°N latitude.

Month	1	2	3	4	5	6
Hours of Sunshine	261	279	363	407	471	482

Month	7	8	9	10	11	12
Hours of Sunshine	486	442	374	329	267	246

Source: C. Winter, *Solar Power Plants.*

   (a) The function f given by $f(x) = 122.3 \sin (0.524x - 1.7) + 367$ models this data. Graph f and the data in $[0, 13, 1]$ by $[200, 600, 50]$.
   (b) Estimate graphically any solutions to the inequality $f(x) \geq 350$ on the interval $[1, 12]$. Interpret the result.

68. *Maximum Yearly Sunshine* The maximum number of hours of sunshine in one year is not constant at each latitude. This is due to the eccentricity of

Earth's orbit and the tilt of Earth's axis. The accompanying table lists the total hours of daylight in one year for selected latitudes. (Note that a negative latitude indicates a location in the Southern Hemisphere.)

Latitude	$-90°$	$-80°$	$-60°$	$-40°$	$-20°$	$0°$
Hours of Sunshine	4290	4297	4347	4367	4376	4383

Latitude	$20°$	$40°$	$60°$	$80°$	$90°$
Hours of Sunshine	4390	4399	4419	4469	4476

Source: C. Winter, *Solar Power Plants.*

(a) Plot the data in the window $[-100, 100, 20]$ by $[4200, 4500, 50]$.

(b) Conjecture whether $y = a \sin(b(x - c)) + d$ could be used to model this data.

(c) Estimate the latitude where the annual hours of sunshine equals 4320 hours.

Exercises 69–70: *Equation of Kepler* (Refer to Example 9.) *Solve Kepler's equation to within two significant digits graphically and numerically.*

69. $\theta = 0.26 + 0.017 \sin \theta$

70. $\theta = 0.18 + 0.249 \sin \theta$

71. *Music and Pure Tones*   A pure tone can be described by a sinusoidal graph. The graph of $P = 0.004 \sin(100 \pi x)$ shown in the accompanying figure represents the pressure of a pure tone on an eardrum in pounds per square foot at time t in seconds. (Source: J. Roederer, *Introduction to the Physics and Psychophysics of Music.*)

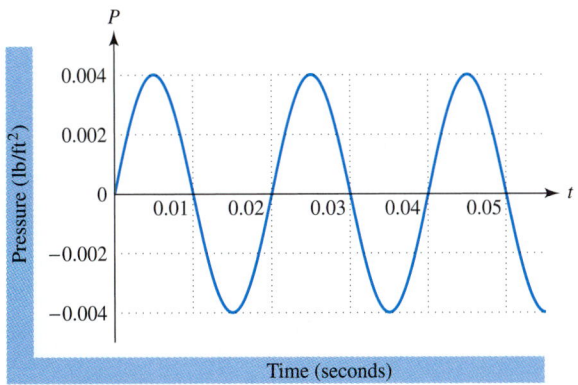

Time (seconds)

(a) Estimate all solutions to the equation $P = 0.004$ on the interval $[0, 0.05]$.

(b) Interpret these solutions.

72. *Music and Pure Tones*   The following graphs show the pressure wave resulting from a musical tone, where t is time in seconds and P is pressure in pounds per square foot. (Source: J. Roederer.)

(a) For each graph describe what a person might hear.

(b) Count the number of solutions to the equation $P = 0.003$ in each graph for $0 \le t \le 0.2$.

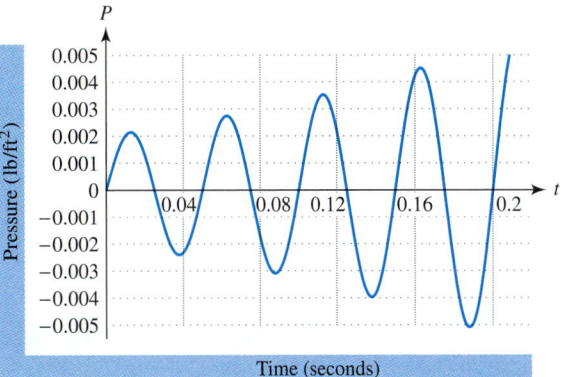

Time (seconds)

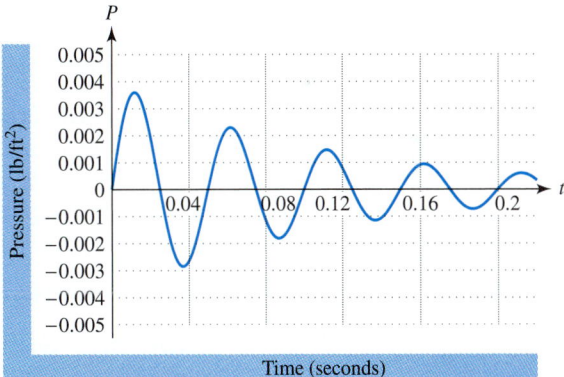

Time (seconds)

73. *Sums of Pure Tones*   If two loud speakers located at different positions produce the same pure tone, the human ear may hear one sound that is equal to the sum of the individual tones. Since the sources are at different locations, their sinusoidal waves will have different phase angles. (Source: N. Fletcher and T. Rossing, *The Physics of Musical Instruments.*)

(a) Let two different musical tones be represented by

$$P_1 = 0.003 \sin(880 \pi t - 0.7) \quad \text{and}$$
$$P_2 = 0.002 \sin(880 \pi t + 0.6).$$

Graph P_1, P_2, and their sum $P = P_1 + P_2$ separately in $[0, 0.01, 0.005]$ by $[-0.005, 0.005, 0.001]$.

(b) Determine the maximum pressure for P.

(c) Is the maximum pressure for P equal to the sum of the maximums of P_1 and P_2? Explain.

74. (Refer to the previous exercise.) Suppose that two loud speakers located at different positions produce pure tones given by

$$P_1 = A_1 \sin (2\pi Ft + \alpha) \quad \text{and}$$

$$P_2 = A_2 \sin (2\pi Ft + \beta),$$

where F is their common frequency. Then the resulting tone heard by a listener may be written as $P = A \sin (2\pi Ft + \theta)$, where

$$A = \sqrt{(A_1 \cos \alpha + A_2 \cos \beta)^2 + (A_1 \sin \alpha + A_2 \sin \beta)^2}$$

and

$$\theta = \arctan \left[\frac{A_1 \sin \alpha + A_2 \sin \beta}{A_1 \cos \alpha + A_2 \cos \beta} \right].$$

(Source: N. Fletcher.)

(a) Find A and θ for P if $F = 440$, $A_1 = 0.003$, $\alpha = -0.7$, $A_2 = 0.002$, and $\beta = 0.6$.

(b) Graph $P = A \sin (2\pi Ft + \theta)$ and $y = P_1 + P_2$ in $[0, 0.01, 0.005]$ by $[-0.005, 0.005, 0.001]$. Do the graphs appear to be identical?

Writing about Mathematics

1. Explain the difference between a conditional equation and an identity. Give one example of each.

2. Explain why knowledge of algebra is important when solving trigonometric equations. Give one example of how knowledge of algebra can be applied to solving a trigonometric equation.

6.4 Sum and Difference Identities

Sum and Difference Identities for Cosine • Other Sum and Difference Identities

Introduction

Music is made up of vibrations that create pressure on our eardrums. Musical tones can sometimes be modeled with sinusoidal graphs. When more than one tone is played, the resulting pressure is equal to the sum of the individual pressures. Sum and difference identities are sometimes helpful in the analysis of music. In this section we are introduced to several trigonometric identities and some of their applications.

Sum and Difference Identities for Cosine

The graph of $f(t) = \cos (t - \pi/2)$ is translated right $\pi/2$ units compared to the graph of $g(t) = \cos t$ as shown in Figure 6.76.

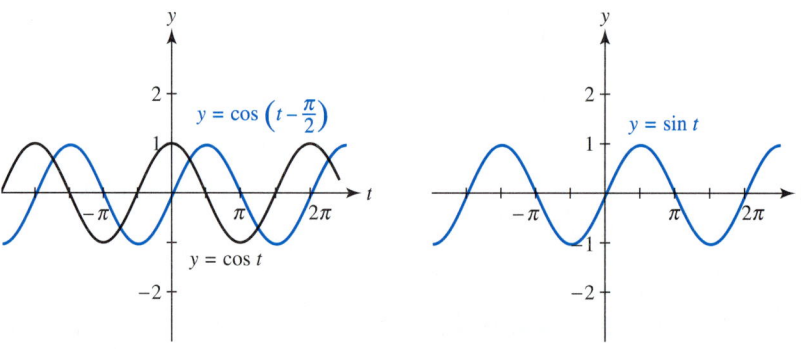

FIGURE 6.76　　　　　　　　**FIGURE 6.77**

X	Y₁	Y₂
0	0	0
.5236	.5	.5
1.0472	.86603	.86603
1.5708	1	1
2.0944	.86603	.86603
2.618	.5	.5
3.1416	0	0

X=0

FIGURE 6.78 Radian Mode

If the graph of $y = \cos t$ is translated right $\frac{\pi}{2}$ units, it coincides with the graph of $y = \sin t$, which is shown in Figure 6.77. We can support this numerically if we table $Y_1 = \cos(X - \pi/2)$ and $Y_2 = \sin(X)$. Notice that $Y_1 = Y_2$ for each x-value in Figure 6.78, where the table increment is $\pi/6$.

It is important to notice that

$$\cos\left(t - \frac{\pi}{2}\right) \neq \cos t - \cos\frac{\pi}{2}$$

$$= \cos t - 0$$

$$= \cos t.$$

Verbal, graphical, and numerical support suggests that

$$\cos\left(t - \frac{\pi}{2}\right) = \sin t.$$

To verify this symbolically, a new identity is needed. Suppose that α and β represent any two angles. Then the following identity can be used to calculate the cosine of their difference. (Its proof is given at the end of this section.)

$$\cos(\alpha - \beta) = \cos\alpha \cos\beta + \sin\alpha \sin\beta$$

The next example demonstrates how to apply this identity.

EXAMPLE 1 *Using the cosine difference identity*

Verify the identity $\cos\left(t - \frac{\pi}{2}\right) = \sin t$.

Solution

Start by letting $\alpha = t$ and $\beta = \frac{\pi}{2}$ in the difference identity

$$\cos(\alpha - \beta) = \cos\alpha \cos\beta + \sin\alpha \sin\beta.$$

Then,

$$\cos\left(t - \frac{\pi}{2}\right) = \cos t \cos\frac{\pi}{2} + \sin t \sin\frac{\pi}{2}$$

$$= \cos t\,(0) + \sin t\,(1)$$

$$= \sin t. \qquad \blacksquare$$

In the next example we use the difference identity for cosine to find the exact value of $\cos 15°$.

EXAMPLE 2 *Applying the cosine difference identity*

Find the exact value of $\cos 15°$. Support your result numerically.

Solution

Since $45° - 30° = 15°$, we proceed as follows.

$$\cos 15° = \cos(45° - 30°)$$

$$= \cos 45° \cos 30° + \sin 45° \sin 30° \quad \text{Difference identity for cosine}$$

$$= \frac{\sqrt{2}}{2} \cdot \frac{\sqrt{3}}{2} + \frac{\sqrt{2}}{2} \cdot \frac{1}{2} \quad \text{Evaluate each function.}$$

$$= \frac{\sqrt{6} + \sqrt{2}}{4} \quad \text{Simplify the } \textit{exact} \text{ value.}$$

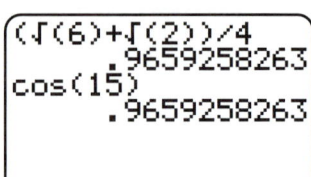

FIGURE 6.79 Degree Mode

In Figure 6.79 we see that the value of $\cos 15°$ agrees with the symbolic result. ∎

With the aid of the difference identity for cosine we can derive a sum identity for cosine.

$$\cos(\alpha + \beta) = \cos(\alpha - (-\beta))$$

$$= \cos\alpha \cos(-\beta) + \sin\alpha \sin(-\beta) \quad \text{Difference identity for cosine}$$

$$= \cos\alpha \cos\beta - \sin\alpha \sin\beta \quad \begin{array}{l}\cos(-\beta) = \cos\beta; \\ \sin(-\beta) = -\sin(\beta)\end{array}$$

Sum and difference identities for cosine are as follows.

Cosine of sum or difference

$$\cos(\alpha + \beta) = \cos\alpha \cos\beta - \sin\alpha \sin\beta$$

$$\cos(\alpha - \beta) = \cos\alpha \cos\beta + \sin\alpha \sin\beta$$

In Section 5.2 we introduced the cofunction identities for an acute angle θ. These identities are true for any angle θ. We verify one of these identities.

EXAMPLE 3   *Verifying a cofunction identity*

Verify that $\cos(90° - \theta) = \sin\theta$ symbolically. Give graphical and numerical support.

Solution

Symbolic Verification   Let $\alpha = 90°$ and $\beta = \theta$ in the cosine difference identity.

$$\cos(90° - \theta) = \cos 90° \cos\theta + \sin 90° \sin\theta$$

$$= (0)\cos\theta + (1)\sin\theta$$

$$= \sin\theta$$

Graphical Support   In Figures 6.80 and 6.81 graphical support is shown, where $Y_1 = \cos(90 - X)$ and $Y_2 = \sin(X)$. Their graphs appear to be identical.

Numerical Support   In Figure 6.82 numerical support is shown. Notice that $Y_1 = Y_2$ for each x-value.

$[0°, 360°, 90°]$ by $[-2, 2, 1]$

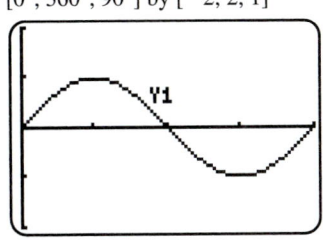

FIGURE 6.80

$[0°, 360°, 90°]$ by $[-2, 2, 1]$

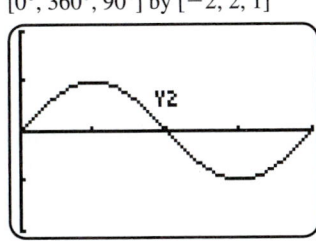

FIGURE 6.81

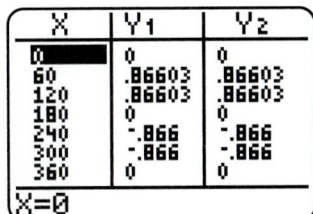

FIGURE 6.82 Degree Mode ∎

The following cofunction identities are valid for any angle θ.

Cofunction identities for any angle θ

$$\sin\theta = \cos(90° - \theta) \qquad \cos\theta = \sin(90° - \theta)$$

$$\tan\theta = \cot(90° - \theta) \qquad \cot\theta = \tan(90° - \theta)$$

$$\sec\theta = \csc(90° - \theta) \qquad \csc\theta = \sec(90° - \theta)$$

Other Sum and Difference Identities

There are also sum and difference identities for sine.

$$\sin(\alpha + \beta) = \cos(90° - (\alpha + \beta)) \qquad \text{Cofunction identity}$$

$$= \cos((90° - \alpha) - \beta) \qquad \text{Associative property}$$

$$= \cos(90° - \alpha)\cos\beta + \sin(90° - \alpha)\sin\beta \qquad \text{Difference identity for cosine}$$

$$= \sin\alpha\cos\beta + \cos\alpha\sin\beta \qquad \text{Cofunction identities}$$

In a similar manner the difference identity for sine can be derived. We now give the sum and difference identities for sine.

Sine of sum or difference

$$\sin(\alpha + \beta) = \sin\alpha\cos\beta + \cos\alpha\sin\beta$$

$$\sin(\alpha - \beta) = \sin\alpha\cos\beta - \cos\alpha\sin\beta$$

EXAMPLE 4   *Analyzing an identity graphically, numerically, verbally, and symbolically*

Give graphical, numerical, and verbal support for $\sin(\theta + \pi) = -\sin\theta$.
Then verify the identity symbolically.

Solution

Graphical Support   Graph $Y_1 = \sin(X + \pi)$ and $Y_2 = -\sin(X)$ as shown in Figures 6.83 and 6.84 on the next page. Their graphs appear to be identical.

Numerical Support Table Y_1 and Y_2 as shown in Figure 6.85. Notice that $Y_1 = Y_2$ for each value of x in the table, where the increment equals $\pi/6$.

Verbal Support The graph of $y = \sin(\theta + \pi)$ is similar to the graph of $y = \sin\theta$ except that it is translated left π units. The graph of $y = -\sin\theta$ is similar to the graph of $y = \sin\theta$ except that it is reflected across the x-axis. Translating the sine graph left π units or reflecting the sine graph across the x-axis results in the same graph. Therefore, we suspect that $\sin(\theta + \pi) = -\sin\theta$ is an identity.

Symbolic Verification Let $\alpha = \theta$ and $\beta = \pi$ in the sum identity for sine.

$$\sin(\theta + \pi) = \sin\theta\cos\pi + \cos\theta\sin\pi$$
$$= \sin\theta\,(-1) + \cos\theta\,(0)$$
$$= -\sin\theta$$

$[-2\pi, 2\pi, \pi/2]$ by $[-2, 2, 1]$

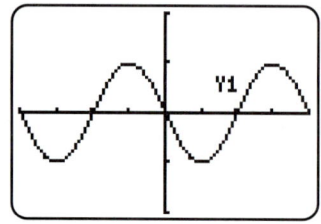

FIGURE 6.83

$[-2\pi, 2\pi, \pi/2]$ by $[-2, 2, 1]$

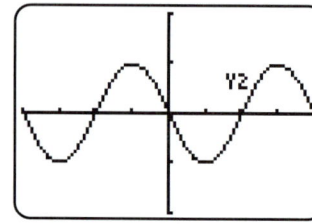

FIGURE 6.84

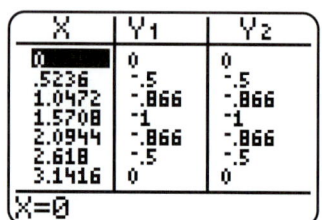

FIGURE 6.85 Radian Mode ■

Sum and difference identities can be used to find exact trigonometric values.

EXAMPLE 5 *Applying sum identities for sine and cosine*

Let $\sin\alpha = \dfrac{4}{5}$ and $\cos\beta = \dfrac{3}{5}$. If α is in quadrant II and β is in quadrant IV, find each of the following.
(a) $\sin(\alpha + \beta)$
(b) $\cos(\alpha + \beta)$
(c) $\tan(\alpha + \beta)$
(d) The quadrant containing $\alpha + \beta$

Solution
(a) First sketch possible angles for α and for β as shown in Figures 6.86 and 6.87. From these figures we can see that $\cos\alpha = -\dfrac{3}{5}$ and $\sin\beta = -\dfrac{4}{5}$. Now apply the sum identities for sine.

$$\sin(\alpha + \beta) = \sin\alpha\cos\beta + \cos\alpha\sin\beta$$
$$= \left(\frac{4}{5}\right)\left(\frac{3}{5}\right) + \left(-\frac{3}{5}\right)\left(-\frac{4}{5}\right)$$
$$= \frac{24}{25}$$

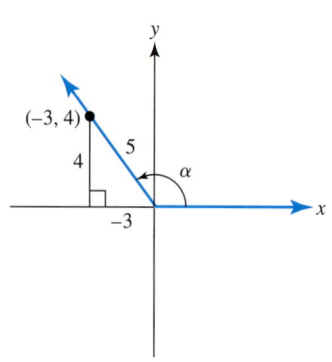

FIGURE 6.86

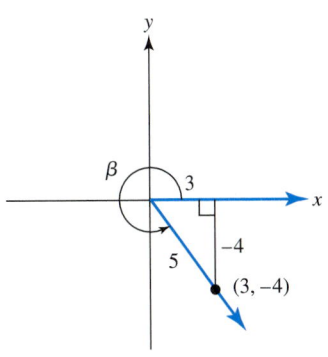

FIGURE 6.87

(b) To find $\cos(\alpha + \beta)$ apply the sum identity for cosine.

$$\cos(\alpha + \beta) = \cos\alpha\cos\beta - \sin\alpha\sin\beta$$

$$= \left(-\frac{3}{5}\right)\left(\frac{3}{5}\right) - \left(\frac{4}{5}\right)\left(-\frac{4}{5}\right)$$

$$= \frac{7}{25}$$

(c) $\tan(\alpha + \beta) = \dfrac{\sin(\alpha + \beta)}{\cos(\alpha + \beta)} = \dfrac{24/25}{7/25} = \dfrac{24}{7}$

(d) Since both $\sin(\alpha + \beta)$ and $\cos(\alpha + \beta)$ are positive, $\alpha + \beta$ is in quadrant I. ∎

Because joints both bend and rotate, trigonometry frequently is applied to human physiology. The next example shows how to calculate the force exerted by a person's back muscles and gives a rather amazing result.

EXAMPLE 6   *Analyzing stress on a person's back*

If a person with weight W bends at the waist with a straight back, then the force F exerted by the lower back muscles may be approximated using $F = 2.89W \sin(\theta + 90°)$, where θ is the angle between a person's torso and the horizontal. See Figure 6.88. (Source: H. Metcalf, *Topics in Classical Biophysics.*)

(a) Let $W = 155$ and use degree mode to graph F in $[0°, 90°, 30°]$ by $[0, 500, 100]$. Interpret the graph.

(b) For what value of θ does F equal 400 pounds?

(c) Show that $F = 2.89W \cos\theta$. Use this equation to solve part (b) symbolically.

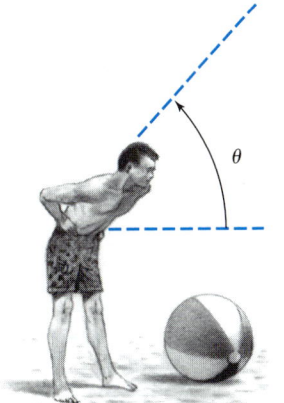

FIGURE 6.88

Solution

(a) Graph $Y_1 = 2.89(155)\sin(X + 90)$ as shown in Figure 6.89. When $\theta = 0°$, the person's back is parallel to the ground and force F exerted by the back muscles has a maximum of about 450 pounds. This is nearly three times the person's weight! As the person straightens up, θ increases, while F decreases. When $\theta = 90°$, the person is standing straight up and $F = 0$.

(b) Graph Y_1 and $Y_2 = 400$. Their graphs intersect near $(26.8, 400)$ as shown in Figure 6.90. The force is 400 pounds when $\theta \approx 26.8°$.

$[0°, 90°, 30°]$ by $[0, 500, 100]$

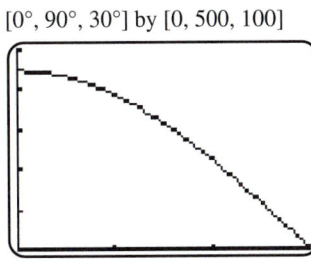

FIGURE 6.89

$[0°, 90°, 30°]$ by $[0, 500, 100]$

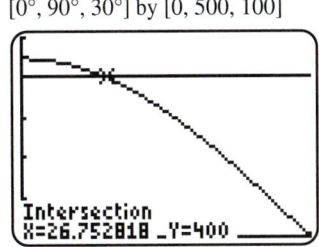

Intersection
X=26.752818 Y=400

FIGURE 6.90

(c) Apply a sum identity for sine.

$$F = 2.89W \sin(\theta + 90°)$$
$$= 2.89W(\sin\theta \cos 90° + \cos\theta \sin 90°)$$
$$= 2.89W(\sin\theta (0) + \cos\theta (1))$$
$$= 2.89W \cos\theta$$

To solve part (b) symbolically, let $F = 400$, $W = 155$, and solve for θ.

$$400 = 2.89(155)\cos\theta \qquad \text{\color{blue}$F = 400$}$$

$$\cos\theta = \frac{400}{2.89(155)} \qquad \text{\color{blue}Solve for $\cos\theta$.}$$

$$\theta = \cos^{-1}\left(\frac{400}{2.89(155)}\right) \qquad \text{\color{blue}Solve for θ.}$$

$$\theta \approx 26.8° \qquad \text{\color{blue}Approximate θ.} \qquad ∎$$

Music is composed of tones with various frequencies. Pressure exerted on the eardrum by a pure tone may be modeled by either $P(t) = a \cos bt$ or $P(t) = a \sin bt$, where a and b are constants and t represents time. When two tuning forks produce the same pure tone, the human ear hears only one sound that is equal to the sum of the individual tones. Trigonometry can be used to model this situation. (*Source: N. Fletcher and T. Rossing, The Physics of Musical Instruments.*)

EXAMPLE 7 *Modeling musical tones*

Let the pressure P in grams per square meter exerted on the eardrum by two different sources be modeled by

$$P_1(t) = 5 \cos(440\pi t) \quad \text{and}$$
$$P_2(t) = 3 \sin(440\pi t),$$

where t is time in seconds.

(a) Graph the total pressure, $P = P_1 + P_2$, on the eardrum in $[0, 0.01, 0.001]$ by $[-8, 8, 1]$.

(b) Use the graph to estimate values for a and k such that $P = a \sin(440\pi t + k)$.

(c) Use a sum or difference identity for sine to verify that $P \approx P_1 + P_2$.

Solution

(a) Let $Y_1 = 5 \cos(440\pi X)$, $Y_2 = 3 \sin(440\pi X)$, and $Y_3 = Y_1 + Y_2$. The graph of Y_3 is shown in Figure 6.91.

(b) A maximum y-value occurs near $(0.00039116, 5.831)$. (Note that the x- and y-value shown in Figure 6.91 may vary slightly.) Thus, let $a = 5.831$. Since $\sin\theta$ is maximum when $\theta = \pi/2$, we let $t = 0.00039116$ and solve the following equation for k.

$$440\pi(0.00039116) + k = \frac{\pi}{2}$$

$$k = \frac{\pi}{2} - 440\pi(0.00039116)$$

$$\approx 1.0301$$

Thus, let $P = 5.831 \sin(440\pi t + 1.0301)$. (Other values for k are possible.)

$[0, 0.01, 0.001]$ by $[-8, 8, 1]$

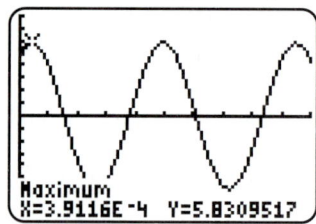

FIGURE 6.91

(c) Apply the difference identity for sine.

$$P = 5.831 \sin(440\pi t + 1.0301)$$
$$= 5.831(\sin(440\pi t)\cos(1.0301) + \cos(440\pi t)\sin(1.0301))$$
$$\approx 5.831(\sin(440\pi t)(0.5147) + \cos(440\pi t)(0.8574))$$
$$\approx 3.00 \sin(440\pi t) + 5.00 \cos(440\pi t))$$
$$\approx P_1 + P_2$$

Sum and difference identities can also be found for the tangent function.

Tangent of sum or difference

$$\tan(\alpha + \beta) = \frac{\tan\alpha + \tan\beta}{1 - \tan\alpha\tan\beta}$$

$$\tan(\alpha - \beta) = \frac{\tan\alpha - \tan\beta}{1 + \tan\alpha\tan\beta}$$

These identities are a result of the sum and difference identities for sine and cosine. For example, the difference identity can be verified as follows.

$$\tan(\alpha - \beta) = \frac{\sin(\alpha - \beta)}{\cos(\alpha - \beta)}$$ Use $\tan\theta = \frac{\sin\theta}{\cos\theta}$ with $\theta = \alpha - \beta$.

$$= \frac{\sin\alpha\cos\beta - \cos\alpha\sin\beta}{\cos\alpha\cos\beta + \sin\alpha\sin\beta}$$ Apply difference identities.

$$= \frac{\dfrac{\sin\alpha\cos\beta}{\cos\alpha\cos\beta} - \dfrac{\cos\alpha\sin\beta}{\cos\alpha\cos\beta}}{\dfrac{\cos\alpha\cos\beta}{\cos\alpha\cos\beta} + \dfrac{\sin\alpha\sin\beta}{\cos\alpha\cos\beta}}$$ Divide each term by $\cos\alpha\cos\beta$.

$$= \frac{\tan\alpha - \tan\beta}{1 + \tan\alpha\tan\beta}$$ Simplify.

EXAMPLE 8   *Using the tangent difference identity*

Use Figure 6.92 to find $\tan\gamma$ if $\tan\alpha = \dfrac{4}{3}$ and $\tan\beta = \dfrac{3}{4}$.

Solution

From Figure 6.92, α and $\beta + \gamma$ are both supplements of angle BAC. Thus, $\alpha = \beta + \gamma$ or $\gamma = \alpha - \beta$.

$$\tan\gamma = \tan(\alpha - \beta)$$ $\gamma = \alpha - \beta$

$$= \frac{\tan\alpha - \tan\beta}{1 + \tan\alpha\tan\beta}$$ Tangent difference identity.

$$= \frac{(4/3) - (3/4)}{1 + (4/3)(3/4)}$$ $\tan\alpha = \dfrac{4}{3};\ \tan\beta = \dfrac{3}{4}$

$$= \frac{7}{24}$$ Simplify.

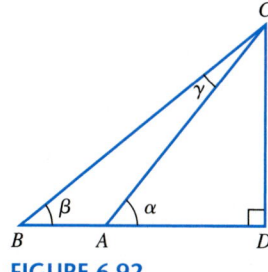

FIGURE 6.92

Musicians sometimes tune instruments by playing the same tone on two different instruments and listening for a phenomenon known as *beats*. Beats occur when two musical tones vary slightly in frequency. When the two instruments are in tune, the beats will disappear. The human ear hears beats because the sound pressure slowly rises and falls as a result of this slight variation in the frequency. The pressure P on an eardrum can be modeled by $P = a \sin(2\pi Ft)$, where F is the frequency of the tone, t is time in seconds, and P is in pounds per square foot. This phenomenon can be modeled with a graphing calculator. (Source: J. Pierce, *The Science of Musical Sound.*)

EXAMPLE 9 *Modeling musical beats*

Consider two tones with similar frequencies of 440 and 443 cycles per second and pressures

$$P_1 = 0.006 \sin(880\pi t) \quad \text{and} \quad P_2 = 0.004 \sin(886\pi t),$$

respectively.

[0.15, 1.15, 0.05] by
[−0.01, 0.01, 0.001]

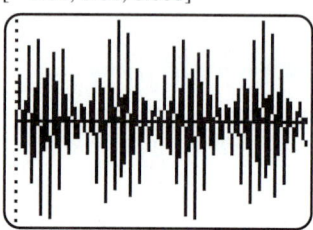

FIGURE 6.93

(a) Graph $P = P_1 + P_2$ in [0.15, 1.15, 0.05] by [−0.01, 0.01, 0.001], where P is the total pressure exerted by the tones on an eardrum. How many beats are there in this one-second interval?

(b) Repeat part (a) with frequencies of 220 and 224.

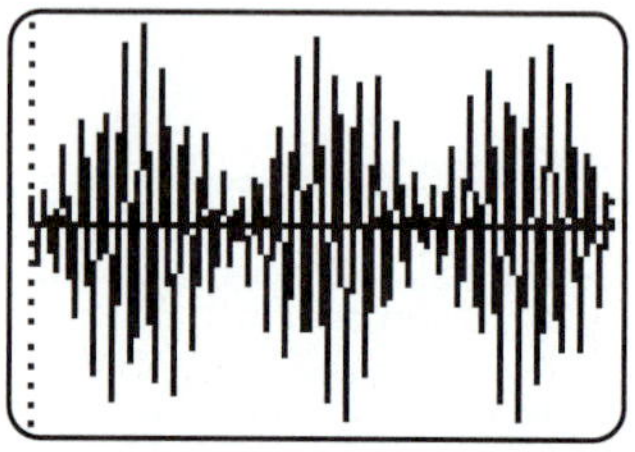

Solution

(a) Graph $Y_1 = 0.006 \sin(880\pi X) + 0.004 \sin(886\pi X)$ as shown in Figure 6.93. There appears to be 3 beats per second.

(b) Graph $Y_1 = 0.006 \sin(440\pi X) + 0.004 \sin(448\pi X)$ as shown in Figure 6.94. There appears to be 4 beats per second. ■

[0.15, 1.15, 0.05] by
[−0.01, 0.01, 0.001]

Critical Thinking

Determine a way to find the number of beats per second if the frequencies of the tones are F_1 and F_2.

FIGURE 6.94

We conclude this section by deriving the difference identity for cosine. Begin by considering the angles α and β in standard position and the unit circle as shown in Figure 6.95. The terminal side of α intersects the unit circle at $(\cos\alpha, \sin\alpha)$ and the terminal side of β intersects the unit circle at the point $(\cos\beta, \sin\beta)$. The angle formed between the terminal sides of α and β equals $\alpha - \beta$. Now consider angle $\alpha - \beta$ in standard position. Its terminal side intersects the unit circle at the point $(\cos(\alpha - \beta), \sin(\alpha - \beta))$. This is shown in Figure 6.96.

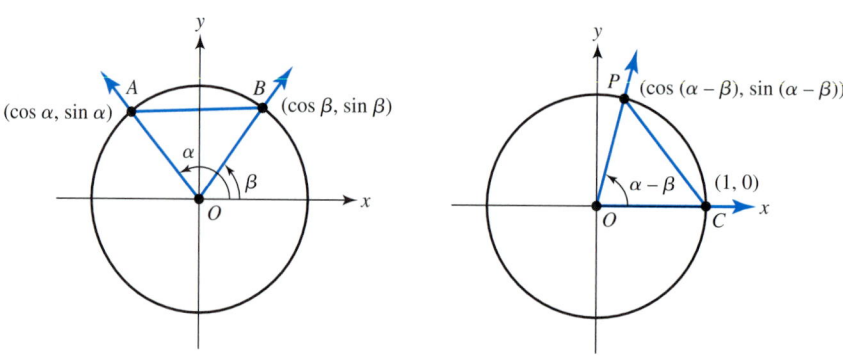

FIGURE 6.95   **FIGURE 6.96**

Since triangles *ABO* and *PCO* are congruent, the distance from *A* to *B* in Figure 6.95 equals the distance from *P* to *C* in Figure 6.96.

$$\sqrt{(\cos\alpha - \cos\beta)^2 + (\sin\alpha - \sin\beta)^2} =$$

$$\sqrt{(\cos((\sigma - \beta) - 1)^2 + (\sin((\alpha - \beta) - 0)^2}$$

Squaring both sides and clearing parentheses produces

$$\cos^2\alpha - 2\cos\alpha\cos\beta + \cos^2\beta + \sin^2\alpha - 2\sin\alpha\sin\beta + \sin^2\beta =$$

$$\cos^2(\alpha - \beta) - 2\cos(\alpha - \beta) + 1 + \sin^2(\alpha - \beta).$$

Since $\sin^2\theta + \cos^2\theta = 1$ for any θ, the above equation simplifies to

$$2 - 2\cos\alpha\cos\beta - 2\sin\alpha\sin\beta = 2 - 2\cos(\alpha - \beta).$$

Solving this equation for $\cos(\alpha - \beta)$ gives the cosine difference identity

$$\cos(\alpha - \beta) = \cos\alpha\cos\beta + \sin\alpha\sin\beta.$$

6.4 PUTTING IT ALL TOGETHER

Sum and difference identities occur in a variety of applications such as music, physiology, and electricity. The sum and difference identities are also used in calculus. The following table lists the important identities in this section.

Cosine Sum and Difference Identities
$\cos(\alpha + \beta) = \cos\alpha\cos\beta - \sin\alpha\sin\beta$
$\cos(\alpha - \beta) = \cos\alpha\cos\beta + \sin\alpha\sin\beta$
Sine Sum and Difference Identities
$\sin(\alpha + \beta) = \sin\alpha\cos\beta + \cos\alpha\sin\beta$
$\sin(\alpha - \beta) = \sin\alpha\cos\beta - \cos\alpha\sin\beta$
Tangent Sum and Difference Identites
$\tan(\alpha + \beta) = \dfrac{\tan\alpha + \tan\beta}{1 - \tan\alpha\tan\beta} \qquad \tan(\alpha - \beta) = \dfrac{\tan\alpha - \tan\beta}{1 + \tan\alpha\tan\beta}$
Confunction Identities
$\sin\theta = \cos(90° - \theta) \qquad \cos\theta = \sin(90° - \theta)$
$\tan\theta = \cot(90° - \theta) \qquad \cot\theta = \tan(90° - \theta)$
$\sec\theta = \csc(90° - \theta) \qquad \csc\theta = \sec(90° - \theta)$

6.4 EXERCISES

Tape 11

Sum and Difference Identities

Exercises 1–10: Find the exact value for each expression. Support your result numerically.

1. $\sin 15°$

2. $\sin 105°$ *(Hint: $105° = 45° + 60°$)*

3. $\tan 15°$

4. $\sin 75°$

5. $\cos 75°$

6. $\cos 105°$

7. $\sin \dfrac{\pi}{12}$　$\left(Hint: \dfrac{\pi}{3} - \dfrac{\pi}{4} = \dfrac{\pi}{12} \right)$

8. $\cos \dfrac{5\pi}{12}$

9. $\sin \dfrac{5\pi}{12}$

10. $\tan \dfrac{\pi}{12}$

Exercises 11–16: Complete the following for the identity.
  (a) *Give graphical, numerical, and verbal support for the identity.*
  (b) *Verify the identity symbolically.*

11. $\sin \left(t + \dfrac{\pi}{2} \right) = \cos t$

12. $\sin \left(t + \dfrac{3\pi}{2} \right) = -\cos t$

13. $\cos (t + \pi) = -\cos t$

14. $\cos \left(t + \dfrac{3\pi}{2} \right) = \sin t$

15. $\sec \left(t - \dfrac{\pi}{2} \right) = \csc t$

16. $\tan \left(t + \dfrac{\pi}{2} \right) = -\cot t$

Exercises 17–20: Use a difference identity to verify the cofunction identity. Give graphical or numerical support.

17. $\cos (90° - \theta) = \sin \theta$

18. $\tan (90° - \theta) = \cot \theta$

19. $\sec (90° - \theta) = \csc \theta$

20. $\csc (90° - \theta) = \sec \theta$

Exercises 21–26: (Refer to Example 5.) Find the following.
  (a) $\sin (\alpha + \beta)$
  (b) $\cos (\alpha + \beta)$
  (c) $\tan (\alpha + \beta)$
  (d) *The quadrant containing $\alpha + \beta$*

21. $\sin \alpha = \dfrac{3}{5}$ and $\sin \beta = \dfrac{5}{13}$, α and β in quadrant I

22. $\cos \alpha = -\dfrac{12}{13}$ and $\cos \beta = -\dfrac{5}{13}$, α and β in quadrant II

23. $\sin \alpha = \dfrac{-8}{17}$ and $\cos \beta = \dfrac{11}{61}$, α in quadrant III and β in quadrant I

24. $\cos \alpha = -\dfrac{24}{25}$ and $\sin \beta = \dfrac{4}{5}$, α in quadrant II and β in quadrant I

25. $\cos \alpha = \dfrac{-3}{5}$ and $\cos \beta = \dfrac{12}{13}$, α in quadrant III and β in quadrant IV

26. $\tan \alpha = \dfrac{3}{4}$ and $\cos \beta = -\dfrac{4}{5}$, α in quadrant I and β in quadrant III

Exercises 27–34: Verify the identity symbolically.

27. $\cos \left(t - \dfrac{\pi}{4} \right) = \dfrac{\sqrt{2}}{2} (\cos t + \sin t)$

28. $\sin \left(t + \dfrac{\pi}{4} \right) = \dfrac{\sqrt{2}}{2} (\cos t + \sin t)$

29. $\tan \left(t + \dfrac{\pi}{4} \right) = \dfrac{1 + \tan t}{1 - \tan t}$

30. $\dfrac{\sin (x - y)}{\sin (x + y)} = \dfrac{\tan x - \tan y}{\tan x + \tan y}$

31. $\sin 2t = 2 \sin t \cos t$

32. $\cos 2t = \cos^2 t - \sin^2 t$

33. $\sin (\alpha + \beta) + \sin (\alpha - \beta) = 2 \sin \alpha \cos \beta$

34. $\cos (\alpha + \beta) + \cos (\alpha - \beta) = 2 \cos \alpha \cos \beta$

Exercises 35–36: Solve Example 8 using the given information.

35. $\tan \alpha = \dfrac{6}{7}$ and $\tan \beta = \dfrac{5}{7}$

36. $\cot \alpha = \dfrac{8}{13}$ and $\cot \beta = \dfrac{11}{13}$

Lines and Slopes

Exercises 37–40: Suppose two lines, l_1 and l_2, intersect the x-axis making angles α and β as shown in the accompanying figure. Then the slopes of l_1 and l_2 satisfy $m_1 = \tan \alpha$ and $m_2 = \tan \beta$, respectively. If l_1 and l_2 intersect with angle θ as shown, then it follows that $\beta = \alpha + \theta$, or equivalent, $\theta = \beta - \alpha$.

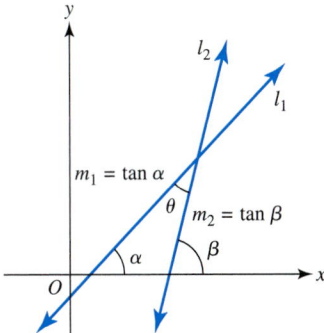

37. Use a difference identity for tangent to show that
$$\tan \theta = \frac{m_2 - m_1}{1 + m_1 m_2}.$$

38. Is the formula in the previous exercise valid if β is an obtuse angle? Explain.

39. Find θ for the two intersecting lines given by $y = 2x - 3$ and $y = \frac{3}{5}x + 1$.

40. Find θ for the two intersecting lines given by $y = \frac{1}{2}x + 1$ and $y = 3 - x$.

Applications

41. *Back Stress*   (Refer to Example 6.)
   (a) Suppose a 200-pound person bends at the waist so that $\theta = 45°$. Estimate the force exerted by the person's back muscles.
   (b) Approximate the value of θ that results in the back muscles exerting a force of 400 pounds.

42. *Sound Waves*   Sound is a result of waves applying pressure to a person's ear drum. For a particular sound wave radiating outward, the trigonometric function $P(r) = \frac{a}{r} \cos(\pi r - 1000t)$ can be used to express the pressure at a radius of r feet from the source after t seconds. In this formula, a is the maximum sound pressure at the source measured in pounds per square foot. (Source: L. Beranek, *Noise and Vibration Control.*)

   (a) Let $a = 0.4$, $t = 1$, and graph the sound pressure for $0 \le r \le 20$. What happens to the pressure P as the radius r increases?
   (b) Use a difference identity to simplify the expression for $P(r)$ when r is an even integer.

43. *Modeling Musical Tones*   (Refer to Example 7.) Let the pressure exerted by two sound waves in grams per square meter be given by
$$P_1(t) = 4 \cos(220\pi t) \quad \text{and}$$
$$P_2(t) = 3 \sin(220\pi t)$$
where t is in seconds.
   (a) Graph the total pressure $P = P_1 + P_2$ in $[0, 0.02, 0.001]$ by $[-6, 6, 1]$.
   (b) Use the graph to estimate values for a and k such that $P = a \sin(220\pi t + k)$.
   (c) Use a sum or difference identity for sine to verify that
$$a \sin(220\pi t + k) \approx$$
$$3 \sin(220\pi t) + 4 \cos(220\pi t).$$

44. *Electricity*   When voltages $V_1 = 50 \sin(120\pi t)$ and $V_2 = 120 \cos(120\pi t)$ are applied to the same circuit, the resulting voltage V is equal to their sum. (Source: D. Bell, *Fundamentals of Electric Circuits.*)
   (a) Graph $V = V_1 + V_2$ in $[0, 0.05, 0.01]$ by $[-160, 160, 40]$.
   (b) Use the graph to estimate values for a and k so that $V = a \sin(120\pi t + k)$.
   (c) Use a sum or difference identity for sine to verify part (b).

Exercises 45–46: Music and Beats (Refer to Example 9.) Given two musical tones P_1 and P_2, graph their sum in $[0.2, 1.2, 0.05]$ by $[-0.01, 0.01, 0.001]$. Count the number of beats in one second.

45. $P_1 = 0.007 \sin(450\pi t)$, $P_2 = 0.005 \sin(454\pi t)$

46. $P_1 = 0.004 \cos(830\pi t)$, $P_2 = 0.005 \sin(836\pi t)$

Writing about Mathematics

1. Are $\sin(45° + 30°)$ and $\sin 45° + \sin 30°$ equivalent expressions? Explain your answer.

2. Are the expressions $\cos(\alpha - \beta)$ and $\cos \alpha - \cos \beta$ equal? Give numerical support for your answer.

CHECKING BASIC CONCEPTS FOR SECTIONS 6.3 AND 6.4

1. Find the reference angle of each angle.

    (a) $225°$ **(b)** $\dfrac{5\pi}{6}$

2. Solve each equation for θ in $[0°, 360°)$ symbolically. Give graphical and numerical support.

    (a) $\cos\theta = \dfrac{1}{2}$ **(b)** $\sin\theta = -\dfrac{\sqrt{3}}{2}$

3. Find all solutions to the given equation.

    (a) $\sin t = -\cos t$ **(b)** $2\sin^2 t = 1 - \cos t$

4. Use a sum or difference identity to find $\cos\dfrac{\pi}{12}$.

5. Verify the identity $\sin(t - \pi) = -\sin t$ symbolically. Give graphical and numerical support.

6.5 Multiple-Angle Identities

Double-Angle Identities • Half-Angle Formulas • Solving Equations • Product-to-Sum and Sum-to-Product Identities (Optional)

Introduction

In 1831 Michael Faraday discovered that when a wire is passed near a magnet, a small electric current is produced in the wire. This property is used to generate electric current for homes, schools, and businesses throughout the world. By rotating thousands of wires near large electromagnets, massive amounts of electricity can be produced. In one year utilities in the United States generate enough electricity to power a 100-watt light bulb for over 3 billion years!

    Voltage, amperage, and wattage are quantities that may be modeled by sinusoidal graphs and functions. To model electricity and other phenomena, trigonometric functions and identities are used. In this section we are introduced to several important multiple-angle identities. (Sources: R. Weidner and R. Sells, *Elementary Classical Physics,* Vol. 2; J. Wright, *The Universal Almanac 1997.*)

Double-Angle Identities

The sum identities can be used to derive double-angle identities for sine, cosine, and tangent.

Sine Double-Angle Identity

$$\begin{aligned}\sin 2\theta &= \sin(\theta + \theta)\\ &= \sin\theta\cos\theta + \cos\theta\sin\theta\\ &= 2\sin\theta\cos\theta\end{aligned}$$

Cosine Double-Angle Identities

$$\begin{aligned}\cos 2\theta &= \cos(\theta + \theta)\\ &= \cos\theta\cos\theta - \sin\theta\sin\theta\\ &= \cos^2\theta - \sin^2\theta\end{aligned}$$

Applying the identity $\sin^2 \theta + \cos^2 \theta = 1$, the expression $\cos^2 \theta - \sin^2 \theta$ can be written as

$$\cos^2 \theta - \sin^2 \theta = \cos^2 \theta - (1 - \cos^2 \theta)$$
$$= 2 \cos^2 \theta - 1$$

or as

$$\cos^2 \theta - \sin^2 \theta = (1 - \sin^2 \theta) - \sin^2 \theta$$
$$= 1 - 2 \sin^2 \theta.$$

Tangent Double-Angle Identity

$$\tan 2\theta = \tan (\theta + \theta)$$
$$= \frac{\tan \theta + \tan \theta}{1 - \tan \theta \tan \theta}$$
$$= \frac{2 \tan \theta}{1 - \tan^2 \theta}$$

A summary of these double-angle identities is given.

Double-angle identities

$$\sin 2\theta = 2 \sin \theta \cos \theta$$
$$\cos 2\theta = \cos^2 \theta - \sin^2 \theta = 2 \cos^2 \theta - 1 = 1 - 2 \sin^2 \theta$$
$$\tan 2\theta = \frac{2 \tan \theta}{1 - \tan^2 \theta}$$

The next example illustrates that $\sin 2\theta \neq 2 \sin \theta$.

EXAMPLE 1   *Using double-angle identities*

Verify numerically, graphically, and symbolically that the expressions $\sin 2\theta$ and $2 \sin \theta$ are *not* equivalent.

Solution

Numerical Verification   Table $Y_1 = \sin (2X)$ and $Y_2 = 2 \sin (X)$, starting at $x = 0$ and incrementing by $\pi/12 \approx 0.2618$ as shown in Figure 6.97. Notice that except for $x = 0$, $Y_1 \neq Y_2$. Therefore, the expressions $\sin 2\theta$ and $2 \sin \theta$ are not equivalent.

X	Y1	Y2
0	0	0
.2618	.5	.51764
.5236	.86603	1
.7854	1	1.4142
1.0472	.86603	1.7321
1.309	.5	1.9319
1.5708	0	2

X=0

FIGURE 6.97 Radian Mode

$[-2\pi, 2\pi, \pi/2]$ by $[-4, 4, 1]$

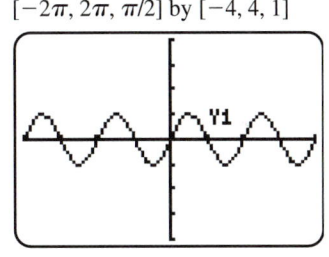

FIGURE 6.98

$[-2\pi, 2\pi, \pi/2]$ by $[-4, 4, 1]$

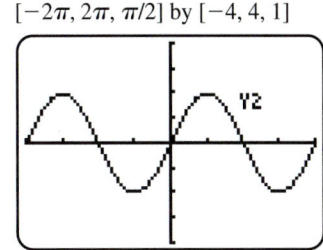

FIGURE 6.99

Graphical Verification Graph $Y_1 = \sin(2X)$ and $Y_2 = 2\sin(X)$ as shown in Figures 6.98 and 6.99. Notice that the graphs are different.

Symbolic Verification The expressions $\sin 2\theta$ and $2\sin\theta$ are not equivalent, rather

$$\sin 2\theta = 2\sin\theta\cos\theta \neq 2\sin\theta$$

unless $\cos\theta = 1$. ∎

Critical Thinking

Verify numerically, graphically, and symbolically that $\cos 2\theta$ and $2\cos\theta$ are not equivalent expressions.

EXAMPLE 2 *Using double-angle identities*

Given $\cos\theta = -\dfrac{12}{13}$ and $\sin\theta > 0$, find $\sin 2\theta$, $\cos 2\theta$, and $\tan 2\theta$. Support your results numerically.

Solution

Since $\cos\theta < 0$ and $\sin\theta > 0$, θ is contained in quadrant II. One possibility for θ is shown in Figure 6.100. We see that $\sin\theta = \dfrac{5}{13}$.

Using double-angle identities, we obtain the following results.

$$\sin 2\theta = 2\sin\theta\cos\theta$$
$$= 2\cdot\frac{5}{13}\cdot\left(-\frac{12}{13}\right)$$
$$= -\frac{120}{169}$$

$$\cos 2\theta = \cos^2\theta - \sin^2\theta$$
$$= \left(-\frac{12}{13}\right)^2 - \left(\frac{5}{13}\right)^2$$
$$= \frac{119}{169}$$

$$\tan 2\theta = \frac{\sin 2\theta}{\cos 2\theta}$$
$$= \frac{-120/169}{119/169}$$
$$= -\frac{120}{119}$$

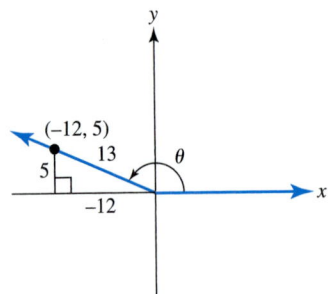

FIGURE 6.100

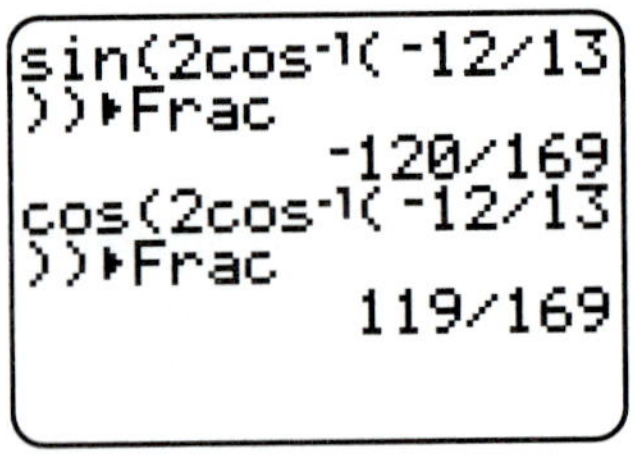

FIGURE 6.101

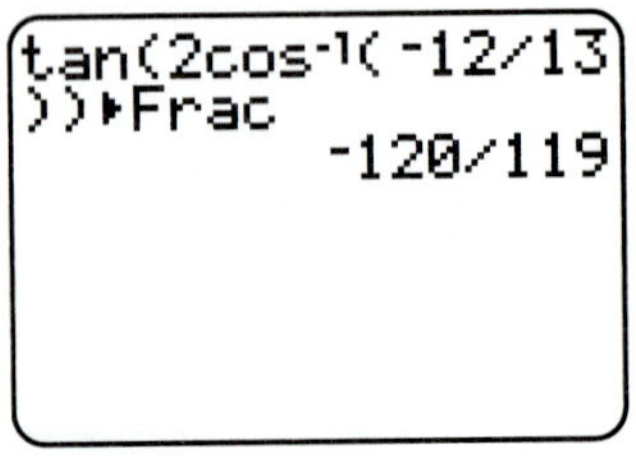

FIGURE 6.102

Numerical Support Since θ is in quadrant II, we can support these results numerically by letting $\theta = \arccos\left(-\dfrac{12}{13}\right) \approx 157.38°$ and performing the calculations shown in Figures 6.101 and 6.102. ∎

EXAMPLE 3   *Deriving a triple-angle identity*

Write $\cos 3\theta$ in terms of $\cos \theta$. Give graphical and numerical support.

Solution

$$\cos 3\theta = \cos (2\theta + \theta)$$

$$= \cos 2\theta \cos \theta - \sin 2\theta \sin \theta \qquad \text{Sum identity for cosine}$$

$$= (2 \cos^2\theta - 1) \cos \theta - (2 \sin \theta \cos \theta) \sin \theta \qquad \text{Double-angle identities}$$

$$= 2 \cos^3 \theta - \cos \theta - 2 \sin^2 \theta \cos \theta \qquad \text{Multiply.}$$

$$= 2 \cos^3 \theta - \cos \theta - 2 (1 - \cos^2 \theta) \cos \theta \qquad \text{Apply } \sin^2 \theta + \cos^2 \theta = 1.$$

$$= 2 \cos^3 \theta - \cos \theta - 2 \cos \theta + 2 \cos^3 \theta \qquad \text{Distributive property}$$

$$= 4 \cos^3 \theta - 3 \cos \theta \qquad \text{Combine like terms.}$$

Thus, $\cos 3\theta = 4 \cos^3 \theta - 3 \cos \theta$. This identity is supported in Figures 6.103–6.105, where $Y_1 = \cos (3X)$ and $Y_2 = 4 (\cos(X))^3 3 - 3 \cos (X)$.

$[-2\pi, 2\pi, \pi/2]$ by $[-2, 2, 1]$ $[-2\pi, 2\pi, \pi/2]$ by $[-2, 2, 1]$

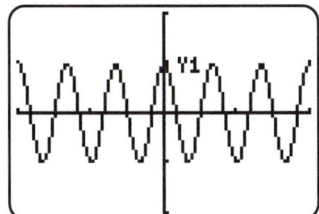

FIGURE 6.103

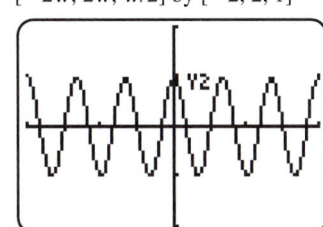

FIGURE 6.104

X	Y1	Y2
0	1	1
.2618	.70711	.70711
.5236	0	0
.7854	-.7071	-.7071
1.0472	-1	-1
1.309	-.7071	-.7071
1.5708	0	0

X=0

FIGURE 6.105 Radian Mode ∎

Half-Angle Formulas

Power-reducing identities for sine, cosine, and tangent can be derived using the double-angle identities.

Sine Power-Reducing Identity   Since

$$\cos 2\theta = 1 - 2 \sin^2 \theta,$$

we can solve for $\sin^2 \theta$ to obtain

$$\sin^2 \theta = \frac{1 - \cos 2\theta}{2}.$$

Cosine Power-Reducing Identity   Solving the identity

$$\cos 2\theta = 2 \cos^2 \theta - 1$$

for $\cos^2 \theta$ gives

$$\cos^2 \theta = \frac{1 + \cos 2\theta}{2}.$$

Tangent Power-Reducing Identity We can use the power-reducing identities

$$\sin^2\theta = \frac{1 - \cos 2\theta}{2} \quad \text{and} \quad \cos^2\theta = \frac{1 + \cos 2\theta}{2}$$

to derive

$$\tan^2\theta = \frac{\sin^2\theta}{\cos^2\theta} = \frac{1 - \cos 2\theta}{1 + \cos 2\theta}.$$

A summary of these identities is now given.

Power-reducing identities

$$\sin^2\theta = \frac{1 - \cos 2\theta}{2} \qquad \cos^2\theta = \frac{1 + \cos 2\theta}{2} \qquad \tan^2\theta = \frac{1 - \cos 2\theta}{1 + \cos 2\theta}$$

EXAMPLE 4   *Using a power-reducing identity*

Find the exact value of $\sin^2(22.5°)$. Support your result numerically.

Solution
Let $\theta = 22.5°$ and $2\theta = 45°$.

$$\sin^2 22.5° = \frac{1 - \cos 45°}{2}$$

$$= \frac{1 - \sqrt{2}/2}{2}$$

$$= \frac{2 - \sqrt{2}}{4}$$

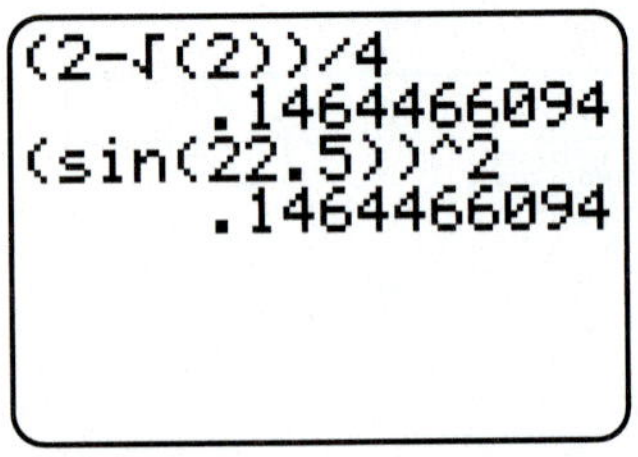

```
(2-√(2))/4
       .1464466094
(sin(22.5))^2
       .1464466094
```

FIGURE 6.106 Degree Mode   Numerical support is shown in Figure 6.106.   ■

In the next example we apply a power-reducing identity to electrical circuits.

EXAMPLE 5   *Using a power-reducing identity to analyze wattage*

Amperage I is a measure of the amount of electricity passing through a wire and voltage V is a measure of the force "pushing" the electricity. The wattage W consumed by an electrical device may be calculated using the equation $W = VI$. (Source: G. Wilcox and C. Hesselberth, *Electricity For Engineering Technology.*)
(a) Voltage in a household circuit is given by $V = 160 \sin(120\pi t)$, where t is in seconds. Suppose the amperage flowing through a toaster is given by $I = 12 \sin(120\pi t)$. Graph the wattage W consumed by the toaster in $[0, 0.04, 0.01]$ by $[-200, 2200, 200]$.
(b) Write the wattage as $W = a \cos(k\pi t) + d$, where a, k, and d are constants.
(c) Compare the periods of the voltage, amperage, and wattage.
(d) The wattage of this toaster equals half the maximum of W. Find the wattage.

Solution
(a) Since $W = VI$, graph $Y_3 = Y_1 * Y_2$ where $Y_1 = 160 \sin(120\pi X)$ and $Y_2 = 12 \sin(120\pi X)$ as shown in Figure 6.107.

(b) $W = VI$

$\qquad = 160 \sin{(120\pi t)}\ 12 \sin{(120\pi t)}$ Substitute for V and I.

$\qquad = 1920 \sin^2{(120\pi t)}$ Multiply.

$\qquad = 1920 \cdot \dfrac{1 - \cos{(240\pi t)}}{2}$ Power-reducing identity

$\qquad = 960 - 960 \cos{(240\pi t)}$ Simplify.

Thus, let $a = -960$, $k = 240$, and $d = 960$. Then, the wattage can be written as

$$W = -960 \cos{(240\pi t)} + 960.$$

(c)   The period for both V and I is $\dfrac{2\pi}{120\pi} = \dfrac{1}{60}$ second, while the period for W is

$\dfrac{2\pi}{240\pi} = \dfrac{1}{120}$ second. This is supported in Figure 6.108, where the graph of V

requires twice as much time as W to complete one oscillation.

(d)   The maximum wattage is 1920 watts. Half this amount is 960 watts, which is the wattage rating for the toaster.

$[0, 0.04, 0.01]$ by $[-200, 2200, 200]$ $[0, 0.04, 0.01]$ by $[-200, 200, 50]$

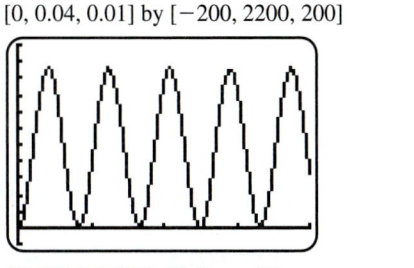   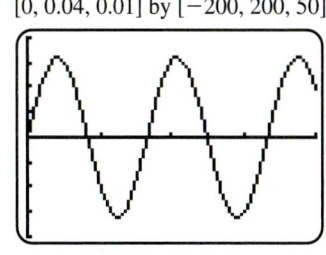

FIGURE 6.107 Wattage W **FIGURE 6.108** Voltage W ■

We can obtain half-angle formulas using the power-reducing identities.

$$\sin^2 x = \dfrac{1 - \cos 2x}{2} \qquad \text{Power-reducing identity}$$

$$\sin x = \pm \sqrt{\dfrac{1 - \cos 2x}{2}} \qquad \text{Square root property}$$

$$\sin \dfrac{\theta}{2} = \pm \sqrt{\dfrac{1 - \cos \theta}{2}} \qquad \text{Let } x = \dfrac{\theta}{2} \text{ and } 2x = \theta.$$

The following gives some half-angle formulas. (Verification of the second and third half-angle formulas for tangent is done in Exercises 65 and 66.)

Half-angle formulas

$$\sin \dfrac{\theta}{2} = \pm \sqrt{\dfrac{1 - \cos \theta}{2}} \qquad \cos \dfrac{\theta}{2} = \pm \sqrt{\dfrac{1 + \cos \theta}{2}}$$

$$\tan \dfrac{\theta}{2} = \pm \sqrt{\dfrac{1 - \cos \theta}{1 + \cos \theta}} \qquad \tan \dfrac{\theta}{2} = \dfrac{1 - \cos \theta}{\sin \theta} \qquad \tan \dfrac{\theta}{2} = \dfrac{\sin \theta}{1 + \cos \theta}$$

To decide whether a positive or negative sign should be used in a half-angle formula, we must determine the quadrant containing $\theta/2$. This is illustrated in the next example.

EXAMPLE 6   *Finding an exact value using a half-angle formula*

Find the exact values of $\sin(-15°)$. Support your results numerically.

Solution

We can use the fact that $\cos(-30°) = \dfrac{\sqrt{3}}{2}$ to find $\sin(-15°)$. We choose the negative sign in the half-angle formula since the sine function is negative in Quadrant IV.

$$\sin(-15°) = \sin\left(\frac{-30°}{2}\right) \qquad \text{Let } \frac{\theta}{2} = -15° \text{ and } \theta = -30°.$$

$$= -\sqrt{\frac{1 - \cos(-30°)}{2}} \qquad \text{Half-angle formula}$$

$$= -\sqrt{\frac{1 - \sqrt{3}/2}{2}} \qquad \cos(-30°) = \frac{\sqrt{3}}{2}$$

$$= -\sqrt{\frac{2 - \sqrt{3}}{4}} \qquad \text{Multiply numerator and denominator by 2.}$$

$$= -\frac{\sqrt{2 - \sqrt{3}}}{2} \qquad \sqrt{4} = 2$$

Numerical support is given in Figure 6.109. ∎

```
-√(2-√(3))/2
           -.2588190451
sin(-15)
           -.2588190451
```

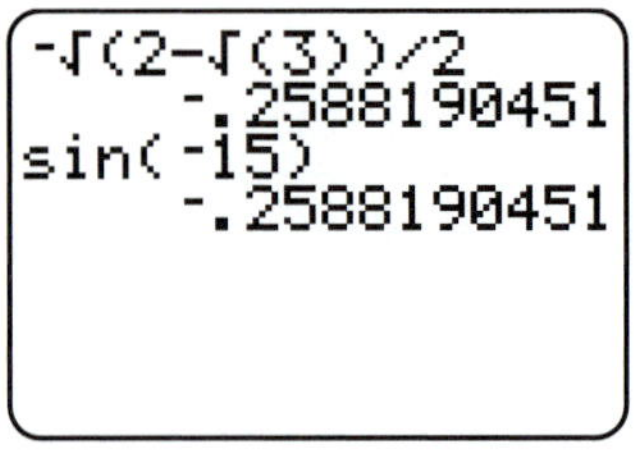

FIGURE 6.109 Degree Mode

Critical Thinking ———————————————
Find the exact value of $\tan(-15°)$. Support your result numerically.

EXAMPLE 7   *Finding exact values using half-angle formulas*

If $\cos\theta = -\dfrac{3}{5}$ and $90° \le \theta \le 180°$, find $\sin\dfrac{\theta}{2}$, $\cos\dfrac{\theta}{2}$, and $\tan\dfrac{\theta}{2}$. Support your results numerically.

Solution

Since $90° \le \theta \le 180°$, it follows that $45° \le \dfrac{\theta}{2} \le 90°$. In quadrant I the values for $\sin\dfrac{\theta}{2}$, $\cos\dfrac{\theta}{2}$, and $\tan\dfrac{\theta}{2}$ are all positive.

$$\sin\frac{\theta}{2} = \sqrt{\frac{1 - \cos\theta}{2}} = \sqrt{\frac{1 + 3/5}{2}} = \sqrt{\frac{4}{5}} \text{ or } \frac{2\sqrt{5}}{5}$$

$$\cos\frac{\theta}{2} = \sqrt{\frac{1 + \cos\theta}{2}} = \sqrt{\frac{1 - 3/5}{2}} = \sqrt{\frac{1}{5}} \text{ or } \frac{\sqrt{5}}{5}$$

$$\tan\frac{\theta}{2} = \sqrt{\frac{1 - \cos\theta}{1 + \cos\theta}} = \sqrt{\frac{1 + 3/5}{1 - 3/5}} = 2$$

The value for $\tan\dfrac{\theta}{2}$ could also be obtained using $\tan\dfrac{\theta}{2} = \dfrac{\sin(\theta/2)}{\cos(\theta/2)}$.

Numerical support can be obtained by finding the exact value of θ. If θ satisfies $\cos\theta = -3/5$ and $90° \le \theta \le 180°$, then $\theta = \cos^{-1}(-3/5) \approx 126.87°$.

$$\sin\frac{126.87°}{2} \approx 0.8944 \approx \frac{2\sqrt{5}}{5}$$

$$\cos\frac{126.87°}{2} \approx 0.4472 \approx \frac{\sqrt{5}}{5}$$

$$\tan\frac{126.87°}{2} \approx 2 \qquad\blacksquare$$

Solving Equations

When trigonometric functions are used in modeling, it is common to use these functions to make predictions. This often results in trigonometric equations. In Example 5, voltage in a household outlet was modeled by

$$f(t) = 160\sin(120\pi t),$$

where t is time in seconds. If we wanted to find the times when the voltage equals 80 volts, we could solve the equation $f(t) = 80$ as follows.

$$160\sin(120\pi t) = 80 \qquad f(t) = 80$$

$$\sin(120\pi t) = \frac{1}{2} \qquad \text{Divide by 160.}$$

The equation $\sin\theta = \dfrac{1}{2}$ is satisfied whenever

$$\theta = \frac{\pi}{6} + 2\pi n \qquad\text{or}\qquad \theta = \frac{5\pi}{6} + 2\pi n,$$

where n is an integer. The expression $2\pi n$ is included because $\sin\theta$ is periodic with 2π. Substituting $120\pi t$ for θ results in

$$120\pi t = \frac{\pi}{6} + 2\pi n \qquad\text{or}\qquad 120\pi t = \frac{5\pi}{6} + 2\pi n.$$

Dividing these equations by 120π produces the following solutions.

$$t = \frac{1}{720} + \frac{n}{60} \qquad\text{or}\qquad t = \frac{5}{720} + \frac{n}{60}$$

Each value of n gives different solutions. For example, if $n = 1$ then

$$t = \frac{1}{720} + \frac{1}{60} = \frac{13}{720} \qquad\text{and}\qquad t = \frac{5}{720} + \frac{1}{60} = \frac{17}{720}.$$

The voltage equals 80 volts after $\dfrac{13}{720}$ second and after $\dfrac{17}{720}$ second.

EXAMPLE 8 *Solving a trigonometric equation*

Solve the equation $\cos\theta - \sin 2\theta = 0$ symbolically, graphically, and numerically for $0° \le \theta < 360°$.

$[0°, 352.5°, 30°]$ by $[-2, 2, 1]$

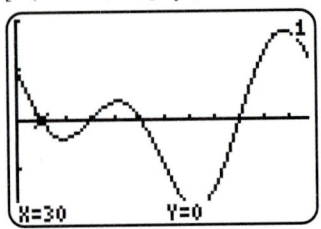

FIGURE 6.110

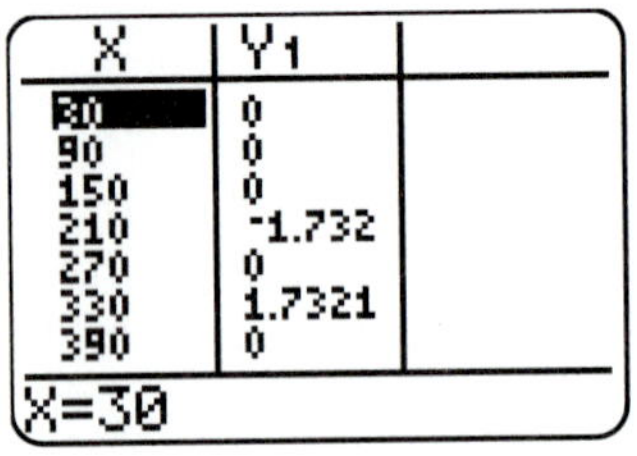

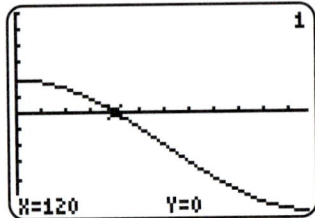

FIGURE 6.111 Degree Mode

Solution

Symbolic Solution We begin by applying the double-angle identity for sine.

$\cos\theta - \sin 2\theta = 0$	Given equation
$\cos\theta - 2\sin\theta\cos\theta = 0$	Double-angle identity
$\cos\theta(1 - 2\sin\theta) = 0$	Factor out $\cos\theta$.
$\cos\theta = 0$ or $1 - 2\sin\theta = 0$	Zero-product property
$\cos\theta = 0$ or $\sin\theta = \dfrac{1}{2}$	Solve for $\sin\theta$.
$\theta = 90°, 270°$ or $\theta = 30°, 150°$	Solve for θ.

On the interval $[0°, 360°)$, the solutions are $30°$, $90°$, $150°$, and $270°$.

Graphical Solution In Figure 6.110 graphical support is given, where $Y_1 = \cos(X) - \sin(2X)$. Notice that there are four x-intercepts that correspond to the four symbolic solutions.

Numerical Solution In Figure 6.111 numerical support is shown. ∎

EXAMPLE 9 *Solving a trigonometric equation*

Solve the equation $4\cos\dfrac{x}{2} - 2 = 0$ symbolically, graphically, and numerically for $0° \le \theta < 360°$.

$[0°, 352.5°, 30°]$ by $[-6, 6, 1]$

FIGURE 6.112

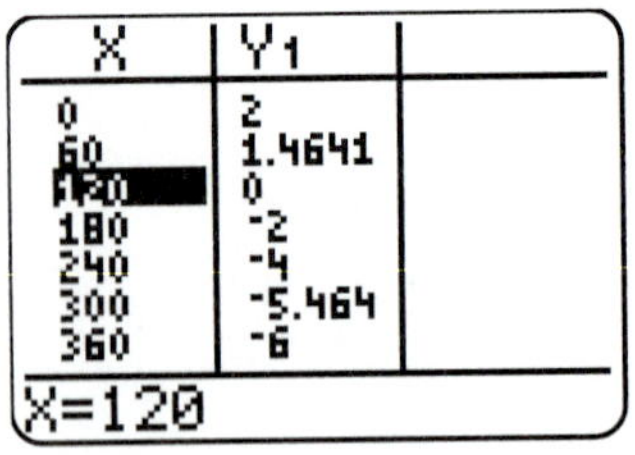

FIGURE 6.113 Degree Mode

Solution

Symbolic Solution Begin by solving for $\cos\dfrac{x}{2}$.

$4\cos\dfrac{x}{2} - 2 = 0$	Given equation
$\cos\dfrac{x}{2} = \dfrac{1}{2}$	Solve for $\cos\dfrac{x}{2}$.
$\dfrac{x}{2} = 60°$ or $\dfrac{x}{2} = 300°$	Solve for $\dfrac{x}{2}$.
$x = 120°$ or $x = 600°$	Multiply by 2.

The only solution on $[0°, 360°)$ is $120°$.

Graphical Solution In Figure 6.112 graphical support is shown, where $Y_1 = 4\cos(X/2) - 2$. There is one x-intercept at $x = 120°$ that agrees with the symbolic solution.

Numerical Solution In Figure 6.113 numerical support is shown. ∎

Product-to-Sum and Sum-to-Product Identities (Optional)

The difference identities for sine and cosine can be used to derive several identities that make it possible to rewrite a product as a sum. For example, adding the identities for $\cos(\alpha + \beta)$ and $\cos(\alpha - \beta)$ results in the following.

$$\cos(\alpha + \beta) = \cos\alpha\cos\beta - \sin\alpha\sin\beta$$
$$\cos(\alpha - \beta) = \cos\alpha\cos\beta + \sin\alpha\sin\beta$$
$$\overline{\cos(\alpha + \beta) + \cos(\alpha - \beta) = 2\cos\alpha\cos\beta}$$

Rewriting gives

$$\cos\alpha\cos\beta = \frac{1}{2}(\cos(\alpha + \beta) + \cos(\alpha - \beta)).$$

Four product-to-sum identities are given. The other three identities are derived in a similar manner.

Product-to-sum identities

$$\cos\alpha\cos\beta = \frac{1}{2}(\cos(\alpha + \beta) + \cos(\alpha - \beta))$$

$$\sin\alpha\sin\beta = \frac{1}{2}(\cos(\alpha - \beta) - \cos(\alpha + \beta))$$

$$\sin\alpha\cos\beta = \frac{1}{2}(\sin(\alpha + \beta) + \sin(\alpha - \beta))$$

$$\cos\alpha\sin\beta = \frac{1}{2}(\sin(\alpha + \beta) - \sin(\alpha - \beta))$$

EXAMPLE 10   *Using a product-to-sum identity*

Write $\cos 5\theta \cos 3\theta$ as a sum.

Solution

We begin by applying the first product-to-sum identity with $\alpha = 5\theta$ and $\beta = 3\theta$.

$$\cos 5\theta \cos 3\theta = \frac{1}{2}(\cos(5\theta + 3\theta) + \cos(5\theta - 3\theta))$$

$$= \frac{1}{2}(\cos 8\theta + \cos 2\theta) \qquad \blacksquare$$

By rewriting the product-to-sum identities, we can derive four sum-to-product identities. Sum-to-product identities have applications in areas such as musical sounds and touch-tone phones. If we let $a = \alpha + \beta$ and $b = \alpha - \beta$, then the identity

$$\cos\alpha\cos\beta = \frac{1}{2}(\cos(\alpha + \beta) + \cos(\alpha - \beta)),$$

reduces to

$$2\cos\frac{a + b}{2}\cos\frac{a - b}{2} = \cos a + \cos b.$$

Note that

$$\frac{a + b}{2} = \frac{\alpha + \beta + \alpha - \beta}{2} = \alpha \text{ and}$$

$$\frac{a - b}{2} = \frac{(\alpha + \beta) - (\alpha - \beta)}{2} = \beta.$$

The other three sum-to-product identities may be derived in a similar manner.

Sum-to-product identities

$$\cos a + \cos b = 2\cos\frac{a+b}{2}\cos\frac{a-b}{2}$$

$$\cos a - \cos b = -2\sin\frac{a+b}{2}\sin\frac{a-b}{2}$$

$$\sin a + \sin b = 2\sin\frac{a+b}{2}\cos\frac{a-b}{2}$$

$$\sin a - \sin b = 2\cos\frac{a+b}{2}\sin\frac{a-b}{2}$$

Touch-tone phones are often used to register for classes or make business transactions. One reason this can be done is that each number has its own unique pair of frequencies. For example, when 1 is pressed two frequencies of 697 hertz and 1209 hertz are simultaneously transmitted through the phone line. A *hertz* (Hz) is equal to one cycle per second. On the other hand, when 2 is pressed the pair of frequencies transmitted is 697 hertz and 1336 hertz. As a result, 2 has a different tone than 1. Table 6.1 shows the frequency pairs for the numbers 0 through 9. (Source: World Wide Web.)

TABLE 6.1   Frequencies Used in Touch-Tone Phones

Number	0	1	2	3	4	5	6	7	8	9
Frequency 1 (Hz)	941	697	697	697	770	770	770	852	852	852
Frequency 2 (Hz)	1336	1209	1336	1477	1209	1336	1477	1209	1336	1477

A tone with frequencies F_1 and F_2 can be modeled by

$$a_1\cos(2\pi F_1 t) + a_2\cos(2\pi F_2 t).$$

If both tones have the same intensity then we can let $a_1 = a_2 = 1$.

EXAMPLE 11   *Analyzing touch-tone phones*

(a) Write an expression modeling the sound of a 5 on a touch-tone phone.
(b) Rewrite the expression in part (a) as a product of trigonometric expressions.
(c) Give graphical and numerical support for your answer in part (b).

Solution

(a) From Table 6.1 we can see that $F_1 = 770$ and $F_2 = 1336$. Thus,

$$y = \cos(1540\pi t) + \cos(2672\pi t).$$

(b) Let $a = 1540\pi t$ and $b = 2672\pi t$ in the first sum-to-product identity.

$$\cos 1540\pi t + \cos 2672\pi t$$

$$= 2\cos\frac{1540\pi t + 2672\pi t}{2}\cos\frac{1540\pi t - 2672\pi t}{2}$$

$$= 2\cos(2106\pi t)\cos(-566\pi t)$$

$$= 2\cos(2106\pi t)\cos(566\pi t) \qquad\qquad \cos(-\theta) = \cos\theta$$

(c) This can be supported by graphing and tabling

$$Y_1 = \cos(1540\pi X) + \cos(2672\pi X) \qquad \text{and}$$
$$Y_2 = 2\cos(2106\pi X)\cos(566\pi X)$$

as shown in Figures 6.114–6.116. Notice that the graphs of Y_1 and Y_2 appear to be identical.

[0, 0.005, 0.001] by [−2, 2, 1]

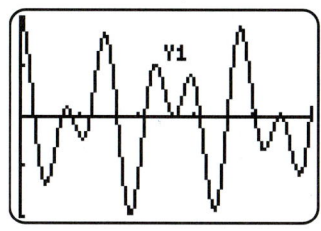

FIGURE 6.114

[0, 0.005, 0.001] by [−2, 2, 1]

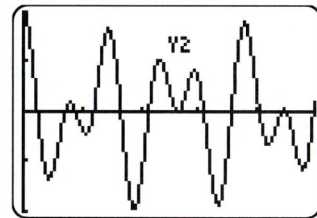

FIGURE 6.115

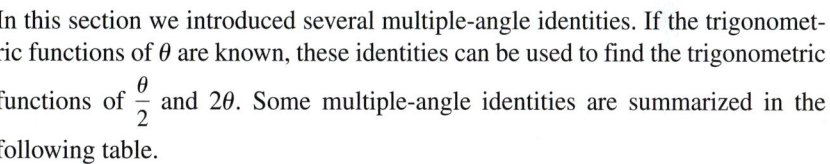

FIGURE 6.116 Radian Mode ∎

6.5 PUTTING IT ALL TOGETHER

In this section we introduced several multiple-angle identities. If the trigonometric functions of θ are known, these identities can be used to find the trigonometric functions of $\frac{\theta}{2}$ and 2θ. Some multiple-angle identities are summarized in the following table.

Double-Angle Identities
$\sin 2\theta = 2\sin\theta\cos\theta$
$\cos 2\theta = \cos^2\theta - \sin^2\theta = 2\cos^2\theta - 1 = 1 - 2\sin^2\theta$
$\tan 2\theta = \dfrac{2\tan\theta}{1 - \tan^2\theta}$

Power-Reducing Identities
$\sin^2\theta = \dfrac{1 - \cos 2\theta}{2} \qquad \cos^2\theta = \dfrac{1 + \cos 2\theta}{2} \qquad \tan^2\theta = \dfrac{1 - \cos 2\theta}{1 + \cos 2\theta}$

Half-Angle Formulas
$\sin\dfrac{\theta}{2} = \pm\sqrt{\dfrac{1 - \cos\theta}{2}} \qquad \cos\dfrac{\theta}{2} = \pm\sqrt{\dfrac{1 + \cos\theta}{2}} \qquad \tan\dfrac{\theta}{2} = \pm\sqrt{\dfrac{1 - \cos\theta}{1 + \cos\theta}}$
$\tan\dfrac{\theta}{2} = \dfrac{1 - \cos\theta}{\sin\theta} \qquad\qquad \tan\dfrac{\theta}{2} = \dfrac{\sin\theta}{1 + \cos\theta}$

6.5 EXERCISES

   Tape 11

Double-Angle Identities

Exercises 1–6: If possible, evaluate each expression and compare their values.

1. (a) $\sin 30° + \sin 30°$
   (b) $\sin 60°$

2. (a) $\sin 45° + \sin 45°$
   (b) $\sin 90°$

3. (a) $\cos 60° + \cos 60°$
   (b) $\cos 120°$

4. (a) $\cos 90° + \cos 90°$
   (b) $\cos 180°$

5. (a) $\tan 45° + \tan 45°$
   (b) $\tan 90°$

6. (a) $\tan 30° + \tan 30°$
   (b) $\tan 60°$

Exercises 7 and 8: (Refer to Example 1.) Verify graphically, numerically, and symbolically that the two expressions are not equivalent.

7. $\tan 2\theta$, $2 \tan \theta$
8. $\cos 3\theta$, $3 \cos \theta$

Exercises 9–16: (Refer to Example 2.) Complete the following.

(a) *Find $\sin 2\theta$, $\cos 2\theta$, and $\tan 2\theta$.*
(b) *Support your results numerically.*

9. $\cos \theta = \dfrac{4}{5}$ and $\sin \theta = \dfrac{3}{5}$

10. $\sin \theta = \dfrac{12}{13}$ and $\cos \theta = \dfrac{5}{13}$

11. $\sin \theta = -\dfrac{24}{25}$ and $\cos \theta > 0$

12. $\cos \theta = -\dfrac{7}{25}$ and $\tan \theta > 0$

13. $\sin \theta = -\dfrac{11}{61}$ and $\sec \theta > 0$

14. $\csc \theta = -2$ and $\sec \theta > 0$

15. $\tan \theta = \dfrac{7}{24}$ and $\cos \theta < 0$

16. $\cot \theta = \dfrac{5}{12}$ and $\sin \theta > 0$

Exercises 17–22: Rewrite the expression using a double-angle identity.

17. $2 \cos \theta \sin \theta$
18. $2 \sin 2\theta \cos 2\theta$
19. $\sin \theta \cos \theta$
20. $(\sin \theta - \cos \theta)(\sin \theta + \cos \theta)$
21. $2 \cos^2 2\theta - 1$
22. $1 - 2 \sin^2 3\theta$

Exercises 23–26: Use a fundamental identity to write the expression as one term.

23. $\sin^2 3\theta + \cos^2 3\theta$
24. $1 + \tan^2 2\theta$
25. $\csc^2 5x - 1$
26. $\sin^2 8x + \cos^2 8x$

Power-Reducing Identities and Half-Angle Formulas

Exercises 27–30: Use a power-reducing identity to find the exact value of the expression. Support your result numerically.

27. $\cos^2 (22.5°)$
28. $\sin^2 (15°)$
29. $\tan^2 (75°)$
30. $\csc^2 (105°)$

Exercises 31–34: Use a half-angle formula to find the exact value of the expression. Support your result numerically.

31. (a) $\cos 15°$ (b) $\tan(-15°)$
32. (a) $\sin 67.5°$ (b) $\cos(-67.5°)$
33. (a) $\tan \dfrac{\pi}{8}$ (b) $\sin\left(-\dfrac{\pi}{8}\right)$
34. (a) $\cos \dfrac{5\pi}{12}$ (b) $\cos\left(-\dfrac{5\pi}{12}\right)$

Exercises 35–40: Use a half-angle formula to simplify the expression. Check your result numerically.

35. $\sqrt{\dfrac{1 - \cos 60°}{2}}$ 36. $\sqrt{\dfrac{1 + \cos 60°}{2}}$

37. $\sqrt{\dfrac{1 + \cos 50°}{2}}$ 38. $\sqrt{\dfrac{1 - \cos 50°}{2}}$

39. $\sqrt{\dfrac{1 - \cos 40°}{1 + \cos 40°}}$ 40. $\sqrt{\dfrac{1 + \cos 26°}{1 - \cos 26°}}$

Exercises 41–46: (Refer to Example 7.) Complete the following.

   (a) *Find* $\sin \dfrac{\theta}{2}$, $\cos \dfrac{\theta}{2}$, *and* $\tan \dfrac{\theta}{2}$.

   (b) *Support your results numerically.*

41. $\cos \theta = \dfrac{4}{5}$ and $0° < \theta < 90°$

42. $\cos \theta = \dfrac{1}{3}$ and $0° < \theta < 90°$

43. $\tan \theta = -\dfrac{5}{12}$ and $-90° < \theta < 0°$

44. $\sec \theta = -2$ and $90° < \theta < 180°$

45. $\csc \theta = \dfrac{25}{24}$ and $90° < \theta < 180°$

46. $\sin \theta = \dfrac{4}{5}$ and $0° < \theta < 90°$

Verifying Identities

Exercises 47–56: Verify the identity. Give graphical and numerical support.

47. $4 \sin 2x = 8 \sin x \cos x$

48. $\cos 4\theta = 1 - 2 \sin^2 2\theta$

49. $\dfrac{2 - \sec^2 x}{\sec^2 x} = \cos 2x$

50. $(\sin x + \cos x)^2 = \sin 2x + 1$

51. $\sec 2x = \dfrac{1}{1 - 2 \sin^2 x}$

52. $2 \csc 2t = \csc t \sec t$

53. $\sin 3\theta = 3 \sin \theta - 4 \sin^3 \theta$

54. $\dfrac{2 \tan x}{1 + \tan^2 x} = \sin 2x$

55. $\sin 4\theta = 4 \sin \theta \cos \theta \cos 2\theta$

56. $\cos 4t = 8 \cos^4 t - 8 \cos^2 t + 1$

Exercises 57–66: Verify the identity.

57. $\dfrac{\sin 2\theta}{\sin \theta} = 2 \cos \theta$

58. $2 \sin^2 4\theta = 1 - \cos 8\theta$

59. $2 \cos^2 \left(\dfrac{\theta}{2} \right) = 1 + \cos \theta$

60. $\dfrac{\sin^2 2\theta}{1 + \cos 2\theta} = 2 \sin^2 \theta$

61. $\cos^4 \theta - \sin^4 \theta = \cos 2\theta$

62. $\dfrac{1 - \tan^2 x}{1 + \tan^2 x} = \cos 2x$

63. $\csc 2t = \dfrac{\csc t}{2 \cos t}$

64. $\tan \theta + \cot \theta = \dfrac{2}{\sin 2\theta}$

65. $\tan \dfrac{x}{2} = \dfrac{\sin x}{1 + \cos x}$

   $\left(Hint: \text{Let } \tan \dfrac{x}{2} = \dfrac{\sin (x/2)}{\cos (x/2)}. \right)$

66. $\tan \dfrac{x}{2} = \dfrac{1 - \cos x}{\sin x}$ (*Hint:* Use the identity in Exercise 65.)

Product-to-Sum and Sum-to-Product Identities

Exercises 67–70: Write each expression as a sum or difference of trigonometric functions.

67. **(a)** $\cos 50° \sin 20°$
      (b) $\cos 40° \cos 20°$

68. **(a)** $2 \sin 74° \sin 24°$
      (b) $8 \sin 144° \cos 104°$

69. **(a)** $\sin 7\theta \cos 3\theta$
      (b) $\sin 8x \sin 4x$

70. **(a)** $2 \cos 5x \cos 7x$
      (b) $4 \cos 9\theta \sin 2\theta$

Exercises 71–74: Write each expression as a product of trigonometric functions.

71. **(a)** $\sin 40° + \sin 30°$
      (b) $\cos 45° + \cos 35°$

72. **(a)** $\cos 104° - \cos 24°$
      (b) $\sin 32° - \sin 64°$

73. **(a)** $\cos 6\theta + \cos 4\theta$
      (b) $\sin 7x + \sin 4x$

74. **(a)** $\sin 3x - \sin 5x$
      (b) $\cos 3\theta - \cos \theta$

Solving Equations

Exercises 75–80: Find the solutions to the equation on $[0°, 360°)$

   (a) *symbolically,*
   (b) *graphically, and*
   (c) *numerically.*

75. $\cos 2\theta = 1$

76. $\sin 2\theta = \dfrac{1}{2}$

77. $\sin 2\theta = 0$

78. $\cos \dfrac{\theta}{2} = -\dfrac{\sqrt{3}}{2}$

79. $\sin \dfrac{\theta}{2} = 1$

80. $\cos 2\theta + \cos \theta = 0$

Exercises 81–84: Find all solutions expressed in radians.

81. $\sin 2t + \sin t = 0$

82. $\sin t - \cos 2t = 0$

83. $2 \sin \dfrac{t}{2} - 1 = 0$

84. $\sin 2t = 2 \cos^2 t$

Applications

85. *Electricity* (Refer to Example 5.) Suppose the voltage in a 220-volt electrical circuit is modeled by $V(t) = 310 \sin(120\pi t)$ and the amperage flowing through a heater is given by $I = 7 \sin(120\pi t)$. (Source: G. Wilcox.)
  (a) Graph the wattage $W = VI$ consumed by the heater in $[0, 0.04, 0.01]$ by $[-500, 2500, 500]$.
  (b) Find values for the constants a, k, and d so that $W = a \cos(k\pi t) + d$

86. *Electricity* If a toaster is plugged into a common household outlet, the wattage W used varies according to the equation $W = \dfrac{V^2}{R}$, where V is the voltage and R is a constant that measures the resistance of the toaster in ohms. (Source: D. Bell, *Fundamentals of Electric Circuits*.)
  (a) Graph W if $R = 15$ and $V = 163 \sin(120\pi t)$ in $[0, 0.05, 0.01]$ by $[-500, 2000, 500]$.
  (b) Approximate the maximum wattage consumed by the toaster.
  (c) Use a power-reducing identity to express the wattage as $W = a \cos(240\pi t) + d$, where a and d are constants.

87. *Highway Curves* When an automobile travels along a circular curve, objects like trees and buildings situated on the inside of the curve can obstruct a driver's vision. If the cars in the figure at the top of the next column are a safe stopping distance apart, then the distance d that should be cleared on the inside of the curve is $d = r\left(1 - \cos\dfrac{\beta}{2}\right)$, where r is the radius of the curve and β is the central angle between the cars. (Source: F. Mannering and W. Kilareski, *Principles of Highway Engineering and Traffic Analysis*.)
  (a) Find d if $\beta = 80°$ and $r = 600$ feet.
  (b) Use the figure to justify this formula.
  (c) Is the given formula equivalent to the formula $d = r\left(1 - \dfrac{1}{2}\cos\beta\right)$? Explain.

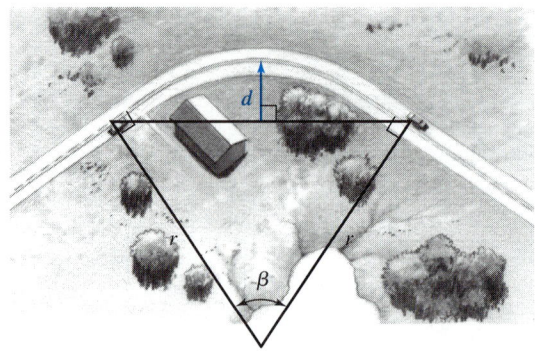

88. *Highway Curves* The figure below represents a circular curve with radius r and central angle θ. The tangent length T is an important distance used by surveyors. (Source: F. Mannering.)
  (a) Show that $T = r \tan\dfrac{\theta}{2}$.
  (b) Find T for a curve with 1500-foot radius and $\theta = 80°$.

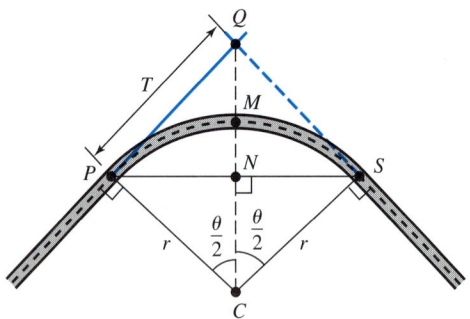

89. *Silver Box* (Refer to Example 11.) A *silver box* contains the standard touch-tone phone keys along with special keys for A, B, C, and D. It has 16 different tones rather than the typical 12 tones. These extra tones are sometimes used to enter security codes. Their frequency pairs in hertz are listed in the table. (Source: World Wide Web.)

Letter	A	B	C	D
Frequency 1 (Hz)	697	770	852	941
Frequency 2 (Hz)	1633	1633	1633	1633

  (a) Write a function f expressed as the *sum* of two trigonometric expressions that models the tone generated by A.
  (b) Write f using a *product* of two trigonometric functions.
  (c) Graph f in $[0, 0.005, 0.001]$ by $[-2, 2, 1]$.

90. *Touch-Tone Phones* (Refer to the previous exercise.) The sound generated by the letter D can be modeled by

$$f(t) = \cos(1882\pi t) + \cos(3266\pi t).$$

(a) Graph f in [0, 0.005, 0.001] by [−2, 2, 1].
(b) Does the graph of f appear to be periodic?
(c) Conjecture whether f is periodic.

91. *Touch-Tone Phones* (Refer to Example 11.) For the numbers 3 and 4 on a touch-tone phone complete the following.
(a) Write an expression that is the *sum* of two trigonometric functions that models the tone generated by each number.
(b) Write the expression for each tone written as a *product* of two trigonometric functions.
(c) Graph the expression for each number in [0, 0.005, 0.001] by [−2, 2, 1]. Do the two tones for 3 and 4 appear to be different? Explain why they should be different.

92. *Musical Tones and Beats* If two musical tones with nearly the same frequency are played simultaneously, a phenomenon called *beats* occurs. Let $P_1 = 0.04\cos(110\pi t)$ and $P_2 = 0.04\cos(116\pi t)$ represent these tones.
(a) Graph $P = P_1 + P_2$ in [0.2, 1.2, 0.2] by [−0.08, 0.08, 0.01]. How many beats are there?
(b) Use a sum-to-product identity to write $P_1 + P_2$ as a product of trigonometric expression.
(c) Give numerical support for your result in part (b).

Writing about Mathematics

1. Suppose a student conjectures that an equation is an identity but cannot verify it symbolically. Discuss techniques that the student could use to support this conjecture.

2. Does the equation $\sin^2 3\theta + \cos^2 2\theta = 1$ represent an identity? Explain your reasoning and give numerical and graphical support.

CHECKING BASIC CONCEPTS FOR SECTION 6.5

1. Find values for $\sin 2\theta$ and $\cos 2\theta$ if $\cos\theta = -\dfrac{7}{25}$ and $\sin\theta = \dfrac{24}{25}$.

2. Use a half-angle formula to find the exact value of $\sin 22.5°$.

3. Find $\sin\dfrac{\theta}{2}$ and $\cos\dfrac{\theta}{2}$ if $\sin\theta = \dfrac{4}{5}$ and θ is acute. Support your results numerically.

4. Verify that $\dfrac{\cos 2\theta}{\cos^2\theta} = 2 - \sec^2\theta$.

5. Solve $\sin 2\theta = 2\cos\theta$ for $0° \le \theta < 360°$. Support your results graphically and numerically.

CHAPTER 6 Summary

An identity is an equation that is true for all meaningful values of a variable. The fundamental trigonometric identities include the reciprocal, quotient, Pythagorean, and negative-angle identities. See Putting It All Together for Section 6.1. Because there is an infinite number of trigonometric identities, it is impossible to know them all. Instead, fundamental identities and other basic identities are used to verify trigonometric identities when they are needed.

Identities are verified by transforming one side of an equation into the other. There is frequently more than one way to verify an identity. Becoming proficient at verifying identities takes practice and effort. Suggestions for verifying identities are given in Putting It All Together for Section 6.2.

Trigonometric equations can be solved symbolically, graphically, and numerically. Many of the algebraic techniques that are used to solve polynomial equations can be applied to trigonometric equations. However, there is one important difference. Trigonometric equations frequently have an infinite number of solutions, whereas polynomial equations have a finite number of solutions. For example, the polynomial equation

$$4x - 3 = 1$$

has one solution of $x = 1$. On the other hand, the trigonometric equation

$$4 \sin \theta - 3 = 1$$

has an infinite number of solutions, since $\sin \theta = 1$ when $\theta = 90° + 360° \cdot n$, where n is an integer. Reference angles, graphs, and tables can all be aids when solving trigonometric equations.

Sum, difference, and multiple-angle identities are trigonometric identities used in applications involving electricity and music. They can also be used to find the exact value of some trigonometric expressions. However, numerical approximations must be used in many instances to evaluate trigonometric functions.

Review Exercises

Exercises 1 and 2: Determine the quadrant containing θ.

1. $\sec \theta < 0$ and $\sin \theta > 0$

2. $\cot \theta > 0$ and $\cos \theta < 0$

Exercises 3–6: Use the given information to find the other four trigonometric functions of θ.

3. $\sin \theta = \dfrac{3}{5}$ and $\cos \theta = -\dfrac{4}{5}$

4. $\sec \theta = -\dfrac{13}{12}$ and $\csc \theta = -\dfrac{13}{5}$

5. $\tan \theta = -\dfrac{7}{24}$ and $\cos \theta = \dfrac{24}{25}$

6. $\cot \theta = -\dfrac{1}{2}$ and $\sin \theta = \dfrac{2}{\sqrt{5}}$

Exercises 7–10: Use a negative-angle identity to write an equivalent trigonometric expression involving a positive angle.

7. $\sin(-13°)$ 8. $\cos(-106°)$

9. $\sec\left(-\dfrac{3\pi}{7}\right)$ 10. $\tan\left(-\dfrac{5\pi}{11}\right)$

11. Explain the difference between an identity and a conditional equation. Give an example of each.

12. Discuss one application where a trigonometric equation or identity occurs.

13. Is there a quadrant where one of the trigonometric functions is positive and the other five are negative? Explain your reasoning.

14. Is it possible for $\sin \theta < 0$, $\cos \theta > 0$, and $\tan \theta > 0$ for some value of θ? Explain.

Exercises 15–20: Simplify each expression. Give numerical support for your result.

15. $\sec \theta \cot \theta \sin \theta$

16. $\sin \theta \csc \theta$

17. $(\sec^2 t - 1)(\csc^2 t - 1)$

18. $\dfrac{\sec \theta}{\csc \theta} + \dfrac{\sin \theta}{\cos \theta}$

19. $\dfrac{\csc \theta \sin \theta}{\sec \theta}$

20. $\dfrac{\cos^2 \theta}{1 - \sin \theta}$

Exercises 21–24: Use a calculator to approximate the other trigonometric functions of θ to four decimal places.

21. $\tan \theta = 1.2367$ and θ is acute

22. $\sin \theta = -0.3434$ and θ is in Quadrant IV

23. $\cos \theta = -0.4544$ and θ is in Quadrant III

24. $\tan \theta = -0.8595$ and θ is in Quadrant II

Exercises 25–28: Factor the trigonometric expression.

25. $\sin^2 \theta + 2 \sin \theta + 1$

26. $2 \cos^2 t - 3 \cos t + 1$

27. $\tan^2 \theta - 9$

28. $2 \sec^2 \theta - 3 \sec \theta - 5$

Exercises 29–42: Verify the identity.

29. $(\sec \theta - 1)(\sec \theta + 1) = \tan^2 \theta$

30. $(\cos \theta + \sin \theta)^2 + (\cos \theta - \sin \theta)^2 = 2$

31. $(1 + \tan t)^2 = \sec^2 t + 2 \tan t$

32. $(1 - \cos^2 t)(1 + \tan^2 t) = \tan^2 t$

33. $\sin (x - \pi) = -\sin x$

34. $\cos (\pi + x) = -\cos x$

35. $\sin 8x = 2 \sin 4x \cos 4x$

36. $\cos^4 x - \sin^4 x = \cos 2x$

37. $\sec 2x = \dfrac{1}{1 - 2 \sin^2 x}$

38. $\dfrac{1 + \tan^2 x}{\sin^2 x + \cos^2 x} = \sec^2 x$

39. $\cos^4 x \sin^3 x = (\cos^4 x - \cos^6 x) \sin x$

40. $\sin^4 x = \dfrac{3}{8} - \dfrac{1}{2} \cos 2x + \dfrac{1}{8} \cos 4x$

41. $\sec^4 \theta - \tan^4 \theta = 1 + 2 \tan^2 \theta$

42. $\dfrac{1 + \cos \theta}{\sin \theta} + \dfrac{\sin \theta}{1 + \cos \theta} = 2 \csc \theta$

Exercises 43–46: Find the reference angle for θ.

43. $\theta = 240°$

44. $\theta = 320°$

45. $\theta = \dfrac{9\pi}{7}$

46. $\theta = -\dfrac{7\pi}{6}$

Exercises 47 and 48: Use the graph to estimate any solutions to the trigonometric equation for $0 \le \theta < 2\pi$. Then use factoring to solve the equation symbolically.

47. $2 \sin t \cos t - \cos t = 0$

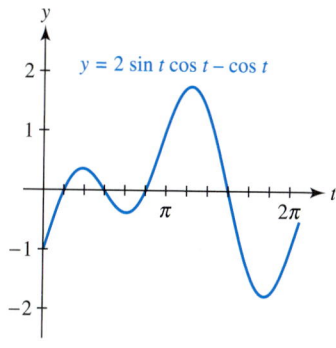

48. $\cos^2 \theta - 2 \cos \theta = 0$

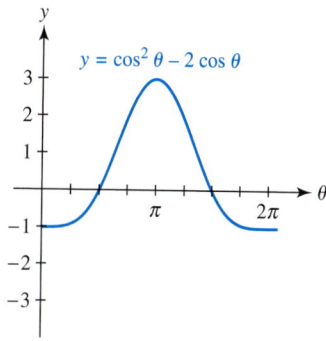

Exercises 49 and 50: Solve the equation for $0° \le \theta < 360°$.

49. **(a)** $\tan \theta = \sqrt{3}$

　　　(b) $\cot \theta = -\sqrt{3}$

50. **(a)** $\sin \theta = 1$

　　　(b) $\cos \theta = -1$

Exercises 51–54: Solve the equation symbolically for $0° \le \theta < 360°$. Support your results graphically and numerically.

51. $2 \cos \theta - 1 = 0$

52. $\cot^2 \theta = 1$

53. $2 \sin^2 \theta + \sin \theta - 3 = 0$

54. $\sin^2 \theta + 2 \cos \theta = 1$

Exercises 55 and 56: Solve the equation for t on the interval $[0, 2\pi)$ symbolically. Support your answer either graphically or numerically.

55. $\tan^2 t - 2 \tan t + 1 = 0$

56. $2 \sin t = \sqrt{3}$

Exercises 57–60: Find all solutions to the equation. Express your results in both degrees and radians.

57. $3 \tan^2 t - 1 = 0$

58. $2 \sin^2 t - \sin t - 1 = 0$

59. $\sin 2t + 3 \cos t = 0$

60. $\cos 2t = 1$

Exercises 61 and 62: Use a half-angle formula to find the exact value for each expression. Support your result numerically.

61. $\cos 105°$

62. $\sin \dfrac{\pi}{12}$

Exercises 63 and 64: Estimate any solutions in $[-2\pi, 2\pi]$ *graphically to three significant digits.*

63. $\tan x = x + 1$ **64.** $\sin(\cos x) = \tan x$

Exercises 65 and 66: Complete the following for the identity.
  (a) *Justify the identity verbally using transformations of graphs.*
  (b) *Verify the identity symbolically.*
  (c) *Give graphical and numerical support.*

65. $\sin\left(t - \dfrac{\pi}{2}\right) = -\cos t$

66. $\cos\left(t - \dfrac{\pi}{2}\right) = \sin t$

Exercises 67 and 68: Find the following.
  (a) $\sin(\alpha + \beta)$
  (b) $\cos(\alpha + \beta)$
  (c) $\tan(\alpha + \beta)$
  (d) *The quadrant containing* $\alpha + \beta$

67. $\cos\alpha = \dfrac{3}{5}$ and $\cos\beta = \dfrac{12}{13}$, α and β are in quadrant I

68. $\sin\alpha = -\dfrac{12}{13}$ and $\tan\beta = -\dfrac{4}{3}$, α and β are in quadrant IV

Exercises 69 and 70: Complete the following.
  (a) *Find* $\sin 2\theta$, $\cos 2\theta$, *and* $\tan 2\theta$.
  (b) *Support your results numerically.*

69. $\sin\theta = \dfrac{4}{5}$ and $\tan\theta = -\dfrac{4}{3}$

70. $\sin\theta = -\dfrac{12}{13}$ and $\cos\theta = -\dfrac{5}{13}$

Exercises 71 and 72: Complete the following.
  (a) *Find* $\sin\dfrac{\theta}{2}$, $\cos\dfrac{\theta}{2}$, *and* $\tan\dfrac{\theta}{2}$.
  (b) *Support your results numerically.*

71. $\cos\theta = \dfrac{1}{4}$ and $0° < \theta < 90°$

72. $\tan\theta = -\dfrac{8}{15}$ and $-90° < \theta < 0°$

73. *Daylight Hours*   The number of daylight hours y at 20° S latitude can be modeled by

$$y = 1.2\cos\left(\frac{\pi}{6}(x - 0.7)\right) + 12.1,$$

where $x = 1$ corresponds to January 1, $x = 2$ to February 1, and so on. Estimate graphically and numerically when the number of daylight hours equals 11.5 hours. (Source: J. Williams.)

74. *Music and Pure Tones*   A pure tone is modeled by the graph of $P(t) = 0.006\cos(50\pi t)$, where P represents the pressure on an eardrum in pounds per square foot at time t in seconds. (Source: J. Roederer, *Introduction to the Physics and Psychophysics of Music*.)
  (a) Graph P in the window $[0, 0.1, 0.01]$ by $[-0.008, 0.008, 0.001]$.
  (b) Estimate all solutions to the equation $P = 0$ on the interval $0 \le t \le 0.1$.

75. *Modeling Musical Tones*   Let the pressure exerted on the eardrum by two sound waves in pounds per square foot be given by

$$P_1(t) = 0.006\cos(100\pi t) \quad \text{and}$$
$$P_2(t) = 0.008\sin(100\pi t).$$

  (a) Graph the pressure on the eardrum $P = P_1 + P_2$ in the window $[0, 0.06, 0.01]$ by $[-0.012, 0.012, 0.002]$.
  (b) Use the graph to find values for a and k such that $P = a\sin(100\pi t + k)$.
  (c) Use a sum or difference identity for sine to verify your result in part (b).

76. *Electricity*   Suppose the voltage in an electrical circuit is given by $V(t) = 17\sin(120\pi t)$ and the amperage flowing through a heater is given by $I(t) = 2\sin(120\pi t)$. (Source: G. Wilcox and C. Hesselberth, *Electricity for Engineering Technology*.)
  (a) Graph the wattage $W = VI$ consumed by the heater in $[0, 0.04, 0.01]$ by $[-40, 40, 10]$.
  (b) Use a power-reducing identity to express the wattage in the form $W = a\cos(k\pi t) + d$, where a, k, and d are constants.
  (c) Support your result in part (b) numerically.

Extended and Discovery Exercises

1. *Piano Strings*   A piano string vibrates at more than one frequency when it is struck. It produces a complex wave that can be modeled by a sum of several pure tones. If a piano key with a frequency of f_1 is played, then the corresponding string will not only vibrate at f_1 but it will also vibrate at the higher frequencies of $2f_1$, $3f_1$, $4f_1$, and so on. The *fundamental* frequency of the string is f_1 and the higher frequencies are called the *upper harmonics*. The human ear will hear the sum of these frequencies as one complex tone. (Source: J. Roederer, *Introduction to the Physics and Psychophysics of Music*.)
  (a) If the A note above middle C is played, its fundamental frequency is $f_1 = 440$ hertz.

(One hertz equals one cycle per second.) The piano string also vibrates at frequencies of $f_2 = 2(440) = 880$, $f_3 = 3(440) = 1320$, $f_4 = 4(440) = 1760$, and so on. Let the pressure for each frequency in pounds per square foot be modeled by

$$P_1 = 0.002 \sin(2\pi(440)t),$$

$$P_2 = \frac{0.002}{2} \sin(2\pi(880)t),$$

$$P_3 = \frac{0.002}{3} \sin(2\pi(1320)t),$$

$$P_4 = \frac{0.002}{4} \sin(2\pi(1760)t), \quad \text{and}$$

$$P_5 = \frac{0.002}{5} \sin(2\pi(2200)t),$$

where t is in seconds. Graph each of the following expressions for P in $[0, 0.01, 0.002]$ by $[-0.005, 0.005, 0.001]$.

i. $P = P_1$
ii. $P = P_1 + P_2$
iii. $P = P_1 + P_2 + P_3$
iv. $P = P_1 + P_2 + P_3 + P_4$
v. $P = P_1 + P_2 + P_3 + P_4 + P_5$.

(b) The final graph of P models what the human ear hears. Describe this graph.
(c) Estimate the maximum pressure of $P = P_1 + P_2 + P_3 + P_4 + P_5$.
(d) A pure tone with a frequency of 440 hertz is modeled by $P = P_1$, whereas a piano generates the graph of $P = P_1 + P_2 + P_3 + P_4 + P_5$. Compare and contrast these two graphs.

2. *Plucking a String* (Refer to the previous exercise.) If a string with a fundamental frequency of 110 hertz is *plucked in the middle,* it will vibrate at the odd harmonics or frequencies of 110, 330, and 550, but not at the even harmonics of 220, 440, and 660. The resulting pressure P caused by this sound wave may be modeled by

$$P = 0.002 \sin 220\pi t + \frac{0.002}{3} \sin 660\pi t$$

$$+ \frac{0.002}{5} \sin 1100\pi t + \frac{0.002}{7} \sin 1540\pi t.$$

(a) Graph P in the window $[0, 0.03, 0.01]$ by $[-0.004, 0.004, 0.001]$.
(b) Describe the graph of P.
(c) At lower frequencies, the inner ear hears a tone only when the eardrum is moving outward or when $P < 0$. Estimate the times in the interval $0 \le t \le 0.03$ when this occurs.

(Source: A. Benade, *Fundamentals of Musical Acoustics.*)

3. *Low Tones and Small Speakers* Small speakers found in older radios and telephones often cannot vibrate at frequencies that are less than 200 hertz. Thirty-five keys on a piano have frequencies less than 200 hertz. Nonetheless these notes can still be heard on these speakers. When a piano string creates a tone of 110 hertz, it also creates tones at 220, 330, 440, 550, and 660 hertz. A small speaker cannot reproduce the 110 hertz vibration but it can reproduce the higher frequencies, which are called the upper harmonics. The low tones can still be heard because the speaker produces *difference tones* of the upper harmonics. The difference between consecutive frequencies is 110 hertz and this difference tone will be heard on a small speaker even though the speaker cannot vibrate at 110 hertz. This phenomenon can be visualized with a graphing calculator. (Source: A. Benade.)
(a) Graph the upper harmonics represented by the pressure wave

$$P = \frac{1}{2} \sin(2\pi(220)t) + \frac{1}{3} \sin(2\pi(330)t)$$

$$+ \frac{1}{4} \sin(2\pi(440)t)$$

in $[0, 0.03, 0.01]$ by $[-1.2, 1.2, 0.5]$.
(b) Estimate the t-values on the interval $0 \le t \le 0.03$ where P is maximum.
(c) Approximate the frequency of these maximum values. What does a person hear in addition to the frequencies of 220, 330, and 440 hertz?
(d) Discuss the advantage of having large speakers instead of smaller ones. (*Hint:* Try graphing the pressure produced by a speaker that can vibrate both at 110 hertz and at the upper harmonics.)

4. *Piano Strings* When a string is set into vibration by striking it, the amplitude A of the vibrations decreases over time, while the frequency of the vibration remains constant. This phenomenon is called *exponential decay* and can be modeled by $A = A_0 e^{-kt} \sin(2\pi F t)$, where F is the frequency, t is time in seconds, and k and A_0 are positive constants. (Source: J. Roederer.)
(a) Graph A when $A_0 = 0.1$, $F = 15$, and $k = 1.2$ in $[0, 1, 0.1]$ by $[-0.15, 0.15, 0.05]$.
(b) Now graph the equations $y_1 = -0.1e^{-1.2t}$ and $y_2 = 0.1e^{-1.2t}$ with A. Describe how the graphs of y_1 and y_2 relate to the graph of A.
(c) The *decay half-time* is the time it takes for the maximum amplitude of A to decrease to $\frac{1}{2}A_0$.

Estimate this time graphically. (The decay half-time for a typical piano string is about 0.4 second.)

CHAPTER 7 *Further Topics in Trigonometry*

*T*rigonometry plays an important role in our society. Almost any phenomenon that involves angles, triangles, or circular motion requires trigonometric functions to model it. For example, aerial photography has become essential for predicting weather, surveying land, promoting national security, and even making archaeological discoveries. Aerial photography began in 1858 when Gaspard Tournachon, a French photographer, took pictures of Paris from a hot air balloon that had a makeshift darkroom. The first archaeological aerial photographs were taken of Stonehenge in 1906. By searching these photographs for unusual soil markings, caused by structures lying below the ground, Stonehenge Avenue was discovered. Today, trigonometry is used extensively in aerial photography, and hot air balloons have been replaced by airplanes and satellites.

In this chapter we use trigonometry in a variety of applications, which include aerial photography, computer graphics, robotics, navigation, GPS, highway

*T*he best way to understanding is a few good examples.

— **Isaac Newton**

design, physics, weather, solar energy, art, electronics, construction, astronomy, surveying, and simulating motion.

Sources: R. Brooks and J. Dieter, *Phytoarchaeology;* F. Moffitt, *Photogrammetry.*

7.1 Law of Sines

Oblique Triangles • Solving Triangles • The Ambiguous Case

Introduction

In Chapter 5 we solved right triangles. Solving a triangle involves finding the measure of each side and each angle in the triangle. In many areas of study, triangles without right angles often occur. To solve these triangles, we will use the law of sines and the law of cosines. In this section the law of sines is derived and used to solve several problems.

Oblique Triangles

If a triangle is not a right triangle then it is called an **oblique triangle.** There are four different situations or cases that can occur when attempting to solve an oblique triangle.

1. All three sides are given. This situation determines a unique triangle and is referred to as SSS. See Figure 7.1. Note that the length of any side must be less than the sum of the lengths of the other two sides.
2. Two sides and the angle included are given. This situation determines a unique triangle and is referred to as SAS. See Figure 7.2.
3. One side and two angles are given. This situation determines a unique triangle and is referred to as ASA or AAS. See Figure 7.3. Note that whenever two angles of a triangle are known, then the third angle can be found using the fact that the sum of the measures of the angles equals 180°.
4. Two sides and an angle opposite one of the sides are given. This situation does *not always* determine a unique triangle and is referred to as SSA. There may be 0, 1, or 2 triangles that can satisfy SSA. As a result we call SSA the **ambiguous case.** See Figures 7.4–7.6.

Cases 1 and 2 are solved in the next section using the *law of cosines,* while Cases 3 and 4 are solved using the *law of sines.*

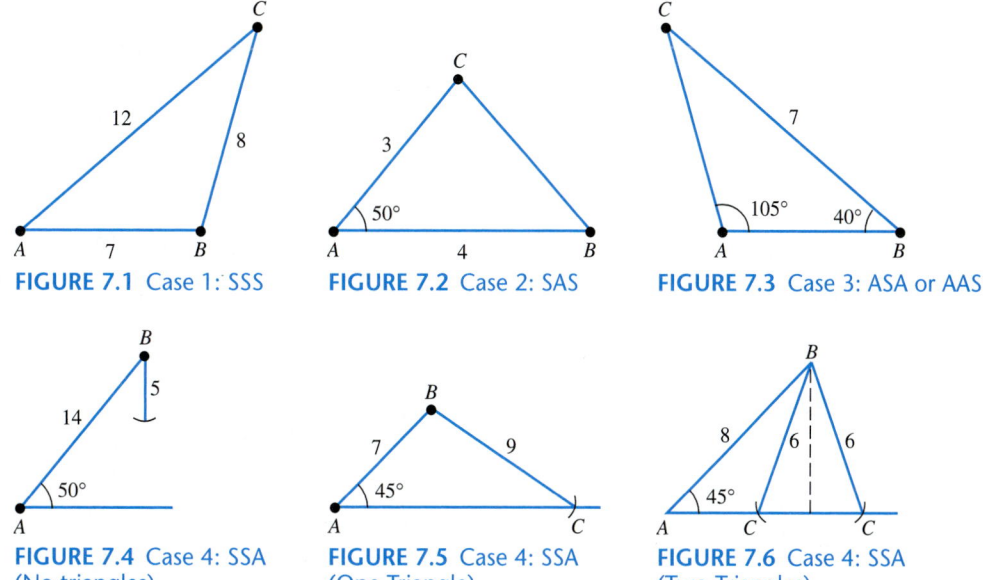

FIGURE 7.1 Case 1: SSS

FIGURE 7.2 Case 2: SAS

FIGURE 7.3 Case 3: ASA or AAS

FIGURE 7.4 Case 4: SSA (No triangles)

FIGURE 7.5 Case 4: SSA (One Triangle)

FIGURE 7.6 Case 4: SSA (Two Triangles)

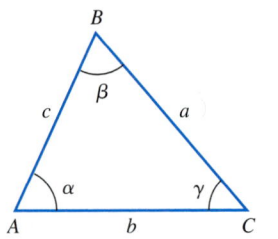

FIGURE 7.7 Standard Labeling

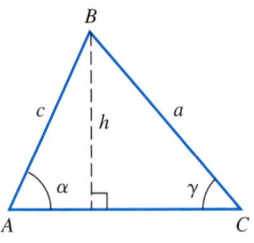

FIGURE 7.8

Solving Triangles

We will label triangles as shown in Figure 7.7 and refer to this labeling as the **standard labeling.** For example, angle α is located at vertex A and side a is opposite angle α. Note that angle γ need not be a right angle.

The law of sines may be derived using the oblique triangle shown in Figure 7.8.

$$\sin \alpha = \frac{h}{c} \quad \text{or} \quad h = c \sin \alpha$$

$$\sin \gamma = \frac{h}{a} \quad \text{or} \quad h = a \sin \gamma$$

Since $h = c \sin \alpha$ and $h = a \sin \gamma$, it follows that

$$c \sin \alpha = a \sin \gamma,$$

or dividing both sides by ac gives

$$\frac{\sin \alpha}{a} = \frac{\sin \gamma}{c}.$$

In a similar manner it can be shown that

$$\frac{\sin \alpha}{a} = \frac{\sin \beta}{b}.$$

This discussion supports the following result.

Law of sines

Any triangle with standard labeling satisfies

$$\frac{\sin \alpha}{a} = \frac{\sin \beta}{b} = \frac{\sin \gamma}{c}, \quad \text{or equivalently,} \quad \frac{a}{\sin \alpha} = \frac{b}{\sin \beta} = \frac{c}{\sin \gamma}.$$

The next example illustrates how the law of sines is used in aerial photography.

EXAMPLE 1 *Finding distances using aerial photography (ASA)*

Trigonometry is used extensively in aerial photography. In Figure 7.9 a camera lens has an angular coverage of 75°. As a picture is taken over level ground, the airplane's distance is 4800 feet from a house located on the edge of the photograph and the angle of elevation of the airplane from the house is 48°. Find the distance across the photograph. (Source: F. Moffit, *Photogrammetry.*)

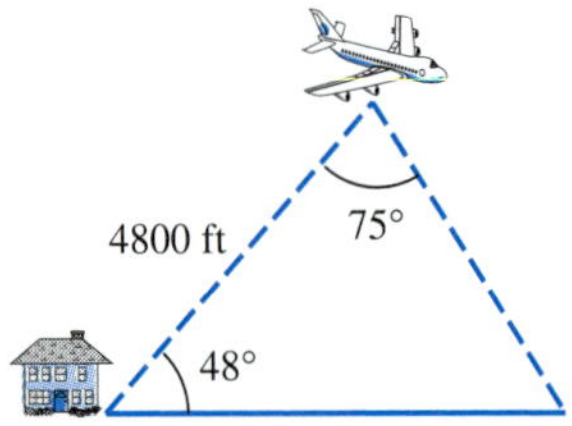

Solution

We are given two angles and the side included (ASA), so let $\alpha = 75°$, $\beta = 48°$, and $c = 4800$ as shown in Figure 7.10. We can find the third angle γ using the fact that the angles sum to 180°.

$$\gamma = 180° - \alpha - \beta$$
$$= 180° - 75° - 48°$$
$$= 57°$$

FIGURE 7.9

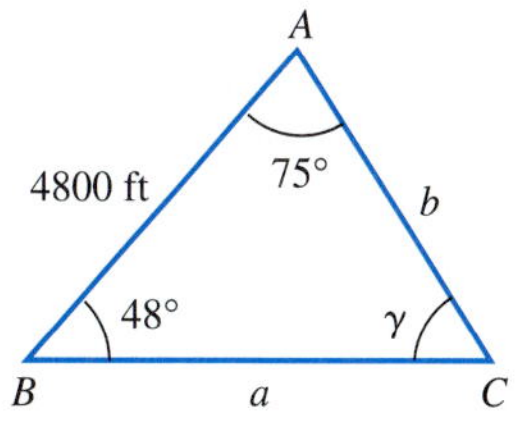

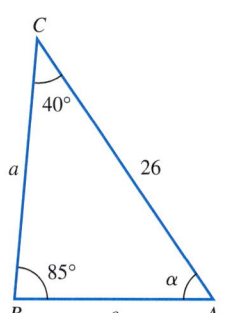

FIGURE 7.10

Side a corresponds to the distance across the photograph and can be found using the law of sines.

$$\frac{a}{\sin \alpha} = \frac{c}{\sin \gamma} \qquad \text{Law of sines}$$

$$\frac{a}{\sin 75°} = \frac{4800}{\sin 57°} \qquad \text{Substitute.}$$

$$a = \frac{4800 \sin 75°}{\sin 57°} \qquad \text{Solve for } a.$$

$$a \approx 5528 \qquad \text{Approximate } a.$$

The picture will show about 5500 feet of ground distance from one edge of the photograph to the other. ∎

EXAMPLE 2 *Solving a triangle (AAS)*

If $\beta = 85°$, $\gamma = 40°$, and $b = 26$, solve triangle ABC.

Solution

Sketch triangle ABC as shown in Figure 7.11.

$$\alpha = 180° - \beta - \gamma$$
$$= 180° - 85° - 40°$$
$$= 55°$$

FIGURE 7.11

Side a may be found using the law of sines.

$$\frac{a}{\sin \alpha} = \frac{b}{\sin \beta} \qquad \text{Law of sines}$$

$$\frac{a}{\sin 55°} = \frac{26}{\sin 85°} \qquad \text{Substitute.}$$

$$a = \frac{26 \sin 55°}{\sin 85°} \qquad \text{Solve for } a.$$

$$a \approx 21.4 \qquad \text{Approximate } a.$$

Side c may be found in a similar manner.

$$\frac{c}{\sin \gamma} = \frac{b}{\sin \beta} \qquad \text{Law of sines}$$

$$\frac{c}{\sin 40°} = \frac{26}{\sin 85°} \qquad \text{Substitute.}$$

$$c = \frac{26 \sin 40°}{\sin 85°} \qquad \text{Solve for } c.$$

$$c \approx 16.8 \qquad \text{Approximate } c. \qquad ∎$$

EXAMPLE 3 *Estimating the distance to the moon*

Since the moon is a relatively close celestial object, its distance can be measured directly using trigonometry. To find this distance, two different photographs of the moon were taken at precisely the same time from two different locations. On April

29, 1976 at 11:35 A.M. the lunar angles of elevation during a partial solar eclipse at Bochum in upper Germany and at Donaueschingen in lower Germany were measured as $\alpha = 52.6997°$ and $\theta = 52.7430°$, respectively. See Figure 7.12. If the two cities are 398.02 kilometers apart, approximate the distance to the moon. Disregard the curvature of the earth in this calculation. (Source: W. Scholosser, *Challenges of Astronomy.*)

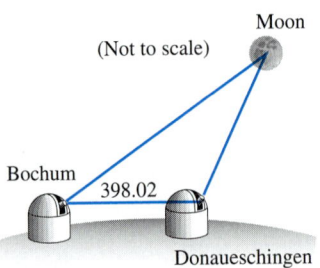

(Not to scale)

Moon

Bochum 398.02

Donaueschingen

FIGURE 7.12

Solution

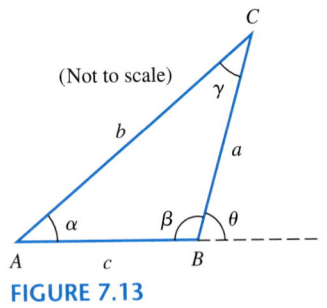

FIGURE 7.13

Consider triangle ABC in Figure 7.13, where $\alpha = 52.6997°$ and $\theta = 52.7430°$. Since $\alpha + \gamma$ and θ are supplements of angle β, it follows that $\alpha + \gamma = \theta$ and so

$$\gamma = \theta - \alpha = 52.7430° - 52.6997° = 0.0433°.$$

The distance to the moon can be approximated by finding either a or b. Why? We find a by applying the law of sines.

$$\frac{a}{\sin \alpha} = \frac{c}{\sin \gamma} \qquad \text{Law of sines}$$

$$\frac{a}{\sin 52.6997°} = \frac{398.02}{\sin 0.0433°} \qquad \text{Substitute.}$$

$$a = \frac{398.02 \sin 52.6997°}{\sin 0.0433°} \qquad \text{Solve for } a.$$

$$a \approx 419{,}000 \text{ km} \qquad \text{Approximate } a.$$

The distance to the moon on that day was about 419,000 kilometers. ∎

Bearings are used both in surveying and aerial navigation to determine directions. If a single angle is used for a **bearing,** then it is understood that the bearing is measured in a *clockwise* direction from due north. Some examples of bearings are shown in Figure 7.14.

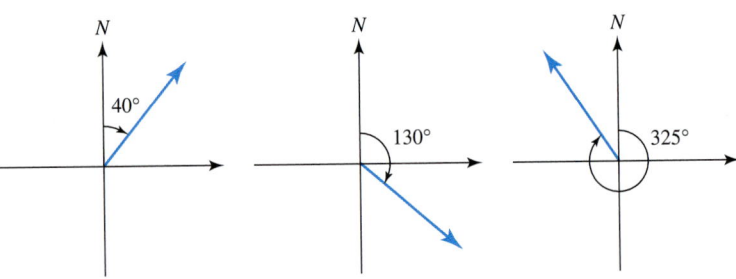

FIGURE 7.14 Examples of Bearings

In the next example we locate a fire using bearings. Locating fires precisely is particularly important for firefighters in rugged areas with limited access. (Source: I. Mueller and K. Ramsayer, *Introduction to Surveying.*)

EXAMPLE 4   *Determining the location of a forest fire*

A fire is spotted from two ranger stations that are 4 miles apart as illustrated in Figure 7.15. From Station A the bearing of the fire is 35° and from Station B the bearing of the fire is 335°. Find the distance between the fire and each ranger station if Station A lies directly west of Station B.

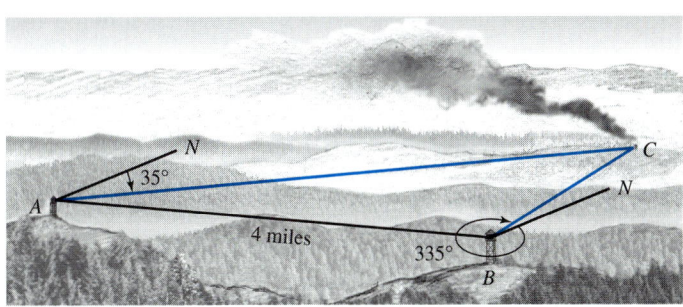

FIGURE 7.15

Solution

Consider triangle *ABC* in Figure 7.16, where $\alpha = 55°$, $\beta = 65°$, and $c = 4$. Thus,

$$\gamma = 180° - 55° - 65° = 60°.$$

Using the law of sines we can find *a*.

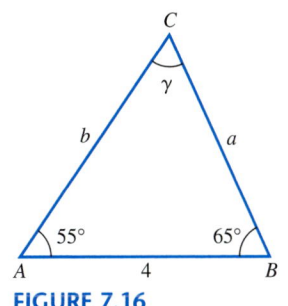

FIGURE 7.16

$$\frac{a}{\sin \alpha} = \frac{c}{\sin \gamma} \qquad \text{Law of sines}$$

$$\frac{a}{\sin 55°} = \frac{4}{\sin 60°} \qquad \text{Substitute.}$$

$$a = \frac{4 \sin 55°}{\sin 60°} \qquad \text{Solve for } a.$$

$$a \approx 3.78 \text{ mi} \qquad \text{Approximate } a.$$

In a similar manner we can find *b*.

$$\frac{b}{\sin \beta} = \frac{c}{\sin \gamma} \qquad \text{Law of sines}$$

$$\frac{b}{\sin 65°} = \frac{4}{\sin 60°} \qquad \text{Substitute.}$$

$$b = \frac{4 \sin 65°}{\sin 60°} \qquad \text{Solve for } b.$$

$$\approx 4.19 \text{ mi} \qquad \text{Approximate } b.$$

The fire is 4.19 miles from Station A and 3.78 miles from Station B. ∎

The Ambiguous Case

If we are given two sides and an angle opposite one of the sides (SSA), there may be 0, 1, or 2 triangles that satisfy these conditions. Refer to Figures 7.4–7.6. For this reason SSA is called the *ambiguous case.* We will discuss how the law of sines can be used to solve each of these situations.

EXAMPLE 5 *Solving the ambiguous case (no solutions)*

Let $\alpha = 62°$, $a = 6$, and $b = 10$. If possible, solve the triangle.

Solution

We begin by attempting to find β using the law of sines.

$$\frac{\sin \beta}{b} = \frac{\sin \alpha}{a} \qquad \text{Law of sines}$$

$$\frac{\sin \beta}{10} = \frac{\sin 62°}{6} \qquad \text{Substitute.}$$

$$\sin \beta = \frac{10 \sin 62°}{6} \qquad \text{Solve for } \sin \beta.$$

$$\sin \beta \approx 1.47 > 1 \qquad \text{Approximate } \sin \beta.$$

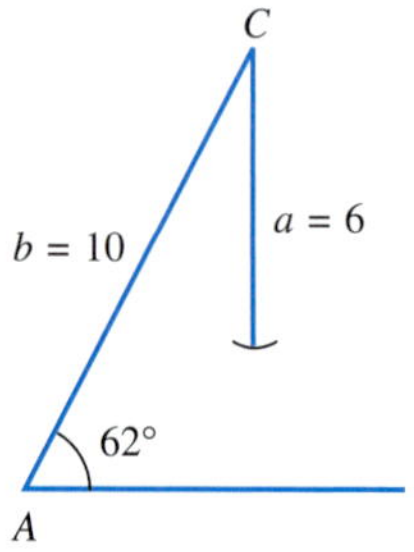

FIGURE 7.17

Since the sine function is never greater than 1, there are no solutions for β. No such triangle exists. See Figure 7.17. ∎

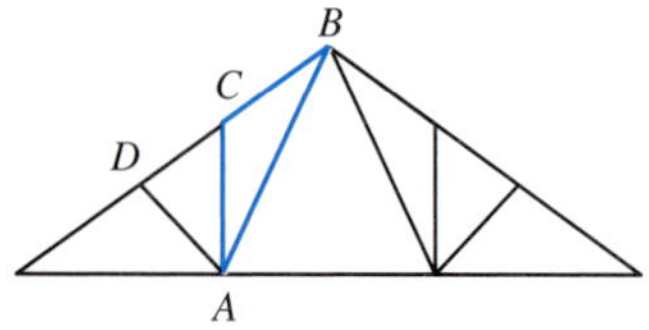

FIGURE 7.18 A Bridge Truss

Trusses are used in construction to support roofs, radio towers, bridges, and aircraft frames. A truss is composed of straight segments joined together at their ends. Many times the smaller shapes within a truss are triangles. See Figure 7.18. If trusses are designed properly, they can be relatively light and very strong. In the next example we determine that a particular truss design results in a unique truss. (Source: W. Riley, *Statics and Mechanics of Materials.*)

EXAMPLE 6 *Solving the ambiguous case (one solution)*

Suppose that an engineer has designed the truss in Figure 7.18 and specified that in triangle *ABC*, *BC* = 17 feet, *AC* = 22 feet, and angle *ABC* = 32°. Determine the length of *AB*. Is this value for *AB* unique?

Solution

Using standard labeling, let $a = 17$, $b = 22$, and $\beta = 32°$. Then find angle *CAB* or α.

$$\frac{\sin \alpha}{a} = \frac{\sin \beta}{b} \qquad \text{Law of sines}$$

$$\frac{\sin \alpha}{17} = \frac{\sin 32°}{22} \qquad \text{Substitute.}$$

$$\sin \alpha = \frac{17 \sin 32°}{22} \qquad \text{Solve for } \sin \alpha.$$

$$\sin \alpha \approx 0.4095 \qquad \text{Approximate } \sin \alpha.$$

There are two values possible for angle α if $\sin\alpha \approx 0.4094$. Angle α lies in quadrants I or II with reference angle $\alpha_R \approx \sin^{-1}(0.4095) \approx 24.2°$. Thus,

$$\alpha \approx 24.2° \quad \text{or} \quad \alpha \approx 180° - 24.2° = 155.8°.$$

However, if $\alpha \approx 155.8°$ then $\alpha + \beta = 155.8° + 32° > 180°$, which is impossible in a triangle. Therefore, $\alpha \approx 24.2°$ is the only possibility and

$$\gamma \approx 180° - 24.2° - 32° = 123.8°.$$

The law of sines allows us to find AB, or side c.

$$\frac{c}{\sin\gamma} = \frac{b}{\sin\beta} \qquad \text{Law of sines}$$

$$\frac{c}{\sin 123.8°} \approx \frac{22}{\sin 32°} \qquad \text{Substitute.}$$

$$c \approx \frac{22\sin 123.8°}{\sin 32°} \qquad \text{Solve for } c.$$

$$c \approx 34.5 \text{ ft} \qquad \text{Approximate } c.$$

Thus, AB has length 34.5 feet and this value is unique. ∎

Critical Thinking

Suppose we are given a, b, and α, and calculate $\sin\beta = 1$. Discuss the number of solutions for triangle ABC.

EXAMPLE 7 *Solving the ambiguous case (two solutions)*

Let $\beta = 55°$, $a = 8.5$, and $b = 7.3$. Solve the triangle.

Solution

Begin by finding α.

$$\frac{\sin\alpha}{a} = \frac{\sin\beta}{b} \qquad \text{Law of sines}$$

$$\frac{\sin\alpha}{8.5} = \frac{\sin 55°}{7.3} \qquad \text{Substitute.}$$

$$\sin\alpha = \frac{8.5\sin 55°}{7.3} \qquad \text{Solve for } \sin\alpha.$$

$$\sin\alpha \approx 0.9538 \qquad \text{Approximate } \sin\alpha.$$

Two angles satisfying $\sin\alpha \approx 0.9538$ in quadrants I and II are

$$\alpha \approx \sin^{-1}(0.9538) \approx 72.5° \quad \text{and} \quad \alpha \approx 180° - 72.5° = 107.5°.$$

Both of these values for α are valid since they do not result in the sum of the angles exceeding 180°.

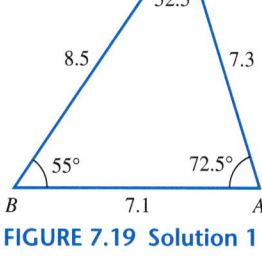

FIGURE 7.19 Solution 1

Solution 1 Let $\alpha \approx 72.5°$. Then,

$$\gamma \approx 180° - 72.5° - 55° = 52.5°.$$

Side c can then be found.

$$\frac{c}{\sin \gamma} = \frac{b}{\sin \beta} \qquad \text{Law of sines}$$

$$c \approx \frac{7.3 \sin 52.5°}{\sin 55°} \qquad \text{Substitute and solve for } c.$$

$$c \approx 7.1 \qquad \text{Approximate } c.$$

A sketch of triangle ABC is shown in Figure 7.19.

Solution 2 Let $\alpha \approx 107.5°$. Then,

$$\gamma \approx 180° - 107.5° - 55° = 17.5°.$$

Then side c can be found.

$$\frac{c}{\sin \gamma} = \frac{b}{\sin \beta} \qquad \text{Law of sines}$$

$$c \approx \frac{7.3 \sin 17.5°}{\sin 55°} \qquad \text{Substitute and solve for } c.$$

$$c \approx 2.7 \qquad \text{Approximate } c.$$

FIGURE 7.20 Solution 2

A sketch of triangle ABC is shown in Figure 7.20. ■

Critical Thinking

Suppose you are given a, b, and α. If $a > b$, what can be said about the number of solutions? Explain your reasoning.

7.1 PUTTING IT ALL TOGETHER

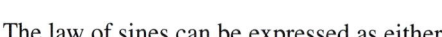

The law of sines can be expressed as either

$$\frac{\sin \alpha}{a} = \frac{\sin \beta}{b} = \frac{\sin \gamma}{c} \qquad \text{or} \qquad \frac{a}{\sin \alpha} = \frac{b}{\sin \beta} = \frac{c}{\sin \gamma}.$$

When using the law of sines a good strategy is to select an equation with the unknown variable in the numerator. The law of sines can be used to solve triangles when we are given ASA, AAS, or SSA. The case where we are given SSA is called the ambiguous case, since there may be 0, 1, or 2 triangles that satisfy the conditions. Several of these situations are shown in the accompanying table.

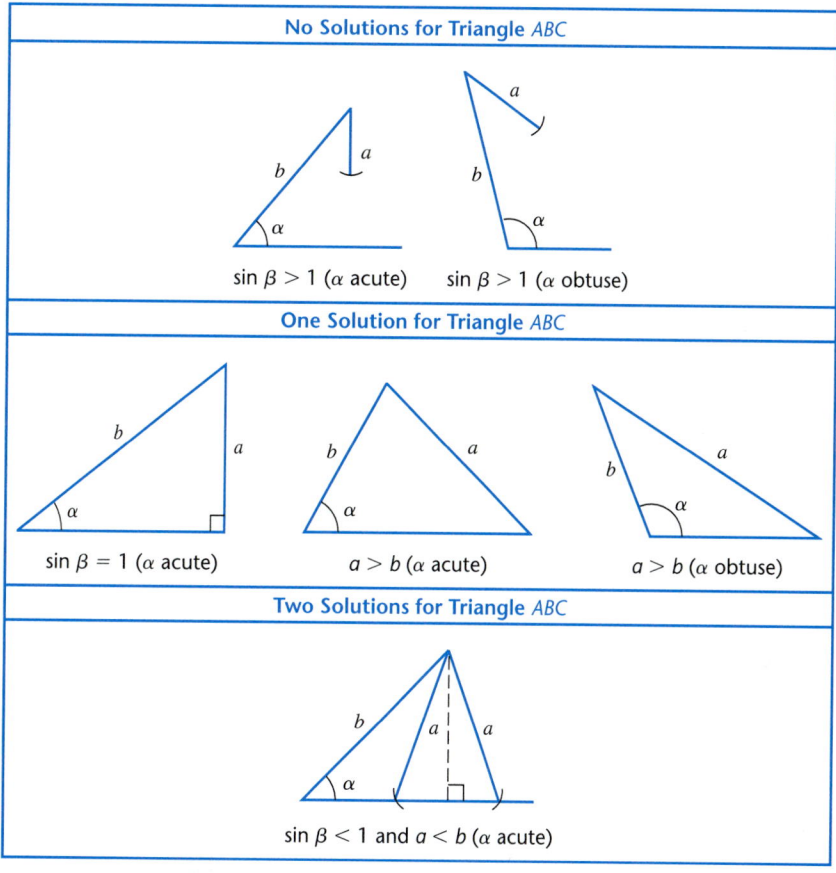

No Solutions for Triangle *ABC*

$\sin \beta > 1$ (α acute) $\sin \beta > 1$ (α obtuse)

One Solution for Triangle *ABC*

$\sin \beta = 1$ (α acute) $a > b$ (α acute) $a > b$ (α obtuse)

Two Solutions for Triangle *ABC*

$\sin \beta < 1$ and $a < b$ (α acute)

7.1 EXERCISES   Tape 12

Recognizing the Ambiguous Case

Exercises 1–8: Let triangle ABC have standard labeling. Given the following angles and sides, decide if solving the triangle results in the ambiguous case.

1. α, β, and a **2.** α, γ, and c

3. a, b, and c **4.** α, a, and b

5. β, b, and c **6.** α, b, and c

7. γ, a, and c **8.** β, b, and α

Solving Triangles

Exercises 9–16: Solve the triangle, if possible. If the triangle represents the ambiguous case solve for all possible triangles. Approximate values to three significant digits.

9.

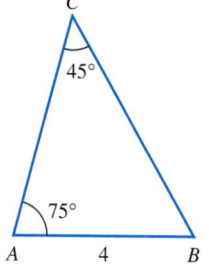

10.

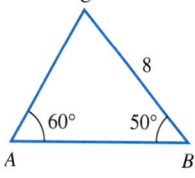

11.

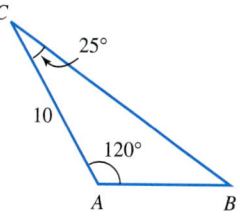

12.

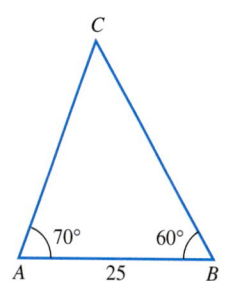

13.

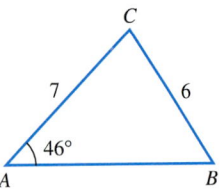

14.

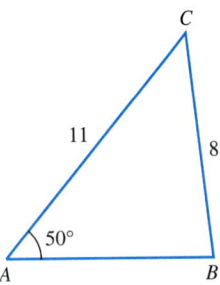

15.

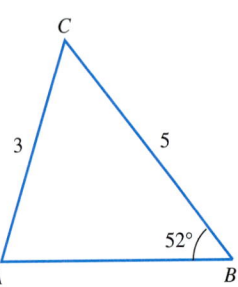

16.

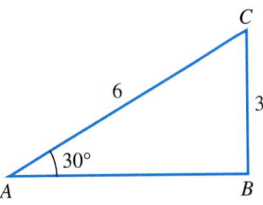

Exercises 17–28: Solve the triangle if possible. Approximate values to three significant digits.

17. $\alpha = 32°, \beta = 55°, b = 12$

18. $\beta = 20°, \gamma = 67°, c = 9$

19. $\alpha = 20°, b = 9, a = 7$

20. $\alpha = 20°, b = 7, a = 9$

21. $b = 10, \beta = 30°, c = 20$

22. $a = 13.5, \alpha = 46°, c = 27.8$

23. $\gamma = 102°, c = 51.6, a = 42.1$

24. $\beta = 43°, b = 22.1, c = 30.7$

25. $\alpha = 55.2°, \gamma = 114.8°, b = 19.5$

26. $c = 225, \alpha = 103.2°, \beta = 62.5°$

27. $b = 6.2, c = 7.4, \beta = 73°$

28. $\alpha = 45°, a = 5, b = 5\sqrt{2}$

Applications

29. *Aerial Photography*   (Refer to Example 1.) In the accompanying figure, a plane takes an aerial photograph with a camera lens that has an angular coverage of 70°. The ground below is inclined at 7°. If the angle of elevation of the plane at B is 52° and distance BC is 3500 feet, estimate the ground distance AB that appears in the picture. (Source: F. Moffit.)

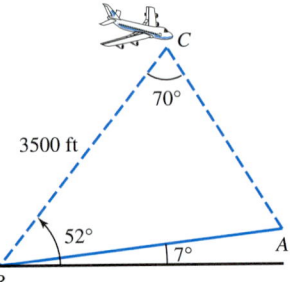

30. *Distance to the Moon*   (Refer to Example 3.) Suppose that the lunar angle at Bochum in upper Germany had been measured as $\alpha = 52.6901°$ instead of $52.6997°$. Determine the effect that this would have on the estimation of the distance to the moon. Interpret the result.

31. *Locating a Ship*   The accompanying figure shows the bearings of a ship on Lake Ontario from two different observation points located on shore. The bearing from the first observation point is 54.3° and the bearing from the second observation point is 325.2°. If the distance between these points is 15 miles, how far is it from the ship to shore? Assume that the first observation point is directly west of the second observation point.

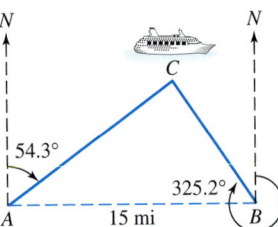

32. *Airplane Navigation*   An airplane takes off with a bearing of 55° and flies 480 miles after which it changes its course to a bearing of 285°. Finally the airplane flies back to its starting point with a bearing of 180°. Find the total mileage flown by the airplane.

33. *Surveying* To find the distance between two points A and B on the opposite sides of a small pond a surveyor determines that AC is 97.3 feet, angle ACB is 55.1°, and angle CAB is 75.7° as illustrated in the accompanying figure. Find the distance between A and B.

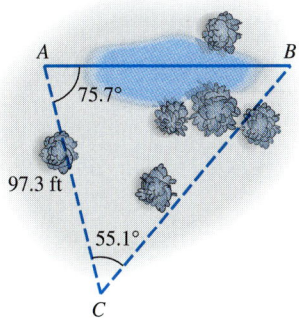

34. *Trigonometric Leveling* In surveying it is often necessary to determine the height of an inaccessible point P as illustrated in the accompanying figure. Points A, B, and C lie on level ground. (Source: P. Kissam, *Surveying Practice*.)
  (a) If angle ABP is 50°, angle PAB is 53.3°, and AB is 102 feet, find PB.
  (b) If angle PBC is 47° find PC.

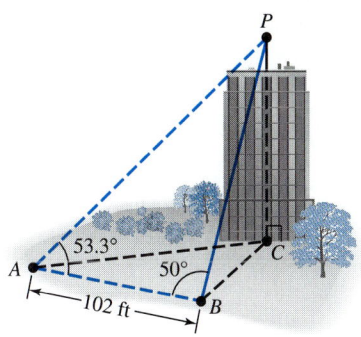

35. *Distance* An ore ship is traveling west toward Duluth on Lake Superior at 18 miles per hour. The bearing of Split Rock Lighthouse is 285°. After one hour the bearing of the lighthouse is 340°. Find the distance between the ship and the lighthouse when the second bearing was determined.

36. *Locating a Ship* From two observation points A and B, a sinking ship is spotted at point C. Angle CAB and angle ABC are measured as 28° and 60° and the distance AB is 4.12 miles as illustrated in the figure.
  (a) How far is the ship from point A?

(b) If the coordinates of A are $(0, 0)$ and the coordinates of B are $(4, 1)$, find the bearing of the ship from point A.

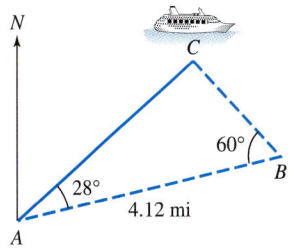

37. *Truss Construction* In the accompanying figure a truss is shown, where AB is 24.2 feet, angle ABD is 118°, and angle BDF is 28°. Find the length of BD. (Source: F. Riley.)

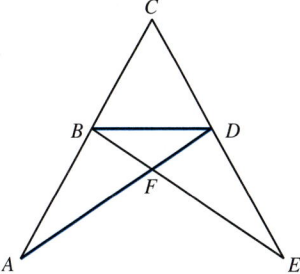

38. *Truss Construction* Use the results of Example 6 to solve triangle ACD in Figure 7.18 if angle BAD is 90°.

39. *Highway Construction* A *reverse curve* or S-curve is used to connect two straight portions of highway that are offset as illustrated in the accompanying figure. Angles α and β will not be equal if the two straight portions of highway have different directions. Typically the same radius r is used for both portions of the reverse curve. If $r = 480$ feet, $\alpha = 38°$, $\beta = 15°$, and $\theta = 75°$, find the distance between A and B. (Source: P. Kissam.)

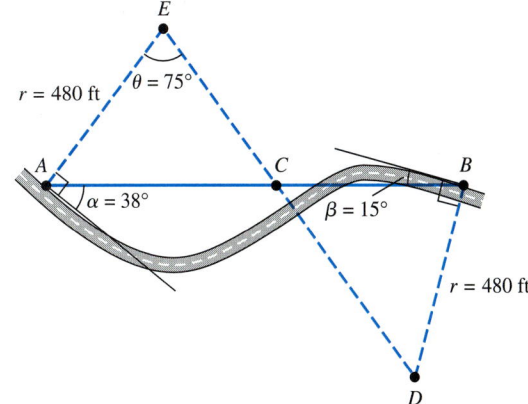

40. *Height of a Hot-Air Balloon* Two observation points *A* and *B* are 1500 feet apart. From these points the angles of elevation of a hot-air balloon are 43° and 47° as illustrated in the accompanying figure. Find the height of the balloon.

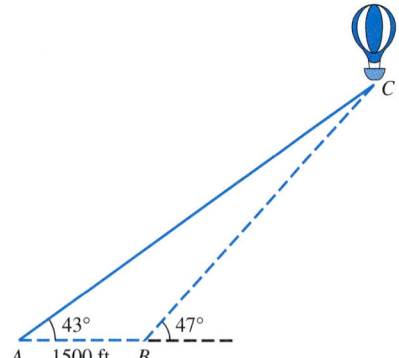

41. *Height of the Gateway Arch* The tallest monument in the world is the Gateway to the West Arch in St. Louis. From point *A* the top of the arch has an angle of elevation of 64.91° and from point *B* the angle of elevation is 60.81°. See the accompanying figure. If distance *AB* is 57 feet, find the height of this monument. (Source: *The Guinness Book of Records,* 1995.)

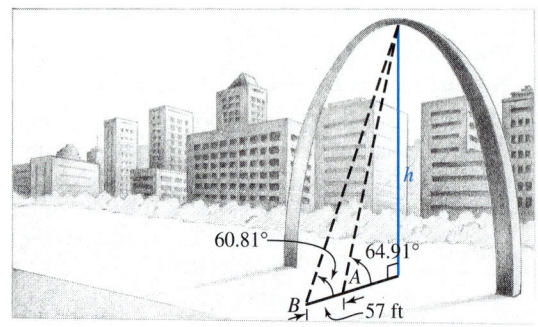

42. *Height of a Tower* A vertical tower supporting a cable for gondolas to transport skiers up a mountain is located on a ski-slope inclined at 28° as illustrated in the accompanying figure. If the length of the tower's shadow is 21 feet along the mountain side when the angle of the sun is 57° with respect to the ski-slope, calculate the height of the tower.

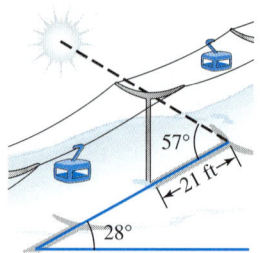

Writing about Mathematics

1. In your own words describe two different situations where the law of sines may be applied. Give an example of each situation.

2. Suppose that you are given α, a, and b for triangle *ABC*. If α is obtuse, what is the maximum number of triangles that could satisfy these conditions? What is the maximum number if α is acute? Explain your reasoning and give examples.

7.2 Law of Cosines

Derivation of the Law of Cosines • Solving Triangles • Area Formulas

Introduction

Surveying has been used for centuries in construction and in the determination of boundaries. Today the Global Positioning System (GPS) is being used to determine distances on Earth. The signal from a GPS satellite contains information necessary for hand-held receivers to calculate both the position of a GPS satellite and its distance. This information may be used to accurately calculate distances and angles between points on the ground. The law of cosines is a generalization of the Pythagorean theorem and is used in GPS calculations. In this section we introduce the law of cosines and use it to solve several applications. (Source: J. Sickle, *GPS for Land Surveyors*.)

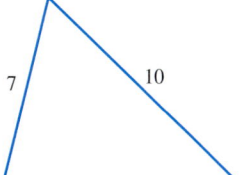

FIGURE 7.21 Given SSS

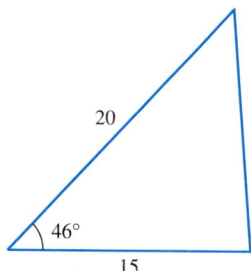

FIGURE 7.22 Given SAS

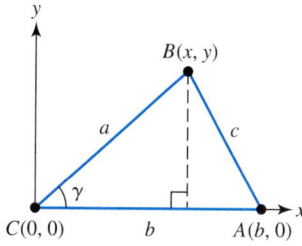

FIGURE 7.23

Derivation of the Law of Cosines

The law of cosines may be used to solve a triangle given either all three sides (SSS) or two sides and the angle included (SAS). In both cases a unique triangle is formed as illustrated in Figures 7.21 and 7.22.

Consider triangle *ABC* shown in Figure 7.23 with point *B* having coordinates (x, y). Using definitions of sine and cosine,

$$\cos \gamma = \frac{x}{a} \quad \text{and} \quad \sin \gamma = \frac{y}{a},$$

or equivalently,

$$x = a \cos \gamma \quad \text{and} \quad y = a \sin \gamma.$$

As a result, the coordinates of *B* are $(a \cos \gamma, a \sin \gamma)$. Since the coordinates of point *A* are $(b, 0)$, the distance *c* between points *A* and *B* may be found.

$$c = \sqrt{(a \cos \gamma - b)^2 + (a \sin \gamma - 0)^2} \qquad \text{Distance formula}$$

$$c^2 = (a \cos \gamma - b)^2 + (a \sin \gamma - 0)^2 \qquad \text{Square both sides.}$$

$$= a^2 \cos^2 \gamma - 2ab \cos \gamma + b^2 + a^2 \sin^2 \gamma \qquad \text{Expand each expression.}$$

$$= a^2(\cos^2 \gamma + \sin^2 \gamma) - 2ab \cos \gamma + b^2 \qquad \text{Distributive property}$$

$$= a^2 + b^2 - 2ab \cos \gamma \qquad \cos^2 \gamma + \sin^2 \gamma = 1$$

This result is valid for any triangle *ABC* and is known as the *law of cosines*. Since the vertices in Figure 7.23 could be rearranged, three possible equations are associated with the law of cosines.

Law of cosines

Any triangle with standard labeling satisfies

$$a^2 = b^2 + c^2 - 2bc \cos \alpha$$
$$b^2 = a^2 + c^2 - 2ac \cos \beta$$
$$c^2 = a^2 + b^2 - 2ab \cos \gamma.$$

Critical Thinking

Let $\gamma = 90°$ in the formula

$$c^2 = a^2 + b^2 - 2ab \cos \gamma$$

and simplify. Discuss the relationship between the law of cosines and the Pythagorean theorem.

Solving Triangles

A common problem in surveying is to find the distance between two points, *A* and *B*, situated on opposite sides of a building as illustrated in Figure 7.24. This distance can be found by applying the law of cosines. (Source: P. Kissam, *Surveying Practice.*)

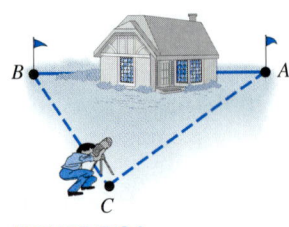

FIGURE 7.24

EXAMPLE 1 | *Finding the distance between two points (SAS)*

In Figure 7.24 a surveyor determines that *CA* is 75 feet, *CB* is 58 feet, and angle *ACB* is 83°. Find distance *AB*.

Solution

Let $\gamma = 83°$, $a = 58$, and $b = 75$. To find c, apply the law of cosines.

$$
\begin{array}{ll}
c^2 = a^2 + b^2 - 2ab \cos \gamma & \text{Law of cosines} \\
\quad = (58)^2 + (75)^2 - 2(58)(75) \cos(83°) & \text{Substitute.} \\
\quad \approx 7929 & \text{Approximate.} \\
c \approx 89 & \text{Take the square root.}
\end{array}
$$

The points *A* and *B* are about 89 feet apart. ∎

Trusses are frequently used to support roofs on buildings as illustrated in Figure 7.25. The simplest type of roof truss is a triangle as shown in Figure 7.26. One basic task when constructing a roof truss is to cut the ends of the rafters so that the roof has the correct slope. (Source: W. Riley, *Statics and Mechanics of Materials.*)

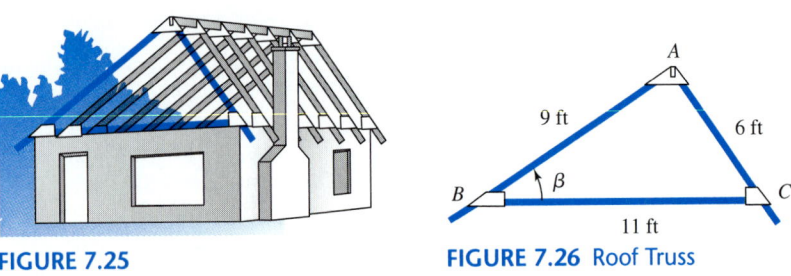

FIGURE 7.25   **FIGURE 7.26** Roof Truss

EXAMPLE 2 | *Designing a roof truss (SSS)*

Find β for the truss shown in Figure 7.26.

Solution

Begin by letting $a = 11$, $b = 6$, and $c = 9$, and then use the law of cosines to find β.

$$b^2 = a^2 + c^2 - 2ac \cos\beta \qquad \text{Law of cosines}$$

$$2ac \cos\beta = a^2 + c^2 - b^2 \qquad \text{Transpose terms.}$$

$$\cos\beta = \frac{a^2 + c^2 - b^2}{2ac} \qquad \text{Divide by } 2ac.$$

$$= \frac{11^2 + 9^2 - 6^2}{2(11)(9)} \qquad \text{Substitute.}$$

$$\approx 0.8384 \qquad \text{Approximate.}$$

Thus, $\beta \approx \cos^{-1}(0.8384) \approx 33°$.   ■

In the next example the distance between two GPS receivers is found. Finding the distance between two points might be important to surveyors or to search parties looking for lost hikers. The distance between two GPS receivers is sometimes called the *baseline*.

EXAMPLE 3   *Using GPS to find a baseline distance (SAS)*

A search party and an injured hiker both have hand-held GPS receivers as illustrated in Figure 7.27. The distance from the satellite to the search party is $b = 20{,}231.15$ kilometers, while the distance from the satellite to the hiker is $c = 20{,}231.57$ kilometers. If it is determined that $\alpha = 0.01456°$, find the baseline a between the search party and the hiker.

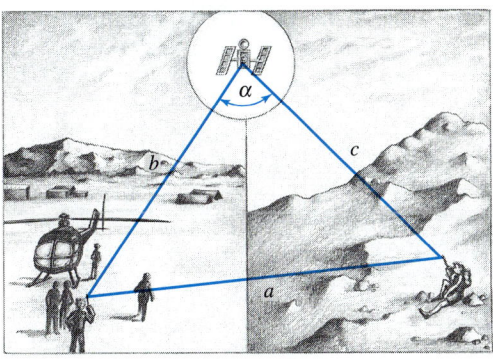

FIGURE 7.27

Solution

We can use the law of cosines to find a.

$$a^2 = b^2 + c^2 - 2bc \cos\alpha$$

$$= (20{,}231.15)^2 + (20{,}231.57)^2 - 2(20{,}231.15)(20{,}231.57)\cos(0.01456°)$$

$$\approx 26.61$$

$$a \approx 5.16$$

The distance between the search party and the hiker is about 5.16 kilometers.   ■

Area Formulas

One task that is frequently performed by surveyors is to find the acreage of a lot using a technique called triangulation. *Triangulation* divides a parcel of land into triangles. The area of the lot equals the sum of the areas of the triangles. We begin our discussion by developing some area formulas for triangles.

The area K of any triangle is given by $K = \frac{1}{2}bh$, where b is its base and h is its height. Using the law of sines we can find a formula for the area of the triangle shown in Figure 7.28.

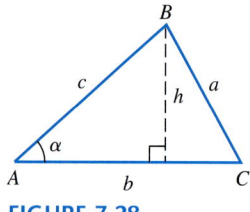

FIGURE 7.28

$$\sin\alpha = \frac{h}{c} \qquad \text{or} \qquad h = c\sin\alpha$$

Thus, the area equals

$$K = \frac{1}{2}bh$$

$$= \frac{1}{2}bc\sin\alpha.$$

Since the labels for the vertices in triangle ABC could be rearranged, three different area formulas can be written. Notice that these formulas can be applied when given SAS.

Area of a triangle

In any triangle with standard labeling, the area K is given by

$$K = \frac{1}{2}ab\sin\gamma, \qquad K = \frac{1}{2}ac\sin\beta, \qquad K = \frac{1}{2}bc\sin\alpha.$$

EXAMPLE 4 *Finding the area of a triangle (SAS)*

Find the area of triangle ABC in Figure 7.29.

Solution

We are given $\beta = 55°$, $a = 34$ feet, and $c = 42$ feet. Thus, the area K is given by the following.

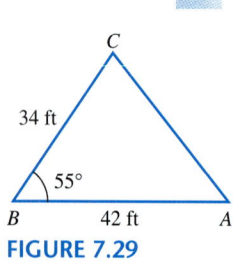

FIGURE 7.29

$$K = \frac{1}{2}ac\sin\beta \qquad \text{Area formula}$$

$$= \frac{1}{2}(34)(42)\sin 55° \qquad \text{Substitute.}$$

$$= 584.9 \text{ square feet} \qquad \text{Approximate.} \qquad \blacksquare$$

The next formula can be used to find the area of a triangle when the lengths of three sides are known. It is named after the Greek mathematician Heron.

Heron's formula

If a triangle has sides with lengths a, b, and c, then its area K is given by

$$K = \sqrt{s(s-a)(s-b)(s-c)},$$

where $s = \dfrac{1}{2}(a+b+c)$.

EXAMPLE 5 *Finding the area of a triangle (SSS)*

Approximate the area of triangle ABC with sides $a = 4$, $b = 5$, and $c = 7$.

Solution

Begin by calculating s.

$$s = \frac{1}{2}(4 + 5 + 7) = 8$$

Then the area is

$$K = \sqrt{8(8-4)(8-5)(8-7)} = \sqrt{96} \approx 9.80 \qquad \blacksquare$$

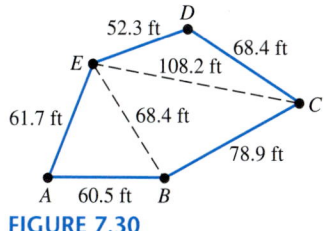

FIGURE 7.30

One method for finding the area of a lot is called the *distance method*. This method can be used to find the area of an irregular lot such as the one shown in Figure 7.30. The distance method does not measure angles, rather it only measures distances between points. Using triangulation and Heron's formula, the area of the lot can be found. (Source: I. Mueller and K. Ramsayer, *Introduction to Surveying.*)

EXAMPLE 6 *Applying the distance method to find the area of a lot*

Find the area of the parcel of land determined by *ABCDE* in Figure 7.30.

Solution

In triangle *ABE*, $s = \dfrac{1}{2}(60.5 + 68.4 + 61.7) = 95.3$ and its area is

$$K_1 = \sqrt{95.3(95.3 - 60.5)(95.3 - 68.4)(95.3 - 61.7)} \approx 1731.$$

In triangle *BCE*, $s = \dfrac{1}{2}(78.9 + 108.2 + 68.4) = 127.75$ and its area is

$$K_2 = \sqrt{127.75(127.75 - 78.9)(127.75 - 108.2)(127.75 - 68.4)} \approx 2691.$$

In triangle *CDE*, $s = \dfrac{1}{2}(68.4 + 52.3 + 108.2) = 114.45$ and its area is

$$K_3 = \sqrt{114.45(114.45 - 68.4)(114.45 - 52.3)(114.45 - 108.2)} \approx 1431.$$

The area of the lot is

$$K_1 + K_2 + K_3 \approx 1731 + 2691 + 1431 = 5853 \text{ square feet.} \qquad \blacksquare$$

7.2 PUTTING IT ALL TOGETHER

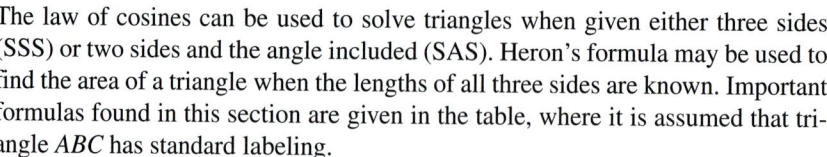

The law of cosines can be used to solve triangles when given either three sides (SSS) or two sides and the angle included (SAS). Heron's formula may be used to find the area of a triangle when the lengths of all three sides are known. Important formulas found in this section are given in the table, where it is assumed that triangle ABC has standard labeling.

Law of Cosines
$a^2 = b^2 + c^2 - 2bc\cos\alpha$
$b^2 = a^2 + c^2 - 2ac\cos\beta$
$c^2 = a^2 + b^2 - 2ab\cos\gamma$
Area Formulas
$K = \dfrac{1}{2}ab\sin\gamma \qquad K = \dfrac{1}{2}ac\sin\beta \qquad K = \dfrac{1}{2}bc\sin\alpha$
$K = \sqrt{s(s-a)(s-b)(s-c)} \qquad$ (Heron's formula),
where $s = \dfrac{1}{2}(a + b + c)$.

7.2 EXERCISES

   Tape 12

Determining a Method to Solve a Triangle

Exercises 1–8: Assume triangle ABC has standard labeling and complete the following.

  (a) *Determine if* AAS, ASA, SSA, SAS, *or* SSS *is given.*

  (b) *Decide if the law of sines or the law of cosines should be used to solve the triangle.*

  1. a, b, and γ **2.** α, γ, and c

  3. a, b, and α **4.** a, b, and c

  5. α, β, and c **6.** a, c, and α

  7. β, a, and γ **8.** b, c, and α

Solving Triangles

Exercises 9–14: Solve the triangle. Approximate values to three significant digits.

9.

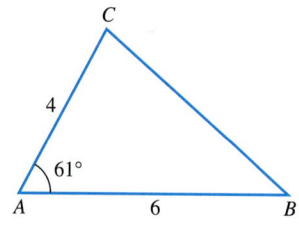

10.

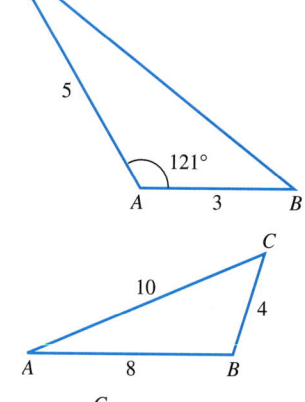

11.

12.

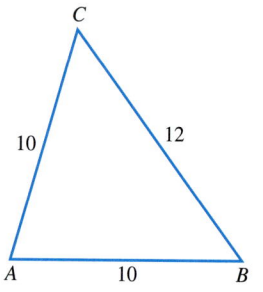

13.

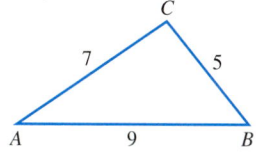

14.

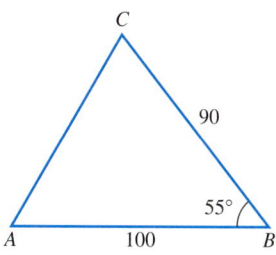

Exercises 15–22: Solve the triangle. Round values to the nearest tenth.

15. $a = 45, \gamma = 35°, b = 24$

16. $c = 7.9, \beta = 52°, a = 9.6$

17. $a = 2.4, b = 1.7, c = 1.4$

18. $a = 43, b = 41, c = 34$

19. $\alpha = 10°30', b = 24.1, c = 15.8$

20. $a = 12.8, b = 15.8, \gamma = 36°$

21. $a = 10.6, b = 25.8, c = 20.6$

22. $a = 104, b = 121, c = 111$

Does This Triangle Exist?

Exercises 23–28: Decide if a triangle exists that satisfies the conditions. Justify your answer.

23. $a = 10, b = 12, c = 25$

24. $a = 10, \beta = 51°, c = 5$

25. $\alpha = 89°, b = 63, \gamma = 112°$

26. $a = 2, b = 10, \alpha = 50°$

27. $\gamma = 54°, b = 63, \alpha = 63°$

28. $a = 5, b = 6, c = 8$

Area of Triangles

Exercises 29–32: Approximate the area of the triangle to three significant digits.

29.

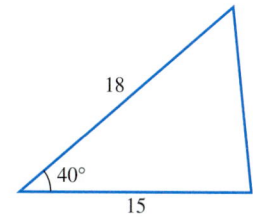

30.

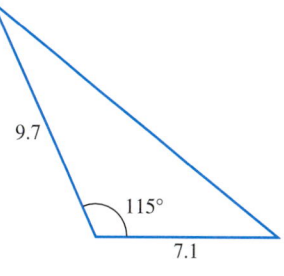

31.

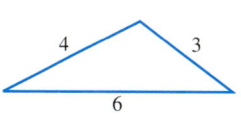

32.

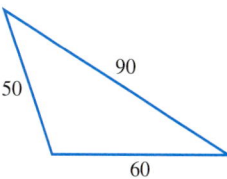

Exercises 33–40: Approximate the area of the triangle to three significant digits.

33. $a = 10, b = 12, \gamma = 58°$

34. $\alpha = 40°, b = 5.8, c = 8.8$

35. $\beta = 78°, a = 5.5, c = 6.8$

36. $\alpha = 23°, \gamma = 47°, b = 53$

37. $\beta = 31°, \alpha = 54°, a = 2.6$

38. $a = 7, b = 8, c = 9$

39. $a = 5.5, b = 6.7, c = 9.2$

40. $a = 104, b = 98, c = 112$

Applications

41. *Obstructed View* (Refer to Example 1.) In the accompanying figure, a surveyor is attempting to find the distance between two points A and B. A grove of trees is obstructing the view so the surveyor determines that AC is 143 feet, BC is 123 feet, and angle ACB is $78°35'$. Find the distance between A and B.

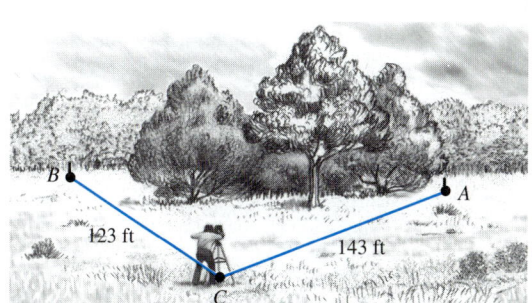

42. *Surveyor* A surveyor measures the sides of a triangular lot to be $a = 145.2$, $b = 136.8$, and $c = 95.3$, where measurements are in feet.
  (a) Approximate angles α, β, and γ.
  (b) What is the area of the lot?

43. *Curvature of the Earth* In times past when sailing ships came into port, people would see the top of the sails before they saw the entire ship. This is one reason why people knew the earth was not flat. Because of the curvature of the earth, surveyors must correct height measurements made over large distances. For example, a surveyor might measure the height of a hill to be too small unless a correction is taken. If an object is x miles away, then $f(x) = 0.585x^2$ computes the number of feet that should be added to the measured height. (Source: W. Rayner and M. Schmidt, *Elementary Surveying*.)
  (a) In the accompanying figure a surveyor measures a small hill 3 miles away to be 96 feet high. Find the actual height of the hill.
  (b) Approximate the correction angle θ in the figure by assuming that triangle ABC is isosceles.

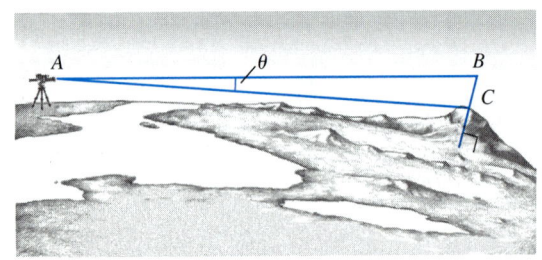

44. *Curvature and Surveying* Suppose that the hill in the previous exercise were 5 miles away. Find the correction angle θ and compare it with the value found in the previous exercise.

45. *Ship Navigation* Two ships set sail with bearings of 52° and 121°, traveling at 20 miles per hour and 14 miles per hour, respectively. Approximate the distance between the ships after 1.5 hours.

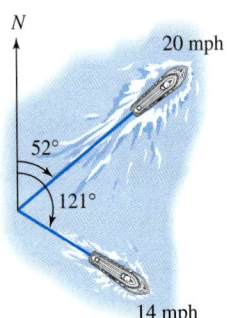

46. *Air Navigation* An airplane flies in a triangular course shown in the figure. Find the bearings of the plane while traveling from A to B and from B to C.

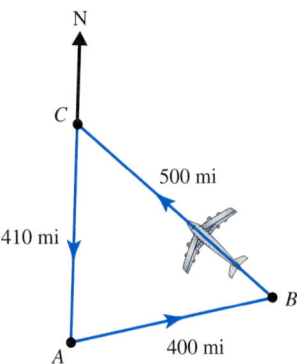

47. *Area of Regular Polygons* If a regular polygon has n sides of equal length L, then its area A is computed by

$$A = \frac{nL^2}{4} \cot\left(\frac{\pi}{n}\right).$$

(Source: M. Mortenson, *Computer Graphics*.)
  (a) Find the area of an equilateral triangle with sides of 6 inches using this formula.
  (b) Find the area of this triangle using Heron's formula. Compare answers.

48. *Area of Regular Polygons* (Refer to the previous exercise.) The measure of an interior angle in a regular polygon with n sides is given by $180°\left(1 - \frac{2}{n}\right)$. For example, a square is a regular polygon with $n = 4$ and each interior angle equals $180°\left(1 - \frac{2}{4}\right) = 90°$.
  (a) Find the area of a regular pentagon with sides of length 8 inches using the formula given in the previous exercise. See the accompanying figure.
  (b) Find the area of this regular pentagon using triangulation and Heron's formula.

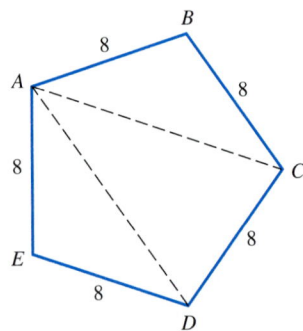

49. *Truss Construction* (Refer to Example 2.) A triangular truss is shown in the figure. Find angle θ.

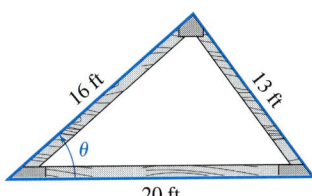

50. *Robotics* The figure illustrates the MIT Scheinman robotic arm. Suppose the length of the upper arm is 20 centimeters and the combined length of the forearm and hand is 30 centimeters. If the arm is positioned so that $\theta = 126°$, find the distance between the hand at point A and the shoulder joint at point B. (Sources: G. Beni and S. Hackwood, *Recent Advances in Robotics.*)

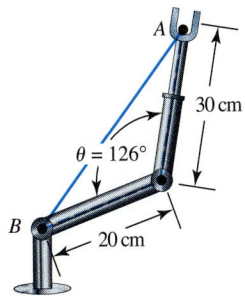

51. *Area of a Lot* Find the area of the lot in the figure.

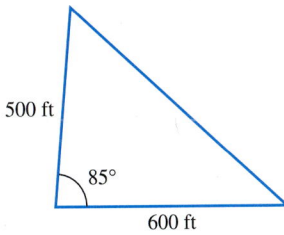

52. *Area of a Lot* Find the area of the quadrangular lot shown in the figure.

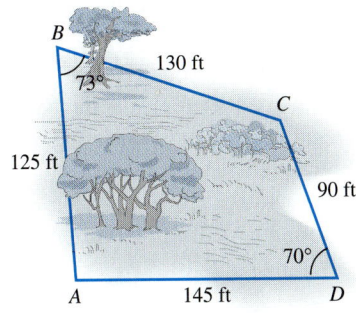

53. *Area of a Lot* Apply the distance method discussed in Example 6 to find the area of the lot.

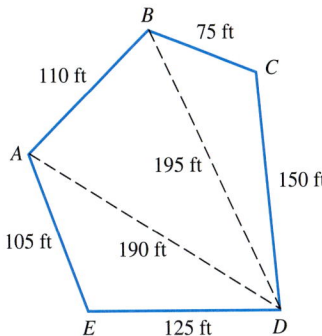

54. *Highway Curve* The most common highway curve consists of a circular arc connecting two sections of straight road as illustrated in the figure. Find the distance between PC (point of curve) and PT (point of tangency). (Source: P. Kissam.)

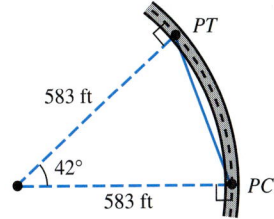

55. *Computer Graphics* When shading a triangle on a computer screen, three pixels are used to determine the vertices of a triangle. Many computer screens are 1024 pixels by 800 pixels and each pixel represents 0.015 inch by 0.015 inch as illustrated in the figure, where pixel $(7, 3)$ is shown. If three pixels $A(100, 300)$, $B(500, 200)$, and $C(320, 600)$ represent vertices of a triangle, approximate the actual area of triangle ABC on the computer screen in square inches. (Source: J. Foley, *Introduction to Computer Graphics.*)

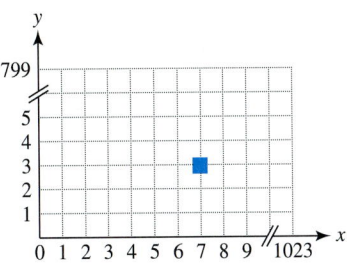

56. *Computer Graphics* Scaling is used in computer graphics to change the size of an object. For example, if a triangle has vertices $A(1, 2)$, $B(4, 0)$, and $C(2, 4)$, then scaling factors of s_x and s_y multiply the x-coordinates of the vertices by s_x and the y-coordinates by s_y. Scaling factors of $s_x = 2$ and $s_y = 3$ result in a new triangle with vertices $D(2, 6)$, $E(8, 0)$, and $F(4, 12)$. (Source: J. Foley.)

(a) Calculate the area of triangles ABC and DEF. How do they compare?

(b) If triangle ABC is scaled using $s_x = 2$ and $s_y = \frac{1}{2}$, find its area.

(c) Conjecture how the area A of a triangle changes for scaling factors of s_x and s_y.

Writing about Mathematics

1. Describe two different situations where the law of cosines may be applied. Give an example of each situation.

2. Describe two methods to find the area of a triangle. What information do you need to apply each method? Give an example of each situation.

CHECKING BASIC CONCEPTS FOR SECTIONS 7.1 AND 7.2

1. Solve triangle ABC if $\alpha = 44°$, $\gamma = 62°$, and $a = 12$.

2. Solve triangle ABC if $\alpha = 32°$, $a = 6$, and $b = 8$. How many solutions are there?

3. Solve the triangles using the law of cosines.

(a)

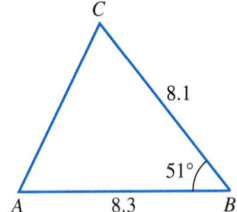

(b)

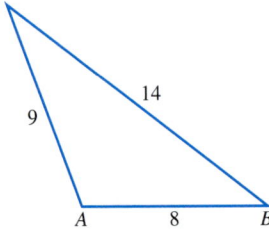

4. Find the area of triangle ABC.
(a) $a = 4.5$, $b = 5.2$, $\gamma = 55°$
(b) $a = 6$, $b = 7$, $c = 9$

7.3 Vectors

Basic Concepts • Operations on Vectors • The Dot Product • Work (Optional)

Introduction

The beginnings of vectors go back centuries to the notion of a directed line segment, but their formal development occurred during the nineteenth and twentieth centuries, after the invention of complex numbers. However, it was not until Einstein used vectors in his theory of general relativity that their importance became readily accepted.

Vectors are a profound invention. They provide a simple model for science and technology to visualize difficult concepts such as force, velocity, and electric fields. One of the most important mathematical devices in creating today's amazing computer graphics is a vector. In this section we discuss some of the important properties and applications of vectors. (Sources: *Historical Topics for the Mathematical Classroom, Thirty-first Yearbook*, NCTM; M. Mortenson, *Computer Graphics*.)

Basic Concepts

Many quantities in mathematics can be described using real numbers or **scalars.** Examples include a person's weight, the cost of a CD player, and the gas mileage of a car. Other quantities must be represented using vector quantities. A **vector quantity** involves both *magnitude* and *direction*. Magnitude may be interpreted as size or length. For example, if a car is traveling north at 50 miles per hour, then the direction *north* coupled with a *speed* of 50 miles per hour represents a vector quantity called *velocity*. In science a distinction is made between speed and velocity—speed is the magnitude of velocity.

A vector quantity may be represented by a directed line segment called a **vector.** A vector **v** representing a velocity of a car traveling 50 miles per hour north is shown in Figure 7.31, while the vector **u** represents a velocity of 25 miles per hour east. Notice that the length of **u** is half the length of **v**. Vectors do *not* have position, rather they only have magnitude and direction. A vector may be translated, provided its direction and magnitude do not change. Two vectors are **equal** if they have the same magnitude and direction. In Figure 7.32 each directed line segment represents the same vector **v.**

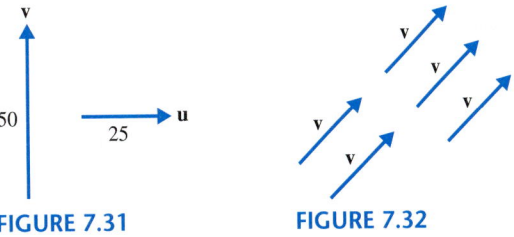

FIGURE 7.31   **FIGURE 7.32**

A vector is usually represented symbolically by a letter printed in boldface type, such as **a, b, v,** or **F.** A second way to denote a vector is using two points. If the **initial point** of a vector **v** is P and its **terminal point** is Q, then $\mathbf{v} = \overrightarrow{PQ}$ as illustrated in Figure 7.33.

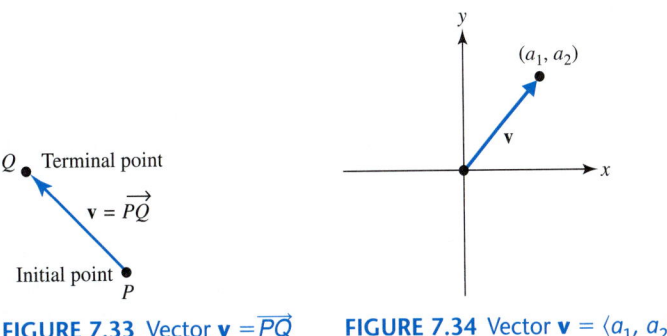

FIGURE 7.33 Vector $\mathbf{v} = \overrightarrow{PQ}$ **FIGURE 7.34** Vector $\mathbf{v} = \langle a_1, a_2 \rangle$

Operations on Vectors

If we place the initial point of vector **v** at the origin as in Figure 7.34, then its terminal point (a_1, a_2) may be used to determine **v.** To distinguish the *point* (a_1, a_2) from the *vector* **v,** we use the notation $\mathbf{v} = \langle a_1, a_2 \rangle$. The **horizontal component** of **v** is a_1 and the **vertical component** of **v** is a_2.

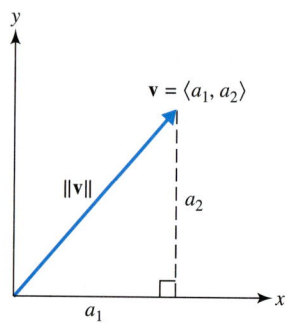

FIGURE 7.35

The length of a vector equals its magnitude. If $\mathbf{v} = \langle a_1, a_2 \rangle$ then the *magnitude* of $\mathbf{v}$ is denoted by $\|\mathbf{v}\|$. By applying the Pythagorean theorem to Figure 7.35, $\|\mathbf{v}\| = \sqrt{a_1^2 + a_2^2}$.

> ### Magnitude of a vector
>
> If $\mathbf{v} = \langle a_1, a_2 \rangle$, then the **magnitude** (or length) of $\mathbf{v}$ is given by
> $$\|\mathbf{v}\| = \sqrt{a_1{}^2 + a_2{}^2}.$$
> If $\|\mathbf{v}\| = 1$, then $\mathbf{v}$ is a **unit vector.**

If a vector has initial point P with coordinates (a_1, b_1) and terminal point Q with coordinates (a_2, b_2) then vector $\overrightarrow{PQ}$ is given by $\overrightarrow{PQ} = \langle a_2 - a_1, b_2 - b_1 \rangle$. This is illustrated in the next example.

EXAMPLE 1   *Finding a vector graphically and symbolically*

Let P have coordinates $(-1, 2)$ and Q have coordinates $(3, 4)$. Find vector $\overrightarrow{PQ}$ graphically and symbolically. Calculate the magnitude of $\overrightarrow{PQ}$.

Solution

To graph $\overrightarrow{PQ}$, plot the points P and Q. Then sketch a directed line segment from P to Q as shown in Figure 7.36. We can see that the horizontal component is 4 and the vertical component is 2. A symbolic representation of $\overrightarrow{PQ}$ is given by

$$\overrightarrow{PQ} = \langle 3 - (-1), 4 - 2 \rangle = \langle 4, 2 \rangle.$$

The magnitude or length of $\overrightarrow{PQ}$ is

$$\|\overrightarrow{PQ}\| = \sqrt{4^2 + 2^2} = \sqrt{20} \approx 4.47. \qquad \blacksquare$$

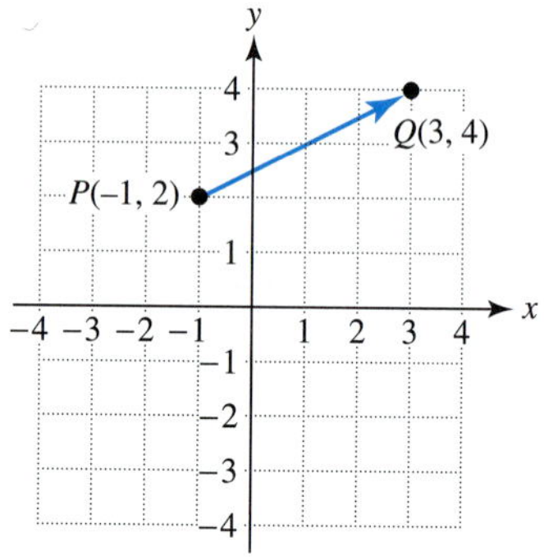

FIGURE 7.36

Vector Addition.   Suppose that a swimmer heads directly across a river at 3 miles per hour. If the current is 4 miles per hour, then the person will be carried a distance downstream before reaching the other side as illustrated in Figure 7.37. We can use vectors to visually find the direction and speed that the swimmer will travel across the river.

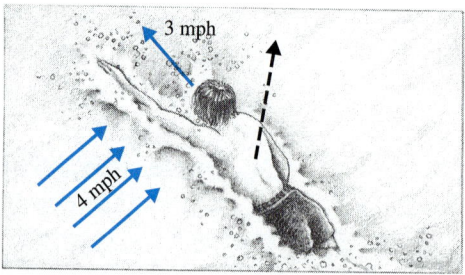

FIGURE 7.37 A Swimmer in a Current

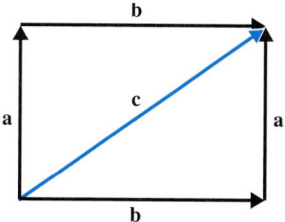

FIGURE 7.38 Vector Addition

Let vector **a** represent the speed and direction of the swimmer with no current, vector **b** represent the direction and speed of the current, and vector **c** represent the final direction and speed of the swimmer. We can find the length and direction of **c** by applying the **parallelogram rule** as shown in Figure 7.38. The speed and direction of the swimmer is represented by the diagonal **c** of the parallelogram (rectangle), which is determined by **a** and **b**. Vector **c** is called the **sum** or **resultant** of vectors **a** and **b**.

Symbolically, we can represent the velocity of the swimmer with no current by $\mathbf{a} = \langle 0, 3 \rangle$, the velocity of the current by $\mathbf{b} = \langle 4, 0 \rangle$, and the velocity of the swimmer in the current by $\mathbf{c} = \langle 4, 3 \rangle$. Vector **c** is the sum of vectors **a** and **b** and can be found as follows.

$$\mathbf{a} + \mathbf{b} = \langle 0, 3 \rangle + \langle 4, 0 \rangle = \langle 0 + 4, 3 + 0 \rangle = \langle 4, 3 \rangle = \mathbf{c}$$

Since $\|\mathbf{c}\| = \sqrt{4^2 + 3^2} = 5$, the swimmer moves 5 miles per hour in the direction of **c**.

Figure 7.39 illustrates graphically how to find $\mathbf{c} = \mathbf{a} + \mathbf{b}$ using the parallelogram rule. The following defines vector addition symbolically.

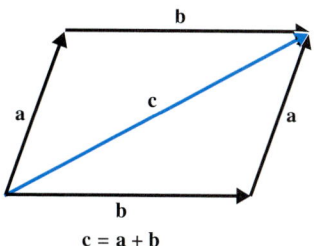

$\mathbf{c} = \mathbf{a} + \mathbf{b}$

FIGURE 7.39 The Parallelogram Rule: $\mathbf{c} = \mathbf{a} + \mathbf{b}$

Vector addition

If $\mathbf{a} = \langle a_1, a_2 \rangle$ and $\mathbf{b} = \langle b_1, b_2 \rangle$, then the **sum** of **a** and **b** is given by

$$\mathbf{a} + \mathbf{b} = \langle a_1, a_2 \rangle + \langle b_1, b_2 \rangle = \langle a_1 + b_1, a_2 + b_2 \rangle.$$

Suppose that vector **a** represents a force of 80 pounds pulling on a water-ski towrope and **b** represents a force of 60 pounds pulling on a second towrope. See Figure 7.40. The resultant force $\mathbf{c} = \mathbf{a} + \mathbf{b}$ is given by the diagonal of the parallelogram shown in Figure 7.41. Vector **c** represents the net force exerted by the two water skiers.

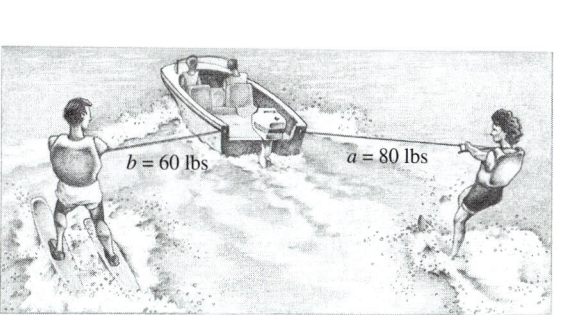

$b = 60$ lbs $a = 80$ lbs

FIGURE 7.40

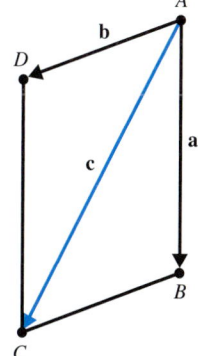

FIGURE 7.41

EXAMPLE 2   *Applying the parallelogram rule*

Find the magnitude of the resultant force on the ski boat in the preceding discussion if the angle between the towropes is $25°$.

Solution

The magnitude of the force equals the length of the diagonal AC in Figure 7.41. Since angle ABC is $180° - 25° = 155°$, we find AC by applying the law of cosines.

$$AC^2 = 60^2 + 80^2 - 2(60)(80)\cos 155°$$

$$\approx 18{,}701$$

$$AC \approx 137 \text{ pounds} \qquad \blacksquare$$

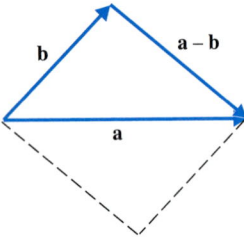

FIGURE 7.42

Vector Subtraction. Subtraction can be defined both graphically and symbolically. The difference $\mathbf{a} - \mathbf{b}$ is shown graphically in Figure 7.42. Notice that by the parallelogram rule $\mathbf{b} + (\mathbf{a} - \mathbf{b}) = \mathbf{a}$.

Vector subtraction can be defined symbolically as follows.

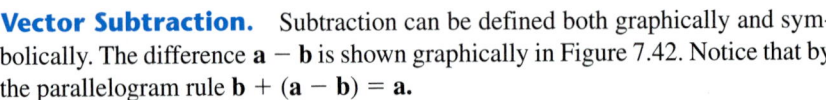

Vector subtraction

If $\mathbf{a} = \langle a_1, a_2 \rangle$ and $\mathbf{b} = \langle b_1, b_2 \rangle$, then the **difference** of $\mathbf{a}$ and $\mathbf{b}$ is given by

$$\mathbf{a} - \mathbf{b} = \langle a_1, a_2 \rangle - \langle b_1, b_2 \rangle = \langle a_1 - b_1, a_2 - b_2 \rangle.$$

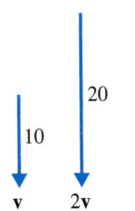

FIGURE 7.43 North Winds

Scalar Multiplication. Scalar multiplication results when a vector $\mathbf{v}$ is multiplied by a real number or *scalar* k to form $k\mathbf{v}$. Vectors $\mathbf{v}$ and $k\mathbf{v}$ are parallel if $k \neq 0$. Vector $k\mathbf{v}$ points in the *same* direction as $\mathbf{v}$ if $k > 0$, and $k\mathbf{v}$ points in the *opposite* direction of $\mathbf{v}$ if $k < 0$. The magnitude of $k\mathbf{v}$ is $|k|$ times the magnitude of $\mathbf{v}$. For example, suppose that a 10-mile-per-hour wind is blowing from the north. Since the wind is blowing toward the south, the wind may be represented by $\mathbf{v} = \langle 0, -10 \rangle$. If the wind speed doubles, but does not change direction, then the *scalar product* $2\mathbf{v}$ models this situation.

$$2\mathbf{v} = 2\langle 0, -10 \rangle = \langle 2 \cdot 0, 2 \cdot (-10) \rangle = \langle 0, -20 \rangle$$

The wind is now blowing at 20 miles per hour toward the south. See Figure 7.43.

Next, suppose that a wind from the southwest is modeled by $\mathbf{v} = \langle 4, 4 \rangle$. Then

$$-\frac{1}{2}\mathbf{v} = -\frac{1}{2}\langle 4, 4 \rangle = \left\langle -\frac{1}{2} \cdot 4, -\frac{1}{2} \cdot 4 \right\rangle = \langle -2, -2 \rangle$$

represents a wind from the northeast in the opposite direction of $\mathbf{v}$ with half the speed. See Figure 7.44. This discussion motivates the following definition of scalar multiplication.

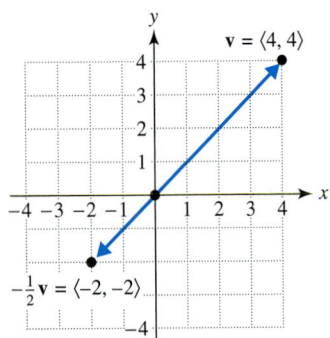

FIGURE 7.44

Scalar multiplication

If $\mathbf{v} = \langle v_1, v_2 \rangle$ and k is a real number, then the **scalar product** $k\mathbf{v}$ is given by

$$k\mathbf{v} = k\langle v_1, v_2 \rangle = \langle kv_1, kv_2 \rangle.$$

Sums, differences, and scalar products can be calculated graphically and symbolically as illustrated in the next example.

EXAMPLE 3 *Performing operations on vectors*

Find each of the following graphically and symbolically if $\mathbf{a} = \langle -3, 4 \rangle$ and $\mathbf{b} = \langle -1, -2 \rangle$.

(a) $\|\mathbf{a}\|$

(b) $-2\mathbf{b}$

(c) $\mathbf{a} + 2\mathbf{b}$

Solution

(a) Graph $\mathbf{a} = \langle -3, 4 \rangle$ as shown in Figure 7.45. The length of $\mathbf{a}$ appears to be about 5. This can be verified symbolically.

$$\|\mathbf{a}\| = \sqrt{(-3)^2 + (4)^2} = 5$$

(b) Graph $\mathbf{b} = \langle -1, -2 \rangle$. The scalar product $-2\mathbf{b}$ points in the opposite direction of $\mathbf{b}$ with twice the length. See Figure 7.46, where $-2\mathbf{b} = \langle 2, 4 \rangle$. Symbolically this is given by

$$-2\mathbf{b} = -2\langle -1, -2 \rangle = \langle -2 \cdot (-1), -2 \cdot (-2) \rangle = \langle 2, 4 \rangle.$$

(c) Graph $\mathbf{a} = \langle -3, 4 \rangle$ and $2\mathbf{b} = \langle -2, -4 \rangle$ as shown in Figure 7.47. Using the parallelogram rule, the diagonal represents $\mathbf{a} + 2\mathbf{b} = \langle -5, 0 \rangle$. This can be verified symbolically.

$$\begin{aligned} \mathbf{a} + 2\mathbf{b} &= \langle -3, 4 \rangle + 2\langle -1, -2 \rangle \\ &= \langle -3, 4 \rangle + \langle -2, -4 \rangle \\ &= \langle -5, 0 \rangle \end{aligned}$$

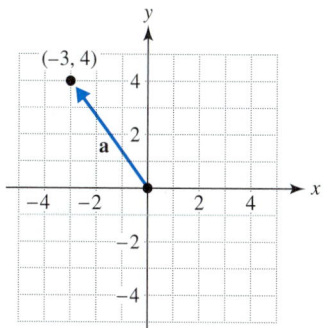

FIGURE 7.45

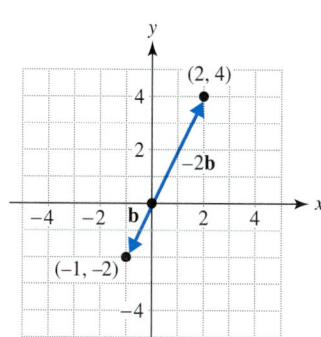

FIGURE 7.46

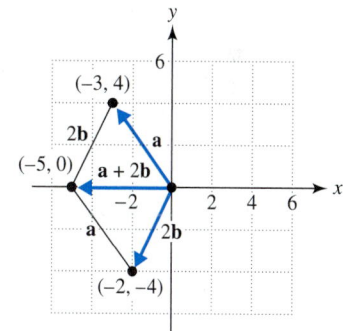

FIGURE 7.47 ■

```
√((-3)²+(4)²)
                5
-2(-1, -2)
            (2 4)
(-3,4)+2(-1,-2)
            (-5 0)
```

FIGURE 7.48

Technology Note *Operations on Vectors*

On some graphing calculators, such as the TI-83, the list feature may be used to perform operations on vectors. In Figure 7.48 this calculator has been used to evaluate the expressions in Example 3. On other calculators, such as the TI-85, vectors may be represented using ordered pairs with parentheses.

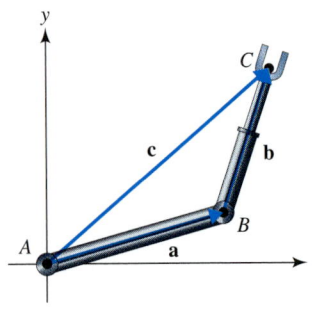

FIGURE 7.49

An Application in Robotics.   Robotic arms are sometimes modeled using vectors. Consider the *planar two-arm manipulator* in Figure 7.49. If $\overrightarrow{AB}$ = **a** and $\overrightarrow{BC}$ = **b,** then the position of the hand is given by $\overrightarrow{AC}$ = **c.** Since **c** = **a** + **b,** we can easily locate the position of the hand if **a** and **b** are known. (Source: J. Craig, *Introduction to Robotics*.)

EXAMPLE 4   *Locating a robotic arm using vectors*

Let **a** = $\langle 3.1, 1.5 \rangle$ and **b** = $\langle 1.4, 2.4 \rangle$ in Figure 7.49.
(a) Find the position of the robotic hand.
(b) Suppose the upper arm represented by **a** doubles its length and the forearm represented by **b** reduces its length by half. Find the new position of the hand.

Solution
(a) To find the position of the hand evaluate **a** + **b.**

$$\mathbf{a} + \mathbf{b} = \langle 3.1, 1.5 \rangle + \langle 1.4, 2.4 \rangle = \langle 4.5, 3.9 \rangle$$

The hand is located at the point (4.5, 3.9).
(b) The new position is represented by

$$2\mathbf{a} + \frac{1}{2}\mathbf{b} = 2\langle 3.1, 1.5 \rangle + \frac{1}{2}\langle 1.4, 2.4 \rangle$$
$$= \langle 6.2, 3.0 \rangle + \langle 0.7, 1.2 \rangle$$
$$= \langle 6.9, 4.2 \rangle.$$

The new coordinates of the robotic hand are (6.9, 4.2).   ∎

Vector Notation Using i and j.   A second type of vector notation involves the vectors **i** = $\langle 1, 0 \rangle$ and **j** = $\langle 0, 1 \rangle$. Given a vector **a** = $\langle a_1, a_2 \rangle$, it can be expressed as

$$\mathbf{a} = \langle a_1, a_2 \rangle = a_1 \langle 1, 0 \rangle + a_2 \langle 0, 1 \rangle = a_1 \mathbf{i} + a_2 \mathbf{j}.$$

For example, $\langle 3, -4 \rangle$ and $3\mathbf{i} - 4\mathbf{j}$ represent the same vector.

---------- **MAKING CONNECTIONS** ----------

In Section 3.5 we discussed the *complex number i*, where $i = \sqrt{-1}$ and $i^2 = -1$. In this section we introduced the *vector* **i** = $\langle 1, 0 \rangle$. Each represents a different mathematical concept.

EXAMPLE 5   *Using vectors in navigation*

An airplane is flying with an airspeed of 300 miles per hour and a bearing of 40° in a 30-mile-per-hour west wind.
(a) Find vectors **v** and **u** that model the velocity of the airplane and the velocity of the wind, respectively.
(b) Use vectors to determine the groundspeed of the plane.

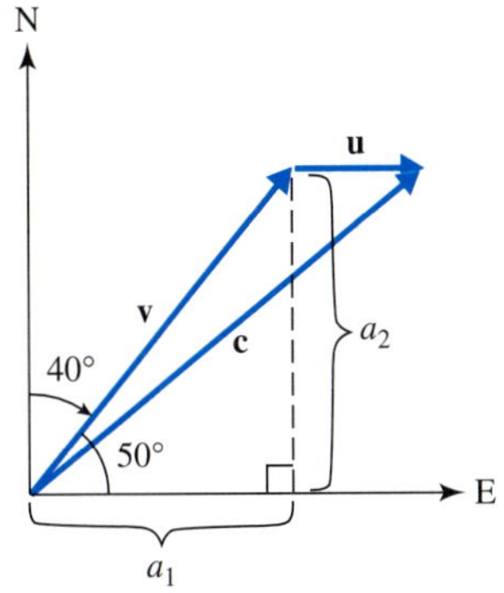

Solution

(a) Consider Figure 7.50, which shows vectors **v** and **u** graphically. Since **u** models a west wind, it points to the right with length 30 and is represented symbolically by $\mathbf{u} = \langle 30, 0 \rangle$. Let a_1 be the horizontal component and a_2 be the vertical component of **v**. Since $\|\mathbf{v}\| = 300$,

$$\cos 50° = \frac{a_1}{300} \quad \text{and} \quad \sin 50° = \frac{a_2}{300}.$$

It follows that $a_1 = 300 \cos 50°$ and $a_2 = 300 \sin 50°$. Thus,

$$\mathbf{v} = \langle 300 \cos 50°, 300 \sin 50° \rangle.$$

(b) The true course of the plane is given by $\mathbf{c} = \mathbf{v} + \mathbf{u}$. As a result, the ground-speed of the plane equals $\|\mathbf{c}\|$.

$$\begin{aligned} \mathbf{c} &= \mathbf{v} + \mathbf{u} \\ &= \langle 300 \cos 50°, 300 \sin 50° \rangle + \langle 30, 0 \rangle \\ &= \langle 300 \cos 50° + 30, 300 \sin 50° \rangle \end{aligned}$$

Then,

$$\begin{aligned} \|\mathbf{c}\| &= \sqrt{(300 \cos 50° + 30)^2 + (300 \sin 50°)^2}, \\ &\approx 320.1 \end{aligned}$$

The groundspeed of the airplane is approximately 320 miles per hour. ∎

N

40°

50°

E

a_1

a_2

v

c

u

FIGURE 7.50

The Dot Product

Thus far we have discussed addition, subtraction, and scalar multiplication of vectors. Another operation on vectors is called the *dot product,* which is important because it can be used to find angles between vectors. The dot product has applications in computer graphics, solar energy, and physics. We begin by defining the dot product and then discuss some of its applications.

> ### Dot product
> Let $\mathbf{a} = \langle a_1, a_2 \rangle$ and $\mathbf{b} = \langle b_1, b_2 \rangle$. The **dot product** of **a** and **b**, denoted by $\mathbf{a} \cdot \mathbf{b}$, is a *real number* given by
>
> $$\mathbf{a} \cdot \mathbf{b} = a_1 b_1 + a_2 b_2.$$

In the next example we calculate dot products. Notice that the dot product of two vectors is a real number, rather than a vector.

EXAMPLE 6 — *Calculating dot products*

Calculate $\mathbf{a} \cdot \mathbf{b}$.

(a) $\mathbf{a} = \langle 4, -3 \rangle$, $\mathbf{b} = \langle -1, 2 \rangle$

(b) $\mathbf{a} = 2\mathbf{i} + 5\mathbf{j}$, $\mathbf{b} = -3\mathbf{i} + 2\mathbf{j}$

Solution

(a) $\mathbf{a} \cdot \mathbf{b} = \langle 4, -3 \rangle \cdot \langle -1, 2 \rangle = (4)(-1) + (-3)(2) = -10$

(b) $\mathbf{a} \cdot \mathbf{b} = (2\mathbf{i} + 5\mathbf{j}) \cdot (-3\mathbf{i} + 2\mathbf{j}) = (2)(-3) + (5)(2) = 4$ ∎

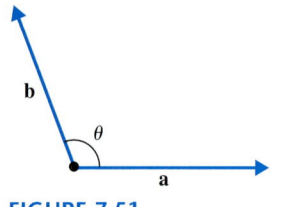

FIGURE 7.51

In Figure 7.51 the *angle between vectors* **a** *and* **b** is θ, where $0° \leq \theta \leq 180°$. If $\theta = 90°$ the vectors are **perpendicular** and if $\theta = 0°$ or $180°$ the vectors are **parallel.** If $\theta = 0°$ the vectors point in the *same* direction, and if $\theta = 180°$ they point in *opposite* directions.

It is shown in the Extended Exercises that for any two nonzero vectors **a** and **b,**

$$\mathbf{a} \cdot \mathbf{b} = \|\mathbf{a}\| \, \|\mathbf{b}\| \cos \theta.$$

This result can be used to find the angle θ between **a** and **b.**

Angle between two vectors

If **a** and **b** are nonzero vectors, then the **angle θ between a and b** is given by

$$\theta = \cos^{-1}\left(\frac{\mathbf{a} \cdot \mathbf{b}}{\|\mathbf{a}\| \, \|\mathbf{b}\|}\right).$$

Vectors **a** and **b** are perpendicular if and only if $\mathbf{a} \cdot \mathbf{b} = 0$.

EXAMPLE 7   *Finding the angle between two vectors*

Sketch the vectors **a** and **b.** Then find the angle θ between **a** and **b.**
(a) $\mathbf{a} = 2\mathbf{i} - 3\mathbf{j}$, $\mathbf{b} = 3\mathbf{i} + 2\mathbf{j}$
(b) $\mathbf{a} = \langle -4, 3 \rangle$, $\mathbf{b} = \langle 1, -2 \rangle$

Solution

(a) Vectors **a** and **b** appear to be perpendicular in Figure 7.52. Since

$$\mathbf{a} \cdot \mathbf{b} = (2)(3) + (-3)(2) = 0,$$

the vectors are perpendicular and $\theta = 90°$.

(b) A sketch of the vectors is shown in Figure 7.53. They are neither perpendicular nor parallel. Since

$$\|\mathbf{a}\| = \sqrt{(-4)^2 + (3)^2} = 5 \quad \text{and} \quad \|\mathbf{b}\| = \sqrt{(1)^2 + (-2)^2} = \sqrt{5},$$

it follows that

$$\theta = \cos^{-1}\left(\frac{(-4)(1) + (3)(-2)}{5\sqrt{5}}\right) \approx 153.4°$$

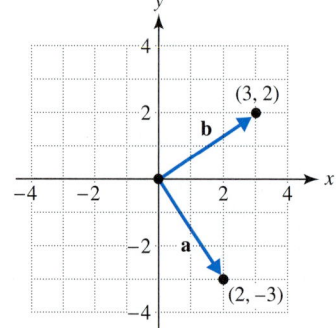

FIGURE 7.52

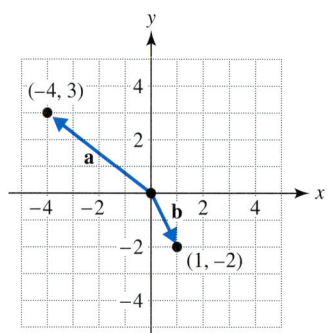

FIGURE 7.53

An Application in Computer Graphics. Vectors frequently are used to determine the color of pixels on computer screens. For example, suppose that we would like to color the right side of the screen blue and the left side gray, where the boundary is vector $\overrightarrow{PQ}$. See Figure 7.54.

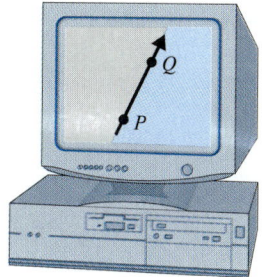

FIGURE 7.54

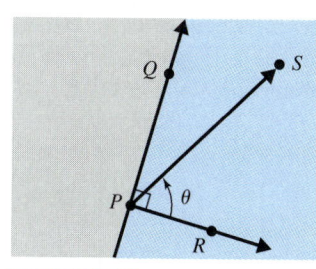

FIGURE 7.55

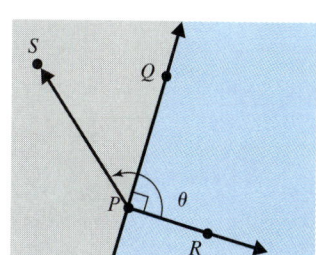

FIGURE 7.56

First, we find $\overrightarrow{PR}$ perpendicular to $\overrightarrow{PQ}$. Then to determine if a pixel at point S should be blue or gray consider the angle θ between vectors $\overrightarrow{PS}$ and $\overrightarrow{PR}$. If θ is acute then S must be on the same side of $\overrightarrow{PQ}$ as R and is colored blue as illustrated in Figure 7.55. If angle θ is obtuse then S is on the opposite side of $\overrightarrow{PQ}$ as R and is colored gray as shown in Figure 7.56. We can determine whether θ is acute or obtuse by calculating the dot product $\overrightarrow{PS} \cdot \overrightarrow{PR}$. If the dot product is positive then

$$\theta = \cos^{-1}\left(\frac{\overrightarrow{PS} \cdot \overrightarrow{PR}}{\|\overrightarrow{PS}\|\,\|\overrightarrow{PR}\|}\right)$$

is acute and S should be blue. If the dot product is negative, then θ is obtuse and S should be gray. (Source: J. Foley, *Introduction to Computer Graphics*.)

EXAMPLE 8 *Using the dot product in computer graphics*

In the previous discussion let P be $(-1, 2)$, S be $(-2, -5)$, $\overrightarrow{PQ} = \mathbf{i} + 5\mathbf{j}$, and $\overrightarrow{PR} = 5\mathbf{i} - \mathbf{j}$. Determine the appropriate color at point S.

Solution

First, find vector $\overrightarrow{PS}$.

$$\overrightarrow{PS} = \langle -2 - (-1), -5 - 2 \rangle = \langle -1, -7 \rangle \text{ or } -\mathbf{i} - 7\mathbf{j}$$

Then compute the dot product $\overrightarrow{PS} \cdot \overrightarrow{PR}$.

$$\overrightarrow{PS} \cdot \overrightarrow{PR} = (-\mathbf{i} - 7\mathbf{j}) \cdot (5\mathbf{i} - \mathbf{j}) = 2 > 0$$

Thus, θ is acute and the point $(-2, -5)$ is in the blue region. See Figure 7.57. ∎

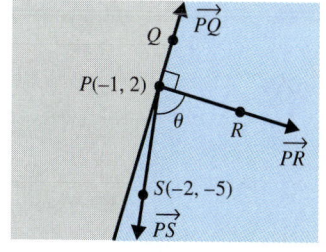

FIGURE 7.57

Work (Optional)

In science a force does work only when an object moves. For example, a person pushing against a brick wall does no work, whereas a person lifting a 20-pound weight does work. Work equals force times distance, *provided* the force is in the same direction as the movement of the object. If a 150-pound person climbs up a 20-foot rope, then the work W done is

$$W = 150 \times 20 = 3000 \text{ foot-pounds.}$$

A **foot-pound** equals the work required to lift one pound a vertical distance of one foot. If the force is not in the same direction as the movement, then we must use trigonometry to determine work.

Consider a person pulling a wagon as shown in Figure 7.58, where **F** represents the force on the handle and **D** represents the distance that the wagon is pulled. The force **F** can be expressed as the sum of a horizontal vector in the direction of **D** and a vertical vector perpendicular to **D** as illustrated in Figure 7.59. Using the right triangle in Figure 7.60, the horizontal component of **F** is given by $\|\mathbf{F}\| \cos \theta$ and the vertical component of **F** is given by $\|\mathbf{F}\| \sin \theta$.

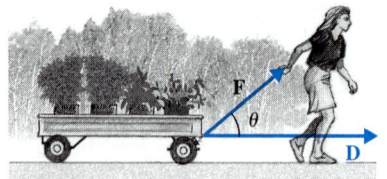

FIGURE 7.58

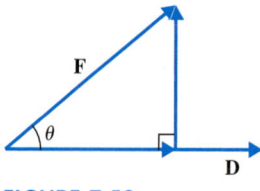

FIGURE 7.59

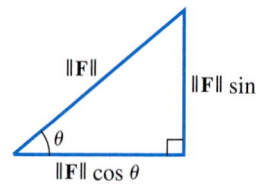

FIGURE 7.60

The work done pulling the wagon is equal to the horizontal component $\|\mathbf{F}\| \cos \theta$ times the distance, which is given by $\|\mathbf{D}\|$. That is,

$$W = \|\mathbf{F}\| \|\mathbf{D}\| \cos \theta = \mathbf{F} \cdot \mathbf{D}.$$

Work equals the dot product of the force vector **F** and the displacement vector **D**.

Work

If a constant force **F** is applied to an object that moves along a vector **D,** then the work W done is

$$W = \mathbf{F} \cdot \mathbf{D}.$$

EXAMPLE 9 *Calculating work*

A 150-pound person walks 500 feet up a hiking trail that is inclined at 20°. Use vectors to compute the work done by the person as illustrated in Figure 7.61.

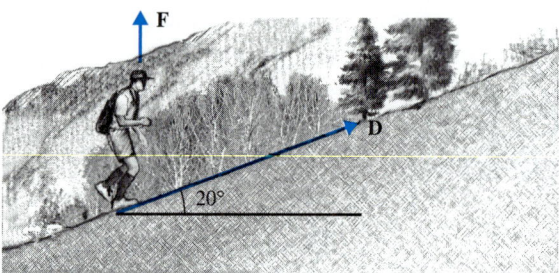

FIGURE 7.61

Solution

Vector **D** is given by $\mathbf{D} = \langle 500 \cos 20°, \ 500 \sin 20° \rangle$. Since gravity pulls downward, the force exerted by the person against gravity is $\mathbf{F} = \langle 0, 150 \rangle$. The work done is

$$W = \mathbf{F} \cdot \mathbf{D} = (0)(500 \cos 20°) + (150)(500 \sin 20°) \approx 25{,}650 \text{ foot-pounds.} \ \blacksquare$$

Critical Thinking

Suppose a force vector **F** is perpendicular to **D**. How much work is done? Interpret your answer.

7.3 PUTTING IT ALL TOGETHER

Vectors are an important invention of the nineteenth and twentieth centuries that have enabled science and technology to model a wide variety of phenomena. The following table summarizes some basic concepts regarding vectors.

Vectors
Vector quantities denote both magnitude and direction. A vector **a** may be expressed as either $\mathbf{a} = \langle a_1, a_2 \rangle$ or $\mathbf{a} = a_1\mathbf{i} + a_2\mathbf{j}$. Its magnitude is given by $\|\mathbf{a}\| = \sqrt{a_1{}^2 + a_2{}^2}$. The numbers a_1 and a_2 are called the horizontal and vertical components of **a**, respectively.
Operations on Vectors
Let $\mathbf{a} = \langle a_1, a_2 \rangle$, $\mathbf{b} = \langle b_1, b_2 \rangle$, and k be a real number. $\mathbf{a} + \mathbf{b} = \langle a_1, a_2 \rangle + \langle b_1, b_2 \rangle = \langle a_1 + b_1, a_2 + b_2 \rangle$ (sum) $\mathbf{a} - \mathbf{b} = \langle a_1, a_2 \rangle - \langle b_1, b_2 \rangle = \langle a_1 - b_1, a_2 - b_2 \rangle$ (difference) $k\mathbf{a} = k\langle a_1\ a_2 \rangle = \langle ka_1, ka_2 \rangle$ (scalar product) $\mathbf{a} \cdot \mathbf{b} = a_1b_1 + a_2b_2$ (dot product)
Work
If a constant force **F** is applied to an object that moves along a vector **D**, then the work done is $W = \mathbf{F} \cdot \mathbf{D}$.

7.3 EXERCISES   Tape 12

Representing Vectors and Their Magnitudes

Exercises 1–4: Use the graphical representation of **v** *to complete the following.*

  (a) *Estimate a_1 and a_2 so that* $\mathbf{v} = \langle a_1, a_2 \rangle$.
  (b) *Calculate* $\|\mathbf{v}\|$.

1.

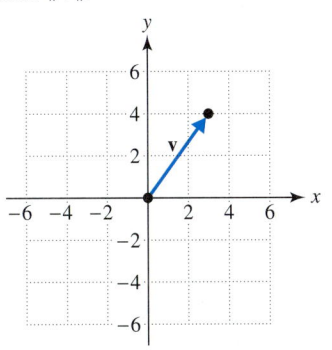

2.

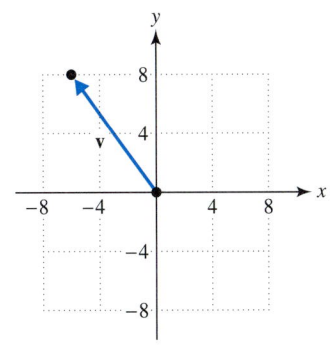

3.

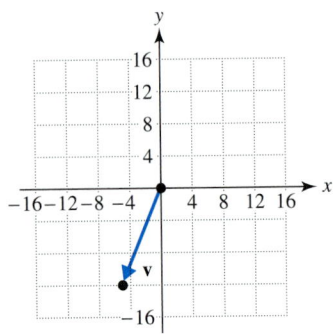

4.

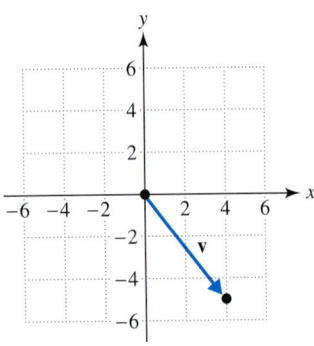

Exercises 5–10: Complete the following.
(a) *Sketch a vector* **v** *that models the situation.*
(b) *Express* **v** *as* $\langle a_1, a_2 \rangle$.
(c) *Find* $2\mathbf{v}$ *and* $-\frac{1}{2}\mathbf{v}$. *Interpret each result.*

5. A 20-mile-per-hour north wind

6. A 10-mile-per-hour west wind

7. A 5-mile-per-hour northwest wind

8. A 7-mile-per-hour southeast wind

9. A 30-pound force upward

10. A 15-pound force pulling at a $45°$ angle in standard position

Exercises 11–18: Complete the following for vector **v**.
(a) *Find the horizontal and vertical components.*
(b) *Calculate* $\|\mathbf{v}\|$ *and decide if* **v** *is a unit vector.*
(c) *Graph* **v** *and interpret* $\|\mathbf{v}\|$.

11. $\mathbf{v} = \langle 1, 1 \rangle$

12. $\mathbf{v} = \langle -1, 0 \rangle$

13. $\mathbf{v} = \langle 3, -4 \rangle$

14. $\mathbf{v} = \langle -2, -2 \rangle$

15. $\mathbf{v} = \mathbf{i}$

16. $\mathbf{v} = -3\mathbf{j}$

17. $\mathbf{v} = 5\mathbf{i} + 12\mathbf{j}$

18. $\mathbf{v} = -\dfrac{3}{5}\mathbf{i} - \dfrac{4}{5}\mathbf{j}$

Exercises 19–24: A vector **v** *has initial point P and terminal point Q.*
(a) *Graph* $\overrightarrow{PQ}$.
(b) *Write* $\overrightarrow{PQ}$ *as* $\mathbf{v} = a_1\mathbf{i} + a_2\mathbf{j}$.

19. $P = (0, 0)$, $\quad Q = (-1, 2)$

20. $P = (0, 0)$, $\quad Q = (4, -6)$

21. $P = (1, 2)$, $\quad Q = (3, 6)$

22. $P = (-1, -2)$, $\quad Q = (4, 4)$

23. $P = (-2, 4)$, $\quad Q = (3, -2)$

24. $P = (0, -4)$, $\quad Q = (1, 3)$

Operations on Vectors

Exercises 25–30: Use the figure to evaluate each of the following.
(a) $\mathbf{a} + \mathbf{b}$
(b) $\mathbf{a} - \mathbf{b}$
(c) $-\mathbf{a}$

25.

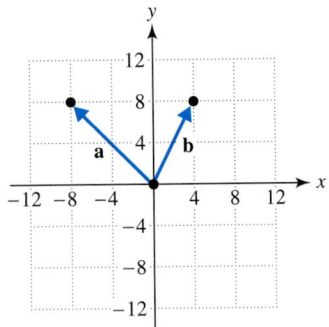

26.

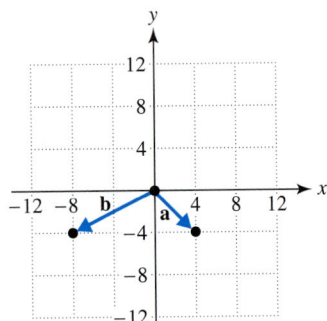

27.

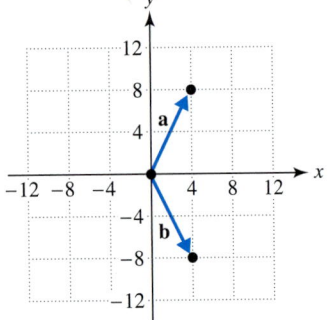

28.

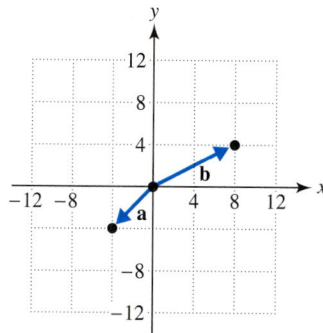

37. $\mathbf{a} = \langle -1, 2 \rangle, \quad \mathbf{b} = \langle 3, 0 \rangle$
38. $\mathbf{a} = \langle -2, -1 \rangle, \quad \mathbf{b} = \langle -3, 2 \rangle$

Exercises 39–42: Approximate the horizontal and vertical components of $\mathbf{v}$ *shown in the figure.*

39.

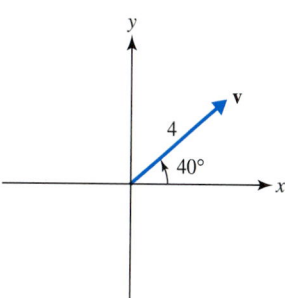

29.

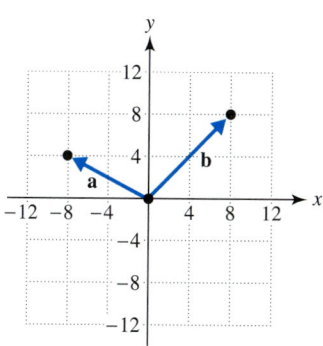

40.

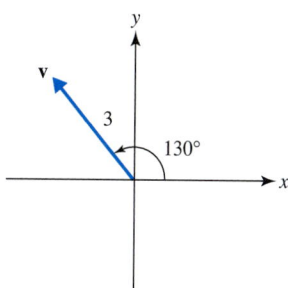

30.

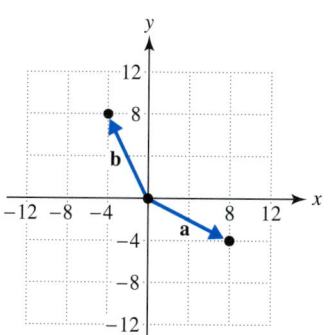

41.

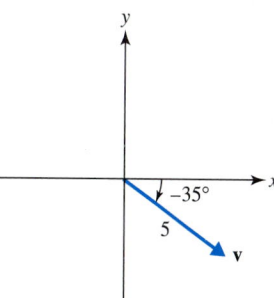

Exercises 31–34: Evaluate each of the following graphically and symbolically.
  (a) $\mathbf{a} + \mathbf{b}$
  (b) $\mathbf{a} - \mathbf{b}$
31. $\mathbf{a} = \langle 0, 2 \rangle, \quad \mathbf{b} = \langle 3, 0 \rangle$
32. $\mathbf{a} = \langle 1, 1 \rangle, \quad \mathbf{b} = \langle -2, 3 \rangle$
33. $\mathbf{a} = 2\mathbf{i} + \mathbf{j}, \quad \mathbf{b} = \mathbf{i} - 2\mathbf{j}.$
34. $\mathbf{a} = \mathbf{i} + 2\mathbf{j}, \quad \mathbf{b} = -2\mathbf{i} + 3\mathbf{j}$

Exercises 35–38: Evaluate each of the following graphically and symbolically.
  (a) $2\mathbf{a}$
  (b) $2\mathbf{a} + 3\mathbf{b}$
35. $\mathbf{a} = 2\mathbf{i}, \quad \mathbf{b} = \mathbf{i} + \mathbf{j}$
36. $\mathbf{a} = -\mathbf{i} + 2\mathbf{j}, \quad \mathbf{b} = \mathbf{i} - \mathbf{j}$

42.

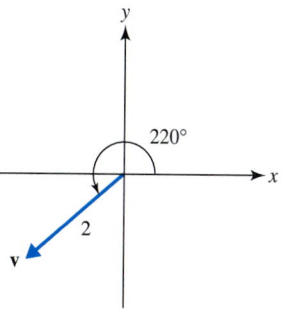

Exercise 43–46: Use the parallelogram rule to find the magnitude of the resultant force for the two forces shown in the figure.

43.

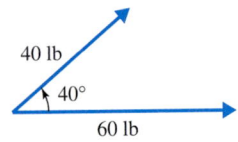

40 lb
40°
60 lb

44.

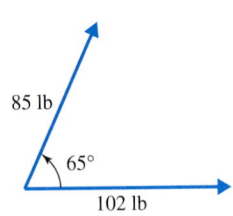

85 lb
65°
102 lb

45.

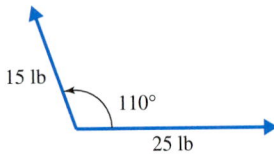

15 lb
110°
25 lb

46.

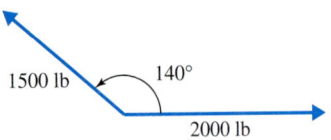

1500 lb
140°
2000 lb

Dot Product and Work

*Exercises 47–54: Complete the following for vectors **a** and **b**.*
   (a) *Find **a** · **b**.*
   (b) *Approximate the angle θ between **a** and **b** to within a tenth of a degree.*
   (c) *State if vectors **a** and **b** are perpendicular, parallel, or neither. If **a** and **b** are parallel, state whether they point in the same direction or in opposite directions.*

47. $\mathbf{a} = \langle 1, -2 \rangle$, $\mathbf{b} = \langle 3, 1 \rangle$

48. $\mathbf{a} = \langle 4, -5 \rangle$, $\mathbf{b} = \langle 2, -2 \rangle$

49. $\mathbf{a} = \langle 6, 8 \rangle$, $\mathbf{b} = \langle -4, 3 \rangle$

50. $\mathbf{a} = \langle 1, -2 \rangle$, $\mathbf{b} = \langle -2, 4 \rangle$

51. $\mathbf{a} = 5\mathbf{i} + 6\mathbf{j}$, $\mathbf{b} = 10\mathbf{i} + 12\mathbf{j}$

52. $\mathbf{a} = -2\mathbf{i} + 6\mathbf{j}$, $\mathbf{b} = 3\mathbf{i} + \mathbf{j}$

53. $\mathbf{a} = \mathbf{i} + 3\mathbf{j}$, $\mathbf{b} = 0.5\mathbf{i} - 1.5\mathbf{j}$

54. $\mathbf{a} = -12\mathbf{i} + 16\mathbf{j}$, $\mathbf{b} = -3\mathbf{i} + 4\mathbf{j}$

Exercises 55–58: Find the work done in each situation.

55. Lifting a 30-pound weight 5 feet into the air

56. Lifting a 15-pound bucket 8 feet into the air

57. Pushing a stalled car with a force of 100 pounds for 1000 feet

58. A 150-pound person running up 5 flights of steps with 10 feet between floors

*Exercises 59–62: Find the work done when a constant force **F** is applied to an object that moves along the vector **D**, where units are in pounds and feet. Find the magnitude of **F**.*

59. $\mathbf{F} = \langle 10, 20 \rangle$, $\mathbf{D} = \langle 15, 22 \rangle$

60. $\mathbf{F} = \langle 64, 36 \rangle$, $\mathbf{D} = \langle 22, -33 \rangle$

61. $\mathbf{F} = 5\mathbf{i} - 3\mathbf{j}$, $\mathbf{D} = 3\mathbf{i} - 4\mathbf{j}$

62. $\mathbf{F} = 7\mathbf{i} - 24\mathbf{j}$, $\mathbf{D} = -2\mathbf{i} - 5\mathbf{j}$

*Exercises 63–66: Find the work done when force **F** = 5**i** + 3**j** moves an object from P to Q.*

63. $P = (-2, 3)$, $Q = (1, 6)$

64. $P = (-2, -1)$, $Q = (1, 3)$

65. $P = (2, -3)$, $Q = (4, -5)$

66. $P = (1, 1)$, $Q = (-1, 6)$

Applications

67. *Swimming in a Current* A swimmer heads directly north across a river at 3 miles per hour in a current that is 2 miles per hour as illustrated in the figure. Find a vector that models the direction and speed of the swimmer. With what speed is the swimmer moving in the river?

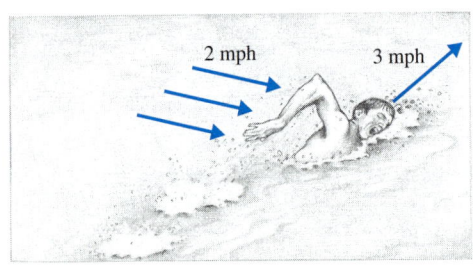

2 mph   3 mph

68. *Air Navigation* An airplane is flying west at 400 miles per hour in a 50-mile-per-hour northwest wind. Find a vector that models the direction and speed of the airplane. Find the groundspeed and bearing of the airplane.

69. *Wind and Vectors* A wind can be described by $\mathbf{v} = 6\mathbf{i} + 8\mathbf{j}$, where vector **j** points north and represents a south wind of 1 mile per hour.
   (a) What is the speed of the wind?
   (b) Find 3**v**. Interpret the result.
   (c) Interpret the wind if it switches to $\mathbf{u} = -8\mathbf{i} + 8\mathbf{j}$.

70. *Force and Water-Ski Towropes* (Refer to Example 2.) Forces of 65 pounds and 110 pounds are exerted by two water-ski towropes. If the angle between the towropes is 19°, find the magnitude of the resultant force.

71. *Measuring Rainfall* Suppose that vector **R** models the amount of rainfall in inches and the direction it falls, while vector **A** models the area in square inches and orientation of the opening of a rain gauge as illustrated in the figure. The total volume V of water collected in the rain gauge is given by $V = |\mathbf{R} \cdot \mathbf{A}|$. This formula calculates the volume of water collected even if the wind is blowing the rain in a slanted direction or the rain gauge is not exactly vertical. Let $\mathbf{R} = \mathbf{i} - 2\mathbf{j}$ and $\mathbf{A} = 0.5\mathbf{i} + \mathbf{j}.$

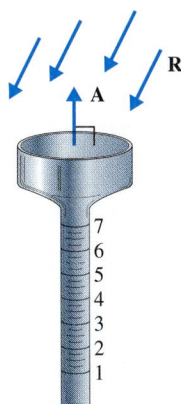

(a) Find $\|\mathbf{R}\|$ and $\|\mathbf{A}\|$. Interpret your results.
(b) Calculate V and interpret this result.
(c) For the rain gauge to collect the maximum amount of water, what should be true about vectors **R** and **A**?

72. *Solar Panels* Suppose that the sun's intensity in watts per square centimeter and direction is given by vector **I,** and a solar panel's area in square centimeters and orientation is given by vector **A** as illustrated in the figure. Then the total number of watts W that are collected by the solar panel is given by $W = |\mathbf{I} \cdot \mathbf{A}|$. Let $\mathbf{I} = 0.01\mathbf{i} - 0.02\mathbf{j}$ and $\mathbf{A} = 400\mathbf{i} + 300\mathbf{j}.$

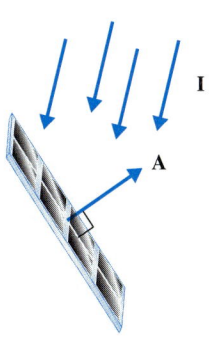

(a) Find $\|\mathbf{I}\|$ and $\|\mathbf{A}\|$. Interpret your results.
(b) Calculate W and interpret this result.
(c) For the solar panel to absorb maximum wattage, what must be true about vectors **I** and **A?**

73. *Robotics* (Refer to Example 4.) Consider the planar two-arm manipulator shown in the accompanying figure. Let the upper arm be modeled by $\mathbf{a} = 3\mathbf{i} + 2\mathbf{j}$ and the forearm be modeled by $\mathbf{b} = -2\mathbf{i} + 2\mathbf{j}$, where units are in feet. (Source: J. Craig.)
(a) Find a vector **c** that represents the position of the hand.
(b) How far is the hand from the origin?
(c) Find the position of the hand if the length of the upper arm triples and the length of the forearm is reduced by half.

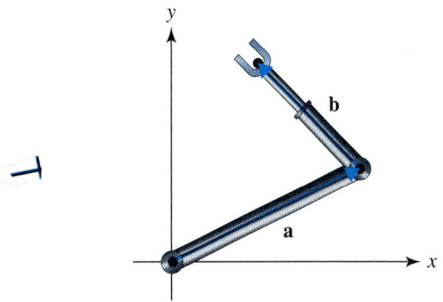

74. *Robotics* A planar three-arm manipulator is shown in the accompanying figure with joint angles measured relative to a positive horizontal axis. (Source: R. Murray, *A Mathematical Introduction to Robotic Manipulation.*)
(a) Find vectors **a, b,** and **c** that represent each part of the robotic arm. (*Hint:* Find the horizontal and vertical components for each part.)
(b) Find a vector **d** that represents the position of the hand. How far is the hand from the origin?

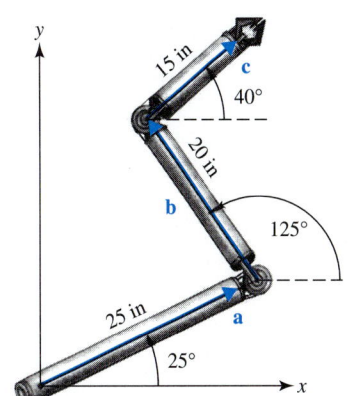

75. *Dot Products in Computer Graphics* (Refer to Example 8.) A computer screen is gray to the left of vector $\overrightarrow{PQ} = 2\mathbf{i} - 3\mathbf{j}$ and blue to the right, where vector $\overrightarrow{PR} = 3\mathbf{i} + 2\mathbf{j}$ is perpendicular to $\overrightarrow{PQ}$. Let point P be $(2, 1)$. Determine the color of a pixel located at S.
(a) $S = (3, -2)$
(b) $S = (2, 2)$

76. *Dot Products in Computer Graphics* A computer screen is to be blue above vector $\overrightarrow{PQ} = 5\mathbf{i} - \mathbf{j}$ and white below, where point P is $(2, 1)$. Determine the color of a pixel located at S.
(a) $S = (100, -10)$
(b) $S = (-500, 50)$

77. *Translations in Computer Graphics* Vectors are used in computer graphics to compute translations of points. For example, suppose we would like to translate the point $(-1, 2)$ by $\mathbf{v} = \langle 2, 1 \rangle$ as illustrated in the accompanying figure, where the *point* $(-1, 2)$ has been represented by the *vector* $\mathbf{a} = \langle -1, 2 \rangle$. The new location of $(-1, 2)$ is modeled by

$$\mathbf{b} = \mathbf{a} + \mathbf{v} = \langle -1, 2 \rangle + \langle 2, 1 \rangle = \langle 1, 3 \rangle.$$

Thus, $(-1, 2)$ has been translated by $\mathbf{v}$ to $(1, 3)$. (Source: J. Foley, *Introduction to Computer Graphics*.)

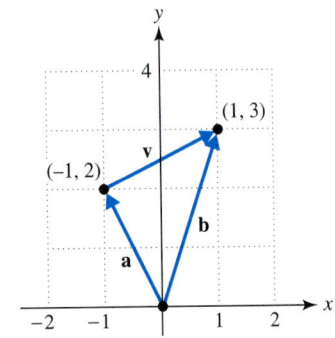

(a) Find the new coordinates of $(-2, 4)$ if it is translated by $\mathbf{v} = \langle 4, -2 \rangle$.
(b) Represent this situation graphically.

78. *Translations in Computer Graphics* (Refer to the previous exercise.) Let triangle ABC have vertices $(1, 1)$, $(3, 0)$, and $(4, 3)$.
(a) Find the new vertices if the triangle ABC is translated by $\mathbf{v} = \langle -2, 1 \rangle$.
(b) Describe the change in triangle ABC if it were translated by $-2\mathbf{v}$.

79. *Work* (Refer to Example 9.) A 145-pound person walks 1.5 miles up a hiking trail inclined at $15°$. Use a dot product to calculate the work done.

80. *Work* A wagon is pulled 500 feet using a force of 10 pounds applied to the handle, which makes a $40°$ angle with the horizontal. See Figure 7.58. Use a dot product to calculate the work done.

81. *Air Navigation* A pilot would like to fly to a city that is 200 miles away and has a bearing of $135°$. The wind is blowing from the north at 30 miles per hour and the trip is to take one hour. Find the direction and speed that the pilot should head the plane to accomplish this.

82. *Work* Calculate the work required to push a 1800-pound car up a $7°$ incline for 0.1 mile.

Writing about Mathematics

1. State the properties of a vector. Does a vector have position? Explain. Give two examples of vectors.

2. State one application of vectors. Give a specific example of a vector for this application and explain how the vector models that application.

7.4 Parametric Equations

Basic Concepts • Applications of Parametric Equations

Introduction

We have used functions to model curves in the *xy*-plane. Sometimes a curve cannot be modeled by a function. For example, a circle cannot be described by a single function because a circle fails the vertical line test. Parametric equations represent a different approach to describing curves in the *xy*-plane. They are used in industry to draw complicated curves and surfaces, such as the hood of an automobile. Parametric equations are also used in computer graphics, engineering, and physics. (Source: F. Hill, *Computer Graphics*.)

Basic Concepts

In previous chapters we graphed curves in the xy-plane determined by $y = f(x)$. Some curves cannot be represented by $y = f(x)$, but they can be represented by parametric equations. Sample graphs of parametric equations are shown in Figures 7.62–7.64.

$[-6, 6, 1]$ by $[-4, 4, 1]$

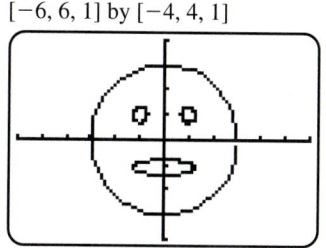

FIGURE 7.62

$[-6, 6, 1]$ by $[-4, 4, 1]$

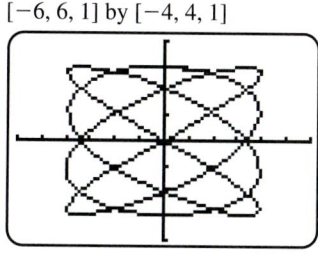

FIGURE 7.63

$[-6, 6, 1]$ by $[-4, 4, 1]$

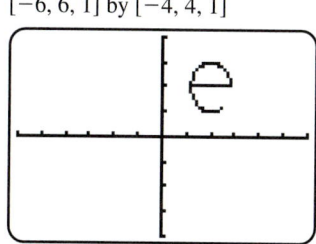

FIGURE 7.64

We now define parametric equations of a plane curve.

Parametric equations of a plane curve

A **plane curve** is a set of points (x, y) such that $x = f(t)$ and $y = g(t)$, where f and g are continuous functions on an interval $a \leq t \leq b$. The equations $x = f(t)$ and $y = g(t)$ are **parametric equations** with **parameter** t.

Parametric equations can be represented symbolically, numerically, graphically, and described verbally. This is illustrated in the next example.

EXAMPLE 1   Representing parametric equations numerically, graphically, and verbally

Let $x = t + 3$ and $y = t^2$ for $-3 \leq t \leq 3$.
(a) Make a table of values for x and y with $t = -3, -2, -1, \ldots, 3$.
(b) Plot the points in the table and graph the curve.
(c) Describe the curve.

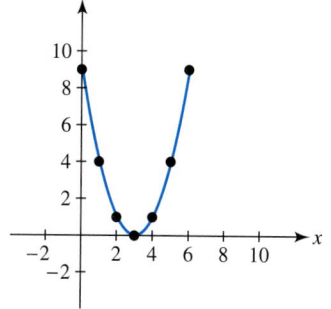

FIGURE 7.65

Solution

(a) *Numerical Representation*   A partial numerical representation of the parametric equations is shown in Table 7.1. For example, if $t = 2$, then $x = 2 + 3 = 5$ and $y = 2^2 = 4$.

TABLE 7.1

t	-3	-2	-1	0	1	2	3
x	0	1	2	3	4	5	6
y	9	4	1	0	1	4	9

(b) *Graphical Representation*   In Figure 7.65 each ordered pair (x, y) in Table 7.1 is plotted and then the points are connected to obtain the graph of the curve. Notice that the curve starts at $(0, 9)$ and ends at $(6, 9)$.
(c) *Verbal Representation*   The curve in Figure 7.65 appears to be the lower portion of a parabola opening upward with vertex $(3, 0)$. ∎

Technology Note   *Parametric Equations*

Graphing calculators are capable of using parametric equations to make tables and graphs. In addition to setting the standard values for the *x*- and *y*-axis, the interval for *t* must also be specified. A window setting, table, and graph for the parametric equations in Example 1 are shown in Figures 7.66–7.68. The variable Tstep represents the increment in the parameter *t* and has a default value of 0.1 on this calculator.

$[-2, 10, 1]$ by $[-2, 10, 1]$

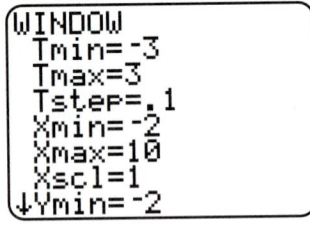

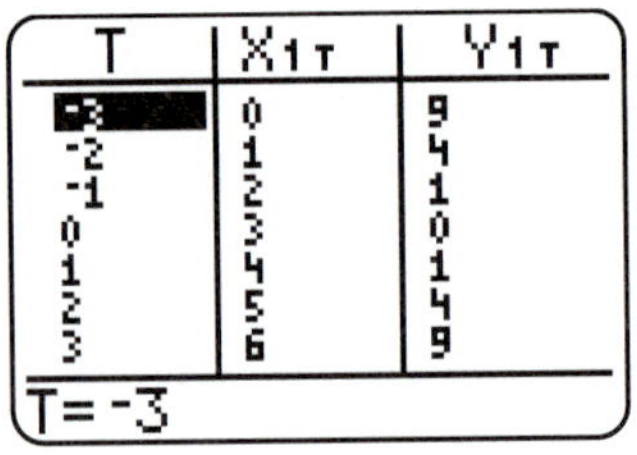

FIGURE 7.66

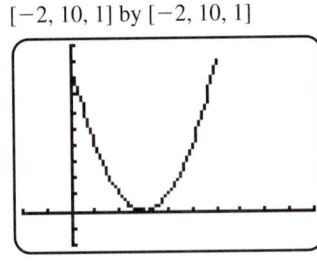

FIGURE 7.67

FIGURE 7.68

We can verify symbolically that the curve in Figure 7.65 is indeed a portion of a parabola.

EXAMPLE 2   *Finding an equivalent rectangular equation*

Find an equivalent rectangular equation for $x = t + 3$ and $y = t^2$, where $-3 \le t \le 3$.

Solution
Begin by solving $x = t + 3$ for t to obtain $t = x - 3$. Substituting for t in $y = t^2$, results in

$$y = (x - 3)^2,$$

which represents a parabola with vertex $(3, 0)$. When $t = -3$ then $x = 0$, and when $t = 3$ then $x = 6$. Thus, the domain is restricted to $0 \le x \le 6$. ■

Parametric equations can model a circle as shown in the next example.

EXAMPLE 3   *Graphing a circle with parametric equations*

Graph $x = 2 \cos t$ and $y = 2 \sin t$ for $0 \le \theta \le 2\pi$. Find an equivalent equation using rectangular coordinates.

Solution
Let $X_1 = 2 \cos (T)$ and $Y_1 = 2 \sin (T)$ and graph these parametric equations as shown in Figures 7.69 and 7.70 (*Note:* Be sure to have the mode of the calculator set for parametric equations. A square viewing rectangle is necessary for the curve to appear circular rather than elliptical.)

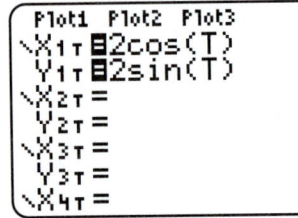

FIGURE 7.69

[−3, 3, 1] by [−2, 2, 1]

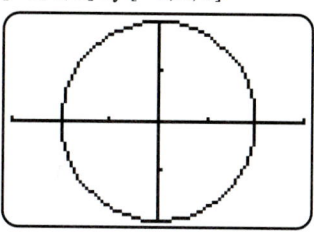

FIGURE 7.70

To verify that this is a circle consider the following.

$$
\begin{aligned}
x^2 + y^2 &= (2 \cos t)^2 + (2 \sin t)^2 \quad &x = 2\cos t,\, y = 2\sin t\\
&= 4 \cos^2 t + 4 \sin^2 t \quad &\text{Properties of exponents}\\
&= 4(\cos^2 t + \sin^2 t) \quad &\text{Distributive property}\\
&= 4 \quad &\cos^2 t + \sin^2 t = 1
\end{aligned}
$$

The parametric equations are equivalent to $x^2 + y^2 = 4$, which is a circle with center $(0, 0)$ and radius 2. ∎

Applications of Parametric Equations

Parametric equations are used frequently in computer graphics to design a variety of figures and letters. Computer fonts are sometimes designed using parametric equations. In the next example, we use parametric equations to design a "smiley" face consisting of a head, two eyes, and a mouth. (Source: F. Hill.)

EXAMPLE 4   *Creating drawings with parametric equations*

Graph a "smiley" face using parametric equations.

Solution

Head   We can use a circle centered at the origin for the head. If the radius is 2, then let $x = 2 \cos t$ and $y = 2 \sin t$ for $0 \le t \le 2\pi$. This is graphed in Figure 7.71.

Eyes   For the eyes we can use two small circles. The eye in the first quadrant can be modeled by $x = 1 + 0.3 \cos t$ and $y = 1 + 0.3 \sin t$. This represents a circle centered at $(1, 1)$ with radius 0.3. The eye in the second quadrant can be modeled by $x = -1 + 0.3 \cos t$ and $y = 1 + 0.3 \sin t$ for $0 \le t \le 2\pi$, which is a circle centered at $(-1, 1)$ with radius 0.3. This is shown in Figure 7.72.

Mouth   For the smile, we can use the lower half of a circle. Using trial and error we might arrive at $x = 0.5 \cos \frac{1}{2}t$ and $y = -0.5 - 0.5 \sin \frac{1}{2}t$. This is a semicircle centered at $(0, -0.5)$ with radius 0.5. Since we are letting $0 \le t \le 2\pi$, the term $\frac{1}{2}t$ ensures that only half a circle (semicircle) is drawn. The minus sign before $0.5 \sin \frac{1}{2}t$ in the y-equation results in the lower half of the semicircle being drawn rather than the upper half. The final result is shown in Figure 7.73. The pupils have been added by plotting the points $(1, 1)$ and $(-1, 1)$ and the coordinate axes have been turned off.

[−3, 3, 1] by [−2, 2, 1]

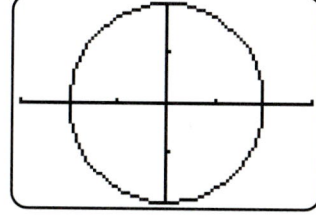

FIGURE 7.71

[−3, 3, 1] by [−2, 2, 1]

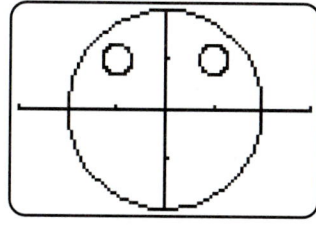

FIGURE 7.72

[−3, 3, 1] by [−2, 2, 1]

FIGURE 7.73   ∎

Critical Thinking ――――――――――――――――――――――――

Modify the face in Example 4 so that it is frowning. Try to find a way to make the right eye shut rather than open.

Parametric equations are used to simulate motion. If a ball is thrown with a velocity of v feet per second at an angle θ with the horizontal, its flight can be modeled by the parametric equations

$$x = (v \cos \theta)t \quad \text{and} \quad y = (v \sin \theta)t - 16t^2,$$

where t is in seconds. The term $-16t^2$ occurs because gravity is pulling downward. See Figure 7.74. These equations ignore air resistance.

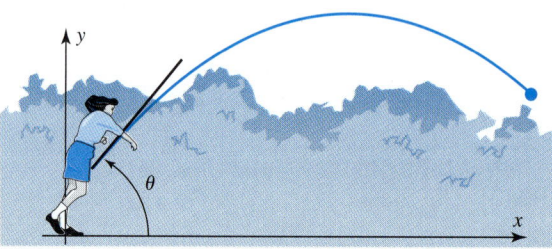

FIGURE 7.74

EXAMPLE 5   *Simulating motion with parametric equations*

Three baseballs are thrown simultaneously into the air at 132 feet per second (90 miles per hour) making angles of 30°, 50°, and 70° with the horizontal as they leave the hand of each player.

(a) Assuming the ground is level, determine graphically which ball travels the farthest. Estimate this distance.

(b) Which ball reaches the greatest height? Estimate this height.

[0, 600, 50] by [0, 400, 50]

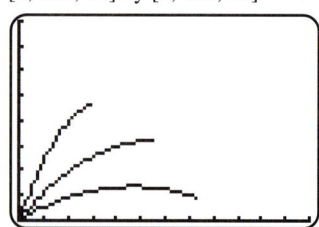

FIGURE 7.75

[0, 600, 50] by [0, 400, 50]

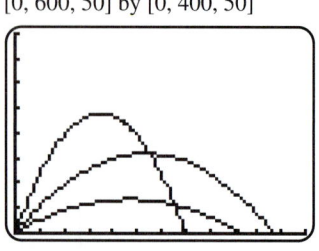

FIGURE 7.76

Solution

(a) The three sets of parametric equations determined by the three baseballs are as follows,

$$X_1 = 132 \cos(30)T, \quad Y_1 = 132 \sin(30)T - 16T\char`^2$$
$$X_2 = 132 \cos(50)T, \quad Y_2 = 132 \sin(50)T - 16T\char`^2$$
$$X_3 = 132 \cos(70)T, \quad Y_3 = 132 \sin(70)T - 16T\char`^2$$

The graphs of the three sets of parametric equations are shown in Figures 7.75 and 7.76, where $0 \le t \le 9$. A graphing calculator in simultaneous mode has been used so that we can view all three balls in flight at the same time. From the second graph we can see that the ball thrown at 50° travels the farthest distance. Using the trace feature, we estimate this distance to be about 540 feet.

(b) The ball thrown at 70° reaches the greatest height of about 240 feet.   ■

Critical Thinking ――――――――――――――――――――――――

If a ball is thrown at 88 feet per second (60 mph), use trial and error to find the angle θ that results in a maximum distance for the ball.

7.4 PUTTING IT ALL TOGETHER

Parametric equations may be used to model a wide variety of curves in the xy-plane that cannot be represented by a single function. Important concepts about parametric equations are summarized in the following table.

Parametric Equations

Let $x = f(t)$, $y = g(t)$, and graph the ordered pairs (x, y) for all t-values in a specified interval. For example, if $x = 2t$ and $y = t^2$ for $-1 \le t \le 2$, then the resulting graph is a portion of a parabola.

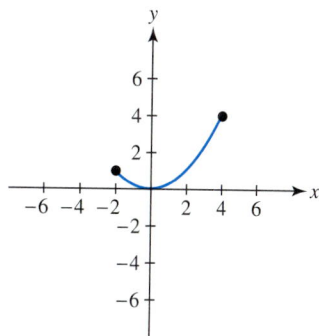

Writing Parametric Equations in Terms of x and y

Solve one of the parametric equations for t and substitute into the second equation. For example, if $x = t^3$ and $y = t^2 - 2$, solve $x = t^3$ for t to obtain $t = x^{1/3}$. Substituting gives $y = (x^{1/3})^2 - 2$ or $y = x^{2/3} - 2$.

7.4 EXERCISES

   Tape 13

Graphs of Parametric Equations

Exercises 1–6: Use the parametric equations to complete the following.

   (a) *Make a table of values for $t = 0, 1, 2, 3$.*
   (b) *Plot the points from the table and graph the curve for $0 \le t \le 3$.*
   (c) *Describe the curve.*

1. $x = t - 1, \quad y = 2t$

2. $x = t + 1, \quad y = t - 2$

3. $x = t + 2, \quad y = (t - 2)^2$

4. $x = \dfrac{1}{3}t^2, \quad y = t - 1$

5. $x = \sqrt{9 - t^2}, \quad y = t$

6. $x = t^2, \quad y = 2t + 1$

Exercises 7–12: Find a rectangular equation for each curve and describe the curve. Support your result by graphing the parametric equations.

7. $x = t, \quad y = \dfrac{1}{2}t^2; \qquad -2 \le t \le 2$

8. $x = \sqrt[3]{t}, \quad y = t; \qquad -2 \le t \le 2$

9. $x = t - 2, \quad y = t^2 + 1; \quad -1 \le t \le 2$

10. $x = 2t, \quad y = t^2 + 1; \qquad -1 \le t \le 2$

11. $x = 3 \sin t, \quad y = 3 \cos t; \ -\pi \le t \le \pi$

12. $x = 2 \cos^2 t, \quad y = 2 \sin^2 t; \ \ 0 \le t \le \pi/2$

Exercises 13–18: Graph each pair of parametric equations for $0 \le t \le 2\pi$. Describe any differences in the two graphs.

13. (a) $x = 3\cos t$, $y = 3\sin t$
    (b) $x = 3\cos 2t$, $y = 3\sin 2t$

14. (a) $x = 2\cos t$, $y = 2\sin t$
    (b) $x = 2\cos t$, $y = -2\sin t$

15. (a) $x = 3\cos t$, $y = 3\sin t$
    (b) $x = 3\sin t$, $y = 3\cos t$

16. (a) $x = t$, $y = t^2$
    (b) $x = t^2$, $y = t$

17. (a) $x = -1 + \cos t$, $y = 2 + \sin t$
    (b) $x = 1 + \cos t$, $y = 2 + \sin t$

18. (a) $x = 2\cos \frac{1}{2}t$, $y = 2\sin \frac{1}{2}t$
    (b) $x = 2\cos t$, $y = 2\sin t$

Exercises 19–30: Graph the parametric equations.

19. $x = t^3$, $y = t^2$; $-2 \le t \le 2$
20. $x = e^t$, $y = t^2 - 1$; $-2 \le t \le 2$
21. $x = t^2$, $y = \ln t$; $0 < t \le 2$
22. $x = t^3 - t$, $y = e^t$; $-1.5 \le t \le 1.5$
23. $x = (t - 1)^2$, $y = t$; $0 \le t \le 2$
24. $x = t^3 + 3t$, $y = 2\cos t$; $-1 \le t \le 1$
25. $x = 2 + \cos t$, $y = \sin t - 1$; $0 \le t \le 2\pi$
26. $x = -2 + \cos t$, $y = \sin t + 1$; $0 \le t \le 2\pi$
27. $x = \cos^3 t$, $y = \sin^3 t$; $0 \le t \le 2\pi$
28. $x = \cos^5 t$, $y = \sin^5 t$; $0 \le t \le 2\pi$
29. $x = |3\sin t|$, $y = |3\cos t|$; $0 \le t \le \pi$
30. $x = 3\sin 2t$, $y = 3\cos t$; $0 \le t \le 2\pi$

Designing Shapes and Figures

Exercises 31–34: Graph the following set of parametric equations for $0 \le t \le 2\pi$ in the viewing rectangle $[0, 6, 1]$ by $[0, 4, 1]$. Identify the letter of the alphabet that is being graphed.

31. $x_1 = 1$, $y_1 = 1 + t/\pi$
    $x_2 = 1 + t/(3\pi)$, $y_2 = 2$
    $x_3 = 1 + t/(2\pi)$, $y_3 = 3$

32. $x_1 = 1$, $y_1 = 1 + t/\pi$
    $x_2 = 1 + t/(3\pi)$, $y_2 = 2$
    $x_3 = 1 + t/(2\pi)$, $y_3 = 3$
    $x_4 = 1 + t/(2\pi)$, $y_4 = 1$

33. $x_1 = 1$, $y_1 = 1 + t/\pi$
    $x_2 = 1 + 1.3\sin(0.5t)$, $y_2 = 2 + \cos(0.5t)$

34. $x_1 = 2 + 0.8\cos(0.85t)$, $y_1 = 2 + \sin(0.85t)$
    $x_2 = 1.2 + t/(1.3\pi)$, $y_2 = 2$

Exercises 35–38: Designing Letters. Find a set of parametric equations that results in a letter similar to the one shown in the figure. Use the viewing rectangle $[-4.7, 4.7, 1]$ by $[-3.1, 3.1, 1]$ and turn off the coordinate axes.

35.

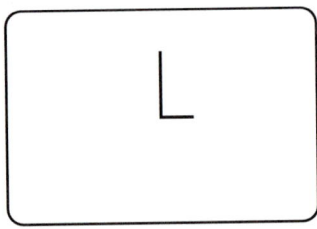

36.

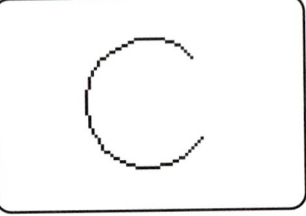

37.

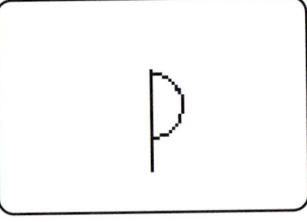

38.

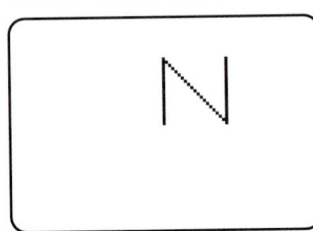

39. *Designing a Face*   (Refer to Example 4.) Use parametric equations to create your own "smiley" face. This face should have a head, a mouth, and eyes.

40. *Designing a Face*   Add a nose to the face that you designed in the previous example.

Applications

Exercises 41–44: Lissajous Figures Lissajous figures occur in electronics and may be used to find the frequency of an unknown voltage. (See Exercises 45–48.) Graph the lissajous figure for $0 \le t \le 6.5$ in the viewing rectangle $[-6, 6, 1]$ by $[-4, 4, 1]$.

41. $x = 2\cos t$, $y = 3\sin 2t$
42. $x = 3\cos 2t$, $y = 3\sin 3t$
43. $x = 3\sin 4t$, $y = 3\cos 3t$
44. $x = 4\sin 4t$, $y = 3\sin 5t$

Exercises 45–48: Electronic Technology Parametric equations have applications in electricity. If two sinusoidal voltages, denoted by $x = V_1(t)$ and $y = V_2(t)$, are applied to an oscilloscope, a stationary pattern called a lissajous figure may appear as shown in the accompanying figure. If the frequency F_1 of V_1 is known and the frequency F_2 of V_2 is unknown, then a lissajous figure may be used to find F_2. The ratio $\dfrac{F_1}{F_2}$ is equal to the ratio of the corresponding number of tangents to the enclosing rectangle. The number of tangents along a vertical side of the rectangle corresponds to F_1 and the number of tangents along a horizontal side corresponds to F_2. In this figure $\dfrac{F_1}{F_2} = \dfrac{3}{2}$. Therefore, $F_2 = \dfrac{2}{3}F_1$. Determine F_2 given the lissajous figure and the frequency F_1 in cycles per second. (Source: R. Smith and R. Dorf. Circuits, Devices, and Systems.)

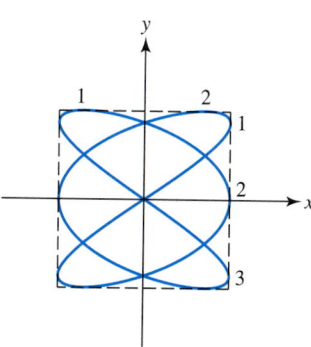

45. $F_1 = 150$

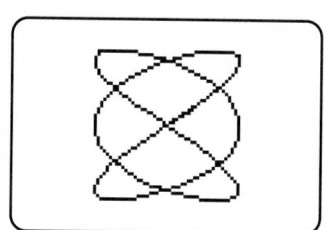

46. $F_1 = 60$

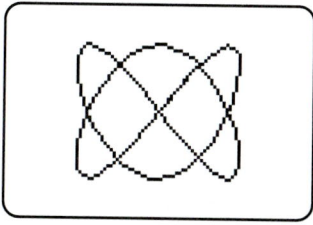

47. $F_1 = 400$

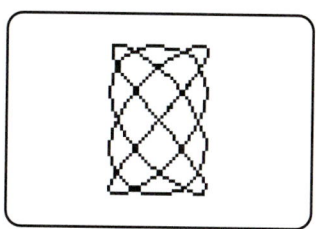

48. $F_1 = 1200$

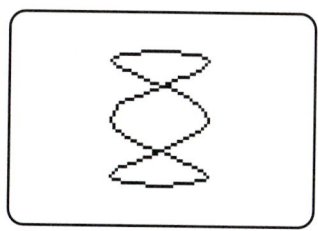

49. *Flight of a Baseball* (Refer to Example 5.) Two baseballs are thrown into the air at 66 feet per second (45 mph) making angles of 35° and 50° with the horizontal as they leave the hand of each player. If the ground is level, estimate graphically the horizontal distance traveled by each baseball.

50. *Flight of a Baseball* Solve the previous example if, instead of being level, the ground is inclined with a slope of $m = 0.1$.

51. *Flight of a Baseball* If a baseball is thrown at 88 feet per second (60 mph) making an angle of 45° with the horizontal, will it go over a fence 10 feet high that is 200 feet away on level ground?

52. *Simulating Gravity on the Moon* (Refer to Example 5.) If an object is thrown on the moon then the parametric equations of flight are

$$x = (v\cos\theta)t \quad \text{and} \quad y = (v\sin\theta)t - 2.66t^2.$$

Estimate graphically the distance that a ball thrown 88 feet per second (60 mph) at an angle of 45° with the horizontal travels on the moon if the moon's surface is level.

Writing about Mathematics

1. Describe the basic form of parametric equations. Give an example. Explain how graphs of parametric equations can differ from graphs of functions.

2. Suppose that a function is defined by $y = f(x)$, where the domain of f is $a \le x \le b$. Explain how we could represent f with parametric equations. Apply your method to $f(x) = x^2 + 1$, where $-2 \le x \le 2$.

CHECKING BASIC CONCEPTS FOR SECTIONS 7.3 AND 7.4

1. Let point P be $(-1, 3)$ and Q be $(3, 7)$. Find the following.
   (a) $\mathbf{v} = \overrightarrow{PQ}$
   (b) $\|\mathbf{v}\|$
   (c) $\overrightarrow{PQ} + \overrightarrow{QP}$

2. Let $\mathbf{v} = 2\mathbf{i} - \mathbf{j}$ and $\mathbf{u} = -3\mathbf{i} + 2\mathbf{j}$. Find the following graphically and symbolically.
   (a) $2\mathbf{v} + \mathbf{u}$
   (b) $2\mathbf{v}$
   (c) $\mathbf{v} - 3\mathbf{u}$

3. Let $\mathbf{a} = \langle 3, -2 \rangle$ and $\mathbf{b} = \langle -1, 3 \rangle$. Find the following.
   (a) $\mathbf{a} \cdot \mathbf{b}$
   (b) The angle θ between $\mathbf{a}$ and $\mathbf{b}$ rounded to a tenth of a degree

4. Represent the parametric equations $x = t + 1$ and $y = (t - 1)^2$ graphically and numerically for $-1 \le t \le 5$. For the numerical representation let $t = -1, 0, 1, \ldots, 5$.

7.5 Polar Equations

The Polar Coordinate System • Graphs of Polar Equations • Solving Polar Equations (Optional)

Introduction

Many times a change in a frame of reference can have a profound effect on the solution of a problem. Thus far we have graphed functions only in the xy-plane. Many interesting curves cannot be represented by a function since these curves fail the vertical line test. By creating a new coordinate system, some types of equations become simpler. For example, the equation $x^2 + y^2 = 1$ describes the unit circle in the rectangular coordinate system. Every point lying on the unit circle is 1 unit from the origin. If we specify a new variable r that represents the radius of the circle, then $r = 1$ also models the unit circle. A change of variable has resulted in a simpler equation.

In this section we introduce a new coordinate system called the *polar coordinate system* that originated with Jakob Bernoulli (1654–1705). This new system allows us to express many curves using relatively simple equations. (Source: H. Eves, *An Introduction to the History of Mathematics.*)

FIGURE 7.77

The Polar Coordinate System

In Figure 7.77 point A is located at $x = -2$. One number completely determines the location of a point on a one-dimensional line. In two dimensions, two numbers are necessary to locate a point. For example, two numbers are needed to locate point B at $(2, 3)$ in Figure 7.78. In the xy-plane we are accustomed to identifying points using (x, y), where x and y are real numbers. However, the xy-plane is not the only way to locate a point in a plane.

The **polar coordinate system** uses r and θ instead of x and y to locate a point P as shown in Figure 7.79. The distance OP between P and the origin is represented by $|r|$. Angle θ can be expressed in either degrees or radians and is assumed to be in standard position. If $r > 0$, point P lies on the terminal side of θ and if $r < 0$, point P lies on the ray pointing in the opposite direction of the terminal side of θ, a distance of $|r|$ from the origin. See Figure 7.80. In the polar coordinate system, the origin is called the **pole** and the positive x-axis corresponds to the **polar axis**.

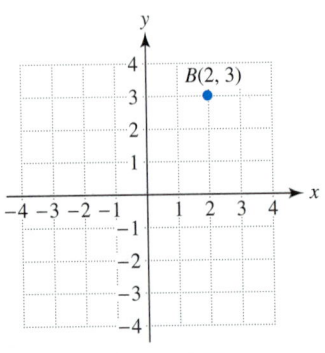

FIGURE 7.78

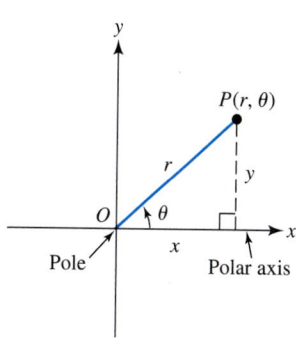

FIGURE 7.79 $r > 0$

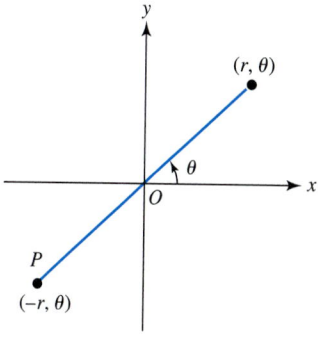

FIGURE 7.80

Using Figure 7.79 and trigonometry, the following relationships can be established between rectangular and polar coordinates.

Rectangular and polar coordinates

If a point has rectangular coordinates (x, y) and polar coordinates (r, θ), then these coordinates are related as follows.

$$x = r\cos\theta, \qquad y = r\sin\theta$$

$$r^2 = x^2 + y^2, \qquad \tan\theta = \frac{y}{x} \; (x \neq 0)$$

EXAMPLE 1   *Plotting points in polar coordinates*

Plot the points (r, θ) on a polar grid.

(a) $(2, 45°)$ **(b)** $(-3, 150°)$ **(c)** $\left(3.5, -\dfrac{\pi}{3}\right)$

Solution

(a) Let $r = 2$ and $\theta = 45°$. Plot a point 2 units from the pole on the terminal side of $\theta = 45°$ as shown in Figure 7.81.

(b) Since $r = -3 < 0$ and $\theta = 150°$ begin by locating the terminal side of θ in the second quadrant. Next plot a point 3 units from the pole in the opposite direction of the terminal side of θ as shown in Figure 7.82.

(c) Since $r = 3.5$ and $\theta = -\dfrac{\pi}{3}$ (radians), the point is located as shown in Figure 7.83.

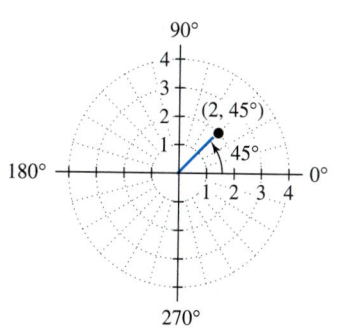

FIGURE 7.81

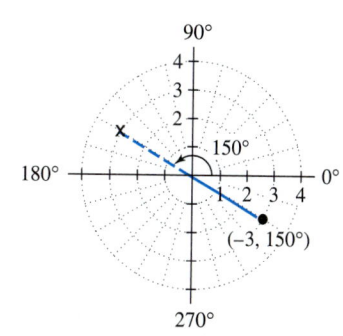

FIGURE 7.82

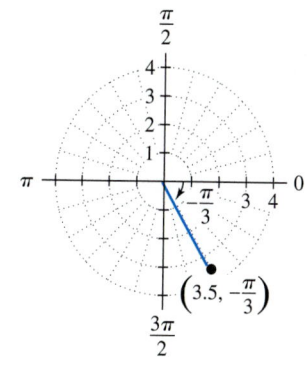

FIGURE 7.83

It is important to note that unlike the xy-coordinate system, polar coordinates are *not* unique. For example, the $r\theta$-coordinates of $(2, 0°)$, $(2, 360°)$, $(2, -360°)$, and $(-2, 180°)$ all represent the same point.

Critical Thinking

What can be said about a point (r, θ) if $r = 0$?

Graphs of Polar Equations

When we graph $y = f(x)$ in the xy-coordinate system, we are graphing a function. A vertical line can intersect the graph of a function at most once. As a result, many shapes such as circles, hearts, and leaves cannot be represented by a function. Like parametric equations, polar equations can be valuable when representing curves.

EXAMPLE 2   *Representing polar equations numerically, graphically, and verbally*

Represent each equation numerically and graphically. Then describe the curve.
(a) $r = 3$
(b) $r = 2 + 2 \cos \theta$

Solution

(a) *Numerical Representation*   A partial numerical representation is given in Table 7.2.

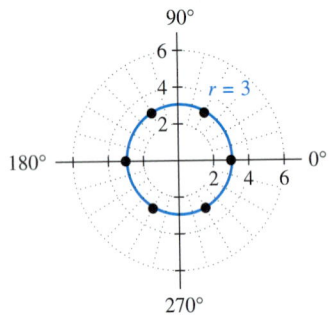

FIGURE 7.84 A Circle

TABLE 7.2

θ	0°	60°	120°	180°	240°	300°
r	3	3	3	3	3	3

Graphical Representation   A graph and these points are plotted in Figure 7.84.

Verbal Representation   The polar equation represents a circle with radius 3.

(b) *Numerical Representation*   A table of values for $r = 2 + 2 \cos \theta$ is shown in Table 7.3. For example, when $\theta = 60°$ then

$$r = 2 + 2 \cos 60° = 2 + 2(0.5) = 3.$$

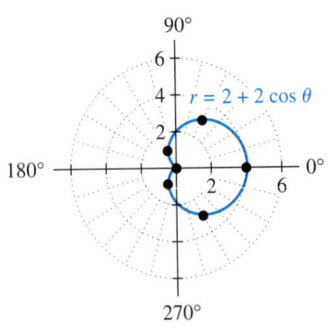

FIGURE 7.85 A Cardioid

TABLE 7.3

θ	0°	60°	120°	180°	240°	300°	360°
r	4	3	1	0	1	3	4

Graphical Representation   To help graph the equation we can plot the points in Table 7.3. See Figure 7.85. Notice that since the cosine function is periodic with 360°, the graph repeats after 360°.

Verbal Representation   The polar equation represents a heart-shaped graph that is called a **cardioid.** ∎

Technology Note   *Polar Coordinates*

Technology can be used to make tables and graphs in polar coordinates. The table and graph in Example 2(b) are shown in Figures 7.86 and 7.87.

As is the case with rectangular and parametric equations, a viewing rectangle must be selected before graphing a polar equation. First choose whether the graph should be plotted in radian or degree mode. Next determine an interval for θ. Many times the interval $0° \leq \theta \leq 360°$ is sufficient to have the entire graph generated. Then select a square viewing rectangle if possible.

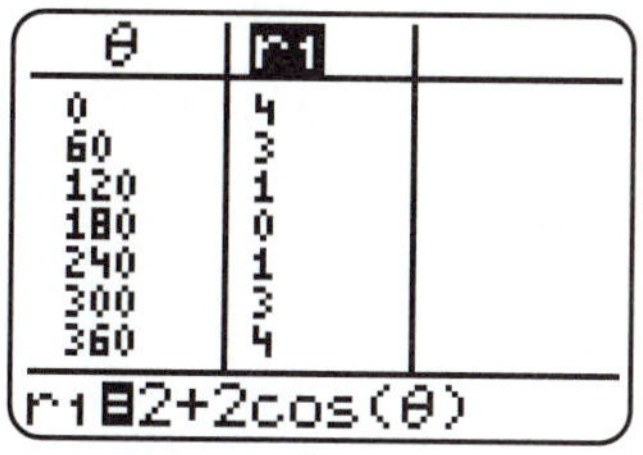

$[-6, 6, 1]$ by $[-4, 4, 1]$

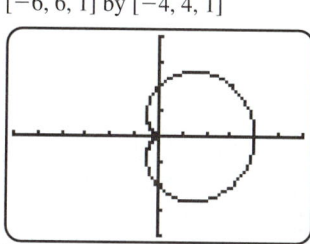

FIGURE 7.86

FIGURE 7.87

EXAMPLE 3   *Representing polar equations*

Represent each equation numerically and graphically. Then describe the curve.
(a) $r = 3 \cos 2\theta$
(b) $r = 1 - 2 \sin \theta$

Solution

(a) Let $r_1 = 3 \cos (2\theta)$ and table r_1 starting at $\theta = 0$ incrementing by 15° as shown in Figure 7.88. A graph of r_1 is shown in Figure 7.89. It is called a **four-leaved rose.** Turn on the polar grid and try tracing the graph. This shows how each leaf of the rose is generated by the polar equation. Notice for example, the location of the point $(-1.5, 60°)$ in Figure 7.89.

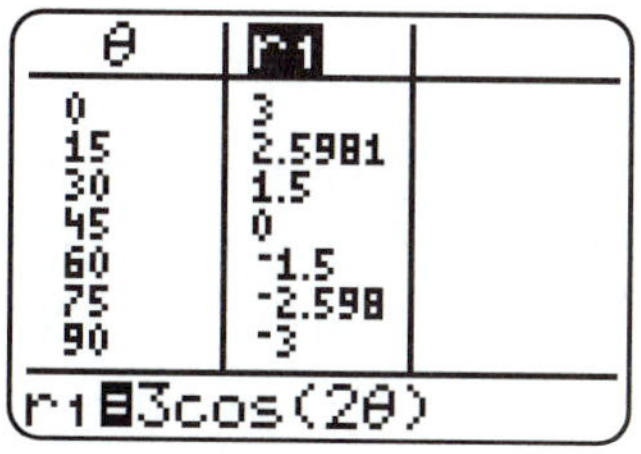

$[-6, 6, 1]$ by $[-4, 4, 1]$

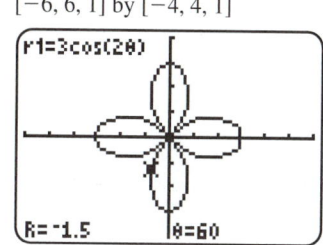

FIGURE 7.88

FIGURE 7.89 A Four-Leaved Rose

(b) Let $r_1 = 1 - 2 \sin (\theta)$ and table r_1 starting at $\theta = 0$ incrementing by 30° as shown in Figure 7.90 on the next page. A graph of r_1 is shown in Figure 7.91. It is called a **limaçon.** A graph with better resolution is shown in Figure 7.92. Notice that the inner loop occurs when $30° \leq \theta \leq 150°$ and $r \leq 0$.

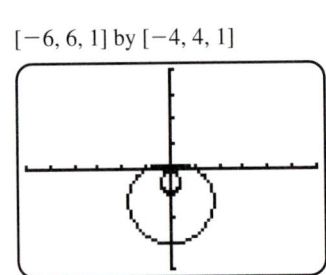

FIGURE 7.90

$[-6, 6, 1]$ by $[-4, 4, 1]$

FIGURE 7.91

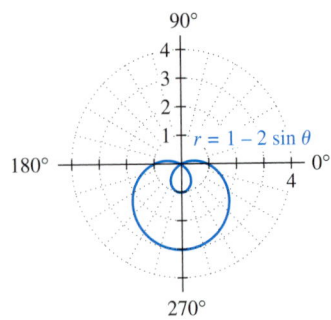

FIGURE 7.92 A Limaçon

Note: It can be shown that the graphs of $r = a \sin n\theta$ or $r = a \cos n\theta$ are rose curves with n leaves when n is odd and $2n$ leaves when n is even.

The polar equation $r = \dfrac{a(1 - e^2)}{1 + e \cos \theta}$ can be used to model the orbits of planets and comets, where a is the average distance of the celestial body from the sun in astronomical units and e is a constant called the *eccentricity*. (One astronomical unit equals 93 million miles.) The sun is located at the pole. Table 7.4 lists a and e for the outer planets. (Source: H. Kartunen, *Fundamentals of Astronomy.*)

TABLE 7.4	Distances and Eccentricities of the Outer Planets	
Planet	a	e
Jupiter	5.20	0.048
Saturn	9.54	0.056
Uranus	19.2	0.047
Neptune	30.1	0.009
Pluto	39.4	0.249

EXAMPLE 4   *Determining the farthest planet from the sun*

Use graphing to determine if Pluto is always the farthest planet from the sun.

Solution

We will compare the orbits of Pluto and Neptune since Neptune is closest to Pluto. Their orbital equations are given and then graphed in Figure 7.93.

$$\text{Neptune: } r_1 = \frac{30.1(1 - 0.009^2)}{1 + 0.009 \cos \theta}, \qquad \text{Pluto: } r_2 = \frac{39.4(1 - 0.249^2)}{1 + 0.249 \cos \theta}$$

The graph shows that their orbits pass near each other. By zooming in we can determine that the orbit of Pluto actually passes inside the orbit of Neptune. See Figure 7.94. Therefore, there are times when Neptune—not Pluto—is the farthest planet from the sun. However, Pluto's average distance from the sun is considerably greater than Neptune's average distance. Neptune was the farthest planet from the sun for a 20-year period ending in 1999.

[−60, 60, 10] by [−40, 40, 10]

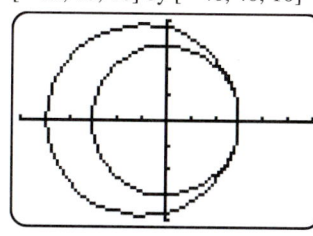

FIGURE 7.93

[27, 33, 1] by [−2, 2, 1]

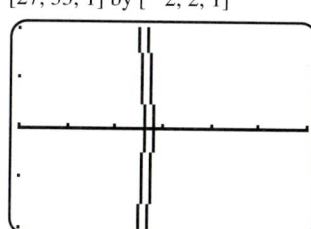

FIGURE 7.94

■

In the next example we convert a polar equation to an equation in x and y.

EXAMPLE 5   *Graphing polar equations*

Write the polar equation in terms of x and y. Graph each equation.

(a) $r = 3 \csc \theta$

(b) $r = \dfrac{2}{4 \cos \theta - 3 \sin \theta}$

[−6, 6, 1] by [−4, 4, 1]

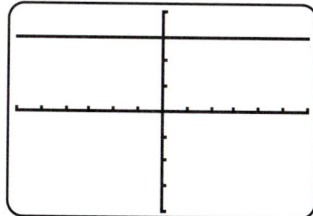

FIGURE 7.95 Polar plane

[−6, 6, 1] by [−4, 4, 1]

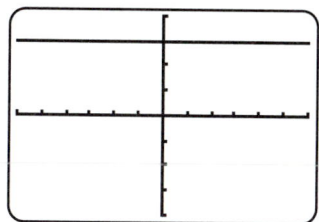

FIGURE 7.96 xy-plane

Solution

(a) Begin by applying a reciprocal identity.

$$r = 3 \csc \theta \qquad \text{Given equation}$$

$$r = \frac{3}{\sin \theta} \qquad \text{Reciprocal identity}$$

$$r \sin \theta = 3 \qquad \text{Multiply by } \sin \theta.$$

$$y = 3 \qquad y = r \sin \theta$$

Graphs of $r_1 = 3/\sin(\theta)$ and $Y_1 = 3$ appear to be identical in Figures 7.95 and 7.96.

(b) Start by cross-multiplying.

$$r = \frac{2}{4 \cos \theta - 3 \sin \theta} \qquad \text{Given equation}$$

$$4r \cos \theta - 3r \sin \theta = 2 \qquad \text{Cross-multiply.}$$

$$4x - 3y = 2 \qquad \text{Substitute.}$$

$$y = \frac{4}{3}x - \frac{2}{3} \qquad \text{Solve for } y.$$

Graphs of $r_1 = 2/(4 \cos(\theta) - 3 \sin(\theta))$ and $Y_1 = (4/3)X - (2/3)$ are shown in Figures 7.97 and 7.98.

[−6, 6, 1] by [−4, 4, 1]

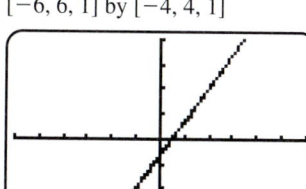

FIGURE 7.97 Polar plane

[−6, 6, 1] by [−4, 4, 1]

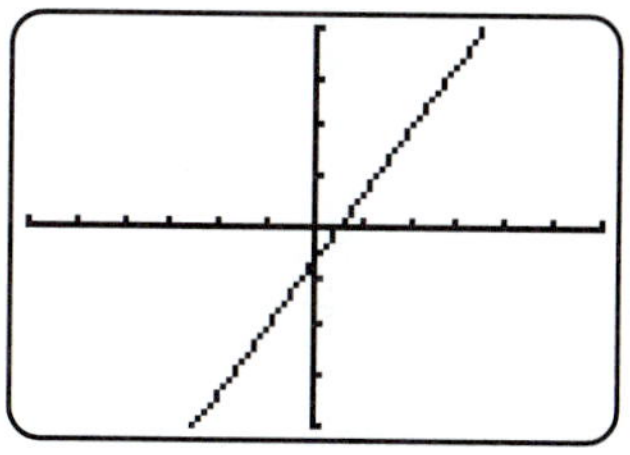

FIGURE 7.98 xy-plane

■

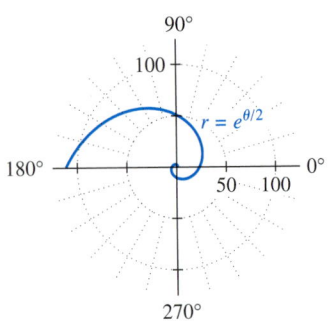

FIGURE 7.99 A Logarithmic Spiral

In 1638 René Descartes described a *logarithmic spiral* using the complicated equation $y = x \tan(\ln(x^2 + y^2))$. This curve, shown in Figure 7.99, cannot be represented by a function. Using polar coordinates, this rectangular equation reduces to the much simpler equation $r = e^{\theta/2}$. Johann Bernoulli was so entranced by this remarkable curve that he ordered it carved on his tombstone. (Source: H. Resnikoff and R. Wells, *Mathematics in Civilization.*)

Critical Thinking

Show that the equation $y = x \tan(\ln(x^2 + y^2))$ reduces to $r = e^{\theta/2}$.

(*Hint:* Let $\tan\theta = \dfrac{y}{x}$ and $r^2 = x^2 + y^2$.)

Solving Polar Equations (Optional)

We can solve polar equations graphically, numerically, and symbolically.

EXAMPLE 6   *Solving a polar equation graphically, numerically, and symbolically*

Find values for θ where the circle $r = 3$ intersects the cardioid $r = 2 + 2\cos\theta$. Let $0° \le \theta \le 360°$.

Solution

Graphical Solution   Using the intersection-of-graphs method, let $r_1 = 3$ and $r_2 = 2 + 2\cos(\theta)$. Their graphs intersect when $\theta = 60°$ and $300°$ as shown in Figures 7.100 and 7.101.

Numerical Solution   Numerical support is shown in Figure 7.102, where $r_1 = r_2 = 3$ when $\theta = 60°$ and $300°$.

Symbolic Solution   Begin by setting the two equations equal.

$$3 = 2 + 2\cos\theta \qquad \text{Set equations equal.}$$

$$\cos\theta = \frac{1}{2} \qquad \text{Solve for } \cos\theta.$$

$$\theta = 60° \text{ or } 300° \qquad \text{Solve for } \theta.$$

$[-6, 6, 1]$ by $[-4, 4, 1]$

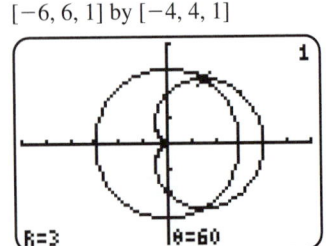

FIGURE 7.100

$[-6, 6, 1]$ by $[-4, 4, 1]$

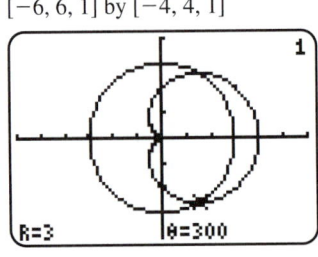

FIGURE 7.101

θ	r_1	r_2
0	3	4
60	3	3
120	3	1
180	3	0
240	3	1
300	3	3
360	3	4

$r_2 \blacksquare 2 + 2\cos(\theta)$

FIGURE 7.102

7.5 PUTTING IT ALL TOGETHER

Polar equations and graphs can be used to describe curves that are not easily represented by equations involving x and y. The polar coordinates (r, θ) are not unique for a given point, whereas in rectangular coordinates a point is determined uniquely by x and y. Polar equations can be solved graphically, numerically, and symbolically. Some basic concepts regarding polar coordinates and graphs of polar equations are summarized in the following table.

Polar Coordinates

A point is determined by r and θ. To convert between rectangular and polar coordinates use $x = r \cos \theta$, $y = r \sin \theta$, $\tan \theta = \dfrac{y}{x}$, and $r^2 = x^2 + y^2$.

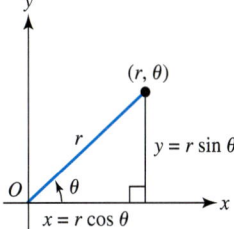

Examples of Graphs in Polar Coordinates

Circle
$r = a$

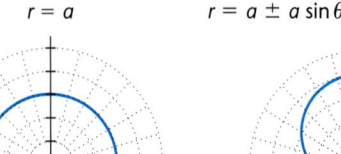

Cardioid
$r = a \pm a \sin \theta$ or $r = a \pm a \cos \theta$

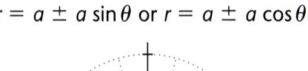

Limaçon ($b \neq a$)
$r = a \pm b \sin \theta$ or $r = a \pm b \cos \theta$

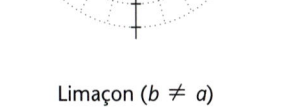
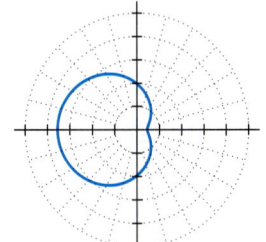

Rose Curve
$r = a \sin n\theta$ or $r = a \cos n\theta$

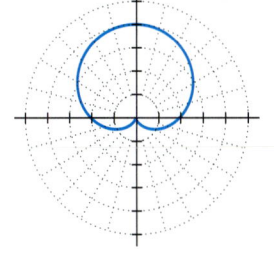
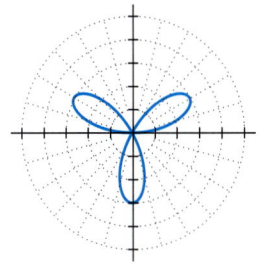

7.5 EXERCISES

Polar Coordinates

Exercises 1–4: Plot the points (r, θ).

1. (a) $(2, 0°)$
   (b) $(3, 120°)$
   (c) $(-1, 135°)$

2. (a) $(-2, 60°)$
   (b) $(1, 120°)$
   (c) $(2, 270°)$

3. (a) $(2, \pi/3)$
   (b) $(-3, -\pi/6)$
   (c) $(0, 3\pi/4)$

4. (a) $(4, \pi)$
   (b) $(1, \pi/4)$
   (c) $(-3, -3\pi/2)$

Exercises 5–10: Determine if the pair of polar coordinates represents the same point.

5. $(2, 180°), (2, -180°)$
6. $(1, 90°), (-1, -90°)$
7. $(3, 45°), (3, -45°)$
8. $(-2, 135°), (2, -135°)$
9. $(0, 40°), (0, 50°)$
10. $(-3, 30°), (3, 210°)$

Exercises 11–16: Change the polar coordinates (r, θ) to rectangular coordinates (x, y).

11. $(3, 45°)$
12. $(-4, 225°)$
13. $(10, 90°)$
14. $(-1, \pi/3)$
15. $(5, 2\pi)$
16. $(3, -\pi/2)$

Graphs in Polar Coordinates

Exercises 17–22: Use the polar equation to complete the following.

(a) Make a table of values with $\theta = 0°, 90°, 180°, 270°$.

(b) Plot the points from the table and graph the curve for $0° \le \theta \le 360°$.

17. $r = 2$
18. $r = \cos \theta$
19. $r = 3 \sin \theta$
20. $r = 2 + 3 \sin \theta$
21. $r = 2 + 2 \sin \theta$
22. $r = 3 - 2 \cos \theta$

Exercises 23–30: (Refer to Example 5.) Write the polar equation in terms of x and y. Graph each equation to support your result.

23. $r = 3$
24. $r = 5$
25. $r = 2 \sec \theta$
26. $r = 2 \csc \theta$
27. $r = \dfrac{3}{2 \cos \theta + 4 \sin \theta}$
28. $r = \dfrac{2}{5 \sin \theta - \sin \theta}$

Tape 13

29. $r = \cos \theta$ (*Hint:* Multiply both sides by *r.*)
30. $r = 2 \sin \theta$

Exercises 31–42: (Refer to Example 3.) Represent each equation numerically and graphically.

31. $r = 3 + 3 \cos \theta$ (cardioid)
32. $r = 2 - 2 \sin \theta$ (cardioid)
33. $r = 3 - 2 \sin \theta$ (limaçon)
34. $r = 4 + \cos \theta$ (limaçon)
35. $r = 2 - 4 \cos \theta$ (limaçon with a loop)
36. $r = 1 + 2 \sin \theta$ (limaçon with a loop)
37. $r = 4 \sin \theta$ (circle)
38. $r = 2 \cos 3\theta$ (three-leaved rose)
39. $r = 2 \cos 5\theta$ (five-leaved rose)
40. $r = 3 \sin 2\theta$ (four-leaved rose)
41. $r = \dfrac{\theta}{2}$ (spiral)
42. $r = e^{\theta/4}$ (logarithmic spiral)
43. $r^2 = 2 \sin 2\theta$ (lemniscate)
44. $r^2 = 4 \cos 2\theta$ (lemniscate)

Solving Equations in Polar Coordinates

Exercises 45–46: Find values for θ that satisfy the equation $r_1 = r_2$, where $0° \le \theta \le 360°$. Check any solutions.

45. $r_1 = 3, r_2 = 2 + 2 \sin \theta$

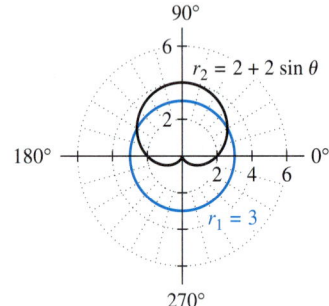

46. $r_1 = 1, r_2 = 2 \cos \theta$

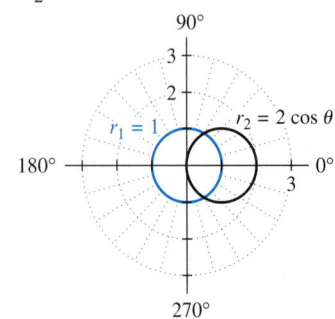

Exercises 47–50: Solve the equation $r_1 = r_2$, where $0° \leq \theta \leq 360°$,
   (a) *graphically,*
   (b) *numerically, and*
   (c) *symbolically.*

47. $r_1 = 3, r_2 = 2 - 2\sin\theta$

48. $r_1 = 3, r_2 = 2 - \sin\theta$

49. $r_1 = 1, r_2 = 2\sin\theta$

50. $r_1 = 2 - \sin\theta, r_2 = 2 + \cos\theta$

Applications

Exercises 51–52: Planetary Orbits   (Refer to Example 4 and Table 7.4.) Graph the planetary orbits in a square viewing rectangle using polar coordinates.

51.   Saturn and Uranus

52.   Jupiter and Neptune

Exercises 53–54: Planetary Orbits   (Refer to Example 4.) Use the table to complete the following.

Planet	a	e
Mercury	0.39	0.206
Venus	0.78	0.007
Earth	1.00	0.017
Mars	1.52	0.093

Source: H. Kartunen.

53.   Graph the orbits of the four inner planets in the same square viewing rectangle.

54.   NASA is planning future missions to Mars. Estimate graphically the closest distance possible between Earth and Mars.

Exercises 55–56: Radio Towers and Broadcasting Patterns   Many times radio stations do not broadcast in all directions with the same intensity. To avoid interference with an existing station to the north, a new station may only be licensed to broadcast east and west. To create an east-west signal, two radio towers are sometimes used as illustrated in the accompanying figure. Locations where the radio signal is received correspond to the interior of the curve defined by $r^2 = 40,000\cos 2\theta$, where the polar axis (or positive x-axis) points east. (Source: R. Weidner and R. Sells, Elementary Classical Physics. Vol 2.)

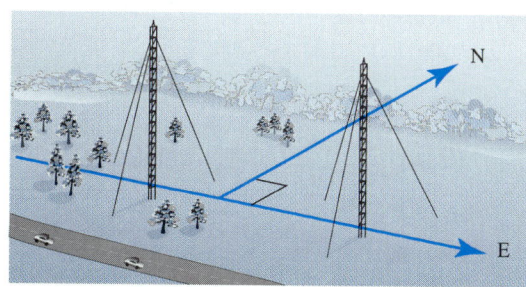

55.   Graph $r^2 = 40,000\cos 2\theta$ for $0° \leq \theta \leq 360°$, where units are in miles. Assuming the radio towers are located near the origin, use the graph to describe the regions where the signal can be received and where the signal cannot be received.

56.   (Refer to the previous exercise.) Suppose a radio signal pattern is given by $r^2 = 22,500\sin 2\theta$. Graph this pattern and interpret the results.

Writing about Mathematics

1.   Explain why (r, θ) and $(-r, \theta + 180°)$ represent the same point in polar coordinates. Give two examples.

2.   Give an example of a curve other than a circle that is more convenient to express in polar coordinates than in rectangular coordinates. Give an example of a curve that is more convenient to express using rectangular coordinates. Explain your reasoning.

7.6 Trigonometric Form and Roots of Complex Numbers

Trigonometric Form • Products and Quotients of Complex Numbers • De Moivre's Theorem • Roots of Complex Numbers

Introduction

One of the earliest encounters with the square root of a negative number was in 50 A.D. when Heron of Alexandria derived the expression $\sqrt{81 - 144}$. Square roots of negative numbers resulted in the invention of complex numbers. As late as the sixteenth and seventeenth centuries, mathematicians felt uneasy about negative numbers and square roots of negative numbers. The famous mathematician René Descartes rejected complex numbers and coined the term "imaginary" numbers.

The historical development of our present day number system was often met with resistance to the introduction of new numbers. Today, complex numbers are readily accepted and play an important role in the design of airplanes, ships, electrical circuits, and fractals. (Sources: M. Kline, *Mathematics: The Loss of Certainty; Historical Topics for the Mathematics Classroom, Thirty-first Yearbook,* NCTM.)

Trigonometric Form

Complex numbers were introduced in Section 3.5. Any complex number can be expressed in *standard form* as $a + bi$, where a and b are real numbers. The *real part* is a and the *imaginary part* is b.

Real numbers may be plotted on a number line. Since complex numbers are determined by both the real part a and the imaginary part b, we use the **complex plane** to plot complex numbers. The horizontal axis is the **real axis** and the vertical axis is the **imaginary axis.** For example, $2 + 3i$ may be plotted in the complex plane as the point $(2, 3)$ and $3 - 4i$ may be plotted as the point $(3, -4)$. See Figure 7.103.

A second form for complex numbers is called *trigonometric form,* which uses the variables r and θ to locate a complex number as illustrated in Figure 7.104. We can see that

$$\cos \theta = \frac{a}{r} \quad \text{and} \quad \sin \theta = \frac{b}{r}.$$

Solving for a and b gives

$$a = r \cos \theta \quad \text{and} \quad b = r \sin \theta.$$

As a result, we can write the complex number $a + bi$ as follows.

$$a + bi = r \cos \theta + (r \sin \theta)i$$
$$= r(\cos \theta + i \sin \theta)$$

By the Pythagorean theorem, $r = \sqrt{a^2 + b^2}$. It also follows that

$$\tan \theta = \frac{b}{a} \quad (a \neq 0).$$

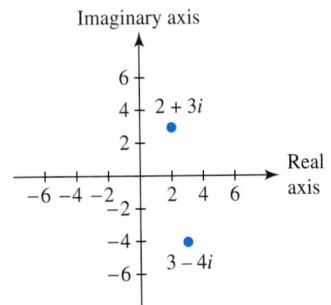

FIGURE 7.103 The Complex Plane

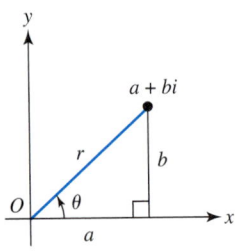

FIGURE 7.104 Trigonometric Form

> ## Trigonometric form of a complex number
>
> The expression
>
> $$r(\cos\theta + i\sin\theta)$$
>
> is called a **trigonometric form** of the complex number $a + bi$, where $a = r\cos\theta$ and $b = r\sin\theta$. The number $r = \sqrt{a^2 + b^2}$ is the **modulus** of $a + bi$ and θ is the **argument** of $a + bi$.

Note: The expression $\cos\theta + i\sin\theta$ may be written as cis θ. The expression $|a + bi|$ is sometimes used to denote the *modulus* of the complex number $a + bi$.

EXAMPLE 1   *Converting standard form to trigonometric form*

Find the trigonometric form for each complex number, where $0° \leq \theta \leq 360°$.
(a) $1 + i$ **(b)** $-1 - i\sqrt{3}$

Solution

(a) Plot $1 + i$ in the complex plane as shown in Figure 7.105. The modulus r is

$$r = \sqrt{1^2 + 1^2} = \sqrt{2}.$$

We can see that $\tan\theta = \dfrac{b}{a} = \dfrac{1}{1}$. Therefore, $\theta = \tan^{-1} 1 = 45°$. The trigonometric form is

$$\sqrt{2}(\cos 45° + i\sin 45°).$$

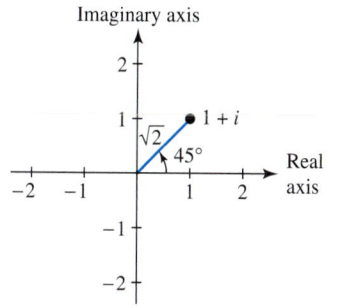

FIGURE 7.105   FIGURE 7.106

(b) Plot $-1 - i\sqrt{3}$ as shown in Figure 7.106. The modulus r is

$$r = \sqrt{(-1)^2 + (-\sqrt{3})^2} = 2.$$

The argument θ is in quadrant III and satisfies $\tan\theta = \dfrac{-\sqrt{3}}{-1} = \sqrt{3}$. The reference angle for θ is $\theta_R = \tan^{-1}(\sqrt{3}) = 60°$. Thus, $\theta = 240°$ and the trigonometric form is $2(\cos 240° + i\sin 240°)$. ∎

Technology Note *Trigonometric Form*

Some calculators have the capability to convert the complex number $a + bi$ to trigonometric or **polar form.** This is illustrated in Figures 7.107 and 7.108, where the first computation gives the modulus and the second gives the argument. Notice that angles of 240° and $-120°$ are coterminal angles. The value of θ is *not unique* in a trigonometric form. If θ_1 and θ_2 are coterminal angles, then

$$r(\cos\theta_1 + i\sin\theta_1) \quad \text{and} \quad r(\cos\theta_2 + i\sin\theta_2)$$

represent equivalent trigonometric forms.

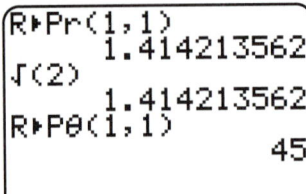

FIGURE 7.107 **FIGURE 7.108**

EXAMPLE 2 *Converting trigonometric form to standard form*

Write the complex number as $a + bi$, where a and b are real numbers.

(a) $4\left(\cos\dfrac{\pi}{2} + i\sin\dfrac{\pi}{2}\right)$ **(b)** $\sqrt{3}(\cos 150° + i\sin 150°)$

Solution

(a)

$$4\left(\cos\frac{\pi}{2} + i\sin\frac{\pi}{2}\right) = 4(0 + i(1))$$

$$= 4i$$

(b)

$$\sqrt{3}(\cos 150° + i\sin 150°) = \sqrt{3}\left(-\frac{\sqrt{3}}{2} + \frac{1}{2}i\right)$$

$$= -\frac{3}{2} + \frac{\sqrt{3}}{2}i$$

∎

Products and Quotients of Complex Numbers

If two complex numbers, z_1 and z_2, are expressed in trigonometric form, it is straightforward to find either their product or quotient. Let

$$z_1 = r_1(\cos\theta_1 + i\sin\theta_1) \quad \text{and} \quad z_2 = r_2(\cos\theta_2 + i\sin\theta_2).$$

Then,

$$z_1 z_2 = r_1(\cos\theta_1 + i\sin\theta_1) \cdot r_2(\cos\theta_2 + i\sin\theta_2)$$

$$= r_1 r_2(\cos\theta_1\cos\theta_2 + i\cos\theta_1\sin\theta_2 + i\sin\theta_1\cos\theta_2 + i^2\sin\theta_1\sin\theta_2)$$

$$= r_1 r_2((\cos\theta_1\cos\theta_2 - \sin\theta_1\sin\theta_2) + i(\cos\theta_1\sin\theta_2 + \sin\theta_1\cos\theta_2)).$$

Using the sum identities for cosine and sine, the above expression reduces to

$$z_1 z_2 = r_1 r_2 (\cos (\theta_1 + \theta_2) + i \sin (\theta_1 + \theta_2)).$$

Using similar reasoning it can be shown that

$$\frac{z_1}{z_2} = \frac{r_1}{r_2} (\cos (\theta_1 - \theta_2) + i \sin (\theta_1 - \theta_2)).$$

These results are summarized in the following.

> ### Product and quotient of complex numbers
>
> Let $z_1 = r_1(\cos \theta_1 + i \sin \theta_1)$ and $z_2 = r_2 (\cos \theta_2 + i \sin \theta_2)$. Then
>
> $$z_1 z_2 = r_1 r_2 (\cos (\theta_1 + \theta_2) + i \sin (\theta_1 + \theta_2))$$
>
> $$\frac{z_1}{z_2} = \frac{r_1}{r_2} (\cos (\theta_1 - \theta_2) + i \sin (\theta_1 - \theta_2)), \; z_2 \neq 0.$$

EXAMPLE 3   *Finding products and quotients*

Find the product and quotient of

$$z_1 = 4(\cos 45° + i \sin 45°) \qquad \text{and} \qquad z_2 = 2(\cos 135° + i \sin 135°).$$

Express the answer in standard form.

Solution

$$
\begin{aligned}
z_1 z_2 &= 4(\cos 45° + i \sin 45°) \cdot 2(\cos 135° + i \sin 135°) \\
&= (4 \cdot 2)(\cos (45° + 135°) + i \sin (45° + 135°)) \\
&= 8(\cos 180° + i \sin 180°) \\
&= 8(-1 + 0) \\
&= -8
\end{aligned}
$$

$$
\begin{aligned}
\frac{z_1}{z_2} &= \frac{4(\cos 45° + i \sin 45°)}{2(\cos 135° + i \sin 135°)} \\
&= 2(\cos (45° - 135°) + i \sin (45° - 135°)) \\
&= 2(\cos (-90°) + i \sin (-90°)) \\
&= 2(0 + -1i) \\
&= -2i
\end{aligned}
$$

De Moivre's Theorem

If a complex number z is expressed in trigonometric form, then z^n for any positive integer n can be computed easily. Let $z = r(\cos \theta + i \sin \theta)$ and consider the following.

$$
\begin{aligned}
z^2 &= r(\cos \theta + i \sin \theta) \cdot r(\cos \theta + i \sin \theta) \\
&= r^2(\cos (\theta + \theta) + i \sin (\theta + \theta)) \\
&= r^2(\cos 2\theta + i \sin 2\theta) \\
z^3 &= z z^2 \\
&= r(\cos \theta + i \sin \theta) \cdot r^2(\cos 2\theta + i \sin 2\theta) \\
&= r^3(\cos 3\theta + i \sin 3\theta)
\end{aligned}
$$

In general it can be shown that

$$z^n = r^n(\cos n\theta + i \sin n\theta).$$

This result is summarized in the following theorem, which is due to Abraham De Moivre (1667–1754), a French Huguenot, who was a close friend of Isaac Newton. This theorem has become a keystone of analytic trigonometry. (Source: H. Eves, *An Introduction to the History of Mathematics.*)

De Moivre's theorem

Let $z = r(\cos \theta + i \sin \theta)$ and n be a positive integer. Then

$$z^n = r^n(\cos n\theta + i \sin n\theta).$$

EXAMPLE 4   *Finding a power of a complex number*

Use De Moivre's theorem to evaluate $(1 + i)^8$ and express the result in standard form.

Solution

From Example 1(a), the trigonometric form of $z = 1 + i$ is

$$z = \sqrt{2}(\cos 45° + i \sin 45°).$$

By De Moivre's theorem

$$\begin{aligned}
z^8 &= (\sqrt{2})^8(\cos(8 \cdot 45°) + i \sin(8 \cdot 45°)) \\
&= 16(\cos 360° + i \sin 360°) \\
&= 16(1 + 0i) \\
&= 16.
\end{aligned}$$

■

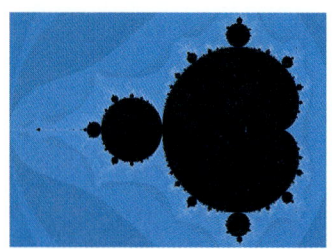

FIGURE 7.109

During the past 20 years, computer graphics and complex numbers have made it possible to produce many beautiful fractals. In 1977 Benoit B. Mandelbrot first used the term *fractal.* Largely because of his efforts, fractal geometry has become a new field of study. A fractal is an enchanting geometric figure with an endless self-similarity property that repeats itself infinitely with ever decreasing dimensions. If you look at smaller and smaller portions of the figure, you will continue to see the whole—much like looking into two parallel mirrors that are facing each other. Fractals not only have aesthetic appeal, they also have applications in science. An example of a fractal is the *Mandelbrot set* shown in Figure 7.109. (Source: F. Hill, *Computer Graphics.*)

EXAMPLE 5   *Analyzing the Mandelbrot set*

The fractal called the Mandelbrot set is shown in Figure 7.109. To determine if a complex number $z = a + bi$ is in the Mandelbrot set, we can perform the following sequence of calculations. Let

$$\begin{aligned}
z_0 &= z \\
z_1 &= z_0{}^2 + z_0 \\
z_2 &= z_1{}^2 + z_0 \\
z_3 &= z_2{}^2 + z_0
\end{aligned}$$

and so on. If the modulus of any z_k ever exceeds 2, then z is not in the Mandelbrot set, otherwise z is in the Mandelbrot set. Determine if the complex number belongs to the Mandelbrot set. (Source: F. Hill.)

(a) $z = 1 + i$

(b) $z = 0.5i$

Solution

(a) Let $z_0 = 1 + i$. Then,

$$z_1 = (1 + i)^2 + (1 + i) = 1 + 3i.$$

Since the modulus of z_1 is

$$|1 - 3i| = \sqrt{1^2 + 3^2} = \sqrt{10} > 2,$$

the complex number $1 + i$ is not in the Mandelbrot set.

(b) Let $z_0 = 0.5i$. Then,

$$z_1 = (0.5i)^2 + 0.5i = -0.25 + 0.5i$$

$$z_2 = (-0.25 + 0.5i)^2 + 0.5i = -0.1875 + 0.25i$$

$$z_3 = (-0.1875 + 0.25i)^2 + 0.5i \approx -0.0273 + 0.406i$$

The modulus of each consecutive z_k never exceeds 2. Thus, $0.5i$ is in the Mandelbrot set. You may find it helpful to use a calculator to perform these calculations. See Figure 7.110.   ∎

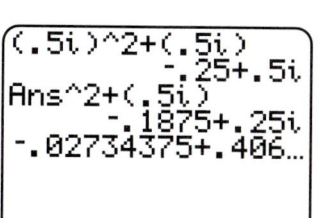

FIGURE 7.110

Roots of Complex Numbers

A number w is the **nth root** of a number z if $w^n = z$. De Moivre's theorem can be used to find roots of complex numbers. To see this let the trigonometric forms of w and z be

$$w = s(\cos \alpha + i \sin \alpha) \quad \text{and} \quad z = r(\cos \theta + i \sin \theta).$$

Then by De Moivre's theorem $w^n = z$ implies that

$$s^n(\cos n\alpha + i \sin n\alpha) = r(\cos \theta + i \sin \theta).$$

It follows that $s^n = r$ or $s = \sqrt[n]{r}$. Furthermore, the following two equations must be satisfied.

$$\cos n\alpha = \cos \theta$$

$$\sin n\alpha = \sin \theta$$

Since the cosine and sine functions are periodic with 360°,

$$n\alpha = \theta + 360° \cdot k$$

for some integer k, or

$$\alpha = \frac{\theta + 360° \cdot k}{n}.$$

Substituting these results in the trigonometric form for w gives

$$w_k = \sqrt[n]{r}\left(\cos\frac{\theta + 360° \cdot k}{n} + i \sin\frac{\theta + 360° \cdot k}{n}\right).$$

We obtain a unique value of w_k for $k = 0, 1, 2, \ldots , n - 1$. This discussion is summarized in the following.

Roots of a complex number

Let $z = r(\cos \theta + i \sin \theta)$ be a nonzero complex number and n be any positive integer. Then z has exactly n distinct nth roots given by

$$w_k = \sqrt[n]{r}\left(\cos \frac{\theta + 360° \cdot k}{n} + i \sin \frac{\theta + 360° \cdot k}{n}\right),$$

where $k = 0, 1, 2, \ldots , n - 1$. If radian measure is used then let

$$w_k = \sqrt[n]{r}\left(\cos \frac{\theta + 2\pi k}{n} + i \sin \frac{\theta + 2\pi k}{n}\right).$$

EXAMPLE 6   *Finding cube roots of a complex number*

Find the three cube roots of $8i$. Check your results with a calculator.

Solution
First write the complex number $8i$ in trigonometric form.

$$8i = 8(\cos 90° + i \sin 90°)$$

The three cube roots of $8i$ can be found by letting $n = 3$, $r = 8$, $\theta = 90°$, and $k = 0, 1, 2$.

$$w_0 = \sqrt[3]{8}\left(\cos \frac{90° + 360° \cdot 0}{3} + i \sin \frac{90° + 360° \cdot 0}{3}\right)$$

$$= 2(\cos 30° + i \sin 30°)$$

$$= 2\left(\frac{\sqrt{3}}{2} + \frac{1}{2}i\right)$$

$$= \sqrt{3} + i$$

$$w_1 = \sqrt[3]{8}\left(\cos \frac{90° + 360° \cdot 1}{3} + i \sin \frac{90° + 360° \cdot 1}{3}\right)$$

$$= 2(\cos 150° + i \sin 150°)$$

$$= 2\left(-\frac{\sqrt{3}}{2} + \frac{1}{2}i\right)$$

$$= -\sqrt{3} + i$$

$$w_2 = \sqrt[3]{8}\left(\cos \frac{90° + 360° \cdot 2}{3} + i \sin \frac{90° + 360° \cdot 2}{3}\right)$$

$$= 2(\cos 270° + i \sin 270°)$$

$$= 2(0 - i)$$

$$= -2i$$

```
(-√(3)+i)^3
                 8i
(√(3)+i)^3
                 8i
(-2i)^3
                 8i
```

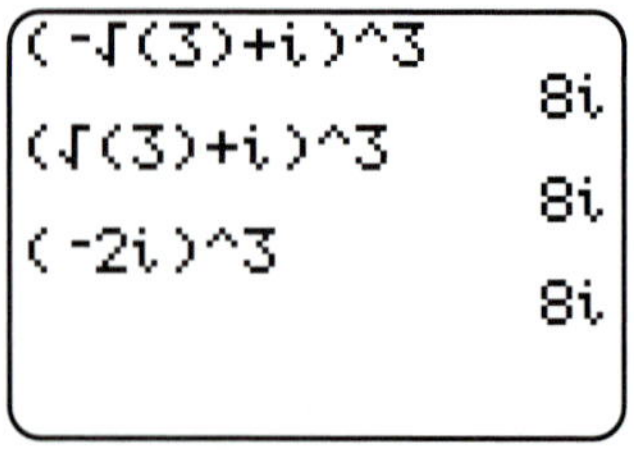

FIGURE 7.111

The three cube roots of $8i$ are $\sqrt{3} + i$, $-\sqrt{3} + i$, and $-2i$. These can be checked using a calculator as shown in Figure 7.111. ∎

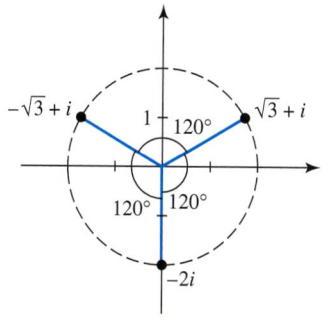

FIGURE 7.112 The Cube Roots of $8i$

If the three cube roots of $8i$ are plotted in the complex plane, they lie on a circle of radius 2, equally spaced 120° apart as shown in Figure 7.112. In general, the n roots of a complex number $z = r(\cos\theta + i\sin\theta)$ will lie on a circle of radius $\sqrt[n]{r}$ equally spaced.

Critical Thinking

One cube root of $z = -1$ is $w = -1$. Find the other two cube roots of z graphically.

EXAMPLE 7   *Finding square roots of a complex number*

Find the two square roots of $1 + i\sqrt{3}$.

Solution

First write the complex number $1 + i\sqrt{3}$ in trigonometric form.

$$1 + i\sqrt{3} = 2\left(\cos\frac{\pi}{3} + i\sin\frac{\pi}{3}\right)$$

The two square roots of $1 + i\sqrt{3}$ can be found by letting $n = 2$, $r = 2$, $\theta = \pi/3$, and $k = 0, 1$.

$$w_0 = \sqrt{2}\left(\cos\frac{\pi/3 + 2\pi \cdot 0}{2} + i\sin\frac{\pi/3 + 2\pi \cdot 0}{2}\right)$$

$$= \sqrt{2}\left(\cos\frac{\pi}{6} + i\sin\frac{\pi}{6}\right)$$

$$= \sqrt{2}\left(\frac{\sqrt{3}}{2} + \frac{1}{2}i\right)$$

$$= \frac{\sqrt{6}}{2} + \frac{\sqrt{2}}{2}i$$

$$w_1 = \sqrt{2}\left(\cos\frac{\pi/3 + 2\pi \cdot 1}{2} + i\sin\frac{\pi/3 + 2\pi \cdot 1}{2}\right)$$

$$= \sqrt{2}\left(\cos\frac{7\pi}{6} + i\sin\frac{7\pi}{6}\right)$$

$$= \sqrt{2}\left(-\frac{\sqrt{3}}{2} - \frac{1}{2}i\right)$$

$$= -\frac{\sqrt{6}}{2} - \frac{\sqrt{2}}{2}i$$

Thus, the two square roots of $1 + i\sqrt{3}$ are $\dfrac{\sqrt{6}}{2} + \dfrac{\sqrt{2}}{2}i$ and $-\dfrac{\sqrt{6}}{2} - \dfrac{\sqrt{2}}{2}i$. ∎

7.6 PUTTING IT ALL TOGETHER

The following table summarizes some of the important topics in this section

Trigonometric Form
If $z = a + bi$, then its trigonometric form is $z = r(\cos\theta + i\sin\theta)$, where $r = \sqrt{a^2 + b^2}$ and $\tan\theta = \dfrac{b}{a}$. The modulus is r and the argument is θ.

Products and Quotients
Let $z_1 = r_1(\cos\theta_1 + i\sin\theta_1)$ and $z_2 = r_2(\cos\theta_2 + i\sin\theta_2)$. Then $$z_1 z_2 = r_1 r_2(\cos(\theta_1 + \theta_2) + i\sin(\theta_1 + \theta_2)) \qquad \text{and}$$ $$\frac{z_1}{z_2} = \frac{r_1}{r_2}(\cos(\theta_1 - \theta_2) + i\sin(\theta_1 - \theta_2)), z_2 \neq 0.$$

De Moivre's Theorem
Let $z = r(\cos\theta + i\sin\theta)$. Then, $z^n = r^n(\cos n\theta + i\sin n\theta)$.

Roots of Complex Numbers
Let $z = r(\cos\theta + i\sin\theta)$ and n be any positive integer. Then the nth roots of z are given by $$w_k = \sqrt[n]{r}\left(\cos\frac{\theta + 360° \cdot k}{n} + i\sin\frac{\theta + 360° \cdot k}{n}\right),$$ where $k = 0, 1, 2, \ldots, n - 1$.

7.6 EXERCISES

The Complex Plane

Exercises 1–4: Plot the numbers in the complex plane.

1. (a) $3 + 2i$
   (b) $-1 + i$
   (c) $3i$

2. (a) $-2i$
   (b) $2 + 2i$
   (c) $2 - 2i$

3. (a) -3
   (b) $4 - 2i$
   (c) $-1 - 3i$

4. (a) $-1 - i$
   (b) $4 + 3i$
   (c) 4

Trigonometric Form

Exercises 5–12: Write the number in standard form.

5. $5(\cos 180° + i\sin 180°)$

6. $3(\cos 90° + i\sin 90°)$

7. $2(\cos 45° + i\sin 45°)$

8. $\cos 150° + i\sin 150°$

9. $4\left(\cos\dfrac{3\pi}{2} + i\sin\dfrac{3\pi}{2}\right)$

10. $2\left(\cos\dfrac{\pi}{6} + i\sin\dfrac{\pi}{6}\right)$

11. $3(\cos 2\pi + i\sin 2\pi)$

Tape 13

12. $5\left(\cos\dfrac{3\pi}{4} + i\sin\dfrac{3\pi}{4}\right)$

Exercises 13–20: Find the modulus of the number.

13. $1 + i$

14. $3 - 4i$

15. $12 - 5i$

16. $-24 + 7i$

17. -6

18. $15i$

19. $2 - 3i$

20. $11 - 60i$

Exercises 21–30: Write the number in trigonometric form. Let $0° \leq \theta < 360°$.

21. $1 + i$

22. $1 - i$

23. 5

24. -3

25. $4i$

26. $-i$

27. $-1 + i\sqrt{3}$

28. $-\sqrt{2} - i\sqrt{2}$

29. $\sqrt{3} + i$

30. $-\dfrac{\sqrt{3}}{2} + \dfrac{1}{2}i$

Exercises 31–34: Write the number in trigonometric form. Let $0 \leq \theta < 2\pi$.

31. -2

32. $4i$

33. $-2 + 2i$

34. $1 + i\sqrt{3}$

Exercises 35–40: Find $z_1 z_2$ and $\dfrac{z_1}{z_2}$. Express your answer in standard form.

35. $z_1 = 9(\cos 45° + i \sin 45°)$,
$z_2 = 3(\cos 15° + i \sin 15°)$

36. $z_1 = 5(\cos 90° + i \sin 90°)$,
$z_2 = 2(\cos 30° + i \sin 30°)$

37. $z_1 = 6\left(\cos \dfrac{3\pi}{4} + i \sin \dfrac{3\pi}{4} \right)$,

$z_2 = \cos \dfrac{\pi}{4} + i \sin \dfrac{\pi}{4}$

38. $z_1 = 4(\cos 300° + i \sin 300°)$,
$z_2 = 2(\cos 60° + i \sin 60°)$

39. $z_1 = \cos 15° + i \sin 15°$,

$z_2 = \cos \left(-\dfrac{\pi}{4} \right) + i \sin \left(-\dfrac{\pi}{4} \right)$

40. $z_1 = 11 \left(\cos \dfrac{2\pi}{3} + i \sin \dfrac{2\pi}{3} \right)$,

$z_2 = 22(\cos 30° + i \sin 30°)$

Powers of Complex Numbers

Exercises 41–46: Use De Moivre's theorem to evaluate the expression. Write the result in standard form.

41. $(2(\cos 30° + i \sin 30°))^3$

42. $(3(\cos 45° + i \sin 45°))^4$

43. $(\cos 10° + i \sin 10°)^{36}$

44. $(\cos 1° + i \sin 1°)^{90}$

45. $(5(\cos 60° + i \sin 60°))^2$

46. $(2(\cos 90° + i \sin 90°))^5$

Exercises 47–50: Use De Moivre's theorem to evaluate the expression. Write the result in standard form and check it using a calculator.

47. $(1 + i)^3$ **48.** $(3i)^4$

49. $(\sqrt{3} + i)^5$ **50.** $(2 - 2i)^6$

Roots of Complex Numbers

Exercises 51–62: Find the following roots. Check your results with a calculator.

51. The square roots of $4(\cos 120° + i \sin 120°)$

52. The cube roots of $27(\cos 180° + i \sin 180°)$

53. The cube roots of $\cos 180° + i \sin 180°$

54. The fourth roots of $16(\cos 240° + i \sin 240°)$

55. The square roots of i

56. The cube roots of 1

57. The cube roots of -8

58. The square roots of $-4i$

59. The cube roots of $64i$

60. The fourth roots of -1

61. The fourth roots of 81

62. The square roots of $-1 + i\sqrt{3}$

Fractals

Exercises 63–66: Mandelbrot Set (Refer to Example 5.) Determine if the complex number belongs to the Mandelbrot set.

63. $-0.4i$ **64.** $0.5 + i$

65. $1 + i$ **66.** $-0.2 + 0.2i$

Applications

67. *Electrical Circuits*   *Impedance* is a measure of the opposition to the flow of current in an electrical circuit. It consists of two parts called the *resistance* and *reactance*. Light bulbs add resistance to an electrical circuit, while reactance occurs when electricity passes through coils of wire like those found in electric motors. Impedance Z in ohms (Ω) may be expressed as a complex number, where the real part represents the resistance and the imaginary part represents the reactance. For example, if the resistive part is 3 ohms and the reactive part is 4 ohms, then the impedance could be described by the complex number $Z = 3 + 4i$. The modulus of Z gives the total impedance in ohms. In a series circuit like the one shown in the figure, the total impedance is the sum of the individual impedances. (Source: R. Smith and R. Dorf. *Circuits, Devices, and Systems.*)
(a) The circuit contains two light bulbs and two electric motors. If it is assumed that the light bulbs represent resistance and the motors represent reactance, express impedance as $Z = a + bi$.
(b) Find total impedance in ohms by calculating the modulus of Z.

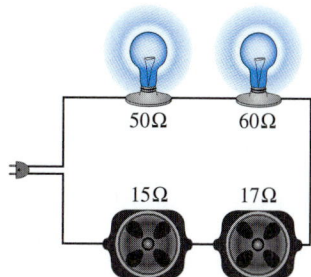

68. (Continuation of the previous exercise.) In the parallel electrical circuit shown in the figure, impedance Z is given by $Z = (1/Z_1 + 1/Z_2)^{-1}$, where Z_1 and Z_2 represent impedances for each branch of the circuit. (Source: G. Wilcox and C. Hesselberth, *Electricity For Engineering Technology.*)

(a) Find Z.
(b) Find total impedance in ohms by calculating
the modulus of Z.

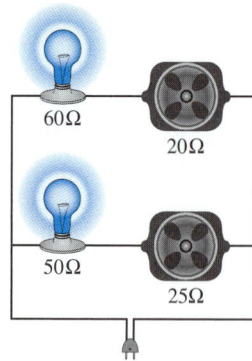

60Ω
20Ω
50Ω
25Ω

Writing about Mathematics

1. Explain how to find a trigonometric form of a complex
number $a + bi$. Give an example. Is trigonometric
form unique for a given complex number z? Explain.

2. Suppose that one square root w of a complex number z is known. Explain how to graphically find the
second square root of z.

CHECKING BASIC CONCEPTS FOR SECTIONS 7.5 AND 7.6

1. Plot the following points (r, θ) on a polar grid.
   (a) $(2, 30°)$
   (b) $(3, -60°)$
   (c) $(-4, 120°)$

2. Represent the polar equations graphically and
numerically.
   (a) $r = 3 + 3\cos\theta$
   (b) $r = 3\cos 2\theta$

3. Plot the numbers in the complex plane.
   (a) $-3 + 2i$
   (b) $-4 - 3i$

4. Find the trigonometric form of $1 + i\sqrt{3}$.

5. Find the three cube roots of i.

CHAPTER 7 Summary

The *law of sines* is given by either

$$\frac{\sin\alpha}{a} = \frac{\sin\beta}{b} = \frac{\sin\gamma}{c} \quad \text{or} \quad \frac{a}{\sin\alpha} = \frac{b}{\sin\beta} = \frac{c}{\sin\gamma},$$

and can be used to solve triangles when given either two angles and a side or two
sides and an angle opposite one of the sides. These triangles are identified by ASA,
AAS, or SSA. The SSA case is called the *ambiguous case* because it can result in
0, 1, or 2 solutions.

The *law of cosines* is given by any of the following formulas.

$$a^2 = b^2 + c^2 - 2bc\cos\alpha$$
$$b^2 = a^2 + c^2 - 2ac\cos\beta$$
$$c^2 = a^2 + b^2 - 2ab\cos\gamma$$

These formulas can be used to solve triangles when given either three sides or two
sides and the included angle. These triangles are identified by SSS and SAS.

A *vector quantity* has both magnitude and direction. A *vector* is a directed line segment with an initial point and a terminal point. A vector **v** may be denoted by

$$\mathbf{v} = \langle a_1, a_2 \rangle, \qquad \mathbf{v} = a_1\mathbf{i} + a_2\mathbf{j}, \qquad \text{or} \qquad \mathbf{v} = \overrightarrow{PQ}.$$

The *horizontal component* of **v** is a_1 and its *vertical component* is a_2. Sums, differences, and scalar products of vectors may be found. The dot product of two vectors, denoted by $\mathbf{a} \cdot \mathbf{b}$, is a real number. If the dot product equals 0, then the vectors are perpendicular.

In *parametric equations* the variables x and y are defined by $x = f(t)$ and $y = g(t)$, where t is a *parameter*. An example is $x = \cos t$ and $y = \sin t$ for $0 \le t \le 2\pi$, which describes the unit circle. Parametric equations can be represented symbolically, graphically, and numerically. They are particularly important to describe curves that cannot be modeled by a function.

Polar equations are graphed in the polar plane in which the *pole* corresponds to the origin and the *polar axis* corresponds to the positive x-axis. In polar coordinates a point is identified by giving (r, θ) rather than (x, y). Like parametric equations, polar equations can be used to describe curves that cannot be modeled by a function. Many equations in rectangular coordinates become simpler in polar coordinates, such as the equation of a circle.

Complex numbers are plotted in the *complex plane*. A complex number z can be represented using either *standard form*, $z = a + bi$, or *trigonometric form*, $z = r(\cos\theta + i\sin\theta)$. In trigonometric form $r = \sqrt{a^2 + b^2}$ is called the *modulus*, and θ is called the *argument*, where $\tan\theta = \dfrac{b}{a}$. Products, quotients, and powers of complex numbers can be calculated easily if z is expressed in trigonometric form. Trigonometric form can also be used to calculate roots of complex numbers.

Review Exercises

Exercises 1–4: Solve the triangle. Approximate values to three significant digits.

1.

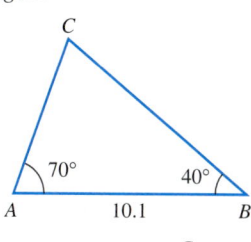

2.

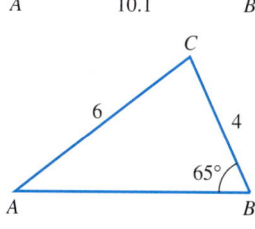

3.

4.

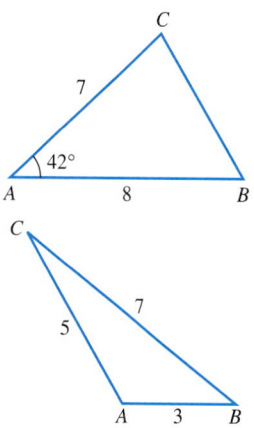

Exercises 5–10: Solve the triangle. Approximate values to the nearest tenth.

5. $\alpha = 19°, \beta = 46°, b = 13$

6. $\alpha = 30°, b = 10, a = 8$

7. $\gamma = 20°, b = 8, c = 11$

8. $\alpha = 70°, b = 17, a = 5$

9. $b = 23, \gamma = 35°, a = 18$

10. $a = 65, b = 45, c = 32$

Exercises 11–14: Approximate the area of each triangle to three significant digits.

11. $a = 12.3, b = 13.7, \gamma = 39°$

12. $\alpha = 40°, \beta = 55°, c = 67$

13. $a = 34, b = 67, c = 53$

14. $a = 2.1, b = 1.7, c = 2.2$

*Exercises 15 and 16: Complete the following for vector **v**.*
   (a) *Give the horizontal and vertical components.*
   (b) *Find* $\|\mathbf{v}\|$.
   (c) *Graph* **v** *and interpret* $\|\mathbf{v}\|$.

15. $\mathbf{v} = \langle 3, 4 \rangle$ **16.** $\mathbf{v} = -5\mathbf{i} + 12\mathbf{j}$

*Exercises 17 and 18: A vector **v** has initial point P and terminal point Q.*
   (a) *Graph* $\overrightarrow{PQ}$.
   (b) *Write* $\overrightarrow{PQ}$ *as* $\mathbf{v} = a_1\mathbf{i} + a_2\mathbf{j}$.
   (c) *Find* $\|\overrightarrow{PQ}\|$.

17. $P = (0, 0), Q = (-2, -4)$

18. $P = (3, 2), Q = (-3, -1)$

Exercises 19–22: Find each of the following.
   (a) $2\mathbf{a}$
   (b) $\mathbf{a} - 3\mathbf{b}$
   (c) $\mathbf{a} \cdot \mathbf{b}$
   (d) *The angle* θ *between* **a** *and* **b** *rounded to a tenth of a degree*

19. $\mathbf{a} = \langle 3, -2 \rangle, \mathbf{b} = \langle 1, 1 \rangle$

20. $\mathbf{a} = \langle 3, 2 \rangle, \mathbf{b} = \langle -2, -3 \rangle$

21. $\mathbf{a} = 2\mathbf{i} + 2\mathbf{j}, \mathbf{b} = \mathbf{i} + \mathbf{j}$

22. $\mathbf{a} = \mathbf{i} - 2\mathbf{j}, \mathbf{b} = 2\mathbf{i} + \mathbf{j}$

23. *Resultant Force* Use the parallelogram rule to find the magnitude of the resultant force of the two forces shown in the figure.

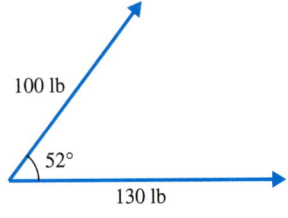

24. *Work* Find the work done when a constant force $\mathbf{F} = 300\mathbf{i} + 400\mathbf{j}$ is applied to an object that moves along the vector $\mathbf{D} = 10\mathbf{i} - 2\mathbf{j}$, where units are in pounds and feet. Approximate the magnitude of $\mathbf{F}$ and interpret the result.

Exercises 25–28: Graph the parametric equations.

25. $x = t + 2, \quad y = t^2 - 3; \quad -2 \le t \le 2$

26. $x = t^3 - 4, \quad y = t - 1; \quad 0 \le t \le 2$

27. $x = 2\cos t, \quad y = -2\sin t; \quad 0 \le t \le 2\pi$

28. $x = 3\sin t, \quad y = 2\cos t; \quad -\pi \le t \le \pi$

29. *Designing Letters* Find parametric equations whose graphs resemble the given letter.
   (a) O   **(b)** H

30. Change the polar coordinates (r, θ) to rectangular coordinates (x, y).
   (a) $(2, 135°)$
   (b) $(-1, 60°)$

Exercises 31 and 32: Use the polar equation to complete the following.
   (a) *Make a table of values with* $\theta = 0°, 90°, 180°, 270°$.
   (b) *Plot the points from the table and graph the curve.*

31. $r = 1 + \cos\theta$ **32.** $r = \sin\theta$

Exercises 33–36: Graph each polar equation.

33. $r = 3\sin 3\theta$ **34.** $r = 2 - \cos\theta$

35. $r = 3 + 3\sin\theta$ **36.** $r = 1 - 2\sin\theta$

37. Plot the numbers in the complex plane.
   (a) $4 - i$ **(b)** $-2 + 2i$
   (c) $-2i$ **(d)** -4

38. Write the number in trigonometric form. Let $0° \le \theta < 360°$.
   (a) $-2 + 2i$ **(b)** $\sqrt{3} + i$
   (c) $5i$ **(d)** -6

39. Find $z_1 z_2$ and $\dfrac{z_1}{z_2}$, if

$$z_1 = 4(\cos 150° + i\sin 150°),$$
$$z_2 = 2(\cos 30° + i\sin 30°).$$

Write the results in standard form.

40. Use De Moivre's theorem to evaluate z^4 if $z = 2(\cos 45° + i\sin 45°)$. Write the result in standard form.

Exercises 41 and 42: Find the following roots. Check your results with a calculator.

41. The square roots of $4(\cos 60° + i\sin 60°)$

42. The cube roots of $27i$

43. *Airplane Navigation*   An airplane takes off with a bearing of 130° and flies 350 miles. Then it changes its course to a bearing of 60° and flies for 500 miles. Determine how far the plane is from its takeoff point.

44. *Obstructed View*   To find the distance between two points *A* and *B* on the opposite sides of a small building, a surveyor measures *AC* as 63.15 feet, angle *ACB* as 43.56°, and *CB* as 103.53 feet. Find the distance between *A* and *B*.

45. *Height of an Airplane*   Two observation points *A* and *B* are 950 feet apart. From these points the angles of elevation of an airplane in the distance are 52° and 57° as illustrated in the figure. Find the height of the airplane.

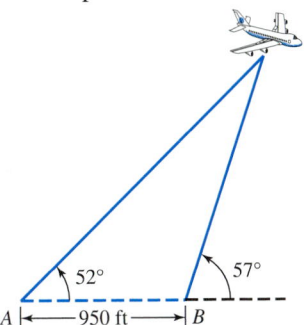

46. *Area of a Lot*   A surveyor measures two sides and the included angle of a triangular lot as *a* = 93.6 feet, *b* = 110.6 feet, and γ = 51.8°. Find the area of the lot.

47. *Area of a Lot*   Find the area of the quadrangular lot shown in the figure.

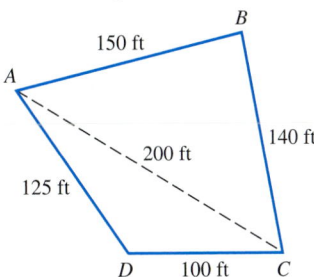

48. *Interpreting a Vector*   A boat is heading west at 20 miles per hour in a current that is flowing at 6 miles per hour toward the south. Find a vector **v** that models the direction and speed of the boat. What does ‖**v**‖ represent?

49. *Solar Panels*   Suppose that the sun's intensity in watts per square meter and direction is given by vector **I**, and a solar panel's area in square meters and orientation is given by vector **S.** Then the total number of watts *W* that can be absorbed by the panel is given by $W = |\mathbf{I} \cdot \mathbf{S}|$. Suppose **I** = 40**i** − 180**j** and **S** = 2**i** + 5**j.**

(a) Find ‖**I**‖ and ‖**S**‖. Interpret your results.
(b) Calculate *W* and interpret the results.

50. *Robotics*   Consider the planar two-arm manipulator shown in the accompanying figure, where units are in centimeters. Let the upper arm be modeled by **a** = 40**i** − 20**j** and the forearm be modeled by **b** = 20**i** + 30**j.** (Source: J. Craig, *Introduction of Robotics.*)

(a) Find a vector **c** that represents the position of the hand.
(b) How far is the hand from the origin?
(c) Find the position of the hand if the length of the forearm doubles.

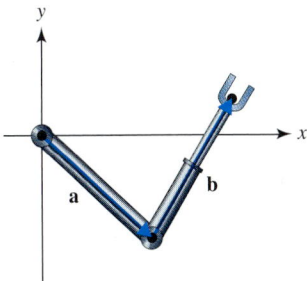

51. *Work*   A 200-pound person walks 0.75 mile up a hiking trail inclined at 15°. Use a dot product to compute the work done in foot-pounds.

52. *Flight of a Football*   A football is passed at 50 feet per second making an angle of 45° with the horizontal as it leaves the quarterback's hand. If the ground is level, estimate graphically the distance traveled by the football.

53. *Aerial Photography*   The distance covered by an aerial photograph is determined by both the focal length of the camera lens and the tilt of the camera from the perpendicular to the ground. Although the tilt is usually small, archaeological and Canadian aerial photographs often use larger tilts. A camera lens with a 12-inch focal length will have an angular coverage of 60°. If an aerial photograph is taken with this camera tilted θ = 35° at an altitude of 5000 feet, calculate the ground distance *d* that will be shown in this photograph. See the accompanying figure. (Sources: R. Brooks and J. Dieter, *Phytoarchaeology;* F. Moffitt, *Photogrammetry.*)

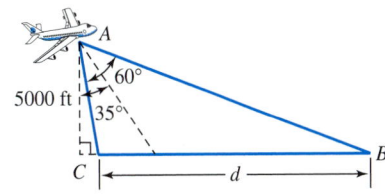

54. *Aerial Photography* A camera lens with a 6-inch focal length has an angular coverage of 86°. Suppose an aerial photograph is taken vertically with no tilt at an altitude of 3500 feet over ground with an increasing slope of 5° as shown in the figure. Calculate the ground distance *CB* that would appear in the resulting photograph. (Source: F. Moffitt.)

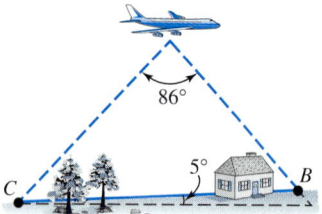

Extended and Discovery Exercises

1. *Velocity of a Star* The velocity vector **v** of a star relative to the sun can be expressed as the resultant vector of two perpendicular vectors—the radial velocity $\mathbf{v}_r$ and the tangential velocity $\mathbf{v}_t$, where $\mathbf{v} = \mathbf{v}_r + \mathbf{v}_t$ as illustrated in the figure. If a star is located near the sun and its velocity is large, then its motion across the sky will also be large. Barnard's Star is relatively close to the sun with a distance of 35 trillion miles. It moves across the sky through an angle of 10.34″ per year, which is the largest of any known star. Its radial velocity is $\mathbf{v}_r = 67$ miles per second toward the sun. (Sources: A. Acker and C. Jaschek, *Astronomical Methods and Calculations;* M. Zelik, *Introductory to Astronomy and Astrophysics.*)

(a) Approximate $\|\mathbf{v}_t\|$ for Barnard's Star.
  (*Hint:* Use $s = r\theta$.)
(b) Compute $\|\mathbf{v}\|$.

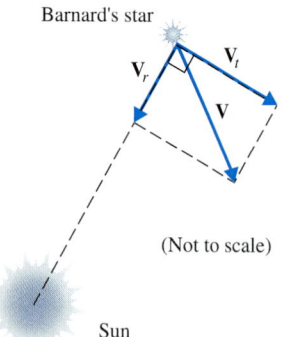

Barnard's star

(Not to scale)

Sun

2. *Fractals* The fractal called the *Julia set* is shown in the figure. To determine if a complex number $z = a + bi$ belongs to this set, repeatedly compute the sequence of values,

$$z_1 = z^2 - 1, \qquad z_2 = z_1^2 - 1, \qquad z_3 = z_2^2 - 1,$$

and so on. If the modulus of any of the resulting complex numbers exceeds 2, then the complex number z is not in the Julia set. Otherwise z is in this set. Determine if the complex numbers belong to the Julia set. (Source: R. Crownover, *Introduction to Fractals and Chaos,© 1995 by Jones and Bartlett Publisher, www.jbpub.com. Reprinted with permission.*)

(a) $z = 0 + 0i$
(b) $z = 1 + i$
(c) $z = -0.2i$

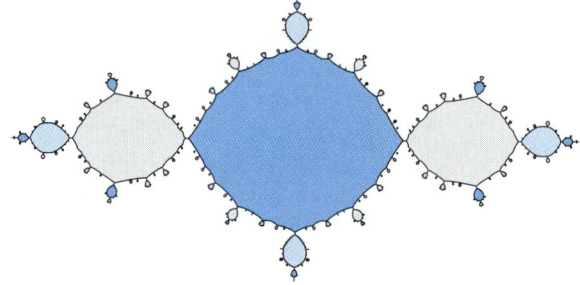

3. *Aerial Photography* Aerial photography from satellites and planes has become important to many applications such as map-making, national security, and surveying. If a photograph is taken from a plane so that the camera is tilted at an angle θ, then trigonometry may be used to find the ground coordinates of an object as illustrated in the accompanying figure on the following page. If an object's photographic coordinates in inches are (x, y), then its ground coordinates (X, Y) in feet can be computed using the formulas

$$X = \frac{ax}{f \sec \theta - y \sin \theta},$$

$$Y = \frac{ay \cos \theta}{f \sec \theta - y \sin \theta},$$

where f is the focal length of the camera in inches and a is the altitude in feet of the airplane. Suppose the photographic coordinates of a house and nearby forest fire are $(x_H, y_H) = (0.9, 3.5)$ and $(x_F, y_F) = (2.1, -2.4)$, respectively. (Source: F. Moffitt, *Photogrammetry*.)

(a) Find the distance in inches between the house and fire on the photograph.

(b) If the photograph was taken at 7400 feet by a camera with a focal length of 6 inches and a tilt of $\theta = 4.1°$, find the ground distance in feet between the house and the fire.

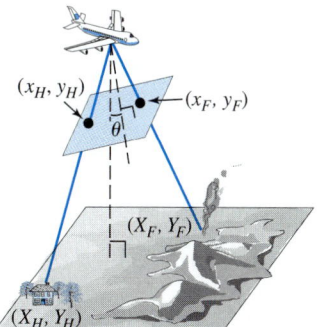

4. *Aerial Photographs and Surveying* To locate important points in aerial photographs, surveyors use basic control monuments established by both the U.S. Coast and Geodetic Survey and the U.S. Geological Survey. These monuments have published *xy*-coordinates called *state plane coordinates*. Using these monuments, coordinates of other important points can be determined. Two basic control monuments A and B have *xy*-coordinates of $x_A = 2{,}101{,}345.1$, $y_A = 998{,}764.3$ and $x_B = 2{,}131{,}667.8$, $y_B = 923{,}541.7$, respectively, where units are in feet. The coordinates of an unknown point P are needed. If angles PAB and PBA are measured as $37°41' \ 37''$ and $57°52' \ 04''$, respectively, determine the state plane coordinates of P. See the accompanying figure. (Source: F. Moffitt.)

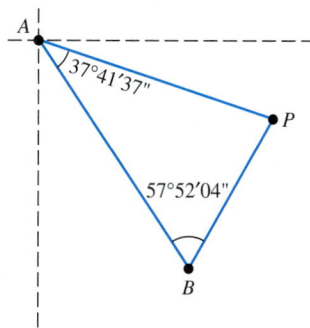

5. *Shadows in Computer Graphics* Vectors are used frequently in computer graphics to simulate realis-

tic shadows. For example, suppose an airplane is taking off from a runway as illustrated in the figure. Let the length and direction of an airplane at take-off be given by vector **L**. If the sunlight is assumed to be perpendicular to the runway, then the length of the airplane's shadow cast on the runway equals $\|\mathbf{L}\| \cos \theta$. From previous work we know that if vector **R** points in the direction of the runway, then

$$\mathbf{L} \cdot \mathbf{R} = \|\mathbf{L}\|\|\mathbf{R}\| \cos \theta.$$

Solving for $\|\mathbf{L}\| \cos \theta$ results in

$$\|\mathbf{L}\| \cos \theta = \frac{\mathbf{L} \cdot \mathbf{R}}{\|\mathbf{R}\|}.$$

The expression $\dfrac{\mathbf{L} \cdot \mathbf{R}}{\|\mathbf{R}\|}$ represents the **component of L in the direction of R.** Find the length of the shadow on the runway for each **L** and **R**. Assume units are in feet. (Source: C. Pokorny and C. Gerald, *Computer Graphics*.)

(a) $\mathbf{L} = 40\mathbf{i} + 10\mathbf{j},$ $\mathbf{R} = \mathbf{i}$
(b) $\mathbf{L} = 35\mathbf{i} + 5\mathbf{j},$ $\mathbf{R} = 10\mathbf{i} + \mathbf{j}$
(c) $\mathbf{L} = 100\mathbf{i} + 8\mathbf{j},$ $\mathbf{R} = 30\mathbf{i} + 2\mathbf{j}$

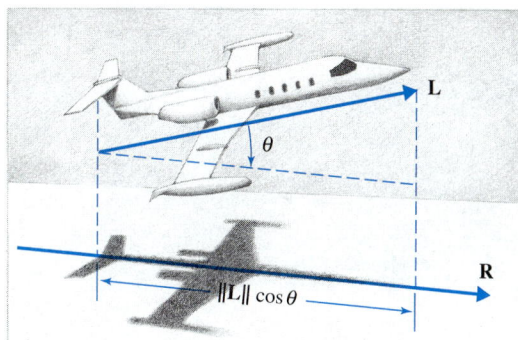

6. *The Dot Product*   In the accompanying figure

$$\mathbf{a} = \langle a_1, a_2 \rangle, \quad \mathbf{b} = \langle b_1, b_2 \rangle, \quad \text{and}$$
$$\mathbf{a} - \mathbf{b} = \langle a_1 - b_1, a_2 - b_2 \rangle.$$

Apply the law of cosines to the triangle and derive the equation

$$\mathbf{a} \cdot \mathbf{b} = \|\mathbf{a}\|\|\mathbf{b}\| \cos \theta.$$

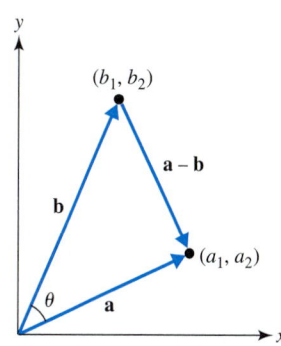

CHAPTER **8** *Systems of Equations and Inequalities*

*O*ne of the most important inventions of the twentieth century was the electronic digital computer. In 1940 John Atanasoff, a physicist from Iowa State University, needed to find the solution to a system of 29 equations. Since this task was too difficult to solve by hand, it led him to invent the first fully electronic digital computer. Today modern supercomputers compute billions of arithmetic operations in a single second and solve over 600,000 equations simultaneously. The need to solve a *mathematical problem* resulted in one of the most profound inventions of our time.

A recent breakthrough in visualizing our world is the invention of digital photography. By the year 2006, TV stations may be required by the Federal Communications Commission (FCC) to transmit digital signals. Digital pictures are represented by numbers rather than film. They are crystal clear without interference. One of the first digital pictures was done by James Blinn of NASA in 1981. It was a simulation of the Voyager-Saturn flyby and required substantial mathematics to create.

Mathematics plays a central role in new technology. In this chapter mathematics is used to solve systems of equations, represent digital pictures, compute movement in computer graphics, and even represent color on computer monitors. Throughout history,

> *T*he future belongs to those who believe in the beauty of their dreams.
>
> — **Eleanor Roosevelt**

many important discoveries and inventions have been based on the creative insights of a few people. These insights have had a profound impact on our society.

References: A. Tucker, *Fundamentals of Computing I; USA Today;* NASA/JPL.

8.1 Functions and Equations of More Than One Variable

Representations of Functions of More Than One Input • Equations of More Than One Variable • The Method of Substitution • Graphical and Numerical Methods • Joint Variation (Optional)

Introduction

Many quantities in everyday life are dependent on more than one variable. Calculating the area of a rectangular room requires both its length and width. The heat index is a function of temperature and humidity, while windchill is determined using temperature and wind speed. Information about grades and credit hours is necessary to compute a grade point average.

In earlier chapters we discussed functions of one input. Quantities determined by more than one variable often are computed by a function of more than one input. The mathematical concepts concerning functions of one input also apply to functions of more than one input. Functions of more than one input can be computed by algorithms and have verbal, graphical, numerical, and symbolic representations. One unifying concept about any function is that it produces *at most one output* each time it is evaluated.

Representations of Functions of More than One Input

The arithmetic operations of addition, subtraction, multiplication, and division are computed by functions of two inputs. In order to perform addition, two numbers must be provided. The addition of x and y results in one output z. The addition function f can be represented symbolically by $f(x, y) = x + y$, where $z = f(x, y)$. For example, the addition of 3 and 4 can be written as $z = f(3, 4) = 3 + 4 = 7$. In this case, $f(x, y)$ is a **function of two inputs** or a **function of two variables.** The **independent variables** are x and y, while z is the **dependent variable.** The output z depends on the inputs x and y. In a similar manner, a division function can be defined by $g(x, y) = \dfrac{x}{y}$, where $z = g(x, y)$.

MAKING CONNECTIONS
Independent and Dependent Variables

In Chapters 1–4 the input for a function f was usually represented by x and the output by y. This was expressed as $y = f(x)$. For functions of two inputs, it is common to use x and y as inputs and z as the output. This is expressed as $z = f(x, y)$.

The arithmetic operations involve three variables: inputs x, y, and output z. Functions of more than one input have verbal, numerical, symbolic, and graphical representations. As an example, we consider the function that computes the average of two numbers.

Verbal Representations. To find the average of two numbers x and y, we compute their sum and divide the result by 2. This is a verbal description of the function f that calculates the average of two numbers. An algorithm can also be used to describe f.

ALGORITHM 8.1 Finding the average of two numbers
Step 1: Input two numbers, x and y.
Step 2: Compute the sum of x and y.
Step 3: Divide the result in STEP 2 by 2.
Step 4: Let the result in STEP 3 be z.
Step 5: Output z, the average of x and y.

This algorithm receives two inputs x and y and computes their average z. Each computation can be expressed as the ordered triple (x, y, z).

Numerical Representations. Data involving three variables can be represented by a three-column list. A partial numerical representation of the averaging function f is given in Table 8.1, where z is the average of x and y.

TABLE 8.1

x	y	z
1	1	1
1	2	1.5
1	3	2
2	1	1.5
2	2	2
2	3	2.5

This data is in the form of ordered triples (x, y, z), such as $(1, 1, 1)$ and $(1, 3, 2)$. Because of the many different possibilities for x and y, a three-column table quickly becomes unmanageable. A slightly more efficient way to represent this data is to use a chart. See Table 8.2. To find the average of $x = 3$ and $y = 5$, locate 3 in the left column and 5 in the top row. The intersection of this row and column gives $z = 4$, which is the average of 3 and 5.

TABLE 8.2 Average of x and y

$x \backslash y$	1	2	3	4	5
1	1.0	1.5	2.0	2.5	**3.0**
2	1.5	2.0	2.5	3.0	**3.5**
3	**2.0**	**2.5**	**3.0**	**3.5**	4.0
4	2.5	3.0	3.5	4.0	4.5
5	3.0	3.5	4.0	4.5	5.0

Symbolic Representations. The averaging function of two numbers could be represented symbolically by $f(x, y) = \dfrac{x + y}{2}$, where $z = f(x, y)$. To compute the average of $x = 6$ and $y = 12$, evaluate

$$f(6, 12) = \frac{6 + 12}{2} = 9.$$

This result could be expressed by the ordered triple $(6, 12, 9)$.

Graphical Representations. To graphically represent one-variable data, the x-axis or real number line is used. To represent two-variable data in the form (x, y), the x- and y-axes are used. Graphing three-variable data in the form (x, y, z) requires three axes, which often are labeled x, y, and z. See Figure 8.1. It can be difficult to graph three-variable data by hand. However, technology is capable of creating excellent three-dimensional graphs. Figure 8.2 shows a three-dimensional graph of the averaging function f. The average z of x and y is shown by the vertical height on the z-axis. Notice that the graph of f is a flat surface or plane.

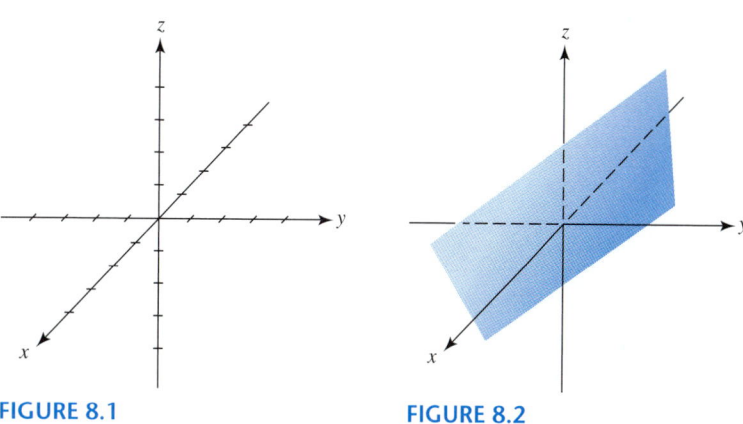

FIGURE 8.1   FIGURE 8.2

EXAMPLE 1   *Evaluating a function of more than one input*

For each function evaluate the expression and interpret the result.
(a) $f(3, -4)$, where $f(x, y) = xy$ represents the multiplication function
(b) $f(120, 5)$, where $f(m, g) = \dfrac{m}{g}$ computes the gas mileage when traveling m miles on g gallons of gasoline
(c) $f(0.5, 2)$, where $f(r, h) = \pi r^2 h$ calculates the volume of a cylindrical barrel with a radius r and height h

Solution
(a) $f(3, -4) = (3)(-4) = -12$. The product of 3 and -4 is -12.
(b) $f(120, 5) = \dfrac{120}{5} = 24$. If a car travels 120 miles on 5 gallons of gasoline, its mileage is 24 miles per gallon.
(c) $f(0.5, 2) = \pi(0.5^2)(2) = 0.5\pi \approx 1.57$. If a barrel has a radius of 0.5 foot and a height of 2 feet, it holds about 1.57 cubic feet of liquid. ■

The surface area of the skin covering the human body is a function of more than one variable. A taller person tends to have a larger surface area, as does a heavier person. Both height and weight influence the surface area of a person's body. The following example demonstrates a function that can be used to estimate this surface area.

EXAMPLE 2   *Estimating the surface area of the human body*

A formula to determine the surface area of a person's body in square meters is computed by $S(w, h) = 0.007184(w^{0.425})(h^{0.725})$, where w is weight in kilograms and h is height in centimeters. Use S to estimate the surface area of a human body that is 65 inches tall and weighs 154 pounds. (Source: H. Lancaster, *Quantitative Methods in Biological and Medical Sciences.*)

Solution

First, convert 65 inches to centimeters. There are approximately 2.54 centimeters in 1 inch, so this person's height is $65 \cdot 2.54 = 165.1$ centimeters. There are about 2.2 pounds in 1 kilogram, so the individual weighs $\dfrac{154}{2.2} = 70$ kilograms.

$$S(70, 165.1) = 0.007184(70^{0.425})(165.1^{0.725}) \approx 1.77$$

Thus, a person weighing 70 kilograms (154 pounds) with a height of 165.1 centimeters (65 inches) has a surface area of approximately 1.77 square meters. ■

Equations of More than One Variable

Any linear equation in one variable can be written as $f(x) = k$, where f is a linear function and k is a constant. Similarly, any nonlinear equation in one variable can be written as $g(x) = k$, where g is a nonlinear function. This mathematical concept can be extended to functions and equations of more than one variable.

A **linear function of two inputs** has the form $f(x, y) = ax + by + c$, where $a, b,$ and c are constants. The subtraction function, represented by $f(x, y) = x - y$, is a linear function with $a = 1, b = -1,$ and $c = 0$. The division function, given by $g(x, y) = \dfrac{x}{y} = xy^{-1}$, is not a linear function.

Any **linear equation in two variables** can be written in the form $ax + by = k$, where a and b are not both equal to zero and k is a constant. A linear equation in two variables can be written in the form $f(x, y) = k$, where f is a linear function of two inputs. Since

$$f(x, y) = \frac{x + y}{2} = \frac{1}{2}x + \frac{1}{2}y,$$

the averaging function f is linear. To find two numbers whose average is 10, we can solve the linear equation

$$f(x, y) = \frac{x + y}{2} = 10.$$

The numbers 5 and 15 average to 10, as do the numbers 0 and 20. In fact, any pair of numbers whose sum is 20 will satisfy this equation. There are an infinite number of solutions to this linear equation. For a unique solution, another restriction must be placed on the variables x and y.

Many situations involving two variables result in the need to determine values for x and y that satisfy two different equations. For example, suppose that we would like to find a pair of numbers whose average is 10, and whose difference is 2. The function $f(x, y) = \dfrac{x + y}{2}$ calculates the average of two numbers and $g(x, y) = x - y$ computes their difference. The solution could be found by solving two linear equations.

$$f(x, y) = 10$$

$$g(x, y) = 2$$

These equations can be written as follows.

$$\frac{x + y}{2} = 10$$

$$x - y = 2$$

This pair of equations is called a **system of linear equations** because we are solving more than one linear equation at once. A *solution* to a system of equations consists of an x-value and a y-value that satisfy *both* equations simultaneously. The set of all solutions is called the *solution set*. Using trial and error, we see that $x = 11$ and $y = 9$ satisfy both equations. This is the only solution and can be expressed as the ordered pair $(11, 9)$.

Critical Thinking

A linear function f of one input x is given by $f(x) = ax + b$. A linear function g of two inputs x and y is given by $g(x, y) = ax + by + c$. Define a linear function h, having three inputs x, y, and z. Give an example.

EXAMPLE 3 *Determining a system of equations*

In 1996 hamburger sales totaled $23.9 billion at McDonald's and Burger King. McDonald's sales exceeded Burger King's by $8.9 billion. (Source: Technomic.)
(a) Write a system of equations involving two variables, whose solution gives the individual sales of each company.
(b) Is the resulting system of equations linear or nonlinear?

Solution

(a) When setting up a system of equations, it is essential that we identify what each variable represents. Let x represent the hamburger sales of McDonald's and y the hamburger sales of Burger King. Their combined sales are $23.9 billion, so $x + y = 23.9$. McDonald's sales exceed Burger King's by $8.9 billion, so $x - y = 8.9$. Thus, the system of equations is as follows.

$$x + y = 23.9$$

$$x - y = 8.9$$

(b) Both equations are written in the form $ax + by = k$. Therefore, it is a system of linear equations. ∎

Systems of equations that are not linear are called **nonlinear systems of equations.** The next example shows how nonlinear functions can lead to a system of nonlinear equations.

EXAMPLE 4 *Determining a nonlinear system of equations*

The volume V of a cylindrical container with a radius r and height h is computed by $V(r, h) = \pi r^2 h$. See Figure 8.3. The lateral surface area S of the container, excluding the circular top and bottom, is computed by $S(r, h) = 2\pi rh$.

(a) Write a system of equations, whose solution is the dimensions for a cylinder with a volume of 38 cubic inches and a lateral surface area of 63 square inches.

(b) Determine if $r = 1.5$ and $h = 4$ represent a solution.

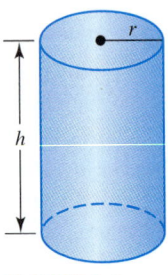

FIGURE 8.3

Solution

(a) The equations $V(r, h) = 38$ and $S(r, h) = 63$ must be satisfied. This results in the following nonlinear system.

$$\pi r^2 h = 38$$
$$2\pi rh = 63$$

(b) To determine if $r = 1.5$ and $h = 4$ represent a solution, substitute their values into V and S.

$$V(1.5, 4) = \pi(1.5^2)(4) = 9\pi \approx 28.3 \neq 38$$
$$S(1.5, 4) = 2\pi(1.5)(4) = 12\pi \approx 37.7 \neq 63$$

Thus, $r = 1.5$ and $h = 4$ do not represent a solution.   ∎

The Method of Substitution

The next example describes the *method of substitution.* It often is used to solve systems of equations involving two variables.

EXAMPLE 5 *Solving a linear system of equations*

Use the method of substitution to solve the system of equations in Example 3. Interpret the result.

Solution

In this system x represents 1996 hamburger sales for McDonald's in billions of dollars, while y represents the corresponding sales for Burger King.

$$x + y = 23.9$$
$$x - y = 8.9$$

Begin by solving an equation for a convenient variable. Then substitute the result into the other equation. For example, if we solve the first equation for y, the result is $y = 23.9 - x$. Then, substitute $(23.9 - x)$ into the second equation for y.

$$x - (23.9 - x) = 8.9$$

This equation can be solved for x.

$$
\begin{aligned}
x - (23.9 - x) &= 8.9 \\
x - 23.9 + x &= 8.9 && \text{Distributive property} \\
2x &= 32.8 && \text{Combine } x\text{-terms and add 23.9.} \\
x &= 16.4 && \text{Divide by 2.}
\end{aligned}
$$

Thus, McDonald's sold \$16.4 billion in hamburgers in 1996. Since $y = 23.9 - x$, it follows that $y = 23.9 - 16.4 = 7.5$. Burger King's share of the market was \$7.5 billion. ∎

Nonlinear systems also can be solved using the method of substitution.

EXAMPLE 6 *Solving a nonlinear system of equations*

A circle with radius r, centered at the origin, has an equation $x^2 + y^2 = r^2$. Use the method of substitution to determine the points where the graph of $y = 2x$ intersects this circle when $r = \sqrt{5}$. Sketch a graph that illustrates the solutions.

Solution

Since $r^2 = 5$, solve the following system.

$$x^2 + y^2 = 5$$
$$y = 2x$$

Substitute $y = 2x$ into the first equation.

$$
\begin{aligned}
x^2 + y^2 &= 5 \\
x^2 + (2x)^2 &= 5 && y = 2x \\
x^2 + 4x^2 &= 5 && \text{Square the expression.} \\
5x^2 &= 5 && \text{Add like terms.} \\
x^2 &= 1 && \text{Divide by 5.} \\
x &= \pm 1 && \text{Square root property}
\end{aligned}
$$

Since $y = 2x$ we see that when $x = 1$, $y = 2$ and when $x = -1$, $y = -2$. The graphs of $x^2 + y^2 = 5$ and $y = 2x$ intersect at the points $(1, 2)$ and $(-1, -2)$. This nonlinear system has two solutions, which are shown in Figure 8.4 on the next page.

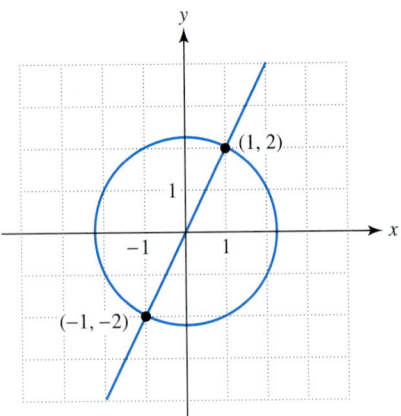

FIGURE 8.4 ■

Graphical and Numerical Methods

The next example illustrates how a linear system with two variables can be solved graphically, numerically, and symbolically.

EXAMPLE 7 *Modeling roof trusses*

Linear systems occur in the design of roof trusses for homes and buildings. See Figure 8.5. One of the simplest types of roof trusses is an equilateral triangle. If a 200-pound force is applied to the peak of a truss as shown in Figure 8.6, then the weights W_1 and W_2 exerted on each rafter of the truss are determined by the following system of linear equations. (Source: R. Hibbeler, *Structural Analysis.*)

$$W_1 - W_2 = 0$$

$$\frac{\sqrt{3}}{2}(W_1 + W_2) = 200$$

(a) Estimate the solution graphically and numerically.
(b) Check your solution by solving the system symbolically.

FIGURE 8.5

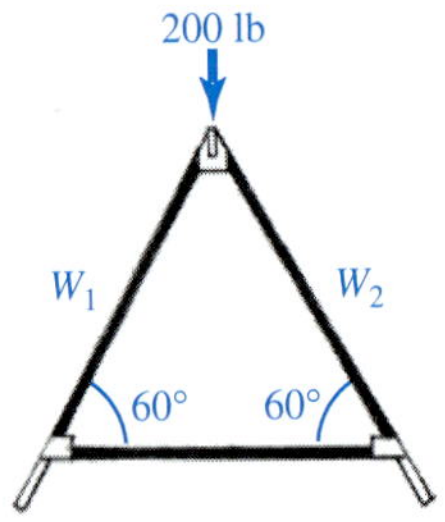

FIGURE 8.6

Solution

(a) *Graphical Solution* Begin by solving each equation for the variable W_2.

$$W_2 = W_1$$

$$W_2 = \frac{400}{\sqrt{3}} - W_1$$

Graph the equations $Y_1 = X$ and $Y_2 = (400/\sqrt{3}) - X$. Their graphs intersect near the point $(115.47, 115.47)$ as shown in Figure 8.7. This means that each rafter supports a weight of approximately 115 pounds. (*Remark: W_1 and W_2* represent the forces parallel to the rafters. Most of the weight is pushing downward, while a smaller portion is attempting to spread the two rafters apart.)

Numerical Solution Numerical support is shown in Figure 8.8, where $Y_1 \approx Y_2$ when $x = 115$.

[0, 200, 50] by [0, 200, 50]

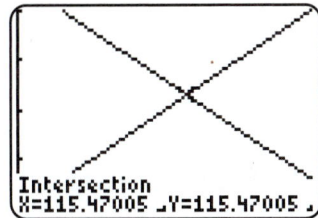

FIGURE 8.7　　　　　**FIGURE 8.8**

(b) *Symbolic Solution* We can solve this system using the method of substitution. From the first equation we see that $W_1 = W_2$.

$$\frac{\sqrt{3}}{2}(W_1 + W_2) = 200 \qquad \text{Second equation}$$

$$\frac{\sqrt{3}}{2}(W_1 + W_1) = 200 \qquad \text{Substitute } W_2 = W_1.$$

$$\sqrt{3}W_1 = 200 \qquad \text{Simplify.}$$

$$W_1 = \frac{200}{\sqrt{3}} \qquad \text{Divide by } \sqrt{3}.$$

Thus, $W_2 = W_1 = \dfrac{200}{\sqrt{3}} \approx 115.47$ lb. This verifies the results from part (a).　■

EXAMPLE 8　*Solving a nonlinear system graphically and numerically*

Determine the radius and height of the cylindrical container in Example 4 graphically. Support the result numerically.

Solution

Graphical Solution　The system of nonlinear equations is the following.

$$\pi r^2 h = 38$$

$$2\pi r h = 63$$

To find the solution graphically, we can solve each equation for h, and then apply the intersection-of-graphs method.

$$h = \frac{38}{\pi r^2}$$

$$h = \frac{63}{2\pi r}$$

Let r correspond to x and h to y. Graph $Y_1 = 38/(\pi X^2)$ and $Y_2 = 63/(2\pi X)$. Their graphs intersect near the point $(1.206, 8.312)$ as shown in Figure 8.9. Therefore, a cylinder with a radius of $r \approx 1.206$ inches and height of $h \approx 8.312$ inches has a volume of 38 cubic inches and lateral surface area of 63 square inches.

Numerical Solution   Numerical support is shown in Figure 8.10, where $Y_1 \approx Y_2$ when $x = 1.2$.

[0, 4, 1] by [0, 20, 5]

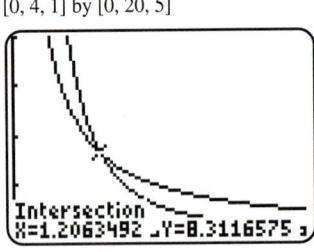

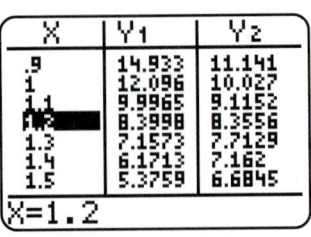

FIGURE 8.9   **FIGURE 8.10**   ■

Joint Variation (Optional)

A quantity may depend on more than one variable. For example, the volume V of a cone is given by $V = \frac{1}{3}\pi r^2 h$. We say that V *varies jointly* as h and the square of r. The *constant of variation* is $\frac{1}{3}\pi$.

Joint variation

Let m and n be real numbers. Then z **varies jointly** as the nth power of x and the mth power of y if a nonzero real number k exists such that

$$z = kx^n y^m.$$

EXAMPLE 9   *Modeling the amount of wood in a tree*

In forestry it is common to estimate the volume of timber in a given area of forest. To do this, formulas are developed to find the amount of wood contained in a tree with height h and diameter d. One study concluded that the volume V of wood in a tree varies jointly as the 1.12 power of h and the 1.98 power of d. (The diameter is measured 4.5 feet above the ground.) (Source: B. Ryan, B. Joiner, and T. Ryan, *Minitab Handbook.*)

(a) Write an equation that relates V, h, and d.

(b) A tree with a 13.8-inch diameter and a 64-foot height has a volume of 25.14 cubic feet. Estimate the constant of variation k.

(c) Estimate the volume of wood in a tree with $d = 11$ inches and $h = 47$ feet.

Solution

(a) $V = kh^{1.12}d^{1.98}$, where k is the constant of variation.

(b) Substitute $d = 13.8$, $h = 64$, and $V = 25.14$ into the equation and solve for k.

$$25.14 = k(64^{1.12})(13.8^{1.98})$$

$$k = \frac{25.14}{(64^{1.12})(13.8^{1.98})} \approx 0.00132$$

Thus, let $V = 0.00132h^{1.12}d^{1.98}$.

(c) $V = 0.00132(47^{1.12})(11^{1.98}) \approx 11.4$ cubic feet.   ∎

8.1 PUTTING IT ALL TOGETHER

Many mathematical concepts that were presented in earlier chapters can be applied to functions and equations of more than one variable. Like functions of one input, functions of more than one input can be either linear or nonlinear. They have verbal, numerical, graphical, and symbolic representations. Functions of more than one variable often lead to equations involving more than one variable. Systems of equations can be solved symbolically, graphically, or numerically.

The following table summarizes some mathematical concepts involved with functions and equations of two variables.

Concept	Comments	Example
Function of two inputs or variables	$z = f(x, y)$, where x and y are inputs and z is the output.	$f(x, y) = x^2 + 5y$ $f(2, 3) = 2^2 + 5(3) = 19$
Linear function of two inputs or variables	$f(x, y) = ax + by + c$, where a, b, and c are constants.	$f(x, y) = 3x + 2y - 1$ $f(4, -3) = 3(4) + 2(-3) - 1 = 5$
Linear system of two equations	Equations written as $ax + by = k$.	$2x - 3y = 6$ $5x + 4y = -8$
Nonlinear system of two equations	A system of equations that is not linear is nonlinear.	$5x^2 - 4xy = 20$ $\dfrac{5}{x} - 2y = 18$
Method of substitution	Solve one equation for a variable and substitute the result into the second equation.	$x - y = 1$ $x + y = 5$ If $x - y = 1$, then $x = 1 + y$. Substitute this into the second equation, $(1 + y) + y = 5$. This results in $y = 2$ and $x = 3$.
Graphical method for two equations	Solve both equations for the same variable. Then apply the intersection-of-graphs method.	If $x + y = 3$ then $y = 3 - x$. If $4x - y = 2$ then $y = 4x - 2$. Graph and locate the point of intersection at $(1, 2)$.

8.1 EXERCISES

   Tape 6

Functions of More than One Input

Exercises 1–4: Evaluate f for the indicated inputs and interpret the result.

1. $f(5, 8)$, where $f(b, h) = \frac{1}{2}bh$ (f computes the area of a triangle with base b and height h.)

2. $f(2, 6)$, where $f(r, h) = \frac{1}{3}\pi r^2 h$ (f computes the volume of a cone with radius r and height h.)

3. $f(20, 35)$, where $f(w, l) = wl$ (f computes the area of a rectangle with width w and length l.)

4. $f(5, 9, 4)$, where $f(x, y, z) = \frac{x + y + z}{3}$ (f computes the average of three numbers x, y, and z.)

Exercises 5 and 6: (Refer to Example 2.) Estimate the surface area of a human body with weight w and height h.

5. $w = 132$ pounds, $h = 62$ inches

6. $w = 220$ pounds, $h = 75$ inches

Exercises 7 and 8: Write an algorithm that computes the specified quantity for the given inputs.

7. The time T that it takes to travel x miles in a car moving at y miles per hour

8. The volume V of a rectangular box with dimensions x, y, and z

Exercises 9–12: Write the symbolic representation for f(x, y), if the function f computes the following quantity.

9. The sum of y and twice x

10. The product of x^2 and y^2

11. The product of x and y divided by $1 + x$

12. The square root of the sum of x and y

Exercises 13–16: Life Expectancy The table lists life expectancy by birth year and sex. Let this table be a partial numerical representation of a function f, where $f(x, y)$ computes the life expectancy at birth of someone born in the year x whose sex is y. (Source: Department of Health and Human Services.)

Year	Male	Female
1920	53.6	54.6
1940	60.8	65.2
1960	66.6	73.1
1980	70.0	77.4
1993	72.1	78.9

13. Evaluate $f(1960, \text{Male})$. Interpret the result.

14. Evaluate $f(1993, \text{Female}) - f(1993, \text{Male})$. Interpret the result.

15. Estimate the value of $f(1970, \text{Female})$.

16. What other variables might affect life expectancy?

Exercises 17–20: Degrees Awarded The table represents a function f, where $f(x, y)$ computes the number of degrees awarded in year x of type y. Let B denote a bachelor's degree, M a master's degree, and D a doctorate degree. (Source: Department of Education.)

Year	Bachelor's	Master's	Doctorate
1970	792,317	208,291	29,912
1980	929,417	298,081	32,615
1990	1,051,344	324,301	38,371
1996	1,198,000	393,000	43,000

17. Evaluate $f(1970, B)$. Interpret the result.

18. Evaluate $\dfrac{f(1996, B)}{f(1996, M) + f(1996, D)}$. Interpret the result.

19. Estimate the value of $f(1985, B)$.

20. Assume that the table is a complete numerical representation of a function f. Does $f(x, y)$ increase as x increases for a fixed y? Interpret your answer.

Exercises 21 and 22: The area of a rectangle with length l and width w is computed by $A(l, w) = lw$, while its perimeter is calculated by $P(l, w) = 2l + 2w$. Use the method of substitution to solve the system of equations for l and w. Interpret the solution.

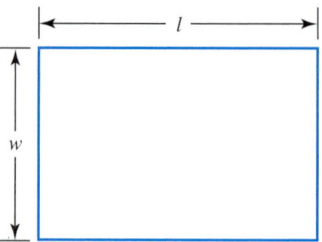

21. $A(l, w) = 35$
    $P(l, w) = 24$

22. $A(l, w) = 300$
    $P(l, w) = 70$

Exercises 23–28: Colors on Computer Monitors Although it might not seem possible to describe color using mathematics, it is done every day on computer screens. Colored light can be broken down into three basic colors: red, green, and blue. Some computer monitors are capable of creating 256 different intensities for each of these three colors, which may be numbered from 0 to 255. The number 0 indicates the absence of a color, while 255 represents the brightest intensity of a color. The color displayed on a screen can be computed by $f(r, g, b)$, where r represents the brightness of red light, g of green light, and b of blue light. The expression $f(255, 0, 0)$ indicates the brightest intensity of red with no green or blue light. Therefore, $f(255, 0, 0) = red$. In a similar manner, $f(0, 255, 0) = green$ and $f(0, 0, 255) = blue$. By mixing intensities of these three basic colors, thousands of colors can be generated. For example, $f(255, 255, 0) = yellow$, since red and green light combine to make yellow light. (Source: I. Kerlow, The Art of 3-D Computer Animation and Imaging.)

23. Conjecture what color $f(255, 0, 255)$ represents.

24. Conjecture what inputs create white light. (*Hint:* White light is the presence of all colors.)

25. Turquoise is a greenish-blue color. Conjecture inputs that create turquoise light.

26. Conjecture whether $f(141, 43, 17)$ represents rust or cream color.

27. If there are 256 different intensities for each basic color, how many different colors can be created on a computer screen by mixing different intensities of these colors?

28. Some computers are capable of generating 65,536 different intensities for each of the three basic colors. Approximate the number of different colors that could be generated by this computer.

Systems of Equations

Exercises 29–32: Determine which ordered pairs are solutions to the given system of equations.

29. $(2, 1), (-2, 1), (1, 0)$
$2x + y = 5$
$x + y = 3$

30. $(3, 2), (3, -4), (5, 0)$
$x - y = 5$
$2x + y = 10$

31. $(4, -3), (0, 5), (4, 3)$
$x^2 + y^2 = 25$
$2x + 3y = -1$

32. $(4, 8), (8, 4), (-4, -8)$
$xy = 32$
$x + y = 12$

Exercises 33 and 34: The figure shows the graphs of a system of two linear equations. Use the graph to estimate the solution to this system of equations. Then solve the system symbolically.

33.

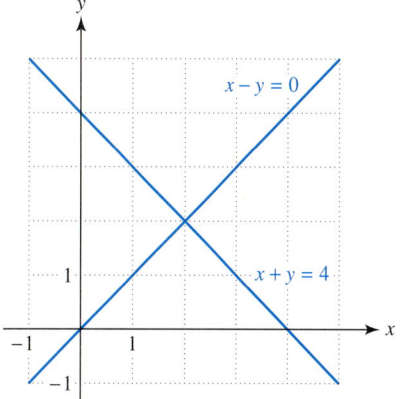

34.

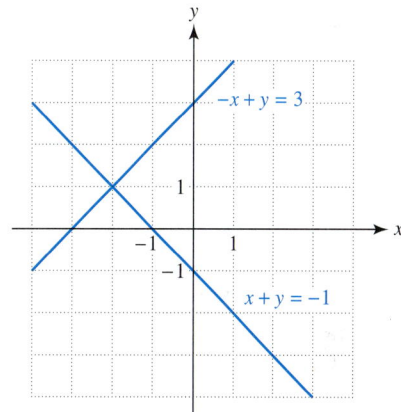

Exercises 35–38: Solve the system of linear equations
(a) *graphically,*
(b) *numerically, and*
(c) *symbolically.*

35. $2x + y = 1$
$x - 2y = 3$

36. $3x + 2y = -2$
$2x - y = -6$

37. $-2x + y = 0$
$7x - 2y = 3$

38. $x - 4y = 15$
$3x - 2y = 15$

Exercises 39–42: Solve the system of nonlinear equations
(a) *graphically,*
(b) *numerically, and*
(c) *symbolically.*

39. $x^2 + y^2 = 16$
$x - y = 0$

40. $x^2 - y = 1$
$3x + y = -1$

41. $xy = 63$
    $x - y = 2$

42. $x^2 + y^2 = 2$
    $x^2 - y = 0$

Exercises 43 and 44: Approximate any solutions to the system of nonlinear equations graphically.

43. $x^3 - 3x + y = 1$
    $x^2 + 2y = 3$

44. $x^2 + y = 5$
    $x + y^2 = 6$

Applications

45. *Bank Robberies* The total number of bank robberies in 1994 and 1995 was 13,787. From 1994 to 1995 the number of bank robberies declined by 271. (Source: Federal Bureau of Investigation.)
    (a) Write a system of equations whose solution represents the number of robberies committed in each of these years.
    (b) Solve the system graphically.
    (c) Solve the system symbolically.

46. *Medical Waste* In 1996 hospital incinerators caused significant toxic pollution. The two states producing the largest medical waste pollution were New York and California. Together they accounted for 185,000 tons. New York's total exceeded California's by 32,900 tons. (Source: Environmental Working Group.)
    (a) Set up a system of linear equations whose solution represents the incinerator pollution from each state.
    (b) Solve the system graphically.
    (c) Solve the system symbolically.

47. *Tourists* In 1996 a total of 6.2 million tourists visited Fort Lauderdale, Florida. There were four times as many American tourists as foreign tourists. Determine the number of foreign and American tourists that visited Fort Lauderdale in 1996. Round your answers to the nearest tenth of a million. (Source: Greater Fort Lauderdale Convention & Visitor Bureau.)

48. *Card Catalogs* Libraries are moving toward computer catalogs. From 1968 to 1996 the number of 3 × 5-inch cards sold to libraries by the Library of Congress declined by 78.19 million. The number of cards sold in 1996 was only 0.72% of the 1968 number. How many cards were sold in 1968 and in 1996? Round your answers to the nearest hundredth of a million. (Source: Library of Congress.)

49. (Refer to Examples 4 and 8.) Approximate graphically the radius and height of a cylindrical container with a volume of 50 cubic inches and a lateral surface area of 65 square inches.

50. (Refer to Examples 4 and 8.) Determine graphically if it is possible to construct a cylindrical container, including the top and bottom, with a volume of 38 cubic inches and a surface area of 38 square inches. (*Hint:* The surface area is $2\pi r^2 + 2\pi rh$.)

51. *Heart Rate* In one study the maximum heart rates of conditioned athletes were examined. A group of athletes was exercised to exhaustion. Let x represent an athlete's heart rate five seconds after stopping exercise and y this rate after ten seconds. It was found that the maximum heart rate H for these athletes satisfied the following two equations.

$$H = 0.491x + 0.468y + 11.2$$
$$H = -0.981x + 1.872y + 26.4$$

If an athlete had a maximum heart rate of $H = 180$, determine x and y graphically. Interpret your answer. (Source: V. Thomas, *Science and Sport.*)

52. *Heart Rate* Repeat the previous exercise for an athlete with a maximum heart rate of 195.

53. *Roof Truss* (Refer to Example 7.) The forces or weights W_1 and W_2 exerted on each rafter for the roof truss shown in the figure are determined by the system of linear equations. Solve the system graphically, numerically, and symbolically.

$$W_1 + \sqrt{2}W_2 = 300$$
$$\sqrt{3}W_1 - \sqrt{2}W_2 = 0$$

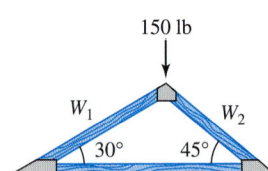

150 lb

W_1 W_2

30°   45°

54. *Height and Weight* The relationship between a professional basketball player's height h in inches and weight w in pounds was modeled using two different samples of players. The resulting equations that modeled each sample were $w = 7.46h - 374$ and $w = 7.93h - 405$. Assume that $65 \le h \le 85$.
    (a) Use each equation to predict the weight of a professional basketball player who is 6'11".

(b) Determine graphically the height where the two models give the same weight.

(c) For each model, what change in weight is associated with a one-inch increase in height?

Joint Variation

Exercises 55 and 56: Approximate the constant of variation to three significant digits.

55. The variable z varies jointly as the second power of x and the third power of y. When $x = 2$ and $y = 2.5$, $z = 31.9$.

56. The variable z varies jointly as the 1.5 power of x and the 2.1 power of y. When $x = 4$ and $y = 3.5$, $z = 397$.

57. *Wind Power*   The electrical power generated by a windmill varies jointly as the square of the diameter of the blades and the cube of the wind velocity. If a windmill with 8-foot blades and a 10 mile-per-hour wind generates 2405 watts, how much power would be generated if the blades were 6 feet and the wind was 20 miles per hour?

58. *Strength of a Beam*   The strength of a rectangular beam varies jointly as its width and the square of its thickness. If a beam 5.5 inches wide and 2.5 inches thick supports 600 pounds, how much can a similar beam that is 4 inches wide and 1.5 inches thick support?

59. *Volume of Wood*   (Refer to Example 9.) One cord of wood contains 128 cubic feet. Estimate the number of cords in a tree that is 105 feet tall and has a diameter of 38 inches.

60. *Carpeting*   The cost of carpet for a rectangular room varies jointly as its width and length. If a room 10 feet wide and 12 feet long costs $1560 to carpet, find the cost to carpet a room that is 11 feet by 23 feet. Interpret the constant of variation.

61. *Surface Area*   Use the results of Example 2 to find a formula for $S(w, h)$ that calculates the surface area of a person's body if w is given in pounds and h is given in inches.

62. *Surface Area*   Use the results of the previous exercise to solve Exercises 5 and 6.

Writing about Mathematics

1. Give an example of a quantity occurring in everyday life that can be computed by a function of more than one input. Identify the inputs and the output.

2. Give an example of a system of linear equations with two variables. Explain how to solve the system graphically, numerically, and symbolically.

8.2 Linear Systems of Equations and Inequalities in Two Variables

Types of Linear Systems in Two Variables • The Elimination Method
• Systems of Inequalities • Linear Programming (Optional)

Introduction

Large systems of equations and inequalities involving thousands of variables are solved by economists, scientists, engineers, and mathematicians every day. Computers and sophisticated numerical methods are essential for finding their solutions. In this section we focus on systems involving two variables. Graphical, numerical, and symbolic techniques can be applied to equations in two variables. Many of the concepts presented in this section apply to larger systems of linear equations.

Types of Linear Systems in Two Variables

A linear equation in two variables can be written in the form

$$ax + by = k$$

with a and b both not equal to zero. The graph of a linear equation in two variables is a line. A system of two linear equations in two variables can be represented graphically by two lines. Two lines in the xy-plane can intersect, coincide, or be parallel. See Figures 8.11–8.13. **Coincident lines** are the same line—they are identical and represent equivalent equations. A system of two linear equations in two variables that has at least one solution is **consistent.** Both intersecting and coincident lines represent consistent linear systems. An **inconsistent** system of equations has no solutions. Parallel lines represent an inconsistent system. These three situations are illustrated in the next examples.

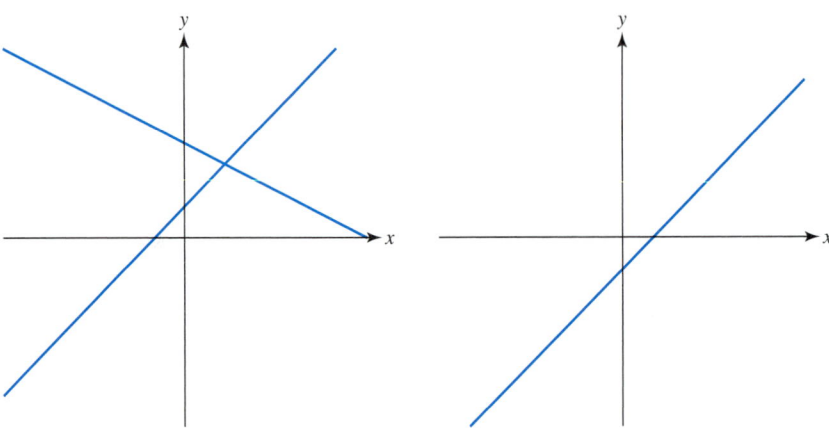

FIGURE 8.11 Intersecting Lines   FIGURE 8.12 Coincident Lines

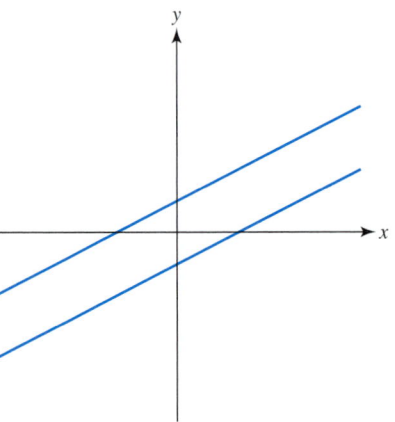

FIGURE 8.13 Parallel Lines

EXAMPLE 1   *Solving a consistent linear system*

Two groups of students go to a movie. The first group buys 2 soft drinks and 2 boxes of popcorn for $7, while the second group purchases 1 soft drink and 3 boxes of popcorn for $7.50. Determine the individual costs of a soft drink and a box of popcorn.

Solution

Let x represent the cost of a soft drink and y the cost of a box of popcorn. Since 2 soft drinks and 2 boxes of popcorn are $7, the equation $2x + 2y = 7$ must be satisfied. In a similar manner, $x + 3y = 7.5$ models the second purchase. Solving each equation for y results in the following.

$$y = \frac{7 - 2x}{2}$$

$$y = \frac{7.5 - x}{3}$$

The graphs of $Y_1 = (7 - 2X)/2$ and $Y_2 = (7.5 - X)/3$ are lines that intersect at $(1.5, 2)$ as shown in Figure 8.14. Thus, the cost for a soft drink is $1.50 and a box of popcorn is $2. The system of equations is consistent and has one solution, $(1.5, 2)$. This result can be checked as follows.

$$2(1.5) + 2(2) = 7\checkmark$$

$$1.5 + 3(2) = 7.5\checkmark$$

This verifies that $(1.5, 2)$ is a solution.

[0, 5, 1] by [0, 5, 1]

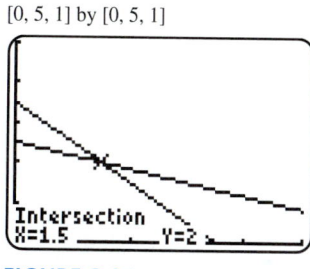

FIGURE 8.14   ∎

Not all consistent systems have a unique solution, as the next example demonstrates.

EXAMPLE 2   *Solving dependent equations*

Suppose the two groups of students in Example 1 go to a different movie theater. The first group buys 1 soft drink and 2 boxes of popcorn for $5, while the second group purchases 2 soft drinks and 4 boxes of popcorn for $10. If possible, determine the individual costs of each item.

Solution

The system of equations describing this situation is

$$x + 2y = 5$$
$$2x + 4y = 10.$$

Solving these equations for y gives

$$y = \frac{5 - x}{2}$$

$$y = \frac{10 - 2x}{4}.$$

When these two equations are graphed, only one line appears in the viewing rectangle. See Figure 8.15. The two lines are coincident—their equations are equivalent.

The two equations contain the same information. It is logical that 2 soft drinks and 4 boxes of popcorn would cost twice as much as 1 soft drink and 2 boxes of popcorn. For this reason, these equations are dependent. If x and y satisfy $x + 2y = 5$, then they also satisfy the equation $2x + 4y = 10$. There are an infinite number of solutions, which can be expressed as

$$\{(x, y) \mid x + 2y = 5\}.$$

For example, two solutions to the system of linear equations are $(1, 2)$ and $(2, 1.5)$. Soft drinks could be \$1 and popcorn \$2, or soft drinks could be \$2 and popcorn \$1.50. Both satisfy the conditions in the problem.

[0, 5, 1] by [0, 5, 1]

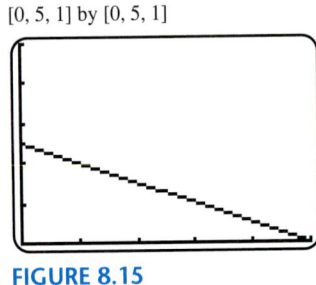

FIGURE 8.15   ∎

EXAMPLE 3 — *Recognizing an inconsistent linear system*

Now suppose the first group of students from Examples 1 and 2 bought 2 soft drinks and 3 boxes of popcorn for \$9 and the second group bought 4 soft drinks and 6 boxes of popcorn for \$16. If possible, determine the individual costs of each item.

Solution

With these conditions the system of equations becomes

$$2x + 3y = \ 9$$
$$4x + 6y = 16.$$

Solving each equation for y provides the following.

$$y = \frac{9 - 2x}{3}$$

$$y = \frac{16 - 4x}{6}$$

Their graphs are parallel lines with slope $-\frac{2}{3}$ as shown in Figure 8.16 on the next page. They do not intersect. This means that the pricing is *inconsistent*. To be consistent, 4 soft drinks and 6 boxes of popcorn should have twice the cost of 2 soft drinks and 3 boxes of popcorn. This system of linear equations is inconsistent—there are no solutions.

[0, 5, 1] by [0, 5, 1]

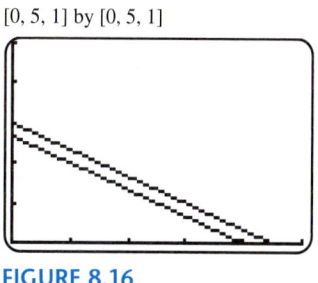

FIGURE 8.16

■

> ## Possible graphs of a system of two linear equations in two variables
>
> **1.** The graphs of the two equations are intersecting lines. The system is *consistent.* There is one solution, which is given by the coordinates of the point of intersection. In this case the equations are *independent.*
> **2.** The graphs of the two equations are the same line. The system is *consistent.* There is an infinite number of solutions, and the equations are *dependent.*
> **3.** The graphs of the two equations consist of distinct parallel lines. The system is *inconsistent.* There are no solutions.

The Elimination Method

The substitution method, presented in Section 8.1, is a symbolic method for solving small systems of equations. A second symbolic method is the **elimination method.** The next example demonstrates this technique.

EXAMPLE 4   *Using the elimination method*

Title IX is landmark legislation that prohibits sex discrimination in sports programs. In 1997 the national average spent on two varsity athletes, one female and one male, was $6050 for Division I-A schools. However, average expenditures for a male athlete exceeded those for a female athlete by $3900. Determine how much was spent per varsity athlete for each sex. Give graphical and numerical support. (Source: *USA Today.*)

Solution

Symbolic Solution   Let x represent average expenditures per male athlete and y average expenditures per female athlete in 1997. Since the average amount spent on one female and one male athlete was $6050, the equation $(x + y)/2 = 6050$ must be satisfied. The expenditures for a male athlete exceed a female athlete by $3900. Thus, $x - y = 3900$.

$$\frac{x + y}{2} = 6050$$

$$x - y = 3900$$

Multiply the first equation by 2 and then add the resulting equations.

$$x + y = 12,100$$
$$\underline{x - y = \quad 3900}$$
$$2x \quad = 16,000$$

By adding the equations, we have *eliminated* the y-variable. The solution to $2x = 16,000$ is $x = 8000$. Thus, the average expenditure per male athlete in 1997 was $8000. To determine y, we can substitute $x = 8000$ into either equation.

$$8000 - y = 3900 \qquad \text{implies} \qquad y = 4100.$$

The average amount spent on each female athlete was $4100.

Graphical Solution Solve each equation for y and graph $Y_1 = 12100 - X$ and $Y_2 = X - 3900$. Their graphs intersect at $(8000, 4100)$. See Figure 8.17. This agrees with the symbolic solution.

Numerical Solution Table Y_1 and Y_2. When $x = 8000$, $Y_1 = Y_2 = 4100$. See Figure 8.18.

[0, 15,000, 5000] by [0, 15,000, 5000]

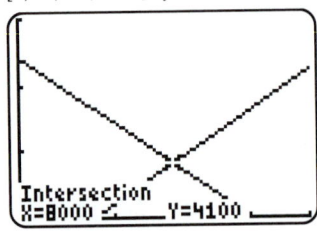

X	Y1	Y2
7700	4400	3800
7800	4300	3900
7900	4200	4000
8000	4100	4100
8100	4000	4200
8200	3900	4300
8300	3800	4400

X=8000

FIGURE 8.17   **FIGURE 8.18**   ■

Linear systems occur frequently when modeling data. For example, suppose that we would like to determine the equation of a line $y = ax + b$ that models a set of data. In this situation we can think of a and b as unknown constants, whose values must be found.

EXAMPLE 5 *Determining the equation of a line*

In 1980, average tuition and fees at four-year public colleges was $804 and in 1995 it was $2860. This growth can be modeled by $y = ax + b$, where y represents the tuition in the year x. Determine appropriate values for a and b. (Source: The College Board.)

Solution
When $x = 1980$, $y = 804$ and when $x = 1995$, $y = 2860$. Since $y = ax + b$, the following system of equations must be satisfied by a and b.

$$804 = a(1980) + b$$
$$2860 = a(1995) + b$$

Subtracting the equations eliminates b.

$$804 = 1980a + b$$
$$\underline{2860 = 1995a + b}$$
$$-2056 = -15a$$

Thus, $a = \dfrac{2056}{15}$. Next, solve the first equation for b and substitute for a.

$$b = 804 - 1980a$$
$$= 804 - 1980\left(\frac{2056}{15}\right)$$
$$= -270{,}588$$

The equation $y = \dfrac{2056}{15}x - 270{,}588$ models the data. In Figure 8.19 the two data points (1980, 804) and (1995, 2860) are graphed with this line. Numerical support is shown in Figure 8.20.

[1978, 1998, 2] by [0, 3000, 1000]

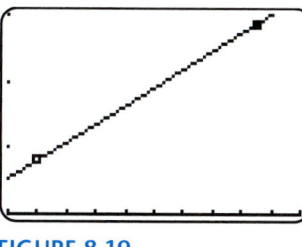

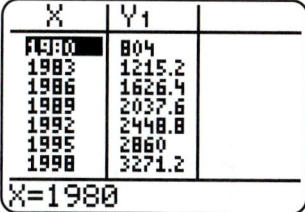

FIGURE 8.19   **FIGURE 8.20**   ■

MAKING CONNECTIONS
Equations of Lines

The equation of the line passing through (1980, 804) and (1995, 2860) also can be found using the point-slope form, $y = m(x - h) + k$.

EXAMPLE 6   *Using elimination on an inconsistent system*

Two trials that measure stopping distance are performed on a car traveling at 20 miles per hour. On the first trial the car takes 35 feet to stop, and on the second trial it requires 30 feet. Suppose we attempt to model this data using $f(x) = ax + b$, where x represents the speed of the car. Find a and b, if possible. (Source: F. Mannering and W. Kilareski, *Principles of Highway Engineering and Traffic Analysis.*)

Solution

Since $f(20) = 35$ and $f(20) = 30$, the system of equations is

$$a(20) + b = 35$$
$$a(20) + b = 30.$$

Rewriting and subtracting gives the following result.

$$\begin{aligned} 20a + b &= 35 \\ \underline{20a + b} &= \underline{30} \\ 0 &= 5 \end{aligned} \quad \text{(Contradiction)}$$

Since $0 \neq 5$, the equations are inconsistent—there are no solutions. The points (20, 35) and (20, 30) lie on a vertical line. See Figure 8.21 on the next page. This data cannot be modeled by a *function*.

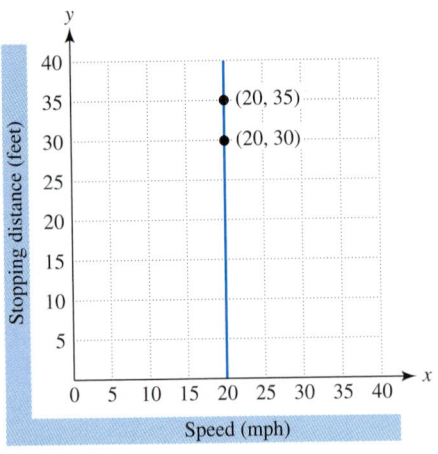

FIGURE 8.21 ■

Sometimes multiplication is performed before elimination is used, as illustrated in the next example.

EXAMPLE 7 *Multiplying before using elimination*

Solve each system of equations using elimination. Support your results graphically and numerically.

(a) $2x - 3y = 18$ (b) $5x + 10y = 10$
    $5x + 2y = 7$ $x + 2y = 2$

Solution

(a) If we multiply the first equation by 2 and the second equation by 3, then the y-coefficients become -6 and 6. Addition eliminates the y-variable.

$$4x - 6y = 36$$
$$\underline{15x + 6y = 21}$$
$$19x = 57 \quad \text{or} \quad x = 3$$

Substituting $x = 3$ into $2x - 3y = 18$ results in

$$2(3) - 3y = 18 \quad \text{or} \quad y = -4.$$

The solution is $(3, -4)$.

    To show graphical and numerical support let $Y_1 = (2X - 18)/3$ and $Y_2 = (7 - 5X)/2$. Their graphs intersect at $(3, -4)$. See Figure 8.22. Numerical support is shown in Figure 8.23.

$[-10, 10, 1]$ by $[-10, 10, 1]$

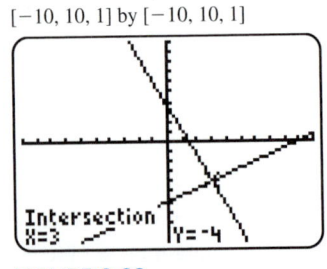

FIGURE 8.22

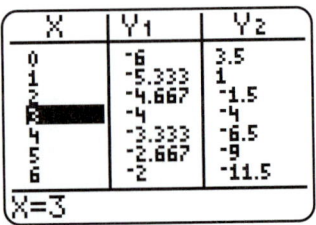

FIGURE 8.23

(b) If the second equation is multiplied by 5, subtraction eliminates both variables.

$$5x + 10y = 10$$
$$\underline{5x + 10y = 10}$$
$$0 = 0$$

The statement $0 = 0$ is always true. This means that the equations are dependent. There is an infinite number of solutions. The solution set is $\{(x, y) \,|\, x + 2y = 2\}$.

To give graphical and numerical support, let $Y_1 = (10 - 5X)/10$ and $Y_2 = (2 - X)/2$. Their graphs are identical. See Figure 8.24. Numerical support is shown in Figure 8.25, where $Y_1 = Y_2$.

$[-10, 10, 1]$ by $[-10, 10, 1]$

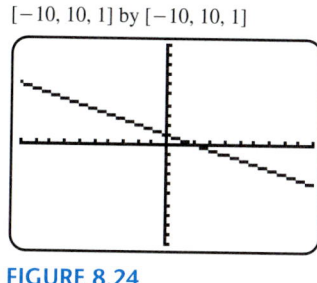

FIGURE 8.24

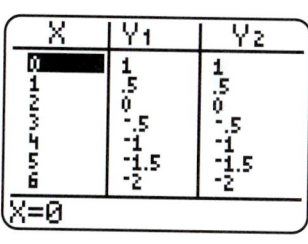

FIGURE 8.25   ∎

Systems of Inequalities

A linear inequality in two variables can be written as

$$ax + by \leq c,$$

where a, b, and c are constants with a and b both not equal to zero. (The symbol $\leq$ can be replaced by $\geq$, $<$, or $>$.) If an ordered pair (x, y) makes the inequality a true statement, then (x, y) is a solution. The set of all solutions is called the *solution set*. The graph of an inequality includes all points (x, y) in the solution set.

The graph of a linear inequality is a *half-plane,* which may include the boundary. For example, to graph the linear inequality $3x - 2y \leq 6$, solve the inequality for y. This results in $y \geq \frac{3}{2}x - 3$. Then graph the line $y = \frac{3}{2}x - 3$. The inequality includes the line and the half-plane above the line as shown in Figure 8.26 on the next page.

A second approach is to write the equation $3x - 2y = 6$ as $y = \frac{3}{2}x - 3$ and graph. To determine which half-plane to shade, select a *test point.* For example, the point $(0, 0)$ satisfies the inequality $3x - 2y \leq 6$, so shade the region containing $(0, 0)$.

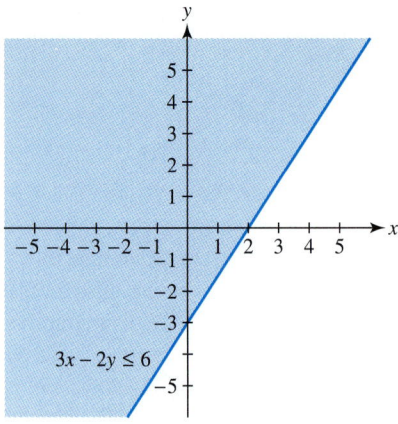

FIGURE 8.26

EXAMPLE 8   *Modeling plant growth*

Two factors that have a critical effect on plant growth are temperature and precipitation. If a region has too little precipitation, it will be a desert. Forests tend to grow in regions where trees can exist at relatively low temperatures and there is sufficient rainfall. At other levels of precipitation and temperature, grasslands may prevail. Figure 8.27 illustrates the relationship between forests, grasslands, and deserts, as suggested by annual average temperature T in degrees Fahrenheit and precipitation P in inches. (Source: A. Miller and J. Thompson, *Elements of Meteorology.*)

(a) Determine a system of linear inequalities that describes where grasslands are likely to occur.

(b) Bismarck, North Dakota, has annual average temperature of 40°F and precipitation of 15 inches. According to the graph, what type of plant growth would you expect near Bismarck? Do these values satisfy the system of inequalities from part (a)?

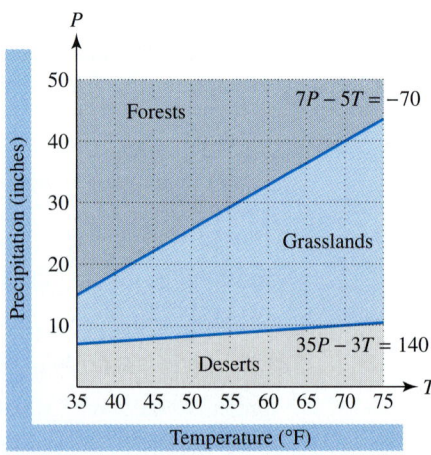

FIGURE 8.27 Effect of Temperature and Precipitation on Plant Growth

Solution

(a) Grasslands occur for ordered pairs (T, P) lying between the two lines in Figure 8.27. The boundary between deserts and grassland is determined by the equation $35P - 3T = 140$. Solving for P results in

$$P = \frac{3}{35}T + \frac{140}{35}.$$

Grasslands grow where values of P are above the line. This is described by $P > \frac{3}{35}T + \frac{140}{35}$, or equivalently, $35P - 3T > 140$. In a similar manner, the region below the boundary between grasslands and forests is represented by $7P - 5T < -70$. Thus, grasslands satisfy the following system of inequalities.

$$35P - 3T > 140$$

$$7P - 5T < -70$$

(b) For Bismarck, $T = 40$ and $P = 15$. From Figure 8.27, it appears that the point $(40, 15)$ lies between the two lines, so the graph predicts that grasslands will exist around Bismarck. Substituting these values for T and P into the system of inequalities results in the following true statements.

$$35(15) - 3(40) = 405 > 140$$

$$7(15) - 5(40) = -95 < -70$$

The temperature and precipitation for Bismarck satisfies the system of inequalities for grasslands. ∎

Critical Thinking

Use Example 8 to find an inequality that describes temperature and precipitation levels in deserts.

In Section 8.1, we learned how systems of equations could be either linear or nonlinear. In an analogous manner, systems of inequalities can be linear or nonlinear. The next example illustrates a system of each type. Both are solved graphically.

EXAMPLE 9   *Solving systems of inequalities graphically*

Solve each system of inequalities by shading the solution set. Use the graph to identify one solution.

(a) $y > x^2$
    $x + y < 4$

(b) $x + 3y \leq 9$
    $2x - y \leq -1$

Solution

(a) This is a nonlinear system. Graph the parabola $y = x^2$ and the line $y = 4 - x$. The region satisfying the system lies above the parabola and below the line. It does not include the boundary. This is shown using dashed graphs. See Figure 8.28 on the next page.

Any point in the shaded region represents a solution. For example, $(0, 2)$ lies in the shaded region and is a solution since $x = 0$ and $y = 2$ satisfy both inequalities.

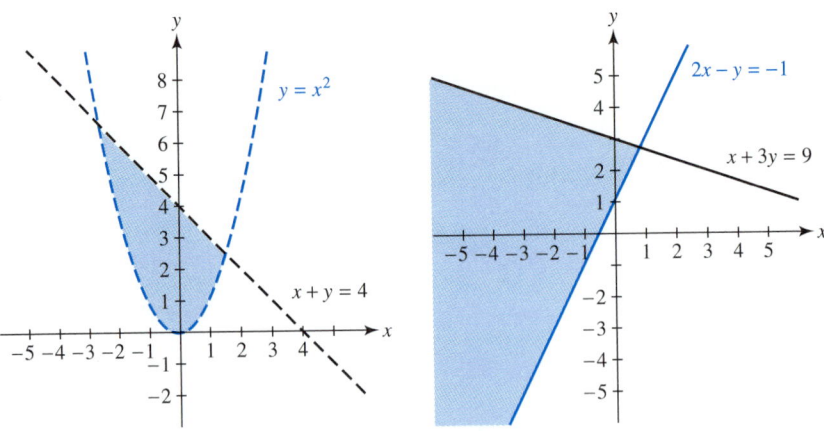

FIGURE 8.28   **FIGURE 8.29**

(b) Begin by solving each linear inequality for y.

$$y \leq -\frac{1}{3}x + 3$$

$$2x + 1 \leq y$$

Graph $y = -\frac{1}{3}x + 3$ and $y = 2x + 1$. The region satisfying the system is below the line $x + 3y = 9$ and above the line $2x - y = -1$. Because equality is included, the boundaries are part of the region, which are shown as solid lines in Figure 8.29. The point $(-3, 0)$ is a solution since it lies in the shaded region and satisfies both inequalities.   ∎

Graphing calculators can be used to shade regions in the xy-plane. The solution set shown in Figure 8.28 is also shown in Figure 8.30, where a graphing calculator has been used. However, the boundary is not dashed.

$[-5, 5, 1]$ by $[-2, 8, 1]$

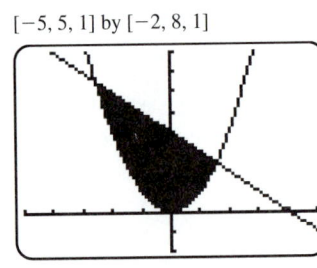

FIGURE 8.30

Linear Programming (Optional)

An important application of mathematics to business and social sciences is linear programming. *Linear programming* is an algorithm used to optimize quantities such as cost and profit. It was developed during World War II as a method of efficiently allocating supplies. Linear programming applications frequently contain thousands of variables. However, here we focus on problems involving two variables.

A linear programming problem consists of an **objective function** and a system of linear inequalities called **constraints.** The solution set for the system of linear inequalities is called the set of **feasible solutions.** The objective function describes a quantity that is to be optimized. For example, linear programming often is used to maximize profit or minimize cost. The following example illustrates these concepts.

EXAMPLE 10

Finding maximum profit

Suppose a small company manufactures two products—radios and CD players. Each radio results in a profit of $15 and each CD player provides a profit of $35. Due to demand, the company must produce at least 5 radios and not more than 25 radios per day. The number of radios cannot exceed the number of CD players, and the number of CD players cannot exceed 30. How many of each should the company manufacture to obtain maximum profit?

Solution

Let x represent the number of radios produced daily and y the number of CD players produced daily. Since the profit from x radios is $15x$ dollars and the profit from y CD players is $35y$ dollars, the total daily profit P is given by

$$P = 15x + 35y.$$

The company produces from 5 to 25 radios per day, so the inequalities

$$x \geq 5 \quad \text{and} \quad x \leq 25$$

must be satisfied. The requirements that the number of radios cannot exceed the number of CD players and the number of CD players cannot exceed 30 indicates that

$$x \leq y \quad \text{and} \quad y \leq 30.$$

Since the number of radios and CD players cannot be negative, we have

$$x \geq 0 \quad \text{and} \quad y \geq 0.$$

Listing all the constraints on production gives

$$x \geq 5, \quad x \leq 25, \quad y \leq 30, \quad x \leq y, \quad x \geq 0, \quad \text{and} \quad y \geq 0.$$

Graphing these constraints results in the shaded region shown in Figure 8.31. This shaded region is the set of feasible solutions. The vertices (or corners) of this region are (5, 5), (25, 25), (25, 30), and (5, 30).

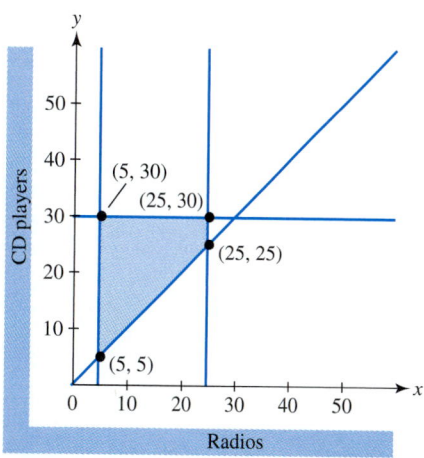

FIGURE 8.31

It can be shown that maximum profit occurs at a vertex of the region of feasible solutions. Thus, we evaluate P at each vertex as shown in Table 8.3.

TABLE 8.3	
Vertex	$P = 15x + 35y$
(5, 5)	$15(5) + 35(5) = 250$
(25, 25)	$15(25) + 35(25) = 1250$
(25, 30)	$15(25) + 35(30) = 1425$
(5, 30)	$15(5) + 35(30) = 1125$

The maximum value of P is 1425 at vertex (25, 30). Thus, the maximum profit is \$1425 and it occurs when 25 radios and 30 CD players are manufactured. ∎

The following holds for *every* linear programming problem.

Fundamental theorem of linear programming

The optimal value for a linear programming problem occurs at a vertex of the region of feasible solutions.

The next example illustrates how to minimize an objective function.

EXAMPLE 11   *Finding the minimum of an objective function*

Find the minimum value of $C = 2x + 3y$ subject to the following constraints.

$$x \geq 0$$
$$y \geq 0$$
$$x + y \geq 4$$
$$2x + y \leq 8$$

Solution

Step 1:   Sketch the region determined by the constraints and find all vertices. See Figure 8.32.

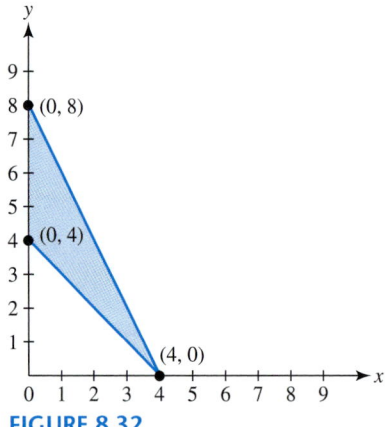

FIGURE 8.32

Step 2: Evaluate the objective function C at each vertex as shown in Table 8.4.

TABLE 8.4

Vertex	$C = 2x + 3y$
(4, 0)	$2(4) + 3(0) = 8$
(0, 8)	$2(0) + 3(8) = 24$
(0, 4)	$2(0) + 3(4) = 12$

Step 3: The minimum value for C is 8 and it occurs at vertex (4, 0), or when $x = 4$ and $y = 0$. ∎

8.2 PUTTING IT ALL TOGETHER

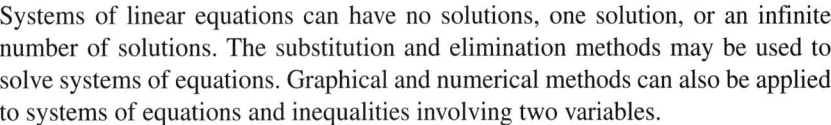

Systems of linear equations can have no solutions, one solution, or an infinite number of solutions. The substitution and elimination methods may be used to solve systems of equations. Graphical and numerical methods can also be applied to systems of equations and inequalities involving two variables.

The following table summarizes some mathematical concepts involved with systems of linear equations in two variables.

Concept	Comments	Example
Consistent linear system of equations in two variables	A consistent linear system has either one or an infinite number of solutions. Its graph is either intersecting lines or identical lines.	$x + y = 10$ $x - y = 4$ $\overline{2x = 14}$ Solution is $x = 7$, $y = 3$. Since the equations have one solution, they are independent.
Dependent linear system of equations in two variables	A dependent linear system has an infinite number of solutions. The graph consists of two identical lines.	$2x + 2y = 2$ and $x + y = 1$ are equivalent equations. The solution set is $\{(x, y) \mid x + y = 1\}$.
Inconsistent linear system of equations in two variables	An inconsistent linear system has no solutions. The graph is two distinct parallel lines.	$x + y = 1$ $x + y = 2$ (Subtract) $\overline{0 = -1}$ (Contradiction) The solution set is empty.
Elimination method	By performing arithmetic operations on a system, a variable is eliminated.	$2x + y = 5$ $x - y = 1$ (y is eliminated.) $\overline{3x = 6}$ so $x = 2$ and $y = 1$.
Linear inequality in two variables	$ax + by \le c$ ($\le$ may be replaced by $<$, $>$, or $\ge$.)	$2x - 3y > 25$ $-2x + y < 4$

8.2 EXERCISES

   Tape 6

Consistent and Inconsistent Linear Systems

Exercises 1–4: The figure represents a system of linear equations. Classify each system as consistent or inconsistent. Solve the system graphically and symbolically, if possible.

1.

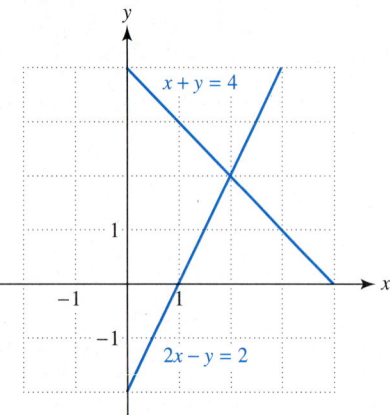

2.

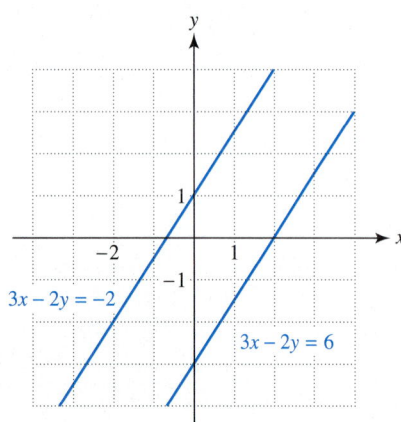

3.

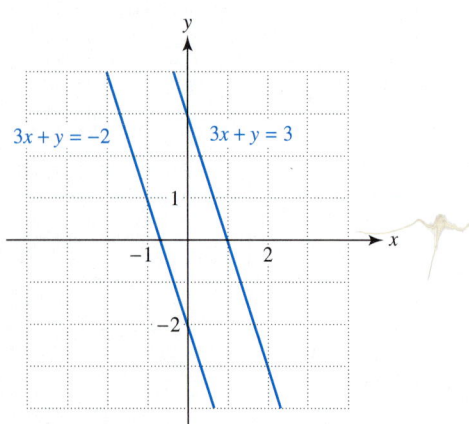

4.

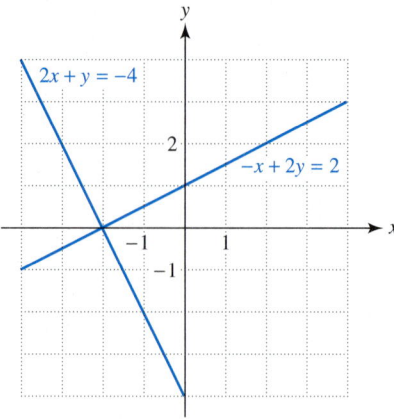

The Elimination Method

Exercises 5–16: Use elimination to solve each system of equations, if possible. Identify the system as consistent or inconsistent. If the system is consistent, state whether the equations are dependent or independent. Support your results graphically and numerically.

5. $x + y = 20$
$x - y = 8$

6. $2x + y = 15$
$x - y = 0$

7. $x + 3y = 10$
$x - 2y = -5$

8. $4x + 2y = 10$
$-2x - y = 10$

9. $x + y = 500$
$0.05x + 0.05y = 25$

10. $2x + 3y = 4$
$x - 2y = -5$

11. $5x + y = -5$
$7x - 3y = -29$

12. $2x + 3y = 5$
$5x - 2y = 3$

13. $2x + 4y = 7$
$-x - 2y = 5$

14. $4x - 3y = 5$
$3x + 4y = 2$

15. $2x + 3y = 2$
$x - 2y = -5$

16. $x - 3y = 1$
$2x - 6y = 2$

Exercises 17–20: Write a system of linear equations in two variables whose solution satisfies the problem. State what each variable represents. Then solve the system
  (a) *graphically, and*
  (b) *symbolically using elimination.*

17. The screen of a rectangular television set is 2 inches wider than it is high. If the perimeter of the screen is 38 inches, find its dimensions.

18. The sum of two numbers is 300 and their difference is 8. Find the two numbers.

19. Admission prices to a movie are $4 for children and $7 for adults. If 75 tickets were sold for $456, how many of each type of ticket were sold?

20. A sample of 16 dimes and quarters has a value of $2.65. How many of each type of coin are there?

Exercises 21–24: Use the method shown in Example 5 to find the line $y = ax + b$ that passes through the given points. Use elimination to solve the system. Then graph the line and the points.

21. $(-2, -2), (2, 6)$ **22.** $(-1, 4), (3, 2)$

23. $(-1, 3), (3, -5)$ **24.** $(-7, 1), (3, -3)$

Exercises 25–28: Modeling Real Data   Use the method shown in Example 6 to find $f(x) = ax + b$ so that f models the data, where x represents the year. Then graph f and the data in the same viewing rectangle.

25. The number of cable TV networks has increased. In 1985 there were 56 cable networks and in 1996 there were 162. (Sources: National Cable Television Association and Warren Publishing.)

26. In 1984 domestic car sales were 76% of total car sales in the United States, while in 1996 they were 59%. (Source: Autodata.)

27. In 1996 the percentage of ABC, CBS, and NBC network evening news stories presented by women was 16%, whereas in 1994 it was 25%. (Source: Center for Media and Public Affairs Analysis.)

28. In 1996 the percentage of ABC, CBS, and NBC network evening news stories presented by minorities was 9%, whereas in 1994 it was 14%. (Source: Center for Media and Public Affairs Analysis.)

Inequalities

Exercises 29–32: Match the system of inequalities with the appropriate graph. Use the graph to identify a solution.

29. $x + y \geq 2$
$\quad\ \ x - y \leq 1$

30. $2x - y > 0$
$\quad\ \ x - 2y \leq 1$

31. $\frac{1}{2}x^3 - y > 0$
$\quad\ \ 2x - y \leq 1$

32. $x^2 + y \leq 4$
$\quad\ \ x^2 - y \leq 2$

a.

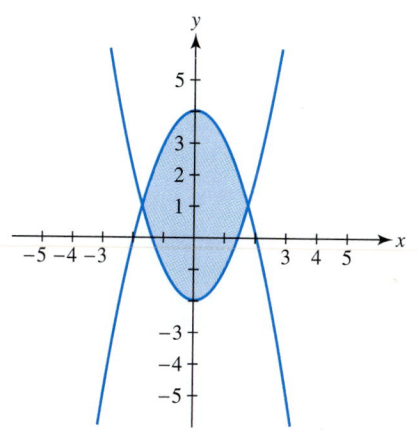

b.

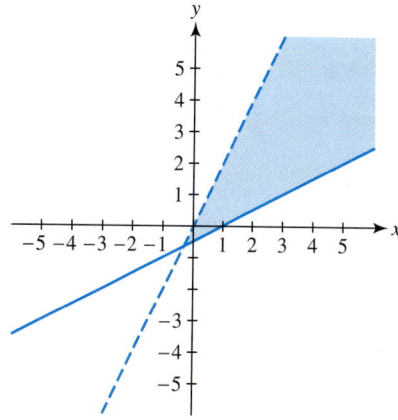

c.

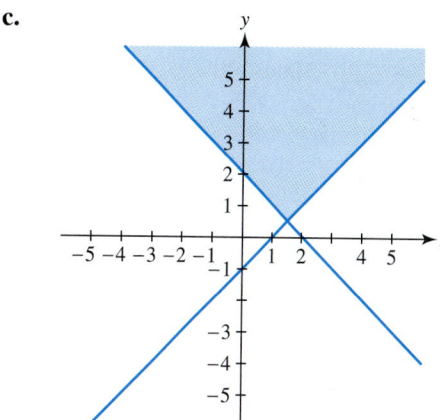

d.

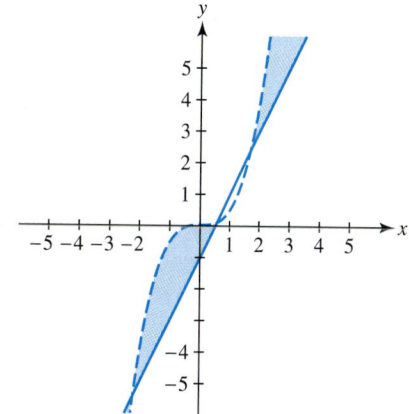

Exercises 33–38: Graph each system of inequalities. Shade the region that represents the solution set. Use the graph to identify a solution.

33. $\quad y \geq x^2$
$\quad\ \ x + y \leq 6$

34. $y \leq \sqrt{x}$
$\quad\ y \geq 1$

35. $x + 2y > -2$
$\quad\ \ x + 2y < 5$

36. $x - y \leq 3$
$\quad\ \ x + y \leq 3$

37. $x^2 + y^2 \leq 16$
$\quad\ \ x + y < 2$

38. $x^2 + y \leq 4$
$\quad\ \ x^2 - y \leq 3$

Applications

39. *Student Loans*   A student takes out two loans totaling $3000 to help pay for college expenses. One
loan is at 8% interest, while the other is at 10%.
Interest for both loans is compounded annually.
  (a) If the first-year interest is $264, determine a
  system of linear equations whose solution is
  the amount of each loan.
  (b) Find the amount of each loan.

40. (Refer to the previous exercise.) Suppose that both
loans have an interest rate of 10% and the total
first-year interest is $300. If possible, determine the
amount of each loan. Interpret your results.

41. (Refer to the previous two exercises.) Suppose that
both loans are at 10% and the total annual interest
is $264. If possible, determine the amount of each
loan. Interpret your results.

42. *Maximizing Area*   Suppose a rectangular pen for a
pet is to be made using 40 feet of fence. Let l represent its length and w its width.
  (a) Find the dimensions that result in an area of
  91 square feet.
  (b) Write a formula for the area A in terms of w.
  Graph A.
  (c) What is the maximum area possible for the
  pen? Interpret this result.

43. *Road Rage*   Aggressive driving incidents occurring in 1994 and 1995 that involved injury or death
totaled 3377. The number in 1995 exceeded the
1994 number by 39. Use elimination to determine
how many aggressive driving incidents occurred
each year. Support your result graphically and
numerically. (Source: American Automobile
Association.)

44. *Charitable Giving*   In 1995 corporations and individuals gave $123.6 billion to charitable organiza-
tions. Individuals gave 15.7 times more than
corporations. Use elimination to determine how
much individuals gave and how much corporations
gave. Round answers to the nearest tenth of a billion dollars. (Source: Giving USA '95 AAFRC
Trust for Philanthropy.)

45. *Meat Consumption*   In 1994 the average person in
the United States ate 4.4 times as much beef as
turkey. The average consumption of beef and
turkey was 77.8 pounds per person. Use elimination to determine the average annual consumption
per person for each type of meat. Round your
answers to the nearest tenth of a pound.

46. *Cable TV and the Internet*   In 1994 consumer
spending on basic cable TV and on-line Internet
averaged $117 per person. Basic cable TV spend-
ing exceeded on-line Internet spending by $103.

Determine the amount spent for basic cable and
for on-line Internet by writing a system of equations and solving it using elimination. (Source:
Statistical Abstract of the United States.)

47. *Air Speed*   A jet airliner travels 1680 miles in
3 hours with a tail wind. The return trip, into the
wind, takes 3.5 hours. Find both the air speed of the
jet and the wind speed. (*Hint:* First find the ground
speed of the airplane in each direction.)

48. *River Current*   A tugboat can push a barge 60 miles
upstream in 15 hours. The same tugboat and barge
can make the return trip downstream in 6 hours.
Determine the speed of the current in the river.

49. *Traffic Control*   The figure shows two intersections labeled A and B that involve one-way streets.
The numbers and variables represent the average
traffic flow rates measured in vehicles per hour. For
example, an average of 500 vehicles per hour enter
intersection A from the west, whereas 150 vehicles
per hour enter this intersection from the north. A
stoplight will control the unknown traffic flow
denoted by the variables x and y. Use the fact that
the number of vehicles entering an intersection
must equal the number leaving to determine x and y.

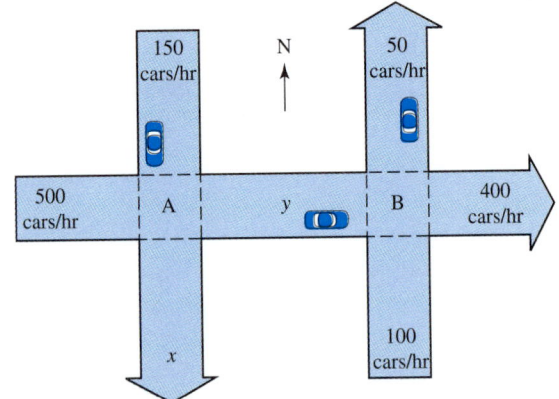

50. (Refer to the previous exercise and accompanying
figure.) Suppose that the number of vehicles entering intersection A from the west varies between
400 and 600. If all other traffic flows remain the
same as in the figure, what effect does this have on
the ranges of the values for x and y?

Exercises 51–54: Weight and Height   *The following graph shows a weight and height chart. The weight w is listed in pounds and the height h in inches. The shaded area is a recommended region.* (Source: Department of Agriculture.)

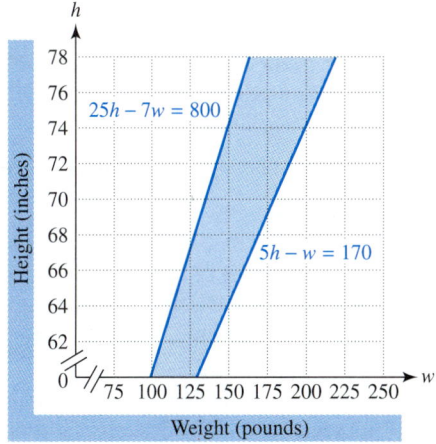

51. What does this chart indicate about an individual weighing 125 pounds with a height of 70 inches?

52. If a person is 74 inches tall, use the graph to estimate the recommended weight range.

53. Use the graph to find a system of linear inequalities that describes the recommended region.

54. Explain why inequalities, rather than equalities, are more appropriate to describe recommended weight and height combinations.

Linear Programming

Exercises 55 and 56: The graph shows a region of feasible solutions. Find the maximum and minimum values of P over this region.

55. $P = 3x + 5y$

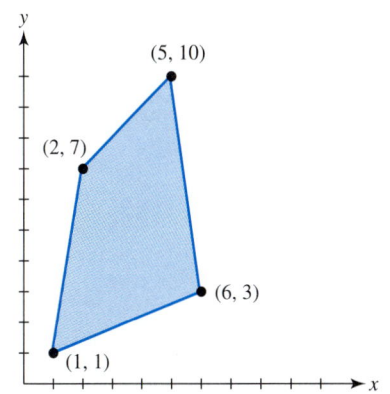

56. $P = 6x + y$

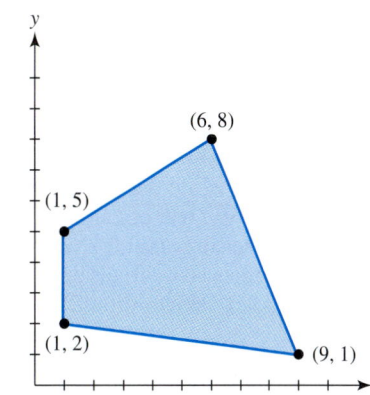

Exercises 57–60: The graph shows a region of feasible solutions. Find the maximum and minimum of C over this region.

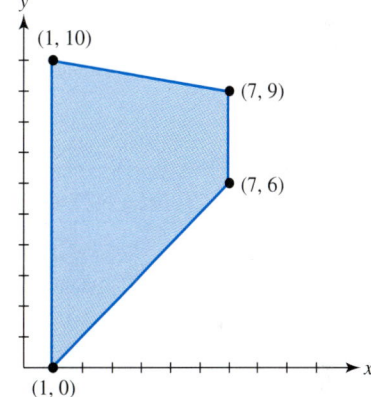

57. $C = 3x + 5y$

58. $C = 5x + 5y$

59. $C = 10y$

60. $C = 3x - y$

61. Find the minimum value of $C = 4x + 2y$ subject to the following constraints.

$$x \geq 0$$
$$y \geq 0$$
$$x + y \geq 3$$
$$2x + 3y \leq 12$$

62. Find the maximum value of $P = 3x + 5y$ subject to the following constraints.

$$x \geq 0$$
$$y \geq 0$$
$$3x + y \leq 8$$
$$x + 3y \leq 8$$

63. Rework Example 10 if the profit from each radio is $20 and the profit from each CD player is $15.

64. *Maximizing Storage*   An office manager wants to buy filing cabinets. Cabinet X costs $100, requires 6 square feet of floor space, and holds 8 cubic feet. Cabinet Y costs $200, requires 8 square feet of floor space, and holds 12 cubic feet. No more than $1400 can be spent, while the office has room for no more than 72 square feet of cabinets. The office manager wants the maximum storage capacity within the limits imposed by funds and space. How many of each type of cabinet should be bought?

65. *Maximizing Profit*   A business manufactures two parts, X and Y. Machines A and B are needed to make each part. To make part X, machine A is needed 4 hours and machine B is needed 2 hours. To make part Y, machine A is needed 1 hour and machine B is needed 3 hours. Machine A is available for 40 hours each week and machine B is available for 30 hours. The profit from part X is $500 and the profit from part Y is $600. How many parts of each type should be made to maximize weekly profit?

66. *Minimizing Cost*   Two substances, X and Y, are found in pet food. Each substance contains the ingredients A and B. Substance X is 20% ingredient A and 50% ingredient B. Substance Y is 50% ingredient A and 30% ingredient B. The cost of substance X is $2 per pound and the cost of substance Y is $3 per pound. The pet store needs at least 251 pounds of ingredient A and at least 200 pounds of ingredient B. If cost is to be minimal, how many pounds of each substance should be ordered? Find the minimum cost.

Writing about Mathematics

1. Give the general form of a system of linear equations in two variables. Discuss what distinguishes a system of linear equations from a system of nonlinear equations.

2. Discuss what the words *consistent* and *inconsistent* mean with regards to linear systems. Give an example of each type of system.

CHECKING BASIC CONCEPTS FOR SECTIONS 8.1 AND 8.2

1. Evaluate $f(13, 18)$ if $f(x, y) = \sqrt{(x-1)^2 + (y-2)^2}$. (The function f computes the distance that the point (x, y) is from the point $(1, 2)$.)

2. Solve the nonlinear system of equations using the method of substitution.

$$2x^2 - y = 0$$
$$3x + 2y = 7$$

3. Solve the system of inequalities graphically. Use the graph to identify one solution.

$$x^2 - y < 3$$
$$x - y \geq 1$$

4. In 1996 the United States and the former Soviet Union had a combined total of 13,820 strategic nuclear warheads. The United States had 480 more warheads than the former Soviet Union. (Sources: Arms Control Association, Natural Resources Defense Council.) Find a system of linear equations whose solution gives the number of nuclear warheads for each country. Solve the system
  (a) graphically,
  (b) numerically, and
  (c) symbolically.

8.3 Solutions of Linear Systems Using Matrices

Representing Systems of Linear Equations with Matrices • Row-Echelon Form • Gaussian Elimination • Solving Systems of Linear Equations with Technology

Introduction

The methods of substitution and elimination can be applied to linear systems involving more than two variables. As the number of variables increases, these methods become more cumbersome and less efficient. However, by combining these two methods into one, they provide a state-of-the-art numerical method, capable of solving systems of linear equations containing thousands of variables. This numerical method is called *Gaussian elimination with backward substitution.* Even though this method dates back to Carl Friedrich Gauss (1777–1855), it continues to be one of the most efficient methods for solving systems of linear equations.

Representing Systems of Linear Equations with Matrices

Arrays of numbers occur frequently in many different situations. Spreadsheets often make use of arrays, where data is displayed in a tabular format. A **matrix** is a rectangular array of elements. The following are examples of matrices whose elements are real numbers.

$$\begin{bmatrix} 4 & -7 \\ -2 & 9 \end{bmatrix} \quad \begin{bmatrix} -1 & -5 & 3 \\ 1.2 & 0 & -1.3 \\ 4.1 & 5 & 7 \end{bmatrix} \quad \begin{bmatrix} -3 & -6 & 9 & 5 \\ \sqrt{2} & -8 & -8 & 0 \\ 3 & 0 & 19 & -7 \\ -11 & -3 & 7 & 8 \end{bmatrix} \quad \begin{bmatrix} 5 & -2 \\ -2 & \pi \\ 1 & -1 \end{bmatrix} \quad \begin{bmatrix} 1 & -0.5 & 9 \\ 5 & 0.4 & -3 \end{bmatrix}$$

$$2 \times 2 \qquad\qquad 3 \times 3 \qquad\qquad\qquad 4 \times 4 \qquad\qquad\qquad 3 \times 2 \qquad\qquad 2 \times 3$$

The dimension of a matrix is given much like the dimensions of a rectangular room. We might say a room is n feet long and m feet wide. The **dimension** of a matrix is $n \times m$ (n by m), if it has n rows and m columns. For example, the last matrix has a dimension of 2×3, because it has 2 rows and 3 columns. If the number of rows and columns are equal, the matrix is a **square matrix.** The first three matrices are square matrices.

Matrices are frequently used to represent systems of linear equations. A linear system with three equations and three variables can be written as follows.

$$a_1 x + b_1 y + c_1 z = d_1$$
$$a_2 x + b_2 y + c_2 z = d_2$$
$$a_3 x + b_3 y + c_3 z = d_3$$

The a_k, b_k, c_k, and d_k are constants and x, y, and z are variables. The coefficients of the variables can be represented using a 3×3 matrix. This matrix is called the **coefficient matrix** of the linear system.

$$\begin{bmatrix} a_1 & b_1 & c_1 \\ a_2 & b_2 & c_2 \\ a_3 & b_3 & c_3 \end{bmatrix}$$

If we enlarge the matrix to include the constants d_k, the system may be written as

$$\begin{bmatrix} a_1 & b_1 & c_1 & | & d_1 \\ a_2 & b_2 & c_2 & | & d_2 \\ a_3 & b_3 & c_3 & | & d_3 \end{bmatrix}.$$

Because this matrix is an enlargement of the coefficient matrix, it is commonly called an **augmented matrix.** The vertical line between the third and fourth column corresponds to where the equals sign occurs in each equation.

Critical Thinking

Give a general form of a system of linear equations with four equations and four variables. Write its augmented matrix.

EXAMPLE 1   *Representing a linear system with an augmented matrix*

Express each linear system with an augmented matrix. State the dimension of the matrix.

(a) $\begin{aligned} 3x - 4y &= 6 \\ -5x + y &= -5 \end{aligned}$

(b) $\begin{aligned} 2x - 5y + 6z &= -3 \\ 3x + 7y - 3z &= 8 \\ x + 7y &= 5 \end{aligned}$

Solution

(a) This system has two equations in two variables. It can be represented by an augmented matrix having dimension 2×3.

$$\left[\begin{array}{cc|c} 3 & -4 & 6 \\ -5 & 1 & -5 \end{array}\right]$$

(b) This system has three equations in three variables. Note that variable z does not appear in the third equation. A zero is input for its coefficient.

$$\left[\begin{array}{ccc|c} 2 & -5 & 6 & -3 \\ 3 & 7 & -3 & 8 \\ 1 & 7 & 0 & 5 \end{array}\right]$$

Since this matrix has 3 rows and 4 columns, its dimension is 3×4. ∎

EXAMPLE 2   *Converting an augmented matrix into a linear system*

Write the linear system represented by the augmented matrix. Let the variables be x, y, and z.

(a) $\left[\begin{array}{ccc|c} 1 & 0 & 2 & -3 \\ 2 & 2 & 10 & 3 \\ -1 & 2 & 3 & 5 \end{array}\right]$

(b) $\left[\begin{array}{ccc|c} 1 & 2 & 3 & -4 \\ 0 & 1 & -6 & 7 \\ 0 & 0 & 1 & 8 \end{array}\right]$

Solution

(a) The first column corresponds to x, the second to y, and the third to z. When a zero appears, the variable for that column does not appear in the equation. The vertical line corresponds to the location of the equals sign. The last column represents the constant terms.

$$\begin{aligned} x \qquad + 2z &= -3 \\ 2x + 2y + 10z &= 3 \\ -x + 2y + 3z &= 5 \end{aligned}$$

(b) The augmented matrix represents the following linear system.

$$\begin{aligned} x + 2y + 3z &= -4 \\ y - 6z &= 7 \\ z &= 8 \end{aligned}$$

∎

Row-Echelon Form

When solving linear systems with augmented matrices, a convenient form is row-echelon form. The following matrices are in row-echelon form.

$$\begin{bmatrix} 1 & 3 & 0 & -1 \\ 0 & 1 & -6 & 1 \\ 0 & 0 & 1 & -2 \end{bmatrix} \quad \begin{bmatrix} 1 & 2 & 0 \\ 0 & 1 & 4 \end{bmatrix} \quad \begin{bmatrix} 1 & 3 & -1 & 5 \\ 0 & 1 & -1 & 3 \\ 0 & 0 & 1 & 0 \end{bmatrix} \quad \begin{bmatrix} 1 & 3 & -1 & 5 \\ 0 & 0 & 1 & 3 \\ 0 & 0 & 0 & 0 \end{bmatrix} \quad \begin{bmatrix} 1 & 3 & 5 \\ 0 & 0 & 1 \end{bmatrix}$$

The elements of the **main diagonal** are highlighted in each matrix. As one moves down the main diagonal of a matrix in row-echelon form, it first contains only 1's, and then possibly 0's. The first element in each nonzero row is 1. This 1 is called a **leading 1.** If two rows contain a leading 1, then the row with the left-most leading 1 is listed first in the matrix. Rows containing only 0's occur last in the matrix. Any element below the main diagonal is always equal to zero.

The next example demonstrates a technique called **backward substitution.** It can be used to solve linear systems represented by an augmented matrix in row-echelon form.

EXAMPLE 3   *Solving a linear system with backward substitution*

Solve the system of linear equations.

(a) $\begin{bmatrix} 1 & 1 & 3 & | & 12 \\ 0 & 1 & -2 & | & -4 \\ 0 & 0 & 1 & | & 3 \end{bmatrix}$ (b) $\begin{bmatrix} 1 & -1 & 5 & | & 5 \\ 0 & 1 & 3 & | & 3 \\ 0 & 0 & 0 & | & 0 \end{bmatrix}$

Solution

(a) The linear system is represented by

$$\begin{aligned} x + y + 3z &= 12 \\ y - 2z &= -4 \\ z &= 3. \end{aligned}$$

Since $z = 3$, substitute this value into the second equation to find y.

$$y - 2(3) = -4 \quad \text{or} \quad y = 2$$

Then $y = 2$ and $z = 3$ can be substituted into the first equation to determine x.

$$x + 2 + 3(3) = 12 \quad \text{or} \quad x = 1$$

The solution is $x = 1$, $y = 2$, and $z = 3$, and can be expressed as the ordered triple $(1, 2, 3)$.

(b) The linear system is represented by

$$\begin{aligned} x - y + 5z &= 5 \\ y + 3z &= 3 \\ 0 &= 0. \end{aligned}$$

The last equation $0 = 0$ is always true. Its presence usually indicates an infinite number of solutions. Use the second equation to write y in terms of z.

$$y = 3 - 3z$$

Next, substitute $(3 - 3z)$ for y in the first equation and write x in terms of z.

$$x - (3 - 3z) + 5z = 5$$
$$x - 3 + 3z + 5z = 5 \qquad \text{Distributive property}$$
$$x = 8 - 8z \qquad \text{Solve for } x.$$

The solution can be written as the ordered triple $(8 - 8z, 3 - 3z, z)$, where z is any real number. There is an infinite number of solutions. For example, if we let $z = 1$, then $y = 3 - 3(1) = 0$ and $x = 8 - 8(1) = 0$. Thus, one solution is $(0, 0, 1)$. ■

Gaussian Elimination

If an augmented matrix is not in row-echelon form, it can be transformed into row-echelon form using *Gaussian elimination*. This algorithm utilizes three basic matrix row transformations.

Matrix row transformations

For any augmented matrix representing a system of linear equations, the following row transformations result in an equivalent system of linear equations.

1. Any two rows may be interchanged.
2. The elements of any row may be multiplied by a nonzero constant.
3. Any row may be changed by adding to (or subtracting from) its elements a multiple of the corresponding elements of another row.

When we transform a matrix into row-echelon form, we also are transforming a system of linear equations. The next example illustrates how Gaussian elimination with backward substitution is performed.

EXAMPLE 4 *Transforming a matrix into row-echelon form*

Use Gaussian elimination with backward substitution to solve the linear system of equations.

$$2x + 4y + 4z = 4$$
$$x + 3y + z = 4$$
$$-x + 3y + 2z = -1$$

Solution

The linear system is included to illustrate how each row transformation affects the corresponding system of linear equations. The initial linear system and augmented matrix are written first.

Linear System   Augmented Matrix

$$2x + 4y + 4z = 4$$
$$x + 3y + z = 4$$
$$-x + 3y + 2z = -1$$

$$\begin{bmatrix} 2 & 4 & 4 & | & 4 \\ 1 & 3 & 1 & | & 4 \\ -1 & 3 & 2 & | & -1 \end{bmatrix}$$

First we obtain a 1 where the coefficient of x in the first row is highlighted. This can be accomplished by either multiplying the first equation by $\frac{1}{2}$, or interchanging rows 1 and 2. We multiply row 1 by $\frac{1}{2}$. This operation is denoted by $\frac{1}{2}R_1$, to indicate that row 1 in the previous augmented matrix is multiplied by $\frac{1}{2}$. The result becomes the new row 1. It is important to write down each row operation, so that we can check our work more easily.

$$
\begin{aligned}
x + 2y + 2z &= 2 \\
x + 3y + z &= 4 \\
-x + 3y + 2z &= -1
\end{aligned}
\qquad
\frac{1}{2}R_1 \to
\begin{bmatrix}
1 & 2 & 2 & 2 \\
1 & 3 & 1 & 4 \\
-1 & 3 & 2 & -1
\end{bmatrix}
$$

The next step is to eliminate the x-variable in rows 2 and 3 by obtaining zeros in the highlighted positions. To do this, subtract row 1 from row 2, and add row 1 to row 3.

$$
\begin{aligned}
x + 2y + 2z &= 2 \\
y - z &= 2 \\
5y + 4z &= 1
\end{aligned}
\qquad
\begin{aligned}
&R_2 - R_1 \to \\
&R_3 + R_1 \to
\end{aligned}
\begin{bmatrix}
1 & 2 & 2 & 2 \\
0 & 1 & -1 & 2 \\
0 & 5 & 4 & 1
\end{bmatrix}
$$

Since we have a 1 for the coefficient of y in the second row, the next step is to eliminate the y-variable in row 3, and obtain a zero where the y-coefficient of 5 is highlighted. Multiply row 2 by -5, and add the result to row 3.

$$
\begin{aligned}
x + 2y + 2z &= 2 \\
y - z &= 2 \\
9z &= -9
\end{aligned}
\qquad
R_3 + (-5)R_2 \to
\begin{bmatrix}
1 & 2 & 2 & 2 \\
0 & 1 & -1 & 2 \\
0 & 0 & 9 & -9
\end{bmatrix}
$$

Finally, make the coefficient of z in the third row equal 1 by multiplying row 3 by $\frac{1}{9}$.

$$
\begin{aligned}
x + 2y + 2z &= 2 \\
y - z &= 2 \\
z &= -1
\end{aligned}
\qquad
\frac{1}{9}R_3 \to
\begin{bmatrix}
1 & 2 & 2 & 2 \\
0 & 1 & -1 & 2 \\
0 & 0 & 1 & -1
\end{bmatrix}
$$

The final matrix is in row-echelon form. Backward substitution may be applied to find the solution. Substituting $z = -1$ into the second equation gives

$$
y - (-1) = 2 \qquad \text{or} \qquad y = 1.
$$

Next, substitute $y = 1$ and $z = -1$ into the first equation to determine x.

$$
x + 2(1) + 2(-1) = 2 \qquad \text{or} \qquad x = 2
$$

The solution of the system is $x = 2$, $y = 1$, and $z = -1$, or $(2, 1, -1)$.   ■

MAKING CONNECTIONS

A Geometric Interpretation of Systems of Linear Equations

Solving a linear equation in one variable reduces to finding the x-value on the number line that satisfies the equation. Solving two linear equations in two variables is often equivalent to finding the xy-coordinates where two lines intersect.

Solving linear equations in three variables also has a geometric interpretation. The graph of a linear equation in the three variables x, y, and z is a flat plane. Finding a unique solution is equivalent to locating a point where three planes in space intersect, as illustrated in Figure 8.33.

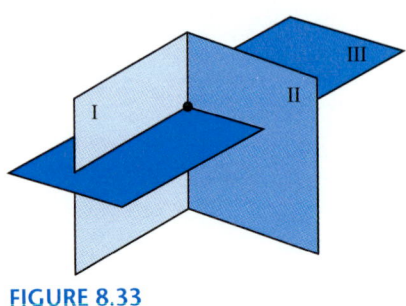

FIGURE 8.33

Sometimes it is convenient to express a matrix in reduced row-echelon form. A matrix in row-echelon form is in **reduced row-echelon form** if every element above and below a leading 1 in a column is zero. The following matrices are examples of reduced row-echelon form.

$$\begin{bmatrix} 1 & 0 \\ 0 & 1 \end{bmatrix} \quad \begin{bmatrix} 1 & 0 \\ 0 & 0 \end{bmatrix} \quad \begin{bmatrix} 1 & 0 & 0 \\ 0 & 1 & 0 \\ 0 & 0 & 1 \end{bmatrix}$$

$$\begin{bmatrix} 1 & 0 & 3 \\ 0 & 1 & -2 \end{bmatrix} \quad \begin{bmatrix} 1 & 0 & 0 & 3 \\ 0 & 1 & 0 & 1 \\ 0 & 0 & 1 & -1 \end{bmatrix} \quad \begin{bmatrix} 1 & 0 & 4 & 8 \\ 0 & 1 & -1 & 2 \\ 0 & 0 & 0 & 0 \end{bmatrix}$$

If an augmented matrix is in reduced row-echelon form, it usually is straightforward to solve the system of linear equations.

EXAMPLE 5   *Determining a solution from a matrix in reduced row-echelon form*

Each matrix represents a system of linear equations. Find the solution.

(a) $\begin{bmatrix} 1 & 0 & 0 & | & 3 \\ 0 & 1 & 0 & | & -1 \\ 0 & 0 & 1 & | & 2 \end{bmatrix}$ (b) $\begin{bmatrix} 1 & 0 & | & 6 \\ 0 & 1 & | & -5 \end{bmatrix}$

Solution

(a) The top row represents the equation $1x + 0y + 0z = 3$ or $x = 3$. Using similar reasoning for the second and third rows: $y = -1$ and $z = 2$. The solution is $(3, -1, 2)$.

(b) The system involves two equations and two unknowns. The solution is $(6, -5)$. ∎

Gaussian elimination may be used to transform an augmented matrix into reduced row-echelon form. It requires more effort than transforming a matrix into row-echelon form, but often eliminates the need for backward substitution.

EXAMPLE 6 *Transforming a matrix into reduced row-echelon form*

Use Gaussian elimination to transform the augmented matrix of the linear system into reduced row-echelon form. State the solution.

$$2x + y + 2z = 10$$
$$x \qquad + 2z = 5$$
$$x - 2y + 2z = 1$$

Solution

The linear system has been included at the left for illustrative purposes.

Linear System	Augmented Matrix

$$2x + y + 2z = 10$$
$$x \qquad + 2z = 5$$
$$x - 2y + 2z = 1$$

$$\begin{bmatrix} 2 & 1 & 2 & | & 10 \\ 1 & 0 & 2 & | & 5 \\ 1 & -2 & 2 & | & 1 \end{bmatrix}$$

Start by obtaining a leading 1 in the first row by interchanging rows 1 and 2.

$$x \qquad + 2z = 5 \qquad R_2 \rightarrow$$
$$2x + y + 2z = 10 \qquad R_1 \rightarrow$$
$$x - 2y + 2z = 1$$

$$\begin{bmatrix} 1 & 0 & 2 & | & 5 \\ 2 & 1 & 2 & | & 10 \\ 1 & -2 & 2 & | & 1 \end{bmatrix}$$

Next subtract 2 times row 1 from row 2. Then, subtract row 1 from row 3. This eliminates the *x*-variable from the second and third equations.

$$x \qquad + 2z = 5$$
$$y - 2z = 0 \qquad R_2 - 2R_1 \rightarrow$$
$$-2y \qquad = -4 \qquad R_3 - R_1 \rightarrow$$

$$\begin{bmatrix} 1 & 0 & 2 & | & 5 \\ 0 & 1 & -2 & | & 0 \\ 0 & -2 & 0 & | & -4 \end{bmatrix}$$

To eliminate the *y*-variable in row 3, add two times row 2 to row 3.

$$x \qquad + 2z = 5$$
$$y - 2z = 0$$
$$-4z = -4 \qquad R_3 + 2R_2 \rightarrow$$

$$\begin{bmatrix} 1 & 0 & 2 & | & 5 \\ 0 & 1 & -2 & | & 0 \\ 0 & 0 & -4 & | & -4 \end{bmatrix}$$

To obtain a leading 1 in row 3, multiply by $-\dfrac{1}{4}$.

$$x \qquad + 2z = 5$$
$$y - 2z = 0$$
$$z = 1 \qquad -\frac{1}{4}R_3 \rightarrow$$

$$\begin{bmatrix} 1 & 0 & 2 & | & 5 \\ 0 & 1 & -2 & | & 0 \\ 0 & 0 & 1 & | & 1 \end{bmatrix}$$

The matrix is in row-echelon form. It can be transformed into reduced row-echelon form by subtracting 2 times row 3 from row 1, and adding 2 times row 3 to row 2.

$$
\begin{array}{c}
x = 3 \\
y = 2 \\
z = 1
\end{array}
\qquad
\begin{array}{c}
R_1 - 2R_3 \rightarrow \\
R_2 + 2R_3 \rightarrow \\
{}
\end{array}
\left[
\begin{array}{ccc|c}
1 & 0 & 0 & 3 \\
0 & 1 & 0 & 2 \\
0 & 0 & 1 & 1
\end{array}
\right]
$$

This matrix is in reduced row-echelon form. The solution is (3, 2, 1). ■

Solving Systems of Linear Equations with Technology

Gaussian elimination is a numerical method. If the arithmetic at each step is done exactly, then it may be thought of as an exact symbolic procedure. However, in real applications the augmented matrix is usually quite large and its elements are not all integers. As a result, calculators and computers often are used to solve systems of equations. Their solutions usually are approximate. The next two examples use a graphing calculator to solve systems of linear equations.

EXAMPLE 7   *Transforming a matrix into reduced row-echelon form*

Three food shelters are operated by a charitable organization. Three different quantities are computed. They include monthly food costs F in dollars, number of people served per month N, and monthly charitable receipts R in dollars. The data is shown in Table 8.5. (Source: D. Sanders, *Statistics: A First Course.*)

TABLE 8.5

Food Costs (F)	Number Served (N)	Charitable Receipts (R)
3000	2400	8000
4000	2600	10,000
8000	5900	14,000

(a) Model this data using the equation $F = aN + bR + c$, where a, b, and c are constants.

(b) Predict the food costs for a shelter that serves 4000 people and receives charitable receipts of $12,000. Round your answer to the nearest hundred dollars.

Solution

(a) Since $F = aN + bR + c$, the constants a, b, and c satisfy the following three equations.

$$3000 = a(2400) + b(8000) + c$$
$$4000 = a(2600) + b(10{,}000) + c$$
$$8000 = a(5900) + b(14{,}000) + c$$

This can be rewritten as

$$2400a + \quad 8000b + c = 3000$$
$$2600a + 10{,}000b + c = 4000$$
$$5900a + 14{,}000b + c = 8000.$$

The resulting augmented matrix is

$$A = \begin{bmatrix} 2400 & 8000 & 1 & 3000 \\ 2600 & 10{,}000 & 1 & 4000 \\ 5900 & 14{,}000 & 1 & 8000 \end{bmatrix}.$$

Figure 8.34 shows the matrix A. The fourth column of A may be viewed using the arrow keys. In Figure 8.35, A has been transformed into reduced row-echelon form. From Figure 8.35, we see that $a \approx 0.6897$, $b \approx 0.4310$, and $c \approx -2103$. Thus, let $F = 0.6897N + 0.431R - 2103$.

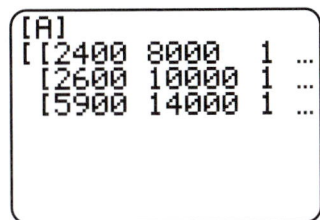

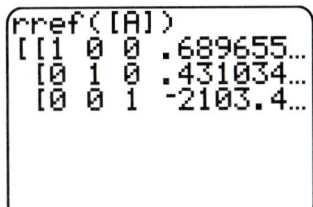

FIGURE 8.34   **FIGURE 8.35**

(b) To predict the food costs for a shelter that serves 4000 people and receives charitable receipts of $12,000, let $N = 4000$ and $R = 12{,}000$ and evaluate F.

$$F = 0.6897(4000) + 0.431(12{,}000) - 2103 = 5827.8.$$

This model predicts monthly food costs of approximately $5800.   ■

The next example shows how a system of linear equations can be used to determine a quadratic function.

EXAMPLE 8   *Determining a quadratic function*

More than half of private-sector employees cannot carry vacation days into a new year. The average number y of paid days off for full-time workers at medium to large companies after x years is listed in Table 8.6. (Source: Bureau of Labor Statistics.)

TABLE 8.6

x (yr)	1	15	30
y (days)	9.4	18.8	21.9

(a) Determine the coefficients for $f(x) = ax^2 + bx + c$ so that f models this data.
(b) Graph f with the data in $[-4, 32, 5]$ by $[8, 23, 2]$.
(c) Estimate the number of paid days off after 3 years of experience. Compare it to the actual value of 11.2 days.

Solution

(a) For f to model the data, the three equations $f(1) = 9.4$, $f(15) = 18.8$, and $f(30) = 21.9$ must be satisfied.

$$f(1) = a(1)^2 + b(1) + c = 9.4$$

$$f(15) = a(15)^2 + b(15) + c = 18.8$$

$$f(30) = a(30)^2 + b(30) + c = 21.9$$

The associated augmented matrix is

$$\begin{bmatrix} 1^2 & 1 & 1 & | & 9.4 \\ 15^2 & 15 & 1 & | & 18.8 \\ 30^2 & 30 & 1 & | & 21.9 \end{bmatrix}.$$

Figure 8.36 shows a portion of the matrix represented in reduced row-echelon form.

$[-4, 32, 5]$ by $[8, 23, 2]$

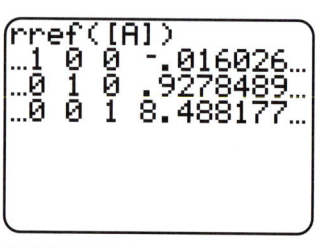

FIGURE 8.36   **FIGURE 8.37**

The solution is $a \approx -0.016026$, $b \approx 0.92785$, and $c \approx 8.4882$.

(b) Graph $Y_1 = -0.016026X^2 + 0.92785X + 8.4882$ with the points $(1, 9.4)$, $(15, 18.8)$, and $(30, 21.9)$. The graph of f passes through the points as expected. See Figure 8.37.

(c) To estimate the number of paid days after 3 years, evaluate $f(3)$.

$$f(3) = -0.016026(3)^2 + 0.92785(3) + 8.4882 \approx 11.1$$

This is quite close to the actual value of 11.2 days.   ■

8.3 PUTTING IT ALL TOGETHER

A matrix is a rectangular array of elements. The elements are often real numbers. An augmented matrix may be used to represent a system of linear equations. One of the most common methods for solving a system of linear equations is Gaussian elimination with backward substitution. Through a sequence of matrix row operations, an augmented matrix is transformed into row-echelon form or reduced row-echelon form. Backward substitution is frequently used to find the solution when a matrix is in row-echelon form. Technology can be used to solve systems of linear equations.

The following table summarizes some mathematical concepts involved with solving systems of linear equations in three variables.

Augmented Matrix

A linear system can be represented using an augmented matrix.

$$\begin{bmatrix} 2 & 0 & -3 & 2 \\ -1 & 2 & -2 & -5 \\ 1 & -2 & -1 & 7 \end{bmatrix} \qquad \begin{array}{rcr} 2x & -3z = & 2 \\ -x + 2y & - 2z = & -5 \\ x - 2y & -z = & 7 \end{array}$$

Row-Echelon Form

The following matrices are in row-echelon form. They represent the three situations where there are no solutions, one solution, and an infinite number of solutions.

$$\begin{bmatrix} 1 & -2 & 1 & 0 \\ 0 & 1 & 2 & 3 \\ 0 & 0 & 0 & 1 \end{bmatrix} \qquad \begin{bmatrix} 1 & -2 & 1 & 0 \\ 0 & 1 & 2 & 3 \\ 0 & 0 & 1 & 1 \end{bmatrix} \qquad \begin{bmatrix} 1 & -2 & 1 & 0 \\ 0 & 1 & 2 & 3 \\ 0 & 0 & 0 & 0 \end{bmatrix}$$

No solutions | One solution | Infinite Number of Solutions
| $(1, 1, 1)$ | $(6 - 5z, 3 - 2z, z)$

Backward Substitution

Backward substitution can be used to solve a system of linear equations represented by an augmented matrix in row-echelon form.

$$\begin{bmatrix} 1 & -2 & 1 & 3 \\ 0 & 1 & -2 & -3 \\ 0 & 0 & 1 & 2 \end{bmatrix}$$

From the last row, $z = 2$.
Substituting into the second row, $y - 2(2) = -3$ or $y = 1$.
Let $z = 2$ and $y = 1$ in the first row, $x - 2(1) + 2 = 3$ or $x = 3$.
The solution is $(3, 1, 2)$.

8.3 EXERCISES

   Tape 6

Dimensions of Matrices and Augmented Matrices

Exercises 1 and 2: State the dimension of each matrix.

1. (a) $\begin{bmatrix} 1 \\ 2 \\ 3 \end{bmatrix}$ **(b)** $\begin{bmatrix} a & b & c \\ d & e & b \end{bmatrix}$ **(c)** $\begin{bmatrix} 3 & 0 \\ 1 & -4 \end{bmatrix}$

2. (a) $\begin{bmatrix} -1 & 1 \end{bmatrix}$ **(b)** $\begin{bmatrix} 1 & -1 \\ 7 & 5 \\ -4 & 0 \end{bmatrix}$

(c) $\begin{bmatrix} 1 & 3 & 8 & -3 \\ 1 & -1 & 1 & -2 \\ 4 & 5 & 0 & -1 \end{bmatrix}$

Exercises 3–6: Represent each linear system by an augmented matrix, and state the dimension of the matrix.

3. $\begin{array}{l} 5x - 2y = 3 \\ -x + 3y = -1 \end{array}$ **4.** $\begin{array}{l} 3x + y = 4 \\ -x + 4y = 5 \end{array}$

5. $\begin{array}{l} -3x + 2y + z = -4 \\ 5x \quad\;\; - z = 9 \\ x - 3y - 6z = -9 \end{array}$ **6.** $\begin{array}{l} x + 2y - z = 2 \\ -2x + y - 2z = -3 \\ 7x + y - z = 7 \end{array}$

Exercises 7–10: Write the system of linear equations that the augmented matrix represents.

7. $\begin{bmatrix} 3 & 2 & 4 \\ 0 & 1 & 5 \end{bmatrix}$ **8.** $\begin{bmatrix} -2 & 1 & 5 \\ 7 & 9 & 2 \end{bmatrix}$

9. $\begin{bmatrix} 3 & 1 & 4 & 0 \\ 0 & 5 & 8 & -1 \\ 0 & 0 & -7 & 1 \end{bmatrix}$

10. $\begin{bmatrix} 1 & -1 & 3 & 2 \\ -2 & 1 & 1 & -2 \\ -1 & 0 & -2 & 1 \end{bmatrix}$

Row-Echelon Form

Exercises 11 and 12: Determine if the matrix is in row-echelon form.

11. (a) $\begin{bmatrix} 1 & 3 & 2 \\ 0 & 1 & -1 \end{bmatrix}$ **(b)** $\begin{bmatrix} 1 & 4 & -1 & 0 \\ 0 & -1 & 1 & 3 \\ 0 & 2 & 1 & 7 \end{bmatrix}$

(c) $\begin{bmatrix} 1 & 6 & -8 & 5 \\ 0 & 1 & 7 & 9 \\ 0 & 0 & 1 & 11 \end{bmatrix}$

12. **(a)** $\begin{bmatrix} 1 & 3 & | & 2 \\ 0 & -1 & | & -1 \end{bmatrix}$

(b) $\begin{bmatrix} 1 & 3 & -1 & | & 8 \\ 0 & 1 & 5 & | & 3 \\ 0 & 0 & 0 & | & 0 \end{bmatrix}$

(c) $\begin{bmatrix} 0 & 0 & 1 & | & 1 \\ 0 & 1 & 7 & | & 9 \\ 1 & 2 & -1 & | & 11 \end{bmatrix}$

Exercises 13–20: The augmented matrix is in row-echelon form and represents a linear system. Solve the system using backward substitution, if possible. Write the solution as either an ordered pair or an ordered triple.

13. $\begin{bmatrix} 1 & 2 & | & 3 \\ 0 & 1 & | & -1 \end{bmatrix}$ **14.** $\begin{bmatrix} 1 & -5 & | & 6 \\ 0 & 0 & | & 1 \end{bmatrix}$

15. $\begin{bmatrix} 1 & -1 & | & 2 \\ 0 & 1 & | & 0 \end{bmatrix}$ **16.** $\begin{bmatrix} 1 & 4 & | & -2 \\ 0 & 1 & | & 3 \end{bmatrix}$

17. $\begin{bmatrix} 1 & 1 & -1 & | & 4 \\ 0 & 1 & -1 & | & 2 \\ 0 & 0 & 1 & | & 1 \end{bmatrix}$

18. $\begin{bmatrix} 1 & -2 & -1 & | & 0 \\ 0 & 1 & -3 & | & 1 \\ 0 & 0 & 1 & | & 2 \end{bmatrix}$

19. $\begin{bmatrix} 1 & 2 & -1 & | & 5 \\ 0 & 1 & -2 & | & 1 \\ 0 & 0 & 0 & | & 0 \end{bmatrix}$

20. $\begin{bmatrix} 1 & -1 & 2 & | & 8 \\ 0 & 1 & -4 & | & 2 \\ 0 & 0 & 1 & | & -1 \end{bmatrix}$

Solving Systems with Gaussian Elimination

Exercises 21–28: Use Gaussian elimination with backward substitution to solve each system of linear equations. Write the solution as an ordered pair or an ordered triple, whenever possible.

21. $\begin{aligned} x + 2y &= 3 \\ -x - y &= 7 \end{aligned}$ **22.** $\begin{aligned} 2x + 4y &= 10 \\ x - 2y &= -3 \end{aligned}$

23. $\begin{aligned} x + 2y + z &= 3 \\ x + y - z &= 3 \\ -x - 2y + z &= -5 \end{aligned}$ **24.** $\begin{aligned} x + y + z &= 6 \\ 2x + 3y - z &= 3 \\ x + y + 2z &= 10 \end{aligned}$

25. $\begin{aligned} 3x + y + 3z &= 14 \\ x + y + z &= 6 \\ -2x - 2y + 3z &= -7 \end{aligned}$ **26.** $\begin{aligned} x + 3y - 2z &= 3 \\ -x - 2y + z &= -2 \\ 2x - 7y + z &= 1 \end{aligned}$

27. $\begin{aligned} x + 2y - z &= 2 \\ 2x + 5y + z &= 8 \\ 3x + 7y &= 5 \end{aligned}$ **28.** $\begin{aligned} x + y + z &= 3 \\ x + y + 2z &= 4 \\ 2x + 2y + 3z &= 7 \end{aligned}$

Exercises 29–32: Reduced Row-Echelon Form   Use Gaussian elimination to transform the augmented matrix of the linear system into reduced row-echelon form. Find the solution.

29. $\begin{aligned} x - y &= 1 \\ x + y &= 5 \end{aligned}$ **30.** $\begin{aligned} 2x + 3y &= 1 \\ x - 2y &= -3 \end{aligned}$

31. $\begin{aligned} x + 2y + z &= 3 \\ y - z &= -2 \\ -x - 2y + 2z &= 6 \end{aligned}$ **32.** $\begin{aligned} x + z &= 2 \\ x - y - z &= 0 \\ -2x + y &= -2 \end{aligned}$

Exercises 33–36: Technology   Use technology to find the solution. Approximate values to four significant digits.

33. $\begin{aligned} 2.1x + 0.5y + 1.7z &= 4.9 \\ -2x + 1.5y - 1.7z &= 3.1 \\ 5.8x - 4.6y + 0.8z &= 9.3 \end{aligned}$

34. $\begin{aligned} 53x + 95y + 12z &= 108 \\ 81x - 57y - 24z &= -92 \\ -9x + 11y - 78z &= 21 \end{aligned}$

35. $\begin{aligned} 0.1x + 0.3y + 1.7z &= 0.6 \\ 0.6x + 0.1y - 3.1z &= 6.2 \\ 2.4y + 0.9z &= 3.5 \end{aligned}$

36. $\begin{aligned} 103x - 886y + 431z &= 1200 \\ -55x + 981y &= 1108 \\ -327x + 421y + 337z &= 99 \end{aligned}$

Applications

37. *Food Shelters*   (Refer to Example 7.) Three food shelters have monthly food costs F in dollars, number of people served per month N, and monthly charitable receipts R in dollars as shown in the table.

Food Costs (F)	Number Served (N)	Charitable Receipts (R)
1300	1800	5000
5300	3200	12,000
6500	4500	13,000

(a) Model this data using $F = aN + bR + c$, where a, b, and c are constants.

(b) Predict the food costs for a shelter that serves 3500 people and receives charitable receipts of $12,500. Round your answer to the nearest hundred dollars.

38. *Computing Time* When computers are programmed to solve large linear systems involved in applications, such as designing aircraft or large electrical circuits, they frequently use Gaussian elimination with backward substitution. Solving a linear system with n equations and n variables requires a computer to perform a total of $T(n) = \frac{2}{3}n^3 + \frac{3}{2}n^2 - \frac{7}{6}n$ arithmetic operations (additions, subtractions, multiplications, and divisions). (Source: R. Burden and J. Faires, *Numerical Analysis.*)

(a) John Atanasoff, the inventor of the modern digital computer, needed to solve a system of 29 linear equations. Evaluate $T(29)$ to find the number of arithmetic operations this would require. Would it be too many to do by hand?

(b) Compute T for $n = 10, 100, 1000, 10,000$, and $100,000$. List the results in a table.

(c) If the number of equations and variables increases by a factor of 10, does the number of arithmetic operations also increase by a factor of 10? Explain.

(d) Discuss why supercomputers are needed to solve large systems of linear equations.

39. *Pumping Water* Three pumps are being used to empty a small swimming pool. The first pump is twice as fast as the second pump. The first two pumps can empty the pool in 8 hours, while all three pumps can empty it in 6 hours. How long would it take each pump to empty the pool individually? (*Hint:* Let x represent the fraction of the pool that the first pump can empty in 1 hour. Let y and z represent this fraction for the second and third pumps, respectively.)

40. *Pumping Water* Suppose in the previous exercise the first pump is three times as fast as the third pump, the first and second pumps can empty the pool in 6 hours, and all three pumps can empty the pool in 8 hours.

(a) Is this data realistic? Explain your reasoning.

(b) Conjecture a mathematical solution to this data.

(c) Test your conjecture by solving the problem.

Exercises 41 and 42: Traffic Flow   The accompanying figure shows three one-way streets with intersections A, B, and C. Numbers indicate the average traffic flow in vehicles per minute. The variables x, y, and z denote unknown traffic flows that need to be determined for timing of stoplights.

(a) *If the number of vehicles per minute entering an intersection must equal the number exiting an*

intersection, verify that the accompanying system of linear equations describes the traffic flow.

(b) *Rewrite the system and solve.*

(c) *Interpret your solution.*

41. A: $x + 5 = y + 7$
B: $z + 6 = x + 3$
C: $y + 3 = z + 4$

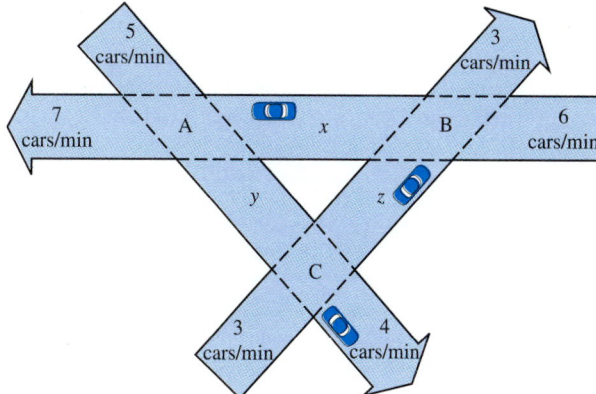

42. A: $x + 7 = y + 4$
B: $4 + 5 = x + z$
C: $y + 8 = 9 + 4$

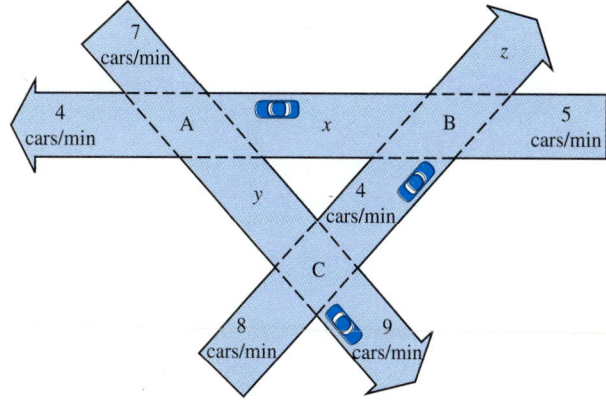

Exercises 43–46: Each set of data can be modeled by $f(x) = ax^2 + bx + c$, where x represents the year.

(a) *Find a linear system whose solution represents values of a, b, and c.*

(b) *Use technology to find the solution.*

(c) *Graph f and the data in the same viewing rectangle.*

(d) *Make your own prediction using f.*

43. *Chronic Health Care*   A large percentage of the U.S. population will require chronic health care in the coming decades. The average caregiving age is 50–64, while the typical person needing chronic care is 85 or older. The ratio of potential caregivers to those needing chronic health care will shrink in the coming years as shown in the table. (Source: Robert Wood Johnson Foundation Chronic Care in America: A 21st Century Challenge.)

Year	1990	2010	2030
Ratio	11	10	6

44. *Home Health Care* The table shows the cost of Medicare home health care y in billions of dollars during the year x. In this table $x = 0$ corresponds to 1990, while $x = 6$ corresponds to 1996. (Source: Health Care Financing Administration.)

x	0	3	6
y	3.9	10.5	18.1

45. *Women in the Military* The table shows the percentage y of the enlisted people in the military who are women. In this table $x = 3$ corresponds to 1973 and $x = 26$ to 1996. (Source: Department of Defense.)

x	3	18	26
y	2.2	10.4	12.8

46. *Carbon Dioxide Levels* Carbon dioxide (CO_2) is a greenhouse gas. Its concentrations in parts per million (ppm) have been measured at Mauna Loa, Hawaii, during the past 30 years. The table lists measurements for three selected years. (Source: A. Nilsson, *Greenhouse Earth.*)

Year	1958	1973	1992
CO_2 (ppm)	315	325	354

Writing about Mathematics

1. A linear equation in three variables can be represented by a flat plane. Describe geometrically possible situations that can occur when a system of three linear equations has either no solutions or an infinite number of solutions.

2. Give an example of an augmented matrix that represents a system of linear equations that has no solutions. Explain your reasoning.

8.4 Properties and Applications of Matrices

Matrix Notation • Sums, Differences, and Scalar Multiples of Matrices • Matrix Products

Introduction

Matrices occur in many fields of study and have a wide variety of applications. New technologies, such as digital photography and computer graphics, frequently utilize matrices to achieve their goals. Many of these technologies have become specialized fields that require substantial interdisciplinary skills, including mathematics. In this section we discuss properties of matrices and some of their applications.

Matrix Notation

The following notation is used to denote elements in a matrix.

$$\begin{bmatrix} a_{11} & a_{12} \\ a_{21} & a_{22} \end{bmatrix} \quad \begin{bmatrix} a_{11} & a_{12} & a_{13} \\ a_{21} & a_{22} & a_{23} \\ a_{31} & a_{32} & a_{33} \end{bmatrix} \quad \begin{bmatrix} a_{11} & a_{12} & a_{13} & a_{14} \\ a_{21} & a_{22} & a_{23} & a_{24} \\ a_{31} & a_{32} & a_{33} & a_{34} \\ a_{41} & a_{42} & a_{43} & a_{44} \end{bmatrix} \quad \begin{bmatrix} a_{11} & a_{12} \\ a_{21} & a_{22} \\ a_{31} & a_{32} \end{bmatrix} \quad \begin{bmatrix} a_{11} & a_{12} & a_{13} \\ a_{21} & a_{22} & a_{23} \end{bmatrix}$$

A general element in a matrix A is denoted by a_{ij}. This refers to the element in the ith row, jth column. For example, a_{23} would be the element of A located in the second row, third column. Two m by n matrices A and B are **equal** if corresponding elements are equal. If A and B have different dimensions, they cannot be equal.

EXAMPLE 1   *Determining matrix elements*

Let a_{ij} denote an element in A and b_{ij} an element in B, where

$$A = \begin{bmatrix} 3 & -3 & 7 \\ 1 & 6 & -2 \\ 4 & 2 & 5 \end{bmatrix} \quad \text{and} \quad B = \begin{bmatrix} 3 & x & 7 \\ 1 & 6 & -2 \\ 4 & 5 & 2 \end{bmatrix}.$$

(a) Identify a_{12}, b_{32}, and a_{13}.
(b) Compute $a_{31}b_{13} + a_{32}b_{23} + a_{33}b_{33}$.
(c) Is there a value for x that will make the statement $A = B$ true?

Solution

(a) The element a_{12} is located in the first row, second column of A. Thus, $a_{12} = -3$. In a similar manner, $b_{32} = 5$ and $a_{13} = 7$.
(b) $a_{31}b_{13} + a_{32}b_{23} + a_{33}b_{33} = (4)(7) + (2)(-2) + (5)(2) = 34$
(c) No, since $a_{32} = 2 \neq 5 = b_{32}$ and $a_{33} = 5 \neq 2 = b_{33}$. Even if we let $x = -3$, there are other corresponding elements in A and B that are not equal. ■

Sums, Differences, and Scalar Multiples of Matrices

The FCC has mandated all television stations to transmit digital signals by the year 2006. Digital photography is a new technology in which matrices play an important role. Figure 8.38 shows a black-and-white photograph. In Figure 8.39 a grid has been laid over this photograph. Each square in the grid represents a pixel in a digitized photograph. A gray scale is illustrated in Figure 8.40 on the next page. In this scale 0 corresponds to white, while 11 represents black. As the scale increases from 0 to 11, the shade of gray darkens.

FIGURE 8.38 Photograph to Be Digitized

FIGURE 8.39 Placing a Grid over the Photograph

Source: *Visualization* by Friedhoff and Benzon © 1991 by W. H. Freeman and Company.

In Figure 8.41 the different shades of gray are stored as numbers in a matrix with dimension 20×20. Notice how the dark pole in the picture is represented by the fifth and sixth columns in the matrix. Both these columns contain the numbers 10 and 11, which indicates a dark portion in the photograph. The matrix is a numerical or digital representation of the photograph, which can be stored and transmitted over the Internet. In real applications, low resolution, digital photographs utilize a grid with 512 rows and 512 columns. The gray scale typically contains 64 levels instead of the 12 in our example. High resolution photography utilizes 2048 rows and columns in the grid with 256 levels of gray.

1	1	1	1	10	11	1	1	1	1	1	1	1	1	1	1	1	1	9	8
1	1	1	1	10	11	1	1	1	1	1	1	1	1	1	1	1	2	8	9
1	1	1	1	10	11	1	1	1	1	1	1	1	1	1	1	1	4	8	10
1	1	1	1	10	11	1	1	1	1	1	1	1	1	1	1	1	4	8	10
1	1	1	1	10	11	1	1	1	1	1	1	1	1	1	1	1	5	8	10
1	1	1	1	10	11	1	1	1	1	1	1	1	1	1	1	1	5	8	8
1	0	0	0	10	11	0	0	0	0	0	1	1	0	0	0	0	5	8	8
0	2	2	2	10	11	2	3	3	4	4	5	5	7	7	7	7	6	8	8
3	5	9	8	10	11	7	7	7	8	8	8	7	6	6	6	6	8	8	8
11	9	9	6	10	11	7	7	8	8	8	8	6	5	5	5	6	8	8	8
10	9	9	6	10	11	7	8	8	8	8	8	3	5	6	4	6	8	8	8
10	7	6	6	10	11	6	7	7	6	6	6	4	3	2	2	5	8	8	8
4	3	8	7	10	11	5	4	3	2	2	2	2	2	2	4	8	8	8	8
6	4	2	2	10	11	2	2	2	2	3	2	2	2	7	9	8	8	8	8
2	2	2	2	10	11	2	2	2	2	2	2	5	7	9	9	7	8	8	8
2	2	2	2	10	11	2	2	2	4	2	8	9	9	10	9	8	7	8	8
2	2	2	2	10	11	2	2	2	2	5	7	9	10	11	9	10	9	7	7
2	2	2	2	10	11	2	2	2	2	7	9	9	9	10	11	10	9	7	7
2	2	5	7	9	8	2	2	2	2	7	9	9	9	10	11	11	11	10	7
7	9	7	5	2	2	2	2	2	2	8	9	9	10	11	11	11	11	11	10

0	1	2	3	4	5	6	7	8	9	10	11

FIGURE 8.40 Gray Levels   **FIGURE 8.41** Digitized Photograph

Source: *Visualization* by Friedhoff and Benzon © 1991 by W. H. Freeman and Company. Used with permission.

Matrix Addition and Subtraction. In order to simplify the above concept of digital photography, we reduce the grid to 3×3 and have four gray levels, where 0 represents white, 1 light gray, 2 dark gray, and 3 black. Suppose that we would like to digitize the letter T shown in Figure 8.42. The gray levels are shown in Figure 8.43.

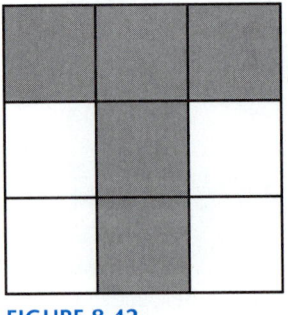

FIGURE 8.42

0	1	2	3

FIGURE 8.43

Since the T is dark gray and the background is white, Figure 8.42 can be represented by

$$A = \begin{bmatrix} 2 & 2 & 2 \\ 0 & 2 & 0 \\ 0 & 2 & 0 \end{bmatrix}.$$

Suppose that we want to make the entire picture darker. If we change every element in A to 3, the entire picture would be black. A more acceptable solution would be to darken each pixel by one gray level. This corresponds to adding 1 to each element in the matrix A. This can be accomplished efficiently using matrix notation.

$$\begin{bmatrix} 2 & 2 & 2 \\ 0 & 2 & 0 \\ 0 & 2 & 0 \end{bmatrix} + \begin{bmatrix} 1 & 1 & 1 \\ 1 & 1 & 1 \\ 1 & 1 & 1 \end{bmatrix} = \begin{bmatrix} 2+1 & 2+1 & 2+1 \\ 0+1 & 2+1 & 0+1 \\ 0+1 & 2+1 & 0+1 \end{bmatrix} = \begin{bmatrix} 3 & 3 & 3 \\ 1 & 3 & 1 \\ 1 & 3 & 1 \end{bmatrix}$$

To add two matrices of equal dimension, add corresponding elements. The result is shown as a picture in Figure 8.44. Notice that the background is now light gray and the T is black. The entire picture is darker.

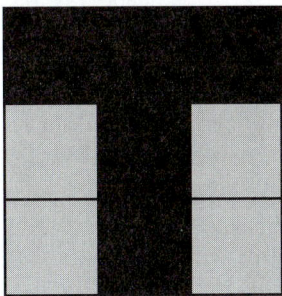

FIGURE 8.44

To lighten the picture in Figure 8.44, subtract 1 from each element. To subtract two matrices of equal dimension, subtract corresponding elements.

$$\begin{bmatrix} 3 & 3 & 3 \\ 1 & 3 & 1 \\ 1 & 3 & 1 \end{bmatrix} - \begin{bmatrix} 1 & 1 & 1 \\ 1 & 1 & 1 \\ 1 & 1 & 1 \end{bmatrix} = \begin{bmatrix} 3-1 & 3-1 & 3-1 \\ 1-1 & 3-1 & 1-1 \\ 1-1 & 3-1 & 1-1 \end{bmatrix} = \begin{bmatrix} 2 & 2 & 2 \\ 0 & 2 & 0 \\ 0 & 2 & 0 \end{bmatrix}$$

Increasing the contrast in a picture causes a light area to become lighter and a dark area to become darker. As a result, there are fewer pixels with intermediate gray levels. Changing contrast is different from making the entire picture lighter or darker.

The digital picture in Figure 8.45 on the next page was taken by *Voyager 1* and faintly shows the rings of Saturn. By increasing the contrast in the picture, as shown in Figure 8.46 on the next page, the rings are much clearer. Notice that light gray features have become lighter, while dark gray areas have become darker. One exciting aspect about digital photography is that contrast enhancement could be performed on Earth after *Voyager 1* sent back the original picture.

FIGURE 8.45 Rings of Saturn
Source: NASA.

FIGURE 8.46 Enhanced Photograph

EXAMPLE 2 *Applying addition of matrices to digital photography*

Increase the contrast of the + sign in Figure 8.47 by changing light gray to white and dark gray to black. Use matrices to represent this computation.

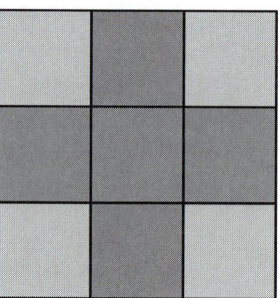

FIGURE 8.47

Solution

Figure 8.47 can be represented by the matrix A.

$$A = \begin{bmatrix} 1 & 2 & 1 \\ 2 & 2 & 2 \\ 1 & 2 & 1 \end{bmatrix}$$

To change the contrast we can reduce each 1 in the matrix A to 0 and increase each 2 to 3. The addition of the matrix B can accomplish this task.

$$A + B = \begin{bmatrix} 1 & 2 & 1 \\ 2 & 2 & 2 \\ 1 & 2 & 1 \end{bmatrix} + \begin{bmatrix} -1 & 1 & -1 \\ 1 & 1 & 1 \\ -1 & 1 & -1 \end{bmatrix} = \begin{bmatrix} 0 & 3 & 0 \\ 3 & 3 & 3 \\ 0 & 3 & 0 \end{bmatrix}$$

The picture corresponding to $A + B$ is shown in Figure 8.48.

FIGURE 8.48 ∎

Critical Thinking

Discuss how the negative image of a digital picture might be represented.

Multiplication of a Matrix by a Scalar. The matrix

$$B = \begin{bmatrix} 1 & 1 & 1 \\ 1 & 1 & 1 \\ 1 & 1 & 1 \end{bmatrix}$$

can be used to lighten or darken a picture. Suppose that a photograph is represented by a matrix A with gray levels 0 through 11. Every time the matrix B is added to A, the picture becomes slightly darker. For example, if

$$A = \begin{bmatrix} 0 & 5 & 0 \\ 5 & 5 & 5 \\ 0 & 5 & 0 \end{bmatrix}$$

then the addition of $A + B + B$ would darken the picture by two gray levels, and could be computed by

$$A + B + B = \begin{bmatrix} 0 & 5 & 0 \\ 5 & 5 & 5 \\ 0 & 5 & 0 \end{bmatrix} + \begin{bmatrix} 1 & 1 & 1 \\ 1 & 1 & 1 \\ 1 & 1 & 1 \end{bmatrix} + \begin{bmatrix} 1 & 1 & 1 \\ 1 & 1 & 1 \\ 1 & 1 & 1 \end{bmatrix} = \begin{bmatrix} 2 & 7 & 2 \\ 7 & 7 & 7 \\ 2 & 7 & 2 \end{bmatrix}.$$

A simpler way to write the expression $A + B + B$ is $A + 2B$. Multiplying B by 2 to obtain $2B$ is called **scalar multiplication.**

$$2B = 2\begin{bmatrix} 1 & 1 & 1 \\ 1 & 1 & 1 \\ 1 & 1 & 1 \end{bmatrix} = \begin{bmatrix} 2(1) & 2(1) & 2(1) \\ 2(1) & 2(1) & 2(1) \\ 2(1) & 2(1) & 2(1) \end{bmatrix} = \begin{bmatrix} 2 & 2 & 2 \\ 2 & 2 & 2 \\ 2 & 2 & 2 \end{bmatrix}$$

Each element of B is multiplied by the real number 2.

Sometimes a matrix B is denoted by $B = [b_{ij}]$, where b_{ij} represents the element in the ith row, jth column. In this way, we could write $2B$ as $2[b_{ij}] = [2b_{ij}]$. This indicates that to calculate $2B$, multiply each b_{ij} by 2. In a similar manner, a matrix A is sometimes denoted by $[a_{ij}]$.

Some operations on matrices are now summarized.

Operations on matrices

Matrix Addition The sum of two $m \times n$ matrices A and B is the $m \times n$ matrix $A + B$, in which each element is the sum of the corresponding elements of A and B. This is written as $A + B = [a_{ij}] + [b_{ij}] = [a_{ij} + b_{ij}]$. If A and B have different dimensions, then $A + B$ is undefined.

Matrix Subtraction The difference of two $m \times n$ matrices A and B is the $m \times n$ matrix $A - B$, in which each element is the difference of the corresponding elements of A and B. This is written as $A - B = [a_{ij}] - [b_{ij}] = [a_{ij} - b_{ij}]$. If A and B have different dimensions, then $A - B$ is undefined.

Multiplication of a Matrix by a Scalar The product of a scalar (real number) k and an $m \times n$ matrix A is the $m \times n$ matrix kA, in which each element is k times the corresponding element of A. This is written as $kA = k[a_{ij}] = [ka_{ij}]$.

EXAMPLE 3   *Performing operations on matrices*

If possible, perform the indicated operations using

$$A = \begin{bmatrix} 4 & -2 \\ 3 & 5 \end{bmatrix}, B = \begin{bmatrix} 0 & 1 \\ -2 & 3 \end{bmatrix}, C = \begin{bmatrix} 1 & -1 \\ 0 & 7 \\ -4 & 2 \end{bmatrix}, \text{ and } D = \begin{bmatrix} -1 & -3 \\ 9 & -7 \\ 1 & 8 \end{bmatrix}.$$

(a) $A + 3B$

(b) $A - C$

(c) $-2C - 3D$

Solution

(a) $A + 3B = \begin{bmatrix} 4 & -2 \\ 3 & 5 \end{bmatrix} + 3\begin{bmatrix} 0 & 1 \\ -2 & 3 \end{bmatrix} = \begin{bmatrix} 4 & -2 \\ 3 & 5 \end{bmatrix} + \begin{bmatrix} 0 & 3 \\ -6 & 9 \end{bmatrix} = \begin{bmatrix} 4 & 1 \\ -3 & 14 \end{bmatrix}$

(b) $A - C$ is undefined because the dimension of A is 2×2 and unequal to the dimension of C, which is 3×2.

(c) $-2C - 3D = -2\begin{bmatrix} 1 & -1 \\ 0 & 7 \\ -4 & 2 \end{bmatrix} - 3\begin{bmatrix} -1 & -3 \\ 9 & -7 \\ 1 & 8 \end{bmatrix} =$

$\begin{bmatrix} -2 & 2 \\ 0 & -14 \\ 8 & -4 \end{bmatrix} - \begin{bmatrix} -3 & -9 \\ 27 & -21 \\ 3 & 24 \end{bmatrix} = \begin{bmatrix} 1 & 11 \\ -27 & 7 \\ 5 & -28 \end{bmatrix}$ ■

Matrix Products

Addition, subtraction, and multiplication can be performed on numbers, variables, and functions. The same operations apply to matrices. Matrix multiplication is different than scalar multiplication.

Suppose two students are taking day classes at one college and night classes at another, in order to graduate on time. Tables 8.7 and 8.8 list the number of credits taken by the students and the cost per credit at each college.

TABLE 8.7		
Credits	**College A**	**College B**
Student 1	10	7
Student 2	11	4

TABLE 8.8	
Cost	**Tuition**
College A	$60
College B	$80

The cost of tuition is computed by multiplying the number of credits times the cost of each credit. Student 1 is taking 10 credits at $60 each and 7 credits at $80 each. The total tuition for Student 1 is 10($60) + 7($80) = $1160. In a similar manner, the tuition for Student 2 is given by 11($60) + 4($80) = $980.

The information in these tables can be represented by matrices. Let A represent Table 8.7 and B represent Table 8.8.

$$A = \begin{bmatrix} 10 & 7 \\ 11 & 4 \end{bmatrix} \quad \text{and} \quad B = \begin{bmatrix} 60 \\ 80 \end{bmatrix}$$

The matrix product AB calculates tuition cost for each student.

$$AB = \begin{bmatrix} 10 & 7 \\ 11 & 4 \end{bmatrix} \begin{bmatrix} 60 \\ 80 \end{bmatrix} = \begin{bmatrix} 10(60) + 7(80) \\ 11(60) + 4(80) \end{bmatrix} = \begin{bmatrix} 1160 \\ 980 \end{bmatrix}$$

Generalizing from this example provides the following definition of matrix multiplication.

Matrix multiplication

The **product** of an $m \times n$ matrix A and an $n \times k$ matrix B is the $m \times k$ matrix AB, which is computed as follows. To find the element of AB in the ith row and jth column, multiply each element in the ith row of A by the corresponding element in the jth column of B. The sum of these products will give the element of row i, column j in AB.

Note: In order to compute the product of two matrices, the number of columns in the first matrix must equal the number of rows in the second matrix.

EXAMPLE 4   *Multiplying matrices*

If possible, compute each product using

$$A = \begin{bmatrix} 1 & -1 \\ 0 & 3 \\ 4 & -2 \end{bmatrix}, B = \begin{bmatrix} -1 \\ -2 \end{bmatrix}, C = \begin{bmatrix} 1 & 2 & 3 \\ 4 & 5 & 6 \end{bmatrix}, \text{and } D = \begin{bmatrix} 1 & -1 & 2 \\ 0 & 3 & -2 \\ -3 & 4 & 5 \end{bmatrix}.$$

(a) AB **(b)** CA **(c)** DC **(d)** CD

Solution

(a) The dimension of A is 3×2 and the dimension of B is 2×1. The dimension of AB is 3×1 and can be found as follows.

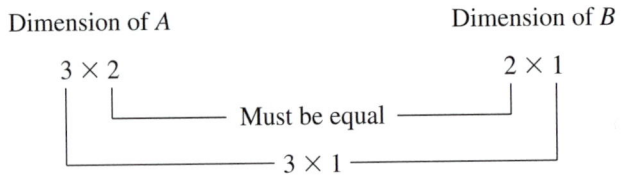

$$AB = \begin{bmatrix} 1 & -1 \\ 0 & 3 \\ 4 & -2 \end{bmatrix} \begin{bmatrix} -1 \\ -2 \end{bmatrix} = \begin{bmatrix} (1)(-1) + (-1)(-2) \\ (0)(-1) + (3)(-2) \\ (4)(-1) + (-2)(-2) \end{bmatrix} = \begin{bmatrix} 1 \\ -6 \\ 0 \end{bmatrix}$$

(b) The dimension of C is 2×3, and the dimension of A is 3×2. Thus, CA has dimension 2×2.

$$\begin{aligned} CA &= \begin{bmatrix} 1 & 2 & 3 \\ 4 & 5 & 6 \end{bmatrix} \begin{bmatrix} 1 & -1 \\ 0 & 3 \\ 4 & -2 \end{bmatrix} \\ &= \begin{bmatrix} 1(1) + 2(0) + 3(4) & 1(-1) + 2(3) + 3(-2) \\ 4(1) + 5(0) + 6(4) & 4(-1) + 5(3) + 6(-2) \end{bmatrix} \\ &= \begin{bmatrix} 13 & -1 \\ 28 & -1 \end{bmatrix} \end{aligned}$$

(c) The dimension of D is 3×3 and the dimension of C is 2×3. Therefore, DC is undefined. Note that D has 3 columns and C has only 2 rows.

(d) The dimension of C is 2×3 and the dimension of D is 3×3. Thus, CD has dimension 2×3.

$$\begin{aligned} CD &= \begin{bmatrix} 1 & 2 & 3 \\ 4 & 5 & 6 \end{bmatrix} \begin{bmatrix} 1 & -1 & 2 \\ 0 & 3 & -2 \\ -3 & 4 & 5 \end{bmatrix} \\ &= \begin{bmatrix} 1(1) + 2(0) + 3(-3) & 1(-1) + 2(3) + 3(4) & 1(2) + 2(-2) + 3(5) \\ 4(1) + 5(0) + 6(-3) & 4(-1) + 5(3) + 6(4) & 4(2) + 5(-2) + 6(5) \end{bmatrix} \\ &= \begin{bmatrix} -8 & 17 & 13 \\ -14 & 35 & 28 \end{bmatrix} \quad \blacksquare \end{aligned}$$

MAKING CONNECTIONS

The Commutative Property and Matrix Multiplication

In Example 4 it was shown that $CD \neq DC$. Unlike multiplication of numbers, variables, and functions, matrix multiplication is *not* commutative. Instead, matrix multiplication is similar to function composition, where for a general pair of functions $f \circ g \neq g \circ f$.

Matrices and Computer Graphics. Matrices with dimension 3×3 are used to describe rotation in two-dimensional computer graphics. The point (x, y) in the xy-plane is represented by the 3×1 **column matrix**

$$X = \begin{bmatrix} x \\ y \\ 1 \end{bmatrix}.$$

The last element in X always equals 1. (This element of 1 is necessary to implement translations in the next section.) An example of a rotation of an object is shown in Figure 8.49, where the hand on a gauge rotates clockwise 45° from a reading of 0 to a reading of 1.

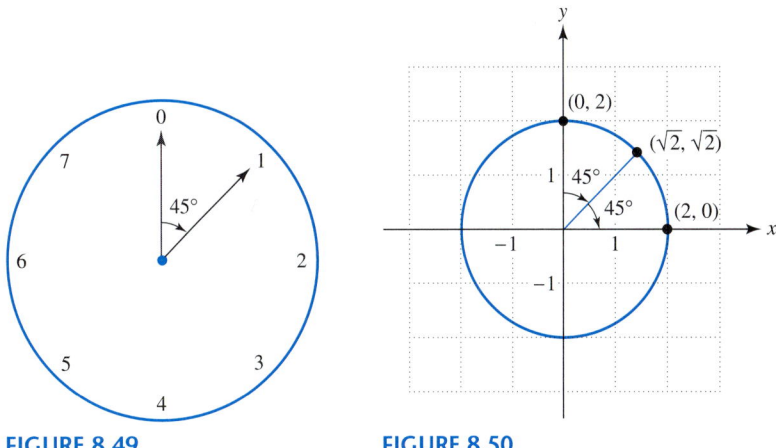

FIGURE 8.49   **FIGURE 8.50**

The new position of the tip of the hand can be computed using matrices. It can be shown that a 45° clockwise rotation of the point $(0, 2)$ is computed by the product AX.

$$AX = \begin{bmatrix} \dfrac{1}{\sqrt{2}} & \dfrac{1}{\sqrt{2}} & 0 \\[2mm] -\dfrac{1}{\sqrt{2}} & \dfrac{1}{\sqrt{2}} & 0 \\[2mm] 0 & 0 & 1 \end{bmatrix} \begin{bmatrix} 0 \\ 2 \\ 1 \end{bmatrix} = \begin{bmatrix} 0 + \dfrac{2}{\sqrt{2}} + 0 \\[2mm] 0 + \dfrac{2}{\sqrt{2}} + 0 \\[2mm] 0 + 0 + 1 \end{bmatrix} = \begin{bmatrix} \sqrt{2} \\ \sqrt{2} \\ 1 \end{bmatrix}$$

Thus, when the point $(0, 2)$ is rotated 45° clockwise about the origin, its new location is $(\sqrt{2}, \sqrt{2})$. See Figure 8.50. If the matrix is applied again to the point $(\sqrt{2}, \sqrt{2})$, it will be rotated another 45° about the origin to the position $(2, 0)$.

$$\begin{bmatrix} \dfrac{1}{\sqrt{2}} & \dfrac{1}{\sqrt{2}} & 0 \\[2mm] -\dfrac{1}{\sqrt{2}} & \dfrac{1}{\sqrt{2}} & 0 \\[2mm] 0 & 0 & 1 \end{bmatrix} \begin{bmatrix} \sqrt{2} \\ \sqrt{2} \\ 1 \end{bmatrix} = \begin{bmatrix} 1 + 1 + 0 \\ -1 + 1 + 0 \\ 0 + 0 + 1 \end{bmatrix} = \begin{bmatrix} 2 \\ 0 \\ 1 \end{bmatrix}$$

This agrees with intuition. If we rotate the point $(0, 2)$ through two 45° angles or 90°, it moves to the point $(2, 0)$. The product AAX performs two 45° rotations clockwise about the origin. (Source: C. Pokorny and C. Gerald, *Computer Graphics*.)

Critical Thinking

Conjecture the result of the product $AAAAAX$ or A^5X. Test your conjecture.

Technology and Matrices. Computing arithmetic operations on large matrices by hand can be a difficult task, prone to errors. Many graphing calculators have the capability to perform addition, subtraction, multiplication, and scalar multiplication with matrices, as this next example demonstrates.

EXAMPLE 5 *Using technology to evaluate a matrix expression*

Evaluate the expression $2A + 3B^3$, where

$$A = \begin{bmatrix} 3 & -1 & 2 \\ -1 & 6 & -1 \\ 2 & -1 & 9 \end{bmatrix} \quad \text{and} \quad B = \begin{bmatrix} 1 & -2 & 5 \\ 3 & 1 & -1 \\ 5 & 2 & 1 \end{bmatrix}.$$

Solution

In the expression $2A + 3B^3$, B^3 is equal to *BBB*. Enter each matrix into a calculator and evaluate the expression. Figure 8.51 shows the result of this computation.

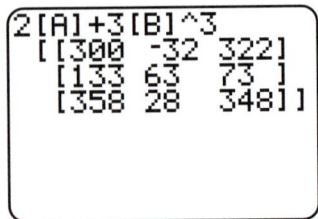

FIGURE 8.51

8.4 PUTTING IT ALL TOGETHER

Addition, subtraction, and multiplication can be performed on numbers, variables, and functions. In this section we learned how these operations also apply to matrices. The following table provides examples of these operations.

<table>
<tr><td colspan="2" align="center">**Matrix Addition**</td></tr>
<tr><td colspan="2">$$\begin{bmatrix} 1 & 2 & 3 \\ 5 & 6 & 7 \end{bmatrix} + \begin{bmatrix} -1 & 0 & 8 \\ 9 & -2 & 10 \end{bmatrix} = \begin{bmatrix} 1+(-1) & 2+0 & 3+8 \\ 5+9 & 6+(-2) & 7+10 \end{bmatrix} = \begin{bmatrix} 0 & 2 & 11 \\ 14 & 4 & 17 \end{bmatrix}$$</td></tr>
<tr><td colspan="2">Both matrices must have the same dimension for their sum to be defined.</td></tr>
<tr><td colspan="2" align="center">**Matrix Subtraction**</td></tr>
<tr><td colspan="2">$$\begin{bmatrix} 1 & -4 \\ -3 & 4 \\ 2 & 7 \end{bmatrix} - \begin{bmatrix} 5 & 1 \\ 3 & 6 \\ 8 & -9 \end{bmatrix} = \begin{bmatrix} 1-5 & -4-1 \\ -3-3 & 4-6 \\ 2-8 & 7-(-9) \end{bmatrix} = \begin{bmatrix} -4 & -5 \\ -6 & -2 \\ -6 & 16 \end{bmatrix}$$</td></tr>
<tr><td colspan="2">Both matrices must have the same dimension for their difference to be defined.</td></tr>
<tr><td colspan="2" align="center">**Scalar Multiplication**</td></tr>
<tr><td colspan="2">$$3\begin{bmatrix} 3 & -2 \\ 0 & 1 \end{bmatrix} = \begin{bmatrix} 3(3) & 3(-2) \\ 3(0) & 3(1) \end{bmatrix} = \begin{bmatrix} 9 & -6 \\ 0 & 3 \end{bmatrix}$$</td></tr>
<tr><td colspan="2" align="center">**Matrix Multiplication**</td></tr>
<tr><td colspan="2">$$\begin{bmatrix} 0 & 1 \\ 2 & -3 \end{bmatrix}\begin{bmatrix} 3 & -5 \\ 4 & 6 \end{bmatrix} = \begin{bmatrix} 0(3)+1(4) & 0(-5)+1(6) \\ 2(3)+(-3)(4) & 2(-5)+(-3)(6) \end{bmatrix} = \begin{bmatrix} 4 & 6 \\ -6 & -28 \end{bmatrix}$$</td></tr>
<tr><td colspan="2">For a matrix product to be defined, the number of columns in the first matrix must equal the number of rows in the second matrix. Matrix multiplication is not commutative. That is, $AB \neq BA$ in general.</td></tr>
</table>

8.4 EXERCISES

   Tape 6

Elements of Matrices

Exercises 1 and 2: Determine the following for the matrix A, if possible.

(a) a_{12}, a_{21}, and a_{32}
(b) $a_{11} a_{22} + 3a_{23}$

1. $A = \begin{bmatrix} 1 & 2 & 3 \\ 4 & 5 & 6 \end{bmatrix}$

2. $A = \begin{bmatrix} 1 & 2 & 3 & 4 \\ 5 & 6 & 7 & 8 \\ 9 & 10 & 11 & 12 \end{bmatrix}$

Exercises 3–8: Matrices and Road Maps   Matrices can be used to represent distances between cities. Consider the following road map illustrating freeway distances in miles between four cities. Each city has been assigned a number. For example, there is a direct route from Denver (City 1) to Colorado Springs (City 2) of approximately 60 miles. Therefore, $a_{12} = 60$ in the accompanying matrix A. The distance from Colorado Springs to Denver is also 60 miles, so $a_{21} = 60$. Since there is no direct freeway between Las Vegas (City 4) and Colorado Springs (City 2), we let $a_{24} = a_{42} = *$. The matrix A is called an **adjacency matrix.** (Reference: S. Baase, *Computer Algorithms: Introduction to Design and Analysis.*)

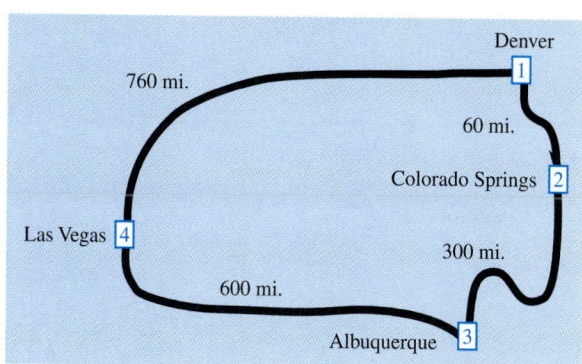

$$A = \begin{bmatrix} 0 & 60 & * & 760 \\ 60 & 0 & 300 & * \\ * & 300 & 0 & 600 \\ 760 & * & 600 & 0 \end{bmatrix}$$

3. Explain how to use *A* to find the freeway distance from Denver to Las Vegas.

4. Explain how to use *A* to find the freeway distance from Denver to Albuquerque.

5. Why are there only zeros on the main diagonal of *A*?

6. If we draw a line along the main diagonal, then the upper triangular portion is a reflection of the lower triangular portion of the matrix *A*. A matrix with this characteristic is said to be **symmetric.** Explain why adjacency matrices that represent highway distances between cities are typically symmetric.

7. If a map has 20 cities, what would the dimension be of the adjacency matrix? How many elements would there be in this matrix?

8. Given an adjacency matrix, can we draw the corresponding road map?

Exercises 9 and 10: (Refer to Exercises 3–8.) Determine an adjacency matrix A for the given road map.

9.

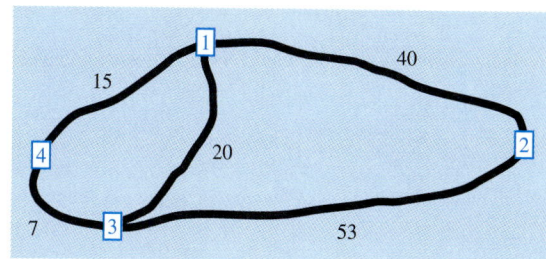

10.

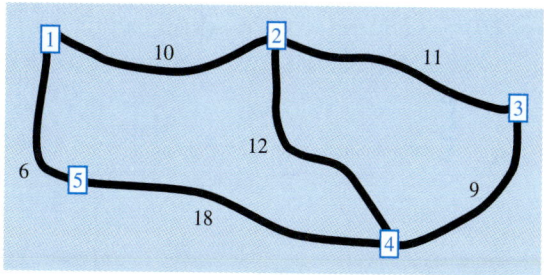

Exercises 11 and 12: (Refer to Exercises 3–8.) Sketch a road map represented by the adjacency matrix A.

11. $A = \begin{bmatrix} 0 & 30 & 20 & 5 \\ 30 & 0 & 15 & * \\ 20 & 15 & 0 & 25 \\ 5 & * & 25 & 0 \end{bmatrix}$

12. $A = \begin{bmatrix} 0 & 5 & * & 13 & 20 \\ 5 & 0 & 5 & * & * \\ * & 5 & 0 & 13 & * \\ 13 & * & 13 & 0 & 10 \\ 20 & * & * & 10 & 0 \end{bmatrix}$

Exercises 13 and 14: (Refer to Exercises 3–8.) Use the adjacency matrix A to calculate the following distance without sketching a road map.

$$A = \begin{bmatrix} 0 & 7 & 9 & * & 10 \\ 7 & 0 & 12 & * & * \\ 9 & 12 & 0 & 30 & * \\ * & * & 30 & 0 & * \\ 10 & * & * & * & 0 \end{bmatrix}$$

13. Find the shortest distance from City 2 to City 5.

14. Find the shortest distance from City 1 to City 4.

Addition, Subtraction, and Scalar Multiples of Matrices

Exercises 15–20: If possible, find each of the following.
  (a) $A + B$ (b) $3A$ (c) $2A - 3B$

15. $A = \begin{bmatrix} 2 & -6 \\ 3 & 1 \end{bmatrix}$, $B = \begin{bmatrix} -1 & 0 \\ -2 & 3 \end{bmatrix}$

16. $A = \begin{bmatrix} 1 & -2 & 5 \\ 3 & -4 & -1 \end{bmatrix}$, $B = \begin{bmatrix} 0 & -1 & -5 \\ -3 & 1 & 2 \end{bmatrix}$

17. $A = \begin{bmatrix} 1 & -1 & 0 \\ 1 & 5 & 9 \\ -4 & 8 & -5 \end{bmatrix}$, $B = \begin{bmatrix} 2 & 8 & -1 \\ 6 & -1 & 3 \end{bmatrix}$

18. $A = \begin{bmatrix} 6 & 2 & 9 \\ 3 & -2 & 0 \\ -1 & 4 & 8 \end{bmatrix}$, $B = \begin{bmatrix} 1 & 0 & -1 \\ 3 & 0 & 7 \\ 0 & -2 & -5 \end{bmatrix}$

19. $A = \begin{bmatrix} -2 & -1 \\ -5 & 1 \\ 2 & -3 \end{bmatrix}$, $B = \begin{bmatrix} 2 & -1 \\ 3 & 1 \\ 7 & -5 \end{bmatrix}$

20. $A = \begin{bmatrix} 0 & 1 \\ 3 & 2 \\ 4 & -9 \end{bmatrix}$, $B = \begin{bmatrix} 5 & 2 & -7 \\ 8 & -2 & 0 \end{bmatrix}$

Exercises 21 and 22: Evaluate the matrix expression.

21. $2\begin{bmatrix} 2 & -1 \\ 5 & 1 \\ 0 & 3 \end{bmatrix} + \begin{bmatrix} 5 & 0 \\ 7 & -3 \\ 1 & 1 \end{bmatrix} - \begin{bmatrix} 9 & -4 \\ 4 & 4 \\ 1 & 6 \end{bmatrix}$

22. $-3\begin{bmatrix} 3 & 8 \\ -1 & -9 \end{bmatrix} + 5\begin{bmatrix} 4 & -8 \\ 1 & 6 \end{bmatrix}$

Matrices and Digital Photography

Exercises 23–26: Digital Photography (Refer to the discussion of digital photography in this section.) Consider the following simplified digital photograph that has a 3 × 3 grid with four gray levels numbered from 0 to 3. It shows the number 1 in dark gray on a light gray background. Let A be the 3 × 3 matrix that represents this figure digitally.

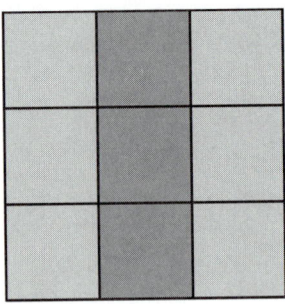

23. Find the matrix A.

24. Find a matrix B such that $A + B$ represents the entire picture becoming one gray level darker. Evaluate $A + B$.

25. Find matrix B such that $A - B$ represents the entire picture becoming lighter by one gray level. Evaluate $A - B$.

26. Find a matrix B such that $A + B$ will result in a contrast enhancement of A by one gray level. Evaluate $A + B$.

27. *Negative Image*   The negative image of a picture interchanges black and white. The number 1 is represented by the matrix A. Determine a matrix B such that $B - A$ represents the negative image of the picture represented by A. Evaluate $B - A$.

$$A = \begin{bmatrix} 0 & 3 & 0 \\ 0 & 3 & 0 \\ 0 & 3 & 0 \end{bmatrix}$$

28. (Refer to the previous exercise.) Consider the matrix A representing a digital photograph. Conjecture a matrix B that represents the negative image of this picture.

$$A = \begin{bmatrix} 0 & 3 & 0 \\ 1 & 3 & 1 \\ 2 & 3 & 2 \end{bmatrix}$$

Exercises 29 and 30: Digital Photography   The accompanying digital photograph represents the letter F using 20 pixels in a 5 × 4 grid. Assume that there are four gray levels from 0 to 3.

29. Find a matrix A that represents a digital photograph of this letter F.

30. (Continuation of the previous exercise)
    (a) Find a matrix B such that $B - A$ represents the negative image of A.
    (b) Find a matrix C where $A + C$ represents a decrease in the contrast of A by one gray level.

Exercises 31–34: Digitizing Letters   Complete the following.
    (a) *Design a matrix A with dimension 4×4 that represents a digital photograph of the given letter. Assume that there are four gray levels from 0 to 3.*
    (b) *Find a matrix B such that $B - A$ represents the negative image of the picture represented by the matrix A from part (a).*

31. Z   **32.** N
33. L   **34.** O

Matrix Multiplication

Exercises 35–38: Tuition Costs   (Refer to the discussion before Example 4.)
    (a) *Find a matrix A and a column matrix B that describe the following tables involving credits and tuition costs.*
    (b) *Find the matrix product AB, and interpret the result.*

35.

Credits	College A	College B
Student 1	12	4
Student 2	8	7

Cost	Tuition
College A	$55
College B	$70

36.

Credits	College A	College B
Student 1	15	2
Student 2	12	4

Cost	Tuition
College A	$90
College B	$75

37.

Credits	College A	College B
Student 1	10	5
Student 2	9	8
Student 3	11	3

Cost	Tuition
College A	$60
College B	$70

38.

Credits	College A	College B	College C
Student 1	6	0	3
Student 2	11	3	0
Student 3	0	12	3

Cost	Tuition
College A	$50
College B	$65
College C	$60

Exercises 39–48: If possible, find the matrix products AB and BA.

39. $A = \begin{bmatrix} 1 & -1 \\ 2 & 0 \end{bmatrix}, \quad B = \begin{bmatrix} -2 & 3 \\ 1 & 2 \end{bmatrix}$

40. $A = \begin{bmatrix} -3 & 5 \\ 2 & 7 \end{bmatrix}, \quad B = \begin{bmatrix} -1 & 2 \\ 0 & 7 \end{bmatrix}$

41. $A = \begin{bmatrix} 5 & -7 & 2 \\ 0 & 1 & 5 \end{bmatrix}, \quad B = \begin{bmatrix} 9 & 8 & 7 \\ 1 & -1 & -2 \end{bmatrix}$

42. $A = \begin{bmatrix} 2 & 1 & -1 \\ 0 & 2 & 1 \\ 3 & 2 & -1 \end{bmatrix}, \quad B = \begin{bmatrix} 1 & 0 \\ 2 & -1 \\ 3 & 1 \end{bmatrix}$

43. $A = \begin{bmatrix} 2 & -3 \\ 5 & 3 \end{bmatrix}, \quad B = \begin{bmatrix} -3 \\ 4 \\ 1 \end{bmatrix}$

44. $A = \begin{bmatrix} 3 & -1 \\ 2 & -2 \\ 0 & 4 \end{bmatrix}, \quad B = \begin{bmatrix} 1 & -4 & 0 \\ -1 & 3 & 2 \end{bmatrix}$

45. $A = \begin{bmatrix} 2 & -1 & 3 \\ 0 & 1 & 0 \\ 2 & -2 & 3 \end{bmatrix}$, $B = \begin{bmatrix} 1 & 5 & -1 \\ 0 & 1 & 3 \\ -1 & 2 & 1 \end{bmatrix}$

46. $A = \begin{bmatrix} 1 & -2 & 5 \\ 1 & 0 & -2 \\ 1 & 3 & 2 \end{bmatrix}$, $B = \begin{bmatrix} -1 & 4 & 2 \\ -3 & 0 & 1 \\ 5 & 1 & 0 \end{bmatrix}$

47. $A = \begin{bmatrix} 2 & -1 \\ 3 & 1 \end{bmatrix}$, $B = \begin{bmatrix} 1 \\ 3 \end{bmatrix}$

48. $A = \begin{bmatrix} 5 & -3 \end{bmatrix}$, $B = \begin{bmatrix} 1 \\ 3 \end{bmatrix}$

Matrices and Computer Graphics

Exercises 49–52: Rotations (Refer to the discussion about rotation.) Make a sketch and conjecture the new location of the point (x, y) when it is rotated 45° clockwise about the origin. Then calculate its position by computing matrix product AX, where

$$A = \begin{bmatrix} \dfrac{1}{\sqrt{2}} & \dfrac{1}{\sqrt{2}} & 0 \\ -\dfrac{1}{\sqrt{2}} & \dfrac{1}{\sqrt{2}} & 0 \\ 0 & 0 & 1 \end{bmatrix} \quad \text{and} \quad X = \begin{bmatrix} x \\ y \\ 1 \end{bmatrix}.$$

49. $(1, 1)$
50. $(-2, -2)$
51. $(0, -2)$
52. $(1, 0)$

Exercises 53–56: Rotations The matrix B rotates a point with coordinates (x, y) about the origin 30° counterclockwise. Use B to calculate the new location of the point by computing BX, where

$$B = \begin{bmatrix} \dfrac{\sqrt{3}}{2} & -\dfrac{1}{2} & 0 \\ \dfrac{1}{2} & \dfrac{\sqrt{3}}{2} & 0 \\ 0 & 0 & 1 \end{bmatrix} \quad \text{and} \quad X = \begin{bmatrix} x \\ y \\ 1 \end{bmatrix}.$$

53. $(0, -2)$
54. $(1, \sqrt{3})$
55. $(\sqrt{3}, 1)$
56. $(2, 2)$

Exercises 57–60: Combining Rotations (Refer to Exercises 49–56 for the matrices A and B.)

57. The matrix A rotates a point 45° clockwise about the origin, while the matrix B rotates a point 30° counterclockwise about the origin. Compute the matrix product ABX, where X represents the point $(3, 0)$. Interpret the result.

58. (Continuation of the previous exercise) Although matrices do not commute in general, conjecture the result of the computation BAX. Test your conjecture by computing this matrix product.

59. (Continuation of the previous two exercises) Use A and B to approximate a matrix C that computes a 15° clockwise rotation of a point about the origin.

60. (Continuation of the previous exercise)
  (a) Use the matrix C to find an approximation for a matrix D that rotates a point 30° clockwise.
  (b) Conjecture the result of the product of DB, where the matrix B rotates a point 30° counterclockwise. Test your conjecture.

Exercises 61–64: Representing Colors (Refer to Exercises 23–28 in Section 8.1.) Colors for computer monitors are often described using ordered triples. One model, called the RGB system, uses red, green, and blue to generate all colors. The accompanying figure describes the relationship between these colors in this system. For example, red is (1, 0, 0), green is (0, 1, 0), and blue is (0, 0, 1). Since equal amounts of red and green combine to form yellow, yellow is represented by (1, 1, 0). Similarly, magenta (a deep reddish purple) is a mixture of blue and red and represented by (1, 0, 1). Cyan is (0, 1, 1), since it is a mixture of blue and green.

Another color model uses cyan, magenta, and yellow. It is referred to as the CMY model and is used in the four-color printing process for textbooks. In this system, cyan is (1, 0, 0), magenta is (0, 1, 0), and yellow is (0, 0, 1). In the CMY model, red is created by mixing magenta and yellow. Thus, red is (0, 1, 1) in this system. To convert ordered triples in the RGB model to ordered triples in the CMY model, we can use the following matrix equation. In both of these systems, color intensities vary between 0 and 1. (Sources: I. Kerlow, The Art of 3-D Computer Animation and Imaging; R. Wolff.)

$$\begin{bmatrix} C \\ M \\ Y \end{bmatrix} = \begin{bmatrix} 1 \\ 1 \\ 1 \end{bmatrix} - \begin{bmatrix} R \\ G \\ B \end{bmatrix}$$

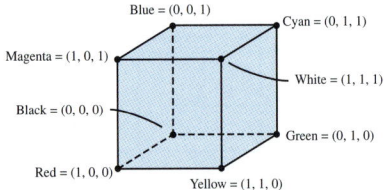

61. In the RGB model, aquamarine is $(0.631, 1, 0.933)$. Use the matrix equation to determine the mixture of cyan, magenta, and yellow that makes aquamarine in the CMY model.

62. In the RGB model, rust is (0.552, 0.168, 0.066). Use the matrix equation to determine the mixture of cyan, magenta, and yellow that makes rust in the CMY model. Explain your results.

63. Use the given matrix equation to find a matrix equation that changes colors represented by ordered triples in the CMY model into ordered triples in the RGB model.

64. In the CMY model, (0.012, 0, 0.597) is a cream color. Use the matrix equation from the previous exercise to determine the mixture of red, green, and blue that makes a cream color in the RGB model.

Exercises 65–68: Properties of Matrices   *Use a graphing calculator to evaluate the expression with the given matrices A, B, and C. Compare your answers for parts (a) and (b). Then, explain the results.*

$$A = \begin{bmatrix} 2 & -1 & 3 \\ 1 & 3 & -5 \\ 0 & -2 & 1 \end{bmatrix}, \quad B = \begin{bmatrix} 6 & 2 & 7 \\ 3 & -4 & -5 \\ 7 & 1 & 0 \end{bmatrix},$$

$$\text{and } C = \begin{bmatrix} 1 & 4 & -3 \\ 8 & 1 & -1 \\ 4 & 6 & -2 \end{bmatrix}$$

65. (a) $A(B + C)$
    (b) $AB + AC$

66. (a) $(A - B)C$
    (b) $AC - BC$

67. (a) $(A - B)^2$
    (b) $A^2 - AB - BA + B^2$

68. (a) $(AB)C$
    (b) $A(BC)$

Writing about Mathematics

1. Discuss whether matrix multiplication is more like multiplication of functions or composition of functions. Explain your reasoning.

2. Describe one application of matrices.

CHECKING BASIC CONCEPTS FOR SECTIONS 5.3 AND 5.4

1. Solve the system of linear equations using Gaussian elimination and backward substitution.

$$\begin{aligned} x \quad\quad + z &= 2 \\ x + y - z &= 1 \\ -x - 2y - z &= 0 \end{aligned}$$

2. Solve the system of linear equations in the previous exercise using technology.

3. Perform the following operations on the given matrices A and B.

$$A = \begin{bmatrix} 1 & 0 & 1 \\ -1 & 1 & 2 \\ 1 & 3 & 0 \end{bmatrix}, \quad B = \begin{bmatrix} -1 & 1 & 2 \\ 0 & 4 & 1 \\ 1 & -2 & 0 \end{bmatrix}$$

  (a) $A + B$
  (b) $2A - B$
  (c) AB

8.5 Inverses of Matrices

Matrix Inverses • Representing Linear Systems with Matrix Equations • Solving Linear Systems with Inverses • Finding Inverses Symbolically

Introduction

In Section 4.2 we discussed how the inverse function f^{-1} will undo or cancel the computation performed by the function f. Like functions, some matrices have inverses. The inverse of a matrix A will undo or cancel the computation performed by A. For example, matrices play an important role in computer graphics. If a matrix A is capable of rotating a figure on a screen 90° clockwise, then the inverse matrix would cause the figure to rotate 90° counterclockwise. This section discusses matrix inverses and some of their applications.

Matrix Inverses

In computer graphics the matrix

$$A = \begin{bmatrix} 1 & 0 & h \\ 0 & 1 & k \\ 0 & 0 & 1 \end{bmatrix}$$

is used to translate a point (x, y), horizontally h units and vertically k units. The translation is to the right if $h > 0$ and left if $h < 0$. Similarly, the translation is up if $k > 0$ and down if $k < 0$. A point (x, y) is represented using the 3×1 column matrix

$$X = \begin{bmatrix} x \\ y \\ 1 \end{bmatrix}.$$

(Source: C. Pokorny and C. Gerald, *Computer Graphics.*)

The third element in X is always equal to 1. For example, the point $(-1, 2)$ could be translated 3 units right and 4 units down by computing the matrix product,

$$AX = \begin{bmatrix} 1 & 0 & 3 \\ 0 & 1 & -4 \\ 0 & 0 & 1 \end{bmatrix} \begin{bmatrix} -1 \\ 2 \\ 1 \end{bmatrix} = \begin{bmatrix} 2 \\ -2 \\ 1 \end{bmatrix} = Y.$$

Its new location is $(2, -2)$. In the matrix A, $h = 3$ and $k = -4$. See Figure 8.52.

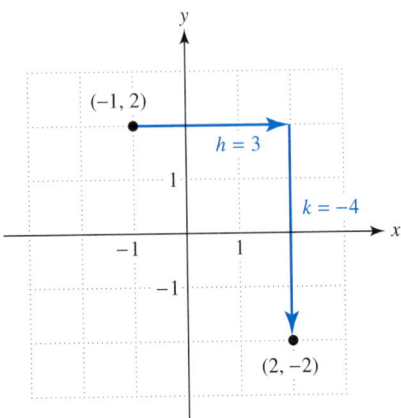

FIGURE 8.52 Translation of a Point

If A translates a point 3 units right and 4 units down, then the inverse matrix translates a point 3 units left and 4 units up. This would return a point to its original position after being translated by A. Therefore, the inverse matrix of A, denoted by A^{-1}, is given by

$$A^{-1} = \begin{bmatrix} 1 & 0 & -3 \\ 0 & 1 & 4 \\ 0 & 0 & 1 \end{bmatrix}.$$

In A^{-1}, $h = -3$ and $k = 4$. The matrix product $A^{-1}Y$ results in

$$A^{-1}Y = \begin{bmatrix} 1 & 0 & -3 \\ 0 & 1 & 4 \\ 0 & 0 & 1 \end{bmatrix} \begin{bmatrix} 2 \\ -2 \\ 1 \end{bmatrix} = \begin{bmatrix} -1 \\ 2 \\ 1 \end{bmatrix} = X.$$

The matrix A^{-1} translates the point located at $(2, -2)$ to its original coordinates of $(-1, 2)$. The two translations acting on the point $(-1, 2)$ can be represented by the following computation.

$$A^{-1}AX = \begin{bmatrix} 1 & 0 & -3 \\ 0 & 1 & 4 \\ 0 & 0 & 1 \end{bmatrix} \begin{bmatrix} 1 & 0 & 3 \\ 0 & 1 & -4 \\ 0 & 0 & 1 \end{bmatrix} \begin{bmatrix} -1 \\ 2 \\ 1 \end{bmatrix}$$

$$= \begin{bmatrix} 1 & 0 & 0 \\ 0 & 1 & 0 \\ 0 & 0 & 1 \end{bmatrix} \begin{bmatrix} -1 \\ 2 \\ 1 \end{bmatrix}$$

$$= \begin{bmatrix} -1 \\ 2 \\ 1 \end{bmatrix} = X$$

That is, the action of A followed by A^{-1} on the point $(-1, 2)$ results in $(-1, 2)$. In a similar manner, if we reverse the order of A^{-1} and A to compute $AA^{-1}X$ the result is again X.

$$AA^{-1}X = \begin{bmatrix} 1 & 0 & 3 \\ 0 & 1 & -4 \\ 0 & 0 & 1 \end{bmatrix} \begin{bmatrix} 1 & 0 & -3 \\ 0 & 1 & 4 \\ 0 & 0 & 1 \end{bmatrix} \begin{bmatrix} -1 \\ 2 \\ 1 \end{bmatrix}$$

$$= \begin{bmatrix} 1 & 0 & 0 \\ 0 & 1 & 0 \\ 0 & 0 & 1 \end{bmatrix} \begin{bmatrix} -1 \\ 2 \\ 1 \end{bmatrix}$$

$$= \begin{bmatrix} -1 \\ 2 \\ 1 \end{bmatrix} = X$$

Notice that the matrix product of $A^{-1}A$ and AA^{-1} resulted in a matrix with 1's on its main diagonal and 0's elsewhere. This matrix is called the **identity matrix.**

The $n \times n$ identity matrix

The **$n \times n$ identity matrix,** denoted by I_n, has only 1's on its main diagonal and 0's elsewhere.

Some examples of identity matrices are

$$I_2 = \begin{bmatrix} 1 & 0 \\ 0 & 1 \end{bmatrix}, \qquad I_3 = \begin{bmatrix} 1 & 0 & 0 \\ 0 & 1 & 0 \\ 0 & 0 & 1 \end{bmatrix}, \qquad \text{and} \qquad I_4 = \begin{bmatrix} 1 & 0 & 0 & 0 \\ 0 & 1 & 0 & 0 \\ 0 & 0 & 1 & 0 \\ 0 & 0 & 0 & 1 \end{bmatrix}.$$

If A is any $n \times n$ matrix, then $I_n A = A$ and $AI_n = A$. For instance, if

$$A = \begin{bmatrix} 2 & 3 \\ 4 & 5 \end{bmatrix}$$

then

$$I_2 A = \begin{bmatrix} 1 & 0 \\ 0 & 1 \end{bmatrix} \begin{bmatrix} 2 & 3 \\ 4 & 5 \end{bmatrix} = \begin{bmatrix} 2 & 3 \\ 4 & 5 \end{bmatrix} = A, \qquad \text{and}$$

$$AI_2 = \begin{bmatrix} 2 & 3 \\ 4 & 5 \end{bmatrix} \begin{bmatrix} 1 & 0 \\ 0 & 1 \end{bmatrix} = \begin{bmatrix} 2 & 3 \\ 4 & 5 \end{bmatrix} = A.$$

Next we formally define the inverse of an $n \times n$ matrix A, whenever it exists.

Inverse of a square matrix

Let A be an $n \times n$ matrix. If there exists an $n \times n$ matrix, denoted by A^{-1}, that satisfies

$$A^{-1}A = I_n \qquad \text{and} \qquad AA^{-1} = I_n,$$

then A^{-1} is the **inverse** of A.

If A^{-1} exists, then A is called **invertible** or **nonsingular.** On the other hand, if a matrix A is not invertible then it is **singular.** Not every matrix has an inverse. For example, the **zero matrix** with dimension 3×3 is given by

$$O_3 = \begin{bmatrix} 0 & 0 & 0 \\ 0 & 0 & 0 \\ 0 & 0 & 0 \end{bmatrix}.$$

The matrix O_3 does not have an inverse. The product of O_3 with any 3×3 matrix B would again be O_3, rather than the identity matrix I_3.

EXAMPLE 1   *Verifying an inverse*

Determine if B is the inverse of A, where

$$A = \begin{bmatrix} 5 & 3 \\ -3 & -2 \end{bmatrix} \qquad \text{and} \qquad B = \begin{bmatrix} 2 & 3 \\ -3 & -5 \end{bmatrix}.$$

Solution
For B to be the inverse of A, it must satisfy the equations $AB = I_2$ and $BA = I_2$.

$$AB = \begin{bmatrix} 5 & 3 \\ -3 & -2 \end{bmatrix} \begin{bmatrix} 2 & 3 \\ -3 & -5 \end{bmatrix} = \begin{bmatrix} 1 & 0 \\ 0 & 1 \end{bmatrix} = I_2$$

$$BA = \begin{bmatrix} 2 & 3 \\ -3 & -5 \end{bmatrix} \begin{bmatrix} 5 & 3 \\ -3 & -2 \end{bmatrix} = \begin{bmatrix} 1 & 0 \\ 0 & 1 \end{bmatrix} = I_2$$

Thus, B is the inverse of A. That is, $B = A^{-1}$. ∎

In Section 8.4 we saw how a matrix can be used to rotate points about the origin. The next example discusses the significance of an inverse matrix in computer graphics.

EXAMPLE 2   *Interpreting an inverse matrix*

The matrix A can be used to rotate a point 90° clockwise about the origin, where

$$A = \begin{bmatrix} 0 & 1 & 0 \\ -1 & 0 & 0 \\ 0 & 0 & 1 \end{bmatrix} \quad \text{and} \quad A^{-1} = \begin{bmatrix} 0 & -1 & 0 \\ 1 & 0 & 0 \\ 0 & 0 & 1 \end{bmatrix}.$$

(a) Use A to rotate the point $(-2, 0)$ clockwise 90° about the origin.
(b) Conjecture the effect of A^{-1} on the resulting point.
(c) Test this conjecture.

Solution

(a) First, let the point $(-2, 0)$ be represented by the column matrix

$$X = \begin{bmatrix} -2 \\ 0 \\ 1 \end{bmatrix}.$$

Then compute

$$AX = \begin{bmatrix} 0 & 1 & 0 \\ -1 & 0 & 0 \\ 0 & 0 & 1 \end{bmatrix} \begin{bmatrix} -2 \\ 0 \\ 1 \end{bmatrix} = \begin{bmatrix} 0 \\ 2 \\ 1 \end{bmatrix} = Y.$$

If the point $(-2, 0)$ is rotated 90° clockwise about the origin, its new location is $(0, 2)$. See Figure 8.53.

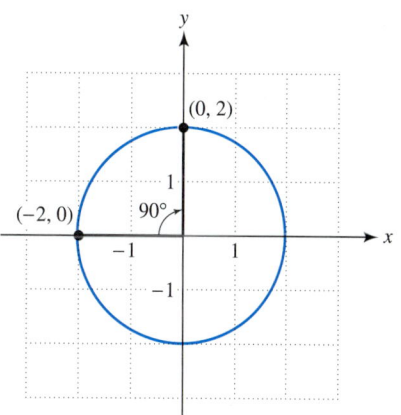

FIGURE 8.53 Rotating a Point about the Origin

(b) Since A^{-1} represents the inverse operation of A, A^{-1} will rotate the point located at $(0, 2)$ counterclockwise 90°, back to $(-2, 0)$.
(c) This conjecture is correct since

$$A^{-1}Y = \begin{bmatrix} 0 & -1 & 0 \\ 1 & 0 & 0 \\ 0 & 0 & 1 \end{bmatrix} \begin{bmatrix} 0 \\ 2 \\ 1 \end{bmatrix} = \begin{bmatrix} -2 \\ 0 \\ 1 \end{bmatrix} = X. \qquad \blacksquare$$

Critical Thinking ————————————————————————————————————

What will the results be of the computations AAX and $A^{-1} A^{-1} X$?

Representing Linear Systems with Matrix Equations

In Section 8.3 linear systems were solved using Gaussian elimination with backward substitution. This method used an augmented matrix to represent a system of linear equations. A system of linear equations can also be represented by a matrix equation.

$$3x - 2y + 4z = 5$$
$$2x + y + 3z = 9$$
$$-x + 5y - 2z = 5$$

Let A, X, and B be matrices defined as

Coefficient Matrix   Variable Matrix   Constant Matrix

$$A = \begin{bmatrix} 3 & -2 & 4 \\ 2 & 1 & 3 \\ -1 & 5 & -2 \end{bmatrix}, \quad X = \begin{bmatrix} x \\ y \\ z \end{bmatrix}, \quad \text{and} \quad B = \begin{bmatrix} 5 \\ 9 \\ 5 \end{bmatrix}.$$

The matrix product AX is given by

$$AX = \begin{bmatrix} 3 & -2 & 4 \\ 2 & 1 & 3 \\ -1 & 5 & -2 \end{bmatrix} \begin{bmatrix} x \\ y \\ z \end{bmatrix} = \begin{bmatrix} 3x + (-2)y + 4z \\ 2x + 1y + 3z \\ (-1)x + 5y + (-2)z \end{bmatrix} = \begin{bmatrix} 3x - 2y + 4z \\ 2x + y + 3z \\ -x + 5y - 2z \end{bmatrix}.$$

Thus, the matrix equation $AX = B$ simplifies to

$$\begin{bmatrix} 3x - 2y + 4z \\ 2x + y + 3z \\ -x + 5y - 2z \end{bmatrix} = \begin{bmatrix} 5 \\ 9 \\ 5 \end{bmatrix}.$$

This matrix equation $AX = B$ is equivalent to the original system of linear equations. Any linear system of equations can be represented using a matrix equation in the form $AX = B$.

EXAMPLE 3   *Representing linear systems with matrix equations*

Represent each system of linear equations in the form $AX = B$.

(a) $3x - 4y = 7$
     $-x + 6y = -3$

(b) $x - 5y = 2$
     $-3x + 2y + z = -7$
     $4x + 5y + 6z = 10$

Solution

(a) This linear system is comprised of two equations and two variables. The equivalent matrix equation is

$$AX = \begin{bmatrix} 3 & -4 \\ -1 & 6 \end{bmatrix} \begin{bmatrix} x \\ y \end{bmatrix} = \begin{bmatrix} 7 \\ -3 \end{bmatrix} = B.$$

(b) The equivalent matrix equation is

$$AX = \begin{bmatrix} 1 & -5 & 0 \\ -3 & 2 & 1 \\ 4 & 5 & 6 \end{bmatrix} \begin{bmatrix} x \\ y \\ z \end{bmatrix} = \begin{bmatrix} 2 \\ -7 \\ 10 \end{bmatrix} = B.$$

■

Solving Linear Systems with Inverses

The matrix equation $AX = B$ can be solved using A^{-1}, if it exists.

$$\begin{aligned} AX &= B && \text{Linear system} \\ A^{-1}AX &= A^{-1}B && \text{Multiply both sides by } A^{-1}. \\ I_nX &= A^{-1}B && A^{-1}A = I_n \\ X &= A^{-1}B && I_nX = X \text{ for any } n \times 1 \text{ matrix } X. \end{aligned}$$

To solve a linear system, multiply both sides of the matrix equation $AX = B$ by A^{-1}, if it exists. The solution to the system can be written as $X = A^{-1}B$.

Note: Since matrix multiplication is not commutative, it is essential to multiply both sides of the equation on the *left* by A^{-1}. That is, $X = A^{-1}B \neq BA^{-1}$ in general.

EXAMPLE 4 *Solving a linear system using the inverse of a 2 × 2 matrix*

Write the linear system as a matrix equation in the form $AX = B$. Find A^{-1} and solve for X.

$$\begin{aligned} 4x - 5y &= 8.1 \\ 7x - 9y &= -4.7 \end{aligned}$$

Solution

The linear system can be written as

$$AX = \begin{bmatrix} 4 & -5 \\ 7 & -9 \end{bmatrix} \begin{bmatrix} x \\ y \end{bmatrix} = \begin{bmatrix} 8.1 \\ -4.7 \end{bmatrix} = B.$$

The matrix A^{-1} can be found with a calculator as shown in Figure 8.54. The solution to the system, given by the product $A^{-1}B$, is $x = 96.4$ and $y = 75.5$. See Figure 8.55.

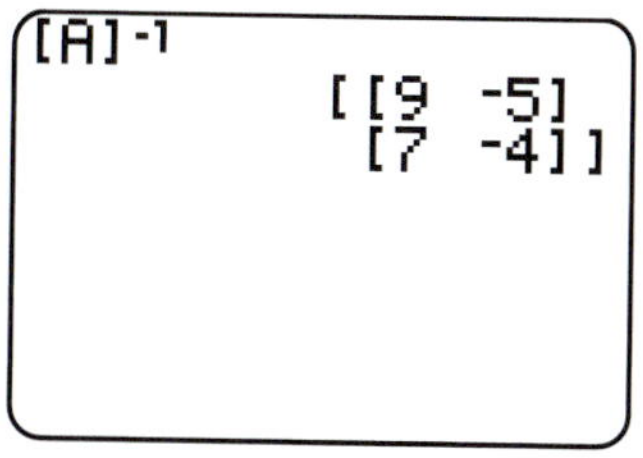
[A]-1
[[9 -5]
[7 -4]]

FIGURE 8.54

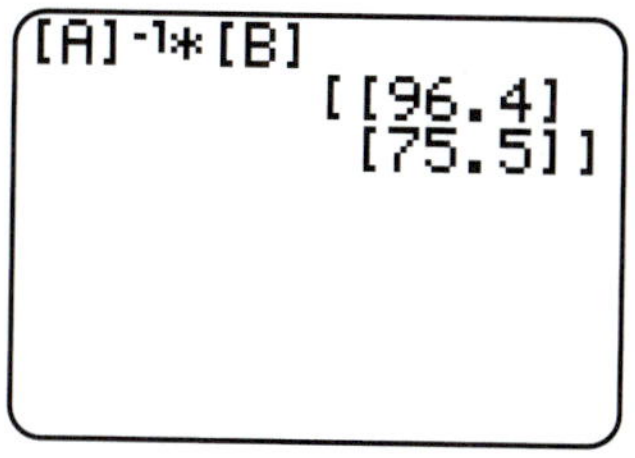
[A]-1*[B]
[[96.4]
[75.5]]

FIGURE 8.55

■

EXAMPLE 5 *Solving a linear system using the inverse of a 3 × 3 matrix*

In one study of adult males, it was believed that systolic blood pressure was affected by both age A in years and weight W in pounds. This was modeled by

$P(A, W) = a + bA + cW$, where a, b, and c are constants. Table 8.9 lists three individuals with representative blood pressures for the group. (Reference: C. H. Brase and C. P. Brase, *Understandable Statistics.*)

(a) Use Table 8.9 to approximate values for the constants a, b, and c.

(b) Estimate a typical systolic blood pressure for an individual that is 55 years old and has a weight of 175 pounds.

TABLE 8.9

P	A	W
113	39	142
138	53	181
152	65	191

Solution

(a) Determine the constants a, b, and c in $P(A, W) = a + bA + cW$ by solving the following three equations.

$$P(39, 142) = a + b(39) + c(142) = 113$$
$$P(53, 181) = a + b(53) + c(181) = 138$$
$$P(65, 191) = a + b(65) + c(191) = 152$$

These three equations can be rewritten as follows.

$$a + 39b + 142c = 113$$
$$a + 53b + 181c = 138$$
$$a + 65b + 191c = 152$$

This system can be represented by the matrix equation $AX = B$.

$$AX = \begin{bmatrix} 1 & 39 & 142 \\ 1 & 53 & 181 \\ 1 & 65 & 191 \end{bmatrix} \begin{bmatrix} a \\ b \\ c \end{bmatrix} = \begin{bmatrix} 113 \\ 138 \\ 152 \end{bmatrix} = B$$

The solution, $X = A^{-1}B$, is shown in Figure 8.56. The values for the constants are $a \approx 32.78$, $b \approx 0.9024$, and $c \approx 0.3171$. Thus, P is given by $P(A, W) = 32.78 + 0.9024A + 0.3171W$.

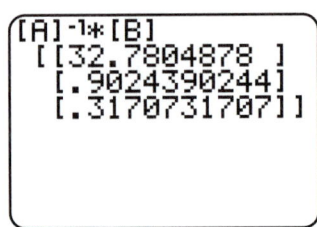

FIGURE 8.56

(b) Evaluate $P(55, 175) = 32.78 + 0.9024(55) + 0.3171(175) \approx 137.9$. This model predicts that a typical (male) individual 55 years old, weighing

175 pounds, has a systolic blood pressure of approximately 138. Clearly, this could vary greatly between individuals. ∎

MAKING CONNECTIONS
Solving Equations with Functions and Matrices

Chapter 4 demonstrated that if a function f has an inverse, then the solution to the equation $f(x) = b$ is given by $x = f^{-1}(b)$. A similar situation occurs with a linear system represented by $AX = B$. If A is invertible, then the solution is given by $X = A^{-1}B$.

Finding Inverses Symbolically

In Examples 4 and 5, technology was used to find A^{-1}. The inverse matrix can be found symbolically by first forming the augmented matrix $\left[A \mid I_n \right]$, and then performing matrix row operations, until the left side of the augmented matrix becomes the identity matrix. The resulting augmented matrix can be written as $\left[I_n \mid A^{-1} \right]$, where the right side of the matrix is A^{-1}. The next example illustrates this method.

EXAMPLE 6 *Finding the inverse of a matrix symbolically*

Find A^{-1} if

$$A = \begin{bmatrix} 1 & 0 & 1 \\ 2 & 1 & 3 \\ -1 & 1 & 1 \end{bmatrix}.$$

Solution
Begin by forming the following 3×6 augmented matrix. Perform matrix row operations to obtain the identity on the left side, while performing the same operation on the right side of this matrix.

$$\begin{bmatrix} 1 & 0 & 1 & 1 & 0 & 0 \\ 2 & 1 & 3 & 0 & 1 & 0 \\ -1 & 1 & 1 & 0 & 0 & 1 \end{bmatrix}$$

$$\begin{matrix} R_2 - 2R_1 \to \\ R_3 + R_1 \to \end{matrix} \begin{bmatrix} 1 & 0 & 1 & 1 & 0 & 0 \\ 0 & 1 & 1 & -2 & 1 & 0 \\ 0 & 1 & 2 & 1 & 0 & 1 \end{bmatrix}$$

$$R_3 - R_2 \to \begin{bmatrix} 1 & 0 & 1 & 1 & 0 & 0 \\ 0 & 1 & 1 & -2 & 1 & 0 \\ 0 & 0 & 1 & 3 & -1 & 1 \end{bmatrix}$$

$$\begin{matrix} R_1 - R_3 \to \\ R_2 - R_3 \to \end{matrix} \begin{bmatrix} 1 & 0 & 0 & -2 & 1 & -1 \\ 0 & 1 & 0 & -5 & 2 & -1 \\ 0 & 0 & 1 & 3 & -1 & 1 \end{bmatrix}$$

The right side is equal to A^{-1}. That is,

$$A^{-1} = \begin{bmatrix} -2 & 1 & -1 \\ -5 & 2 & -1 \\ 3 & -1 & 1 \end{bmatrix}.$$

It can be verified that $A^{-1}A = I_3 = AA^{-1}$. ∎

Note: If it is not possible to obtain the identity matrix on the left side of the augmented matrix using matrix row operations, then A^{-1} does not exist.

8.5 PUTTING IT ALL TOGETHER

The inverse of a matrix A is denoted by A^{-1}, if it exists. The inverse matrix will undo or cancel the operations performed by A. Inverse matrices frequently are used in computer graphics. They also can be used to solve systems of linear equations.

The following table summarizes some of the mathematical concepts presented in this section.

Identity Matrix
The $n \times n$ identity matrix I_n has only 1's on the main diagonal and 0's elsewhere. When it is multiplied by any $n \times n$ matrix A, the result is A.
Matrix Inverse
If an $n \times n$ matrix A has an inverse it is unique, is denoted by A^{-1}, and satisfies the equations $AA^{-1} = I_n$ and $A^{-1} A = I_n$. Matrix inverses can be found using technology. They can also be found using pencil and paper by performing matrix row operations on the augmented matrix $[A \mid I_n]$ until it is reduced to $[I_n \mid A^{-1}]$.
Solutions to Systems of Linear Equations
Systems of linear equations can be written using the matrix equation $AX = B$. If A is invertible, then there will be a unique solution given by $X = A^{-1}B$. If A is not invertible then there could be either no solutions or an infinite number of solutions. In this case, Gaussian elimination should be applied.

8.5 EXERCISES

   Tape 6

Identifying Inverse Matrices

Exercises 1–4: Determine if B is the inverse matrix of A.

1. $A = \begin{bmatrix} 4 & 3 \\ 5 & 4 \end{bmatrix}$, $B = \begin{bmatrix} 4 & -3 \\ -5 & 4 \end{bmatrix}$

2. $A = \begin{bmatrix} -1 & 2 \\ -3 & 8 \end{bmatrix}$, $B = \begin{bmatrix} -4 & 1 \\ -2 & 0.5 \end{bmatrix}$

3. $A = \begin{bmatrix} 1 & -1 & 2 \\ 0 & 1 & -1 \\ 1 & 0 & 2 \end{bmatrix}$, $B = \begin{bmatrix} 2 & 2 & -1 \\ -1 & 0 & 1 \\ -1 & -1 & 1 \end{bmatrix}$

4. $A = \begin{bmatrix} 2 & 1 & 1 \\ -1 & 0 & -1 \\ 0 & 2 & -1 \end{bmatrix}$, $B = \begin{bmatrix} 2 & 3 & -1 \\ -1 & -2 & 1 \\ -2 & -4 & 1 \end{bmatrix}$

Exercises 5–8: Determine the value of the constant k in the matrix B so that $B = A^{-1}$.

5. $A = \begin{bmatrix} 1 & 1 \\ 1 & 2 \end{bmatrix}$, $B = \begin{bmatrix} 2 & -1 \\ -1 & k \end{bmatrix}$

6. $A = \begin{bmatrix} -2 & 2 \\ 1 & -2 \end{bmatrix}$, $B = \begin{bmatrix} -1 & k \\ -0.5 & -1 \end{bmatrix}$

7. $A = \begin{bmatrix} 1 & 3 \\ -1 & -5 \end{bmatrix}$, $B = \begin{bmatrix} k & 1.5 \\ -0.5 & -0.5 \end{bmatrix}$

8. $A = \begin{bmatrix} -2 & 5 \\ -3 & 4 \end{bmatrix}$, $B = \begin{bmatrix} \frac{4}{7} & -\frac{5}{7} \\ k & -\frac{2}{7} \end{bmatrix}$

Exercises 9–12: Predict the results of I_nA and AI_n. Then, verify your prediction.

9. $I_2 = \begin{bmatrix} 1 & 0 \\ 0 & 1 \end{bmatrix}$, $\quad A = \begin{bmatrix} 1 & -2 \\ 4 & 3 \end{bmatrix}$

10. $I_3 = \begin{bmatrix} 1 & 0 & 0 \\ 0 & 1 & 0 \\ 0 & 0 & 1 \end{bmatrix}$, $\quad A = \begin{bmatrix} 1 & -4 & 3 \\ 1 & 9 & 5 \\ 3 & -5 & 0 \end{bmatrix}$

11. $I_3 = \begin{bmatrix} 1 & 0 & 0 \\ 0 & 1 & 0 \\ 0 & 0 & 1 \end{bmatrix}$, $\quad A = \begin{bmatrix} 0 & 0 & 0 \\ 0 & 0 & 0 \\ 0 & 0 & 0 \end{bmatrix}$

12. $I_4 = \begin{bmatrix} 1 & 0 & 0 & 0 \\ 0 & 1 & 0 & 0 \\ 0 & 0 & 1 & 0 \\ 0 & 0 & 0 & 1 \end{bmatrix}$, $\quad A = \begin{bmatrix} 5 & -2 & 6 & -3 \\ 0 & 1 & 4 & -1 \\ -5 & 7 & 9 & 8 \\ 0 & 0 & 3 & 1 \end{bmatrix}$

Interpreting Inverses

Exercises 13 and 14: Translations   (Refer to the discussion in this section about translating a point.) The matrix product AX performs a translation on the point (x, y), where

$$A = \begin{bmatrix} 1 & 0 & h \\ 0 & 1 & k \\ 0 & 0 & 1 \end{bmatrix} \quad \text{and} \quad X = \begin{bmatrix} x \\ y \\ 1 \end{bmatrix}.$$

(a) Predict the new location of the point (x, y), when it is translated by A. Compute $Y = AX$ to verify your prediction.

(b) Conjecture what $A^{-1}Y$ represents. Find A^{-1} and calculate $A^{-1}Y$ to test your conjecture.

(c) What will AA^{-1} and $A^{-1}A$ equal?

13. $A = \begin{bmatrix} 1 & 0 & 2 \\ 0 & 1 & 3 \\ 0 & 0 & 1 \end{bmatrix}$ $\quad$ and $\quad X = \begin{bmatrix} 0 \\ 1 \\ 1 \end{bmatrix}$

14. $A = \begin{bmatrix} 1 & 0 & -4 \\ 0 & 1 & 5 \\ 0 & 0 & 1 \end{bmatrix}$ $\quad$ and $\quad X = \begin{bmatrix} 4 \\ 2 \\ 1 \end{bmatrix}$

Exercises 15 and 16: Translations   (Refer to the discussion in this section about translating a point.) Find a 3×3 matrix A that performs the following translation of a point (x, y) represented by X. Find A^{-1} and describe what it computes.

15. 3 units left and 5 units down

16. 6 units right and 1 unit up

Exercises 17 and 18: Scaling Matrix   The given matrix A can be used to change the distance between the point

*(x, y) and the origin without any rotation. In computer graphics this is called **scaling**. (Source: C. Pokorny.)*

(a) Find the distance between the given point and the origin.

(b) Let X represent the given point. Compute $Y = AX$, and then compute the distance between the point represented by Y and the origin. Describe how A changes the distance between the point (x, y) and the origin.

(c) Calculate $A^{-1}Y$. Explain the computation performed by A^{-1}.

17. $(3, 4)$; $\quad A = \begin{bmatrix} 3 & 0 & 0 \\ 0 & 3 & 0 \\ 0 & 0 & 1 \end{bmatrix}$ and

$$A^{-1} = \begin{bmatrix} \frac{1}{3} & 0 & 0 \\ 0 & \frac{1}{3} & 0 \\ 0 & 0 & 1 \end{bmatrix}$$

18. $(-5, 12)$; $\quad A = \begin{bmatrix} \frac{1}{2} & 0 & 0 \\ 0 & \frac{1}{2} & 0 \\ 0 & 0 & 1 \end{bmatrix}$ and

$$A^{-1} = \begin{bmatrix} 2 & 0 & 0 \\ 0 & 2 & 0 \\ 0 & 0 & 1 \end{bmatrix}$$

19. *Rotation*   The matrix B rotates (x, y) clockwise about the origin $45°$, where

$$B = \begin{bmatrix} \frac{1}{\sqrt{2}} & \frac{1}{\sqrt{2}} & 0 \\ -\frac{1}{\sqrt{2}} & \frac{1}{\sqrt{2}} & 0 \\ 0 & 0 & 1 \end{bmatrix} \text{ and } B^{-1} = \begin{bmatrix} \frac{1}{\sqrt{2}} & -\frac{1}{\sqrt{2}} & 0 \\ \frac{1}{\sqrt{2}} & \frac{1}{\sqrt{2}} & 0 \\ 0 & 0 & 1 \end{bmatrix}.$$

(a) Let X represent the point $(-\sqrt{2}, -\sqrt{2})$. Compute $Y = BX$.

(b) Find $B^{-1}Y$. Interpret the computation performed by B^{-1}.

20. (Continuation of the previous exercise.) Predict the result of the computations $BB^{-1}X$ and $B^{-1}BX$ for any point (x, y) represented by X. Explain this result geometrically.

21. *Translations*   The matrix A translates a point to the right 4 units and down 2 units, and the matrix B translates a point left 3 units and up 3 units, where

$$A = \begin{bmatrix} 1 & 0 & 4 \\ 0 & 1 & -2 \\ 0 & 0 & 1 \end{bmatrix} \quad \text{and} \quad B = \begin{bmatrix} 1 & 0 & -3 \\ 0 & 1 & 3 \\ 0 & 0 & 1 \end{bmatrix}.$$

(a) Let X represent the point $(1, 1)$. Predict the result of $Y = ABX$. Check your prediction.

(b) Predict the form of the matrix product AB, and then compute AB.

(c) In this exercise, would you expect $AB = BA$? Verify your answer.

(d) Conjecture the form of $(AB)^{-1}$. Explain your reasoning.

22. (Refer to Exercises 13 and 19 for matrices A and B.) Use a calculator to find any matrix products.

(a) Let X represent the point $(0, \sqrt{2})$. If this point is rotated about the origin $45°$ clockwise, and then translated 2 units right and 3 units up, determine its new coordinates geometrically.

(b) Compute the matrix product $Y = ABX$, and explain the result.

(c) Is your answer in part (b) equal to BAX? Interpret your answer.

(d) Find a matrix that translates Y back to X. Test your conjecture.

Calculating Inverses

Exercises 23–26: Use a graphing calculator to calculate the inverse of A.

23. $A = \begin{bmatrix} 0.5 & -1.5 \\ 0.2 & -0.5 \end{bmatrix}$ 24. $A = \begin{bmatrix} -0.5 & 0.5 \\ 3 & 2 \end{bmatrix}$

25. $A = \begin{bmatrix} 1 & 2 & 0 \\ -1 & 4 & -1 \\ 2 & -1 & 0 \end{bmatrix}$

26. $A = \begin{bmatrix} -2 & 0 & 1 \\ 5 & -4 & 1 \\ 1 & -2 & 0 \end{bmatrix}$

Exercises 27–34: (Refer to Example 6.) Find A^{-1} without a calculator.

27. $A = \begin{bmatrix} 1 & 2 \\ 1 & 3 \end{bmatrix}$ 28. $A = \begin{bmatrix} 1 & 0 \\ 1 & -1 \end{bmatrix}$

29. $A = \begin{bmatrix} -1 & 2 \\ 3 & -5 \end{bmatrix}$ 30. $A = \begin{bmatrix} 1 & 3 \\ 2 & 5 \end{bmatrix}$

31. $A = \begin{bmatrix} 0 & 0 & 1 \\ 1 & 0 & 0 \\ 0 & 1 & 0 \end{bmatrix}$

32. $A = \begin{bmatrix} 1 & 0 & 0 \\ 1 & 1 & 0 \\ 0 & 1 & 1 \end{bmatrix}$

33. $A = \begin{bmatrix} 1 & 0 & 1 \\ 2 & 1 & 3 \\ -1 & 1 & 1 \end{bmatrix}$

34. $A = \begin{bmatrix} -2 & 1 & 0 \\ 1 & 0 & 1 \\ -1 & 1 & 0 \end{bmatrix}$

Solving Linear Systems

Exercises 35–42: Complete the following for the given system of linear equations.

(a) Write the system in the form $AX = B$.

(b) Solve the linear system by computing $X = A^{-1}B$ with a calculator. Approximate the solution to four significant digits when appropriate.

35. $\begin{aligned} 1.5x + 3.7y &= 0.32 \\ -0.4x - 2.1y &= 0.36 \end{aligned}$

36. $\begin{aligned} 31x + 18y &= 64.1 \\ 5x - 23y &= -59.6 \end{aligned}$

37. $\begin{aligned} 0.08x - 0.7y &= -0.504 \\ 1.1x - 0.05y &= 0.73 \end{aligned}$

38. $\begin{aligned} -231x + 178y &= -439 \\ 525x - 329y &= 2282 \end{aligned}$

39. $\begin{aligned} 3.1x + 1.9y - z &= 1.99 \\ 6.3x \quad\quad - 9.9z &= -3.78 \\ -x + 1.5y + 7z &= 5.3 \end{aligned}$

40. $\begin{aligned} 17x - 22y - 19z &= -25.2 \\ 3x + 13y - 9z &= 105.9 \\ x - 2y + 6.1z &= -23.55 \end{aligned}$

41. $\begin{aligned} 3x - y + z &= 4.9 \\ 5.8x - 2.1y &= -3.8 \\ -x + 2.9z &= 3.8 \end{aligned}$

42. $\begin{aligned} 1.2x - 0.3y - 0.7z &= -0.5 \\ -0.4x + 1.3y + 0.4z &= 0.9 \\ 1.7x + 0.6y + 1.1z &= 1.3 \end{aligned}$

Applications

43. *Cost of CDs*   A music store has compact discs that sell for three different prices marked A, B, and C. Each row in the table shows the total cost of a purchase. Use this information to determine the cost of one CD of each type by setting up a matrix equation and solving it with an inverse.

A	B	C	Total
2	3	4	$120.91
1	4	0	$62.95
2	1	3	$79.94

44. *Determining Prices* A group of students bought 3 soft drinks and 2 boxes of popcorn at a movie for $8.50. A second group bought 4 soft drinks and 3 boxes of popcorn for $12.
  (a) Find a matrix equation $AX = B$ whose solution gives the individual prices of a soft drink and a box of popcorn. Solve this matrix equation using A^{-1}.
  (b) Could these prices be determined if both groups had bought 3 soft drinks and 2 boxes of popcorn for $8.50? Explain by trying to calculate A^{-1}.

45. *Traffic Flow* (Refer to Exercises 41 and 42 in Section 8.3.) The accompanying figure shows four one-way streets with intersections A, B, C, and D. Numbers indicate the average traffic flow in vehicles per minute. The variables x_1, x_2, x_3, and x_4 denote unknown traffic flows.
  (a) The number of vehicles per minute entering an intersection equals the number exiting an intersection. Verify that the given system of linear equations describes the traffic flow.
  (b) Write the system as $AX = B$ and solve using A^{-1}.
  (c) Interpret your results.

$$A: x_1 + 5 = 4 + 6$$
$$B: x_2 + 6 = x_1 + 3$$
$$C: x_3 + 4 = x_2 + 7$$
$$D: 6 + 5 = x_3 + x_4$$

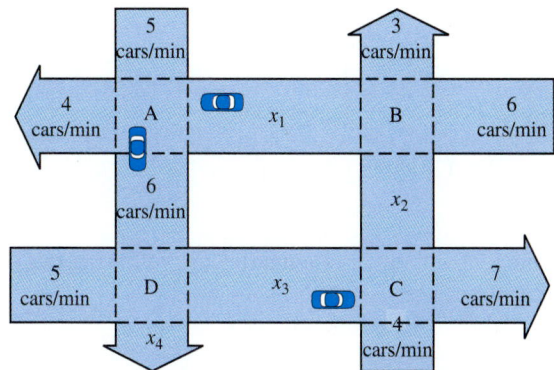

46. *Tire Sales* In one study the relationship between annual tire sales T in thousands, automobile registrations A in millions, and personal disposable

income I in millions of dollars was investigated. Representative data for three different years is shown in the table. (Source: J. Jarrett, *Business Forecasting Methods.*)

T	A	I
10,170	113	308
15,305	133	622
21,289	155	1937

The data was modeled by $T = aA + bI + c$, where a, b, and c are constants.
  (a) Use the data to write a system of linear equations, whose solution gives a, b, and c.
  (b) Solve this linear system. Write a formula for T.
  (c) If $A = 118$ and $I = 311$, predict T. (The actual value for T was 11,314.)

47. *Home Prices* The selling price of a home can depend on several factors. Real estate companies sometimes study how the selling price of a home is related to its size and condition. The accompanying table contains representative data for sales of three homes. Price P is measured in thousands of dollars, home size S in square feet, and condition C is rated on a scale from 1 to 10, where 10 represents excellent condition. The variables were found to be related by the linear equation $P = a + bS + cC$.

P	S	C
122	1500	8
130	2000	5
158	2200	10

  (a) Use the table to write a system of linear equations whose solution gives a, b, and c.
  (b) Estimate the selling price of a home with 1800 square feet and a condition of 7.

48. *Plate Glass Sales* The amount of plate glass sales G can be affected by the number of new building contracts B issued and automobiles A produced, since plate glass is used in buildings and cars. A plate glass company in California wanted to forecast future sales. The table on the next page contains sales data for three consecutive years. All units are in millions. (Source: S. Makridakis and S. Wheelwright, *Forecasting Methods for Management.*)

G	A	B
603	5.54	37.1
657	6.93	41.3
779	7.64	45.6

The data was modeled by $G = aA + bB + c$, where a, b, and c are constants.
(a) Write a system of linear equations whose solution gives a, b, and c.
(b) Solve this linear system. Write a formula for G.

(c) For the following year, it was estimated that $A = 7.75$ and $B = 47.4$. Predict G. (The actual value for G was \$878 million.)

Writing about Mathematics

1. Discuss how to solve the equation $f(x) = k$ if f is one-to-one. Then discuss how to solve the matrix equation $AX = B$ if A^{-1} exists.

2. Give an example of a 2×2 matrix A that does not have an inverse. Explain what happens if one attempts to find the inverse of A symbolically.

8.6 Determinants

Definition and Calculation of Determinants • Area of Regions • Cramer's Rule • Limitations on the Method of Cofactors and Cramer's Rule

Introduction

Determinants are used in mathematics for theoretical purposes. However, they also are used to test if a matrix is invertible and to find the area of certain geometric figures. A *determinant* is a real number associated with a square matrix. We begin our discussion by defining a determinant for a 2×2 matrix.

Definition and Calculation of Determinants

The determinant of a matrix with dimension 2×2 is a straightforward arithmetic calculation.

Determinant of a 2 × 2 matrix

The **determinant** of

$$A = \begin{bmatrix} a & b \\ c & d \end{bmatrix}$$

is a real number defined by

$$\det A = ad - cb.$$

Later we will define determinants for any $n \times n$ matrix. The following can be used to determine if a matrix is invertible.

Invertible matrix

A square matrix A is invertible if and only if $\det A \neq 0$.

EXAMPLE 1 *Determining if a 2 × 2 matrix is invertible*

Determine if A^{-1} exists by computing the determinant of the matrix A.

(a) $A = \begin{bmatrix} 3 & -4 \\ -5 & 9 \end{bmatrix}$ **(b)** $A = \begin{bmatrix} 52 & -32 \\ 65 & -40 \end{bmatrix}$

Solution

(a) The determinant of the 2×2 matrix A is calculated as follows.

$$\det A = \det \begin{bmatrix} 3 & -4 \\ -5 & 9 \end{bmatrix} = (3)(9) - (-5)(-4) = 7$$

Since $\det A = 7 \neq 0$, the matrix A is invertible and A^{-1} exists.

(b) In a similar manner,

$$\det A = \det \begin{bmatrix} 52 & -32 \\ 65 & -40 \end{bmatrix} = (52)(-40) - (65)(-32) = 0$$

Since $\det A = 0$, A^{-1} does not exist. Try finding A^{-1}. What happens? ∎

We can use determinants of 2×2 matrices to find determinants of larger square matrices. In order to do this, we first define the concepts of a *minor* and a *cofactor*.

Minors and cofactors

The **minor,** denoted by M_{ij}, for element a_{ij} in the square matrix A is the real number computed by performing the following steps.

1. Delete the ith row and jth column from the matrix A.
2. M_{ij} is equal to the determinant of the resulting matrix.

The **cofactor,** denoted by A_{ij}, for a_{ij} is defined by $A_{ij} = (-1)^{i+j} M_{ij}$.

EXAMPLE 2 *Calculating minors and cofactors*

Find the following minors and cofactors for the matrix A.

$$A = \begin{bmatrix} 2 & -3 & 1 \\ -2 & 1 & 0 \\ 0 & -1 & 4 \end{bmatrix}$$

(a) M_{11} and M_{21}
(b) A_{11} and A_{21}

Solution

(a) To obtain the minor M_{11}, begin by crossing out the first row and first column of A.

$$A = \begin{bmatrix} \cancel{2} & \cancel{-3} & \cancel{1} \\ -\cancel{2} & 1 & 0 \\ \cancel{0} & -1 & 4 \end{bmatrix}$$

The remaining elements form the 2 × 2 matrix

$$B = \begin{bmatrix} 1 & 0 \\ -1 & 4 \end{bmatrix}.$$

The minor M_{11} is equal to det $B = (1)(4) - (-1)(0) = 4$.

M_{21} is found by crossing out the second row and first column of A.

$$A = \begin{bmatrix} 2 & -3 & 1 \\ -2 & 1 & 0 \\ 0 & -1 & 4 \end{bmatrix}$$

The resulting matrix is

$$B = \begin{bmatrix} -3 & 1 \\ -1 & 4 \end{bmatrix}.$$

Thus, $M_{21} = \det B = (-3)(4) - (-1)(1) = -11$.

(b) Since $A_{ij} = (-1)^{i+j}M_{ij}$, A_{11} and A_{21} can be computed as follows.

$$A_{11} = (-1)^{1+1}M_{11} = (-1)^2(4) = 4$$
$$A_{21} = (-1)^{2+1}M_{21} = (-1)^3(-11) = 11 \qquad ■$$

Using the concept of a cofactor, we can calculate the determinant of any square matrix.

Determinant of a matrix using the method of cofactors

Multiply each element in any row or column of the matrix by its cofactor. The sum of the products is equal to the determinant.

To compute the determinant of a 3 × 3 matrix A, begin by selecting a row or column.

$$A = \begin{bmatrix} a_{11} & a_{12} & a_{13} \\ a_{21} & a_{22} & a_{23} \\ a_{31} & a_{32} & a_{33} \end{bmatrix}$$

For example, if the second row of A is selected, the elements are a_{21}, a_{22}, and a_{23}. Then,

$$\det A = a_{21}A_{21} + a_{22}A_{22} + a_{23}A_{23}.$$

On the other hand, utilizing the elements of a_{11}, a_{21}, and a_{31} in the first column results in

$$\det A = a_{11}A_{11} + a_{21}A_{21} + a_{31}A_{31}.$$

Regardless of the row or column selected, the value of det A is the same. The calculation is easier if some elements in the selected row or column equal 0.

EXAMPLE 3 *Evaluating the determinant of a 3 × 3 matrix*

Find det A, if

$$A = \begin{bmatrix} 2 & -3 & 1 \\ -2 & 1 & 0 \\ 0 & -1 & 4 \end{bmatrix}.$$

Solution

To find the determinant of A, we can select any row or column. If we begin expanding about the first column of A, then

$$\det A = a_{11}A_{11} + a_{21}A_{21} + a_{31}A_{31}.$$

In the first column, $a_{11} = 2$, $a_{21} = -2$, and $a_{31} = 0$. In Example 2, the cofactors A_{11} and A_{21} were computed as 4 and 11, respectively. Since A_{31} is multiplied by $a_{31} = 0$, its value is unimportant. Thus,

$$\begin{aligned} \det A &= a_{11}A_{11} + a_{21}A_{21} + a_{31}A_{31} \\ &= 2(4) + (-2)(11) + (0)A_{31} \\ &= -14. \end{aligned}$$

We could have also expanded about the second row.

$$\begin{aligned} \det A &= a_{21}A_{21} + a_{22}A_{22} + a_{23}A_{23} \\ &= (-2)A_{21} + (1)A_{22} + (0)A_{23} \end{aligned}$$

To complete this computation we only need to determine A_{22}, since A_{21} is known to be 11 and A_{23} is multiplied by 0. To compute A_{22}, delete the second row and column of A to obtain M_{22}.

$$M_{22} = \det \begin{bmatrix} 2 & 1 \\ 0 & 4 \end{bmatrix} = 8 \quad \text{and} \quad A_{22} = (-1)^{2+2}(8) = 8$$

Thus, $\det A = (-2)(11) + (1)(8) + (0)A_{23} = -14$. The same value for $\det A$ is obtained in both calculations. ∎

Critical Thinking

If a row or column in a matrix A contains only zeros, what is $\det A$?

Instead of calculating $(-1)^{i+j}$ for each cofactor, the following sign matrix can be utilized to find determinants of 3×3 matrices. The checkerboard pattern can be expanded to include larger matrices.

$$\begin{bmatrix} + & - & + \\ - & + & - \\ + & - & + \end{bmatrix}$$

For example, if

$$A = \begin{bmatrix} 2 & 3 & 7 \\ -3 & -2 & -1 \\ 4 & 0 & 2 \end{bmatrix},$$

we can compute det A by expanding about the second column to take advantage of the 0. The second column contains $-$, $+$, and $-$ signs. Therefore,

$$\det A = (3) \det \begin{bmatrix} -3 & -1 \\ 4 & 2 \end{bmatrix} \quad (-2) \det \begin{bmatrix} 2 & 7 \\ 4 & 2 \end{bmatrix} \quad (0) \det \begin{bmatrix} 2 & 7 \\ -3 & -1 \end{bmatrix}$$

$$= -3(-2) + (-2)(-24) - (0)(19)$$

$$= 54.$$

Many graphing calculators can evaluate the determinant of a matrix, as illustrated in the next example.

EXAMPLE 4   *Using technology to find a determinant*

Find the determinant of A.

(a) $A = \begin{bmatrix} 2 & -3 & 1 \\ -2 & 1 & 0 \\ 0 & -1 & 4 \end{bmatrix}$ (b) $A = \begin{bmatrix} 2 & -3 & 1 & 5 \\ 7 & 1 & -8 & 0 \\ 5 & 4 & 9 & 7 \\ -2 & 3 & 3 & 0 \end{bmatrix}$

Solution

(a) The determinant of this matrix was calculated in Example 3 by hand. To use technology enter the matrix and evaluate its determinant as shown in Figure 8.57. The result is det $A = -14$, which agrees with our earlier calculation.

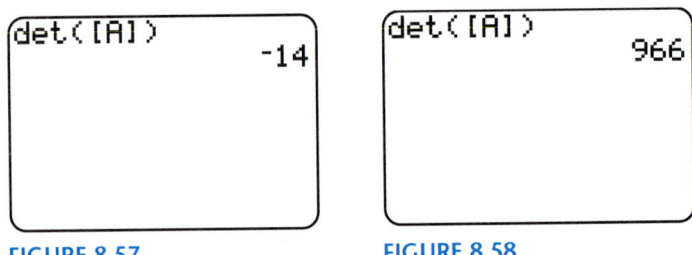

FIGURE 8.57   FIGURE 8.58

(b) The determinant of a 4×4 matrix can be computed using cofactors. However, it is considerably easier to use technology. From Figure 8.58 we see that det $A = 966$. ∎

Area of Regions

Determinants may be used to find the area of a triangle. If a triangle has vertices (a_1, b_1), (a_2, b_2), and (a_3, b_3), as shown in Figure 8.59, then its area is equal to the absolute value of D, where

$$D = \frac{1}{2} \det \begin{bmatrix} a_1 & a_2 & a_3 \\ b_1 & b_2 & b_3 \\ 1 & 1 & 1 \end{bmatrix}.$$

If the vertices are entered into the columns of D in a counterclockwise direction, then D will be positive. (Source: W. Taylor, *The Geometry of Computer Graphics.*)

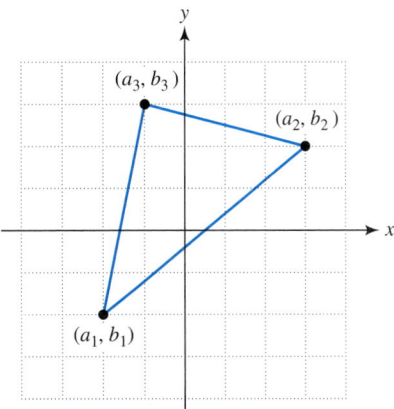

FIGURE 8.59

EXAMPLE 5 *Computing area of a parallelogram*

Calculate the area of the parallelogram in Figure 8.60 using determinants.

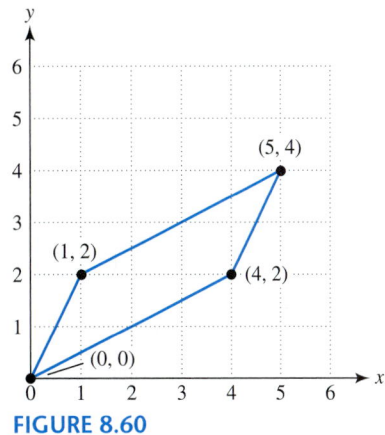

FIGURE 8.60

Solution

To find the area of the parallelogram, we view the parallelogram as being comprised of two triangles. One possible triangle has vertices at $(0, 0)$, $(4, 2)$, and $(1, 2)$, while the other triangle has vertices at $(4, 2)$, $(5, 4)$, and $(1, 2)$. The area of the parallelogram is equal to the sum of the areas of the two triangles. Since these triangles are congruent, we can calculate the area of one triangle and double it. The area of one triangle is equal to D, where

$$D = \frac{1}{2} \det \begin{bmatrix} 0 & 4 & 1 \\ 0 & 2 & 2 \\ 1 & 1 & 1 \end{bmatrix} = 3.$$

Since the vertices were entered in a counterclockwise direction, D is positive. The area of one triangle is equal to 3. Therefore, the area of the parallelogram is twice this value or 6.

Critical Thinking ———————————————————————

Suppose we are given three distinct vertices and $D = 0$. What must be true about the three points?

Cramer's Rule

Determinants can be used to solve linear systems by employing a method called *Cramer's rule*.

Cramer's rule for linear systems in two variables

The solution to the linear system

$$a_1 x + b_1 y = c_1$$
$$a_2 x + b_2 y = c_2$$

is given by $x = \dfrac{E}{D}$ and $y = \dfrac{F}{D}$, where

$$E = \det \begin{bmatrix} c_1 & b_1 \\ c_2 & b_2 \end{bmatrix}, \quad F = \det \begin{bmatrix} a_1 & c_1 \\ a_2 & c_2 \end{bmatrix}, \quad \text{and} \quad D = \det \begin{bmatrix} a_1 & b_1 \\ a_2 & b_2 \end{bmatrix} \neq 0.$$

EXAMPLE 6   *Using Cramer's rule to solve a linear system in two variables*

The height of a falling object above the ground after x seconds can be modeled by a quadratic function given by $f(x) = ax^2 + b$.
(a) If the object is 146 feet above the ground after 2 seconds and 66 feet after 3 seconds, find values for the constants a and b.
(b) Use f to find the height of the object after 1 second.

Solution

(a) The constants a and b must satisfy the following two equations.

$$f(2) = a(2)^2 + b = 146$$
$$f(3) = a(3)^2 + b = 66$$

These can be expressed as a linear system in two variables.

$$4a + b = 146$$
$$9a + b = 66$$

In this system of equations the unknowns are a and b. By Cramer's rule, the solution can be found as follows.

$$E = \det \begin{bmatrix} c_1 & b_1 \\ c_2 & b_2 \end{bmatrix} = \det \begin{bmatrix} 146 & 1 \\ 66 & 1 \end{bmatrix} = (146)(1) - (66)(1) = 80$$

$$F = \det \begin{bmatrix} a_1 & c_1 \\ a_2 & c_2 \end{bmatrix} = \det \begin{bmatrix} 4 & 146 \\ 9 & 66 \end{bmatrix} = (4)(66) - (9)(146) = -1050$$

$$D = \det \begin{bmatrix} a_1 & b_1 \\ a_2 & b_2 \end{bmatrix} = \det \begin{bmatrix} 4 & 1 \\ 9 & 1 \end{bmatrix} = (4)(1) - (9)(1) = -5$$

The solution is

$$a = \frac{E}{D} = \frac{80}{-5} = -16 \quad \text{and} \quad b = \frac{F}{D} = \frac{-1050}{-5} = 210.$$

Thus, the function f is given by $f(x) = -16x^2 + 210$.

(b) After 1 second the height of the object was 194 feet, since $f(1) = 194$. ■

Limitations on the Method of Cofactors and Cramer's Rule

Systems of linear equations involving more than two variables can be solved using Cramer's rule. If a linear system has n equations, then Cramer's rule requires the computation of $n + 1$ determinants with dimension $n \times n$. Cramer's rule is seldom employed in real applications because of the substantial number of arithmetic operations needed to compute determinants of large matrices.

It can be shown that the cofactor method of calculating the determinant of an $n \times n$ matrix, $n > 2$, generally involves more than $n!$ multiplication operations. For example, a 4×4 determinant requires over $4! = 4 \cdot 3 \cdot 2 \cdot 1 = 24$ multiplication operations, while a 10×10 determinant would involve over

$$10! = 10 \cdot 9 \cdot 8 \cdot 7 \cdot 6 \cdot 5 \cdot 4 \cdot 3 \cdot 2 \cdot 1 = 3{,}628{,}800$$

multiplication operations. Factorials grow rapidly and are discussed in Chapter 6.

In real-life applications, it is not uncommon to solve linear systems involving thousands of equations. Suppose that we were to solve a modest linear system involving 20 equations using Cramer's rule, and expand each determinant by the method of cofactors. Then, the calculation of one 20×20 determinant would require over $20!$ multiplication operations. The fastest supercomputers can perform 1 trillion (10^{12}) multiplication operations per second. On this supercomputer, just one 20×20 determinant would require over

$$\frac{20!}{10^{12}} \approx 2{,}432{,}902 \text{ seconds} \approx 28.2 \text{ days}$$

to compute. We would need to compute 21 of these determinants. This would require approximately $21 \times 28.2 = 592.2$ days, which is the best part of two years. Also, it is not uncommon for the cost of supercomputer time to exceed $2000 per hour. So, besides the time element, these computations would cost over $28 million to perform. On a typical personal computer, this task could not be completed in a lifetime.

It becomes obvious why this method is not implemented even on a modest linear system, regardless of the technology available. Modern software packages are more efficient than Cramer's rule on linear systems involving as little as three variables. (Source: Minnesota Supercomputer Institute.)

8.6 PUTTING IT ALL TOGETHER

The determinant of a square matrix A is a real number, denoted by det A. If det $A \neq 0$, then the matrix A is invertible. The computation of a determinant frequently involves a large number of arithmetic operations. Graphing calculators can be a valuable aid in computing determinants. Cramer's rule is a method for solving systems of linear equations that involves determinants.

The following table summarizes the calculation of 2×2 and 3×3 determinants by hand.

Determinants of 2 × 2 Matrices

The determinant of a 2×2 matrix A is given by

$$\det A = \det \begin{bmatrix} a & b \\ c & d \end{bmatrix} = ad - cb.$$

Determinants of 3 × 3 Matrices

The determinant of a 3×3 matrix A can be reduced to calculating the determinants of three 2×2 matrices. This calculation can be performed using cofactors.

$$\det A = \det \begin{bmatrix} a_1 & b_1 & c_1 \\ a_2 & b_2 & c_2 \\ a_3 & b_3 & c_3 \end{bmatrix}$$

$$= a_1 \det \begin{bmatrix} b_2 & c_2 \\ b_3 & c_3 \end{bmatrix} - a_2 \det \begin{bmatrix} b_1 & c_1 \\ b_3 & c_3 \end{bmatrix} + a_3 \det \begin{bmatrix} b_1 & c_1 \\ b_2 & c_2 \end{bmatrix}$$

$$= a_1 (b_2 c_3 - b_3 c_2) - a_2 (b_1 c_3 - b_3 c_1) + a_3 (b_1 c_2 - b_2 c_1)$$

8.6 EXERCISES

   Tape 7

Calculating Determinants

Exercises 1–4: Determine if the matrix A is invertible by calculating det A.

1. $A = \begin{bmatrix} 4 & 3 \\ 5 & 4 \end{bmatrix}$ **2.** $A = \begin{bmatrix} 1 & -3 \\ 2 & 6 \end{bmatrix}$

3. $A = \begin{bmatrix} -4 & 6 \\ -8 & 12 \end{bmatrix}$ **4.** $A = \begin{bmatrix} 10 & -20 \\ -5 & 10 \end{bmatrix}$

Exercises 5–8: Find the specified minor and cofactor for the matrix A.

5. M_{12} and A_{12} if $A = \begin{bmatrix} 1 & -1 & 3 \\ 2 & 3 & -2 \\ 0 & 1 & 5 \end{bmatrix}$

6. M_{23} and A_{23} if $A = \begin{bmatrix} 1 & 2 & -1 \\ 4 & 6 & -3 \\ 2 & 3 & 9 \end{bmatrix}$

7. M_{22} and A_{22} if $A = \begin{bmatrix} 7 & -8 & 1 \\ 3 & -5 & 2 \\ 1 & 0 & -2 \end{bmatrix}$

8. M_{31} and A_{31} if $A = \begin{bmatrix} 0 & 0 & -1 \\ 6 & -7 & 1 \\ 8 & -9 & -1 \end{bmatrix}$

Exercises 9–12: Find det A by expanding about the first column. State whether A^{-1} exists.

9. $A = \begin{bmatrix} 1 & 4 & -7 \\ 0 & 2 & -3 \\ 0 & -1 & 3 \end{bmatrix}$

10. $A = \begin{bmatrix} 0 & 2 & 8 \\ -1 & 3 & 5 \\ 0 & 4 & 1 \end{bmatrix}$

11. $A = \begin{bmatrix} 5 & 1 & 6 \\ 0 & -2 & 0 \\ 0 & 4 & 0 \end{bmatrix}$

12. $A = \begin{bmatrix} 3 & 2 & 3 \\ 2 & 2 & 2 \\ 1 & 3 & 1 \end{bmatrix}$

Exercises 13–20: Find det A using the method of cofactors.

13. $A = \begin{bmatrix} 2 & 0 & 0 \\ 0 & 3 & 0 \\ 0 & 0 & 5 \end{bmatrix}$ **14.** $A = \begin{bmatrix} 0 & 0 & 2 \\ 0 & 3 & 0 \\ 5 & 0 & 0 \end{bmatrix}$

15. $A = \begin{bmatrix} 0 & 0 & 0 \\ -8 & 3 & -9 \\ 15 & 5 & 9 \end{bmatrix}$ **16.** $A = \begin{bmatrix} 1 & 1 & 5 \\ -3 & -3 & 0 \\ 7 & 0 & 0 \end{bmatrix}$

17. $A = \begin{bmatrix} 3 & -1 & 2 \\ 0 & 5 & 7 \\ 1 & 0 & -1 \end{bmatrix}$ **18.** $A = \begin{bmatrix} 3 & 0 & -1 \\ 2 & 3 & -4 \\ 6 & -5 & 1 \end{bmatrix}$

19. $A = \begin{bmatrix} 1 & -5 & 2 \\ -7 & 1 & 3 \\ 0 & 4 & -2 \end{bmatrix}$

20. $A = \begin{bmatrix} 1 & -1 & 2 \\ -2 & 0 & 1 \\ 1 & 1 & -1 \end{bmatrix}$

Exercises 21–24: Use technology to calculate det A.

21. $A = \begin{bmatrix} 11 & -32 \\ 1.2 & 55 \end{bmatrix}$

22. $A = \begin{bmatrix} 17 & -4 & 3 \\ 11 & 5 & -15 \\ 7 & -9 & 23 \end{bmatrix}$

23. $A = \begin{bmatrix} 2.3 & 5.1 & 2.8 \\ 1.2 & 4.5 & 8.8 \\ -0.4 & -0.8 & -1.2 \end{bmatrix}$

24. $A = \begin{bmatrix} 1 & -1 & 3 & 7 \\ 9 & 2 & -7 & -4 \\ 5 & -7 & 1 & -9 \\ 7 & 1 & 3 & 6 \end{bmatrix}$

Calculating Area

Exercises 25–28: Find the area of the figure.

25.

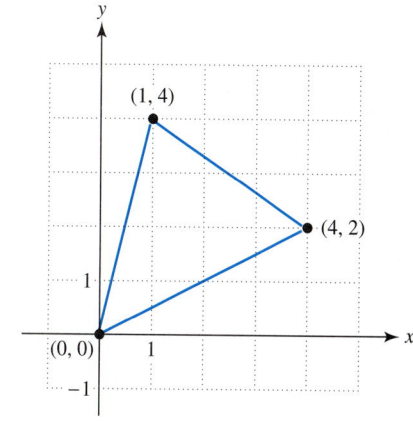

26.

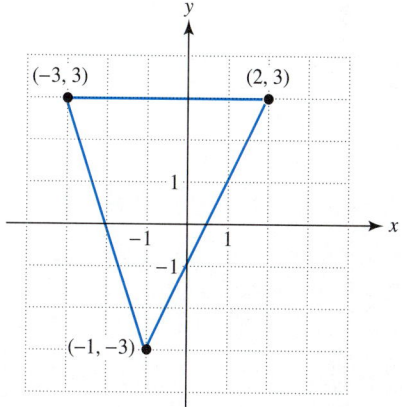

27.

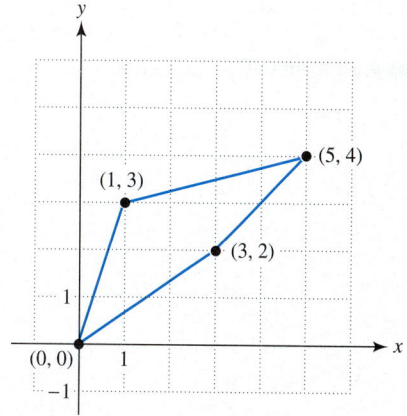

28.

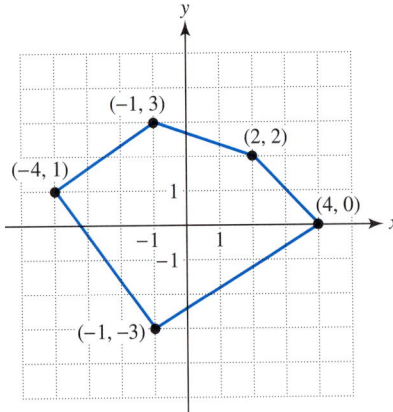

Cramer's Rule

Exercises 29–32: Use Cramer's rule to solve each system of linear equations.

29. $7x + 4y = 23$
$11x - 5y = 70$

30. $-7x + 5y = 8.2$
$6x + 4y = -0.4$

31. $1.7x - 2.5y = -0.91$
$-0.4x + 0.9y = 0.423$

32. $-2.7x + 1.5y = -1.53$
$1.8x - 5.5y = -1.68$

Applying a Concept

Exercises 33–36: Use the concept of the area of a triangle to determine if the three points are collinear.

33. $(1, 3), (-3, 11), (2, 1)$

34. $(3, 6), (-1, -6), (5, 11)$

35. $(-2, -5), (4, 4), (2, 3)$

36. $(4, -5), (-2, 10), (6, -10)$

Writing about Mathematics

1. Choose two matrices A and B with dimension 2×2. Calculate det A, det B, and det (AB). Repeat this process until you are able to discover how these three determinants are related. Summarize your results.

2. Calculate det A and det A^{-1} for different matrices. Compare the determinants. Try to generalize your results.

CHECKING BASIC CONCEPTS FOR SECTIONS 8.5 AND 8.6

1. Find the inverse of the matrix A by hand. Use technology to check your answer.

$$A = \begin{bmatrix} 0 & 0 & 1 \\ 1 & 1 & 0 \\ 1 & 0 & 1 \end{bmatrix}$$

2. Write the following system of linear equations as a matrix equation $AX = B$. Use technology to solve the system utilizing A^{-1}.

$$3.1x - 5.3y = -2.682$$
$$-0.1x + 1.8y = 0.787$$

3. Find the determinant of the matrix A using the method of cofactors. Use technology to check your answer. Is A invertible?

$$A = \begin{bmatrix} 1 & -1 & 2 \\ 2 & 3 & 1 \\ 0 & -2 & 5 \end{bmatrix}$$

CHAPTER **8** Summary

Many quantities in everyday life can be computed by a function of more than one input. Calculation of windchill, gas mileage, grade point average, and loan payments are examples. Functions of more than one input have verbal, numerical, graphical, and symbolic representations.

Functions of more than one input often lead to systems of equations of more than one variable. For example, if the function A computes the area of a rectangle with length l and width w and P calculates its perimeter, then the equations $A(l, w) = lw = 12$ and $P(l, w) = 2l + 2w = 14$ result in a system of two equations in two variables. The solution of $l = 4$ and $w = 3$ gives a rectangle with an area of 12 and a perimeter of 14. This solution satisfies both equations.

Systems of linear equations can have no solutions, one solution, or an infinite number of solutions. If the linear system has a solution, then it is consistent—otherwise the system is inconsistent. An efficient method for solving linear systems is Gaussian elimination with backward substitution. Substitution, elimination, and graphing also are methods to solve systems of equations. The inverse of a matrix, if it exists, can be used to solve a system of linear equations.

A matrix is a rectangular array of elements. Matrices often are written as augmented matrices to solve linear systems. Matrices have many applications in areas such as digital photography and computer graphics. Like numbers, variables, and functions—matrices can be added, subtracted, and multiplied. There is no division of matrices. However, square matrices sometimes have inverses. The inverse matrix A^{-1} will undo or cancel the computation performed by the matrix A.

A determinant is a real number associated with a square matrix. If the determinant of a matrix A is equal to 0, then A^{-1} does not exist. Determinants can be used to compute solutions to linear systems and find areas of certain types of regions. The computation of a determinant of a large matrix is usually avoided because of the enormous number of arithmetic computations it involves.

Review Exercises

Exercises 1 and 2: Evaluate f for the indicated inputs.

1. $f(3, 6)$, where $f(b, h) = \frac{1}{2}bh$

2. $f(2, 5)$, where $f(r, h) = \pi r^2 h$

3. *Ultraviolet Light*   There are two types of ultraviolet light: UV-A and UV-B. UV-B is responsible for

both skin and eye damage. The table on the next page lists the maximum doses of UV-B in the northern hemisphere for selected latitudes L and dates D of the year. (Sources: Orbital Sciences Corporation; J. Williams, *The Weather Almanac 1995.*)

L/D	Mar 21	June 21	Sept 21	Dec 21
0°	325	254	325	272
10°	311	275	280	220
20°	249	292	256	143
30°	179	248	182	80
40°	99	199	127	34
50°	57	143	75	13

(a) A person living at Topeka, Kansas, (39°N) is planning spring break in Honolulu, Hawaii, (21°N) during March 19–23. Approximate how many times stronger the UV-B light is in Hawaii than in Topeka.

(b) Compare the sun's UV-B strength at Glasgow, Montana, (48°N) on June 21 with Honolulu on December 21.

4. *Wave Height*   The table lists the wave heights produced on the ocean for various wind speeds and durations. (Source: J. Navarra, *Atmosphere, Weather and Climate.*)

Wind Speed (mph)	Duration of the Wind			
	10 hr	20 hr	30 hr	40 hr
11.5	2 ft	2 ft	2 ft	2 ft
17.3	4 ft	5 ft	5 ft	5 ft
23.0	7 ft	8 ft	9 ft	9 ft
34.5	13 ft	17 ft	18 ft	19 ft
46.0	21 ft	28 ft	31 ft	33 ft
57.5	29 ft	40 ft	45 ft	48 ft

(a) What is the expected wave height if a 46 mile-per-hour wind blows for 30 hours?

(b) If the wave height is 5 feet, can the speed of the wind and its duration be determined?

(c) Describe the relationship between wave height, wind speed, and duration of the wind.

(d) If the duration of the wind doubles, does the wave height double?

(e) Estimate the wave height if a 46 mile-per-hour wind blows for 15 hours.

Exercises 5 and 6: Solve the system of equations
(a) *graphically, and*
(b) *symbolically using substitution.*

5. $3x + y = 1$
   $2x - 3y = 8$

6. $x^2 - y = 1$
   $x + y = 1$

7. *Area and Perimeter*   Let l represent the length of a rectangle and w its width. Then, its area can be computed by $A(l, w) = lw$ and its perimeter by $P(l, w) = 2l + 2w$. Use substitution to solve the system of equations determined by $A(l, w) = 77$ and $P(l, w) = 36$.

8. *Cylinder*   Graphically approximate the radius r and height h of a cylindrical container with a volume V of 30 cubic inches and a lateral surface area S of 45 square inches.

Exercises 9 and 10: Use the elimination method to solve each system of linear equations, if possible. Identify the system as consistent or inconsistent.

9. $2x + y = 7$
   $x - 2y = -4$

10. $3x + 3y = 15$
    $-x - y = -4$

11. *Student Loans*   A student takes out two loans totaling $2000 to help pay for college expenses. One loan is at 7% interest, while the other is at 9%. Interest for both loans is compounded annually.
(a) If the total first-year interest for both loans is $156, find the amount of each loan symbolically.
(b) Determine the amount of each loan graphically and numerically.

12. *Dimensions of a Screen*   The screen of a rectangular television set is 3 inches wider than it is high. If the perimeter of the screen is 42 inches, find its dimensions by writing a system of linear equations and using elimination.

Exercises 13 and 14: Graph the system of inequalities. Shade the region that represents the solution set. Use the graph to find a solution.

13. $x^2 + y^2 < 9$
    $x + y > 3$

14. $y \le 4 - x^2$
    $y \ge 2 - x$

Exercises 15 and 16: Each augmented matrix represents a system of linear equations. Solve the system.

15. $\begin{bmatrix} 1 & 5 & | & 6 \\ 0 & 1 & | & 3 \end{bmatrix}$

16. $\begin{bmatrix} 1 & 2 & -2 & | & 8 \\ 0 & 1 & 1 & | & 5 \\ 0 & 0 & 0 & | & 0 \end{bmatrix}$

Exercises 17 and 18: Use Gaussian elimination with backward substitution to solve each system of linear equations.

17. $x + 3y = 8$
 $-x + \ y = 4$

18. $x + \qquad z = \ \ 4$
 $x + y - 2z = -3$
 $-x + y + \ z = \ \ 4$

Exercises 19 and 20: Evaluate the following.
 (a) $A + 2B$
 (b) $A - B$

19. $A = \begin{bmatrix} 1 & -3 \\ 2 & -1 \end{bmatrix}$, $B = \begin{bmatrix} 3 & 2 \\ -5 & 1 \end{bmatrix}$

20. $A = \begin{bmatrix} 4 & 0 & 1 \\ -2 & 8 & 9 \end{bmatrix}$, $B = \begin{bmatrix} -5 & 3 & 2 \\ -4 & 0 & 7 \end{bmatrix}$

Exercises 21–24: If possible, find AB and BA.

21. $A = \begin{bmatrix} 2 & 0 \\ -5 & 3 \end{bmatrix}$, $B = \begin{bmatrix} -1 & -2 \\ 4 & 7 \end{bmatrix}$

22. $A = \begin{bmatrix} 1 & -2 \\ 2 & 3 \end{bmatrix}$, $B = \begin{bmatrix} 1 & 0 & 2 \\ -1 & 3 & 4 \end{bmatrix}$

23. $A = \begin{bmatrix} 2 & -1 & 3 \\ 2 & 4 & 0 \end{bmatrix}$, $B = \begin{bmatrix} 1 & 0 \\ -1 & 2 \\ 0 & 3 \end{bmatrix}$

24. $A = \begin{bmatrix} 1 & -1 & 2 \\ 0 & 3 & 4 \\ 1 & 0 & 2 \end{bmatrix}$, $B = \begin{bmatrix} -1 & 0 & 0 \\ 2 & 0 & -1 \\ 1 & 4 & 2 \end{bmatrix}$

Exercises 25 and 26: Use technology to find det A. State if A is invertible.

25. $A = \begin{bmatrix} 13 & 22 \\ 55 & -57 \end{bmatrix}$

26. $A = \begin{bmatrix} 6 & -7 & -1 \\ -7 & 3 & -4 \\ 23 & 54 & 77 \end{bmatrix}$

Exercises 27 and 28: Determine if B is the inverse matrix of A.

27. $A = \begin{bmatrix} 8 & 5 \\ 6 & 4 \end{bmatrix}$, $B = \begin{bmatrix} 2 & -2.5 \\ -3 & 4 \end{bmatrix}$

28. $A = \begin{bmatrix} -1 & 1 & 2 \\ 1 & 0 & -1 \\ 0 & 1 & 2 \end{bmatrix}$,
 $B = \begin{bmatrix} -1 & 0 & 1 \\ 2 & 2 & -1 \\ -1 & -1 & -1 \end{bmatrix}$

Exercises 29 and 30: Complete the following.
 (a) *Write the system of linear equations in the form* $AX = B$.
 (b) *Solve the linear system by computing* $X = A^{-1}B$.

29. $11x + 31y = -27.6$
 $37x - 19y = \ \ 240$

30. $12x + \ 7y - \ 3z = \ \ 14.6$
 $8x - 11y + 13z = -60.4$
 $-23x \qquad + \ 9z = -14.6$

31. If possible, graphically approximate the solution of each system of equations to four significant digits. Identify each system as consistent or inconsistent. If the system is consistent, determine if the equations are dependent or independent.
 (a) $3.1x + 4.2y = 6.4$
 $1.7x - 9.1y = 1.6$
 (b) $6.3x - 5.1y = 9.3$
 $4.2x - 3.4y = 6.2$
 (c) $0.32x - 0.64y = \ \ 0.96$
 $-0.08x + 0.16y = -0.72$

32. *CD Prices*   A music store sells compact discs at two different prices marked A and B. Each row in the table represents a purchase. Determine the cost of each type of CD using a matrix inverse.

A	B	Total
1	2	$37.47
2	3	$61.95

Exercises 33 and 34: Find A^{-1}.

33. $A = \begin{bmatrix} 1 & -2 \\ -1 & 1 \end{bmatrix}$

34. $A = \begin{bmatrix} 1 & 0 & 1 \\ 1 & 1 & 1 \\ 0 & 1 & -1 \end{bmatrix}$

Exercises 35 and 36: Find det A using the method of cofactors.

35. $A = \begin{bmatrix} 2 & 1 & 3 \\ 0 & 3 & 4 \\ 1 & 0 & 5 \end{bmatrix}$

36. $A = \begin{bmatrix} 3 & 0 & 2 \\ 1 & 3 & 5 \\ -5 & 2 & 0 \end{bmatrix}$

37. *Digital Photography*   Design a 3×3 matrix A that represents a digital photograph of the letter T in black on a white background. Find a matrix B, such that $A + B$ darkens only the white background by one gray level.

38. *Area*   Find the area of the triangle whose vertices are $(0, 0)$, $(5, 2)$, and $(2, 5)$.

39. *Voter Turnout*   The table on the following page shows the percent y of voter turnout in the United States for three different presidential elections in year x, where $x = 0$ corresponds to 1900. Find a quadratic function defined by $f(x) = ax^2 + bx + c$ that models this data. Approximate a, b, and c to five significant digits. Graph f together with the data. (Source: Committee for the Study of the American Electorate.)

x	24	60	96
y	48.9	62.8	48.8

40. *Geometry* Complete the following.
    (a) Write a system of inequalities that describes possible dimensions for a cylinder with a volume V greater than or equal to 30 cubic inches, and a lateral surface area S less than or equal to 45 square inches.
    (b) Graph and identify the region of solutions.
    (c) Use the graph to estimate one solution to the system of inequalities.

41. *Snowmobile Fatalities* Through February during the 1996–1997 winter, snowmobile fatalities in Michigan exceeded the average number for an entire winter by 5 fatalities. This was a 19.2% increase over the yearly average, with two months still left in the winter. Approximate graphically the average number of snowmobile fatalities in Michigan annually, and the number of fatalities through February during the winter of 1996–1997. (Source: *USA Today.*)

42. *Flood Control* The spillway capacity of a dam is important in flood control. Spillway capacity Q is measured in cubic feet of water per second and depends on the width W in feet of the spillway and the depth D in feet of the water over the spillway. See the figure. Spillway capacity is computed by $Q = f(W, D)$. The table is a partial numerical representation of f. (Source: D. Callas, Project Director, *Snapshots of Applications in Mathematics.*)

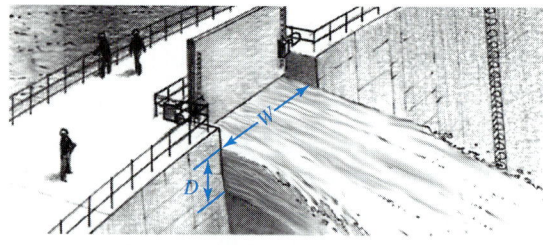

$W \backslash D$	1.0	1.5	2.0	2.5	3.0
10	33	61	94	131	173
20	66	122	188	262	345
30	99	183	282	394	518
40	133	244	376	525	690
50	166	305	470	656	863

$W \backslash D$	3.5	4.0	4.5	5.0	5.5
10	217	266	317	371	428
20	434	531	634	742	856
30	652	797	951	1114	1285
40	870	1062	1268	1485	1713
50	1087	1328	1585	1856	2141

    (a) Evaluate $f(40, 2)$ and interpret the result.
    (b) Solve the equation $f(W, 4) = 1062$. Interpret the result.
    (c) If the depth D doubles, does the spillway capacity Q roughly double? Explain.
    (d) If the width W doubles, does the spillway capacity Q roughly double? Explain.

43. *Joint Variation* Suppose P varies jointly as the square of x and the cube of y. If $P = 432$ when $x = 2$ and $y = 3$, find P when $x = 3$ and $y = 5$.

44. *Linear Programming* Find the maximum value of $P = 3x + 4y$ subject to the following constraints.

$$x \geq 0$$
$$y \geq 0$$
$$x + 3y \leq 12$$
$$3x + y \leq 12$$

Extended and Discovery Exercises

1. To form the **transpose** of a matrix A, denoted by A^T, let the first row of A be the first column of A^T, the second row of A be the second column of A^T, and so on, for each row of A. The following are examples of A and A^T. If A has dimension $m \times n$, then A^T has dimension $n \times m$.

$$A = \begin{bmatrix} 3 & -3 & 7 \\ 1 & 6 & -2 \\ 4 & 2 & 5 \end{bmatrix} \quad A^T = \begin{bmatrix} 3 & 1 & 4 \\ -3 & 6 & 2 \\ 7 & -2 & 5 \end{bmatrix}$$

$$A = \begin{bmatrix} 1 & 2 \\ 3 & 4 \\ 5 & 6 \end{bmatrix} \quad A^T = \begin{bmatrix} 1 & 3 & 5 \\ 2 & 4 & 6 \end{bmatrix}$$

Find the transpose of each matrix A.

(a) $A = \begin{bmatrix} 3 & -3 \\ 2 & 6 \\ 4 & 2 \end{bmatrix}$

(b) $A = \begin{bmatrix} 0 & 1 & -2 \\ 2 & 5 & 4 \\ -4 & 3 & 9 \end{bmatrix}$

(c) $A = \begin{bmatrix} 5 & 7 \\ 1 & -7 \\ 6 & 3 \\ -9 & 2 \end{bmatrix}$

Least-Squares Fit and Matrices

The table shows the average cost of tuition and fees y in dollars at four-year public colleges. In this table x = 0 represents 1980 and x = 15 corresponds to 1995. (Source: The College Board.)

x	0	5	10	15
y	804	1318	1908	2860

The data is modeled using a line in the accompanying figure.

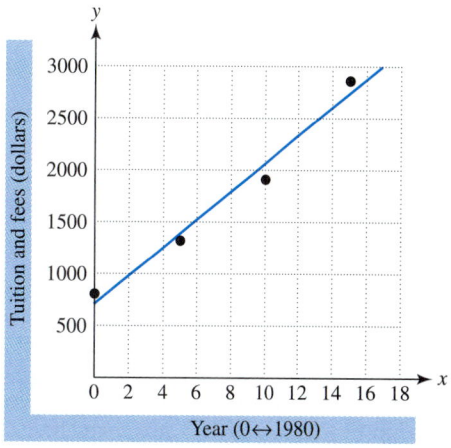

Year (0↔1980)

The equation of this line can be found using linear regression based on least squares. (Refer to Section 2.5.) If possible, we would like $f(x) = ax + b$ to satisfy the following four equations.

$$f(0) = a(0) + b = 804$$
$$f(5) = a(5) + b = 1318$$
$$f(10) = a(10) + b = 1908$$
$$f(15) = a(15) + b = 2860$$

Since the data points are not collinear, it is impossible for the graph of f to pass through all four points. These four equations can be written as

$$AX = \begin{bmatrix} 0 & 1 \\ 5 & 1 \\ 10 & 1 \\ 15 & 1 \end{bmatrix} \begin{bmatrix} a \\ b \end{bmatrix} = \begin{bmatrix} 804 \\ 1318 \\ 1908 \\ 2860 \end{bmatrix} = B.$$

The least-squares solution is found by solving the **normal equations**

$$A^T A X = A^T B$$

for X. The solution is $X = (A^T A)^{-1} A^T B$. Using technology, $a = 135.16$ and $b = 708.8$. Thus, f is given by $f(x) = 135.16x + 708.8$. The function f and the data can be graphed. See the accompanying figures.

```
([A]ᵀ[A])⁻¹[A]ᵀ[B
]
          [[135.16]
           [708.8 ]]
```

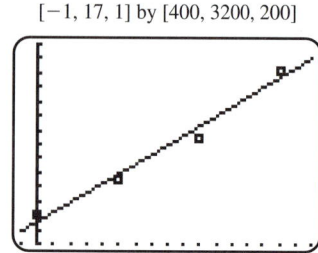

[−1, 17, 1] by [400, 3200, 200]

Exercises 2–4: Least-Squares Models Solve the normal equations to model the data with $f(x) = ax + b$. Plot the data and f in the same viewing rectangle.

2. *Tuition and Fees* The table shows average cost of tuition and fees y in dollars at private four-year colleges. In this table $x = 0$ corresponds to 1980 and $x = 15$ to 1995. (Source: The College Board.)

x	0	5	10	15
y	3617	6121	9340	12,432

3. *Satellite TV* The table lists the estimated number of satellite television subscribers y in millions. In this table $x = 0$ corresponds to 1995 and $x = 5$ to the year 2000. (Source: USA Today.)

x	0	1	2	3	4	5
y	2.2	4.5	7.9	10.5	13	15

4. *Federal Deficit* The table lists the annual federal deficit y in billions of dollars from 1992 to 1996, where $x = 2$ corresponds to 1992 and $x = 6$ to 1996. (Source: Department of the Treasury.)

x	2	3	4	5	6
y	290	255	203	164	107

Exercises 5 and 6: Nonlinear Least-Squares Normal equations can be used to solve problems involving nonlinear regression. Solve the normal equations so that f models the given data.

5. $f(x) = \dfrac{a}{x} + \dfrac{b}{x^2}$

x	1	2	3	4
y	6.30	1.88	0.967	0.619

6. $f(x) = a \log x + bx + c$ (*Hint:* The matrix A has dimension 5×3.)

x	2	3	4	5	6
y	2.45	3.12	3.70	4.25	4.77

Cryptography

7. Businesses and government agencies frequently send classified messages in code. One cryptographic technique that involves matrices is the **polygraphic system.** In this system each letter in the alphabet is associated with a number between 1 and 26. The following table gives a common example. (Source: A. Sinkov, *Elementary Cryptanalysis: A Mathematical Approach.*)

A	B	C	D	E	F	G	H	I	J	K	L	M
1	2	3	4	5	6	7	8	9	10	11	12	13

N	O	P	Q	R	S	T	U	V	W	X	Y	Z
14	15	16	17	18	19	20	21	22	23	24	25	26

For example, the word MATH is coded as 13 1 20 8. Enter these numbers in a matrix B.

$$B = \begin{bmatrix} 13 & 20 \\ 1 & 8 \end{bmatrix}$$

To code these letters a 2×2 matrix, such as

$$A = \begin{bmatrix} 2 & 1 \\ -5 & -2 \end{bmatrix},$$

is multiplied times B to form the product AB.

$$AB = \begin{bmatrix} 2 & 1 \\ -5 & -2 \end{bmatrix} \begin{bmatrix} 13 & 20 \\ 1 & 8 \end{bmatrix} = \begin{bmatrix} 27 & 48 \\ -67 & -116 \end{bmatrix}$$

Since the resulting elements of AB are less than 1 or greater than 26, they may be scaled between 1 and 26 by adding or subtracting multiples of 26.

$$27 - 1(26) = 1 \qquad 48 - 1(26) = 22$$
$$-67 + 3(26) = 11 \qquad -116 + 5(26) = 14$$

Thus, the word MATH is coded as 1 11 22 14 or AKVN. The advantage of this coding technique is that a particular letter is not always coded the same each time. Use the matrix A to code the following words.

(a) HELP

(b) LETTER (*Hint: B* has dimension 2×3.)

8. (Refer to the previous exercise.) To decode a message, A^{-1} is used. For example, to decode AKVN, write 1 11 22 14. This is stored as

$$C = \begin{bmatrix} 1 & 22 \\ 11 & 14 \end{bmatrix}.$$

To decode the message, evaluate $A^{-1}C$, where

$$A^{-1} = \begin{bmatrix} -2 & -1 \\ 5 & 2 \end{bmatrix}.$$

$$A^{-1}C = \begin{bmatrix} -2 & -1 \\ 5 & 2 \end{bmatrix} \begin{bmatrix} 1 & 22 \\ 11 & 14 \end{bmatrix} = \begin{bmatrix} -13 & -58 \\ 27 & 138 \end{bmatrix}$$

Scaling the matrix elements of $A^{-1}C$ between 1 and 26 results in the following.

$$-13 + (1)26 = 13 \qquad -58 + 3(26) = 20$$
$$27 - 1(26) = 1 \qquad 138 - 5(26) = 8$$

The number 13 1 20 8 represents MATH. Use A^{-1} to decode each of the following.

(a) UBNL

(b) QNABMV

CHAPTER 9 *Further Topics in Algebra*

$\mathcal{M}$athematics permeates the fabric of modern society. At times its presence is subtle and frequently missed by the average person, but its influence is nonetheless profound. Mathematics is the language of technology—it allows society to quantify its experiences.

In previous chapters we saw hundreds of examples where mathematics is used to describe physical phenomena and events. These included computers, CD players, cars, electricity, light, road construction, government data, telephones, medicine, ecology, business, sports, and psychology. Mathematics is diverse in its ability to adapt to new situations and solve complex problems. If any subject area is studied in enough detail, mathematics usually appears. Mathematics even describes aspects of seemingly intangible concepts such as color or photography.

This last chapter introduces further topics in mathematics. It represents only a small fraction of the topics found in mathematics. Although it may be difficult to

> $\mathcal{T}$he art of asking the right questions
> in mathematics is more important
> than the art of solving them.
> — **Georg Cantor**

predict exactly what the twenty-first century will bring, one thing is certain—mathematics will play an increasingly important role.

9.1 Sequences

Sequences and Functions • Representations of Sequences
• Arithmetic Sequences • Geometric Sequences

Introduction

Sequences are a fundamental concept in mathematics with many applications. A sequence is a function that computes an ordered list. For example, the average person in the United States uses 100 gallons of water each day. The function $f(n) = 100n$ generates the terms of the sequence

$$100, 200, 300, 400, 500, 600, 700, \ldots,$$

when $n = 1, 2, 3, 4, 5, 6, 7, \ldots$. This list represents the gallons of water used by the average person after n days.

Sequences and Functions

A second example of a sequence involves investing money. If \$100 is deposited into a savings account, paying 5% interest compounded annually, then the function defined by $g(n) = 100(1.05)^n$ calculates the account balance after n years, which is given by

$$g(1), g(2), g(3), g(4), g(5), g(6), g(7), \ldots.$$

These terms can be approximated as

$$105, 110.25, 115.76, 121.55, 127.63, 134.01, 140.71, \ldots.$$

We now define a sequence formally.

Sequence

An **infinite sequence** is a function that has the set of natural numbers as its domain. A **finite sequence** is a function with domain $D = \{1, 2, 3, \ldots, n\}$, for some fixed natural number n.

Since sequences are functions, many of the concepts discussed in previous chapters apply to sequences. Instead of letting y represent the output, it is common to write $a_n = f(n)$, where n is a natural number in the domain of the sequence. The **terms** of a sequence are

$$a_1, a_2, a_3, \ldots, a_n, \ldots$$

The first term is $a_1 = f(1)$, the second term is $a_2 = f(2)$, and so on. The **nth term** or **general term** of a sequence is $a_n = f(n)$.

EXAMPLE 1 *Computing terms of a sequence*

The average distances of the planets from the sun display a pattern that was first described by Johann Bode in 1772. This relationship is called *Bode's Law,* and was proposed even before Uranus, Neptune, and Pluto were discovered. It is a sequence defined by

$$f(1) = 0.4$$
$$f(n) = 0.3(2)^{n-2} + 0.4, \qquad \text{for } n = 2, 3, 4, \dots , 10,$$

where $a_n = f(n)$. In this sequence, a distance of one unit corresponds to the average Earth-sun distance. The number n represents the nth planet. The actual distances of the planets, including an average distance for the asteroids, are listed in Table 9.1. (*Source: M. Zeilik, Introductory Astronomy and Astrophysics.*)

(a) Find $a_4 = f(4)$, and interpret the result.

(b) Calculate the terms of the sequence defined by f. Compare them with the values in Table 9.1.

(c) If there is another planet beyond Pluto, use Bode's Law to predict its distance from the sun.

TABLE 9.1

Planet	Distance
Mercury	0.39
Venus	0.72
Earth	1.00
Mars	1.52
Asteroids	2.8
Jupiter	5.20
Saturn	9.54
Uranus	19.2
Neptune	30.1
Pluto	39.5

Solution

(a) $a_4 = f(4) = 0.3(2)^{4-2} + 0.4 = 1.6$. Bode's Law predicts that the fourth planet, Mars, is located 1.6 times farther from the sun than Earth.

(b) We will calculate $f(n)$ for $n = 1, 2, 3, \dots , 10$. For example,

$$f(1) = 0.4 \qquad \text{and}$$
$$f(2) = 0.3(2)^{2-2} + 0.4 = 0.7.$$

Other values are found in a similar manner. The ten terms of the sequence are

$$0.4, \quad 0.7, \quad 1, \quad 1.6, \quad 2.8, \quad 5.2, \quad 10, \quad 19.6, \quad 38.8, \quad 77.2.$$

Bode's Law works quite well for the seven inner planets and asteroids, but fails to predict the location of Neptune at 30.1. It appears to locate planet Pluto.

(c) Since Pluto's distance is 39.5, Bode's Law predicts the next planet at 77.2. ∎

EXAMPLE 2 *Finding terms of a sequence*

Write the first four terms a_1, a_2, a_3, and a_4 of each sequence, where $a_n = f(n)$.
(a) $f(n) = 2n - 5$
(b) $f(n) = 4(2)^{n-1}$
(c) $f(n) = (-1)^n \left(\dfrac{n}{n + 1} \right)$

Solution
(a) Evaluate the following.

$$a_1 = f(1) = 2(1) - 5 = -3$$
$$a_2 = f(2) = 2(2) - 5 = -1$$

In a similar manner, $a_3 = f(3) = 1$ and $a_4 = f(4) = 3$.

(b) Since $f(n) = 4(2)^{n-1}$,

$$a_1 = f(1) = 4(2)^{1-1} = 4.$$

Similarly, $a_2 = 8$, $a_3 = 16$, and $a_4 = 32$.

(c) Let $f(n) = (-1)^n \left(\dfrac{n}{n + 1} \right)$, and substitute $n = 1, 2, 3$, and 4.

$$a_1 = f(1) = (-1)^1 \left(\frac{1}{1 + 1} \right) = -\frac{1}{2}$$

$$a_2 = f(2) = (-1)^2 \left(\frac{2}{2 + 1} \right) = \frac{2}{3}$$

$$a_3 = f(3) = (-1)^3 \left(\frac{3}{3 + 1} \right) = -\frac{3}{4}$$

$$a_4 = f(4) = (-1)^4 \left(\frac{4}{4 + 1} \right) = \frac{4}{5}$$

Notice that the factor $(-1)^n$ causes the terms of the sequence to alternate sign. ∎

Some sequences are not defined using a general term. Instead they are defined *recursively*. With a recursive formula, we must find terms a_1 through a_{n-1} before we can find a_n.

A population model for an insect with a life span of one year can be described using a sequence. Suppose each adult female insect produces r female offspring that survive to reproduce the following year. Let $f(n)$ calculate the insect population during year n. Then, the number of female insects is given recursively by

$$f(n) = rf(n - 1) \qquad \text{for } n > 1.$$

The number of female insects in the year n is equal to r times the number of female insects in the previous year $n - 1$. This symbolic representation of a function is fundamentally different than any other equation we have encountered thus far. The reason is that the function f is defined symbolically in terms involving itself. To evaluate $f(n)$ we evaluate $f(n - 1)$. To evaluate $f(n - 1)$ we evaluate $f(n - 2)$, and so on. If we know the number of adult female insects during the first year, then we can determine the sequence. That is, if $f(1)$ is given, we can determine $f(n)$ by first computing

$$f(1), f(2), f(3), \ldots, f(n - 1).$$

The next example illustrates this recursively defined sequence. (Source: D. Brown and P. Rothery, *Models in Biology: Mathematics, Statistics and Computing.*)

EXAMPLE 3   *Finding terms of a sequence defined recursively*

Suppose that the initial density of adult female insects is 1000 per acre and $r = 1.1$. Then, the density of female insects is described by

$$f(1) = 1000$$
$$f(n) = rf(n-1), \qquad n > 1.$$

(a) Rewrite this symbolic representation in terms of a_n.
(b) Find a_4 and interpret the result. Is the density of female insects increasing or decreasing?
(c) A general term for this sequence is given by $f(n) = 1000(1.1)^{n-1}$. Use this representation to find a_4.

Solution

(a) Since $a_n = f(n)$ for all n, $a_{n-1} = f(n-1)$. The sequence can be expressed as

$$a_1 = 1000$$
$$a_n = ra_{n-1}, \qquad n > 1.$$

(b) In order to calculate the fourth term, a_4, we must first determine a_1, a_2, and a_3. Since $r = 1.1$, $a_n = 1.1a_{n-1}$.

$$a_1 = 1000$$
$$a_2 = 1.1a_1 = 1.1(1000) = 1100$$
$$a_3 = 1.1a_2 = 1.1(1100) = 1210$$
$$a_4 = 1.1a_3 = 1.1(1210) = 1331$$

The fourth term is $a_4 = 1331$. The female population density is increasing and reaches 1331 per acre during the fourth year.

(c) Since $a_4 = f(4)$,

$$a_4 = 1000(1.1)^{4-1} = 1331.$$

It is less work to find a_n using a formula for a general term than a recursive formula—particularly if n is large.   ■

Critical Thinking

How does the value of *r* in Example 3 affect the population density in future years?

Representations of Sequences

Sequences are functions. Therefore, they have graphical, numerical, and symbolic representations. The next example illustrates numerical and graphical representations for a sequence involving population growth.

EXAMPLE 4   *Representing sequences numerically and graphically*

Frequently the population of a particular insect does not continue to grow indefinitely, as it does in Example 3. Instead, its population grows rapidly at first, and then levels off because of competition for limited resources. In one study, the behavior of the winter moth was modeled using a sequence similar to the following, where a_n represents the population density in thousands per acre during year n. (Source: G. Varley and G. Gradwell, "Population models for the winter moth.")

$$a_1 = 1$$
$$a_n = 2.85a_{n-1} - 0.19a_{n-1}^2, \qquad n \geq 2$$

(a) Give a partial numerical representation for $n = 1, 2, 3, \ldots, 10$. Approximate each term to three significant digits. Describe what happens to the population density of the winter moth.
(b) Use the numerical representation to graph the sequence.

Solution
(a) Evaluate $a_1, a_2, a_3, \ldots, a_{10}$ recursively. Since $a_1 = 1$,

$$a_2 = 2.85a_1 - 0.19a_1^2 = 2.85(1) - 0.19(1)^2 = 2.66, \qquad \text{and}$$
$$a_3 = 2.85a_2 - 0.19a_2^2 = 2.85(2.66) - 0.19(2.66)^2 \approx 6.24.$$

Other terms are shown in Table 9.2. Figure 9.1 shows the computation of the sequence using a calculator. The sequence is denoted by $u(n)$ rather than u_n.

TABLE 9.2

n	1	2	3	4	5	6	7	8	9	10
a_n	1	2.66	6.24	10.4	9.11	10.2	9.31	10.1	9.43	9.98

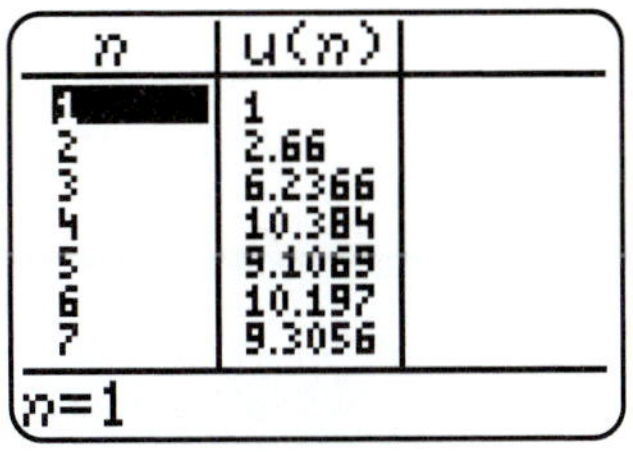

n	u(n)
1	1
2	2.66
3	6.2366
4	10.384
5	9.1069
6	10.197
7	9.3056

$n=1$

FIGURE 9.1

(b) The graph of a sequence is a set of discrete points. Plot the points

$$(1, 1), (2, 2.66), (3, 6.24), \ldots, (10, 9.98)$$

as shown in Figure 9.2 on the next page. At first the insect population increases rapidly, and then oscillates about the line $y = 9.7$. (See the following Critical Thinking Exercise.) The oscillations become smaller as n increases, indicating that the population density may stabilize near 9.7 thousand per acre. Some calculators can plot sequences as shown in Figure 9.3. In this figure the first 20 terms have been plotted.

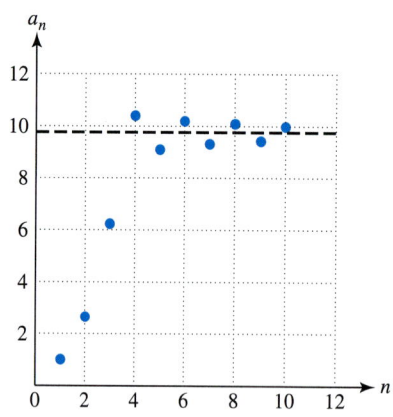

FIGURE 9.2 Insect Population

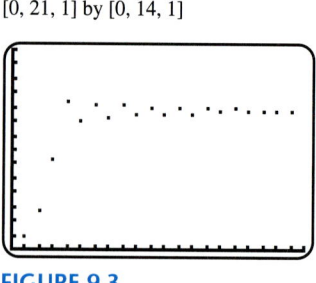

[0, 21, 1] by [0, 14, 1]

FIGURE 9.3

Critical Thinking

In Example 4, the insect population stabilizes near the value $k = 9.74$ thousand. This value of k can be found by solving the quadratic equation $k = 2.85k - 0.19k^2$. Try to explain why this is true.

Arithmetic Sequences

Suppose that a person's starting salary was $20,000 per year. Thereafter, this person receives a $1000 raise each year. The salary after the nth year is represented by

$$f(n) = 1000n + 20{,}000,$$

where f is a linear function. After 10 years of experience, the annual salary would be

$$f(10) = 1000(10) + 20{,}000 = \$30{,}000.$$

If a sequence can be defined by a linear function, it is an *arithmetic sequence.*

Infinite arithmetic sequence

An **infinite arithmetic sequence** is a linear function whose domain is the set of natural numbers.

An arithmetic sequence can be defined recursively by $a_n = a_{n-1} + d$, where d is a constant. Since $d = a_n - a_{n-1}$ for each valid n, d is called the **common difference.** If $d = 0$, then the sequence is a **constant sequence.** A **finite arithmetic sequence** is similar to an infinite arithmetic sequence except its domain is $D = \{1, 2, 3, \ldots, n\}$, where n is a fixed natural number.

EXAMPLE 5   *Determining arithmetic sequences*

Determine if f is an arithmetic sequence.
(a) $f(n) = n^2 + 3n$
(b) A graph of f is shown in Figure 9.4 on the following page.

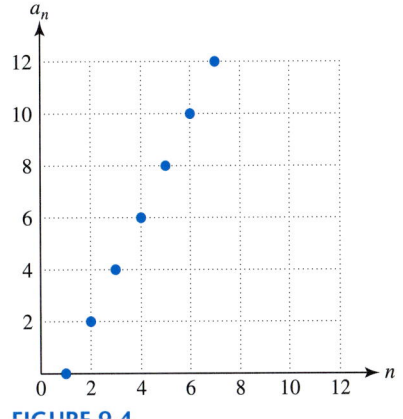

FIGURE 9.4

(c) The function f is represented numerically in Table 9.3.

TABLE 9.3							
n	1	2	3	4	5	6	7
$f(n)$	-1.5	0	1.5	3	4.5	6	7.5

Solution

(a) This sequence is not arithmetic, because $f(n) = n^2 + 3n$ is nonlinear.

(b) The sequence in Figure 9.4 represents an arithmetic sequence because the points lie on a line. A linear function could generate these points. Notice that the slope between points is always equal to 2. This slope represents the common difference d of the sequence.

(c) The successive terms

$$-1.5, 0, 1.5, 3, 4.5, 6, 7.5$$

increase by precisely 1.5. Therefore, the common difference is $d = 1.5$. Since $a_n = a_{n-1} + 1.5$ for each valid n, the sequence is arithmetic. ∎

Since an arithmetic sequence is a linear function, it can always be represented by $f(n) = dn + c$, where d is the common difference and c is a constant.

EXAMPLE 6   *Finding a symbolic representation for an arithmetic sequence*

Find a general term $a_n = f(n)$ for each arithmetic sequence.

(a) $a_1 = 3$ and $d = -2$

(b) $a_3 = 4$ and $a_9 = 17$

Solution

(a) Let $f(n) = dn + c$. Since $d = -2, f(n) = -2n + c$.

$$a_1 = f(1) = -2(1) + c = 3 \quad \text{or} \quad c = 5$$

Thus, $a_n = -2n + 5$.

(b) Since $a_3 = 4$ and $a_9 = 17$, we find a linear function $f(n) = dn + c$ that satisfies the equations $f(3) = 4$ and $f(9) = 17$. The common difference is equal to the slope between the points $(3, 4)$ and $(9, 17)$.

$$d = \frac{17 - 4}{9 - 3} = \frac{13}{6}$$

It follows that $f(n) = \dfrac{13}{6} n + c.$

$$a_3 = f(3) = \frac{13}{6}(3) + c = 4 \qquad \text{or} \qquad c = -\frac{5}{2}$$

Thus, $a_n = \dfrac{13}{6} n - \dfrac{5}{2}.$ ∎

MAKING CONNECTIONS

Linear Functions and Arithmetic Sequences

In Chapter 2 we discussed several techniques for finding symbolic representations of linear functions. These methods can be applied to finding symbolic representations of arithmetic sequences. It is important to realize that the mathematical concept of a linear function is simply being applied to the new topic of sequences.

Critical Thinking

Explain why the nth term of an arithmetic sequence is given by

$$a_n = a_1 + (n - 1)d.$$

Geometric Sequences

Suppose that a person with a starting salary of $20,000 per year receives a 5% raise each year. If $a_n = f(n)$ computes this salary at the beginning of the nth year, then

$$f(1) = 20{,}000$$
$$f(2) = 20{,}000(1.05) = 21{,}000$$
$$f(3) = 21{,}000(1.05) = 22{,}050$$
$$f(4) = 22{,}050(1.05) = 23{,}152.5$$

are salaries for the first four years. Each salary results from multiplying the previous salary by 1.05. A general term in the sequence can be written as

$$f(n) = 20{,}000(1.05)^{n-1}.$$

During the tenth year, the annual salary is

$$f(10) = 20{,}000(1.05)^{10-1} \approx \$31{,}027.$$

This type of sequence is an example of a *geometric sequence* given by $f(n) = cr^{n-1}$, where c and r are constants. Geometric sequences are capable of either rapid growth or decay. The first five terms from some geometric sequences are shown in Table 9.4 on the following page. The corresponding values of c and r have been included.

TABLE 9.4	**Terms of Geometric Sequences**	
c	r	a_1, a_2, a_3, a_4, a_5
1	2	1, 2, 4, 8, 16
1	$\dfrac{1}{2}$	$1, \dfrac{1}{2}, \dfrac{1}{4}, \dfrac{1}{8}, \dfrac{1}{16}$
2	-4	2, -8, 32, -128, 512
3	$\dfrac{1}{10}$	3, 0.3, 0.03, 0.003, 0.0003

The terms of a geometric sequence can be found by multiplying the previous term by r. We now define a geometric sequence formally.

Infinite geometric sequence

An **infinite geometric sequence** is a function defined by $f(n) = cr^{n-1}$, where c and r are nonzero constants. The domain of f is the set of natural numbers.

A geometric sequence can be defined recursively by $a_n = ra_{n-1}$, where $a_n = f(n)$ and the first term is $a_1 = c$. Since $r = \dfrac{a_n}{a_{n-1}}$ for each valid n, r is referred to as the **common ratio.**

The next example illustrates how to recognize symbolic, graphical, and numerical representations of geometric sequences.

EXAMPLE 7   *Representing geometric sequences*

Decide which of the following represents a geometric sequence, where $a_n = f(n)$.
(a) $f(n) = 4(0.5)^n$
(b) The function f is represented graphically in Figure 9.5.

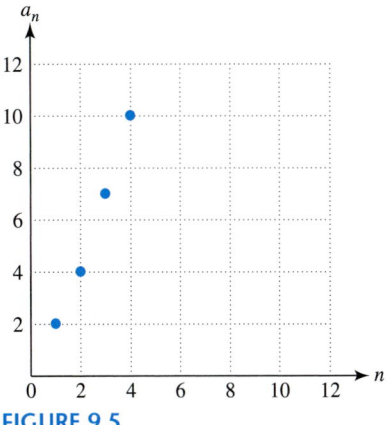

FIGURE 9.5

(c) Table 9.5 is a numerical representation of f.

TABLE 9.5

n	1	2	3	4	5	6
$f(n)$	1	-3	9	-27	81	-243

Solution

(a) The symbolic representation of f can be written as

$$f(n) = 4(0.5)^n = 4(0.5)(0.5)^{n-1} = 2(0.5)^{n-1}.$$

Thus, $f(n)$ represents a geometric sequence with $c = 2$ and $r = 0.5$.

(b) The points on the graph are $(1, 2)$, $(2, 4)$, $(3, 7)$, and $(4, 10)$. Thus, $a_1 = 2$, $a_2 = 4$, $a_3 = 7$, and $a_4 = 10$. Taking ratios of successive terms results in

$$\frac{a_2}{a_1} = 2, \qquad \frac{a_3}{a_2} = \frac{7}{4}, \qquad \text{and} \qquad \frac{a_4}{a_3} = \frac{10}{7}.$$

Since these ratios are not equal, there is no *common* ratio. The sequence is not geometric.

(c) Notice that

$$1, -3, 9, -27, 81, -243$$

result from multiplying the previous term by -3. This sequence can be expressed as

$$a_n = -3a_{n-1} \qquad \text{with} \qquad a_1 = 1.$$

Therefore, the sequence is geometric. ∎

EXAMPLE 8 *Finding a symbolic representation for a geometric sequence*

Find a general term a_n for each geometric sequence.
(a) $a_1 = 5$ and $r = 1.12$
(b) $a_2 = 8$ and $a_5 = 512$

Solution

(a) Since the first term is 5 and the common ratio is 1.12, $a_n = 5(1.12)^{n-1}$.

(b) We must find $a_n = cr^{n-1}$ so that $a_2 = 8$ and $a_5 = 512$. Start by determining the common ratio r. Since

$$\frac{a_5}{a_2} = \frac{cr^{5-1}}{cr^{2-1}} = \frac{r^4}{r^1} = r^3 \qquad \text{and} \qquad \frac{a_5}{a_2} = \frac{512}{8} = 64,$$

$r^3 = 64$ or $r = 4$. So, $a_n = c(4)^{n-1}$. Now,

$$a_2 = c(4)^{2-1} = 8 \qquad \text{or} \qquad c = 2.$$

Thus, $a_n = 2(4)^{n-1}$. ∎

9.1 PUTTING IT ALL TOGETHER

An infinite sequence is a function whose domain is the natural numbers. A finite sequence has the finite domain $D = \{1, 2, 3, \ldots, n\}$ for some fixed natural number n. The graphs of sequences are not continuous. They consist of discrete points.

Sequences are functions and have graphical, numerical, and symbolic representations. A symbolic representation for a sequence can be defined either recursively or with standard function notation in the form $a_n = f(n)$. A recursive definition requires that the terms a_1 through a_{n-1} be found before finding a_n.

Two important types of sequences are arithmetic and geometric. The following table summarizes concepts related to arithmetic and geometric sequences.

Arithmetic Sequence

Recursive Definition: $a_n = a_{n-1} + d$, where d is the common difference.
Function Definition: $f(n) = dn + c$, or equivalently, $f(n) = a_1 + d(n - 1)$, where $a_n = f(n)$ and d is the common difference.
Example: $a_n = a_{n-1} + 3$, $a_1 = 4$, and $f(n) = 3n + 1$ describe the same sequence. The common difference is $d = 3$. The terms of the sequence are

$$4, 7, 10, 13, 16, 19, 22, \ldots$$

Geometric Sequence

Recursive Definition: $a_n = ra_{n-1}$, where r is the common ratio.
Function Definition: $f(n) = cr^{n-1}$, where $c = a_1$ and r is the common ratio.
Example: $a_n = -2a_{n-1}$, $a_1 = 3$, and $f(n) = 3(-2)^{n-1}$ describe the same sequence. The common ratio is $r = -2$. The terms of the sequence are

$$3, -6, 12, -24, 48, -96, 192, \ldots$$

9.1 EXERCISES

   Tape 7

Finding Terms of Sequences

Exercises 1–8: Find the first four terms of each sequence.

1. $a_n = 2n + 1$

2. $a_n = 3(n - 1) + 5$

3. $a_n = 4(-2)^{n-1}$

4. $a_n = 2(3)^n$

5. $a_n = \dfrac{n}{n^2 + 1}$

6. $a_n = 5 - \dfrac{1}{n^2}$

7. $a_n = (-1)^n \left(\dfrac{1}{2}\right)^n$

8. $a_n = (-1)^n \left(\dfrac{1}{n}\right)$

Exercises 9 and 10: Use the graphical representation to write the terms of the sequence.

9.

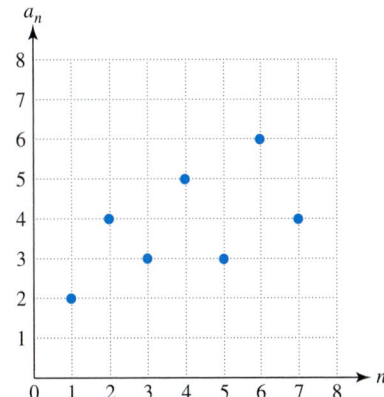

10.

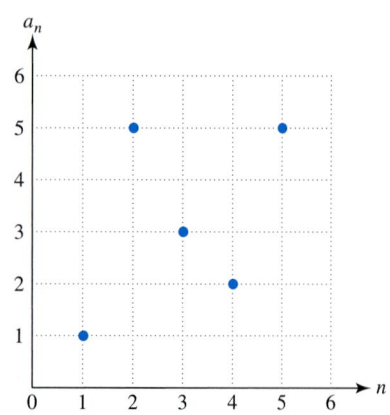

Exercises 11–16: Complete the following for the recursively defined sequence.

(a) *Find the first four terms.*

(b) *Graph these terms.*

11. $a_n = 2a_{n-1}; a_1 = 1$

12. $a_n = a_{n-1} + 5; a_1 = -4$

13. $a_n = a_{n-1} - a_{n-2}; a_1 = 2, a_2 = 5$

14. $a_n = 2a_{n-1} + a_{n-2}; a_1 = 0, a_2 = 1$

15. $a_n = a_{n-1}^2; a_1 = 2$

16. $a_n = \frac{1}{2}a_{n-1}^3 + 1; a_1 = 0$

Modeling Insect and Bacteria Populations

Exercises 17 and 18: *Insect Population* *The annual population density of a species of insect after n years is modeled by a sequence. Use the graph to discuss trends in the insect population.*

17.

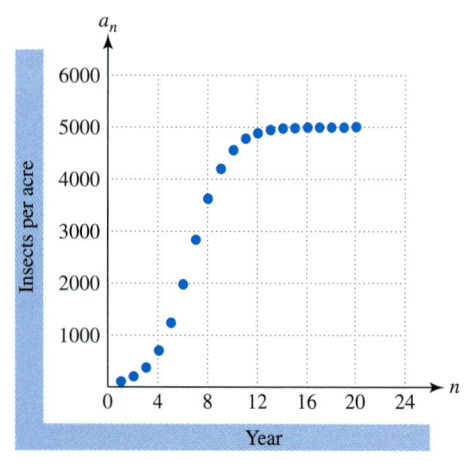

18.

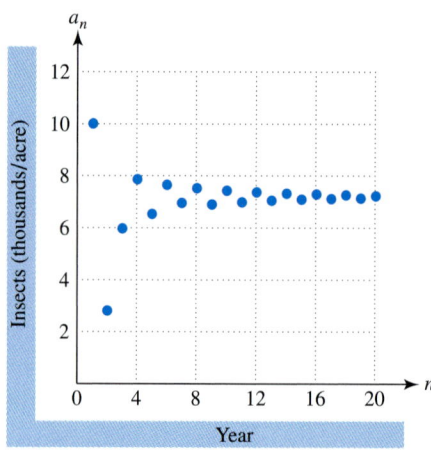

19. *Insect Population* (Refer to Example 3.) Suppose the initial density of female insects is 500 per acre with $r = 0.8$.

(a) Write a recursive sequence that describes this data, where a_n denotes the female insect density during year n.

(b) Find the terms $a_1, a_2, a_3, \ldots, a_6$. Interpret the results.

(c) Find a formula for a_n.

20. *Insect Population* (Refer to Example 4.) Suppose an insect population density during year n can be modeled by the recursively defined sequence

$$a_1 = 8$$
$$a_n = 2.9a_{n-1} - 0.2a_{n-1}^2, \qquad n > 1.$$

(a) Find the population for $n = 1, 2, 3$.

(b) Use technology to graph the sequence for $n = 1, 2, 3, \ldots, 20$. Interpret the graph.

21. *Bacteria Growth* Some strains of bacteria are incapable of producing an amino acid necessary for cell division, called *histidine*. If such bacteria are cultured in a medium with sufficient histidine, they can double their size and divide every 40 minutes. (Source: F. Hoppensteadt and C. Peskin, *Mathematics in Medicine and the Life Sciences*.)

(a) Write a recursive sequence that describes this growth where each value of n represents a 40-minute interval. Let $a_1 = 300$ represent the initial number of bacteria per milliliter. Find the first five terms.

(b) Determine the number of bacteria after 10 hours have elapsed.

(c) Is this sequence arithmetic or geometric? Explain.

22. *Bacteria Growth* (Refer to the previous exercise.) If bacteria are cultured in a medium with limited nutrients, competition ensues and growth slows. According to *Verhulst's model*, the number of bacteria at 40-minute intervals is given by

$$a_n = \left(\frac{2}{1 + a_{n-1}/K} \right) a_{n-1},$$ where K is a constant.

(a) Let $a_1 = 200$ and $K = 10{,}000$. Use technology to graph the sequence for $n = 1, 2, 3, \ldots, 20$.

(b) Describe the growth of this bacteria.

(c) Trace the graph of the sequence. Make a conjecture as to why K is called the *saturation constant*. Test your conjecture by changing the value of K.

Representations of Sequences

Exercises 23–26: The first five terms of an arithmetic sequence are given. Find

(a) *numerical,*

(b) *graphical, and*

(c) *symbolic*

representations of the sequence. Include at least eight terms of the sequence.

23. $1, 3, 5, 7, 9$

24. $4, 1, -2, -5, -8$

25. $7.5, 6, 4.5, 3, 1.5$

26. $5.1, 5.5, 5.9, 6.3, 6.7$

Exercises 27–30: The first five terms of a geometric sequence are given. Find

(a) *numerical,*

(b) *graphical, and*

(c) *symbolic*

representations of the sequence. Include at least eight terms of the sequence.

27. $8, 4, 2, 1, \dfrac{1}{2}$

28. $32, -8, 2, -\dfrac{1}{2}, \dfrac{1}{8}$

29. $\dfrac{3}{4}, \dfrac{3}{2}, 3, 6, 12$

30. $\dfrac{1}{27}, \dfrac{1}{9}, \dfrac{1}{3}, 1, 3$

Exercises 31–36: Find a general term a_n for each arithmetic sequence.

31. $a_1 = 5, \quad d = -2$

32. $a_1 = -3, \qquad d = 5$

33. $a_3 = 1, \qquad d = 3$

34. $a_4 = 12, \qquad d = -10$

35. $a_2 = 5, \qquad a_6 = 13$

36. $a_3 = 22, \qquad a_{17} = -20$

Exercises 37–42: Find a general term a_n for each geometric sequence.

37. $a_1 = 2, \qquad r = \dfrac{1}{2}$

38. $a_1 = 0.8, \qquad r = -3$

39. $a_3 = \dfrac{1}{32}, \qquad r = -\dfrac{1}{4}$

40. $a_4 = 3, \qquad r = 3$

41. $a_3 = 2, \qquad a_6 = \dfrac{1}{4}$

42. $a_2 = 6, \qquad a_4 = 24$

Identifying Types of Sequences

Exercises 43–48: Given the terms of a finite sequence, classify it as arithmetic, geometric, or neither.

43. $-5, 2, 9, 16, 23, 30$

44. $5, 2, -2, -6, -11$

45. $2, 8, 32, 128, 512$

46. $5.75, 5.5, 5.25, 5, 4.75, 4.5$

47. $100, 110, 130, 160, 200$

48. $0.7, 0.21, 0.063, 0.0189, 0.00567$

Exercises 49–52: Use the graph to determine if the sequence is arithmetic or geometric. If the sequence is arithmetic, state the sign of the common difference d and estimate its value. If the sequence is geometric, give the sign of the common ratio r and state if $|r| < 1$.

49.

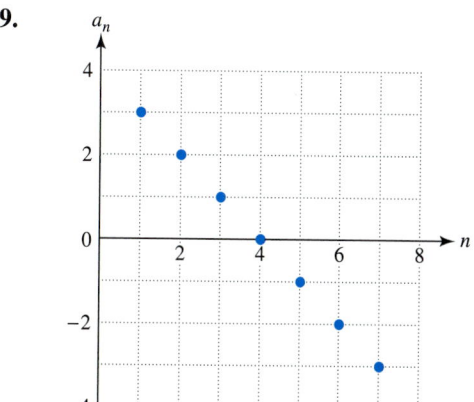

50.

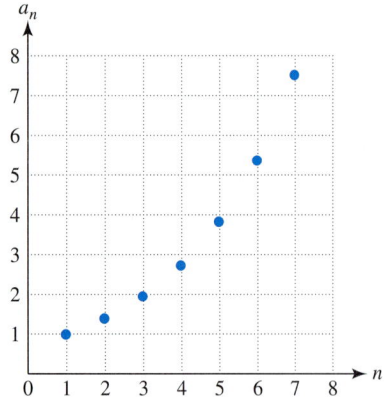

51.

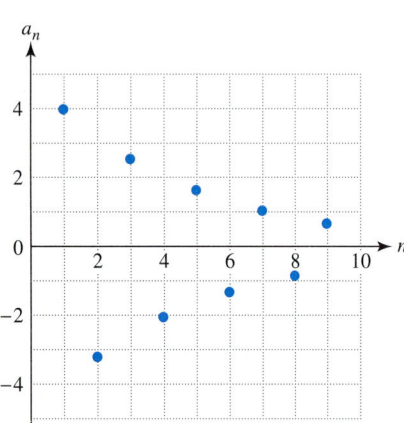

52.

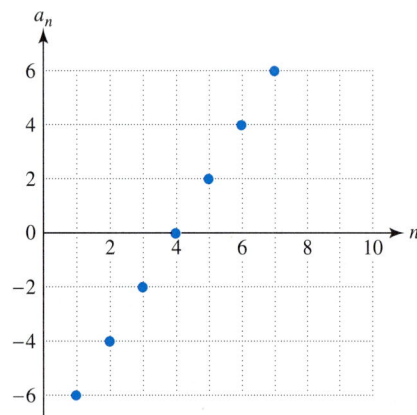

53. The *Fibonacci Sequence* dates back to 1202. It is one of the most famous sequences in mathematics and can be defined recursively by

$$a_1 = 1, a_2 = 1$$
$$a_n = a_{n-1} + a_{n-2} \quad \text{for } n > 2.$$

(a) Find the first 12 terms of this sequence.

(b) Compute $\dfrac{a_n}{a_{n-1}}$ when $n = 2, 3, 4, \ldots , 12$. What happens to this ratio?

(c) Show that for $n = 2, 3,$ and 4 the terms of the Fibonacci Sequence satisfy the equation

$$a_{n-1} \cdot a_{n+1} - a_n^2 = (-1)^n.$$

54. *Bouncing Ball*   If a tennis ball is dropped, it bounces or rebounds to 80% of its initial height.

(a) Write the first five terms of a sequence that gives the maximum height attained by the tennis ball on each rebound when it is dropped from an initial height of five feet. Let $a_1 = 5$. What type of sequence is this?

(b) Give a graphical representation of these terms.

(c) Find a general term a_n.

55. *Salary Increases*   Suppose an employee's initial salary is $30,000.

(a) If this person receives a $2000 raise for each year of experience, determine a sequence that gives the salary at the beginning of the nth year. What type of sequence is this?

(b) Suppose another employee has the same starting salary and receives a 5% raise after each year. Find a sequence that computes the salary at the beginning of the nth year. What type of sequence is this?

(c) Which salary is larger at the beginning of the tenth year, and the twentieth year?

(d) Graph both sequences in the same viewing rectangle. Compare the two salaries.

56. *Area*   A sequence of smaller squares is formed by connecting the midpoints of the sides of a larger square as shown in the figure. If the area of the largest square is one square unit, give the first five terms of a sequence that describes the area of each successive square. What type of sequence is this? Write an expression for the area of the nth square.

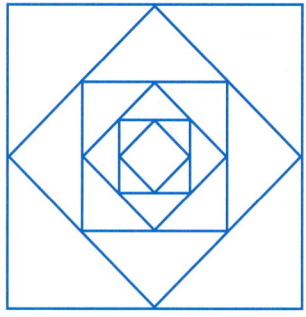

Exercises 57 and 58: Computing Square Roots   The following recursively defined sequence can be used to compute $\sqrt{k}$ for any positive number k.

$$a_1 = k$$
$$a_n = \frac{1}{2}\left(a_{n-1} + \frac{k}{a_{n-1}}\right)$$

This sequence was known to Sumerian mathematicians 4000 years ago, but it is still used today. Utilize this sequence to approximate the given square root by finding a_6. Compare your result with the actual value. (Source: P. Heinz-Otto, *Chaos and Fractals.*)

57. $\sqrt{2}$

58. $\sqrt{11}$

Writing about Mathematics

1. Explain how we can distinguish between an arithmetic and a geometric sequence. Give examples.

2. Compare and contrast a sequence whose nth term is given by $a_n = f(n)$, and a sequence that is defined recursively. Give examples. Which symbolic representation for defining a sequence is usually more convenient? Explain why.

9.2 Series

Basic Concepts • Arithmetic Series • Geometric Series • Summation Notation

Introduction

Although the terms *sequence* and *series* are sometimes used interchangeably in everyday English, they represent different mathematical concepts. In mathematics, a sequence is a function whose domain is the set of natural numbers, whereas a series is a summation of the terms in a sequence. Series have played a central role in the development of modern mathematics. Today series are often used to approximate functions that are too complicated to have simple symbolic representations. Series are also instrumental in calculating approximations of numbers like π and e.

Basic Concepts

Suppose a person has a starting salary of $30,000 per year and receives a $2000 raise each year. Then,

$$30,000, 32,000, 34,000, 36,000, 38,000$$

are terms of the sequence that describe this person's salaries over a five-year period. The total earned is given by the finite series

$$30,000 + 32,000 + 34,000 + 36,000 + 38,000,$$

whose sum is $170,000. Any sequence can be used to define a series. For example, the infinite sequence

$$1, \frac{1}{3}, \frac{1}{9}, \frac{1}{27}, \frac{1}{81}, \frac{1}{243}, \cdots$$

defines the terms of the infinite series

$$1 + \frac{1}{3} + \frac{1}{9} + \frac{1}{27} + \frac{1}{81} + \frac{1}{243} + \cdots.$$

We now define the concept of a series, where $a_1, a_2, a_3, \ldots, a_n, \ldots$ represent terms of a sequence.

> ## Series
>
> A **finite series** is an expression of the form
>
> $$a_1 + a_2 + a_3 + \cdots + a_n,$$
>
> and an **infinite series** is an expression of the form
>
> $$a_1 + a_2 + a_3 + \cdots + a_n + \cdots.$$

An infinite series contains an infinite number of terms. We must define what is meant by the sum of an *infinite* series. Let the following be a **sequence of partial sums.**

$$S_1 = a_1$$
$$S_2 = a_1 + a_2$$
$$S_3 = a_1 + a_2 + a_3$$
$$\vdots$$
$$S_n = a_1 + a_2 + a_3 + \cdots + a_n$$

If S_n approaches a real number S as $n \to \infty$, then the sum of the infinite series is S. For example, let $S_1 = 0.3$, $S_2 = 0.3 + 0.03$, $S_3 = 0.3 + 0.03 + 0.003, \ldots$. Then, as $n \to \infty$, $S_n \to \dfrac{1}{3}$. We say that the infinite series

$$0.3 + 0.03 + 0.003 + 0.0003 + \cdots = 0.\overline{3}$$

has sum $\dfrac{1}{3}$. Some infinite series do not have a sum S.

EXAMPLE 1   *Computing total reported cases of AIDS*

Table 9.6 is a numerical representation of a finite sequence that computes the number of AIDS cases diagnosed from 1991 to 1995, where $a_n = f(n)$ and $n = 1$ corresponds to 1991. (Source: Department of Health and Human Services.)

TABLE 9.6

n	1	2	3	4	5
$f(n)$	58,991	76,898	75,534	64,026	40,051

(a) Let a_n represent the number of AIDS cases diagnosed in year n. Write a series whose sum represents the total number of AIDS cases diagnosed from 1991 to 1995. Find its sum.

(b) Interpret the sum $a_1 + a_2 + a_3 + \cdots + a_{10}$.

Solution

(a) The terms of the finite sequence in Table 9.6 are

$$a_1 = 58{,}991,\ a_2 = 76{,}898,\ a_3 = 75{,}534,\ a_4 = 64{,}026,\ \text{and}\ a_5 = 40{,}051.$$

The series that represents the total numbers of AIDS cases diagnosed is

$$a_1 + a_2 + a_3 + a_4 + a_5 = 58{,}991 + 76{,}898 + 75{,}534 + 64{,}026 + 40{,}051$$
$$= 315{,}500.$$

Thus, there were 315,500 AIDS cases diagnosed from 1991 to 1995.

(b) The series $a_1 + a_2 + a_3 + \cdots + a_{10}$ would represent the total number of AIDS cases diagnosed from 1991 to 2000. ∎

The techniques used to calculate π throughout history is a fascinating story. Since π is an irrational number, it cannot be represented exactly by a fraction. Its decimal expansion neither repeats nor does it have a discernible pattern. The ability to compute π was essential to the development of a society, because π appears in formulas used in construction, surveying, and geometry. In early historical records, π was given the value of 3. Later the Egyptians used a value of

$$\frac{256}{81} \approx 3.1605.$$

It was not until the discovery of series that exceedingly accurate decimal approximations of π were possible. In 1989, after 100 hours of supercomputer time, π was computed to 1,073,740,000 digits. Why would anyone want to compute π to so many decimal places? One practical reason is to test electrical circuits in new computers. If a computer has a small defect in its hardware, there is a good chance that an error will appear after performing trillions of arithmetic calculations during the computation of π. (Sources: P. Beckmann, *A History of PI;* P. Heinz-Otto, *Chaos and Fractals.*)

EXAMPLE 2   *Computing π with a series*

The infinite series given by

$$\frac{\pi^4}{90} = \frac{1}{1^4} + \frac{1}{2^4} + \frac{1}{3^4} + \frac{1}{4^4} + \frac{1}{5^4} + \cdots + \frac{1}{n^4} + \cdots$$

can be used to estimate π.

(a) Approximate π by finding the sum of the first four terms.

(b) Use technology to approximate π by summing the first 50 terms. Compare the result to the actual value of π.

Solution

(a) Summing the first four terms results in the following.

$$\frac{\pi^4}{90} \approx \frac{1}{1^4} + \frac{1}{2^4} + \frac{1}{3^4} + \frac{1}{4^4} \approx 1.078751929$$

This approximation can be solved for π by multiplying by 90 and then taking the fourth root. Thus,

$$\pi \approx \sqrt[4]{90(1.078751929)} \approx 3.139.$$

(b) Some calculators are capable of summing the terms of a sequence as shown in Figure 9.6 on the next page. (Summing the terms of a sequence is equivalent to finding the sum of a series.) The first 50 terms of the series provides an approximation of $\pi \approx 3.141590776$. This computation matches the actual value of π to six significant digits.

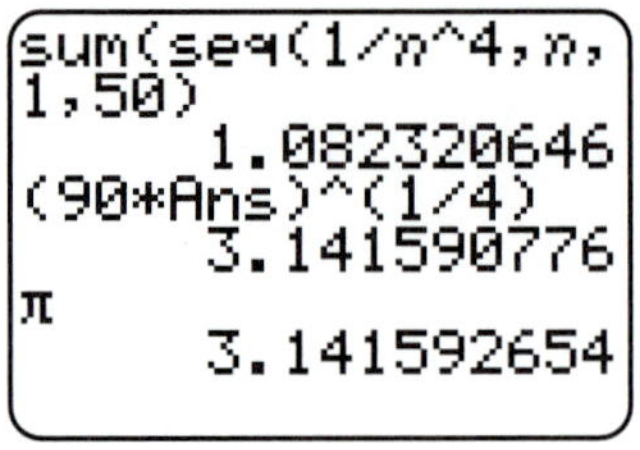

```
sum(seq(1/n^4,n,
1,50)
          1.082320646
(90*Ans)^(1/4)
          3.141590776
π
          3.141592654
```

FIGURE 9.6   ■

Arithmetic Series

Summing the terms of an arithmetic sequence results in an **arithmetic series.** For example, the sequence defined by $a_n = 2n - 1$ for $n = 1, 2, 3, \ldots, 7$ is the arithmetic sequence

$$1, 3, 5, 7, 9, 11, 13.$$

The corresponding arithmetic series is

$$1 + 3 + 5 + 7 + 9 + 11 + 13.$$

The following formula can be used to sum the first n terms of an arithmetic sequence. (For a proof see Exercise 3 in the Extended Exercises.)

Sum of the first n terms of an arithmetic sequence

The **sum of the first n terms of an arithmetic sequence,** denoted by S_n, is found by averaging the first and nth terms and then multiplying by n. That is,

$$S_n = a_1 + a_2 + a_3 + \cdots + a_n = n\left(\frac{a_1 + a_n}{2}\right).$$

Since $a_n = a_1 + (n - 1)d$, S_n can also be written as follows.

$$S_n = n\left(\frac{a_1 + a_n}{2}\right)$$

$$= \frac{n}{2}(a_1 + a_1 + (n - 1)d)$$

$$= \frac{n}{2}(2a_1 + (n - 1)d)$$

EXAMPLE 3   *Finding the sum of a finite arithmetic series*

A person has a starting annual salary of $30,000 and receives a $1500 raise each year.
(a) Calculate the total amount earned after 10 years.
(b) Verify this value using a calculator.

Solution
(a) The sequence describing the salary during year n is computed by the arithmetic sequence

$$a_n = 30{,}000 + 1500(n - 1).$$

The first and tenth year's salaries are:

$$a_1 = f(1) = 30,000 + 1500(1 - 1) = 30,000$$
$$a_{10} = f(10) = 30,000 + 1500(10 - 1) = 43,500.$$

Thus, the total amount earned during this ten-year period is

$$S_{10} = 10\left(\frac{30,000 + 43,500}{2}\right) = \$367,500.$$

This can also be found using

$$S_{10} = \frac{10}{2}(2 \cdot 30,000 + (10 - 1)1500) = \$367,500.$$

(b) To verify this with a calculator compute the sum

$$a_1 + a_2 + a_3 + \cdots + a_{10},$$

where $a_n = 30,000 + 1500(n - 1)$. This calculation is shown in Figure 9.7. The result of 367,500 agrees with part (a).

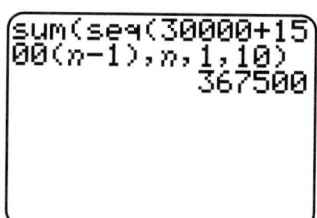

FIGURE 9.7   ∎

For tax purposes, businesses frequently depreciate equipment. Two different methods of depreciation are called *straight-line depreciation* and *sum-of-the-years'-digits*.

Suppose a college student buys a $2000 computer to start a business that provides Internet searches. This student estimates the life of the computer at five years, after which its value will be $200. The difference between $2000 and $200 is $1800, which may be deducted from his or her taxable income over a five-year period.

In straight-line depreciation, equal portions of $1800 are deducted over five years. The *sum-of-the-years'-digits* method calculates depreciation differently. A computer with a useful life of five years has a sum-of-the-years computed by

$$1 + 2 + 3 + 4 + 5 = 15.$$

With this method, $\frac{5}{15}$ of $1800 is deducted the first year, $\frac{4}{15}$ the second year, and so on, until $\frac{1}{15}$ is deducted the fifth year. Both depreciation methods deduct a total of $1800 over five years. (Source: Sharp Electronics Corporation, *Conquering the Sciences.*)

EXAMPLE 4   *Calculating depreciation*

For each of the previous depreciation methods, complete the following.
(a) Find an arithmetic sequence that gives the amount depreciated each year.
(b) Write a series whose sum is the amount depreciated over five years.

Solution

(a) With straight-line depreciation, $\frac{1}{5}(1800) = \$360$ is depreciated each year. The sequence describing this depreciation is the constant arithmetic sequence

$$360, 360, 360, 360, 360.$$

For sum-of-the-years'-digits, the following depreciation schedule is computed.

$$\frac{5}{15} \text{ of } \$1800 = \$600$$

$$\frac{4}{15} \text{ of } \$1800 = \$480$$

$$\frac{3}{15} \text{ of } \$1800 = \$360$$

$$\frac{2}{15} \text{ of } \$1800 = \$240$$

$$\frac{1}{15} \text{ of } \$1800 = \$120$$

The arithmetic sequence describing these amounts is

$$600, 480, 360, 240, 120.$$

(b) The arithmetic series that describes the total amount deducted for straight-line depreciation is

$$360 + 360 + 360 + 360 + 360.$$

Although we know that the amount depreciated is equal to $1800, it can also be verified using the following formula.

$$S_5 = 5\left(\frac{360 + 360}{2}\right) = \$1800$$

For the sum-of-the-years'-digits, the total depreciation is calculated by the series

$$600 + 480 + 360 + 240 + 120.$$

Notice that this series also sums to $1800.

$$S_5 = 5\left(\frac{600 + 120}{2}\right) = \$1800 \qquad \blacksquare$$

EXAMPLE 5 *Finding a term of an arithmetic series*

The sum of an arithmetic series with 15 terms is 285. If $a_{15} = 40$, find a_1.

Solution

To find a_1, we apply the sum formula

$$S_n = n\left(\frac{a_1 + a_n}{2}\right) = a_1 + a_2 + a_3 + \cdots + a_n,$$

with $n = 15$ and $a_n = 40$.

$$15\left(\frac{a_1 + 40}{2}\right) = 285$$

$$15(a_1 + 40) = 570 \qquad \text{Multiply by 2.}$$

$$a_1 + 40 = 38 \qquad \text{Divide by 15.}$$

$$a_1 = -2 \qquad \text{Subtract 40.} \qquad\blacksquare$$

Critical Thinking

Explain why a formula for the sum of an infinite arithmetic series is not given.

Geometric Series

What will happen if we attempt to find the sum of the terms of an infinite geometric sequence? The mathematical concept of infinity dates back to at least the paradoxes of Zeno (450 B.C.). The following illustrates a difficulty that occurred in attempting to comprehend infinity.

Suppose that a person walks 1 mile on the first day, $\frac{1}{2}$ mile the second day, $\frac{1}{4}$ mile the third day, and so on. How far down the road would this person travel? This distance is described by the infinite series

$$1 + \frac{1}{2} + \frac{1}{4} + \frac{1}{8} + \frac{1}{16} + \frac{1}{32} + \frac{1}{64} + \cdots.$$

Does the sum of an infinite number of positive values always become infinitely large? The solution to problems like this took centuries for mathematicians to answer. (Source: H. Eves, *An Introduction to the History of Mathematics.*)

Finite Geometric Series. In an analogous manner to how an arithmetic series was defined, a **geometric series** is the sum of the terms of a geometric sequence. In order to calculate sums of infinite geometric series, we begin by finding sums of finite geometric series.

Any finite geometric sequence can be written as

$$a_1, a_1 r, a_1 r^2, a_1 r^3, \ldots, a_1 r^{n-1}.$$

The summation of these n terms is a finite geometric series. Its sum S_n is expressed by

$$S_n = a_1 + a_1 r + a_1 r^2 + a_1 r^3 + \cdots + a_1 r^{n-1}.$$

To find the value of S_n, multiply this equation by r.

$$r S_n = a_1 r + a_1 r^2 + a_1 r^3 + \cdots + a_1 r^{n-1} + a_1 r^n$$

Subtracting this equation from the previous equation results in

$$S_n - r S_n = a_1 - a_1 r^n$$

$$S_n(1 - r) = a_1(1 - r^n)$$

$$S_n = a_1\left(\frac{1 - r^n}{1 - r}\right), \qquad \text{provided } r \neq 1.$$

This formula can be used to find the sum of the first n terms of a geometric sequence.

> ### Sum of the first n terms of a geometric sequence
>
> If a geometric sequence has first term a_1 and common ratio r, then the sum of the first n terms is given by
>
> $$S_n = a_1\left(\frac{1 - r^n}{1 - r}\right),$$
>
> provided $r \neq 1$.

EXAMPLE 6 *Finding the sum of a finite geometric series*

Approximate the sum for the given values of n.

(a) $1 + \dfrac{1}{2} + \dfrac{1}{4} + \cdots + \left(\dfrac{1}{2}\right)^{n-1}$; $n = 5$, 10, and 20

(b) $3 - 6 + 12 - 24 + 48 - \cdots + 3(-2)^{n-1}$; $n = 3$, 8, and 13

Solution

(a) This geometric series has $a_1 = 1$ and $r = \dfrac{1}{2} = 0.5$. Thus,

$$S_5 = 1\left(\frac{1 - 0.5^5}{1 - 0.5}\right) = 1.9375$$

$$S_{10} = 1\left(\frac{1 - 0.5^{10}}{1 - 0.5}\right) \approx 1.998047$$

$$S_{20} = 1\left(\frac{1 - 0.5^{20}}{1 - 0.5}\right) \approx 1.999998.$$

(b) This geometric series has $a_1 = 3$ and $r = -2$.

$$S_3 = 3\left(\frac{1 - (-2)^3}{1 - (-2)}\right) = 9$$

$$S_8 = 3\left(\frac{1 - (-2)^8}{1 - (-2)}\right) = -255$$

$$S_{13} = 3\left(\frac{1 - (-2)^{13}}{1 - (-2)}\right) = 8193$$ ■

Annuities.   A sequence of deposits made at equal periods of time is called an **annuity**. Suppose A_0 dollars is deposited into an account at the end of each year paying an annual interest rate i compounded annually. At the end of the first year the account contains A_0 dollars. At the end of the second year A_0 dollars would be deposited again. In addition, the first deposit of A_0 dollars would have received interest during the second year. Therefore, the value of the annuity after two years is

$$A_0 + A_0(1 + i).$$

After three years the balance is

$$A_0 + A_0(1 + i) + A_0(1 + i)^2,$$

and after n years this amount is given by

$$A_0 + A_0(1 + i) + A_0(1 + i)^2 + \cdots + A_0(1 + i)^{n-1}.$$

This is a geometric series with first term $a_1 = A_0$ and common ratio $r = (1 + i)$. The sum of the first n terms is given by

$$S_n = a_1 \left(\frac{1 - (1 + i)^n}{1 - (1 + i)} \right) = a_1 \left(\frac{(1 + i)^n - 1}{i} \right).$$

EXAMPLE 7   *Finding the future value of an annuity*

Suppose that a 20-year-old worker deposits $1000 into an account at the end of each year until age 65. If the interest rate is 12%, find the future value of the annuity.

Solution

Let $a_1 = 1000$, $i = 0.12$, and $n = 45$. The future value of the annuity is given by

$$S_n = a_1 \left(\frac{(1 + i)^n - 1}{i} \right)$$

$$= 1000 \left(\frac{(1 + 0.12)^{45} - 1}{0.12} \right)$$

$$\approx \$1{,}358{,}230.$$ ∎

Infinite Geometric Series.   In Example 6 the value of r affects the sum of a finite geometric series. If $|r| > 1$, as in part (b), then r^n becomes large in absolute value for increasing n. As a result, the sum of the series also becomes large in absolute value. On the other hand, in part (a) the common ratio satisfies $|r| < 1$. The values of r^n become closer to 0 as n increases. For large values of n,

$$S_n = a_1 \left(\frac{1 - r^n}{1 - r} \right) \approx a_1 \left(\frac{1 - 0}{1 - r} \right) = \frac{a_1}{1 - r}.$$

This result is summarized in the following.

Sum of an infinite geometric sequence

The sum of the infinite geometric sequence with first term a_1 and common ratio r is given by

$$S = \frac{a_1}{1 - r},$$

provided $|r| < 1$. If $|r| \geq 1$, then this sum does not exist.

Infinite series can be used to describe repeating decimals. For example, the fraction $\frac{5}{9}$ can be written as the repeating decimal $0.555555 \ldots$. This decimal can be expressed as an infinite series.

$$\frac{5}{9} = 0.5 + 0.05 + 0.005 + 0.0005 + \cdots + 0.5(0.1)^{n-1} + \cdots$$

In this series $a_1 = 0.5$ and $r = 0.1$. Since $|r| < 1$, the sum exists and is given by

$$S = \frac{0.5}{1 - 0.1} = \frac{5}{9},$$

as expected.

We are now able to answer the question concerning how far a person will walk if he or she travels 1 mile on the first day, $\frac{1}{2}$ mile on the second day, $\frac{1}{4}$ mile the third day, and so on.

EXAMPLE 8   *Finding the sum of an infinite geometric series*

Find the sum of the infinite geometric series

$$1 + \frac{1}{2} + \frac{1}{4} + \cdots + \left(\frac{1}{2}\right)^{n-1} + \cdots.$$

Solution

In this series, the first term is $a_1 = 1$ and the common ratio is $\frac{1}{2} = 0.5$. Its sum is

$$S = \frac{a_1}{1 - r} = \frac{1}{1 - 0.5} = 2.$$

If it were possible to walk in the prescribed manner, the total distance traveled after each day would always be less than 2 miles. ∎

Summation Notation

Summation notation is used to write series efficiently. The symbol Σ, the upper-case Greek letter *sigma*, is used to indicate a sum.

Summation notation

$$\sum_{k=1}^{n} a_k = a_1 + a_2 + a_3 + \cdots + a_n.$$

The letter k is called the **index of summation.** The numbers 1 and n represent the subscripts of the first and last terms in the series. They are called the **lower limit** and **upper limit** of the summation, respectively.

EXAMPLE 9   *Using summation notation*

Evaluate each series.

(a) $\displaystyle\sum_{k=1}^{5} k^2$

(b) $\displaystyle\sum_{k=1}^{4} 5$

(c) $\displaystyle\sum_{k=3}^{6} (2k - 5)$

Solution

(a) $\displaystyle\sum_{k=1}^{5} k^2 = 1^2 + 2^2 + 3^2 + 4^2 + 5^2 = 55$

(b) $\displaystyle\sum_{k=1}^{4} 5 = 5 + 5 + 5 + 5 = 20$

(c) $\displaystyle\sum_{k=3}^{6} (2k - 5) = \underset{k=3}{(2(3) - 5)} + \underset{k=4}{(2(4) - 5)} + \underset{k=5}{(2(5) - 5)} + \underset{k=6}{(2(6) - 5)}$

$$= 1 + 3 + 5 + 7 = 16$$ ∎

Summation notation is used frequently in statistics, when there is a large number of data items. The next example demonstrates how averages can be expressed using summation notation.

EXAMPLE 10 *Applying summation notation*

Express the average of n numbers $x_1, x_2, x_3, \ldots, x_n$ using summation notation.

Solution

The average of n numbers can be written as

$$\frac{x_1 + x_2 + x_3 + \cdots + x_n}{n}.$$

This is equivalent to $\displaystyle\frac{1}{n}\left(\sum_{k=1}^{n} x_k\right).$ ∎

Series play an essential role in mathematics and its applications, as illustrated by the next example.

EXAMPLE 11 *Expressing a series in summation notation*

Suppose that an air filter removes 90% of the impurities entering it.
(a) Find a series that represents the amount of impurities removed by a sequence of n air filters. Express this answer in summation notation.
(b) How many air filters would be necessary to remove 99.99% of the impurities? Could 100% of the impurities be removed from the air?

Solution

(a) The first filter removes 90% of the impurities so 10%, or 0.1, passes through it. Of the 0.1 that passes through the first filter, 90% is removed by the second filter, while 10% of 10%, or 0.01, passes through. Then, 10% of 0.01, or 0.001, passes through the third filter. See Figure 9.8 on the next page. From this we can establish a pattern. Let 100% or 1 represent the amount of impurities entering the first air filter. The amount removed by n filters would equal

$$(0.9)(1) + (0.9)(0.1) + (0.9)(0.01) + (0.9)(0.001) + \cdots + (0.9)(0.1)^{n-1}.$$

In summation notation, this series can be written as $\displaystyle\sum_{k=1}^{n} 0.9(0.1)^{k-1}.$

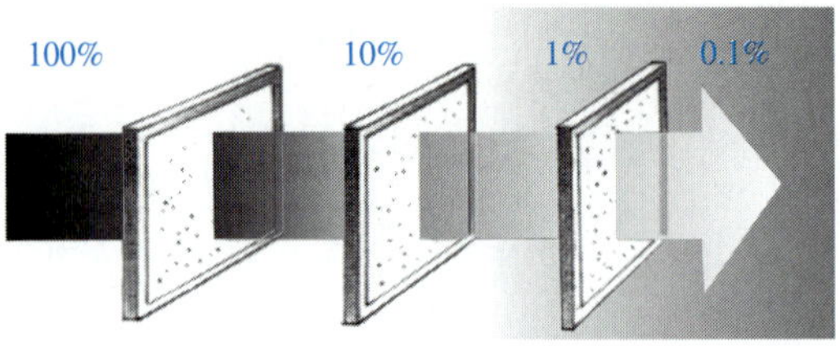

FIGURE 9.8 Percent of Impurities Passing through Air Filters

(b) To remove 99.99% or 0.9999 of the impurities requires 4 air filters, since

$$\sum_{k=1}^{4} 0.9(0.1)^{k-1} = (0.9)(1) + (0.9)(0.1) + (0.9)(0.01) + (0.9)(0.001)$$

$$= 0.9 + 0.09 + 0.009 + 0.0009$$

$$= 0.9999.$$

Intuition tells us that it is impossible to remove 100% of the impurities. If k increases without a maximum value, then the fraction of impurities removed is

$$\sum_{k=1}^{\infty} (0.9)(0.1)^{k-1} = (0.9)(1) + (0.9)(0.1) + (0.9)(0.1)^2 + \cdots$$

$$= 0.99999 \ldots .$$

Notice that since the upper limit has no maximum, the symbol ∞ is used. If k is allowed to increase without bound, then the series equals the repeating decimal $0.\overline{9}$. This is an infinite geometric series with $a_1 = 0.9$ and $r = 0.1$. Its sum is

$$S = \frac{a_1}{1 - r} = \frac{0.9}{1 - 0.1} = \frac{0.9}{0.9} = 1.$$

The mathematics of this problem is telling us that it would require an infinite number of air filters to remove 100% of the impurities. ■

9.2 PUTTING IT ALL TOGETHER

A sequence is a function f whose domain is the natural numbers. Its terms can be listed as

$$a_1, a_2, a_3, a_4, a_5, \ldots, a_n, \ldots,$$

where $a_n = f(n)$. A series is the summation of the terms of a sequence, and can be expressed as

$$a_1 + a_2 + a_3 + a_4 + a_5 + \cdots + a_n + \cdots.$$

Sequences and series can either be infinite or finite. The sum of a finite series always exists. However, the sum of an infinite series may not exist.

The following table summarizes concepts related to arithmetic and geometric series.

Arithmetic Series

Finite Arithmetic Series: $\displaystyle\sum_{k=1}^{n} a_k = a_1 + a_2 + a_3 + \cdots + a_n$, where $a_k = dk + c$
for some constants c and d.

Sum of the First n Terms: $S_n = n\left(\dfrac{a_1 + a_n}{2}\right)$ or $S_n = \dfrac{n}{2}(2a_1 + (n-1)d)$.

Example: The series $4 + 7 + 10 + 13 + 16 + 19 + 22$ is defined by $a_k = 3k + 1$.
Its sum is $S_7 = 7\left(\dfrac{4+22}{2}\right) = 91$.

Geometric Series

Infinite Geometric Series: $\displaystyle\sum_{k=1}^{n} a_k = a_1 + a_2 + a_3 + \cdots + a_n + \cdots$
where $a_k = a_1 r^{k-1}$ for some nonzero constants a_1 and r.

Sum of First n Terms: $S_n = a_1\left(\dfrac{1-r^n}{1-r}\right)$, where a_1 is the first term and r is the common
ratio.

Example: The series $3 + 6 + 12 + 24 + 48 + 96$ has $a_1 = 3$ and $r = 2$.
Its sum is $S_6 = 3\left(\dfrac{1-2^6}{1-2}\right) = 189$.

Sum of Infinite Geometric Series: $S = \dfrac{a_1}{1-r}$, if $|r| < 1$. S does not exist if $|r| \geq 1$.

Example: The series $4 + 1 + \dfrac{1}{4} + \dfrac{1}{16} + \dfrac{1}{64} + \cdots$ has sum $S = \dfrac{4}{1 - \frac{1}{4}} = \dfrac{16}{3}$.

9.2 EXERCISES

   Tape 7

Finding Sums of Series

1. *Prison Escapees* The table lists the number of escapees from state prisons each year. (Source: Bureau of Justice Statistics.)

Year	1990	1991	1992	1993	1994	1995
Escapees	8518	9921	10,706	14,035	14,307	12,249

   (a) Write a series whose sum is the total number of escapees from 1990 to 1995. Find its sum.
   (b) Is the series finite or infinite?

2. *Captured Prison Escapees* (Refer to the previous exercise.) The table lists the number of escapees from state prisons who were captured, including inmates who may have escaped during a previous year. (Source: Bureau of Justice Statistics.)

Year	1990	1991	1992	1993	1994	1995
Captured	9324	9586	10,031	12,872	13,346	12,166

   (a) Write a series whose sum is the total number of escapees captured from 1990 to 1995. Find its sum.

   (b) Compare the number of escapees to the number captured during this time period.

Exercises 3–8: Find the sum of the arithmetic series using a formula.

3. $3 + 5 + 7 + 9 + 11 + 13 + 15 + 17$

4. $7.5 + 6 + 4.5 + 3 + 1.5 + 0 + (-1.5)$

5. $1 + 2 + 3 + 4 + \cdots + 50$

6. $1 + 3 + 5 + 7 + \cdots + 97$

7. $-7 + (-4) + (-1) + 2 + 5 + \cdots + 98 + 101$

8. $89 + 84 + 79 + 74 + \cdots + 9 + 4$

Exercises 9–12: Find the sum of the finite geometric series using $S_n = a_1\left(\dfrac{1-r^n}{1-r}\right)$.

9. $1 + 2 + 4 + 8 + 16 + 32 + 64 + 128$

10. $2 + \dfrac{1}{2} + \dfrac{1}{8} + \dfrac{1}{32} + \dfrac{1}{128} + \dfrac{1}{512}$

11. $0.5 + 1.5 + 4.5 + 13.5 + 40.5 + 121.5 + 364.5$

12. $0.6 + 0.3 + 0.15 + 0.075 + 0.0375$

Exercises 13–16: Find the sum of the infinite geometric series.

13 $1 + \dfrac{1}{3} + \dfrac{1}{9} + \dfrac{1}{27} + \dfrac{1}{81} + \cdots$

14. $5 + \dfrac{5}{2} + \dfrac{5}{4} + \dfrac{5}{8} + \dfrac{5}{16} + \cdots$

15. $6 - 4 + \dfrac{8}{3} - \dfrac{16}{9} + \dfrac{32}{27} - \dfrac{64}{81} + \cdots$

16. $-2 + \dfrac{1}{2} - \dfrac{1}{8} + \dfrac{1}{32} - \dfrac{1}{128} + \cdots$

17. *Area* (Refer to Exercise 56, Section 9.1.) Use a geometric series to sum the area of the squares if they continue indefinitely.

18. *Perimeter* (Refer to Exercise 56, Section 9.1.) Use a geometric series to find the sum of the perimeters of the squares if they continue indefinitely.

Exercises 19–22: Annuities (Refer to Example 7.) Find the future value of each annuity. Interpret the result.

19. $A_0 = \$2000, \quad i = 0.08, n = 20$

20. $A_0 = \$500, \quad\ i = 0.15, n = 10$

21. $A_0 = \$10,000, i = 0.11, n = 5$

22. $A_0 = \$3000, \quad i = 0.19, n = 45$

Decimal Numbers and Geometric Series

Exercises 23–26: Write each rational number in the form of an infinite geometric series.

23. $\dfrac{2}{3}$

24. $\dfrac{1}{9}$

25. $\dfrac{9}{11}$

26. $\dfrac{14}{33}$

Exercises 27–30: Write the sum of each geometric series as a rational number.

27. $0.8 + 0.08 + 0.008 + 0.0008 + \cdots$

28. $0.9 + 0.09 + 0.009 + 0.0009 + \cdots$

29. $0.45 + 0.0045 + 0.000045 + \cdots$

30. $0.36 + 0.0036 + 0.000036 + \cdots$

Summation Notation

Exercises 31–38: Write out the terms of the series.

31. $\displaystyle\sum_{k=1}^{4} (k + 1)$

32. $\displaystyle\sum_{k=1}^{6} (3k - 1)$

33. $\displaystyle\sum_{k=1}^{8} 4$

34. $\displaystyle\sum_{k=2}^{6} (5 - 2k)$

35. $\displaystyle\sum_{k=1}^{7} k^3$

36. $\displaystyle\sum_{k=1}^{4} 5(2)^{k-1}$

37. $\displaystyle\sum_{k=4}^{5} (k^2 - k)$

38. $\displaystyle\sum_{k=1}^{5} \log k$

Exercises 39–42: Write each series using summation notation.

39. $1^4 + 2^4 + 3^4 + 4^4 + 5^4 + 6^4$

40. $1 + \dfrac{1}{5} + \dfrac{1}{25} + \dfrac{1}{125} + \dfrac{1}{625}$

41. $1 + \dfrac{1}{2^2} + \dfrac{1}{3^2} + \dfrac{1}{4^2} + \dfrac{1}{5^2} + \cdots$

42. $1 + \dfrac{1}{10} + \dfrac{1}{100} + \dfrac{1}{1000} + \dfrac{1}{10,000} + \cdots$

43. Verify the formula $\displaystyle\sum_{k=1}^{n} k = \dfrac{n(n + 1)}{2}$ using the formula for the sum of the first n terms of a finite arithmetic sequence.

44. Use the previous exercise to find the sum of the series $\displaystyle\sum_{k=1}^{200} k.$

45. *Stacking Logs* A stack of logs is made in layers, with one less log in each layer. See the accompanying figure. If the top layer has 7 logs and the bottom layer has 15 logs, what is the total number of logs in the pile? Use a formula to find the sum.

46. (Refer to the previous exercise.) Suppose a stack of logs has 13 logs in the top layer and a total of 7 layers. How many logs are in the stack?

Exercises 47 and 48: Depreciation (Refer to Example 4.) Let the total depreciation be T over n years.
  (a) *Find an arithmetic sequence that gives the amount depreciated each year using straight-line and sum-of-the-years'-digits depreciation.*
  (b) *Write a series for each method whose sum is the total amount depreciated over n years.*

47. $T = \$10,000, n = 4$

48. $T = \$42,000, n = 7$

Exercises 49 and 50: The Natural Exponential Function
The following series can be used to estimate the value of e^a for any real number a.

$$e^a \approx 1 + a + \frac{a^2}{2!} + \frac{a^3}{3!} + \cdots + \frac{a^n}{n!},$$

where $n! = 1 \cdot 2 \cdot 3 \cdot 4 \cdots \cdots n$. Use the first eight terms of this series to approximate the given expression. Compare this estimate with the actual value.

49. e

50. e^{-1}

Computing Partial Sums

Exercises 51–54: Use a_k and n to find $S_n = \sum_{k=1}^{n} a_k$. Then, evaluate the infinite geometric series $S = \sum_{k=1}^{\infty} a_k$. Compare S to the values for S_n.

51. $a_k = \left(\frac{1}{3}\right)^{k-1}$; $n = 2, 4, 8, 16$

52. $a_k = 3\left(\frac{1}{2}\right)^{k-1}$; $n = 5, 10, 15, 20$

53. $a_k = 4\left(-\frac{1}{10}\right)^{k-1}$; $n = 1, 2, 3, 4, 5, 6$

54. $a_k = 2(-0.02)^{k-1}$; $n = 1, 2, 3, 4, 5, 6$

Writing about Mathematics

1. Discuss the difference between a sequence and a series. Give examples of each.

2. Under what circumstances can we find the sum of a geometric series? Give examples to illustrate your answer.

CHECKING BASIC CONCEPTS FOR SECTIONS 9.1 AND 9.2

1. Give graphical and numerical representations of the sequence defined by $a_n = -2n + 3$, where $a_n = f(n)$. Include the first six terms.

2. Determine if the sequence is arithmetic or geometric. If it is arithmetic state the common difference—if it is geometric give the common ratio.
(a) $2, -4, 8, -16, 32, -64, 128, \ldots$
(b) $-3, 0, 3, 6, 9, 12, \ldots$
(c) $4, 2, 1, \frac{1}{2}, \frac{1}{4}, \frac{1}{8}, \frac{1}{16}, \ldots$

3. Determine if the series is arithmetic or geometric. Use a formula to find its sum.
(a) $1 + 5 + 9 + 13 + \cdots + 37$
(b) $3 + 1 + \frac{1}{3} + \frac{1}{9} + \frac{1}{27} + \frac{1}{81}$
(c) $2 + \frac{1}{2} + \frac{1}{8} + \frac{1}{32} + \cdots$
(d) $0.9 + 0.09 + 0.009 + 0.0009 + \cdots$

4. Write each series in the previous exercise using summation notation.

9.3 Conic Sections

Types of Conic Sections • Parabolas • Ellipses • Hyperbolas • Translation of Axes

Introduction

Throughout history, people have been fascinated by the universe around them and compelled to understand it. Conic sections have played an important role in gaining this understanding. Although conic sections were described and named by the Greek astronomer Apollonius in 200 B.C., it was not until much later that they were used to model motion in the universe. In the sixteenth century Tycho Brahe, the greatest observational astronomer of the age, recorded precise data on planetary movement in the sky. Using Brahe's data in 1619, Johannes Kepler determined that planets move in elliptical orbits around the sun. In 1686 Newton used Kepler's work to show that elliptical orbits are the result of his famous theory of gravitation. We now know that all celestial objects—including planets, comets, asteroids, and satellites—travel in paths described by conic sections. Today scientists search the sky for information about the universe with enormous radio telescopes in the shape of parabolic dishes.

The search to understand the universe has been directed toward both the infinite and the infinitesimal. In 1911 Ernest Rutherford determined the basic structure of the atom. Small atomic particles are capable of traveling in trajectories described by conic sections.

Parabolas, ellipses, and hyperbolas have had a profound influence on our understanding of ourselves and the cosmos around us. In this section we learn about these age-old curves. (Source: *Historical Topics for the Mathematics Classroom, Thirty-first Yearbook,* NCTM.)

Types of Conic Sections

Conic sections are named after the different ways that a plane can intersect a cone. See Figure 9.9. The three basic curves are parabolas, ellipses, and hyperbolas. A circle is an example of an ellipse.

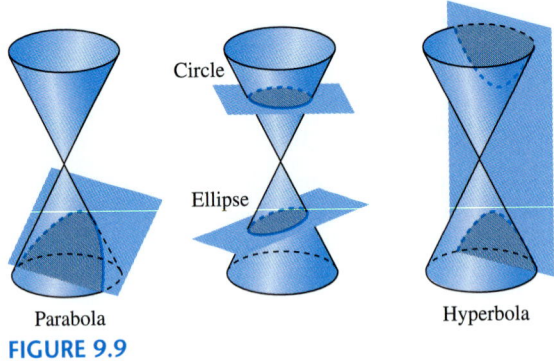

FIGURE 9.9

Figures 9.10–9.12 show examples of conic sections.

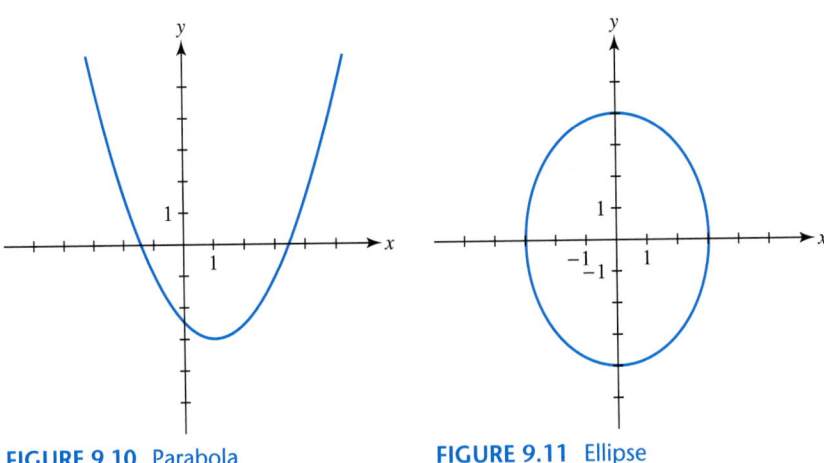

FIGURE 9.10 Parabola   FIGURE 9.11 Ellipse

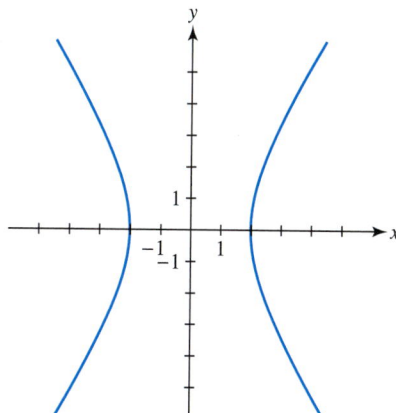

FIGURE 9.12 Hyperbola

Parabolas

In Chapter 1 we learned that a parabola with vertex $(0, 0)$ can be represented symbolically by the equation $y = ax^2$. With this representation a parabola can either open upward when $a > 0$ or downward when $a < 0$. The following definition of a parabola allows it to open in any direction.

> ### Parabola
>
> A **parabola** is the set of points in a plane equidistant from a fixed point and a fixed line. The fixed point is called the **focus** and the fixed line is called the **directrix** of the parabola.

Figures 9.13 and 9.14 show two parabolas with vertex V and focus F. The first parabola has a vertical axis and directrix $y = -p$, while the second has a horizontal axis and directrix $x = -p$. For any point P located at (x, y) on the parabola, the distance d_1 from F to P is equal to the perpendicular distance d_2 from P to the directrix. By the vertical line test, the parabola in Figure 9.14 cannot be represented by a function.

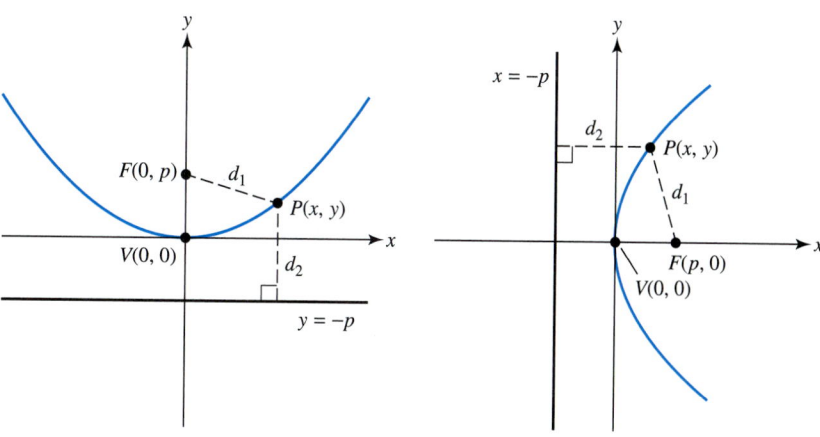

FIGURE 9.13   **FIGURE 9.14**

MAKING CONNECTIONS
Functions and Points

In Figures 9.13 and 9.14, the point P is labeled as $P(x, y)$. This resembles function notation involving two inputs, since the point P is determined by x and y.

If the value of p is known, then the equation of a parabola with vertex $(0, 0)$ can be found using one of the following equations.

Equation of a parabola with vertex (0, 0)

Vertical Axis
The parabola with a focus at $(0, p)$ and directrix $y = -p$ has equation

$$x^2 = 4py.$$

The parabola opens upward if $p > 0$ and downward if $p < 0$.

Horizontal Axis
The parabola with a focus at $(p, 0)$ and directrix $x = -p$ has equation

$$y^2 = 4px.$$

The parabola opens to the right if $p > 0$ and to the left if $p < 0$.

EXAMPLE 1   *Sketching graphs of parabolas*

Sketch a graph of each parabola. Label the vertex, focus, and directrix.
(a) $x^2 = 8y$
(b) $y^2 = -2x$

Solution

(a) The equation $x^2 = 8y$ is in the form $x^2 = 4py$, where $8 = 4p$. Therefore, the parabola has a vertical axis with $p = 2$. Since $p > 0$, the parabola opens upward. The focus is located at $(0, p)$ or $(0, 2)$, and the directrix is $y = -p$ or $y = -2$. The graph of the parabola is shown in Figure 9.15.

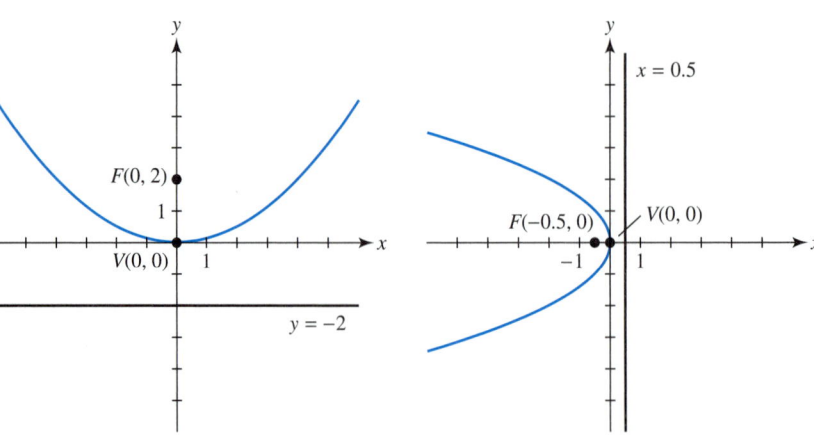

FIGURE 9.15   FIGURE 9.16

(b) The equation $y^2 = -2x$ has the form $y^2 = 4px$, where $-2 = 4p$. Therefore, the parabola has a horizontal axis with $p = -0.5$. Since $p < 0$, the parabola opens to the left. The focus is located at $(-0.5, 0)$, and the directrix is $x = 0.5$. See Figure 9.16. ∎

When a parabola is rotated about its axis, it sweeps out a shape called a **paraboloid** as shown in Figure 9.17. Paraboloids have a special reflective property. When incoming rays of light from the sun or distant stars strike the surface of a paraboloid, each ray is reflected toward the focus. See Figure 9.18. If the rays are sunlight, intense heat is produced, which can be used to generate solar heat. Radio signals from distant space also concentrate at the focus. Scientists can measure these signals by placing a receiver at the focus. This property of a paraboloid can also be used in reverse. If a light source is placed at the focus, then the light is reflected straight ahead as shown in Figure 9.19. Searchlights, flashlights, and car headlights make use of this property.

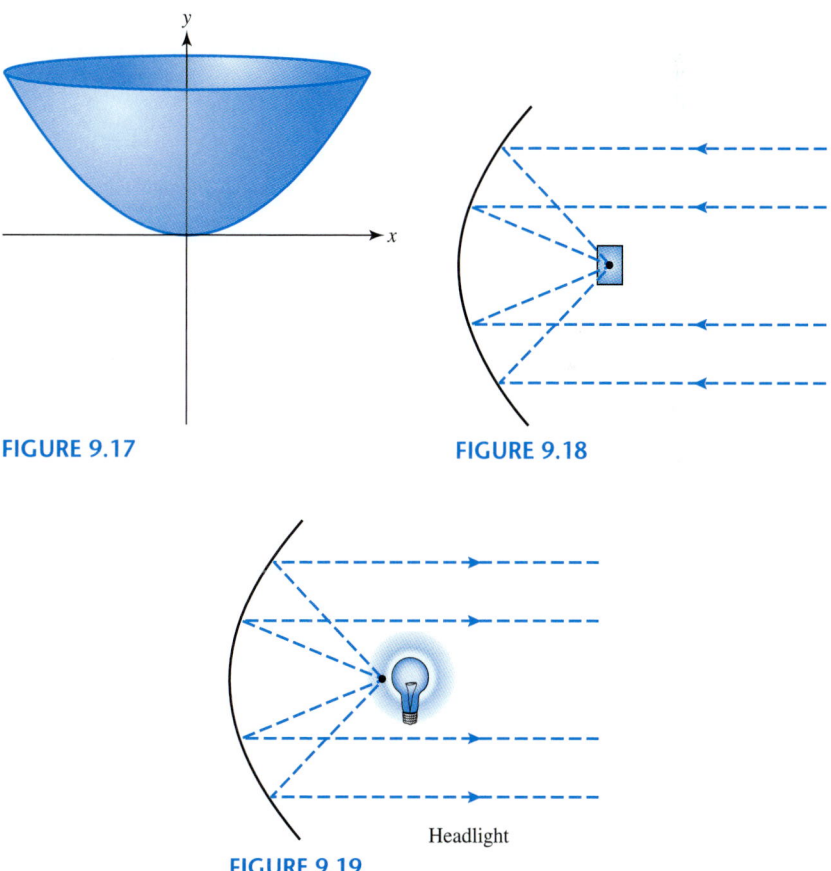

FIGURE 9.17   **FIGURE 9.18**

Headlight

FIGURE 9.19

The next example illustrates this property for radio telescopes.

EXAMPLE 2   *Locating the receiver for a radio telescope*

The U.S. Naval Research Laboratory designed a giant radio telescope weighing 3450 tons. Its parabolic dish has a diameter of 300 feet and a depth of 44 feet. See

Figure 9.20. (Source: J. Mar, *Structure Technology for Large Radio and Radar Telescope Systems.*)

(a) Find an equation in the form $y = ax^2$ that describes a cross section of this dish.

(b) If the receiver is located at the focus, how far should it be from the vertex?

FIGURE 9.20

Solution

(a) Locate a parabola that passes through $(-150, 44)$ and $(150, 44)$ as in Figure 9.21. Substitute either point into $y = ax^2$.

$$y = ax^2$$

$$44 = a(150)^2$$

$$a = \frac{44}{150^2} = \frac{11}{5625}$$

The equation of the parabola is $y = \frac{11}{5625}x^2$, where $-150 \le x \le 150$.

(b) The value of p represents the distance from the vertex to the focus. To determine p, write the equation in the form $x^2 = 4py$. Then,

$$y = \frac{11}{5625}x^2 \quad \text{is equivalent to} \quad x^2 = \frac{5625}{11}y.$$

It follows that $4p = \frac{5625}{11}$ or $p = \frac{5625}{44} \approx 127.84$. Therefore, the receiver should be located about 128 feet from the vertex.

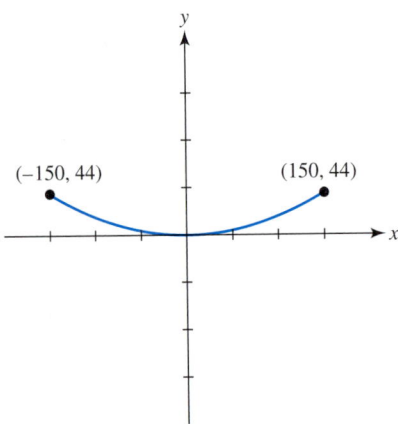

FIGURE 9.21

Ellipses

One method to sketch an ellipse is to tie a string to two nails driven into a flat board. If a pencil is placed inside the loop, the resulting curve shown in Figure 9.22 is an ellipse. The sum of the distances d_1 and d_2 between the pencil and each of the nails is always fixed by the string. The locations of the nails correspond to the *foci* of the ellipse. If the two nails coincide, the ellipse becomes a circle. As the nails spread farther apart, the ellipse becomes more elongated or eccentric.

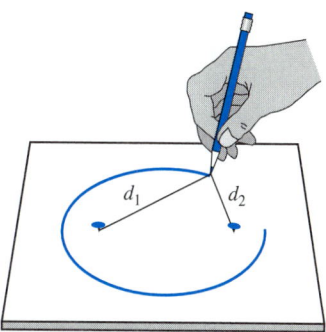

FIGURE 9.22

This method of sketching an ellipse suggests the following definition.

Ellipse

An **ellipse** is the set of points in a plane, the sum of whose distances from two fixed points is constant. Each fixed point is called a **focus** (plural **foci**) of the ellipse.

In Figures 9.23 and 9.24 the **major axis** and **minor axis** are labeled for each ellipse. Figure 9.23 shows an ellipse with a *horizontal major axis,* while Figure 9.24 illustrates an ellipse with a *vertical major axis.* The **vertices,** V_1 and V_2, of each ellipse are located at the endpoints of the major axis.

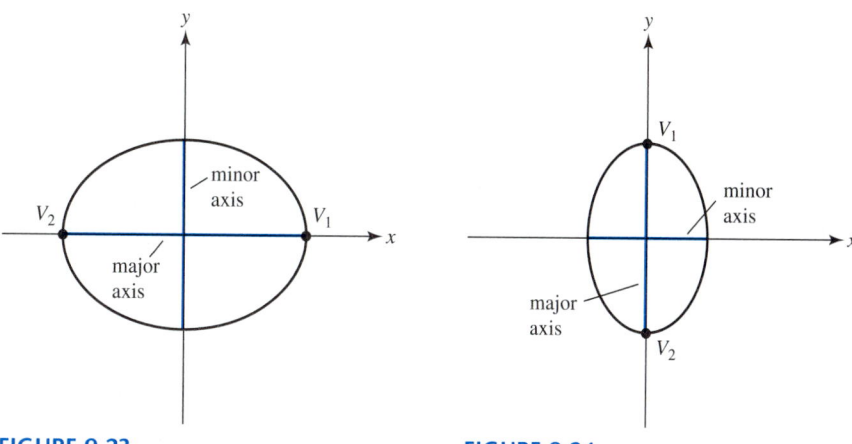

FIGURE 9.23   **FIGURE 9.24**

Since a vertical line can intersect the graph of an ellipse more than once, an ellipse cannot be modeled by a single function. However, some ellipses can be represented by the following equations.

Standard equations for ellipses centered at (0, 0)

The ellipse with center at the origin, *horizontal* major axis, and equation

$$\frac{x^2}{a^2} + \frac{y^2}{b^2} = 1 \qquad (a > b > 0)$$

has vertices $(\pm a, 0)$, endpoints of the minor axis $(0, \pm b)$, and foci $(\pm c, 0)$, where $c^2 = a^2 - b^2$ and $c \geq 0$.

The ellipse with center at the origin, *vertical* major axis, and equation

$$\frac{x^2}{b^2} + \frac{y^2}{a^2} = 1 \qquad (a > b > 0)$$

has vertices $(0, \pm a)$, endpoints of the minor axis $(\pm b, 0)$, and foci $(0, \pm c)$, where $c^2 = a^2 - b^2$ and $c \geq 0$.

Figures 9.25 and 9.26 show two ellipses. The first has a horizontal major axis and the second has a vertical major axis. The coordinates of the vertices V_1 and V_2, foci F_1 and F_2, and endpoints of the minor axis U_1 and U_2 are labeled.

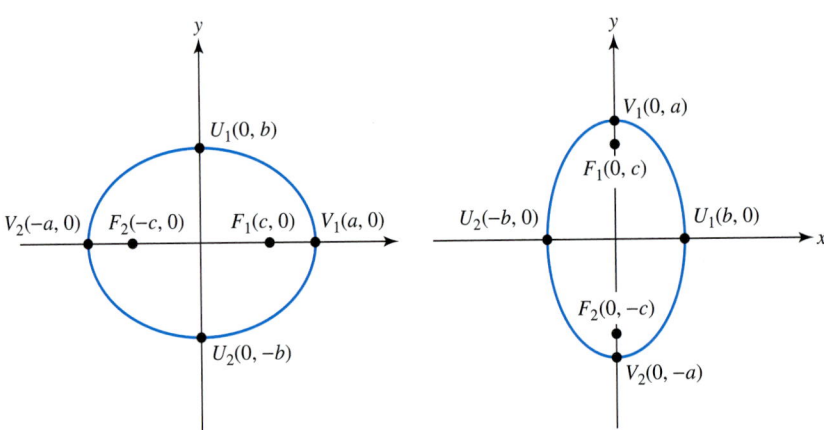

FIGURE 9.25   **FIGURE 9.26**

EXAMPLE 3   *Sketching graphs of ellipses*

Sketch a graph of each ellipse. Label the vertices, foci, and endpoints of the minor axes.

(a) $\dfrac{x^2}{9} + \dfrac{y^2}{4} = 1$

(b) $25x^2 + 16y^2 = 400$

Solution

(a) The equation $\dfrac{x^2}{9} + \dfrac{y^2}{4} = 1$ describes an ellipse with $a = 3$ and $b = 2$. The ellipse has a horizontal major axis with vertices $(\pm 3, 0)$. The endpoints of the minor axis are $(0, \pm 2)$. To locate the foci, find c.

$$c^2 = a^2 - b^2 = 9 - 4 = 5 \qquad \text{or}$$
$$c = \sqrt{5} \approx 2.236.$$

The foci are located on the major axis with coordinates $(\pm \sqrt{5}, 0)$. See Figure 9.27.

(b) The equation $25x^2 + 16y^2 = 400$ can be put into standard form by dividing both sides by 400.

$$25x^2 + 16y^2 = 400$$

$$\frac{25x^2}{400} + \frac{16y^2}{400} = \frac{400}{400} \qquad \textcolor{blue}{\text{Divide by 400.}}$$

$$\frac{x^2}{16} + \frac{y^2}{25} = 1 \qquad \textcolor{blue}{\text{Simplify.}}$$

This ellipse has a vertical major axis with $a = 5$ and $b = 4$. The value of c is given by

$$c^2 = 5^2 - 4^2 = 9 \qquad \text{or} \qquad c = 3.$$

This ellipse has foci $(0, \pm 3)$, vertices $(0, \pm 5)$, and endpoints of the minor axis located at $(\pm 4, 0)$. See Figure 9.28.

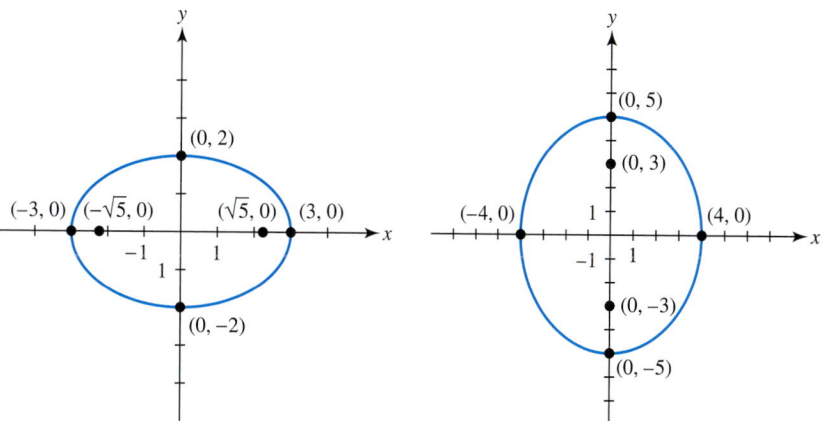

FIGURE 9.27   **FIGURE 9.28**   ■

The planets travel around the sun in elliptical orbits. Although their orbits are nearly circular, many planets have a slight eccentricity to them. The **eccentricity** e of an ellipse is defined by

$$e = \frac{\sqrt{a^2 - b^2}}{a} = \frac{c}{a}.$$

Since the foci of an ellipse lie inside the ellipse, $0 \le c < a$ and $0 \le \dfrac{c}{a} < 1$. Therefore, the eccentricity e of an ellipse satisfies $0 \le e < 1$. If $e = 0$, then $a = b$

and the ellipse is a circle. See Figure 9.29. As e increases, the foci spread apart and the ellipse becomes more elongated. See Figures 9.30 and 9.31.

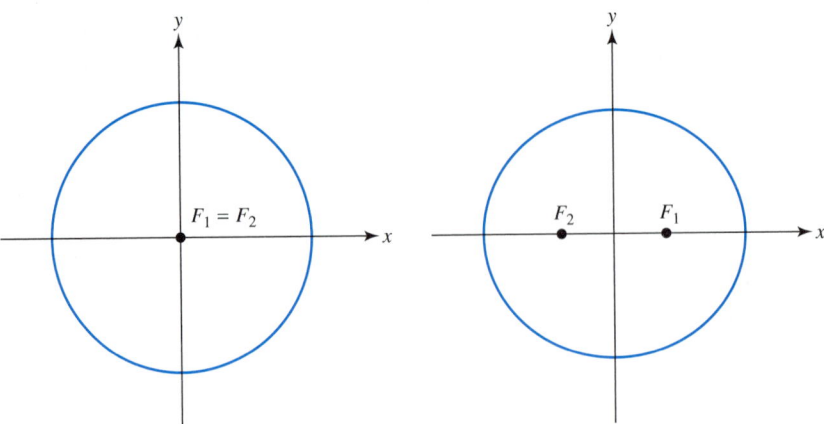

FIGURE 9.29 $e = 0$ **FIGURE 9.30** $e = 0.4$

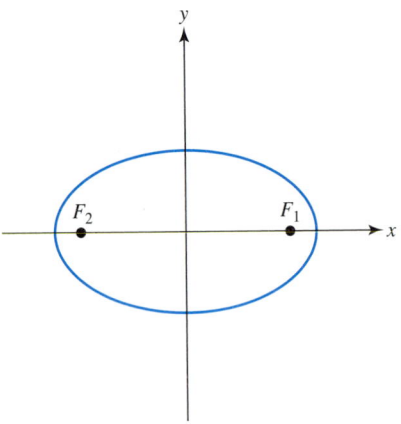

FIGURE 9.31 $e = 0.8$

Astronomers have measured values of e and a for each planet. Utilizing this information and the fact that the sun is located at one focus of the ellipse, the equation of a planet's orbit can be found.

EXAMPLE 4   *Finding the orbital equation of the planet Pluto*

The planet Pluto has the greatest eccentricity of any planet with $e = 0.249$ and $a = 39.44$. (For Earth $a = 1$.) Graph the orbit of Pluto and the position of the sun in $[-60, 60, 10]$ by $[-40, 40, 10]$. (*Source: M. Zeilik, Introductory Astronomy and Astrophysics.*)

Solution

Let the orbit of Pluto be given by $\dfrac{x^2}{a^2} + \dfrac{y^2}{b^2} = 1$. Then,

$$e = \frac{c}{a} = 0.249 \qquad \text{implies} \qquad c = 0.249a = 0.249(39.44) \approx 9.821.$$

To find b, solve the equation $c^2 = a^2 - b^2$ for b.

$$b = \sqrt{a^2 - c^2}$$
$$= \sqrt{39.44^2 - 9.821^2} \approx 38.20$$

Pluto's orbit is modeled by $\dfrac{x^2}{39.44^2} + \dfrac{y^2}{38.20^2} = 1$. Since $c \approx 9.821$, the foci are $(\pm 9.821, 0)$. The sun could be located at either one. We locate the sun at $(9.821, 0)$. See Figures 9.32 and 9.33.

$[-60, 60, 10]$ by $[-40, 40, 10]$

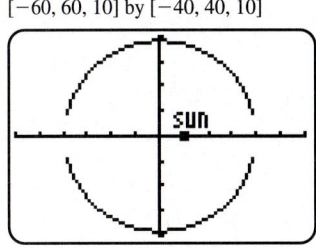

FIGURE 9.32   **FIGURE 9.33**   ■

Note: To graph an ellipse on some types of graphing calculators, the equation must be solved for y. This results in two equations to graph. For example,

$$\frac{x^2}{39.44^2} + \frac{y^2}{38.20^2} = 1$$

$$\frac{y^2}{38.20^2} = 1 - \frac{x^2}{39.44^2}$$

$$\frac{y}{38.20} = \pm\sqrt{1 - \frac{x^2}{39.44^2}}$$

$$y = \pm 38.20\sqrt{1 - \frac{x^2}{39.44^2}}.$$

Hyperbolas

The third type of conic section is a hyperbola.

Hyperbola

A **hyperbola** is the set of points in a plane, the difference of whose distances from two fixed points is constant. Each fixed point is called a **focus** of the hyperbola.

In Figure 9.34 on the following page a point $P(x, y)$ is shown on a hyperbola with distance d_1 from focus $F_1(c, 0)$ and distance d_2 from focus $F_2(-c, 0)$.

Regardless of the location of the point $P(x, y)$, $|d_2 - d_1| = 2a$. The **transverse axis** is the line segment connecting the **vertices** $V_1(a, 0)$ and $V_2(-a, 0)$.

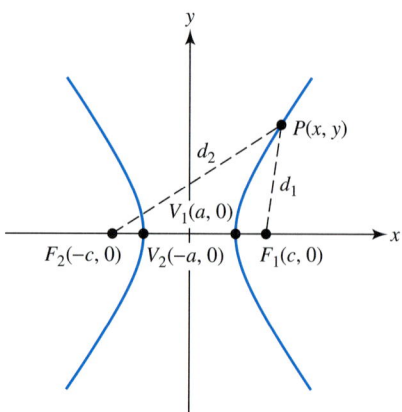

FIGURE 9.34

By the vertical line test, a hyperbola cannot be represented by a function, but many can be described by the following equations. The constants a, b, and c are positive.

Standard equations for hyperbolas centered at (0, 0)

The hyperbola with center at the origin, *horizontal* transverse axis, and equation

$$\frac{x^2}{a^2} - \frac{y^2}{b^2} = 1$$

has vertices $(\pm a, 0)$ and foci $(\pm c, 0)$, where $c^2 = a^2 + b^2$.

The hyperbola with center at the origin, *vertical* transverse axis, and equation

$$\frac{y^2}{a^2} - \frac{x^2}{b^2} = 1$$

has vertices $(0, \pm a)$ and foci $(0, \pm c)$, where $c^2 = a^2 + b^2$.

Two hyperbolas are shown in Figures 9.35 and 9.36 on the next page. The coordinates of the vertices and foci are labeled. The two parts of the hyperbola in Figure 9.35 are the **left branch** and **right branch**, whereas in Figure 9.36 the hyperbola has an **upper branch** and a **lower branch.** A line segment connecting the points $(0, \pm b)$ in Figure 9.35 and $(\pm b, 0)$ in Figure 9.36 is the **conjugate axis.** The lines $y = \pm \frac{b}{a} x$ and $y = \pm \frac{a}{b} x$ are **asymptotes** for each respective hyperbola. They can be used as an aid in graphing.

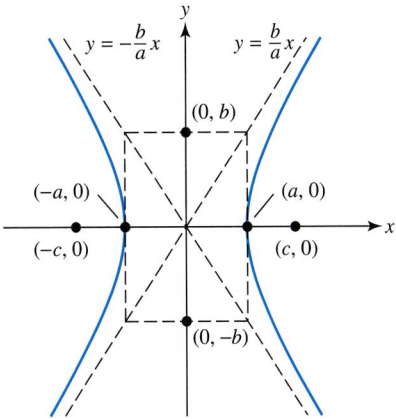

FIGURE 9.35

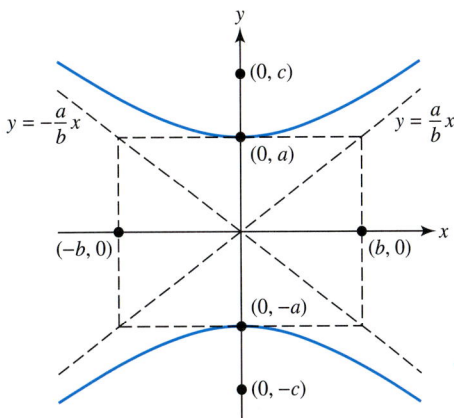

FIGURE 9.36

Note: A hyperbola consists of two solid curves or branches. The asymptotes, foci, transverse axis, conjugate axis, and dashed rectangle are not part of the hyperbola, but are aids for sketching its graph.

One interpretation of an asymptote can be made using trajectories of comets as they approach the sun. Comets travel in parabolic, elliptic, or hyperbolic trajectories. If the speed of a comet is too slow, the gravitational pull of the sun captures the comet in an elliptical orbit. See Figure 9.37. If the speed of the comet is too fast, the sun's gravity is too weak and the comet passes by the sun in a hyperbolic trajectory. Near the sun the gravitational pull is stronger and the comet's trajectory is curved. Farther from the sun, gravity becomes weaker and the comet eventually returns to a straight-line trajectory that is determined by the *asymptote* of the hyperbola. See Figure 9.38. Finally, if the speed is neither too slow nor too fast, the comet will travel in a parabolic path. See Figure 9.39. In all three cases, the sun is located at a focus of the conic section.

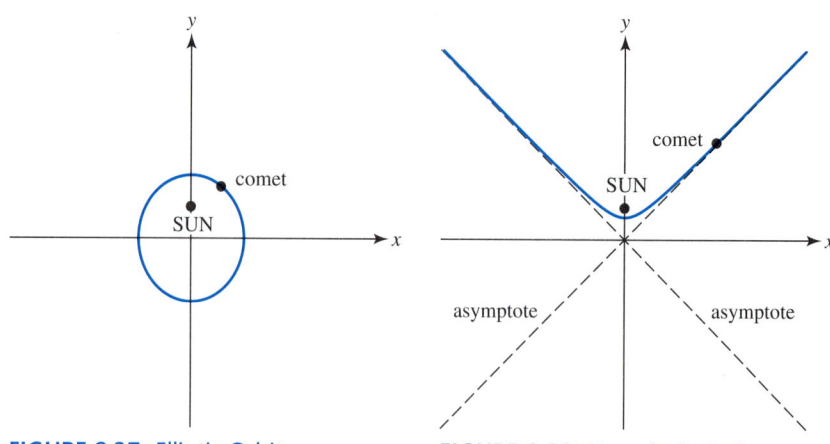

FIGURE 9.37 Elliptic Orbit   **FIGURE 9.38** Hyperbolic Path

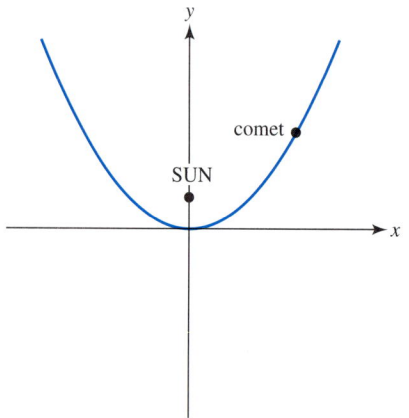

FIGURE 9.39 Parabolic Path

Critical Thinking

If a comet is seen at regular intervals, what type of path must it follow?

EXAMPLE 5   *Sketching the graph of a hyperbola*

Sketch a graph of $\dfrac{x^2}{4} - \dfrac{y^2}{9} = 1$. Label the vertices, foci, and asymptotes.

Solution

The equation is in standard form with $a = 2$ and $b = 3$. It has a horizontal transverse axis with vertices $(\pm 2, 0)$. The endpoints of the conjugate axis are $(0, \pm 3)$. To locate the foci find c.

$$c^2 = a^2 + b^2 = 4 + 9 = 13 \qquad \text{or}$$
$$c = \sqrt{13} \approx 3.61.$$

The foci are $(\pm\sqrt{13}, 0)$. The asymptotes are $y = \pm\dfrac{b}{a}x$ or $y = \pm\dfrac{3}{2}x$. See Figure 9.40.

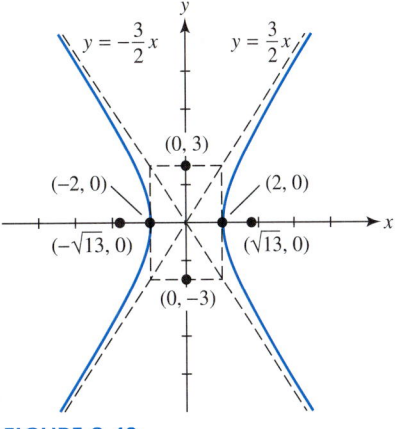

FIGURE 9.40 ∎

EXAMPLE 6

Finding the equation of a hyperbola

Find the equation of the hyperbola centered at the origin with a vertical transverse axis of length 6 and focus (0, 5). Sketch a graph of the hyperbola.

Solution

Since the hyperbola is centered at the origin with a vertical axis, its equation is $\dfrac{y^2}{a^2} - \dfrac{x^2}{b^2} = 1$. The transverse axis has length $6 = 2a$, so $a = 3$. Since one focus is located at (0, 5), $c = 5$. We can find b using the following equation.

$$b^2 = c^2 - a^2$$
$$b = \sqrt{c^2 - a^2}$$
$$b = \sqrt{5^2 - 3^2} = 4$$

The equation of this hyperbola is $\dfrac{y^2}{9} - \dfrac{x^2}{16} = 1$. Its asymptotes are $y = \pm\dfrac{a}{b}x$ or $y = \pm\dfrac{3}{4}x$. Its graph is shown in Figure 9.41.

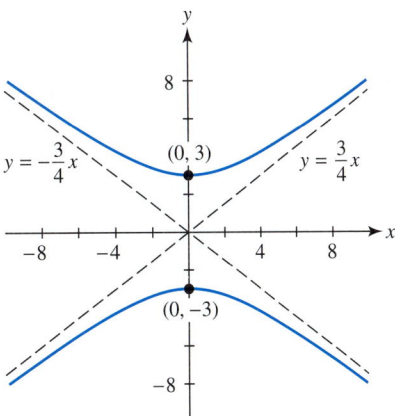

FIGURE 9.41 ∎

EXAMPLE 7   *Graphing a hyperbola with technology*

Graph $\dfrac{y^2}{4.2} - \dfrac{x^2}{8.4} = 1$ using a graphing calculator.

Solution

Begin by solving the given equation for y.

$$\frac{y^2}{4.2} = 1 + \frac{x^2}{8.4} \qquad\qquad \text{Add } \frac{x^2}{8.4}.$$

$$y^2 = 4.2\left(1 + \frac{x^2}{8.4}\right) \qquad\qquad \text{Multiply by 4.2.}$$

$$y = \pm\sqrt{4.2\left(1 + \frac{x^2}{8.4}\right)} \qquad\qquad \text{Square root property}$$

Graph $Y_1 = \sqrt{(4.2(1 + X^2/8.4))}$ and $Y_2 = -\sqrt{(4.2(1 + X^2/8.4))}$. See Figures 9.42 and 9.43.

$[-10, 10, 1]$ by $[-10, 10, 1]$

FIGURE 9.42   **FIGURE 9.43**   ■

Translation of Axes

All three types of conic sections can be translated so that they are centered at a point (h, k), rather than at the origin. The techniques from Section 1.6 can be applied to conic sections.

EXAMPLE 8   *Translating an ellipse*

Translate the ellipse with equation $\dfrac{x^2}{9} + \dfrac{y^2}{4} = 1$ so that it is centered at $(-1, 2)$. Find this equation and sketch its graph.

Solution

To translate the center to $(-1, 2)$, replace x with $(x + 1)$ and y with $(y - 2)$. This new equation is

$$\frac{(x + 1)^2}{9} + \frac{(y - 2)^2}{4} = 1.$$

This ellipse is congruent to the given ellipse, except that it is centered at $(-1, 2)$. The given ellipse is shown in Figure 9.44 on the next page and the translated ellipse is shown in Figure 9.45.

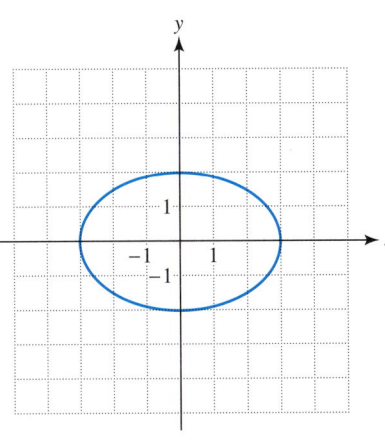

FIGURE 9.44

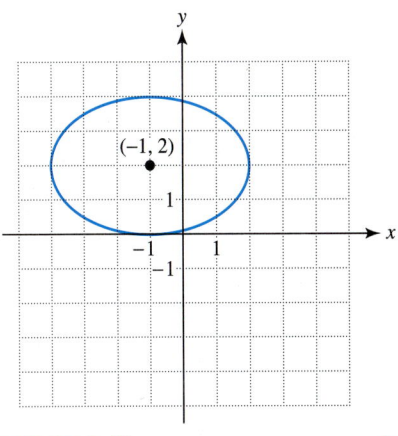

FIGURE 9.45   ■

A circle is an ellipse where $a = b$. If a circle has radius r and center (h, k), then an equation for the circle is $\dfrac{(x - h)^2}{r^2} + \dfrac{(y - k)^2}{r^2} = 1$. Multiplying the equation by r^2 provides the following result.

> ### Standard equation of a circle
>
> The **standard equation of a circle** with center (h, k) and radius r is
> $$(x - h)^2 + (y - k)^2 = r^2.$$

EXAMPLE 9

Finding the standard equation of a circle

Find the standard equation of a circle with radius 4 and center $(5, -3)$.

Solution

Let $h = 5$, $k = -3$, and $r = 4$. The standard equation is
$$(x - 5)^2 + (y + 3)^2 = 16.$$
■

EXAMPLE 10

Finding the center and radius of a circle

Find the center and radius of the circle given by $x^2 + 6x + y^2 - 2y = -6$.

Solution

Use completing the square to write the standard equation of the circle.

$x^2 + 6x + y^2 - 2y = -6$	Given equation
$(x^2 + 6x + 9) + (y^2 - 2y + 1) = -6 + 9 + 1$	Complete the square.
$(x + 3)^2 + (y - 1)^2 = 4$	Factor.

The center is $(-3, 1)$ and the radius is 2. ■

9.3 PUTTING IT ALL TOGETHER

The three types of conic sections are parabolas, ellipses, and hyperbolas. The following table summarizes concepts related to conic sections.

Parabola with Vertex (0, 0)

Standard Forms
$x^2 = 4py$ (vertical axis)   or $y^2 = 4px$ (horizontal axis)
Meaning of p
Both the vertex-focus distance and the vertex-directrix distance are p. The sign of p determines if the parabola opens upward or downward—left or right.

Ellipse Centered at (0, 0)

Standard Forms with $a > b > 0$
$\dfrac{x^2}{a^2} + \dfrac{y^2}{b^2} = 1$ (horizontal major axis)   or $\dfrac{y^2}{a^2} + \dfrac{x^2}{b^2} = 1$ (vertical major axis)
Meaning of a, b, and c
The distance from the center to a vertex is a, the distance from the center to an endpoint of the minor axis is b, and the distance from the center to a focus is c. The ratio $\dfrac{c}{a}$ equals the eccentricity e. They are related by $c^2 = a^2 - b^2$.

Hyperbola Centered at (0, 0)

Standard Forms with $a, b > 0$
$\dfrac{x^2}{a^2} - \dfrac{y^2}{b^2} = 1$ (horizontal transverse axis)   or $\dfrac{y^2}{a^2} - \dfrac{x^2}{b^2} = 1$ (vertical transverse axis)
Meaning of a, b, and c
The distance from the center to a vertex is a, the distance from the center to a focus is c. The asymptotes are $y = \pm\dfrac{b}{a}x$ if the transverse axis is horizontal, and $y = \pm\dfrac{a}{b}x$ if the transverse axis is vertical. They are related by $c^2 = a^2 + b^2$.

9.3 EXERCISES

 Tape 7

Parabolas

Exercises 1–6: Graph the parabola. Label the vertex, focus, and directrix.

1. $16y = x^2$

2. $y = -2x^2$

3. $x = \dfrac{1}{8}y^2$

4. $-y^2 = 6x$

5. $-4x = y^2$

6. $\dfrac{1}{2}y^2 = 3x$

Exercises 7–10: Sketch a graph of a parabola with focus and directrix as shown in the figure. Find an equation of the parabola.

7.

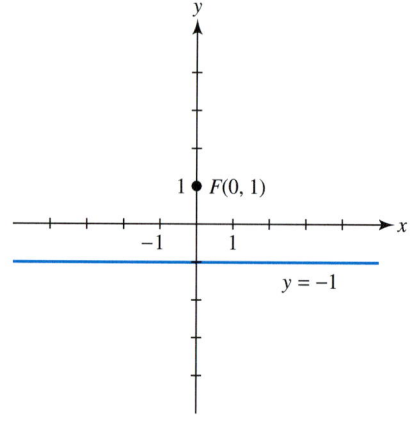

8.

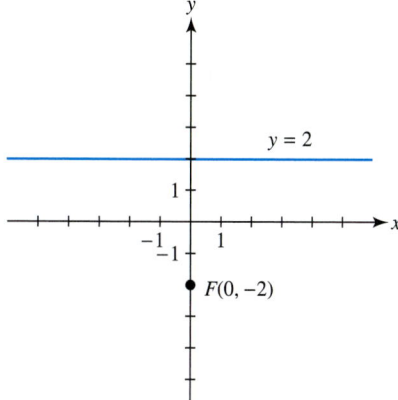

9.

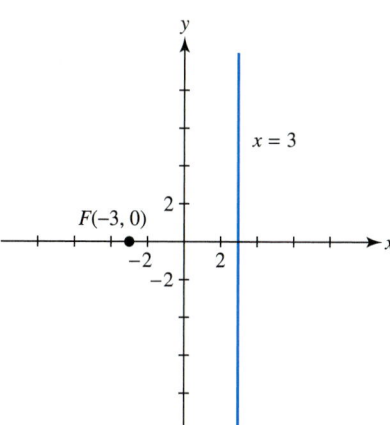

10.

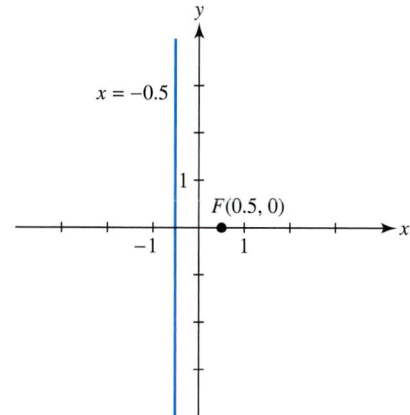

Exercises 11–14: Find an equation of the parabola with vertex V(0, 0) that satisfies the given conditions. Sketch its graph.

11. Focus $\left(0, \frac{3}{4}\right)$

12. Directrix $y = 2$

13. Directrix $x = 2$

14. Focus $(-1, 0)$

Exercises 15 and 16: (Refer to Example 2.) Use the dimensions of a television satellite dish in the shape of a

paraboloid to calculate how far from the vertex the receiver should be located.

15. Six-foot diameter, nine inches deep

16. Nine-inch radius, two inches deep

Ellipses

Exercises 17–22: Graph the ellipse. Label the foci and the endpoints of each axis.

17. $\dfrac{x^2}{4} + \dfrac{y^2}{9} = 1$

18. $\dfrac{x^2}{9} + \dfrac{y^2}{4} = 1$

19. $\dfrac{x^2}{36} + \dfrac{y^2}{16} = 1$

20. $x^2 + \dfrac{y^2}{4} = 1$

21. $9x^2 + 5y^2 = 45$

22. $x^2 + 4y^2 = 400$

Exercises 23–26: The foci F_1 and F_2, vertices V_1 and V_2 and endpoints U_1 and U_2 of the minor axis of an ellipse are labeled in the figure. Graph the ellipse and find its standard equation.

23.

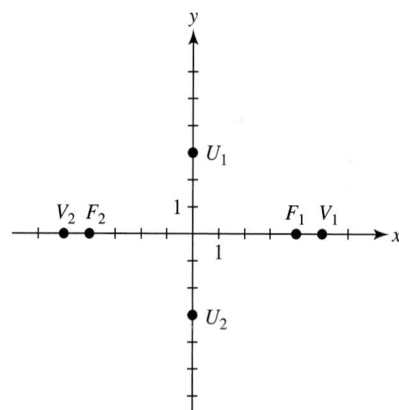

24.

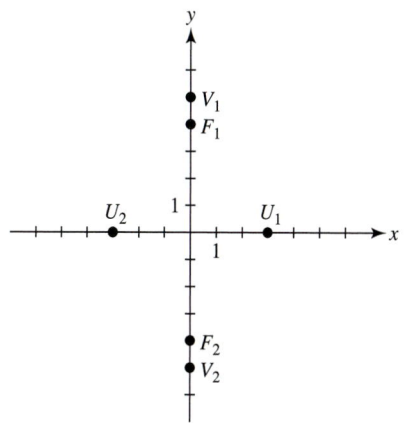

25.

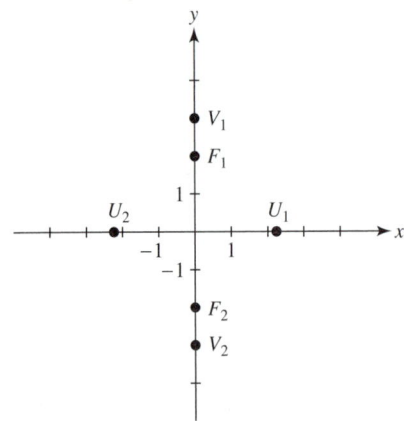

26.

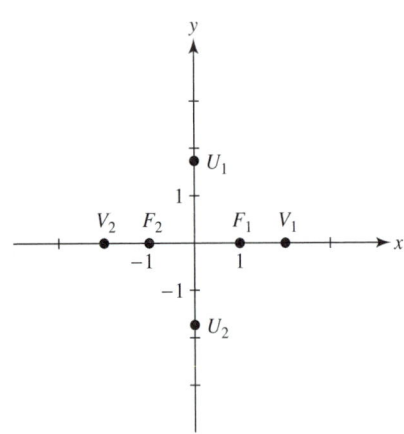

Exercises 27–32: Find an equation of the ellipse, centered at the origin, satisfying the conditions. Sketch its graph.

27. Foci $(0, \pm 2)$, vertices $(0, \pm 4)$

28. Foci $(0, \pm 3)$, vertices $(0, \pm 5)$

29. Foci $(\pm 5, 0)$, vertices $(\pm 6, 0)$

30. Vertical major axis of length 12, minor axis of length 8

31. Eccentricity $\frac{2}{3}$, horizontal major axis of length 6

32. Eccentricity $\frac{3}{4}$, vertices $(0, \pm 8)$

Exercises 33 and 34: (Refer to Example 4.) Find an equation of the orbit for the planet. Graph its orbit and the location of the sun at a focus.

33. Mercury: $e = 0.206$, $a = 0.387$

34. Mars: $e = 0.093$, $a = 1.524$

Hyperbolas

Exercises 35–40: Sketch a graph of the hyperbola, including the asymptotes. Give the coordinates of the foci.

35. $\dfrac{x^2}{9} - \dfrac{y^2}{49} = 1$

36. $\dfrac{x^2}{16} - \dfrac{y^2}{4} = 1$

37. $\dfrac{y^2}{36} - \dfrac{x^2}{16} = 1$

38. $\dfrac{y^2}{4} - \dfrac{x^2}{4} = 1$

39. $x^2 - y^2 = 9$

40. $49y^2 - 25x^2 = 1225$

Exercises 41–44: Sketch a graph of a hyperbola, centered at the origin, with the given foci, vertices, and asymptotes shown in the figure. Find an equation of the hyperbola.

41.

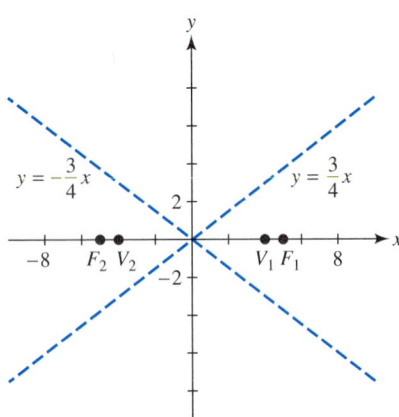

42.

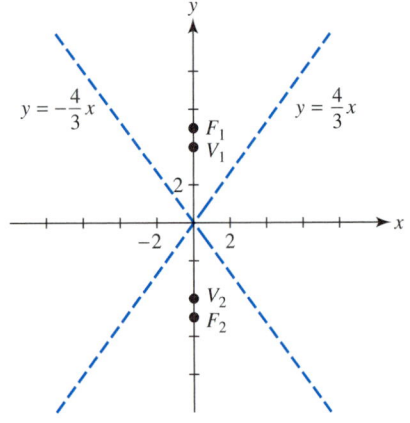

43.

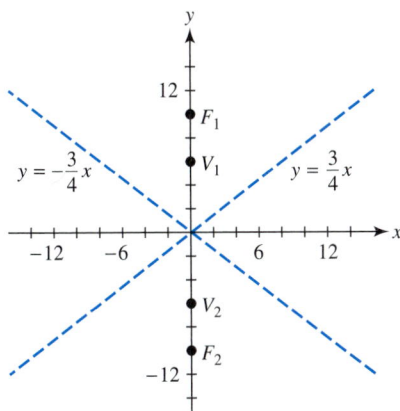

Translation of Axes

Exercises 51–54: Find an (approximate) equation of the conic section shown in the figure.

51.

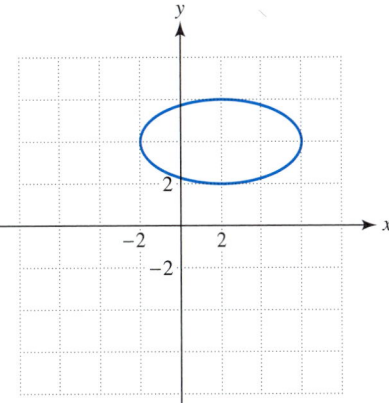

44.

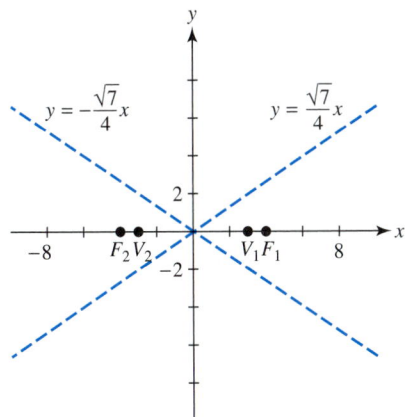

52.

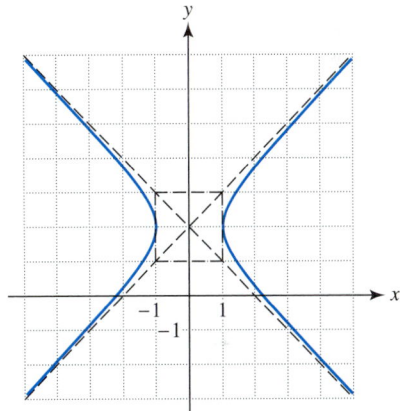

Exercises 45–50: Determine an equation of the hyperbola, centered at the origin, satisfying the conditions. Sketch its graph, including the asymptotes.

45. Foci $(0, \pm 13)$, vertices $(0, \pm 12)$

46. Foci $(\pm 13, 0)$, vertices $(\pm 5, 0)$

47. Vertical transverse axis of length 4, foci $(0, \pm 5)$

48. Horizontal transverse axis of length 12, foci $(\pm 10, 0)$

49. Vertices $(\pm 3, 0)$, asymptotes $y = \pm \frac{2}{3} x$

50. Endpoints of conjugate axis $(\pm 4, 0)$, vertices $(0, \pm 2)$

53.

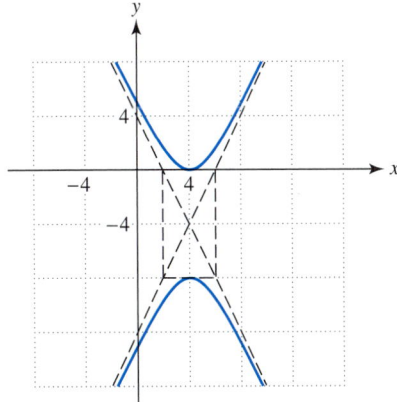

54.

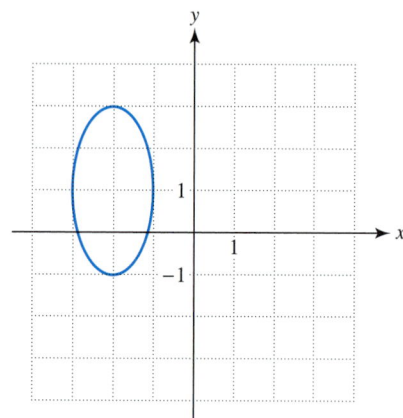

Exercises 55–58: Find the standard equation of the circle that satisfies the conditions. Graph the circle.

55. Center $(0, 0)$, radius of 4

56. Center $(1, -2)$, radius of 3

57. Center $(3, -4)$, radius of 1

58. Center $(-1, -3)$, passing through the point $(3, 0)$

Exercises 59–64: Sketch a graph of the conic section.

59. $(x - 1)^2 = 8(y + 1)$

60. $(y + 3)^2 = 4(x - 2)$

61. $\dfrac{(y + 2)^2}{25} + \dfrac{(x + 1)^2}{16} = 1$

62. $\dfrac{(x - 4)^2}{9} + \dfrac{y^2}{4} = 1$

63. $\dfrac{(y - 2)^2}{36} - \dfrac{(x + 2)^2}{4} = 1$

64. $\dfrac{(x + 1)^2}{4} - \dfrac{(y - 1)^2}{4} = 1$

Exercises 65–68: Find the center and radius of the circle.

65. $x^2 - 4x + y^2 - 2y = 11$

66. $x^2 + 6x + y^2 - 4y = 12$

67. $x^2 + y^2 + 10y = 0$

68. $x^2 - 2x + y^2 + 8y = 19$

Applications

69. *Satellite Orbits* The trajectory of a satellite near Earth can trace a hyperbola, parabola, or ellipse. If the satellite follows either a hyperbolic or parabolic path, it escapes Earth's gravitational influence after a single pass. The path that a satellite travels near Earth depends both on its velocity V in meters per second and its distance D in meters from the center

of Earth. Its path is hyperbolic if $V > \dfrac{k}{\sqrt{D}}$, parabolic if $V = \dfrac{k}{\sqrt{D}}$, and elliptic if $V < \dfrac{k}{\sqrt{D}}$, where $k = 2.82 \times 10^7$ is a constant. (Sources: W. Loh, *Dynamics and Thermodynamics of Planetary Entry;* W. Thomson, *Introduction to Space Dynamics.*)

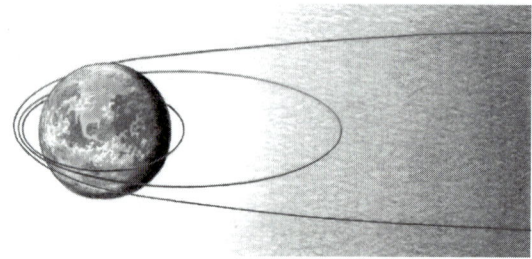

(a) When Explorer IV was at a maximum distance of 42.5×10^6 meters from Earth's center, it had a velocity of 2090 m/sec. Determine the shape of its trajectory.

(b) If an orbiting satellite is scheduled to escape Earth's gravity, so that it can travel to another planet, its velocity must be increased so its trajectory changes from elliptic to hyperbolic. What range of velocities would allow Explorer IV to leave Earth's influence when it is at a maximum distance?

(c) Explain why it is easier to change a satellite's trajectory from an ellipse to a hyperbola when D is maximum rather than minimum.

70. *Planet Velocity* The maximum and minimum velocities in kilometers per second of a planet moving in an elliptical orbit can be calculated by

$$v_{max} = \frac{2\pi a}{P}\sqrt{\frac{1 + e}{1 - e}} \quad \text{and} \quad v_{min} = \frac{2\pi a}{P}\sqrt{\frac{1 - e}{1 + e}}.$$

In these equations, a is the length of the semimajor axis of the orbit in kilometers, P is its orbital period in seconds, and e is the eccentricity of the orbit. (Source: M. Zeilik.)

(a) Calculate v_{max} and v_{min} for Pluto if $a = 5.913 \times 10^9$ kilometers, $P = 2.86 \times 10^{12}$ seconds, and $e = 0.249$.

(b) If a planet has a circular orbit, what can be said about its orbital velocity?

71. *Satellite Orbit* The orbit of Explorer VII and the outline of Earth's surface are shown in the figure on the next page. This orbit can be described by the equation $\dfrac{x^2}{a^2} + \dfrac{y^2}{b^2} = 1$, where $a = 4464$ and $b = 4462$. The surface of Earth can be described by the equation $(x - 164)^2 + y^2 = 3960^2$. Find the maximum and minimum heights of the satellite above Earth's surface if all units are in miles. (Sources: W. Loh; W. Thomson.)

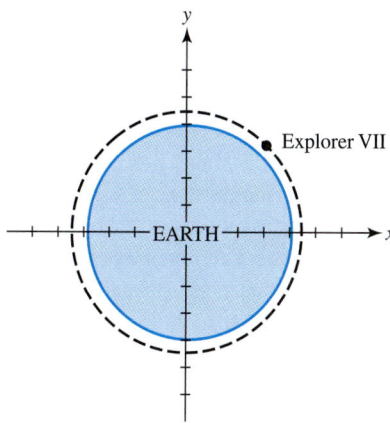

72. *Radio Telescope* (Refer to Example 2.) The Parkes radio telescope has the shape of a parabolic dish with a diameter of 210 feet and a depth of 32 feet. The dish is shown in the figure. (Source: J. Mar.)

(a) Determine an equation of the form $y = ax^2$ describing a cross section of the dish.
(b) The receiver is placed at the focus. How far from the vertex is the receiver located?

Graphing Conic Sections with Technology

Exercises 73–78: Graph the conic section.

73. $x = 2.3(y + 1)^2$

74. $(y - 2.5)^2 = 4.1(x + 1)$

75. $\dfrac{y^2}{10} + \dfrac{x^2}{15} = 1$

76. $\dfrac{(x - 1.2)^2}{7.1} + \dfrac{y^2}{3.5} = 1$

77. $\dfrac{(y - 1)^2}{11} - \dfrac{x^2}{5.9} = 1$

78. $\dfrac{x^2}{5.3} - \dfrac{y^2}{6.7} = 1$

Writing about Mathematics

1. Outline the distinguishing features of parabolas and ellipses. Discuss the location of any foci.

2. Outline the distinguishing features of hyperbolas. Discuss the location of any foci.

CHECKING BASIC CONCEPTS FOR SECTION 9.3

1. Graph the parabola defined by $x = \dfrac{1}{2}y^2$. Include the focus and directrix.

2. Graph the ellipse defined by $\dfrac{x^2}{36} + \dfrac{y^2}{100} = 1$. Include the foci and label the major and minor axes.

3. Graph the hyperbola defined by $\dfrac{x^2}{9} - \dfrac{y^2}{16} = 1$. Include the foci and asymptotes.

4. A parabolic reflector for a searchlight has a diameter of four feet and a depth of one foot. How far from the vertex of the reflector should the filament of the light bulb be located?

9.4 Counting

Fundamental Counting Principle • Permutations • Combinations • The Binomial Theorem

Introduction

The notion of counting in mathematics includes much more than simply counting from 1 to 100. It also includes determining the number of ways that an event can occur. For example, how many ways are there to answer a true-false quiz with ten questions? The answer involves counting the different ways that a student could answer such a quiz. Counting is an important concept that is used to calculate probabilities. Probability is discussed in the next section.

Fundamental Counting Principle

Suppose a quiz has only two questions. The first is a multiple-choice question with four choices: A, B, C, or D, and the second is a true-false (T-F) question. The tree diagram in Figure 9.46 can be utilized to count the different ways that this quiz can be answered.

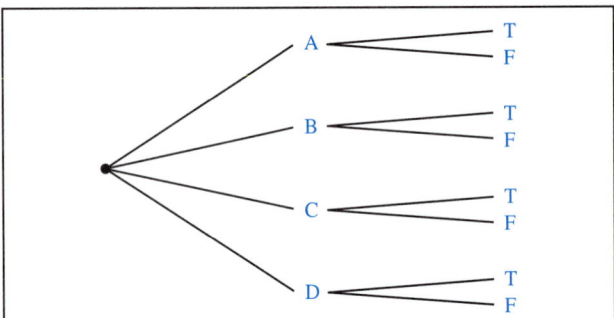

FIGURE 9.46 Different Ways to Answer a Quiz

A tree diagram is a systematic way of listing every possibility. From Figure 9.46, we can see that there are eight different ways to answer the test. They are

AT, AF, BT, BF, CT, CF, DT, and DF.

For instance, CF indicates a quiz with answers of C on the first question and F on the second question.

A tree diagram is not always practical, because it can quickly become very large. For this reason mathematicians have developed more efficient ways of counting. Since the multiple-choice question has four possible answers, after which the true-false question has two possible choices, there are $4 \cdot 2 = 8$ possible ways of answering the test. This is an application of the *fundamental counting principle,* which applies to independent events. Two events are **independent** if neither event influences the outcome of the other.

> **Fundamental counting principle**
>
> Let $E_1, E_2, E_3, \ldots, E_n$ be a sequence of n independent events. If event E_k can occur m_k different ways for $k = 1, 2, 3, \ldots, n$, then there are
>
> $$m_1 \cdot m_2 \cdot m_3 \cdots \cdots m_n$$
>
> different ways for all n events to occur.

EXAMPLE 1   *Counting ways to answer an exam*

An exam contains four true-false questions and six multiple-choice questions. Each multiple-choice question has five possible answers. Count the number of different ways that the exam can be answered.

Solution

Answering these ten questions can be thought of as a sequence of ten independent events. There are two ways to answer each of the first four questions, and five ways of answering the next six questions. The number of ways to answer the exam is

$$2 \cdot 2 \cdot 2 \cdot 2 \cdot 5 \cdot 5 \cdot 5 \cdot 5 \cdot 5 \cdot 5 = 2^4 5^6 = 250{,}000. \qquad \blacksquare$$

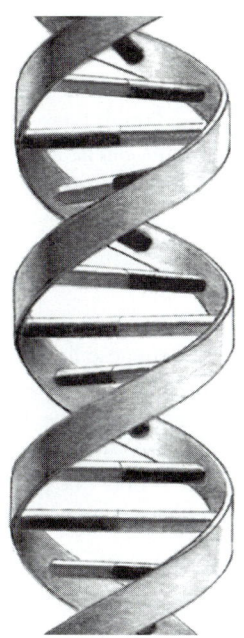

One of the most important discoveries of our time was made in 1953 when the double helical structure of DNA (deoxyribonucleic acid) was identified. DNA is the material of heredity. It is composed of four bases called adenine (A), cytosine (C), guanine (G), and thymine (T). The genetic code of a person is composed of approximately six billion of these bases stored in a long list. The nucleus of every cell contains a copy of this genetic code, tightly coiled in the shape of a double helix. The DNA of a single cell has the potential to store 30 encyclopedic volumes three times over. Any one of the four bases can occur anytime in the genetic code sequence.

Small changes in this code can have dramatic effects. In a normal red blood cell, hemoglobin contains the genetic code

ACTCCTGAGGAGGAGT,

whereas a person with sickle cell anemia has the exact same sequence except thymine (T) has been substituted for adenine (A):

ACTCCTGTGGAGGAGT.

Thus, a single error in a list of six billion letters causes this serious disease. This unique list represented by the four letters A, C, G, and T defines the characteristics of an individual. Approximately 99–99.9% of this genetic code is identical for all humans. Only 0.1–1% is unique to a particular individual. (Source: S. Easteal, *DNA Profiling: Principles, Pitfalls and Potential.*)

EXAMPLE 2   *Counting genetic codes in DNA*

Write a numeric expression that represents the total number of genetic codes that are possible using the four letters A, C, G, and T.

Solution

A human genetic code is composed of a list of six billion letters. For each letter there are four choices. By the fundamental counting principle there are $4^{6,000,000,000}$ different genetic codes of this type. (Many of these would not be valid for human beings.) This number is too large to approximate with a calculator. ∎

EXAMPLE 3   *Counting toll-free 800 telephone numbers*

The number of available 800 telephone numbers for new businesses and individuals is decreasing rapidly. Count the total number of 800 numbers, if the local portion of a telephone number does not start with a 0 or 1. (Source: Database Services Management.)

Solution

A toll-free 800 number assumes the following form.

Local Number

We can think of choosing the remaining digits for the local number as seven independent events. Since the local number cannot begin with a 0 or 1, there are eight possibilities (2 to 9) for the first digit. The remaining six digits can be any number from 0 to 9, so there are ten possibilities for each of these digits. The total is given by

$$8 \cdot 10 \cdot 10 \cdot 10 \cdot 10 \cdot 10 \cdot 10 = 8 \cdot 10^6 = 8,000,000.$$ ∎

Permutations

A permutation is an *ordering* or *arrangement*. For example, suppose three students are scheduled to give a speech in a class. The different arrangements of how these speeches can be ordered are called *permutations*. Initially, any one of the three students could give the first speech. After the first speech, there are two students remaining for the second speech. For the third speech there is only one possibility. By the fundamental counting principle, the total number of permutations is equal to

$$3 \cdot 2 \cdot 1 = 6.$$

If the students are denoted as A, B, and C, then these six permutations are ABC, ACB, BAC, BCA, CAB, and CBA. In a similar manner, if there were ten students scheduled to give speeches, the number of permutations would increase to

$$10 \cdot 9 \cdot 8 \cdot 7 \cdot 6 \cdot 5 \cdot 4 \cdot 3 \cdot 2 \cdot 1 = 3,628,800.$$

A more efficient way of writing the previous two products is to use *factorial notation*. The number $n!$ (read "*n*-factorial") is defined as follows.

---

n-factorial

For any natural number n,

$$n! = n(n-1)(n-2) \cdots (3)(2)(1)$$

and

$$0! = 1.$$

The reason for the definition of 0! will become apparent later. Factorials grow rapidly and can be computed on most calculators.

EXAMPLE 4   *Calculating factorials*

Compute $n!$ for $n = 0, 1, 2, 3, 4,$ and 5 by hand. Use a calculator to find 8!, 13!, and 25!.

Solution

The values for $n!$ can be calculated as

$$0! = 1, \quad 1! = 1, \quad 2! = 2 \cdot 1 = 2, \quad 3! = 3 \cdot 2 \cdot 1 = 6,$$
$$4! = 4 \cdot 3 \cdot 2 \cdot 1 = 24, \quad \text{and} \quad 5! = 5 \cdot 4 \cdot 3 \cdot 2 \cdot 1 = 120.$$

Figure 9.47 shows the values of 8!, 13!, and 25!. The value for 25! is an approximation. Notice how dramatically $n!$ increases.

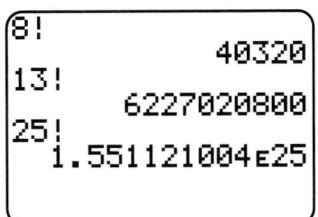

```
8!
              40320
13!
        6227020800
25!
     1.551121004E25
```

FIGURE 9.47   ■

One of the most famous unanswered questions in computing today is called the *traveling salesperson problem.* It is a relatively simple problem to state, but if someone could design an algorithm to solve this problem *efficiently,* he or she would not only become famous, but would also provide a valuable method for businesses to save millions of dollars on scheduling problems, such as bus routes and truck deliveries.

One instance of the traveling salesperson problem can be stated as follows. A salesperson must begin and end at home and travel to three different cities. Assuming that the salesperson can travel between any pair of cities, what route would minimize the salesperson's mileage? In Figure 9.48, the four cities are labeled A, B, C, and D. Let the salesperson live in city A. There are six different routes that could be tried. They are listed in Table 9.7 on the next page with the appropriate mileage for each.

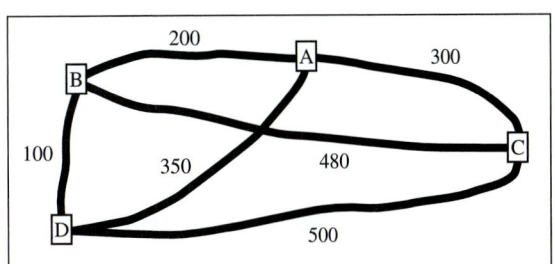

FIGURE 9.48 Traveling Salesperson Problem

TABLE 9.7

Route	Mileage
A B C D A	200 + 480 + 500 + 350 = 1530
A B D C A	200 + 100 + 500 + 300 = 1100
A C B D A	300 + 480 + 100 + 350 = 1230
A C D B A	300 + 500 + 100 + 200 = 1100
A D B C A	350 + 100 + 480 + 300 = 1230
A D C B A	350 + 500 + 480 + 200 = 1530

The shortest route of 1100 miles occurs when the salesperson either starts at A and travels through B, D, and C, and back to A, or reverses this route. Currently, this method of listing all possible routes to find the minimum distance is the only known way to consistently find the optimal solution for any general map containing n cities. In fact, people have not been able to determine whether or not a *significantly* faster method even exists. (Source: J. Smith, *Design and Analysis of Algorithms.*)

Counting the number of routes involves the fundamental theorem of counting. At the first step, the salesperson can travel to any one of three cities. Once this city is selected, there are then two possible cities to choose, and then one. Finally the salesperson returns home. The total number of routes is given by

$$3! = 3 \cdot 2 \cdot 1 = 6.$$

If the salesperson must travel to 30 cities, then there are

$$30! \approx 2.7 \times 10^{32}$$

routes to check, far too many to check even with the largest supercomputers.

Next suppose that a salesperson must visit three of eight possible cities. At first there are eight cities to choose. After the first city has been visited, there are seven cities to select. Since the salesperson only travels to three of the eight cities, there are

$$8 \cdot 7 \cdot 6 = 336$$

possible routes. This number of permutations is denoted by $P(8, 3)$. It represents the number of arrangements that can be made using three elements taken from a sample of eight.

Permutations of n elements taken r at a time

If $P(n, r)$ denotes the number of permutations of n elements taken r at a time, with $r \leq n$, then

$$P(n, r) = \frac{n!}{(n - r)!} = n(n - 1)(n - 2) \cdots (n - r + 1).$$

EXAMPLE 5   *Calculating permutations*

If there is a class of 30 students, in how many different orderings can four people give a speech?

Solution

The number of permutations of 30 elements taken 4 at a time is given by

$$P(30, 4) = 30 \cdot 29 \cdot 28 \cdot 27 = 657{,}720.$$

Thus, there are 657,720 orderings or arrangements of the four speeches. ∎

EXAMPLE 6   *Calculating P(n, r)*

Calculate each of the following by hand. Then, support your answers using a calculator.

(a) $P(7, 3)$
(b) $P(100, 2)$

Solution

(a) $P(7, 3) = 7 \cdot 6 \cdot 5 = 210$
(b) $P(100, 2) = 100 \cdot 99 = 9900$. One also can compute this number as follows.

$$P(100, 2) = \frac{100!}{(100 - 2)!} = \frac{100 \cdot 99 \cdot 98!}{98!} = 100 \cdot 99 = 9900$$

In this case, it is helpful to cancel 98! before performing the arithmetic. Both of these computations are performed by a calculator in Figure 9.49.

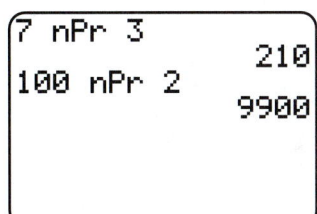

FIGURE 9.49   ∎

Critical Thinking

Count the number of arrangements of 52 cards in a standard deck. Is it likely that there are arrangements that no one has ever shuffled at any time in the history of the world? Explain.

Combinations

Unlike a permutation, a combination is not an ordering or arrangement, but rather a subset of a set of elements. Order is unimportant when finding combinations. For example, suppose we want to select a tennis team of two players from four people. The order in which the selection is made does not affect the final team of two players. From a set of four people, we select a subset of two players. This number of possible subsets or combinations is denoted by either $C(4, 2)$ or $\binom{4}{2}$.

To calculate $C(4, 2)$, we first consider $P(4, 2)$. Denote the four players by the letters A, B, C, and D. There are $P(4, 2) = 4 \cdot 3 = 12$ permutations given by

$$AB, \; BA, \; AC, \; CA, \; AD, \; DA, \; BC, \; CB, \; BD, \; DB, \; CD, \; DC.$$

However, the team comprised of person A and person B is equivalent to the team with person B and person A. The sets {AB} and {BA} are equal. The valid combinations are the following two-element subsets of {A, B, C, D}.

$$\{AB\} \quad \{AC\} \quad \{AD\} \quad \{BC\} \quad \{BD\} \quad \{CD\}$$

That is, $C(4, 2) = \dfrac{P(4, 2)}{2!} = 6$. The relationship between $P(n, r)$ and $C(n, r)$ is now given.

> ### Combinations of *n* elements taken *r* at a time
>
> If $C(n, r)$ denotes the number of combinations of n elements taken r at a time, with $r \leq n$, then
>
> $$C(n, r) = \frac{P(n, r)}{r!} = \frac{n!}{(n - r)!\, r!}.$$

EXAMPLE 7   *Counting combinations*

A college student has five courses left in his or her major and plans to take two courses this semester. Assuming that this student has the prerequisites for all five courses, determine how many different ways these two courses can be selected.

Solution

The order in which the courses are selected is unimportant. From a set of five courses, the student selects a subset of two courses. This number of subsets is given by $C(5, 2)$.

$$C(5, 2) = \frac{5!}{(5 - 2)!\, 2!} = \frac{5!}{3!\, 2!} = 10$$

There are ten different ways to select two courses from a set of five.   ■

EXAMPLE 8   *Calculating C(n, r)*

Calculate each of the following. Support your answers using a calculator.
(a) $C(7, 3)$

(b) $\dbinom{50}{47}$

Solution

(a) $C(7, 3) = \dfrac{7!}{(7 - 3)!\, 3!} = \dfrac{7!}{4!\, 3!} = \dfrac{7 \cdot 6 \cdot 5 \cdot 4!}{4!\, 3!} = \dfrac{7 \cdot 6 \cdot 5}{3!} = \dfrac{210}{6} = 35$

(b) The notation $\dbinom{50}{47}$ is equivalent to $C(50, 47)$.

$$\binom{50}{47} = \frac{50!}{(50 - 47)!\, 47!} = \frac{50!}{3!\, 47!} = \frac{50 \cdot 49 \cdot 48 \cdot 47!}{3!\, 47!}$$

$$= \frac{50 \cdot 49 \cdot 48}{3!} = \frac{117{,}600}{6} = 19{,}600$$

These computations are performed by a calculator in Figure 9.50 on the next page.

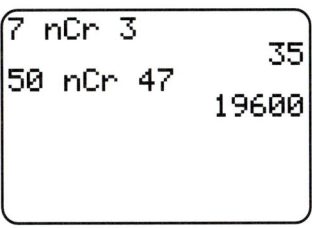

FIGURE 9.50

EXAMPLE 9   *Calculating the number of ways to play the lottery*

To win the jackpot in a lottery, a person must select five different numbers from 1 to 49 and then pick the powerball, which is numbered from 1 to 42. Count the different ways to play the game. (Source: Minnesota State Lottery.)

Solution

From 49 numbers a player picks 5 numbers. There are $C(49, 5)$ different ways of doing this. There are 42 different ways to choose the powerball. Using the fundamental counting principle, the number of ways to play the game equals $C(49, 5) \cdot 42 = 80,089,128$.

The Binomial Theorem

Combinations play a central role in the development of the binomial theorem. The binomial theorem can be used to expand expressions in the form $(a + b)^n$. Before stating the binomial theorem, we begin by first counting the number of "words" or strings of a given length that can be formed using only the letters a and b.

EXAMPLE 10   *Calculating distinguishable strings*

Count the number of distinguishable strings that can be formed with the given number of a's and b's. List these strings.
(a) Two a's, one b
(b) Two a's, three b's
(c) Four a's, no b's

Solution
(a) Using two a's and one b, we can form strings of length three. Once the b has been positioned, the string is determined. For example, if the b is placed in the middle position,

$$\square \boxed{b} \square$$

then the string must be *aba*. From a set of three slots, we choose one slot to place the b. This is computed by $C(3, 1) = 3$. The strings are *aab, aba,* and *baa*.

(b) With two a's and three b's we can form strings of length five. Once the locations of the three b's have been selected, the string is determined. For instance, if the b's are placed in the first, third, and fifth positions,

$$\boxed{b} \square \boxed{b} \square \boxed{b}$$

then the string becomes *babab*. From a set of five slots, we select three to place the b's. This is computed by $C(5, 3) = 10$. The ten strings are

bbbaa, bbaba, bbaab, babba, babab, baabb, abbba, abbab, ababb, and *aabbb*.

(c) There is only one string of length four containing no b's. This is *aaaa* and is computed by

$$C(4, 0) = \frac{4!}{(4-0)!\,0!} = \frac{24}{24(1)} = 1.$$ ∎

Next we expand $(a+b)^n$ for a few values of n, without simplifying.

$$(a+b)^1 = a + b$$

$$(a+b)^2 = (a+b)(a+b)$$

$$= aa + ab + ba + bb$$

$$(a+b)^3 = (a+b)(a+b)^2$$

$$= (a+b)(aa + ab + ba + bb)$$

$$= aaa + aab + aba + abb + baa + bab + bba + bbb$$

Notice that $(a+b)^1$ is the sum of all possible strings of length one that can be formed using a and b. The only possibilities are a and b. The expression $(a+b)^2$ is the sum of all possible strings of length two using a and b. The strings are *aa*, *ab*, *ba*, and *bb*. In a similar manner, $(a+b)^3$ is the sum of all possible strings of length three using a and b. This pattern continues for higher powers of $(a+b)$.

Strings with equal numbers of a's and equal numbers of b's can be combined into one term. For example, in $(a+b)^2$ the terms *ab* and *ba* can be combined as $2ab$. Notice that there are $C(2, 1) = 2$ distinguishable strings of length two containing one b. Similarly, in the expansion of $(a+b)^3$, the terms containing one a and two b's can be combined as

$$abb + bab + bba = 3ab^2.$$

There are $C(3, 2) = 3$ strings of length three that contain two b's.

We can use these concepts to expand $(a+b)^4$. The expression $(a+b)^4$ consists of the sum of all strings of length four using only the letters a and b. There is $C(4, 0) = 1$ string containing no b's, $C(4, 1) = 4$ strings containing one b, and so on, until there is $C(4, 4) = 1$ string containing four b's. Thus,

$$(a+b)^4 = \binom{4}{0}a^4b^0 + \binom{4}{1}a^3b^1 + \binom{4}{2}a^2b^2 + \binom{4}{3}a^1b^3 + \binom{4}{4}a^0b^4$$

$$= a^4 + 4a^3b + 6a^2b^2 + 4ab^3 + b^4.$$

These results are summarized by the binomial theorem.

Binomial theorem

For any positive integer n and numbers a and b,

$$(a+b)^n = \binom{n}{0}a^n + \binom{n}{1}a^{n-1}b^1 + \cdots + \binom{n}{n-1}a^1b^{n-1} + \binom{n}{n}b^n.$$

Expanding $(a+b)^n$ for various n results in the following triangle.

$$(a+b)^0 = \qquad\qquad\qquad 1$$

$$(a+b)^1 = \qquad\qquad\qquad 1a + 1b$$

$$(a+b)^2 = \qquad\qquad\quad 1a^2 + 2ab + 1b^2$$

$$(a+b)^3 = \qquad\qquad 1a^3 + 3a^2b + 3ab^2 + 1b^3$$

$$(a+b)^4 = \qquad\quad 1a^4 + 4a^3b + 6a^2b^2 + 4ab^3 + 1b^4$$

$$(a+b)^5 = \quad 1a^5 + 5a^4b + 10a^3b^2 + 10a^2b^3 + 5ab^4 + 1b^5$$

The triangle formed by the highlighted numbers is called **Pascal's triangle.** It can be used to efficiently compute the binomial coefficients, $C(n, r)$. The triangle consists of 1's along the sides. Each element inside the triangle is the sum of the two numbers above it. Pascal's triangle is usually written without variables as follows. It can be extended to include as many rows as needed.

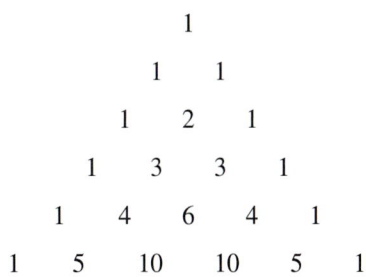

```
              1
            1   1
          1   2   1
        1   3   3   1
      1   4   6   4   1
    1   5   10  10  5   1
```

EXAMPLE 11

Expanding expressions with the binomial theorem

Expand each of the following.
(a) $(2x + 1)^5$
(b) $(3x - y)^3$

Solution

(a) To expand $(2x + 1)^5$, let $a = 2x$ and $b = 1$ in the binomial theorem. We can use Pascal's triangle to obtain the coefficients 1, 5, 10, 10, 5, and 1.

$$(2x + 1)^5 = 1(2x)^5 + 5(2x)^4(1)^1 + 10(2x)^3(1)^2 + 10(2x)^2(1)^3 + 5(2x)^1(1)^4 + 1(1)^5$$
$$= 32x^5 + 80x^4 + 80x^3 + 40x^2 + 10x + 1$$

(b) Let $a = 3x$ and $b = -y$ in the binomial theorem. Use the coefficients 1, 3, 3, and 1 from Pascal's triangle.

$$(3x - y)^3 = 1(3x)^3 + 3(3x)^2(-y)^1 + 3(3x)^1(-y)^2 + 1(-y)^3$$
$$= 27x^3 - 27x^2y + 9xy^2 - y^3$$ ∎

9.4 PUTTING IT ALL TOGETHER

The fundamental counting principle may be used to determine the number of ways a sequence of independent events can occur. Permutations are arrangements or orderings, whereas combinations are a subset of a set of events. The binomial theorem can be used to expand expressions written in the form $(a + b)^n$.

The following table summarizes some concepts related to counting in mathematics.

Notation	Meaning	Examples
n-factorial $n!$	$n!$ represents the product $n(n - 1) \cdots (3)(2)(1)$.	$6! = 6 \cdot 5 \cdot 4 \cdot 3 \cdot 2 \cdot 1 = 720$ $0! = 1$
$P(n, r) = \dfrac{n!}{(n - r)!}$	$P(n, r)$ represents the number of permutations of n elements taken r at a time.	The number of two-letter "words" that can be formed using the four letters A, B, C, and D exactly once is given by $P(4, 2) = \dfrac{4!}{(4 - 2)!} = \dfrac{4!}{2!} = 12$ or $P(4, 2) = 4 \cdot 3 = 12$.
$C(n, r) = \dfrac{n!}{(n - r)! \, r!}$	$C(n, r)$ represents the number of combinations of n elements taken r at a time.	The number of committees of three that can be formed from five people is given by $C(5, 3) = \dfrac{5!}{(5 - 3)! \, 3!} = \dfrac{5!}{2!3!} = 10$.

9.4 EXERCISES

    Tape 7

Counting

Exercises 1–4: Exam Questions   Count the number of ways that the questions on an exam could be answered.

1. Ten true-false questions

2. Ten multiple-choice questions with five choices each

3. Five true-false questions and ten multiple-choice questions with four choices each

4. One question involving matching ten items in one column with ten items in another column, using a one-to-one correspondence

Exercises 5–8: License Plates   Count the number of possible license plates with the given constraints.

5. Three digits followed by three letters

6. Two letters followed by four digits

7. Three letters followed by three digits or letters

8. Two letters followed by either three or four digits

Exercises 9 and 10: Counting Strings   Count the number of five-letter strings that can be formed with the given letters, assuming a letter can be used more than once.

9. A, B, C

10. W, X, Y, Z

Exercises 11–14: Counting Strings   Count the number of strings that can be formed with the given letters, assuming each letter is used exactly once.

11. A, B

12. A, B, C

13. W, X, Y, Z

14. V, W, X, Y, Z

15. *Combination Lock*   A briefcase has two separate locks. The combination to each lock consists of a three-digit number, where digits may be repeated. See the accompanying figure. How many combinations are possible in all? (*Hint:* The word *combination* is a misnomer. Mathematically, it is a permutation where the arrangement of the numbers is important.)

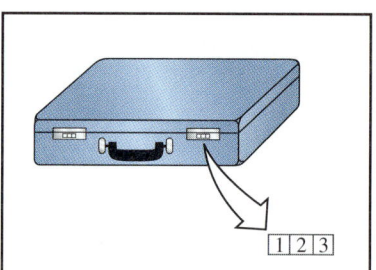

16. *Combination Lock*   A typical combination for a padlock consists of three numbers from 0 to 39. Count the number of combinations that are possible with this type of lock, if a number may be repeated.

17. *Garage Door Openers* The code for some garage door openers consists of 12 electrical switches that can be set to either 0 or 1 by the owner. With this type of opener, how many different codes are possible? (Source: Promax.)

18. *Lottery* To win the jackpot in a lottery game, a person must pick three numbers from 0 to 9 in the correct order. If a number can be repeated, how many different ways are there to play the game?

19. *Radio Stations* Call letters for a radio station usually begin with either a K or W, followed by three letters. In 1995 there were 11,834 radio stations on the air. Is there any shortage of call letters for new radio stations? (Source: M. Street Corporation.)

20. *Access Codes* ATM access codes often consist of a four-digit number. How many different codes are possible without giving two different accounts the same access code?

21. *Computer Sale* A computer sale offers a choice of two monitors, three printers, and four types of software. How many different packages can be purchased?

22. *Dice* A red die and a blue die are thrown. How many ways are there for both dice to show an even number?

Permutations

Exercises 23–28: Find the number.

23. $P(5, 3)$ **24.** $P(10, 2)$

25. $P(8, 1)$ **26.** $P(6, 6)$

27. $P(7, 3)$ **28.** $P(12, 3)$

29. In how many different ways can four people stand in a line?

30. How many arrangements are there of six different books on a shelf?

31. How many ways could five basketball players be introduced at a game?

32. How many ways can three students from a class of 15 be selected to give a speech?

33. *Traveling Salesperson* (Refer to the discussion before Example 5.) A salesperson must travel to three of seven cities. Direct travel is possible between every pair of cities. How many different orderings are there for the salesperson to visit these three cities? Assume that traveling a route in reverse order constitutes a different route.

34. *Keys* In how many distinguishable ways can four keys be put on a key ring?

Combinations

Exercises 35–40: Find the number.

35. $C(3, 1)$ **36.** $C(4, 3)$

37. $C(6, 3)$ **38.** $C(7, 5)$

39. $C(5, 0)$ **40.** $C(10, 2)$

41. *Lottery* To win the jackpot in a lottery, one must select five different numbers from 1 to 39. How many ways are there to play this game?

42. *Selecting a Committee* In how many ways can a committee of five be selected from eight people?

43. *Selecting a Committee* How many committees of four people can be selected from five women and three men, if a committee must have two people of each sex on it?

44. *Essay Questions* On a test involving six essay questions, students are asked to answer four questions. How many ways can the essay questions be selected?

Binomial Theorem

Exercises 45–48: (Refer to Example 10.) Calculate the number of distinguishable strings that can be formed with the given number of a's and b's.

45. Three a's, two b's

46. Five a's, three b's

47. Four a's, four b's

48. One a, five b's

Exercises 49–56: Use the binomial theorem to expand each expression.

49. $(x + y)^2$ **50.** $(x + y)^4$

51. $(m + 2)^3$ **52.** $(m + 2n)^5$

53. $(2x - 3)^3$ **54.** $(x + y^2)^3$

55. $(p - q)^6$ **56.** $(p^2 - 3)^4$

Writing about Mathematics

1. Explain the difference between a permutation and a combination. Give examples.

2. Explain what counting is, as presented in this section.

9.5 Probability

Definition of Probability • Compound Events • Independent and Dependent Events

Introduction

Questions of chance have no doubt engaged the minds of people since antiquity. However, the mathematical treatment of probability did not occur until the fifteenth century. The birth of probability theory as a mathematical discipline began in the seventeenth century with the work of Blaise Pascal and Pierre Fermat. Today probability pervades modern society. It is not only used to determine outcomes in gambling, but is also used to predict weather, genetic outcomes, and the risk involved with various types of substances and behaviors.

Risk is the chance or probability that a harmful event will actually occur. The following activities carry an annual increased risk of death by one chance in a million: flying 1000 miles in a jet, traveling 300 miles in a car, riding 10 miles on a bicycle, smoking 1.4 cigarettes, living two days in New York, having one chest x-ray, or living two months with a cigarette smoker. In the previous section we saw that the likelihood of winning the jackpot in a lottery was one chance in 80,089,128.

Ideally, we would like to live in a risk-free world. However, almost every action or substance exposes people to some risk. Recognizing relative risks is important for long life. Probability provides us with a measure of the likelihood that an event will occur. Knowledge about probability allows individuals to make informed decisions about their lives. (Sources: *Historical Topics for the Mathematics Classroom, Thirty-first Yearbook, NCTM;* J. Rodricks, *Calculated Risk.*)

Definition of Probability

In the study of probability, experiments often are performed. An experiment might involve tossing a coin or measuring the cholesterol level of a heart attack patient. A result from an experiment is called an **outcome.** The set of all possible outcomes is the **sample space.** Any subset of a sample space is an **event.**

For example, if an experiment involves rolling a die, then the sample space consists of $S = \{1, 2, 3, 4, 5, 6\}$. The event $E = \{1, 6\}$ contains the outcomes of 1 or 6 showing on the die. If $n(E)$ and $n(S)$ denote the number of outcomes in E and S, then $n(E) = 2$ and $n(S) = 6$. The probability of rolling a 1 or 6 is given by $P(E) = \frac{2}{6}$. That is, the likelihood of event E occurring is two chances in six. These concepts are summarized in the following.

Probability of an event

If the outcomes of a finite sample space S are equally likely, and if E is an event in S, then the **probability of E** is given by

$$P(E) = \frac{n(E)}{n(S)},$$

where $n(E)$ and $n(S)$ represent the number of outcomes in E and S, respectively.

Since $n(E) \leq n(S)$, the probability of an event E satisfies $0 \leq P(E) \leq 1$. If $P(E) = 1$, then event E is certain to occur. If $P(E) = 0$, then event E is impossible.

EXAMPLE 1   *Drawing a card*

One card is drawn from a standard deck of 52 cards. Find the probability that the card is an ace.

Solution

The sample space S consists of 52 outcomes that correspond to drawing any one of 52 cards. Each outcome is equally likely. Let E represent the event of drawing an ace. There are four aces in the deck, so event E contains four outcomes. Therefore, $n(S) = 52$ and $n(E) = 4$. The probability of drawing an ace is given by

$$P(E) = \frac{n(E)}{n(S)} = \frac{4}{52} = \frac{1}{13}.$$ ■

EXAMPLE 2   *Estimating probability of organ transplants*

In 1997 there were 51,277 people waiting for an organ transplant. Table 9.8 lists the number of patients waiting for the most common types of transplants. Assuming none of these people need two or more transplants, approximate the probability of a transplant patient needing
(a) a kidney or a heart;
(b) neither a kidney nor a heart.

TABLE 9.8	
Heart	3774
Kidney	35,025
Liver	7920
Lung	2340

Source: Coalition on Organ and Tissue Donation.

Solution

(a) Let each patient represent an outcome in a sample space S. The event E of a transplant patient needing either a kidney or a heart contains $35,025 + 3774 = 38,779$ outcomes. The desired probability is

$$P(E) = \frac{n(E)}{n(S)} = \frac{38,779}{51,277} \approx 0.76.$$

In 1997, about 76% of transplant patients needed either a kidney or a heart.
(b) Let F be the event of a patient waiting for an organ other than a kidney or a heart. Then,

$$n(F) = n(S) - n(E) = 51,277 - 38,779 = 12,498.$$

The probability of F is

$$P(F) = \frac{n(F)}{n(S)} = \frac{12{,}498}{51{,}277} \approx 0.24 \qquad \text{or} \qquad 24\%. \qquad \blacksquare$$

Notice that $P(E) + P(F) = 1$ in Example 2. The events E and F are **complements** because $E \cap F = \varnothing$ and $E \cup F = S$. That is, a transplant patient is either waiting for a kidney or a heart (event E), or not waiting for a kidney or a heart (event F). The complement of E may be denoted by E'.

Probability concepts can be illustrated using **Venn diagrams.** In Figure 9.51 the sample space S of an experiment is composed of event E and its complement E'. Notice that $E \cup E' = S$ and $E \cap E' = \varnothing$.

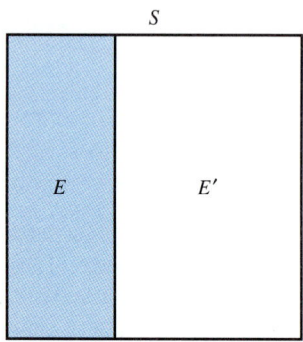

FIGURE 9.51

If $P(E)$ is known, then $P(E')$ can be calculated as follows.

$$P(E') = \frac{n(E')}{n(S)} = \frac{n(S) - n(E)}{n(S)} = 1 - \frac{n(E)}{n(S)} = 1 - P(E)$$

In part (b) of Example 2, the probability of $F = E'$ could have also been calculated using

$$P(F) = 1 - P(E) \approx 1 - 0.76 = 0.24.$$

Probability of a complement

Let E be an event and E' be its complement. If the probability of E is $P(E)$, then the probability of its complement is given by

$$P(E') = 1 - P(E).$$

EXAMPLE 3   *Finding probabilities of human eye color*

In 1865 Johann Mendel performed important research in genetics. His work led to a better understanding of dominant and recessive genetic traits. One example is human eye color, which is determined by a pair of genes called a *genotype*. Brown eye color B is dominant over blue eye color b. If a person has the genotype of BB, Bb, or bB, he or she will have brown eyes. A genotype of bb will result in blue eyes. A person receives one gene (B or b) from each parent. Table 9.9 on the next page shows how these two genes can be paired together. (Source: H. Lancaster, *Quantitative Methods in Biology and Medical Sciences.*)

TABLE 9.9

	B	b
B	BB	Bb
b	bB	bb

(a) Assuming that each genotype is equally likely, find the probability of blue eyes.

(b) What is the probability that a person has brown eyes?

Solution

(a) The sample space S consists of four equally likely outcomes denoted by BB, Bb, bB, and bb. The event E of blue eye color (bb) occurs once. Therefore,

$$P(E) = \frac{n(E)}{n(S)} = \frac{1}{4} = 0.25.$$

(b) In this chart, brown eyes are the complement of blue eyes. The probability of brown eyes is

$$P(E') = 1 - P(E) = 1 - 0.25 = 0.75.$$

This also could be computed as

$$P(E') = \frac{n(E')}{n(S)} = \frac{3}{4} = 0.75,$$

since there are three genotypes that result in brown eye color. ■

Compound Events

Frequently the probability of more than one event is needed. For example, suppose a college has a total of 2500 students with 225 students enrolled in college algebra, 75 in trigonometry, and 30 in both. Let E_1 denote the event that a student is enrolled in college algebra, and E_2 the event that a student is enrolled in trigonometry. Then, the Venn diagram in Figure 9.52 visually describes this situation.

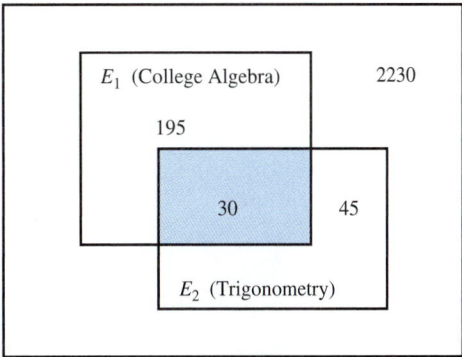

FIGURE 9.52

In this Venn diagram, it is important not to count the 30 students taking both courses twice. The set E_1 has a total of $195 + 30 = 225$ students, while set E_2 contains $45 + 30 = 75$ students.

EXAMPLE 4 *Calculating the probability of a union*

In the preceding discussion, suppose a student is selected at random. What is the probability that this student is enrolled in college algebra, trigonometry, or both?

Solution

We would like to find the probability $P(E_1 \text{ or } E_2)$, which is denoted by $P(E_1 \cup E_2)$. Since $n(E_1 \cup E_2) = 195 + 30 + 45 = 270$ and $n(S) = 2500$,

$$P(E_1 \cup E_2) = \frac{n(E_1 \cup E_2)}{n(S)} = \frac{270}{2500} = 0.108.$$

There is a 10.8% chance that a student will be taking algebra, trigonometry, or both. ■

In the previous example, it would have been incorrect to simply add the probability of a student taking algebra and the probability of a student taking trigonometry. Their sum would be

$$P(E_1) + P(E_2) = \frac{n(E_1)}{n(S)} + \frac{n(E_2)}{n(S)}$$

$$= \frac{225}{2500} + \frac{75}{2500}$$

$$= \frac{300}{2500}$$

$$= 0.12 \quad \text{or} \quad 12\%.$$

This sum is greater than $P(E_1 \cup E_2)$ because this calculation counts the 30 students taking both courses twice. In order to find the correct probability for $P(E_1 \cup E_2)$, we must subtract out the probability of the intersection $E_1 \cap E_2$.

$$P(E_1 \cup E_2) = P(E_1) + P(E_2) - P(E_1 \cap E_2)$$

$$= \frac{n(E_1)}{n(S)} + \frac{n(E_2)}{n(S)} - \frac{n(E_1 \cap E_2)}{n(S)}$$

$$= \frac{225}{2500} + \frac{75}{2500} - \frac{30}{2500}$$

$$= \frac{270}{2500}$$

$$= 0.108 \quad \text{or} \quad 10.8\%$$

This is the same result obtained in Example 4 and suggests the following property.

Probability of the union of two events

For any two events E_1 and E_2,

$$P(E_1 \cup E_2) = P(E_1) + P(E_2) - P(E_1 \cap E_2).$$

EXAMPLE 5 *Rolling dice*

Suppose two dice are rolled. Find the probability that the dice show a sum of eight or a pair.

Solution

In Table 9.10 the roll of the dice is represented by an ordered pair. For example, the ordered pair (3, 6) represents the first die showing 3 and the second die 6.

TABLE 9.10

(1, 1)	(1, 2)	(1, 3)	(1, 4)	(1, 5)	(1, 6)
(2, 1)	**(2, 2)**	(2, 3)	(2, 4)	(2, 5)	**(2, 6)**
(3, 1)	(3, 2)	**(3, 3)**	(3, 4)	**(3, 5)**	(3, 6)
(4, 1)	(4, 2)	(4, 3)	**(4, 4)**	(4, 5)	(4, 6)
(5, 1)	(5, 2)	**(5, 3)**	(5, 4)	**(5, 5)**	(5, 6)
(6, 1)	**(6, 2)**	(6, 3)	(6, 4)	(6, 5)	**(6, 6)**

Since each die can show six different outcomes, there is a total of $6 \cdot 6 = 36$ outcomes in the sample space S. Let E_1 denote the event of rolling a sum of eight, and E_2 the event of rolling a pair. Then,

$$E_1 = \{(6, 2), (5, 3), (4, 4), (3, 5), (2, 6)\} \quad \text{and}$$
$$E_2 = \{(1, 1), (2, 2), (3, 3), (4, 4), (5, 5), (6, 6)\}.$$

The intersection of E_1 and E_2 is

$$E_1 \cap E_2 = \{(4, 4)\}.$$

Since $n(S) = 36$, $n(E_1) = 5$, $n(E_2) = 6$, and $n(E_1 \cap E_2) = 1$, the following can be computed.

$$P(E_1 \cup E_2) = P(E_1) + P(E_2) - P(E_1 \cap E_2)$$
$$= \frac{n(E_1)}{n(S)} + \frac{n(E_2)}{n(S)} - \frac{n(E_1 \cap E_2)}{n(S)}$$
$$= \frac{5}{36} + \frac{6}{36} - \frac{1}{36}$$
$$= \frac{10}{36} \quad \text{or} \quad \frac{5}{18}$$

This result can be verified by counting the number of highlighted outcomes in Table 9.10. There are 10 outcomes that satisfy the conditions out of a total of 36, so the probability is $\frac{10}{36}$ or $\frac{5}{18}$. ∎

If $E_1 \cap E_2 = \varnothing$, then the events E_1 and E_2 are **mutually exclusive.** Mutually exclusive events have no outcomes in common so $P(E_1 \cap E_2) = 0$. Thus, $P(E_1 \cup E_2) = P(E_1) + P(E_2)$.

EXAMPLE 6 *Drawing cards*

Find the probability of drawing either an ace or a king from a standard deck of 52 playing cards.

Solution

The event E_1 of drawing an ace and the event E_2 of drawing a king are mutually exclusive. No card can be both an ace and a king. Therefore, $P(E_1 \cap E_2) = 0$. The probability of drawing an ace is $P(E_1) = \dfrac{4}{52}$, since there are four aces in 52 cards. Similarly, the probability of drawing a king is $P(E_2) = \dfrac{4}{52}$.

$$P(E_1 \cup E_2) = P(E_1) + P(E_2)$$
$$= \frac{4}{52} + \frac{4}{52}$$
$$= \frac{8}{52} \quad \text{or} \quad \frac{2}{13}$$

Thus, the probability of drawing either an ace or a king is $\dfrac{2}{13}$. ∎

Independent and Dependent Events

Two events are **independent** if they do not influence one another. Otherwise they are **dependent.** An example of independent events would be if the same coin is tossed twice. The result of the first toss does not affect the second.

Probability of independent events

If E_1 and E_2 are independent events, then
$$P(E_1 \cap E_2) = P(E_1) \cdot P(E_2).$$

EXAMPLE 7 *Tossing a coin*

Suppose a coin is tossed twice. Determine the probability that the result is two heads.

Solution

Let E_1 be the event of a head on the first toss, and E_2 the event of a head on the second toss. Then, $P(E_1) = P(E_2) = \dfrac{1}{2}$. The two events are independent. The probability of two heads occurring is

$$P(E_1 \cap E_2) = \frac{1}{2} \cdot \frac{1}{2} = \frac{1}{4}.$$

This probability of $\dfrac{1}{4}$ also can be found using a tree diagram as shown in Figure 9.53 on the next page. There are four equally likely outcomes. Tosses resulting in two heads occur once.

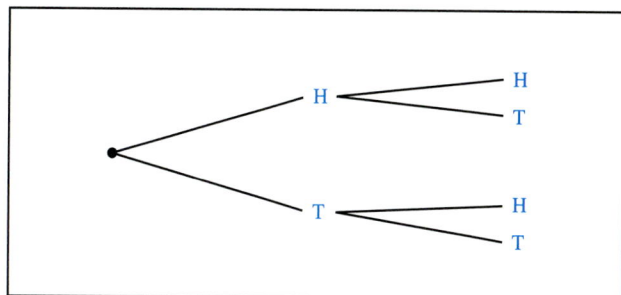

FIGURE 9.53

EXAMPLE 8   *Rolling dice*

What is the probability of rolling a sum of 12 with two dice?

Solution

The roll of one die does not influence the roll of the other. They are independent events. To obtain a sum of 12, both dice must show a six. Let E_1 be the event of rolling a six with the first die, and E_2 the event of rolling a six with the second die. Then, $P(E_1) = P(E_2) = \dfrac{1}{6}$. The probability of rolling a 12 is

$$P(E_1 \cap E_2) = P(E_1) \cdot P(E_2) = \frac{1}{6} \cdot \frac{1}{6} = \frac{1}{36}.$$

This can be verified by using Table 9.10. There is only one outcome out of 36 that results in a sum of 12.

Critical Thinking

When rolling a pair of dice, what sum is most likely to appear?

If events E_1 and E_2 influence each other, they are *dependent*. The probability of dependent events is given as follows.

Probability of dependent events

If E_1 and E_2 are dependent events, then

$$P(E_1 \cap E_2) = P(E_1) \cdot P(E_2, \text{ given that } E_1 \text{ occurred}).$$

EXAMPLE 9   *Drawing cards*

Find the probability of drawing two hearts from a standard deck of 52 cards, when the first card is
(a) replaced before drawing the second card;
(b) not replaced.

Solution

(a) Let E_1 denote the event of the first card being a heart, and E_2 the event of the second card being a heart. If the first card is replaced before the second card is drawn, the two events are independent. Since there are 13 hearts in a standard deck of 52 cards, the probability of two hearts being drawn is

$$P(E_1 \cap E_2) = P(E_1) \cdot P(E_2) = \frac{13}{52} \cdot \frac{13}{52} = \frac{1}{16}.$$

(b) If the first card is not replaced, then the outcome for the second card is influenced by the first card. Therefore, the events E_1 and E_2 are dependent. The probability of drawing a second heart, given that the first card is a heart, is represented by $P(E_2$, given E_1 has occurred$) = \frac{12}{51}$. That is, if the first card drawn were a heart and removed from the deck, then there would be 12 hearts in a sample space of 51 cards.

$$P(E_1 \cap E_2) = P(E_1) \cdot P(E_2, \text{ given that } E_1 \text{ occurred}) = \frac{13}{52} \cdot \frac{12}{51} = \frac{1}{17}$$

Thus, the probability of drawing two hearts is slightly less if the first card is not replaced. ∎

EXAMPLE 10 Analyzing a polygraph test

Suppose there is a 6% chance that a polygraph test will incorrectly say a person is lying when he or she is actually telling the truth. If a person tells the truth 95% of the time, what percentage of the story will the polygraph test incorrectly indicate is a lie?

Solution

Let E_1 be the event that the person is telling the truth and E_2 be the event that the polygraph test is incorrect. Then,

$$P(E_1 \cap E_2) = P(E_1) \cdot P(E_2, \text{ given the person is telling the truth})$$
$$= (0.95)(0.06)$$
$$= 0.057 \text{ or } 5.7\%. \qquad \blacksquare$$

9.5 PUTTING IT ALL TOGETHER

Probability is a real number between 0 and 1 that measures the likelihood or chance of an event occurring. A probability of 0 indicates an event is impossible, and a probability of 1 indicates that an event is certain to occur. Compound events involve more than one event. If E_1 and E_2 are two events, then the probability of both events occurring is denoted by the expression $P(E_1 \cap E_2)$. Independent events do not influence each other, whereas dependent events do. The probability of either E_1 or E_2 occurring is denoted by $P(E_1 \cup E_2)$. If E_1 and E_2 have no outcomes in common, then they are mutually exclusive events.

The following table summarizes concepts related to probability if the sample space is finite.

Probability

$P(E) = \frac{n(E)}{n(S)}$, where $n(E)$ is the number of outcomes in event E and

$n(S)$ is the number of outcomes in the sample space S.

Compound Events

Probability of either E_1 or E_2 (or both) occurring:

$$P(E_1 \cup E_2) = P(E_1) + P(E_2) - P(E_1 \cap E_2).$$

If E_1 and E_2 are *mutually exclusive,* then $E_1 \cap E_2 = \varnothing$, and

$$P(E_1 \cup E_2) = P(E_1) + P(E_2).$$

Probability of *both* E_1 and E_2 occurring:
  If E_1 and E_2 are *independent,* then

$$P(E_1 \cap E_2) = P(E_1) \cdot P(E_2).$$

If E_1 and E_2 are *dependent,* then

$$P(E_1 \cap E_2) = P(E_1) \cdot P(E_2, \text{given that } E_1 \text{ has occurred}).$$

9.5 EXERCISES      Tape 7

Probability of an Event

Exercises 1–8: Determine if the number could represent a probability.

1. $\dfrac{11}{13}$

2. 0.995

3. 2.5

4. 1

5. 0

6. 110%

7. −0.375

8. $\dfrac{9}{8}$

Exercises 9–18: Find the probability of each event.

9. Tossing a head with a fair coin

10. Tossing a tail with a fair coin

11. Rolling a 2 with a fair die

12. Rolling a 5 or 6 with a fair die

13. Guessing the correct answer for a true-false question

14. Guessing the correct answer for a multiple-choice question with five choices

15. Drawing an ace from a standard deck of 52 cards

16. Drawing a club from a standard deck of 52 cards

17. Randomly guessing a four-digit ATM access code

18. Randomly picking the winning team at a basketball game

19. The table shows some of the favorite pizza toppings. (Source: *USA Today.*)

Pepperoni	43%
Sausage	19%
Mushrooms	14%
Vegetables	13%

  (a) If a person is selected at random, what is the probability that pepperoni is not his or her favorite topping?

  (b) Find the probability that a person's favorite topping is either mushrooms or sausage.

20. (Refer to Example 2.) Find the probability that a transplant patient in 1997 was waiting for the following.

  (a) A lung

  (b) A lung or a liver

Probability of Compound Events

Exercises 21–30: Find the probability of each compound event.

21. Tossing a coin twice with the outcomes of two tails

22. Tossing a coin three times with the outcomes of three heads

23. Rolling a die three times and obtaining a 5 or 6 on each roll

24. Rolling a sum of 2 with two dice

25. Rolling a sum of 7 with two dice

26. Rolling a sum other than 7 with two dice

27. Rolling a die four times without obtaining a 6

28. Rolling a die four times and obtaining at least one 6

29. Drawing four consecutive aces from a standard deck of 52 cards without replacement

30. Drawing a pair (two cards with the same value) from a standard deck of 52 cards without replacement

31. *Quality Control*   A quality control experiment involves selecting one string of decorative lights from a box of 20. If the string is defective, the entire box of 20 is rejected. Suppose the box contains four defective strings of lights. What is the probability of rejecting the box?

32. *Quality Control*   (Refer to the previous exercise.) Suppose three strings of lights are tested. If any of the strings are defective, the entire box of 20 is rejected. What is the probability of rejecting the box if there are four defective strings of lights in a box? (*Hint:* Start by finding the probability that the box is not rejected.)

33. *Entrance Exams*   A group of students are preparing for college entrance exams. It is estimated that 50% need help with mathematics, 45% with English, and 25% with both.
    (a) Draw a Venn diagram representing this data.
    (b) Use this diagram to find the probability that a student needs help with mathematics, English, or both.
    (c) Solve part (b) symbolically by applying a probability formula.

34. *College Classes*   In a college of 5500 students, 950 students are enrolled in English classes, 1220 in business classes, and 350 in both. If a student is picked at random, find the probability that he or she is enrolled in an English class, a business class, or both.

35. *New Books Published*   In 1994 there was a total of 51,863 new books and editions published. The table lists the number of books published in specific areas. If a new book or edition is selected at random, find the probability that its subject area satisfies the following.

Art	1621
Business	1616
History	2507
Music	364
Religion	2730
Science	3021

(a) Art or music
(b) Neither science nor religion

36. *Death Rates*   In 1994, the U.S. death rate per 100,000 people was 876.9. What is the probability that a person selected at random in 1994 died? (Source: Department of Health and Human Services.)

37. *Death Rates*   In 1994, the death rate per 100,000 people between the ages of 15 and 24 was 99.6. What is the probability that a person selected at random from this age group died during 1994? (Source: Department of Health and Human Services.)

38. *AIDS*   By 1994, a total of 473,435 cases of AIDS had been diagnosed. The table lists AIDS cases diagnosed in certain cities. Estimate the probability that a person diagnosed with AIDS satisfied the following conditions. (Source: Department of Health and Human Services.)

New York	81,604
Los Angeles	31,085
San Francisco	22,835
Miami	16,372

(a) Resided in New York
(b) Did not reside in New York
(c) Resided in Los Angeles or Miami

39. *Tossing a Coin*   Find the probability of tossing a coin n times and obtaining n heads. What happens to this probability as n increases? Does this agree with your intuition? Explain.

40. *Rolling Dice*   Find the probability of rolling a die five times and obtaining a 6 on the first two rolls, a 5 on the third roll, and a 1, 2, 3, or 4 on the last two rolls.

41. *Unfair Coin*   Suppose a coin is not fair, but instead the probability of obtaining a head (H) is $\frac{3}{4}$ and a tail (T) is $\frac{1}{4}$. What is the probability of the following events?
    (a) HT
    (b) HH
    (c) HHT
    (d) THT

42. *Unfair Die* Suppose a die is not fair, but instead the probability P of each number n is as listed in the table. Find the probability of each event occurring.

n	1	2	3	4	5	6
P	0.1	0.1	0.1	0.2	0.2	0.3

(a) Rolling a number that is 4 or higher
(b) Rolling a 6 twice on consecutive rolls

43. *Dice* Suppose there are two dice, one red and one blue, having the probabilities shown in the table from the previous exercise. If both dice are rolled, find the probability of the given sum appearing.
(a) 12   **(b)** 11

44. *Garage Door Code* The code for some garage door openers consists of 12 electrical switches that can be set to either 0 or 1 by the owner. Each setting represents a different code. What is the probability of guessing someone's code at random? (Source: Promax.)

45. *Lottery* To win a lottery, a person picks three numbers from 0 to 9 in the correct order. If a number may be repeated, what is the probability of winning this game with one play?

46. *Lottery* To win the jackpot in a lottery, a person picks five numbers from 1 to 49, and then picks the powerball number from 1 to 42. If the numbers are picked at random, what is the probability of winning this game with one play?

47. *Marbles* A jar contains 22 red marbles, 18 blue marbles, and 10 green marbles. If a marble is drawn from the jar at random, find the probability that the ball is the following.
(a) Red
(b) Not red
(c) Blue or green

48. *Marbles* A jar contains 55 red marbles and 45 blue marbles. If two marbles are drawn from the jar at random without replacement, find the probability that the marbles satisfy the following.
(a) Both are blue.
(b) Neither is blue.

(c) The first marble is red and the second marble is blue.

49. *Cancer and Saccharin* Saccharin is the *least* potent carcinogen ever detected in an animal study. The dose-risk curve is shown in the figure for saccharin-induced bladder tumors in rats. Doses on the x-axis represent the percentage of the animals' diets consisting of saccharin. The associated lifetime risk R or probability of the animal developing bladder cancer is shown on the y-axis. (Source: J. Rodricks.)

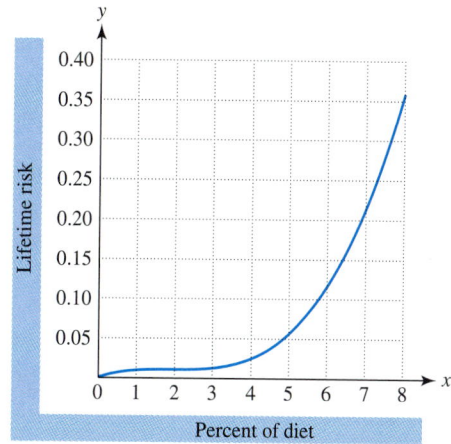

(a) If the diet of a rat consists of 6% saccharin, estimate the risk of developing bladder cancer.
(b) Discuss the information shown in this graph.

50. *Horse Racing* The favorite to win the Kentucky Derby has won the race 48 out of 122 times. If a past favorite is selected at random, what is the probability that the horse won the Kentucky Derby? (Source: Churchill Downs.)

Writing about Mathematics

1. What values are possible for a probability? Interpret different probabilities and give examples.

2. Discuss the difference between dependent and independent events. How are their probabilities calculated?

CHECKING BASIC CONCEPTS FOR SECTIONS 9.4 AND 9.5

1. Count the ways to answer a quiz consisting of eight true-false questions.

2. Count the number of five-card poker hands that are possible using a standard deck of 52 cards.

3. In 1995 there were 2.6 million high school graduates, of which 1.2 million were male. If a 1995 high school graduate is selected at random, estimate the probability that this graduate is female. (Source: The American College Testing Program.)

4. Find the probability of tossing a coin four times and obtaining a head every time.

CHAPTER 9 Summary

An infinite *sequence* is a function whose domain is the natural numbers. The output of a sequence is an ordered list, denoted by the terms $a_1, a_2, a_3, a_4, \ldots$. Two basic types of sequences are arithmetic and geometric. An *arithmetic sequence* is defined by $f(n) = dn + c$. Successive terms are found by adding a *common difference* d to the previous term. A *geometric sequence* is defined by $g(n) = a_1 r^{n-1}$, where a_1 is the first term and r is a *common ratio*. Successive terms are found by multiplying r times the previous term.

A *series* is a summation of the terms of a sequence. The sum of an infinite series may not exist. Two basic types of series are arithmetic and geometric. *Arithmetic series* are the summation of the terms of an arithmetic sequence. *Geometric series* are the summation of the terms of a geometric sequence.

Three types of conic sections are parabolas, ellipses, and hyperbolas. A line called a *directrix* and a point called a *focus* can be used to define a *parabola*. If a parabola has vertex $(0, 0)$ then its equation is either $x^2 = 4py$ or $y^2 = 4px$, where p is the distance between the vertex and the focus. Ellipses and hyperbolas have two foci. *Ellipses* centered at the origin are given by either $\dfrac{x^2}{a^2} + \dfrac{y^2}{b^2} = 1$ or $\dfrac{y^2}{a^2} + \dfrac{x^2}{b^2} = 1$, where $a > b > 0$. Their foci are $(\pm c, 0)$ or $(0, \pm c)$, respectively, with $c^2 = a^2 - b^2$. *Hyperbolas* centered at the origin are defined by either $\dfrac{x^2}{a^2} - \dfrac{y^2}{b^2} = 1$ or by $\dfrac{y^2}{a^2} - \dfrac{x^2}{b^2} = 1$. Their foci are $(\pm c, 0)$ or $(0, \pm c)$, respectively, where the constants a, b, and c satisfy $a^2 + b^2 = c^2$. Hyperbolas have asymptotes, which are lines that approximate the hyperbola large distances from its center.

Counting in mathematics often refers to determining the number of outcomes that can occur for a particular event, and frequently makes use of the *fundamental counting principle*. A *permutation* is an ordering or arrangement. On the other hand, a *combination* is not an arrangement, but rather a subset of a set of elements.

Probability is a real number from 0 to 1. A probability of 0 indicates an event is impossible, whereas a probability of 1 means that an event is certain to occur. If an *event* E can occur $n(E)$ ways and the entire *sample space* S contains $n(S)$ equally likely outcomes, then the probability of event E in sample space S is given by $\dfrac{n(E)}{n(S)}$.

Review Exercises

Exercises 1–4: Find the first four terms of the sequence.

1. $a_n = -3n + 2$

2. $a_n = n^2 + n$

3. $a_n = 2a_{n-1} + 1; a_1 = 0$

4. $a_n = a_{n-1} + 2a_{n-2}; a_1 = 1, a_2 = 4$

Exercises 5 and 6: Use the graphical representation to identify the terms of the finite sequence.

5.
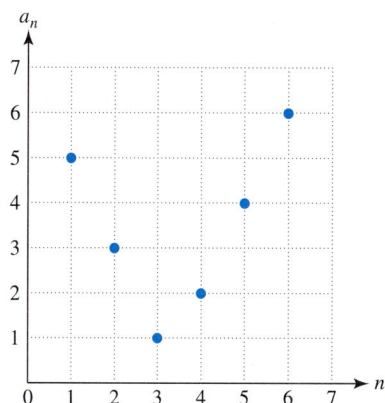

6.
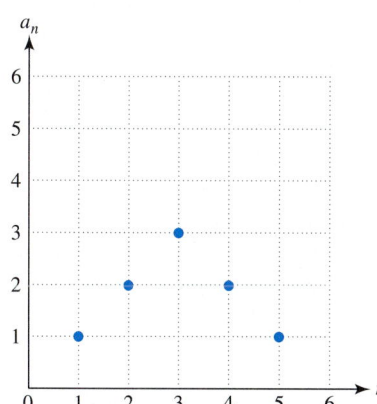

Exercises 7–10: The first five terms of an infinite arithmetic or geometric sequence are given. Find

  (a) *numerical,*

  (b) *graphical, and*

  (c) *symbolic*

representations of the sequence. Include at least the first eight terms of the sequence.

7. 2, 4, 6, 8, 10

8. 3, 1, −1, −3, −5

9. 81, 27, 9, 3, 1

10. 1.5, −3, 6, −12, 24

11. Find a general term a_n for the arithmetic sequence with $a_3 = -3$ and $d = 4$.

12. Find a general term a_n for the geometric sequence with $a_1 = 2.5$ and $a_6 = -80$.

13. *Height of a Ball* When a ping-pong ball is dropped it rebounds to 90% of its initial height.

  (a) Write the first five terms of a sequence that gives the maximum height attained by the ball on each rebound when it is dropped from an initial height of four feet. Let $a_1 = 4$. What type of sequence describes these maximums?

  (b) Give a graphical representation of this sequence for the first five terms.

  (c) Find a formula for a_n.

14. *Falling Object* If air resistance is ignored, an object falls 16, 48, 80, and 112 feet during each successive one-second interval.

  (a) What type of sequence describes these distances?

  (b) Determine how far an object falls during the sixth second.

  (c) Find a formula for the nth term of this sequence.

Exercises 15–18: Find the sum of the series using a formula.

15. $-2 + 1 + 4 + 7 + 10 + 13 + 16 + 19 + 22$

16. $2 + 4 + 6 + 8 + \cdots + 98 + 100$

17. $1 + 3 + 9 + 27 + 81 + 243 + 729 + 2187$

18. $64 + 16 + 4 + 1 + \dfrac{1}{4} + \dfrac{1}{16}$

Exercises 19–22: Find the sum of the infinite geometric series.

19. $2 + 1 + \dfrac{1}{2} + \dfrac{1}{4} + \dfrac{1}{8} + \dfrac{1}{16} + \cdots$

20. $4 - \dfrac{4}{3} + \dfrac{4}{9} - \dfrac{4}{27} + \dfrac{4}{81} - \dfrac{4}{243} + \cdots$

21. $0.2 + 0.02 + 0.002 + 0.0002 + 0.00002 + \cdots$

22. $0.25 + 0.0025 + 0.000025 + 0.00000025 + \cdots$

Exercises 23 and 24: Write out the terms of the series.

23. $\displaystyle\sum_{k=1}^{5} (5k + 1)$

24. $\displaystyle\sum_{k=1}^{4} (2 - k^2)$

Exercises 25 and 26: Write each series using summation notation.

25. $1^3 + 2^3 + 3^3 + 4^3 + 5^3 + 6^3$

26. $1 + \dfrac{1}{10} + \dfrac{1}{100} + \dfrac{1}{1000} + \dfrac{1}{10,000}$

Exercises 27–30: Determine an equation of the conic section that satisfies the given conditions. Sketch its graph.

27. Parabola with focus $(2, 0)$ and vertex $(0, 0)$

28. Ellipse with foci $(\pm 4, 0)$ and vertices $(\pm 5, 0)$

29. Ellipse centered at the origin with vertical major axis of length 14 and minor axis of length 8

30. Hyperbola with foci $(0, \pm 10)$ and endpoints of the conjugate axis $(\pm 6, 0)$

Exercises 31–36: Sketch a graph of the conic section. Give the coordinates of any foci.

31. $-4y = x^2$

32. $y^2 = 8x$

33. $\dfrac{x^2}{25} + \dfrac{y^2}{4} = 1$

34. $49x^2 + 36y^2 = 1764$

35. $\dfrac{x^2}{16} - \dfrac{y^2}{9} = 1$

36. $\dfrac{y^2}{4} - x^2 = 1$

Exercises 37 and 38: Sketch a graph of the conic section. Identify the coordinates of its center.

37. $\dfrac{(y - 2)^2}{4} + \dfrac{(x + 1)^2}{16} = 1$

38. $\dfrac{(x - 2)^2}{16} - \dfrac{y^2}{4} = 1$

39. *Exam Questions*   Count the ways that an exam, consisting of 20 multiple-choice questions with four choices each, could be answered.

40. *License Plates*   Count the different license plates having four numeric digits followed by two letters.

41. *Combination Lock*   A combination lock consists of four numbers from 0 to 49. If a number may be repeated, find the number of possible combinations.

42. *Dice*   A red die and a blue die are rolled. How many different ways are there for the sum to be 4?

Exercises 43 and 44: Find the number.

43. $P(6, 3)$ **44.** $C(7, 4)$

45. *Standing in Line*   In how many arrangements can five people stand in a line?

46. *Giving a Speech*   How many orderings are possible for four students to give a speech out of a class of 15?

47. *Committees*   In how many different ways can a committee of three be selected from six people?

48. *Committees*   Find the number of different committees with three women and three men that can be selected from a group of seven women and five men.

49. *Test Questions*   On a test involving ten essay questions, students are asked to answer six questions. How many different ways can the essay questions be selected?

50. *Binomial Theorem*   Use the binomial theorem to expand the expression $(2x - y)^4$.

Exercises 51 and 52: Find the probability of each event.

51. A 1, 2, or 3 appears when rolling a die

52. Tossing three heads in a row using a fair coin

53. *Quality Control*   A quality control experiment involves selecting two batteries from a pack of 16. If either battery is defective, the entire pack of 16 is rejected. Suppose a pack contains two defective batteries. What is the probability of not rejecting this pack?

54. *Venn Diagram*   From a group of 82 students, 19 are enrolled in music, 22 in art, and 10 in both.
   (a) Draw a Venn diagram representing this data.
   (b) Use this diagram to determine the probability that a student selected at random is enrolled in music, art, or both.
   (c) Solve part (b) symbolically by applying a probability formula.

55. *Marbles*   A jar contains 13 red, 27 blue, and 20 green marbles. If a marble is drawn from the jar at random, find the probability that the marble is the following.
   (a) Blue
   (b) Not blue
   (c) Red

56. *Cards*   Find the probability of drawing two diamonds from a standard deck of 52 cards without replacing the first card.

57. *Insect Populations*   The monthly density of an insect population, measured in thousands per acre, is described by the recursive sequence

$$a_1 = 100$$

$$a_n = \frac{2a_{n-1}}{1 + (a_{n-1}/4000)}, \qquad n > 1.$$

Use your calculator to graph the sequence in $[0, 16, 1]$ by $[0, 5000, 1000]$. Include the first 15 terms and discuss any trends illustrated by the graph.

Extended and Discovery Exercises

*Exercises 1 and 2: Antibiotic Resistance   Due to the frequent use of antibiotics in society, many strains of bacteria are becoming resistant. Some types of **haploid** bacteria contain genetic material called **plasmids**. Plasmids are capable of making a strain of bacteria resistant to antibiotic drugs. Genetic engineers want to predict the resistance of bacteria after many generations.*

1. Suppose a strain of bacteria contains two plasmids R_1 and R_2. Plasmid R_1 is resistant to the antibiotic ampicillin, whereas plasmid R_2 is resistant to the antibiotic tetracycline. When bacteria reproduce through cell division, the type of plasmids passed on to each new cell is random. For example, a daughter cell could have two plasmids of type R_1 and no plasmid of type R_2, one of each type, or no plasmid of type R_1 and two plasmids of type R_2. The probability $P_{k,j}$ that a mother cell with k plasmids of type R_1 produces a daughter cell with j plasmids of type R_1 can be calculated by the formula

$$P_{k,j} = \frac{\binom{2k}{j}\binom{4-2k}{2-j}}{\binom{4}{2}}.$$

   (Source: F. Hoppensteadt and C. Peskin, *Mathematics in Medicine and the Life Sciences.*)

   (a) Compute $P_{k,j}$ for $0 \le k, j \le 2$. Assume that $\binom{0}{0} = 1$ and $\binom{k}{j} = 0$ whenever $k < j$.

   Record your results in the matrix

$$P = \begin{bmatrix} P_{00} & P_{01} & P_{02} \\ P_{10} & P_{11} & P_{12} \\ P_{20} & P_{21} & P_{22} \end{bmatrix}.$$

   (b) Which elements in P are the greatest? Interpret the result.

2. (Continuation of the previous exercise.) The genetic makeup of future generations of the haploid bacteria can be modeled using matrices. Let $A = [a_1, a_2, a_3]$ be a 1×3 matrix containing three probabilities. The value of a_1 is the probability that a cell has two R_1 plasmids and no R_2 plasmid; a_2 is the probability that it has one R_1 plasmid and one R_2 plasmid; a_3 is the probability that a cell has no R_1 plasmid and two R_2 plasmids. If an entire generation of bacteria has one plasmid of each type then $A_1 = [0, 1, 0]$. In this case the bacteria is resistant to both antibiotics. The probabilities A_n for plasmids R_1 and R_2 in nth generation of bacteria can be calculated with the matrix recurrence equation $A_n = A_{n-1}P$, where $n > 1$ and P is the 3×3 matrix determined in the previous exercise. The resulting phenomena was not well understood until quite recently. It is now used in the genetic engineering of plasmids.

   (a) If an entire strain of bacteria is resistant to both the antibiotics ampicillin and tetracycline, make a conjecture as to the drug resistance of future generations of bacteria.

   (b) Test your conjecture by repeatedly computing the matrix product $A_n = A_{n-1}P$ with your graphing calculator. Let $A_1 = [0, 1, 0]$ and $n = 2, 3, \ldots, 12$. Interpret the result. (It may surprise you.)

3. The sum of the first n terms of an arithmetic series is given by $S_n = n\left(\dfrac{a_1 + a_n}{2}\right)$. Justify this formula using a geometric discussion. (*Hint:* Start by graphing an arithmetic sequence where n is an odd number.)

BIBLIOGRAPHY

Acker, A., and C. Jaschek. *Astronomical Methods and Calculations*. New York: John Wiley and Sons, 1986.

Andre-Pascal, R. *Global Energy: The Changing Outlook*. Paris, France: Organization for Economic Cooperation and Development/International Energy Agency, 1992.

Baase, S. *Computer Algorithms: Introduction to Design and Analysis*. 2nd ed. Reading, Mass.: Addison-Wesley Publishing Company, 1988.

Battan, L. *Weather in Your Life*. San Francisco: W. H. Freeman, 1983.

Beckmann, P. *A History of PI*. New York: Barnes and Noble, Inc., 1993.

Bell, D. *Fundamentals of Electric Circuits*. Reston, Va.: Reston Publishing Company, 1981.

Benade, A. *Fundamentals of Musical Acoustics*. New York: Oxford University Press, 1976.

Beni, G., and S. Hackwood. *Recent Advances in Robotics*. New York: John Wiley and Sons, 1985.

Beranek, L. *Noise and Vibration Control*. Washington: Institute of Noise Control Engineering, 1988.

Brase, C. H., and C. P. Brase. *Understandable Statistics*. 5th ed. Lexington, Mass.: D. C. Heath and Company, 1995.

Brearley, J., and A. Nicholas. *This Is the Bichon Frise*. Hong Kong: TFH Publication, 1973.

Brooks, R. and J. Dieter. *Phytoarchaeology*. Portland, Or.: Dioscorides Press, 1990.

Brown, D., and P. Rothery. *Models in Biology: Mathematics, Statistics and Computing*. West Sussex, England: John Wiley and Sons Ltd, 1993.

Brown, F., J. Hastings, and J. Palmer. *The Biological Clock*. New York: Academic Press, 1970.

Bünning, E. *The Physiological Clock*. New York: Springer-Verlag, 1967.

Burden, R., and J. Faires. *Numerical Analysis*. 5th ed. Boston: PWS-KENT Publishing Company, 1993.

Callas, D. *Snapshots of Applications in Mathematics*. Deli, New York: State University College of Technology, 1994.

Carlson, T. "Über Geschwindigkeit und Grösse der Hefevermehrung in Würze." *Biochem. A*. 57: 313–334.

Cheney, W., and D. Kincaid. *Numerical Mathematics and Computing*. 3rd ed. Pacific Grove, Calif.: Brooks/Cole Publishing Company, 1994.

Clime, W. *The Economics of Global Warming*. Washington, D.C.: Institute for International Economics, 1992.

Conquering the Sciences. Sharp Electronics Corporation, 1986.

Cooper, J., and R. Glassow. *Kinesiology*. 2nd ed. St. Louis: The C. V. Mosby Company, 1968.

Cotton, W., and R. Pielke. *Human Impacts on Weather and Climate*. Geophysical Science Series, vol. 2. Fort Collins, Colo.: *ASTeR Press, 1992.

Craig, J. *Introduction to Robotics: Mechanics and Control.* Reading, Mass.: Addison-Wesley Publishing Company, 1989.

Crone, D. *Elementary Photogrammetry.* New York: Frederick Ungar Publishing Company, 1968.

Crownover, R. *Introduction to Fractals and Chaos.* Boston: Jones and Bartlett, 1995.

Duffet-Smith, P. *Practical Astronomy with your Calculator.* New York: Cambridge University Press, 1988.

Easteal, S., N. McLeod, and K. Reed. *DNA Profiling: Principles, Pitfalls and Potential.* Philadelphia: Harwood Academic Publishers, 1991.

Elton, C. S., and M. Nicholson. "The ten year cycle in numbers of lynx in Canada." *J. Anim. Ecol.* 11, (1942): 215–244.

Eves, H. *An Introduction to the History of Mathematics,* 5th ed. Philadelphia: Saunders College Publishing, 1983.

Eysenck, H. *Smoking, Personality, and Stress: Psychosocial Factors in the Prevention of Cancer and Coronary Heart Disease.* New York: Springer-Verlag, 1991.

Fletcher, N. and T. Rossing. *The Physics of Musical Instruments.* New York: Springer-Verlag, 1991.

Foley, J., A. van Dam, S. Feiner, J. Hughes, and R. Phillips. *Introduction to Computer Graphics.* Reading Mass.: Addison-Wesley Publishing Company, 1994.

Foster, R., and J. Bates. "Use of mussels to monitor point source industrial discharges." *Environ. Sci. Technol.* 12: 958–962.

Freebury, H. *A History of Mathematics.* New York: MacMillan, 1961.

Freedman, B. *Environmental Ecology: The Ecological Effects of Pollution, Disturbance, and Other Stresses.* 2nd ed. San Diego: Academic Press, 1995.

Friedhoff, M., and W. Benzon. *The Second Computer Revolution: Visualization.* New York: W. H. Freeman, 1991.

Glass, L., and M. Mackey. *From Clocks to Chaos.* Princeton, New Jersey: Princeton University Press, 1988.

Glassner, A. *An Introduction to Ray Tracing.* San Diego: Academic Press, 1993.

Goldstein, M., and J. Larson. *Jackie Joyner-Kersee: Superwoman.* Minneapolis: Lerner Publications Company, 1994.

Good, I. J. "What is the most amazing approximate integer in the universe?" *Pi Mu Epsilon Journal* 5 (1972): 314–315.

Grigg, D. *The World Food Problem.* Oxford: Blackwell Publishers, 1993.

Haber-Schaim, U., J. Cross, G. Abegg, J. Dodge, and J. Walter. *Introductory Physical Science.* Englewood Cliffs, N.J.: Prentice Hall, Inc., 1972.

Haefner, L. *Introduction to Transportation Systems.* New York: Holt, Rinehart and Winston, 1986.

Harker, J. *The Physiology of Diurnal Rhythms.* New York: Cambridge University Press, 1964.

Harrison, F., F. Hills, J. Paterson, and R. Saunders. "The measurement of liver blood flow in conscious calves." *Quarterly Journal of Experimental Physiology* 71: 235–247.

Hartman, D. *Global Physical Climatology.* San Diego: Academic Press, 1994.

Heinz-Otto, P., H. Jürgens, and D. Saupe. *Chaos and Fractals: New Frontiers in Science.* New York: Springer-Verlag, 1993.

Hibbeler, R. *Structural Analysis.* Englewood Cliffs, N.J.: Prentice Hall, 1995.

Hill, F. *Computer Graphics.* New York: Macmillan Publishing Company, 1990.

Hines, A., T. Ghosh, S. Loyalka, and R. Warder, Jr. *Indoor Air Quality and Control.* Englewood Cliffs, N.J.: Prentice Hall, 1993.

Historical Topics for the Mathematics Classroom, Thirty-first Yearbook. National Council of Teachers of Mathematics, 1969.

Hoggar, S. *Mathematics for Computer Graphics*. New York: Cambridge University Press, 1993.

Hoppensteadt, F., and C. Peskin. *Mathematics in Medicine and the Life Sciences*. New York: Springer-Verlag, 1992.

Hosmer, D., and S. Lemeshow. *Applied Logistic Regression*. New York: John Wiley and Sons, 1989.

Howatson, A. *Electrical Circuits and Systems*. New York: Oxford University Press, 1996.

Howells, G. *Acid Rain and Acid Waters*. 2nd ed. New York: Ellis Horwood, 1995.

Huffman, R. *Atmospheric Ultraviolet Remote Sensing*. San Diego: Academic Press, 1992.

Huxley, J. *Problems of Relative Growth*. London: Methuen and Co. Ltd., 1932.

Jarrett, J. *Business Forecasting Methods*. Oxford: Basil Blackwell Ltd., 1991.

Karttunen, H., P. Kroger, H. Oja, M. Poutanen, K. Donner, eds. *Fundamental Astronomy*. 2nd ed. New York: Springer-Verlag, 1994.

Kerlow, I. *The Art of 3-D Computer Animation and Imaging*. New York: Van Nostrand Riehold, 1996.

Kincaid, D., and W. Cheney. *Numerical Analysis*. Pacific Grove, Calif.: Brooks/Cole Publishing Company, 1991.

Kissam, P. *Surveying Practice*. 3rd ed. New York: McGraw-Hill, 1978.

Kline, M. *The Loss of Certainty*. New York: Oxford University Press, 1980.

Kraljic, M. *The Greenhouse Effect*. New York: The H. W. Wilson Company, 1992.

Kress, S. *Bird Life—A Guide to the Behavior and Biology of Birds*. Racine, Wisc.: Western Publishing Company, 1991.

Lack, D. *The Life of a Robin*. London: Collins, 1965.

Lancaster, H. *Quantitative Methods in Biological and Medical Sciences: A Historical Essay*. New York: Springer-Verlag, 1994.

Leder, J. *Martina Navratilova*. Mankato, Minn.: Crestwood House, 1985.

Leick, A. *GPS Satellite Surveying*. New York: Wiley, 1990.

Loh, W. *Dynamics and Thermodynamics of Planetary Entry*. Englewood Cliffs, N.J.: Prentice-Hall, 1963.

Makridakis, S., and S. Wheelwright. *Forecasting Methods for Management*. New York: John Wiley and Sons, 1989.

Mannering, F., and W. Kilareski. *Principles of Highway Engineering and Traffic Analysis*. New York: John Wiley and Sons, 1990.

Mar, J., and H. Liebowitz. *Structure Technology for Large Radio and Radar Telescope Systems*. Cambridge, Mass.: The MIT Press, 1969.

Mason, C. *Biology of Freshwater Pollution*. New York: Longman and Scientific and Technical, John Wiley and Sons, 1991.

Medley, G., D. Cox, and L. Billard. "Incubation period of AIDS in patients infected via blood transfusions." *Nature* 328: 719–721.

Meeus, J. *Astronomical Algorithms*. Richmond, Va.: Willman-Bell, 1991.

Mehrotra, A. *Cellular Radio: Analog and Digital Systems*. Boston: Artech House, 1994.

Metcalf, H. *Topics in Classical Biophysics*. Englewood Cliffs, N.J.: Prentice-Hall, 1980.

Miller, A., and J. Thompson. *Elements of Meteorology*. 2nd ed. Columbus, Ohio: Charles E. Merrill Publishing Company, 1975.

Miller, A., and R. Anthes. *Meteorology*. 5th ed. Columbus, Ohio: Charles E. Merrill Publishing Company, 1985.

Moffitt, F. *Photogrammetry*. Scranton, Pa.: International Textbook Company, 1967.

Monroe, J. *Steffi Graf*. Mankato, Minn.: Crestwood House, 1988.

Mortenson, M. *Computer Graphics: An Introduction to Mathematics and Geometry*. New York: Industrial Press Inc., 1989.

Motz, L., and J. Weaver. *The Story of Mathematics.* New York: Plenum Press, 1993.

Mueller, I., and K. Ramsayer. *Introduction to Surveying.* New York: Frederick Ungar Publishing Company, 1979.

Navarra, J. *Atmosphere, Weather and Climate.* Philadelphia: W. B. Saunders Company, 1979.

Nemerow, N., and A. Dasgupta. *Industrial and Hazardous Waste Treatment.* New York: Van Nostrand Reinhold, 1991.

Nicholson, A. J. "An Outline of the dynamics of animal populations." *Austr. J. Zool.* 2 (1935): 9–65.

Nielson, G., and B. Shriver, eds. *Visualization in Scientific Computing.* Los Alamitos, Calif.: IEEE Computer Society Press, 1990.

Nilsson, A. *Greenhouse Earth.* New York: John Wiley and Sons, 1992.

Paetsch, M. *Mobile Communications in the U.S. and Europe: Regulation, Technology, and Markets.* Norwood, Mass.: Artech House, Inc., 1993.

Parmar and Grosjean. "Removal of Air Pollutants from Museum Display Cases." Marina del Rey, Calif.: Getty Conservation Institute, 1989.

Payne, A. "Responses of the three test Algae of the algal assay procedure: bottle test." *Water Res.* 9: 437–445.

Pearl, R., T. Edwards, and J. Miner. "The growth of *Cucumis melo* seedlings at different temperatures." *J. Gen. Physiol.* 17: 687–700.

Pielou, E. *Population and Community Ecology: Principles and Methods.* New York: Gordon and Breach Science Publishers, 1974.

Pierce, J. *The Science of Musical Sound.* New York: W. H. Freeman, 1992.

Pokorny, C., and C. Gerald. *Computer Graphics: The Principles behind the Art and Science.* Irvine, Calif.: Franklin, Beedle, and Associates, 1989.

Pugh, J. *Surveying for Field Scientists.* Pittsburgh: University of Pittsburgh Press, 1975.

Raggett, G. "Modeling the Eyam plague." *The Institute of Mathematics and Its Applications* 18: 221–226.

Rayner, W., and M. Schmidt. *Elementary Surveying.* 4th ed. New York: D. Van Nostrand Company, Inc., 1963.

Resnikoff H., and R. Wells, Jr. *Mathematics in Civilization.* New York: Dover Publications, Inc., 1984.

Rezvan, R. L. "Effectiveness of Local Ventilation in Removing Simulated Pollutants from Point Sources." *Proceedings of the Third International Conference on Indoor Air Quality and Climate,* 1984.

Riley, W., L. Sturges, and D. Morris. *Statics and Mechanics of Materials: An Integrated Approach.* New York: John Wiley & Sons, Inc., 1995.

Rodricks, J. *Calculated Risk.* New York: Cambridge University Press, 1992.

Roederer, J. *Introduction to the Physics and Psychophysics of Music.* New York: Springer-Verlag, 1973.

Ronan, C. *The Natural History of the Universe.* New York: MacMillan Publishing Company, 1991.

Ryan, B., B. Joiner, and T. Ryan. *Minitab Handbook.* Boston: Duxbury Press, 1985.

Sanders, D. *Statistics: A First Course.* 5th ed. New York: McGraw-Hill, 1995.

Schaufele, C., and N. Zumoff. *Earth Algebra.* New York: HarperCollins, 1995.

Schlosser, W. *Challenges of Astronomy.* New York: Springer-Verlag, 1991.

Schneier, B. *Applied Cryptography: Protocols, Algorithms, and Source Code in C.* New York: John Wiley and Sons, 1994.

Schulz, J. *The Economics of Aging.* 6th ed. Westport, Conn.: Auburn House, 1995.

Sharov, A., and I. Novikov. *Edwin Hubble, The Discoverer of the Big Bang Universe.* New York: Cambridge University Press, 1993.

Sinkov, A. *Elementary Cryptanalysis: A Mathematical Approach.* New York: Random House, 1968.

Smith, J. *Design and Analysis of Algorithms.* Boston: PWS Publishing Company, 1989.

Smith, R., and R. Dorf. *Circuits, Devices and Systems.* 5th ed. New York: John Wiley and Sons, Inc., 1992.

Stadler, W. *Analytical Robotics and Mechatronics.* New York: McGraw-Hill, Inc., 1995.

Stent, G. S. *Molecular Biology of Bacterial Viruses.* San Francisco: W. H. Freeman, 1963.

Taylor, W. *The Geometry of Computer Graphics.* Pacific Grove, Calif.: Wadsworth and Brooks/Cole, 1992.

Teutsch, S., and R. Churchill. *Principles and Practice of Public Health Surveillance.* New York: Oxford University Press, 1994.

Thomas, D. *Swimming Pool Operators Handbook.* National Swimming Pool Foundation of Washington, D.C., 1972.

Thomas, V. *Science and Sport.* London: Faber and Faber, 1970.

Thomson, W. *Introduction to Space Dynamics.* New York: John Wiley and Sons, 1961.

Toffler, A., and H. Toffler. *Creating a New Civilization: The Politics of the Third Wave.* Kansas City, Mo.: Turner Publication, 1995.

Tucker, A., A. Bernat, W. Bradley, R. Cupper, and G. Scragg. *Fundamentals of Computing I Logic: Problem Solving, Programs, and Computers.* New York: McGraw-Hill, 1995.

Turner, R. K., D. Pierce, and I. Bateman. *Environmental Economics, An Elementary Approach.* Baltimore: The Johns Hopkins University Press, 1993.

Van Sickle, J. *GPS for Land Surveyors.* Chelsey, Mich.: Ann Arbor Press, 1996.

Varley, G., and G. Gradwell. "Population models for the winter moth." *Symposium of the Royal Entomological Society of London* 4: 132–142.

Walker, A. *Observation and Inference: An Introduction to the Methods of Epidemiology.* Newton Lower Falls, Mass.: Epidemiology Resources Inc., 1991.

Wang, T. *ASHRAE Trans.* 81, Part 1 (1975): 32.

Watt, A. *3D Computer Graphics.* Reading, Mass.: Addison-Wesley Publishing Company, 1993.

Webb, T. *Celestial Objects for Common Telescopes.* New York: Dover Publications Inc., 1962.

Weidner, R., and R. Sells. *Elementary Classical Physics,* Vol. 1, Vol. 2. Boston: Allyn and Bacon, 1965.

West, G. *"Differential GPS—how accurate is it?" Trailer Boats.* 25 (10): 72–73.

Wilcox, G. and C. Hesselberth. *Electricity for Engineering Technology.* Boston: Allyn and Bacon, 1970.

Williams, J. *The Weather Almanac 1995.* New York: Vintage Books, 1994.

Winter, C. *Solar Power Plants.* New York: Springer-Verlag, 1991.

Wolff, R., and L. Yaeger. *Visualization of Natural Phenomena.* New York: Springer-Verlag, 1993.

Wright, J. *The Universal Almanac 1997.* Kansas City: Andrews and McMeel, 1997.

Wuebbles, D., and J. Edmonds. *Primer on Greenhouse Gases.* Chelsea, Mich.: Lewis Publishers, 1991.

Zeilik, M., S. Gregory, and D. Smith. *Introductory Astronomy and Astrophysics.* 3rd ed. Philadelphia: Saunders College Publishers, 1992.

Zhoa, Y. *Vehicle Location and Navigation Systems.* Boston, Mass.: Artech House, 1997.

ANSWERS TO SELECTED EXERCISES

CHAPTER 1   INTRODUCTION TO FUNCTIONS AND GRAPHS

1.1   ANSWERS

1. Natural number, integer, rational number, real number

3. Integer, rational number, real number

5. Rational number, real number

7. Real number

9. Natural number, $\sqrt{9}$; integers, -3, $\sqrt{9}$; rational numbers, -3, $\frac{2}{9}$, $\sqrt{9}$, $1.\overline{3}$; irrational numbers: π, $-\sqrt{2}$

11. Rational numbers   **13.** Rational numbers

15. Integers   **17.** 1.863×10^5

19. 5.333×10^{-2} **21.** 0.000001

23. $200{,}000{,}000$ **25.** 3.4×10^{19}

27. 1.8×10^{-1} **29.** 5.769

31. 0.058 **33.** 0.419

35. Tuition and fees: 91.3%, CPI: 37.7%

37. (a) Approximately $1820 per person
(b) Approximately $18,690 per person
(c) Estimates will vary.

39. 2.9×10^{-4} cm

41. (a) 16,666,667 ft   (b) Yes

43. (a) Increased significantly
(b) $527 billion, which is quite close to the actual amount

45. (a) Increased from 1940 to 1980, decreased from 1980 to 1992
(b) 4.9×10^8 gal

47. 68°F

49. -40°F; the Celsius and Fahrenheit temperatures have the same numeric value.

51. 2.50   **53.** 29.3   **55.** 22.0

57. 14,083   **59.** 409,722

61. The length of a side

63. **ALGORITHM Tuition and Fees**
Step 1: Input the number of credits x.
Step 2: Multiply x by 95.25.
Step 3: Add 31.50 to the answer in STEP 2. Let the result be y.
Step 4: Output y, the tuition and fees in dollars.

65. **ALGORITHM Sale Price**
Step 1: Input the regular price P and the percentage discount x.
Step 2: Compute $100 - x$.
Step 3: Divide the answer in STEP 2 by 100.
Step 4: Multiply the answer in STEP 3 by P. Call this result y.
Step 5: Output y, the sale price.

1.2   ANSWERS

1.

-30	-30	-10	5	15	25	45	55	61

(a) Max: 61; min: -30
(b) Avg: $\frac{136}{9} \approx 15.11$; median: 15; range: 91

3. $\sqrt{15} \approx 3.87$, $2^{2.3} \approx 4.92$, $\sqrt[3]{69} \approx 4.102$, $\pi^2 \approx 9.87$, $2^{\pi} \approx 8.82$, 4.1

$\sqrt{15}$	4.1	$\sqrt[3]{69}$	$2^{2.3}$	2^{π}	π^2

(a) Max: π^2; min: $\sqrt{15}$
(b) Avg: 5.95; median: 4.51; range: $\pi^2 - \sqrt{15} \approx 6.00$

5. (a)

(b) Avg: 273; median: 212; range: 756
The average area of the eight largest islands in the world is 273,000 square miles. Half of the islands have areas under 212,000 square miles and half are over. The largest difference in area between any two islands is 756,000 square miles.
(c) Greenland

A-1

7. **(a)**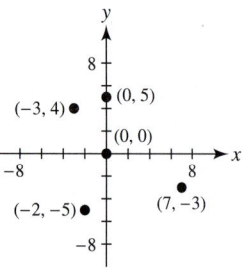

        4 8 12 16 20 24 28 32

     (b) Avg: 19.0; median: 19.3; range: 21.7

          The average of the maximum elevations for the seven continents is 19,000 feet. About half of these elevations are below 19,300 feet and about half are above. Their largest difference is 21,700 feet.

     (c) Mount Everest

9. **(a)** \$1,168,333

     (b) \$109,000

     (c) One large salary can raise the average considerably, while the majority of the players have considerably lower salaries.

11. **(a)** 2.13

     (b) Max: 5910; min: 3780

13. **(a)** $D = \{0, -3, -2, 7, 0\}, R = \{5, 4, -5, -3, 0\}$

     (b) x-min: -3; x-max: 7; y-min: -5; y-max: 5

     (c) & (d)

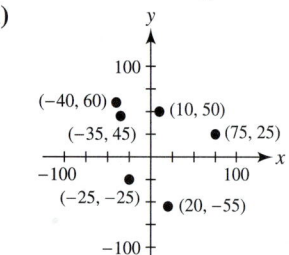

15. **(a)** $D = \{10, -35, 20, 75, -40, -25\},$
$R = \{50, 45, -55, 25, 60, -25\}$

     (b) x-min: -40; x-max: 75; y-min: -55; y-max: 60

     (c) & (d)

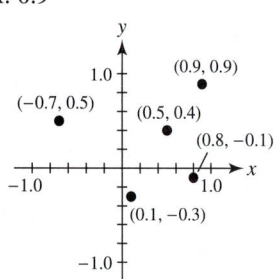

17. **(a)** $D = \{0.1, 0.5, -0.7, 0.8, 0.9\},$
$R = \{-0.3, 0.4, 0.5, -0.1, 0.9\}$

     (b) x-min: -0.7; x-max: 0.9; y-min: -0.3; y-max: 0.9

     (c) & (d)

19. x-axis: 10; y-axis: 10   **21.** x-axis: 10; y-axis: 5

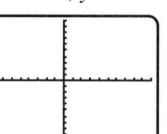   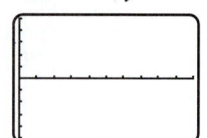

23. x-axis: 16; y-axis: 5   **25.** d   **27.** c

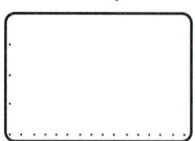

29. $[-6, 6, 1]$ by $[-4, 4, 1]$

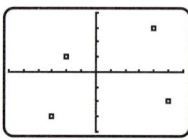

31. $[-50, 50, 10]$ by $[-50, 50, 10]$

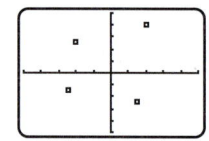

33. $[-6, 6, 2]$ by $[-12, 12, 2]$

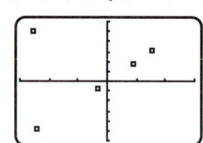

35. **(a)** x-min: 1988; x-max: 1994; y-min: 2916; y-max: 3472

     (b) $[1986, 1996, 1]$ by $[2500, 3500, 100]$. Answers may vary.

     (c) 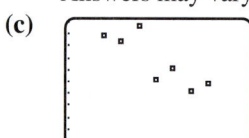   **(d)**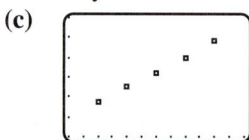

37. **(a)** x-min: 1996; x-max: 2004; y-min: 9.7; y-max: 12.8

     (b) $[1994, 2006, 1]$ by $[8, 14, 1]$. Answers may vary.

     (c) 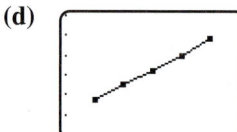   **(d)**

39. CD sales have increased, while cassette tape sales have remained flat. Cassette sales may decline and CD sales may increase because it is a newer technology.

41. Answers may vary.

Year	1950	1960	1970	1980	1990	1995
Doctorate Degrees	6000	10,000	30,000	33,000	38,000	41,000

43. **(a)** No. The number manufactured has both increased and decreased.
   (b) 1986 to 1989

1.3   ANSWERS

1.

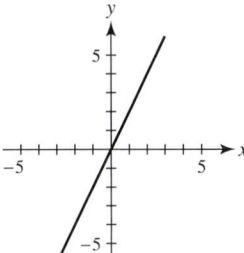

3.

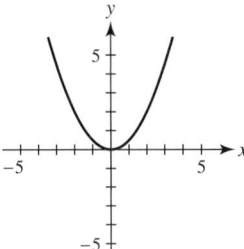

5.

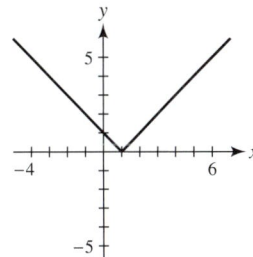

7. $f(x) = 3.785x$ [0, 100, 10] by [0, 400, 100]

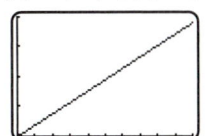

9. $f(x) = \dfrac{x}{1.609}$ [0, 100, 10] by [0, 60, 10]

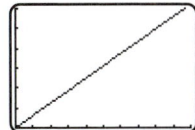

11. Verbal: Square the input x.
   Graphical: Graph $Y_1 = X^2$.
   $[-10, 10, 1]$ by $[-10, 10, 1]$

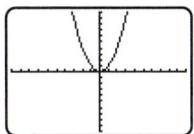

Numerical:

−2	−1	0	1	2
4	1	0	1	4

13. Verbal: Divide 1 by x, or equivalently, compute the reciprocal of x.
   Graphical: Graph $Y_1 = 1/X$.
   $[-6, 6, 1]$ by $[-4, 4, 1]$

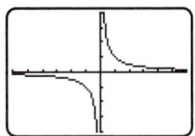

Numerical:

−2	−1	0	1	2
−0.5	−1	—	1	0.5

15. Verbal: Compute the absolute value of the input x.
   Graphical: Graph $Y_1 = \text{abs}(X)$.
   $[-6, 6, 1]$ by $[-4, 4, 1]$

Numerical:

−2	−1	0	1	2
2	1	0	1	2

17.

Bills (millions)	0	1	2	3	4	5	6
Counterfeit Bills	0	9	18	27	36	45	54

19. **(a)**

Number of Lines	1	2	3	4	5	6
Cost	$9.60	$9.60	$9.60	$12.80	$16.00	$19.20

[0, 7, 1] by [0, 20, 1]

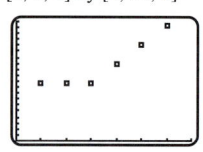

(b) $16

21. (a) 11.0
(b) $D = \{1990, 1991, 1992, 1993\}$,
$R = \{9.2, 6.1, 7.6, 11.0\}$
(c) $D = \{1991, 1992, 1993, 1994\}$,
$R = \{6.1, 7.6, 11.0, 11.2\}$

23. (a) The average temperature is least in January and greatest in July.

[0.5, 12.5, 1] by [65, 95, 5]

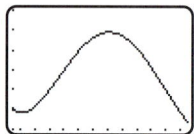

(b)

x	1	2	3	4	5	6
$f(x)$	69	71	75	80	85	88

x	7	8	9	10	11	12
$f(x)$	89	89	85	81	75	69

(c) 75°F

25. (a) $f(0) = -2$, $f(2) = 2$ **(b)** $x = 1$
27. (a) [−4.7, 4.7, 1] by [−3.1, 3.1, 1] **(b)** $f(2) = 1$

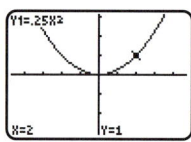

29. (a) [−4.7, 4.7, 1] by [−3.1, 3.1, 1] **(b)** $f(2) = 2$

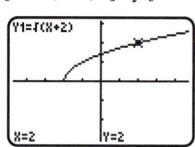

31. (a) $f(-2) = -8$, $f(5) = 125$
(b) All real numbers
33. (a) $f(4) = 2$, $f(a) = \sqrt{a}$
(b) Nonnegative real numbers
35. (a) $f(-1) = -\dfrac{1}{2}$, $f(a + 1) = \dfrac{1}{a}$
(b) $x \neq 1$
37. (a) $f(6) = -7$, $f(a) = -7$
(b) All real numbers

39. (a) $f(4) = \dfrac{1}{16}$, $f(-7) = \dfrac{1}{49}$ **(b)** $x \neq 0$

41. $D: -3 \leq x \leq 3$; $R: 0 \leq y \leq 3$

43. $D: -2 \leq x \leq 4$; $R: -2 \leq y \leq 2$

45. (a) 0.2 in.
(b) Yes. For each input x, there is one output P.
(c) $x = 2, 3, 7, 11$

47. Yes. Domain and range include all real numbers.

49. No, 2

51. Yes. $D: -4 \leq x \leq 4$; $R: 0 \leq y \leq 4$

53. Yes. Each real number input x has exactly one cube root.

55. No. Usually, several students pass a given exam.

57. No **59.** Yes **61.** No

1.4 ANSWERS
1. (a) $f(x) = 68$

[0, 24, 4] by [0, 100, 10]

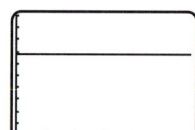

(b)

x	0	4	8	12	16	20	24
$f(x)$	68	68	68	68	68	68	68

(c) Constant

3. (a) No **(b)** $f(x) = 7$
(c) [0, 15, 3] by [0, 10, 1]

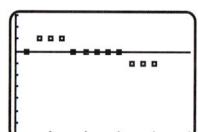

5. (a) f: linear; g: nonlinear; h: linear
(b) f: circumference; g: area; h: diameter

7. The velocity of the car is zero.

9. $f(x) = \dfrac{x}{16}$ **11.** $f(x) = 50x$ **13.** $f(x) = 500$

15. $\dfrac{1}{2}$ **17.** -8 **19.** $-\dfrac{23}{30} \approx 0.7667$

21. Undefined

23. (a) [1820, 1995, 20] by [0, 40, 10]

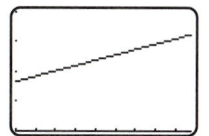

(b) $f(1820) = 16.7$ yr; $f(1995) = 32.45$ yr
(c) $m = 0.09$. The median age increased by 0.09 year, on average, each year from 1820 to 1995.

25. (a) [1948, 1996, 4] by [21, 25, 1]

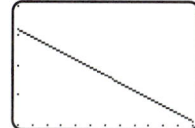

(b) $m = -0.0635$. The time to run the 200-meter dash is *decreasing* by 0.0635 second per year on average.
(c) -0.254 sec

27. (a) Linear   **(b)** Decreasing for all x

[−10, 10, 1] by [−10, 10, 1]

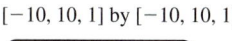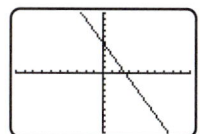

29. (a) Nonlinear   **(b)** Increasing for $x \geq 0$

[−10, 10, 1] by [−10, 10, 1]

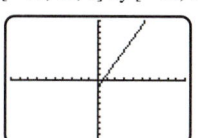

31. (a) Constant (and linear)
(b) Neither increasing nor decreasing

[−10, 10, 1] by [−10, 10, 1]

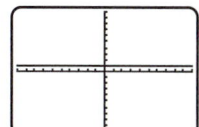

33. (a) Nonlinear
(b) Increasing for $x \leq -k$ or $x \geq k$ and decreasing for $-k \leq x \leq k$, where $k \approx 1.8$

[−10, 10, 1] by [−10, 10, 1]

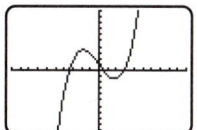

35. (a) Nonlinear

(b) Increasing for $x \leq 2.5$, decreasing for $x \geq 2.5$
[−10, 10, 1] by [−10, 10, 1]

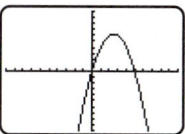

37. (a) Nonlinear
(b) Decreasing for $x \leq -k$ or $0 \leq x \leq k$, increasing for $-k \leq x \leq 0$ and $x \geq k$, where $k \approx 1.6$
[−10, 10, 1] by [−10, 10, 1]

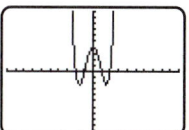

39. (a) [0, 4000, 500] by [0, 280, 35]

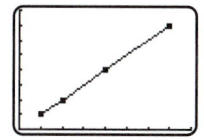

(b) Increasing for given x-values
(c) Linear

41. (a) [1965, 1995, 5] by [0, 25,000, 5000]

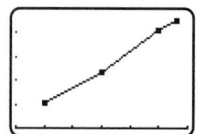

(b) Increasing for given x-values
(c) Nonlinear

43. f_1: decreasing; f_2: neither; f_3: increasing; f_4: neither

For Exercises 45–50, answers may vary.

45.

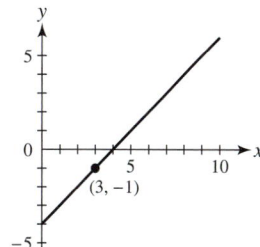

47. The function g must both increase and decrease.

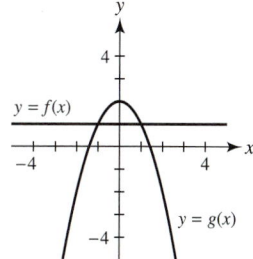

49. Average rates of change are positive.

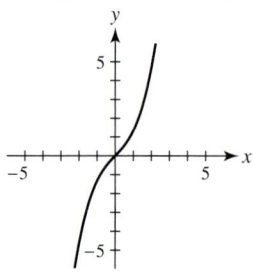

51. Always equal to 0

53. For negative x-values, the average rates of change are positive; for positive x-values, the average rates of change are negative.

55. 7; the slope of the line passing through $(1, f(1))$ and $(4, f(4))$ is 7.

57. 20.75; the slope of the line passing through $(0.5, f(0.5))$ and $(4.5, f(4.5))$ is 20.75.

59. 0.62; the slope of the line passing through $(1, f(1))$ and $(3, f(3))$ is approximately 0.62.

61. **(a)** $[-10, 90, 10]$ by $[0, 80, 10]$

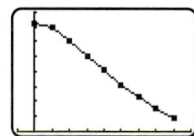

     (b) $-0.24, -0.98, -0.97, -0.95, -0.93, -0.85,$ $-0.76, -0.63$; as a woman becomes older, her *remaining* life expectancy is reduced each year by the average rate of change.

     (c) 80.1 yr, 85.5 yr; there is risk associated with living from 20 years old to 70 years old that does not apply to someone who is already 70 years old.

63. **(a)** 1705, 4864.2, 2899.2

     (b) From 1981 to 1985, AIDS deaths increased by 1705 per year. Other values can be interpreted similarly.

     (c) The rate of increase in deaths may be starting to slow down after increasing during the 1980s.

1.5 ANSWERS

1.

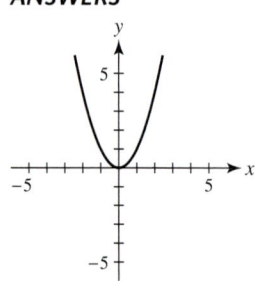

3.

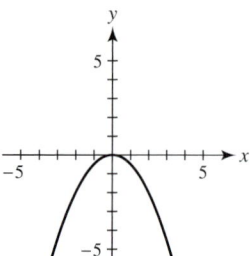

5.

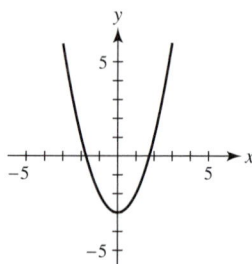

7. **(a)** Quadratic   **(b)** 13

9. **(a)** Linear   **(b)** 6.5

11. **(a)** Quadratic   **(b)** -1

13. -3.3 **15.** 45

17. **(a)** $(0, 6)$

     (b) Increasing $x \le 0$, decreasing $x \ge 0$

19. **(a)** $(3, -9)$

     (b) Increasing $x \ge 3$, decreasing $x \le 3$

21. **(a)** $(1.9, -5.61)$

     (b) Increasing $x \ge 1.9$, decreasing $x \le 1.9$

23. **(a)** $(-0.25, 1.875)$

     (b) Increasing $x \le -0.25$, decreasing $x \ge -0.25$

25. **(a)** $[-1, 20, 2]$ by $[-50, 800, 100]$

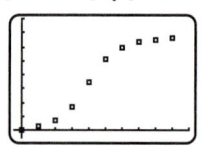

     (b) From 6 to 8 hours

     (c) The population was initially small, then started to grow faster, and finally leveled off.

27. **(a)** The optimum temperature is about 30°C. If the temperature is too low or too high, the seedlings do not grow as well.

     (b) f_1 appears to model the data better.

        $[15, 40, 2]$ by $[10, 30, 2]$

     (c) 28.4°C

29. **(a)** Yes. Systolic blood pressure appears to increase with age.

(b) f_1 appears to model the data better.

[30, 80, 5] by [110, 170, 5]

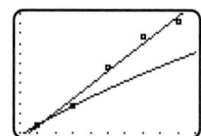

31. d   **33.** a

35. 146 ft after 2.75 sec

[0, 6, 1] by [0, 160, 10]

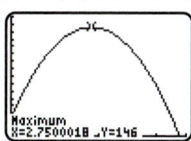

37. 322.8 ft after 6.77 sec

[0, 15, 5] by [0, 400, 100]

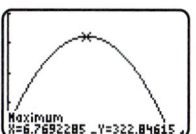

39. **(a)** With stress 7%, without stress 3%

(b) Increases the likelihood

(c) Adding stress to smoking increases the likelihood.

41. **(a)** 10 mg, 110 mg, 155 mg; answers may vary.

(b) Hatching to birth: 16.7 mg per week; 6 weeks to 12 weeks: 7.5 mg per week. The fish grow the fastest during the first 6 weeks.

43. **(a)** Increased with small, but regular fluctuations

(b) The amounts of carbon dioxide oscillate.

(c) Overall carbon dioxide levels have increased, possibly due to the burning of fossil fuels and deforestation. The small oscillations are due to the seasons. Vegetation stores carbon dioxide through photosynthesis during the summer and releases it through respiration during the winter.

1.6 ANSWERS

1.

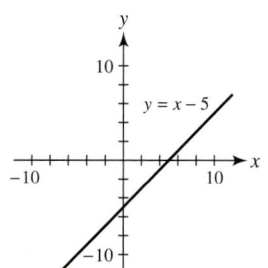

3.

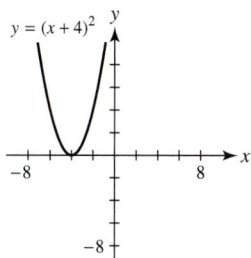

5.

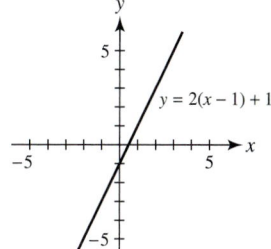

7.

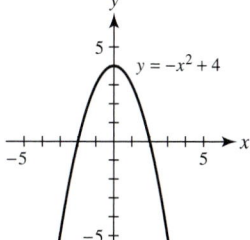

9.

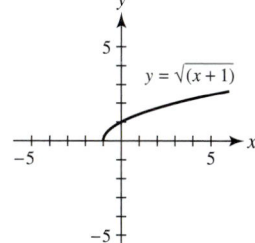

11.

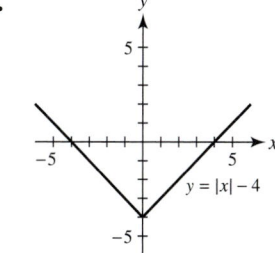

13.

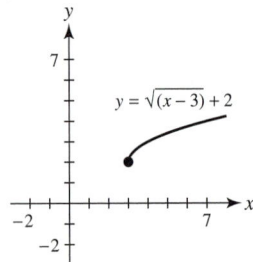

$y = \sqrt{(x-3)} + 2$

15. (a)

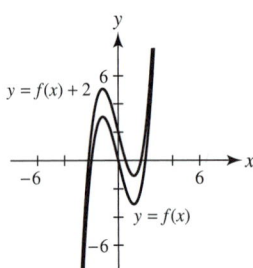

$y = f(x) + 2$, $y = f(x)$

(b)

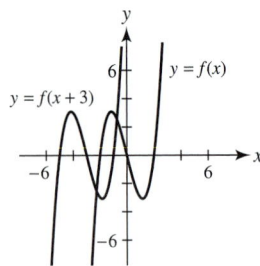

$y = f(x+3)$, $y = f(x)$

17. (a)

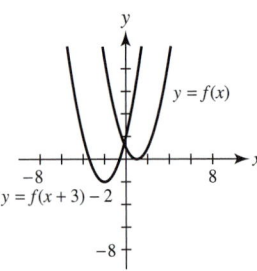

$y = f(x)$, $y = f(x+3) - 2$

(b)

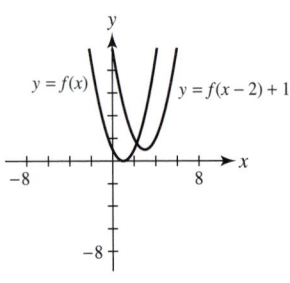

$y = f(x)$, $y = f(x-2) + 1$

19. (a)

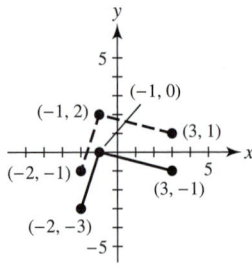

(b)

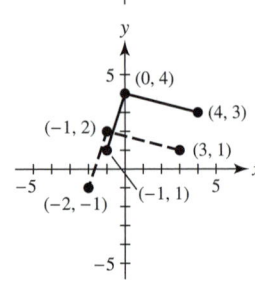

21.

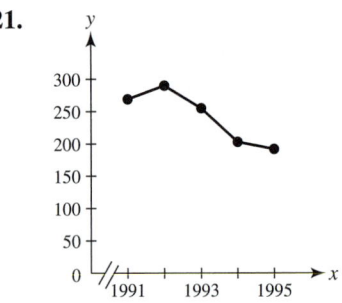

23. $f(x+3) = 2(x+3)$ **25.** $f(x) - 7 = |x| - 6$

27. $f(x+1) - 3 = (x+1)^3 - 5$

29. $y = (x-2)^3 - 3$
$[-10, 10, 1]$ by $[-10, 10, 1]$

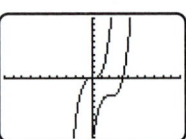

31. $y = (x+6)^2 - 4(x+6) + 5$
$[-10, 10, 1]$ by $[-10, 10, 1]$

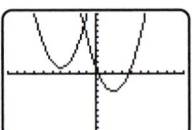

33. (a)

x	0	1	2	3	4	5	6
$g(x)$	\$360	\$380	\$410	\$430	\$460	\$490	\$520

(b) $D_f = \{1990, 1991, 1992, 1993, 1994, 1995, 1996\}$, $D_g = \{0, 1, 2, 3, 4, 5, 6\}$

(c) $g(x - 1990) = f(x)$ where $1990 \leq x \leq 1996$, or $g(x) = f(x + 1990)$ where $0 \leq x \leq 6$

35.

x	1	2	3	4	5
$g(x)$	12	8	13	9	14

37.

x	-2	0	2	4	6
$g(x)$	5	2	-3	-5	-9

39.

x	0	1	2	3	4
$g(x)$	0	2	1	5	6

41. $y = \dfrac{1}{2}(x - 1) - 3$

$[-10, 10, 1]$ by $[-10, 10, 1]$

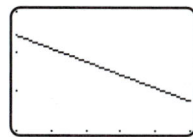

43. $y = -165.1(x - 1995) + 4582$

$[1990, 2000, 2]$ by $[3000, 6000, 1000]$

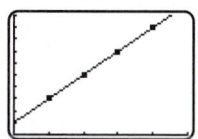

45.   **(b)** $f(x) = 2(x - 1) + 3$

$[0, 5, 1]$ by $[0, 10, 1]$

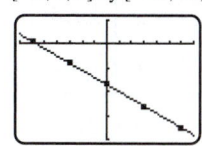

   (c)

47.   **(b)** $f(x) = -3(x + 6) + 1$

$[-7, 7, 1]$ by $[-40, 10, 10]$

   (c)

49.   **(b)** $f(x) = 3.4(x - 1960) + 87.8$. Answers may vary.

$[1955, 1995, 5]$ by $[70, 220, 5]$

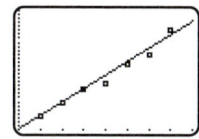

   (c)   Municipal solid waste is increasing by approximately 3.4 million tons each year.
   (d) $f(1995) = 206.8$ million tons.

51.   **(b)** $f(x) = 0.38(x - 1996) + 9.7$. Answers may vary.

$[1995, 2003, 1]$ by $[9, 13, 1]$

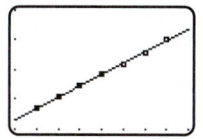

   (c)   The Asian-American population is predicted to increase by about 0.38 million each year.
   (d) $f(2005) = 13.12$ million

53. $y = 2(x - 2)^2 - 3$

$[-10, 10, 1]$ by $[-10, 10, 1]$

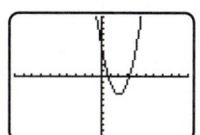

55. $y = -2.59(x - 1990)^2 + 35$

$[1980, 2000, 5]$ by $[-100, 100, 50]$

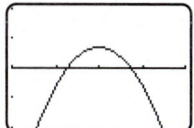

57.   **(b)** $f(x) = 2(x - 2)^2 + 3$

$[0, 6, 1]$ by $[0, 22, 1]$

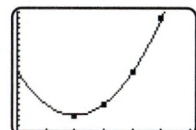

   (c)

59. **(b)** $f(x) = -3(x + 5)^2 + 3$

[−8, 1, 2] by [−25, 5, 2]

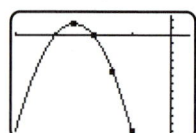

(c)

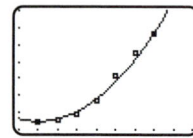

61. **(b)** $f(x) = 5.8(x − 1960)^2 + 1490$. Answers may vary.

[1955, 2000, 5] by [1000, 8000, 1000]

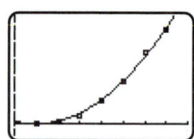

(c) $f(1995) = \$8595$

63. **(b)** $f(x) = 3100(x − 1982)^2 + 1586$. Answers may vary.

[1980, 1996, 2] by [−50,000, 500,000, 10,000]

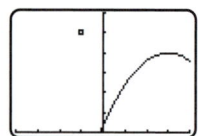

(c) $f(1998) = 795{,}186$ cases

65. $f(x) = -0.26(x − 13)^2 + 95$. Answers may vary.

67. $y = -0.4(x − 3)^2 + 4$ (mountain)

[−4, 4, 1] by [0, 6, 1]

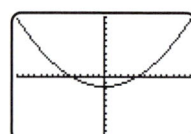

69. **(a)** $y = \dfrac{1}{20}x^2 − 1.6$

[−15, 15, 1] by [−10, 10, 1]

(b) $y = \dfrac{1}{20}(x − 2.1)^2 − 2.5$. The front reaches Columbus by midnight.

[−15, 15, 1] by [−10, 10, 1]

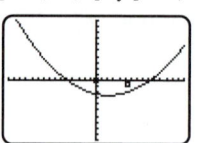

CHAPTER 1 REVIEW ANSWERS

1. Natural number: $\sqrt{16}$; integer: $-2, 0, \sqrt{16}$; rational number: $-2, \dfrac{1}{2}, 0, 1.23, \sqrt{16}$, real number: $-2, \dfrac{1}{2}$, $0, 1.23, \sqrt{7}, \sqrt{16}$

3. 1.891×10^6 **5.** 4.39×10^{-5}

7. 15,200 **9.** 760 sec **11.** 23.5

13.

−23	−5	8	19	24

(a) Max: 24; min: −23

(b) Avg: 4.6; median: 8; range: 47

15. **(a)** $D = \{-1, 4, 0, -5, 1\}, R = \{-2, 6, -5, 3, 0\}$

(b) x-max: 4; x-min: −5, y-max: 6, y-min: −5

(c) & (d)

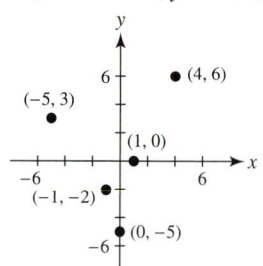

17. Not a function

[−50, 50, 10] by [−50, 50, 10]

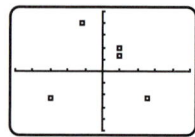

19. **(a)** $f(1993) \approx 16$

(b) $x = 1992$. Sales reached 4 million in 1992.

(c) From 1992 to 1993, the increase was about 12 million units.

(d)

1991	1992	1993	1994
1	4	16	24

21. $f(x) = 16x$

[0, 100, 10] by [0, 1800, 300]

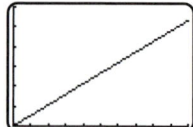

x	0	25	50	75	100
$f(x)$	0	400	800	1200	1600

23. (a) $f = \{(1991, 20.6), (1992, 20.6), (1993, 20.7),$
       $(1994, 20.8)\}$
   (b) $D = \{1991, 1992, 1993, 1994\}$,
       $R = \{20.6, 20.7, 20.8\}$
   (c) $x = 1991$ and 1992

25. (a) $f(-8) = -2, f(1) = 1$
   (b) All real numbers

27. (a) $f(-3) = \dfrac{1}{5}; f(2)$ is undefined.
   (b) $D = \{x \,|\, x \neq \pm 2\}$

29. No. Input $x = 1$ results in outputs $y = \pm 1$. Answers may vary.

31. It is a function.

33. $-\dfrac{3}{4}$ **35.** 0   **37.** Linear

39. Nonlinear   **41.** Constant (and linear)

43. (a) $[-10, 10, 1]$ by $[-10, 10, 1]$

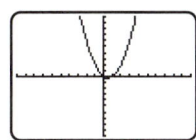

   (b) Decreasing when $x \leq 0.5$, increasing when $x \geq 0.5$

45. f: increasing; g: decreasing; h: neither

47. (a) The data decreases rapidly, indicating a very high mortality rate during the first year.
       $[-1, 5, 1]$ by $[0, 110, 10]$

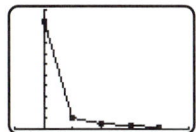

   (b) Yes

49. (a) 1   (b) $(2, 5)$

51.

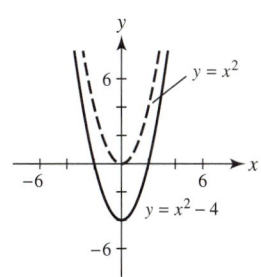

53.

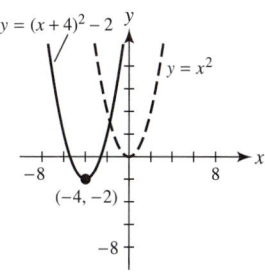

55. $f(x - 2) = 2(x - 2)$

57. $f(x + 1) - 2 = (x + 1)^2 - 5$

59.

x	1	2	3	4
$g(x)$	-1	5	6	9

61. $f(x) = -3(x - 1) + 7$

63. (a) The data appears to be approximately linear.
       $[1994, 2001, 1]$ by $[0, 30, 5]$

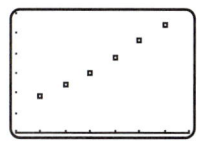

   (b) $f(x) = 3.5(x - 1995) + 9$. Answers may vary.
   (c) The number of children with access to the Internet increased by about 3.5 million per year.

65. (a) $f(1985) \approx 4.7$. In 1985 there were about 4.7 billion people in the world.
   (b) 6.1 billion

CHAPTER 2 LINEAR FUNCTIONS AND EQUATIONS

2.1 ANSWERS

1. (a) $3x - 8.5 = 0, h(x) = 3x - 8.5$
   (b) Both are linear.
   (c) The graph of h is linear.
       $[-10, 10, 1]$ by $[-10, 10, 1]$

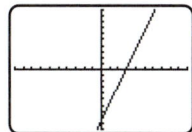

3. (a) $0.2x^2 - 3x - 1 = 0, h(x) = 0.2x^2 - 3x - 1$
   (b) Both are nonlinear.
   (c) The graph of h is nonlinear.
       $[-20, 20, 5]$ by $[-20, 20, 5]$

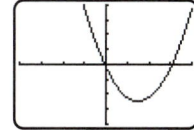

5. **(a)** $3x - 31 = 0$, $h(x) = 3x - 31$
     (b) Both are linear.
     (c) The graph of h is linear.
         $[-60, 60, 10]$ by $[-40, 40, 10]$

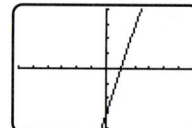

7. $x = 3$ **9.** $x = 1.3$ **11.** $x = 0.675$

13. $x \approx 3.621$ **15.** $x \approx 2.294$

17. $x \approx -2.405$ **19.** $x \approx -3$

21. $x = 8.6$ **23.** $x = 3.5$ **25.** $x \approx -0.2$

27. **(a)** $x = 5$ **(c)** Conditional

29. **(a)** No solutions   **(c)** Contradiction

31. **(a)** $x = \dfrac{8}{3}$ **(c)** Conditional

33. **(a)** All real numbers   **(c)** Identity

35. **(a)** $x = 3$ **(c)** Conditional

37. **(a)** All real numbers   **(c)** Identity

39. **(a)** $x = 1990$ **(c)** Conditional

41. $x = 2$ **43.** $x = 7.5$ **45.** $x \approx 1985$

47. **(a)** $V(x) = 900x$
     (b) $V(50) = 45{,}000$ cubic feet per hour
     (c) 4.5 **(d)** $3\frac{1}{3}$ times

49. $\dfrac{1000}{36.5} \approx 27.4$ yr   **51.** 41.25 ft

53. 189 **55.** 259 million

57. No. If there were 100 people surveyed then each person represents 1%. 13.7% is not possible.

59. **(a)** Approximately 264,800 tons
     (b) Approximately 174,700 tons
     (c) Approximately 1.40×10^9 orders
     (d) Approximately 11.2 orders per person

61. $f(x) = 0.75x$, $\$42.18$

63. The point $(4, 3.5)$ lies on the graph of f. The graph of f is continuous, so there must be a point where it crosses the line $y = 3.5$.

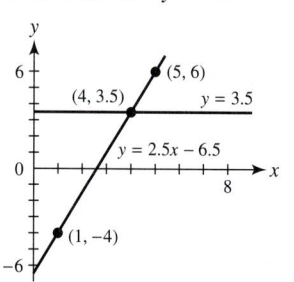

2.2 ANSWERS

1. **(a)** A loan amount of \$1000 results in annual interest of \$100.
     (b) A loan amount greater than \$1000 results in annual interest of more than \$100.
     (c) A loan amount less than \$1000 results in annual interest of less than \$100.
     (d) A loan amount greater than or equal to \$1000 results in annual interest of \$100 or more.

3. **(a)** $f(x) = 100$ when $x = 1000$
     (b) $f(x) > 100$ when $x > 1000$
     (c) $f(x) < 100$ when $x < 1000$
     (d) $f(x) \geq 100$ when $x \geq 1000$

5. **(a)** Car A is traveling faster since its graph has the greater slope.
     (b) 2.5 hr, 225 mi   **(c)** $0 \leq x < 2.5$

7. **(a)** $x = 8$ **(b)** $x < 8$

9. $x > 2.8$ **11.** $x \leq 1987.5$

13. $x > k$, where $k \approx -1.82$

15. $4 \leq x < 6.4$ **17.** $4.6 \leq x \leq 15.2$

19. $Y_1 > 0$ when $x < 4$, and $Y_1 \leq 0$ when $x \geq 4$

21. $Y_1 \geq Y_2$ when $x \leq 50$, and $Y_1 < Y_2$ when $x > 50$

23. **(a)** $-4x - 6 > 0$, where $h(x) = -4x - 6$. Answers may vary.
     (b) $x = -1.5$, $x < -1.5$

25. **(a)** $0.6x - 15.2 < 0$, where $h(x) = 0.6x - 15.2$. Answers may vary.
     (b) $x \approx 25.3$, $x < k$, where $k \approx 25.3$

27. $[5, \infty)$ **29.** $[4, 19)$ **31.** $(-\infty, -37]$

33. $[-1, \infty)$ **35.** $(-3, 5]$ **37.** $(-\infty, -2)$

39. $[2, \infty)$ **41.** $(-\infty, 10.5)$ **43.** $[13, \infty)$

45. $\left[-\dfrac{1}{2}, 2\right]$ **47.** $[-16, 1]$ **49.** $(-\infty, -4)$

51. $x > 4$ **53.** $x < 3$

55. $0 \leq x < k$, where $k \approx 0.86$

57. $\left(\dfrac{13}{2}, \infty\right)$

59. $[k, \infty)$, where $k \approx 0.717$

61. **(a)** $k < x \leq 6$, where $k \approx 1.8$
     (b) The x-intercept represents the altitude where the temperature is $0°F$.
     (c) $\dfrac{53}{29} < x \leq 6$

63. **(a)** The median price of a single-family home has increased, on average, by \$3421 per year.
     (b) & (c) From 1983 to 1988 (approximately)

65. **(a)** $1900 + \dfrac{331}{36.1} \approx 1909$

(b) The x-intercept represents the last year when there were no surfaced roads.

67. (a) $[10, 75, 5]$ by $[0, 100, 5]$

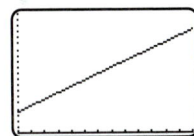

(b) Increasing the ventilation increases the percentage of pollutants removed. A slope of 1.06 means that for each additional liter of air removed per second, the amount of pollutants decreases by 1.06%.

(c) From approximately 40.4 to 59.3 liters per second

69. (a) 1981

(b) From approximately 1981 to 2005

71. $x < -\dfrac{b}{a}$ **73.** $ac \geq bc$

75. $a^2 < b^2$ **77.** $\dfrac{1}{a} > \dfrac{1}{b}$

2.3 ANSWERS

1. (a) Away from home

(b) The point $(1, 35)$ indicates that after 1 hour the distance is 35 miles. The point $(3, 95)$ indicates that after 3 hours the distance is 95 miles.

(c) $y = 30x + 5$. Since $m = 30$, the car is traveling at 30 mph.

(d) $y = 125$

3. e   **5.** d   **7.** f   **9.** d   **11.** a

13. $y = -2.4(x - 4) + 5$

15. $y = -\dfrac{1}{2}(x - 1) - 2$ or $y = -\dfrac{1}{2}(x + 9) + 3$

17. $y = 2(x - 1980) + 5$ or $y = 2(x - 1990) + 25$

19. $y = -7.8x + 5$

21. $y = -\dfrac{1}{2}x + 45$ or $y = -\dfrac{1}{2}(x - 90)$

23. $y = 4(x + 4) - 7$ **25.** $y = \dfrac{3}{2}(x - 1980) + 10$

27. $x = -5$ **29.** $y = 6$

31. $x = 4$ **33.** $x = 19$

35. x-intercept: $\dfrac{5}{8}$; y-intercept: -5

37. x-intercept: $\dfrac{11}{3}$; y-intercept: -11

39. (a) In 1984 the average cost of tuition and fees was $1225. In the year 1987, the average cost of tuition and fees was $1621.

(b) $y = 132(x - 1984) + 1225$. Tuition and fees increased, on average, by $132 per year.

(c) 1990

41. (a) $k = 0.01$ **(b)** 1.1 mm

(c) No, it was a substantial thinning of the ozone layer.

43. (a) $f(x) = 0.29x + 4200$

(b) It represents the annual fixed costs.

45. (a) $y = -\dfrac{11}{3}(x - 1984) + 30$

(b) During 1989

47. (a) No, the slope is not zero.

$[0, 3, 1]$ by $[-2, 2, 1]$

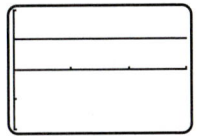

(b) The resolution of most calculator screens is not high enough to show the slight increase in the y-values.

49. (a) They do not appear to be perpendicular in the standard viewing rectangle. (Answers may vary for different calculators.)

(b) In $[-15, 15, 1]$ by $[-10, 10, 1]$ and $[-3, 3, 1]$ by $[-2, 2, 1]$ they appear to be perpendicular.

$[-10, 10, 1]$ by $[-10, 10, 1]$ $[-15, 15, 1]$ by $[-10, 10, 1]$

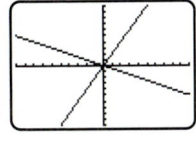   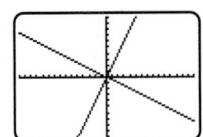

$[-10, 10, 1]$ by $[-3, 3, 1]$ $[-3, 3, 1]$ by $[-2, 2, 1]$

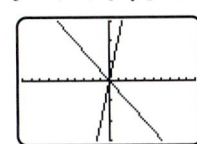   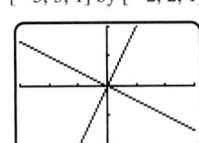

(c) The lines appear perpendicular when the distance shown along the x-axis is approximately 1.5 times farther than the distance along the y-axis.

51. Graph $Y_1 = X$, $Y_2 = -X$, $Y_3 = X + 2$, $Y_4 = -(X - 2) + 2$.

$[-6, 6, 1]$ by $[-4, 4, 1]$

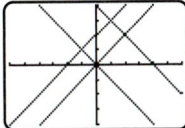

53. Graph $Y_1 = X + 4$, $Y_2 = X - 4$, $Y_3 = 4 - X$, $Y_4 = -X - 4$.

[−9, 9, 1] by [−6, 6, 1]

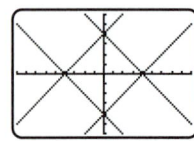

55. $k = 2.5$, $y = 20$ when $x = 8$

57. $k = 0.06$, $x = \$85$ when $y = \$5.10$

59. $\$1048$, $k = 65.5$

61. **(a)** $k = \dfrac{8}{15}$ **(b)** $13\dfrac{1}{3}$ in

63. No. Doubling the speed more than doubles the stopping distance.

2.4 ANSWERS

1. **(a)** Max: 55 mph; min: 30 mph
(b) 12 mi
(c) $f(4) = 40$, $f(12) = 30$, $f(18) = 55$
(d) $x = 4, 6, 8, 12,$ and 16. The speed limit changes at each discontinuity.

3. **(a)** Initial: 50,000 gal; final: 30,000 gal
(b) $0 \le x \le 1$ or $3 \le x \le 4$
(c) $f(2) = 45$, $f(4) = 40$
(d) 5000 gal/day

5. $f(-3) = -10$, $f(1) = 4$, $f(2) = 4$, $f(5) = 1$

7. f is not continuous

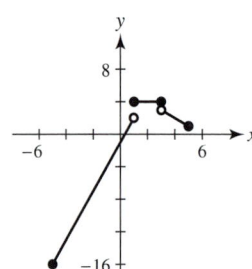

9. $g(-8) = 10$, $g(-2) = -2$, $g(2) = 2$, $g(8) = 5$

11. g is continuous.

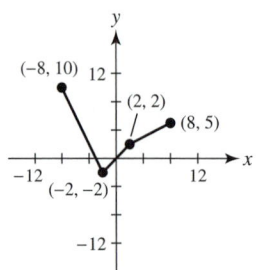

13. **(a)** [−10, 10, 1] by [−10, 10, 1]

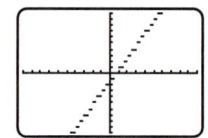

(b) $f(-3.1) = -8$, $f(1.7) = 2$

15. **(a)** [−10, 10, 1] by [−10, 10, 1]

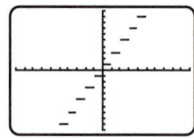

(b) $f(-3.1) = -7$, $f(1.7) = 3$

17. **(a)** $f(x) = 0.8[\![x/2]\!]$ for $6 \le x \le 18$
(b) [6, 18, 1] by [0, 8, 1]

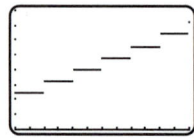

(c) $f(8.5) = \$3.20$, $f(15.2) = \$5.60$

19. **(a)** [−10, 10, 1] by [−10, 10, 1]

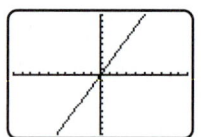

(b) [−10, 10, 1] by [−10, 10, 1] **(c)** $x = 0$

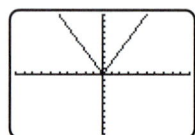

21. **(a)** [−10, 10, 1] by [−10, 10, 1]

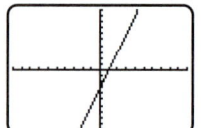

(b) [−10, 10, 1] by [−10, 10, 1] **(c)** $x = 1$

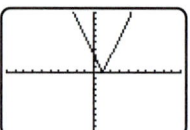

23. **(a)** [−10, 10, 1] by [−10, 10, 1]

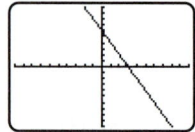

(b) $[-10, 10, 1]$ by $[-10, 10, 1]$ **(c)** $x = 3$

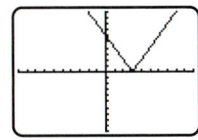

25. **(a)** $x = -1$ or 7 **(b)** $-1 < x < 7$
    (c) $x < -1$ or $x > 7$

27. $x = -5$ or 5; $-5 \le x \le 5$

29. $x = -2.5$ or 7.5; $-2.5 < x < 7.5$

31. $x = \dfrac{7}{3}$ or 1; $x < 1$ or $x > \dfrac{7}{3}$

33. $x = -\dfrac{17}{21}$ or $\dfrac{31}{21}$; $x \le -\dfrac{17}{21}$ or $x \ge \dfrac{31}{21}$

35. $x = -\dfrac{1}{3}$ or $\dfrac{1}{3}$; $x < -\dfrac{1}{3}$ or $x > \dfrac{1}{3}$

37. There are no solutions for the equation or inequality.

39. $\left(-\dfrac{7}{3}, 3\right)$ **41.** $\left[-1, \dfrac{9}{2}\right]$ **43.** $\left(-\dfrac{5}{2}, \dfrac{11}{2}\right)$

45. $(-\infty, 1) \cup (2, \infty)$ **47.** $\left(-\infty, \dfrac{5}{3}\right] \cup \left[\dfrac{11}{3}, \infty\right)$

49. $(-\infty, -8) \cup (16, \infty)$

51.

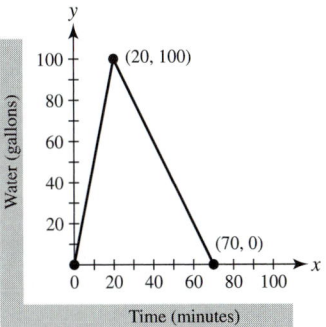

53.

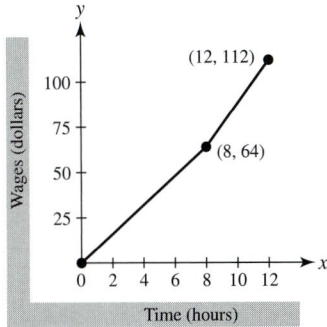

55. **(a)** $19 \le T \le 67$
    (b) The monthly average temperatures in Marquette vary between a low of 19°F and a high of 67°F. The monthly averages are always within 24 degrees of 43°F.

57. **(a)** $28 \le T \le 72$
    (b) The monthly average temperatures in Boston vary between a low of 28°F and a high of 72°F. The monthly averages are always within 22 degrees of 50°F.

59. **(a)** $49 \le T \le 74$
    (b) The monthly average temperatures in Buenos Aires vary between a low of 49°F (possibly in July) and a high of 74°F (possibly in January). The monthly averages are always within 12.5 degrees of 61.5°F.

61. **(a)** $y = x - 1$
    (b)

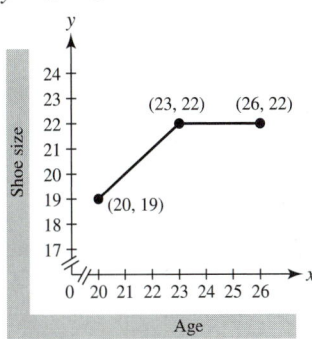

63. **(a)**

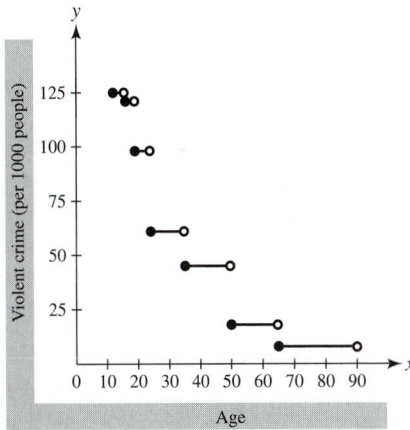

    (b) The likelihood of being a victim of violent crime decreases with age.

65. **(a)** $[1968, 1997, 5]$ by $[460, 490, 5]$

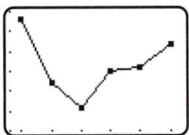

    (b) It appears that scores fell dramatically from 1970 to 1980 and then recovered slightly from 1980 to 1995.
    (c) In this viewing rectangle, it appears that the scores have not changed significantly. Altering the y-scale can affect how a person interprets the data.

[1968, 1997, 5] by [200, 800, 100]

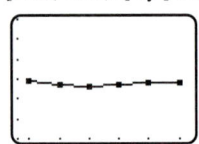

67. (a) Cesarean births increased at a constant rate of 6% every five-year period between 1970 and 1985. It then leveled off at 23%.

[1965, 1995, 5] by [0, 25, 5]

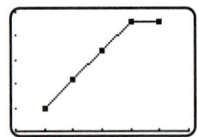

(b) $a = 1.2, b = 5, c = 23$

(c) $f(1978) = 14.6\%, f(1988) = 23\%$. In 1978, 14.6% of births were Cesarean births, while in 1988 the percentage was 23%.

69. (a) Make a line graph using the points (1994, 6), (1997, 18), and (2000, 18).

[1993, 2001, 1] by [0, 20, 5]

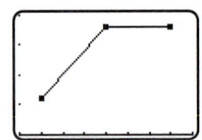

(b) $f(x) = \begin{cases} 4(x - 1994) + 6 & \text{if } 1994 \leq x \leq 1997 \\ 18 & \text{if } 1997 < x \leq 2000 \end{cases}$

2.5 ANSWERS

1. $349 billion **3.** 543,949 inmates

5. 10 sec **7.** $(3, -0.5)$ **9.** $(-2.1, -0.35)$

11. $(\sqrt{2}, 0)$ **13.** $(0, 2b)$ **15.** 5

17. $\sqrt{29} \approx 5.39$ **19.** $\sqrt{133.37} \approx 11.55$

21. $2\sqrt{2} + \sqrt{10} \approx 5.991$

[-4.5, 4.5, 1] by [-1, 5, 1]

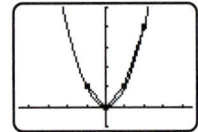

23. $2\sqrt{1.25} + \sqrt{13.25} \approx 5.876$

[-4.5, 4.5, 1] by [0, 6, 1]

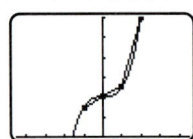

25. 9.747

27. (a) $y = 1.5x - 3.2$

(b) When $x = -2.7$, $y = -7.25$ (interpolation); when $x = 6.3$, $y = 6.25$ (extrapolation)

29. (a) $y = -2.1x + 105.2$

(b) When $x = -2.7$, $y = 110.87$ (extrapolation); when $x = 6.3$, $y = 91.97$ (interpolation)

31. (a) $a \approx 3.0929, b \approx -2.2143$

(b) $y \approx 5.209$

33. (a) $a \approx -3.8857, b \approx 9.3254$

(b) $y \approx -0.00028$. Answers may vary slightly.

35. (a) [1982, 1994, 1] by [200, 500, 100]

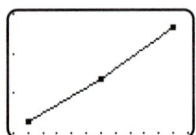

(b) High **(c)** 347

37. 13%

39. (a) Not exactly linear

(b) $f(x) = 263.25(x - 1985) + 1828$

(c) $f(1987) = 2354.5$, which equals the midpoint estimate. The midpoint lies on the graph of f.

41. (a) Yes **(b)** $57.9 billion

43. (a) 1973

(b) The function f will not continue to be a good model far into the future. Once the water is no longer polluted, the number of species in the river will reach its natural level.

45. (a) [-100, 1800, 100] by [-1000, 28,000, 1000]

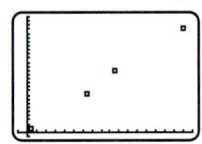

(b) $y = ax + b$, where $a \approx 14.680$ and $b \approx 277.82$

(c) 2500 light years

47. (a) [1973, 1997, 2] by [0, 25, 1]

(b) $y = ax + b$, where $a \approx 0.66045$ and $b \approx -1296.5$

(c) The percentage of women in state legislatures has increased by about 0.66% per year.

(d) 23.1%

49. (a) The graph begins to look like a straight line.

(b) A linear approximation will provide a good approximation over a small interval.

CHAPTER 2   REVIEW ANSWERS

1. $x = 6.4$ **3.** $x = \dfrac{15}{7} \approx 2.143$ **5.** $x \approx -2.9$

7. **(a)** $x \approx 1983$. In 1983 the median U.S. family income was about \$25,000.

(b) $x = \dfrac{3847}{1321.7} + 1980 \approx 1983$

9. **(a)** $x = 2$ **(b)** $x > 2$ **(c)** $x < 2$

11. $(-\infty, 3]$ **13.** $(-1, 3.5]$

15. From 2001 to 2006   **17.** $y = 7x + 30$

19. $y = -3x + 2$ **21.** $x = 6$ **23.** $y = 3$

25. Initially, the car is at home. After traveling 30 mph for 1 hour, the car is 30 miles away from home. During the second hour the car travels 20 mph until it is 50 miles away. During the third hour the car travels toward home at 30 mph until it is 20 miles away. During the fourth hour the car travels away from home at 40 mph until it is 60 miles away from home. During the last hour, the car travels 60 miles at 60 mph until it arrives home.

27. **(a)** $f(-2) = 4, f(-1) = 6, f(2) = 3, f(3) = 4$
(b) f is continuous.

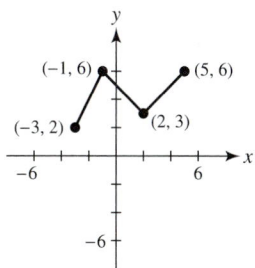

(c) $x = -2.5$ or 2

29.

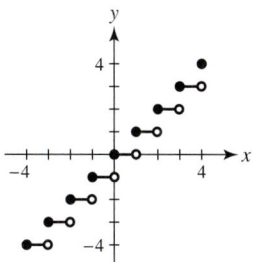

31. $x = \pm 3; x < -3$ or $x > 3$

33. $x = \dfrac{17}{3}$ or $x = -1; x < -1$ or $x > \dfrac{17}{3}$

35. **(a)** [1988, 1995, 1] by [20.5, 20.9, 0.1]

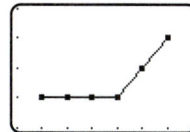

(b) $f(x) = \begin{cases} 20.6 & \text{if } 1989 \le x \le 1992 \\ 0.1(x - 1992) + 20.6 & \text{if } 1992 < x \le 1994 \end{cases}$

37. 155,590

39. **(a)** $f(x) = 40{,}000(x - 1996) + 900{,}000$,
$g(x) = 40{,}000(x - 1996) + 650{,}000$
(b) They both have slope 40,000.

[1996, 2000, 1] by [6×10^5, 1.1×10^6, 10^5]

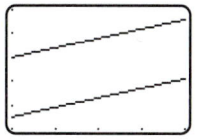

(c) No. This distance equals 250,000 and represents the difference between the maximum and minimum number of cases in a given year x.

41. **(a)** $y = 1.5(x - 1991) + 422$ or $y = 1.5(x - 1995) + 428$. Scores rose, on average, by 1.5 points per year.
(b) 429.5 (extrapolation)

43. The tank initially contains 25 gallons. When $0 \le x \le 4$, only the 5 gal/min inlet pipe is open. When $4 < x \le 8$, only the outlet pipe is open. When $8 < x \le 12$, both inlet pipes are open. When $12 < x \le 16$, all pipes are open. When $16 < x \le 24$, the 2 gal/min inlet pipe and the outlet pipe are open. When $24 < x \le 28$, all pipes are closed.

CHAPTER 3 NONLINEAR FUNCTIONS AND EQUATIONS

3.1   ANSWERS

1. **(a)** $a > 0$ **(b)** $x = -6$ or 2 **(c)** Positive

3. **(a)** $a > 0$ **(b)** $x = -4$ **(c)** Zero

5. $x = -2$ or 3 **7.** $x = \pm\sqrt{3} \approx \pm 1.7$

9. $x = 1.5$ **11.** $x = 2 \pm \sqrt{2} \approx 3.4, 0.6$

13. No real solutions   **15.** $x = -0.75$ or 0.2

17. $x = 0.7$ or 1.2 **19.** $x = -96$ or 72

21. **(a)** $3x^2 - 12 = 0$
(b) $b^2 - 4ac = 144 > 0$. There are two real solutions.
(c) $x = \pm 2$

23. **(a)** $x^2 - 2x + 1 = 0$
(b) $b^2 - 4ac = 0$. There is one real solution.
(c) $x = 1$

25. **(a)** $2x^2 + 5x - 12 = 0$
(b) $b^2 - 4ac = 121 > 0$. There are two real solutions.
(c) $x = -4$ or 1.5

27. (a) $\frac{1}{4}x^2 + 2x + 4 = 0$

    (b) $b^2 - 4ac = 0$. There is one real solution.

    (c) $x = -4$

29. (a) $\frac{1}{2}x^2 + x + \frac{13}{2} = 0$

    (b) $b^2 - 4ac = -12 < 0$. There are no real solutions.

31. (a) $3x^2 + x - 1 = 0$

    (b) $b^2 - 4ac = 13 > 0$. There are two real solutions.

    (c) $x = -\frac{1}{6} \pm \frac{1}{6}\sqrt{13} \approx 0.4343, -0.7676$

33. $-2 \pm \sqrt{10}$ **35.** $-\frac{5}{2} \pm \frac{1}{2}\sqrt{41}$

37. (a) R quadruples   (b) $a = 0.5$

    (c) $x = \sqrt{1000} \approx 31.6$. The safe speed for a curve with radius 500 feet is about 31 mph or less.

39. (a) $[-5, 45, 5]$ by $[0, 1500, 100]$

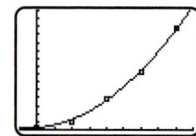

    (b) and (c) $x \approx 35.2$ or in 1985

41. (a) $f(x) = -0.00184x^2 + 50$. Answers may vary.

    (b) $x \approx \pm 165$. Only the positive solution has meaning. It represents the number of days (165) for all the worms to die.

43. 8.5 by 11

45. (a) $-3 < x < 2$ (b) $x \le -3$ or $x \ge 2$

47. (a) $x = -2$ (b) $x \ne -2$

49. $-1 < x < 4$ **51.** $x < -3$ or $x > 2$

53. $-2 \le x \le 2$ **55.** All real numbers

57. $x \le -2$ or $x \ge 3$ **59.** $-\frac{1}{3} < x < \frac{1}{2}$

61. $-\sqrt{5} \le x \le \sqrt{5}$

63. $x \le k$ or $x \ge 22.4$, where $k \approx 3.29$

65. From 1991 to 1993

67. (a) $163x^2 - 146x + 205 > 2000$ or $163x^2 - 146x - 1795 > 0$, where x is positive

    (b) $x > k$, where $k \approx 3.796$. This corresponds to 1989 or later.

69. (a) The height does not change by the same amount each 15-second interval.

    (b) From 43 to 95 sec (approximately)

(c) $[-25, 200, 25]$ by $[-5, 20, 5]$

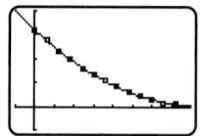

(d) From 43.8 to 93.1 sec (approximately)

3.2 ANSWERS

1. Power and root function

3. Polynomial with degree 2

5. Power but not a root function

7. (a) The data is linear. The linear function f might model the data better.

    (b) The graph of f fits the data better than g.

    $[0, 10, 1]$ by $[0, 15, 1]$

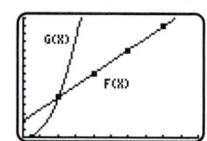

9. Local maxima: approximately 800 and 900; local minima: approximately 600 and 700

11. (a) Local max: approximately 5.5; local min: approximately -5.5

    (b) No absolute extrema

13. (a) Local maxima: approximately 17 and 27; local minima: approximately -10 and 24

    (b) No absolute extrema

15. (a) Local max: 5.25   (b) Absolute max: 5.25

17. (a) Local min: -8; local max: 4.5

    (b) Absolute min: -8

19. (a) Local max: 8   (b) Absolute max: 8

21. (a) The energy consumption increased, reached a maximum value, and then started to decrease.

    $[0, 30, 5]$ by $[6, 16, 1]$ $[0, 30, 5]$ by $[6, 16, 1]$

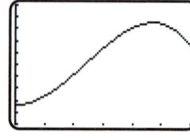

    (b) Local max: approximately 14.5. In 1973 maximum energy use peaked at 14.5 quadrillion Btu.

23. (a) Possible absolute maximum in January and absolute minimum in July

    (b) Absolute maximum of $140 in January and an absolute minimum of $15 in July

[1, 12, 1] by [0, 150, 10] [1, 12, 1] by [0, 150, 10]

25. *f* is odd and the point (8, −6) lies on its graph.

27. *f* is odd.

29. Note that *f*(0) can be any number.

x	−3	−2	−1	0	1	2	3
f(*x*)	21	−12	−25	1	−25	−12	21

31. Answers may vary.

x	−2	−1	0	1	2
f(*x*)	5	1	2	3	4

33.

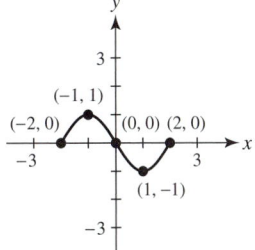

35. **(a)** Even **(b)** 83°F **(c)** They are equal.
(d) Monthly average temperatures are symmetric about July. July has the highest average and January the lowest. The pairs June-August, May-September, April-October, March-November, and February-December have approximately the same average temperatures.

37. **(a)** *h*(−2) = *h*(2) = 336. Two seconds before and two seconds after the time when the maximum height is reached, the projectile's height is 336 feet.
(b) *h*(−5) = *h*(5) = 0. Five seconds before and five seconds after the time when the maximum height is reached, the projectile is on the ground.
(c) *h* is an even function. [−5, 5, 1] by [0, 500, 100]

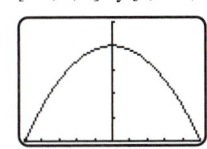

(d) *h*(−*x*) = *h*(*x*) when −5 ≤ *x* ≤ 5. This means that the projectile is at the same height either *t* seconds before or after the time when the maximum height is attained.

39. Even **41.** Odd **43.** Neither

45. Even **47.** Even

49. Answers may vary.

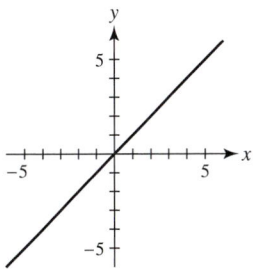

51. No. If (2, 5) is on the graph of an odd function *f*, then so is (−2, −5). Since *f* would pass through (−3, −4) and then (−2, −5), it could not always be increasing.

53. 4 **55.** $\frac{1}{8}$ **57.** 9 **59.** 2

61. **(a)** $f(x) = \sqrt{x^3}$
(b) All nonnegative real numbers
(c) *x* = 9

63. **(a)** $f(x) = \sqrt[3]{x^4}$
(b) All real numbers
(c) $x = \pm 10^{3/4} = \pm\sqrt[4]{1000} \approx \pm5.6234$

65. 2 **67.** 81 **69.** 7 **71.** −1

73. $f(1.2) = 1.2^{1.62} \approx 1.3436$

75. $f(50) = 50^{3/2} - 50^{1/2} \approx 346.48$

77. b

79. **(a)** *b* = 1960
(b) *a* ≈ −1.2

[0, 5, 1] by [0, 5000, 1000]

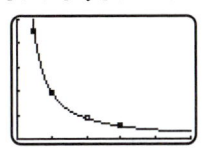

(c) $f(4) = 1960(4)^{-1.2} \approx 371$. If the zinc ion concentration reaches 371 mg/l, a rainbow trout will survive, on average, 4 minutes.

81. **(a)** *f*(2) ≈ 1.06 g
(b) & **(c)** Approximately 1.1 g

3.3 ANSWERS

1. (a) The turning points are approximately (1.6, 3.6), (3, 1.2), (4.4, 3.6).
   (b) After 1.6 minutes the runner is 360 feet from the starting line. The runner turns and jogs toward the starting line. After 3 minutes the runner is 120 feet from the starting line, turns, and jogs away from the starting line. After 4.4 minutes the runner is again 360 feet from the starting line. The runner turns and jogs back to the starting line.

3. (a) 0.5   (b) Positive   (c) 1

5. (a) $-6, -1$, and 6   (b) Negative
   (c) 4

7. (a) $-3, -1, 0, 1$, and 2
   (b) Positive
   (c) 5

9. (a) $[-10, 10, 1]$ by $[-10, 10, 1]$

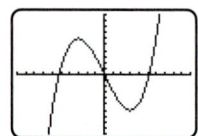

   (b) $(-3, 6), (3, -6)$
   (c) Local max: 6; local min: -6

11. (a) $[-10, 10, 1]$ by $[-10, 10, 1]$

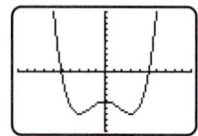

   (b) There are three turning points located at $(-3, -7.025), (0, -5)$, and $(3, -7.025)$.
   (c) Local min: -7.025; local max: -5

13. (a) Degree: 3; leading coefficient: -1
   (b) Up on left side, down on right side
   (c) $[-10, 10, 1]$ by $[-10, 10, 1]$

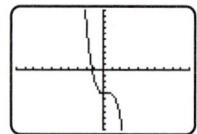

15. (a) Degree: 5; leading coefficient: 0.1
   (b) Down on left side, up on right side
   (c) $[-10, 10, 1]$ by $[-10, 10, 1]$

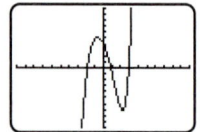

17. (a) Degree: 2; leading coefficient: $-\dfrac{1}{2}$
   (b) Down on both sides
   (c) $[-10, 10, 1]$ by $[-10, 10, 1]$

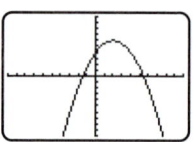

19. As the viewing rectangle increases in size, the graphs begin to look alike. Each formula contains the term $2x^4$, which determines the end behavior of the graph for large values of $|x|$.

$[-4, 4, 1]$ by $[-4, 4, 1]$ $[-10, 10, 1]$ by $[-100, 100, 10]$

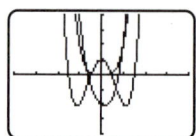   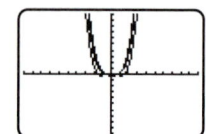

$[-100, 100, 10]$ by $[-10^6, 10^6, 10^5]$

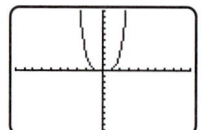

21. (a) $[-5, 5, 1]$ by $[-2, 12, 1]$

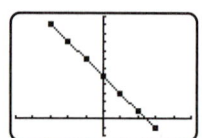

   (b) Degree 1

23. (a) $[-5, 5, 1]$ by $[-5, 15, 5]$

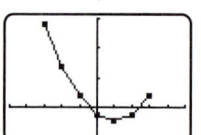

   (b) Degree 2

25. Answers may vary.

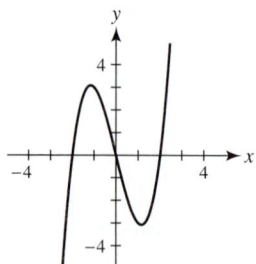

27. Answers may vary.

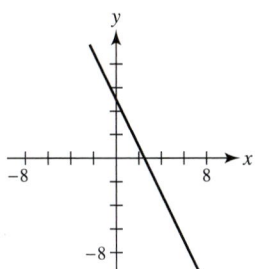

29. Answers may vary.

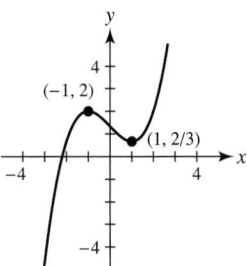

31.

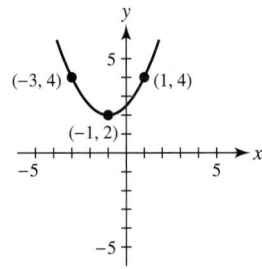

33. **(a)** Marijuana use among seniors peaked in 1978 at 37%. There was a decline until 1990 when marijuana use was at a low of 14%. Marijuana use then increased significantly from 1990 to 1995.
[−2, 22, 1] by [10, 40, 10]

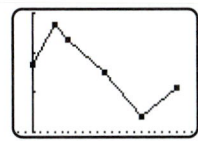

(b) The cubic function h may model the data best.
(c) h models the data best.
[−2, 22, 1] by [10, 40, 10]   [−2, 22, 1] by [10, 40, 10]

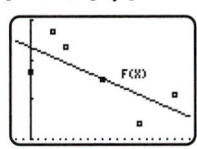 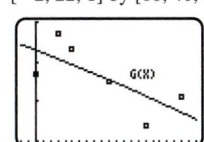

[−2, 22, 1] by [10, 40, 10]

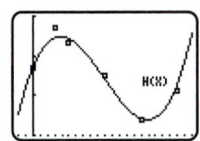

35. **(a)** As the speed x increases, so does the minimum sight distance y.
(b) Degree: 1; $D(x) = 34x + 130$. Answers may vary.

[15, 75, 5] by [500, 2800, 100]

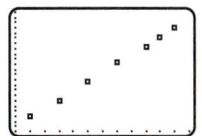

(d) $D(43) = 1592$ ft. Answers may vary.

37. **(a)** Approximately $(1, 13)$ and $(7, 72)$
(b) The monthly average low temperature of 13°F occurs in January. The monthly average high temperature of 72°F occurs in July.

39. The graph of $g(x) = -x^3 + 1$ is a reflection of the graph of $f(x)$ across the x-axis.

[−10, 10, 1] by [−10, 10, 1]

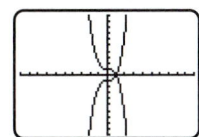

41. The graph of $g(x) = x^2 - 2x - 2$ is a reflection of the graph of $f(x)$ across the x-axis.

[−10, 10, 1] by [−10, 10, 1]

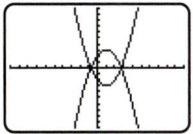

43. The graph of $g(x) = -2x - 6$ is a reflection of the graph of $f(x)$ across the y-axis.

[−10, 10, 1] by [−20, 10, 1]

45. The graph of $g(x) = -x^2 - 7x - 4$ is a reflection of the graph of $f(x)$ across the y-axis.

[−10, 10, 1] by [−10, 10, 1]

47.

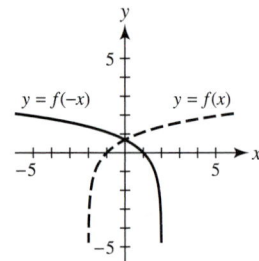

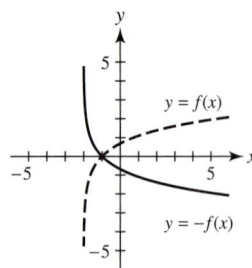

49. $[-9, 9, 1]$ by $[-6, 6, 1]$ $[-9, 9, 1]$ by $[-6, 6, 1]$

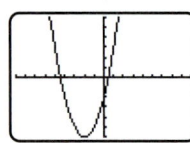   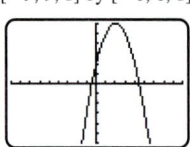

51. $f(-2) \approx 5, f(1) \approx 0$

53. $f(-2) = 6, f(1) = 7, f(2) = 9$

55.   **(a)**

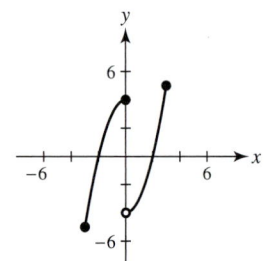

    (b) f is not continuous.

57.   **(a)**

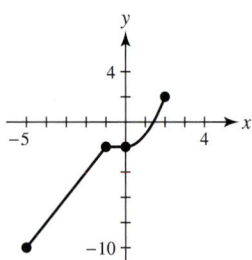

    (b) f is continuous.

59.   **(a)**   4 sec
          $[-1, 8, 1]$ by $[-10, 170, 10]$

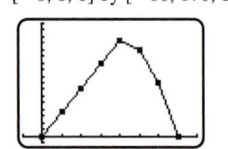

(b)   From 0 to 4 sec; from 4 to 7 sec

(c) $m = 36, a = -16, b = 144$

3.4   ANSWERS

1.   Quotient: $x^2 + x - 2$; remainder: 0

3.   Quotient: $2x^3 - 9x^2 + 4x - 23$; remainder: 40

5.   Quotient: $3x^2 + 3x - 4$; remainder: 6

7.   Quotient: $x^2 - 3x + 2$; remainder: 0

9. $x^3 - 8x^2 + 15x - 6$

11. $f(x) = 2\left(x - \dfrac{11}{2}\right)(x - 7)$

13. $f(x) = (x + 2)(x - 1)(x - 3)$

15. $f(x) = -2(x + 5)\left(x - \dfrac{1}{2}\right)(x - 6)$

17. $f(x) = 2(x - 1)(x - 2)(x - 3)\left(x - \dfrac{7}{2}\right)$

19. $f(x) = 2(x + 3)(x - 2)(x - 10)$

21. $f(x) = -7(x + 10)\left(x + \dfrac{7}{3}\right)(x + \sqrt{3})(x - \sqrt{3})$

23. $f(x) = (x + 4)(x - 2)(x - 8)$

25. $f(x) = -1(x + 8)(x + 4)(x + 2)(x - 4)$

For Exercises 27–32, answers may vary slightly.

27. $f(x) \approx (x + 2.0095)(x - 0.11639)(x - 2.9931)$

29. $f(x) \approx -0.7(x + 4.0503)(x + 0.51594)(x - 1.7091)$

31. $f(x) \approx 2(x + 2.6878)(x + 1.0957)(x - 0.55475)$
          $(x - 3.9787)$

33. -2 (odd), 4 (even)

35. -6 (even), -1 (even), 4 (odd)

37. $f(x) = (x + 1)^2(x - 6)$

39. $f(x) = (x - 2)^3(x - 6)$

41. $f(x) = (x + 2)^2(x - 4)$

43. $f(x) = -1(x + 3)^2(x - 3)^2$

45. $f(x) = (x - 1)(x - 3)(x - 5)$

47. $f(x) = -4(x + 4)\left(x - \dfrac{3}{4}\right)(x - 3)$

49. $x = -6$ **51.** $x \approx -0.6823$

53.   **(a)** $-3, \dfrac{1}{2}, 1$

          (b) $f(x) = 2(x + 3)\left(x - \dfrac{1}{2}\right)(x - 1)$

55.   **(a)** $-2, -1, 1, \dfrac{3}{2}$

          (b) $f(x) = 2(x + 2)(x + 1)(x - 1)\left(x - \dfrac{3}{2}\right)$

57.   **(a)** $\dfrac{1}{3}, 1, 4$

          (b) $f(x) = 3\left(x - \dfrac{1}{3}\right)(x - 1)(x - 4)$

59. (a) 1

(b) $f(x) = (x + \sqrt{7})(x - 1)(x - \sqrt{7})$

61. $-3, 0, 2$ **63.** $-1, 1$ **65.** $-2, 0, 2$

67. $x = -5, 0,$ or 5 **69.** $x = \pm 2$

71. $x = -3, 0,$ or 6 **73.** $x = 0$ or 1

75. (a) As x increases, C decreases.

(b) $[0, 70, 10]$ by $[0, 22, 5]$

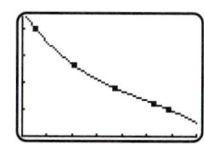

(c) $0 \le x < k$, where $k \approx 32.1$

77. (a) $f(x) \approx -0.184(x + 6.01)(x - 2.15)(x - 11.7)$

(b) The zero of -6.01 has no significance. The zeros of $2.15 \approx 2$ and $11.7 \approx 12$ indicate that during February and December the average temperature is $0°F$.

79. (a) During 1986

(b) $x \approx 6.53$ or during 1986

3.5 ANSWERS

1. $2i$ **3.** $10i$ **5.** $i\sqrt{23}$

7. $i\sqrt{12} = 2i\sqrt{3}$ **9.** $8i$ **11.** $-2 - i$

13. $13 - 16i$ **15.** $-1 + 6i$ **17.** $4 + 8i$

19. $5 - i$ **21.** $4 - 7i$ **23.** $\dfrac{1}{2} - \dfrac{1}{2}i$

25. $\dfrac{19}{26} + \dfrac{9}{26}i$ **27.** $-\dfrac{2}{25} + \dfrac{4}{25}i$

29. $-18.5 + 87.4i$ **31.** $8.7 - 6.7i$

33. $-117.27 + 88.11i$ **35.** $-0.9207 - 0.2364i$

37. $10 + 6i$

39. (a) Two real zeros   (b) $x = -1$ or $\dfrac{3}{2}$

41. (a) Two imaginary zeros

(b) $x = -\dfrac{1}{2} \pm \dfrac{\sqrt{7}}{2}i$

43. (a) Two imaginary zeros

(b) $x = -\dfrac{2}{5} \pm \dfrac{1}{5}i$

45. One real zero; two imaginary zeros

47. Two real zeros; two imaginary zeros

49. Three real zeros; two imaginary zeros

51. (a) $f(x) = (x - 6i)(x + 6i)$

(b) $f(x) = x^2 + 36$

53. (a) $f(x) = -1(x + 1)(x - 2i)(x + 2i)$

(b) $f(x) = -x^3 - x^2 - 4x - 4$

55. (a) $f(x) = 10(x - 1)(x + 1)(x - 3i)(x + 3i)$

(b) $f(x) = 10x^4 + 80x^2 - 90$

57. $f(x) = (x - 5i)(x + 5i)$

59. $f(x) = 3(x - 0)(x - i)(x + i)$ or $3x(x - i)(x + i)$

61. $f(x) = (x - i)(x + i)(x - 2i)(x + 2i)$

63. $f(x) = (x - 1)(x + 3)(x - 2i)(x + 2i)$

65. $x = 2 \pm i$ **67.** $x = \dfrac{3}{2} \pm \dfrac{\sqrt{11}}{2}i$

69. $x = -1 \pm i\sqrt{3}$ **71.** $x = 0, \pm i$

73. $x = 2, \pm i\sqrt{7}$ **75.** $x = 0, \pm i\sqrt{5}$

77. $x = 0, \dfrac{1}{2} \pm \dfrac{\sqrt{15}}{2}i$

3.6 ANSWERS

1. Yes   **3.** Yes   **5.** No

7. Yes, since $f(x) = \dfrac{4 + x}{x}$

9. Horizontal: $y = 4$; vertical: $x = 2$

11. Horizontal: $y = -4$; vertical: $x = \pm 2$

13. Horizontal: $y = 2$; vertical: $x = 3$

15. Horizontal: $y = 0$; vertical: $x = \pm\sqrt{5}$

17. Horizontal: none; vertical: $x = -5$ or 2

19. Horizontal: none; vertical: none, since $f(x) = x - 3$ for $x \ne 3$

21. b   **23.** d

25. Horizontal: $y = 2$; vertical: $x = 0$

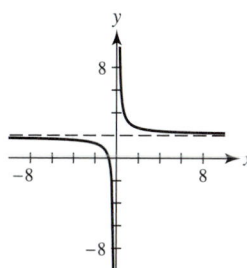

27. Horizontal: $y = -1$; vertical: $x = -3$

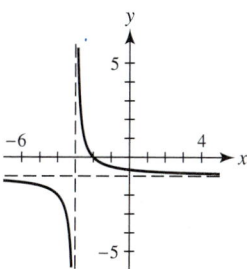

29. (a) $D = \{x \mid x \ne 2\}$

(c) Horizontal: $y = 1$; vertical: $x = 2$

(d)

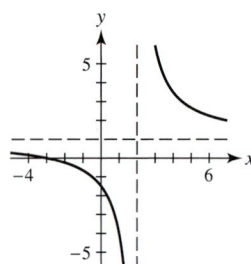

31. (a) $D = \{x \mid x \neq \pm 2\}$
(c)   Horizontal: $y = 0$; vertical: $x = \pm 2$
(d)

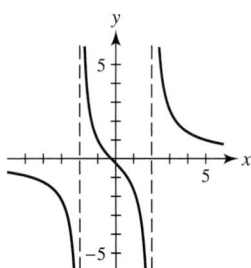

33. (a) $D = \{x \mid x \neq \pm 2\}$
(c)   Horizontal: $y = 0$; vertical: $x = \pm 2$
(d)

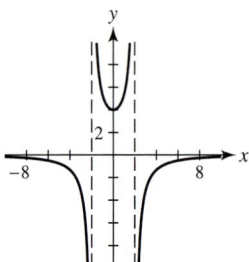

35. (a) $D = \{x \mid x \neq 2\}$
(c)   Horizontal: none; vertical: none, since $f(x) = x + 2$ for $x \neq 2$
(d)

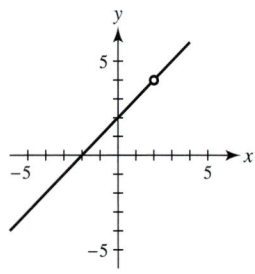

37. $y = 3$

39.   Answers may vary.

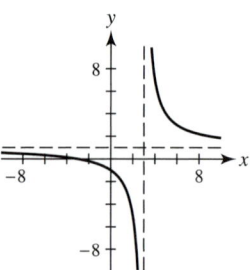

41.   Answers may vary.

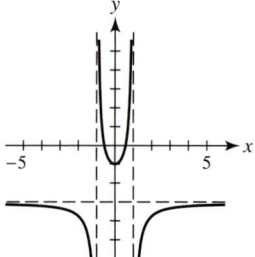

43. $f(x) = \dfrac{1}{x - 5}$. Answers may vary.

45. $f(x) = \dfrac{x + 1}{x + 3}$. Answers may vary.

47. $f(x) = \dfrac{1}{x^2 - 9}$. Answers may vary.

49.   Slant: $y = x - 1$; vertical: $x = -1$

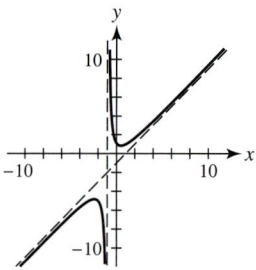

51.   Slant: $y = \dfrac{1}{2}x - 3$; vertical: $x = -2$

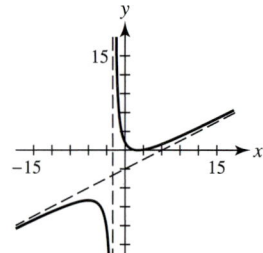

53. **(a)** $y = 10$
[0, 14, 1] by [0, 14, 1]

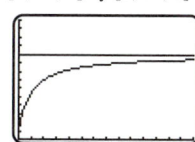

(b) When $x = 0$, there are 1 million insects.
(c) It starts to level off at 10 million.
(d) The horizontal asymptote $y = 10$ represents the limiting population after a long time.

55. $x = -3$ **57.** $x = \frac{1}{2}$ or 2 **59.** $x = \frac{1}{2}$

61. $x = \pm\sqrt{2}$ **63.** No real solutions

65. **(a)** About 12.4 cars per minute
(b) 3

67. **(a)** $f(400) = \frac{2540}{400} = 6.35$ in. A curve designed for 60 mph with a radius of 400 feet should have the outer rail elevated 6.35 inches.
(b) As the radius x of the curve increases, the elevation of the outer rail decreases.
[0, 600, 100] by [0, 50, 5]

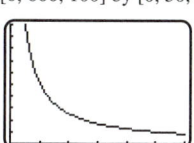

(c) The horizontal asymptote is $y = 0$. As the radius of the curve increases without a bound $(x \to \infty)$, the tracks become straight and no elevation or banking $(y \to 0)$ is necessary.

69. **(a)** $P(9) = \frac{9 - 1}{9} \approx 0.89 = 89\%$

(b) $P(1.9) = \frac{1.9 - 1}{1.9} \approx 0.47 = 47\%$

71. **(a)**

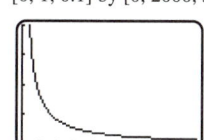

(b) As the uphill grade x increases, the stopping distance decreases, which agrees with intuition.

73. **(a)** [0, 1, 0.1] by [0, 2000, 500]

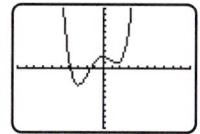

(b) The vertical asymptote is $x = 0$. As the coefficient of friction x decreases to 0, the stopping distances become larger without a maximum.

75. $k = 6$ **77.** $k = 8$

79. $T = 160$ **81.** $y = 2$

83. Becomes half as much

85. Becomes 27 times as much

87. $k = 0.5, n = 2$

89. $k = 3, n = 1$ **91.** 1.18 g

93. $\sqrt{50} \approx 7$ times farther **95.** $\frac{2}{9}$ ohm

CHAPTER 3 REVIEW ANSWERS

1. **(a)** $a > 0$ **(b)** $-3, 2$ **(c)** Positive

3. $x = -4$ or 5 **5.** $x < 1$ or $x > 2$

7. Neither **9.** Even **11.** Odd

13. $f(x) = (x - 1)(x - 2)(x - 3)$

15. $f(x) \approx (x + 3.0938)(x + 0.83704)(x - 1.9308)$

17. **(a)** Local min: $-4.5, -0.5$; local max: 0
(b) Absolute min: -4.5; absolute max: none

19. **(a)** Local min: -60.08; local max: 15, 75.12
(b) Absolute min: none; absolute max: 75.12
(c) Increases on $(-\infty, -3.996] \cup [0.9995, 5.007]$; decreases on $[-3.996, 0.9995] \cup [5.007, \infty)$, where endpoints are rounded to four significant digits

21. **(a)** One local maximum, two local minima, two x-intercepts
(b) Two local maxima, one local minimum, three x-intercepts
(c) One local maximum, two local minima, four x-intercepts

[-10, 10, 1] by [-100, 100, 10]

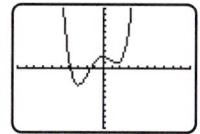

[-10, 10, 1] by [-10, 10, 1]

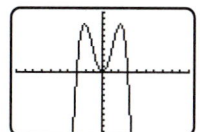

[-10, 10, 1] by [-100, 100, 10]

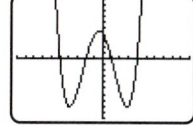

23. Answers may vary.

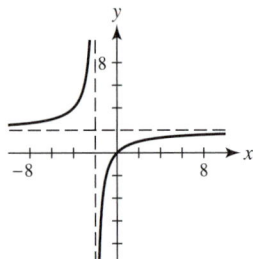

25. $x = 0, \pm\sqrt{3}$

27. $f(x) = (x + 2)^2(x - 2)^3$. The leading coefficient may vary.

29. Answers may vary.

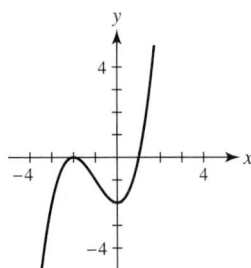

31. (a) 1   (b) 2   (c) 3

33. -1 **35.** $-10 - 11i$ **37.** $x = \pm\frac{3}{2}i$

39. One real zero, two imaginary zeros

41. $f(x) = 2(x - i\sqrt{2})(x + i\sqrt{2})$

43. Zeros: $-3, \frac{1}{2}, 2; f(x) = 2(x + 3)\left(x - \frac{1}{2}\right)(x - 2)$

45. $x = 4$ **47.** $x = 6$ **49.** $x = 4$

51. (a) Dog: 148; person: 69
   (b) 6.4 in

53. (a) The change in debt is not constant each 4-year period.
   (b) $x = 1990$. Visa and MasterCard debt reached \$212 billion in 1990.

55. (a) f is odd. From September to March 21, daylight hours are less than 12 hours.
   (b)

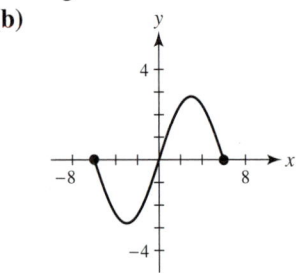

57. (a) $a \approx 0.4$. Answers may vary.
   (b) $x \approx 2001$. The number of accounts may reach 10 million in 2001.

CHAPTER 4   INVERSE FUNCTIONS

4.1   ANSWERS

1. (a) 5   (b) 5   (c) 0   (d) Undefined

3.

x	-2	0	2	4
$(f + g)(x)$	6	5	5	15
$(f - g)(x)$	-6	5	9	5
$(fg)(x)$	0	0	-14	50
$(f/g)(x)$	0	—	-3.5	2

5. (a) 2   (b) 4   (c) 0   (d) $-\frac{1}{3}$

7. (a) -5 (b) -5 (c) -3 (d) $-\frac{1}{3}$

9. (a) $(f + g)(x) = 2x + x^2$; all real numbers
   (b) $(f - g)(x) = 2x - x^2$; all real numbers
   (c) $(fg)(x) = 2x^3$; all real numbers
   (d) $(f/g)(x) = \frac{2}{x}; D = \{x \mid x \neq 0\}$

11. (a) $(f + g)(x) = 2x; D = \{x \mid x \geq 1\}$
   (b) $(f - g)(x) = -2\sqrt{x - 1}; D = \{x \mid x \geq 1\}$
   (c) $(fg)(x) = x^2 - x + 1; D = \{x \mid x \geq 1\}$
   (d) $(f/g)(x) = \frac{x - \sqrt{x - 1}}{x + \sqrt{x - 1}}; D = \{x \mid x \geq 1\}$

13. (a) $(f + g)(x) = \frac{4}{x + 1}; D = \{x \mid x \neq -1\}$
   (b) $(f - g)(x) = -\frac{2}{x + 1}; D = \{x \mid x \neq -1\}$
   (c) $(fg)(x) = \frac{3}{(x + 1)^2}; D = \{x \mid x \neq -1\}$
   (d) $(f/g)(x) = \frac{1}{3}; D = \{x \mid x \neq -1\}$

15. (a) $Y_1 = \sqrt{(X)}, Y_2 = X + 1$, and $Y_3 = Y_1 + Y_2$
      $[0, 9, 1]$ by $[0, 15, 1]$

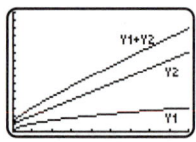

   (b) To determine the graph $f + g$ add corresponding y-coordinates on the graphs of f and g.

17. (a) 5   (b) Undefined   (c) 4

19. (a) -4 (b) 2

21. (a) 3   (b) 4

23. (a) $(f \circ g)(x) = (x^2 + 3x - 1)^3$; all real numbers
   (b) $(g \circ f)(x) = x^6 + 3x^3 - 1$; all real numbers
   (c) $(f \circ f)(x) = x^9$; all real numbers

25. (a) $(f \circ g)(x) = 1 - x; D = \{x \mid x \leq 1\}$

(b) $(g \circ f)(x) = \sqrt{1 - x^2}; D = \{x \mid -1 \le x \le 1\}$
(c) $(f \circ f)(x) = x^4$; all real numbers
27. (a) $(f \circ g)(x) = 2 - 3x^3$; all real numbers
(b) $(g \circ f)(x) = (2 - 3x)^3$; all real numbers
(c) $(f \circ f)(x) = 9x - 4$; all real numbers

Answers may vary for Exercises 29–36.
29. $f(x) = x - 2, g(x) = \sqrt{x}$
31. $f(x) = x + 2, g(x) = 5x^2 - 4$
33. $f(x) = 2x + 1, g(x) = 4x^3$
35. $f(x) = x^3 - 1, g(x) = x^2$
37. (a)

x (yr)	1991	1992	1993	1994	1995	1996
h(x) ($ million)	38	42	44	50	54	61

(b) $h(x) = f(x) + g(x)$
39. (a) 50
(b) $(f + g)(x)$ computes the total SO_2 emissions from coal and oil during year x.
(c)

x	1860	1900	1940	1970	2000
(f + g)(x)	2.4	12.8	26.5	50.0	78.0

41. (a)

x	1990	2000	2010	2020	2030
h(x)	32	35.5	39	42.5	46

(b) $h(x) = f(x) + g(x)$
43. $h(x) = 0.35x - 664.5$
45. (a) $h(x) = g(x) - f(x)$
(b) $h(1995) = 200, h(2000) = 250$
(c) $h(x) = 10(x - 1995) + 200$ or $h(x) = 10x - 19{,}750$
47. (a)

x	C	Si	N	K	P
h(x)	5400	2000	30,400	1300	80,000

(b) $h(x)$ computes how many times greater the concentration of element x is in aquatic plants than in freshwater.
49. (a)

x	4	8	12
g(x)	1.96	1.80	1.39

(b) The function g represents the factor by which the new cases are increasing each 4-week period. The numbers 1.96, 1.80, and 1.39 are decreasing. When this ratio is less than one, the number of new cases decreases.
51. (a) 30,000 gal
(b) $(g \circ f)(x)$ computes the gallons of water in the pool after x days.
53. (a) $(g \circ f)(1960) = 1.98$
(b) $(g \circ f)(x) = 0.165(x - 1948)$
(c) $f, g,$ and $g \circ f$ are all linear functions.

4.2 ANSWERS
1. Closing a window
3. Closing a book, standing up, and walking out of the classroom
5. Subtract 2 from x; $x + 2$ and $x - 2$
7. Divide x by 3 and then add 2; $3(x - 2)$ and $\frac{x}{3} + 2$
9. Subtract 1 from x and cube the result; $\sqrt[3]{x} + 1$ and $(x - 1)^3$
11. Take the reciprocal of x; $\frac{1}{x}$ and $\frac{1}{x}$
13. One-to-one   **15.** Not one-to-one
17. Not one-to-one   **19.** One-to-one
21. One-to-one   **23.** Not one-to-one
25. Not one-to-one   **27.** Not one-to-one
29. No
31. Yes, the total debt has increased each fiscal year.
33.

x	5	7	9
$f^{-1}(x)$	1	2	3

For f: $D = \{1, 2, 3\}, R = \{5, 7, 9\}$;
for f^{-1}: $D = \{5, 7, 9\}, R = \{1, 2, 3\}$
35.

x	0	1	4	9	16
$f^{-1}(x)$	0	1	2	3	4

For f: $D = \{0, 1, 2, 3, 4\}, R = \{0, 1, 4, 9, 16\}$;
for f^{-1}: $D = \{0, 1, 4, 9, 16\}, R = \{0, 1, 2, 3, 4\}$
37.

x	−3	0	3	6
$f^{-1}(x)$	−8	−5	−2	1

39.

x	−8	−1	8	27
$f^{-1}(x)$	−2	−1	2	3

41. 5
43. 1

45. **(a)** $f(1) \approx \$110$ **(b)** $f^{-1}(110) \approx 1$ year
(c) $f^{-1}(160) \approx 5$ years
$f^{-1}(x)$ computes the years necessary for the account to accumulate x dollars.

47.

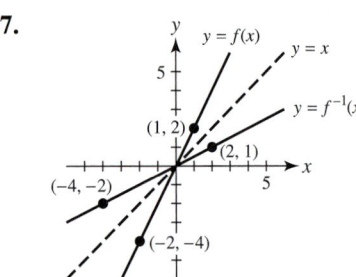

49.

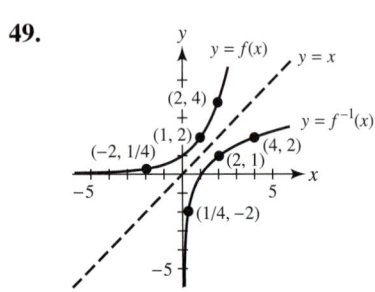

51. $f^{-1}(x) = x^3$ **53.** $f^{-1}(x) = -\dfrac{1}{2}x + 5$

55. $f^{-1}(x) = \dfrac{x + 1}{3}$

57. $f^{-1}(x) = \dfrac{x + 15}{5}$. D and R are all real numbers.

59. $f^{-1}(x) = \dfrac{1}{x} - 3. D = \{x \mid x \neq 0\}$ and
$R = \{y \mid y \neq -3\}$

61. $f^{-1}(x) = \sqrt[3]{\dfrac{x}{2}}$. D and R are all real numbers.

63. $f^{-1}(x) = \sqrt{x}$. D and R include all nonnegative real numbers.

65. $(f \circ g)(x) = f(g(x)) = f\left(\dfrac{x}{5}\right) = 5\left(\dfrac{x}{5}\right) = x$

$(g \circ f)(x) = g(f(x)) = g(5x) = \dfrac{5x}{5} = x$

67. $(f \circ g)(x) = f(g(x)) = f\left(\dfrac{x-1}{4}\right) = 4\left(\dfrac{x-1}{4}\right) + 1$
$= (x - 1) + 1 = x$
$(g \circ f)(x) = g(f(x)) = g(4x + 1) =$
$\dfrac{(4x + 1) - 1}{4} = \dfrac{4x}{4} = x$

69. $(f \circ g)(x) = f(g(x)) = f(x^3 - 7) =$
$\sqrt[3]{(x^3 - 7) + 7} = \sqrt[3]{x^3} = x$
$(g \circ f)(x) = g(f(x)) = g(\sqrt[3]{x + 7}) =$
$(\sqrt[3]{x + 7})^3 - 7 = (x + 7) - 7 = x$

71. **(a)** The hours it takes to travel x miles
(b) The hours it takes to travel 200 miles
(c) Evaluate $f^{-1}(200)$.

73. $x = 3$ **75.** $x = \dfrac{10}{11}$

77. **(a)** f^{-1} computes the year when x cases of AIDS were reported.
(b) The year when 260,000 AIDS cases were reported
(c) $f^{-1}(260,000) \approx 1991$

79. **(a)** $f(1995) = 1$ **(b)** 1995
(c) $f^{-1}(1) = 1995$
In 1995 a 30-second Super Bowl commercial cost $1 million.

81. **(a)** $[1885, 1975, 5]$ by $[-5, 35, 5]$

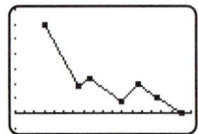

(b) f is not one-to-one, so f does not have an inverse.

83. **(a)** $(f \circ g)(2) = 10{,}560$ represents the number of feet in 2 miles.
(b) $f^{-1}(26{,}400) = 8800$ represents the number of yards in 26,400 feet.
(c) $(g^{-1} \circ f^{-1})(21{,}120) = 4$ represents the number of miles in 21,120 feet.

85. **(a)** $(g \circ f)(96) = 1.5$ represents the number of quarts in 96 tablespoons.
(b) $g^{-1}(2) = 8$ represents the number of cups in 2 quarts.
(c) $(f^{-1} \circ g^{-1})(1.5) = 96$ represents the number of tablespoons in 1.5 quarts.

87. **(a)** $f(1930) = 62.5, f(1980) = 65.5$; increased by 3%.
(b) $f^{-1}(x)$ computes the year when the cloud cover was x percent.
(c) $f^{-1}(62.5) = 1930, f^{-1}(65.5) = 1980$
(d) $f^{-1}(x) = \dfrac{50}{3}(x - 62.5) + 1930$

89. $[-9, 9, 1]$ by $[-6, 6, 1]$

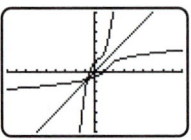

4.3 ANSWERS

1. $\dfrac{1}{8}$ **3.** 2   **5.** 9   **7.** 8

9. 2   **11.** e^{2x} **13.** $\$841.53$

15. $1730.97   **17.** $4451.08

19. $A_n = 500e^{0.072n}$

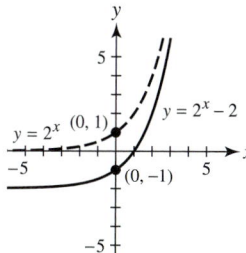

21. The account with $200 will have double the money since twice the money was deposited.

23. Approximately $778.7 billion

25. $19,870.65

27. (i) b   (ii) d   (iii) a   (iv) c

29. (a)

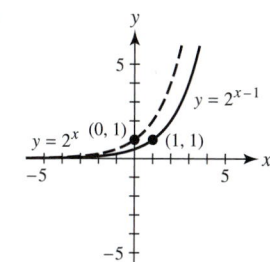

(b)

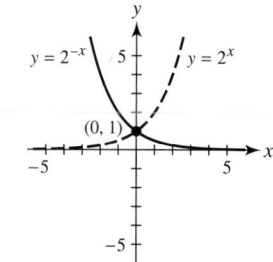

(c)

(d)

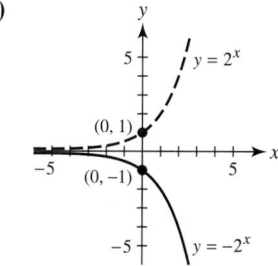

31. 22.20   **33.** 71.26   **35.** -0.7586

37. (a) New York: 18.2 million; Florida: 14 million
    (b) 2011, 18.5 million

39. (a) 2.5 ppm   (b) 1.4 days

41. (a) $[1968, 1998, 2]$ by $[-1, 6, 1]$ (b) 39%

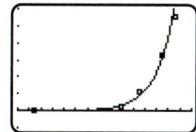

43. (a) $P(2) \approx 0.20$ and $P(20) \approx 0.90$. There is a 20% chance that at least one tree is located within a circle having radius 2 feet and a 90% chance for a circle having radius 20 feet.

(b) The larger the circle, the more likely it is to contain a tree.

$[0, 25, 5]$ by $[0, 1, 0.1]$

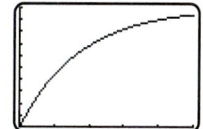

(c) $x \approx 6.1$. A circle of radius 6.1 feet has a 50-50 chance of containing at least one tree.

45. (a) 31.7 mg. The half-life is less than 50 years.
    (b) 30.2 yr

47.

x (in)	6	7.5	9	10.5	12
y (lb $\times$ 1000)	80	116	168	242	350

49. (a) H is increasing.

$[0, 100, 10]$ by $[0, 5, 1]$

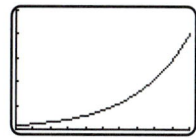

(b) $H(30) \approx 0.42$. About 0.42 horsepower is required for each ton pulled at 30 mph.

(c) About 2100 hp   (d) 2

51. (a) $f(25) \approx 0.065$ and $f(65) \approx 0.82$. For people age 25, 6.5% have some CHD, whereas for people age 65, 82% have some CHD.

(b) 48 yr (approximately)

4.4 ANSWERS

1. (a) 1   (b) 4   (c) -20 (d) -2

3. (a) $n = 1, \log 79 \approx 1.898$
    (b) $n = 2, \log 500 \approx 2.699$
    (c) $n = 0, \log 5 \approx 0.6990$
    (d) $n = -1, \log 0.5 \approx -0.3010$

5. (a) 3   (b) 6   (c) -2 (d) -4

7. (a) $\log 250 \approx 2.398$
    (b) $\log 4 \approx 0.6021$
    (c) $\log 0.5 \approx -0.3010$

9. 6   **11.** $\dfrac{1}{2}$ **13.** 0   **15.** -1

17.

x	$\dfrac{1}{16}$	1	8	32
$f(x)$	-4	0	3	5

19.

x	5	6	7	21
$f(x)$	—	0	2	8

   $f(5)$ is undefined.

21. $\dfrac{\log 25}{\log 2} \approx 4.644$ **23.** $\dfrac{\log 130}{\log 5} \approx 3.024$

25. $x = \log_2 72 = \dfrac{\log 72}{\log 2} \approx 6.170$

27. $x = \log_5 0.25 = \dfrac{\log 0.25}{\log 5} \approx -0.8614$

29. $x = \ln 25 \approx 3.219$ **31.** $x = -\ln 3 \approx -1.099$

33. $x = 10^{2.3} \approx 199.5$ **35.** $x = 2^{1.2} \approx 2.297$

37. $x = e^{-2} \approx 0.1353$ **39.** $\log 6$

41. $-\dfrac{3}{2}\ln 5$ **43.** $\dfrac{3}{2}\log x$ **45.** $\ln \dfrac{2}{e}$

47. $2 \log x - \log 3$

49. $\ln 2 + 7 \ln x - \ln 3 - \ln k$

51. $2 + 2 \log_2 k + 3 \log_2 x$

53. **(a)** Yes

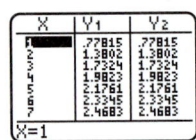

   (b) By Property 2: $\log 3x + \log 2x =$
   $\log (3x \cdot 2x) = \log 6x^2$

55. **(a)** No

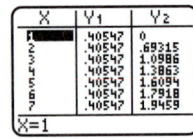

57. **(a)** Yes

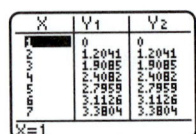

   (b) By Property 4: $\log (x^4) = 4 \log x$

59. **(a)** Yes

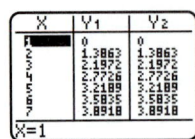

   (b) By Property 4: $\ln x^4 - \ln x^2 =$
   $4 \ln x - 2 \ln x = 2 \ln x$

61. $f^{-1}(x) = \ln x$

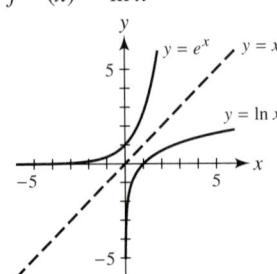

63. d   **65.** c   **67.** $x \approx 4.714$

69. $x \approx \pm 2.035$

71. **(a)** Since L is increasing, heavier planes generally require longer runways.
   $[0, 50, 10]$ by $[0, 6, 1]$

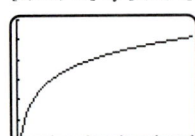

   (b) No. It increases by 3000 feet.
   (c) If the weight increases tenfold, the runway
   length increases by 3000 feet.

73. **(a)** $10^{-4.92} \approx 0.000012$
   (b) $10^{-3.9} \approx 0.000126$

75. **(a)** Since the data is leveling off, f_2 may model
   the data better.
   (b) Let $Y_1 = f_1(x)$ and $Y_2 = f_2(x)$
   $[1996, 2007, 1]$ by $[8, 12, 1]$

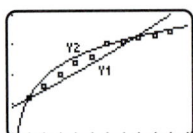

77. **(a)** $f(0) = 27, f(100) \approx 29.2$ inches. At the eye,
   the barometric air pressure is 27 inches, while
   100 miles away it is 29.2 inches.
   (b) The air pressure rises rapidly at first and then
   starts to level off.
   $[0, 250, 50]$ by $[25, 30, 1]$

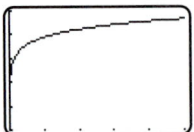

79. **(a)** $T(C(x)) = 6.5 \ln (1.3(1.005)^x)$ and $T(100) \approx$ 4.95. This model predicts an average global temperature increase of about 5°F in the year 2100.

(b) The graph of Y_1 is exponential and increasing; the graph of Y_2 is linear and increasing.

[0, 200, 50] by [0, 1000, 100]

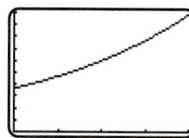

[0, 200, 50] by [0, 10, 1]

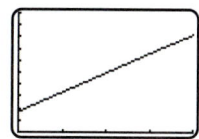

(c) If carbon dioxide levels increase exponentially, the average global temperatures increase linearly.

81. **(a)** Yugoslavia: $x = 1,000,000$;
Indonesia: $x = 100,000,000$

(b) 100

4.5 ANSWERS

1. **(a)** $x \approx 2$ **(b)** $x = \ln 7.5 \approx 2.015$

3. **(a)** $x \approx 2$ **(b)** $x = 5 \log 2.5 \approx 1.990$

5. $x = \ln 1.25 \approx 0.2231$ **7.** $x = \log 20 \approx 1.301$

9. $x = -\dfrac{5}{6} \ln 0.4 \approx 0.7636$

11. $x = \dfrac{\ln 0.5}{\ln 0.9} \approx 6.579$

13. $x = 1 + \dfrac{\log 4}{\log 1.1} \approx 15.55$

15. $x = 1$ or 2

17. No solutions

19. $x = 1 - \log 7 \approx 0.1549$

21. **(a)** $x = \dfrac{\ln 1.5}{0.09} \approx 4.505$

(b) $500 invested at 9% compounded continuously results in $750 after 4.5 years.

23. $x = 10^{2/3} \approx 4.642$ **25.** $x = \dfrac{1}{2}e^5 \approx 74.21$

27. $x = \pm\sqrt{50} = \pm 5\sqrt{2} \approx \pm 7.071$

29. $x = 10^{-11}$ **31.** $x = e \approx 2.718$

33. $x = 2^{2.1} \approx 4.287$

35. $x = \sqrt{50} = 5\sqrt{2} \approx 7.071$

37. $x = 1 + e \approx 3.718$

39. $x \approx 3$ or 1990   **41.** $100 - 10^{1.16} \approx 85.5\%$

43. $\dfrac{\ln (12/3.4)}{\ln 1.07} + 1995 \approx 2014$

45. $\dfrac{\log (20,000/2329)}{\log 1.2406} + 1988 \approx 1998$

47. $x = \dfrac{\ln (0.6)}{-0.000121} \approx 4222$ yr

49. **(a)** $f(x) = 364(1.005)^x$. Answers may vary.

(b) $x \approx 86$ or in 2086

51. $P = 1 - e^{-0.3214} \approx 0.275$. If a $60 tax is placed on each ton of carbon burned, carbon dioxide emissions could decrease by 27.5%.

53. $a \approx 1.10$, $b \approx 2.10$, or $f(x) = 2.10(1.10)^x$

[0, 6, 1] by [0, 4, 1]

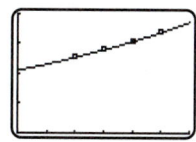

55. $a \approx 0.20$, $b \approx 3.20$, or $f(x) = 3.20x^{0.20}$

[1, 9, 1] by [0, 6, 1]

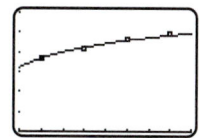

57. **(a)** $a \approx 0.99986$, $b \approx 1.4734$, or $f(x) = 1.4734(0.99986)^x$

(b) Approximately 0.55 kg/m³

59. **(a)** The data is not linear.

[−2, 32, 5] by [0, 80, 10]

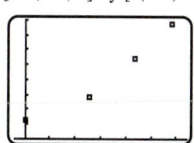

(b) $a \approx 1.066$, $b \approx 12.42$, or $f(x) = 12.42(1.066)^x$

(c) $f(39) \approx 150$. Chemical fertilizer use increased but at a slower rate than predicted by f.

CHAPTER 4   REVIEW ANSWERS

1. **(a)** 8   **(b)** 0
(c) −6   **(d)** Undefined

3. **(a)** 7   **(b)** 1   **(c)** 0   **(d)** $-\dfrac{9}{2}$

5. **(a)** 2   **(b)** 1

7. **(a)** $\sqrt{6}$ **(b)** 12

9. Subtract 6 from x and then multiply the result by 10; $\dfrac{x}{10} + 6$ and $10(x - 6)$.

11. f is one-to-one.   **13.** f is not one-to-one.

15. (a) $f(1) \approx \$1050$

(b) $f^{-1}(1200) \approx 4$ yr f^{-1} computes the number of years it takes to accumulate x dollars.

17. $f^{-1}(x) = \dfrac{x+5}{3}$ **19.** $f^{-1}(x) = \dfrac{1}{x} - 7$

21. $1562.71   **23.** 11.023

25. $\ln 19 \approx 2.9444$ **27.** 3   **29.** -21

31. 2   **33.** 1   **35.** $\dfrac{\log 18}{\log 3} \approx 2.631$

37. $x = \log 125 \approx 2.097$

39. $x = 10 \ln 5.2 \approx 16.49$

41. $x = -\dfrac{1}{\log 5} \approx -1.431$

43. $x = 10^{1.5} \approx 31.62$ **45.** $x = e^{3.4} \approx 29.96$

47. $\log 30x$ **49.** $\ln 4 - 2 \ln x$

51. $x = 10^{1/4} \approx 1.778$ **53.** $x = \dfrac{100{,}000}{3} \approx 33{,}333$

55. The x-intercept is b. If $(0, b)$ is on the graph of f, then $(b, 0)$ is on the graph of f^{-1}.

57. $x \approx 13$ or 2008

59. (a) $(g \circ f)(32) = 1$ represents the number of quarts in 32 fluid ounces.

(b) $f^{-1}(1) = 16$ represents the number of fluid ounces in 1 pint.

(c) $(f^{-1} \circ g^{-1})(1) = 32$ represents the number of fluid ounces in 1 quart.

61. (a) $W(1) \approx 19.5$

(b) $x \approx 2.74$. The fish weighs 50 mg at about three weeks.

63. (a)

x	0	15	30	45	60	75	90	125
$g(x)$	7	21	57	111	136	158	164	178

(b) $g(x) = 261 - f(x)$

(c) y_1 models $g(x)$ better.
[$-10, 140, 10$] by [$-20, 200, 10$]

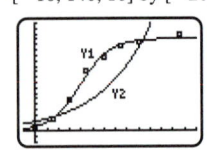

(d) $f(x) = 261 - \dfrac{171}{1 + 18.6e^{-0.0747x}}$

65. $a \approx 0.74, b \approx 3.50$, or $f(x) = 3.50(0.74)^x$
[$0, 5, 1$] by [$0, 3, 1$]

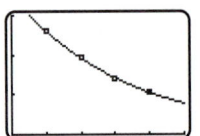

CHAPTER 5   TRIGONOMETRIC FUNCTIONS

5.1   ANSWERS

1. (a)

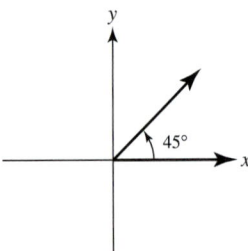

(b)

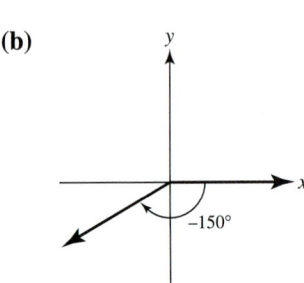

(c)

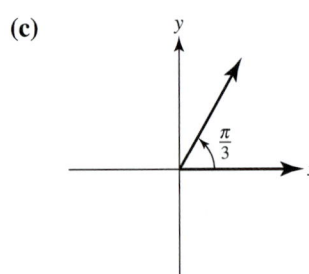

(d)

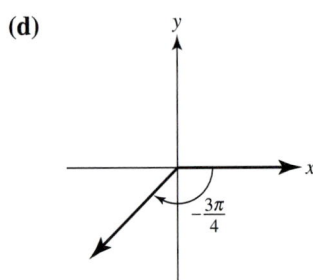

3. Answers may vary.

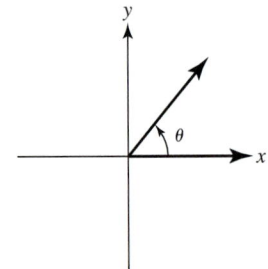

5.

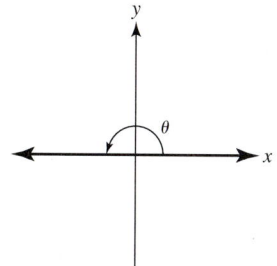

7. Answers may vary.

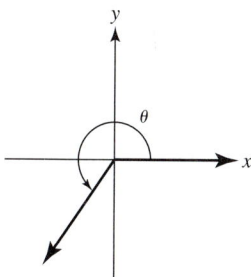

9. Answers may vary.

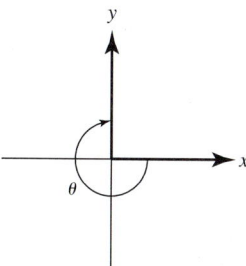

11. (a) $\dfrac{1}{4}$ (b) $\dfrac{1}{12}$ (c) $\dfrac{1}{6}$ (d) $\dfrac{1}{8}$

13. $\theta = 2$ radians or $\theta \approx 114.6°$

15. $\theta = 1.3$ radians or $\theta \approx 74.5°$

17. $\alpha = 34.1°, \beta = 124.1°$

19. $\alpha = 4°36'15'', \beta = 94°36'15''$

21. $\alpha = 66°19'25'', \beta = 156°19'25''$

23. $125.25°$ **25.** $108.76°$

27. $125°18'$ **29.** $51°21'36''$

31. (a) $30°$ (b) $12°$
(c) $-300°$ (d) $-210°$

33. (a) $\dfrac{\pi}{4}$ (b) $\dfrac{3\pi}{4}$
(c) $-\dfrac{2\pi}{3}$ (d) $-\dfrac{7\pi}{6}$

35. $510°, -210°$; Answers may vary.

37. $288°, -432°$; Answers may vary.

39. $\dfrac{5\pi}{2}, -\dfrac{3\pi}{2}$; Answers may vary.

41. $\dfrac{9\pi}{5}, -\dfrac{11\pi}{5}$; Answers may vary.

43. $s = \dfrac{2\pi}{3}$ in   **45.** $\theta = \dfrac{12}{5}$ radians

47. $r = \dfrac{5}{\pi}$ ft   **49.** $\dfrac{\pi}{4}$ m

51. π ft   **53.** $\dfrac{7\pi}{360}$ mi

55. 2π in, $\dfrac{2\pi}{15}$ in/min   **57.** 10π in, $\dfrac{2\pi}{15}$ in/min

59. 16.25 ft/sec, about 11.1 mi/hr

61. 1.5π in^2 **63.** 4.5π in^2

65. $\dfrac{17{,}161\pi}{3000} \approx 5.72\pi$ cm^2

67. $\dfrac{3\pi}{16}$ ft^2 **69.** 240π in^2 **71.** 4292π cm^2

73. 810 mi   **75.** 14.5 in

77. (a) 2.5 revolutions
(b) $\dfrac{65\pi}{6} \approx 34$ ft/sec

79. (a) 78,370 mi/hr (b) 66,630 mi/hr
(c) 29,250 mi/hr (d) 12,160 mi/hr
Planets farther from the sun have slower orbital velocities.

81. 137.2 m   **83.** 0.069 mi or 364 ft

85. $s = r\theta\left(\dfrac{\pi}{180°}\right)$, where θ is in degrees. The formula for radian measure is simpler.

87. (a) 388.8 m   (b) 881.8 m

89. Radius $\approx$ 3947 mi, circumference $\approx$ 24,800 mi

5.2 ANSWERS

1.

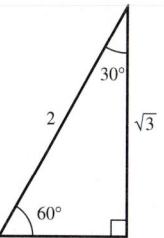

3.

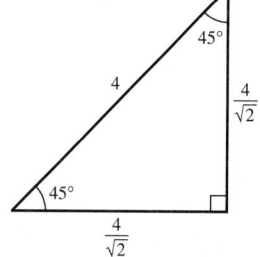

5. $\sin\theta = \dfrac{4}{5}$, $\cos\theta = \dfrac{3}{5}$, $\tan\theta = \dfrac{4}{3}$

   $\csc\theta = \dfrac{5}{4}$, $\sec\theta = \dfrac{5}{3}$, $\cot\theta = \dfrac{3}{4}$

7. $\sin\theta = \dfrac{12}{13}$, $\cos\theta = \dfrac{5}{13}$, $\tan\theta = \dfrac{12}{5}$

   $\csc\theta = \dfrac{13}{12}$, $\sec\theta = \dfrac{13}{5}$, $\cot\theta = \dfrac{5}{12}$

9. $a = 8\sqrt{3}$, $b = 8$

11. $b = \dfrac{6}{\tan 50°} \approx 5.03$, $c = \dfrac{6}{\sin 50°} \approx 7.83$

13. $a = 8\sin 41° \approx 5.25$, $b = 8\cos 41° \approx 6.04$

15. $a = 16.1\tan 46°15' \approx 16.8$,

   $c = \dfrac{16.1}{\cos 46°15'} \approx 23.3$

17. $\sec\theta = 3$ **19.** $\csc\theta = \dfrac{13}{12}$ **21.** $\tan\theta = \dfrac{4}{3}$

23. $\sin 25° \approx 0.4226$, $\cos 25° \approx 0.9063$,
   $\tan 25° \approx 0.4663$, $\csc 25° \approx 2.366$,
   $\sec 25° \approx 1.103$, $\cot 25° \approx 2.145$

25. $\sin 5°35' \approx 0.09729$, $\cos 5°35' \approx 0.9953$,
   $\tan 5°35' \approx 0.09776$, $\csc 5°35' \approx 10.28$,
   $\sec 5°35' \approx 1.005$, $\cot 5°35' \approx 10.23$

27. $\sin 13°45'30'' \approx 0.2378$, $\cos 13°45'30'' \approx 0.9713$,
   $\tan 13°45'30'' \approx 0.2449$, $\csc 13°45'30'' \approx 4.205$,
   $\sec 13°45'30'' \approx 1.030$, $\cot 13°45'30'' \approx 4.084$

29. $\sin 1.05° \approx 0.01832$, $\cos 1.05° \approx 0.9998$,
   $\tan 1.05° \approx 0.01833$, $\csc 1.05° \approx 54.57$,
   $\sec 1.05° \approx 1.000$, $\cot 1.05° \approx 54.56$

31. (a) $\cos 20° \approx 0.9397$ **(b)** $\sin 50° \approx 0.7660$

33. (a) $\sec 41° \approx 1.3250$ **(b)** $\csc 27° \approx 2.2027$

35. $1500\tan 37°30' \approx 1151$ ft

37. $100\tan 35° \approx 70$ ft

39. Barnard's Star: 3.5×10^{13} mi, 5.9 light-years
   Sirius: 5.1×10^{13} mi, 8.6 light-years
   61 Cygni: 6.6×10^{13} mi, 11.1 light-years
   Procyon: 6.7×10^{13} mi, 11.3 light-years

41. Min: 29×10^6 mi; max: 44×10^6 mi

43. 6306 mi. As θ increases, the altitude also increases.

45. 13,500 ft   **47.** $PQ \approx 194.5$ ft

49. (a) $d = \cot\dfrac{\theta}{2}$

   (b) $\cot\left(\dfrac{1°45'15''}{2}\right) \approx 65.32$ m

51. (a) 704 ft   **(b)** 595 ft

(c) Increasing θ, decreases r.

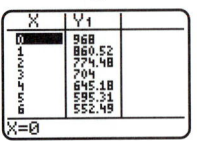

53. $d = 625\left(\dfrac{1}{\cos 54°} - 1\right) \approx 438$ ft

55. (a) $\sin^2\theta + \cos^2\theta = \left(\dfrac{a}{c}\right)^2 + \left(\dfrac{b}{c}\right)^2$

$$= \dfrac{a^2}{c^2} + \dfrac{b^2}{c^2}$$

$$= \dfrac{a^2 + b^2}{c^2}$$

$$= \dfrac{c^2}{c^2}$$

$$= 1$$

(b)

X	Y₁	
0	1	
10	1	
20	1	
30	1	
40	1	
50	1	
60	1	
X=0		

5.3   ANSWERS

1. (a) 13

   (b) $\sin\theta = \dfrac{5}{13}$, $\cos\theta = \dfrac{12}{13}$

3. (a) 17

   (b) $\sin\theta = \dfrac{8}{17}$, $\cos\theta = -\dfrac{15}{17}$

5. $\sin\theta = \dfrac{3}{5}$, $\cos\theta = \dfrac{4}{5}$

7. $\sin\theta = -\dfrac{2}{\sqrt{5}}$, $\cos\theta = \dfrac{1}{\sqrt{5}}$

9. $\sin\theta = -\dfrac{8}{17}$, $\cos\theta = -\dfrac{15}{17}$

11. $\sin 45° = \dfrac{1}{\sqrt{2}}$, $\cos 45° = \dfrac{1}{\sqrt{2}}$

13. $\sin(-30°) = -\dfrac{1}{2}$, $\cos(-30°) = \dfrac{\sqrt{3}}{2}$

15. $\sin\dfrac{\pi}{3} = \dfrac{\sqrt{3}}{2}$, $\cos\dfrac{\pi}{3} = \dfrac{1}{2}$

17. $\sin\left(-\dfrac{\pi}{2}\right) = -1$, $\cos\left(-\dfrac{\pi}{2}\right) = 0$

19. $\sin\theta = \dfrac{2}{\sqrt{5}}$, $\cos\theta = \dfrac{1}{\sqrt{5}}$

21. $\sin\theta = -\dfrac{3}{\sqrt{10}}$, $\cos\theta = \dfrac{1}{\sqrt{10}}$

23. $\sin 93.2° \approx 0.9984$, $\cos 93.2° \approx -0.0558$

25. $\sin 123°50' \approx 0.8307$, $\cos 123°50' \approx -0.5568$

27. $\sin(-4) \approx 0.7568$, $\cos(-4) \approx -0.6536$

29. $\sin\dfrac{11\pi}{7} \approx -0.9749$, $\cos\dfrac{11\pi}{7} \approx 0.2225$

31. $\sin\theta = \dfrac{3}{5}$, $\cos\theta = \dfrac{4}{5}$

33. $\sin\theta = -\dfrac{5}{13}$, $\cos\theta = \dfrac{12}{13}$

35. $\sin 0 = 0$ **37.** $\sin\dfrac{\pi}{2} = 1$

39. $\sin\left(-\dfrac{\pi}{6}\right) = -\dfrac{1}{2}$ **41.** $\cos 0 = 1$

43. $\cos\dfrac{\pi}{3} = \dfrac{1}{2}$ **45.** $\cos\left(-\dfrac{3\pi}{2}\right) = 0$

47. $\sin(-\pi) = 0$ **49.** $\cos\left(\dfrac{3\pi}{4}\right) = -\dfrac{1}{\sqrt{2}}$

51. $\sin\left(-\dfrac{\pi}{2}\right) = -1$

53. $[-2\pi, 2\pi, \pi/2]$ by $[-2, 2, 1]$

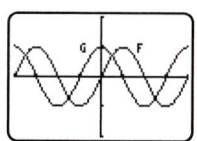

The translated graph and the graph of g are identical.

55. **(a)** $f(t) = \cos(2t)$

  (b) $[-2\pi, 2\pi, \pi/2]$ by $[-2, 2, 1]$

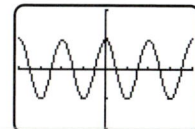

  (c)

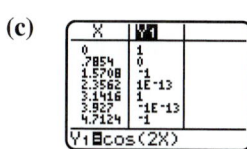

57. ALGORITHM $f(t) = 2\sin t$

  Step 1: Input a real number t

  Step 2: Let θ be an angle of t radians in standard position.

  Step 3: Determine the point (x, y) where the terminal side of θ intersects the unit circle.

  Step 4: Output $2y$.

  $f(\pi/2) = 2\sin(\pi/2) = 2$

59. No, their graphical and numerical representations are different.

$[-2\pi, 2\pi, \pi/2]$ by $[-4, 4, 1]$

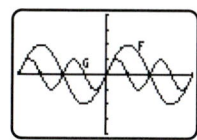

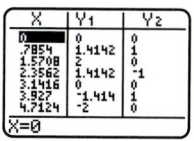

61. **(a)** $R = \{y \mid -3 \le y \le 3\}$

  $[-2\pi, 2\pi, \pi/2]$ by $[-4, 4, 1]$

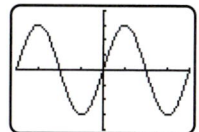

  (b) $R = \{y \mid -1 \le y \le 1\}$

  $[-2\pi, 2\pi, \pi/2]$ by $[-4, 4, 1]$

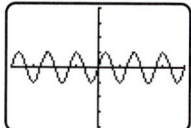

63. **(a)** $R = \{y \mid -1 \le y \le 3\}$

  $[-2\pi, 2\pi, \pi/2]$ by $[-4, 4, 1]$

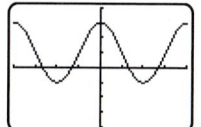

  (b) $R = \{y \mid -2 \le y \le 0\}$

  $[-2\pi, 2\pi, \pi/2]$ by $[-4, 4, 1]$

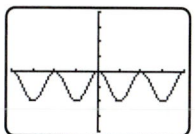

65. **(a)** The graph is similar to the sine graph except that it is reflected about the x-axis.

  $[-2\pi, 2\pi, \pi/2]$ by $[-4, 4, 1]$

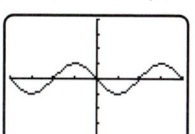

  (b) The graph is similar to the cosine graph except that it is reflected about the x-axis.

  $[-2\pi, 2\pi, \pi/2]$ by $[-4, 4, 1]$

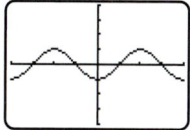

67. (a) The graph is similar to the sine graph except that it is shifted upward 1 unit.

$[-2\pi, 2\pi, \pi/2]$ by $[-4, 4, 1]$

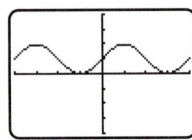

(b) The graph is similar to the cosine graph except that it is shifted downward 1 unit.

$[-2\pi, 2\pi, \pi/2]$ by $[-4, 4, 1]$

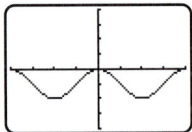

69. (a) 3%   **(b)** $25{,}000\left(\dfrac{3}{\sqrt{10{,}009}}\right) \approx 750$ lb

71. (a) V is sinusoidal and varies between -310 volts and 310 volts.

$[0, 1/15, 1/60]$ by $[-400, 400, 100]$

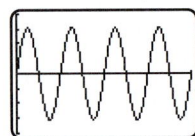

(b) $V(1/120) = 0$. After 1/120 second, the voltage is 0.

(c) $310/\sqrt{2} \approx 219$ volts

73. (9.6, 3.5)

75. (a) 31.4 ft   **(b)** 78.3 ft

(c) When the speed limit is higher, more land needs to be cleared on the inside of the curve.

77. (a) 201 ft   **(b)** 258 ft

(c) It is easier to stop going uphill ($\theta > 0$) than downhill ($\theta < 0$).

79. (a) $[0, 1/100, 1/880]$ by $[-1.5, 1.5, 0.5]$

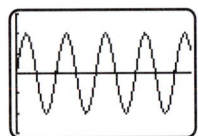

(b) $\dfrac{1}{440} \approx 0.00227$ sec

(c) 440 cycles per second

5.4 ANSWERS

1. (a) 5

(b) $\sin\theta = -\dfrac{3}{5}$, $\cos\theta = \dfrac{4}{5}$, $\tan\theta = -\dfrac{3}{4}$

$\csc\theta = -\dfrac{5}{3}$, $\sec\theta = \dfrac{5}{4}$, $\cot\theta = -\dfrac{4}{3}$

3. $\sin\theta = \dfrac{12}{13}$, $\cos\theta = \dfrac{5}{13}$, $\tan\theta = \dfrac{12}{5}$

$\csc\theta = \dfrac{13}{12}$, $\sec\theta = \dfrac{13}{5}$, $\cot\theta = \dfrac{5}{12}$

5. $\sin\theta = -\dfrac{24}{25}$, $\cos\theta = \dfrac{7}{25}$, $\tan\theta = -\dfrac{24}{7}$

$\csc\theta = -\dfrac{25}{24}$, $\sec\theta = \dfrac{25}{7}$, $\cot\theta = -\dfrac{7}{24}$

7. $\sin\theta = 0$, $\cos\theta = 1$, $\tan\theta = 0$
$\csc\theta$ is undefined, $\sec\theta = 1$, $\cot\theta$ is undefined

9. Symmetric with respect to the origin.

$[-2\pi, 2\pi, \pi/2]$ by $[-4, 4, 1]$

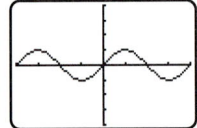

11. Symmetric with respect to the origin

$[-2\pi, 2\pi, \pi/2]$ by $[-4, 4, 1]$

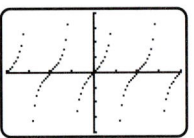

13. Symmetric with respect to the y-axis

$[-2\pi, 2\pi, \pi/2]$ by $[-4, 4, 1]$

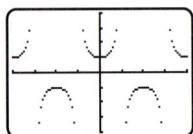

15. $\sin 90° = 1$, $\cos 90° = 0$, $\tan 90°$ is undefined
$\csc 90° = 1$, $\sec 90°$ is undefined, $\cot 90° = 0$

17. $\sin(-45°) = -\dfrac{1}{\sqrt{2}}$, $\cos(-45°) = \dfrac{1}{\sqrt{2}}$,

$\tan(-45°) = -1$, $\csc(-45°) = -\sqrt{2}$,

$\sec(-45°) = \sqrt{2}$, $\cot(-45°) = -1$

19. $\sin\pi = 0$, $\cos\pi = -1$, $\tan\pi = 0$
$\csc\pi$ is undefined, $\sec\pi = -1$, $\cot\pi$ is undefined

21. $\sin\left(-\dfrac{\pi}{3}\right) = -\dfrac{\sqrt{3}}{2}$, $\cos\left(-\dfrac{\pi}{3}\right) = \dfrac{1}{2}$,

$\tan\left(-\dfrac{\pi}{3}\right) = -\sqrt{3}$, $\csc\left(-\dfrac{\pi}{3}\right) = -\dfrac{2}{\sqrt{3}}$,

$\sec\left(-\dfrac{\pi}{3}\right) = 2$, $\cot\left(-\dfrac{\pi}{3}\right) = -\dfrac{1}{\sqrt{3}}$

23. $\sin \theta = \dfrac{4}{\sqrt{17}}$, $\cos \theta = -\dfrac{1}{\sqrt{17}}$, $\tan \theta = -4$

$\csc \theta = \dfrac{\sqrt{17}}{4}$, $\sec \theta = -\sqrt{17}$, $\cot \theta = -\dfrac{1}{4}$

The slope of the line equals $\tan \theta$.

25. $\sin \theta = -\dfrac{6}{\sqrt{37}}$, $\cos \theta = -\dfrac{1}{\sqrt{37}}$, $\tan \theta = 6$

$\csc \theta = -\dfrac{\sqrt{37}}{6}$, $\sec \theta = -\sqrt{37}$, $\cot \theta = \dfrac{1}{6}$

The slope of the line equals $\tan \theta$.

27. **(a)** 0.9984   **(b)** 1.0016

29. **(a)** 1.4045   **(b)** 0.7120

31. **(a)** −0.8637   **(b)** −1.1578

33. **(a)** 0.2225   **(b)** 4.4940

35. $\sin \theta = \dfrac{1}{\sqrt{2}}$, $\cos \theta = \dfrac{1}{\sqrt{2}}$, $\tan \theta = 1$,

$\csc \theta = \sqrt{2}$, $\sec \theta = \sqrt{2}$, $\cot \theta = 1$

37. $\sin \theta = -\dfrac{12}{13}$, $\cos \theta = \dfrac{5}{13}$, $\tan \theta = -\dfrac{12}{5}$,

$\csc \theta = -\dfrac{13}{12}$, $\sec \theta = \dfrac{13}{5}$, $\cot \theta = -\dfrac{5}{12}$

39. $\tan(-90°)$ is undefined, $\tan(-45°) = -1$, $\tan(0°) = 0$, $\tan(45°) = 1$, $\tan(90°)$ is undefined

$[-352.5°, 352.5°, 90°]$ by $[-4, 4, 1]$

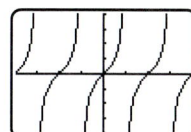

41. $\csc(-90°) = -1$, $\csc(-45°) = -\sqrt{2}$, $\csc(0°)$ is undefined, $\csc(45°) = \sqrt{2}$, $\csc(90°) = 1$

$[-352.5°, 352.5°, 90°]$ by $[-4, 4, 1]$

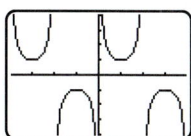

43. $D = $ all real numbers, $R = \{y \mid -1 \le y \le 1\}$, period $= 2\pi$

$[-2\pi, 2\pi, \pi/2]$ by $[-4, 4, 1]$

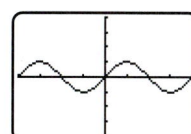

45. $D = \left\{ t \mid t \ne \pm \dfrac{\pi}{2}, \pm \dfrac{3\pi}{2}, \pm \dfrac{5\pi}{2}, \ldots \right\}$,

$R = $ all real numbers, period $= \pi$

$[-2\pi, 2\pi, \pi/2]$ by $[-4, 4, 1]$

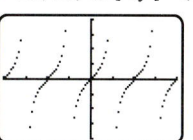

47. $D = \left\{ t \mid t \ne \pm \dfrac{\pi}{2}, \pm \dfrac{3\pi}{2}, \pm \dfrac{5\pi}{2}, \ldots \right\}$,

$R = \{y \mid |y| \ge 1\}$, period $= 2\pi$

$[-2\pi, 2\pi, \pi/2]$ by $[-4, 4, 1]$

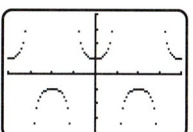

49. The zeros of $\sin t$ correspond to the asymptotes on the graph of $\csc t$.

If $\sin t = \pm 1$, then $\csc t = \pm 1$.

$[-2\pi, 2\pi, \pi/2]$ by $[-4, 4, 1]$

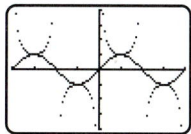

51. **(a)** $t = -\pi, 0, \pi, 2\pi$ **(b)** $t = \pm \dfrac{\pi}{2}, \pm \dfrac{3\pi}{2}$

   (c) $t = -\pi, 0, \pi, 2\pi$ **(d)** No solutions

   (e) No solutions   **(f)** $t = \pm \dfrac{\pi}{2}, \pm \dfrac{3\pi}{2}$

53. $\tan \theta = \dfrac{3}{4}$, $\cot \theta = \dfrac{4}{3}$, $\csc \theta = \dfrac{5}{3}$, $\sec \theta = \dfrac{5}{4}$

55. $\sin \theta = -\dfrac{15}{17}$, $\cos \theta = -\dfrac{8}{17}$, $\tan \theta = \dfrac{15}{8}$, $\cot \theta = \dfrac{8}{15}$

57. $\sin \theta = \dfrac{5}{13}$, $\csc \theta = \dfrac{13}{5}$, $\cot \theta = \dfrac{12}{5}$, $\sec \theta = \dfrac{13}{12}$

59. **(a)** The period of g is half the period of f.
   (b) The period of g is one-third the period of f.
   (c) The period of g is one-fourth the period of f.
   (d) The period of g is half the period of f.

61. Twice   **63.** None

65. $\tan 0 = 0$, $\cot 0$ is undefined.

67. $\tan\left(-\dfrac{\pi}{4}\right) = -1$, $\cot\left(-\dfrac{\pi}{4}\right) = -1$

69. $\sec \pi = -1$, $\csc \pi$ is undefined.

71. $\sec\left(-\dfrac{\pi}{2}\right)$ is undefined, $\csc\left(-\dfrac{\pi}{2}\right) = -1$

73. **(a)** $f(t) = \cot(2t)$

(b) $[-\pi, \pi, \pi/2]$ by $[-4, 4, 1]$

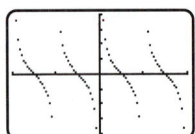

(c)

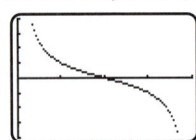

75. **ALGORITHM** Computing $f(t) = \tan t - 5$

Step 1:   Input a real number t.

Step 2:   Let θ be an angle of t radians in standard position.

Step 3:   Determine the point (x, y) where the terminal side of θ intersects the unit circle.

Step 4:   If $x \neq 0$, output $\dfrac{y}{x} - 5$.

   $f(0) = -5$

77. $y = \sin x$ **79.** $y = \cot x$

81. $y = \sec x$ **83.** $f(t) = 3 \cos t$

85. $f(t) = -3 \tan t$ **87.** $f(t) = 4 \csc t$

89. **(a)** $[0, \pi, \pi/4]$ by $[-4, 4, 1]$

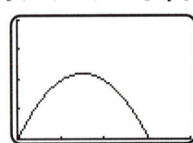

(b)   Just after sunrise ($\theta = 0$) the shadow is very long. As the elevation of the sun increases, the shadow decreases in length until it is 0 when $\theta = \pi/2$. In the afternoon the shadow increases in length in the opposite direction until sunset ($\theta = \pi$).

91. About 41%

93. $[0, 20, 000, 5000]$ by $[0, 4000, 1000]$

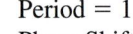

(a)   Approximately (7612, 2197)

(b)   About 15,223 ft

5.5 ANSWERS

1. c   **3.** b

5. Amplitude = 3   **7.** Amplitude = 4
   Period = 4π Period = 2
   Phase Shift = 0   Phase Shift = 1

9. c   **11.** d   **13.** a

15.

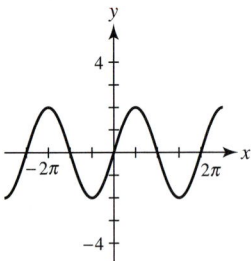

17.

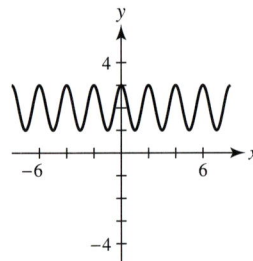

19.

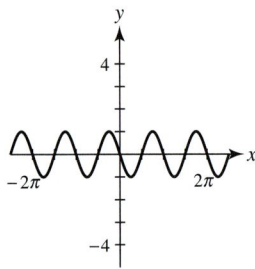

21. Amplitude = 2 $[-2\pi, 2\pi, \pi/2]$ by $[-4, 4, 1]$
   Period = π
   Phase Shift = 0

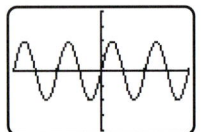

23. Amplitude = $\dfrac{1}{2}$ $[-2\pi, 2\pi, \pi/2]$ by $[-4, 4, 1]$

   Period = $\dfrac{2\pi}{3}$

   Phase Shift = $-\dfrac{\pi}{3}$

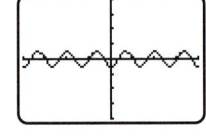

25. Amplitude = 1 $[-2\pi, 2\pi, \pi/2]$ by $[-4, 4, 1]$

   Period = $\dfrac{\pi}{2}$

   Phase Shift = 0

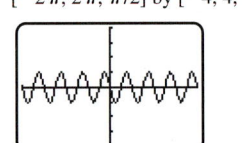

27. Amplitude = 1 $[-2\pi, 2\pi, \pi/2]$ by $[-4, 4, 1]$
   Period = 1
   Phase Shift = 0

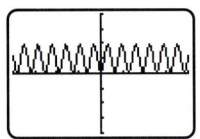

29. $a = 3, b = 2$ **31.** $a = -2, b = \dfrac{1}{2}$

33. $y = 3 \sin \left(\dfrac{1}{2} x \right)$ **35.** $y = 4 \sin (\pi(x - 1))$

37. Period $= \dfrac{\pi}{2}$

Phase Shift $= 0$

Asymptotes: $x = \pm \dfrac{\pi}{4}, \pm \dfrac{3\pi}{4}, \pm \dfrac{5\pi}{4}, \pm \dfrac{7\pi}{4}$

$[-2\pi, 2\pi, \pi/2]$ by $[-4, 4, 1]$

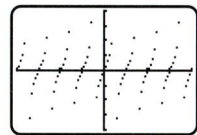

39. Period $= \pi$

Phase Shift $= \dfrac{\pi}{2}$

Asymptotes: $x = 0, \pm \pi, \pm 2\pi$

$[-2\pi, 2\pi, \pi/2]$ by $[-4, 4, 1]$

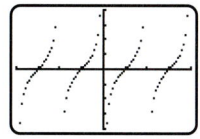

41. Period $= \dfrac{\pi}{2}$

Phase Shift $= 0$

Asymptotes: $x = 0, \pm \dfrac{\pi}{2}, \pm \pi, \pm \dfrac{3\pi}{2}, \pm 2\pi$

$[-2\pi, 2\pi, \pi/2]$ by $[-4, 4, 1]$

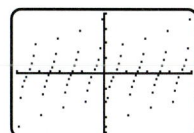

43. Period $= 4\pi$ $[-2\pi, 2\pi, \pi/2]$ by $[-4, 4, 1]$
Phase Shift $= 0$
Asymptotes: $x = \pm \pi$

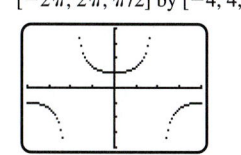

45. Period $= 2\pi$
Phase Shift $= \pi$
Asymptotes: $x = 0, \pm \pi, \pm 2\pi$
$[-2\pi, 2\pi, \pi/2]$ by $[-4, 4, 1]$

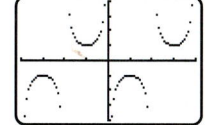

47. Period $= 6\pi$

Phase Shift $= \dfrac{\pi}{6}$

Asymptotes: $x = -\dfrac{4\pi}{3}, \dfrac{5\pi}{3}$

$[-2\pi, 2\pi, \pi/2]$ by $[-4, 4, 1]$

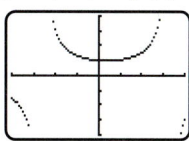

49. (a) $40°F, -40°F$

(b) Amplitude $= 40$, Period $= 12$

(c) The months when the average temperature is $0°F$

51. (a) Amplitude $= 34$
Period $= 12$
Phase Shift $= 4.3$

$[0, 25, 2]$ by $[-50, 50, 10]$

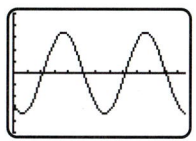

(b) $f(5) \approx 12.2°F$
$f(12) \approx -26.4°F$

(c) About $0°F$

53. (a) $[0, 25, 2]$ by $[0, 80, 10]$

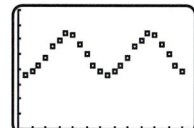

(b) $f(x) = 14 \sin \left(\dfrac{\pi}{6}(x - 4) \right) + 50$

(c) $[0, 25, 2]$ by $[0, 80, 10]$

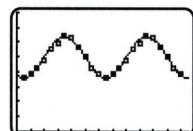

55. (a) $f(x) = 17 \cos \left(\dfrac{\pi}{6}(x - 7) \right) + 75$

(b) Yes. The period of the graph is 12 so, for example, the phase shift could be $c = 7 + 12 = 19$ or $c = 7 - 12 = -5$.

57. About 1300 ft

59. (a) About 18.5 hr; June 21

(b) About 6 hr; December 21

(c) The amplitude represents half the difference in daylight between the longest and shortest day. The period represents one year. Answers may vary.

61. (a) Max $\approx$ 8 in; min $\approx$ 0.5 in

    (b) Amp $\approx$ 3.75. The amplitude represents half the difference between the maximum and minimum monthly average precipitations.

    (c) Yes. An x-intercept would correspond to no precipitation.

    (d) About 51 in (Answers may vary.)

63. (a) 11 hours 10 minutes

    (b) There is no consistent amplitude because high tides vary.

65. (a) Max $\approx$ 87°F; min $\approx$ 62°F

    (b) Increase

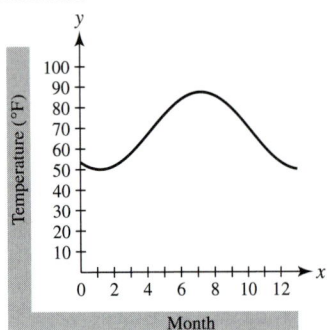

67. (a) [0, 25, 2] by [60, 90, 5]

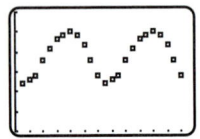

    (b) $y = 6.5 \sin\left(\dfrac{\pi}{6}(x - 4)\right) + 78.5$

69. (a) The general trend in carbon dioxide levels has been increasing. During the year there are seasonal variations.

    [20, 35, 5] by [320, 370, 10]

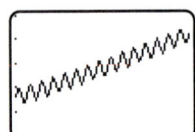

    (b) The quadratic function models the general trend in carbon dioxide levels, while the sine function models the seasonal fluctuations in carbon dioxide levels.

71. (a) $s(t) = 2 \cos(4\pi t)$

    (b) $s(1) = 2$. The weight is neither moving upward nor downward. At $t = 1$ the motion of the weight is changing from up to down.

73. (a) $s(t) = -3 \cos(2.5\pi t)$

    (b) $s(1) = 0$. The weight is moving upward.

75. $Y_1 = 0.21 \cos(55\pi X)$

[0, 0.05, 0.01] by [−0.3, 0.3, 0.1]

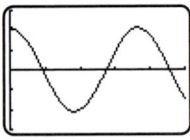

77. $Y_1 = 0.14 \cos(110\pi X)$

[0, 0.05, 0.01] by [−0.3, 0.3, 0.1]

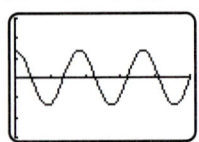

79. $y \approx 13.2 \sin(0.524x - 2.18) + 49.7$

[0, 25, 2] by [30, 80, 10]

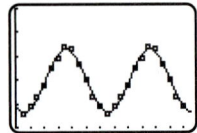

81. $y \approx 16.9 \sin(0.522x - 2.09) + 75.4$

[0, 25, 2] by [50, 100, 10]

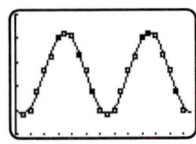

5.6 ANSWERS

1. one-to-one   **3.** π

5. 0   **7.** $f^{-1}(x) = \dfrac{x}{3}$

9. $f^{-1}(x) = x^3$ **11.** $f^{-1}(x) = \sqrt{x} + 1$

13. *Verbal:* Subtract 6 from x.

    Numerical:

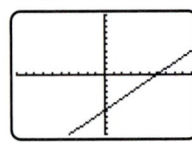

    Graphical: [−10, 10, 1] by [−10, 10, 1]

    Symbolic: $f^{-1}(x) = x - 6$

15. *Verbal:* Add 1 to x and divide the result by 2.
Numerical:

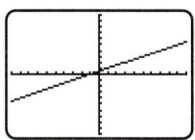

Graphical: $[-10, 10, 1]$ by $[-10, 10, 1]$

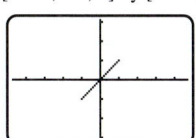

Symbolic: $f^{-1}(x) = \dfrac{x+1}{2}$

17. $[-2, 2]$ **19.** $\dfrac{\pi}{2}$ **21.** $-\dfrac{\pi}{4}$ **23.** $\dfrac{2\pi}{3}$

25. **(a)** $\dfrac{\pi}{2}$ or $90°$ **(b)** 0 or $0°$

   (c) $-\dfrac{\pi}{3}$ or $-60°$

27. **(a)** $\dfrac{\pi}{2}$ or $90°$ **(b)** π or $180°$

   (c) $\dfrac{\pi}{3}$ or $60°$

29. **(a)** $\dfrac{\pi}{4}$ or $45°$ **(b)** $-\dfrac{\pi}{4}$ or $-45°$

   (c) $\dfrac{\pi}{3}$ or $60°$

31. **(a)** Error   **(b)** 1.47 or $84.3°$
   (c) 2.42 or $138.6°$

33. **(a)** β **(b)** α **(c)** β

35. 1   **37.** $\dfrac{3\pi}{4}$ **39.** -3

41. **(a)** The graph matches $y = x$ for $-1 \le x \le 1$.
   $[-4.7, 4.7, 1]$ by $[-3.1, 3.1, 1]$

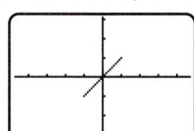

   (b) The graph matches $y = x$ for $-1 \le x \le 1$.
   $[-4.7, 4.7, 1]$ by $[-3.1, 3.1, 1]$

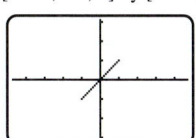

   (c) The graph matches $y = x$.

$[-4.7, 4.7, 1]$ by $[-3.1, 3.1, 1]$

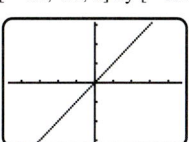

43. **(a)** $90°$ **(b)** $90°$ **(c)** $90°$
   $\sin^{-1}x + \cos^{-1}x = 90°$ whenever $-1 \le x \le 1$.

45. $\alpha = \tan^{-1}\dfrac{7}{24} \approx 16.3°$

   $\beta = \tan^{-1}\dfrac{24}{7} \approx 73.7°$

   $c = 25$

47. $\alpha = \sin^{-1}\dfrac{6}{10} \approx 36.9°$

   $\beta = \cos^{-1}\dfrac{6}{10} \approx 53.1°$

   $b = 8$

49. $\beta = 35°$

   $b = \dfrac{5}{\tan 55°} \approx 3.5$

   $c = \dfrac{5}{\sin 55°} \approx 6.1$

51. $\tan\theta = \dfrac{x}{\sqrt{1-x^2}}$

53. $\cos\theta = \dfrac{x}{\sqrt{1+x^2}}$

55. $90°$ **57.** $45°$ **59.** $90°$ **61.** $82.8°$

63. $80.5°$ **65.** $68.9°$ **67.** -0.197

69. 1.10 **71.** -0.282 **73.** $1.37, 1.77$

75. $\tan^{-1}\dfrac{50}{85} \approx 30.5°$

77. **(a)** $\tan^{-1}\dfrac{11}{5} \approx 65.6°$

   (b) $\tan^{-1}\left(\dfrac{-3}{1}\right) \approx -71.6°$

79. $\tan^{-1}\dfrac{4}{10} \approx 21.8°$

81. **(a)** $60°$ **(b)** $120°$

83. $\theta \approx 41.9°$

85. **(a)** 14.9 hr   **(b)** 13.8 hr   **(c)** 14.6 hr

87. $\theta_2 \approx 28.9°$ **89.** $x \approx 6.2$ ft

$[0, 50, 10]$ by $[0, 50, 10]$

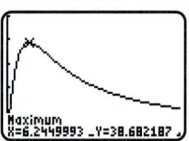

CHAPTER 5   REVIEW ANSWERS

1. **(a)**

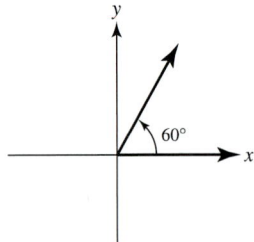

(b)

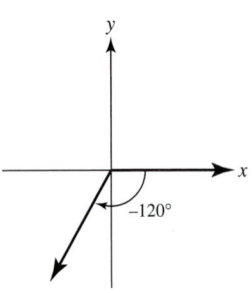

(c)

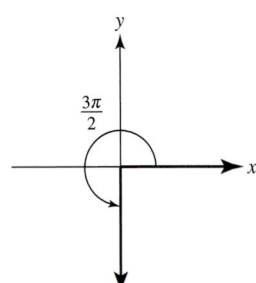

(d)

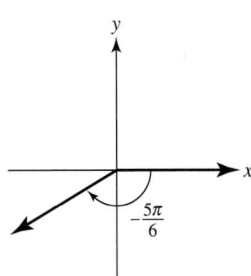

3. **(a)** $60°$ **(b)** $5°$
   (c) $-150°$ **(d)** $-315°$

5. 2π ft

7. $\sin\theta = \dfrac{8}{\sqrt{145}},\ \cos\theta = \dfrac{9}{\sqrt{145}},\ \tan\theta = \dfrac{8}{9},$
   $\csc\theta = \dfrac{\sqrt{145}}{8},\ \sec\theta = \dfrac{\sqrt{145}}{9},\ \cot\theta = \dfrac{9}{8}$

9. 3

11. $\sin 25° \approx 0.4226,\ \cos 25° \approx 0.9063,$
   $\tan 25° \approx 0.4663,\ \csc 25° \approx 2.366,$
   $\sec 25° \approx 1.103,\ \cot 25° \approx 2.145$

13. $\sin\theta = -\dfrac{2}{\sqrt{5}},\ \cos\theta = \dfrac{1}{\sqrt{5}},\ \tan\theta = -2,$
   $\csc\theta = -\dfrac{\sqrt{5}}{2},\ \sec\theta = \sqrt{5},\ \cot\theta = -\dfrac{1}{2}$

15. $\sin(-45°) = -\dfrac{1}{\sqrt{2}},\ \cos(-45°) = \dfrac{1}{\sqrt{2}},$
   $\tan(-45°) = -1,\ \csc(-45°) = -\sqrt{2},$
   $\sec(-45°) = \sqrt{2},\ \cot(-45°) = -1$

17. $\sin\theta = \dfrac{\sqrt{3}}{2},\ \cos\theta = -\dfrac{1}{2},\ \tan\theta = -\sqrt{3},$
   $\csc\theta = \dfrac{2}{\sqrt{3}},\ \sec\theta = -2,\ \cot\theta = -\dfrac{1}{\sqrt{3}}$

19. -1 **21.** 0

23. $\tan\theta = -\dfrac{4}{3},\ \csc\theta = -\dfrac{5}{4},\ \sec\theta = \dfrac{5}{3},\ \cot\theta = -\dfrac{3}{4}$

25. $y = \sin x$

27. Amplitude $= 3$ $[-2\pi, 2\pi, \pi/2]$ by $[-4, 4, 1]$
   Period $= \pi$
   Phase Shift $= 0$

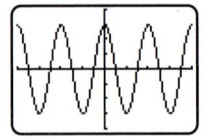

29. Amplitude $= 1$ $[-2\pi, 2\pi, \pi/2]$ by $[-4, 4, 1]$
   Period $= \dfrac{2\pi}{3}$
   Phase Shift $= \dfrac{\pi}{3}$

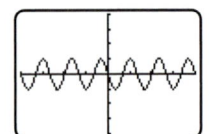

31. $a = 2,\ b = 3$

33. Period $= \dfrac{\pi}{2}$ $[-2\pi, 2\pi, \pi/2]$ by $[-4, 4, 1]$
   Phase Shift $= 0$

35. Period $= 2\pi$ $[-2\pi, 2\pi, \pi/2]$ by $[-4, 4, 1]$
   Phase Shift $= -\pi$

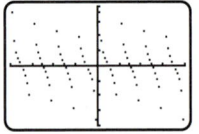

37. **(a)** $-\dfrac{\pi}{2}$ or $-90°$ **(b)** $\dfrac{\pi}{3}$ or $60°$

   (c) $\dfrac{\pi}{4}$ or $45°$

39. **(a)** -0.64 or $-36.9°$ **(b)** 1.37 or $78.7°$
   (c) 1.45 or $83.1°$

41. $\alpha = \tan^{-1}\dfrac{5}{3} \approx 59.0°$

   $\beta = \tan^{-1}\dfrac{3}{5} \approx 31.0°$
   $c = \sqrt{34}$

43. 30°   **45.** 30°   **47.** 78.5°   **49.** 71.6°

51. −0.6435   **53.** 0.4510   **55.** 89 ft

57. **(a)** $\sin^{-1}\dfrac{350}{6000} \approx 3.3°$

   (b) $\sin^{-1}\dfrac{160}{4500} \approx 2.0°$

59. 161 mi

61. **(a)** [0, 25, 2] by [0, 70, 10]

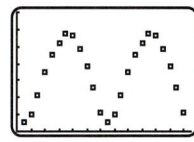

   (b) $f(x) = 26\cos\left(\dfrac{\pi}{6}(x-7)\right) + 32$

   (c) [0, 25, 2] by [0, 70, 10]

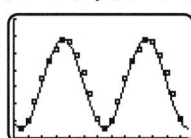

63. $y = 6\sin(x - 220°)$

CHAPTER 6   TRIGONOMETRIC IDENTITIES AND EQUATIONS

6.1   ANSWERS

1. $\cot\theta = 2$ **3.** $\sec\theta = \dfrac{7}{2}$ **5.** $\cos\theta = -\dfrac{1}{4}$

7. Quadrant IV   **9.** Quadrant III

11. Quadrant II

13. $\tan\theta = -\dfrac{3}{4}$, $\csc\theta = \dfrac{5}{3}$, $\sec\theta = -\dfrac{5}{4}$, $\cot\theta = -\dfrac{4}{3}$

15. $\cos\theta = -\dfrac{7}{25}$, $\tan\theta = \dfrac{24}{7}$,
   $\csc\theta = -\dfrac{25}{24}$, $\sec\theta = -\dfrac{25}{7}$

17. $\sin\theta = -\dfrac{\sqrt{3}}{2}$, $\tan\theta = -\sqrt{3}$, $\csc\theta = -\dfrac{2}{\sqrt{3}}$,
   $\sec\theta = 2$, $\cot\theta = -\dfrac{1}{\sqrt{3}}$

19. $\sin\theta = -\dfrac{\sqrt{8}}{3}$, $\cos\theta = \dfrac{1}{3}$, $\tan\theta = -\sqrt{8}$,
   $\csc\theta = -\dfrac{3}{\sqrt{8}}$, $\cot\theta = -\dfrac{1}{\sqrt{8}}$

21. $\sin\theta = \dfrac{3}{7}$, $\cos\theta = -\dfrac{\sqrt{40}}{7}$, $\tan\theta = -\dfrac{3}{\sqrt{40}}$,
   $\sec\theta = -\dfrac{7}{\sqrt{40}}$, $\cot\theta = -\dfrac{\sqrt{40}}{3}$

23. $-\sin 13°$ **25.** $-\tan\dfrac{\pi}{11}$

27. $\sec\dfrac{2\pi}{5}$ **29.** 0.5

31. 4   **33.** Yes   **35.** No

37. $1 + \tan^2 x = \sec^2 x$ **39.** $\sec x = \dfrac{1}{\cos x}$

41. No. For example, $\tan 30° = \dfrac{1}{\sqrt{3}} \neq \sqrt{3} = \cot 30°$.

43. No. For example, $\sin 30° = \dfrac{1}{2} \neq -\dfrac{1}{2} = \sin(-30°)$.

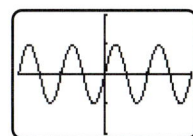

45. This is incorrect. The variable should be included. For example, $\cos^2(2\theta) + \sin^2(\theta) \neq 1$, but $\cos^2\theta + \sin^2\theta = 1$.

47. 1   **49.** 1   **51.** 1

53. $\cos\theta = \sqrt{1 - x^2}$, 0.8586

55. $\sin\theta = -\dfrac{1}{\sqrt{1 + x^2}}$, −0.8899

57. **(a)** $[-2\pi, 2\pi, \pi/2]$ by $[-1, 1, 0.5]$ **(b)** Odd

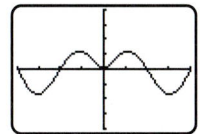

   (c) The product of an even function times an odd function is an odd function. This does not agree with products of integers.

59. **(a)** $[-2\pi, 2\pi, \pi/2]$ by $[-4, 4, 1]$ **(b)** Even

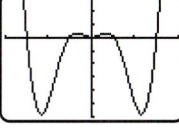

   (c) The product of an odd function times an odd function is an even function. This does not agree with products of integers.

61. **(a)** $[-2\pi, 2\pi, \pi/2]$ by $[-12, 6, 2]$ **(b)** Even

(c) The product of an even function times an even function is an even function. This agrees with products of integers.

63. $d = 93{,}000{,}000 \csc \theta$

65. **(a)** $P = 16k \cos^2 (2\pi t)$

(b) Let $Y_1 = 32(\cos(2\pi X))^{\wedge}2$. Y_1 has a maximum value of 32 when $t = 0, 0.5, 1, 1.5, 2.0$, and Y_1 has a minimum value of 0 when $t = 0.25, 0.75, 1.25, 1.75$. The spring is either stretched or compressed the most when Y_1 is maximum.

[0, 2, 0.5] by [−1, 40, 8]

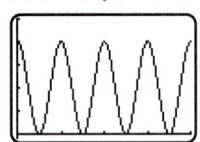

(c) $P = 16k(1 - \sin^2(2\pi t))$

67. **(a)** The sum of L and C equals 3.

[0, 10^{-6}, 10^{-7}] by [−1, 4, 1]

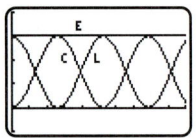

(b) Let $Y_1 = L(t)$, $Y_2 = C(t)$, and $Y_3 = E(t)$. $Y_3 = 3$ for all inputs.

X	Y2	Y3
0	0	3
1E-7	.95646	3
2E-7	2.6061	3
3E-7	2.8451	3
4E-7	1.3688	3
5E-7	.05974	3
6E-7	.58747	3

Y3◗Y1+Y2

(c) $E(t) = L(t) + C(t)$
$= 3 \cos^2 (6{,}000{,}000t) +$
$3 \sin^2 (6{,}000{,}000t)$
$= 3(\cos^2 (6{,}000{,}000t) +$
$\sin^2 (6{,}000{,}000t))$
$= 3(1)$
$= 3$

6.2   ANSWERS

1. **(a)** $1 - x^2$ **(b)** $\cos^2 \theta$

3. **(a)** $x^2 - x$ **(b)** $\sec^2 \theta - \sec \theta$

5. **(a)** $(x + 1)(x + 1)$ **(b)** $(\cos \theta + 1)(\cos \theta + 1)$

7. **(a)** $x(x - 2)$ **(b)** $\sec t(\sec t - 2)$

9. **(a)** $\dfrac{2}{1 - x^2}$ **(b)** $2 \csc^2 \theta$

11. **(a)** $\dfrac{x^2 + y^2}{xy}$ **(b)** $\sec t \csc t$

13. **(a)** $y^2 + x^2$ **(b)** 1

15. **(a)** x **(b)** $\cos \theta$

17. $\sin \theta$ **19.** $\sin \theta - \sec \theta$ **21.** $2 \tan t + \sec^2 t$

23. $\sin^2 \theta$ **25.** $(1 - \tan \theta)(1 + \tan \theta)$

27. $(\sec t - 3)(\sec t + 2)$ **29.** $\sec^2 \theta(\tan^2 \theta + 2)$

31. $\cot \theta \sin \theta = \dfrac{\cos \theta}{\sin \theta} \cdot \sin \theta$
$= \cos \theta$

33. $(1 - \cos^2 \theta)(1 + \tan^2 \theta) = \sin^2 \theta \sec^2 \theta$
$= \dfrac{\sin^2 \theta}{\cos^2 \theta}$
$= \tan^2 \theta$

35. $\cos t(\tan t - \sec t) = \cos t \left(\dfrac{\sin t}{\cos t} - \dfrac{1}{\cos t} \right)$
$= \sin t - 1$

37. $\dfrac{\tan(-\theta)}{\sin(-\theta)} = \dfrac{-\tan \theta}{-\sin \theta}$
$= \dfrac{1}{\cos \theta}$
$= \sec \theta$

39. Not an identity

41. $\dfrac{\sin^2 t}{\cos t} = \dfrac{1 - \cos^2 t}{\cos t}$
$= \dfrac{1}{\cos t} - \cos t$
$= \sec t - \cos t$

43. Not an identity

45. $\sec^2 \theta - 1 = (1 + \tan^2 \theta) - 1$
$= \tan^2 \theta$

47. $\dfrac{\tan^2 t}{\sec t} = \dfrac{\sec^2 t - 1}{\sec t}$
$= \sec t - \dfrac{1}{\sec t}$
$= \sec t - \cos t$

49. $\sec^2 t + \csc^2 t = \dfrac{1}{\cos^2 t} + \dfrac{1}{\sin^2 t}$
$= \dfrac{\sin^2 t + \cos^2 t}{\cos^2 t \sin^2 t}$
$= \dfrac{1}{\cos^2 t \sin^2 t}$
$= \sec^2 t \csc^2 t$

51. $(\sec t - 1)(\sec t + 1) = \sec^2 t - 1$
$= \tan^2 t$

53. $\dfrac{1 - \sin^2 \theta}{\cos \theta} = \dfrac{\cos^2 \theta}{\cos \theta}$
$= \cos \theta$

55. $\dfrac{\sin^4 t - \cos^4 t}{\sin^2 t - \cos^2 t} = \dfrac{(\sin^2 t - \cos^2 t)(\sin^2 t + \cos^2 t)}{\sin^2 t - \cos^2 t}$
$= \sin^2 t + \cos^2 t$
$= 1$

57.
$$\frac{\cot^2 t}{\csc t + 1} = \frac{\csc^2 t - 1}{\csc t + 1}$$
$$= \frac{(\csc t + 1)(\csc t - 1)}{\csc t + 1}$$
$$= \csc t - 1$$

59.
$$\frac{\cot t}{\cot t + 1} = \frac{\cot t}{\cot t + 1} \cdot \frac{\tan t}{\tan t}$$
$$= \frac{\cot t \tan t}{\cot t \tan t + \tan t}$$
$$= \frac{1}{1 + \tan t}$$

61.
$$\frac{1}{1 - \sin t} + \frac{1}{1 + \sin t} = \frac{(1 + \sin t) + (1 - \sin t)}{(1 - \sin t)(1 + \sin t)}$$
$$= \frac{2}{1 - \sin^2 t}$$
$$= \frac{2}{\cos^2 t}$$
$$= 2 \sec^2 t$$

63.
$$\frac{\csc t + \cot t}{\csc t - \cot t} = \frac{\csc t + \cot t}{\csc t - \cot t} \cdot \frac{\csc t + \cot t}{\csc t + \cot t}$$
$$= \frac{(\csc t + \cot t)^2}{\csc^2 t - \cot^2 t}$$
$$= \frac{(\csc t + \cot t)^2}{\csc^2 t - (\csc^2 t - 1)}$$
$$= (\csc t + \cot t)^2$$

65. Apply the Pythagorean theorem on triangle RST.

67. Using the similar triangles ABC and OPQ,
$$\frac{\cot \theta}{\csc \theta} = \frac{\cos \theta}{1}.$$

69. Applying the Pythagorean theorem to triangles ABC and OPQ, $\sin^2 \theta + \cos^2 \theta = 1$ and $\csc^2 \theta - \cot^2 \theta = 1$.

71. **(a)** $W(t) = 5(1 - \sin^2 (120 \pi t)); V = 25 \sin (120 \pi t)$ is maximum or minimum when $\sin(120 \pi t) = \pm 1$ and so $W = 0$.

(b) Whenever there is a peak or valley on the graph of V, the graph of W intersects the x-axis, which corresponds to a zero of W.

[0, 1/15, 1/60] by [−30, 30, 10]

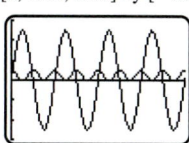

6.3 ANSWERS

1. $60°$ **3.** $85°$ **5.** $65°$ **7.** $\dfrac{\pi}{6}$

9. $\dfrac{\pi}{3}$ **11.** $\dfrac{\pi}{4}, \dfrac{5\pi}{4}$ **13.** $\dfrac{\pi}{3}, \dfrac{5\pi}{3}$

15. $60°, 120°, 240°, 300°; 60° + 180° \cdot n, 120° + 180° \cdot n$

17. $30°, 210°; 30° + 180° \cdot n$

19. **(a)** $\theta = 90°$ **(b)** $\theta = 270°$

21. **(a)** $60°, 240°$ **(b)** $120°, 300°$

23. **(a)** $60°, 300°$ **(b)** $120°, 240°$

25. **(a)** No solutions   **(b)** No solutions

27. **(a)** $x = \dfrac{1}{2}$ **(b)** $\theta = 30°, 150°$

29. **(a)** $x = 0, 1$ **(b)** $\theta = 0°, 90°, 180°$

31. **(a)** $x = \pm 1$ **(b)** $\theta = 45°, 135°, 225°, 315°$

33. **(a)** $x = -2, 1$ **(b)** $\theta = 0°$

35. $\dfrac{\pi}{3}, \dfrac{2\pi}{3}, \dfrac{4\pi}{3}, \dfrac{5\pi}{3}$ **37.** No solutions

39. $\dfrac{\pi}{2}, \dfrac{3\pi}{2}$ **41.** $\dfrac{\pi}{4}, \dfrac{5\pi}{4}$

43. $\dfrac{\pi}{4} + \dfrac{\pi n}{2}; 45° + 90° \cdot n$

45. No solutions   **47.** $\pi n; 180° \cdot n$

49. $\dfrac{\pi}{4} + \pi n; 45° + 180° \cdot n$ **51.** $\dfrac{\pi n}{2}; 90° \cdot n$

53. $2\pi n, \dfrac{\pi}{2} + 2\pi n; 360° \cdot n, 90° + 360° \cdot n$

55. $x = 0, x \approx 4.5$ **57.** $x \approx 0.39, 2.0$

59. $x \approx 3.6$

61. $x = 180° + 360° \cdot n \ (F = 1)$
$$x = 60° + 360° \cdot n, 300° + 360° \cdot n \ \left(F = \frac{1}{4}\right)$$

63. $x \approx 2.6, 10.7$; near February 17 and October 22

65. $x \approx 2.6, 10.7$; near February 17 and October 22

67. **(a)** [0, 13, 1] by [200, 600, 50]

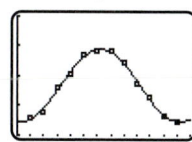

(b) $k_1 \leq x \leq k_2$, where $k_1 \approx 2.98$ and $k_2 \approx 9.51$. At 50°N latitude the maximum monthly hours of sunshine is greater than or equal to 350 hours roughly from March through September.

69. $\theta \approx 0.26$

71. **(a)** $0.005, 0.025, 0.045$

(b) At these times the pressure on an eardrum is maximum.

73. **(a)** [0, 0.01, 0.005] by [−0.005, 0.005, 0.001]

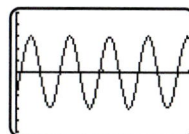

[0, 0.01, 0.005] by [−0.005, 0.005, 0.001]

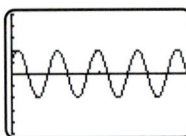

[0, 0.01, 0.005] by [−0.005, 0.005, 0.001]

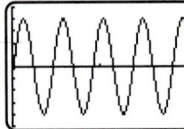

(b) Maximum: $P \approx 0.004$

(c) No. The maximum of P_1 is 0.003 and the maximum of P_2 is 0.002.

6.4 ANSWERS

1. $\dfrac{\sqrt{6} - \sqrt{2}}{4}$ **3.** $2 - \sqrt{3}$ **5.** $\dfrac{\sqrt{6} - \sqrt{2}}{4}$

7. $\dfrac{\sqrt{6} - \sqrt{2}}{4}$ **9.** $\dfrac{\sqrt{6} + \sqrt{2}}{4}$

11. (a) *Graphical:* Graphs of $y = \sin\left(t + \dfrac{\pi}{2}\right)$ and

$y = \cos t$ are the same.

Numerical: Tables of $y = \sin\left(t + \dfrac{\pi}{2}\right)$ and

$y = \cos t$ are the same.

Verbal: If the sine graph is translated $\dfrac{\pi}{2}$ units

left it coincides with the cosine graph.

(b) $\sin\left(t + \dfrac{\pi}{2}\right) = \sin t \cos\dfrac{\pi}{2} + \cos t \sin\dfrac{\pi}{2}$

$\qquad = \sin t\,(0) + \cos t\,(1)$

$\qquad = \cos t$

13. (a) *Graphical:* Graphs of $y = \cos(t + \pi)$ and
$y = -\cos t$ are the same.
Numerical: Tables of $y = \cos(t + \pi)$ and
$y = -\cos t$ are the same.
Verbal: If the cosine graph is translated π
units left it coincides with the cosine graph
reflected about the x-axis.

(b) $\cos(t + \pi) = \cos t \cos \pi - \sin t \sin \pi$
$\qquad = \cos t\,(-1) - \sin t\,(0)$
$\qquad = -\cos t$

15. (a) *Graphical:* Graphs of $y = \sec\left(t - \dfrac{\pi}{2}\right)$ and

$y = \csc t$ are the same.

Numerical: Tables of $y = \sec\left(t - \dfrac{\pi}{2}\right)$ and

$y = \csc t$ are the same.

Verbal: If the secant graph is translated $\dfrac{\pi}{2}$
units right it coincides with the cosecant
graph.

(b) $\sec\left(t - \dfrac{\pi}{2}\right) = \dfrac{1}{\cos(t - \pi/2)}$

$\qquad = \dfrac{1}{\cos t \cos(\pi/2) + \sin t \sin(\pi/2)}$

$\qquad = \dfrac{1}{\cos t\,(0) + \sin t\,(1)}$

$\qquad = \dfrac{1}{\sin t}$

$\qquad = \csc t$

17. $\cos(90° - \theta) = \cos 90° \cos \theta + \sin 90° \sin \theta$
$\qquad = (0) \cos \theta + (1) \sin \theta$
$\qquad = \sin \theta$

19. $\sec(90° - \theta) = \dfrac{1}{\cos(90° - \theta)}$

$\qquad = \dfrac{1}{\cos 90° \cos \theta + \sin 90° \sin \theta}$

$\qquad = \dfrac{1}{(0) \cos \theta + (1) \sin \theta}$

$\qquad = \dfrac{1}{\sin \theta}$

$\qquad = \csc \theta$

21. (a) $\dfrac{56}{65}$ **(b)** $\dfrac{33}{65}$

(c) $\dfrac{56}{33}$ **(d)** Quadrant I

23. (a) $-\dfrac{988}{1037}$ **(b)** $\dfrac{315}{1037}$

(c) $-\dfrac{988}{315}$ **(d)** Quadrant IV

25. (a) $-\dfrac{33}{65}$ **(b)** $-\dfrac{56}{65}$

(c) $\dfrac{33}{56}$ **(d)** Quadrant III

27. $\cos\left(t - \dfrac{\pi}{4}\right) = \cos t \cos\dfrac{\pi}{4} + \sin t \sin\dfrac{\pi}{4}$

$\qquad = \cos t\left(\dfrac{\sqrt{2}}{2}\right) + \sin t\left(\dfrac{\sqrt{2}}{2}\right)$

$\qquad = \dfrac{\sqrt{2}}{2}(\cos t + \sin t)$

29. $\tan\left(t + \dfrac{\pi}{4}\right) = \dfrac{\tan t + \tan(\pi/4)}{1 - \tan t \tan(\pi/4)}$

$\qquad = \dfrac{\tan t + 1}{1 - \tan t\,(1)}$

$\qquad = \dfrac{1 + \tan t}{1 - \tan t}$

31. $\sin 2t = \sin(t+t)$
$= \sin t \cos t + \cos t \sin t$
$= 2 \sin t \cos t$

33. $\sin(\alpha+\beta) + \sin(\alpha-\beta)$
$= \sin\alpha\cos\beta + \cos\alpha\sin\beta + \sin\alpha\cos\beta -$
$\quad \cos\alpha\sin\beta$
$= \sin\alpha\cos\beta + \sin\alpha\cos\beta$
$= 2\sin\alpha\cos\beta$

35. $\tan\gamma = \tan(\alpha-\beta)$
$= \dfrac{\tan\alpha - \tan\beta}{1 + \tan\alpha\tan\beta}$
$= \dfrac{(6/7) - (5/7)}{1 + (6/7)(5/7)}$
$= \dfrac{7}{79}$

37. $\tan\theta = \tan(\beta-\alpha)$
$= \dfrac{\tan\beta - \tan\alpha}{1 + \tan\alpha\tan\beta}$
$= \dfrac{m_2 - m_1}{1 + m_1 m_2}$

39. $\theta = \tan^{-1}\dfrac{7}{11} \approx 32.5°$

41. **(a)** $F \approx 409$ lb   **(b)** $46.2°$

43. **(a)** $[0, 0.02, 0.001]$ by $[-6, 6, 1]$

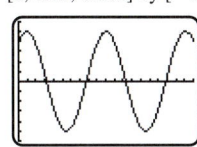

(b) $a = 5, k \approx 0.9272$
(c) $5\sin(220\pi t + 0.9272)$
$= 5(\sin(220\pi t)\cos(0.9272) +$
$\quad \cos(220\pi t)\sin(0.9272))$
$\approx 5(0.6\sin(220\pi t) + 0.8\cos(220\pi t))$
$= 3\sin(220\pi t) + 4\cos(220\pi t)$

45. Two beats
$[0.2, 1.2, 0.05]$ by $[-0.01, 0.01, 0.001]$

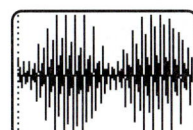

6.5 ANSWERS

1. **(a)** 1 **(b)** $\dfrac{\sqrt{3}}{2}$

3. **(a)** 1 **(b)** $-\dfrac{1}{2}$

5. **(a)** 2 **(b)** Undefined

7. *Graphical:* Graphs of $y = \tan 2\theta$ and $y = 2\tan\theta$ are not the same.
Numerical: Tables of $y = \tan 2\theta$ and $y = 2\tan\theta$ are not the same.
Symbolical: $\tan 2\theta = \dfrac{2\tan\theta}{1 - \tan^2\theta} \neq 2\tan\theta$,
unless $\tan\theta = 0$.

9. $\sin 2\theta = \dfrac{24}{25}, \cos 2\theta = \dfrac{7}{25}, \tan 2\theta = \dfrac{24}{7}$

11. $\sin 2\theta = -\dfrac{336}{625}, \cos 2\theta = -\dfrac{527}{625}, \tan 2\theta = \dfrac{336}{527}$

13. $\sin 2\theta = -\dfrac{1320}{3721}, \cos 2\theta = \dfrac{3479}{3721}, \tan 2\theta = -\dfrac{1320}{3479}$

15. $\sin 2\theta = \dfrac{336}{625}, \cos 2\theta = \dfrac{527}{625}, \tan 2\theta = \dfrac{336}{527}$

17. $\sin 2\theta$ **19.** $\dfrac{1}{2}\sin 2\theta$ **21.** $\cos 4\theta$ **23.** 1

25. $\cot^2 5x$ **27.** $\dfrac{2+\sqrt{2}}{4}$ **29.** $\dfrac{2+\sqrt{3}}{2-\sqrt{3}}$

31. **(a)** $\dfrac{\sqrt{2+\sqrt{3}}}{2}$

(b) $-\sqrt{\dfrac{2-\sqrt{3}}{2+\sqrt{3}}}$ or $-\dfrac{1}{2+\sqrt{3}}$ or $\sqrt{3}-2$

33. **(a)** $\sqrt{\dfrac{2-\sqrt{2}}{2+\sqrt{2}}}$ or $\dfrac{\sqrt{2}}{2+\sqrt{2}}$ or $\sqrt{2}-1$

(b) $-\dfrac{\sqrt{2-\sqrt{2}}}{2}$

35. $\sin 30°$ **37.** $\cos 25°$ **39.** $\tan 20°$

41. $\sin\dfrac{\theta}{2} = \dfrac{1}{\sqrt{10}}, \cos\dfrac{\theta}{2} = \dfrac{3}{\sqrt{10}}, \tan\dfrac{\theta}{2} = \dfrac{1}{3}$

43. $\sin\dfrac{\theta}{2} = -\dfrac{1}{\sqrt{26}}, \cos\dfrac{\theta}{2} = \dfrac{5}{\sqrt{26}}, \tan\dfrac{\theta}{2} = -\dfrac{1}{5}$

45. $\sin\dfrac{\theta}{2} = \dfrac{4}{5}, \cos\dfrac{\theta}{2} = \dfrac{3}{5}, \tan\dfrac{\theta}{2} = \dfrac{4}{3}$

47. $4\sin 2x = 4(2\sin x\cos x)$
$= 8\sin x\cos x$

49. $\dfrac{2 - \sec^2 x}{\sec^2 x} = \dfrac{2}{\sec^2 x} - 1$
$= 2\cos^2 x - 1$
$= \cos 2x$

51. $\sec 2x = \dfrac{1}{\cos 2x}$
$= \dfrac{1}{1 - 2\sin^2 x}$

53. $\sin 3\theta = \sin(2\theta + \theta)$
$= \sin 2\theta \cos\theta + \cos 2\theta \sin\theta$
$= (2\sin\theta\cos\theta)\cos\theta + (1 - 2\sin^2\theta)\sin\theta$
$= 2\sin\theta\cos^2\theta + \sin\theta - 2\sin^3\theta$
$= 2\sin\theta(1 - \sin^2\theta) + \sin\theta - 2\sin^3\theta$
$= 2\sin\theta - 2\sin^3\theta + \sin\theta - 2\sin^3\theta$
$= 3\sin\theta - 4\sin^3\theta$

55. $\sin 4\theta = 2\sin 2\theta \cos 2\theta$
$= 2(2\sin\theta\cos\theta)\cos 2\theta$
$= 4\sin\theta\cos\theta\cos 2\theta$

57. $\dfrac{\sin 2\theta}{\sin\theta} = \dfrac{2\sin\theta\cos\theta}{\sin\theta}$
$= 2\cos\theta$

59. $2\cos^2\dfrac{\theta}{2} = 2\left(\dfrac{1 + \cos\theta}{2}\right)$
$= 1 + \cos\theta$

61. $\cos^4\theta - \sin^4\theta = (\cos^2\theta - \sin^2\theta)(\cos^2\theta + \sin^2\theta)$
$= (\cos 2\theta)(1)$
$= \cos 2\theta$

63. $\csc 2t = \dfrac{1}{\sin 2t}$
$= \dfrac{1}{2\sin t\cos t}$
$= \dfrac{\csc t}{2\cos t}$

65. $\tan\dfrac{x}{2} = \dfrac{\sin(x/2)}{\cos(x/2)}$
$= \dfrac{2\sin(x/2)\cos(x/2)}{2\cos(x/2)\cos(x/2)}$
$= \dfrac{\sin x}{2\cos^2(x/2)}$
$= \dfrac{\sin x}{1 + \cos x}$

67. **(a)** $\dfrac{1}{2}(\sin 70° - \sin 30°)$

(b) $\dfrac{1}{2}(\cos 60° + \cos 20°)$

69. **(a)** $\dfrac{1}{2}(\sin 10\theta + \sin 4\theta)$

(b) $\dfrac{1}{2}(\cos 4x - \cos 12x)$

71. **(a)** $2\sin 35°\cos 5°$ **(b)** $2\cos 40°\cos 5°$

73. **(a)** $2\cos 5\theta\cos\theta$ **(b)** $2\sin\dfrac{11x}{2}\cos\dfrac{3x}{2}$

75. $0°, 180°$ **77.** $0°, 90°, 180°, 270°$ **79.** $180°$

81. $\pi n, \dfrac{2\pi}{3} + 2\pi n, \dfrac{4\pi}{3} + 2\pi n$

83. $\dfrac{\pi}{3} + 4\pi n, \dfrac{5\pi}{3} + 4\pi n$

85. **(a)** $[0, 0.04, 0.01]$ by $[-500, 2500, 500]$

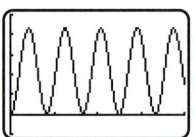

(b) $a = -1085, k = 240, d = 1085$

87. **(a)** $d = 600\left(1 - \cos\dfrac{80°}{2}\right) \approx 140.4$ ft

(b) *Hint:* First show that $r = d + r\cos\dfrac{\beta}{2}$.

(c) No, since $\cos\dfrac{\beta}{2} \neq \dfrac{1}{2}\cos\beta$ in general.

89. **(a)** $f(t) = \cos(1394\pi t) + \cos(3266\pi t)$
(b) $2\cos(2330\pi t)\cos(936\pi t)$
(c) $[0, 0.005, 0.001]$ by $[-2, 2, 1]$

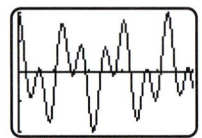

91. Let $f(t)$ model the tone for the number 3 and $g(t)$ model the tone for number 4.
(a) $f(t) = \cos(1394\pi t) + \cos(2954\pi t)$
$g(t) = \cos(1540\pi t) + \cos(2418\pi t)$
(b) $f(t) = 2\cos(2174\pi t)\cos(780\pi t)$
$g(t) = 2\cos(1979\pi t)\cos(439\pi t)$
(c) They are different. These two numbers sound different so their graphs should be different.

$[0, 0.005, 0.001]$ by $[-2, 2, 1]$ $[0, 0.005, 0.001]$ by $[-2, 2, 1]$

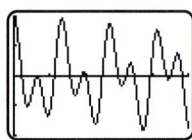   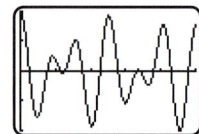

Graph of Number 3   Graph of Number 4

CHAPTER 6   REVIEW EXERCISES
1. Quadrant II

3. $\tan\theta = -\dfrac{3}{4}, \csc\theta = \dfrac{5}{3}, \sec\theta = -\dfrac{5}{4}, \cot\theta = -\dfrac{4}{3}$

5. $\sin\theta = -\dfrac{7}{25}, \csc\theta = -\dfrac{25}{7}, \sec\theta = \dfrac{25}{24},$
$\cot\theta = -\dfrac{24}{7}$

7. $-\sin(13°)$

9. $\sec\left(\dfrac{3\pi}{7}\right)$

11. An identity is an equation that is valid for all meaningful values of the variable. An example is $\sin^2\theta + \cos^2\theta = 1$. A conditional equation is valid only for certain values of the variable. An example is $\tan\theta = 1$.

13. No. If one trigonometric function is positive then its reciprocal must also be positive.

15. 1   **17.** 1   **19.** $\cos\theta$

21. $\sin\theta \approx 0.7776$, $\cos\theta \approx 0.6288$, $\csc\theta \approx 1.2860$, $\sec\theta \approx 1.5904$, $\cot\theta \approx 0.8086$

23. $\sin\theta \approx -0.8908$, $\tan\theta = 1.9604$, $\csc\theta \approx -1.1226$, $\sec\theta \approx -2.2007$, $\cot\theta \approx 0.5101$

25. $(\sin\theta + 1)(\sin\theta + 1)$

27. $(\tan\theta + 3)(\tan\theta - 3)$

29.
$$(\sec\theta - 1)(\sec\theta + 1) = \sec^2\theta - 1$$
$$= (1 + \tan^2\theta) - 1$$
$$= \tan^2\theta$$

31.
$$(1 + \tan t)^2 = 1 + 2\tan t + \tan^2 t$$
$$= \sec^2 t + 2\tan t$$

33.
$$\sin(x - \pi) = \sin x \cos\pi - \cos x \sin\pi$$
$$= \sin x (-1) - \cos x (0)$$
$$= -\sin x$$

35.
$$\sin 8x = \sin(2 \cdot 4x)$$
$$= 2\sin 4x \cos 4x$$

37.
$$\sec 2x = \frac{1}{\cos 2x}$$
$$= \frac{1}{1 - 2\sin^2 x}$$

39.
$$\cos^4 x \sin^3 x = \cos^4 x \sin^2 x \sin x$$
$$= \cos^4 x (1 - \cos^2 x) \sin x$$
$$= (\cos^4 x - \cos^6 x) \sin x$$

41.
$$\sec^4\theta - \tan^4\theta = (\sec^2\theta - \tan^2\theta)(\sec^2\theta + \tan^2\theta)$$
$$= (1)(1 + \tan^2\theta + \tan^2\theta)$$
$$= 1 + 2\tan^2\theta$$

43. $60°$

45. $\dfrac{2\pi}{7}$

47. $\dfrac{\pi}{6}, \dfrac{\pi}{2}, \dfrac{5\pi}{6}, \dfrac{3\pi}{2}$

49. (a) $60°, 240°$
     (b) $150°, 330°$

51. $60°, 300°$ **53.** $90°$

55. $\dfrac{\pi}{4}, \dfrac{5\pi}{4}$

57. $\dfrac{\pi}{6} + \pi n, -\dfrac{\pi}{6} + \pi n$; $30° + 180° \cdot n$, $-30° + 180° \cdot n$

59. $\dfrac{\pi}{2} + \pi n$; $90° + 180° \cdot n$

61. $-\dfrac{\sqrt{2 - \sqrt{3}}}{2}$

63. $-4.43, 1.13, 4.53$

65. (a) If the sine graph is translated $\dfrac{\pi}{2}$ units right, it corresponds to the cosine graph reflected across the x-axis.

  (b)
$$\sin\left(t - \frac{\pi}{2}\right) = \sin t \cos\frac{\pi}{2} - \cos t \sin\frac{\pi}{2}$$
$$= \sin t (0) - \cos t (1)$$
$$= -\cos t$$

  (c) Graphs and tables of $y = \sin(t - \pi/2)$ and $y = -\cos t$ are the same.

67. $\sin(\alpha + \beta) = \dfrac{63}{65}$, $\cos(\alpha + \beta) = \dfrac{16}{65}$

  $\tan(\alpha + \beta) = \dfrac{63}{16}$; Quadrant I

69. $\sin 2\theta = -\dfrac{24}{25}$, $\cos 2\theta = -\dfrac{7}{25}$, $\tan 2\theta = \dfrac{24}{7}$

71. $\sin\dfrac{\theta}{2} = \sqrt{\dfrac{3}{8}}$, $\cos\dfrac{\theta}{2} = \sqrt{\dfrac{5}{8}}$, $\tan\dfrac{\theta}{2} = \sqrt{\dfrac{3}{5}}$

73. $x = 4.7, 8.7$ or about April 21 and August 22

75. (a) $[0, 0.06, 0.01]$ by $[-0.012, 0.012, 0.002]$

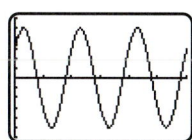

  (b) $\alpha = 0.01$, $k \approx 0.6435$
  (c)
$$0.01\sin(100\pi t + 0.6435)$$
$$= 0.01(\sin(100\pi t)\cos(0.6435) + \cos(100\pi t)\sin(0.6435))$$
$$\approx 0.01(0.8\sin(100\pi t) + 0.6\cos(100\pi t))$$
$$\approx 0.008\sin(100\pi t) + 0.006\cos(100\pi t)$$

CHAPTER 7   FURTHER TOPICS IN TRIGONOMETRY

7.1   ANSWERS

1. No

3. No

5. Yes

7. Yes

9. $\beta = 60°, a \approx 5.46, b \approx 4.90$

11. $\beta = 35°, a \approx 15.1, c \approx 7.37$

13. $\beta_1 \approx 57.1°, \gamma_1 \approx 76.9°, c_1 \approx 8.12$
$\beta_2 \approx 122.9°, \gamma_2 \approx 11.1°, c_2 \approx 1.61$

15. There are no solutions.

17. $\gamma = 93°, a \approx 7.76, c \approx 14.6$

19. $\beta_1 \approx 26.1°, \gamma_1 \approx 134°, c_1 \approx 14.7$
$\beta_2 \approx 154°, \gamma_2 \approx 6.1°, c_2 \approx 2.17$

21. $\gamma = 90°, \alpha = 60°, a = 10\sqrt{3} \approx 17.3$

23. $\alpha \approx 52.9°, \beta \approx 25.1°, b \approx 22.4$

25. $\beta = 10°, a \approx 92.2, c \approx 102$

27. There are no solutions.

29. 3629 ft

31. $d \approx 7.2$ mi

33. $AB \approx 105.4$ ft

35. 5.69 mi

37. 28.8 ft

39. 1054 ft

41. 630 ft

7.2 ANSWERS

1. (a) SAS   (b) Law of cosines

3. (a) SSA   (b) Law of sines

5. (a) ASA   (b) Law of sines

7. (a) ASA   (b) Law of sines

9. $a \approx 5.36, \beta \approx 40.7°, \gamma \approx 78.3°$

11. $\alpha \approx 22.3°, \beta \approx 108°, \gamma \approx 49.5°$

13. $\alpha \approx 33.6°, \beta \approx 50.7°, \gamma \approx 95.7°$

15. $c \approx 28.8, \alpha \approx 116.5°, \beta \approx 28.5°$

17. $\alpha \approx 101.0°, \beta \approx 44.0°, \gamma \approx 34.9°$
Angles do not sum to 180° due to rounding.

19. $a \approx 9.0, \beta \approx 150.9°, \gamma \approx 18.6°$

21. $\alpha \approx 23.1°, \beta \approx 107.2°, \gamma \approx 49.7°$

23. No, since $a + b < c$.

25. No, since $89° + 112° > 180°$.

27. Yes, since we are given ASA and $\alpha + \gamma < 180°$.

29. 86.8

31. 5.33

33. 50.9

35. 18.3

37. 2.14

39. 18.3

41. 169 ft

43. (a) 101.3 ft
(b) $\theta \approx 0.019°$

45. 29.8 mi

47. (a) $9\sqrt{3} \approx 15.6$ in^2
(b) The results are equal.

49. $\theta \approx 40.5°$ **51.** 149,429 ft^2

53. 21,309 ft^2 **55.** About 16.0 in^2

7.3 ANSWERS

1. (a) $a_1 \approx 3, a_2 \approx 4$
(b) $\|\mathbf{v}\| = 5$

3. (a) $a_1 \approx -5, a_2 \approx -12$
(b) $\|\mathbf{v}\| = 13$

5. (a)

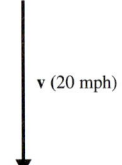

v (20 mph)

(b) $\mathbf{v} = \langle 0, -20 \rangle$
(c) $2\mathbf{v} = \langle 0, -40 \rangle$; this represents a 40-mph north wind.
$-\frac{1}{2}\mathbf{v} = \langle 0, 10 \rangle$; this represents a 10-mph south wind.

7. (a) v (5 mph)

(b) $\mathbf{v} = \langle 5/\sqrt{2}, -5/\sqrt{2} \rangle$ or $\left\langle \frac{5}{4}\sqrt{2}, -\frac{5}{2}\sqrt{2} \right\rangle$
(c) $2\mathbf{v} = \langle 10/\sqrt{2}, -10/\sqrt{2} \rangle$ or $\langle 5\sqrt{2}, -5\sqrt{2} \rangle$; this represents a 10-mph northwest wind.
$-\frac{1}{2}\mathbf{v} = \left\langle -\frac{5}{4}\sqrt{2}, \frac{5}{4}\sqrt{2} \right\rangle$; this represents a 2.5-mph southeast wind.

9. (a)

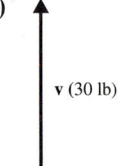

v (30 lb)

(b) $\mathbf{v} = \langle 0, 30 \rangle$
(c) $2\mathbf{v} = \langle 0, 60 \rangle$; this represents a 60 lb force upward.
$-\frac{1}{2}\mathbf{v} = \langle 0, -15 \rangle$; this represents a 15 lb force downward.

11. (a) Horizontal = 1, vertical = 1
(b) $\|\mathbf{v}\| = \sqrt{2}$; $\mathbf{v}$ is not a unit vector.
(c) $\|\mathbf{v}\|$ represents the length of $\mathbf{v}$.

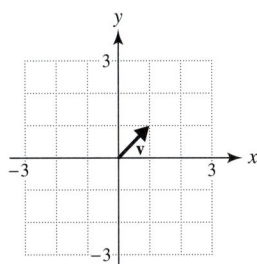

13. (a) Horizontal = 3, vertical = −4
(b) $\|\mathbf{v}\| = 5$; $\mathbf{v}$ is not a unit vector.
(c) $\|\mathbf{v}\|$ represents the length of $\mathbf{v}$.

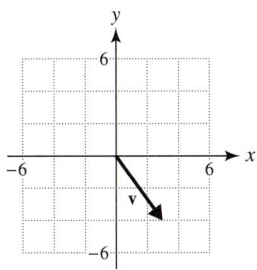

15. (a) Horizontal = 1, vertical = 0
(b) $\|\mathbf{v}\| = 1$; $\mathbf{v}$ is a unit vector.
(c) $\|\mathbf{v}\|$ represents the length of $\mathbf{v}$.

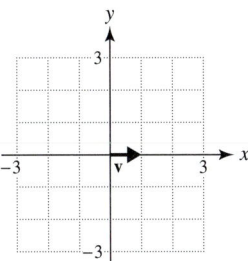

17. (a) Horizontal = 5, vertical = 12
(b) $\|\mathbf{v}\| = 13$; $\mathbf{v}$ is not a unit vector.
(c) $\|\mathbf{v}\|$ represents the length of $\mathbf{v}$.

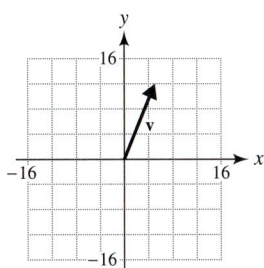

19. (a)

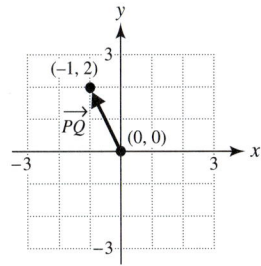

(b) $\overrightarrow{PQ} = -\mathbf{i} + 2\mathbf{j}$

21. (a)

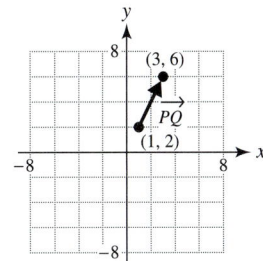

(b) $\overrightarrow{PQ} = 2\mathbf{i} + 4\mathbf{j}$

23. (a)

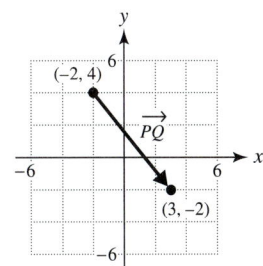

(b) $\overrightarrow{PQ} = 5\mathbf{i} - 6\mathbf{j}$

25. (a) $\mathbf{a} + \mathbf{b} = \langle -4, 16 \rangle$
(b) $\mathbf{a} - \mathbf{b} = \langle -12, 0 \rangle$
(c) $-\mathbf{a} = \langle 8, -8 \rangle$

27. (a) $\mathbf{a} + \mathbf{b} = \langle 8, 0 \rangle$
(b) $\mathbf{a} - \mathbf{b} = \langle 0, 16 \rangle$
(c) $-\mathbf{a} = \langle -4, -8 \rangle$

29. (a) $\mathbf{a} + \mathbf{b} = \langle 0, 12 \rangle$
(b) $\mathbf{a} - \mathbf{b} = \langle -16, -4 \rangle$
(c) $-\mathbf{a} = \langle 8, -4 \rangle$

31. (a) $\mathbf{a} + \mathbf{b} = \langle 3, 2 \rangle$
(b) $\mathbf{a} - \mathbf{b} = \langle -3, 2 \rangle$

33. (a) $\mathbf{a} + \mathbf{b} = 3\mathbf{i} - \mathbf{j}$
(b) $\mathbf{a} - \mathbf{b} = \mathbf{i} + 3\mathbf{j}$

35. (a) $2\mathbf{a} = 4\mathbf{i}$
(b) $2\mathbf{a} + 3\mathbf{b} = 7\mathbf{i} + 3\mathbf{j}$

37. (a) $2\mathbf{a} = \langle -2, 4 \rangle$
(b) $2\mathbf{a} + 3\mathbf{b} = \langle 7, 4 \rangle$

39. Horizontal $= 4 \cos 40° \approx 3.06$
Vertical $= 4 \sin 40° \approx 2.57$

41. Horizontal $= 5 \cos(-35°) \approx 4.10$
Vertical $= 5 \sin(-35°) \approx -2.87$

43. 94.2 lb

45. 24.4 lb

47. (a) 1
(b) 81.9°
(c) Neither

49. (a) 0
(b) 90°
(c) Perpendicular

51. (a) 122
(b) 0°
(c) Parallel, same direction

53. (a) -4
(b) 143.1°
(c) Neither

55. 150 ft-lb

57. 100,000 ft-lb

59. Work $= 590$ ft–lb, $\|\mathbf{F}\| = 10\sqrt{5} \approx 22.4$ lb

61. Work $= 27$ ft–lb, $\|\mathbf{F}\| = \sqrt{34} \approx 5.83$ lb

63. 24

65. 4

67. $\mathbf{v} = \langle 2, 3 \rangle$, speed $= \sqrt{13} \approx 3.6$ mph

69. (a) 10 mph
(b) $3\mathbf{v} = 18\mathbf{i} + 24\mathbf{j}$; this represents a 30-mph wind in the direction of $\mathbf{v}$.
(c) $\mathbf{u}$ represents a southeast wind of $\sqrt{128} \approx 11.3$ mph.

71. (a) $\|\mathbf{R}\| = \sqrt{5} \approx 2.2$, $\|\mathbf{A}\| = \sqrt{1.25} \approx 1.1$
About 2.2 inches of rain fell. The area of the opening of the rain gauge is about 1.1 square inches.
(b) $V = 1.5$; the volume of rain was 1.5 cubic inches.
(c) $\mathbf{R}$ and $\mathbf{A}$ should be parallel and point in opposite directions.

73. (a) $\mathbf{c} = \mathbf{a} + \mathbf{b} = \mathbf{i} + 4\mathbf{j}$
(b) $\sqrt{17} \approx 4.1$ ft
(c) $3\mathbf{a} + \dfrac{1}{2}\mathbf{b} = 8\mathbf{i} + 7\mathbf{j}$

75. (a) Gray
(b) Blue

77. (a) $\langle 2, 2 \rangle$
(b) $\mathbf{b} = \mathbf{a} + \mathbf{v}$

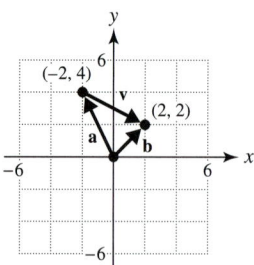

79. $W = \langle 0, 145 \rangle \cdot \langle 7920 \cos 15°, 7920 \sin 15° \rangle \approx$ 297,228 ft–lb

81. Speed $= 180$ mph, bearing $= 128.2°$

7.4 ANSWERS

1. (a)

t	0	1	2	3
x	-1	0	1	2
y	0	2	4	6

(b) (Plot the points in the table.)

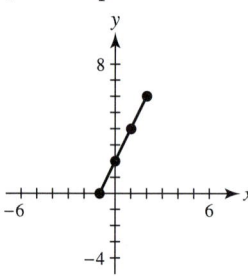

(c) Line segment

3. (a)

t	0	1	2	3
x	2	3	4	5
y	4	1	0	1

(b)

(c) Lower portion of a parabola

5. (a)

t	0	1	2	3
x	3	$\sqrt{8}$	$\sqrt{5}$	0
y	0	1	2	3

(b)

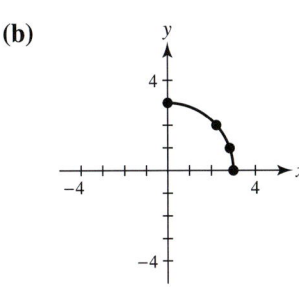

(c)   Portion of a circle with radius 3

7. $y = \dfrac{1}{2}x^2$; portion of a parabola

$[-4.7, 4.7, 1]$ by $[-3.1, 3.1, 1]$

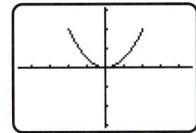

9. $y = x^2 + 4x + 5$; portion of a parabola

$[-4.7, 4.7, 1]$ by $[0, 6.2, 1]$

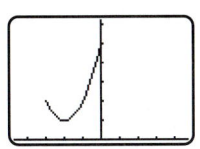

11. $x^2 + y^2 = 9$; circle with radius 3

$[-4.7, 4.7, 1]$ by $[-3.1, 3.1, 1]$

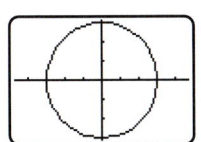

13. (a)   Traces a circle of radius 3 once

$[-4.7, 4.7, 1]$ by $[-3.1, 3.1, 1]$

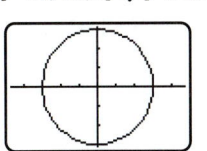

(b)   Traces a circle of radius 3 twice

$[-4.7, 4.7, 1]$ by $[-3.1, 3.1, 1]$

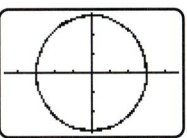

15. (a)   Traces a circle of radius 3 once starting at $(3, 0)$

$[-4.7, 4.7, 1]$ by $[-3.1, 3.1, 1]$

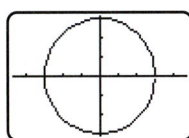

(b)   Traces a circle of radius 3 once starting at $(0, 3)$

$[-4.7, 4.7, 1]$ by $[-3.1, 3.1, 1]$

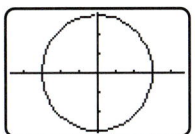

17. (a)   Circle of radius 1 centered at $(-1, 2)$

$[-4.7, 4.7, 1]$ by $[-3.1, 3.1, 1]$

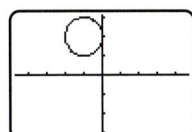

(b)   Circle of radius 1 centered at $(1, 2)$

$[-4.7, 4.7, 1]$ by $[-3.1, 3.1, 1]$

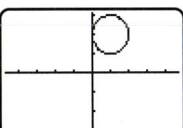

19. $[-9.4, 9.4, 1]$ by $[-6.2, 6.2, 1]$

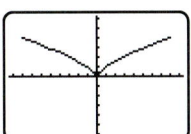

21. $[0, 6, 1]$ by $[-2, 2, 1]$

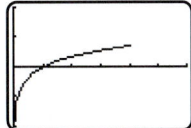

23. $[-1, 5, 1]$ by $[-1, 3, 1]$

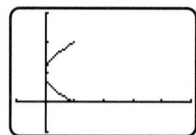

25. $[-4.7, 4.7, 1]$ by $[-3.1, 3.1, 1]$

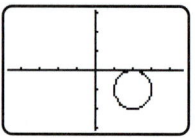

27. $[-1.5, 1.5, 0.5]$ by $[-1, 1, 0.5]$

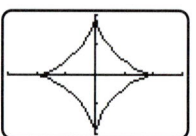

29. $[-4.7, 4.7, 1]$ by $[-3.1, 3.1, 1]$

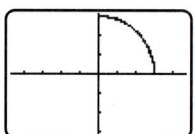

31.   F

$[0, 6, 1]$ by $[0, 4, 1]$

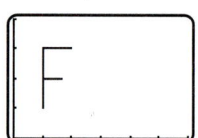

33.   D

$[0, 6, 1]$ by $[0, 4, 1]$

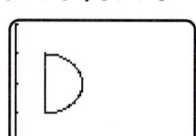

35. $x_1 = 0, y_1 = 2t; x_2 = t, y_2 = 0; 0 \le t \le 1$
Answers may vary.

37. $x_1 = \sin t, y_1 = \cos t; x_2 = 0, y_2 = t - 2; 0 \le t \le \pi$
Answers may vary.

39.   Answers will vary.

41. $[-6, 6, 1]$ by $[-4, 4, 1]$

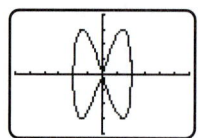

43. $[-6, 6, 1]$ by $[-4, 4, 1]$

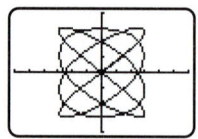

45. $F_2 = 100$

47. $F_2 = 300$

49.   The ball thrown at $35°$ travels about 128 feet.
The ball thrown at $50°$ travels about 134 feet.

51.   Yes

7.5   ANSWERS

1.

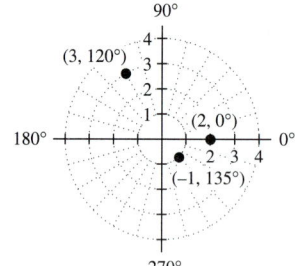

3.

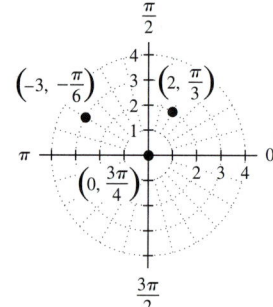

5.   Yes

7.   No

9.   Yes

11. $\left(\dfrac{3}{\sqrt{2}}, \dfrac{3}{\sqrt{2}}\right)$ or $\left(\dfrac{3\sqrt{2}}{2}, \dfrac{3\sqrt{2}}{2}\right)$

13. $(0, 10)$

15. $(5, 0)$

17. (a)

θ	0°	90°	180°	270°
r	2	2	2	2

(b)

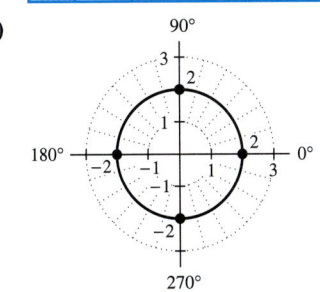

19. (a)

θ	0°	90°	180°	270°
r	0	3	0	−3

(b)

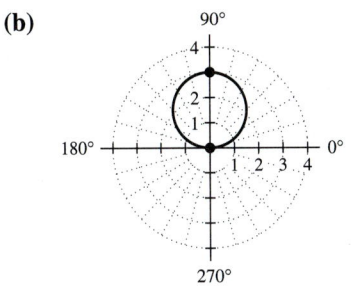

21. (a)

θ	0°	90°	180°	270°
r	2	4	2	0

(b)

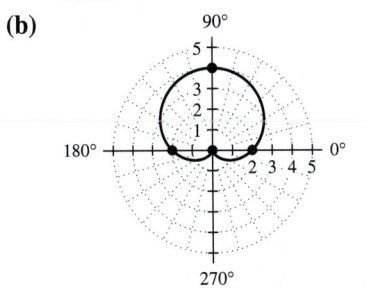

23. $x^2 + y^2 = 9$

25. $x = 2$

27. $2x + 4y = 3$

29. $x^2 + y^2 = x$

31.

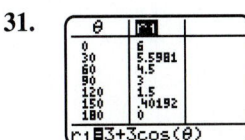

$[-9.4, 9.4, 1]$ by $[-6.2, 6.2, 1]$

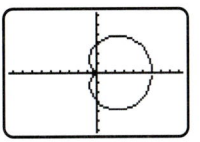

33.

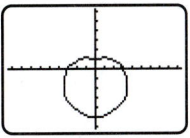

$[-9.4, 9.4, 1]$ by $[-6.2, 6.2, 1]$

35.

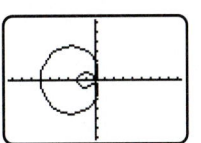

$[-9.4, 9.4, 1]$ by $[-6.2, 6.2, 1]$

37.

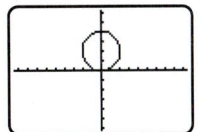

$[-9.4, 9.4, 1]$ by $[-6.2, 6.2, 1]$

39.

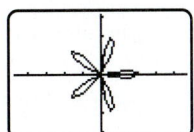

$[-4.7, 4.7, 1]$ by $[-3.1, 3.1, 1]$

41. Use radian mode with $0 \le \theta \le \dfrac{9\pi}{2}$.

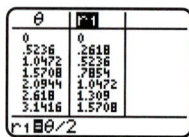

$[-9.4, 9.4, 1]$ by $[-6.2, 6.2, 1]$

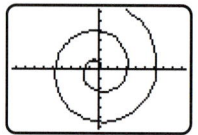

43. Let $r_1 = \sqrt{2 \sin(2\theta)}$ and $r_2 = -\sqrt{2 \sin(2\theta)}$.

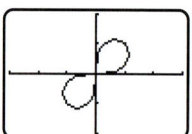

$[-3, 3, 1]$ by $[-2, 2, 1]$

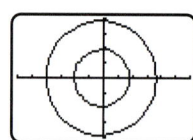

45. $30°, 150°$

47. $210°, 330°$

49. $30°, 150°$

51. $[-30, 30, 10]$ by $[-20, 20, 10]$

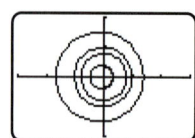

53. $[-3, 3, 1]$ by $[-2, 2, 1]$

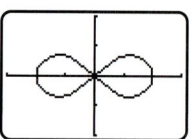

55. Inside the "figure eight" the radio signal can be received. This region is generally in an east–west direction from the two radio towers with a maximum distance of 200 miles.

$[-300, 300, 100]$ by $[-200, 200, 100]$

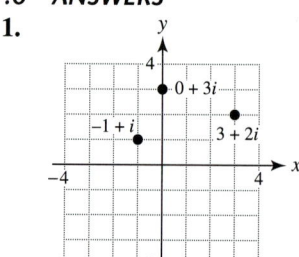

7.6 ANSWERS

1.

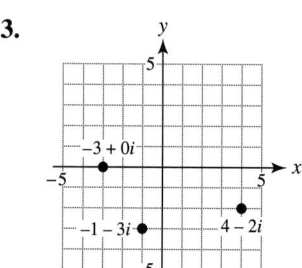

3.

5. -5

7. $\sqrt{2} + i\sqrt{2}$

9. $-4i$

11. 3

13. $\sqrt{2}$

15. 13

17. 6

19. $\sqrt{13}$

21. $\sqrt{2}(\cos 45° + i \sin 45°)$

23. $5(\cos 0° + i \sin 0°)$

25. $4(\cos 90° + i \sin 90°)$

27. $2(\cos 120° + i \sin 120°)$

29. $2(\cos 30° + i \sin 30°)$

31. $2(\cos \pi + i \sin \pi)$

33. $\sqrt{8}\left(\cos \dfrac{3\pi}{4} + i \sin \dfrac{3\pi}{4}\right)$

35. $z_1 z_2 = \dfrac{27}{2} + \dfrac{27\sqrt{3}}{2} i$

$\dfrac{z_1}{z_2} = \dfrac{3\sqrt{3}}{2} + \dfrac{3}{2} i$

37. $z_1z_2 = -6$

$\dfrac{z_1}{z_2} = 6i$

39. $z_1z_2 = \dfrac{\sqrt{3}}{2} - \dfrac{1}{2}i$

$\dfrac{z_1}{z_2} = \dfrac{1}{2} + \dfrac{\sqrt{3}}{2}i$

41. $8i$

43. 1

45. $-\dfrac{25}{2} + \dfrac{25\sqrt{3}}{2}i$

47. $-2 + 2i$

49. $-16\sqrt{3} + 16i$

51. $1 + i\sqrt{3}, -1 - i\sqrt{3}$

53. $-1, \dfrac{1}{2} + \dfrac{\sqrt{3}}{2}i, \dfrac{1}{2} - \dfrac{\sqrt{3}}{2}i$

55. $\dfrac{\sqrt{2}}{2} + \dfrac{\sqrt{2}}{2}i, -\dfrac{\sqrt{2}}{2} - \dfrac{\sqrt{2}}{2}i$

57. $-2, 1 + i\sqrt{3}, 1 - i\sqrt{3}$

59. $-4i, 2\sqrt{3} + 2i, -2\sqrt{3} + 2i$

61. $\pm 3, \pm 3i$

63. Yes

65. No

67. (a) $z = 110 + 32i$
(b) $\sqrt{13{,}124} \approx 114.6$ ohms

CHAPTER 7   REVIEW EXERCISES

1. $\gamma = 70°, a = 10.1, b \approx 6.91$

3. $a \approx 5.46, \beta \approx 59.1°, \gamma \approx 78.9°$

5. $\gamma = 115°, a \approx 5.9, c \approx 16.4$

7. $\beta \approx 14.4°, \alpha \approx 145.6°, a \approx 18.2$

9. $c \approx 13.2, \beta \approx 93.6°, \alpha \approx 51.4°$

11. 53.0

13. 891

15. (a) Horizontal = 3, vertical = 4
(b) $\|\mathbf{v}\| = 5$
(c) $\|\mathbf{v}\|$ represents the length of **v**.

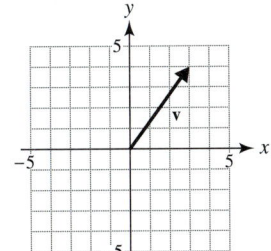

17. (a)

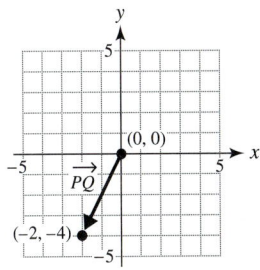

(b) $\overrightarrow{PQ} = -2\mathbf{i} - 4\mathbf{j}$
(c) $\|\overrightarrow{PQ}\| = \sqrt{20}$

19. (a) $2\mathbf{a} = \langle 6, -4\rangle$
(b) $\mathbf{a} - 3\mathbf{b} = \langle 0, -5\rangle$
(c) $\mathbf{a} \cdot \mathbf{b} = 1$
(d) $\theta \approx 78.7°$

21. (a) $2\mathbf{a} = 4\mathbf{i} + 4\mathbf{j}$
(b) $\mathbf{a} - 3\mathbf{b} = -\mathbf{i} - \mathbf{j}$
(c) $\mathbf{a} \cdot \mathbf{b} = 4$
(d) $\theta = 0°$

23. 207.1 lb

25. $[-4.7, 4.7, 1]$ by $[-3.1, 3.1, 1]$

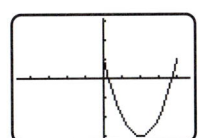

27. $[-4.7, 4.7, 1]$ by $[-3.1, 3.1, 1]$

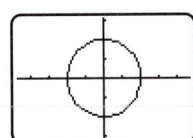

29. Answers will vary. (Graph with the axes turned off.)
(a) $x = \cos t, y = 2\sin t; 0 \le t \le 2\pi$
(b) $x_1 = -1 + \dfrac{t}{\pi}, y_1 = 0; 0 \le t \le 2\pi$

$x_2 = -1, y_2 = -1 + \dfrac{t}{\pi}; 0 \le t \le 2\pi$

$x_3 = 1, y_3 = -1 + \dfrac{t}{\pi}; 0 \le t \le 2\pi$

31. (a)

θ	0°	90°	180°	270°
r	2	1	0	1

(b)

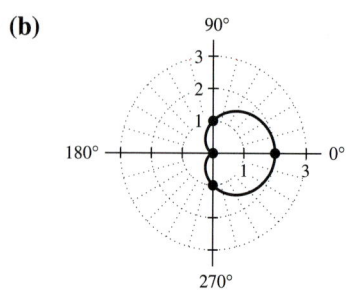

33. $[-4.7, 4.7, 1]$ by $[-3.1, 3.1, 1]$

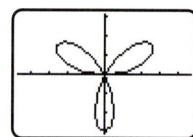

35. $[-9.4, 9.4, 1]$ by $[-6.2, 6.2, 1]$

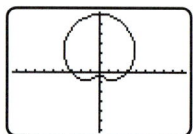

37.

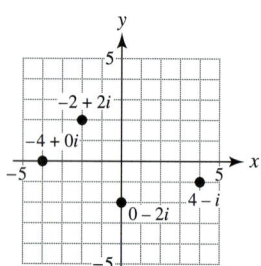

39. $z_1z_2 = -8, \dfrac{z_1}{z_2} = -1 + i\sqrt{3}$

41. $\sqrt{3} + i, -\sqrt{3} - i$

43.   701.6 mi

45.   7204 ft

47.   15,600 ft^2

49.   **(a)** $\|\mathbf{I}\| = \sqrt{34,000} \approx 184.4, \|\mathbf{S}\| = \sqrt{29} \approx 5.4$
    The intensity of the sun is about 184.4 watts per square meter. The area of the solar panel is about 5.4 square meters.
   (b) $W = |\mathbf{I} \cdot \mathbf{S}| = 820$. The solar panel is collecting 820 watts.

51.   204,985 ft–lb

53.   10,285 ft

CHAPTER 8   SYSTEMS OF EQUATIONS AND INEQUALITIES

8.1   ANSWERS

1. $f(5, 8) = 20$. The area of a triangle with base 5 and height 8 is 20.

3. $f(20, 35) = 700$. The area of a rectangle with width 20 and length 35 is 700.

5. $S(60, 157.48) \approx 1.60m^2$

7.   **ALGORITHM Time to Travel**
   Step 1:   Input the number of miles x and the speed y in mph.
   Step 2:   Evaluate $\dfrac{x}{y}$.
   Step 3:   Let the answer from STEP 2 be T.
   Step 4:   Output T, the time to travel x miles at y miles per hour.

9. $f(x, y) = y + 2x$

11. $f(x, y) = \dfrac{xy}{1 + x}$

13. $f(1960, \text{Male}) = 66.6$. A male born in 1960 had a life expectancy at birth of 66.6 years.

15. $f(1970, \text{Female}) \approx 75.3$. Answers may vary.

17. $f(1970, B) = 792{,}317$. In 1970 there were 792,317 bachelor degrees awarded.

19. $f(1985, B) \approx 990{,}381$. Answers may vary.

21. $l = 7, w = 5$

23. $f(255, 0, 255) = $ purple (or magenta)

25. $f(0, 255, 255) = $ cyan or turquoise. Answers may vary.

27. $256 \cdot 256 \cdot 256 = 16{,}777{,}216$ colors

29. $(2, 1)$

31. $(4, -3)$

33. $(2, 2)$

35. $(1, -1)$

37. $(1, 2)$

39. $(-\sqrt{8}, -\sqrt{8}), (\sqrt{8}, \sqrt{8})$

41. $(-7, -9), (9, 7)$

43. $(-1.5878, 0.2395), (0.1637, 1.4866),$
    $(1.9241, -0.3511)$

45.   **(a)**   Let x represent the number of bank robberies in 1994 and y the number in 1995.
      $x + y = 13{,}787$
      $x - y = 271$
   (b) & (c) $(7029, 6758)$

47.   1.2 million foreign tourists, 5.0 million American tourists

49. $r \approx 1.538, h \approx 6.724$

51. $x \approx 177.1, y \approx 174.9$; if an athlete's maximum heart rate is 180 bpm, then it will be about 177 bpm after 5 seconds and 175 bpm after 10 seconds.

53. $W_1 = \dfrac{300}{1 + \sqrt{3}} \approx 109.8$ lb, $W_2 = \dfrac{300\sqrt{3}}{\sqrt{6} + \sqrt{2}} \approx$ 134.5 lb

55. 0.510

57. Approximately 10,823 watts

59. Approximately 2.54 cords

61. $S(w, h) = 0.0101(w^{0.425})(h^{0.725})$

8.2   ANSWERS

1. The system is consistent with solution $(2, 2)$.

3. The system is inconsistent.

5. $(14, 6)$; consistent and independent

7. $(1, 3)$; consistent and independent

9. $\{(x, y) \mid x + y = 500\}$; consistent and dependent

11. $(-2, 5)$; consistent and independent

13. Inconsistent

15. $\left(-\dfrac{11}{7}, \dfrac{12}{7}\right)$; consistent and independent

17. 10.5 inches wide, 8.5 inches high

19. 23 child tickets, 52 adult tickets

21. $y = 2x + 2$

$[-8, 8, 1]$ by $[-8, 8, 1]$

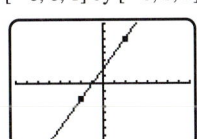

23. $y = -2x + 1$

$[-8, 8, 1]$ by $[-8, 8, 1]$

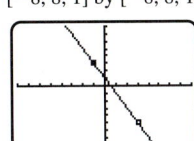

25. $f(x) = \dfrac{106}{11}x - \dfrac{209{,}794}{11} \approx 9.6364x - 19{,}072$

$[1983, 1998, 1]$ by $[40, 180, 10]$

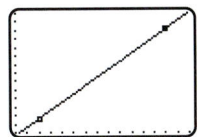

27. $f(x) = -4.5x + 8998$

$[1993, 1997, 1]$ by $[12, 28, 5]$

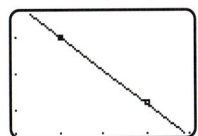

29. c; one solution is $(2, 3)$. Answers may vary.

31. d; one solution is $(-1, -1)$. Answers may vary.

33. One solution is $(0, 2)$. Answers may vary.

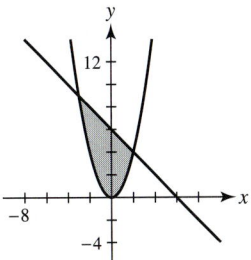

35. One solution is $(0, 0)$. Answers may vary.

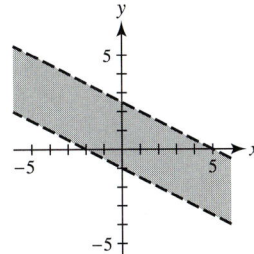

37. One solution is $(-1, 1)$. Answers may vary.

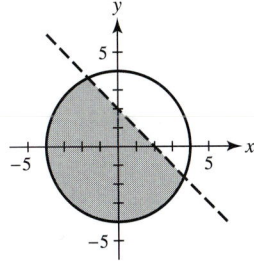

39. **(a)** $x + y = 3000, \quad 0.08x + 0.10y = 264$
(b) \$1800 at 8% and \$1200 at 10%

41. There are no solutions. If loans totaling \$3000 are at 10%, then the interest must be \$300.

43. 1669 incidents in 1994 and 1708 incidents in 1995

45. 63.4 lb of beef, 14.4 lb of turkey

47. Air speed: 520 mph; wind speed: 40 mph

49. $x = 300, y = 350$

51. This individual has a weight that is less than recommended for his or her height.

53. $25h - 7w \leq 800$
$5h - w \geq 170$

55. Max: 65; min: 8

57. Max: 66; min: 3

59. Max: 100; min: 0

61. Min: 6

63. 25 radios, 30 CD players

65. Part X: 9; part Y: 4

8.3 ANSWERS

1. (a) 3×1
(b) 2×3
(c) 2×2

3. Dimension: 2×3
$$\begin{bmatrix} 5 & -2 & | & 3 \\ -1 & 3 & | & -1 \end{bmatrix}$$

5. Dimension: 3×4
$$\begin{bmatrix} -3 & 2 & 1 & | & -4 \\ 5 & 0 & -1 & | & 9 \\ 1 & -3 & -6 & | & -9 \end{bmatrix}$$

7. $3x + 2y = 4$
$y = 5$

9. $3x + y + 4z = 0$
$5y + 8z = -1$
$-7z = 1$

11. (a) Yes
(b) No
(c) Yes

13. $(5, -1)$

15. $(2, 0)$

17. $(2, 3, 1)$

19. $\{(3 - 3z, 1 + 2z, z \mid z \text{ is a real number})\}$

21. $(-17, 10)$

23. $(0, 2, -1)$

25. $(3, 2, 1)$

27. No solutions

29. $(3, 2)$

31. $(-2, 1, 3)$

33. $(5.211, 3.739, -4.655)$

35. $(7.993, 1.609, -0.4011)$

37. (a) $F = 0.5714N + 0.4571R - 2014$
(b) \$5700

39. Pump 1: 12 hr; pumps 2 and 3: 24 hr

41. (a) At intersection A incoming traffic is equal to $x + 5$. The outgoing traffic is given by $y + 7$.

Therefore, $x + 5 = y + 7$. The other equations can be justified in a similar way.

(b) The three equations can be written as:
$$x - y = 2$$
$$x - z = 3$$
$$y - z = 1$$
The solution can be written as
$\{(z + 3, z + 1, z) \mid z \geq 0\}$.

(c) There are an infinite number of solutions since some cars could be driving around the block continually.

43. (a) $1990^2 a + 1990b + c = 11$
$2010^2 a + 2010b + c = 10$
$2030^2 a + 2030b + c = 6$
(b) $f(x) = -0.00375x^2 + 14.95x - 14{,}889.125$
(c) [1985, 2035, 5] by [5, 12, 1]

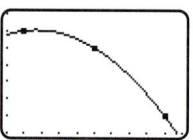

(d) Answers may vary. For example, in 2015 the ratio could be $f(2015) \approx 9.3$.

45. (a) $3^2 a + 3b + c = 2.2$
$18^2 a + 18b + c = 10.4$
$26^2 a + 26b + c = 12.8$
(b) $f(x) = -0.010725x^2 + 0.77188x - 0.019130$
(c) [0, 30, 5] by [0, 14, 1]

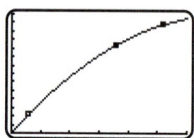

(d) Answers may vary. For example, in 1990 this percentage was $f(20) \approx 11.1$.

8.4 ANSWERS

1. (a) $a_{12} = 2$, $a_{21} = 4$, a_{32} is undefined.
(b) $a_{11}a_{22} + 3a_{23} = 23$

3. Denver is City 1, while Las Vegas is City 4. Since $a_{14} = 760$, the distance is 760 miles.

5. Diagonal elements are denoted by a_{11}, a_{22}, a_{33}, and a_{44}. They represent the distance from City k to City k, which is always zero.

7. Dimension: 20×20; 400 elements

9. $A = \begin{bmatrix} 0 & 40 & 20 & 15 \\ 40 & 0 & 53 & * \\ 20 & 53 & 0 & 7 \\ 15 & * & 7 & 0 \end{bmatrix}$

11. Format of map may vary slightly.

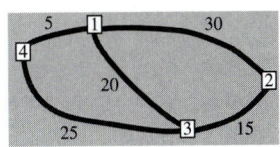

13. 17

15. (a) $A + B = \begin{bmatrix} 1 & -6 \\ 1 & 4 \end{bmatrix}$

(b) $3A = \begin{bmatrix} 6 & -18 \\ 9 & 3 \end{bmatrix}$

(c) $2A - 3B = \begin{bmatrix} 7 & -12 \\ 12 & -7 \end{bmatrix}$

17. (a) $A + B$ is undefined.

(b) $3A = \begin{bmatrix} 3 & -3 & 0 \\ 3 & 15 & 27 \\ -12 & 24 & -15 \end{bmatrix}$

(c) $2A - 3B$ is undefined.

19. (a) $A + B = \begin{bmatrix} 0 & -2 \\ -2 & 2 \\ 9 & -8 \end{bmatrix}$

(b) $3A = \begin{bmatrix} -6 & -3 \\ -15 & 3 \\ 6 & -9 \end{bmatrix}$

(c) $2A - 3B = \begin{bmatrix} -10 & 1 \\ -19 & -1 \\ -17 & 9 \end{bmatrix}$

21. $\begin{bmatrix} 0 & 2 \\ 13 & -5 \\ 0 & 1 \end{bmatrix}$

23. $A = \begin{bmatrix} 1 & 2 & 1 \\ 1 & 2 & 1 \\ 1 & 2 & 1 \end{bmatrix}$

25. $B = \begin{bmatrix} 1 & 1 & 1 \\ 1 & 1 & 1 \\ 1 & 1 & 1 \end{bmatrix}, A - B = \begin{bmatrix} 0 & 1 & 0 \\ 0 & 1 & 0 \\ 0 & 1 & 0 \end{bmatrix}$

27. $B = \begin{bmatrix} 3 & 3 & 3 \\ 3 & 3 & 3 \\ 3 & 3 & 3 \end{bmatrix}$,

$B - A = \begin{bmatrix} 3 & 0 & 3 \\ 3 & 0 & 3 \\ 3 & 0 & 3 \end{bmatrix}$

29. $A = \begin{bmatrix} 3 & 3 & 3 & 3 \\ 3 & 0 & 0 & 0 \\ 3 & 3 & 3 & 0 \\ 3 & 0 & 0 & 0 \\ 3 & 0 & 0 & 0 \end{bmatrix}$

31. (a) One possibility is $A = \begin{bmatrix} 3 & 3 & 3 & 3 \\ 0 & 0 & 3 & 0 \\ 0 & 3 & 0 & 0 \\ 3 & 3 & 3 & 3 \end{bmatrix}$.

(b) $B = \begin{bmatrix} 3 & 3 & 3 & 3 \\ 3 & 3 & 3 & 3 \\ 3 & 3 & 3 & 3 \\ 3 & 3 & 3 & 3 \end{bmatrix}$

33. (a) One possibility is $A = \begin{bmatrix} 3 & 0 & 0 & 0 \\ 3 & 0 & 0 & 0 \\ 3 & 0 & 0 & 0 \\ 3 & 3 & 3 & 3 \end{bmatrix}$.

(b) $B = \begin{bmatrix} 3 & 3 & 3 & 3 \\ 3 & 3 & 3 & 3 \\ 3 & 3 & 3 & 3 \\ 3 & 3 & 3 & 3 \end{bmatrix}$

35. (a) $A = \begin{bmatrix} 12 & 4 \\ 8 & 7 \end{bmatrix}, B = \begin{bmatrix} 55 \\ 70 \end{bmatrix}$

(b) $AB = \begin{bmatrix} 940 \\ 930 \end{bmatrix}$. Tuition for Student 1 is $940, and tuition is $930 for Student 2.

37. (a) $A = \begin{bmatrix} 10 & 5 \\ 9 & 8 \\ 11 & 3 \end{bmatrix}, B = \begin{bmatrix} 60 \\ 70 \end{bmatrix}$

(b) $AB = \begin{bmatrix} 950 \\ 1100 \\ 870 \end{bmatrix}$. The product AB calculates tuition for each student.

39. $AB = \begin{bmatrix} -3 & 1 \\ -4 & 6 \end{bmatrix}, BA = \begin{bmatrix} 4 & 2 \\ 5 & -1 \end{bmatrix}$

41. AB and BA are undefined.

43. AB and BA are undefined.

45. $AB = \begin{bmatrix} -1 & 15 & -2 \\ 0 & 1 & 3 \\ -1 & 14 & -5 \end{bmatrix}, BA = \begin{bmatrix} 0 & 6 & 0 \\ 6 & -5 & 9 \\ 0 & 1 & 0 \end{bmatrix}$

47. $AB = \begin{bmatrix} -1 \\ 6 \end{bmatrix}$. BA is undefined.

49. If the point $(1, 1)$ rotates $45°$ clockwise, it is located on the positive x-axis with coordinates $(\sqrt{2}, 0)$. Its distance from the origin equals $\sqrt{2}$.

51. If the point $(0, -2)$ rotates $45°$ clockwise, it will be located in the third quadrant, a distance of 2 from the origin. Its coordinates are $(-\sqrt{2}, -\sqrt{2})$.

53. $(1, -\sqrt{3})$

55. $(1, \sqrt{3})$

57. $ABX \approx \begin{bmatrix} 2.8978 \\ -0.77646 \\ 1 \end{bmatrix}$.

If the point $(3, 0)$ is rotated $30°$ counterclockwise and then $45°$ clockwise, the net result is a $15°$ rotation clockwise. Its new coordinates are approximately $(2.8978, -0.77646)$.

59. $C = AB = \begin{bmatrix} 0.96593 & 0.25882 & 0 \\ -0.25882 & 0.96593 & 0 \\ 0 & 0 & 1 \end{bmatrix}$

61. $(0.369, 0, 0.067)$

63. $\begin{bmatrix} R \\ G \\ B \end{bmatrix} = \begin{bmatrix} 1 \\ 1 \\ 1 \end{bmatrix} - \begin{bmatrix} C \\ M \\ Y \end{bmatrix}$

65. They both equal $\begin{bmatrix} 36 & 36 & 8 \\ -15 & -38 & -4 \\ -11 & 13 & 10 \end{bmatrix}$. The distributive property appears to hold for matrices.

67. They both equal $\begin{bmatrix} 50 & 3 & 12 \\ -6 & 55 & 8 \\ 27 & -3 & 29 \end{bmatrix}$. Matrices appear to conform to rules of algebra except that $AB \neq BA$.

8.5 ANSWERS

1. B is the inverse of A.

3. B is the inverse of A.

5. $k = 1$

7. $k = 2.5$

9. A

11. A

13. (a) $(2, 4)$
(b) It will translate $(2, 4)$ back to $(0, 1)$;
$A^{-1} = \begin{bmatrix} 1 & 0 & -2 \\ 0 & 1 & -3 \\ 0 & 0 & 1 \end{bmatrix}$
(c) I_3

15. $A = \begin{bmatrix} 1 & 0 & -3 \\ 0 & 1 & -5 \\ 0 & 0 & 1 \end{bmatrix}$ and $A^{-1} = \begin{bmatrix} 1 & 0 & 3 \\ 0 & 1 & 5 \\ 0 & 0 & 1 \end{bmatrix}$.

A^{-1} will translate a point 3 units to the right and 5 units up.

17. (a) 5
(b) $AX = \begin{bmatrix} 9 \\ 12 \\ 1 \end{bmatrix} = Y$ The distance from $(9, 12)$ to the origin is 15. A increases the distance by a factor of 3.
(c) $A^{-1}Y = \begin{bmatrix} 3 \\ 4 \\ 1 \end{bmatrix} = X$. A^{-1} reduces the distance by a factor of $\frac{1}{3}$.

19. (a) $BX = \begin{bmatrix} -2 \\ 0 \\ 1 \end{bmatrix} = Y$
(b) $B^{-1}Y = \begin{bmatrix} -\sqrt{2} \\ -\sqrt{2} \\ 1 \end{bmatrix} = X$
B^{-1} rotates the point represented by Y $45°$ counterclockwise.

21. (a) $ABX = \begin{bmatrix} 2 \\ 2 \\ 1 \end{bmatrix} = Y$.
(b) The net result of A and B is to translate a point 1 unit right and 1 unit up.
$AB = \begin{bmatrix} 1 & 0 & 1 \\ 0 & 1 & 1 \\ 0 & 0 & 1 \end{bmatrix}$
(c) Yes
(d) Since AB translates a point 1 unit right and 1 unit up, the inverse of AB would translate a point 1 unit left and 1 unit down. Therefore,
$(AB)^{-1} = \begin{bmatrix} 1 & 0 & -1 \\ 0 & 1 & -1 \\ 0 & 0 & 1 \end{bmatrix}$

23. $A^{-1} = \begin{bmatrix} -10 & 30 \\ -4 & 10 \end{bmatrix}$

25. $A^{-1} = \begin{bmatrix} 0.2 & 0 & 0.4 \\ 0.4 & 0 & -0.2 \\ 1.4 & -1 & -1.2 \end{bmatrix}$

27. $A^{-1} = \begin{bmatrix} 3 & -2 \\ -1 & 1 \end{bmatrix}$

29. $A^{-1} = \begin{bmatrix} 5 & 2 \\ 3 & 1 \end{bmatrix}$

31. $A^{-1} = \begin{bmatrix} 0 & 1 & 0 \\ 0 & 0 & 1 \\ 1 & 0 & 0 \end{bmatrix}$

33. $A^{-1} = \begin{bmatrix} -2 & 1 & -1 \\ -5 & 2 & -1 \\ 3 & -1 & 1 \end{bmatrix}$

35. (a) $AX = \begin{bmatrix} 1.5 & 3.7 \\ -0.4 & -2.1 \end{bmatrix}\begin{bmatrix} x \\ y \end{bmatrix} = \begin{bmatrix} 0.32 \\ 0.36 \end{bmatrix} = B$

   (b) $X = \begin{bmatrix} 1.2 \\ -0.4 \end{bmatrix}$

37. (a) $AX = \begin{bmatrix} 0.08 & -0.7 \\ 1.1 & -0.05 \end{bmatrix}\begin{bmatrix} x \\ y \end{bmatrix} = \begin{bmatrix} -0.504 \\ 0.73 \end{bmatrix} = B$

   (b) $X = \begin{bmatrix} 0.7 \\ 0.8 \end{bmatrix}$

39. (a) $AX = \begin{bmatrix} 3.1 & 1.9 & -1 \\ 6.3 & 0 & -9.9 \\ -1 & 1.5 & 7 \end{bmatrix}\begin{bmatrix} x \\ y \\ z \end{bmatrix} = \begin{bmatrix} 1.99 \\ -3.78 \\ 5.3 \end{bmatrix} = B$

   (b) $X = \begin{bmatrix} 0.5 \\ 0.6 \\ 0.7 \end{bmatrix}$

41. (a) $AX = \begin{bmatrix} 3 & -1 & 1 \\ 5.8 & -2.1 & 0 \\ -1 & 0 & 2.9 \end{bmatrix}\begin{bmatrix} x \\ y \\ z \end{bmatrix}$
   $= \begin{bmatrix} 4.9 \\ -3.8 \\ 3.8 \end{bmatrix} = B$

   (b) $X \approx \begin{bmatrix} 9.262 \\ 27.39 \\ 4.504 \end{bmatrix}$

43. Type A: \$10.99; type B: \$12.99; type C: \$14.99

45. (a) Intersection A: incoming traffic is $x_1 + 5$ and outgoing traffic is $4 + 6$, so $x_1 + 5 = 4 + 6$. Intersection B: incoming traffic is $x_2 + 6$ and outgoing traffic is $x_1 + 3$, so $x_2 + 6 = x_1 + 3$. Intersection C: incoming traffic is $x_3 + 4$ and outgoing traffic is $x_2 + 7$, so $x_3 + 4 = x_2 + 7$. Intersection D: incoming traffic is $6 + 5$ and outgoing traffic is $x_3 + x_4$, so $6 + 5 = x_3 + x_4$.

   (b) $AX = \begin{bmatrix} 1 & 0 & 0 & 0 \\ -1 & 1 & 0 & 0 \\ 0 & -1 & 1 & 0 \\ 0 & 0 & 1 & 1 \end{bmatrix}\begin{bmatrix} x_1 \\ x_2 \\ x_3 \\ x_4 \end{bmatrix} = \begin{bmatrix} 5 \\ -3 \\ 3 \\ 11 \end{bmatrix} = B$

   The solution is $x_1 = 5$, $x_2 = 2$, $x_3 = 5$, and $x_4 = 6$.

   (c) The traffic traveling west from intersection B to intersection A has a rate of $x_1 = 5$ cars per minute. The other values for $x_2, x_3,$ and x_4 can be interpreted in a similar manner.

47. (a) $\begin{bmatrix} 1 & 1500 & 8 \\ 1 & 2000 & 5 \\ 1 & 2200 & 10 \end{bmatrix}\begin{bmatrix} a \\ b \\ c \end{bmatrix} = \begin{bmatrix} 122 \\ 130 \\ 158 \end{bmatrix}$

   (b) \$130,000

8.6 ANSWERS

1. $\det A = 1 \neq 0.$ A is invertible.

3. $\det A = 0.$ A is not invertible.

5. $M_{12} = 10, A_{12} = -10$

7. $M_{22} = -15, A_{22} = -15$

9. $\det A = 3 \neq 0.$ A^{-1} exists.

11. $\det A = 0.$ A^{-1} does not exist.

13. 30

15. 0

17. -32

19. 0

21. 643.4

23. -4.484

25. 7

27. 6.5

29. $(5, -3)$

31. $(0.45, 0.67)$

33. The points are collinear.

35. The points are not collinear.

CHAPTER 8   REVIEW ANSWERS

1. $f(3, 6) = 9$

3. (a) $\dfrac{249}{99} \approx 2.5$

   (b) They are approximately equal.

5. $(1, -2)$

7. $l = 11, w = 7$

9. $(2, 3)$, consistent

11. \$1,200 at 7%, \$800 at 9%

13. One solution is (2, 2).

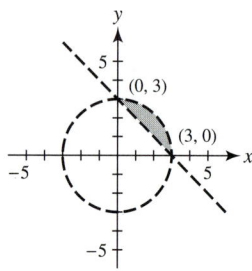

15. $(-9, 3)$

17. $(-1, 3)$

19. **(a)** $A + 2B = \begin{bmatrix} 7 & 1 \\ -8 & 1 \end{bmatrix}$

(b) $A - B = \begin{bmatrix} -2 & -5 \\ 7 & -2 \end{bmatrix}$

21. $AB = \begin{bmatrix} -2 & -4 \\ 17 & 31 \end{bmatrix}$, $BA = \begin{bmatrix} 8 & -6 \\ -27 & 21 \end{bmatrix}$

23. $AB = \begin{bmatrix} 3 & 7 \\ -2 & 8 \end{bmatrix}$, $BA = \begin{bmatrix} 2 & -1 & 3 \\ 2 & 9 & -3 \\ 6 & 12 & 0 \end{bmatrix}$

25. $\det A = -1951 \neq 0$. A is invertible.

27. B is the inverse of A.

29. **(a)** $AX = \begin{bmatrix} 11 & 31 \\ 37 & -19 \end{bmatrix} \begin{bmatrix} x \\ y \end{bmatrix} = \begin{bmatrix} -27.6 \\ 240 \end{bmatrix} = B$

(b) $X = \begin{bmatrix} 5.1 \\ -2.7 \end{bmatrix}$

31. **(a)** $x \approx 1.838$, $y \approx 0.1675$; consistent and independent

(b) All points (x, y) satisfying $y = \dfrac{4.2x - 6.2}{3.4}$, where x is a real number; consistent and dependent

(c) No solutions; inconsistent

33. $A^{-1} = \begin{bmatrix} -1 & -2 \\ -1 & -1 \end{bmatrix}$

35. $\det A = 25$

37. $A = \begin{bmatrix} 3 & 3 & 3 \\ 0 & 3 & 0 \\ 0 & 3 & 0 \end{bmatrix}$, $B = \begin{bmatrix} 0 & 0 & 0 \\ 1 & 0 & 1 \\ 1 & 0 & 1 \end{bmatrix}$

39. $a \approx -0.010764$, $b \approx 1.2903$, $c \approx 24.133$

[20, 100, 20] by [45, 65, 5]

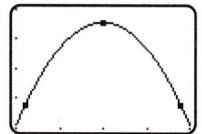

41. There were 31 fatalities through February of 1996–1997. The annual average is 26 fatalities.

43. 4500

CHAPTER 9   FURTHER TOPICS IN ALGEBRA

9.1   ANSWERS

1. $a_1 = 3, a_2 = 5, a_3 = 7, a_4 = 9$

3. $a_1 = 4, a_2 = -8, a_3 = 16, a_4 = -32$

5. $a_1 = \dfrac{1}{2}, a_2 = \dfrac{2}{5}, a_3 = \dfrac{3}{10}, a_4 = \dfrac{4}{17}$

7. $a_1 = -\dfrac{1}{2}, a_2 = \dfrac{1}{4}, a_3 = -\dfrac{1}{8}, a_4 = \dfrac{1}{16}$

9. 2, 4, 3, 5, 3, 6, 4

11. **(a)** $a_1 = 1, a_2 = 2, a_3 = 4, a_4 = 8$

(b) [0, 5, 1] by [0, 9, 1]

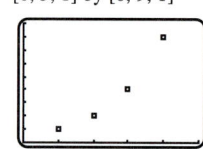

13. **(a)** $a_1 = 2, a_2 = 5, a_3 = 3, a_4 = -2$

(b) [0, 5, 1] by [−3, 6, 1]

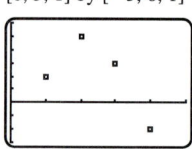

15. **(a)** $a_1 = 2, a_2 = 4, a_3 = 16, a_4 = 256$

(b) [0, 5, 1] by [0, 300, 50]

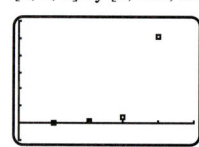

17. The insect population density increases rapidly and then levels off near 5000 per acre.

19. **(a)** $a_n = 0.8a_{n-1}, a_1 = 500$

(b) $a_1 = 500, a_2 = 400, a_3 = 320, a_4 = 256, a_5 = 204.8$, and $a_6 = 163.84$. The population density decreases by 20% each year.

(c) $a_n = 500(0.8)^{n-1}$.

21. **(a)** $a_1 = 300$
$a_n = 2a_{n-1}, n > 1$
$a_1 = 300, a_2 = 600, a_3 = 1200, a_4 = 2400, a_5 = 4800$

(b) $a_{16} = 9{,}830{,}400$ bacteria per milliliter

(c) Geometric, since each term is found by multiplying the previous term by 2

23. (a)

n	1	2	3	4	5	6	7	8
a_n	1	3	5	7	9	11	13	15

(b) [0, 10, 1] by [0, 16, 1]

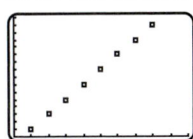

(c) $a_n = 2n - 1$

25. (a)

n	1	2	3	4	5	6	7	8
a_n	7.5	6	4.5	3	1.5	0	-1.5	-3

(b) [0, 12, 1] by [-4, 8, 1]

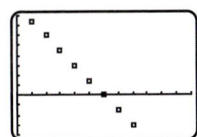

(c) $a_n = -1.5n + 9$

27. (a)

n	1	2	3	4	5	6	7	8
a_n	8	4	2	1	$\frac{1}{2}$	$\frac{1}{4}$	$\frac{1}{8}$	$\frac{1}{16}$

(b) [0, 10, 1] by [-1, 9, 1]

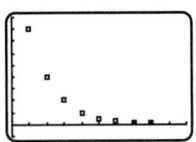

(c) $a_n = 8\left(\dfrac{1}{2}\right)^{n-1}$

29. (a)

n	1	2	3	4	5	6	7	8
a_n	$\frac{3}{4}$	$\frac{3}{2}$	3	6	12	24	48	96

(b) [0, 10, 1] by [-10, 110, 10]

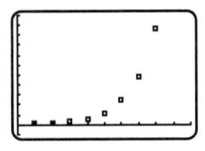

(c) $a_n = \dfrac{3}{4}(2)^{n-1}$

31. $a_n = -2n + 7$

33. $a_n = 3n - 8$

35. $a_n = 2n + 1$

37. $a_n = 2\left(\dfrac{1}{2}\right)^{n-1}$

39. $a_n = \dfrac{1}{2}\left(-\dfrac{1}{4}\right)^{n-1}$

41. $a_n = 8\left(\dfrac{1}{2}\right)^{n-1}$

43. Arithmetic

45. Geometric

47. Neither

49. Arithmetic, $d < 0$, $d = -1$

51. Geometric, $r < 0$, $|r| < 1$

53. (a) 1, 1, 2, 3, 5, 8, 13, 21, 34, 55, 89, 144

(b) $\dfrac{a_2}{a_1} = 1, \dfrac{a_3}{a_2} = 2, \dfrac{a_4}{a_3} = 1.5, \dfrac{a_5}{a_4} = \dfrac{5}{3} \approx 1.6667,$

$\dfrac{a_6}{a_5} = \dfrac{8}{5} = 1.6, \dfrac{a_7}{a_6} = \dfrac{13}{8} = 1.625, \dfrac{a_8}{a_7} = \dfrac{21}{13} \approx$ 1.6154,

$\dfrac{a_9}{a_8} = \dfrac{34}{21} \approx 1.6190, \dfrac{a_{10}}{a_9} = \dfrac{55}{34} \approx 1.6176,$

$\dfrac{a_{11}}{a_{10}} = \dfrac{89}{55} = 1.6182,$ and $\dfrac{a_{12}}{a_{11}} = \dfrac{144}{89} \approx 1.6180.$

These ratios appear to approach a number near 1.618.

(c) $n = 2$:
$a_1 \cdot a_3 - a_2^2 = (1)(2) - (1)^2 = 1 = (-1)^2$
$n = 3$:
$a_2 \cdot a_4 - a_3^2 = (1)(3) - (2)^2 = -1 = (-1)^3$
$n = 4$:
$a_3 \cdot a_5 - a_4^2 = (2)(5) - (3)^2 = 1 = (-1)^4$

55. (a) $a_n = 2000n + 28{,}000$ or $a_n = 30{,}000 + 2000(n - 1)$, arithmetic

(b) $b_n = 30{,}000(1.05)^{n-1}$, geometric

(c) Since $a_{10} = \$48{,}000 > b_{10} \approx \$46{,}540$, the first salary is larger after 10 years. Since $a_{20} = \$68{,}000 < b_{20} \approx \$75{,}809$, the second salary is larger after 20 years.

(d) With time the geometric sequence with $r > 1$ overtakes the arithmetic sequence.

[0, 30, 10] by [0, 150,000, 50,000]

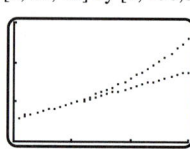

57. $a_6 \approx 1.414213562$, $\sqrt{2} \approx 1.414213562$

9.2 ANSWERS

1. **(a)** $8518 + 9921 + 10,706 + 14,035 + 14,307 + 12,249 = 69,736$

   (b) Finite

3. $S_8 = 8\left(\dfrac{3 + 17}{2}\right) = 80$

5. $S_{50} = 50\left(\dfrac{1 + 50}{2}\right) = 1275$

7. $S_{37} = 37\left(\dfrac{-7 + 101}{2}\right) = 1739$

9. $S_8 = 1\left(\dfrac{1 - 2^8}{1 - 2}\right) = 255$

11. $S_7 = 0.5\left(\dfrac{1 - 3^7}{1 - 3}\right) = 546.5$

13. $S = \dfrac{1}{1 - \dfrac{1}{3}} = \dfrac{3}{2}$

15. $S = \dfrac{6}{1 - \left(-\dfrac{2}{3}\right)} = \dfrac{18}{5}$

17. 2

19. \$91,523.93

21. \$62,278.01

23. $\dfrac{2}{3} = 0.6 + 0.06 + 0.006 + 0.0006 + 0.00006 + \cdots$

25. $\dfrac{9}{11} = 0.81 + 0.0081 + 0.000081 + 0.00000081 + \cdots$

27. $\dfrac{8}{9}$

29. $\dfrac{5}{11}$

31. $\displaystyle\sum_{k=1}^{4}(k + 1) = 2 + 3 + 4 + 5$

33. $\displaystyle\sum_{k=1}^{8}4 = 4 + 4 + 4 + 4 + 4 + 4 + 4 + 4$

35. $\displaystyle\sum_{k=1}^{7}k^3 = 1 + 8 + 27 + 64 + 125 + 216 + 343$

37. $\displaystyle\sum_{k=4}^{5}(k^2 - k) = 12 + 20$

39. $\displaystyle\sum_{k=1}^{6}k^4$

41. $\displaystyle\sum_{k=1}^{\infty}\left(\dfrac{1}{k^2}\right)$

43. $\displaystyle\sum_{k=1}^{n}k = 1 + 2 + 3 + 4 + \cdots + n$ is an arithmetic series with $a_1 = 1$ and $a_n = n$. Its sum equals $S_n = n\left(\dfrac{a_1 + a_n}{2}\right) = n\left(\dfrac{1 + n}{2}\right) = \dfrac{n(n + 1)}{2}$.

45. $S_9 = 9\left(\dfrac{7 + 15}{2}\right) = 99$ logs

47. **(a)** 2500, 2500, 2500, 2500
   4000, 3000, 2000, 1000

   (b) $2500 + 2500 + 2500 + 2500$
   $4000 + 3000 + 2000 + 1000$

49. $1 + 1 + \dfrac{1}{2} + \dfrac{1}{6} + \dfrac{1}{24} + \dfrac{1}{120} + \dfrac{1}{720} + \dfrac{1}{5040} \approx 2.718254$; $e \approx 2.718282$

51. $S_2 = \dfrac{4}{3} \approx 1.3333$, $S_4 = \dfrac{40}{27} \approx 1.4815$, $S_8 \approx 1.49977$, $S_{16} \approx 1.49999997$; $S = 1.5$
   As n increases, the partial sums approach S.

53. $S_1 = 4$, $S_2 = 3.6$, $S_3 = 3.64$, $S_4 = 3.636$, $S_5 = 3.6364$, $S_6 = 3.63636$; $S = \dfrac{40}{11} = 3.\overline{63}$.
   As n increases, the partial sums approach S.

9.3 ANSWERS

1. Vertex: $V(0, 0)$; focus: $F(0, 4)$; directrix: $y = -4$

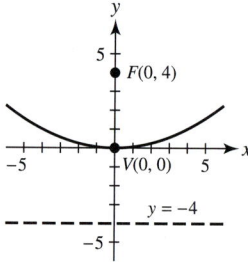

3. Vertex: $V(0, 0)$; focus: $F(2, 0)$; directrix: $x = -2$

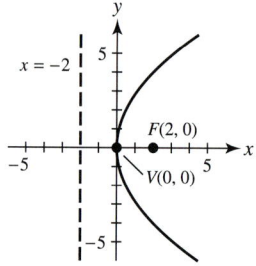

5. Vertex: $V(0, 0)$; focus: $F(-1, 0)$; directrix: $x = 1$

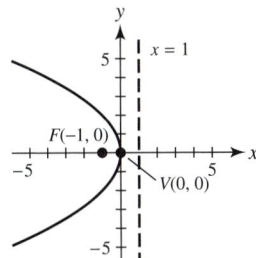

7. $x^2 = 4y$

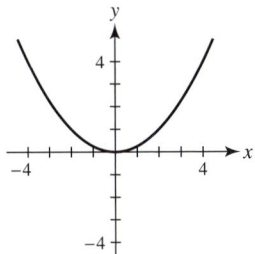

9. $y^2 = -12x$

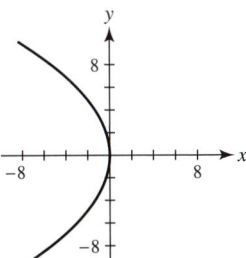

11. $x^2 = 3y$

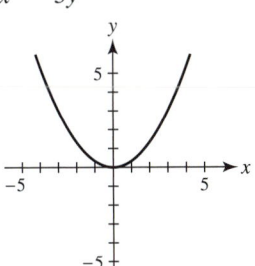

13. $y^2 = -8x$

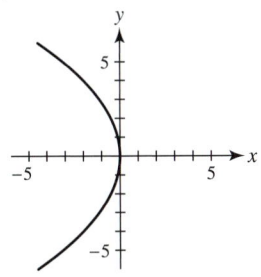

15. $p = 3$ ft

17. Foci: $F(0, \pm\sqrt{5})$; vertices: $V(0, \pm3)$; endpoints of the minor axis $U(\pm2, 0)$

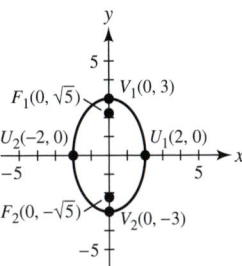

19. Foci: $F(\pm\sqrt{20}, 0)$; vertices: $V(\pm6, 0)$; endpoints of the minor axis $U(0, \pm4)$

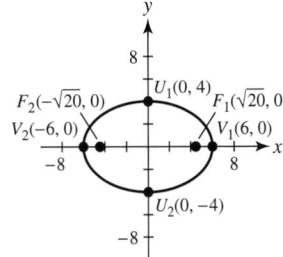

21. Foci: $F(0, \pm2)$; vertices: $V(0, \pm3)$; endpoints of the minor axis $U(\pm\sqrt{5}, 0)$

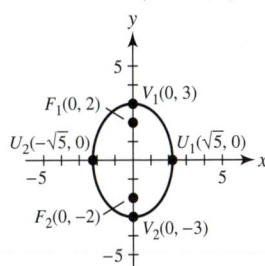

23. $\dfrac{x^2}{25} + \dfrac{y^2}{9} = 1$

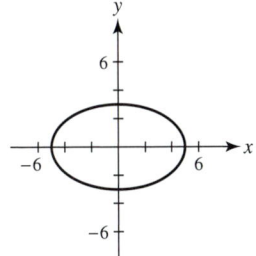

25. $\dfrac{x^2}{5} + \dfrac{y^2}{9} = 1$

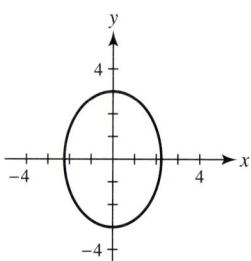

27. $\dfrac{x^2}{12} + \dfrac{y^2}{16} = 1$

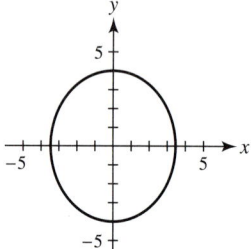

29. $\dfrac{x^2}{36} + \dfrac{y^2}{11} = 1$

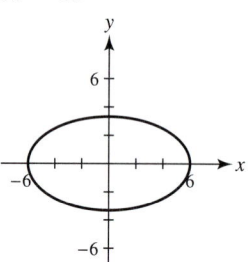

31. $\dfrac{x^2}{9} + \dfrac{y^2}{5} = 1$

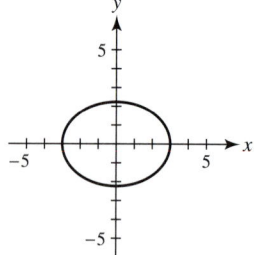

33. $\dfrac{x^2}{0.387^2} + \dfrac{y^2}{0.379^2} = 1$; sun: $(0.0797, 0)$

$[-0.6, 0.6, 0.1]$ by $[-0.4, 0.4, 0.1]$

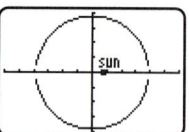

35. Asymptotes: $y = \pm\dfrac{7}{3}x$; $F(\pm\sqrt{58}, 0)$

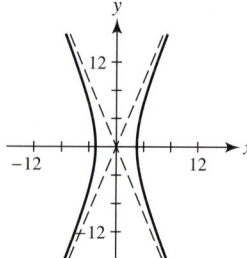

37. Asymptotes: $y = \pm\dfrac{3}{2}x$; $F(0, \pm\sqrt{52})$

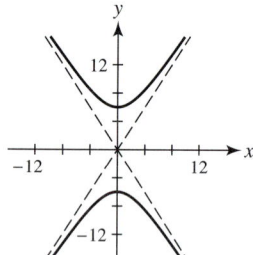

39. Asymptotes: $y = \pm x$; $F(\pm\sqrt{18}, 0)$

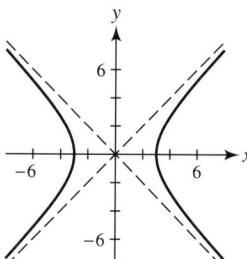

41. $\dfrac{x^2}{16} - \dfrac{y^2}{9} = 1$

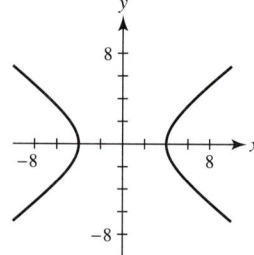

43. $\dfrac{y^2}{36} - \dfrac{x^2}{64} = 1$

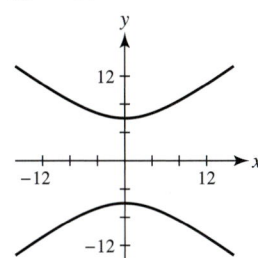

45. $\dfrac{y^2}{144} - \dfrac{x^2}{25} = 1$; asymptotes: $y = \pm\dfrac{12}{5}x$

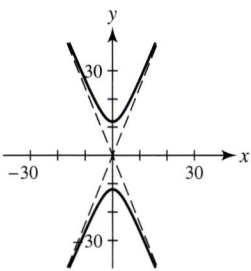

47. $\dfrac{y^2}{4} - \dfrac{x^2}{21} = 1$; asymptotes: $y = \pm\dfrac{2}{\sqrt{21}}x$

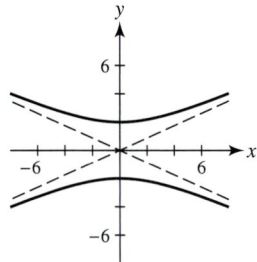

49. $\dfrac{x^2}{9} - \dfrac{y^2}{4} = 1$; asymptotes: $y = \pm\dfrac{2}{3}x$

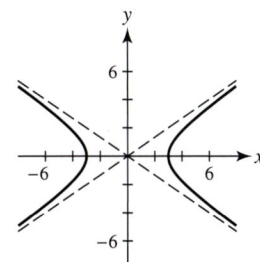

51. $\dfrac{(x-2)^2}{16} + \dfrac{(y-4)^2}{4} = 1$

53. $\dfrac{(y+4)^2}{16} - \dfrac{(x-4)^2}{4} = 1$

55. $x^2 + y^2 = 16$

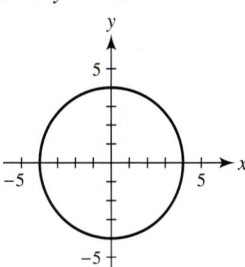

57. $(x - 3)^2 + (y + 4)^2 = 1$

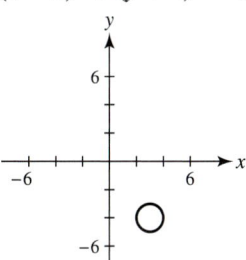

59.

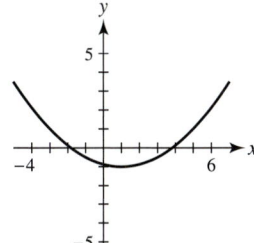

61.

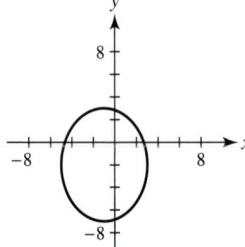

63.

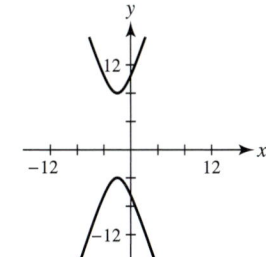

65. Center: $(2, 1)$; radius: $r = 4$
67. Center: $(0, -5)$; radius: $r = 5$
69. **(a)** Elliptic
    (b) Its speed should be 4326 m/sec or greater.

(c) If D is larger then $\dfrac{k}{\sqrt{D}}$ is smaller, so smaller values for V satisfy $V > \dfrac{k}{\sqrt{D}}$.

71. Max: 668 mi; min: 340 mi

73. $y = -1 \pm \sqrt{\dfrac{x}{2.3}}$

$[-6, 6, 1]$ by $[-4, 4, 1]$

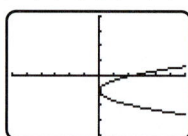

75. $y = \pm\sqrt{10\left(1 - \dfrac{x^2}{15}\right)}$

$[-6, 6, 1]$ by $[-4, 4, 1]$

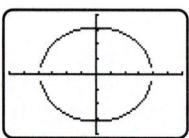

77. $y = 1 \pm \sqrt{11\left(1 + \dfrac{x^2}{5.9}\right)}$

$[-15, 15, 5]$ by $[-10, 10, 5]$

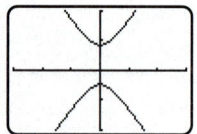

9.4   ANSWERS
1. $2^{10} = 1024$
3. $2^5 4^{10} = 33,554,432$
5. $10^3 \cdot 26^3 = 17,576,000$
7. $26^3 \cdot 36^3 = 820,025,856$
9. $3^5 = 243$
11. 2
13. 24
15. 1,000,000
17. $2^{12} = 4096$
19. No, there are 35,152 call letters possible.
21. 24
23. $P(5, 3) = 60$
25. $P(8, 1) = 8$
27. $P(7, 3) = 210$
29. $P(4, 4) = 4! = 24$
31. $P(5, 5) = 5! = 120$
33. $7 \cdot 6 \cdot 5 = 210$
35. $C(3, 1) = 3$

37. $C(6, 3) = 20$
39. $C(5, 0) = 1$
41. $C(39, 5) = 575,757$
43. $C(5, 2) \cdot C(3, 2) = 30$
45. $C(5, 2) = 10$
47. $C(8, 4) = 70$
49. $x^2 + 2xy + y^2$
51. $m^3 + 6m^2 + 12m + 8$
53. $8x^3 - 36x^2 + 54x - 27$
55. $p^6 - 6p^5q + 15p^4q^2 - 20p^3q^3 + 15p^2q^4 - 6pq^5 + q^6$

9.5   ANSWERS
1. Yes
3. No
5. Yes
7. No
9. $\dfrac{1}{2}$
11. $\dfrac{1}{6}$
13. $\dfrac{1}{2}$
15. $\dfrac{4}{52} = \dfrac{1}{13}$
17. $\dfrac{1}{10,000}$
19. **(a)** 0.57 or 57%
    (b) 0.33 or 33%
21. $\dfrac{1}{4}$
23. $\dfrac{1}{27}$
25. $\dfrac{6}{36} = \dfrac{1}{6}$
27. $\dfrac{625}{1296} \approx 0.482$
29. $\dfrac{1}{270,725}$
31. $\dfrac{4}{20} = 0.2$
33. **(a)**

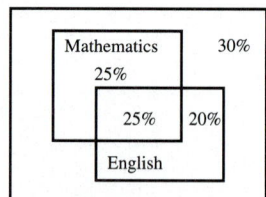

(b) 0.7 or 70%
(c) Let M denote the event of needing help with math and E the event of needing help with

English. Then, $P(M \cup E) = P(M) + P(E) - P(M \cap E) = 0.5 + 0.45 - 0.25 = 0.7$

35. (a) $\dfrac{1985}{51{,}863} \approx 0.038$

(b) $\dfrac{46{,}112}{51{,}863} \approx 0.889$

37. $\dfrac{99.6}{100{,}000} = 0.000996.$

39. The probability is $\left(\dfrac{1}{2}\right)^{n}$. As n increases the probability decreases. This agrees with intuition. The probability of tossing a long string of consecutive heads is small. The longer the string, the less chance there is of it happening.

41. (a) $\dfrac{3}{16}$

(b) $\dfrac{9}{16}$

(c) $\dfrac{9}{64}$

(d) $\dfrac{3}{64}$

43. (a) 0.09

(b) 0.12

45. $\dfrac{1}{1000}$

47. (a) $\dfrac{22}{50} = 0.44$

(b) $\dfrac{28}{50} = 0.56$

(c) $\dfrac{28}{50} = 0.56$

49. (a) 0.1 or 10%

(b) There is a smaller lifetime risk of bladder cancer when the percentage is under 4%. Higher percentages result in a dramatic increase in risk.

CHAPTER 9   REVIEW ANSWERS

1. $-1, -4, -7, -10$

3. $0, 1, 3, 7$

5. $5, 3, 1, 2, 4, 6$

7. (a)

n	1	2	3	4	5	6	7	8
a_n	2	4	6	8	10	12	14	16

(b) [0, 17, 1] by [0, 17, 1]

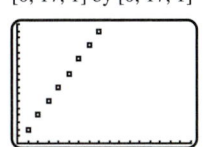

(c) $a_n = 2n$

9. (a)

n	1	2	3	4	5	6	7	8
a_n	81	27	9	3	1	$\dfrac{1}{3}$	$\dfrac{1}{9}$	$\dfrac{1}{27}$

(b) [0, 10, 1] by [−10, 90, 10]

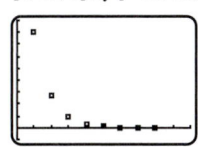

(c) $a_n = 81\left(\dfrac{1}{3}\right)^{n-1}$

11. $a_n = 4n - 15$

13. (a) $4, 3.6, 3.24, 2.916, 2.6244$; geometric

(b) [0, 6, 1] by [0, 6, 1]

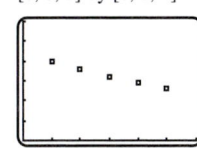

(c) $a_n = 4(0.9)^{n-1}$

15. $S_9 = 9\left(\dfrac{-2 + 22}{2}\right) = 90$

17. $S_8 = 1\left(\dfrac{1 - 3^8}{1 - 3}\right) = 3280$

19. $S = \dfrac{2}{1 - \dfrac{1}{2}} = 4$

21. $S = \dfrac{0.2}{1 - 0.1} = \dfrac{2}{9}$

23. $6 + 11 + 16 + 21 + 26$

25. $\displaystyle\sum_{k=1}^{6} k^3$

27. $y^2 = 8x$

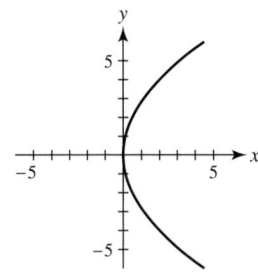

29. $\dfrac{x^2}{16} + \dfrac{y^2}{49} = 1$

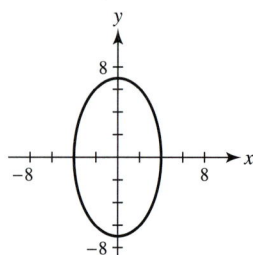

31. $F(0, -1)$

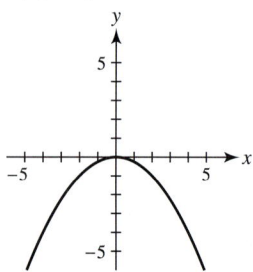

33. $F(\pm\sqrt{21}, 0)$

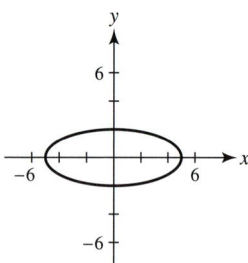

35. $F(\pm 5, 0)$

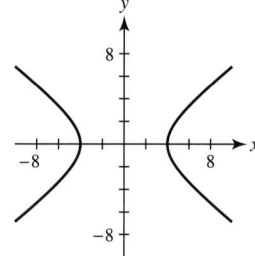

37. Center: $(-1, 2)$

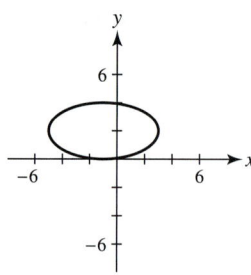

39. $4^{20} \approx 1.1 \times 10^{12}$

41. $50^4 = 6,250,000$

43. $P(6, 3) = 120$

45. $P(5, 5) = 120$

47. $C(6, 3) = 20$

49. $C(10, 6) = 210$

51. $\dfrac{1}{2}$

53. $\dfrac{91}{120} \approx 0.758$

55. (a) $\dfrac{27}{60} = 0.45$

     (b) $\dfrac{33}{60} = 0.55$

     (c) $\dfrac{13}{60} \approx 0.217$

57. Initially, the population density grows slowly, then increases rapidly, and finally levels off near 4,000,000 per acre.

[0, 16, 1] by [0, 5000, 1000]

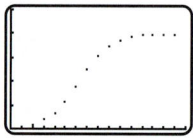

INDEX OF APPLICATIONS

INDEX

GEOMETRIC FORMULAS

Figure	Formulas	Examples
Trapezoid	Area: $A = \frac{1}{2}h(b_1 + b_2)$ Perimeter: $\quad P = a + b_1 + c + b_2$	
Sphere	Volume: $V = \frac{4}{3}\pi r^3$ Surface area: $S = 4\pi r^2$	
Cone	Volume: $V = \frac{1}{3}\pi r^2 h$ Surface area: $\quad S = \pi r \sqrt{r^2 + h^2}$	
Cube	Volume: $V = e^3$ Surface area: $\quad S = 6e^2$	
Rectangular Solid	Volume: $V = LWH$ Surface area: $\quad S = 2HW + 2LW + 2LH$	
Right Circular Cylinder	Volume: $V = \pi r^2 h$ Surface area: $\quad S = 2\pi rh + 2\pi r^2$	
Right Pyramid	Volume: $V = \frac{1}{3}Bh$ $\quad B = $ area of the base	

GEOMETRIC FORMULAS

Figure	Formulas	Examples
Square	Perimeter: $P = 4s$ Area: $A = s^2$	
Rectangle	Perimeter: $P = 2L + 2W$ Area: $A = LW$	
Triangle	Perimeter: $P = a + b + c$ Area: $A = \frac{1}{2}bh$	
Pythagorean Theorem (for Right Triangles)	$c^2 = a^2 + b^2$	
Sum of the Angles of a Triangle	$A + B + C = 180°$	
Circle	Diameter: $d = 2r$ Circumference: $\quad C = 2\pi r = \pi d$ Area: $A = \pi r^2$	
Parallelogram	Area: $A = bh$ Perimeter: $P = 2a + 2b$	